FEATURES AND BENEFITS

Algebra 1

■ The lessons and exercises were written by an **experienced author team,** led by Dr. Mary P. Dolciani, to provide honors students with a rigorous, comprehensive course of study in first-year algebra. See pp. iii–viii (Table of Contents).

■ Algebra concepts and skills are developed using these time-tested and effective varieties of **exercises:**

Oral Exercises check students' understanding of basic lesson ideas. See p. 73.

Written Exercises, graded A, B, and C, range from straightforward exercises reinforcing lesson concepts to thought-provoking exercises that challenge better students. See pp. 235–236.

Numerous **tests and reviews** check students' mastery of algebra skills. See pp. 374–379.

Problems give students a chance to apply the problem solving methods discussed in the lessons. See pp. 132, 189–194.

NEW optional **Computer Exercises** following selected sections enable students to use a computer to help them learn algebra. See p. 129.

NEW Mixed Reviews test students' ability to solve problems encountered out of context. See p. 319.

■ **NEW** student learning aids:

More **worked-out examples** in the lessons and exercise sets help students apply lesson concepts to solve exercises and problems. See pp. 350–351.

Reading Algebra features help students read mathematical explanations more effectively. See p. 349.

Preparing for College Entrance Exams gives students opportunities to practice on test items like those on college entrance exams. See pp. 109, 211.

■ Realistic **Applications** enhance students' knowledge of the applications of mathematics. See pp. 540–541.

■ Challenging **Contest Problems** motivate students to sharpen their logical reasoning skills. See p. 321.

■ **On the Calculator** and **Programming in BASIC** features extend algebra lesson concepts. See pp. 28, 301.

■ Teaching aids include the following:

Comprehensive **Teacher's Edition** provides teacher's commentary followed by annotated pupil book pages with side-column notes. See Contents, p. T3.

Tests on duplicating masters provide tests on groups of sections, chapter tests, and cumulative tests.

Solution Key contains worked solutions to pupil book exercises.

Teacher's Edition

Algebra 1

Mary P. Dolciani
Richard A. Swanson
John A. Graham

Editorial Adviser

Andrew M. Gleason

Teacher Consultants

Bruce Brombacher

Elizabeth B. Jayne

Carla J. Lund

HOUGHTON MIFFLIN COMPANY · Boston

Atlanta   Dallas   Geneva, Ill.   Lawrenceville, N.J.   Palo Alto   Toronto

Authors

Mary P. Dolciani Professor of Mathematical Sciences, Hunter College of the City of New York

Richard A. Swanson Supervisor of Mathematics, Liverpool Central Schools, Liverpool, New York

John A. Graham Mathematics Teacher, Buckingham Browne and Nichols School, Cambridge, Massachusetts

Editorial Adviser

Andrew M. Gleason Hollis Professor of Mathematics and Natural Philosophy, Harvard University, Cambridge, Massachusetts

Teacher Consultants

Bruce Brombacher Mathematics Teacher, Jones Middle School, Westerville, Ohio

Elizabeth B. Jayne Mathematics Teacher, Paul Blazer High School, Ashland, Kentucky

Carla J. Lund Mathematics Teacher, Washington High School, Tulsa, Oklahoma

Printed in U.S.A.

ISBN: 0-395-34374-7

ABCDEFGHIJ-RM-93210/898765

Contents

Using the Teacher's Edition

This Teacher's Edition includes nearly full-sized textbook pages annotated with answers. Time-saving references, suggestions, and extra examples and exercises appear in the adjacent side-columns, where they will be most useful. The simulated Teacher's Edition pages below illustrate the material in the side columns.

For each section in the textbook, page references are given for the corresponding **Teaching Suggestions** and **Related Activities** in the Lesson Commentary at the front of the Teacher's Edition.

Supplementary Materials provide references to the Tests that accompany the textbook.

Key Ideas state the major concepts presented in the lesson.

Chalkboard Examples provide additional examples to use in presenting the lesson to your students.

Reading Algebra sections suggest ways you may help your students read algebra more easily and with greater comprehension.

The **Common Errors** sections alert you to errors that students often make and suggest how you can help your students avoid these errors.

Teaching Suggestions p. T58

Related Activities p. T58

Supplementary Material Test 3

Key Ideas
Translate numerical relationships stated in words into mathematical sentences.

Chalkboard Examples
Write a mathematical sentence for the word sentence.
1. The cube of a number n is three times the square of the same number. $n^3 = 3n^2$

Reading Algebra
When reading word problems, students should look for the words that indicate an operation. In the problem "Find the total cost of c pounds of coffee and t pounds of tea," students should see that the word *total* indicates addition.

Common Errors
Students often interpret the phrase "five less than p" as "$5 - p$" rather than "$p - 5$." Encourage them to substitute numerical examples (What is 5 less than 12?) to test.

1–9 Words into Symbols: Sentences

Just as you can translate a word phrase into a mathematical expression, you also can translate a *word sentence* into a mathematical sentence.

EXAMPLE 1 Write a mathematical sentence for each word sentence.
  a. The sum of seven and twice a number n is twenty-five.
  b. The quotient of a number y divided by three is greater than four times y, decreased by eleven.

SOLUTION **a.** $7 + 2n = 25$ SIMULATED PAGE

  b. $\frac{y}{3} > 4y - 11$

A word sentence may involve an unknown number or quantity without specifying a variable. In such cases, you may choose any variable to represent the number. Then write a mathematical sentence that relates the facts in the given situation.

EXAMPLE 2 Write a mathematical sentence that represents the given information.
  a. The sum of a number and forty-nine is ninety-three.
  b. The perimeter of a square is less than or equal to sixty centimeters.

SOLUTION **a.** Let n represent the unknown number.
      $n + 49 = 93$

  b. Let s represent the length of one *side* of the square.
      $4s \leq 60$

Oral Exercises

Translate each word sentence into a mathematical sentence.

1. The sum of five and nineteen is twenty-four. $5 + 19 = 24$
2. The difference when seven is subtracted from thirty-two is greater than twenty. $34 - 7 > 20$
3. The product of a number a and three is less than or equal to forty-seven. $3a \leq 47$
4. The total of five and a number b is thirty-nine. $5 + b = 39$
5. The quotient when a number c is divided by six is two less than the product of c and six. $\frac{c}{6} = 6c - 2$
6. Fifteen less than twice a number d is greater than fifty more than one half d. $2d - 15 > 50 + \frac{1}{2}d$

One of the features of the Teacher's Edition not illustrated here is **Problem Solving Strategies** (see p. 1). **Permission-to-reproduce tests** and **reviews** and a number of other useful features are located in the front of the Teacher's Edition. See Contents, p. T3, for a complete listing.

---

7. Thirty-five is not the product when the number that is one greater than a number e is multiplied by the number that is one less than e.
8. The quotient when the sum of a number f and twelve is divided by seven is sixteen more than f. $\frac{f + 12}{7} = 16 + f$ SIMULATED PAGE
9. The cube of the difference when a number h is subtracted from a number g is greater than the sum of g and h. $(g - h)^3 > g + h$
10. The square of the sum of a number j and a number k is not equal to the sum of the square of j and the square of k. $(j + k)^2 \neq j^2 + k^2$

7. $35 \neq (e + 1)(e - 1)$

Translate each mathematical sentence into a word sentence.

11. $r - 3 = 10$ 12. $4s = 12$ 13. $22 < t + 8$ 14. $\frac{u}{5} \geq 5$

15. $7 - 2v \neq v^2$ 16. $\frac{5}{w} = \frac{1}{2}w$ 17. $8(9 - x) = 4$ 18. $(6 + y)^3 \neq 6 + y^3$

Written Exercises

Write a mathematical sentence for each word sentence.

A 1. The total cost of x tickets at four dollars per ticket is seventy-six dollars. $4x = 76$
2. After a deposit of y dollars in an account containing fifty dollars, the new balance is ninety-five dollars. $y + 50 = 95$
3. A customer used a twenty-dollar bill to pay for an item that cost z dollars and received less than four dollars in change. $20 - z < 4$
4. A thousand-dollar profit was divided among w partners, and each partner received more than three hundred dollars. $\frac{1000}{w} > 300$
5. The winner of the election received twenty-one votes, which is one more than two thirds of the total number, v, of votes cast. $21 = 1 + \frac{2}{3}v$
6. In today's game the team scored nine runs, which is one less than twice the number of runs, r, that they scored in yesterday's game. $9 = 2r - 1$
7. Mark is m years old, and his age in eight years will be three times his present age. $m + 8 = 3m$
8. Donna is d years old, and her age one year ago was half her age six years from now. $d - 1 = \frac{1}{2}(d + 6)$
9. The area of a rectangle that is six meters long and w meters wide is twenty-one square meters. $6w = 21$
10. The perimeter of an equilateral triangle in which each side measures s centimeters is not fifty-four centimeters. $3s \neq 54$
11. The total value of d dimes and q quarters is not ninety cents. $10d + 25q \neq 90$
12. The total value of p pennies and n nickels is one dollar forty-six cents. $0.01p + 0.05n = 1.46$

Suggested Assignments for minimum, average, and maximum courses are given with each lesson and also in the **Assignment Guide** at the front of the Teacher's Edition.

Answers that cannot be annotated on the textbook page appear in the side column.

You can use the **Additional A Exercises** to check whether your students are ready to start the Written Exercises on their own.

Mixed Reviews provide practice in skills taught in previous lessons and help keep these skills alive.

For each Self-Test in the textbook, there is a corresponding **Quick Quiz**.

Teaching the Course

The Teacher's Edition and Solution Key have been designed to help you teach algebra. For each chapter in the textbook, the Teacher's Edition provides Lesson Commentary and slightly reduced reproductions of student pages with annotated answers. Side columns next to the student pages present additional teaching aids. The Teacher's Edition also includes extra tests and reviews, an assignment guide, term projects, and special Reading Algebra, Problem Solving Strategies, and Error Analysis sections.

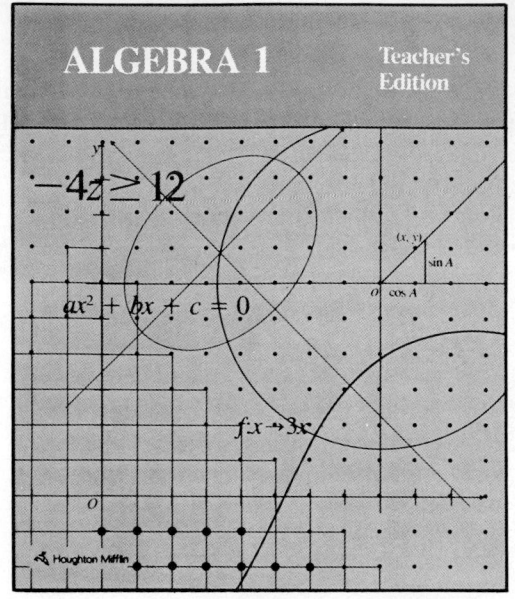

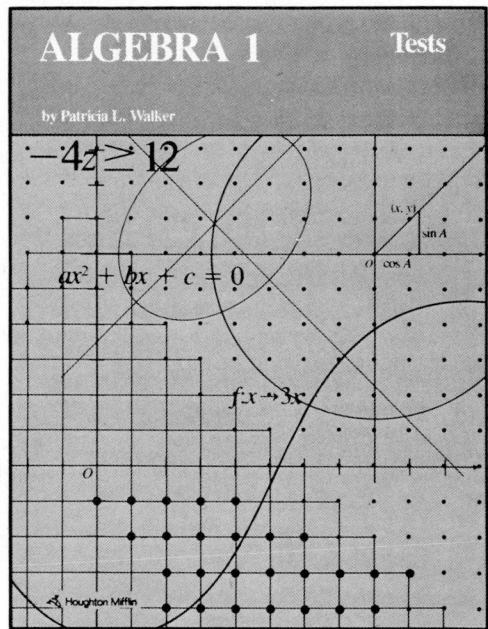

The optional Tests on duplicating masters contain tests on chapter subdivisions, chapter tests, and cumulative tests. The masters are keyed to the student textbook, and a separate Answer Key with answers annotated on reduced facsimiles of the tests is provided.

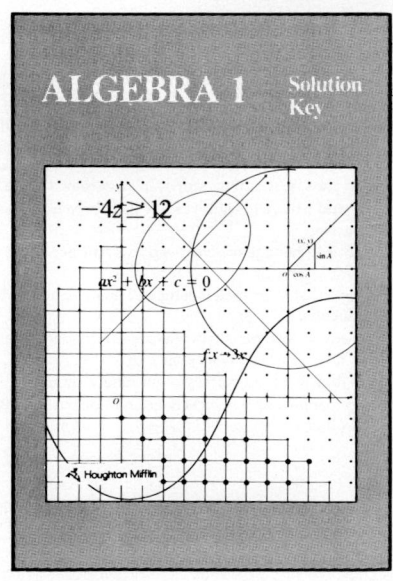

The Solution Key provides step-by-step solutions, including diagrams, for all written exercises in the student book.

Alternate Chapter Tests

(The starred problems may be used in place of the unstarred ones to individualize the tests or make them more challenging.)

Chapter 1 Test

1. On a number line, what is the coordinate of the point that is one third of the way from $^-6$ to 10?

2. List these numbers in order from least to greatest:
$^-5$, $^-2$, $^-3\frac{1}{2}$, 7.75, 0, $^-3.75$

Tell whether each statement is true or false.

3. $0 \in$ {the negative real numbers}
4. {3, 6, 9, 12} $\not\subset$ {the rational numbers}
5. {the positive integers} $\subset$ {the natural numbers}
6. {the rational numbers} $\neq$ {the irrational numbers}

Evaluate each expression when $x = 3$ and $y = \frac{1}{6}$.

7. $3xy$

8. $\dfrac{y}{x}$

9. $\dfrac{x + y}{4}$

Simplify.

10. $\{2[6 + 7(9)] \div 3\}4$

11. $3\left\{\dfrac{3[15 - 3(4)]}{3 + 9(2)}\right\} - 1$

12. $11 + (8 - 2)8 \div 3(1 + 3)$

13. $4(3)^2 + (4 + 3^2)3 \div 3$

Solve each open sentence if $z \in$ {the whole numbers}.

14. $\dfrac{z}{7} = 5z$

15. $z < 6.9$

16. $5z + 1 \geq 36$

Write a variable expression for each word phrase.

17. the total number of seconds in m minutes
18. the total value in cents of d dimes and q quarters

Write an open sentence for each word sentence.

19. The square of the difference when a number b is subtracted from a number a is greater than the quotient when the sum of a and b is divided by the product of a and b.

20. Johanna is p years old, and her age seven years ago was one-fifth her age twenty-one years from now.

★ 1. On a number line, what is the coordinate of the point that is one half the distance from $^-2\frac{1}{3}$ to $3\frac{1}{2}$?

★ 7. Evaluate the expression $\dfrac{y - \frac{1}{x}}{x - \frac{1}{y}}$ when $x = 2$ and $y = 3$.

★15. Solve $w \geq 3w - 2$ if $w \in$ {the whole numbers}.

Chapter 2 Test

1. Find the value of a variable that makes this statement true: "There exists a whole number w such that $10w - 2 = 28$."

2. Name the axiom that is illustrated by the following:
$(a + 5) + 2a = (5 + a) + 2a$

Simplify.

3. $17 + 39 + 13 + 11$

4. $(4y)(10y)(3x)$

5. $4m^2 + 3m + m^2 + 6m + 7$

6. $4(3a + 1) + 5(a + 2)$

7. $-[-(71 - 43)]$

8. $|-15| - |5 - 4|$

9. $4 + (-21)$

10. $[4 + (-15)] + (-12)$

11. Solve $m + 9 = 4$.

12. Simplify $-[-19 + 9] + 29 + (-15)$.

13. Three months ago, Roy sold 11 records from his collection. A month later, he bought 6 new records, and last month, he bought 9 new records. This month, he sold 16 records. If he has 114 records now, how many did he have three months ago?

14. Give the reason that justifies the following statement:
$a + [b + (-b)] = a + 0$

Simplify.

15. $-47 - (-17 + 3)$

16. $14y - (2y + 1) - (3y - 1)$

17. $(6m)(-5m)(4m)$

18. $-5(10a - 7b) - 20a$

19. $\frac{1}{7}(-28)\left(-\frac{1}{2}\right)$

20. $(-5x - 45y)\left(-\frac{1}{5}\right)$

21. $\frac{-39m^2n}{13mn}$

22. $49x^2y \div \left(-\frac{7}{10}\right)$

★ **6.** Simplify $\frac{1}{4}\left[3t + 6\left(\frac{3}{2} + \frac{1}{3}t\right)\right] + \frac{3}{4}$.

★ **11.** Solve $-\frac{1}{2} + \left(-\frac{3}{4}\right) = -g + \left(-\frac{1}{3}\right)$.

★ **15.** Evaluate $-[x - (y + z)] - [-(w - x)]$ when $x = -3$, $y = -\frac{1}{2}$, $z = \frac{2}{3}$, and $w = 2$.

Chapter 3 Test

1. Name the property that is illustrated by the following:
"If $\frac{2x}{3} = 12$, then $\left(-\frac{3}{2}\right)\left(\frac{2x}{3}\right) = \left(-\frac{3}{2}\right)(12)$."

Solve.

2. $m + 6 = 2$

3. $7 = 4 - y$

4. $|-14| = x - 8$

5. $21z = -189$

6. $-\frac{2c}{7} = -20$

7. $-5|t| = 75$

8. $4y - 1 = -9$

9. $\frac{4y + 6}{7} = 2$

10. $9y - 4(y + 8) = 18$

11. One number is four fifths another number. If the sum of the numbers is 18, what is the lesser number?

12. Alec's age is twelve years greater than twice Rachel's age. If the sum of their ages is 72, how old is Alec?

Solve.

13. $14 - 5k = 10 - 3k$

14. $7(m - 12) = 6(m + 1)$

15. When six is subtracted from twice a certain number, the result is the number's additive inverse. Find the number.

16. Kim is ten years older than Kenny. Two years ago Kim was twice as old as Kenny. How old is Kim now?

17. Solve for y if $B = \dfrac{a(c + y)}{5}$.

18. Use the formula $A = \pi r^2$ to find the radius of a circle having area 154 cm². Use $\pi = \frac{22}{7}$.

★ **4.** Solve $-3|1 - w| + 2 = -7$.

★ **6.** Solve $\frac{1}{5}a = -2\frac{1}{5}$.

★**10.** Solve $-5(x - 3) - x = 2(1 - 2x) + 5$.

★**15.** Find the number that is three greater than one-half the sum of its additive inverse and six.

★**18.** The length of a field is 12 m greater than twice its width. It cost $1228.50 to put a fence around the field, at a cost of $5.25 a meter.
   a. Using the formula $P = 2(l + w)$ for perimeter, express the width w in terms of the perimeter P and the length l.
   b. What is the width of the field?

Chapter 4 Test

1. Name the property that is illustrated by the following:
   "If $a + 3 < 5$ and $5 < b + 1$, then $a + 3 < b + 1$."

Solve.

2. $11 - 3r > 20$

3. $f + 45 < 3f + 10 + 5f$

Let $A = \{$the integers greater than $-2\}$ and $B = \{$the natural numbers less than or equal to 6$\}$. Specify the following.

4. $A \cup B$

5. $A \cap B$

Solve.

6. $-5 < 3 - 4y \leq 7$

7. $2t + 7 < 6$ or $3t - 2 > 13$

8. $|5 - 4s| < 7$

9. $3|2 + 3p| - 1 > 11$

10. Find four consecutive odd integers such that the sum of the least integer and the greatest integer is -36.

11. The measure of an angle is 24° less than twice the measure of its complement. Find the measure of the angle.

12. Allison drove with her friends from her house to a park at an average speed of 36 km/h. She bicycled back home at an average speed of 9 km/h. It took her 3 hours longer to cycle back than it took her to drive the same distance. How far does Allison live from the park?

13. Chris has $3.55 in dimes and quarters. If he has 11 more dimes than quarters, how many of each coin does he have?

14. Explain why the following problem has no solution: A zoo charges $2.50 admission for adults and $1.50 for children. On a certain day, the zoo collected a total of $172, from 116 paid admissions. How many adults and how many children entered the zoo that day?

★10. Find all sets of three consecutive multiples of 7 whose sum is between −41 and −85.

★12. Hugh left home at 7:15 A.M. to drive to work at an average speed of 50 km/h. His mother noticed he'd forgotten his wallet, and left at 7:30 A.M. to get the wallet to him. If Hugh's mother drove an average of 80 km/h, at what time did she overtake him on the highway?

Chapter 5 Test

Name the coordinates of each point plotted on the coordinate plane at the right.

1. *A*   2. *B*   3. *C*   4. *D*

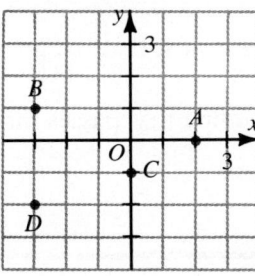

Plot each point on a coordinate plane.

5. *E* (4, 0)   6. *F* (−3, −3)
7. *G*(2, −2)   8. *H* (0, −1)

Given the relation {(−2, 0), (0, −2), (2, 0), (1, 1)}:

9. Draw a mapping diagram that represents the relation.

10. Determine whether or not the relation is a function.

Given *f*: *x* → 4 − 3*x* with domain *D* = {−2, −1, 0, 1, 2}:

11. Graph the function.   12. Compute $2f(0) + f(2)$.

Solve when $x \in \{-1, 0, 1\}$ and $y \in \{-2, -1, 0, 1, 2\}$.

13. $y = -2 + x$ **14.** $y > -2x$

Graph on a coordinate plane.

15. $4x - 4y = 8$ **16.** $x = -1$ **17.** $y = -x$

18. $x - y < 3$ **19.** $x \geq 2$ **20.** $y \geq 2$

21. Graph the line that passes through the point $(-2, 1)$ and has slope $\frac{1}{2}$.

22. Determine the slope of the line that passes through the points $(-4, 5)$ and $(3, -1)$.

Determine an equation of the line that satisfies the given requirements. The equation should be in the form $ax + by = c$, where a, b, and c are integers.

23. has slope $\frac{2}{3}$ and y-intercept -1

24. has zero slope and passes through $(2, -4)$

25. passes through $(-2, -1)$ and has slope $-\frac{1}{3}$

26. passes through the points $(-1, -1)$ and $(3, -2)$

★11. Graph the function $f: x \rightarrow 2x - x^2$ with domain $\{-3, -2, -1, 0, 1, 2, 3\}$.

★22. Find the slope and y-intercept of the line whose equation is $c(x + 2y) = d$, where $c, d \neq 0$.

★23. Determine the value of a such that the graph of the equation $ay = 2x - 1$ has slope 4.

★26. Find an equation of the line that passes through the points $(x_1, 0)$ and $(0, y_1)$.

Chapter 6 Test

1. Solve the system $\begin{array}{l} x + 2y = 4 \\ x = 3 - 2y \end{array}$, using graphs.

2. Is the system $\begin{array}{l} 5x + 5y = -25 \\ 3x - 3y = 15 \end{array}$ consistent or inconsistent?

Solve each system using addition or subtraction.

3. $a + 5b = 18$
$a - 2b = -3$

4. $5r + 5s = 12 + 7r$
$-3r + 3 = 5s - 6r$

Solve using the linear combination method.

5. $3c - 2d = 11$
   $4c + 5d = 7$

6. $3x + 2y = -10$
   $-8x + 5y = 6$

Solve each system using the substitution method.

7. $m + 3n = 4$
   $2m - 5n = 8$

8. $3(x + y) = 2(x + 2)$
   $2(x - 1) = y - 1$

Solve.

9. Five years ago Toni was three-fifths as old as Cindy was then. Seven years from now, Cindy will be twice as old as Toni is now. How old are Cindy and Toni now?

10. Jim swam 4 km upstream in 4 h, but his return trip downstream took 2 h less. Find his swimming rate in still water and the rate of the current.

11. The units' digit of a three-digit number is twice the tens' digit, and the hundreds' digit is one less than the units' digit. The sum of the digits is 19. What is the number?

12. Graph the solution set of the system: $x + y \leq 5$
    $2x - 2 \leq 3y$
    $x \geq -1$

13. Graph the solution set of the system at the right and label the corner points with their coordinates.
    $x + y \leq 4$
    $x - y \geq -2$
    $y \geq -1$
    $x \leq 3$

14. Find the maximum value of the expression $3x - 4y$ over the region graphed in Exercise 13.

15. Solve the system: $x + y + z = 1$
    $2x - y + 3z = 10$
    $2x - y - z = 2$

★ 2. Solve for x and y in terms of n: $x + 2y = 4n$
    $4x - 6n = 2y$

★10. Indira walked home 3 km/h faster than she walked to the shopping mall, and it took 20 minutes less time. Write an equation that expresses her rate r as she walked to the mall in terms of the time t (in hours) it took her to walk there.

★11. The tens' digit of a three-digit number is one more than three times the hundreds' digit, and the tens' digit is half the units' digit. The units' digit is seven more than the hundreds' digit. Find the number.

Chapter 7 Test

1. Simplify $(-3ab + 2a - 1) - (ab - 5a + 1)$.
2. Simplify $(6f^2 + 4f - 5) + (f^2 - 6f - 7)$.
3. Solve $(6a - 1) + (a + 3) = -12$.

Simplify.

4. $(-2m^3n)^3(-3m^2n^2)^2$
5. $(3b^2)(3b^6) - (3b^4)^2$
6. $-3z^2(3z^2 - z - 1)$
7. $(5g + 9)(g - 4)$
8. $(e + 1)(3e^2 - e - 1)$
9. $3a(a^2 - 5) - 2a^2(4a + 1)$
10. Solve $2p(3p - 1) = 6p(p + 3) - 20$.

Simplify.

11. $(5i + 3j)^2$
12. $8(2r^2 - 5s)(2r^2 + 5s)$

13. A rectangular garden is 5 times as long as it is wide. A second rectangular garden has the same area as the first garden, but is 4 m wider and 12 m shorter. What are the dimensions of the second garden?
14. Find the GCF of $135x^2y^3z$ and $90x^3y^3z^3$.

Factor completely.

15. $25a^5b^4 + 15a^4b^3 - 75a^3b^3$
16. $t(t - 9) + 4(9 - t)$
17. $64p^3 + 1$
18. $6x^2 - 31x + 25$
19. $2q^4 - 98z^2$
20. $3d^2 + 30d + 72$
21. $9r^3 - 6r^2 - 35r$
22. $30x^3 - 33x^2 - 18x$

Solve.

23. $-5t(t + 3)(3t - 2) = 0$
24. $10x^2 + 21 = 47x - 10x^2$

25. Find two consecutive positive even integers such that the difference of their squares is 52.

Simplify. Assume that n is a positive integer.

★ 4. $(e^{n+1})^2(-2e^nf)^2$
★ 7. $(x^{6n} - y^{4n})(x^{6n} + y^{4n})$

Factor completely.

★15. $6a^4b^2 - 10a^3b^3 - 4a^2b^4$
★22. $2x^8 - 33x^4 + 16$

★24. Write a quadratic equation whose solution set is $\{\frac{4}{3}, -3\}$. The equation should be in the form $ax^2 + bx + c = 0$.

Chapter 8 Test

Simplify. Assume that no denominator equals zero.

1. $\dfrac{7m}{-56m^2}$

2. $\dfrac{35a^6bc^4}{0.7ab^3c^3}$

3. $\dfrac{(-p^3)^5}{p^6}$

Express each quotient as a sum.

4. $\dfrac{5b^4 + 20b^6}{5b^5}$

5. $\dfrac{30c^3d^2 - 60c^2d - 75c^2d^3}{15c^2d}$

Divide the first polynomial by the second. Assume that no divisor equals zero.

6. $9x^2 - 9x - 4;\ 3x + 1$

7. $t^3 - 7t^2 + 13t - 4;\ t - 2$

Simplify using only positive exponents.

8. $\dfrac{12s^3t^3}{4s^2t^2 + 8s^3t^2}$

9. $\dfrac{k^2 + 7k + 10}{k^2 - 25}$

10. $\dfrac{m^{-1}y^{-3}}{z^0}$

11. $\dfrac{(-5)^0 a^{-3}b^2}{a^2b^{-2}c^{-1}}$

12. Express the product $3000 \times 20{,}000 \times 0.00003$ in scientific notation.

Simplify using only positive exponents.

13. $\dfrac{5w^3}{6w} \cdot \dfrac{12w}{25w^2}$

14. $\dfrac{2h - 4}{h^3} \cdot \dfrac{h^2 + 2h}{h^2 - 4}$

15. $\dfrac{8b + 3}{8b^3} \div \dfrac{8b - 3}{8b^2}$

16. $\dfrac{d^2 - 3d + 2}{d^3} \div (d - 2)$

17. $\dfrac{a^2 - 7a + 12}{a^2 - 8a + 16} \cdot \dfrac{a^2 - 2a}{a^2 - 3a}$

18. $\dfrac{2x - 3y}{12xy} + \dfrac{4x + 2y}{12xy} - \dfrac{6x - y}{12xy}$

19. $\dfrac{v}{v^2 - 16} - \dfrac{1}{4v + 16}$

20. $\dfrac{g}{f^2 - g^2} + \dfrac{1}{f - g}$

21. $\dfrac{\dfrac{15rs^2}{8t}}{\dfrac{5rs}{2t}}$

22. $\dfrac{(j + 2)^{-1}(j + 2)}{(j^2 - 4)^{-1}(3j - 6)}$

★ **2.** $\left[\dfrac{-2rs^{-2}(3r)^0}{(r^{-3}s^2)^{-2}}\right]^{-1}$

★ **8.** $-\left(1 + \dfrac{m}{n}\right)\left(1 - \dfrac{m}{n}\right) \div \left(\dfrac{2m}{n} - \dfrac{2n}{m}\right)$

Chapter 9 Test

1. Solve $\dfrac{z-3}{12} \geq \dfrac{9-2z}{36} + \dfrac{1}{18}$.

2. Solve $\dfrac{m+5}{2} - \dfrac{4m-12}{8} = \dfrac{m+1}{4}$.

3. A suit was on sale for $105. If the original selling price was $150, what was the percent discount?

4. How many grams of water must be evaporated from 80 grams of a solution that is 5% salt, to leave a solution that is 25% salt?

5. Solve $\dfrac{a}{a+5} - \dfrac{1}{a+2} = 1$.

6. Solve $\dfrac{1}{x^2-4} = \dfrac{2}{x+2} - \dfrac{1}{5}$.

7. Two thirds of a number is 5 more than one fourth of the number. Find the number.

8. The difference of two numbers is 3, and the sum of their reciprocals is $\frac{3}{20}$. Find the numbers.

9. Working alone, Ms. Martin can weed the garden in 6 hours. It takes her daughter 8 hours to weed the garden if she works alone. How long will it take them to weed the garden if they work together?

10. Leah rows her boat 8 km downstream in the same time it takes her to row 4 km upstream. If the current in the river flows at 4 km/h, how fast does Leah's boat travel in still water?

11. Solve $\dfrac{n-8}{n} = \dfrac{3}{7}$.

12. Find the ratio of a to b when $\dfrac{b-2a}{7} = \dfrac{a+3b}{5}$.

13. How many gallons of heavy cream are needed to produce 36 lb of butter, if $6\frac{2}{9}$ gallons of cream yield 4 lb of butter?

14. If v varies directly as w^2, and if $v = 1200$ when $w = 2$, find v when $w = 1.5$.

15. The distance which a freely falling body falls varies directly as the square of the time it falls. If an object falls 400 ft in 5 seconds, how far will it fall in 2 seconds?

16. If x varies inversely as y, and if $x = 40$ when $y = 30$, find y when $x = 20$.

17. If a is inversely proportional to b, and if $a = 18$ when $b = \frac{2}{3}$, find a when $b = \frac{1}{6}$.

18. If x varies jointly as y and z, and if $x = 30$ when $y = 15$ and $z = 4$, find x when $y = 6$ and $z = 11$.

★ **1.** Solve $\dfrac{3x-1}{3} - \dfrac{x+5}{6} \le \dfrac{4-x}{4}$.

★ **4.** Kendra invested twice as much money at an annual rate of 7% as she did at an annual rate of $5\frac{1}{2}\%$. How much did she invest at the higher rate if her income from these investments for one year totaled $175.50?

★ **5.** Solve $\dfrac{3}{x-2} - \dfrac{5}{x+2} = \dfrac{2}{x-1}$.

★ **9.** A pipe will fill a tank in 6 h and a second pipe will fill it in 10 h. After the first pipe has been operating for 1 h, the second pipe is opened as well. If the two pipes, working together, finish filling the tank, how long has each pipe been operating?

Chapter 10 Test

1. Write $\dfrac{48}{90}$, $\dfrac{56}{80}$, and $\dfrac{28}{112}$ in order from least to greatest.

2. Find the rational number that is one fifth of the way from 2 to $4\frac{2}{3}$.

3. Express $\dfrac{41}{12}$ as a decimal.

4. Express $0.2\overline{45}$ as a fraction in simplest form.

5. Express $(1.\overline{2})(0.\overline{39})$ as a fraction in simplest form.

Simplify.

6. $-\sqrt{\dfrac{49}{25}}$

7. $\sqrt{(-25)^2}$

8. $\sqrt{(5-x)^2}$

9. $\sqrt{800}$

10. $\sqrt{384}$

11. $\sqrt{\dfrac{32}{81}}$

12. Solve $4v^2 - 9 = 19$.

13. Given that the length of the hypotenuse of a right triangle is $\sqrt{42}$ cm and that the length of a second side is $\sqrt{7}$ cm, find the length of the third side.

14. Bess rode her bike from the store due west for 12 km and then due south for 8 km. How far was she from the store? Approximate the answer to the nearest hundredth.

15. Find the distance between the points $(5, 1)$ and $(2, 10)$.

16. Determine whether or not the triangle with vertices $A(-1, 1)$, $B(2, 5)$, and $C(4, -4)$ is a right triangle.

Simplify. Assume that all variables represent positive real numbers.

17. $\sqrt{45} \cdot \sqrt{24}$

18. $\dfrac{\sqrt{80}}{\sqrt{15}}$

19. $\sqrt{6m^7} \cdot \sqrt{12m^2}$

20. $\sqrt{180} - 2\sqrt{48} + \dfrac{1}{3}\sqrt{27}$

21. $(3\sqrt{3} - \sqrt{5})^2$

22. $\dfrac{5\sqrt{2}}{3 + \sqrt{10}}$

23. $\sqrt[5]{64}$

24. $\sqrt[4]{32} + \sqrt[4]{162}$

25. $\sqrt[3]{\dfrac{9}{4}}$

Solve.

26. $\sqrt{\dfrac{4 - 2x}{3}} = 4$

27. $5\sqrt{m} - \dfrac{2}{5} = \dfrac{13}{5}$

28. $\sqrt{y} + 6 = y$

★**15.** Find the distance between $(a, -2b)$ and $(-4a, 2b)$.

★**18.** Simplify $\dfrac{\sqrt{6a^2b^3}\sqrt{12a^3b^2}}{\sqrt{2ab^6}}$.

★**20.** Simplify $\sqrt{8y^2} + 2\sqrt{18y^3} - \sqrt{32y^2} - y\sqrt{50y}$.

★**21.** Simplify $(\sqrt{3} - 2\sqrt{2})^3$.

Chapter 11 Test

Solve by completing the square. Express irrational solutions in simplest form.

1. $n^2 - 6n = 27$

2. $3z^2 = z + 1$

Solve by using the quadratic formula. Express irrational solutions in simplest form.

3. $2b^2 = 4 - 3b$

4. $4y^2 = 6y + 1$

5. Determine the number of real roots of the equation $x = 3x^2 - 4$.

Solve. Approximate irrational solutions to the nearest hundredth.

6. Starting at his house, Mark can ride his bike due east to the store and then due north to the park. The distance between the park and the store is 10 km more than the distance between the store and Mark's house. If the park is 15 km from Mark's house, find the distance between the park and the store.

7. The sum of a number and twice its reciprocal is $\frac{9}{2}$. Find the number.

Graph each function on a coordinate plane.

8. $h: x \rightarrow 4 - x^2$

9. $f: x \rightarrow x^2 - 2x + 4$

For each function, determine its zeros, if any, and its maximum or minimum value.

10. $F: x \rightarrow -x^2 - 5x$

11. $G: x \rightarrow 4x^2 - 4x + 1$

12. Tell whether the graph of the function $f: x \rightarrow x^2 - 5x - 1$ opens upward or downward and whether the function has a maximum or a minimum value.

Sketch the graph of each function and estimate its zeros, if any.

13. $f: x \rightarrow 4 - x^3$

14. $g: x \rightarrow 2x^3 - 8x$

15. Graph $y < x^2 - 2x$ on a coordinate plane.

16. Graph the solution set of the system $\begin{array}{l} y \geq x^2 + 1 \\ y < 5 \end{array}$ on a coordinate plane.

★ **1.** Solve $x^2 - 6cx = d$ by completing the square.

★ **4.** Solve $a^2x^2 = 1 - 4bx$ by using the quadratic formula.

★ **14.** Sketch the graph of $f: x \rightarrow x^3 - 3x^2 - x + 3$.

Chapter 12 Test

1. If A is a quadrantal angle and $-290° \geq m\angle A \geq -200°$ find $m\angle A$.

2. If A is in standard position, $90° \leq m\angle A \leq 180°$, and $\angle A$ is co-terminal with an angle measuring $-570°$, find $m\angle A$.

3. Evaluate $\sin 450°$.

4. Angle A is in standard position and its terminal side contains the point $\left(\frac{4}{5}, -\frac{3}{5}\right)$ on the unit circle. Find $\tan A$.

5. If $1° \leq m\angle A \leq 90°$ and $\sin A = 0.4368$, find $m\angle A$ to the nearest degree.

6. In right $\triangle ABC$, $c = 20$ and $b = 12$. Find $m\angle A$ to the nearest degree.

7. After take-off, an airplane maintains a flight angle of $12°$ with the ground. Find its elevation to the nearest meter after it has covered a ground distance of 2500 m.

8. If a vector in standard position has direction $210°$ and y-component 48, find its norm.

9. If vectors $\mathbf{s}$ and $\mathbf{t}$ are in standard position with terminal points $(3, 7)$ and $(3, 3)$ respectively, and $\mathbf{v} = \mathbf{s} + \mathbf{t}$, find $\|\mathbf{v}\|$.

10. Frank has attached a rope to his bicycle, which is stuck in the mud. What is the magnitude of the force applied vertically to the bike if Frank pulls on the rope with a force of 60 N and the rope makes an angle of $20°$ with the horizontal? Give the magnitude to the nearest tenth of a newton.

★ 4. Find $\sin A$, $\cos A$, and $\tan A$ for an angle A in standard position whose terminal side contains the point (x^n, y^n).

★ 9. Find the norm of $\mathbf{u} + \mathbf{v}$ and its direction to the nearest degree, given: $\|\mathbf{u}\| = 24, 45°$; $\|\mathbf{v}\| = 10, 315°$.

Chapter 13 Test

1. Make a table showing the frequencies, relative frequencies, and percents for the following set of data.

   9, 20, 19, 15, 17, 20, 16, 15, 19, 19

2. Make a histogram and a frequency polygon for the data at the right. Group them in the intervals 10–20, 20–30, 30–40, 40–50.

11	23	25	32	43
41	21	12	31	27
35	23	50	37	29

3. Make a table showing the frequencies, percents, cumulative frequencies, and cumulative percents for the data in Exercise 2.

4. Find the median, mode, and mean for the following data.

   6, 3, 6, 5, 4, 9, 6, 9, 8, 8, 3, 6, 9, 7, 4

For the list of data 1, 3, 4, 7, 9, 12, find to the nearest tenth:

5. the variance

6. the standard deviation

7. What is the probability of drawing a three at random from a standard deck of 52 cards?

8. A jar contains 7 orange, 4 red, 2 blue, and 5 green marbles. If a marble is drawn at random from the jar, what is the probability that the marble is green or red?

9. A softball team won 15 of their last 27 games. What is the experimental probability that they will win their game next weekend?

10. If a manufacturer finds that 5 batteries in a sample of 1000 batteries are defective, what is the experimental probability that a customer who buys one of this brand of batteries will find it to be defective?

★ 7. What is the probability that a card drawn at random from a standard deck of 52 cards is neither black nor an ace?

Cumulative Tests

Test for Chapters 1–4

1. On a number line, what is the coordinate of the point that is one fifth of the way from −9 to 1?

Tell whether each statement is true or false.

2. $0 \in \emptyset$

3. {the irrational numbers} $\not\subset$ {the real numbers}

Evaluate each expression when $x = 7$ and $y = \frac{1}{4}$.

4. $\dfrac{x}{y}$

5. $\dfrac{xy}{x + y}$

Simplify.

6. $\{6[5 + 3(10)] \div 4\}2$

7. $2^5 - 2^4 \div 2^3 + 2^2$

8. Solve $2p + 5 \leq 25$ if $p \in$ {the whole numbers}.

9. Write a variable expression for the following word phrase:
   the total value in dollars of p pennies, q quarters, and x dollars.

10. Write a mathematical sentence for the following word sentence:
    The quotient when the sum of a number n and 2 is divided by the square of n is less than the square of the sum of n and 3.

11. Find a value of the variable that makes this statement true:
    "There exists a whole number w such that $2w = w$."

12. Name the axiom that is illustrated by the following:
    $2(3 + 7) = 2 \times 3 + 2 \times 7$

Simplify.

13. $5(a + 3) + 2(a + 1) + 7$

14. $8b^2 + 6b + 3b^2 + 5b + 7 + 2b^2$

15. $5|11 - 4| - |-9|$

16. $-[-26 + (-17)] + (-38) + 19$

17. Give the reason that justifies the following statement:
    $-(-x + y) = -(-x) + (-y)$

Simplify.

18. $-22 - (-14) - 35$

19. $6w - (3w + 4) - (7 - 5w)$

20. $(-8p)(3p)(-7p)$

21. $3(5m - 3n) - 2(8m - 5n)$

22. $\dfrac{-72mn}{18m}$, $m \neq 0$

23. $-\dfrac{5}{7}st \div \left(-\dfrac{10}{7}\right)$

24. Name the property that is illustrated by the following:
  "If $2n - 17 = 25$, then $(2n - 17) + 17 = 25 + 17$."

Solve.

25. $7 - t = 8$

26. $-5|b| = -60$

27. $\dfrac{4x - 8}{4} = 7$

28. $-6c + 4(c + 3) = 16$

29. One number is five greater than a second number. If the lesser number is subtracted from three times the greater number, the difference is 9. Find the numbers.

30. The length of a certain rectangle is 3 cm greater than three times its width. If the perimeter of the rectangle is 46 cm, find its length.

31. $5r - 17 = -2r + 18$

32. $6(d - 5) = 7(d + 6) - d$

33. When twenty is subtracted from a certain number, the result is the number's additive inverse. What is the number?

34. Beverly is three years older than her sister. Six years ago, Beverly was twice as old as her sister was then. How old is Beverly now?

35. Solve for a if $x = -\dfrac{b}{2a}$.

36. The formula $A = \frac{1}{2}bh$ gives the area A of a triangle in terms of the length of the base b and the height of the triangle h. Find the length of the base of a triangle that has an area of 35 m^2 if its height is 8 m.

37. Name the property that is illustrated by the following:
  "If $y - 3 < -4$, then $y < -1$."

Solve each inequality and graph its solution set.

38. $3 - a < -4$

39. $4(b - 1) + b < 5b + 1$

Let $A = \{$the natural numbers less than 2$\}$ and $B = \{$the negative integers greater than $-2\}$. Specify the following.

40. $A \cup B$

41. $A \cap B$

Solve each open sentence and graph its solution set.

42. $-4x < -12$ or $x + 6 < 4$

43. $2 \geq x - 2 > -3$

44. $|m - 3| = 3$

45. $|n + 4| \leq 3$

Solve. If a problem has no solution, explain why.

46. Find the greatest three consecutive odd integers whose sum is less than -18.

47. The measure of an angle is $44°$ more than the measure of its supplement. Find the measure of each angle.

48. Two trains leave Fort Point Station at 2:00 P.M. One is traveling due north at an average speed of 45 km/h, and the other is traveling due south at an average speed of 55 km/h. At what time will the trains be 175 km apart?

49. How many liters of acid must be added to 8 L of a solution that is 25% acid in order to produce a solution that is 50% acid?

50. The measure of one angle of a triangle is $60°$ less than the measure of a second angle. The measure of the third angle is four times the measure of the second angle. Find the measure of each angle.

Test for Chapters 5–8

Plot the given points on a coordinate plane.

1. $A(3, -1)$

2. $B(-2, 0)$

3. $C(-4, -1)$

4. $D(0, 3)$

5. Give the domain D and the range R of the relation $\{(-2, 2), (-1, 1), (0, 0), (1, -1)\}$. Is this relation a function?

6. Given $f: x \longrightarrow 2x - 1$, compute $f(-1) + 2f(2)$.

7. Solve $y > x + 1$ when $x \in \{-1, 0, 1\}$ and $y \in \{-2, -1, 0, 1, 2\}$.

Graph on a coordinate plane.

8. $x - 2y = 4$

9. $y = -2$

10. $3x - y < 1$

11. $x \geq -3$

12. Graph the line that passes through the point $(0, -2)$ and has slope $-\frac{3}{4}$.

Determine an equation of the line that satisfies the given requirements. Use the form $ax + by = c$, where a, b, and c are integers.

13. has slope $\frac{3}{5}$ and passes through the point $(-1, 0)$

14. passes through the points $(-2, 3)$ and $(0, 2)$

15. Solve the system $\begin{array}{l} 3y = 2x + 7 \\ 5y = -3x - 1 \end{array}$ using graphs.

Solve each system by any method.

16. $10 = m - 3n$
$\quad -5 = m + 2n$

17. $6r - 4s = 20$
$\quad 5r + 3s = -15$

18. $7k - 16 = 5j$
$\quad j + 4k = 13$

Solve each problem using two equations in two variables.

19. Rob is three years older than Liz. The sum of their ages is 53. How old is Liz?

20. Flying against the wind, a plane makes a trip of 1600 km in 5 h. Flying with the same wind, the plane makes the return trip in 4 h. Find the wind speed and the speed of the plane.

21. The sum of the digits of a two-digit number is 8. If the order of the digits is reversed, the result is a number that is 54 less than the original number. Find the number.

22. Graph the solution set of the system: $\begin{array}{l} x > -2 \\ y < 2x + 3 \end{array}$

23. Find the minimum value of the expression $5x + 3y$ over the region that has corner points $(0, 8)$, $(2, 4)$, $(4, 2)$, and $(8, 0)$.

24. Solve the system: $\begin{array}{l} 2x + y = 5 \\ 4x - z = 2 \\ y + z = 5 \end{array}$

Simplify.

25. $2a^2b - 2ab^2 - 9ab + 3a^2b$

26. $(6n^2 - 3n + 4) - (4n^2 - 2n)$

27. $(-ab^2c)(-a^2b)^3$

28. $(-2m)(3m^7) - (3m^4)^2$

29. $-2j^2(3j^2k - jk + 2jk^2)$

30. $(z + 1)(z^2 + z + 1)$

31. $(x + 8)(x - 9)$

32. $(y - 5)^2$

33. $(z + 2)^3$

34. The difference of the squares of two consecutive negative odd integers is 72. Find the integers.

35. Factor 432 over the set of prime numbers.

36. Find the GCF and the LCM of $-20a^2b^5$ and $30a^3b$.

Factor completely.

37. $38r^2s^7 - 57r^4s^2t$

38. $4n(n - 1) - 3(1 - n)$

39. $49p^2 - 121q^6$

40. $4j^2 - 24jk + 9k^2$

41. $45 + 4c - c^2$

42. $2d^2 + 16d + 24$

43. $4u^2 + 7u - 15$

44. $2v^5 - 32v$

45. $8t^3 - 8$

46. Solve $3m^2 + m = 2$.

47. The length of a certain rectangle is 2 cm greater than twice its width. The area of the rectangle is 112 cm². Find the length of the rectangle.

48. Express $\dfrac{14m^3n + 6m^2n^2 - 4mn^3}{2m^2n}$ as a sum.

49. Divide $2a^2 + 4a + 3$ by $a + 1$, assuming $a + 1 \neq 0$.

Express in scientific notation.

50. 0.00000000593

51. 34,000,000,000,000

Simplify using only positive exponents.

52. $\dfrac{14x^4y^2z^3}{21x^2yz^4}$

53. $\dfrac{2a - 4b}{8b - 4a}$

54. $\dfrac{n^2 + 6n - 16}{n^2 + 12n + 32}$

55. $\dfrac{(-2)^0 a^{-4} b^{-3}}{(-2)^{-2} a^3 b^{-4}}$

56. $\dfrac{x + 1}{2x - 5} \cdot \dfrac{10 - 4x}{3 + 3x}$

57. $(y - 4) \div \dfrac{y^2 + 3y - 28}{y}$

58. $\dfrac{3x - y}{x - 2y} - \dfrac{2x + y}{x - 2y}$

59. $\dfrac{2}{x + 2} + \dfrac{8}{x^2 - 4}$

60. $\dfrac{\dfrac{4}{u} + \dfrac{2}{v}}{\dfrac{1}{u} + \dfrac{2}{v}}$

Test for Chapters 9–11

Solve.

1. $\dfrac{x + 3}{2} - \dfrac{4}{3} = \dfrac{3x - 7}{3}$

2. $11 - \dfrac{2n}{5} < \dfrac{n}{3}$

3. $\dfrac{7}{t} - \dfrac{3}{4} = \dfrac{11}{2t}$

4. $\dfrac{v}{v - 2} - 7 = \dfrac{2}{v - 2}$

5. $\dfrac{6}{w - 4} = \dfrac{3}{w}$

6. $\dfrac{3 - 2a}{4a - 1} = -\dfrac{2}{3}$

7. What is 7% of 86?

8. What percent of 325 is 195?

9. 51 is 75% of what number?

10. A buyer for a sporting goods store paid $120 each for some sets of golf clubs. The amount of the markup is to be 25% of the selling price. What should be the selling price of each set?

11. Two less than four fifths of a number is five more than one third the number. Find the number.

12. It takes Tim 5 h to load a furniture truck, while Henry can do the same job in 4 h. How long would it take the two of them working together to load the truck?

13. It takes Tina and Angela twice as long to row 8 km up the Muddy River as it takes them to make the return trip. Given that the speed of the current is 1 km/h, how fast can they row in still water?

14. The measures of the three angles of a triangle are in the ratio $2:3:4$. Find the measure of each angle in degrees.

15. If w varies directly as t^3, and if $w = 24$ when $t = 2$, find w when $t = 3$.

16. If x varies inversely as y, and if $x = 6$ when $y = 20$, find y when $x = 15$.

17. If x varies jointly as a and b, and if $x = 36$ when $a = 2$ and $b = 3$, find a when $x = 90$ and $b = 5$.

18. If r varies directly as s and inversely as t^2, and if $r = 2$ when $s = 16$ and $t = 2$, find r when $s = 72$ and $t = 3$.

19. Write $-\frac{5}{6}$, $-\frac{19}{24}$, and $-\frac{11}{12}$ in order from least to greatest.

20. Find the rational number that is one fourth of the way from $-\frac{2}{3}$ to $\frac{2}{5}$.

21. Express $\frac{11}{12}$ as a decimal.

22. Express $0.81\overline{3}$ as a fraction in simplest form.

Simplify.

23. $-\sqrt{(-25)^2}$ **24.** $\sqrt{81m^4n^8}$ **25.** $\sqrt{(x + 3)^2}$ **26.** $\sqrt{340}$

Solve.

27. The length of the hypotenuse of a right triangle is 29 cm and the length of a second side is 21 cm. Find the length of the third side.

28. A city park is in the shape of a rectangle that measures 15 m by 20 m. A concrete walkway crosses the park along a diagonal of the rectangle. Find the length of the walkway.

29. Find the distance between the points $A(5, -2)$ and $B(8, 3)$.

Simplify. Assume that all variables represent positive real numbers.

30. $\sqrt{7} \cdot \sqrt{14}$

31. $\sqrt{2a^2} \cdot \sqrt{6a^5}$

32. $\dfrac{10}{\sqrt{5}}$

33. $\sqrt{\dfrac{7}{3}}$

34. $\sqrt{8} - \sqrt{12} + \sqrt{18}$

35. $\sqrt{5}(\sqrt{10} - \sqrt{15})$

36. $(4 + \sqrt{5})(4 - \sqrt{5})$

37. $(2\sqrt{3} + \sqrt{5})^2$

38. $\dfrac{\sqrt{6}}{2 - \sqrt{2}}$

39. $\sqrt[3]{-64}$

40. $\sqrt[4]{48}$

41. $\dfrac{6}{\sqrt[3]{2}}$

Solve. Simplify irrational solutions. If the equation has no real roots, so state.

42. $a^2 = 24$

43. $6 - b^2 = 42$

44. $8c^2 - 14 = 4$

45. $\sqrt{x} = 16$

46. $2\sqrt{y} + 8 = 12$

47. $z = \sqrt{z} + 2$

Solve by completing the square.

48. $t^2 + 4t = 21$

49. $2s^2 + 3s + 1 = 0$

50. $x = \dfrac{x + 7}{x - 5}$

Solve by using the quadratic formula.

51. $2y^2 - 8y + 3 = 0$

52. $4m^2 + 7m = 1$

53. $\dfrac{1}{z + 2} + \dfrac{1}{z + 6} = 1$

Use the discriminant to determine the number of real roots of each equation.

54. $4c^2 - c + 2 = 0$

55. $5d^2 + 7d = 6$

56. $9k^2 + 4 = 12k$

57. The area of a rectangular floor is 60 m², and the length of the floor is 1 m less than four times its width. Find the dimensions of the floor.

Graph each of the following on a coordinate plane.

58. $f: x \longrightarrow x^2 - 3$

59. $g: x \longrightarrow -x^3 + 1$

60. $y > -x^2$

Test for Chapters 12–13

1. If $\angle A$ is a quadrantal angle and $90° < m \angle A < 270°$, find $m \angle A$.

2. If $\angle A$ is in standard position, $90° \le m \angle A \le 180°$, and $\angle A$ is coterminal with an angle of $-585°$, find $m \angle A$.

Give the sine, cosine, and tangent of an angle A in standard position whose terminal side contains the given point on the unit circle.

3. $(1, 0)$

4. $\left(-\dfrac{3}{5}, -\dfrac{4}{5}\right)$

5. $\left(\dfrac{2\sqrt{5}}{5}, -\dfrac{\sqrt{5}}{5}\right)$

Use the table on page 684 to find the required value.

6. $\sin 33°$

7. $\cos 47°$

8. $\tan 90°$

Use the table on page 684 to find the measure of angle A, $1° \le m \angle A \le 90°$.

9. $\cos A = 0.1392$

10. $\tan A = 20.9271$

11. $\sin A = 0.5695$

Solve the given right triangle ABC using the given information and the table on page 684. Give angle measures to the nearest degree and lengths to the nearest tenth of a unit.

12. $m \angle A = 50°$; $a = 4$

13. $m \angle B = 22°$; $c = 10$

14. $b = 4$; $c = 8$

15. A road rises 12 m vertically over a horizontal distance of 180 m. What is the angle of elevation of the road, to the nearest degree?

Find the norm of v to the nearest tenth and the direction of v to the nearest degree.

16. **v** is in standard position; terminal point is $(1, 7)$.

17. **v** is in standard position; terminal point is $(-2, -3)$.

18. **v** has initial point $(1, 2)$ and terminal point $(4, 6)$.

19. **v** has initial point $(3, -1)$ and terminal point $(-1, 3)$.

20. If vectors **s** and **t** are in standard position with terminal points $(4, -2)$ and $(-3, 5)$ respectively, and **v** = **s** + **t**, find the norm of **v** to the nearest tenth and the direction of **v** to the nearest degree.

21. A ship sails 12 km due north from a harbor, then turns and sails 5 km due east. How far from the harbor is the ship at that point, and what is its bearing with respect to the harbor?

Exercises 22 and 23 refer to the set of data at the right.

22. Draw a dot frequency diagram for the given data.

23. Make a table showing the frequencies and relative frequencies, as fractions and percents, for the given data.

12	11	14	12	13
15	12	14	15	13

Exercises 24 and 25 refer to the set of data at the right.

24. Make a table showing the frequencies and relative frequencies, as fractions and percents, for the given data. Group the data in the intervals 0–3, 3–6, 6–9, and 9–12.

25. Draw a histogram and a frequency polygon for the given data, using the table in Exercise 24.

1	7	4	5	1
8	6	4	4	9
8	1	7	4	8
11	5	7	5	4
2	6	3	11	11

Exercises 26 and 27 refer to the set of data at the right.

26. Make a table showing the frequencies, percents, cumulative frequencies, and cumulative percents for the given data.

27. Draw a cumulative frequency polygon for the given data.

7	9	10	7	7
8	5	10	8	6
10	7	9	8	6
8	6	7	9	8

Exercises 28–33 refer to the set of data at the right.
Find each of the following to the nearest tenth.

28. the median
29. the mode(s)
30. the mean
31. the range
32. the variance
33. the standard deviation

101	101	100	104
102	103	105	100

Suppose that a card is drawn at random from a standard deck of 52 cards. What is the probability that it will be:

34. a red card?
35. an eight?
36. a red ace?
37. not a king?

Suppose that the results of 50 successive random drawings of one marble from a jar containing an unknown number of colored marbles are as follows: 8 red, 18 blue, 14 white, and 10 black. What is the experimental probability that the next marble drawn will be:

38. blue?
39. green?
40. not black or white?

Topical Reviews

Review of Factoring Polynomials and Simplifying Expressions

Factor completely.

1. $6a + 12b$

2. $15c^2 + 10c$

3. $13j^2k^3 - 26j^3k^2$

4. $18m^3 + 42m^2 - 90m$

5. $49p^5 - 63p^3 + 77p^2$

6. $14s^4t^2 - 18s^2t^5 + 18s^3t^3$

7. $121 - 16v^2$

8. $169x^4 - 225y^2$

9. $a^2 + 18a + 81$

10. $25b^2 + 60bc + 36c^2$

11. $d^3 - 64$

12. $8j^6 + 27k^9$

13. $4k^2 - 64$

14. $20m^2 - 60m + 45$

15. $16p^3 - 64p$

16. $7s^2 + 42s + 63$

17. $t^2 + 14t + 49 - v^2$

18. $w^2 + 7w + xw + 7x$

19. $a^2 + 9a + 20$

20. $b^2 + 4b - 21$

21. $c^2 + 2c - 24$

22. $2d^3 - 12d^2 - 54d$

23. $2j^3 - 14j^2k - 60jk^2$

24. $3a^2 + 5a + 2$

25. $5m^2 + 16m + 3$

26. $6p^2 + 11p - 10$

27. $20r^2 - 19r - 28$

28. $12s^2 - 27t^2$

29. $18x^2 - 45x + 25$

30. $28y^2 + 19y - 20$

31. $18b^5 + 39b^3 + 18b$

32. $12c^5 - 46c^3 + 40c$

33. $24d^3 - 34d^2 - 80d$

34. $4j^4 - k^2$

35. $4m^4 + 3m^2 - 1$

36. $p^4 - 17p^2 + 16$

Simplify. Use only positive exponents. Assume that all variables represent positive real numbers.

37. $\{3[5 + 4(9)] \div 6\}4$

38. $3\left\{\dfrac{7[12 - 2(3^2)]}{5 + 2(6)}\right\} + 3$

39. $4|-9| - 3|-6|$

40. $(6a + 9) + (7b + 14)$

41. $10c + 6(2c + 3) - (4c - 1)$

42. $d^2 - 2d - 3(2d^2 - d)$

43. $\dfrac{j^6}{j^4}$

44. $\dfrac{k^7}{k^8}$

45. $\dfrac{20m^2n^4}{12m^5n^3}$

46. $\dfrac{r - s}{s - r}$

47. $\dfrac{16r^3 - 12r}{8r}$

48. $\dfrac{14s^5 + 35s^4 - 42s^2}{-7s^2}$

49. $\dfrac{9t^2v^4 + 12v^4w^3 - 3tw^2}{6tvw}$

50. $\dfrac{9x^2 - 3x}{5 - 15x}$

51. $\dfrac{(y + z)^2(y - 2z)}{2(2z + y)(y + z)}$

52. $\dfrac{8a^2 - 2a - 3}{8a^2 + 6a - 9}$

53. $\dfrac{b^2 - 2bc - 8c^2}{2b^2 - 7bc - 4c^2}$

54. $\dfrac{12k^2 - 21k + 24}{32k^2 - 56k + 64}$

Simplify. Use only positive exponents. Assume that all variables represent positive real numbers.

55. $\dfrac{5j^{-1}}{35j^3}$

56. $\left(\dfrac{2m^2 + 3m^2n^0}{15m^2}\right)^{-2}$

57. $\dfrac{27p^3q^3}{2r^2} \cdot \dfrac{8pr^2}{9q^3}$

58. $\dfrac{6c - 1}{4c^2 - c + 7} - \dfrac{4c + 5}{4c^2 - c + 7}$

59. $\dfrac{2d}{d + 1} + \dfrac{d + 2}{2d - 1}$

60. $\dfrac{\dfrac{j}{4k} - \dfrac{k}{2j}}{\dfrac{1}{2k} + \dfrac{1}{4j}}$

61. $\sqrt{480}$

62. $\sqrt{270} + \sqrt{333}$

63. $\sqrt{6p} \cdot \sqrt{8p}$

64. $\sqrt{\dfrac{36m^2}{25n^2}}$

65. $\sqrt{\dfrac{125}{98r}}$

66. $\sqrt{25s} + 2\sqrt{s} - \sqrt{64s}$

67. $(2\sqrt{6} + \sqrt{5})^2$

68. $\dfrac{4}{\sqrt{5} + 2}$

69. $\dfrac{\sqrt{6}}{\sqrt{48}}$

70. $\sqrt[5]{-243}$

71. $\sqrt[3]{81}$

72. $\sqrt[3]{\frac{1}{4}}$

Review of Solving Equations, Inequalities, and Systems

Solve.

1. $-6 + x = 13$

2. $22 = y - 14$

3. $-18z = 12$

4. $\frac{a}{-3} = -15$

5. $23 - b = 31$

6. $-4 = -\frac{1}{7}|c|$

7. $p - 9 > -17$

8. $4 < \frac{g}{3}$

9. $\frac{r + 2}{-4} > -2$

10. $6t - 5 < 4 - t$

11. $2s - 5(s + 3) - 4 < 0$

12. $9(w + 4) > -(12 - 4w)$

13. $-8 \le h + 7 < 3$

14. $14 \ge 2j \ge -6$

15. $6 \ge -3k + 4 \ge -5$

16. $|x - 7| = 9$

17. $|13 - y| \le 5$

18. $|4 - 5z| < 12$

19. $4 + |5 - a| \ge 7$

20. $5 + |b - 3| \le 10$

21. $2|c - 5| - 3 \le 8$

22. $7 - |3p + 4| \ge -5$

23. $6 - |5q - 2| < -5$

24. $3|2r - 5| + 2 > 8$

Solve each system of equations.

25. $2x + y = 6$
    $3x + y = 11$

26. $7s + 4t = -2$
    $4s - 4t = 7$

27. $4n = 5p - 20$
    $4n = 10p - 16$

28. $7a - 4b = -4$
    $3b + 7a = 2$

29. $6j = -24 - 5k$
    $5k = 4j - 34$

30. $4c + 3d = -2$
    $6c - 2d = 4$

31. $s + 3t = 6$
    $2s - 4t = -4$

32. $n = \frac{3}{4}m$
    $3m + 5n = 6$

33. $\frac{a}{2} + \frac{b}{4} = 8$
    $2a + 3b = 10$

34. $x + y + z = 1$
    $x - y + z = -1$
    $2x + y - z = 2$

35. $2x + y + z = 3$
    $2x + y - z = 2$
    $x - 2y = 0$

36. $3x + y - 2z = 5$
    $4x - y + 3z = 4$
    $x + 2y - 4z = -1$

Solve.

37. $\frac{3s}{4} - \frac{2s}{3} = -2$

38. $2k - \frac{k}{6} = \frac{4}{3}$

39. $\frac{3j + 6}{5} > -4$

40. $\frac{m - 3}{5} - \frac{3m + 1}{4} < \frac{3}{10}$

41. $\frac{4}{3b} + \frac{2}{b} = \frac{6}{b} - \frac{1}{9}$

42. $2 + \frac{3}{c + 3} = \frac{c + 5}{c + 3}$

43. $d^2 - 11 = 0$

44. $3t^2 + 5 = 26$

45. $4s^2 - 17 = 63$

46. $\sqrt{r} + 3 = 9$

47. $\sqrt{x - 9} = 7$

48. $\sqrt{\frac{3y - 5}{3}} = 3$

49. $z^2 + 5z + 6 = 0$

50. $a^2 + \frac{3}{2}a - 20 = 0$

51. $12b^2 + 3b - 54 = 0$

52. $c^2 + 6c = -5$

53. $3d^2 = 10d - 7$

54. $6e^2 - 4 = 5e$

55. $2s^2 + 3s - 6 = 0$

56. $3m^2 - 8m + 1 = 0$

57. $k^2 - 2k = 7$

58. $6p^2 = 7p + 9$

59. $x^2 + \frac{2}{3}x - \frac{4}{3} = 0$

60. $y^2 - \frac{3}{4}y = \frac{1}{4}$

Review of Solving Word Problems

Solve. If a problem has no solution, explain why.

1. Janice rode her bicycle from Goldmont to Broomville in 2.1 h. On the return trip, riding uphill, it took her 3.5 h. If her average speed on her return trip was 14 km/h less than her average speed on the way out, find Janice's average speed in each direction.

2. A merchant has two types of tea, one worth $2.70 per kilogram and the other worth $3.00 per kilogram. How many kilograms of each type should the merchant use in order to produce 30 kg of a blend that is worth $2.95 per kilogram?

3. One outlet of a grain elevator will empty the elevator in 3 h. A second outlet will empty the same elevator in 4.5 h. How long will it take both outlets working together to empty the elevator?

4. Find two positive integers that differ by 4 and whose product is 221.

5. A cable from the top of a radio antenna is attached to the ground at a point 35 m from the base of the antenna. If the cable is 145 m long, how high is the pole, to the nearest hundredth of a meter?

6. Ralph invested $6000, part at an annual interest rate of 9% and the rest at an annual interest rate of 12%. If his total income on the investment for one year was $615, how much did he invest at each rate?

7. A ticket to the school play costs $5 for adults and $2 for students. On Saturday night, the total amount collected in ticket sales was $1990. How many adult tickets were sold that night?

8. A rectangular swimming pool is 12 m wide and 14 m long. Surrounding the pool is a concrete walkway of uniform width. The combined area of the pool and walkway is 360 m². Find the width of the walk.

9. The frequency of a microwave is inversely proportional to its length. If a microwave of length 3 m has a frequency of 100 MHz, find the frequency of a microwave whose length is 0.2 m.

10. The vertices of a triangle are $A(-6, 4)$, $B(3, 7)$, and $C(7, -5)$. Determine the slope of each side of the triangle.

11. With a tail wind, a plane flew 490 km in 50 min. With no change in the wind, the return trip took 1 h. Find the wind speed and the air speed of the plane.

12. The area of a wall is 33 m², and the length of the wall is 1 m longer than four times its height. Find the dimensions of the wall.

13. The hundreds' digit of a three-digit number is 3 less than the tens' digit. The tens' digit is one more than twice the units' digit. If the order of digits is reversed, the number obtained is 198 less than the original number. Find the original number.

14. The measure of one angle of a triangle is 30° less than the measure of a second angle. The measure of the third angle is 20° less than the sum of the measures of the other two. Find the measure of each angle of the triangle.

15. The rotational speed of a gear wheel is inversely proportional to the number of teeth on the wheel. How fast is a gear wheel with 20 teeth revolving if it is meshed with a gear wheel with 45 teeth that is revolving at 120 rpm?

16. Find the least three consecutive odd integers whose sum is greater than -32.

17. The formula $V = \frac{1}{3}\pi r^2 h$ gives the volume V of a cone in terms of the radius of the base r and the height h. Find the height of a cone with base of radius 21 cm if its volume is 1848 cm³. Use $\pi = \frac{22}{7}$.

18. Mary's age forty years from now will be the same as three times her age two years ago. How old is she now?

19. John travels at an average speed of 30 km/h from his home to a point that is 120 km away. What must be his average speed on the return trip if he wants his average speed for the entire trip to be 40 km/h?

20. The difference between a number and three times its reciprocal is $\frac{1}{2}$. Find the number.

Review of Graphing

Exercises 1–12 refer to the coordinate plane at the right.

Name the coordinates of each point.

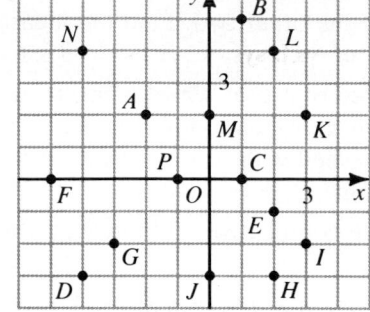

1. A 2. B

3. C 4. D

5. E 6. F

Name the point that has the given coordinates.

7. $(-3, -2)$ 8. $(-1, 0)$

9. $(2, 4)$ 10. $(-4, 4)$

11. $(0, -3)$ 12. $(2, -3)$

Graph each open sentence on a coordinate plane.

13. $y = -3$ 14. $2x - 3y = 9$

15. $1 - y = \frac{x}{2}$ 16. $x < 3$

17. $2x - y < 3$ 18. $2x + 3y \le 6$

Solve each open sentence and graph its solution set on a number line.

19. $8 - 2x > 2x - 4$ 20. $7(t - 2) \ge 4 - 3t$

21. $-3 < 2a - 1 \le 5$ 22. $6 < 2t$ or $-2t > 4$

23. $|3j - 9| = 6$ 24. $5 + |2 - z| \ge 8$

Determine the slope-intercept form of the equation of the line that has the given slope m if any, and that passes through the given point.

25. $m = 2$; $(1, 2)$ 26. $m = -3$; $(-2, 3)$

27. $m = \frac{1}{4}$; $(-4, 0)$ 28. no slope; $(0, 1)$

Determine an equation of the line that passes through the given points. The equation should be in the form $ax + by = c$, where a, b, and c are integers.

29. $(2, 4), (-3, -6)$ 30. $(0, -2), (-6, -4)$

31. $(4, 0), (0, 3)$ 32. $(-3, 2), (-3, 4)$

Solve each system of equations using graphs.

33. $2x - y = 5$
$x - y = 3$

34. $x + y = -4$
$y = 3x$

35. $4x - 5y = 10$
$4x = 20 + 5y$

Graph the solution set of each system of inequalities.

36. $x < 3$
$y > -2$

37. $y \geq x - 1$
$y \leq -x + 2$

38. $y > -2$
$x > -3$
$y < -x$

Graph each function on a coordinate plane.

39. $f\colon x \longrightarrow -x^2 - 1$

40. $g\colon x \longrightarrow x^2 + 2x + 4$

Answers

Alternate Chapter Tests

Chapter 1

1. $\frac{-2}{3}$ 2. $^-5$, $^-3.75$, $^-3\frac{1}{2}$, $^-2$, 0, 7.75 3. F   4. F

5. T   6. T   7. $1\frac{1}{2}$ 8. $\frac{1}{18}$ 9. $\frac{19}{24}$ 10. 184   11. $\frac{2}{7}$

12. 15   13. 49   14. $\{0\}$ 15. $\{0, 1, 2, 3, 4, 5, 6\}$
16. $\{7, 8, 9, \ldots\}$ 17. $60m$ 18. $10d + 25q$
19. $(a - b)^2 > \dfrac{a + b}{ab}$ 20. $p - 7 = \frac{1}{5}(p + 21)$

★1. $\frac{7}{12}$ ★7. $\frac{3}{2}$ ★15. $\{0, 1\}$

Chapter 2

1. $\{3\}$ 2. Commutative axiom for addition   3. 80
4. $120xy^2$ 5. $5m^2 + 9m + 7$ 6. $17a + 14$ 7. 28
8. 14   9. -17 10. -23 11. $\{-5\}$ 12. 24   13. 126
14. Axiom of additive inverses   15. -33 16. $9y$
17. $-120m^3$ 18. $-70a + 35b$ 19. 2   20. $x + 9y$
21. $-3m$ 22. $-70x^2y$

★6. $\frac{5}{4}t + 3$ ★11. $\left\{\frac{11}{12}\right\}$ ★15. $8\frac{1}{6}$

Chapter 3

1. Multiplication property of equality   2. $\{-4\}$
3. $\{-3\}$ 4. $\{22\}$ 5. $\{-9\}$ 6. $\{70\}$ 7. $\varnothing$ 8. $\{-2\}$
9. $\{2\}$ 10. $\{10\}$ 11. 8   12. 52 years old   13. $\{2\}$

14. $\{90\}$ 15. 2   16. 22 years old
17. $y = \dfrac{5B}{a} - c$, $a \neq 0$ 18. 7 cm

★4. $\{-2, 4\}$ ★6. $\{-11\}$ ★10. $\{4\}$ ★15. 4

★18. a. $w = \dfrac{P - 2l}{2}$ b. 35 m

Chapter 4

1. Transitive property of order   2. $r < -3$ 3. $f > 5$

4. $\{-2, -1, 0, 1, 2, 3, 4, 5, 6\}$ 5. $\{1, 2, 3, 4, 5, 6\}$
6. $-1 \leq y < 2$ 7. $t < -\frac{1}{2}$ or $t > 5$ 8. $-\frac{1}{2} < s < 3$

9. $p < -2$ or $p > \frac{2}{3}$ 10. $\{-21, -19, -17, -15\}$

11. $52°$ 12. 36 km   13. 18 dimes, 7 quarters
14. The only solution of an equation that represents the relationships in the problem is a negative integer. The number of adults or children entering the zoo cannot be negative.
★10. $-35, -28, -21$; $-28, -21, -14$; and $-21, -14, -7$
★12. 7:55 A.M.

Chapter 5

1. $(2, 0)$ 2. $(-3, 1)$ 3. $(0, -1)$ 4. $(3, -2)$
5–8.
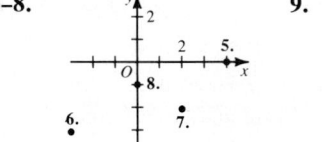
9.
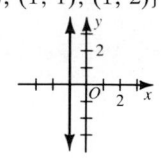

10. Yes   11.
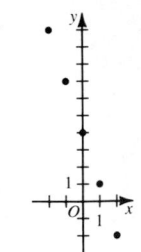

12. 6   13. $\{(0, -2), (1, -1)\}$
14. $\{(0, 1), (0, 2), (0, -1), (1, 0), (1, 1), (1, 2)\}$
15.   16.

17.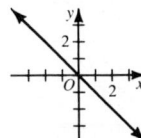

18.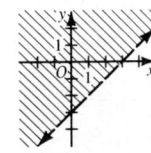

19.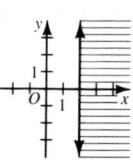

20.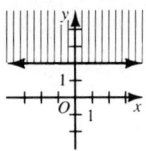

21.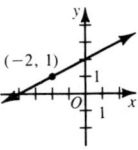

22. $-\dfrac{6}{7}$

23. $-2x + 3y = -3$

24. $y = -4$

25. $x + 3y = -5$

26. $x + 4y = -5$

★**11.** See graph at right.

★**22.** slope: $-\dfrac{1}{2}$; y-intercept: $\dfrac{d}{2c}$

★**23.** $a = \dfrac{1}{2}$ ★**26.** $y = -\dfrac{y_1}{x_1}x + y_1$

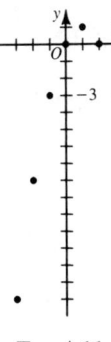

Ex. ★**11**

Chapter 6

1. ∅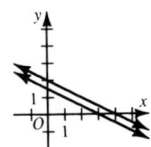

2. consistent   **3.** $(3, 3)$ **4.** $(9, 6)$ **5.** $(3, -1)$
6. $(-2, -2)$ **7.** $(4, 0)$ **8.** $(1, 1)$ **9.** Cindy. 15 yr old;
Toni: 11 yr old   **10.** swimming rate: 1.5 km/h; rate of
current: 0.5 km/h   **11.** 748

12. 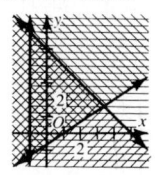   **13.**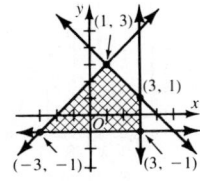

14. 13 at $(3, -1)$ **15.** $\{(1, -2, 2)\}$

★**2.** $\{(2n, n)\}$ ★**10.** $r = (9t - 3)$ km/h   ★**11.** 148

Chapter 7

1. $-4ab + 7a - 2$ **2.** $7f^2 - 2f - 12$ **3.** $\{-2\}$
4. $-72m^{13}n^7$ **5.** 0   **6.** $-9z^4 + 3z^3 + 3z^2$ **7.** $5g^2 -$
$11g - 36$ **8.** $3e^3 + 2e^2 - 2e - 1$ **9.** $-5a^3 - 2a^2 -$
$15a$ **10.** $\{1\}$ **11.** $25i^2 + 30ij + 9j^2$ **12.** $32r^4 - 200s^2$
13. 10 m by 18 m   **14.** $45x^2y^3z$ **15.** $5a^3b^3(5a^2b + 3a -$
$15)$ **16.** $(t - 9)(t - 4)$ **17.** $(4p + 1)(16p^2 - 4p + 1)$
18. $(x - 1)(6x - 25)$ **19.** $2(q^2 + 7z)(q^2 - 7z)$
20. $3(d + 6)(d + 4)$ **21.** $r(3r + 5)(3r - 7)$
22. $3x(2x - 3)(5x + 2)$ **23.** $\left\{0, -3, \dfrac{2}{3}\right\}$ **24.** $\left\{\dfrac{3}{5}, \dfrac{7}{4}\right\}$

25. 12, 14
★**4.** $4e^{4n+2}f^2$ ★**7.** $x^{12n} - y^{8n}$
★**15.** $2a^2b^2(3a + b)(a - 2b)$
★**22.** $(2x^4 - 1)(x^2 + 4)(x + 2)(x - 2)$
★**24.** $3x^2 + 5x - 12 = 0$

Chapter 8

1. $-\dfrac{1}{8m}$ **2.** $\dfrac{50a^5c}{b^2}$ **3.** $-p^9$ **4.** $\dfrac{1}{b} + 4b$ **5.** $2cd - 4 -$

$5d^2$ **6.** $3x - 4$ **7.** $t^2 - 5t + 3 + \dfrac{2}{t - 2}$ **8.** $\dfrac{3st}{1 + 2s}$

9. $\dfrac{k + 2}{k - 5}$ **10.** $\dfrac{1}{my^3}$ **11.** $\dfrac{b^4c}{a^5}$ **12.** 1.8×10^3 **13.** $\dfrac{2w}{5}$

14. $\dfrac{2}{h^2}$ **15.** $\dfrac{8b + 3}{b(8b - 3)}$ **16.** $\dfrac{d - 1}{d^3}$ **17.** $\dfrac{a - 2}{a - 4}$ **18.** 0

19. $\dfrac{3v + 4}{4(v + 4)(v - 4)}$ **20.** $\dfrac{f + 2g}{f^2 - g^2}$ **21.** $\dfrac{3s}{4}$ **22.** $\dfrac{j + 2}{3}$

★**2.** $-\dfrac{r^5}{2s^2}$ ★**8.** $\dfrac{m}{2n}$

Chapter 9

1. $\{z: z \geq 4\}$ **2.** 15   **3.** 30%   **4.** 64 g   **5.** $\left\{-\dfrac{3}{2}\right\}$

6. 7, 3   **7.** 12   **8.** 15 and 12 or $\dfrac{4}{3}$ and $-\dfrac{5}{3}$ **9.** $\dfrac{3}{7}$ h

10. 12 km/h   **11.** 14   **12.** 16 to -17 **13.** 56 lb
14. 675   **15.** 64 ft   **16.** 60   **17.** 72   **18.** 33
★**1.** $\{x: x \leq 2\}$ ★**4.** \$1800   ★**5.** $\left\{\dfrac{1}{2}, 4\right\}$

★**9.** first pipe: $4\dfrac{1}{8}$ h; second pipe: $3\dfrac{1}{8}$ h

Chapter 10

1. $\frac{28}{112}, \frac{48}{90}, \frac{56}{80}$ 2. $2\frac{8}{15}$ 3. 3.4165 4. $\frac{27}{110}$ 5. $\frac{13}{27}$

6. $-\frac{7}{5}$ 7. 25 8. $|5 - x|$ 9. $20\sqrt{2}$ 10. $8\sqrt{6}$

11. $\frac{4}{9}\sqrt{2}$ 12. $\{\sqrt{7}, -\sqrt{7}\}$ 13. $\sqrt{35}$ cm

14. 14.42 km 15. $3\sqrt{10}$ 16. No, not a right triangle

17. $6\sqrt{30}$ 18. $\frac{4\sqrt{3}}{3}$ 19. $6m\sqrt[4]{2m}$ 20. $6\sqrt{5} - 7\sqrt{3}$

21. $32 - 6\sqrt{15}$ 22. $-15\sqrt{2} + 10\sqrt{5}$ 23. $2\sqrt[5]{2}$

24. $5\sqrt[4]{2}$ 25. $\frac{\sqrt[3]{18}}{2}$ 26. $\{-22\}$ 27. $\left\{\frac{9}{25}\right\}$ 28. $\{9\}$

★10. $\sqrt{25a^2 + 16b^2}$ ★15. $\frac{6a^2\sqrt{b}}{b}$ ★16. $y\sqrt{2y} - 2y\sqrt{2}$

★18. $27\sqrt{3} - 34\sqrt{2}$

Chapter 11

1. $\{9, -3\}$ 2. $\left\{\frac{1 + \sqrt{13}}{6}, \frac{1 - \sqrt{13}}{6}\right\}$ 3. $\left\{\frac{-3 + \sqrt{41}}{4},\right.$

$\left.\frac{-3 - \sqrt{41}}{4}\right\}$ 4. $\left\{\frac{3 + \sqrt{13}}{4}, \frac{3 - \sqrt{13}}{4}\right\}$ 5. two

6. 14.36 km 7. 4 or $\frac{1}{2}$

8. 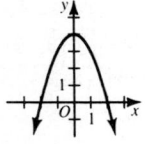 9.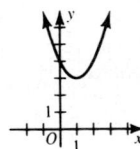

10. The zeros are 0, −5; the maximum value is $\frac{25}{4}$.

11. The zero is $\frac{1}{2}$; the minimum value is 0.

12. upward; minimum

13. The zero is approximately $1\frac{1}{2}$. 14. The zeros are 0, 2, −2.

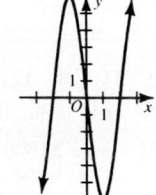 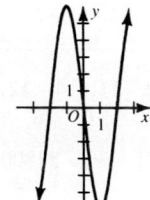

15. 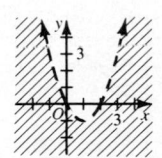 16.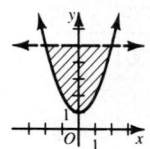

★1. $\{3c + \sqrt{d + 9c^2}, 3c - \sqrt{d + 9c^2}\}$

★4. $\left\{\frac{-2b + \sqrt{4b^2 + a^2}}{a^2}, \frac{-2b - \sqrt{4b^2 + a^2}}{a^2}\right\}$

★14.

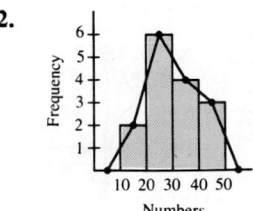

Chapter 12

1. $-270°$ 2. $150°$ 3. 1 4. $-\frac{3}{4}$ 5. $26°$ 6. $53°$

7. 531 m 8. 96 9. 4 10. 20.5 N

★4. $\sin A = \frac{y^n\sqrt{x^{2n} + y^{2n}}}{x^{2n} + y^{2n}}$; $\cos A = \frac{x^n\sqrt{x^{2n} + y^{2n}}}{x^{2n} + y^{2n}}$;

$\tan A = \frac{y^n}{x^n}$ ★9. $\|u + v\| = 26, 0°$

Chapter 13

1.

Number	Freq.	Rel. Freq.	%
15	2	$\frac{2}{10}$	20
16	1	$\frac{1}{10}$	10
17	1	$\frac{1}{10}$	10
19	4	$\frac{4}{10}$	40
20	2	$\frac{2}{10}$	20
Total:	10	$\frac{10}{10}$	100

2.

3.

Interval	Freq.	%	Cum. Freq.	Cum. %
10–20	2	$13\frac{1}{3}$	2	$13\frac{1}{3}$
20–30	6	40	8	$53\frac{1}{3}$
30–40	4	$26\frac{2}{3}$	12	80
40–50	3	20	15	100
Total:	15	100		

4. median: 6; mode: 6; mean: 6.2   **5.** 16.8   **6.** 4.1

7. $\frac{1}{13}$ **8.** $\frac{6}{18}$ or $\frac{1}{3}$ **9.** $\frac{5}{9}$ **10.** $\frac{5}{1000}$ or $\frac{1}{200}$

★**7.** $\frac{6}{13}$

Cumulative Tests

Test for Chapters 1–4

1. −7   **2.** False   **3.** False   **4.** 28   **5.** $\frac{7}{29}$

6. 105   **7.** 34   **8.** {0, 1, 2, . . . , 10}

9. $0.01p + 0.25q + x$ **10.** $\frac{n + 2}{n^2} < (n + 3)^2$

11. $w = 0$ **12.** Distributive axiom   **13.** $7a + 24$
14. $13b^2 + 11b + 7$ **15.** 26   **16.** 24
17. Property of the opposite of a sum   **18.** −43
19. $8w - 11$ **20.** $168p^3$ **21.** $-m + n$ **22.** $-4n$
23. $\frac{st}{2}$ **24.** Addition property of equality   **25.** {−1}
26. {12, −12}   **27.** {9}   **28.** {−2}   **29.** −3 and 2
30. 18 cm   **31.** {5}   **32.** Ø   **33.** 10

34. 12 years   **35.** $a = -\frac{b}{2x}, x \neq 0$ **36.** 8.75 m

37. Addition property of order
38. {$a: a > 7$}   (number line: 0 2 4 6 8, open circle at 7)

39. $\mathcal{R}$ (number line: −4 −2 0 2 4)

40. {−1, 1}
41. Ø
42. {$x: x < -2$ or $x > 3$}   (number line: −4 −2 0 2 4)

43. {$x: -1 < x \leq 4$}   (number line: −4 −2 0 2 4)

44. {0, 6}   (number line: −2 0 2 4 6)

45. {$n: -7 \leq n \leq -1$}   (number line: −8 −6 −4 −2 0)

46. −9, −7, and −5   **47.** 68° and 112°
48. 3:45 P.M.   **49.** 4 L

50. No solution; the only measures that meet the conditions stated in the problem are −20°, 40°, and 160°, but these cannot be the measures of the angles of a triangle.

Test for Chapters 5–8

1-4.

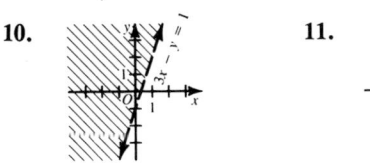

5. $D = \{-2, -1, 0, 1\}$; $R = \{-1, 0, 1, 2\}$; Yes
6. 3   **7.** {(−1, 1), (−1, 2), (0, 2)}
8.   **9.**

10.   **11.**

12.

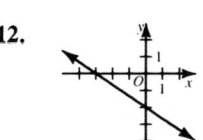

13. $-3x + 5y = 3$ **14.** $x + 2y = 4$

T41

15. $\{(-2, 1)\}$

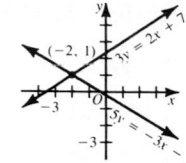

16. $\{(1, -3)\}$ **17.** $\{(0, -5)\}$ **18.** $\{(1, 3)\}$
19. 25 years **20.** wind speed: 40 km/h; speed of plane: 360 km/h **21.** 71

22.

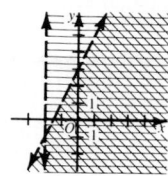

23. 22 **24.** $\{(1, 3, 2)\}$ **25.** $5a^2b - 2ab^2 - 9ab$
26. $2n^2 - n + 4$ **27.** a^7b^5c **28.** $-15m^8$
29. $-6j^4k + 2j^3k - 4j^3k^2$ **30.** $z^3 + 2z^2 + 2z + 1$
31. $x^2 - x - 72$ **32.** $y^2 - 10y + 25$
33. $z^3 + 6z^2 + 12z + 8$ **34.** -19 and -17
35. $2^4 \cdot 3^3$ **36.** $10a^2b$; $60a^3b^5$
37. $19r^2s^2(2s^5 - 3r^2t)$ **38.** $(n - 1)(4n + 3)$
39. $(7p + 11q^3)(7p - 11q^3)$ **40.** $(2j - 3k)^2$
41. $-(c - 9)(c + 5)$ **42.** $2(d + 6)(d + 2)$
43. $(4u - 5)(u + 3)$ **44.** $2v(v^2 + 4)(v + 2)(v - 2)$
45. $8(t - 1)(t^2 + t + 1)$ **46.** $\left\{\dfrac{2}{3}, -1\right\}$ **47.** 16 cm

48. $7m + 3n - \dfrac{2n^2}{m}$ **49.** $2a + 2 + \dfrac{1}{a + 1}$
50. 5.93×10^{-9} **51.** 3.4×10^{13}

52. $\dfrac{2x^2y}{3z}$ **53.** $-\dfrac{1}{2}$ **54.** $\dfrac{n - 2}{n + 4}$ **55.** $\dfrac{4b}{a^7}$

56. $-\dfrac{2}{3}$ **57.** $\dfrac{y}{y + 7}$ **58.** 1 **59.** $\dfrac{2}{x - 2}$

60. $\dfrac{2u + 4v}{2u + v}$

Test for Chapters 9–11

1. $\{5\}$ **2.** $\{n: n > 15\}$ **3.** $\{2\}$ **4.** $\emptyset$
5. $\{-4\}$ **6.** $\left\{-\dfrac{7}{2}\right\}$ **7.** 6.02 **8.** 60%
9. 68 **10.** \$160 **11.** 15 **12.** $2\dfrac{2}{9}$ h
13. 3 km/h **14.** 40°, 60°, and 80° **15.** 81
16. 8 **17.** 3 **18.** 4 **19.** $-\dfrac{11}{12}, -\dfrac{5}{6}, -\dfrac{19}{24}$

20. $-\dfrac{2}{5}$ **21.** $0.91\overline{6}$ **22.** $\dfrac{61}{75}$ **23.** -25
24. $9m^2n^4$ **25.** $|x + 3|$ **26.** $2\sqrt{85}$ **27.** 20 cm
28. 25 m **29.** $\sqrt{34}$ **30.** $7\sqrt{2}$ **31.** $2a^3\sqrt{3a}$
32. $2\sqrt{5}$ **33.** $\dfrac{\sqrt{21}}{3}$ **34.** $5\sqrt{2} - 2\sqrt{3}$
35. $5\sqrt{2} - 5\sqrt{3}$ **36.** 11 **37.** $17 + 4\sqrt{15}$
38. $\sqrt{6} + \sqrt{3}$ **39.** -4 **40.** $2\sqrt[4]{3}$ **41.** $3\sqrt[3]{4}$
42. $\{2\sqrt{6}, -2\sqrt{6}\}$ **43.** $\emptyset$ **44.** $\left\{\dfrac{3}{2}, -\dfrac{3}{2}\right\}$
45. $\{256\}$ **46.** $\{4\}$ **47.** $\{4\}$ **48.** $\{3, -7\}$
49. $\left\{-\dfrac{1}{2}, -1\right\}$ **50.** $\{-1, 7\}$
51. $\left\{\dfrac{4 + \sqrt{10}}{2}, \dfrac{4 - \sqrt{10}}{2}\right\}$
52. $\left\{\dfrac{-7 + \sqrt{65}}{8}, \dfrac{-7 - \sqrt{65}}{8}\right\}$
53. $\{-3 + \sqrt{5}, -3 - \sqrt{5}\}$ **54.** 0 **55.** 2
56. 1 **57.** width: 4 m; length: 15 m
58.

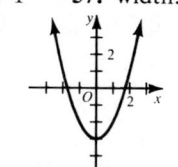

59.

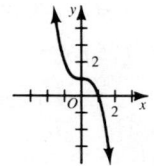

60.

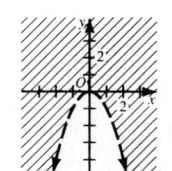

Test for Chapters 12–13

1. 180° **2.** 135° **3.** $\sin A = 0$; $\cos A = 1$; $\tan A = 0$ **4.** $\sin A = -\dfrac{4}{5}$; $\cos A = -\dfrac{3}{5}$; $\tan A = \dfrac{4}{3}$
5. $\sin A = -\dfrac{\sqrt{5}}{5}$; $\cos A = \dfrac{2\sqrt{5}}{5}$; $\tan A = -\dfrac{1}{2}$
6. 0.5446 **7.** 0.6820 **8.** undefined
9. 82° **10.** 87° **11.** 35°
12. $m \angle B = 40°$; $b \approx 3.4$; $c \approx 5.2$
13. $m \angle A = 68°$; $a \approx 3.7$; $b \approx 9.3$
14. $m \angle A = 60°$; $m \angle B = 30°$; $a \approx 6.9$ **15.** 4°
16. $\|\mathbf{v}\| \approx 7.1$; 82° **17.** $\|\mathbf{v}\| \approx 3.6$; 236°
18. $\|\mathbf{v}\| = 5$; 53° **19.** $\|\mathbf{v}\| \approx 5.7$; 135°
20. $\|\mathbf{v}\| \approx 3.2$; 72° **21.** 13 km; 23°

22.

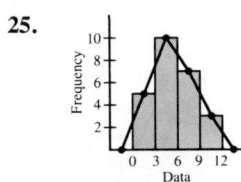

23.

No.	Freq.	Rel. Freq. Frac.	%
11	1	$\frac{1}{10}$	10
12	3	$\frac{3}{10}$	30
13	2	$\frac{2}{10}$	20
14	2	$\frac{2}{10}$	20
15	2	$\frac{2}{10}$	20
Total:	10	1	100

24.

Interval	Freq.	Rel. Freq. Frac.	%
0– 3	5	$\frac{5}{25}$	20
3– 6	10	$\frac{10}{25}$	40
6– 9	7	$\frac{7}{25}$	28
9–12	3	$\frac{3}{25}$	12
Total:	25	1	100

25.

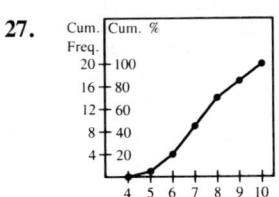

26.

No.	Freq.	%	Cum. Freq.	Cum. %
5	1	5	1	5
6	3	15	4	20
7	5	25	9	45
8	5	25	14	70
9	3	15	17	85
10	3	15	20	100
Total:	20	100		

27.

28. 101.5   **29.** 100, 101   **30.** 102   **31.** 5

32. 3   **33.** 1.7   **34.** $\frac{1}{2}$ **35.** $\frac{1}{13}$ **36.** $\frac{1}{26}$

37. $\frac{12}{13}$ **38.** $\frac{18}{50} = 0.36$ **39.** 0   **40.** $\frac{26}{50} = 0.52$

Topical Reviews

Review of Factoring Polynomials and Simplifying Expressions

1. $6(a + 2b)$
2. $5c(3c + 2)$
3. $13j^2k^2(k - 2j)$
4. $6m(3m^2 + 7m - 15)$
5. $7p^2(7p^3 - 9p + 11)$
6. $2s^2t^2(7s^2 - 9t^3 + 4st)$
7. $(11 + 4v)(11 - 4v)$
8. $(13x^2 + 15y)(13x^2 - 15y)$
9. $(a + 9)^2$
10. $(5b + 6c)^2$
11. $(d - 4)(d^2 + 4d + 16)$
12. $(2j^2 + 3k^3)(4j^4 - 6j^2k^3 + 9k^6)$
13. $4(k + 4)(k - 4)$
14. $5(2m - 3)^2$

15. $4(2p + 4)(2p - 4)$
16. $7(s + 3)^2$
17. $(t + v + 7)(t - v + 7)$
18. $(w + x)(w + 7)$
19. $(a + 4)(a + 5)$
20. $(b + 7)(b - 3)$
21. $(c + 6)(c - 4)$
22. $2d(d + 3)(d - 9)$
23. $2j(j - 10k)(j + 3k)$
24. $(3a + 2)(a + 1)$
25. $(5m + 1)(m + 3)$
26. $(3p - 2)(2p + 5)$
27. $(4r - 7)(5r + 4)$
28. $3(2s + 3t)(2s - 3t)$
29. $(6x - 5)(3x - 5)$
30. $(7y - 4)(4y + 5)$
31. $3b(2b^2 + 3)(3b^2 + 2)$
32. $2c(2c^2 - 5)(3c^2 - 4)$
33. $2d(4d + 5)(3d - 8)$
34. $(2j^2 + k)(2j^2 - k)$

35. $(m^2 + 1)(2m + 1)(2m - 1)$
36. $(p + 4)(p - 4)(p + 1)(p - 1)$

37. 82

38. $-\frac{75}{17}$

39. 18

40. $6a + 7b + 23$

41. $18c + 19$

42. $-5d^2 + d$

43. j^2

44. $\frac{1}{k}$

45. $\frac{5n}{3m^3}$

46. -1

47. $\frac{4r^2 - 3}{2}$

48. $-2s^3 - 5s^2 + 6$

49. $\frac{3tv^3}{2w} + \frac{2v^3w^2}{t} - \frac{w}{2v}$

50. $-\frac{3x}{5}$

51. $\frac{(y + z)(y - 2z)}{2(2z + y)}$

52. $\frac{2a + 1}{2a + 3}$

53. $\frac{b + 2c}{2b + c}$

54. $\frac{3}{8}$

55. $\frac{1}{7j^4}$

56. 9

57. $12p^4$

58. $\frac{2(c + 2)}{4c^2 - c + 7}$

59. $\frac{5d^2 + d + 2}{(d + 1)(2d - 1)}$

60. $\frac{j^2 - 2k^2}{2j + k}$

61. $4\sqrt{30}$

62. $3(\sqrt{30} + \sqrt{37})$

63. $4p\sqrt{3}$

64. $\frac{6m}{5n}$

65. $\frac{5\sqrt{10r}}{14r}$

66. $-\sqrt{s}$

67. $29 + 4\sqrt{30}$

68. $4(\sqrt{5} - 2)$

69. $\frac{\sqrt{2}}{4}$

70. -3

71. $3\sqrt[3]{3}$

72. $\frac{\sqrt[3]{2}}{2}$

21. $\left\{c: -\frac{1}{2} \leq c \leq \frac{21}{2}\right\}$ 22. $\left\{p: -\frac{16}{3} \leq p \leq \frac{8}{3}\right\}$

23. $\left\{q: q > \frac{13}{5} \text{ or } q < -\frac{9}{5}\right\}$

24. $\left\{r: r > \frac{7}{2} \text{ or } r < \frac{3}{2}\right\}$ 25. $\{(5, -4)\}$

26. $\left\{\left(\frac{5}{11}, -\frac{57}{44}\right)\right\}$ 27. $\left\{\left(-6, -\frac{4}{5}\right)\right\}$

28. $\left\{\left(-\frac{4}{49}, \frac{6}{7}\right)\right\}$ 29. $\{(1, -6)\}$ 30. $\left\{\left(\frac{4}{13}, -\frac{14}{13}\right)\right\}$

31. $\left\{\left(\frac{6}{5}, \frac{8}{5}\right)\right\}$ 32. $\left\{\left(\frac{8}{9}, \frac{2}{3}\right)\right\}$ 33. $\left\{\left(\frac{43}{2}, -11\right)\right\}$

34. $\left\{\left(\frac{1}{3}, 1, -\frac{1}{3}\right)\right\}$ 35. $\left\{\left(1, \frac{1}{2}, \frac{1}{2}\right)\right\}$

36. $\left\{\left(\frac{11}{5}, -\frac{72}{5}, -\frac{32}{5}\right)\right\}$ 37. $\{-24\}$ 38. $\left\{\frac{8}{11}\right\}$

39. $\left\{j: j > -\frac{26}{3}\right\}$ 40. $\left\{m: m > -\frac{23}{11}\right\}$ 41. $\{24\}$

42. $\{-4\}$ 43. $\{\sqrt{11}, -\sqrt{11}\}$ 44. $\{\sqrt{7}, -\sqrt{7}\}$

45. $\{2\sqrt{5}, -2\sqrt{5}\}$ 46. $\{36\}$ 47. $\{58\}$

48. $\left\{\frac{32}{3}\right\}$ 49. $\{-3, -2\}$ 50. $\left\{-4, \frac{5}{2}\right\}$

51. $\left\{2, -\frac{9}{4}\right\}$ 52. $\{-1, -5\}$ 53. $\left\{1, \frac{7}{3}\right\}$

54. $\left\{-\frac{1}{2}, \frac{4}{3}\right\}$ 55. $\left\{\frac{-3 + \sqrt{57}}{4}, \frac{-3 - \sqrt{57}}{4}\right\}$

56. $\left\{\frac{4 + \sqrt{13}}{3}, \frac{4 - \sqrt{13}}{3}\right\}$ 57. no real roots

58. $\left\{\frac{7 + \sqrt{265}}{12}, \frac{7 - \sqrt{265}}{12}\right\}$

59. $\left\{\frac{-1 + \sqrt{13}}{6}, \frac{-1 - \sqrt{13}}{6}\right\}$ 60. $\left\{-\frac{1}{4}, 1\right\}$

Review of Solving Equations, Inequalities, and Systems

1. $\{19\}$ 2. $\{36\}$ 3. $\left\{-\frac{2}{3}\right\}$ 4. $\{45\}$

5. $\{-8\}$ 6. $\{28, -28\}$ 7. $\{p: p > -8\}$

8. $\{q: q > 12\}$ 9. $\{r: r < 6\}$ 10. $\left\{t: t < \frac{9}{7}\right\}$

11. $\left\{s: s > -\frac{19}{3}\right\}$ 12. $\left\{w: w > -\frac{48}{5}\right\}$

13. $\{h: -15 \leq h < -4\}$ 14. $\{j: 7 \geq j \geq -3\}$

15. $\left\{k: -\frac{2}{3} \leq k \leq 3\right\}$ 16. $\{-2, 16\}$

17. $\{y: 8 \leq y \leq 18\}$ 18. $\left\{z: \frac{16}{5} > z > -\frac{8}{5}\right\}$

19. $\{a: a \leq 2 \text{ or } a \geq 8\}$ 20. $\{b: -2 \leq b \leq 8\}$

Review of Solving Word Problems

1. 35 km/h to Broomville; 21 km/h to Goldmont
2. 5 kg of the $2.70/kg type; 25 kg of the $3.00/kg type
3. 1.8 h
4. 13 and 17
5. 140.71 m
6. $3500 at 9%; $2500 at 12%
7. no solution; not enough information is given
8. 3 m
9. 1500 MHz
10. slope of $\overline{AB}$: $\frac{1}{3}$; slope of $\overline{BC}$: -3; slope of $\overline{AC}$: $-\frac{9}{13}$
11. wind speed: 49 km/h; air speed: 539 km/h

12. height: 2.75 m; length: 12 m
13. 694
14. 35°; 65°; 80°
15. 270 rpm
16. −11, −9, and −7
17. 4 cm
18. 23 years
19. 60 km/h
20. $-\frac{3}{2}$ or 2

Review of Graphing

1. (−2, 2)   **2.** (1, 5)   **3.** (1, 0)
4. (−4, −3)   **5.** (2, −1)   **6.** (−5, 0)
7. G   **8.** P   **9.** L   **10.** N   **11.** J   **12.** H

13.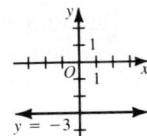

14.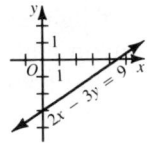

15.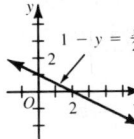

16.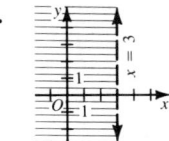

17.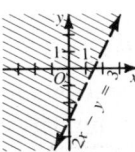

18.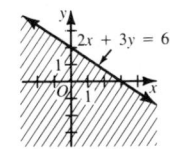

19. {x: x < 3}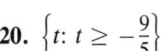

20. $\left\{t: t \geq -\frac{9}{5}\right\}$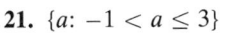

21. {a: −1 < a ≤ 3}   (number line)

22. {t: t < −2 or t > 3}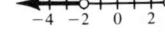

23. {1, 5}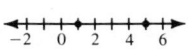

24. {z: z ≤ −1 or z ≥ 5}

25. y = 2x   **26.** y = −3x − 3
27. $y = \frac{1}{4}x + 1$ **28.** x = 0
29. 2x − y = 0   **30.** x − 3y = 6
31. 3x + 4y = 12   **32.** x = −3

33. {(2, −1)}   **34.** {(−1, −3)}

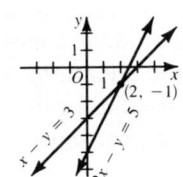   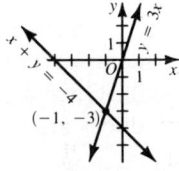

35. ∅   **36.**

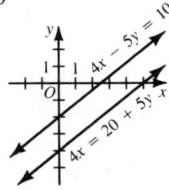   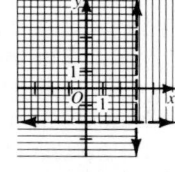

37.   **38.**

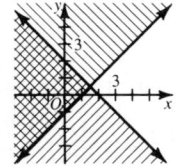   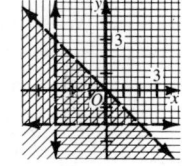

39.   **40.**

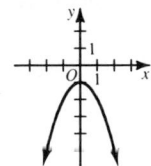   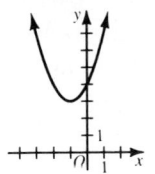

T45

Reading Algebra

Suggestions for helping your students learn how to read an algebra textbook

The question is sometimes asked, "Why do we need to teach reading in an algebra class?" A study of reading algebra reveals that it is integrally tied to the learning of algebra. Difficulties that students encounter in algebra are often the result of difficulties in reading. To improve reading algebra is to improve algebra. To improve reading is to increase the chance of making the student a more independent learner since reading is a learning-to-learn skill. Once a student learns to read proficiently, doors are opened that were previously closed.

The ideas of algebra are expressed in compact form, with extensive use of mathematical symbols, and are applied in the solution of a wide variety of problems. Exposition and examples are illustrated with diagrams that must be read and understood along with the text. Following through the solution of an equation often requires up-and-down eye movements as well as the familiar left-to-right motion, together with pauses to consider how successive steps in the solution are related. For all these reasons, even students who can read other material proficiently may have trouble reading algebra. Throughout the course, students will profit from the teacher's specific suggestions for improving their reading skills.

Many familiar words—for example, *power, variation, imaginary*—take on specialized meanings in algebra. These meanings need to be pointed out to students, and the importance of learning definitions and using mathematical terms correctly needs to be emphasized.

Diagrams are of great value in almost all branches of mathematics. Their importance in algebra should not be overlooked. Students need to realize that diagrams are essential parts of discussions, examples, and often exercises, and that the diagrams must be read along with the words of the text. Identifying and interpreting the information provided by diagrams, graphs, and charts is a skill that has to be learned. Without adequate preparation, students are likely to become frustrated as they attempt to cope with material presented in this way.

Clearly, successful reading calls for practice and patience on the part of both teacher and students. To help the student, this textbook has been organized in a way that is easy to follow, with important material highlighted. Sections titled "Reading Algebra" contain hints that are designed to help the student use the textbook more effectively. Additional suggestions of ways for the teacher to aid students in improving reading are included in appropriate sections of the Lesson Commentary. The following six major reading objectives are emphasized.

Objective 1

Reading and Communicating Orally

This first objective is to ensure that students have a basic understanding of the relationship between what is spoken and what is written. A mathematics textbook contains many symbols, diagrams, charts, and graphs that are not commonly used in other books. It is often difficult to express their content in words. Further, the interrelationship between symbols is often complex, and the method of verbalizing may be unclear. It is not uncommon for a set of symbols to be verbalized in several ways within one class lesson. For example, a^2 may be called the square of a, the second power of a, a squared, or a to the second power. Practice is necessary if the relationship between the spoken and the written is to be understood.

Math teachers need to stress the use of oral language by having students verbalize in sentences, summarize, repeat, say another way, read aloud, work together, and perform other tasks that enhance both their own understanding and their ability to communicate the meaning of algebraic expressions.

Objective 2

Reading Silently

This second objective is to provide students with the necessary skills to be able to read silently, recognizing that one's purpose determines the speed and type of reading. When students read silently it is often beneficial if they do so with a purpose—to find the main idea, to look for a specific detail, to summarize, to answer a question, to study an example.

Different purposes demand different types of reading. For previewing or reviewing a lesson, skimming (rapid reading) is used. By contrast, slow reading is used when the student is trying to learn the main ideas and important details of a lesson. For slow reading, it is beneficial for teachers to provide questions in advance that will be discussed upon completion of the reading. When beginning a section, students may be directed first to skim the page to find new ideas and symbols, then to read carefully, and finally to answer questions or to work exercises.

Objective 3

Using Symbols

The third objective is to assist students in recognizing symbols and in associating symbols with words and diagrams. As was suggested earlier, the verbalization of symbols creates a reading problem for many students. The word forms of symbols often are not obvious to students. Once learned, they must be reinforced to ensure retention.

Other difficulties introduced by the use of symbols are their conciseness and their association with concepts not actually stated. The following example demonstrates these difficulties.

$$7(x + 10) - 4x = (-6 + x)(-5)$$
$$7x + 70 - 4x = 30 - 5x$$
$$3x + 70 = 30 - 5x$$
$$8x = -40$$
$$x = -5$$

One way to recognize the conciseness of the symbolism is to try to write out in words the ideas that appear in symbolic form. While this compactness demands slower reading and greater concentration than reading English sentences, it has the advantage of showing the structure of a complex reasoning process. The reader is expected to mentally supply the property that justifies replacing each equation by the equivalent equation that follows it.

Students will benefit from the teacher's explicit demonstration of the appropriate choice and ordering of symbols in the solution of equations and word problems. When several different symbolic forms can be used to represent the same idea, comparisons of these forms will help students recognize the flexibility of algebraic symbolism.

Objective 4

Using Mathematical Words

This fourth objective is to stress the mathematical meaning of words. A mathematics page contains words of everyday language (*and, when, the*), words that are unique to mathematics (*monomial, abscissa*), and familiar words with specialized meanings in mathematics (*prime, variable, complex*). The most commonly used words are primarily *structure* words, whereas most of the *content* words are mathematical. This implies that the meaning of the content words needs to be taught in the mathematics classroom.

You can help students find clues for words within the words themselves and from content on the page, such as charts, diagrams, and symbols. Call attention to prefixes such as *bi-* and *quadr-* that students may know already and that they can use in figuring out the meaning of new words. Point out also that certain important words—*equivalent,* for example—are used without change of meaning in combination with many other words. You should be aware that students may have been exposed to several different names for the same concept—distributive rule, distributive axiom, distributive property, for instance—and that they may therefore be confused.

Understanding definitions of terms is of primary importance. Use of the glossary and the index should be encouraged. It is also a good idea to have a spelling quiz from time to time.

Objective 5

Reading Word Problems

This fifth objective is to give students confidence in attacking word problems independently and to help them develop their reasoning power in a variety of situations.

If teachers or students were asked how reading skills are used in algebra, they would probably say, "In word problems." Few mathematics teachers would question the importance of careful reading as the first step in solving a problem, but one needs to do more than say, "Read the problem carefully." How does the student "read carefully"? More specific help is needed.

One learns to read carefully by having some specific questions to ask oneself while reading, by taking advantage of all available resources that aid in comprehension, and by thinking about the problem in a well-organized way. The teacher can help students by working with them through the solution of a variety of problems, following the plan of attack suggested in the textbook on page 132. You may need to point out that many problems that *look* very different are basically alike and can be solved by the same method.

Objective 6

Reading Charts, Graphs, and Diagrams

This sixth objective is to make sure that students can relate the reading of charts, graphs, and diagrams to the rest of the exposition and exercises. Tables and graphs play an integral role in the development of many lessons in algebra. Geometric diagrams are used in problems throughout the textbook, and the discussion of a number of important concepts is illustrated by diagrams. Many students fail to use these visual aids to their advantage. Through oral reading and questioning you can assist students in relating these aids to the words and symbols on the page.

Many students may need a good deal of help in understanding how a table or chart is organized and in relating a table to the graphic representation of the data. Construction of a chart on the chalkboard, with a discussion of the procedure, will help to make the structure clearer to students. The best way to check students' understanding is to involve them in constructing and explaining charts and diagrams.

Problem Solving Strategies

Some strategies that your students can use to become better problem solvers

A problem solving strategy is simply a plan or technique for solving a problem. There are a number of well-known problem solving strategies that relate specifically to algebra. For example, applying the quadratic formula is a strategy for solving quadratic equations. One of the goals of an algebra course is to familiarize students with these standard techniques and to give them enough practice with these techniques so that they can use them confidently and successfully to solve algebra problems.

Algebra, however, provides an excellent opportunity to teach not only the standard algorithms but also more general problem solving strategies that are useful in other branches of mathematics and in other subject areas as well as in this course. Since these general strategies provide an *approach* to solving a problem rather than a specific method of solution, they are particularly useful for attacking a problem when the method of solution is not obvious.

Students will learn problem solving skills as they have successful experiences in nonthreatening situations. The teacher should endeavor to expose them to a variety of problems that appeal to their interests and that they can solve with a reasonable amount of effort. Number problems, coin problems, uniform-motion problems, and probability problems, to name but a few, often lend themselves to many different approaches.

Just to understand a problem thoroughly requires careful reading, knowing the vocabulary, being able to recognize necessary, irrelevant, and contradictory data, identifying special conditions and restraints, and visualizing the situation described. Even if the statement of a problem is understood, a correct solution is not guaranteed. The method outlined on page 132 of the textbook is a basic strategy that students can use throughout the course. Some refinements and extensions of the method are suggested in the Reading Algebra features on pages 143 and 504.

In many cases, the problem solving approach to be used will depend on the individual student's ingenuity. The following list suggests some questions that students may ask themselves as they search for a plan of attack on an unfamiliar problem.

- Is it similar to a problem I have seen before?
- Can I write and solve a simpler problem of the same type?
- Is there a recognizable pattern in the data?
- Can I draw a diagram to represent the given conditions?
- Can the data be organized in a table, chart, or graph?
- Is there a theorem or formula that I can apply?
- Is trial and error a suitable approach?
- Can I devise a simulation of the situation or an experiment that will lead to a solution?
- Can I frame and test a hypothesis?
- Can I work the problem backward?
- Can I use logic and reason deductively?
- Is it possible that the problem has no solution?

Once a strategy has been devised, the solution of the problem should be attempted. If a strategy does not work, it should be examined to see whether it should be rejected or be modified and tried again.

In the side column at the beginning of many chapters (see, for example, page 1), there is a short list of problem solving strategies appropriate to types of problems presented in the chapter, together with references to places in the chapter where these strategies might profitably be used and discussed.

Error Analysis

Anticipating common errors and helping students avoid them

Since mathematics builds on previously learned symbols, concepts, and skills, error patterns that are left uncorrected will impede students' progress. Of course, there are many different reasons for errors, but certain types of errors are more common than others. If you are aware of these common errors, you can help students avoid them and you can be better prepared to help students overcome them if they do occur. Throughout the book, in the side columns next to the textbook pages, common errors have been identified and suggestions for avoiding them have been provided. (See, for example, "Common Errors" on pages 72, 328, and 559.) The errors discussed in the side columns are fairly specific. However, many of these errors can be grouped into one or more of the following categories:

Errors in Reading and Translating

(See, for example, pp. 44 and 191.)

Students often have difficulty in translating English phrases and sentences into mathematical expressions and sentences. For example, students may translate the expression "five less than a number" as $5 - n$. Not reading word problems carefully, with concentration on their meaning, is another frequent cause of difficulty. Students may make mistakes because they do not fully understand the meanings of mathematical terms—for example, "maximum value of a function." Words that have a different meaning in mathematics than in everyday speech, such as *or*, may cause confusion.

Failure to Understand Symbols

(See, for example, pp. 64, 216, and 585.)

Students often do not fully understand the meanings of mathematical symbols. As a result, they may make the following errors:

$$\frac{6}{0} = 0 \qquad -x^2 = (-x)^2 \qquad b^n = nb$$

$$2^{-2} = -4 \qquad (a + b)^2 = a^2 + b^2 \qquad -2 > x > 3$$

$$0.\overline{5} = \frac{1}{2} \qquad 2\sqrt{3} = \sqrt{6} \qquad \sqrt{x^2} = x$$

Misunderstanding of Properties

(See, for example, pp. 58 and 365.)

Recurring errors often stem from students' misapplying properties in the ways shown below.

Addition property of equality: $x + 2 = 6$
$$x = 8$$

Multiplication property of equality: $\frac{x}{2} + \frac{x}{3} = 5$
$$3x + 2x = 5$$

Division property of equality: $3x(x + 2) = 0$
$$x + 2 = 0$$

Multiplication property of order: $2x > -8$
$$x < -4$$

Distributive axiom: $-4(x + 1) = 9$
$$-4x + 1 = 9$$

Zero-product property: $(x - 3)(x + 2) = 14$
$$x - 3 = 14 \quad \text{or} \quad x + 2 = 14$$

Errors in Using Standard Forms

(See, for example, pp. 506 and 549.)

Although students may have memorized the Pythagorean theorem or the quadratic formula, they may not understand the importance of using the standard form when they apply these formulas. Consequently, they often try to work with an expression or an equation without first putting it into standard form. This means that they may substitute incorrect values in the formula $a^2 + b^2 = c^2$ or try to solve an equation by the quadratic formula before transforming it so that one side is 0. Other errors that students are liable to make are illustrated below.

$$2x + 3y = 6 \qquad y = 1 - 3x$$
$$\text{slope} = 2 \qquad \text{slope} = 1$$
$$\qquad\qquad\qquad y\text{-intercept} = -3$$

$$x^2 + 1 - 2x = 0$$
$$a = 1, b = 1, c = -2$$

Use of Incorrect Formulas

(See, for example, p. 242.)

Many errors are the result of students using formulas that are incorrect. Some of the more common "impostors" are shown below.

$$p = l + w \qquad C = \pi r^2 \qquad \text{slope} = \frac{x_2 - x_1}{y_2 - y_1}$$

$$x = -b \pm \frac{\sqrt{b^2 - 4ac}}{2a} \qquad \text{discriminant} = \sqrt{b^2 - 4ac}$$

Errors in Simplifying

(See, for example, pp. 390, 394, 408, and 493.)

In addition to some of the reasons already given, students may make errors in simplifying expressions because of incorrect assumptions such as

$\sqrt{a^2 - b^2} = \sqrt{a^2} - \sqrt{b^2}$. They may also make errors because they do not take notice of grouping symbols such as the fraction bar or because they do not follow the prescribed order of operations or because they add unlike terms. Students sometimes confuse the rules of exponents—multiplying exponents when they are multiplying, and dividing exponents when they are dividing. Students may forget to (or not realize that they must) change the sign of every term of a polynomial that they are subtracting. Simplifying fractions seems to be particularly troublesome. Thinking that $\frac{n}{n} = 0$ can lead to errors such as $6x^2 + 8x + 2 = 2(3x^2 + 4x)$. Errors such as $\frac{\cancel{n} + 3}{\cancel{n}} = 3$ and $\frac{\overset{3}{\cancel{6}}\cancel{y}(x + 3)}{2\cancel{y}^2(x + 1)_{1}} = \frac{3(x + 3)}{2(x + 1)}$ are common.

Errors in Checking

(See, for example, p. 133.)

Checking can help students develop self-confidence and alert them to errors. However, a check that is incorrectly performed is not useful. Students often fail to realize that it is not only helpful but necessary to check the roots of fractional and radical equations. The following checking errors may occur: Students may occasionally substitute a value for a variable such as 8 for x, get a true statement such as $4 = 4$, and conclude that the solution is 4. Students may not realize that they must check their answers with the *words* of word problems or that they must check their solutions to systems of linear equations in *both* of the *original* equations. Some students may think that they should always discard negative solutions. Students may not think of checking their answers when the method involves, for example, considering whether or not an answer is reasonable or multiplying to check factoring.

Assignment Guide

This guide may be of some help to you in planning your year's work, but both the Time Schedule and daily assignments should be considered suggestions only. By keeping track of the amount of time your students spend on homework, you can adapt the assignments to conform to the policies of your school.

The following guide offers suggestions for planning separate minimum, average, and maximum courses. Chapters 1-11 comprise a minimum course in elementary algebra that ends with a discussion of quadratic and polynomial functions. In the average course, the addition of Chapter 13 extends the minimum course to include the basic concepts of statistics and probability. In the maximum course, the inclusion of Chapter 12 provides the student with the opportunity to cover the basic concepts of numerical trigonometry and vectors.

Because students' interests and backgrounds differ widely from class to class, most of the optional features are not listed in the Assignment Guide. You will want to choose those features that best suit your individual classes. If you have access to a computer that accepts BASIC, you may wish to allow some time for your students to do the Programming in BASIC features or the Computer Exercises. Please see the note on page ix regarding these features.

All the assignments refer to the Written Exercises, with the letter "P" indicating word problems. The letter "S" indicates the spiraled portion of the assignment, which reviews earlier work. The letter "R" indicates a review built into the text. "EP" refers to the Extra Practice section, which contains extra exercises and problems to be used as needed.

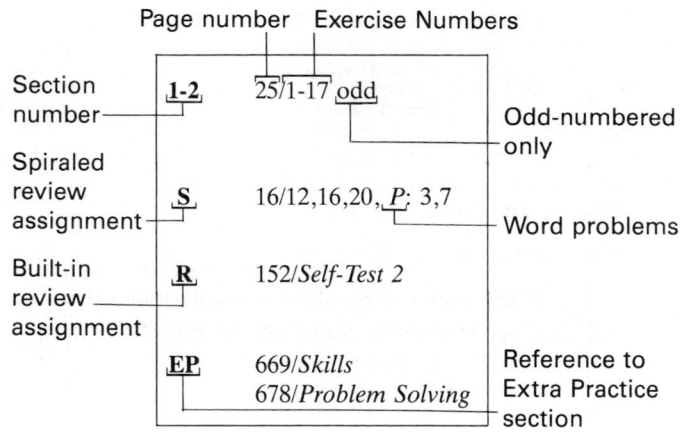

Approximate Time Schedule

Chapter	1	2	3	4	5	6	7	8	9	10	11	12	13	Total
Minimum Course	11	17	17	15	13	14	19	13	14	17	10	0	0	160
Average Course	11	15	15	13	12	14	19	13	13	14	9	0	12	160
Maximum Course	11	15	15	12	11	12	16	12	13	13	9	10	11	160

Trimester   Semester   Trimester

Day	Minimum Course		Average Course		Maximum Course	
1	**1-1**	4/1-32	**1-1**	4/7-43 odd, 44-50	**1-1**	4/7-49 odd, 51-44
2	**1-2**	9/1-30	**1-2**	9/5-33 odd, 35-39	**1-2**	9/5-39 odd, 40-43
3	**1-3** **R**	13/1-19 odd, 20-30 15/*Self-Test 1*	**1-3** **R**	13/1, 7-41 odd, 43-48 15/*Self-Test 1*	**1-3** **R**	13/1, 7-47 odd, 49-52 15/*Self-Test 1*
4	**1-4**	19/1-19 odd, 21-36	**1-4**	19/7-12, 13-49 odd, 50-54	**1-4**	19/7-12, 13-55 odd, 56-60
5	**1-5** **S**	22/2-42 even 13/2-18 even	**1-5** **S**	22/10-46 even 14/28-42 even	**1-5** **S**	22/11-45 odd, 47-54 14/32-48 even
6	**1-6** **R**	26/1-41 odd 27/*Self-Test 2*	**1-6** **R**	26/5-47 odd, 49-54 27/*Self-Test 2*	**1-6** **R**	26/5-33 odd, 55-60 27/*Self-Test 2*
7	**1-7**	32/1-15 odd, 17-32	**1-7**	32/11-31 odd, 33-48	**1-7**	32/11-47 odd, 49-54
8	**1-8** **S**	37/2-18 even, 19-26 26/26-34 even	**1-8** **S**	37/8-26 even, 27-28 27/40-48 even	**1-8** **S**	37/8-38 even, 39-42 27/46-54 even
9	**1-9** **R**	41/1-23 odd 43/*Self-Test 3*	**1-9** **R**	41/5-33 odd 43/*Self-Test 3*	**1-9** **R**	41/5-33 odd, 35-38 43/*Self-Test 3*
10	*Prepare for Chapter Test* **R** **EP**	45/*Chapter Review* 660/*Skills* 675/*Problem Solving*	*Prepare for Chapter Test* **R** **EP**	45/*Chapter Review* 660/*Skills* 675/*Problem Solving*	*Prepare for Chapter Test* **R** **EP**	45/*Chapter Review* 660/*Skills* 675/*Problem Solving*
11	*Administer Chapter Test* 29/*Reading Algebra:* 1-12		*Administer Chapter Test* 29/*Reading Algebra:* 1-12		*Administer Chapter Test* 29/*Reading Algebra:* 1-12	
12	**2-1**	51/1-12	**2-1**	51/1-20	**2-1**	51/1-22
13	**2-2**	56/1-24	**2-2**	56/5-17 odd, 19-32	**2-2**	56/5-35 odd, 36-39
14	**2-3**	60/1-26, 27-35 odd	**2-3**	60/21-47 odd, 49-58	**2-3**	60/25-57 odd, 58-62
15	**2-4** **R**	65/1-30 66/*Self-Test 1*	**2-4** **R**	65/7-41 odd, 43-54 66/*Self-Test 1*	**2-4** **R**	65/7-41 odd, 43-58 66/*Self-Test 1*
16	**2-5**	70/1-15	**2-5** **S**	70/7-53 odd 60/38-48 even	**2-5** **S**	70/7-53 odd, 55-60 60/50-56 even
17	**2-5** **S**	70/16-36 60/28-36 even	**2-6**	73/5-29 odd 75/*P:* 5-11	**2-6**	73/5-33 odd 75/*P:* 7-13
18	**2-6**	73/1-12 75/*P:* 1-5	**2-6**	74/30-36 76/*P:* 12-16	**2-6**	74/34-40 76/*P:* 14-17
19	**2-6**	74/13-28 75/*P:* 6-10	**2-7**	81/1-11 odd, 13-16, 17-23 odd	**2-7**	81/5-11 odd, 13-28

Day	Minimum Course		Average Course		Maximum Course	
20	**2-7**	81/1-13	**2-8**	84/1-35 odd 86/*P*: 1-6	**2-8**	84/1-45 odd 86/*P*: 1-6
21	**2-7**	81/14-18	**2-8** **R**	85/36-48 86/*P*: 7-11 87/*Self-Test 2*	**2-8** **R**	85/47-62 86/*P*: 7-11 87/*Self-Test 2*
22	**2-8**	84/1-20 86/*P*: 1-4	**2-9**	91/7-57 odd, 58-63	**2-9**	91/7-59 odd, 61-68
23	**2-8** **R**	85/21-39 86/*P*: 5-8 87/*Self-Test 2*	**2-10** **S**	97/13-39 odd 85/26-34 even	**2-10** **S**	97/13-49 odd 85/38-46 even
24	**2-9**	91/1-25, 27-39 odd	**2-10** **2-11** **R**	98/40-50 101/11-41 odd 103/*Self-Test 3*	**2-10** **2-11** **R**	98/51-56 101/11-57 odd 103/*Self-Test 3*

25	**2-10**
	S
		97/1-22
		87/9, 10, 11
			Prepare for Chapter Test
			R 105/*Chapter Review*
			EP 661/*Skills*
			      676/*Problem Solving*
					Prepare for Chapter Test
					R 105/*Chapter Review*
					EP 661/*Skills*
					      676/*Problem Solving*	

26	**2-10** **2-11** **R**	97/23-30 101/1-28 103/*Self-Test 3*	*Administer Chapter Test* **R** 107/*Mixed Review*		*Administer Chapter Test* **R** 107/*Mixed Review*	
27	*Prepare for Chapter Test* **R** 105/*Chapter Review* **EP** 661/*Skills* 667/*Problem Solving*		**3-1**	116/1-12, 13-19 odd	**3-1**	117/3-15, 16-22 even
28	*Administer Chapter Test* **R** 107/*Mixed Review*		**3-2**	120/11-49 odd, 50-58	**3-2**	120/11-57 odd, 58-66
29	**3-1**	116/1-7	**3-3**	123/5-37 odd, 39-46	**3-3**	123/5-41 odd, 43-50
30	**3-1**	117/8-12	**3-4**	127/7-35 odd	**3-4**	128/11-49 odd
31	**3-2**	120/1-24, 25-39 odd	**3-4** **R**	128/37-55 odd 129/*Self-Test 1*	**3-4** **R**	128/50-60 129/*Self-Test 1*
32	**3-3**	123/1-32	**3-5** **S**	132/*P*: 1-11 odd, 13-22 120/40-48 even	**3-5** **S**	132/*P*: 5-15 odd, 17-22 120/48-56 even
33	**3-4**	127/1-28	**3-6**	136/1-29 odd, 31-34	**3-6**	136/11-41 odd
34	**3-4** **R**	128/29-36 129/*Self-Test 1*	**3-6**	137/35-50	**3-6**	137/42-56
35	**3-5**	132/*P*: 1-10	**3-7**	139/*P*: 1-12	**3-7**	139/*P*: 1-15

Day	Minimum Course		Average Course		Maximum Course	
36	**3-5** **S**	133/*P*: 11-18 120/26-34 even	**3-7** **S**	140/*P*: 13-20 132/4-12 even	**3-7** **S**	141/*P*: 16-24 132/6-14 even
37	**3-6**	136/1-15	**3-8**	146/1-25	**3-8**	146/1-30
38	**3-6**	136/16-32	**3-8**	146/26-40 147/*P*: 1-3	**3-8**	146/31-49 147/*P*: 1-3
39	**3-7**	139/*P*: 1-8	**3-8** **R**	146/41-55 147/*P*: 4-8 148/*Self-Test 2*	**3-8** **R**	146/50-58 147/*P*: 4-8 148/*Self-Test 2*

40	**3-7**
	 S
		140/*P*: 9-15
		 133/19, 20
			Prepare for Chapter Test
			 R 151/*Chapter Review*
			 EP 662/*Skills*
			      676/*Problem Solving*
					Prepare for Chapter Test
					 R 151/*Chapter Review*
					 EP 662/*Skills*
					      676/*Problem Solving*	

41	**3-8**	146/1-24	*Administer Chapter Test*
			      143/*Reading Algebra:* 1-10
			 R 154/*Cumulative Review*
					Administer Chapter Test
					      143/*Reading Algebra:* 1-10
					 R 154/*Cumulative Review*	

42	**3-8**	146/25-36 147/*P*: 1, 2	**4-1**	161/9-33 odd, 34-38	**4-1**	161/9-37 odd, 38-46
43	**3-8** **R**	146/37, 38 147/*P*: 3-6 148/*Self-Test 2*	**4-2**	166/1-29 odd, 30-36	**4-2**	166/1-33 odd, 34-40
44	*Prepare for Chapter Test* **R** 151/*Chapter Review* **EP** 662/*Skills* 676/*Problem Solving*		**4-3**	169/5-33 odd, 34-39	**4-3**	169/5-37 odd, 39-44
45	*Administer Chapter Test* 143/*Reading Algebra:* 1-10 **R** 154/*Cumulative Review*		**4-4** **S**	173/1-31 odd, 32-38 166/26, 28, 37	**4-4** **S**	173/1-37 odd, 39-46 166/26-32 even
46	**4-1**	161/1-24	**4-5** **R**	177/1-55 odd 178/*Self-Test 1*	**4-5** **R**	177/5-65 odd 178/*Self-Test 1*
47	**4-2**	166/1-24	**4-6**	182/*P*: 1-13 odd, 15-22	**4-6**	182/*P*: 1-13 odd, 15-24
48	**4-3**	169/1-18	**4-7**	186/*P*: 1-13 odd, 15-22	**4-7** **S**	186/*P*: 1-13 odd, 15-24 178/58-66 even
49	**4-4** **S**	173/1-24 166/25-27	**4-8**	192/*P*: 1-10	**4-8**	192/*P*: 1-13 odd, 20-25
50	**4-5** **R**	177/1-10, 12-30 even 178/*Self-Test 1*	**4-8** **S**	193/*P*: 11-22 178/42-50 even	**4-9**	199/*P*: 1-9 odd, 11-18

Day	Minimum Course	Average Course	Maximum Course
51	4-6 182/*P*: 1-14	4-9 199/*P*: 1-14	4-10 202/*P*: 1-18 R 203/*Self-Test 2*
52	4-7 186/*P*: 1-8	4-10 202/*P*: 1-16 R 203/*Self-Test 2*	*Prepare for Chapter Test* R 208/*Chapter Review* EP 663/*Skills* 677/*Problem Solving*
53	4-7 186/*P*: 9-16	*Prepare for Chapter Test* R 208/*Chapter Review* EP 663/*Skills* 677/*Problem Solving*	*Administer Chapter Test* 206/*Extra*: 1-17 odd R 210/*Mixed Review*
54	4-8 192/*P*: 1-9	*Administer Chapter Test* R 210/*Mixed Review*	5-1 218/1-39 odd, 40-44
55	4-8 193/*P*: 10-20 S 177/31-34	5-1 218/1-37 odd	5-2 222/1-25 odd, 26-31
56	4-9 199/*P*: 1-5	5-2 222/1-19 odd, 21-26	5-3 226/13-55 odd R 227/*Self-Test 1*
57	4-9 199/*P*: 6-14	5-3 226/13-31 odd, 33-45 R 227/*Self-Test 1*	5-4 230/5-41 odd, 43-48 S 226/50-56 even
58	4-10 202/*P*: 1-12 R 203/*Self-Test 2*	5-4 230/1-31 odd, 33-42 S 226/26-32 even	5-5 235/5-37 odd, 38-42
59	*Prepare for Chapter Test* R 208/*Chapter Review* EP 663/*Skills* 677/*Problem Solving*	5-5 235/1-35 odd, 36	5-6 239/9-29 odd, 30-36 R 240/*Self-Test 2*
60	*Administer Chapter Test* R 210/*Mixed Review*	5-6 239/5-23 odd, 26-30 R 240/*Self-Test 2*	5-7 246/5-31 odd, 32-34
61	5-1 218/1-24	5-7 246/1-23 odd, 24-32	5-8 251/9-33 odd, 37-39 S 236/43-48
62	5-2 222/1-20	5-8 251/1-35 odd S 236/39-42	5-9 254/7-33 odd, 34-40 R 255/*Self-Test 3*
63	5-3 226/1-24 R 227/*Self-Test 1*	5-9 254/1-23 odd	*Prepare for Chapter Test* R 259/*Chapter Review* EP 664/*Skills*
64	5-4 230/1-24 S 226/25-27	5-9 254/25-38 R 255/*Self-Test 3*	*Administer Chapter Test* 241/*Reading Algebra*: 1-9 257/*Extra*: 1-10 R 262/*Cumulative Review*

Day	Minimum Course		Average Course		Maximum Course	
65	**5-5**	235/1-22	*Prepare for Chapter Test* **R** 259/*Chapter Review* **EP** 664/*Skills*		**6-1**	272/7-31 odd, 32-38
66	**5-6** **R**	239/1-18 240/*Self-Test 2*	*Administer Chapter Test* 241/*Reading Algebra*: 1-9 **R** 262/*Cumulative Review*		**6-2**	276/5-23 odd, 25-28
67	**5-7**	246/1-18	**6-1**	272/7-27 odd, 28-32	**6-3**	279/6-20
68	**5-8**	251/1-23 odd	**6-2**	276/1-23 odd	**6-4** **R**	282/7-29 odd, 31-36 283/*Self-Test 1*
69	**5-8** **S**	251/2-24 even 235/23-26	**6-3**	279/1-15	**6-5** **S**	286/*P*: 5-15 odd, 17-22 276/16-24 even
70	**5-9**	254/1-12	**6-4** **R**	282/1-25 odd, 26-30 283/*Self-Test 1*	**6-6**	290/*P*: 1-7 odd, 9-15
71	**5-9** **R**	254/13-24 255/*Self-Test 3*	**6-5**	286/*P*: 1-8	**6-7** **R**	296/*P*: 9-20 297/*Self-Test 2*
72	*Prepare for Chapter Test* **R** 259/*Chapter Review* **EP** 664/*Skills*		**6-5** **S**	287/*P*: 9-20 276/16-24 even	**6-8** **S**	300/1-21 odd 290/*P*: 2-8 even
73	*Administer Chapter Test* 241/*Reading Algebra*: 1-9 **R** 262/*Cumulative Review*		**6-6**	290/*P*: 1-6	**6-9**	304/1-6, 9-16
74	**6-1**	272/1-23 odd	**6-6** **S**	291/*P*: 7-12 282/20, 22, 24	**6-10** **R**	309/5-17 odd 310/*P*: 1-6 311/*Self-Test 3*
75	**6-2**	276/1-12	**6-7** **R**	296/*P*: 1-14 297/*Self-Test 2*	*Prepare for Chapter Test* **R** 316/*Chapter Review* **EP** 665/*Skills* 677/*Problem Solving*	
76	**6-3** **S**	279/1-12 272/12, 29, 30	**6-8**	300/5-15 odd, 16-20	*Administer Chapter Test* 315/*Extra*: 1-6 **R** 319/*Mixed Review*	
77	**6-4** **R**	282/1-23 odd 283/*Self-Test 1*	**6-9**	305/1-6, 7, 9, 11-14	**7-1**	327/11-49 odd, 50-54
78	**6-5**	286/*P*: 1-7	**6-10** **R**	309/1-8 310/*P*: 1-4 311/*Self-Test 3*	**7-2**	330/1-39 odd, 40-50

Day	Minimum Course		Average Course		Maximum Course	
79	**6-5** **S**	287/*P*: 8-16 276/13-17	*Prepare for Chapter Test* **R** 316/*Chapter Review* **EP** 665/*Skills* 677/*Problem Solving*		**7-3**	334/7-57 odd
80	**6-6**	290/*P*: 1-5	*Administer Chapter Test* **R** 319/*Mixed Review*		**7-4** **R**	338/25-53 odd, 55-60 339/*P*: 5-12 340/*Self-Test 1*
81	**6-6** **S**	291/*P*: 6-10 279/13-15	**7-1**	326/1-39 odd, 40-48	**7-5** **S**	344/1-45 odd, 47-53 334/46-56 even
82	**6-7** **R**	296/*P*: 1-10 297/*Self-Test 2*	**7-2**	330/1-15	**7-6**	347/1-53 odd
83	**6-8**	300/1-15 odd	**7-2**	330/16-38	**7-7**	352/1-47 odd, 49-54
84	**6-9**	304/1-8	**7-3**	334/1-43 odd	**7-8**	357/1-41 odd
85	**6-10** **R**	309/1-8 311/*Self-Test 3*: 1, 2, 4	**7-4** **R**	338/5-53 odd 339/*P*: 1-8 340/*Self-Test 1*	**7-8** **S**	357/43-52 348/46-54 even
86	*Prepare for Chapter Test* **R** 316/*Chapter Review* **EP** 665/*Skills* 677/*Problem Solving*		**7-5** **S**	344/1-39 odd, 41-46 334/36-44 even	**7-9**	360/1-29 odd, 31-34
87	*Administer Chapter Test* **R** 319/*Mixed Review*		**7-6**	347/1-43 odd	**7-9** **R**	360/35-44 361/*Self-Test 2*
88	**7-1**	326/1-30	**7-7**	352/1-29 odd	**7-10**	366/1-53 odd, 54
89	**7-2**	330/1-15	**7-7**	352/31-48	**7-11**	369/*P*: 1-14
90	**7-2**	330/16-30	**7-8**	357/1-29 odd	**7-11** **R**	370/*P*: 15-26 371/*Self-Test 3*
91	**7-3**	334/1-30	**7-8** **S**	357/31-46 348/34-44 even	*Prepare for Chapter Test* **R** 374/*Chapter Review* **EP** 666/*Skills* 678/*Problem Solving*	
92	**7-4** **R**	338/1-25 339/*P*: 1-4 340/*Self-Test 1*	**7-9**	360/1-29 odd	*Administer Chapter Test* 349/*Reading Algebra*: 1-6 373/*Extra*: 1-12 **R** 377/*Cumulative Review*	
93	**7-5** **S**	344/1-34 334/31-34	**7-9** **R**	360/31-42 361/*Self-Test 2*	**8-1**	384/1-39 odd, 41-46

Day	Minimum Course		Average Course		Maximum Course	
94	**7-6**	347/1-32	**7-10**	366/1-31 odd	**8-2**	386/1-29 odd, 30-34
95	**7-7**	352/1-15	**7-10**	367/32-44	**8-3** **R**	391/1-23 odd, 25-30 392/*Self-Test 1*
96	**7-7**	352/16-30	**7-11**	369/*P*: 1-12	**8-4** **S**	395/1-31 odd, 32-36 385/32-40 even
97	**7-8**	357/1-18	**7-11** **R**	370/*P*: 13-24 371/*Self-Test 3*	**8-5** **R**	400/1-43 odd, 45-50 402/*Self-Test 2*
98	**7-8** **S**	357/19-30 339/*P*: 5-8	*Prepare for Chapter Test* **R** 374/*Chapter Review* **EP** 666/*Skills* 678/*Problem Solving*		**8-6**	406/1-45 odd, 46-52
99	**7-9**	360/1-15	*Administer Chapter Test* 349/*Reading Algebra*: 1-6 **R** 377/*Cumulative Review*		**8-7**	409/1-27 odd, 29-32
100	**7-9** **R**	360/16-30 361/*Self-Test 2*	**8-1**	384/1-31 odd, 33-40	**8-8** **S**	412/1-37 odd, 38-44 406/38-44 even
101	**7-10**	366/1-14	**8-2**	386/1-23 odd, 24-28	**8-9**	418/1-22
102	**7-10**	366/15-24	**8-3** **R**	391/1-15 odd, 17-24 392/*Self-Test 1*	**8-9** **R**	418/23-34 419/*Self-Test 3*
103	**7-11** **S**	369/*P*: 1-6 367/25-27	**8-4** **S**	395/1-25 odd, 26-30 384/22-32 even	*Prepare for Chapter Test* **R** 420/*Chapter Review* **EP** 668/*Skills*	
104	**7-11** **R**	370/*P*: 7-14 371/*Self-Test 3*	**8-5** **R**	400/1-31 odd, 33-44 402/*Self-Test 2*	*Administer Chapter Test* **R** 423/*Mixed Review*	
105	*Prepare for Chapter Test* **R** 374/*Chapter Review* **EP** 666/*Skills* 678/*Problem Solving*		**8-6**	406/1-29 odd, 31-42	**9-1**	429/1-31 odd, 33-38
106	*Administer Chapter Test* 349/*Reading Algebra*: 1-6 **R** 377/*Cumulative Review*		**8-7**	409/1-23 odd, 24-28	**9-2** **R**	433/*P*: 1-20 436/*Self-Test 1*
107	**8-1**	384/1-20	**8-8**	412/1-23 odd	**9-3**	438/7-39 odd, 41-45
108	**8-2**	386/1-20	**8-8** **S**	413/25-36 406/22-30 even	**9-4**	442/*P*: 5-11 odd, 13-24
109	**8-3** **R**	391/1-16 392/*Self-Test 1*	**8-9**	418/1-18	**9-5**	445/*P*: 1-10

Day	Minimum Course		Average Course		Maximum Course	
110	**8-4** **S**	395/1-18 384/21-25	**8-9** **R**	418/19-28 419/*Self-Test 3*	**9-5** **S**	446/*P*: 11-16 439/30-40 even
111	**8-5** **R**	400/1-32 402/*Self-Test 2*	*Prepare for Chapter Test* **R** 420/*Chapter Review* **EP** 668/*Skills*		**9-6** **R**	448/*P*: 5-14 449/*Self-Test 2*
112	**8-6**	406/1-27	*Administer Chapter Test* **R** 423/*Mixed Review*		**9-7**	453/5-41 odd 455/*P*: 15-24
113	**8-7**	409/1-18	**9-1**	429/1-27 odd, 29-32	**9-8**	461/*P*: 1-7 odd, 9-22
114	**8-8**	412/1-12	**9-2** **R**	433/*P*: 1-16 436/*Self-Test 1*	**9-9** **S**	465/*P*: 7-20 454/38, 40, 42 456/*P*: 25, 26
115	**8-8** **S**	413/13-24 406/28-32	**9-3**	438/1-39 odd	**9-10** **R**	470/*P*: 6-20 472/*Self-Test 3*
116	**8-9**	418/1-8	**9-4**	442/*P*: 1-11 odd, 13-20	*Prepare for Chapter Test* **R** 473/*Chapter Review* **EP** 669/*Skills* 678/*Problem Solving*	
117	**8-9** **R**	418/9-18 419/*Self-Test 3*	**9-5**	445/*P*: 1-8	*Administer Chapter Test* **R** 477/*Cumulative Review*	
118	*Prepare for Chapter Test* **R** 420/*Chapter Review* **EP** 668/*Skills*		**9-5** **S**	445/*P*: 9-14 439/30-40 even	**10-1**	485/7-29 odd, 30-34
119	*Administer Chapter Test* **R** 423/*Mixed Review*		**9-6** **R**	447/*P*: 1-7 odd, 8-13 449/*Self-Test 2*	**10-2**	490/7-49 odd, 50
120	**9-1**	429/1-24	**9-7**	453/1-31 odd 454/*P*: 1-19 odd	**10-3**	495/1-45 odd, 47-56
121	**9-2** **R**	433/*P*: 1-10 436/*Self-Test 1*	**9-8**	461/*P*: 1-16	**10-4** **S**	500/1-43 odd 490/26-40 even
122	**9-3**	438/1-27 odd	**9-9** **S**	465/*P*: 1-9 odd, 11-16 454/33-36	**10-4** **R**	500/44-61 502/*Self-Test 1*
123	**9-4**	442/*P*: 1-12	**9-10** **R**	470/*P*: 4-18 472/*Self-Test 3*	**10-5**	510/5-41 odd 511/*P*: 1-11 odd, 14-18
124	**9-5**	445/*P*: 1-6	*Prepare for Chapter Test* **R** 473/*Chapter Review* **EP** 669/*Skills* 678/*Problem Solving*		**10-6** **R**	514/1-29 odd, 30-33 516/*Self-Test 2*

Day	Minimum Course		Average Course		Maximum Course	
125	**9-5** S	445/P: 7-11 439/20-28 even	*Administer Chapter Test* R	*477/Cumulative Review*	**10-7**	519/1-43 odd, 45-48
126	**9-6** R	447/P: 1-7 449/Self-Test 2	**10-1**	484/1-23 odd, 25-32	**10-8**	523/9-55 odd
127	**9-7**	453/1-26	**10-2**	490/1-49 odd	**10-9** S	526/1-27 odd, 28-32 520/49-52
128	**9-7**	454/P: 1-14	**10-3**	495/1-33 odd, 34-44	**10-10** R	529/9-37 odd, 38 530/Self-Test 3
129	**9-8**	461/P: 1-8	**10-4** S	500/1-24 490/20-28 even	*Prepare for Chapter Test* R EP	534/Chapter Review 670/Skills 679/Problem Solving
130	**9-9** S	465/P: 1-10 448/P: 8, 9, 10	**10-4** R	500/25-47 502/Self-Test 1	*Administer Chapter Test* R	504/Reading Algebra: 1-5 532/Extra: 1-24 537/Mixed Review
131	**9-10** R	470/P: 1-12 472/Self-Test 3	**10-5**	510/1-37 odd 511/P: 1-12	**11-1**	547/1-43 odd
132	*Prepare for Chapter Test* R EP	473/Chapter Review 669/Skills 678/Problem Solving	**10-6** R	514/1-25 odd, 27-30 516/Self-Test 2	**11-2**	550/1-41 odd, 43-48
133	*Administer Chapter Test* R	*477/Cumulative Review*	**10-7**	519/1-35 odd, 36-40	**11-3** R	554/P: 8-17 556/Self-Test 1
134	**10-1**	484/1-24	**10-8**	523/1-49 odd	**11-4** S	562/7-33 odd 551/38, 40, 42
135	**10-2**	490/1-24	**10-9** S	526/1-15 odd, 17-28 520/41-44	**11-5**	565/2-20 even
136	**10-3**	495/1-15	**10-10**	529/1-18	**11-6**	569/1-12
137	**10-3**	495/16-32	**10-10** R	529/19-30 530/Self-Test 3	**11-6** R	569/13-28 569/Self-Test 2
138	**10-4**	500/1-27 odd	*Prepare for Chapter Test* R EP	534/Chapter Review 670/Skills 679/Problem Solving	*Prepare for Chapter Test* R EP	573/Chapter Review 672/Skills 680/Problem Solving

Day	Minimum Course		Average Course		Maximum Course	
139	**10-4** **R**	500/2-28 even, 29-36 502/*Self-Test 1*	*Administer Chapter Test* **R**	504/*Reading Algebra*: 1-5 537/*Mixed Review*	*Administer Chapter Test* **R**	572/*Extra*: 1-16 575/*Cumulative Review*
140	**10-5**	510/1-8 511/*P*: 1-3	**11-1**	547/1-35 odd	**12-1**	582/1-30
141	**10-5** **S**	510/9-24 511/*P*: 4-8 490/25-28	**11-2**	550/1-35 odd	**12-2**	588/1-23
142	**10-6** **R**	514/1-12 516/*Self-Test 2*	**11-3** **R**	554/*P*: 1-14 556/*Self-Test 1*	**12-2**	589/24-38
143	**10-7**	519/1-19 odd	**11-4** **S**	562/1-27 odd 551/26-32 even	**12-3**	592/1-34
144	**10-7**	519/2-20 even, 21-32	**11-5**	565/1-17 odd	**12-4** **R**	596/3-37 odd 599/*P*: 1-11 odd 601/*Self-Test 1*
145	**10-8**	523/1-10, 11-23 odd	**11-6**	569/1-8, 13-16	**12-5** **S**	606/1-22 599/*P*: 2-10 even
146	**10-9** **S**	526/1-16 515/19-22	**11-6** **R**	569/17-24 569/*Self-Test 2*	**12-6**	610/1-20
147	**10-10**	529/1-12	*Prepare for Chapter Test* **R** **EP**	573/*Chapter Review* 672/*Skills* 680/*Problem Solving*	**12-7** **R**	614/*P*: 1-11 615/*Self-Test 2*
148	**10-10** **R**	529/13-25 530/*Self-Test 3*	*Administer Chapter Test* **R**	575/*Cumulative Review*	*Prepare for Chapter Test* **R** **EP**	616/*Chapter Review* 673/*Skills* 680/*Problem Solving*
149	*Prepare for Chapter Test* **R** **EP**	534/*Chapter Review* 670/*Skills* 679/*Problem Solving*	**13-1**	625/1-10	*Administer Chapter Test* **R**	618/*Mixed Review*
150	*Administer Chapter Test* **R**	504/*Reading Algebra*: 1-5 537/*Mixed Review*	**13-2**	628/1-7	**13-1**	625/1-11
151	**11-1**	547/1-25 odd	**13-3** **R**	632/1-14 636/*Self-Test 1*	**13-2**	628/1-8

Day	Minimum Course		Average Course		Maximum Course	
152	**11-2**	550/1-12	**13-4**	639/1-16	**13-3** **R**	632/1-15 636/*Self-Test 1*
153	**11-2**	550/13-24	**13-5** **S**	642/1-6 634/15	**13-4**	639/1-20
154	**11-3** **R**	554/*P:* 1-10 556/*Self-Test 1*	**13-5**	642/7-11 643/*Self-Test 2*	**13-5** **S**	642/1-6 635/16
155	**11-4** **S**	562/1-6 551/25, 26, 27	**13-6**	647/1-26	**13-5** **R**	642/7-12 643/*Self-Test 2*
156	**11-4**	562/7-11, 12-22 even	**13-6**	648/27-46	**13-6**	647/1-26
157	**11-5**	565/1-12	**13-7**	651/1-8	**13-6**	648/27-48
158	**11-6** **R**	569/1-12 569/*Self-Test 2*	**13-7** **R**	651/9-15 652/*Self-Test 3*	**13-7** **R**	651/1-15 652/*Self-Test 3*
159	*Prepare for Chapter Test* **R** 573/*Chapter Review* **EP** 672/*Skills* 680/*Problem Solving*		*Prepare for Chapter Test* **R** 654/*Chapter Review* **EP** 674/*Skills*		*Prepare for Chapter Test* **R** 654/*Chapter Review* **EP** 674/*Skills*	
160	*Administer Chapter Test* **R** 575/*Cumulative Review*		*Administer Chapter Test* **R** 656/*Cumulative Review*		*Administer Chapter Test* **R** 656/*Cumulative Review*	

Term Projects

Following is a list of topics suitable for written term papers or class presentations by students. A brief description of each topic aimed at generating student interest and a short list of suggested references designed to give the student a starting point are provided.

Semester I

Figurate Numbers

The Pythagoreans of ancient Greece were deeply interested in the geometric and physical representation of numbers by points in a plane and the principles involved with these numbers. In the plane, these points took the shape of polygons and in space, the form of polyhedra. What are some examples of these numbers? How are the numbers generated? Is there symmetry within a pattern?

Gardner, Martin. "On the Patterns and the Unusual Properties of Figurate Numbers." *Scientific American* 231: 116–120; July, 1974.

Hartman, Janet. "Figurate Numbers." *Mathematics Teacher* 69: 47–50.

Historical Topics for the Mathematics Classroom. The Thirty-first Yearbook of the National Council of Teachers of Mathematics. Washington, DC: The Council, 1969.*

Olson, Melfried, et al. "Triangular Numbers: The Building Blocks of Figurate Numbers." *Mathematics Teacher* 76: 624–625.

Weaver, Cloman. "Figurate Numbers." *Mathematics Teacher* 67: 661–666.

*The current address of the National Council of Teachers of Mathematics is 1906 Association Drive, Reston, VA 22091.

Graphing in Three Dimensions

Just as every real number can be associated with a point on a line, and every ordered pair of real numbers can be associated with a point in a plane, every ordered triple of real numbers can be associated with a point in space.

Dolciani, Mary P., et al. *Algebra 2 and Trigonometry.* Boston: Houghton Mifflin Company, 1986.

Jenks, G. F., and Browns, D. A. "Three-Dimensional Map Construction." *Science*, Vol. 54, No. 3750 (18 Nov. 1966), pp. 857–864.

Keller, Jane, and Anderson, Robert. "A Model of 3-Space." *Mathematics Teacher* 74: 350–353.

Königsberg Bridge Problem

Through the city of Königsberg flowed a river. In the river were two islands connected to the mainland and each other by seven bridges. A question was posed—is it possible to walk through the city and cross each bridge only once? The resolution of this simple puzzle was the beginning of the intricate theory of topology.

Hollist, J. Taylor. "A New Look at an Old Puzzle." *Mathematics Teacher* 70: 2–3.

Honsberger, Ross. *Mathematical Gems.* Washington, DC: Mathematical Association of America, 1973. Chapters 2 and 7.

Jacobs, Harold R. *Mathematics—A Human Endeavor.* 2d ed. San Francisco: W. H. Freeman and Company, 1982.

Latin Squares

Many experiments in social science, biology, and medicine make use of Latin Squares. What are Latin Squares? Graeco-Latin Squares? What is

Leonhard Euler's contribution to the study of these squares? What are orthogonal Graeco-Latin Squares?

Alter, Ronald. "How Many Latin Squares Are There?" *American Mathematical Monthly* 82: 632–634; June–July, 1975.

Gardner, Martin. *Martin Gardner's New Mathematical Diversions from "Scientific American."* Chicago, IL: University of Chicago Press, 1984.

Warrington, P. D. "Graeco-Latin Cubes." *Journal of Recreational Mathematics* 6: 47–53; Winter, 1973.

Wynne, Bayard E. "Perfect Magic Cubes of Order Seven." *Journal of Recreational Mathematics* 8: 285–293; 1975–76.

Magic Squares

A square array of numbers such that the sums of the numbers in each row, each column, and each diagonal are all equal is called a *magic square*. How do you form a 3×3, 4×4, 5×5, . . . , $n \times n$ magic square? How is the magic constant found in an $n \times n$ magic square? How do you solve a magic square algebraically? How does multiplication or addition affect the magic square and its constant? What are anti-magic squares?

Benson, William H., and Jacoby, Oswald. *New Recreations with Magic Squares.* New York: Dover Publications, 1976.

Engelmeyer, William J. "Magic Squares." *Mathematics Teacher* 68: 399–402.

Gardner, Martin. "A Breakthrough in Magic Squares, and the First Perfect Magic Cube." *Scientific American* 234: 118–122; Jan., 1976.

Lyon, Betty Clayton. "Using Magic Borders to Generate Magic Squares." *Mathematics Teacher* 77: 223–226.

Munger, Ralph. "An Algebraic Treatment of Magic Squares." *Mathematics Teacher* 66: 101–107.

Modular Arithmetic

What is modular arithmetic? What rules of ordinary arithmetic hold true in modular arithmetic? What are some of the applications of modular arithmetic to everyday life? How can it be used to devise codes?

Boyd, Henry. "Modulo Seven Arithmetic—a Perfect Example of Field Properties." *Mathematics Teacher* 65: 525–528.

Park, Lyman C. *Secret Codes, Remainder Arithmetic and Matrices.* Washington, DC: National Council of Teachers of Mathematics, 1964.

Pomerance, Carl. "The Search for Prime Numbers." *Scientific American,* Vol. 247, pp. 136–147; Dec., 1982.

Singer, Richard. "Modular Arithmetic and Divisibility Criteria." *Mathematics Teacher* 63: 653–656.

Struik, Dirk J. *A Source Book in Mathematics, 1200–1800.* Cambridge, MA: Harvard University Press, 1969.

Random Numbers

When a number is chosen at random, how random is it? Why is there a need for books of random numbers? How are random numbers generated? How are they used?

Gardner, Martin. *Mathematical Carnival.* New York: Alfred A. Knopf, 1975.

Gardner, Martin. "Order and Surprise." *Philosophy of Science*, Jan., 1950, pp. 109–117.

Hull, T. E., and Dobell, A. R. "Random Number Generators." *SIAM Review*, July, 1962, pp. 230–254.

Kimberling, Clark. "Microcomputer-Assisted Discoveries: Generate Your Own Random Numbers." *Mathematics Teacher* 77: 118–123.

Wallis, W. Allen, and Roberts, Henry V. *The Nature of Statistics.* New York: Free Press, 1962. Chapter 6.

Tests of Divisibility

Simple tests of divisibility by 2, 3, 4, 5, 8, 9, 10, 11, and 12 exist. What are these tests? Why do they work? What are some tests of divisibility by 7?

Burton, David M. "Devising Divisibility Tests." *Journal of Recreational Mathematics* 9(4): 258–260; 1976–77.

Engle, Jessie Ann. "A Rediscovered Test for Divisibility by Eleven." *Mathematics Teacher* 69: 669.

Gardner, Martin. *The Unexpected Hanging and Other Mathematical Diversions*. New York: Simon and Schuster, 1972.

Glenn, W. H., and Johnson, Donovan. *Invitation to Mathematics*. New York: Doubleday and Co., 1962.

continued fractions related to the Euclidean algorithm? to repeating decimals? to irrational numbers? to approximations for π and e?

Henle, James M. *Numerous Numerals*. Reston, VA: National Council of Teachers of Mathematics, 1975.

Historical Topics for the Mathematics Classroom. Thirty-first Yearbook of the National Council of Teachers of Mathematics. Washington, DC: The Council, 1969.

Holmes, Joseph E. "Continued Fractions." *Mathematics Teacher* 61: 12–17.

Kasner, Edward, and Newman, James. *Mathematics and the Imagination*. New York: Simon and Schuster, 1961.

Moore, Charles G. *An Introduction to Continued Fractions*. Reston, VA: National Council of Teachers of Mathematics, 1964.

Semester II

Complex Numbers

Complex numbers are of the form $a + bi$ where a and b are real numbers and $i = \sqrt{-1}$. How did Diophantus, Descartes, Euler, and Gauss contribute to the development of complex numbers? Why are complex numbers necessary? How are they related to the coordinate plane?

Bell, E. T. *Men of Mathematics*. New York: Simon and Schuster, 1937.

Green, D. R. "The Historical Development of Complex Numbers." *The Mathematical Gazette,* Vol. 60, No. 412, pp. 99–107; June, 1976.

Continued Fractions

A continued fraction is of the following form: a number plus a fraction whose denominator is a number plus a fraction, and so on. How are

Fibonacci Sequence

Leonardo of Pisa in his book, *Liber abaci* (1202), included a problem about rabbits that was answered in the following sequence: 1, 1, 2, 3, 5, 8, 13, . . . , later known as the *Fibonacci sequence*. How does this sequence appear in nature? What is the significance of the ratio of the numbers? What is its relationship to the Lucas sequence? to the Golden Ratio? How is the sequence related to Pythagorean triples?

Brown, Stephen I. "From the Golden Rectangle and Fibonacci to Pedagogy and Problem Posing." *Mathematics Teacher* 69: 180–188.

Dalton, Leroy C., and Snyder, Henry D., eds. *Topics for Mathematics Clubs*. 2d ed. Reston, VA: National Council of Teachers of Mathematics, 1983.

Gardner, Martin. "The Multiple Fascinations of the Fibonacci Sequence." *Scientific American*, Vol. 220, No. 3, pp. 116–120; March, 1969.

Verno, C. Ralph. "Fibonacci Numbers and Pythagorean Triples." *Mathematics Teacher* 66: 652.

The Number π

The number π is the ratio of the circumference of a circle to the diameter of the circle. Archimedes, in 240 B.C., wrote that the value of π is less than $3\frac{1}{7}$ but greater than $3\frac{10}{71}$. Throughout the centuries, mathematicians have sought a simple, yet elegant, means of calculating π. How is π related to the problem of "squaring the circle"? What are some of the formulas for determining π? How can an approximation for π be derived statistically? Is π an algebraic number? How can π be expressed using continued fractions?

Beckmann, Petr. *A History of π.* 4th ed. Boulder, CO: Golem Press, 1977.

Gardner, Martin. *Martin Gardner's New Mathematical Diversions from "Scientific American."* Chicago, IL: University of Chicago Press, 1984.

Hatcher, Robert S. "Some Little-known Recipes for π." *Mathematics Teacher* 66: 470–474.

Vervoort, Gerardus. "Pie in the Street, or How to Calculate π from the License Plates in the Parking Lot." *Mathematics Teacher* 68: 580–582.

Pascal's Triangle

This triangular array of numbers has many exciting properties and applications. How is it formed? What are some of the patterns found in the triangle? Is there a rule for generating the nth row? How is Pascal's triangle related to probability? to the Fibonacci sequence?

Dalton, LeRoy C., and Snyder, Henry D., eds. *Topics for Mathematics Clubs.* 2d ed. Reston, VA: National Council of Teachers of Mathematics, 1983.

Historical Topics for the Mathematics Classroom. Thirty-first Yearbook of the National Council of Teachers of Mathematics. Washington, DC: The Council, 1969.

Kenney, Margaret J. *The Incredible Pascal's Triangle.* Chestnut Hill, MA: Boston College Press, 1976.

Stanley, Francis W. "Serendipitous Discovery of Pascal's Triangle." *Mathematics Teacher* 68: 95–98.

Struik, Dirk J. *A Concise History of Mathematics.* 3d ed. New York: Dover Publications, 1967.

Prime Number Conjectures

There are many unproved conjectures concerning prime numbers. Are there infinitely many primes of the form n and $n + 2$, i.e., twin primes, such as 3 and 5, or 29 and 31? Can every integer greater than 2 be represented as the sum of two primes (Goldbach's Conjecture)? Is the set of primes determined by $n^2 + 1$ infinite? What are some methods for finding prime numbers?

Card, Leslie E. "Additional Twin Prime Curiosities." *Journal of Recreational Mathematics* 6: 202–203; Summer, 1973.

Enrichment Mathematics for High School. Twenty-eighth Yearbook of the National Council of Teachers of Mathematics. Washington, DC: The Council, 1963.

Eves, Howard. *An Introduction to the History of Mathematics.* 5th ed. Philadelphia: Saunders College Publishing, 1983.

Kennedy, Robert E. "Proofs of the Infinitude of Primes." *Pentagon* 35: 25–27; Fall, 1975.

Ore, Oystein. *Number Theory and Its History.* New York: McGraw-Hill, 1948.

Pomerance, Carl. "The Search for Prime Numbers." *Scientific American,* Vol. 247, pp. 136–147; Dec., 1982.

Pythagorean Triples

For thousands of years, mathematicians have been motivated to find sets of three integers that satisfy Pythagoras's formula, $a^2 + b^2 = c^2$. Properties of these triples and formulas that generate them also have been of prime concern. How did Pythagoras derive these numbers from figurate numbers? Are there formulas for generating the triples?

Cohen, Israel. "Pythagorean Numbers." *Mathematics Teacher* 67: 667–669.

Eves, Howard. *An Introduction to the History of Mathematics.* 5th ed. Philadelphia: Saunders College Publishing, 1983.

Historical Topics for the Mathematics Classroom. Thirty-first Yearbook of the National Council of Teachers of Mathematics. Washington, DC: The Council, 1969.

Roensch, Steve. "Pythagorean Triples." *Mathematics Teacher* 70: 388–389.

Rothbart, Andrea, and Paulsell, Bruce. "Pythagorean Triples: A New Easy-to-derive Formula with Some Geometric Applications." *Mathematics Teacher* 67: 215–218.

Verno, C. Ralph. "Fibonacci Numbers and Pythagorean Triples." *Mathematics Teacher* 66: 652.

Sequences and Series

From the time of the ancient Hindus and Chinese, mathematicians have been exploring the orderly progression of numbers. What is an arithmetic progression? a geometric progression? a harmonic progression? What is the difference between a sequence and a series? How can you find the value of the nth term of a sequence? How can you determine which infinite series have a sum and how can you find the sum?

Ellis, Wade. "Sequences and Progressions." *Mathematics Teacher* 64: 455–458.

Gardner, Martin. *Martin Gardner's Sixth Book of Mathematical Games from "Scientific American."* Chicago: University of Chicago Press, 1984.

Lyng, Merwin J. "Theoretical and Practical Solutions for a Geometric Sequence." *Mathematics Teacher* 61: 393.

Smith, David Eugene. *History of Mathematics,* Vols. I and II. New York: Dover Publications, 1957.

Solving Polynomial Equations of Degree Greater than Two

The quadratic formula provides a simple method for solving any quadratic equation. Is there an equally simple formula for solving cubic or quartic equations? What is Horner's Method? Cardan's Formula? Descartes's Rule of Signs? the method of Double False Position?

Easton, Joy B. "The Rule of Double False Position." *Mathematics Teacher* 60: 56–58.

Eves, Howard. *An Introduction to the History of Mathematics.* 5th ed. Philadelphia: Saunders College Publishing, 1983.

Historical Topics for the Mathematics Classroom. Thirty-first Yearbook of the National Council of Teachers of Mathematics. Washington, DC: The Council, 1969.

Lesson Commentary

1 Numbers and Variables

This chapter introduces basic algebraic concepts while reviewing essential arithmetic properties and operations. The major subsets of the real numbers are reviewed, along with the vocabulary and notation of simple set theory. Numerical expressions, grouping symbols, and order of operations are introduced. Students are then asked to translate word phrases and sentences into algebraic phrases and sentences. Simple equations are solved over specific replacement sets.

Negative signs are raised as in $^-5$, to encourage students to think of the negative sign as part of the number. Do not allow students to read $^-5$ as "minus 5." The negative signs in this book will be lowered to the conventional position when the concept of opposites is introduced on page 62.

Have students practice reading "$<$" and "$>$" statements so they can correctly translate them with ease.

1-1 (pages 1–5)

Teaching Suggestions

Emphasize the one-to-one correspondence between the set of real numbers and the set of points on a number line. (The term *one-to-one correspondence* is defined in Section 1-2.) Point out that the location of the origin and the size of the intervals are arbitrary. A chalkboard diagram like the following should help students visualize that the one-to-one correspondence between the set of real numbers and the points on a number line exists regardless of the size of the interval.

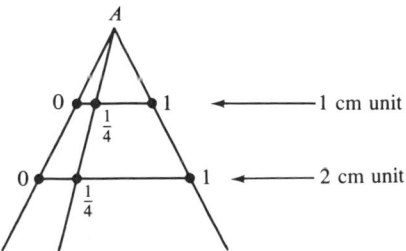

Let the 1 cm unit represent the numbers from 0 to 1. Then a point such as *A* can be used to project every number on the 1 cm unit into the larger 2 cm unit.

1-2 (pages 6–10)

Teaching Suggestions

If your students are already familiar with this material, this section may be completed rapidly. Be sure, however, that students understand the meaning of the ideas and symbols before continuing. Sets are described by roster, rule, and graph throughout mathematics. Point out that every nonempty set has at least two subsets, itself and the empty set.

Caution students that {0} and {∅} are symbols that express different ideas. Both are sets with one element, but {0} has one number as its element and {∅} has one set as its element. The students should also recognize that ∅ and { } are both symbols for the empty set, which has no elements.

Related Activities

Have students create Venn diagrams for sets containing non-numerical members. For example, ask students to make a Venn diagram for the following sets:

a. dogs, cats, mammals, poodles, kittens

b. crops, barley, grain, berries, alfalfa, fruits

c. James Madison, authors, Americans, presidents, Louisa M. Alcott, Andrew Jackson, Mark Twain

d. plants, roses, trees, maples, moss, daisies, flowers, elms, ivy

If students draw different Venn diagrams, ask them to support their cases.

Have students create collections of their own for which their classmates will make Venn diagrams. So that the work can be checked, the creators should make diagram keys of their collections before giving their lists to a classmate.

1-3 (pages 11–15)

Teaching Suggestions

Some students confuse the set of natural numbers and the set of whole numbers and may need extra help differentiating between them. All students should understand the meaning of integers, and of natural, whole, rational, irrational, and real numbers. They should also know the abbreviations used in this text: J (integers), N (natural), Q (rational), W (whole), and R (real).

To enhance understanding of the relationships among the real, rational, and irrational numbers, point out that the set of rationals and the set of irrationals together make up the set of real numbers.

Related Activities

Repeating decimals are discussed in Section 10-2. However, some students will enjoy exploring this topic before it is formally presented to them.

Repeating decimals often develop patterns that interest students. For each of the following fractions, ask students to divide the numerator by the denominator until the quotient digits begin to repeat:

$$\frac{1}{7}, \frac{2}{7}, \frac{3}{7}, \frac{4}{7}, \frac{5}{7}, \frac{6}{7}$$

Have students note that each repeating decimal uses the same six digits and that each digit is followed by the same digit; only the first digit changes.

Now have students repeat the division for these fractions:

$$\frac{1}{9}, \frac{2}{9}, \frac{3}{9}, \frac{4}{9}, \frac{5}{9}, \frac{6}{9}, \frac{7}{9}, \frac{8}{9}$$

Ask them what the decimal representation of $\frac{9}{9}$ would be if it followed the observed pattern. Ask them if they think that $0.999\ldots = 1.0$. Have them give reasons to support their answer. (The statement is true but it will not be proved in Algebra I.)

Have students use calculators to identify other patterns for repeating decimals, first noting that calculators will not show enough places to aid them in all cases.

In Chapter 10, students will learn how to find a common fraction equivalent for a given repeating decimal. You may want to recall this activity then.

1-4 (pages 17–20)

Teaching Suggestions

Point out the difference between a numerical expression and a variable expression. Note that a numerical expression can be evaluated immediately by simplifying, whereas a variable expression can be evaluated only after the replacement set of the variable is given.

To be sure that students understand the substitution principle, ask them to give you its meaning in their own words.

Use the term domain when you describe the replacement set. As students evaluate expressions, remind them that there is a convention for the order of operations. This convention is summarized in Section 1-6.

1-5 (pages 21–24)

Teaching Suggestions

To illustrate the use of grouping symbols, ask students to insert parentheses in the expression $3 + 5 \cdot 4 - 2$ to name as many different numbers as possible. It can represent 30, 16, 21, or 13.

Emphasize that when an expression contains more than one set of grouping symbols, the expression within the innermost symbols must be simplified first. Encourage students to write out the results in each step of the simplifying process as is done in the examples. In this way, they will be less likely to make errors in evaluating expressions.

Have interested students explore recreational mathematics books to find puzzles to fit this concept. For example, how many ways can you show a total of 4, using only 2's?

1-6 (pages 24–27)

Teaching Suggestions

Be sure students understand the meaning of the terms power, base, and exponent. Extend the exponent concept with the generalization, $a^n = a \cdot a \cdot a \cdot \ldots \cdot a$ where there are n factors. Emphasize that the exponent determines the number of repeated factors, not the number of multiplications. Also point out that $2x^3 = 2 \cdot x \cdot x \cdot x$ and not $2 \cdot x \cdot 2 \cdot x \cdot 2 \cdot x$.

Students should begin to memorize common squares (1, 4, 9, . . . , 169, 225, 400, 625), and some common cubes (1, 8, 27, 64, 125). They should be able to mentally compute the lower powers of 2 (2, 4, 8, 16, 32, 64, 128, 256).

Point out that the order of operations stated on page 25 is an arbitrary but standard rule. Use the following examples to demonstrate the need for the convention in evaluating ungrouped expressions.

$$48 \div 6 \times 2 = (48 \div 6) \times 2$$
$$\neq 48 \div (6 \times 2)$$
$$19 - 5 + 7 = (19 - 5) + 7$$
$$\neq 19 - (5 + 7)$$

Related Activities

Have students construct or draw equilateral triangles. As shown in the diagrams that follow, three

of these triangles can form a trapezoid; three trapezoids can form a larger triangle; and so on. Each figure or "level" represents an increasing power of 3. Have students interpret $3^1 = 3$, $3^2 = 9$, $3^3 = 27$, . . . in terms of levels and the number of triangles. (Note: $3^0 = 1$ can be interpreted as Level 0 and would be represented by one triangle. The concept of a zero exponent is discussed in Section 8-5.)

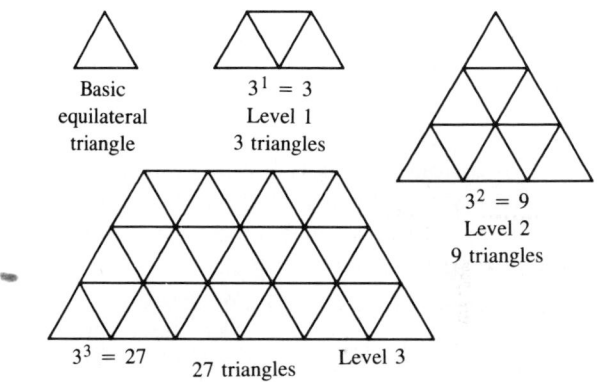

Basic equilateral triangle

$3^1 = 3$
Level 1
3 triangles

$3^2 = 9$
Level 2
9 triangles

$3^3 = 27$
27 triangles
Level 3

1-7 (pages 30–33)

Teaching Suggestions

Mathematical sentences include both equations and inequalities. A sentence with $<$, $\leq$, $>$, or $\geq$ is an inequality and not an equation.

Solving simple sentences over specific domains can often be done by inspection. For instance, in solving $x + 2 = 5$ over the domain {1, 2, 3, 4} it is easy to see that the solution set is {3}. The important idea is to demonstrate that some elements of the domain may make the sentence a true statement, while others may not.

This text is written to accommodate your asking for either the "solution" or "solution set" of mathematical sentences.

Students must be careful to note the domain when solving an open sentence. For example, $2x = 5$ has no solution if the domain is the set of integers. If, however, the domain is the set of rational numbers, it does have a solution.

1-8 (pages 35-39)

Teaching Suggestions

Remind students that several different words can identify the same arithmetic operation. For instance, "increased by," "sum," and "plus" all identify addition. If necessary, hold oral drills to associate various key words with their related arithmetic operations.

A common error occurs when students attempt to translate word by word rather than by phrases (for example, "two less than m" becomes $2 - m$ rather than $m - 2$).

Encourage the use of parentheses to keep connected expressions together. For those having difficulties, write word phrases on the chalkboard and analyze them with boxes and underlining to help students identify the separate parts.

Have students create their own word phrases for other students to translate. Also, have them create as many word phrases as possible from a given variable expression.

The students should become proficient with this skill before they move on to translating complete sentences into equations and inequalities.

1-9 (pages 40-43)

Teaching Suggestions

To aid students in making translations, point out that word sentences are usually written in two parts with the connection being "is" or another verb. This verb shows the relationship between the two parts of the sentence and is translated as either "=" or an inequality. Have students identify and translate the word or words signifying the relationship. Then have them translate the two parts of the sentence that will form the two sides of the equation or inequality.

This section can be difficult for some students because of the different ways of expressing a mathematical sentence in English. Remind them that being a careful reader is an important mathematical skill. Suggest that students read the sentences more than once before they start to translate them. To help students further, use chalkboard demonstrations with boxes and underlining to identify and analyze the component parts.

Remind students that the emphasis in this section is on *writing* mathematical sentences for word sentences, not on *solving* open sentences.

2 Working with Real Numbers

In this chapter the field properties of real numbers are named and stated formally. Practice is given in using these axioms to simplify expressions and prove statements or theorems.

Addition of real numbers and the concept of an additive inverse are introduced through the use of number lines. Thorough practice of arithmetic skills is stressed. Absolute value is introduced through the idea of opposites. The chapter concludes with the presentation of the rules for subtraction, multiplication, and division of real numbers.

2-1 (pages 49-52)
Teaching Suggestions

Point out to students that quantifiers are used in combination with variables to determine if open sentences are true or false. Note that words such as "any," "all," "each," and "every" are used for universal statements. Demonstrate with examples that only one counterexample is needed to prove a universal statement false. Add that other words and phrases, such as "some," "there is," "there exists," and "at least one," are used in existence statements. For such a statement to be true, only one true case needs to be exhibited.

Before assigning problems involving sets of numbers, review the characteristics of natural numbers, whole numbers, integers, rational numbers, and real numbers.

Related Activities

The symbols "$\forall$" and "$\exists$" are called the universal and existential quantifiers, respectively, and are read "for each" (or "for all") and "there exists" (or "there is at least one").

As examples of their use, the true statement "$\forall_x \ x + 1 > x$" is translated "For each real number x, $x + 1$ is greater than x," and the true statement "$\exists_x \forall_y \ x + y = y$" is translated "There exists a real number x such that for all real numbers y, $x + y = y$." (The second statement is true because of the existence of the number zero.)

Decide whether each of the following statements is true or false, if x, y, and z are real numbers. Give a reason for each answer.

1. $\forall_x \exists_y \ x + y = 0$
   True (Axiom of additive inverses)

2. $\exists_x \forall_y \ x + y = x$
   False (For $y = 1$, for example, we have $x + 1 = x$. The solution set is $\emptyset$.)

3. $\forall_x \forall_y \ x + y = y + x$
   True (Commutative axiom of addition)

4. $\exists_x x^2 = 25$ True ($5^2 = 25$)

Have the students write several statements using quantifiers, noting their truth values. Then have them exchange only the statements and repeat the instructions for statements 1-4.

2-2 (pages 52–57)

Teaching Suggestions

The axioms of closure, commutativity, and associativity for addition and multiplication, and the axioms of equality are presented as statements assumed to be true. To demonstrate the need for axioms or postulates, have the students try to define a common idea (e.g., table, coin) in simpler terms. Whatever words are used in the definition, questions could be asked about their meaning. Students should realize that in order to avoid an endless chain of definitions, some words must be accepted without further explanation.

Students should know that the axioms will be used to prove various mathematical statements and to justify steps in simplifying expressions and solving equations. In remembering the axioms, students normally connect "commutative" with "order" and "associative" with "grouping."

2-3 (pages 57–61)

Teaching Suggestions

Students should learn to recognize the various forms of the distributive axiom, some of which are $3a + 4a = (3 + 4)a$, $7(c + d) = 7c + 7d$, and $5 \times 3\frac{2}{5} = (5 \times 3) + (5 \times \frac{2}{5})$. You may wish to generalize the use of the distributive axiom to more than two terms by proving $a(b + c + d) = ab + ac + ad$.

Remind students to distribute over each term; a common error is to rewrite $3(a + b)$ as $3a + b$.

Students sometimes think that multiplication will distribute over multiplication. Use a counterexample to show this is not true, e.g., $5(2 \cdot 3) \neq (5 \cdot 2)(5 \cdot 3)$.

Note that equivalent expressions are ones which are equal for all values of the variables. For example, $3x = 2x + 5$ is true for only one value of x, but $3x = 2x + x$ is true for all values of x. Therefore, $3x$ and $2x + x$ are equivalent expressions, but $3x$ and $2x + 5$ are not.

Related Activities

For all real numbers a and b, consider the operation $\otimes$ defined by $a \otimes b = 2a + b$ and the operation $\oplus$ defined by $a \oplus b = 2ab$.

1. Is $\otimes$ distributive over $\oplus$? No. For example, $3 \otimes (2 \oplus 5) = 3 \otimes (20) = 26$, but $(3 \otimes 2) \oplus (3 \otimes 5) = 8 \oplus 11 = 176$.

2. Is $\oplus$ distributive over $\otimes$? Yes. For example, $2 \oplus (7 \otimes 3) = 2 \oplus (17) = 68$ and $(2 \oplus 7) \otimes (2 \oplus 3) = (28) \otimes (12) = 68$. In variable form, both $a \oplus (b \otimes c)$ and $(a \oplus b) \otimes (a \oplus c)$ are equal to $4ab + 2ac$.

Have the students create their own operations and then determine whether or not axioms such as the commutative, associative, or distributive axioms hold for those operations.

2-4 (pages 62–66)

Teaching Suggestions

In this section the minus sign is moved to the lowered position. Reading $-a$ as the "opposite of a" rather than "negative a" should clarify the meaning of the expression and avoid the incorrect conclusion that $-a$ always represents a negative number.

The interpretation of absolute value as distance on a number line helps students realize that absolute value can never be negative. If students think of $|x|$ as representing the distance from x to the origin, it will not be difficult for them to think of $|x - 3|$ as representing the distance from x to 3. (This extension of the absolute value concept is done in Section 4-5.) They should then be able to solve problems such as $|x - 3| > 4$ by mentally picturing the set of all points which are more than 4 units from 3.

Another aid to understanding absolute value is thinking of real numbers as having both magnitude (the length of the arrow on a number line) and direction (which way the arrow is pointing). Absolute value can be thought of as the magnitude part of a real number.

2-5 (pages 67–70)

Teaching Suggestions

Using arrows on a number line is an effective way to introduce the addition of real numbers. Use of numerous chalkboard examples, with students doing the drawing, will help to establish the mental process.

As a lead-in to the next section, you might ask students if they have discovered shortcuts or rules for the addition of real numbers. For example, the answer will have the sign of the number with the longest arrow; that is, the number whose absolute value is greater.

2-6 (pages 71–76)

Teaching Suggestions

The formal rules for addition should be explained and demonstrated, but do not expect students to memorize them as printed. Instead, stress the gaining of computational skills; suggest they use abbreviated forms of the rules and rely on a mental image of addition on a number line.

In the description of the addition of numbers with opposite signs, avoid the use of the word "subtract," as the concept of subtraction is introduced in Section 2-8. Instead, tell students to "find the difference between the absolute values."

The skill of adding real numbers is well suited to mental computation exercises. Devoting a few minutes of class time in this way will help students master this important basic skill.

2-7 (pages 78–82)

Teaching Suggestions

Formal proof is presented with the conventional two-column pattern of statements and reasons. Proof is a basic mathematical skill, important in algebra as well as in geometry.

Students should realize that each statement in a proof is supported by an axiom or a theorem.

Stress that because variables are used in proofs, the theorems proved can be used in proving other theorems as well as in computational situations.

Many theorems have more than one valid proof. Students should know that all logical proofs are acceptable.

In exercises throughout the text, students will be asked to *show* why a relationship is true. For your most capable students you may wish to change the directions, asking them to give a formal *proof* using statements and reasons.

2-8 (pages 82–87)

Teaching Suggestions

Students should realize that the examples shown at the beginning of this section do not constitute a proof. Instead a definition is given that generalizes an observed pattern.

In the early stages of working with subtraction, students should actually rewrite problems using the definition of subtraction, i.e., adding the opposite of a number. Taking shortcuts too early will likely result in mistakes and the learning of improper techniques.

In Example 2, two approaches are given for simplifying expressions with several sums and differences. Although students will probably prefer one approach to another, they should be able to demonstrate both. Numerous chalkboard examples will help students master the techniques.

With problems involving many subtractions in groupings, students should be able to use the following properties: $-(a + b) = -a + (-b)$ and $-(a - b) = -a + b$.

2-9 (pages 88–93)

Teaching Suggestions

The multiplicative property of -1 and the property of opposites in products are used to lead to the statement of the rules for multiplication of real numbers. In the early problem exercises, students will probably find the sign of the product and the absolute value of the product separately. As they have more practice, these operations will be combined. Mental arithmetic drills are very appropriate for this section.

There are several graphic presentations that reinforce the rules for multiplication of real numbers. One is a hero$(+)$/villain$(-)$ who is entering$(+)$/leaving$(-)$ town. Thus, a hero leaving town has a negative effect, whereas a villain leaving town has a positive effect.

Another image for students is two movies taken, one of a person skating forward and one of a person skating backward. Showing the movies on a projector that can be run either forward or backward produces the following apparent motions:

movie		projector	apparent motion
skating forward,	+	forward, +	skating forward, +
skating forward,	+	reverse, −	skating backward, −
skating backward,	−	forward, +	skating backward, −
skating backward,	−	reverse, −	skating forward, +

Students should realize that these presentations are given as memory aids and not as proofs for the rules for multiplication of real numbers.

2-10 (pages 93–98)

Teaching Suggestions

Techniques that the students have been using for years in arithmetic (for example, $\frac{1}{20} = \frac{1}{4} \cdot \frac{1}{5}$) are now presented and proved for all nonzero real numbers. Because the ideas are familiar to the students, special attention should be given to examples and exercises involving proofs. Students are apt to skip steps when they know a technique. In working with multiplicative inverses, students must be careful to restrict statements to nonzero expressions.

Related Activities

Consider the operation $*$ defined by the table below over the set $T = \{a, b, c, e\}$:

$*$	a	b	c	e
a	b	e	a	e
b	e	c	b	a
c	a	b	c	e
e	e	a	e	b

1. Does this operation have an identity element? If so, what is it?   Yes; c

2. Is this operation commutative?   Yes

3. Is this operation associative?
   No; $a * (e * e) = e$ but $(a * e) * e = b$.

4. Does every member of T have an inverse? No. Since c does not appear in the a or e rows or columns, a and e do not have inverses.

Students may like to make up operations that satisfy all four of the conditions mentioned above, or some other combination of the conditions. Suggest to them that they also try to work with a set of two, three, and five elements.

2-11 (pages 99–103)

Teaching Suggestions

As subtraction was defined in terms of addition, division is defined in terms of multiplication. Rules from the previous work with multiplication will apply to the division of real numbers.

To aid them in simplifications and later transformations, students should be able to interchange $-\frac{1}{a}$, $\frac{1}{-a}$, and $\frac{-1}{a}$.

At the conclusion of this section, oral drills can be conducted with real numbers for all arithmetic operations.

3 Solving Equations and Problems

The four properties of equality are presented as the basis for solving equations. The first equations given involve one step, or transformation, for their solution. More complicated equations are then presented that require two or more steps. These include the solution of formulas. Word problems are then presented and a thorough discussion of some techniques for solving them is given in the *Plan for Solving a Word Problem* on page 132.

3-1 (pages 113–118)

Teaching Suggestions

Students are introduced to the properties of equality that will be used to solve equations throughout their study of mathematics. The proof of the addition property of equality is shown; the others are left as exercises. Completing the exercises will reinforce the axiomatic support of the operations,

showing how they help form the foundation of structure in mathematics.

It is important to discuss the sequence of the steps in the proofs as well as the reasons. Some students will not have memorized the axioms from the previous chapters and will need review.

At this stage of equation work, you should show each step needed for the solution, and justify each with the appropriate reason. Do not take shortcuts. In later work, abbreviated solutions will be allowed, but only after the basic process has been justified and mastered.

3-2 (pages 118–121)

Teaching Suggestions

Although it is important to solve simple equations by inspection, students will sometimes try to solve all equations by inspection.

Seeing a few simple equations (for example, $4x + 5 = -9$) which do not have a simple integral solution will sometimes demonstrate that a logical solution process will produce the answer more efficiently than the inspection method. Stress that showing the steps of the solutions develops a technique, which is more important than the actual numerical answer to any assigned problem. This technique will be invaluable when problems become more complicated.

Students should understand that transformations produce a series of equivalent equations, each with the same solution set. Refer frequently to the "balance-beam" model, and the concept of "balance" between the two sides of the equation. Although the appearance of the equation changes in each step, the solution set remains unchanged.

The most difficult equation form for many students is one like $5 - m = 3$, where the variable term is subtracted from the constant term. For such students, either transform the equation first to an addition statement, $5 + (-m) = 3$, or suggest first adding m to each side. A third approach is to transform the equation so there is only a single constant term: subtract 5 from each side, producing $-m = -2$.

3-3 (pages 121–124)

Teaching Suggestions

For students to solve equations involving multiplication and division, they must be able to identify reciprocals. You may wish to have a short drill on this skill.

When transforming an equation, students need to be careful not to multiply by zero. It is not likely that students will actually multiply the sides of an equation by "0." In future work, however, they may multiply by an expression such as $x - 3$, which has a value of zero when $x = 3$. Similarly, students need to be careful to avoid dividing by an expression that is equal to zero.

In a problem such as $\frac{5}{3} = \frac{c}{12}$, some students may use the "cross-product" from proportion problems. The result of this step here is the equation $5(12) = 3c$, which is easily solved. Show that multiplication by 12, the reciprocal of $\frac{1}{12}$, produces the answer in fewer steps.

Related Activities

The theorem about "cross products" is proved in Section 9-7. You may wish, however, to have some students prove the result at this time.

It is useful to students to see that a test such as the cross-product test is justified using the axioms of algebra. Write this equation on the chalkboard:

$$\frac{a}{b} = \frac{c}{d}$$

Ask students to use axioms of algebra rather than cross multiplication to prove that $ad = bc$. Tell students that a, b, c, and d are real numbers with $b \neq 0$ and $d \neq 0$.

Tell students that all such tests must be able to be proved on the basis of the axioms.

Have pairs of students use the axioms to solve for c in $\frac{a + b}{c} = \frac{d}{e}$, $c \neq 0$ and $e \neq 0$. Ask for volunteer pairs to put their justifications on the chalkboard.

3-4 (pages 124–129)

Teaching Suggestions

In solving equations requiring transformations by both addition and multiplication, some students believe one of these operations must always be done first. To counter this belief, write $\frac{x}{2} + 7 = 3$ twice on the chalkboard. Ask two volunteers to solve the equations, one beginning with a multiplication transformation (multiplication by 2 yields $x + 14 = 6$), and the other with an addition transformation (addition of -7 yields $\frac{x}{2} = -4$). Point out that in equations having fractions, multiplication by the LCD is usually the most efficient approach.

To help students realize there is always more than one approach when solving an equation, write $\frac{x}{2} + 7 = 3$ on the chalkboard again. For the first transformation, add 7. $\left(\frac{x}{2} + 14 = 10\right)$ Ask the students if this is a correct transformation. (Yes) Ask them if it was an efficient transformation to use in finding a solution. (No) Then ask them to help you arrive at a solution using exactly five transformations. In discussing and analyzing assigned problems, stress that efficiency is a goal, but that performing correct transformations is what is most important.

Related Activities

Some students need to be reminded that a check of the solution must be done. To demonstrate this, put this equation on the chalkboard:

$$17 - (3 + a) = 9$$

Tell students that the solution to this equation is 5. (The actual solution is −5.) Then have them use the original equation to check this solution.

Many students will see that 5 is not the correct solution. Encourage these students to explain where and why the error was made.

You may wish to have students set up equations and possible solutions for checking by fellow students.

3-5 (pages 130–134)

Teaching Suggestions

The solution of word problems has always been one of the most difficult areas for beginning algebra students. Even students who had previously demonstrated excellent skills in mechanical algebraic manipulations may experience difficulty with word problems.

These solutions require students to read well, interpret the vocabulary and phrases, translate the statements into algebraic expressions, and, finally, apply their transformational skills to an equation.

To help the students begin to develop an organized approach to solving these problems, do several examples following the steps in the *Plan for Solving a Word Problem* on page 132.

The first step in translating a word problem into a mathematical sentence is to read the problem carefully. Encourage students to be sure they thoroughly understand what a problem says before they try to answer it.

To write an appropriate equation, select a variable, deciding what quantity it should represent. Express the information from the problem in terms of this variable. Then write the equation. Finally, read the word problem again to check that the equation correctly represents the problem.

After the answer to the equation has been found, it should be checked with the original statement of the problem to see that the answer makes sense. Remind students that substituting the solution of the equation back into the equation will not help if the equation is an incorrect translation of the word problem statement.

Related Activities

After students have practiced with the word problems in their text, ask them to write 2 or 3 original word problems complete with a separate step-by-step proof for each problem as set up in their text. The proofs will be the answer key against which work will be judged. Have students exchange problems, solve, and return for evaluation.

Let student writers correct the completed problems, checking for use of the proper steps as well as the answer. If there are errors, have the writers point out where and how they were made.

3-6 (pages 134–137)

Teaching Suggestions

Solving equations with the variable on both sides requires no new transformations. The appropriate addition or subtraction transformations will yield an equivalent equation with the variable appearing on only one side. After that is done, the solution will follow from the steps practiced in Sections 3-2 through 3-4.

Examples 2 and 3 in this section illustrate what happens when transformations completely eliminate the variable from the equation. At first students may find this confusing. To aid them, write 1 = 3 on the chalkboard and ask the students to see if they can make legitimate transformations that will make the equation true. For example, adding c to both sides produces $1 + c = 3 + c$ and there is no substitution that will make this true. Point out that all equations equivalent to the false statement $1 = 3$ have the same solution set, $\emptyset$.

Ask students to think of equations that are equivalent to the true statement $2 = 2$ and write them on the chalkboard. (Examples are $2 + x = 2 + x$ and $x = x$.) Point out that every equation equivalent to the true statement $2 = 2$ has the same solution set, $\mathscr{R}$.

In performing transformations, students should be careful not to divide by a variable, or a variable term, that has a value of zero.

3-7 (pages 139–141)

Teaching Suggestions

Charts are a long-established aid to the organization of information in verbal problems. For many students, they greatly facilitate the task of translating the information into an equation. Students should realize, however, that using charts is no guarantee of automatic solutions. The information must be correctly presented in the chart. Then an equation must be made that shows a true relationship between the entries in the chart.

3-8 (pages 144–147)

Teaching Suggestions

Some students are confused when asked to solve for one variable in terms of other variables. To help them, write the following equation on the chalkboard:

$$\frac{a + b}{c} - d = e + f$$

Then ask students to perform transformations using only the variables shown; tell them that no solution is being sought. (The first one might be to add d or to multiply by c.) After several transformations, remind the students that a formula is simply an equation. Emphasize that the same rules apply when working with a formula as when working with any other equation.

In many mathematical and scientific courses, students will work with formulas and various transformations. Explain that if they know one basic formula and have algebraic skills, they can avoid memorizing all the various transformed variations of the basic formula. Note: similar work may be referred to in other texts as "solving literal equations."

4   Solving Inequalities and Problems

This chapter introduces the properties of order and provides practice in solving inequalities and graphing their solution sets. The set operations of union and intersection are introduced, with emphasis on their application to solving compound sentences involving inequalities and absolute values.

Equations and inequalities are used to solve verbal problems. Emphasis is on consecutive integer, angle relationship, uniform motion, and mixture problems. Problems with no solutions or with an infinite number of solutions are discussed. Students learn techniques to solve these problems.

4-1 (pages 157–163)

Teaching Suggestions

It is useful for students to compare the order properties and the properties of equality. Several numerical examples and the use of a number line will help students visualize the order properties. Students will find it helpful to refer mentally to their own specific numerical examples as they familiarize themselves with these ideas.

Point out that the inequality sign should be changed in the same step as the multiplication or division by a negative number. For example, $2 < 5$ and $-3 \cdot 2 > -3 \cdot 5$. Failure to change the sign at this step results in a statement that is not equivalent, $-3 \cdot 2 < -3 \cdot 5$.

4-2 (pages 163–167)

Teaching Suggestions

Students should be aware that transformations that produce equivalent inequalities are similar to those used in Chapter 3 to produce equivalent equations. Adding and subtracting each side of an inequality by the same real number parallels the technique used in solving equations. Ask students to cite the important exception: the direction of an inequality sign is reversed when each side is multiplied or divided by the same negative real number.

In doing examples and exercises, have students describe some of the solution sets in their own words. Then ask them to pick some specific solutions and test them. Such checking will aid students to clarify their thinking about negative numbers.

Review the correct use of arrows and open circles in graphing sets of numbers on a number line.

Students may be interested in learning symbols that will lessen the amount of writing they must do. One way of writing the solution set for $3m - 2 > 19$ is {all real numbers greater than 7}. Set builder notation for the solution set is $\{m: m > 7\}$, which is read "the set of all m such that m is greater than 7." Set builder notation provides for greater precision and compactness.

4-3 (pages 167–170)

Teaching Suggestions

Some students may confuse the set operations of union and intersection. For those who do confuse the two, an appeal to the ordinary meaning of the words may help to reinforce the distinction.

Some students forget the symbols associated with the operations. To help them remember the symbol for union point out its resemblance to the first letter of union.

Students frequently confuse the symbols $\emptyset$, { }, {0}, and $\{\emptyset\}$. The following argument will help them distinguish among the symbols. In order for sets to be equal, they must (1) contain the same *number* of elements and (2) contain the exact *same* elements. Since $\emptyset$ and { } fulfill both conditions, $\emptyset = \{ \}$; they are just different names for the same idea. On the other hand, {0} and $\{\emptyset\}$ contain 1 element, but 0 is a number and $\emptyset$ is a set. It is not possible for a number to equal a set. Thus, because they do not contain the same elements, $\{0\} \neq \{\emptyset\}$.

Related Activities

Have students investigate the operations of union and intersection with respect to the commutative, associative, and distributive axioms to see if there appear to be similar axioms for operations on sets. Suggest they work with simple sets and then carry out the operations and compare the results. In particular, investigate whether $A \cap (B \cup C) = (A \cap B) \cup (A \cap C)$ and $A \cup (B \cap C) = (A \cup B) \cap (A \cup C)$. (Yes for both.)

4-4 (pages 170–174)

Teaching Suggestions

Students will need explicit connections made between this section and the preceding one. It will help them to understand the term conjunction if they realize that it is the intersection of the solution sets of the two clauses of the compound sentence. They must read carefully to discriminate between 'and' and 'or.' If this is clear, they can easily discern that a disjunction is the union of the solution sets of the two joined sentences. Note that $m \leq 4$ is a disjunction.

Review the use of $\leq$ and solid circles on a number line.

In combined inequalities such as $-3 < x + 2 \leq 5$, the convention is that the inequalities must have the same direction. A combined inequality is a conjunction. It is not correct to write a sentence such as $-3 > x < 1$.

4-5 (pages 174–178)

Teaching Suggestions

Reinforce the idea that absolute value expresses the distance between points. Any point satisfying the equation $|x - 3| = 5$ must be 5 units from 3. Students should be able to extend this idea to finding solutions for $|x - 3| < 5$ or $|x - 3| > 5$.

Students may have difficulty switching between problems of the $|a - b|$ pattern and those of the $|a + b|$ pattern. Have them verbalize the thinking process and demonstrate procedures on the number line. Some may readily be able to rewrite the $|a + b|$ pattern as $|a - (-b)|$.

Expressing equations or inequalities containing absolute values as equivalent conjunctions or disjunctions is useful once the basic ideas are firmly in place.

$|x| = a$ is equivalent to the disjunction $x = a$ or $x = -a$.
$|x| > a$ is equivalent to the disjunction $x > a$ or $x < -a$.
$|x| < a$ is equivalent to the conjunction $-a < x < a$.

These rules may be applied to more complicated inequalities. For example, $|3x - 4| > 2$ can be written as the disjunction $3x - 4 < -2$ or $3x - 4 > 2$; and $|2x - 5| < 1$ can be written as the

conjunction $-1 < 2x - 5$ and $2x - 5 < 1$ or as the combined inequality $-1 < 2x - 5 < 1$.

4-6 (pages 179–183)

Teaching Suggestions

Students need to become familiar with using variables to represent consecutive integers. Allow students to discover the pattern for consecutive integers in the following way. Write x on the chalkboard, tell them it represents an integer, and then ask them to name the next larger integer. When $x + 1$ is named, ask for the next larger integer, and so on. Then ask for the next smaller integer, $x - 1$, and so on. Repeat the process for generating consecutive even integers and consecutive odd integers. Use x to represent the first even or odd integer and have students recognize the pattern of counting by twos: $\ldots, x - 4, x - 2, x, x + 2, x + 4, \ldots$

Consecutive multiples can be developed by presenting students with $5x$ as a multiple of 5 and asking them for the next larger multiple of 5. If $5x + 5$ is given by them, ask that it be rewritten as a product, $5(x + 1)$. Then generate the pattern, $\ldots 5(x - 2), 5(x - 1), 5x, 5(x + 1), 5(x + 2), \ldots$

Before beginning the exercises, have the students review the steps in the *Plan for Solving a Word Problem* on page 132. Encourage them to use these steps in solving word problems.

Related Activities

A common method of using variables for representing even numbers is $2n$ where $n \in$ {integers}. Thus n may be even or odd, but $2n$ has a factor of 2 and is therefore even. Consecutive even numbers would then be represented as $\ldots, 2(n - 1), 2n, 2(n + 1), 2(n + 2), \ldots$, or as $\ldots, 2n - 2, 2n, 2n + 2, 2n + 4, \ldots$

Continuing with this line of reasoning, if $2n$ represents an even number then $2n + 1$ represents an odd number. Consecutive odd numbers then become $\ldots, 2(n - 1) + 1, 2n + 1, 2(n + 1) + 1,$

$2(n + 2) + 1, \ldots$ or $\ldots, 2n - 1, 2n + 1, 2n + 3, 2n + 5, \ldots$.

Have students do some of the exercises involving odd and even numbers using the above method of representing those numbers. Discuss the advantages and disadvantages of each method.

4-7 (pages 184–187)

Teaching Suggestions

Before beginning this section, discuss with students the vocabulary, notation, and angle properties illustrated on page 184. Students who confuse complementary and supplementary angles might associate "c comes before s alphabetically" with "90° comes before 180° numerically." Make sure all students are familiar with the method of marking angles to show equality of measure.

Encourage students to use sketches as an aid in solving problems about angles. For estimating the appropriateness of angle measurement, suggest to students that they sketch a right angle and subdivide for angles of measure 30°, 45°, and 60°.

4-8 (pages 189–194)

Teaching Suggestions

Although students may have had practice in the past with the formula $d = rt$, you may want to discuss its meaning and the different forms it may take: $r = \frac{d}{t}$ and $t = \frac{d}{r}$. Oral practice in deciding which formula to use for a given problem will be useful. Also, having students put charts on the chalkboard during the discussion will help them learn how to use charts with motion problems.

Students frequently need practice in dealing with units of measure. They need to pay particular attention to making sure they are dealing with similar units. They may need practice in converting units. They also need to be sure that the answer they give is in the units called for by the problem.

You might emphasize that students need to return to the problem to make sure they have completed all the steps requested.

Stress the usefulness of making a sketch and a chart for these problems. These visualizations will pull the facts from the text and make it readily apparent to students which facts are still unknown. They also help to establish the basis for the equation. For example, distances are equal; sum of the distances adds up to the whole distance; and so on.

Related Activities

Students sometimes feel it is difficult to solve word problems, but that it must be easy to create them. Have students create word problems similar to those in the exercises, complete with answers. In doing so they will gain a better understanding about the elements of a problem, the numbers and equations involved, and the matching of the problem with reality. Students should come to realize that any equation and solution can describe a great many real life situations.

Once the problems have been written, have the students exchange their problems, find solutions, and then compare them with the answers determined by the designer of the problem.

4-9 (pages 196–200)

Teaching Suggestions

Again, charts may be particularly useful in helping students resolve their difficulties with mixture problems. As with problems of motion, such charts readily show which facts are available and which ones must be expressed in terms of unknowns.

A very simple version of the problem may be memorized and kept in mind as a model. It might be helpful to emphasize general relationships such as: the number of parts times the value of each part is equal to the total value. For mixture problems, note that the value of a mixture is the sum of the values of the materials mixed. The familiar expression, "The whole is equal to the sum of its parts," may be useful for students to apply to mixture problems.

Teaching Suggestions

Students may benefit from a discussion of problems they have encountered which they could not solve. Some problems do not give enough information. Others produce inconsistencies.

You may find it helpful to emphasize that results need to be looked at in the light of common sense. Encouraging students to check their results will help them either correct their errors or reinforce their conclusions that the problems cannot be solved. You may wish to emphasize verbal explanations of the reasons a problem is unsolvable.

5 Graphs and Functions

The function is an important unifying concept throughout mathematics. After the fundamental concepts of ordered pairs and the coordinate plane are developed, the ideas of relations and functions are investigated. Linear functions and their graphs are then studied, including the standard slope and intercept properties. Techniques for linear equations are extended to graphing linear inequalities.

5-1 (pages 215–218)

Teaching Suggestions

The terms "abscissa" and "ordinate" may be new to the students. Remembering the correct order may be made easier by pointing out the alphabetical ordering in each of these ordered pairs: (abscissa, ordinate), (x, y), (first, second), and (horizontal, vertical).

Be sure that students do not confuse ordered pair notation, (a, b), with set notation, $\{a, b\}$ or $\{(a, b)\}$. Remind them that $\{a, b\} = \{b, a\}$, but that $(a, b) \neq (b, a)$.

Have students become familiar with the naming of the quadrants, I, II, III, IV. Then have them realize that points whose coordinates are of the form $(a, 0)$, which lie on the x-axis, and $(0, b)$, which lie on the y-axis, are not assigned to any quadrant.

In graphing points and assigning coordinates, students will see that the process will sometimes involve non-integral numbers. Remind students that every point on the plane has unique coordinates and that every ordered pair names a unique point. Graphing ordered pairs such as $(2\frac{1}{2}, 4\frac{1}{3})$ will aid students in approximating the location of a point. Students might enjoy thinking about the inexactness of a dot as a representation of the idea of a point. For example, given an ordered pair, a dot can be drawn that will cover the place of the point on the plane, but given a dot representing a point, the ordered pair might never be guessed.

When students begin to draw their own graphs, make sure that they use the same scale on both axes whenever possible, or the distances and shapes of the graphs will be distorted.

5-2 (pages 219–223)

Teaching Suggestions

Students should realize the difference between a relation and a function. Saying that a certain set of ordered pairs is a relation is redundant, as a relation is defined as a set of ordered pairs. Saying that a set of ordered pairs is a function, however, says that no two ordered pairs have the same first coordinate.

Examples from everyday life may be used to illustrate the concept of a function: to each person we can assign a height; to each item in a store we can assign a price; and so on. Reinforce the definition of a function by pointing out that it would not make much sense for a person to be assigned two heights or for an item in a store to have two prices.

Note that in this section relations are represented in four ways: as data in tables, as mapping diagrams, as graphs on coordinate planes, and as sets of ordered pairs. In each case, be sure that students can distinguish between those relations which are functions and those which are not. Help them to formulate simple rules for determining whether a relation is a function. For example, in a mapping diagram, if no element has more than one arrow emanating from it, the relation is a function.

5-3 (pages 224–227)

Teaching Suggestions

Students may find the function machine analogy helpful when learning the $f(x)$ notation.

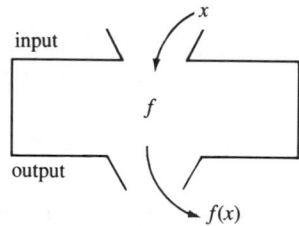

The output of the machine is dependent on, or is a function of, the input. That is, the value $f(x)$ is dependent on the value of x and the rule f that assigns a number in the range to each number in the domain. Point out that $f(x)$ and y play identical roles in function notation.

In your instruction, use function names besides f, like h and F, as well as both forms of notations [$k: x \longrightarrow 3x$ and $k(x) = 2x - 1$].

Emphasize that the rule defines the relation. Students should recognize that if given the rule and a domain, they can determine all the ordered pairs that form the relation. Using a table format will reinforce this and also help minimize errors.

Students should realize the letter F can be used to name a rule in one exercise and can be used to name a different rule in another exercise.

Related Activities

Many different types of information can be communicated by plotting points on coordinate planes. Graphs on coordinate planes may show patterns that are not apparent from a list or table presentation of the information. Drawing the following graph on the chalkboard will demonstrate the rate of increase of the world's population over a 200 year period. Point out that while the y-axis begins at 0, the x-axis begins at the year 1800.

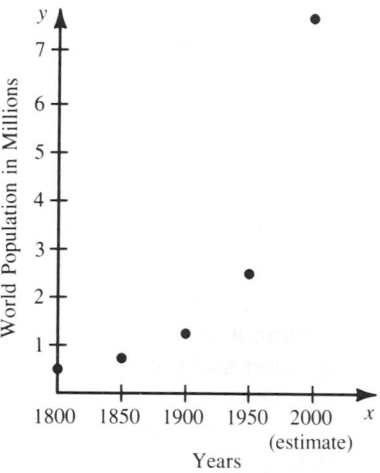

Divide the class into small groups to create graphs of information that they find from reference books like almanacs and encyclopedias. Examples are the population of their city or state over a given period of time; the price of housing over time; and the number of cars manufactured in the United States each year since 1900.

5-4 (pages 228–231)

Teaching Suggestions

Emphasize the fact that the solution set for an open sentence in two variables is a set of ordered pairs and not a set of numbers. Point out that to find the ordered pairs that satisfy an open sentence, one must know which numbers are in the replacement sets. Only those numbers in the replacement sets may be used to form ordered pairs to be tested for membership in the solution set.

Solving an equation for y helps students bridge those exercises in Sections 5-3 and 5-5 that use functional notation. Point out that the ordered pairs in the solution set satisfy the equation regardless of its form.

In this section, students begin graphing equations and inequalities using restricted replacement sets. This allows students to graph equations and inequalities that would otherwise be difficult for them to graph. The solution sets for open sentences containing absolute values and inequalities provide students with good opportunities to distinguish between relations and functions.

5-5   (pages 232–236)

Teaching Suggestions

To draw a line as the graph of a linear equation, students should graph at least three ordered pairs satisfying the equation and then draw the line that passes through the graphs of the ordered pairs. Two points that are often easiest to use are the points of intersection of the line with the coordinate axes. You may wish to introduce the words x-intercept and y-intercept. (The definition of y-intercept is given in Section 5-8.)

Equations in which one variable is absent are difficult for some students. To help, rewrite an equation like $x = 5$ in the form $ax + by = c$ ($1 \cdot x + 0 \cdot y = 5$) and ask students to name ordered pairs that satisfy it. Students should realize that all permissible values for y will work as long as the x-coordinate is 5. The set of all such points is a vertical line. Similar reasoning applies to horizontal lines.

Students may ask why arrowheads are used twice on lines graphed, but only once on coordinate axes. Point out that arrowheads on the lines indicate that the lines extend indefinitely and that the single arrowhead on the coordinate axes indicates the direction of the increasing positive coordinates.

When finding solutions of equations like $\frac{1}{3}x + y = 2$, students should see that selecting values for x which are either positive or negative multiples of 3 will provide ordered pairs that simplify graphing.

5-6   (pages 236–239)

Teaching Suggestions

To graph linear inequalities, students must remember the procedures used to solve inequalities for y. Of particular difficulty are problems involving a negative coefficient in the y term, for example, $4 - 2y < 3x$. Several review problems may be useful.

When discussing the choice of using a solid or dashed boundary line, emphasize the analogy with the "heavy dot" and "open dot" for endpoints used in number line graphing. If the linear inequality symbol is $\leq$ or $\geq$, equality is part of the relationship and the boundary line is solid; if the symbol is $<$ or $>$, equality is *not* part of the relationship and the line is dashed.

Transforming inequalities to the form $y > g(x)$ or $y < R(x)$ greatly aids the decision about which half-plane to shade. In these forms, greater than can be thought of as "above" the boundary line and less than as "below."

To check the graph, an ordered pair for a point in the solution half-plane should be tested in the original inequality. Students should realize that such a test is not a true check as it would be for a statement of equality. Still, if the test results in a false statement, then the steps can be reviewed.

Related Activities

The space shuttle orbits the earth at an approximate speed of 29,000 kilometers per hour. By writing a linear equation to express this fact, $y = 29,000x$, and graphing the equation, students can determine the distance traveled by the shuttle at any point in the mission.

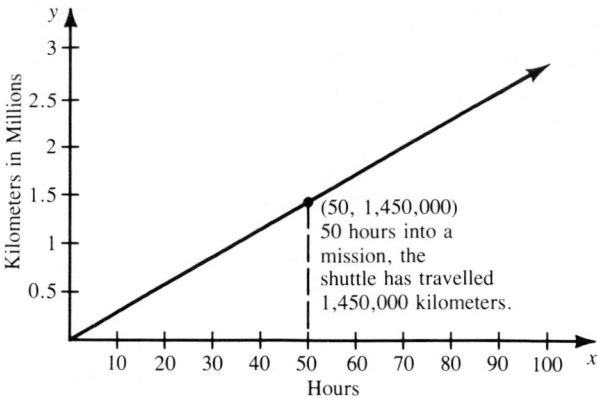

(50, 1,450,000)
50 hours into a
mission, the
shuttle has travelled
1,450,000 kilometers.

Have students do research to find another instance of an object traveling at a constant speed. Have them write a linear equation to express that fact and then have them graph the equation. Examples are the distance traveled by the Explorer satellite over a period of years, the distance traveled by the earth in a year, and the distance sound or light travels in a given period of time.

Encourage students to use calculators in their projects. Have them present their findings or post them on the bulletin board.

5-7 (pages 242–246)

Teaching Suggestions

Once students can present examples of the intuitive notion of slope, they should become familiar with the various forms of the slope given in the definition. The most useful form is

$$\text{slope} = \frac{y_2 - y_1}{x_2 - x_1}, \; x_1 \neq x_2$$

Students may not be familiar with subscripts as in the definition of slope just given. Explain that the subscripts do not affect the numerical value of the coordinates, but are used for purposes of identification. For example, they make it easier to speak about the coordinates of point 1, (x_1, y_1) and point 2, (x_2, y_2)

In graphing a line with a given slope, make sure students start determining a second point by counting from a known point on the line and not from the origin.

In using a negative slope such as $-\frac{3}{2}$, students can use either $\frac{-3}{2}$ or $\frac{3}{-2}$.

5-8 (pages 248–252)

Teaching Suggestions

To help students discover the roles of m and b in $y = mx + b$, draw two sets of coordinate axes and graph $y = 2x + 3$ on each. On one set of axes, graph other linear equations of the form $y = mx + 3$ and vary the values for m. On the other set of axes, graph linear equations of the form $y = 2x + b$ and vary the value for b. Conclude with a summary of statements about m and b. This presentation should help students realize the value of transforming equations to the slope-intercept form when they wish to graph the solution set. Special note should be made of the situations when the slope is 0 ($y = 0 \cdot x + b$) and when the slope is undefined ($0 = mx + b$).

Students may ask why the expression "an equation of a line" is used instead of "the equation of a line." Explain that every linear equation has equivalent equations and that each of these equations has the same graph.

Related Activities

The concept of slope has many applications. One is calculating the height of an object when direct measurement is difficult. For example, to find the height of a flagpole, given that one student has height y_s cm, have another student measure the length of the shadow of the given student, x_s, and of the flagpole, x_f. Because the sun, the source of light, is so distant, and the shadows are measured at approximately the same time, the shadows can be seen as being parallel. The imaginary lines from the sun, one passing over the student to the end of the student's shadow, and one passing over the flagpole to the end of the flagpole's shadow, have the same slope. Thus, to find the height of the flagpole y_f use the following linear equations:

$$y_s = mx_s \text{ and } y_f = mx_f.$$

Solving for m in $y_s = mx_s$ yields $m = \dfrac{y_s}{x_s}$. Substituting for m in $y_f = mx_f$ yields $y_f = \dfrac{y_s}{x_s}x_f$.

For example, a student 160 cm tall casts a shadow 200 cm (or 2 m) long. At the same time, a flagpole casts a shadow 15 m (or 1,500 cm) long. Using the equation above,

$$y_f = \dfrac{y_s}{x_s}x_f = \dfrac{160}{200} \cdot (15) = \dfrac{4}{5} \cdot (15) = 12$$

∴ the flagpole is 12 m tall.

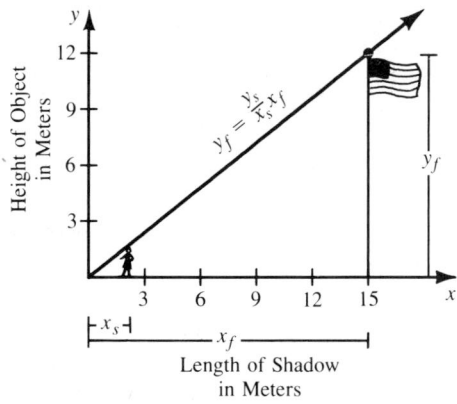

Have students use this technique to calculate the height of various objects either individually or in groups. Encourage them to use calculators in these endeavors.

5-9   (pages 252–255)

Teaching Suggestions

This section describes how the slope-intercept form of an equation can be used to find an equation of a line when you know either the slope of the line and the coordinates of one point through which the line passes or the coordinates of two points through which the line passes.

It is important to note the exception that must be made for a vertical line. Students should always check that the abscissas are different before applying the slope-intercept method.

---

6   Systems of Open Sentences

The chapter introduces methods of solving linear systems in two variables. The graphing method of solution is presented first. This is followed by the algebraic methods of linear combinations and substitution. The chapter then applies the techniques of solving linear systems to solving word problems such as motion and digit problems. In Section 6-9, students learn that linear programming is a direct extension of solving systems of linear equations, and they develop a sense of its power and importance as a problem solving method in science and

industry. In Section 6-10, the methods of solving systems of two linear equations in two variables are extended to solve systems of three linear equations in three variables.

6-1   (pages 267–273)

Teaching Suggestions

Be sure students understand that a system of two linear equations can have no solutions, infinitely many solutions, or one solution. Relate these facts to the geometry of each situation and to the terms *consistent* (coincident lines or lines intersecting at only one point) and *inconsistent* (parallel lines).

Oral drills are recommended on the nature of a system of equations and its possible solution sets. For example, you might state, "What can you conclude about a system of linear equations if you have a point whose coordinates form a solution to both equations?"

Inform students that there is no restriction on the number of equations in a system. Systems of two and three equations are studied in this text, but some applications require systems of many more equations.

Related Activities

Have students draw a triangle on graph paper such that at least one side is parallel to an axis, and the coordinates of all the vertices are integers. Then have them:
a. Write the equation of each side of the triangle.
b. Calculate the actual area of the triangle.

Have students write down only the equations of the sides of their triangle and exchange their equations with a classmate. Then see if the classmates can find the vertices of the triangle and the area of the triangle.

6-2 (pages 274–277)

Teaching Suggestions

When students realize that graphing may provide only approximate solutions, they will perceive the need for methods to find exact solutions. Use Examples 1 and 2 to explain that the algebraic methods use number properties to transform a system of linear equations into an equivalent system that can be solved by inspection. As you discuss these examples, emphasize that the goal of the algebraic method is to develop a system of the form $x = a$ and $y = b$, and to write the solution in the form $\{(a, b)\}$, where there is one solution. Encourage

students to check the solution set in each of the given equations in the system.

The decision of whether to add or subtract is made by inspection of the coefficients. A brief introductory drill of this skill may be made prior to assigning exercises.

Some students will notice that the examples and many of the exercises have simple solutions with integral values and won't see the need to move beyond graphical techniques. Explain that simple problems are used to investigate the basic idea. Point out that even the method of addition or subtraction is limited in its applicability; it is useful only when the coefficients of one of the variables have the same absolute value. Other methods are still needed to solve all linear systems. A system such as

$$7x - \frac{2}{3}y = 5$$

$$-x + 17y = \frac{3}{8}$$

should convince students that it will be helpful to learn additional methods.

Some students may have difficulty when the equations are added and both variables are eliminated.

Case I

$x - 2y = 2 \longrightarrow \quad x - 2y = 2$
$\underline{2y - x = 4} \longrightarrow \underline{-x + 2y = 4}$
$\qquad\qquad\qquad\qquad 0 = 6$

Solution: ∅

Case II

$x = 1 - y \longrightarrow x + y = 1$
$\underline{y = 1 - x} \longrightarrow \underline{x + y = 1}$
$\qquad\qquad\qquad\qquad 0 = 0$

Solution: $\{(x, y): x + y = 1\}$

If the end result is an equation that is obviously false (for example, $0 = 6$), then the solution set is the empty set since there are no substitutions that will make $0 = 6$. The graphs are parallel lines. If an equation that is always true (for example, $0 = 0$) results, then the solution set includes all points that satisfy either equation and the graphs are a pair of coincident lines.

6-3 (pages 277–280)

Teaching Suggestions

Linear combinations provide a method for solving all systems of two linear equations in two variables. The method of addition or subtraction in Section 6-2 can be viewed as a special case of the linear combination method in which the equations are multiplied by 1 or -1. Point out that there are many possible choices of multipliers that will eliminate a given variable.

Before students perform any calculations, encourage them to study the system carefully to determine the simplest and quickest procedure to use. This involves the choice of which variable to eliminate and which multipliers to use in eliminating that variable. Some students will discover that finding the least common multiple of the coefficients of the x- or y-terms is an expedient approach. This method will reward students with speed and ease of solution.

To minimize errors, suggest that students use the following procedure in solving systems of equations:
1. Copy the system.
2. Keep a record of the steps taken in transforming the system of equations.
3. Solve the system.
4. Check the solution in the original equations.

Keeping a record of the steps taken in transforming a system of equations may help some students avoid errors. This will also simplify the later reading of the solution by the student or teacher.

6-4 (pages 280–283)

Teaching Suggestions

The substitution method for solving a system of linear equations is most easily applied when the coefficient of one of the variables is either 1 or -1.

Students sometimes find the substitution method appealing. In using this method, they are solving an equation in terms of a single variable, a task with which they are quite familiar. Although it is often not the most convenient method for solving linear systems, the substitution method is widely applied in later work.

It should be evident that this method demands accurate work with parentheses and the distributive property. Common errors found in Chapter 2 are likely to occur with these exercises.

6-5 (pages 284–288)

There are many problem situations that are best solved by the use of a system of two linear equations. You may wish to create original problems from your own experiences as a consumer, from the news, or from reports.

Emphasize that whenever two variables are used in solving problems, two equations are necessary, each representing a separate relationship. Encourage students to use diagrams, charts, and graphs to help solve problems, and insist that they check their solutions against the wording of the problem. The 5-step *Plan for Solving a Word Problem* is used throughout this and subsequent problem sections.

Point out to students that although each problem in this section can be solved using one equation in one variable, two variables are used in order to learn the technique involved. Note that in most cases, it is easier to use two variables than one because the resulting equations are less complicated.

Related Activities

Have students split up into small groups to research the following problem (or one of their own creation), showing all equations, and arriving at a solution.

The Student Government is sponsoring a film to raise money for the purchase of a VCR for the student lounge. If half the student body attends, and enough non-students attend to fill the auditorium, how much admission should be charged if students pay half the amount that non-students pay?

Have students use newspaper ads to find an appropriate price for the VCR, and have them find the capacity of the school's auditorium. Once the problem is completed, students can discuss the advisibility of calculating for 100% capacity. This can prepare them for problem solving with systems of inequalities later in this chapter.

Students should look for these relationships:

1. $\underset{(n)}{\underset{\text{non-students}}{\text{number of}}} + \underset{(s)}{\underset{\text{students}}{\text{number of}}} = \underset{\text{auditorium*}}{\text{capacity of}}$

2. $\text{admission} \times (n) + \dfrac{\text{admission}}{2} \times (s)$
$= \text{price of VCR*}$

3. $s = \frac{1}{2}$ the student body*

* denotes information to be gathered by students.

6-6 (pages 289–291)

Teaching Suggestions

The physical principle involved in these motion problems is simple. When two forces act together on an object, the resultant or combined effect must be calculated. If the forces act in the same direction, addition is used. If the forces act in opposite directions, subtraction is used. Point out that the approach covered here is concerned with winds and currents acting along the line of travel. Forces acting at an angle require more advanced techniques.

Some of the students may have had personal experiences involving winds or currents. Asking them to remember such times as rowing a boat in a stream or bicycling with or against the wind will help them to visualize these problems.

Stress that a chart may be a valuable tool for organizing the problem information. While many students will find that a chart is a help, some students may be able to write appropriate equations without any aids. Some students may also be able to obtain a solution without using any equations by simply thinking through the problem. Insist that equations be written, reminding students

that the skill being taught is more important than the solution to the particular problem.

Related Activities

Have students research the times recorded at track meets for the 100-meter dash. Have them note if the times were measured in tenths or hundredths of a second. Then have them calculate the effect on the winning time if there had been a head or tail wind of various speeds. Ask them to determine what they would consider to be a maximum allowable wind for a 100-meter dash race.

6-7 (pages 292–297)

Teaching Suggestions

Difficulties with digit problems usually arise when students do not think clearly about the distinction between the digits and the value represented by the digits. Some students may confuse the sum of the digits of a number with the value of the number. Others may write expressions like *tu* and *ut* for a number before and after its digits are reversed. These problems can be anticipated by discussing the basic ideas involved in writing numbers in the base-ten system of numeration. Two ideas that could be reviewed are place value and the expanded form of a number.

6-8 (pages 298–300)

Teaching Suggestions

The use of colored chalk to show the shading of inequalities in a system is effective. If an overhead projector is available, you may want to graph each inequality of a system on a transparency, then place one transparency on top of the other to show the solution set of the system.

You should review the process for graphing inequalities developed in the last chapter. Some students will need to review the solid line or dashed line situations and whether to shade the half-plane above or below the boundary line.

Some students may ask about solving a system of linear inequalities algebraically; that is, by elimination or by substitution as was done with systems of equations. If so, have students experiment to see if it is possible, or suggest they try with the system $x + y \geq 6$ and $x - y \geq 2$. Point out that transformations used for pairs of equations cannot be applied to pairs of inequalities with the guarantee of producing equivalent inequalities. For example, if $a > b$ and $c > d$, it is not necessarily true that $a - c > b - d$. Furthermore, statements like $a < b$ and $a < c$ do not lead to any conclusion about the relationship between b and c.

6-9 (pages 302–306)

Teaching Suggestions

Students may be interested to know that linear programming is a problem solving method that is used widely in industry and science and requires only the methods of elementary algebra to be understood. This is a relatively new field of mathematics, originating in the early 1940's. The techniques learned here can be extended to situations involving linear equations of several variables.

Elaborate slightly on the theorem on page 303 by noting that a linear expression may not have both a maximum and minimum value on a convex polygonal region if the region is not "closed." In the vitamin pill example, no linear expression will have a maximum on the polygonal region graphed on page 302, although some linear expressions will have a minimum value on this region. The important point is that if a linear expression does have a maximum or minimum value, it will occur at a corner point on the region.

You may want to point out to your more able students the conditions under which a maximum or minimum value occurs at more than one point of the region. This happens when the linear expression has the same slope as one of the boundary lines. The maximum or minimum will then occur at any point along this boundary, including the two corner points on it. Exercise 14 on page 306 is an example of this.

6-10 (pages 306–310)

Teaching Suggestions

In solving a system of two equations in two variables, algebraic techniques are used to eliminate one variable and reduce the system to one equation in one variable, which can be solved easily. To solve a system of three equations in three variables, algebraic techniques are again used to eliminate the same variable from two pairs of the original equations, yielding two equations in two variables.

Emphasize to students that they should have a strategy before beginning to perform transformations. As the complexity of the problem increases, so does the need for careful planning.

Related Activities

Remind students that the graph of a linear equation in three variables is a plane in space. Point out planes in the classroom such as a desk top, a wall, the floor, etc. By noticing that the floor and a wall intersect in a line, conclude that two planes may intersect in a line. The corner where the ceiling and two walls meet shows that three planes may intersect in one point.

Have students use pieces of cardboard to demonstrate the different ways in which three planes may intersect.

The eight possibilities of three planes intersecting are as follows:
1. Three planes intersecting in a single point.
2. Three different planes intersecting in a single line.
3. Two coincident planes intersecting a third plane in a line.
4. Three coincident planes.
5. Three parallel planes.
6. Two coincident planes parallel to a third plane.
7. Three planes intersecting in three parallel lines.
8. Two parallel planes intersecting a third plane in two parallel lines.

Representations of these eight possibilities for planes $\mathcal{J}$, $\mathcal{P}$, and $\mathcal{S}$ are as follows:

1.
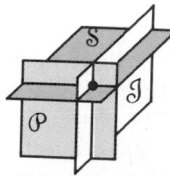

2.
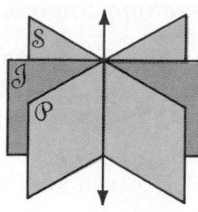

3.
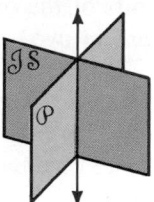

4.

5.
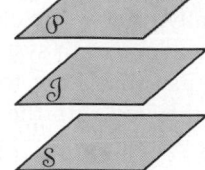

6.
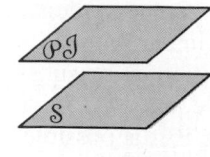

7.
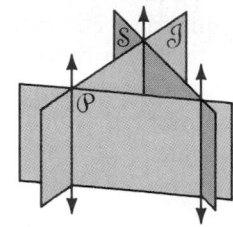

8.
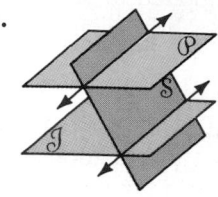

7 Polynomials and Factoring

The arithmetic of polynomials is introduced in this chapter, including finding sums, differences, and products. Exponent rules are extended to cover products of powers. Special attention is given to multiplying binomials. Factoring is developed as an application of the distributive axiom. Finding the greatest common monomial factor of polynomials begins the study of the complete factorization of quadratic trinomials. Factoring skills are applied to the solution of quadratic equations. Finally, quadratic equations are used to solve word problems.

7-1 (pages 323–327)

Teaching Suggestions

Many new vocabulary words are introduced in this section. Use them frequently when discussing the material and insist that the students use them in asking and answering questions.

Check that students' addition of terms such as $3x^2$ and $5x^2$ does not result in the expression $8x^4$.

For addition and subtraction, the column form reinforces the idea that only similar terms can be combined. When subtraction is performed on the same line, care must be taken to apply the subtraction to each term within the parentheses.

7-2 (pages 328–331)

Teaching Suggestions

Demonstrate the product of powers rule with these examples:

$$c^3 \cdot c^4 = (c\ c\ c)(c\ c\ c\ c) = c^7$$
$$(a^2)^3 = (a^2)(a^2)(a^2) = a^6$$
$$(uv)^4 = (uv)(uv)(uv)(uv) = u^4v^4$$

After several examples, ask the students to generalize the rule using variables for exponents. Oral drills may be necessary to insure mastery. Note that at this stage the variables used as exponents must be restricted to the set of positive integers.

Zero and negative exponents are discussed in Chapter 8.

Common errors when multiplying monomials include writing $x^2 \cdot x^3$ as x^6 and $(c^2)^3$ as c^5. Be sure that the students realize that $a^3 \cdot b^2$ cannot be combined as the bases are different. With more complicated expressions like $(-3r^2s^4)^3$, parentheses may be useful in the first steps. This is a good time to review the simple powers of 2, 3, 4, and 5.

7-3 (pages 332–334)

Teaching Suggestions

When multiplying polynomials, the horizontal arrangement is probably the easier arrangement to use when one factor is a monomial or when both factors are binomials. Students may prefer to use the vertical arrangement (as illustrated in Example 3) for multiplying polynomials of three or more terms.

It is important to remind students that the distributive axiom applies in both directions. Thus, $3a(2a - b) = 3a(2a) - 3a(b) = 6a^2 - 3ab$ and $6a^2 - 3ab = 3a(2a) - 3a(b) = 3a(2a - b)$. If students can see this dual nature of the distributive axiom now, it will be easier for them to learn factoring.

Related Activities

Using the chalkboard, have the students write and solve an equation using the information that the difference of the squares of two consecutive positive integers is 15.

$$(x + 1)^2 - x^2 = 15$$
$$x^2 + 2x + 1 - x^2 = 15$$
$$2x + 1 = 15$$
$$2x = 14$$
$$x = 7$$
$$x + 1 = 8$$

Then have the students construct a square of 7 units per side on graph paper. Ask them to extend each side to 8 units and shade in the additional 15 squares. Then have them construct new problems using different squares and different orders of integers.

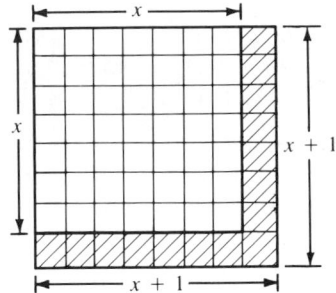

7-4 (pages 335–340)

Teaching Suggestions

A mastery of algebra includes the ability to mentally find the products of binomial expressions. Use of the vertical and horizontal form will demonstrate to students that the initial product of two binomials is the sum of four monomials. These four monomials are the result of multiplying each term of one binomial by each term of the other binomial.

A common and successful technique used to organize and produce these products is the FOIL method (First, Outer, Inner, Last). To demonstrate the use of the FOIL method, expand $(2x + 3y)(4x - 5y)$.

The First terms are $2x$ and $4x$.
The Outer terms are $2x$ and $-5y$.
The Inner terms are $3y$ and $4x$.
The Last terms are $3y$ and $-5y$.

Product of First Terms

$(2x + 3y)(4x - 5y)$ $2x \cdot 4x = 8x^2$

Product of Outer Terms

$(2x + 3y)(4x - 5y)$ $2x(-5y) = -10xy$

Product of Inner Terms

$(2x + 3y)(4x - 5y)$ $3y \cdot 4x = 12xy$

T93

Product of Last Terms

$$(2x + 3y)(4x - 5y) \qquad 3y(-5y) = -15y^2$$

$$\therefore (2x + 3y)(4x - 5y)$$
$$= 8x^2 - 10xy + 12xy - 15y^2$$
$$= 8x^2 + 2xy - 15y^2$$

   Some students avoid errors when multiplying binomials by first drawing in arcs connecting the terms being multiplied, as shown above.
   Making sure students can write with ease the expansions of $(a + b)^2$, $(a - b)^2$, and $(a + b)(a - b)$ will help them with factoring.

___

Related Activity

Have the students construct a trinomial square by cutting from cardboard or paper a square large enough to work with comfortably. Name the length of the side, x. Cut three strips of length x and width 1 inch each. Place them edge to edge along the side of the square. Cut three more strips 1 in. wide and $x + 3$ in. long and place them along the bottom of the square. Shade Section C as shown below. Find the area of each of the three sections and add.

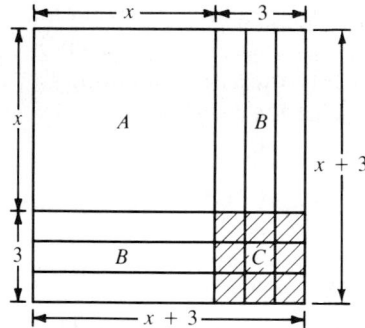

$$\text{area A} = x \cdot x = x^2$$
$$2(\text{area B}) = 2(3x) = 6x$$
$$\text{area C} = 3 \cdot 3 = 9$$
$$\text{area A} + 2(\text{area B}) + \text{area C} = x^2 + 6x + 9$$
$$\therefore (x + 3)^2 = x^2 + 6x + 9$$

Students could then make other strips to continue their exploration. For example, they could use four strips of length x and four of length $x + 4$.

___

7-5 (pages 341–345)

Teaching Suggestions

Most students will be familiar with a method of finding the GCF of two or more numbers. It is worthwhile to review the process here, so that it will be easy to extend the process to finding the GCF of monomials. The work with finding the GCF of monomials will make it easier to recognize common monomial factors of a polynomial in the next section.
   Review with students the fact that 1 is not considered to be a prime number. Note that if numbers are to have unique prime factorizations, 1 cannot be prime. Otherwise 18 could be factored as $2 \cdot 3^2 \cdot 1$ and $2 \cdot 3^2 \cdot 1^2$ and $2 \cdot 3^2 \cdot 1^3$ and so on.

___

7-6 (pages 346–348)

Teaching Suggestions

Review the distributive axiom as you develop the procedure for factoring a common monomial. With an example like $48x^3 - 42x^2$, first identify the GCF as $6x^2$. Then rewrite the binomial:

$$48x^3 - 42x^2 = 6x^2(8x) - 6x^2(7)$$

The right side can then be rewritten as $6x^2(8x - 7)$ by the distributive axiom.
   With practice, students will be able to use mental computations to identify the GCF and write the result directly in factored form. As a check on their work, they should first investigate the binomial in parentheses for any factor they overlooked. (In the example above, do $8x$ and 7 have a factor in common other than 1?) Second, they should mentally multiply the two factors and compare the product with the original expression.
   In working with factors that are opposites of each other, as in Example 3, students must take special care in manipulating minus and opposite signs. As a precaution against making errors, urge students to take extra steps to show the steps they are taking.

7-7 *(pages 350–353)*

Teaching Suggestions

The special factored forms $(a + b)^2$, $(a - b)^2$, $(a + b)(a - b)$, $(a - b)(a^2 + ab + b^2)$, and $(a + b)(a^2 - ab + b^2)$ can be introduced by first simplifying these expressions. The products are, respectively, $a^2 + 2ab + b^2$, $a^2 - 2ab + b^2$, $a^2 - b^2$, $a^3 - b^3$, and $a^3 + b^3$. To demonstrate how the general form applies to specific instances, substitute 5 or some other monomial for b in some of the forms listed above. Students should soon be able to match the form of a given polynomial with one of the patterns and to rewrite the polynomial in factored form. Oral drills of the pattern will aid memorization.

Remind the students that the first step in factoring a polynomial is to determine the greatest common monomial factor (other than 1). This will help to prevent the common error of neglecting the common monomial factor.

With more complicated exercises, such as $8c^6d^3 - 27e^3$, the use of parentheses are recommended for the first step, that is, $(2c^2d)^3 - (3e)^3$.

Students sometimes have difficulty when factoring by grouping when there are no terms to the second degree as in Example 5. Urge them to pay special attention to the numerical factors of the monomials. In Example 5, 3 and 15 share a factor other than 1 and so do 20 and 4. This makes them likely candidates for grouping.

Related Activity

Ask nine students to stand in a three by three pattern. Then ask them to rearrange themselves, keeping a rectangle but making a two by four pattern. One student will have nowhere to go.

. . .   .

. . .   .

. . .

Thus, if the original pattern illustrates $x \cdot x = x^2$, then the second pattern illustrates $(x + 1) \cdot (x - 1) = x^2 - 1$, or one less student in the pattern. The same principle could be shown by a four by four pattern, or by any square pattern, and by using pictures or objects instead of students.

$$x \cdot x = x^2 \qquad\qquad (x + 1)(x - 1) = x^2 - 1$$
$$3 \cdot 3 = 9 \qquad\qquad (3 + 1)(3 - 1) = 8$$

7-8 *(pages 353–357)*

Teaching Suggestions

Giving special attention to the text development of factoring $x^2 + bx + c$ as $(x + p)(x + q)$ will tie the task of factoring quadratic trinomials to the familiar FOIL method of multiplying two binomials. Essentially, students will reverse the FOIL procedure in factoring a quadratic trinomial. In factoring $x^2 + bx + c$, they will discover that the product of the First terms is ax^2, the sum of the product of the Outer terms and the product of the Inner terms is bx, and the product of the Last terms is c.

A certain amount of trial and error is necessary in testing possible factor combinations, and checking by multiplying must be done to determine if the correct factors have been chosen. Still, developing a systematic approach to factoring trinomials will make it easier for students to make efficient guesses about possible factors.

7-9 *(pages 358–361)*

Teaching Suggestions

Following the text development of factoring $ax^2 + bx + c$ as $(px + r)(qx + s)$ will demonstrate to students that the approach is similar to factoring done in the previous section. Difficulties arise from the additional factor possibilities that must now be considered.

Remind students that the first step in factoring a polynomial is to determine if the terms have any factors in common. For example, in attempting to factor $12x^2 + 34x + 14$, noticing the common factor of 2 reduces the polynomial to $2(6x^2 + 17x + 7)$, which has fewer possibilities.

7-10 (pages 363–367)

Teaching Suggestions

Only in the case $ab = 0$ can the disjunction $a = 0$ or $b = 0$ be formed. Students may try to apply the zero-product theorem incorrectly to a statement of the form $ab = k$, $k \neq 0$. Have students compare $(x + 3)(x - 1) = 0$ with $(x + 3)(x - 1) = 12$ by discussing the simpler forms $ab = 0$ and $ab = 12$. By asking how many solutions are possible for $ab = 0$ (2) and for $ab = 12$ (infinite), they should realize the limited and specific use of the zero-product theorem.

The zero-product property can be generalized. If a product of any number of factors is zero, then at least one of the factors is zero. Thus, the theorem can be extended to equations involving polynomials of degree greater than 2.

Students should be cautioned about dividing both sides of an equation by an expression involving the variable. Dividing each side of $3x^2 = 9x$ by x, for instance, results in the equation $3x = 9$. This new equation, however, is not equivalent to the original equation. The solution set of the new equation is {3}, whereas the solution set of the original equation is {0, 3}. This problem is discussed further in Section 9-3.

7-11 (pages 368–371)

Teaching Suggestions

Emphasize the need to check answers. The solution sets of quadratic equations often contain two numbers even though the problem may have only one answer. One member of the solution set can usually be ruled out due to the implied conditions in the problem, such as the presence of physical quantities that cannot be represented by negative numbers. Checking solutions in the original statement of the problem will reveal which solutions are correct and which must be rejected.

8 Polynomials and Rational Expressions

This chapter extends the study of polynomials to the concepts of quotients of polynomials. The laws of exponents for division lead to the definition of zero and negative exponents. This knowledge is used in the arithmetic of scientific notation. Students learn to find and simplify the sum, difference, product, and quotient of rational expressions.

After some practice, students should be able to do most of the A-exercises in this section mentally. Encourage them to do so. Be sure students do not confuse the terms x and x^0; remind them that $x = x^1$.

Be careful not to make such summary statements as "Any number divided by itself is one." The exclusion "Any non-zero number . . ." must be part of such a statement.

8-1 (pages 381–385)

Teaching Suggestions

The laws of exponents for division follow easily from work with reducing fractions. Examples such as those on page 383 will make clear how to use these laws to simplify quotients with monomial dividends and divisors.

8-2 (pages 385–387)

Teaching Suggestions

When students divide a polynomial by a monomial, the most common error they make is to forget to divide each term in the polynomial by the monomial. Stress that the theorem $\frac{a+b}{c} = \frac{a}{c} + \frac{b}{c}$ can be extended to any number of terms in the numerator. $\left(\text{For example, let } b = d + e, \text{ then } \frac{a+b}{c} = \frac{a+d+e}{c} = \frac{a}{c} + \frac{d}{c} + \frac{e}{c}.\right)$

Note that for simplicity throughout the exercises, we assume that no denominator has zero as a value.

It is not advisable to find these quotients with a denominator $\overline{)\text{numerator}}$ format because many have fractional coefficients. In fractional form, $\frac{8x^2 - 2x}{6x}$ is easy to transform to $\frac{4x}{3} - \frac{1}{3}$. Written as $6x\overline{)\,8x^2 - 2x}$, the process is awkward and offers no real advantage.

8-3 (pages 389–391)

Teaching Suggestions

Point out to your students the connection between the division algorithm for integers and that for polynomials. Notice that when dividing one polynomial by another the terms in both the dividend and divisor are arranged in order of decreasing degree. Example 1 illustrates how to insert a term with a zero coefficient for a missing term in the dividend.

Note that for the remainder of the text, it will be assumed that the replacement sets of the variables in a fraction include no numbers for which the denominator is zero.

Related Activities

Computers operate on numbers that have been transformed into binary or base 2 form. In the base 10 numeration system, the place values of numbers are ones (10^0), tens (10^1), hundreds (10^2), thousands (10^3), and so on. In the binary system, the place values are ones (2^0), twos (2^1), fours (2^2), eights (2^3), and so on. Each place value is represented by a zero or a one. The following table shows some equivalent numerals.

Base 10	Expansion in Powers of 2	Base 2
1	$1(2^0)$	1
2	$1(2^1) + 0(2^0)$	10
6	$1(2^2) + 1(2^1) + 0(2^0)$	110
11	$1(2^3) + 0(2^2) + 1(2^1) + 1(2^0)$	1011
15	$1(2^3) + 1(2^2) + 1(2^1) + 1(2^0)$	1111

The computer can perform division with binary numbers in a manner similar to polynomial division. For example,

(1)
$$1(2) + 1\overline{)\,1(2^4) + 1(2^3) + 0(2^2) + 1(2) + 1}$$

quotient: $1(2^3) + 0(2^2) + 0(2) + 1$

$$\underline{1(2^4) + 1(2^3)}$$
$$0(2^3) + 0(2^2) + 1(2) + 1$$
$$\underline{1(2) + 1}$$
$$0$$

(2) The next example is worked using base 2 number facts only.

$$101\overline{)\,110101} \quad \text{quotient } 1010 \text{ R } 11$$
$$\underline{101}$$
$$110$$
$$\underline{101}$$
$$11$$

Check the answers by transforming to base 10.

(1) $1(2) + 1 = 3$
  $1(2^3) + 0(2^2) + 0(2^1) + 1 = 9$
  $1(2^4) + 1(2^3) + 0(2^2) + 1(2^1) + 1 = 27$
  $3\overline{)\,27} \quad \text{quotient } 9 \quad \checkmark$

(2) $1(2^2) + 0(2^1) + 1 = 5$
  $1(2^3) + 0(2^2) + 1(2^1) + 0(1) = 10$
  $1(2) + 1 = 3$
  $1(2^5) + 1(2^4) + 0(2^3) + 1(2^2) + 0(2^1) + 1 = 53$
  $5\overline{)\,53} \quad \text{quotient } 10 \text{ R3} \quad \checkmark$

Have students perform long division on other pairs of binary numbers and check their answers. Challenge them to devise similar methods to divide using different bases. (Students may want to try bases 8 and 16 since these bases are of particular interest in computer work.)

Teaching Suggestions

To simplify or reduce a fraction, we divide the numerator and denominator by the same factor. To help students avoid errors such as

$\dfrac{\cancel{x+3}}{(\cancel{x-4})(\cancel{x+3})} = x - 4$ and $\dfrac{x^2 + \cancel{5}}{\cancel{5}} = x^2 + 1$, have

them write all factors in parentheses. Thus,

$\dfrac{x + 3}{(x - 4)(x + 3)}$ becomes $\dfrac{(1)(x + 3)}{(x - 4)(x + 3)} = \dfrac{1}{x - 4}$

and $\dfrac{x^2 + 5}{5}$ becomes $\dfrac{(1)(x^2 + 5)}{(1)(5)}$ which cannot be

reduced. Emphasize that simplification of fractions occurs when numerators and denominators are factored and there are factors in common.

The term *cancel* should not be used with this simplification process because of its ambiguous

usage. The problem $\dfrac{4x}{4x}$ simplifies to 1 while $3c -$

$3c$ simplifies to 0. Different ideas are employed in each case and should have different terms.

Teaching Suggestions

The definitions and laws of exponents for multiplication and division have thus far been defined only for exponents which are positive integers. These laws are not sufficiently strong to prove that

$a^0 = 1$ and $a^{-n} = \dfrac{1}{a^n}$. Instead, definitions are given

based on the investigation of patterns for exponents. Defining a^{-n} as the reciprocal of a^n can be confirmed by the following exercise.

After establishing that $a^0 = 1$ when $a \neq 0$, show that

$$a^{-n}a^n = a^{-n+n} = a^0 = 1$$

and, from previous work,

$$\frac{1}{a^n} \cdot a^n = \frac{a^n}{a^n} = 1.$$

Since $a^{-n} = \dfrac{1}{a^n}$, then a^{-n} is another form of the

reciprocal of a^n.

Students often think of $(x + y)^a$ as $x^a + y^a$. The operation of raising to a power is not distributive over addition (or subtraction) although it is distributive over multiplication. Thus, for example, since $(x + y)^0 = 1$ and $x^0 + y^0 = 1 + 1 = 2$, $(x + y)^0 \neq x^0 + y^0$, but $(xy)^0 = 1$ and $x^0 \cdot y^0 = 1$.

Remind the students that the topic of scientific notation is valuable for approximations of numbers and computations with measurements in science.

Related Activities

Physicists find the momentum of an object by finding the product of the mass of the object and its velocity. Physicists usually represent momentum by the letter p. Consequently, the formula for the momentum of an object is written $p = mv$, where p is the momentum, m is the mass of the object, and v is the object's velocity. It is easy to compute the momentum and velocity of most objects in our everyday environment. For example, a baseball weighing 150 grams travelling at 4,000 centimeters per second results in a momentum of 600,000 gram-centimeters per second.

$$p = mv$$
$$p = (150 \text{ g})(4,000 \text{ cm/sec})$$
$$= 600,000 \text{ g-cm/sec}$$

Similarly a car weighing 1,000 kg and traveling at 30 m/sec has the following momentum.

$$p = mv$$
$$p = (1,000 \text{ kg})(30 \text{ m/sec})$$
$$= 30,000 \text{ kg-m/sec}$$

For many calculations of momentum, standard notation for numbers becomes cumbersome. Scientific notation is usually used for numbers in a problem such as *Find the momentum of an electron traveling at 20% of the speed of light.*

Mass of an electron $= m_e = 9.1 \times 10^{-18}$ g.

Speed of light $= c = 3 \times 10^{10}$ cm/sec.

The momentum of the given electron is as follows.

$$p = mv$$
$$p = (9.1 \times 10^{-18} \text{ g}) \cdot$$
$$(0.2 \times 3 \times 10^{10} \text{ cm/sec})$$
$$= 5.46 \times 10^{-8} \text{ g-cm/sec}$$

Have students perform this calculation without

using scientific notation. They should then realize the facility of scientific notation. Have students do library research to determine the mass of other subatomic particles. Then have them make similar calculations of momentums at various velocities, using scientific notation. Encourage students to use calculators in their efforts.

8-6 (pages 403–407)

Teaching Suggestions

Introduce multiplication of rational expressions with familiar products of numerical fractions. Show students that the process is made easier by *first* reducing the factors and *then* multiplying. The process of multiplication can then be extended to rational expressions involving variables. The steps to follow are: (1) factor each numerator and denominator completely, (2) divide numerator and denominator by common factors, (3) rewrite the product as a single fraction, and (4) multiply out the numerator and denominator. If (2) is performed correctly, the answer will be in simplest form. Since a common factor might be missed, students should inspect the final product for any further simplification.

8-7 (pages 407–409)

Teaching Suggestions

In problems such as $\frac{a}{x} - \frac{b-c}{x}$ students often incorrectly simplify the expression to $\frac{a-b-c}{x}$. Encourage them to include an additional step, as follows:

$$\frac{a}{x} - \frac{b-c}{x} = \frac{a-(b-c)}{x} = \frac{a-b+c}{x}.$$

When students have transformed the sum or difference to a single fraction, they need to inspect it for possible simplifications. For example,

$$\frac{2c}{4} + \frac{2c+8}{4} = \frac{2c+2c+8}{4} = \frac{4c+8}{4} = \frac{4(c+2)}{4}$$
$$= c + 2.$$

8-8 (pages 410–413)

Teaching Suggestions

This section is the most difficult one in the chapter, requiring the use of many skills developed earlier. Students must factor denominators and then find a common denominator for the rational expressions to be added or subtracted. Each rational expression must be expressed with this common denominator; then the addition (or subtraction) can be performed. After adding (or subtracting), the numerator must be factored (the common denominator has already been factored), and common factors eliminated to reduce to lowest terms. This final step of factoring and reducing will also compensate for the possibility that students may have used a common denominator other than the LCD.

Remind students of the algebraic manipulation $(a - b) = (-1)(b - a)$. This relationship may be useful in some of the Written Exercises.

8-9 (pages 415–418)

Teaching Suggestions

Simplifying complex fractions calls for organizational skills as well as arithmetic skills. Insist that students present their work in a systematic and accurate manner.

Method 1 asks students to simplify the numerator and denominator separately before continuing with simplification, while Method 2 asks them to look at the entire expression at the beginning in order to find the LCD of all the simple fractions. Make sure the students are familiar with both methods, but allow them to choose which is appropriate for a particular problem. Point out that neither method may be used at the beginning of simplifying a fraction where the numerator or denominator is itself a complex fraction. In Example 3, for instance, Method 2 is used but only after the denominator has been simplified to the point of being a simple fractional expression.

Complex fractions can be used to calculate the resistance in parallel electrical circuits. The following diagram represents two resistors in parallel circuits.

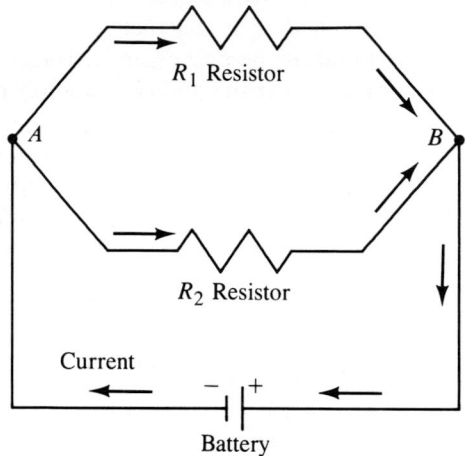

The current flows from the battery to point A, where the current splits to flow through R_1 and R_2. The two parts of the current rejoin at point B and return to the battery, thus completing the circuit. In a parallel circuit, a larger resistance in one branch of a circuit results in a lower current through that branch. To compute the total resistance R through a circuit like the one shown above, use the following formula where resistance is measured in ohms (Ω).

$$R = \cfrac{1}{\dfrac{1}{R_1} + \dfrac{1}{R_2}}$$

For example, find R when $R_1 = 15\ \Omega$ and $R_2 = 25\ \Omega$.

$$R = \cfrac{1}{\dfrac{1}{R_1} + \dfrac{1}{R_2}} = \cfrac{1}{\dfrac{1}{15} + \dfrac{1}{25}} = \cfrac{1}{\dfrac{5+3}{75}} = \cfrac{1}{\dfrac{8}{75}} = \dfrac{75}{8}$$
$$= 9.375\ \Omega$$

Therefore, the total resistance in the circuit is 9.375 ohms. Have students compute values of R for other values of R_1 and R_2.

9 Rational Expressions in Open Sentences

This chapter applies the student's knowledge of the properties of rational expressions to solving open sentences involving such expressions. These open sentences appear in percent, number, work, and motion problems, as well as in problems involving direct, inverse, joint, and combined variation.

9-1 (pages 427–430)

Teaching Suggestions

In solving open sentences containing the sum or difference of fractions, some students will want to first combine the terms. While this is not incorrect, it is sometimes less efficient and provides more opportunities for errors.

Point out that open sentences containing fractions can be easily solved by (1) finding the LCD of the fractions, (2) multiplying each term in the

open sentence by the LCD, and (3) solving the resulting open sentence with no fractions by the customary transformations.

Some students will multiply each term of the open sentence by the LCD but will still have the fractions. For example,

$$\frac{3x}{5} - \frac{x}{2} = 3$$
$$10\left(\frac{3x}{5}\right) - \left(\frac{x}{2}\right) = 10 \cdot 3$$
$$\frac{30x}{5} - \frac{10x}{2} = 30.$$

Explain that the purpose of multiplying each term by the LCD is to *eliminate* the fractions from the sentence, that is,

$$10\left(\frac{3x}{5}\right) - \left(\frac{x}{2}\right) = 10 \cdot 3$$
$$6x - 5x = 30$$
$$x = 30.$$

9-2 (pages 431–435)

Teaching Suggestions

Students must realize that the terms "percentage" and "percent" are not the same. Percentage involves units of measure (dollars, grams, liters, and so on) while percent is a number. Although the "percentage" is frequently described as being a "part of the whole," it may be larger than the whole (or the base) in problems with a rate greater than 100%.

The word problems in this section and Sections 9-5 and 9-6 can all be solved by applying a relationship of the form $x = y \cdot z$. In this section, the relationship is $p = br$.

Many of the problems in this section illustrate by means of practical application the importance of learning how to solve equations involving percents.

Related Activities

Have students collect the data for this exercise from several local banks offering varied term money market accounts. They will need to find out the nominal interest rate, the effective annual yield, and the restrictions on the term.

The investment is $4500. One third must be invested in a short-term account (a 3-month term) and the rest for a term no longer than one year. What combination of accounts at which banks give the best return? Assume that the short-term account will be "rolled over" (renewed) throughout the term of the long-term account. What is the total income from these investments? Note that the nominal interest rate will not give the true income, but the effective annual yield rate will. Thus,

Income = effective annual yield · $3,000 + effective annual yield · $1500.

Have students note also that although the short-term account will be renewed three times, the effective annual yield is computed on an *annual* basis.

9-3 (pages 437–439)

Teaching Suggestions

Inform students that solving fractional equations involves the same process as solving open sentences with fractional coefficients (Section 9-1) except that the LCD of the fractions will contain variables.

Because division by zero is not defined, students must carefully examine a fractional equation prior to solving it to determine what replacement values for the variable would produce a sentence in which division by zero would be shown. If, after solving the equation, any such values appear in the solution set, they must be rejected.

9-4 (pages 440–443)

Teaching Suggestions

Students should experience little or no difficulty with this material. You may want to remind them that consecutive integers can be represented as $n, n + 1, n + 2$, and so on; consecutive odd or consecutive even integers can be represented as $n, n + 2, n + 4$, and so on.

Although the examples in this section are solved using only one variable, some students may prefer to use a system of equations in two variables and should feel free to do so.

9-5 (pages 443–446)

Teaching Suggestions

The relationship of the form $x = y \cdot z$ used in the text to solve work problems is

work done = work rate × time.

We assume that work is additive, that is, the work done by two workers in a given time is the sum of the work each would do in that time if they worked separately.

9-6 *(pages 446–448)*

Teaching Suggestions

You may want to review the three types of uniform motion problems presented in Section 4-8, that is, motion in the same direction, motion in opposite directions, and round trips. In order to avoid fractional equations, the problems in Section 4-8 were solved using the uniform motion formula in the form $d = rt$. In this section, the students will see that it may be more convenient to use one of the other forms of the equation, namely,

$$r = \frac{d}{t} \text{ or } t = \frac{d}{r}.$$

Related Activities

Have students measure the rate of flow of water from two sinks (use cold water only) by using the following procedure.
1. Devise a method for determining when a tap is completely open, half open, and so on.
2. Make collections as suggested below and record the results.

   a. With the tap $\frac{1}{4}$-open, collect water for 5 s.

   b. Repeat for each tap when the taps are $\frac{1}{2}$-open, $\frac{3}{4}$-open, and completely open.

   c. Find the ratio of water flow of Tap 1 to Tap 2 at the different openings. Are the ratios the same?

	Tap 1 (liters)	Tap 2 (liters)	Ratio of Tap 1 to Tap 2
$\frac{1}{4}$-open			
$\frac{1}{2}$-open			
$\frac{3}{4}$-open			
full open			

3. What is the ratio of water flow of Tap 1 when it is $\frac{1}{4}$-open to Tap 1 when it is $\frac{1}{2}$-open? Is this what you would expect it to be?

9-7 *(pages 449–456)*

Teaching Suggestions

It is most likely that the students will have studied ratio and proportion before. Make sure they recognize that a ratio of two numbers is the quotient of one number divided by the other number and that the ratio of x to y, $x:y$, is not the same as the ratio of y to x, $y:x$, except in the case of $x = y$.

In solving fractional equations where each side of the equation is a ratio, a proportion is involved. From previous study, students may refer to "the product of the extremes equals the product of the means" as "cross-multiplication." Use the terms "extremes" and "means" frequently so that students become familiar with them. Students may realize that this method of solving fractional equations is equivalent to multiplying both sides of an equation by the LCD. This will become more evident if they prove the theorem in Exercise 39 on page 454.

9-8 *(pages 458–462)*

Teaching Suggestions

As students work with direct variation, it should become clear to them that every direct variation is a linear function whose graph passes through the origin. An important feature of direct variation is the fact that the proportional change in one variable is the same as that in the other. If a varies directly as b, then when $|b|$ increases or decreases by $p\%$, so does $|a|$.

Point out that although a direct variation has an equation $y = kx$, the proportions derived from this equation, such as $\frac{y_1}{x_1} = \frac{y_2}{x_2}$, are often more useful and efficient in solving problems.

9-9 (pages 462–466)

Teaching Suggestions

An inverse variation can be defined as any function in which the product of the coordinates of its ordered pairs is a nonzero constant; that is, for any ordered pair (x, y) of the function,

$$xy = k \text{ or } y = \frac{k}{x},$$

where k is the nonzero constant. In another form, if (x_1, y_1) and (x_2, y_2) are ordered pairs of an inverse variation, then

(1) $x_1 y_1 = k$ and $x_2 y_2 = k$

and   (2) $x_1 y_1 = x_2 y_2$.

Divide both sides by $x_2 y_1$ to get

(3) $\dfrac{x_1}{x_2} = \dfrac{y_2}{y_1}$.

Using equations (2) or (3) eliminates finding the constant of variation and enables students to find specified values rapidly.

Related Activities

The illumination from a light source varies inversely as the square of the distance. Using the chart below, and a single light source for the whole class, have students estimate the illumination at their desks from that source. Illumination is measured in lumens/m². Therefore, if the light source is a 60 watt bulb, then the illumination at a desk 2 m from the light would be

$$I = \frac{835}{2^2} \approx 209 \text{ lumens/m}^2.$$

Have students use the following chart to determine whether the ratio of watts to lumens is the same for different wattage tungsten lamps and for different wattage fluorescent lamps. Have students propose other relationships using these data.

Tungsten lamps		Fluorescent lamps	
watts	lumens	watts	lumens
10	78	4	73
25	260	6	210
40	465	8	330
60	835	14	490
100	1630	20	960
200	3650	30	1500
500	9950	40	2320
1000	21500	100	4400

9-10 (pages 467–471)

Teaching Suggestions

Be sure students understand that joint variation is *always* direct variation and that combined variation involves *both* direct and inverse variation.

Joint and combined variations are examples of functions of more than one variable. In the case of a function of two variables, the domain is a set of ordered pairs. Functions of the form $y = kxz$ and $y = \dfrac{kx}{z}$ fall into this category. Some other examples of functions of two variables are

(1) the multiplication of real numbers
$$f: (a, b) \longrightarrow ab$$
(2) the addition of real numbers
$$g: (a, b) \longrightarrow a + b$$

Students will find examples of joint and combined variations throughout their studies, in areas such as physics, chemistry, biology, and economics.

10 Irrational Numbers and Radicals

This chapter introduces square roots, a method for approximating square roots, and two applications that involve the use of square roots—the Pythagorean Theorem and the distance formula. Students learn how to convert repeating decimals to fractions, how to simplify square roots and other radicals, how to find sums and products of expressions involving radicals, and how to solve equations involving radicals.

decimal, where the repeating digit is zero. For example, $\frac{1}{5} = 0.2 = 0.2\overline{0}$.

Investigate other repeating decimals with repeating blocks greater than two digits. Students usually enjoy working with fractions with denominator 7 and discovering, for instance, that $0.\overline{142857}$ is the exact decimal equivalent for $\frac{1}{7}$.

10-1 (pages 481–485)

Teaching Suggestions

The definition of a rational number should be reviewed. Point out that rational numbers appear in many different forms such as integers, fractions, mixed numerals, as well as decimals (repeating and terminating decimals are covered in Section 10-2) and some percents, but that all of these forms can be renamed as the quotient of two integers, where the divisor is not zero.

Some students are puzzled by the consequences of the *Density Property of Rational Numbers*. It implies that between any two different rational numbers there are an infinite number of rational numbers. On the drawing that represents a number line, it appears that between any two dots only a finite number of dots may be placed. Remind students that dots represent points; dots have size but points do not.

10-2 (pages 486–491)

Teaching Suggestions

Some students will need to see that repeating decimals renamed as fractions can be translated back to the original repeating decimal.

Some students will notice that a terminating decimal could also be interpreted as a repeating

10-3 (pages 493–496)

Teaching Suggestions

Students will have no difficulty with the fact that $(5)^2$ and $(-5)^2$ both simplify to 25. They may become confused, however, when told that $\sqrt{25}$ names 5 and not -5. In this course, $\sqrt{25}$ will mean the principal square root only, following the definition that $\sqrt{a^2} = |a|$. To aid students here, tell them to read $\sqrt{a}$ as "the positive square root of a" instead of "the square root of a." Likewise, they should read $-\sqrt{a}$ as "the negative square root of a."

Refrain from stating that negative numbers do not have square roots unless you include the phrase "in the set of real numbers." (The Extra on page 530 explains that the square roots of negative real numbers involve *imaginary* numbers.)

This lesson represents the students' first exposure to solving quadratic equations. In solving equations such as $x^2 = 25$, you may want to take the following approach:

$$x^2 = 25$$
$$\sqrt{x^2} = \sqrt{25}$$
$$|x| = 5$$

If $x \geq 0$, then $x = 5$;
if $x < 0$, then $x = -5$.

This approach is consistent with the definition of $\sqrt{x}$ as nonnegative and accounts for the positive and negative roots of the equation $x^2 = 25$.

10-4 (pages 497–501)

Teaching Suggestions

Some of the techniques for approximating square roots include the use of a pocket calculator, a square root table, and the method of successive approximations. Compare the different approaches and discuss their strengths and weaknesses.

When a calculator or square root table is not available, students should know a technique for computing an approximation of a square root. You may wish to discuss the *divide-and-average method* described in Appendix A on page 685. The exercises given in the Appendix provide students with practice with this method.

10-5 (pages 505–512)

Teaching Suggestions

In your presentation of the Pythagorean Theorem, be sure students use this method:
1. Identify the hypotenuse by identifying the right angle.
2. Identify a right triangle when it is part of another figure.
3. Express any of the symbols a^2, b^2, and c^2 in terms of the others when given $c^2 = a^2 + b^2$. (That is, $a^2 = c^2 - b^2$ and $b^2 = c^2 - a^2$.)

Beware of the common misconception that $c^2 = a^2 + b^2$ implies $c = a + b$. Comparing two simple numerical examples like $3^2 + 4^2 = 5^2$ and $3 + 4 \neq 5$ should make clear that the above reasoning is incorrect.

Stress that the equation $x^2 = a$ $(a > 0)$ has two roots, namely $-\sqrt{a}$ and $\sqrt{a}$. The answers to the problems on page 511, however, are all positive because they represent quantities such as length and distance, which are nonnegative.

Students should learn to recognize the most common Pythagorean triples $\{3, 4, 5\}$ and $\{5, 12, 13\}$ and to realize that any multiples such as $\{12, 16, 20\}$ or $\left\{\frac{5}{2}, 6, \frac{13}{2}\right\}$ also work as rational solutions to $c^2 = a^2 + b^2$. $\left(\left\{\frac{5}{2}, 6, \frac{13}{2}\right\}\right.$, however, is not a Pythagorean triple because the members of the set are not all integers.)

Related Activities

Have the students extend a right triangle with sides of length 3 cm, 4 cm, and 5 cm into one with sides 6 cm, 8 cm, and 10 cm, and again to 9 cm, 12 cm, and 15 cm. Ask them to determine the relationships between the triangles. They should see that the measure of the angles does not change, only the lengths of the sides. Also, the ratio of the sides is constant: $\dfrac{AB}{AC} = \dfrac{5}{3} = \dfrac{10}{6} = \dfrac{15}{9}$.

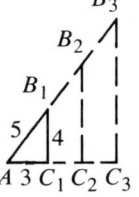 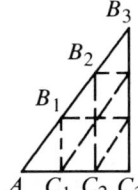

Now have the students determine that the area of the largest triangle is 9 times the area of the smallest triangle by drawing 9 congruent triangles inside the largest triangle. Ask them to explain the relationship $3^2 = 9$ has to the picture. The 3 represents the tripling of the measures of the sides of the small triangle; the 2 represents two dimensions; and the 9 represents the number of resultant small triangles or the number of times the area of the largest triangle is greater than the area of the smallest triangle.

Have students extend the sides to 12 cm, 16 cm, and 20 cm and determine the relationships. They can then try the same investigation with a 5 cm, 12 cm, and 13 cm right triangle and also an equilateral triangle.

10-6 *(pages 512–515)*

Teaching Suggestions

The distance formula is derived directly from the Pythagorean Theorem. Note that if P_1 and P_2 lie on a vertical line, the formula reduces to $P_1P_2 = |y_2 - y_1|$. If P_1 and P_2 lie on a horizontal line, $P_1P_2 = |x_2 - x_1|$. The absolute value signs are needed since distance by definition is nonnegative.

Within the radical sign, students must be careful to use parentheses accurately to reduce simple arithmetic errors. The most common mistakes will result from improper order of operations.

Related Activities

Have the students use the distance formula to prove that the coordinates of the midpoint M of the segment joining $P_1(x_1, y_1)$ and $P_2(x_2, y_2)$ are $\left(\dfrac{x_1 + x_2}{2}, \dfrac{y_1 + y_2}{2}\right)$.

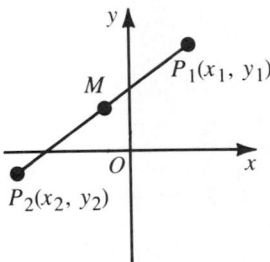

Begin by assuming that $M = \left(\dfrac{x_1 + x_2}{2}, \dfrac{y_1 + y_2}{2}\right)$.

The proof that M is the midpoint of $\overline{P_1P_2}$ requires two steps.

Step (1) Prove that P_1, P_2, and M lie on the same line.

This can be done by showing that the slope of $\overline{P_1M}$ equals the slope of $\overline{P_1P_2}$.

$$\text{Slope of } \overline{P_1M} = \frac{y_1 - \dfrac{y_1 + y_2}{2}}{x_1 - \dfrac{x_1 + x_2}{2}}$$

$$= \frac{\dfrac{y_1 - y_2}{2}}{\dfrac{x_1 - x_2}{2}} = \frac{y_1 - y_2}{x_1 - x_2}.$$

$$\text{Slope of } \overline{P_1P_2} = \frac{y_1 - y_2}{x_1 - x_2}.$$

Slope of $\overline{P_1M}$ = slope of $\overline{P_1P_2}$.

$\therefore P_1$, P_2, and M lie on the same line.

Step (2) Prove that M is the midpoint of $\overline{P_1P_2}$. This can be done by showing that the length of $\overline{P_1M}$ is half the length of $\overline{P_1P_2}$.

$$P_1M = \sqrt{\left(x_1 - \frac{x_1 + x_2}{2}\right)^2 + \left(y_1 - \frac{y_1 + y_2}{2}\right)^2}$$

$$= \sqrt{\left(\frac{x_1 - x_2}{2}\right)^2 + \left(\frac{y_1 - y_2}{2}\right)^2}$$

$$= \frac{1}{2}\sqrt{(x_1 - x_2)^2 + (y_1 - y_2)^2}$$

$$\frac{1}{2}(P_1P_2) = \frac{1}{2}\sqrt{(x_1 - x_2)^2 + (y_1 - y_2)^2}$$

$$P_1M = \frac{1}{2}P_1P_2$$

$\therefore M$ is the midpoint of $\overline{P_1P_2}$.

Thus, the midpoint of $\overline{P_1P_2}$ is the point $\left(\dfrac{x_1 + x_2}{2}, \dfrac{y_1 + y_2}{2}\right)$.

10-7 *(pages 516–520)*

Teaching Suggestions

To help students use the product property of square roots to simplify radicals, have them start the process by supplying two empty radical signs, reserving the first radical sign for the square of a monomial. Thus, to simplify $\sqrt{48x^3}$, first write:

$$\sqrt{48x^3} = \sqrt{} \cdot \sqrt{}$$

Reserved for: square of a   remaining factors
           monomial   of $48x^3$

Then: $\sqrt{48x^3} = \sqrt{16x^2} \cdot \sqrt{3x} = 4x\sqrt{3x}$

Point out to students that in all the exercises, variables are assumed to denote positive real numbers only. To demonstrate the reason for this assumption, square a negative number, zero, and a positive number. In no case is a negative number generated. Thus, in the example above, if x were negative, $\sqrt{3x}$ would not have a value among the real numbers. Also, if the restriction were not stated at the beginning of the exercises, $\sqrt{3a^2}$ would simplify to $|a|\sqrt{3}$ to guarantee a positive result. This form will be seen in later exercises when variables are not restricted to nonnegative values.

10-8 (pages 520–524)

Teaching Suggestions

Stress the fact that since square-root radicals are presently defined as real numbers, algebraic properties and other properties of real numbers apply to them.

Encourage students to do as much mental simplification as they can. For example, to simplify $\sqrt{10}(\sqrt{2} - \sqrt{5})$, students might reason that since $\sqrt{10}$ contains $\sqrt{2}$ as a factor, the first product will result in $2\sqrt{5}$. Similarly, since $\sqrt{10}$ contains $\sqrt{5}$ as a factor, the second product will result in $5\sqrt{2}$. Thus, $\sqrt{10}(\sqrt{2} - \sqrt{5}) = 2\sqrt{5} - 5\sqrt{2}$.

Point out that when conjugates are used to rationalize binomial denominators, both the numerator and denominator are multiplied by the same value. Thus, the original number is being multiplied by 1 so its value is not changed.

10-9 (pages 524–526)

Teaching Suggestions

Using the procedure from Sec. 10-7 will help students simplify expressions involving nth roots. To simplify $\sqrt[3]{48x^3}$, write two empty radical signs:

$$\sqrt[3]{48x^3} = \sqrt[3]{} \cdot \sqrt[3]{}$$
$$\uparrow \qquad\qquad \uparrow$$

Reserved for: nth power of a   remaining factors
monomial   of the expression

Then: $\sqrt[3]{48x^3} = \sqrt[3]{8x^3} \cdot \sqrt[3]{6} = 2x\sqrt[3]{6}$

Stress the fact that negative numbers have no nth roots if n is even and that they have only one root if n is odd. Reaffirm that when n is even and positive, the nth root of a real number is referred to as the principal nth root and is denoted by $\sqrt[n]{a}$. The negative nth root of a is denoted by $-\sqrt[n]{a}$.

This is an appropriate time to hold oral drills on the various powers of 2, -2, 3, -3, 5, and -5 and also to ask for the 4th root if 81, the 5th root of -32, and so on.

10-10 (pages 527–529)

Teaching Suggestions

Students should have little difficulty transforming equations involving a variable in a radicand. Emphasize, however, that raising both sides of an equation to a power should be done only after the radical with the variable is alone on one side of the equation.

Stress the necessity of checking solutions. Point out that squaring often results in extraneous solutions. Cite the following situation: squaring both sides of a false equation can result in a true statement: $3 = -3$, $(3)^2 = (-3)^2$, $9 = 9$.

Students should be reminded that the principal square root of a real number x, $\sqrt{x}$, is positive. Thus, the solution set for $\sqrt{x} = -3$ is $\emptyset$.

Related Activities

Some equations of the form $ax + b\sqrt{x} + c = 0$ can be solved by the same technique used to factor quadratic trinomials. Notice, however, that here the factors will be of the form $(m\sqrt{x} + r)$, rather than $(mx + r)$. Have the students solve each of the following equations.

1. $x - 8\sqrt{x} + 7 = 0$
   $(\sqrt{x} - 7)(\sqrt{x} - 1) = 0$
   $\sqrt{x} = 7$ or $\sqrt{x} = 1$
   $x = 49$ or $x = 1$
   $\therefore$ the solution set is $\{49, 1\}$.
2. $2x - 11\sqrt{x} + 12 = 0$
   $(2\sqrt{x} - 3)(\sqrt{x} - 4) = 0$
   $\sqrt{x} = \frac{3}{2}$ or $\sqrt{x} = 4$
   $x = \frac{9}{4}$ or $x = 16$
   $\therefore$ the solution set is $\left\{\frac{9}{4}, 16\right\}$.

11 Quadratic Equations and Functions

This chapter completes the discussion of quadratic polynomials and thus represents what is traditionally the end of a first-year algebra course. Solving quadratic equations by completing the square or using the quadratic formula is covered as well as graphing quadratic functions, simple polynomial functions, and quadratic inequalities.

11-1 (pages 543–548)

Teaching Suggestions

You may want to begin your presentation by reviewing how to solve a quadratic equation by factoring. Solve $2x^2 - 5x - 12 = 0$ by factoring and then try to solve $a^2 - 9a + 5 = 0$ by the same method. Inform students that the equations in this section are similar to $a^2 - 9a + 5 = 0$ in that they cannot be solved by conventional factoring. Proceed to solve each of the equations by completing the square.

To set the stage for using this technique, you can have your students supply the missing term and rewrite each of the following as the square of a binomial.

$$x^2 + 20x + \underline{\ ?\ }$$ $$\quad y^2 \pm \underline{\ ?\ } + 81$$
$$x^2 - 20x + \underline{\ ?\ }$$ $$\quad m^2 \pm \underline{\ ?\ } + a^2$$
$$x^2 + 5x + \underline{\ ?\ }$$ $$\quad n^2 \pm \underline{\ ?\ } + \frac{9}{4}$$

For completing the square, first practice the technique with several exercises having a quadratic-term coefficient of 1. Then, when $a \neq 0$ and $a \neq 1$ in $ax^2 + bx + c = 0$, remind students to first divide every term by a.

Checking a solution like $\dfrac{-3 + 3\sqrt{13}}{2}$ in a quadratic equation is a demanding task. Checking every such solution is not productive, but some should be checked as reinforcement for the technique of completing the square. The process also serves as a review of the techniques used in simplifying expressions involving radicals discussed in the previous chapter.

11-2 (pages 548–551)

Teaching Suggestions

Since it would be cumbersome to derive the quadratic formula every time it is needed, ask students to memorize it. In an equation such as $x^2 - 5x + 9 = 0$, students will sometimes say b is 5 rather than -5. To minimize such errors, have students state the values of a, b, and c before they begin to use the formula. Such in-class oral drills should be continued until use of the quadratic formula has become a familiar process.

Be sure that students realize the significance of the $\pm$ disjunction in the quadratic formula. Students should be able to rewrite the solution set $\{2 \pm \sqrt{5}\}$ as $\{2 + \sqrt{5}, 2 - \sqrt{5}\}$.

Students should learn to inspect the value of the discriminant to determine the nature of the roots. To approach this another way, have the students develop new quadratic equations which have two different real roots, a double root, or no real roots.

11-3 (pages 552–555)

Teaching Suggestions

Techniques used in previous sections for solving word problems can be used in this section. Be sure to demonstrate with your chalkboard examples that solutions to the quadratic equations are sometimes not solutions to the given problem. In particular, problems that ask for such things as weights, distances, and ages will not have negative values so they must be discarded from the solution set of the quadratic equation. Because of this, it is especially important that tentative solutions be checked in the original problem.

Teaching Suggestions

Be sure your students understand the difference between the quadratic equation in one variable, $ax^2 + bx + c = 0$, and the quadratic function determined by the quadratic equation in two variables, $y = ax^2 + bx + c$. The quadratic function is a set of ordered pairs $\{(x, y): y = ax^2 + bx + c\}$, whereas the solution set of $ax^2 + bx + c = 0$ consists of the set of first coordinates of the ordered pairs mentioned above in which the second coordinate is zero. This fact can be reinforced by pointing out that if and when the graph of a quadratic function crosses the x-axis, the x-coordinates of those points are solutions of $ax^2 + bx + c = 0$. In other words, a zero of a quadratic function is an x-intercept of the graph of the function.

Stress the fact that any equation equivalent to $y = ax^2 + bx + c$, $(a \neq 0)$ defines a quadratic function whose graph is a parabola. Urge students to use the following facts as an aid to graphing: When $a > 0$, the parabola opens upward; when $a < 0$, the parabola opens downward. The relative widths of parabolas can be compared by comparing the coefficients of their quadratic terms. For example, $y = 4x^2$ will be narrower than $y = x^2$ because the y values are 4 times greater, given the same value of x.

If the graph opens upward, the function has a minimum value; if the graph opens downward, the function has a maximum value. Students should understand that if x_1 and x_2 are zeros of a quadratic function, then the equation of the axis of symmetry is $x = \dfrac{x_1 + x_2}{2}$.

If students have memorized the expression for the *sum* of the roots from Section 11-2, $-\dfrac{b}{a}$, the *average* of the roots is $-\dfrac{b}{a} \div 2$, or $-\dfrac{b}{2a}$. Remembering this relationship should help students to avoid confusing the two expressions.

Also, be sure students remember that the *values of a function* are members of the range. Thus, a maximum or minimum value of a function is the greatest or least number that is a member of the range.

Related Activities

Given a sequence of numbers, you can determine by looking at "finite differences" the degree of the polynomial function which determines the sequence. Subtracting successive terms of the sequence from each other determines what are called "first differences." Consider the following sequence:

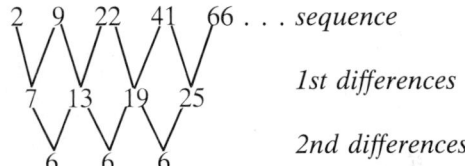

Repeating the process with these *first differences* determines what are called *second differences* and so on. A theorem states that if the *n*th *differences* are constant, then the degree of the polynomial function that determines the sequence is n. In the above example, the function is $f: x \longrightarrow 3x^2 - 2x + 1$.

Two other sequences for which the students can find the degree of the corresponding polynomial function are

$$5, 6, 9, 17, 33, 60, 101, 159 \quad \text{4th}$$

and

$$-12, -4, 11, 33, 62, 98. \quad \text{2nd}$$

Next have students work in small groups to try to build their own sequences. Have them share their ideas with the entire class.

Teaching Suggestions

You can help students in graphing third- and fourth-degree polynomials by discussing the general forms of these graphs.

In developing the general topic of graphing third-degree polynomials, use the graphs shown in Examples 1 and 2 on pages 564–565. In general, when $y = ax^3 + bx^2 + cx + d$, and $a > 0$, the graph will extend upward to the right for sufficiently large values of x, and the graph will extend downward to the left for sufficiently small values of x (see Example 1). If $a < 0$, the graph will extend upward to the left and downward to the right as in Example 2. Emphasize that when there are three distinct real zeros, the graph will cross the x-axis three times.

For the fourth-degree polynomial function
$$y = ax^4 + bx^3 + cx^2 + dx + e,$$
if $a > 0$ and $|x|$ is sufficiently great, the value of the function will be positive so the graph will extend upward both to the left and to the right: if $a < 0$, the graph will extend downward both to the left and to the right.

Even though students know what general form to expect, they will still have to plot many points to be certain their graph is a fairly accurate picture of the given function. For fairly accurate estimation of zeros, students should plot their graphs on $\frac{1}{10}$-in. or 1-mm graph paper. The use of calculators is helpful in making an accurate graph since it is often necessary to determine numerous ordered pairs (see the following Related Activity).

Related Activities

The graphing of polynomial functions can be great-ly facilitated by encouraging students to use calculators in making their tables of values. Once a zero has been approximated from the graph, calculator work could be utilized to improve the accuracy of the approximation with a minimum of effort. Ask the students to make a table of values for $f: x \longrightarrow 2x - x^3$, concentrating on values of x between -1 and 1. Then ask them to find another example for which using the calculator to make a table of values is a decided advantage.

11-6 (pages 567–569)

Teaching Suggestions

You may wish to allow students to use a method similar to the one outlined for solving linear inequalities in *Teaching Suggestions* for Section 5-6. Students should first graph the quadratic function associated with the inequality using a solid or dashed curve, whichever is appropriate. Then one point not on the curve should be chosen; $(0, 0)$ is usually an easy one with which to work. If the coordinates of the point satisfy the inequality, the region in which the point lies should be shaded. If they do not satisfy the inequality, the opposite side of the curve should be shaded.

12 Trigonometry and Vectors

This chapter acquaints students with some algebraic topics that will be covered more extensively in later courses. The trigonometry sections include standard position of an angle, the unit circle, the basic trigonometric functions, and the use of tables in solving right triangles. The vector portion of the chapter deals with the concepts of the standard position and the norm of a vector and the resultant of two vectors. The chapter concludes with a discussion of applications of vectors.

12-1 (pages 579–584)

Teaching Suggestions

For the development of the trigonometric functions on the unit circle, it is important that students extend their concept of an angle to a directed angle as described in the text.

You may wish to present some examples to develop this extended concept of an angle: the turning of a clock hand, the turning of wheels, and so on. Point out that counterclockwise rotation of an angle is positive and that clockwise rotation is negative.

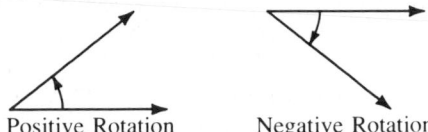

Positive Rotation   Negative Rotation

Students' recognition of the need for angle measures greater than 360° may be motivated by having students consider questions such as:

(1) Through how many degrees does the minute hand of a clock rotate in 3 h and 20 min? −1200°

(2) If a car tire makes $4\frac{5}{9}$ rotations while backing up, through how many degrees does the tire turn? 1640°

For the exercises in this section, it is strongly recommended that students sketch a picture that describes the given information. The measures of the quadrantal angles should be memorized (0°, ±90°, ±180°, ±270°, 360°).

12-2 (pages 585–590)

Teaching Suggestions

A number of new vocabulary words and concepts are introduced in this section. Plan to devote more time on this section than on other sections.

Trigonometry is introduced here as a circular function, with no mention of opposite side, adjacent side, or hypotenuse; these will be brought in later as a special case. If your students can understand the circular-function nature of trigonometry now, they will have a firm foundation for future extensions of the basic ideas.

Many students will have difficulty remembering that the sign of a trigonometric function depends upon the quadrant in which the terminal side of the angle lies. The following diagram may help students recall in which quadrant each trigonometric function assumes positive values.

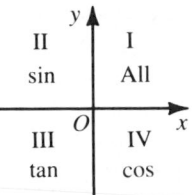

The first letters of the words in "**A**ll **s**tudents **t**ake **c**alculus" may also be used as a memory device.

Some students may like to see a derivation of the equation of the unit circle from the distance formula so they can understand how the relationship $\sin^2 A + \cos^2 A = 1$ is determined.

12-3 (pages 590–592)

Teaching Suggestions

When students are learning how to use trigonometric tables, it may be helpful for them to do examples with you, using the table of trigonometric values on page 684. Since interpolation is not taught at this level, make sure that students know how to approximate the function values which do not appear in the table.

Students may observe that for first-quadrant angles, as the measure of the angle increases, the sine and tangent increase, whereas the cosine decreases.

As you work these exercises, reinforce the functional nature of sine, cosine, and tangent. Stress that the domain for each of the functions consists of angles, whereas the range consists of real numbers.

12-4 (pages 593–600)

Teaching Suggestions

This section develops the relationships between the coordinates of any point on the terminal side of an angle with the three trigonometric functions. With these relationships, the functions can be applied to right triangles not associated with a coordinate system.

Students must memorize the trigonometric formulas. A common mnemonic device is SOH-CAH-TOA (Sine: Opposite, Hypotenuse and so on).

A valuable application of the trigonometric formulas is indirect measurement. Have the students consider how to find the height of a tree without actually measuring it. Suggest that by measuring the shadow of the tree and using the tangent function, a good approximation of the height can be found. Point out the advantages of making a sketch when solving this type of problem.

$$\tan 35° = \frac{TR}{16}$$
$$TR = 16(\tan 35°)$$
$$TR \approx 16(0.7002)$$
$$TR \approx 11 \text{ m}$$

Oral Exercises 11 and 12 on page 596 demonstrate that in solving a triangle it is generally easier to use a trigonometric ratio in which the value to be determined appears in the numerator. You may wish to have students solve these examples using different ratios and compare the ease of solution.

Related Activities

One application for the trigonometric functions is the calculation of the heights of objects when direct measurement is difficult. For example, to find the height of a tall building, a student whose eye level is y_s cm above the ground takes two measurements: (1) the angle the student's line of sight to the top of the building makes with the horizontal ($\angle A$) and (2) the distance to the building (x_b).

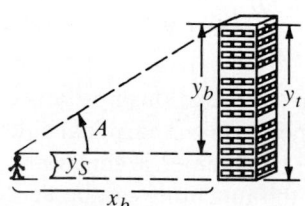

To find the height of the building (y_t), use the following equations.

$$\frac{y_b}{x_b} = \tan A$$
$$y_b = (x_b) \tan A$$

Thus, $y_t = y_b + y_s$.

For a specific example, imagine that a student whose eye level is 150 cm above the ground measures an angle of 68° to the top of the building, and measures a distance of 100 m to the building. Use the equations just given to find the height of the building to the nearest meter.

$$y_b = (x_b) \tan A$$
$$= (100) \tan 68°$$
$$\approx 100(2.4751)$$
$$= 247.51 \text{ m}$$
$$y_t = y_b + y_s$$
$$\approx 247.51 \text{ m} + 1.5 \text{ m}$$
$$= 249.01 \text{ m} \approx 249.0 \text{ m}$$

Have students use this technique to calculate the height of various objects.

(See also the suggested Related Activities in the Lesson Commentary for Sections 12-6 and 12-7.)

12-5 (pages 604–607)

Teaching Suggestions

Since a vector in standard position can be described by the coordinates of the terminal point alone, some students may have difficulty remembering that, in general, both the magnitude and direction of a vector must be given to fully describe it. You may therefore want to include in your discussion a method for finding the norm and direction of several vectors not in standard position.

Make sure that students can picture the x- and y-components of a vector graphically. Point out that if they do this, they will be using the Pythagorean Theorem to find the norm and the right triangle definition of tangent to find the direction.

12-6 (pages 608–611)

Teaching Suggestions

The concept of equivalent vectors is essential to finding the resultant of vectors not in standard position. Make sure students understand that vectors are equivalent when their norms and directions are the same, although their initial and terminal points may differ.

You may wish to present the resultant vector as the diagonal of a parallelogram with the two original vectors as adjacent sides. If you do, make sure that your students are able to position equivalent vectors correctly to form the parallelogram.

Related Activities

Another application of trigonometry is in the field of solar energy. When the sun is shining on a surface above the earth's atmosphere that is perpendicular to the sun's rays, the sun transmits 1,353 W/m². This amount of energy is represented as I_{sc}. However, when the sun shines on a surface that is not perpendicular to its rays, the energy is spread over a larger area, resulting in a reduction of the energy received by a given area. This reduced reception of energy can be expressed by the following equation, $I_{\angle A} = I_{sc}(\cos A)$, where $I_{\angle A}$ is the energy received by a surface that is at an angle A from the perpendicular as shown in the following drawing.

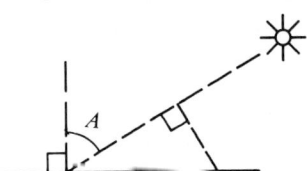

Have students calculate the amount of energy received by flat surfaces in space at various angles to the sun. The same equation can be used to determine the relative amount of solar energy reaching the surface of the earth. Have students research the angle of the sun for zenith in your area at the solstices and at an equinox. Then have them calculate the energy received and compare it to the energy received from direct overhead sunlight.

12-7 (pages 613–614)

Teaching Suggestions

Follow the established procedures for solving word problems with these exercises, giving special emphasis to transferring the information in the problem to a diagram. Such a diagram will help the student choose the correct trigonometric function.

Your students may not realize that vectors can be used to represent any quantity having both magnitude and direction. The following examples can be used to show how one vector could represent a number of different physical situations.

1. An object is being pulled by two forces. One force of 12 kg is pulling due north and the other force of 5 kg is pulling due east. The resultant force is equivalent to a force of 13 kg pulling at an angle with measure of approximately 67° from the direction of the 5 kg force.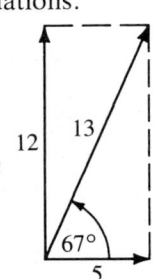

2. A football is kicked from the midfield line due north toward the goal post at a rate of 12 m/s and the wind is blowing due east at a rate of 5 m/s. At the end of 1 s, the football would be located 13 m from where it was kicked at an angle with measure of approximately 67° with the midfield line.

Related Activities

Challenge students to use their knowledge of trigonometry to derive a formula to find the area of any triangle when measures of two sides and the included angle are known. Use the following drawing to help them in this task. Assume that you know lengths a and b and the measure of $\angle C$.

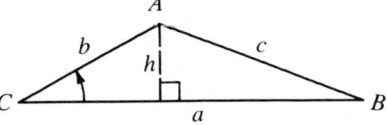

In this example, $h = b \sin C$.
$\therefore$ Area of $\triangle ABC = \frac{1}{2}ah$
$\qquad\qquad\qquad\quad = \frac{1}{2}ab \sin C.$

13 Statistics and Probability

This chapter is an overview of introductory statistics and probability. Various ways of representing data such as dot frequency diagrams, histograms, and relative frequency tables are presented leading to a brief exploration of some measure of central tendency. The chapter concludes with a look at experimental probability.

13-1 (pages 623–626)

Teaching Suggestions

This section and the two that follow deal with various ways of displaying data. Have your students bring in examples of frequency distributions from news magazines and the local paper to stimulate discussion about why different sets of data are displayed in different manners.

The conversion of the relative frequency in fraction form to percent will enable most students to better understand the degree to which the data is distributed. Remind students that with some data, round-off errors will cause the percent total to be greater or less than 100.

13-2 (pages 626–630)

Teaching Suggestions

Students may be confused about when to use a dot frequency diagram versus a histogram or frequency polygon. To help your students get a clearer understanding of the type of data that requires a histogram or frequency polygon, use an example like the following: A quiz is given to a large group of students; the percentage grades range from 50 to 100% with grades such as 87% possible. A histogram made by grouping grades into 5- or 10-point intervals will be much easier to interpret than a frequency diagram, which would have 51 discrete points on one axis.

To reinforce this idea, provide students with a data sheet containing about 100 values. Ask them to prepare a histogram and a frequency polygon. Ask some students to use 2-unit intervals, others 3-unit, and so on up to 20-unit. The comparison of the results will probably reveal that an interval of 5 or 10 units resulting in 10 to 20 intervals was both easy to work with and produced an effective visual display of the data. For most data, number intervals such as multiples of 10 or 100 are usually chosen for interval boundaries for convenience.

A frequent misconception concerning frequency polygons is that there is meaning to the slope of the lines connecting the data points. You may want to discuss this with your class after completing Oral Exercise 10 on page 628.

13-3 (pages 631–635)

Teaching Suggestions

Test scores, demographic data, and longevity data are often displayed in cumulative frequency polygons. The cumulative frequency polygon is well suited for these kinds of data because it shows percentile ranking graphically.

Related Activities

Have students make a frequency distribution table of the sneaker sizes of everyone in class. The table should show frequencies, relative frequencies in percents, cumulative frequencies, and cumulative percent. Using this data have students draw:
a. a dot frequency diagram
b. a histogram (Assume no half sizes are available. Choose appropriate intervals.)
c. a cumulative frequency diagram

Have the class discuss which of the graphical representations of the data would be most helpful for a storekeeper to use when stocking various sizes.

Teaching Suggestions

Explain to your students that the mean, median, and mode are called measures of central tendency in a distribution because each in some sense represents the "middle" of the distribution. Sometimes one of these measures will give a more accurate or useful picture of a distribution than the others. You may want to discuss the following examples with your students.

1. The buyer for a hat shop would be most interested in the *mode*(s) of hat sizes, for this would give a good indication of which size(s) of hat to stock most heavily.

2. A student might be interested in the *median* grade of a test in order to compare his or her grade to the grade the "middle person" in the class received.

3. A director of a tennis tournament would be interested in the *mean* number of tennis balls used in a match. Then, multiplying the mean times the number of matches will provide the director with the number of tennis balls needed for the tournament.

Caution students to order the data before trying to find the mode or median. Remind them that the median of an even number of terms is half the sum of the two middle terms.

13-5 (pages 640–643)

Teaching Suggestions

Students may question why the procedure for determining variance involves finding the squares of the differences from the mean. You may want to use the following line of reasoning as an explanation: Since the variance is a measure of how scattered the data are about the mean, it would appear logical to sum up the differences between each term and the mean and divide by the number of terms n. As your students should realize, this sum is always 0. To counter this, one might consider using the absolute value of the differences to give some measure of the variance. The difficulty with this measure is that a large sample of narrowly scattered data could have a greater variance than a small set of widely scattered data. Squaring the differences makes the degree of variance more apparent. Summing these squares and dividing by the number of terms, n, gives an accurate picture of how scattered the data are and avoids the difficulties involved in other methods.

Related Activities

Many characteristics of large populations can be described by normal distributions. In a normal distribution, measurements are distributed symmetrically above and below the mean. Consider the following example.

A teacher gives a multiple-choice test that has 125 questions worth one point each. The scores received by the students can be described by a normal distribution. The mean x is 100 and the standard deviation s is 10. Approximately 34% of the scores fall within one standard deviation of the mean. (This means that 34% of the scores fall within one standard deviation above the mean and 34% fall within one standard deviation below the mean. Thus, 34% of the scores are between 100 and 110, and 34% are between 90 and 100.) On the same test, approximately 48% of the scores fall within two standard deviations of the mean.

Have the students find the percent of the scores on the test that fall in each of the following regions.

1. between 80 and 100   48%
2. between 90 and 110   68%
3. between 80 and 120   96%
4. less than 100   50%
5. between 90 and 120   82%
6. greater than 120   2%

Give the students the class scores from a test they took early in the year and have them determine if the scores form a normal distribution.

Teaching Suggestions

In general, there are two types of probability, *a priori* and *a posteriori*. The first states that given a sample space of n equally likely, mutually exclusive outcomes, and given an event E consisting of e of these outcomes, the probability of event E is $\frac{e}{n}$.

Experimental, or *a posteriori*, probability is based upon past observations. For example, if over a period of time an experiment has been conducted n times and an event E has occurred e times, the probability that E will occur the next time is $\frac{e}{n}$. *A priori* probability is based on the inherent mathematical properties of an event; *a posteriori* probability is based on testing or sampling.

It is a common misconception that if a coin is tossed 100 times and comes up heads 100 times, then the probability that it will come up heads the 101st time is *less* than $\frac{1}{2}$. Assuming that it is a fair coin, the probability that it will come up heads on the 101st toss is *exactly* $\frac{1}{2}$. One might begin to wonder, however, about the supposed randomness of this event. If in fact the coin is weighted, the probability that it will come up heads may be *greater* than $\frac{1}{2}$.

Related Activities

As a preliminary exercise, have students determine all possible outcomes of rolling two dice and make a table (right, above) showing the probability of each outcome. (Note that there are 36 possible outcomes when two dice are rolled. Thus $n = 36$.)

Next have small groups roll two dice 50 times, and record the results. Each group should then compare the results of their experiment to the expected results.

Finally have all groups pool their experimental results. The students should be able to conclude that as the number of rolls increases, the actual frequency of each outcome approaches the expected probability.

Sum of two dice E	Number of outcomes in E e	$P(E)$ $\frac{e}{n} = \frac{e}{36}$
2	1, 1	$\frac{1}{36}$
3	1, 2 or 2, 1	$\frac{2}{36}$
4	1, 3 or 2, 2 or 3, 1	$\frac{3}{36}$
5	1, 4 or 2, 3 or 3, 2 or 4, 1	$\frac{4}{36}$
6	1, 5 or 2, 4 or 3, 3 or 2, 4 or 5, 1	$\frac{5}{36}$
7	1, 6 or 2, 5 or 3, 4 or 4, 3 or 5, 2 or 6, 1	$\frac{6}{36}$
8	2, 6 or 3, 5 or 4, 4 or 5, 3 or 6, 2	$\frac{5}{36}$
9	3, 6 or 4, 5 or 5, 4 or 6, 3	$\frac{4}{36}$
10	4, 6 or 5, 5 or 6, 4	$\frac{3}{36}$
11	5, 6 or 6, 5	$\frac{2}{36}$
12	6, 6	$\frac{1}{36}$

Teaching Suggestions

If possible, plan to have your students perform some probability experiments. Experimental results from flipping coins, tossing dice, and drawing cards can be compared with the theoretical results. Individual results can be combined to produce an experimental probability for the entire class. Attempting to predict the nature of a complete collection can be done by taking samples from a box of different-colored marbles or from a deck of cards. By recording the results of these experiments and discussing why experimental probability may differ from *a priori* probability, your students should develop a good informal concept of simple probability.

Algebra 1

Mary P. Dolciani
Richard A. Swanson
John A. Graham

Editorial Adviser
Andrew M. Gleason

Teacher Consultants
Bruce Brombacher
Elizabeth B. Jayne
Carla J. Lund

HOUGHTON MIFFLIN COMPANY · **Boston**

Atlanta   Dallas   Geneva, Ill.   Lawrenceville, N.J.   Palo Alto   Toronto

Authors

Mary P. Dolciani Professor of Mathematical Sciences, Hunter College of the City of New York

Richard A. Swanson Supervisor of Mathematics, Liverpool Central Schools, Liverpool, New York

John A. Graham Mathematics Teacher, Buckingham Browne and Nichols School, Cambridge, Massachusetts

Editorial Adviser

Andrew M. Gleason Hollis Professor of Mathematics and Natural Philosophy, Harvard University, Cambridge, Massachusetts

Teacher Consultants

Bruce Brombacher Mathematics Teacher, Jones Middle School, Westerville, Ohio

Elizabeth B. Jayne Mathematics Teacher, Paul Blazer High School, Ashland, Kentucky

Carla J. Lund Mathematics Teacher, Washington High School, Tulsa, Oklahoma

Printed in U.S.A.

ISBN: 0-395-34373-9

ABCDEFGHIJ-RM-93210/89765

Contents

Chapter 6   *Systems of Open Sentences*   267

Chapter 7   *Polynomials and Factoring*   323

Chapter 10 *Irrational Numbers and Radicals* 481

Chapter 11 *Quadratic Equations and Functions* 543

Using a Computer with This Course

There are two types of optional computer material in this text: Programming in BASIC features and Computer Exercises. The Programming in BASIC features can be used by students without previous programming experience. These features usually include a program that students can run to explore an algebra topic covered in the chapter. Some writing of programs may be required in some of these features.

The optional Computer Exercises are designed for students who have some familiarity with programming in BASIC. Students are usually asked to write one or more programs related to the lesson just presented.

An appendix that summarizes BASIC programming may be found beginning on page 686. This appendix can be used as a summary by students who are learning BASIC or as a review by students who are familiar with BASIC.

Symbols

READING ALGEBRA

An algebra textbook requires a different type of reading than a novel or short story. You will need to read every paragraph with great care and concentration. Keep a paper and pencil handy for doing calculations and drawing sketches. It is also a good practice to keep an algebra notebook in which you summarize important ideas.

Vocabulary

Important vocabulary words are printed in red when they are first explained. At the beginning of each Self-Test, there is a list of all such words that have appeared in that part of the chapter. You can also find these words referenced in the Glossary and Index at the back of the book. The Glossary will give you a definition of the word, and the Index will give page references for more information.

Symbols

The language of algebra is primarily symbolic. Thus, in order to understand algebra, you must be able to read symbols. For example, $x \in \{1, 3, 5\}$ is read, "x is an element of the set whose members are 1, 3, and 5." A list of symbols that appear in this book faces this page. If you do not recall what a symbol means, check this list.

Diagrams

Diagrams are often used in this book to illustrate a concept that is difficult to express in words only. Study these diagrams carefully when you read the text that accompanies them.

Displayed Material

Throughout this book, red boxes are used to display axioms, theorems, summaries, and other important material. These boxes will help you identify key concepts as you read and when you study for a test.

Examples and Solutions

Each section of this book contains one or more clearly labeled examples with detailed solutions. Study these carefully, since they will help you in doing many of the exercises and problems that follow.

Reading Aids

Throughout the book, the sections titled *Reading Algebra* contain hints that are designed to help you use your book more effectively.

The design of the Red Rocks Theater, near Denver, Colorado, required skill in electronics and civil engineering as well as architecture. All of these fields involve an extensive use of mathematics.

Chapter 1

Numbers and Variables

The Real Numbers

OBJECTIVES for Sections 1-1 through 1-3:
1. *To graph real numbers as points on a number line.*
2. *To show the order of real numbers.*
3. *To use set notation.*
4. *To specify and graph sets of real numbers.*

1–1 Number Lines

Whenever you count or measure, you use numbers. In order to picture numbers and the relationships among them, numbers can be placed in correspondence with points on a line. Such a picture is called a **number line.**

For example, in your study of mathematics thus far you have encountered *positive numbers* such as 1, 0.5, $\frac{7}{3}$, $\sqrt{2}$, and π. You also may have encountered *negative numbers* such as $^-1$ (read "negative one"), $^-0.5$, $^-\frac{7}{3}$, $^-\sqrt{2}$, and $^-\pi$. The figure below shows how these numbers can be pictured as points on a number line

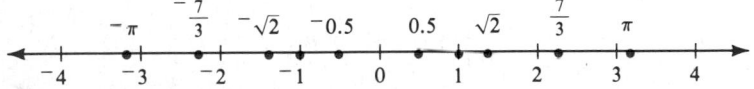

How do you construct a number line? The numbered steps on the following page outline a general procedure.

See p. T49 for a general discussion of problem solving strategies.
Recognizing a Problem Type In Section 1-8, students should look for patterns in translating word phrases into variable expressions. For instance, words or phrases such as *sum, increased by,* or *more than* indicate addition, while *product* or *one sixth of* indicate multiplication.
In Section 1-9, students should look for some form of the verb *to be.* In translating a word sentence into a mathematical sentence, students should make sure that the left side of the equation matches the quantities mentioned before the verb *to be* in the word sentence. Quantities mentioned after the form of the verb *to be* will be on the right side of the equals sign in the equation. Forms of the verb *to be* include *is, was, is not,* and *is greater than.*

Key Ideas

Construct a number line.
Graph real numbers as
points on a number line.
Show the order of real numbers.

Chalkboard Examples

1. State the coordinate of
   each point graphed.

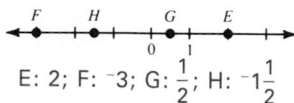

E: 2; F: $^-3$; G: $\frac{1}{2}$; H: $^-1\frac{1}{2}$

2. Use $<$ and $>$ to compare
   the numbers.
   a. $^-7$, 14 $^-7 < 14$
   b. 12, $^-9$ $12 > {^-9}$
   c. $^-6$, $^-4$ $^-6 < {^-4}$

3. What are the directed
   numbers for each?
   a. 5 steps down $^-5$
   b. 5 steps up   5
   c. a $50 loss $^-50$
   d. a $50 gain   50

1. Choose a starting point on a line and label it 0. This point
    corresponds to the number zero and is called the **origin.** The origin
    separates a number line into two opposite sides, the *positive side* and
    the *negative side*. If the number line is horizontal, the side to the right
    of the origin is usually taken to be the positive side.

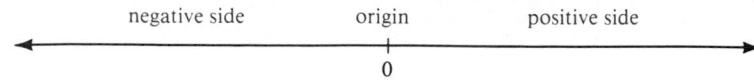

2. Choose any length to be one *unit*. Mark the point that is one unit to
    the positive side of 0 and label it 1. Using the same *unit length*,
    mark and label points to the positive side of 1 as 2, 3, and so on.

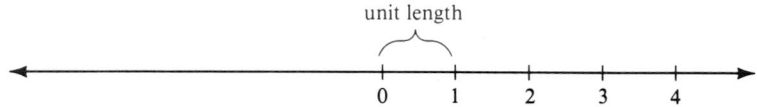

Using the points labeled 0, 1, 2, and so on as a reference, it is also
possible to mark the points that correspond to fractions. For example, to
label the points that correspond to $\frac{2}{3}$, $1\frac{2}{3}$, $2\frac{2}{3}$, and so on, separate each
unit length into three segments of equal length. Then the point that is
two thirds of the distance from 0 to 1 can be labeled $\frac{2}{3}$, the point that is
two thirds of the distance from 1 to 2 can be labeled $1\frac{2}{3}$, and so on.

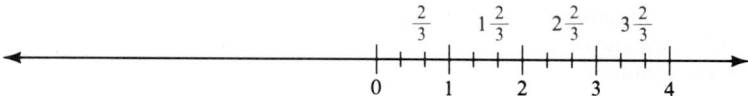

In general, a number that corresponds to a point on the positive side
of a number line is called a **positive number.** A number that corresponds
to a point on the negative side of a number line is called a **negative
number.** To construct the negative side of a number line, continue the
general procedure with the following step.

3. Using the same unit length as before, mark the point that is one unit
    to the negative side of the origin and label it $^-1$. Then mark and label
    points to the negative side of $^-1$ as $^-2$, $^-3$, and so on.

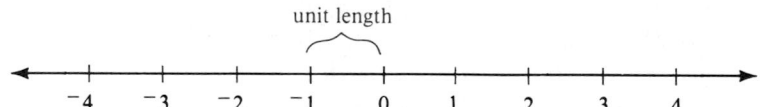

As on the positive side of the number line, the points labeled 0, $^-1$, $^-2$,
and so on may be used as a reference in locating and labeling those
points that correspond to negative fractions.

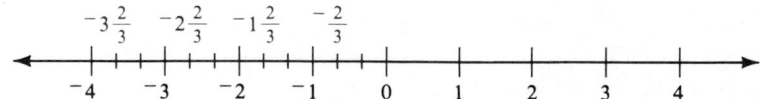

2   *Chapter 1*

Note that the number zero itself is neither positive nor negative.

On a number line, the point that corresponds to a number is called the **graph of the number.** The number that corresponds to a point is called the **coordinate of the point.** A heavy dot is used to indicate the graph of a number.

EXAMPLE   State the coordinate of each point graphed below.

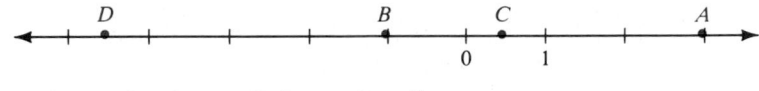

SOLUTION A: 3; B: $^-1$; C: $\frac{1}{2}$; D: $^-4\frac{1}{2}$

Any number that is either a positive number, a negative number, or zero is called a **real number.** When you graph real numbers, you take the following for granted.

1. Each real number corresponds to exactly one point on a number line.
2. Each point on a number line corresponds to exactly one real number.

On a horizontal number line that is marked with positive numbers to the right, the real numbers increase from left to right and decrease from right to left. Thus a number line helps to determine the *order* of two real numbers. The *inequality symbols* $<$ and $>$ are often used when comparing numbers. For example:

$^-5$ is to the left of 1	1 is to the right of $^-5$
$^-5$ *is less than* 1	1 *is greater than* $^-5$
$^-5 < 1$	$1 > ^-5$

To avoid confusing the symbols $<$ and $>$, notice that the greater number is placed at the greater, or open, end of the inequality symbol. The statements $^-5 < 1$ and $1 > ^-5$ give the same information.

A number line also helps to determine if one real number is *between* two others. For example, on a number line you can see that 0 is between $^-5$ and 1 because $^-5$ is less than 0 *and* 0 is less than 1. You can use inequality symbols to write this relationship in the following ways.

$^-5 < 0$ and $0 < 1$	$1 > 0$ and $0 > ^-5$
$^-5 < 0 < 1$	$1 > 0 > ^-5$

Because positive and negative numbers suggest opposite *directions,* they are sometimes called **directed numbers.** You use them for measurements that involve direction as well as size. For example:

a *gain* of \$10: 10	a *loss* of \$10: $^-10$
a temperature *rise* of 3°C: 3	a temperature *drop* of 3°C: $^-3$
12° *east* longitude: 12	12° *west* longitude: $^-12$

Suggested Assignments
Minimum
   4/1-32
Average
   4/7-43 odd, 44-50
Maximum
   4/7-49 odd, 51-54

Sometimes to emphasize that a number like 10 is a positive number, it is called "positive ten" and is denoted by the symbol $^+10$. Note that the small signs $^+$ and $^-$ in the symbols $^+10$ and $^-10$ indicate the directions of the corresponding points from the origin on a number line. They do *not* indicate the operations of addition and subtraction.

Oral Exercises

Exercises 1–16 refer to the number line below.

A B C D E F G H I J K L M N P Q R

$^-8$ $^-7$ $^-6$ $^-5$ $^-4$ $^-3$ $^-2$ $^-1$ 0 1 2 3 4 5 6 7 8

Name the point that is the graph of the given number.

1. $^-5$ D   2. 1 J   3. 0 I   4. 7 Q   5. 3 L   6. $^-3$ F

State the coordinate of the given point.

7. Q 7   8. G $^-2$ 9. B $^-7$ 10. P 6   11. E $^-4$ 12. I 0

13. the point halfway between D and F $^-4$
14. the point halfway between G and K 0
15. the point one third of the way from I to P 2
16. the point one fourth of the way from C to K $^-4$

Translate each statement into words.

17. $^-6 < 1$ 18. $^-2 > ^-5$ 19. $8 > 0 > ^-10$ 20. $^-9 < 4 < 7$

17. $^-6$ is less than 1   18. $^-2$ is greater than $^-5$ 19. 8 is greater than 0 and 0 is greater than $^-10$ 20. $^-9$ is less than 4 and 4 is less than 7

Written Exercises

Write a positive number for each measurement. Then write the opposite of that number and describe the measurement indicated by the opposite.

A   1. thirty wins   2. a gain of five yards
      3. a profit of $5000   4. a deposit of $350
      5. three floors up   6. five steps to the right
      7. 400 m above sea level   8. ten seconds after liftoff
      9. 90 km east   10. 35° north latitude

Graph the given numbers on a horizontal number line. Construct a separate number line for each exercise.

11. $^-5$, $^-2$, 0, 2, 5   12. 0, 1, 6, $^-1$, $^-6$

13. 3, 4, 0, $\frac{^-1}{2}$, $^-3$ 14. $^-2$, $^-1.5$, $^-1$, 2, 3

4   *Chapter 1*

_____

Additional A Exercises

Write a positive number for each measurement. Then write the opposite of that number and describe the measurement indicated by the opposite.

1. two months after birth
   $^+2$; $^-2$; two months before birth
2. a profit of ten dollars
   $^+\$10$; $^-\$10$; a loss of $10
Graph the given numbers on a horizontal number line. Construct a separate number line for each exercise.
3. $^-4$, 1, 3, 4

$^-4$ $^-3$ $^-2$ $^-1$ 0 1 2 3 4

4. $^-5$, $^-2.5$, $^-0.5$, 0, 2

   $^-2.5$ $^-0.5$
$^-5$ $^-4$ $^-3$ $^-2$ $^-1$ 0 1 2 3

Replace each $\underline{\ ?\ }$ with one of the symbols $<$ or $>$ to make a true statement.
5. $^-0.1\ \underline{\ ?\ }\ ^-0.2$ $>$
6. $\frac{^-3}{4}\ \underline{\ ?\ }\ \frac{^-1}{4}\ \underline{\ ?\ }\ 0$ $<$ $<$

15. $^-3, ^-1.5, 0.5, 2, 3.5$

16. $1\frac{1}{2}, 2\frac{1}{2}, 0, \frac{^-1}{2}, ^-3\frac{1}{2}$

17. $^-2\frac{1}{4}, \frac{^-3}{4}, 0, \frac{1}{4}, 3\frac{1}{4}$

18. $\frac{2}{3}, \frac{5}{3}, \frac{^-1}{3}, \frac{^-4}{3}, \frac{^-8}{3}$

Replace each ___?___ with one of the symbols < or > to make a true statement.

19. $^-3$ __?__ 2 $<$

20. 1 __?__ $^-8$ $>$

21. $^-3$ __?__ $^-7$ $>$

22. $^-12$ __?__ $^-4$ $\leq$

23. $\frac{^-1}{5}$ __?__ $\frac{^-2}{5}$ $>$

24. $^-3$ __?__ $^-3.5$ $>$

25. $^-0.25$ __?__ $^-0.75$ $>$

26. $\frac{^-1}{2}$ __?__ $\frac{^-9}{10}$ $>$

27. $^-9$ __?__ 0 __?__ 9 $<, <$

28. $^-15$ __?__ $^-7$ __?__ 0 $<, <$

29. 11 __?__ 0 __?__ $^-2.5$ $>, >$

30. $^-4$ __?__ $^-4.5$ __?__ $^-4.75$ $>, >$

31. $\frac{^-7}{3}$ __?__ $\frac{^-5}{3}$ __?__ $\frac{^-2}{3}$ $<, <$

32. $\frac{^-5}{2}$ __?__ $^-3$ __?__ $\frac{^-7}{2}$ $>, >$

Write the given numbers in order from least to greatest.

B **33.** $5, ^-7, ^-9, 4, 0, ^-3$ $^-9, ^-7, ^-3, 0, 4, 5$

34. $^-12, 1, ^-1, ^-16, 15, ^-8$ $^-16, ^-12, ^-8, ^-1, 1, 15$

35. $\frac{^-1}{4}, \frac{^-1}{6}, \frac{^-1}{2}, \frac{^-1}{10}, \frac{^-1}{3}, \frac{^-1}{5}$ $\frac{^-1}{2}, \frac{^-1}{3}, \frac{^-1}{4}, \frac{^-1}{5}, \frac{^-1}{6}, \frac{^-1}{10}$

36. $\frac{^-2}{7}, \frac{3}{7}, \frac{^-5}{7}, \frac{^-9}{7}, \frac{4}{7}, \frac{^-1}{7}$ $\frac{^-9}{7}, \frac{^-5}{7}, \frac{^-2}{7}, \frac{^-1}{7}, \frac{3}{7}, \frac{4}{7}$

37. $4.5, ^-1.5, ^-2, 5.5, 0, ^-2.5$ $^-2.5, ^-2, ^-1.5, 0, 4.5, 5.5$

38. $^-1.25, ^-1.5, ^-1.15, ^-1.05, ^-1.75, ^-1.1$ $^-1.75, ^-1.5, ^-1.25, ^-1.15, ^-1.1, ^-1.05$

Exercises 39–54 refer to the number line below.

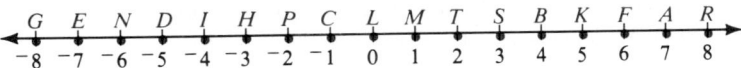

State the coordinate of the given point.

39. the point five eighths of the way from L to R $\qquad$ 5

40. the point three eighths of the way from G to L $\qquad$ $^-5$

41. the point one half of the way from M to T $\qquad$ $1\frac{1}{2}$

42. the point one half of the way from N to D $\qquad$ $^-5\frac{1}{2}$

43. the point one fourth of the way from P to B $\qquad$ $\frac{^-1}{2}$

44. the point three fourths of the way from C to K $\qquad$ $3\frac{1}{2}$

45. the point one third of the way from B to R $\qquad$ $5\frac{1}{3}$

46. the point one third of the way from N to C $\qquad$ $^-4\frac{1}{3}$

47. the point three fifths of the way from T to F $\qquad$ $4\frac{2}{5}$

48. the point two fifths of the way from G to P $\qquad$ $^-5\frac{3}{5}$

49. the point two thirds of the way from D to S $\qquad$ $\frac{1}{3}$

50. the point five eighths of the way from N to B $\qquad$ $\frac{1}{4}$

C **51.** the point between L and F that is twice as far from L as it is from F $\quad$ 4

52. the point between G and B that is three times as far from G as it is from B 1

53. the point to the left of P that is twice as far from C as it is from P $\quad$ $^-3$

54. the point to the right of M that is half as far from M as it is from I 6

Numbers and Variables **5**

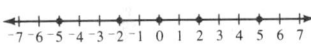
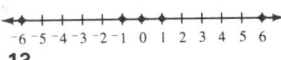

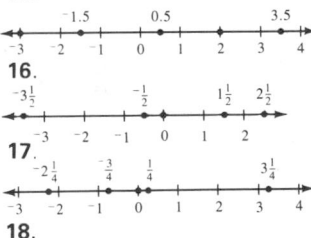

Key Ideas

Use set notation.

Chalkboard Examples

1. Graph: {real numbers

greater than $-2\frac{1}{2}$}

2. Draw a Venn diagram to illustrate A ⊄ B.

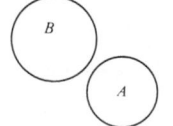 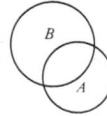

1–2 Sets and Symbols

A **set** is a collection of objects. The objects are called the **members,** or **elements,** of the set. You use braces, { }, to indicate that a set is being named. Within the braces, you separate the members of the set by commas. For example, to indicate "the set whose members are 1, 3, 5, and 7," you write

$$\{1, 3, 5, 7\}.$$

To indicate that 5 is a member of this set, you use the symbol ∈, *is a member of,* and write

$$5 \in \{1, 3, 5, 7\}.$$

If you wish to indicate that 9 is not a member of this set, you use the symbol ∉, *is not a member of,* and write

$$9 \notin \{1, 3, 5, 7\}.$$

Sets that contain exactly the same members are called **equal sets.** You use the symbol =, *is equal to,* to indicate that sets are equal. The order in which you list the members of a set does not matter, and so

$$\{1, 3, 5, 7\} = \{3, 1, 7, 5\}.$$

On the other hand, the sets {1, 3, 5, 7} and {1, 3, 5, 9} do *not* contain exactly the same members. Therefore, you use the symbol ≠, *does not equal*, and write

$$\{1, 3, 5, 7\} \neq \{1, 3, 5, 9\}.$$

Although the sets {1, 3, 5, 7} and {1, 3, 5, 9} are not equal, there is an important relationship between them. The figure that follows shows a pairing that assigns to each member of each set *one and only one* member of the other set.

$$\{1, 3, 5, 7\}$$
$$\updownarrow \updownarrow \updownarrow \updownarrow$$
$$\{1, 3, 5, 9\}$$

Such a pairing of the members of two sets is called a **one-to-one correspondence.** A one-to-one correspondence of great importance in mathematics is that between the set of points on a line and the set of real numbers. (Recall Section 1–1.)

When you list all the members of a set, you *specify* the set by **roster.** You can also specify a set by writing within the braces a **rule,** or description, that identifies the members of the set. For example,

$$\{1, 3, 5, 7\} = \{\text{the odd numbers between 0 and 8}\}.$$

A third way to specify a set of *numbers* is to graph the numbers on a number line. The set of points corresponding to a set of numbers is called the **graph of the set.** When graphing sets of numbers, recall that,

6   *Chapter 1*

on a horizontal number line that is marked with positive numbers to the right, "greater than" means "to the right of," while "less than" means "to the left of."

Often you speak of a set of numbers that are "between" two given numbers. In such cases, note that the given numbers indicate the *boundaries* of the set, but they are not themselves *members* of the set.

EXAMPLE   Graph each set of numbers.

    a. {1, 3, 5, 7}

    b. {the odd numbers between 0 and 8}

    c. {the real numbers between 0 and 8}

    d. {the real numbers greater than 0}

    e. {the real numbers less than or equal to 8}

SOLUTION   **a.**

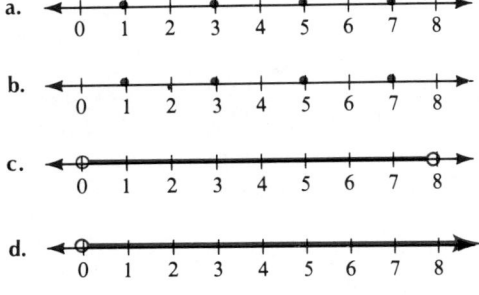

Notice in parts (c), (d), and (e) of the Example that heavy shading is used to show that all points on the indicated portion of the number line belong to the graph. Open dots and portions of the line not shaded show points that do *not* belong to the graph. A heavy arrowhead shows that the graph continues without end in the indicated direction.

Capital letters are often used to name sets. Thus you can write

$$S = \{1, 3, 5, 7\} \quad \text{and} \quad T = \{\text{the real numbers between 0 and 8}\}.$$

If every member of a set S is also a member of a set T, then S is called a **subset** of T. You use the symbol $\subset$, *is a subset of*, to indicate that one set is a subset of another. Thus for the sets S and T just specified,

$$S \subset T.$$

Every set is a subset of itself.

$$S \subset S \qquad T \subset T$$

To indicate that one set is *not* a subset of another set, you use the symbol $\not\subset$, *is not a subset of*.

$$T \not\subset S$$

Numbers and Variables   **7**

Suggested Assignments
Minimum
  9/1-30
Average
  9/5-33 odd, 35-39
Maximum
  9/5-39 odd, 40-43

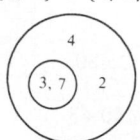

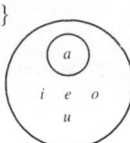
The set that contains no members is called the **empty set,** or the **null set.** You denote the empty set by the symbol Ø. As an example,

$$\varnothing = \{\text{the negative numbers between 0 and 8}\}.$$

The empty set is a subset of every set.

$$\varnothing \subset S \qquad \varnothing \subset T \qquad \varnothing \subset \varnothing$$

Be careful not to confuse the empty set Ø with {0}. The set {0} contains exactly one member, namely 0.

Often it is useful to draw a diagram that shows how certain sets are related. One such type of diagram is the **Venn diagram.** (Venn diagrams are named after John Venn, an English mathematician who was among the first to use them extensively.) For example, to show that a set S is a subset of a set T, you can use a Venn diagram such as the one at the right.

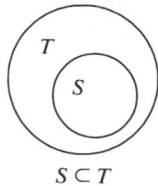

$S \subset T$

Oral Exercises

Specify each set by roster.

1. {the days of the week} {Sunday, Monday, Tuesday, Wednesday, Thursday, Friday, Saturday}
2. {the last five letters of the English alphabet} {v, w, x, y, z}
3. {the months that have thirty-one days} {January, March, May, July, August, October, December}
4. {the months that have thirty-two days} Ø
5. {the positive numbers between ⁻10 and 0} Ø
6. {the even numbers between 1 and 15} {2, 4, 6, 8, 10, 12, 14}

7. 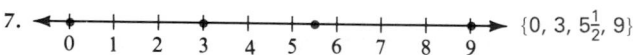 {0, 3, 5½, 9}

8. 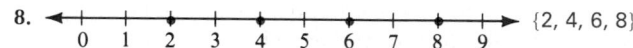 {2, 4, 6, 8}

Specify each set by rule.

9. {summer, fall, winter, spring}   10. {January, June, July}   11. {a, b, c, d, e, f}
12. {a, e, i, o, u}   13. {12, 14, 16, 18}   14. {11, 13, 15, 17, 19}

15.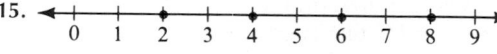

16.

17.

18.

8   Chapter 1

Tell whether each statement is true or false. Exercises 27–30 refer to the Venn diagram at the right.

19. $\{1, 2, 3, 4, 5\} = \{3, 1, 5, 2, 4\}$ true

20. $\{1, 2, 3, 4, 5\} = \{$the real numbers between 0 and 6$\}$ false

21. $3 \in \{1, 2, 3, 4, 5\}$ true

22. $6 \notin \{1, 2, 3, 4, 5\}$ true

23. $\{1, 2\} \subset \{1, 2, 3, 4, 5\}$ true

24. $\{5, 6\} \subset \{1, 2, 3, 4, 5\}$ false

25. $\{1, 2, 3, 4, 5\} \not\subset \varnothing$ true

26. $\varnothing \subset \{1, 2, 3, 4, 5\}$ true

27. $A \subset D$ true

28. $B \subset C$ false

29. $A \not\subset B$ true

30. $C \subset D$ true

Exs. 27–30

Written Exercises

Replace each ___?___ with one of the symbols $\in$ or $\subset$ to make a true statement.

A

1. 2 ___?___ $\{0, 2, 4, 6\}$ $\in$

2. $\{2, 4\}$ ___?___ $\{0, 2, 4, 6\}$ $\subset$

3. $\{0\}$ ___?___ $\{0, 2, 4, 6\}$ $\subset$

4. $\{4, 0, 6, 2\}$ ___?___ $\{0, 2, 4, 6\}$ $\subset$

5. 0 ___?___ $\{0\}$ $\in$

6. $\varnothing$ ___?___ $\{0\}$ $\subset$

7. $\varnothing$ ___?___ $\varnothing$ $\subset$

8. $\varnothing$ ___?___ $\{0, 2, 4, 6\}$ $\subset$

Graph each set of numbers.

9. $\{{}^-5, {}^-3, 0, 3\}$

10. $\{{}^-6, {}^-4, {}^-2, 1, 3, 5\}$

11. $\left\{{}^-3\frac{1}{2}, \frac{{}^-1}{2}, 2\frac{1}{2}, 4\frac{1}{2}\right\}$

12. $\{{}^-4.5, {}^-0.5, 0, 1.5, 2.5\}$

13. $\{$the odd numbers between 4 and 14$\}$

14. $\{$the even numbers between 1 and 11$\}$

15. $\{$the multiples of 3 between ${}^-10$ and 0$\}$

16. $\{$the multiples of 4 between ${}^-10$ and 1$)\}$

17. $\{$the real numbers between ${}^-10$ and 10$\}$

18. $\{$the real numbers between ${}^-4$ and 5$\}$

19. $\{$the real numbers greater than ${}^-2\}$

20. $\{$the real numbers less than or equal to $3\frac{1}{2}\}$

21. $\{$the real numbers greater than ${}^-2$ and less than or equal to 5$\}$

22. $\{$the real numbers greater than or equal to ${}^-5\frac{1}{2}$ and less than ${}^-1\}$

23. $\{$the positive real numbers$\}$

24. $\{$the negative real numbers$\}$

25. $\{$the positive real numbers less than 6$\}$

26. $\{$the negative real numbers greater than ${}^-1\}$

27. $\{$the positive and negative real numbers$\}$

28. $\{$the real numbers that are neither positive nor negative$\}$

Numbers and Variables **9**

9.

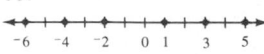

10.

11.

12.

13.

14.

15.

16.

17.

18.

19.

20.

21.

22.

23.

24.

25.

26.

27.

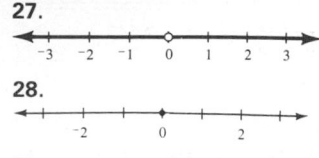

28.

29. **30.**

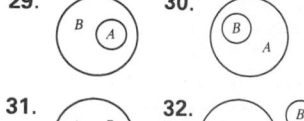

31. **32.**

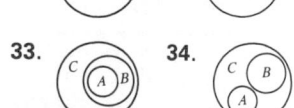

33. **34.**

35. $\{a\}$, $\{b\}$, $\{c\}$
36. $\{a, b\}$, $\{a, c\}$, $\{b, c\}$
37. $\{a, b, c\}$ **38.** $\emptyset$ **39.** 8
40. $\{a\}$, $\{b\}$, $\{c\}$, $\{d\}$, $\{a, b\}$,
$\{a, c\}$, $\{a, d\}$, $\{b, c\}$, $\{b, d\}$,
$\{c, d\}$, $\{a, b, c\}$, $\{a, b, d\}$,
$\{a, c, d\}$, $\{b, c, d\}$,
$\{a, b, c, d\}$, $\emptyset$; 16
41. $\{a\}$, $\{b\}$, $\{c\}$, $\{d\}$, $\{e\}$,
$\{a, b\}$, $\{a, c\}$, $\{a, d\}$, $\{a, e\}$,
$\{b, c\}$, $\{b, d\}$, $\{b, e\}$, $\{c, d\}$,
$\{c, e\}$, $\{d, e\}$, $\{a, b, c\}$,
$\{a, b, d\}$, $\{a, b, e\}$, $\{a, c, d\}$,
$\{a, c, e\}$, $\{a, d, e\}$, $\{b, c, d\}$,
$\{b, c, e\}$, $\{b, d, e\}$, $\{c, d, e\}$,
$\{a, b, c, d\}$, $\{a, b, c, e\}$,
$\{a, b, d, e\}$, $\{a, c, d, e\}$,
$\{b, c, d, e\}$, $\{a, b, c, d, e\}$,
$\emptyset$; 32
42. 64 **43.** 256

Mixed Review

Graph the given numbers on a horizontal number line.
1. 4   **2.** ⁻2   **3.** ⁻3   **4.** 0
5. $3\frac{1}{2}$ **6.** $-1\frac{1}{2}$

Draw a Venn diagram to illustrate each statement.

29. $A \subset B$

30. $B \subset A$

B **31.** $A \subset B$ and $B \subset A$

32. $A \not\subset B$ and $B \not\subset A$

33. $A \subset B$ and $B \subset C$

34. $A \subset C$ and $B \subset C$, but $A \not\subset B$ and $B \not\subset A$

For Exercises 35–39, let $S = \{a, b, c\}$.

35. List all the subsets of S that contain exactly one member.

36. List all the subsets of S that contain exactly two members.

37. List all the subsets of S that contain exactly three members.

38. List all the subsets of S that contain no members.

39. What is the total number of subsets of S?

C **40.** Let $T = \{a, b, c, d\}$. List all the subsets of T. What is the total number of subsets of T?

41. Let $U = \{a, b, c, d, e\}$. List all the subsets of U. What is the total number of subsets of U?

42. If a set V contains six members, what is the total number of subsets of V?

43. If a set W contains eight members, what is the total number of subsets of W?

Marjorie Lee Browne

1914–1979

"I have always, always, always liked mathematics." With these words, Marjorie Lee Browne conveyed the enthusiasm she brought to a teaching career that spanned four decades.

Marjorie Lee Browne was born in Memphis, Tennessee. Although money was scarce in the days of the Great Depression, a combination of scholarships, jobs, and loans saw her through Howard University in Washington, D.C. She pursued graduate studies at the University of Michigan, where she became one of the first two American black women to receive a doctorate in mathematics.

Dr. Browne taught mathematics at North Carolina Central University from 1949 to 1979. Under her leadership, the university became the first predominantly black college to conduct a National Science Foundation Institute for teachers of secondary mathematics.

10 *Chapter 1*

1-3 Sets of Numbers

The set of numbers that is used in counting is called the set of **natural numbers,** or **counting numbers.** Often this set is referred to as N. Since it would be impossible to make a roster of *all* the natural numbers, you can specify the set N as follows.

$$N = \{1, 2, 3, 4, 5, \ldots\}$$

The three dots after the 5 are read "and so on" and indicate that the pattern established by the listed numbers continues without end.

The set N may also be specified by the following graph.

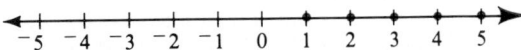

Here the heavy arrowhead indicates that the graph continues indefinitely to the right following the pattern established by the numbers that are graphed with heavy dots.

When you expand the set of natural numbers to include the number 0, you obtain the set W of **whole numbers.**

$$W = \{0, 1, 2, 3, 4, 5, \ldots\}$$

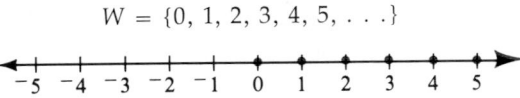

The set J, where

$$J = \{\ldots, {}^-3, {}^-2, {}^-1, 0, 1, 2, 3, \ldots\},$$

is called the set of **integers.** Here the dots at the beginning and end of the list indicate that the pattern of the numbers continues without end in *both* directions. Therefore, to graph the set of integers you need to use a heavy arrowhead at both ends of the number line.

EXAMPLE   Specify each set by roster.
  a. {the natural numbers greater than 9}
  b. {the odd integers between ⁻4 and 4}
  c. {the whole numbers less than 15}

SOLUTION   **a.** {10, 11, 12, 13, 14, . . .}
  b. {⁻3, ⁻1, 1, 3}
  c. {0, 1, 2, 3, . . . , 14}

In part (c) of the Example, the three dots are read "and so on through" and indicate that the pattern established by the first four listed numbers continues until you reach the last listed number.

Numbers and Variables   **11**

Teaching Suggestions
p. T70

Related Activities p. T70

Supplementary Material

Test 1

Key Ideas

Specify and graph sets of real numbers.

Chalkboard Examples

1. Specify {the natural numbers less than 20} by roster.
   {1, 2, 3, 4, . . . , 19}

2. Label rational or irrational.
   a. $\sqrt{8}$ irrational
   b. $1\frac{1}{3}$ rational

A **rational number** is any number that can be expressed as the quotient of two integers, provided that the divisor is not 0. Thus, the following numbers are rational numbers.

$$0 = \frac{0}{1} \qquad 7 = \frac{7}{1} \qquad \frac{2}{9} \qquad 3\frac{1}{2} = \frac{7}{2} \qquad 0.21 = \frac{21}{100}$$

The set of rational numbers is often referred to as Q.

As you might imagine, it would be impossible to label all the rational numbers on a number line. Even if you *could* do so, there would still be many points of the number line that would not be labeled. These points correspond to numbers such as $\sqrt{2}$ and π. Such numbers *cannot* be expressed as the quotient of two integers, and they are called **irrational numbers**.

Together the rational numbers and the irrational numbers make up the set of *real numbers*, which is denoted by the symbol $\mathcal{R}$. Recall that a real number was previously defined as any number that is either a positive number, a negative number, or zero. Thus you can graph the set of real numbers by graphing the entire number line, as follows.

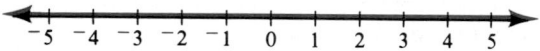

To illustrate the relationships among the sets N, W, J, Q, and $\mathcal{R}$, you can use a Venn diagram such as the one at the right.

Each of the sets N, W, J, Q, $\mathcal{R}$, and the set of irrational numbers is called an **infinite set** because the process of counting its members would never come to an end. A set that is not an infinite set is called a **finite set**.

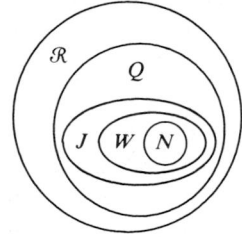

Oral Exercises

Match.

1. $\{1, 2, 3, 4, \ldots\}$ f.
2. $\{\ldots, 1, 2, 3, 4\}$ k.
3. $\{1, 2, 3, 4\}$ b.
4. $\{^-4, ^-3, ^-2, ^-1\}$ no match
5. $\{\ldots, ^-4, ^-3, ^-2, ^-1\}$ a
6. $\{^-4, ^-3, ^-2, ^-1, \ldots\}$ l.
7. $\{^-4, ^-3, ^-2, ^-1, \ldots, 4\}$ d.
8. $\{^-4, ^-2, 0, 2, 4, \ldots\}$ c.
9. $\{\ldots ^-3, ^-1, 1, 3\}$ no match
10. $\{0, 2, 4\}$ g.
11. $\{1, 3\}$ h.
12. $\emptyset$ j.

a. {the negative integers less than 5}
b. {the natural numbers less than 5}
c. {the even integers greater than $^-5$}
d. {the integers between $^-5$ and 5}
e. {the odd integers less than 5}
f. {the natural numbers greater than $^-5$}
g. {the even whole numbers less than 5}
h. {the odd natural numbers less than 5}
i. {the negative integers greater than $^-5$}
j. {the positive integers less than $^-5$}
k. {the integers less than 5}
l. {the integers greater than $^-5$}

12 *Chapter 1*

Suggested Assignments

Minimum
13/1-19 odd, 20-30
R 15/*Self-Test 1*

Average
13/1, 7-41 odd, 43-48
R 15/*Self-Test 1*

Maximum
13/1, 7-47 odd, 49-52
R 15/*Self-Test 1*

List all the sets N, W, J, Q and $\mathcal{R}$ of which each number is a member. Each number may be a member of more than one set.

13. 0
W, J, Q, $\mathcal{R}$

14. 7
N, W, J, Q, $\mathcal{R}$

15. ⁻3 J, Q, $\mathcal{R}$

16. $\frac{5}{6}$ Q, $\mathcal{R}$

17. $\frac{-8}{4}$ J, Q, $\mathcal{R}$

18. ⁻$\sqrt{5}$
$\mathcal{R}$

Tell whether each set is finite or infinite.

19. {the whole numbers greater than 8} *infinite*

20. {the natural numbers less than 8} *finite*

21. {the integers between ⁻9 and 9} *finite*

22. {the integers between 0 and 1} *finite*

23. {the rational numbers between 0 and 1} *infinite*

24. {the negative irrational numbers} *infinite*

Written Exercises

Specify each set by roster.

A

1. 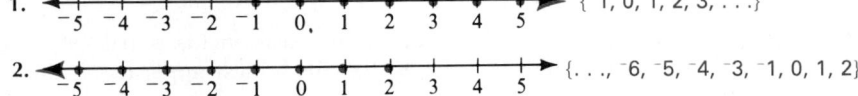 {⁻1, 0, 1, 2, 3, . . .}

2. 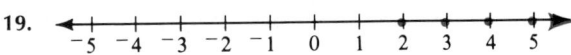 {. . ., ⁻6, ⁻5, ⁻4, ⁻3, ⁻1, 0, 1, 2}

3. {the natural numbers less than 7} {1, 2, 3, 4, 5, 6}

4. {the whole numbers greater than 24} {25, 26, 27, 28, . . .}

5. {the integers greater than ⁻10 and less than 2} {⁻9, ⁻8, ⁻7, . . ., 1}

6. {the positive integers between ⁻10 and 2} {1}

7. {the negative integers greater than or equal to ⁻2} {⁻2, ⁻1}

8. {the even integers less than 3} {. . ., ⁻4, ⁻2, 0, 2}

9. {the natural numbers between 9 and 100} {10, 11, 12, . . ., 99}

10. {the odd integers greater than ⁻50 and less than 50} {⁻49, ⁻47, ⁻45, ⁻43, . . ., 49}

Specify each set by rule.

11. {1, 2, 3, 4, 5}

12. {⁻3, ⁻2, ⁻1, 0, 1, 2}

13. {⁻6, ⁻5, ⁻4, ⁻3, ⁻2, . . .}

14. {. . ., ⁻9, ⁻7, ⁻5, ⁻3, ⁻1}

15. {2, 4, 6, 8, . . . , 100}

16. {⁻50, ⁻49, ⁻48, ⁻47, . . . , 50}

17. {. . . , ⁻6, ⁻3, 0, 3, 6, . . .}

18. {. . . , ⁻10, ⁻5, 0, 5, 10, . . .}

19.
```
 ⁻5  ⁻4  ⁻3  ⁻2  ⁻1   0   1   2   3   4   5
```

20.
```
 ⁻5  ⁻4   3   2  ⁻1   0   1   2   3   4   5
```

21.
```
 ⁻5  ⁻4  ⁻3  ⁻2  ⁻1   0   1   2   3   4   5
```

22.
```
 ⁻5  ⁻4  ⁻3  ⁻2  ⁻1   0   1   2   3   4   5
```

Numbers and Variables **13**

Additional A Exercises

Specify each set by roster.

1. {the odd integers greater than ⁻3} {⁻1, 1, 3, 5, . . .}

2. {the natural numbers between 6 and 7} ∅

Specify each set by rule.

3. {0, 4, 8, 12, . . .} {the nonnegative multiples of 4}

4. {. . . , ⁻3, ⁻2, ⁻1} {the integers less than 0}

Graph each set of numbers.

5. {. . . , ⁻5, 0, 5}

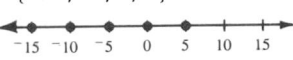

6. {the integers greater than ⁻5}

Additional Answers Written Exercises

11. {the natural numbers less than 6}

12. {the integers greater than ⁻4 and less than 3}

13. {the integers greater than ⁻7}

14. {the odd integers less than 0}

15. {the even integers greater than 1 and less than 101}

16. {the integers greater than ⁻51 and less than 51}

17. {the integers that are multiples of 3}

18. {the integers that are multiples of 5}

19. {the integers greater than 1}

20. {the integers greater than ⁻3 and less than 3}

21. {the natural numbers less than 2}

22. {the integers less than 3}

23.

24.

25.

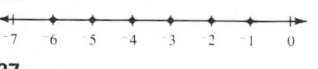

26.

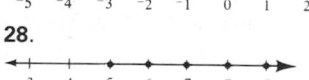

27.

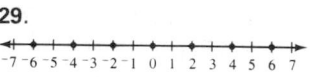

28.

29.

30.

Graph each set of numbers.

23. {. . . , ⁻2, ⁻1, 0, 1, 2, 3}
24. {. . . , ⁻4, ⁻2, 0, 2, 4, . . .}
25. {the natural numbers less than 10}
26. {the negative integers greater than ⁻7}
27. {the integers less than or equal to 1}
28. {the whole numbers greater than 4}
29. {the even integers between ⁻7 and 7}
30. {the natural numbers greater than ⁻4 and less than 4}

Replace each __?__ with one of the words *All*, *Some*, or *No* to make a true statement that has the widest application.

EXAMPLE   __?__ integers are real numbers.

SOLUTION   Replacing the __?__ with either *All* or *Some* yields a true statement. However, the statement with *All* has the widest application. Thus: *All* integers are real numbers.

B 31. __?__ rational numbers are real numbers. All
32. __?__ real numbers are irrational numbers. Some
33. __?__ irrational numbers are integers. No
34. __?__ natural numbers are integers. All
35. __?__ real numbers are whole numbers. Some
36. __?__ rational numbers are integers. Some
37. __?__ integers are rational numbers. All
38. __?__ rational numbers and irrational numbers are real numbers. All
39. __?__ real numbers are rational numbers or irrational numbers. All
40. __?__ whole numbers are natural numbers. Some
41. __?__ natural numbers are irrational numbers. No
42. __?__ rational numbers are negative integers. Some

For Exercises 43–52, $A = \{⁻3, ⁻2, ⁻\frac{1}{2}, 0, 1, 3\}$ and $B = \{⁻2, \frac{1}{2}, 1, \frac{3}{2}, 2, \pi\}$. Specify each of the following sets by roster.

43. {the integers that are members of A but not of B} {⁻3, 0, 3}
44. {the integers that are members of B but not of A} {2}
45. {the integers that are members of either A or B} {⁻3, ⁻2, 0, 1, 2, 3}
46. {the integers that are members of both A and B} {⁻2, 1}
47. {the rational numbers that are members of either A or B} {⁻3, ⁻2, ⁻$\frac{1}{2}$, 0, $\frac{1}{2}$, 1, $\frac{3}{2}$, 2, 3}
48. {the rational numbers that are members of both A and B} {⁻2, 1}

14   *Chapter 1*

C **49.** {the nonintegers that are members of both A and B} $\emptyset$
  50. {the nonintegers that are members of B but not of A} $\{\frac{1}{2}, \frac{3}{2}, \pi\}$
  51. {the positive integers that are members of neither A nor B} $\{4, 5, 6, \ldots\}$
  52. {the positive integers that are not members of both A and B} $\{2, 3, 4, \ldots\}$

■ Self-Test 1

VOCABULARY number line (p. 1) specify a set by rule (p. 6)
origin (p. 2) graph of a set of numbers
positive number (p. 2) (p. 6)
negative number (p. 2) subset (p. 7)
graph of a number (p. 3) empty set or null set (p. 8)
coordinate of a point (p. 3) Venn diagram (p. 8)
real number (p. 3) natural number or counting
directed number (p. 3) number (p. 11)
set (p. 6) whole number (p. 11)
member or element of a set integer (p. 11)
(p. 6) rational number (p. 12)
equal sets (p. 6) irrational number (p. 12)
one-to-one correspondence infinite set (p. 12)
(p. 6) finite set (p. 12)
specify a set by roster (p. 6)

Graph the given numbers on a horizontal number line. Construct a separate number line for each exercise.

  1. $^-4, ^-1, 0, 3, 5$ **2.** $^-4.5, ^-2, ^-0.5, \frac{1}{2}, 2.5$ *Obj. 1, p. 1*

Replace each __?__ with one of the symbols $<$ or $>$ to make a true statement.

  3. $2 \underline{\ ?\ } ^-9 \ >$ **4.** $^-7.5 \underline{\ ?\ } 3 \ <$ **5.** $\frac{^-2}{5} \underline{\ ?\ } \frac{^-3}{5} \underline{\ ?\ } \frac{^-4}{5}$ *Obj. 2, p. 1*
$> \quad >$

Tell whether each statement is true or false.

  6. $0 \in \emptyset$ false   **7.** $\{2, 1\} \subset \{1\}$ false   **8.** $\{0, 1\} = \{1, 0\}$ true   *Obj. 3, p. 1*

Graph each set of numbers.

  9. {the real numbers between $^-4$ and 1}   *Obj. 4, p. 1*
  10. {the positive integers greater than $^-3$ and less than 3}

Check your answers with those at the back of the book.

Numbers and Variables **15**

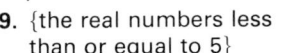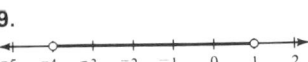

Graph each set of numbers.
Construct a separate number
line for each exercise.

1. {the odd numbers be-
   tween 0 and 6}

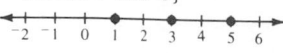

2. {the numbers greater than
   ⁻5 but less than 0}

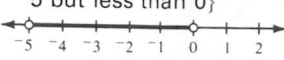

3. {the positive multiples of
   three that are less than
   10}

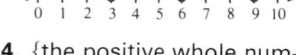

4. {the positive whole numbers less than 2}

Careers
Oceanography

Oceanographers study the characteristics of the ocean in order to predict its behavior and to safely utilize it as a source of food, chemicals, and energy. Since there are so many aspects of the ocean to be studied, many oceanographers choose to specialize in a single field.

Physical oceanographers study tide and wave patterns and the ability of ocean waters to conduct sound, light, and heat waves. *Biological oceanographers* are concerned with the living organisms of the ocean and with the relationship of these organisms to their environment. *Geological oceanographers* concentrate on the formation and physical characteristics of the ocean floor, while *chemical oceanographers* study the composition of the water and sediment.

While some oceanographers are college professors or work in industry, most are involved in research. This research may take place in laboratories or on boats and platforms in the ocean, where easy underwater exploration can take place. Research trips to the ocean may vary in length from a few days to several months, depending on the amount and type of data to be gathered.

EXAMPLE An oceanographer needs to know the depth of a ship that is lying on the ocean floor. *Sonar*, a system of sending out sound waves and monitoring their return, is one method of determining the depth of an object in the ocean. A sound wave is sent down to the ship and returns in 0.076 s. Given that the speed of sound in water is 1490 m/s, what is the depth of the ship?

SOLUTION The total time needed for the sound wave to travel to the sunken ship *and back* is 0.076 s. Thus, the time needed for the sound wave to reach the ship is *half* the total time, or 0.038 s. The depth can be found by multiplying this time by the speed of sound in water, which is given as 1490 m/s.

$$\text{Depth} = \underbrace{\text{Time to reach ship}} \times \underbrace{\text{Speed of sound}}$$
$$= \qquad 0.038 \text{ s} \qquad \times \qquad 1490 \text{ m/s}$$
$$= \qquad\qquad 56.62 \text{ m}$$

Therefore, the depth of the ship is *approximately* 57 m.

16 Chapter 1

Numerical and Variable Expressions

OBJECTIVES for Sections 1-4 through 1-6:
1. To simplify numerical expressions and evaluate variable expressions.
2. To simplify and evaluate expressions that contain grouping symbols.
3. To simplify and evaluate expressions that contain exponents.

1–4 Simplifying and Evaluating Expressions

Teaching Suggestions
p. T70

Key Ideas
Simplify numerical expressions and evaluate variable expressions.

One kind of expression that is used in algebra is called a *numerical expression*. A **numerical expression,** or **numeral,** is simply a name for a number. The number is called the **value of the expression.** For example, since the numerical expressions 4×9 and 36 name the same number, thirty-six, they have the same value. To show that they have the same value, you use the equals sign, $=$. You write

$$4 \times 9 = 36,$$

which is read "four times nine *equals* (or *is equal to*) thirty-six." To show that two numerical expressions do *not* have the same value, you use the symbol $\neq$. For example, you can write

$$4 \times 9 \neq 37,$$

which is read "four times nine *is not equal to* (or *does not equal*) thirty-seven."

Note that a raised dot may also be used as a multiplication symbol.

$$4 \times 9 \text{ may be written } 4 \cdot 9.$$

When you replace a numerical expression with the simplest, or most common, name of its value, you **simplify the expression.** Thus, since 36 is the most common name for the number thirty-six, you simplify the numerical expression 4×9, or $4 \cdot 9$, when you replace it with 36.

In simplifying a numerical expression, you use the following basic principle.

Substitution Principle

Changing the numeral by which a number is named in an expression does not change the value of the expression.

Simplify.

1. $(91 \div 7) - 6$
   $13 - 6$
   7

2. Evaluate the expression
   $51 \div \dfrac{33}{b}$ when $b = 11$.

   $51 \div \dfrac{33}{11} = 51 \div 3 = 17$

3. Find the greatest value of
   the expression $\dfrac{c + 20}{c}$ if
   $c \in \{3, 20\}$.
   When $c = 3$,
   $\dfrac{c + 20}{c} = \dfrac{3 + 20}{3}$
   $= \dfrac{23}{3}$
   $= 7\dfrac{2}{3}$
   When $c = 20$,
   $\dfrac{c + 20}{c} = \dfrac{20 + 20}{20}$
   $= \dfrac{40}{20}$
   $= 2$
   Therefore, the greatest
   value of $\dfrac{c + 20}{c}$ over the
   given domain is $7\dfrac{2}{3}$.

EXAMPLE 1 Simplify.

   a. $7 + 19 + 140$ b. $4 \cdot 17 + 23$

SOLUTION   a. $\underbrace{7 + 19} + 140$ b. $\underbrace{4 \cdot 17} + 23$

   $\underbrace{26 + 140}$ $\underbrace{68 + 23}$

   166 91

Another kind of symbol used in algebra is a *variable*. A **variable** is a symbol that is used to represent one or more numbers. The set of numbers that a variable may represent is called the **domain,** or **replacement set,** of the variable. Each number in the domain is called a **value of the variable.** A variable may be a letter, such as n, or a different type of symbol.

An expression that contains a variable is called a **variable expression.** A variable expression may contain more than one variable, and it may also contain other symbols, including numerals. When you write a product that contains a variable, you usually omit the multiplication symbol.

$3 \times n$ is usually written $3n$.

$y \times z$ is usually written yz.

When you replace each variable in a variable expression with one of its values and simplify the resulting numerical expression, you **evaluate the expression,** or **find the value of the expression.**

EXAMPLE 2 Evaluate each expression when $a = 24$.

   a. $a - 5$ b. $3a$ c. $2a + 9$ d. $40 - \dfrac{a}{3}$

SOLUTION   a. $a - 5 = 24 - 5 = 19$

   b. $3a = 3 \times 24 = 72$

   c. $2a + 9 = 2 \times 24 + 9 = 48 + 9 = 57$

   d. $40 - \dfrac{a}{3} = 40 - \dfrac{24}{3} = 40 - 8 = 32$

EXAMPLE 3 Find the greatest value of the expression $k + 7$ if $k \in \{4, 10, 15\}$.

SOLUTION   When $k = 4$,   When $k = 10$,   When $k = 15$,
   $k + 7 = 4 + 7$ $k + 7 = 10 + 7$ $k + 7 = 15 + 7$
   $= 11$ $= 17$ $= 22$

Therefore, the greatest value of $k + 7$ over the given domain is 22.

Often a numerical expression or a variable expression is referred to as a **mathematical expression.**

Oral Exercises

Tell whether each statement is true or false.

1. $6 \times 9 = 9 \times 6$ true
2. $7 \times 0 = 0 \times 5$ true
3. $8 + 0 \neq 8 - 0$ false
4. $2 \div 1 \neq 1 \div 2$ true
5. $6 \times 3 = 6 \div 3$ false
6. $10 \times 1 = 10 \div 1$ true
7. $\frac{1}{3} \times 12 = 12 \div 3$ true
8. $8 \div 2 \neq \frac{8}{2}$ false
9. $3 \times 5 \neq 5 + 5 + 5$ false
10. $25 - 9 = 2 \times 2 \times 2 \times 2$ true

Simplify.

11. $25 + 92$ 117
12. $53 - 12$ 41
13. $32 \cdot 3$ 96
14. $\frac{144}{12}$ 12

Evaluate each expression when $x = 6$.

15. $x + 18$ 24
16. $9 + x$ 15
17. $x - 5$ 1
18. $30 - x$ 24
19. $12x$ 72
20. $\frac{1}{3}x$ 2
21. $\frac{x}{2}$ 3
22. $\frac{24}{x}$ 4

Find the greatest value of each expression if $y \in \{2, 4, 6, 8, 10\}$.

23. $y + 2$ 12
24. $y - 2$ 8
25. $4y$ 40
26. $\frac{1}{2}y$ 5
27. $16 - y$ 14
28. $2y + 8$ 28
29. $\frac{y - 2}{2}$ 4
30. $\frac{2}{y + 2}$ $\frac{1}{2}$

Written Exercises

Simplify.

A
1. $4392 + 227 + 46$ 4665
2. $1990 - 1776$ 214
3. $3 \times 6 \times 50$ 900
4. $\frac{415}{5}$ 83
5. $12 - 0.49$ 11.51
6. $9.5 + 4 + 8.32$ 21.82
7. $6.5 \div 0.13$ 50
8. 1.2×0.08 0.096
9. $52\frac{3}{4} + 97\frac{1}{2}$ $150\frac{1}{4}$
10. $24 - \frac{7}{8}$ $23\frac{1}{8}$
11. $2\frac{2}{3} \times 3\frac{1}{2}$ $9\frac{1}{3}$
12. $62 \div \frac{1}{2}$ 124

Evaluate each expression when $a = 4$, $b = 8$, and $c = 18$.

13. $a + 19$ 23
14. $20 - c$ 2
15. $10b$ 80
16. $\frac{b}{5}$ $\frac{8}{5}$ or $1\frac{3}{5}$
17. $2b + 5$ 21
18. $13 - \frac{c}{3}$ 7
19. $\frac{3}{c - 3}$ $\frac{1}{5}$
20. $\frac{b + 2}{2}$ 5

Numbers and Variables **19**

Suggested Assignments

Minimum
19/1-19 odd, 21-36

Average
19/7-12, 13-49 odd, 50-54

Maximum
19/7-12, 13-55 odd, 56-60

Additional A Exercises

Simplify.

1. $5 \div \frac{1}{2}$ 10
2. $3\frac{1}{2} \times 10$ 35

Evaluate each expression when $a = 7$ and $b = 3$.

3. $9b$ 27
4. $\frac{56}{a}$ 8

Evaluate each expression when $x = 4$ and $y = 10$.

5. $y - x$ 6
6. $\frac{x}{y}$ $\frac{2}{5}$

Draw a Venn diagram to illustrate the relationship between the following sets of numbers.

1. {real numbers}; {rational numbers}

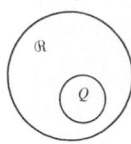

2. {whole numbers}; {natural numbers}

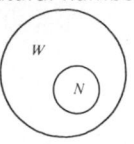

3. {whole numbers that are multiples of three}; {rational numbers}

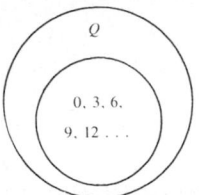

4. {rational numbers}; {integers}

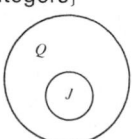

Evaluate each expression when $a = 4$, $b = 8$, and $c = 18$.

21. $a + c$ 22

22. $c - b$ 10

23. bc 144

24. $4ab$ 128

25. $3a + b$ 20

26. $3b + a$ 28

27. $\dfrac{6}{a + b}$ $\frac{1}{2}$

28. $\dfrac{a + b}{6}$ 2

29. $a + c - b$ 14

30. $c - b - a$ 6

31. abc 576

50 32. $ab + c$

33. $\dfrac{bc}{a}$ 36

34. $\dfrac{ac}{b}$ 9

35. $c - \dfrac{b}{a}$ 16

$3\frac{1}{2}$ 36. $b - \dfrac{c}{a}$

Evaluate each expression when the variables have the given values.

B

37. $a + b$; $a = \frac{1}{2}$, $b = \frac{1}{3}$ $\frac{5}{6}$

38. $a - b$; $a = 1\frac{1}{2}$, $b = 1\frac{1}{4}$ $\frac{1}{4}$

39. $c - d + 2$; $c = 0.24$, $d = 0.08$ 2.16

40. $c - d + 1$; $c = 1.04$, $d = 0.6$ 1.44

41. $fg + 1$; $f = \frac{3}{4}$, $g = \frac{1}{2}$ $1\frac{3}{8}$

42. $fg - 1$; $f = 1\frac{2}{3}$, $g = \frac{3}{4}$ $\frac{1}{4}$

43. $\dfrac{m}{n}$; $m = \frac{3}{4}$, $n = \frac{1}{2}$ $1\frac{1}{2}$

44. $\dfrac{m}{n}$; $m = 1\frac{1}{3}$, $n = 1\frac{1}{6}$ $1\frac{1}{7}$

45. $2 + \dfrac{p}{q}$; $p = 0.75$, $q = 0.05$ 17

46. $2 - \dfrac{p}{q}$; $p = 0.6$, $q = 1.5$ 1.6

47. $\dfrac{1}{t}$; $t = \frac{2}{3}$ $1\frac{1}{2}$

48. $\dfrac{1}{t}$; $t = 1\frac{2}{3}$ $\frac{3}{5}$

49. $\dfrac{1}{u} + \dfrac{1}{v}$; $u = \frac{1}{2}$, $v = \frac{1}{3}$ 5

50. $\dfrac{1}{u} - \dfrac{1}{v}$; $u = 2\frac{1}{3}$, $v = 3\frac{1}{2}$ $\frac{1}{7}$

51. xy; $x = \frac{1}{3}$, $y = \frac{3}{4}$ $\frac{1}{4}$

52. $5xy$; $x = 1\frac{1}{5}$, $y = 3$ 18

53. Write four variable expressions that contain the variable z and that have a value of 1 when $z = 3$. $z - 2$, $\frac{z}{3}$, $2z - 5$, $\frac{1}{z} + \frac{2}{3}$ (Answers may vary.)

54. How many variable expressions contain the variable w and have a value of 0 when $w = 3$? an infinite number

Find values of a and b that make each statement true. Answers may vary.

C

55. The expressions $a + b$ and $a - b$ have the same value. a = 1, b = 0

56. The expressions ab and $\dfrac{a}{b}$ have the same value. a = 2, b = 1

57. The expression $a - b$ has the same value as b. a = 2, b = 1

58. The expression $\dfrac{a}{b}$ has the same value as b. a = 4, b = 2

59. What is true of the value of $\dfrac{1}{n}$ and of the value of $n + \dfrac{1}{n}$ as the value of n increases, if $n \in \{1, 2, 3, 4, 5, \ldots\}$? $\frac{1}{n}$ decreases toward 0; n + $\frac{1}{n}$ increases

60. What is true of the value of $\dfrac{1}{n}$ and of the value of $n + \dfrac{1}{n}$ as the value of n decreases, if $n \in \{\frac{1}{1}, \frac{1}{2}, \frac{1}{3}, \frac{1}{4}, \frac{1}{5}, \ldots\}$. $\frac{1}{n}$ increases; n + $\frac{1}{n}$ increases

20   *Chapter 1*

1–5 Grouping Symbols

Simplifying or evaluating an expression frequently involves more than one operation. In such cases parentheses, ( ), are often used to indicate which operations are to be performed first. Different groupings may produce expressions for different numbers. For example,

$$(3 \cdot 7) + 9 = 21 + 9 = 30,$$

while

$$3 \cdot (7 + 9) = 3 \cdot 16 = 48.$$

A **grouping symbol** is a device, such as a pair of parentheses, that is used to enclose an expression. Brackets, [ ], and braces, { }, may also be used as grouping symbols.

A product such as $3 \cdot (7 + 9)$ is usually written without the multiplication symbol simply as $3(7 + 9)$. You read this expression formally as "three times *the quantity* seven plus nine." Similarly, a product such as $3 \cdot 16$ may be written in one of the following ways.

$$3(16) \qquad \text{or} \qquad (3)16 \qquad \text{or} \qquad (3)(16)$$

Variables may be grouped in the same way that numerals are grouped. When you group a product that involves a variable, you usually omit the grouping symbols.

$(5n) + 2$ is usually written $5n + 2$.
$16 \div (4y)$ is usually written $16 \div 4y$.
$(8a) - (5b)$ is usually written $8a - 5b$.

A fraction bar is both a division symbol and a grouping symbol. When operation symbols appear above or below the fraction bar, those operations are to be performed before the division. For example,

$$\frac{6 + 18}{2 + 6} = \frac{24}{8} = 3,$$

and,

$$\text{when } q = 21, \frac{q + 7}{q - 7} = \frac{21 + 7}{21 - 7} = \frac{28}{14} = 2.$$

EXAMPLE 1 Evaluate each expression when $x = 3$.

a. $8(x - 2) - 5$ b. $54 \div 2x$ c. $\dfrac{x + 9}{4x}$

SOLUTION a. $8(x - 2) - 5 = 8(3 - 2) - 5 = 8(1) - 5 = 8 - 5 = 3$

b. $54 \div 2x = 54 \div 2(3) = 54 \div 6 = 9$

c. $\dfrac{x + 9}{4x} = \dfrac{3 + 9}{4(3)} = \dfrac{12}{12} = 1$

Numbers and Variables **21**

Key Ideas

Simplify and evaluate expressions that contain grouping symbols.

Chalkboard Examples

Evaluate each expression when $x = 5$.

1. $\dfrac{x}{5} + x = \dfrac{5}{5} + 5 = 1 + 5 = 6$

2. $4(x + 2) - 6$
$= 4(5 + 2) - 6$
$= 4(7) - 6$
$= 28 - 6$
$= 22$

3. Simplify $10[50 - 8(4 + 2)]$.
$= 10[50 - 8(6)]$
$= 10[50 - 48]$
$= 10(2)$
$= 20$

If an expression contains more than one grouping symbol, first simplify the numeral in the innermost grouping symbol. Then work toward the outermost grouping symbol until you have simplified the entire expression.

EXAMPLE 2   Simplify $9[24 - 3(5 + 1)]$.

SOLUTION
$$9[24 - 3(5 + 1)] = 9[24 - 3(6)]$$
$$= 9[24 - 18]$$
$$= 9(6)$$
$$= 54$$

Suggested Assignments

Minimum
22/2-42 even
S 13/2-18 even

Average
22/10-46 even
S 14/28-42 even

Maximum
22/11-45 odd, 47-54
S 14/32-48 even

Oral Exercises

Tell whether each statement is true or false.

1. $2(3 + 4) \ne 2(3) + 4$ true
2. $(2 + 2)2 = 2 + 2(2)$ false
3. $\frac{10 + 4}{2} = \frac{10}{2} + 4$ false
4. $\frac{12 + 4}{4 + 2} \ne \frac{12}{4} + \frac{4}{2}$ true
5. $12 \div (1 + 1) \ne (12 \div 1) + 1$ true
6. $4 - (2 \div 1) = 4 \div (2 - 1)$ false
7. $8 + (4 + 1) = (8 + 4) + 1$ true
8. $(14 - 6) - 5 = 14 - (6 - 5)$ false
9. $\frac{24}{6 \div 2} \ne \frac{24}{6} \cdot 2$ false
10. $\frac{10 \cdot 3}{2} = \frac{10}{2} \cdot 3$ true
11. $36 \div (6 \div 2) = (36 \div 6) \div 2$ false
12. $(7 \cdot 5) \cdot 2 \ne 7 \cdot (5 \cdot 2)$ false

Simplify.

13. $5(3) + 4$ 19
14. $47 - 2(5)$ 37
15. $7(8 - 2)$ 42
16. $(3 \cdot 3) + (4 \cdot 4)$ 25
17. $\frac{3 \cdot 8}{6 + 2}$ 3
18. $3(7) + \frac{15}{3}$ 26
19. $\frac{3(9) + 1}{10 + 4}$ 2
20. $(8 - 2)(3 + 5)$ 48

Written Exercises

Simplify.

A
1. $21(5) + 4$ 109
2. $85 - 3(9)$ 58
3. $23(4) - 6(13)$ 14
4. $17(3) + 10(3)$ 81
5. $16(25 - 14)$ 176
6. $(26 + 9)7$ 245
7. $\frac{90 - 10}{30 - 10}$ 4
8. $\frac{80 - 62}{15 + 57}$ $\frac{1}{4}$
9. $2(19) + \frac{48}{3}$ 54
10. $\frac{144}{3} - 14(3)$ 6
11. $\frac{3(9) - 7}{7(5) + 5}$ $\frac{1}{2}$
12. $\frac{9(9) + 12(12)}{4(4) + 3(3)}$ 9
13. $[24 - 9(2)]2$ 12
14. $8[3(5) - 4]$ 88
15. $12[2(2) + 4(4)]$ 240
16. $[6(14) + 2(13)]9$ 990
17. $[5(13) - 11] \div 3$ 18
18. $[6(10) + 7(0)] \div 4$ 15

Additional A Exercises
Simplify.

1. $\frac{12(11) + 11}{26 \div 2}$ 11
2. $[3(9) - (4 + 6)]2$ 34
3. $54 - 6(8)$ 6

Evaluate each expression when $d = 10$, $e = 0$, $f = 4$, $g = 3$, and $h = 2$.

4. $\frac{d + e}{f - g}$ 10
5. $\frac{h - e}{g \cdot d}$ $\frac{1}{15}$
6. $f(e \cdot h) + d$ 10

22   Chapter 1

Evaluate each expression when $v = 0$, $w = 1$, $x = 2$, $y = 3$, and $z = 5$.

19. $3x + 5y$ 21
20. $7z - 3y$ 26
21. $2(v + 3) - 2$ 4
22. $16 - 3(z - 5)$ 16

23. $xy - z$ 1
24. $vw + yz$ 15
25. $x(z - v) - y$ 7
26. $(w + y)v + yz$ 15

27. $\dfrac{xz - 1}{x + 1}$ 3
28. $\dfrac{y + z}{x} + w$ 5
29. $xy - \dfrac{w}{z}$ $5\frac{4}{5}$
30. $xyz - \dfrac{v}{w}$ 30

Simplify.

B
31. $[7(3) - 6] + [5 + 3(4)]$ 32
32. $2[2(5) - 7] + [9 + 5(5)]$ 40

33. $6[9(3) - 17][6(6) - 5(7)]$ 60
34. $[6(9) + 7][6 + 9(7)]$ 100

35. $8\{[5(6) - 7(3)] - 9\}$ 0
36. $\{5 + 3[2(5) - 4] - 3\}$ 240

37. $\{[2(25) + 5(70)] \div [80(15) \div 6(10)]\} + [25(12 \div 6)]$ 70

38. $\{[(15 - 3)(7 + 5)] + 6(6)\} - 9\{(15 \div 3)[5(4) \div (12 - 2)]\}$ 90

Replace each __?__ with one of the symbols = or ≠ to make a true statement.

39. $3[(5 + 1)(6 \div 2)]$ __?__ $[18 + (144 \div 8)] + 3$ ≠

40. $[3(5 - 2)]4$ __?__ $96 \div [32 \div (5 - 3)]$ ≠

41. $\dfrac{[9 + 3(13)] \div 12}{4 - (6 \div 2)} + 1$ __?__ $27 - \dfrac{40 + 4}{8 - 6}$ =

42. $\dfrac{[5(10) - 2] - [3(14) - 6(6)]}{(36 \div 2) + 2}$ __?__ $\dfrac{30 - 3}{1 + 2} + 3$ ≠

43. $\dfrac{4(3) + 22}{8 + (27 \div 3)} - \dfrac{8 - 4}{2 + 2}$ __?__ $\dfrac{7(6) + 6}{30 - 2(3)} - 1$ =

44. $\dfrac{\{[3(5) - 7] + 2(2)\} + 40}{\{[2(16) - 5] - 3(3)\} - 5}$ __?__ $\dfrac{(54 \div 3) + 2(3)}{(60 \div 5) - 2(3)}$ =

45. $2\left\{\dfrac{[5 - 2(2)] + 8(3)}{[(24 \div 6) + 4] - 3}\right\}$ __?__ $[(2 + 5) - (30 \div 6)]\left(\dfrac{12 - 2}{5 - 3}\right)$ =

46. $\left\{\dfrac{90 - [13(18) \div 3]}{[7 + 6(4)] - 19}\right\}4$ __?__ $\{[18(3) \div 9] - 5\}\left(\dfrac{24 \div 2}{2(3)}\right)$ ≠

Let $m \in \{1, 2, 3, \ldots, 10\}$. Find the greatest and least values of each expression.

C
47. $4m - 3$ 37; 1
48. $60 - 5m$ 55; 10
49. $m(2m - 1)$ 190; 1
50. $\dfrac{12}{m + 1}$ 6; $1\frac{1}{11}$

Let $m \in \{\frac{1}{1}, \frac{1}{2}, \frac{1}{3}, \ldots, \frac{1}{10}\}$. Find the greatest and least values of each expression.

51. $2m + 3$ 5; $3\frac{1}{5}$
52. $10 - 3m$ $9\frac{7}{10}$; 7
53. $m(1 - m)$ 0; $\frac{9}{100}$
54. $\dfrac{2m + 1}{m}$ 12; 3

Simplify.

1. $\dfrac{3}{4} + \dfrac{5}{4}$ 2

2. $(19 - 8)4$ 44

Evaluate each expression when $a = 3$, $b = 10$, and $c = 15$.

3. $3b. - \dfrac{c}{a}$ 25

4. $\dfrac{a + b + c}{4a}$ $\dfrac{7}{3}$

Evaluate each expression when the variables have the given values.

5. $t + uv$; $t = \dfrac{1}{2}$, $u = \dfrac{1}{4}$, $v = 1\dfrac{3}{4}$

6. xy; $x = 2\dfrac{1}{8}$, $y = 8$ 17

Computer Exercises

The optional Computer Exercises that appear throughout this book can be worked using BASIC or any other programming language. Students using BASIC may find Appendix B at the back of the book summarizing the fundamental concepts and terminology of BASIC useful.

Write a program to evaluate each expression when $n = 6$.

1. $2n + 2$ 14

2. $2(n + 2)$ 16

3. $\dfrac{n}{2}$ 3

4. $\dfrac{n + 2}{2}$ 4

5. $\dfrac{2}{n - 2}$ .5

6. $\dfrac{n + 2}{n - 2}$ 2

7. $\dfrac{2n + 2}{2n - 2}$ 1.4

8. $\dfrac{2n + 2}{2n} - 2$ $-.833333333$

9. Let $n \in \{1, 2, 3, 4, 5, \ldots, 100\}$. Write a program to evaluate the expression $\dfrac{2n - 1}{2n + 1}$ over this domain. What is true of the value of this expression as the value of n increases?

The value of $\frac{2n - 1}{2n + 1}$ increases to 1.

Teaching Suggestions
p. T71

Related Activities p. T71

Supplementary Material
Test 2

Key Ideas

Simplify and evaluate expressions that contain exponents.

1–6 Exponents and Order of Operations

When one factor of a product is used a number of times, you may use an *exponent* to simplify the notation. For example:

$$\underbrace{3 \times 3 \times 3 \times 3 \times 3}_{5 \text{ factors}} = 3^5$$

You read the expression 3^5 as "three to the fifth power" or as "the fifth power of three." In this expression, 3 is called the **base** and 5 is called the **exponent**. The number 3^5, or 243, is called a **power** of 3. The expression 3^5 is called the **exponential form** of the power.

Other powers of 3 can be defined and written as follows.

$3^1 = 3$	"three to the first power"
$3^2 = 3 \times 3 = 9$	"three to the second power" or "three squared" or "the square of three"
$3^3 = 3 \times 3 \times 3 = 27$	"three to the third power" or "three cubed" or "the cube of three"
$3^4 = 3 \times 3 \times 3 \times 3 = 81$	"three to the fourth power"

Notice the special language that is associated with the second and third powers of a number. In fact, the expression 3^2 is *usually* read as "three squared" or as "the square of three" because 3^2 can represent the area of a square with sides of length 3 units. Similarly, the expression 3^3 is usually read as "three cubed" or as "the cube of three" because 3^3 can represent the volume of a cube with edges of length 3 units.

24 *Chapter 1*

EXAMPLE 1 Simplify.

     a. 2^4 **b.** $3^2 \times 4^3$ **c.** 2×5^3 **d.** $(2 \times 5)^3$

SOLUTION   **a.** $2^4 = 2 \times 2 \times 2 \times 2 = 16$

                b. $3^2 \times 4^3 = (3 \times 3) \times (4 \times 4 \times 4) = 9 \times 64 = 576$

                c. $2 \times 5^3 = 2 \times (5 \times 5 \times 5) = 2 \times 125 = 250$

                d. $(2 \times 5)^3 = (10)^3 = 10 \times 10 \times 10 = 1000$

When simplifying expressions, these steps should be followed in order.

Order of Operations

1. First simplify expressions within grouping symbols.
2. Then simplify powers.
3. Then simplify products and quotients in order from left to right.
4. Then simplify sums and differences in order from left to right.

EXAMPLE 2   Simplify $29 + 32 \div (8 - 6)^2$.

SOLUTION $29 + 32 \div (8 - 6)^2 = 29 + 32 \div (2)^2$

                              $= 29 + 32 \div 4$

                              $= 29 + 8$

                              $= 37$

Exponents may also be used in variable expressions. For example:

$$\underbrace{n \times n \times n \times n}_{4 \text{ factors}} = n^4$$

EXAMPLE 3   Evaluate the expression $100 - a(a + 1)^3$ when $a = 2$.

SOLUTION $100 - a(a + 1)^3 = 100 - 2(2 + 1)^3$

                            $= 100 - 2(3)^3$

                            $= 100 - 2(27)$

                            $= 100 - 54$

                            $= 46$

In general, if x denotes any real number and n denotes any positive integer, the expression x^n is defined as follows.

$$x^n = \underbrace{x \cdot x \cdot \ldots \cdot x}_{n \text{ factors}}$$

Numbers and Variables   **25**

Chalkboard Examples

Simplify.

1. $(2 + 3)^3 = 5^3 = 5 \times 5 \times 5 = 125$

2. $183 - 120 \div (9 - 7)^2$
$= 183 - 120 \div 2^2$
$= 183 - 120 \div 4$
$= 183 - 30$
$= 153$

3. Evaluate the expression $59 + c(c - 2)^3$ when $c = 5$.
$= 59 + 5(5 - 2)^3$
$= 59 + 5(3)^3$
$= 59 + 5(27)$
$= 59 + 135$
$= 194$

Suggested Assignments

Minimum
  26/1-41 odd
R 27/Self-Test 2

Average
  26/5-47 odd, 49-54
R 27/Self-Test 2

Maximum
  26/5-53 odd, 55-60
R 27/Self-Test 2

Oral Exercises

State the exponential form of each expression.

1. $7 \times 7 \times 7 \times 7 \times 7$ 7^5
2. $5 \times 5 \times 5 \times 5 \times 5 \times 5 \times 5$ 5^7
3. $2 \times 2 \times 3 \times 3 \times 3 \times 3$ $2^2 \times 3^4$
4. $3 \times 3 \times 11 \times 11$ $3^2 \times 11^2$
5. $x \cdot x$ x^2
6. $y \cdot y \cdot y$ y^3
7. $x \cdot x \cdot y \cdot y \cdot y$ $x^2 y^3$
8. $x \cdot x \cdot x \cdot y \cdot y$ $x^3 y^2$
9. $5 \cdot a \cdot a$ $5a^2$
10. $3 \cdot a \cdot a \cdot a \cdot b \cdot b \cdot c$ $3a^3 b^2 c$
11. $(n + 1)(n + 1)$ $(n + 1)^2$
12. $11(m - n)(m - n)(m - n)$ $11(m - n)^3$

Simplify.

13. 5^2 25
14. 2^3 8
15. $(0.5)^2$ 0.25
16. $(0.1)^2$ 0.01
17. 2×3^2 18
18. $2^2 \times 3$ 12
19. $(2 \times 3)^2$ 36
20. $2^2 \times 3^2$ 36

Evaluate each expression when $a = 3$ and $b = 2$.

21. a^2 9
22. $2a$ 6
23. ab^2 12
24. $(ab)^2$ 36
25. $a + b^2$ 7
26. $(a + b)^2$ 25
27. $a^3 - b^3$ 19
28. $(a - b)^3$ 1

Additional A Exercises

Write each expression in exponential form.

1. one third the cube of p
  $\frac{1}{3}p^3$ or $\frac{p^3}{3}$

2. $b \cdot b \cdot 2 \cdot 2 \cdot 2$ $2^3 b^2$
Simplify.
3. 5^3 125

4. $\frac{(3 + 4)^2}{(7 - 5)^3}$ $\frac{49}{8} = 6\frac{1}{8}$

Evaluate each expression when the variable has the given value.
5. $t^3 - t^2 - t$; $t = 4$ 44

6. $\frac{(s - 1)^2}{(s + 1)^2}$; $s = 5$ $\frac{4}{9}$

Written Exercises

Write each expression in exponential form.

A
1. $x \cdot x \cdot x \cdot x \cdot x$ x^5
2. $y \cdot y \cdot y \cdot y \cdot y \cdot y$ y^6
3. $7 \cdot a \cdot a \cdot a \cdot a$ $7a^4$
4. $2 \cdot b \cdot b \cdot b \cdot b \cdot c$ $2b^4 c$
5. $(x - y)(x - y)(x - y)$ $(x - y)^3$
6. $5 \cdot r \cdot r \cdot r \cdot (s + t)(s + t)$ $5r^3(s + t)^2$
7. m squared m^2
8. the cube of n n^3
9. p plus the square of q $p + q^2$
10. the cube of the quantity r plus s $(r + s)^3$
11. one half the fifth power of t $\frac{1}{2}t^5$
12. the fifth power of one half t $(\frac{1}{2}t)^5$

Simplify.

13. 3^5 243
14. 2^6 64
15. 7^4 2401
16. 11^3 1331
17. $27 + 2^4$ 43
18. $(18 + 5)^2$ 529
19. $2^2 + 3^3 + 4^4$ 287
20. $5^4 - 9^2 - 2^5$
21. $12^2 - 2(5)^2$ 94
22. $4(3^2) + 6^3$ 252
23. $(12^2 - 5^3)2^3$ 152
24. $5(5^3 - 4^3)$
25. $\frac{6^2 - 4^2}{6 - 2^2}$ 10
26. $\frac{7^2 - 5^2}{2^4 + 3^2}$ $\frac{24}{25}$
27. $\frac{(3 + 7)^3}{7^2 - 3^2}$ 25
28. $\frac{10^2 - 5^2}{(10 - 5)^2}$ 3

20. 512 24. 305

Evaluate each expression when the variable has the given value.

29. $2y^2 - 3y$; $y = 5$ 35
30. $z^2 - z - 1$; $z = 3$ 5
31. $10 - 5a + 3a^2$; $a = 1$ 8
32. $b^2 - 8b + 9$; $b = 0$ 9
33. $\frac{c^2 - 2c - 35}{c - 5}$; $c = 7$ 0
34. $\frac{d^2 + d - 20}{d^2 + 2d - 15}$; $d = 5$ $\frac{1}{2}$

26   Chapter 1

Simplify.

B 35. $2^6 \div 2^2 \div 2^3 \div 2$ 1

36. $3^4 - 3^2 \div 3^2 - 3$ 77

37. $4(9 + 2) - 24 \div 2(4)$ 41

38. $(8 - 1)12 \div 3(4) - 6$ 1

39. $75 + 36 \div 3(5 - 3)^2$ 78

40. $84 \div 7 + 5(12 - 2^2)$ 52

41. $3(5^2 - 1) - 10(3^2) \div 5$ 54

42. $(5^3 - 11^2)10 \div 5(2)^2 + 3$ 5

43. $(12 - 3)^2 + (7^2 - 2)^2 - (13 - 2)^2$ 2169

44. $[(8^2 - 5^2) - (7^2 - 6^2) - (3^2 + 4^2)]^2$ 1

45. $[(0.4)^2 \div (0.2)^2] - (0.4 \div 0.2)^2$ 0

46. $[(1.1)^2 - (0.7)^2] \div [(0.2)(0.3)]^2$ 200

47. $\left(\dfrac{1}{2}\right)^2 + \left(\dfrac{5-3}{5}\right)^2 + \left(\dfrac{3+4}{10}\right)^2$ $\dfrac{9}{10}$

48. $\left(\dfrac{2^3}{2^3 + 1}\right)^2 \div \left(\dfrac{3^2 + 1}{3^2}\right)^2$ $\dfrac{16}{25}$

Evaluate each expression when the variable has the given value.

49. $4a - 5a^2;\ a = 0.3$ 0.75

50. $b^2 - 9(b - 1)^2;\ b = 1.1$ 1.12

51. $12x - 9(1 - x)^2;\ x = \dfrac{2}{3}$ 7

52. $y^2 + y + 2;\ y = \dfrac{3}{5}$ $2\frac{24}{25}$

53. $0.5(1 - 2m + 2m^2);\ m = 0.5$ 0.25

54. $(n + 1)^2(n^2 + 2n + 1);\ n = 0.1$ 1.4641

Replace each __?__ with one of the symbols $+$, $-$, $\times$, or $\div$ to make a true statement.

C 55. $12 \underline{\ ?\ } 2 \underline{\ ?\ } 2 \underline{\ ?\ } 8 \underline{\ ?\ } 4 = 18$

56. $36 \underline{\ ?\ } 3 \underline{\ ?\ } 12 \underline{\ ?\ } 6 \underline{\ ?\ } 1 = 36$

57. $64 \underline{\ ?\ } \dfrac{1}{2} \underline{\ ?\ } 8 \underline{\ ?\ } 4 \underline{\ ?\ } 4 \underline{\ ?\ } \dfrac{1}{2} = 8$

58. $18 \underline{\ ?\ } 6 \underline{\ ?\ } \dfrac{1}{3} \underline{\ ?\ } 2 \underline{\ ?\ } 6 \underline{\ ?\ } \dfrac{1}{3} = 4$

59. $24 \underline{\ ?\ } 6 \underline{\ ?\ } 8 \underline{\ ?\ } 0.5 \underline{\ ?\ } 16 \underline{\ ?\ } 0.5 = 12$

60. $12 \underline{\ ?\ } 0.25 \underline{\ ?\ } 16 \underline{\ ?\ } 0.5 \underline{\ ?\ } 0.5 \underline{\ ?\ } 24 = 2$

Self-Test 2

VOCABULARY

numerical expression or numeral (p. 17)

value of a numerical expression (p. 17)

simplify a numerical expression (p. 17)

substitution principle (p. 17)

variable (p. 18)

domain or replacement set of a variable (p. 18)

value of a variable (p. 18)

variable expression (p. 18)

evaluate or find the value of a variable expression (p. 18)

mathematical expression (p. 18)

grouping symbol (p. 21)

base (p. 24)

exponent (p. 24)

power (p. 24)

exponential form of a power (p. 24)

Numbers and Variables **27**

Quick Quiz

Find the least value of each expression if $x \in \{2, 4, 6\}$.

1. $3x + 7$ 13

2. $19 - \dfrac{x}{2}$ 16

3. $\dfrac{168}{x + 2}$ 21

Evaluate each expression when $a = 3$ and $b = 5$.

4. $a + 5(b - 2)$ 18

5. $6b \div 2a$ 5

6. $\dfrac{ab + 9}{ab - 7}$ 3

Simplify.

7. 2^4 16

8. $9^2 - 4(2)^3$ 49

9. $6 \div 3 + 4(15 - 3^2)$ 26

Show whether each statement is true or false.

1. $(29 - 7)3 = 29 - 7(3)$
$(29 - 7)3 = 29 - 7(3)$
$(22)3 = 29 - 21$
$66 \neq 8$ False

2. $[3 + (9 + 5)]2$
$\quad\quad\quad = [(3 + 9) + 5]2$
$[3 + (9 + 5)]2$
$\quad\quad\quad = [(3 + 9) + 5]2$
$(3 + 14)2 = (12 + 5)2$
$(17)2 = (17)2$
$34 = 34$ True

Use parentheses for the following expressions.

3. Multiply the sum of 9 and 6 by the difference when 6 is subtracted from 9.
$(9 + 6) \times (9 - 6)$ or
$(9 + 6)(9 - 6)$

4. Subtract the product of 2 and y from 10 and cube the result. $(10 - 2y)^3$

Find the least value of each expression if $x \in \{1, 2, 3, 4\}$.

1. $6x + 5$ 11

2. $15 - 2x$ 7

3. $\dfrac{60}{x + 1}$ 12

Obj. 1, p. 17

Evaluate each expression when $a = 2$ and $b = 6$.

4. $a + 7(b - 1)$ 37

5. $8b \div 3a$ 8

6. $\dfrac{ab + 4}{ab - 4}$ 2

Obj. 2, p. 17

Simplify.

7. 5^4 625

8. $10^2 - 3(2)^2$ 88

9. $8 \div 2 + 2(8 - 2^3)$ 4

Obj. 3, p. 17

Check your answers with those at the back of the book.

ON THE CALCULATOR

Does your calculator follow the correct order of operations in simplifying expressions? Experiment with your calculator by entering the following example exactly as it appears here:

$$7 + 3 \times 5 =$$

If your calculator displays the answer 22, it evaluated the expression correctly by following the order of operations outlined on page 25. Your calculator has an algebraic operating system.

If your calculator displays the answer 50, it performed the operations in the order in which they were entered. This type of calculator can compute accurately, but you must enter the numbers and operations in the proper sequence. To obtain a correct answer on this type of calculator, you must enter the preceding example as follows:

$$3 \times 5 + 7 =$$

Use a calculator to simplify each expression.

1. $26 + 87 \div 3$ 55

2. $48 - 35 + 17$ 30

3. $95 + 17 \times 4$ 163

4. $9 \times 27 + 12 + 35$ 290

5. $8(46 - 19)$ 216

6. $(67 + 77) \div 12$ 12

7. $172 - 6 \times 19$ 58

8. $\dfrac{532}{19 \times 4}$ 7

9. $278 - 14^2$ 82

10. $16 + 23^2 - 1$ 544

11. $4 + 8 \times 17^2$ 2316

12. $2^3(24^2 - 7)$ 4552

28   *Chapter 1*

READING ALGEBRA Numerical Expressions

Algebra is the primary language of mathematics. As you study algebra, you will learn to read and write expressions in this language.

As you may know, there is often more than one way to translate a given mathematical expression into words. The following are some familiar examples of *numerical expressions*. After each expression you see several different ways that it can be translated into a word phrase.

9 + 2

"nine plus two"
"the sum of nine and two"
"nine increased by two"
"two more than nine"

7 − 4

"seven minus four"
"the difference when four is subtracted from seven"
"seven decreased by four"
"four less than seven"

12 × 3

"twelve multiplied by three"
"the product of twelve and three"
"twelve times three"

10 ÷ 2

"ten divided by two"
"the quotient when ten is divided by two"

$\frac{1}{3} \times 4$

"one third times four"
"one third of four"
"the product of one third and four"

$\frac{4}{3}$

"four thirds"
"four divided by three"
"the quotient when four is divided by three"

Exercises

Write two different word phrases than can represent each numerical expression. Answers will vary.

1. 14 − 8   2. $\frac{16}{7}$ 3. 4 × 15   4. 18 + 9

Translate each word phrase into a numerical expression.

5. six times twenty 6 × 20
6. the sum of fourteen and three 14 + 3
7. one tenth of ninety $\frac{1}{10} \times 90$
8. nine decreased by six 9 − 6
9. four more than twelve 12 + 4
10. the product of sixteen and seven 16 × 7
11. nine thirds $\frac{9}{3}$
12. zero plus six 0 + 6

Numbers and Variables **29**

Mathematical Sentences and Problem Solving

OBJECTIVES for Sections 1-7 through 1-9:
1. *To solve an open sentence in one variable over a specified domain and to graph its solution set.*
2. *To translate numerical relationships stated in word phrases into mathematical expressions.*
3. *To translate numerical relationships stated in word sentences into mathematical sentences.*

Teaching Suggestions
p. T71

Key Ideas

Solve an open sentence in one variable over a specified domain and graph its solution set.

1–7 Equations and Inequalities

A statement that indicates a relationship between two mathematical expressions is called a **mathematical sentence**. If a mathematical sentence states that two expressions name the same number, then the sentence is an **equation**, and the equals sign, =, is used to show the relationship. The two expressions are called the **sides of the equation**. For example,

$$4 \cdot 7 = 28$$

is an equation; $4 \cdot 7$ is the *left* side of the equation, and 28 is the *right* side.

A mathematical sentence that is formed by placing an *inequality symbol* between two mathematical expressions is called an **inequality**. Among the commonly used inequality symbols are $<$, $>$, $\leq$, $\geq$, and $\neq$. The following are examples of inequalities that make use of these symbols.

$3 < 8$	read	*"3 is less than 8"*
$9 > 4$	read	*"9 is greater than 4"*
$6 + 7 \leq 18$	read	*"6 plus 7 is less than or equal to 18"*
$32 - 9 \geq 23$	read	*"32 minus 9 is greater than or equal to 23"*
$48 \div 16 \neq 4$	read	*"48 divided by sixteen is not equal to 4"*

As with equations, the expressions to the left and to the right of an inequality symbol are called the **sides of the inequality**.

If both sides of a mathematical sentence are *numerical* expressions, then the sentence may be classified as either *true* or *false*. For example, the mathematical sentences

$$5 + 9 = 14 \qquad 7 < 9 \qquad 6 \neq 40 \div 8$$

are true, whereas the mathematical sentences

$$5 + 9 = 13 \qquad 7 > 9 \qquad 5 \neq 40 \div 8$$

are false. Note that the mathematical sentences

$$12 \geq 7 \qquad \text{and} \qquad 12 \geq 12$$

are *both* true. A "greater than or equal to" sentence such as these is true *either* if the first number is greater than the second *or* if the first number is equal to the second. Similarly, a "less than or equal to" sentence is true either if the first number is *less than* the second or if the first number is equal to the second.

A mathematical sentence that contains at least one variable is called an **open mathematical sentence,** or simply an **open sentence.** The following are examples of open sentences.

$$2x - 5 = 35 \qquad a > 11 \qquad 24 \neq n \div 3$$

Generally a domain is specified for each variable in an open sentence. Any value of the variables for which the open sentence is true is called a **solution,** or **root,** of the open sentence *over the given domain.* The set of *all* such values is called the **solution set** of the open sentence, and each member of the solution set is said to **satisfy** the open sentence. When you determine the solution set of an open sentence over a given domain, you **solve** the open sentence over that domain.

EXAMPLE 1 Solve $x - 5 = 6$ if $x \in \{10, 11, 12, 13, 14\}$.

SOLUTION Replace x with each of its values in turn. Determine whether the resulting sentence is true or false.

$$
\begin{array}{ll}
x - 5 = 6 & \\
10 - 5 = 6 & \text{False} \\
11 - 5 = 6 & \text{True} \\
12 - 5 = 6 & \text{False} \\
13 - 5 = 6 & \text{False} \\
14 - 5 = 6 & \text{False}
\end{array}
$$

Therefore, the solution set is $\{11\}$.

Sometimes the given domain for a variable in an open sentence is an infinite set. Although in such cases it is impossible to replace the variable with *each* of its values, it still may be possible to determine the solution set of the open sentence.

EXAMPLE 2 Solve $3n = 7$ if $n \in \{\text{the positive integers}\}$.

SOLUTION Since the possible values of n are the positive integers, or 1, 2, 3, 4, . . . , the corresponding values of $3n$ are the positive multiples of 3, or 3, 6, 9, 12, ... Since 7 is not one of these numbers, the equation $3n = 7$ has no solution over the domain of the positive integers.

Therefore, the solution set is $\emptyset$.

Numbers and Variables **31**

1. Solve $x + 5 = 11$ if $x \in \{4, 5, 6, 7\}$.

$$
\begin{array}{ll}
x + 5 = 11 & \\
4 + 5 = 11 & \text{False} \\
5 + 5 = 11 & \text{False} \\
6 + 5 = 11 & \text{True} \\
7 + 5 = 11 & \text{False}
\end{array}
$$
∴ the solution set is $\{6\}$.

2. Solve $5n = 25$ if $n \in \{$the negative integers$\}$.
Since the possible values of n are the negative integers, or $^-1, ^-2, ^-3, ^-4, \ldots$, the corresponding values of $5n$ are negative. Since 25 is positive, the equation $5n = 25$ has no solution over the domain of the negative integers.
∴ the solution set is $\emptyset$.

3. Solve $x - 3 < 5$ over Q and graph its solution set.

$$
\begin{array}{l}
x - 3 < 5 \\
\quad\ \ x < 8
\end{array}
$$
∴ the solution set is $\{x: x < 8\}$.

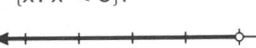

Minimum
32/1-15 odd, 17-32

Average
32/11-31 odd, 33-48

Maximum
32/11-47 odd, 49-54

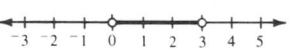
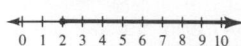

The **graph of an open sentence** is the graph of its solution set.

EXAMPLE 3 Solve $x + 8 > 9$ over $\mathcal{R}$ and graph its solution set.

SOLUTION In order for the sentence $x + 8 > 9$ to be true, x must represent a real number that is greater than 1.
Therefore, the solution set is $\{x: x > 1\}$.
The graph of the solution set is the following.

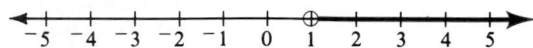

The colon within the brackets in Example 3 is read as "such that." You read the expression

$$\{x: x > 1\}$$

as, "the set of all x such that x is greater than 1." This notation is often called **set-builder notation**.

Oral Exercises

Solve each open sentence if $x \in \{0, 1, 2, 3, 4, 5\}$.

1. $x - 3 = 2$ {5}   **2.** $9 - x = 9$ {0}   **3.** $3x = 12$ {4}   **4.** $\frac{1}{2}x = 8$ ∅

5. $x + 1 > 5$ {5}   **6.** $10 - x < 7$ {4, 5} **7.** $2x > 12$ ∅   **8.** $5x < 20$ {0, 1, 2, 3}

9. $x + 2 = 2x$ {2} **10.** $4x > x + 8$ **11.** $6x < 0$ ∅   **12.** $x = 2x - 1$ {1}
                    {3, 4, 5}

Solve each open sentence if $z \in$ {the positive integers}.

13. $z + 3 = 5$ {2}   **14.** $15 - z = 9$ {6}   **15.** $4z = 28$ {7}   **16.** $\frac{1}{3}z = 5$ {15}

17. $z + 1 < 6$ **18.** $5 + z > 8$ **19.** $3z \geq 18$ **20.** $\frac{1}{2}z \leq 5$

21. $2z = z$ **22.** $z - 1 < z$ **23.** $^-3 < z < 3$ **24.** $5 \leq z < 9$

Written Exercises

Solve each open sentence if $n \in \{0, 1, 2, 3, 4, 5, 6, 7, 8\}$.

A **1.** $2 + n = 6$ {4}   **2.** $12 = n + 9$ {3}   **3.** $5 = n - 2$ {7}   **4.** $10 - n = 7$ {3}

5. $6n = 12$ {2}   **6.** $24 = 3n$ {8}   **7.** $3 = \frac{1}{3}n$ ∅   **8.** $\frac{n}{2} = 4$ {8}

9. $2n + 3 = 11$ {4} **10.** $18 - 2n = 16$ {1} **11.** $2n + 3 < 11$ **12.** $18 - 2n > 16$ {0}
                                        {0, 1, 2, 3}

13. $n^2 = n^3$ **14.** $n^2 > n$ **15.** $2n < n^2$ **16.** $n^2 \neq 4$

32   *Chapter 1*

Solve each open sentence if $y \in$ {the positive real numbers} and graph the solution set.

17. $y \geq 2$

18. $y \leq 5$

19. $0.5 < y$

20. $3\frac{1}{2} > y$

21. $1 < y < 6$

22. $7 \geq y \geq 2$

23. $6\frac{1}{2} > y > 3$

24. $1.5 < y < 5$

25. $y + 3 = 9$

26. $4 + y > 5$

27. $7.5 - y = 2$

28. $y - 1 = 2\frac{1}{2}$

29. $4y + 1 = 17$

30. $4y - 1 \geq 11$

31. $y(y + 3) = 4$

32. $y(y + 2) > 15$

B **33.** $3 \leq y + 1 \leq 7$ **34.** $5 \geq y - 3 \geq 1$ **35.** $8 > y - 2 > 4$ **36.** $1 < 1 + y < 6$

37. $y + 2 \geq y$ **38.** $y \geq y + 1$ **39.** $y + 2 \leq y + 1$ **40.** $y + 1 > y - 1$

Let $a \in$ {1, 2} and $b \in$ {3, 4}. In each open sentence, replace a and b with all possible pairs of values from their respective domains. Tell whether the resulting statements are true or false.

EXAMPLE $2a + 3b = 14$

SOLUTION $2(1) + 3(3) = 14$ False $2(2) + 3(3) = 14$ False
$2(1) + 3(4) = 14$ True $2(2) + 3(4) = 14$ False

41. $a + b = 6$ **42.** $b - a = 2$ **43.** $7 - b = 2a$ **44.** $3a = 2b - 5$

45. $b < 3a$ **46.** $a + 6 \geq 2b$ **47.** $3a + b > 7$ **48.** $8 \leq 3b - a$

Write two different open sentences for which the solution set over $\mathcal{R}$ is the given set. Answers will vary.

C **49.** {5} $\begin{array}{l} x - 1 = 4; \\ x + 1 = 6 \end{array}$ **50.** $\{\frac{1}{3}\}$ $\begin{array}{l} 3x = 1; \\ 6x = 2 \end{array}$ **51.** $\mathcal{R}$ $\begin{array}{l} x + 1 > x; \\ x - 2 < x \end{array}$ **52.** $\emptyset$ **52.** $x + 2 < x; x - 1 > x$

53. {all real numbers except 0}   **54.** {all real numbers except 1}
$2x \neq 0; x^2 > 0$ $x - 1 \neq 0; x(x - 2) > -1$

ON THE CALCULATOR

You can use your calculator to evaluate variable expressions for given values of the variables if you keep in mind the order in which the values must be entered, as discussed on page 28.

Use a calculator to evaluate each expression when $a = 4.5$, $b = 0.9$, and $c = 1.2$.

1. $a + bc$ **5.58**   **2.** $a(b + c)$ **9.45**   **3.** $(a + b)c$ **6.48**   **4.** $a + b^2$ **5.31**

5. $c^2 - b$ **0.54**   **6.** $(8a)(2c)$ **86.4**   **7.** $2b + 5c$ **7.8**   **8.** $a + 3(a - b)$ **15.3**

9. $a + 3b - c - a$ **1.5**   **10.** $5a - 3b - 6b - 2a$ **5.4**

11. $2(a - b + c) - 3(c - b)$ **8.7**   **12.** $3(a^2 + b) + 4(a^2 + b)$ **148.05**

Numbers and Variables   **33**

19. {y: y > 0.5}

20. $\left\{y: 0 < y < 3\frac{1}{2}\right\}$

21. {y: 1 < y < 6}

22. {y: 2 ≤ y ≤ 7}

23. $\left\{y: 3 < y < 6\frac{1}{2}\right\}$

24. {y: 1.5 < y < 5}

25. {6}

26. {y: y > 1}

27. {5.5}

28. $\left\{3\frac{1}{2}\right\}$

29. {4}

30. {y: y ≥ 3}

31. {1}

32. {y: y > 3}

33. {y: 2 ≤ y ≤ 6}

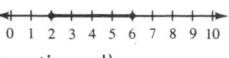

(continued)

Additional Answers
(continued)

34. {y: 4 ≤ y ≤ 8}

0 1 2 3 4 5 6 7 8 9

35. {y: 6 < y < 10}

2 3 4 5 6 7 8 9 10 11 12

36. {y: 0 < y < 5}

0 1 2 3 4 5 6 7 8

37. {y: y > 0}

−4 −2 0 2 4

38. Ø

0 1 2 3 4

39. Ø

0 1 2 3 4 5 6 7 8 9 10

40. {y: y > 0}

0 1 2 3 4

41. 1 + 3 = 6, False; 2 + 3 = 6, False; 1 + 4 = 6, False; 2 + 4 = 6, True.

42. 3 − 1 = 2, True; 4 − 1 = 2, False; 3 − 2 = 2, False; 4 − 2 = 2, True.

43. 7 − 3 = 2(1), False; 7 − 4 = 2(1), False; 7 − 3 = 2(2), True; 7 − 4 = 2(2), False.

44. 3(1) = 2(3) − 5, False; 3(2) = 2(3) − 5, False; 3(1) = 2(4) − 5, True; 3(2) = 2(4) − 5, False.

45. 3 < 3(1), False; 4 < 3(1), False; 3 < 3(2), True; 4 < 3(2), True.

46. 1 + 6 ≥ 2(3), True; 2 + 6 ≥ 2(3), True; 1 + 6 ≥ 2(4), False; 2 + 6 ≥ 2(4), True.

47. 3(1) + 3 > 7, False; 3(2) + 3 > 7, True; 3(1) + 4 > 7, False; 3(2) + 4 > 7, True.

48. 8 ≤ 3(3) − 1, True; 8 ≤ 3(4) − 1, True; 8 ≤ 3(3) − 2, False; 8 ≤ 3(4) − 2, True.

PROGRAMMING IN BASIC

You can use a computer to solve open sentences over a given finite domain by programming the computer to test each value in the domain to see if it satisfies the open sentence. For example, the program that follows will provide a graph of the open sentence X + 8 > 9 over the domain {0, 1, 2, . . . , 15}.

```
10    PRINT "TO GRAPH AN OPEN SENTENCE"
20    PRINT "IN ONE VARIABLE:"
30    PRINT "(SENTENCE IS IN LINE 70.)"
40    PRINT
50    REM *DOMAIN
60    FOR X = 0 TO 15
70    IF X + 8 > 9 THEN 120
80    REM *ELSE
90    PRINT "-";
100   GOTO 130
110   REM *THEN
120   PRINT "*";
130   NEXT X
140   PRINT ">"
150   PRINT "0"
160   END
```

Notice that the program illustrates several of the fundamental structures of the BASIC computer programming language:

       FOR-NEXT loop
       IF-THEN conditional transfer
       GOTO unconditional transfer

The program also utilizes the PRINT and REM (remark) statements. If you need additional information about any of these items, see Appendix B at the back of the book.

Exercises

1. Type in and RUN the program as given. Compare the result with the graph shown on page 32.

2. What is the purpose of lines 140 and 150? Line 140 labels the arrow head and line 150 labels the origin.

Change line 70 as indicated, then RUN the revised program.

3. 70   IF X − 5 = 6 THEN 120

4. 70   IF 3 * X = 7 THEN 120

34   *Chapter 1*

5. 70   IF 3 * X = 12 THEN 120   **6.** 70   IF X + 3 = 5 THEN 120

7. 70   IF 15 − X = 9 THEN 120   **8.** 70   IF 3 * X >= 18 THEN 120

9. 70   IF X + 1 < 6 THEN 120   **10.** 70   IF 5 <= X THEN 120

11. 70   IF 4 * X > X + 8 THEN 120   **12.** 70   IF X > 2 * X − 1 THEN 120

13. 70   IF 3 <= X + 1 AND X + 1 <= 7 THEN 120

14. 70   IF 5 >= X − 3 AND X − 3 >= 1 THEN 120

If possible, save this program for later use.

1–8 Words into Symbols: Expressions

Often you must translate a *word phrase* about numbers into a numerical or variable expression. In order to do so, you must be able to translate each part of the word phrase into an appropriate mathematical symbol. For example, the word phrase

forty-six *decreased by* nineteen

can be translated into the numerical expression

$$46 − 19.$$

Sometimes a single mathematical expression might be used to translate many different word phrases. For example, the variable expression

$$n + 50$$

might represent any of the following word phrases:

the *sum* of a number n and fifty

a number n *increased by* fifty

fifty *more than* a number n

the *total* of a number n and fifty

a number n *plus* fifty

EXAMPLE 1   Write a mathematical expression for each word phrase.
a. the difference when four is subtracted from seventy-two
b. the product of fifteen and a number y
c. seventeen times the quantity ninety-four plus twelve
d. two less than the quotient when a number m is divided by 9

SOLUTION   **a.** $72 − 4$ **b.** $15y$ **c.** $17(94 + 12)$ **d.** $\frac{m}{9} − 2$

Numbers and Variables   **35**

Write a mathematical expression for each word expression.

1. the quotient when 3 is divided by 4 $\frac{3}{4}$

2. half Fred's age three years from now if Fred is a years old now $\frac{a + 3}{2}$

3. the number of ounces in y pounds
One pound consists of 16 · 1 ounces, or 16 ounces.
Two pounds consist of 16 · 2 ounces, or 32 ounces.
Therefore, y pounds consist of 16 · y ounces, or 16y ounces.

Mathematical expressions often represent real-life situations.

EXAMPLE 2   Ann is a years old now. Write a variable expression for each word phrase.
   a. Ann's age five years from now
   b. four years less than half Ann's present age
   c. twice Ann's age three years ago
   d. David's age, if the sum of Ann's age and David's age is thirty

SOLUTION   **a.** $a + 5$ **b.** $\frac{1}{2}a - 4$ **c.** $2(a - 3)$ **d.** $30 - a$

EXAMPLE 3   Write a variable expression for each word phrase.
   a. the number of months in y years
   b. the total value in cents of d dimes and q quarters

SOLUTION   **a.** One year consists of 12 · 1 months, or 12 months.
   Two years consist of 12 · 2 months, or 24 months.
   Therefore, y years consist of 12 · y months, or 12y months.
   b. The value of one dime is 10 cents.
   The value of d dimes is 10 · d cents, or 10d cents.
   The value of one quarter is 25 cents.
   The value of q quarters is 25 · q cents, or 25q cents.
   The *total* value of d dimes and q quarters is the *sum* of the values of the dimes and quarters, or $10d + 25q$ cents.

Oral Exercises

Match.

1. twelve less than a number n i.
2. twelve decreased by a number n c.
3. a number n increased by twelve   k.
4. a number n multiplied by twelve   a.
5. one twelfth of a number n f.
6. twelve more than half a number n l.
7. the difference when twice a number n is subtracted from twelve   g.
8. twice the difference when twelve is subtracted from a number n b.
9. the product of twelve and the sum of a number n and twelve   e.
10. the quotient when the total of a number n and twelve is divided by twelve   d.
11. twelve divided by the sum of twelve and a number n j.
12. the sum of a number n and twelve, divided by the product of n and twelve   h.

a. $12n$
b. $2(n - 12)$
c. $12 - n$
d. $\frac{n + 12}{12}$
e. $12(n + 12)$
f. $\frac{1}{12}n$
g. $12 - 2n$
h. $\frac{n + 12}{12n}$
i. $n - 12$
j. $\frac{12}{12 + n}$
k. $n + 12$
l. $\frac{1}{2}n + 12$

Suggested Assignments

Minimum
   37/2-18 even, 19-26
S 26/26-34 even

Average
   37/8-26 even, 27-38
S 27/40-48 even

Maximum
   37/8-38 even, 39-42
S 27/46-54 even

36   *Chapter 1*

Replace each __?__ with a variable expression to make a true statement.

13. The value in cents of n nickels is __?__. 5n
14. The value in cents of x dollars is __?__. 100x
15. The number of seconds in m minutes is __?__. 60m
16. The number of hours in d days is __?__. 24d
17. If x is a positive integer, the next greater positive integer is __?__. x + 1
18. If y is a positive odd integer, the next greater positive odd integer is __?__. x + 2
19. If you were a years old six years ago, right now you are __?__ years old. a + 6
20. If you will be z years old four years from now, right now you are __?__ years old. z − 4

Written Exercises

Write a variable expression for each word phrase.

A
1. five more than twice a number x 2x + 5
2. eight less than the square of a number y y^2 − 8
3. forty-five decreased by one fourth a number z $45 - \frac{1}{4}z$
4. the cube of a number w, increased by seventy-nine w^3 + 79
5. the product of seventeen and the square of the difference when seven is subtracted from a number m $17(m - 7)^2$
6. the quotient when the fourth power of the sum of a number n and sixteen is divided by five $\frac{(n + 16)^4}{5}$
7. the sum of a number p and eleven, multiplied by the difference when fourteen is subtracted from a number q (p + 11)(q − 14)
8. the quotient when three times a number r is divided by the product of two and a number s $\frac{3r}{2s}$
9. the total of a number a and twice a number b, divided by the total of a number c and the fifth power of a number d $\frac{a + 2b}{c + d^5}$
10. the product when the sum of a number g and the square of a number h is multiplied by the difference when a number k is subtracted from the cube of a number j $(g + h^2)(j^3 - k)$
11. the number of days in w weeks   7w
12. the number of minutes in h hours   60h
13. the number of days in h hours $\frac{h}{24}$
14. the number of minutes in s seconds $\frac{s}{60}$
15. the total value in cents of n nickels and d dimes   5n + 10d
16. the total value in cents of p pennies, d dimes, and q quarters   p + 10d + 25q
17. the total value in dollars of n nickels and q quarters   0.05n + 0.25q
18. the total value in dollars of p pennies, d dimes, and x dollars   0.01p + 0.10d + x

Numbers and Variables   **37**

Additional A Exercises

Write a variable expression for each word phrase.

1. three less than one half a number y $\frac{y}{2} - 3$

2. a quotient when five times a number p is divided by the sum of 9 and 7 $\frac{5p}{9 + 7}$

3. the number of hours in k days   24k

4. the number of pounds in b tons   2000b

5. the difference when a positive integer b is subtracted from the next greater positive integer $(b + 1) - b$

6. your sister's age if the sum of your age d and your sister's age is thirty-one $31 - d$

Mixed Review

Solve each open sentence if $n \in$ {the integers}.

1. $6 = 2n - 18$ {12}
2. $23 > 7n + 3$
   {2, 1, 0, −1, . . . }
3. $4\frac{1}{3} \leq n$ {5, 6, 7, . . . }
4. $n + 1 > -25$
   {−25, −24, −23, . . . }

Write a variable expression for each word phrase.

19. the sum of a positive integer j and the next greater positive integer $j + (j + 1)$
20. the sum of a positive even integer k and the next greater positive even integer $k + (k + 2)$
21. the product of a positive odd integer x and the next greater positive odd integer $x(x + 2)$
22. the quotient when a positive integer y is divided by the next greater positive integer $\frac{y}{y + 1}$
23. your age in years, if your sister is a years old and the sum of your age and your sister's age is twenty-four years $24 - a$
24. your mother's age in years, if you are b years old and your mother is two years older than three times your age $3b + 2$
25. twice your age six years ago, if you are m years old now $2(m - 6)$
26. one fourth your age twenty years from now, if you are n years old now $\frac{1}{4}(n + 20)$

Represent the answer to each question in terms of the given variable(s).

B 27. The sum of two numbers is forty-eight. One number is x. What is the other number? $48 - x$
28. The product of two numbers is thirty-six. One number is y. What is the other number? $\frac{36}{y}$
29. The difference between two numbers is ten. If the lesser number is a, what is the greater number? $a + 10$
30. The quotient when one number is divided by another is four. If the lesser number is z, what is the greater number? $4z$
31. Linda has six more quarters than nickels. If she has n nickels, what is the total value in cents of her quarters and nickels? $5n + 25(n + 6)$
32. Darryl has three fewer dimes than quarters. If he has d dimes, what is the total value in cents of his dimes and quarters? $10d + 25(d + 3)$
33. Ken has nine times as many pennies as dimes. If he has d dimes, what is the total value in cents of his pennies and dimes? $9d + 10d$
34. Alice has six times as many dollar bills as quarters. If she has q quarters, what is the total value in dollars of her dollar bills and quarters? $6q + 0.25q$
35. Sam is m years old. Lisa is three times as old as Sam was five years ago. How old is Lisa? $3(m - 5)$
36. Claudia is n years old. Joel is one third as old as Claudia will be in ten years. How old is Joel? $\frac{1}{3}(n + 10)$
37. Two sides of an isosceles triangle each measure y centimeters. The third side is five centimeters longer. What is the perimeter of the triangle? $y + y + (y + 5)$
38. The width of a rectangle is three centimeters shorter than the length. If the width is x centimeters, what is the area of the rectangle? $x(x - 3)$

38 *Chapter 1*

C **39.** A passenger train travels at a speed of v km/h. A freight train travels at a speed that is x km/h slower. How much farther can the passenger train travel than the freight train in h hours? $vh - (v - x)h$

40. Today Jan drove her truck for h hours at a speed of r km/h. Yesterday she drove x fewer hours at a speed that was y km/h faster. What is the total distance that Jan drove yesterday and today? $rh + (h - x)(r + y)$

41. Green peppers cost n cents per pound. This is z cents per pound less than the cost of red peppers. What is the total cost of g pounds of green peppers and r pounds of red peppers? $gn + r(n + z)$

42. Last week Carl bought m pounds of potatoes for his restaurant at a cost of y cents per pound. This week he bought n more pounds than last week at a cost that was z cents per pound greater. What is the difference between the amount he paid for the potatoes last week and the amount he paid this week? $(m + n)(y + z) - my$

Computer Exercises   For students with computer experience

1. Write a program that will compute the equivalent number of seconds when you input any number of minutes. The output should be in the form: x minutes = y seconds.

2. Write a program that will compute the equivalent number of hours and minutes when you input any number of minutes. The output should be in the form: x minutes = y hours z minutes.

3. Write a program that will compute the total value of a collection of coins when you input the number of pennies, nickels, dimes, quarters, half dollars, and dollars in the collection. The output should display the total value in dollars using the correct $ and . notation.

ON THE CALCULATOR

Solve each equation if $n \in \{1, 2, 3, 4\}$.

1. $\dfrac{n(n + 1)}{2} = 1$ {1}

2. $\dfrac{n(n + 1)}{2} = 1 + 2$ {2}

3. $\dfrac{n(n + 1)}{2} = 1 + 2 + 3$ {3}

4. $\dfrac{n(n + 1)}{2} = 1 + 2 + 3 + 4$ {4}

Let $n \in$ {the natural numbers}. Guess what value of n satisfies each equation, then use a calculator to check your guess.

5. $\dfrac{n(n + 1)}{2} = 1 + 2 + 3 + \ldots + 10$ {10}

6. $\dfrac{n(n + 1)}{2} = 1 + 2 + 3 + \ldots + 24$ {24}

Key Ideas

Translate numerical relationships stated in words into mathematical sentences.

Chalkboard Examples

Write a mathematical sentence for each word sentence.

1. The square of a number n is one half the cube of the same number. $n^2 = \dfrac{n^3}{2}$

2. The area of a square is greater than or equal to 95 square centimeters. Let s represent the length in cm of one side of the square. $s^2 \geq 95$

Suggested Assignments

Minimum
  41/1-23 odd
R 43/Self-Test 3

Average
  41/5-33 odd
R 43/Self-Test 3

Maximum
  41/5-33 odd, 35-38
R 43/Self-Test 3

1–9 Words into Symbols: Sentences

Just as you can translate a word phrase into a mathematical expression, you also can translate a *word sentence* into a mathematical sentence.

EXAMPLE 1 Write a mathematical sentence for each word sentence.

  a. The sum of seven and twice a number n is twenty-five.

  b. The quotient of a number y divided by three is greater than four times y, decreased by eleven.

SOLUTION **a.** $7 + 2n = 25$

  b. $\dfrac{y}{3} > 4y - 11$

A word sentence may involve an unknown number or quantity without specifying a variable. In such cases, you may choose any variable to represent the number. Then write a mathematical sentence that relates the facts in the given situation.

EXAMPLE 2 Write a mathematical sentence that represents the given information.

  a. The sum of a number and forty-nine is ninety-three.

  b. The perimeter of a square is less than or equal to sixty centimeters.

SOLUTION **a.** Let n represent the unknown number.

    $n + 49 = 93$

  b. Let s represent the length of one *side* of the square.

    $4s \leq 60$

Oral Exercises

Translate each word sentence into a mathematical sentence.

1. The sum of five and nineteen is twenty-four. $5 + 19 = 24$

2. The difference when seven is subtracted from thirty-two is greater than twenty. $34 - 7 > 20$

3. The product of a number a and three is less than or equal to forty-seven. $3a \leq 47$

4. The total of five and a number b is thirty-nine. $5 + b = 39$

5. The quotient when a number c is divided by six is two less than the product of c and six. $\dfrac{c}{6} = 6c - 2$

6. Fifteen less than twice a number d is greater than fifty more than one half d. $2d - 15 > 50 + \dfrac{1}{2}d$

7. Thirty-five is not the product when the number that is one greater than a number e is multiplied by the number that is one less than e.

8. The quotient when the sum of a number f and twelve is divided by seven is sixteen more than f. $\frac{f + 12}{7} = 16 + f$

9. The cube of the difference when a number h is subtracted from a number g is greater than the sum of g and h. $(g - h)^3 > g + h$

10. The square of the sum of a number j and a number k is not equal to the sum of the square of j and the square of k. $(j + k)^2 \neq j^2 + k^2$

7. $35 \neq (e + 1)(e - 1)$

Translate each mathematical sentence into a word sentence.

11. $r - 3 = 10$ 12. $4s = 12$ 13. $22 < t + 8$ 14. $\frac{u}{5} \geq 5$

15. $7 - 2v \neq v^2$ 16. $\frac{5}{w} = \frac{1}{2}w$ 17. $8(9 - x) = 4$ 18. $(6 + y)^3 \neq 6 + y^3$

Written Exercises

Write a mathematical sentence for each word sentence.

A
1. The total cost of x tickets at four dollars per ticket is seventy-six dollars. $4x = 76$

2. After a deposit of y dollars in an account containing fifty dollars, the new balance is ninety-five dollars. $y + 50 = 95$

3. A customer used a twenty-dollar bill to pay for an item that cost z dollars and received less than four dollars in change. $20 - z < 4$

4. A thousand-dollar profit was divided among w partners, and each partner received more than three hundred dollars. $\frac{1000}{w} > 300$

5. The winner of the election received twenty-one votes, which is one more than two thirds of the total number, v, of votes cast. $21 = 1 + \frac{2}{3}v$

6. In today's game the team scored nine runs, which is one less than twice the number of runs, r, that they scored in yesterday's game. $9 = 2r - 1$

7. Mark is m years old, and his age in eight years will be three times his present age. $m + 8 = 3m$

8. Donna is d years old, and her age one year ago was half her age six years from now. $d - 1 = \frac{1}{2}(d + 6)$

9. The area of a rectangle that is six meters long and w meters wide is twenty-one square meters. $6w = 21$

10. The perimeter of an equilateral triangle in which each side measures s centimeters is not fifty-four centimeters. $3s \neq 54$

11. The total value of d dimes and q quarters is not ninety cents. $10d + 25q \neq 90$

12. The total value of p pennies and n nickels is one dollar forty-six cents. $0.01p + 0.05n = 1.46$

Numbers and Variables   **41**

13. x; $2x + 2 = 42$

14. y; $\frac{1}{3}y - 8 = 4$

15. z; $3z + z = 32$

16. n; $n + (n + 5) > 25$

17. b; $\frac{b}{5} < 6$

18. c; $c - 5 = 13$

19. s; $4s = 96$

20. s; $s^2 > 100$

21. n; $5n < 90$

22. q; $0.25q = 4.25$

23. n; $n + (n + 1) = 139$

24. m; $(m - 1)m = 650$

25. Let x = the length in centimeters of the shorter piece of balsa wood. $x + (x + 4) = 14$

26. Let l = the number of losses. $5l + l + 1 = 21$

27. Let w = the width in centimeters of the rectangle. $w(w + 3) < 45$

28. Let l = the length in meters of the rectangle. $2l + 2(2l - 4) = 52$

29. Let a = the measure in centimeters of the adjacent side of the parallelogram.
$2a + 2\left(\frac{1}{3}a - 1\right) = 22$

30. Let b = the measure in millimeters of the base of the triangle.
$\frac{1}{2}b(4b) = 32$

31. Let j = Jack's age now.
$\frac{1}{2}(j + 20) = 2(j - 1)$

32. Let d = Donna's age now. $d + 8 = 3d$

33. Let q = the number of quarters Carlos has.
$0.10(q - 5) + 0.25q = 7.55$

Choose a variable to represent the quantity described in parentheses. Then write a mathematical sentence that represents the given information.

13. Two more than twice a number is forty-two. (unknown number)

14. Eight less than one third a number is four. (unknown number)

15. One number is three times a second number, and their sum is thirty-two. (second number)

16. One number is five less than a second number, and their sum is greater than twenty-five. (first number)

17. When the cost of a new basketball is shared equally by five friends, each friend pays less than six dollars. (cost of the basketball)

18. When you receive a five-dollar discount on the original cost of a hair dryer, you pay only thirteen dollars. (original cost of the hair dryer)

19. The perimeter of a square is ninety-six centimeters. (length of one side of the square)

20. The area of a square is greater than one hundred square meters. (length of one side of the square)

21. The total value of a number of nickels is less than ninety cents. (number of nickels)

22. The total value of a number of quarters is four dollars twenty-five cents. (number of quarters)

23. The sum of a positive integer and the next greater positive integer is one hundred thirty-nine. (first integer)

24. The product of a positive integer and the next greater positive integer is six hundred fifty. (greater integer)

Choose a variable to represent one of the unknown quantities. Describe the quantity that the variable represents, then write a mathematical sentence that represents the given information.

B 25. When a strip of balsa wood that is fourteen centimeters long is cut into two pieces, one piece is four centimeters shorter than the other.

26. In a season of twenty-one games, the team won five times as many games as it lost, and one game ended in a tie.

27. The area of a rectangle whose length is three centimeters greater than its width is less than forty-five square centimeters.

28. The perimeter of a rectangle whose width is four meters less than twice its length is fifty-two meters.

29. The perimeter of a parallelogram in which one side is one centimeter shorter than one third the measure of the adjacent side is twenty-two centimeters.

30. The area of a triangle whose height is four times as great as the measure of its base is thirty-two square millimeters.

31. Half Jack's age twenty years from now will be the same as twice his age one year ago.

32. Eight years from now, Donna will be three times as old as she is now.

33. Carlos has five fewer dimes than quarters, and the total value of his dimes and quarters is $7.55.

34. If Leah had twelve more nickels, the total value of her nickels would be three times as great as the total value of the nickels she has now.

Choose a variable to represent one of the unknown quantities and write an open sentence that represents the given information. Then solve the open sentence over $\mathcal{R}$.

C 35. There are three more boys than girls in a class of twenty-five students.

36. Of twelve birds at the bird feeder, there are twice as many sparrows as robins.

37. When the product of five and a number is decreased by four, the result is the same as the result when the number is increased by eight.

38. When the quotient of a number divided by four is increased by five, the result is the same as the result when twice the number is decreased by nine.

34. Let n = the number of nickels Leah has.
$5(n + 12) = 3(5n)$

35. Let g = the number of girls. $g + (g + 3) = 25$, {11}

36. Let r = the number of robins. $2r + r = 12$, {4}

37. Let n = the number.
$5n - 4 = n + 8$, {3}

38. Let x = the number.
$\frac{x}{4} + 5 = 2x - 9$, {8}

▪ Self-Test 3

VOCABULARY mathematical sentence (p. 30)   solution set (p. 31)
equation (p. 30)   satisfy (p. 31)
side of an equation (p. 30)   solve an open sentence (p. 31)
inequality (p. 30)   graph of an open sentence
side of an inequality (p. 30)   (p. 32)
open sentence (p. 31)   set-builder notation (p. 32)
solution or root of an open
   sentence (p. 31)

Solve each open sentence if $x \in$ {the positive real numbers} and graph its solution set.

1. $3x - 5 = 7$ 2. $x + 2 > 8$ 3. $x < x + 3$ *Obj. 1, p. 30*

Write a variable expression for each word phrase.

4. the quotient when the cube of the sum of a number m and one is   *Obj. 2, p. 30*
divided by the sixth power of m

5. the total value in cents of d dimes and x dollars

Numbers and Variables   **43**

Additional Answers
Self-Test 3

1. {4}
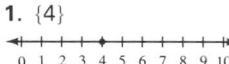

2. $\{x: x > 6\}$
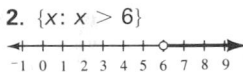

3. $\{x: x > 0\}$
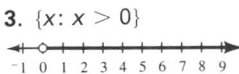

4. $\dfrac{(m + 1)^3}{m^6}$

5. $10d + 100x$

6. $\left(\dfrac{n}{3}\right)^2 < \dfrac{n^2}{3}$

7. $\dfrac{1}{2}(b + 14) = 2(b - 10)$

Quick Quiz

Solve each open sentence if $x \in$ {the positive real numbers} and graph its solution set.

1. $5x - 2 = 13$ {3}

+---+---+---+---+---+---+---+---+
 ⁻3 ⁻2 ⁻1   0   1   2   3   4

2. $x + 3 < 11$ $\{x: 0 < x < 8\}$

+---+---+---+---+---+---+---+---+---+---+
 ⁻1 0   1   2   3   4   5   6   7   8   9

3. $x < x + 7$ $\{x: x > 0\}$

+---+---+---+---+---+---+---+---+
 ⁻2 ⁻1   0   1   2   3   4   5

Write a variable expression for each word phrase.

4. the product of the square of p and the cube of p
$p^2 \times p^3$

5. the total value of q quarters and p pennies
$25q + p$

Write a mathematical sentence that represents the given information.

6. The cube of the quotient when a number t is divided by four is greater than the quotient when the cube of t is divided by

four. $\left(\dfrac{t}{4}\right)^3 > \dfrac{t^3}{4}$

7. Marguerite is c years old now, and one third her age twenty years from now will be double her age ten years ago.
$\dfrac{1}{3}(c + 20) = 2(c - 10)$

Common Errors

Students will often interpret the phrase "five less than p" as "$5 - p$" rather than "$p - 5$." Encourage them to substitute numerical examples (What is 5 less than 12?) to test.

Write a mathematical sentence that represents the given information.

6. The square of the quotient when a number n is divided by three is less than the quotient when the square of n is divided by three. *Obj. 3, p. 30*

7. Bill is b years old now, and half his age in fourteen years will be the same as twice his age ten years ago.

Check your answers with those at the back of the book.

Chapter Summary

1. Any number that is either a positive number, a negative number, or zero is called a *real number*. The real numbers can be pictured as points on a *number line*, thereby showing their order. On a number line, the point that corresponds to a number is called the *graph of the number*, and the number that corresponds to a point is called the *coordinate of the point*.

2. A *set* is a collection of objects that are called the *members*, or *elements*, of the set. Some of the commonly used set symbols include { } (indicates a set), $\in$ (indicates that something is a member of a set), = (indicates that two sets are equal), $\subset$ (indicates that one set is a subset of another), and $\emptyset$ (the *empty set*, or *null set*). If the members of two sets can be paired so that each member of each set is assigned to one and only one member of the other set, the relationship between the sets is said to be a *one-to-one correspondence*.

3. A *rational number* is any number that can be expressed as the quotient of two integers. All other real numbers are called *irrational numbers*.

4. A *numerical expression*, or *numeral*, is a name for a number. The number is called the *value of the expression*. You *simplify* a numerical expression when you replace it with the simplest, or most common, name of its value.

5. A *variable* is a symbol that is used to represent one or more numbers. The set of numbers that a variable may represent is called the *domain*, or *replacement set*, of the variable. An expression that contains a variable is called a *variable expression*. You *evaluate* a variable expression when you replace each variable with one of its. values and simplify the resulting numerical expression.

6. When simplifying or evaluating an expression, *grouping symbols* indicate which operations are to be performed first. After the operations within grouping symbols have been performed, you continue with the *order of operations* outlined on page 25.

7. The product $7 \times 7 \times 7 \times 7$ may be written 7^4. In the expression 7^4, 7 is called the *base* and 4 is called the *exponent*.

44   *Chapter 1*

8. An *equation* is a mathematical sentence which states that two expressions name the same number. An *inequality* is a mathematical sentence that is formed by placing an inequality symbol between two expressions. If a mathematical sentence contains at least one variable, it is called an *open sentence*. The set of all values of the variables for which an open sentence is true is called the *solution set* of the open sentence over the given domain.

Chapter Review

Write the letter of the correct answer.

1. What is the coordinate of the point on a number line that is five sixths of the way from $^-5$ to 7? *1-1*
   a. $^-3$ b. $^-1$ c. 0   d.) 5

2. Which of the following statements is false?
   a. $^-4 < 3$ b. $0 > ^-9.5$ c.) $^-7 > ^-5 > 1$ d. $^-9 < ^-2 < 0$

3. Let $A = \{0, 1\}$ and $B = \{0, 1\}$. Which of the following statements is false? *1-2*
   a. $A \subset B$ b.) $B \in A$ c. $\emptyset \subset A$ d. $0 \in B$

4. Let $S = \{x, y\}$. What is the total number of subsets of S?
   a. 1   b. 2   c. 3   d.) 4

5. Which set of numbers is graphed on the number line below? *1-3*

   a.) {the integers greater than $^-3$}
   b. {the rational numbers greater than $^-3$}
   c. {the real numbers greater than or equal to $^-2$}
   d. {the integers between $^-3$ and 6}

6. Specify {the whole numbers between $^-30$ and 3} by roster.
   a. $\{^-29, ^-28, ^-27, ^-26, \ldots, 2\}$ b. $\{^-29, ^-28, ^-27, ^-26, \ldots, 0\}$
   c.) $\{0, 1, 2\}$ d. $\{1, 2\}$

7. Simplify 2.5×3.24. *1-4*
   a. 5.74   b. 3.49   c. 0.081   d.) 8.1

8. Evaluate the expression $10 - \dfrac{j}{k}$ when $j = \dfrac{2}{3}$ and $k = \dfrac{1}{2}$.
   a. 7   b.) $8\dfrac{2}{3}$ c. $9\dfrac{1}{3}$ d. $9\dfrac{2}{3}$

Numbers and Variables **45**

Write a mathematical sentence for each word sentence.

1. Seventeen is greater than the product of 3 and 5. $17 > 3 \times 5$

2. Nine is greater than 4 but less than g. $4 < 9 < g$

3. The sum of 3 and 9 is equal to the product of 6 and 2. $3 + 9 = 6 \times 2$

4. The difference of Jim's age r and his mother's age t is equal to the square of 5. $t - r = 5^2$

Represent the answer to each question in terms of the given variable.

5. Corn costs z cents per ear. What is the total cost of a dozen ears? $12z$ cents

6. The area of a square is $4s^2$. What is the length of each side? $2s$

9. Simplify $\{6[24 - 6(3)] + 4\}2$. *1-5*
   a. 60   (b.) 80   c. 120   d. 656

10. Evaluate the expression $ab - \dfrac{a}{b}$ when $a = 1.5$ and $b = 3$.
    a. 44.5   b. 43   (c.) 4   d. 2.5

11. Simplify $64 \div 4 + 4(10 - 2^2)$.   *1-6*
    (a.) 40   b. 48   c. 264   d. 512

12. Evaluate the expression $n(n + 1)^2 - n^3$ when $n = 0.1$.
    a. 0.0011   b. 0.003   c. 0.111   (d.) 0.12

13. Solve $z^2 = z + 2$ *if* $z \in \{$the whole numbers$\}$.   *1-7*
    a. $\{0\}$ b. $\{0, 2\}$ (c.) $\{2\}$ d. $\varnothing$

14. Solve $y \leq {}^-4$ if $y \in \{$the positive integers$\}$.
    a. $\{\ldots, {}^-8, {}^-7, {}^-6, {}^-5, {}^-4\}$ b. $\{{}^-4, {}^-3, {}^-2, {}^-1, 0, \ldots\}$
    c. $\{{}^-4, {}^-3, {}^-2, {}^-1\}$ (d.) $\varnothing$

15. Translate the following word phrase into a variable expression: "the quotient when the sum of a number m and five is divided by the sum of five and the square of m"   *1-8*
    (a.) $\dfrac{m + 5}{5 + m^2}$ b. $\dfrac{m + 5}{(5 + m)^2}$ c. $\dfrac{m + 5}{5 + 2m}$ d. $\dfrac{m + 5}{5^2 + m^2}$

16. Translate the following word sentence into an open sentence: "The sum of a positive odd integer n and the next greater positive odd integer is eighty-eight."   *1-9*
    a. $n(n + 1) = 88$ b. $n(n + 2) = 88$
    c. $n + (n + 1) = 88$ (d.) $n + (n + 2) = 88$

Chapter Test

1. On a number line, what is the coordinate of the point that is one fourth of the way from $^-3$ to 7? $2\frac{1}{2}$ *1-1*

2. List these numbers in order from least to greatest:
   $4, {}^-2, {}^-1\frac{1}{2}, 3.25, 0, {}^-2.75$ $^-2.75, {}^-2, {}^-1\frac{1}{2}, 0, 3.25, 4$

Tell whether each statement is true or false.

3. $0 \notin \{$the positive real numbers$\}$ true   *1-2*
4. $\{10, 20, 30, 40, 50\} \not\subset \varnothing$ true
5. $\{$the natural numbers$\} = \{$the positive integers$\}$ true   *1-3*
6. $\{$the rational numbers$\} \subset \{$the irrational numbers$\}$ false

Evaluate each expression when $x = 4$ and $y = \frac{1}{2}$.

7. $5xy$ 10

8. $\dfrac{x}{y}$ 8

9. $\dfrac{x + y}{3}$ $1\frac{1}{2}$

1-4

Simplify.

10. $\{2[4 + 6(10)] \div 4\}5$ 160

11. $2\left\{\dfrac{6[16 - 2(4)]}{6 + 3(2)}\right\} - 8$ 0

1-5

12. $8 + (9 - 5)10 \div 2(7 + 3)$ 10

13. $3(5)^2 + (5 + 2^2)4 \div 2$ 93

1-6

Solve each open sentence if $z \in \{$the whole numbers$\}$.

14. $\dfrac{z}{2} = 2z$ {0}

15. $z < 3.2$ {0, 1, 2, 3}

16. $3z + 1 \geq 16$
{5, 6, 7, . . .}

1-7

Write a variable expression for each word phrase.

17. the total number of hours in d days 24d

1-8

18. the total value in cents of x dollars and q quarters 100x + 25q

Write an open sentence for each word sentence.

19. The square of the sum of a number a and a number b is greater than the quotient when the product of a and b is divided by the sum of a and b. $(a + b)^2 > \frac{ab}{a + b}$

1-9

20. Tina is t years old, and her age five years ago was one fourth her age sixteen years from now. $t - 5 = \frac{1}{4}(t + 16)$

Contest Problems

The following are mathematical problems with a strong emphasis on logical reasoning. They are similar to problems that you might encounter in a mathematical contest or competition.

1. Jim knows that x is an integer greater than 2 and less than 8. Kim knows that x is an integer less than 9 and greater than 4. Tim knows that x is an integer greater than or equal to 2 and less than or equal to 6. How many possible values are there for x? 2 values

2. Starting with a certain number, Jane divided by 2 and then divided by 3. Her answer was $\frac{1}{15}$. However, she should have multiplied by 2 and then multiplied by 3. What is the correct answer? $2\frac{2}{5}$

3. Let a, b, and c be positive real numbers. If $ab = 48$, $bc = 96$, and $ac = 72$, what is the value of abc? 576

4. Let $x \in \{1, 2, 3, . . . , 20\}$. For how many values of x will the value of the expression $\dfrac{3 + x}{5} + \dfrac{12}{x}$ be an integer? 3 values

Numbers and Variables **47**

The icicles hanging from the trees in this photo were formed when the air temperature dropped to 0°C, or below. The body of water in the photo has maintained a higher temperature, however, and has remained unfrozen. Recorded temperatures in the United States have ranged from ⁻62°C to 57°C.

Chapter 2

Working with Real Numbers

Basic Assumptions and Definitions

OBJECTIVES for Sections 2-1 through 2-4:
1. *To determine whether a statement that contains a quantifier is true or false.*
2. *To use basic axioms of addition, multiplication, and equality to simplify expressions.*
3. *To simplify expressions and solve open sentences involving opposites and absolute values.*

2–1 Using Quantifiers

As you have seen, an open sentence such as

$$x + 3 = 3 + x$$

is by itself neither true nor false. If a domain is specified for the variable in an open sentence, however, you can determine whether the sentence is true for any particular value of the variable from that domain. In the open sentence just given, if the domain of x is specified as $\mathcal{R}$, you obtain a true statement when you replace x with *any* value from its domain.

Working with Real Numbers **49**

Directional Reasoning
The Problems on pages 75–76 in Section 2-6 involve quantities that can be expressed as signed numbers, which are then combined. Problems 11–17 on page 76 ask the student to retrace the steps given in the problem to determine the point at which the event began. Students need to add the opposite of each number given to return to the beginning of the event.
Interpreting Answers
In Section 2-8, the Problems on pages 86–87 ask the students to compute differences between positive and negative numbers. The students must deduce from the meaning of the problem whether a quantity should be considered positive or negative. They will also need to interpret their answers to see whether they *make sense*.

Teaching Suggestions
p. T72

Related Activities p. T73

Key Ideas

Determine whether a statement that contains a quantifier is true or false.

Tell whether each statement is true or false.

1. There exists an integer n such that $2n + 5 = 9$.
   True, since $2(2) + 5 = 9$ is true.

2. For all real numbers x, $x + 3 < 7$.
   False; for example, $4 + 3 < 7$ is false.

Find a value of the variable that makes the statement true.

3. For some whole number w, $w = 2w$. The statement $0 = 2(0)$ is true. Therefore, the statement is true when $w = 0$.

Mixed Review

Write a variable expression for each of the following.

1. seven more than three times a number a $3a + 7$

2. twenty-one decreased by one-half of a number b
   $21 - \frac{1}{2}(b)$

3. the number of seconds in m minutes $60m$

4. The total value of p pennies and n nickels is greater than 14. $p + 5n > 14$

5. When a piece of lumber 3 m long is cut into 2 pieces, one piece is twice the length of the other.
   $2w + w = 3$

Minimum
   51/1–12
Average
   51/1–20
Maximum
   51/1–22

For example, each of the following is a true statement.

$$7 + 3 = 3 + 7$$
$$\frac{1}{5} + 3 = 3 + \frac{1}{5}$$
$$\sqrt{2} + 3 = 3 + \sqrt{2}$$

Thus the following statement is also true.

> For any real number x, $x + 3 = 3 + x$.

Other ways to state this same assertion are the following:

> For each real number x, $x + 3 = 3 + x$.
> For every real number x, $x + 3 = 3 + x$.
> For all real numbers x, $x + 3 = 3 + x$.

Now consider the following statement.

> For each integer y, $y + 2 > 3$.

This statement is certainly false, since you obtain a false statement when you replace y with the integer 1:

$$1 + 2 > 3 \text{ is false.}$$

When you replace y with the integer 4, however, you obtain a *true* statement:

$$4 + 2 > 3 \text{ is true.}$$

Because there is *some* integer y for which "$y + 2 > 3$" is a true statement, the following statement is also true.

> For some integer y, $y + 2 > 3$.

Other ways to state this same assertion are the following:

> There is an integer y such that $y + 2 > 3$.
> There exists an integer y such that $y + 2 > 3$.
> For at least one integer y, $y + 2 > 3$.

Words and phrases such as *any, each, every, all, some, there is, there exists,* and *at least one* are used to convey the idea of *how many,* or *quantity.* For this reason, when such a word or phrase is used in combination with a variable in an open sentence, the word or phrase is called a **quantifier.**

EXAMPLE 1   Tell whether each statement is true or false.
   a. There exists a real number m such that $3m + 7 = 16$.
   b. For some real number r, $r + 1 < r + 2$.

SOLUTION   a. True, since $3(3) + 7 = 16$ is true.
   b. True; for example, $7 + 1 < 7 + 2$ is true.

50   *Chapter 2*

EXAMPLE 2 Find a value of the variable that makes the statement true:
For some whole number w, $2 + w = 2w$.

SOLUTION The statement $2 + 2 = 2(2)$ is true.
Therefore, the statement is true when $w = 2$.

Oral Exercises

Tell whether each statement is true or false.

1. There exists a real number a such that $a < 0$. True
2. There is a natural number b such that $b < 0$. False
3. For any whole number w, $w > 0$. False
4. For any negative integer z, $0 > z$. True
5. For at least one positive integer m, $m + 1 > 0$. True
6. For every integer n, $n(n + 1) > 0$. False
7. For some integer c, $c^4 = 0$. True
8. For all positive integers d, $d^2 > d$. True
9. There is a positive integer p such that $2p = p^2$. True
10. For some positive integer q, $3q = 10$. False
11. For each positive integer x, $3x > x$. True
12. For any real number y, $3y > y$. False

Written Exercises

Find a value of the variable that makes each statement true.

A
1. There is a real number u such that $3u + 1 = 16$. $u = 5$
2. For at least one rational number v, $5v = 4$. $v = \frac{4}{5}$
3. For some whole number j, $j + 1 = 2j$. $j = 1$
4. There exists a natural number k such that $k(k - 1) = 0$. $k = 1$
5. For at least one whole number s, $6 > 2s + 5$. $s = 0$
6. There is an integer t such that $14 \leq 3 + 2t \leq 15$. $t = 6$

Find a value of the variable that makes each statement false.

7. For all real numbers p, $p^2 > 0$ $p = 0$
8. For each positive integer q, $q < q^3$. $q = 1$
9. For any natural number a, $a^2 + a > 2$. $a = 1$
10. For any whole number b, $0 < b^2 - b$. $b = 0$ or $b = 1$
11. For each integer m, $5m \neq \frac{m}{5}$. $m = 0$
12. For every real number n, $2n \neq n^2$. $n = 0$ or $n = 2$

Working with Real Numbers **51**

Additional Answers
Written Exercises

(continued)

20. For each positive real number x, there exists a positive real number y such that $\frac{x}{y} = y$.

21. 7. For some real number p, $p^2 > 0$.
 8. There is a positive integer q such that $q < q^3$.
 9. There exists a natural number a such that $a^2 + a > 2$.
 10. For at least one whole number b, $0 < b^2 - b$.
 11. For some integer m, $5m \neq \frac{m}{5}$.
 12. There is a real number n, such that $2n \neq n^2$.

22. 1. For any real number u, $3u + 1 = 16$.
 2. For each rational number v, $5v = 4$.
 3. For every whole number j, $j + 1 = 2j$.
 4. For all natural numbers k, $k(k - 1) = 0$.
 5. For any whole number s, $6 > 2s + 5$.
 6. For each integer t, $14 \leq 3 + 2t \leq 15$.

Teaching Suggestions
p. T73

Key Ideas

Use the basic axioms of closure, commutativity, associativity, and equality to simplify expressions.

Rewrite each statement, using a variable expression to represent the relationship between y and x.

EXAMPLE   For every natural number x, there is a natural number y that is one less than twice x.

SOLUTION   For every natural number x, there is a natural number y such that $y = 2x - 1$.

B 13. For any real number x, there is a real number y such that y is three greater than x.
14. For each real number x, there exists a real number y such that y is one fourth x.
15. For every whole number x, there exists a whole number y that is five more than three times x.
16. For any natural number x, there is a natural number y that exceeds x by one.
17. For all real numbers x and y, the sum when x is increased by y is the same as the sum when y is increased by x.
18. There is a real number y such that, for any real number x, the sum of x and y is x.
19. There exists a nonzero real number y such that, for each real number x, the quotient when x is divided by y is x.
20. For each positive real number x, there exists a positive real number y such that the quotient when x is divided by y is y.

C 21. Rewrite each of the statements in Written Exercises 7–12, changing the quantifier so that the statement becomes true. Answers may vary. Examples are given.
22. Rewrite each of the statements in Written Exercises 1–6, changing the quantifier so that the statement becomes false. Answers may vary. Examples are given.

2-2 Basic Assumptions: Axioms for the Real Numbers

In working with real numbers, you use two basic operations, *addition* and *multiplication*. Each of these operations is called a **binary operation** because it pairs any *two* real numbers with a third real number.

The operation of addition pairs any two real numbers a and b with a real number that is called their **sum**. The sum is denoted as $a + b$, and in this sum the real numbers a and b are called **terms**. Multiplication pairs any two real numbers a and b with a real number that is called their **product**. The product may be denoted as ab, and in this product the real numbers a and b are called **factors**.

The rules that you use when adding or multiplying real numbers are all based on a few basic statements, called **axioms** or **postulates,** that are *assumed* to be true. For example, in describing the sum and the product of two real numbers, it was assumed that the result of adding or multiplying two real numbers is always a real number. That is, the set of real numbers was assumed to be *closed* under the operations of addition and multiplication. Furthermore, it was assumed that any two real numbers have one and only one sum and one and only one product. In other words, the sum and the product of two real numbers were each assumed to be *unique*. These assumptions are stated formally as the following axioms.

Axioms of Closure

For all real numbers a and b:

$a + b$ is a unique real number.

ab is a unique real number.

There are other axioms for the real numbers that are general statements of familiar properties of addition and multiplication. For example, you know that when you add or multiply two real numbers, you get the same sum or product no matter what order you use in performing the operation:

$$5 + 12 = 12 + 5 \quad \text{and} \quad 5 \times 12 = 12 \times 5$$

This fact can be stated formally as follows.

Commutative Axioms

For all real numbers a and b:

$a + b = b + a$

$ab = ba$

Since addition and multiplication are both binary operations, expressions such as $2 + 3 + 5$ and $2 \times 3 \times 5$, in which *more than* two numbers are to be added or multiplied, must be defined. The following pattern is used in defining sums and products of three or more real numbers.

$a + b + c = (a + b) + c,$ $abc = (ab)c,$

$a + b + c + d = (a + b + c) + d,$ $abcd = (abc)d,$

$a + b + c + d + e = (a + b + c + d) + e,$ $abcde = (abcd)e,$

and so on.

1. Simplify.

a. $28 + 160 + 72 + 140$
$= (28 + 72) + (160 + 140)$
$= 100 + 300$
$= 400$

b. $\frac{1}{3} \times 20 \times 18 \times \frac{1}{2}$

$= \left(\frac{1}{3} \times 18\right) \times \left(20 \times \frac{1}{2}\right)$

$= 6 \times 10$

$= 60$

c. $(5m)(14m)$
$= (5 \cdot 14)(m \cdot m)$
$= 70m^2$

2. Solve over $\mathcal{R}$.
$8 + (b + 27) = 8 + (27 + b)$
By the commutative axiom for addition, $8 + (b + 27) = 8 + (27 + b)$ for every real value of b.
$\therefore$ the solution set is $\mathcal{R}$.

3. Name the axiom of equality that is illustrated.
If $-3 + 9 = 6$ and $6 = 10 - 4$, then $-3 + 9 = 10 - 4$.
Transitive Axiom

When you add or multiply three or more real numbers, you get the same sum or product no matter how you group, or *associate*, the numbers. For example:

$$(19 + 47) + 23 = 19 + (47 + 23) \quad \text{and} \quad (9 \times 2) \times 5 = 9 \times (2 \times 5)$$

The following axiom is a formal statement of this fact.

Associative Axioms

For all real numbers a, b, and c:

$$(a + b) + c = a + (b + c)$$
$$(ab)c = a(bc)$$

The commutative and associative axioms permit you to add or multiply numbers *in any order* and *in any groups of two*. Thoughtful use of these axioms can sometimes help you in simplifying numerical expressions.

EXAMPLE 1   Simplify.

        a. $37 + 2\frac{1}{3} + 23 + 5\frac{2}{3}$ **b.** $\frac{2}{3} \times 4 \times 90 \times 75$

SOLUTION   **a.** $37 + 2\frac{1}{3} + 23 + 5\frac{2}{3} = (37 + 23) + \left(2\frac{1}{3} + 5\frac{2}{3}\right)$

$$= 60 + 8$$
$$= 68$$

        b. $\frac{2}{3} \times 4 \times 90 \times 75 = \left(\frac{2}{3} \times 90\right)(4 \times 75)$

$$= (60)(300)$$
$$= 18{,}000$$

Since variables represent real numbers, you can also use the commutative and associative axioms in simplifying variable expressions and in solving open sentences.

EXAMPLE 2   Simplify.

        a. $9 + 6m^3 + 14$ **b.** $(7n)(13n)$

SOLUTION   **a.** $9 + 6m^3 + 14 = 6m^3 + (9 + 14) = 6m^3 + 23$

        b. $(7n)(13n) = (7 \cdot 13)(n \cdot n) = 91n^2$

EXAMPLE 3   Solve each open sentence over $\mathcal{R}$.

        a. $(p + 97) + 86 = p + (97 + 86)$

        b. $8 \cdot 15q < 15 \cdot 8q$

SOLUTION **a.** By the associative axiom for addition,
$(p + 97) + 86 = p + (97 + 86)$ for every real value of p.
Therefore, the solution set is $\mathcal{R}$.

b. By the commutative axiom for multiplication,
$8 \cdot 15q = 15 \cdot 8q$ for every real value of q.
Therefore, the solution set is $\emptyset$.

In Example 3, notice that the domain of the variables was specified as $\mathcal{R}$. *Throughout the rest of this book, the domain of all variables is $\mathcal{R}$ unless otherwise specified.*

The way that the symbol $=$ is used in mathematical sentences is consistent with the following axioms.

Axioms of Equality
For all real numbers a, b, and c:

Reflexive Axiom	$a = a$
Symmetric Axiom	If $a = b$, then $b = a$.
Transitive Axiom	If $a = b$ and $b = c$, then $a = c$.

EXAMPLE 4 Name the axiom of equality that is illustrated.
a. If $8 = x$, then $x = 8$.
b. If $2 + 5 = 7$ and $7 = 4 + 3$, then $2 + 5 = 4 + 3$.
c. If $a = b$, $b = c$, and $c = d$, then $a = d$.

SOLUTION **a.** The symmetric axiom   **b.** The transitive axiom
c. The transitive axiom, in two steps:
Step 1. If $a = b$ and $b = c$, then $a = c$.
Step 2. If $a = c$ and $c = d$, then $a = d$.

Oral Exercises

Name the axiom that is illustrated.

1. $4 + (^-5)$ is a real number.

2. $^-3(^-7)$ is a real number.

3. $2 \times (50 \times 64) = (2 \times 50) \times 64$

4. $3\frac{1}{5} + \left(6\frac{4}{5} + 2\frac{3}{8}\right) = \left(3\frac{1}{5} + 6\frac{4}{5}\right) + 2\frac{3}{8}$

5. $m \cdot 8 = 8 \cdot m$

6. $(n + 12) + 29 = n + (12 + 29)$

7. $\frac{1}{2}(6x) = \left(\frac{1}{2} \cdot 6\right)x$

8. $51 + y = y + 51$

9. $(38 + p) + 46 = (p + 38) + 46$

10. $(25 + 7) + q = q + (25 + 7)$

11. If $35 = 3z$, then $3z = 35$.

12. If $a = 2$ and $2 = b$, then $a = b$.

Working with Real Numbers **55**

Suggested Assignments

Minimum
  56/1–24
Average
  56/5–17 odd, 19–32
Maximum
  56/5–35 odd, 36–39

Additional A Exercises

Simplify.

1. $320 + 76 + 80 + 24$ 500

2. $18 + 579 + 182 + 21$ 800

3. $5 \times 14 \times 20 \times 3$ 4200

4. $25 \times 30 \times 40 \times 15$
    450,000

5. $\frac{1}{7} + \frac{1}{5} + \frac{6}{7} + \frac{4}{5}$ 2

6. $(3y)(2y)$ $6y^2$

Additional Answers
Written Exercises

29. a. 9
   b. True
   c. Commutative
   d. Associative

30. a. 24
   b. True
   c. Commutative
   d. Associative

31. a. 13
   b. True
   c. Not commutative
   d. Not associative

Name the axiom that justifies each lettered step. A check ($\checkmark$) indicates that the step is justified by the substitution principle. Note that the transitive axiom of equality is also used in writing each step after the first.

13.
$$
\begin{aligned}
(79 + 57) + 21 &= (57 + 79) + 21 \\
&= 57 + (79 + 21) \\
&= 57 + 100 \\
&= 157
\end{aligned}
$$
  (a) Commutative axiom for addition
  (b) Associative axiom for addition
  $\checkmark$
  $\checkmark$

14.
$$
\begin{aligned}
(v + 9) + (w + 8) &= v + [9 + (w + 8)] \\
&= v + [9 + (8 + w)] \\
&= v + [(9 + 8) + w] \\
&= v + (17 + w) \\
&= v + (w + 17) \\
&= (v + w) + 17
\end{aligned}
$$
  (a) Associative axiom for addition
  (b) Commutative axiom for addition
  (c) Associative axiom for addition
  $\checkmark$
  (d) Commutative axiom for addition
  (e) Associative axiom for addition

15.
$$
\begin{aligned}
(125r)(8s) &= [(125r)8]s \\
&= [(r \cdot 125)8]s \\
&= [r(125 \cdot 8)]s \\
&= (r \cdot 1000)s \\
&= (1000 \cdot r)s \\
&= 1000(rs)
\end{aligned}
$$
  (a) Associative axiom for multiplication
  (b) Commutative axiom for multiplication
  (c) Associative axiom for multiplication
  $\checkmark$
  (d) Commutative axiom for multiplication
  (e) Associative axiom for multiplication

Written Exercises

Simplify.

A

1. $540 + 37 + 60 + 43$ 680

2. $9 + 38 + 912 + 4391$ 5350

3. $5 \times 23 \times 3 \times 2$ 690

4. $3 \times 4 \times 9 \times 75$ 8100

5. $\frac{1}{2} \times 45 \times 24 \times \frac{1}{5}$ 108

6. $\frac{2}{3} \times \frac{5}{11} \times 44 \times 45$ 600

7. $8.7 + 5.9 + 1.3 + 0.1$ 16

8. $2.24 + 3.3 + 4.51 + 17.76$ 27.81

9. $3\frac{1}{5} + 7\frac{1}{2} + 1\frac{4}{5} + 3\frac{1}{2}$ 16

10. $98\frac{3}{8} + 4\frac{1}{9} + 11\frac{4}{9} + 1\frac{5}{8}$ $115\frac{5}{9}$

11. $\frac{9}{2} \times \frac{3}{7} \times \frac{2}{9} \times 49$ 21

12. $\frac{1}{2} + \frac{1}{3} + \frac{1}{4} + \frac{1}{6}$ $1\frac{1}{4}$

13. $22 + 6z + 95$ $6z + 117$

14. $(2x + 7) + (5y + 9)$ $2x + 5y + 16$

15. $13(4m)$ $52m$ **16.** $(15n)6$ $90n$

17. $(7p)(13q)$ $91pq$ **18.** $(17r)(5s)(3t)$
                                   $255rst$

Solve.

19. $127 + (a + 96) = 47 + 96 + 127$ {47}

20. $(12b)11 = 11 \times 12 \times 13$ {13}

21. $(c + 12) + 23 = c + (12 + 23)$ $\mathcal{R}$

22. $3(5d) < (3 \cdot 5)d$ $\emptyset$

23. $7(22f) \geq 22(7f)$ $\mathcal{R}$

24. $41 + (g + 72) \leq 72 + (g + 41)$ $\mathcal{R}$

B **25.** $25 + (x + 39) < x + 64$ Ø
26. $(11y)18 > 198y$ Ø
27. $(1.4)3w > 7w(0.6)$ Ø
28. $(2.7 + z) + 7.3 \le (z + 6.09) + 3.91$
$\mathcal{R}$

In each of Exercises 29–36, an operation ☆ is defined over the set of natural numbers.
a. Find $3 ☆ 4$.
b. Tell whether the statement "For all natural numbers a and b, $a ☆ b$ is a natural number" is true or false.
c. Tell whether or not ☆ is a commutative operation.
d. Tell whether or not ☆ is an associative operation.

29. $a ☆ b = a + (b + 2)$
30. $a ☆ b = a(2b)$
31. $a ☆ b = 3a + b$
32. $a ☆ b = 3(a + b)$
33. $a ☆ b = (a + b)^2$
34. $a ☆ b = a + b^2$
35. $a ☆ b = \frac{a}{2} + b$
36. $a ☆ b = \frac{ab}{2}$

Tell whether each relationship is reflexive, symmetric, and/or transitive.

C **37.** is greater than
Transitive
38. is less than or equal to
Reflexive, transitive
39. is not equal to
Symmetric

2–3 The Distributive Axiom

In order to simplify the expression $12(10 + 3)$, you may first add 10 and 3, as indicated by the parentheses, and then multiply this sum by 12.

$$12(10 + 3) = 12(13) = 156$$

Another way to simplify this same expression is to first *distribute* 12 as the multiplier of both 10 and 3, then find the sum.

$$(12 \times 10) + (12 \times 3) = 120 + 36 = 156$$

Either way, the result is the same. Thus you can write:

$$12(10 + 3) = (12 \times 10) + (12 \times 3)$$

This last equation illustrates another fact that is often used in working with real numbers: multiplication is *distributive with respect to addition*.

> ### Distributive Axiom of Multiplication with Respect to Addition
> For all real numbers a, b, and c,
> $$a(b + c) = ab + ac \quad \text{and} \quad (b + c)a = ba + ca.$$

Working with Real Numbers **57**

32. a. 21
b. True
c. Commutative
d. Not associative
33. a. 49
b. True
c. Commutative
d. Not associative
34. a. 19
b. True
c. Not commutative
d. Not associative
35. a. $5\frac{1}{2}$
b. False
c. Not commutative
d. Not associative
36. a. 6
b. False
c. Commutative
d. Associative

Teaching Suggestions p. T73

Related Activities p. T73

Key Ideas
Use the distributive axiom of multiplication with respect to addition.

Simplify.

1. $60\left(\frac{1}{4} + \frac{1}{5}\right)$

$\left(60 \times \frac{1}{4}\right) + \left(60 \times \frac{1}{5}\right)$

$= 15 + 12$

$= 27$

2. $4x + (9x^2 + 3x)$
$= (4x + 3x) + 9x^2$
$= (4 + 3)x + 9x^2$
$= 7x + 9x^2$

3. $5(2a + 1) + 4(a + 3)$
$= 10a + 5 + 4a + 12$
$= (10a + 4a) + (5 + 12)$
$= (10 + 4)a + (5 + 12)$
$= 14a + 17$

Common Errors

When distributing multiplication over addition, students often forget to distribute over every term inside the parentheses. For example, they might write $4(a + b) = 4a + b$. To guard against this error, ask students to read and think of $4(a + b)$ as "four times the quantity of a plus b."

By applying the symmetric axiom of equality, you can also state the distributive axiom in the following form.

> For all real numbers a, b, and c,
>
> $$ab + ac = a(b + c) \quad \text{and} \quad ba + ca = (b + c)a.$$

The distributive axiom can be helpful in simplifying both numerical and variable expressions.

EXAMPLE 1 Simplify.

  a. $24\left(\frac{5}{6} + \frac{3}{8}\right)$ **b.** $\left(3\frac{4}{5}\right)10$ **c.** $\frac{1}{9} \times 20 + \frac{1}{9} \times 16$

SOLUTION   **a.** $24\left(\frac{5}{6} + \frac{3}{8}\right) = 24 \times \frac{5}{6} + 24 \times \frac{3}{8} = 20 + 9 = 29$

  b. $\left(3\frac{4}{5}\right)10 = \left(3 + \frac{4}{5}\right)10 = 3 \times 10 + \frac{4}{5} \times 10 = 30 + 8 = 38$

  c. $\frac{1}{9} \times 20 + \frac{1}{9} \times 16 = \frac{1}{9}(20 + 16) = \frac{1}{9} \times 36 = 4$

EXAMPLE 2 Show that, for every real number y, $8y + 7y = 15y$.

SOLUTION $8y + 7y = (8 + 7)y$ by the distributive axiom
        $= 15y$ by the substitution principle

As shown in Example 2, basic properties of the real numbers guarantee that, for all values of the variable, the expressions

$$8y + 7y \qquad \text{and} \qquad 15y$$

represent the same number. Therefore, the two expressions are said to be **equivalent expressions**. Note that the expression $8y + 7y$ has two terms, $8y$ and $7y$, while the expression $15y$ has only one term. When you replace a variable expression with an equivalent expression that has as few terms as possible, you **simplify the expression.**

EXAMPLE 3 Simplify.

  a. $5a + 3 + 8a + 4$ **b.** $(7b^2 + 10b) + 9b^2$

SOLUTION   **a.** $5a + 3 + 8a + 4 = (5a + 8a) + (3 + 4)$
          $= (5 + 8)a + (3 + 4)$
          $= 13a + 7$

  b. $(7b^2 + 10b) + 9b^2 = (7b^2 + 9b^2) + 10b$
          $= (7 + 9)b^2 + 10b$
          $= 16b^2 + 10b$

Frequently you need to use *both* forms of the distributive axiom in simplifying an expression.

EXAMPLE 4   Simplify $9(2m + 3) + 2(m + 5)$.

SOLUTION $9(2m + 3) + 2(m + 5) = 18m + 27 + 2m + 10$
$$= (18m + 2m) + (27 + 10)$$
$$= (18 + 2)m + (27 + 10)$$
$$= 20m + 37$$

Oral Exercises

Name the axiom or principle that justifies each step.

1. $\left(\frac{1}{2} + \frac{1}{3}\right)24 = \frac{1}{2} \times 24 + \frac{1}{3} \times 24$ Distributive

$\qquad = 12 + 8$ Substitution

$\qquad = 20$ Substitution

2. $\frac{1}{7} \times 23 + \frac{1}{7} \times 33 = \frac{1}{7}(23 + 33)$

$\qquad = \frac{1}{7} \times 56$

$\qquad = 8$

3. $35 \times 2\frac{2}{5} = 35\left(2 + \frac{2}{5}\right)$ Substitution

$\qquad = 35 \times 2 + 35 \times \frac{2}{5}$ Distributive

$\qquad = 70 + 14$ Substitution

$\qquad = 84$ Substitution

4. $216 \times 101 = 216(100 + 1)$

$\qquad = 216 \times 100 + 216 \times 1$

$\qquad = 21,600 + 216$

$\qquad = 21,816$

Commutative
5. $3m + (8 + 6m) = 3m + (6m + 8)$

$\qquad = (3m + 6m) + 8$ Associative

$\qquad = (3 + 6)m + 8$ Distributive

$\qquad = 9m + 8$ Substitution

6. $4(n + 6) + 5 = (4 \times n + 4 \times 6) + 5$

$\qquad = (4n + 24) + 5$

$\qquad = 4n + (24 + 5)$

$\qquad = 4n + 29$

Simplify.

7. $4(50 + 7)$ 228

8. $8(60 + 2)$ 496

9. $20 \times 3\frac{1}{5}$ 64

10. $14 \times 2\frac{3}{7}$ 34

11. $6(5.5)$ 33

12. $(3.25)8$ 26

13. $\frac{1}{3} \times 8 + \frac{1}{3} \times 19$ 9

14. $11 \times \frac{5}{8} + 13 \times \frac{5}{8}$ 15

15. $4a + 12a$ 16a

16. $7b + 3b + 9b$ 19b

17. $16c + 14 + 4c$ 20c + 14

18. $(23d^2 + 8) + 17d^2$ 40d^2 + 8

19. $3f + 7g + 8f$ 11f + 7g

20. $12k + (5j + 6k) + 2j$ 7j + 18k

21. $4x + 9x^2 + 6x$ 10x + 9x^2

22. $(6y^2 + 7y) + (2y^2 + 8y)$ 8y^2 + 15y

23. $2(z + 6) + 9$ 2z + 21

24. $8w^2 + 4(w^2 + 5)$ 12w^2 + 20

Working with Real Numbers   **59**

Suggested Assignments

Minimum
  60/1–26, 27–35 odd
Average
  60/21–47 odd, 49–58
Maximum
  60/25–57 odd, 58–62

Additional A Exercises

Simplify.

1. $3(40 + 9)$ 147

2. $\left(\dfrac{1}{2} + \dfrac{4}{3}\right)18$ 33

3. $10 \times 5\dfrac{1}{2}$ 55

4. $14a^2 + 15a^2$ $29a^2$

5. $(1.25)4$ 5

6. $6n - 2m - n$ $5n - 2m$

Written Exercises

Simplify.

A

1. $42\left(\dfrac{1}{3} + \dfrac{5}{7}\right)$ 44

2. $\left(\dfrac{6}{5} + \dfrac{3}{4}\right)40$ 78

3. $80 \times 3\dfrac{9}{10}$ 312

4. $(12.75)4$ 51

5. $(18.25)27 + (1.75)27$ 540

6. $\dfrac{5}{8} \times 21 + \dfrac{5}{8} \times 75$ 60

7. $17x + 64x$ $81x$

8. $49y + 82y$ $131y$

9. $85z^2 + 15z^2$ $100z^2$

10. $5v^3 + 78v^3$ $83v^3$

11. $1.6g + 3g + 4.4g$ $9g$

12. $\dfrac{1}{3}h^2 + 5h^2 + \dfrac{5}{3}h^2$ $7h^2$

13. $12s + 17s + 43$ $29s + 43$

14. $11t^2 + 14 + 27t^2$ $38t^2 + 14$

15. $6p + 11p + 9q$ $17p + 9q$

16. $8m + 15n + 13m$ $21m + 15n$

17. $5a^2 + 6a^2 + 2a$ $11a^2 + 2a$

18. $2b^3 + 5b + 4b^3$ $6b^3 + 5b$

19. $2c^2 + 9c^3 + 14c^3$ $23c^3 + 2c^2$

20. $16d^4 + 8d^2 + 9d^4$ $25d^4 + 8d^2$

21. $4m + 17 + 2m + 8$ $6m + 25$

22. $24n^2 + 8 + 14 + 9n^2$ $33n^2 + 22$

23. $17x + 3y + 15y + 5x$ $22x + 18y$

24. $21z^3 + 4z + 41z + 19z^3$ $40z^3 + 45z$

25. $5p^4 + 8 + 9p^4 + 11p^4 + 2$ $25p^4 + 10$

26. $4q + 6 + 3 + 9q + 7q$ $20q + 9$

27. $3(u + 5) + 6u$ $9u + 15$

28. $8v^2 + 4(1 + v^2)$ $12v^2 + 4$

29. $6(c^2 + 2) + 5(c^2 + 6)$ $11c^2 + 42$

30. $2(d + 5) + 4(d + 9) + 3d$

31. $4(3a + 5) + 8(a + 1)$ $20a + 28$

32. $7(3b + 1) + 3(2b + 7) + 1$

33. $5(y^2 + 2y) + 7(2y^2 + 3y)$ $19y^2 + 31y$

34. $(5z + 3z^2)3 + (z + z^2)6$

35. $12(j + k) + 3(j + k) + 3j$ $18j + 15k$

36. $2(m + n) + 11n + 3(m + n)$

30. $9d + 46$
32. $27b + 29$
34. $15z^2 + 21z$
36. $5m + 16n$

B

37. $(2 + 3g + 2g^2) + (4g^2 + 9g + 7)$ $6g^2 + 12g + 9$

38. $(4 + 10h^2 + 3h) + (2h^2 + 1 + 5h)$ $12h^2 + 8h + 5$

39. $(6a + 8a^3 + 2a^2) + (3a + 9a^2)$ $8a^3 + 11a^2 + 9a$

40. $(11b^4 + 2b^2) + (3b + 5b^2 + b^3 + 6b^4)$ $17b^4 + b^3 + 7b^2 + 3b$

41. $(2 + 7p^2 + 8p + 4p^3) + (9p + 8p^3 + 3p^2 + 9)$ $12p^3 + 10p^2 + 17p + 11$

42. $(6q^3 + 4 + 3q + 7q^2) + (6 + 2q^3 + 5q + 6q^2)$ $8q^3 + 13q^2 + 8q + 10$

43. $2(1 + 3y + 4y^2 + y^3) + (4y^3 + 5y^2 + 8y + 3)$ $6y^3 + 13y^2 + 14y + 5$

44. $(5x^2 + 3x + 4 + 2x^3) + 3(x^2 + 4x + x^3 + 2)$ $5x^3 + 8x^2 + 15x + 10$

45. $5(m + 2m^2) + 3(3m^2 + 4)$ $19m^2 + 5m + 12$

46. $6(n^2 + 2) + 2(3n + 1) + 9n^2$ $15n^2 + 6n + 14$

47. $5(3r + s) + 9(2r + 1) + 7(s + 6)$ $33r + 12s + 51$

48. $2(u + 2v) + 3(v + 3w) + 5(u + w)$ $7u + 7v + 14w$

Write a variable expression for each word phrase. Then simplify the variable expression.

49. three times the sum of a number u and a number v, increased by twice the sum of u and twice v $3(u + v) + 2(u + 2v)$; $5u + 7v$

50. twice the sum when a number b is added to three times a number a, increased by three times the sum when twice b is added to a
  $2(3a + b) + 3(a + 2b)$; $9a + 8b$

60 Chapter 2

51. five times the sum of three and the square of a number w, increased by twice the sum when five is added to the square of w

52. twice the sum when three is added to the square of a number b, increased by triple the sum of one and four times the square of b

53. the product of two and the cube of a number y, increased by six times the sum of three times the square of y and two times the cube of y

54. triple the sum when the square of a number c is added to the fourth power of c, increased by four times the square of c

Simplify.

C 55. $4[2a + 5(2 + 3a)] + 2(a + 1)$ $70a + 42$

56. $3[4b + 2(b + 3) + 2] + 5(2 + 3b)$ $33b + 34$

57. $2[3(3x + 1) + 2(2x + 5)] + 3[1 + 4(x + 2) + 5x]$ $53x + 53$

58. $3[2(1 + 4y) + 2(y + 4) + 2y] + 2[5y + 4(2 + y)]$ $54y + 46$

59. $4[5(m + 1) + 2(m + 3n) + 1] + 5[4 + 2(2m + n)]$ $48m + 34n + 44$

60. $2[3(2u + v) + 2(5 + v)] + 3[2(u + v) + 3(v + 1)]$ $18u + 25v + 29$

61. $4[3p^2 + 2(p^2 + 2p + 1)] + 3[2 + 3(3p^2 + p + 2)] + 2(p^2 + p + 1)$

62. $6[2q^2 + 3(q^2 + q + 4)] + 3[3(q^2 + q + 7) + 4q] + 3(2 + 2q + q^2) + 5q$

$42p^2 + 50q + 141$

61. $49p^2 + 27p + 34$

Karl Friedrich Gauss

1777–1855

Karl Friedrich Gauss was born in Brunswick, Germany, the son of a bricklayer. At the age of three he showed his remarkable computational ability by correcting his father's figuring of the weekly payroll. At ten, he astounded his teacher by mentally computing, within seconds, the sum of the integers from 1 to 100. It was not until he was nineteen, however, that he decided to become a mathematician.

Gauss's mathematical discoveries were amazing in that they dealt not only with the abstract fields of algebra and analytic function theory, but also with mathematical physics, astronomy, geodesy, and electromagnetism. Gauss himself thought that his most significant achievements were in the field of *number theory*, that branch of mathematics which is concerned with the properties of integers. Referring to number theory as *arithmetic*, it was Gauss who proclaimed, "Mathematics is the queen of the sciences, and arithmetic is the queen of mathematics."

Additional Answers Written Exercises

51. $5(3 + w^2) + 2(w^2 + 5)$; $7w^2 + 25$

52. $2(b^2 + 3) + 3(1 + 4b^2)$; $14b^2 + 9$

53. $2y^3 + 6(3y^2 + 2y^3)$; $14y^3 + 18y^2$

54. $3(c^4 + c^2) + 4c^2$; $3c^4 + 7c^2$

Mixed Review

Simplify.

1. $733 + 47 + 67 + 43$ 890

2. $4.1 + 0.9 + 2.9 + 6.9$ 14.8

3. $25 \times 25 \times 8 \times 16$ 80,000

4. $(75q)(5r)(2s)$ $750qrs$

5. $\frac{7}{11} \times \frac{41}{13} \times \frac{11}{21} \times \frac{39}{41}$ 1

6. $(4y + 2) + (2y + 18)$ $6y + 20$

Key Ideas

Use opposites and absolute values of numbers in solving open sentences and simplifying expressions.

2–4 Opposites and Absolute Values

The figure below shows pairings of selected points on a number line. Note that the paired points are at the same distance from the origin, but on *opposite* sides of the origin. The origin is paired with itself.

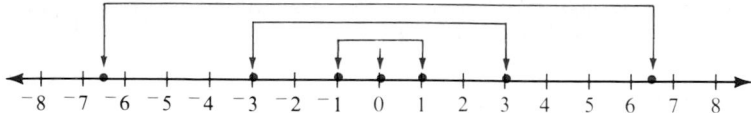

The coordinates of the paired points can also be paired.

0 with 0 $^-$1 with 1 $^-$3 with 3 $^-$6.5 with 6.5

Each number in such a pair is called the **opposite** of the other number. The symbol for the opposite of a number a is $-a$ (note the lowered position of the minus sign). For example:

$-8 = {}^-8$,	read	"The opposite of eight equals negative eight."
$-({}^-4) = 4$,	read	"The opposite of negative four equals four."
$-0 = 0$,	read	"The opposite of zero equals zero."

Notice that the numerals -8 (lowered minus sign) and $^-8$ (raised minus sign) name the same number. Thus you can read -8 as "negative eight" as well as "the opposite of eight." Therefore, in order to simplify notation, *lowered minus signs will be used in the numerals for negative numbers throughout the rest of this book.*

Be sure that you understand the meaning of the variable expression $-a$, the opposite of a. For example, if the value of a is 8, then the value of $-a$ is -8; if the value of a is -8, then the value of $-a$ is 8. In general, the following statements hold true.

1. If a is a positive number, then $-a$ is a negative number.
2. If a is a negative number, then $-a$ is a positive number.
3. If a is zero, then $-a$ is zero.

The following property of opposites is often helpful in simplifying expressions and in solving equations.

Cancellation Property of Opposites

For all real numbers a,

$$-(-a) = a.$$

That is, the opposite of $-a$ is a.

EXAMPLE 1 Simplify.

  a. $-(-1.75)$ **b.** -0 **c.** $-\left(\dfrac{1}{4} + \dfrac{3}{4}\right)$ **d.** $-[-(-2)]$

SOLUTION   **a.** 1.75 **b.** 0 **c.** -1 **d.** -2

EXAMPLE 2 Solve $-n = 12$.

SOLUTION   If $-n = 12$, then $-(-n) = -12$, and $n = -12$.
Therefore, the solution set is $\{-12\}$.

In any pair of nonzero opposites, such as -8 and 8, one number is negative and the other is positive. The positive number of any pair of opposite nonzero real numbers is called the **absolute value** of *each* number in the pair. Thus, 8 is the absolute value of -8; 8 is also the absolute value of 8. The absolute value of a number is denoted by writing the numeral for the number between a pair of vertical bars, $|\ |$. For example,

$$|-8| = 8 \qquad \text{and} \qquad |8| = 8.$$

The absolute value of 0 is defined to be 0.

$$|0| = 0$$

Formally, the absolute value of any real number is defined as follows.

> For any real number a,
>
> $$|a| = a, \text{ if } a \text{ is nonnegative;}$$
> $$|a| = -a, \text{ if } a \text{ is negative.}$$

EXAMPLE 3 Solve $|x| = 12$.

SOLUTION   Since $|12| = 12$ *and* $|-12| = 12$, there are two values of x that make a true statement, 12 and -12.
Therefore, the solution set is $\{12, -12\}$.

In computations involving absolute values, the vertical bars also act as grouping symbols.

EXAMPLE 4 Simplify $7|-8| - |9 + 7|$.

SOLUTION
$$
\begin{aligned}
7|-8| - |9 + 7| &= 7|-8| - |16| \\
&= 7(8) - 16 \\
&= 56 - 16 \\
&= 40
\end{aligned}
$$

Working with Real Numbers **63**

1. Simplify.
  a. $-(2.4 + 1.7)$
    $= -(4.1)$
    $= -4.1$
  b. $-[-(-4)]$
    $= -[4]$
    $= -4$
  c. $|-2| + |-4 - 1|$
    $= |-2| + |-5|$
    $= 2 + 5$
    $= 7$

2. Solve.
  a. $10 = -x$
    If $10 = -x$, then $-10 = -(-x)$, and $-10 = x$.
    $\therefore$ the solution is $\{-10\}$.
  b. $4 = |s|$
    Since $|4| = 4$ and $|-4| = 4$, there are two values of s that make a true statement.
    $\therefore$ the solution is $\{4, -4\}$.

3. Graph $|y| = 5$.
  The graph consists of the two points that are exactly five units from the origin, that is, 5 and -5.

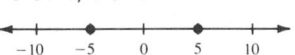

When a negative number is within absolute value bars, students tend to make mistakes when simplifying. For example, for the expression $-|-3|$, they obtain 3 instead of -3. Encourage students to add an intermediate "think" step similar to the following to avoid this type of error.

$$\text{Think: } |-3| = 3.$$
$$\therefore -|-3| = -3.$$

Mixed Review

Simplify.

1. $50\left(\dfrac{1}{2} + \dfrac{1}{25}\right)$ 27

2. $\dfrac{7}{11} \times 71 + \dfrac{7}{11} \times 28$ 63

3. $12.7(25) + 7.3(25)$ 500

4. $2(u + 7) + 4u$ $6u + 14$

5. $(8 + 3h^2 + 2h) + (h^2 - 4 + 3h)$ $4h^2 + 5h + 4$

6. Write a variable expression for the following word phrase: three times the sum when a number x is added to twice a number y. $3(x + 2y)$

Suggested Assignments

Minimum
  65/1–30
R 66/Self-Test 1
Average
  65/7–41 odd, 43–54
R 66/Self-Test 1
Maximum
  65/7–41 odd, 43–58
R 66/Self-Test 1

The absolute value of a number may also be thought of as the distance of the graph of the number from the origin on a number line. For example, the graphs of both -8 and 8 are 8 units from the origin.

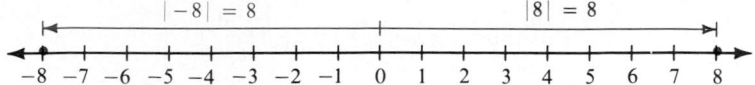

This interpretation of absolute value may be especially helpful in solving inequalities that involve absolute values.

EXAMPLE 5   Graph each open sentence.

       a. $|a| = 2$ **b.** $|x| < 1$ **c.** $|y| \geq 3$

SOLUTION   **a.** The graph consists of the two points that are exactly two units from the origin, that is, the graphs of -2 and 2.

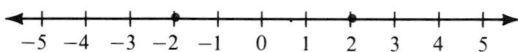

       b. The graph consists of all the points that are *less than* one unit from the origin in either direction.

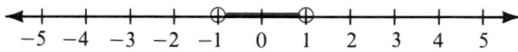

       c. The graph consists of all the points that are *at least* three units from the origin in either direction.

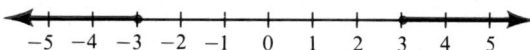

Notice in part (b) of Example 5 that the graph of the open sentence $|x| < 1$ is also the graph of the inequality $-1 < x < 1$.

Oral Exercises

Name the opposite and the absolute value of each number.

  1. 5 $-5, 5$ **2.** -4 4, 4   **3.** $-\dfrac{3}{7}$ $\dfrac{3}{7}, \dfrac{3}{7}$ **4.** $\dfrac{1}{3}$ $-\dfrac{1}{3}, \dfrac{1}{3}$

  5. -6.92 6.92, 6.92   **6.** 6.92 -6.92, 6.92   **7.** 0   0, 0   **8.** 1 -1, 1

Simplify.

  9. $-(-11)$ 11   **10.** $-[-(-15)]$ -15 **11.** $-(9 + 14)$ -23 **12.** $-[-(37 - 12)]$ 25

 13. $|-17|$ 17   **14.** $-|17|$ -17 **15.** $-|-23|$ -23 **16.** $-|0|$ 0

 17. $5|12|$ 60   **18.** $|-8|6$ 48   **19.** $-|11 + 7|$ -18 **20.** $-|19 - 5|$ -14

Tell whether each statement is true or false.

 21. $|-9| = 9$ True   **22.** $|7| = -7$ False   **23.** $|-8| = -|8|$ False   **24.** $-|-2| = -|2|$ True

 25. $|5| > |-5|$ False **26.** $|3| \leq |-3|$ True   **27.** $-1 < |-4|$ True   **28.** $|-10| > 0$ True

29. The opposite of any real number is less than zero. False

30. The absolute value of any real number is greater than zero. False

31. The absolute value of any real number is equal to the absolute value of the opposite of the number. True

32. The absolute value of the opposite of any real number is equal to the opposite of the absolute value of the number. False

33. The opposite of the opposite of any real number is equal to the number itself. True

34. The absolute value of the absolute value of any real number is equal to the number itself. False

Written Exercises

Simplify.

A 1. $-(-3.5)$ 3.5
2. $-[-(-93)]$ −93
3. $-(36 + 49)$ −85

4. $-(8.4 - 1.2)$ −7.2
5. $[-(-79)] - 22$ 57
6. $24 + [-(-137)]$ 161

7. $\left|-6\frac{1}{7}\right|$ $6\frac{1}{7}$
8. $-|0.77|$ −0.77
9. $-|10.5 - 4.3|$ −6.2

10. $|-(223 + 19)|$ 242
11. $6|-(-14)|$ 84
12. $\left|-\frac{1}{5}\right|$ $3\frac{3}{5}$

13. $5|-0.2| + |3|$ 4
14. $6|6| - 7|-5|$ 1
15. $|9 + 22| + |-29|$ 60

16. $6|-4| - |4 + 11|$ 9
17. $3|9 + 5| - 6|-7|$ 0
18. $7|3 + 8| - 4|8 - 3|$ 57

Solve.

19. $-x = 18$ {−18}
20. $2 = -y$ {−2}
21. $-z = -4$ {4}
22. $-9 = -w$ {9}

23. $|a| = 5$ {5, −5}
24. $16 = |b|$ {16, −16}
25. $|c| = 0$ {0}
26. $|d| = -7$ Ø

27. $|-f| = 10$ {10, −10}
28. $-|g| = 1$ Ø
29. $-|j| = -6$ {6, −6}
30. $-|-k| = 3$ Ø

B 31. $-r + 1 = 5$ {−4}
32. $5 + (-s) = 9$ {−4}
33. $-u - 3 = 12$ {−15}

34. $6 = -v - 14$ {−20}
35. $|p| + 2 = 8$ {6, −6}
36. $9 - |q| = 7$ {2, −2}

37. $17 = |s| - 8$ {25, −25}
38. $4 + |t| = 13$ {9, −9}
39. $|-m| + 3 = 4$

40. $16 - |-n| = 5$ {11, −11}
41. $|j| + |3| = 9$ {6, −6}
42. $|-k| - |-7| = 15$

39. {1, −1}
42. {22, −22}

Graph each inequality.

43. $|a| \le 4$
44. $|b| > 5$
45. $|c| \ge 0$
46. $|d| < 0$

47. $|x| < -1$
48. $|y| \ge -4$
49. $|-w| > 5$
50. $|-z| \le 6$

51. $0 \le |j| \le 2$
52. $3 > |k| > 0$
53. $-1 \le |m| < 5$
54. $3 \ge |n| > -7$

Name the set of all values of x for which each inequality is a true statement.

C 55. $|x| > x$
56. $|x| < -x$ Ø
57. $|x| \le -x$
58. $|x| \ge x$ $\mathcal{R}$

Working with Real Numbers **65**

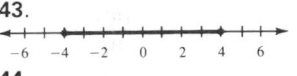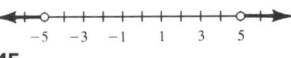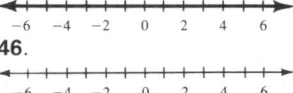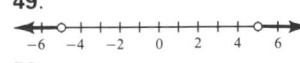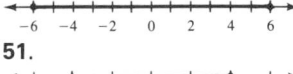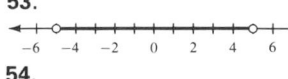

Quick Quiz

Find a value of the variable that makes each statement true.

1. For some whole number w, $4w + 1 = 9$. 2

2. There exists a real number y such that $2y + 5 \leq 3$. Answers will vary. Example: $y \leq -1$.

Simplify.

3. $\dfrac{7}{10} + \dfrac{5}{6} + \dfrac{3}{10} + \dfrac{7}{6}$ 3

4. $\dfrac{1}{2} \times 15 \times 50 \times \dfrac{1}{3}$ 125

5. $4y^2 + 5 + 7y^2 + 6$ $11y^2 + 11$

6. $3(2x + 1) + 4(2x + 1)$ $14x + 7$

7. $-\{-[17 + (-19)]\}$ -2

8. $-2|-5| + (-|4 + 9|)$ -23

Solve.

9. $(-b) + (-17) = 17$ $b = -34$

10. $|m| + (-|-3|) = -2$ $m = 1$ or -1

Computer Exercises   For students with computer experience

Write a program that uses the computer's absolute value function to evaluate each expression when $n \in \{25, -25\}$.

1. $|n|$ 25; 25

2. $-|n|$ −25; −25

3. $|-n|$ 25; 25

4. $-|-n|$ −25; −25

5. $|n| + 2$ 27; 27

6. $|n + 2|$ 27; 23

7. $|2n|$ 50; 50

8. $\dfrac{|n|}{2}$ 12.5; 12.5

9. Write a program that will print the absolute value of a number *without* using the computer's absolute value function. RUN your program for a positive number, a negative number, and zero.

Self-Test 1

VOCABULARY

quantifier (p. 50)
binary operation (p. 52)
sum (p. 52)
terms (p. 52)
product (p. 52)
factors (p. 52)
axiom or postulate (p. 53)
axioms of closure (p. 53)
commutative axioms (p. 53)
associative axioms (p. 54)
reflexive axiom (p. 55)

symmetric axiom (p. 55)
transitive axiom (p. 55)
distributive axiom (p. 57)
equivalent expressions (p. 58)
simplify a variable expression (p. 58)
opposite of a number (p. 62)
cancellation property of opposites (p. 62)
absolute value (p. 63)

Find a value of the variable that makes each statement true.

1. For some natural number n, $2n - 3 = 11$. $n = 7$ *Obj. 1, p. 49*

2. There exists a whole number w such that $3w + 4 < 6$. $w = 0$

Simplify.

3. $9.5 + 6.8 + 0.5 + 3.2$ 20

4. $\dfrac{1}{3} \times 44 \times 21 \times \dfrac{1}{4}$ 77   *Obj. 2, p. 49*

5. $8 + 2x^2 + 7 + 3x^2$ $5x^2 + 15$

6. $2(3y + 8) + 3(4 + 5y)$ $21y + 28$

7. $-[-(41 - 26)]$ 15

8. $9|-7| - |12 + 16|$ 35   *Obj. 3, p. 49*

Solve.

9. $13 + (-a) = 21$ $\{-8\}$

10. $|b| - |-4| = 15$ $\{19, -19\}$

Check your answers with those at the back of the book.

Addition and Subtraction

OBJECTIVES for Sections 2-5 through 2-8:
1. *To picture the addition of real numbers on a number line.*
2. *To add two or more real numbers and to solve word problems involving addition.*
3. *To use basic axioms of addition and equality as reasons for steps in proving theorems.*
4. *To subtract real numbers and to solve word problems involving subtraction.*

2–5 Addition on a Number Line

The addition of real numbers can be pictured as a series of moves, or *displacements*, along a number line. A positive number is represented by a displacement in the positive direction, and a negative number is represented by a displacement in the negative direction. Arrows are used to picture these displacements.

For example, to add 3 and 4 on a horizontal number line that is marked with positive numbers to the right, first start at the origin and move 3 units to the right. In the diagram that follows, the short black arrow represents this displacement, that is, the number 3. Then, starting at the graph of 3, move 4 units *farther* to the right, as represented by the red arrow. Together the two displacements amount to a displacement of 7 units to the right from the origin.

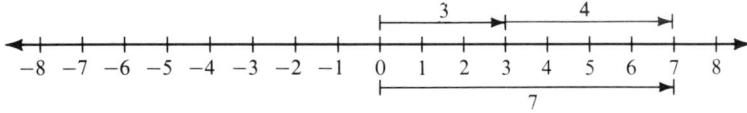

Therefore, this diagram may be used to picture the fact

$$3 + 4 = 7.$$

To find the sum of −3 and −4, first move 3 units to the *left* from the origin, then move 4 units farther left from this point.

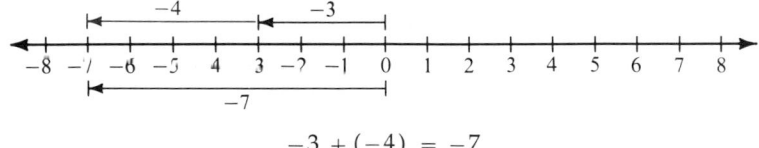

$$-3 + (-4) = -7$$

Notice the use of parentheses in the expression "−3 + (−4)." These parentheses separate the plus sign that means "add" from the minus sign that is part of the numeral for negative four.

Working with Real Numbers **67**

Teaching Suggestions
p. T74

Key Ideas

Use diagrams to picture addition of positive and negative numbers on a number line.

Chalkboard Examples

Find the following sums. Use diagrams to picture the addition process.

1. 7 + (−3)

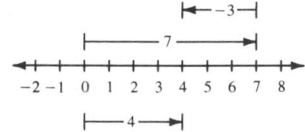

$$7 + (-3) = 4$$

2. −4 + 2

$$-4 + 2 = -2$$

(continued)

3. Solve $5 + m = -1$. Use a number line diagram to help you.

The equation states that 5 plus a number m is equal to -1. On a number line, to go from 5 to -1 you move 6 units to the left.

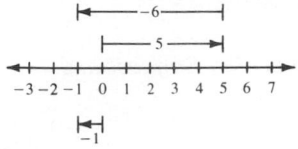

$5 + (-6) = -1$
∴ the solution set is $\{-6\}$.

To find the sum $3 + (-4)$, first move 3 units to the *right* from the origin. Then move 4 units to the *left* from this point.

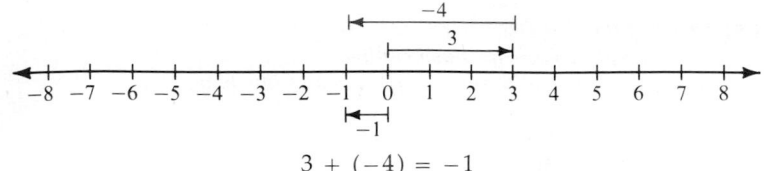

$$3 + (-4) = -1$$

The following diagram shows the displacements that would be used to find the sum $-3 + 4$.

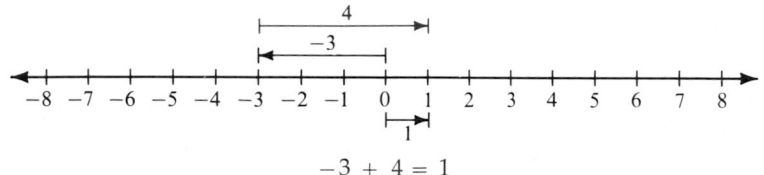

$$-3 + 4 = 1$$

Can you visualize $-3 + 0$ on a number line? If you interpret "add 0" as "no displacement," you can see that

$$-3 + 0 = -3 \qquad \text{and} \qquad 0 + (-3) = -3.$$

These equations illustrate the special property of zero for addition of real numbers: When 0 is added to any given real number, the sum is *identical* to the given number. Thus 0 is called the **identity element for addition,** and the following statement is assumed to be true.

Identity Axiom for Addition

There is a unique real number 0 such that, for every real number a,

$$a + 0 = a \quad \text{and} \quad 0 + a = a.$$

What is the result of adding a pair of opposites, such as -3 and 3, on a number line? As shown in the figures that follow,

$$3 + (-3) = 0 \qquad \text{and} \qquad -3 + 3 = 0.$$

Because the sum of a number and its opposite is always zero, the identity element for addition, the opposite of a number is also called the **additive inverse** of that number. The numeral -3 can then be read "negative three," "the opposite of three," or "the additive inverse of three."

68 *Chapter 2*

The following axiom is a formal way of saying that the sum of a number and its opposite is always zero.

Axiom of Additive Inverses

For every real number a, there is a unique real number $-a$ such that

$$a + (-a) = 0 \quad \text{and} \quad -a + a = 0.$$

A number line is sometimes helpful in solving equations that involve addition.

EXAMPLE   Solve $4 + t = -9$.

SOLUTION   The equation states that 4 plus a number t is equal to -9. On a number line, to go from 4 to -9 you move 13 units to the left.

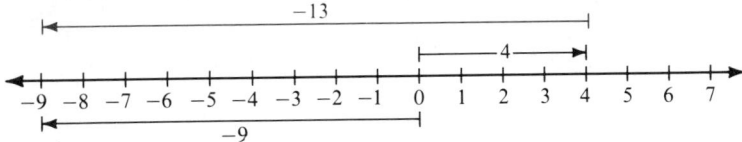

$$4 + (-13) = -9$$

Therefore, the solution set is $\{-13\}$.

Oral Exercises

Give an addition statement pictured by each diagram.

1.   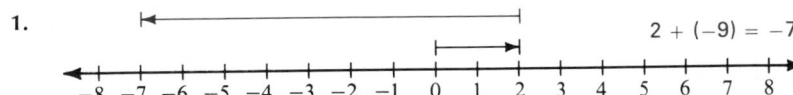 $2 + (-9) = -7$

2.   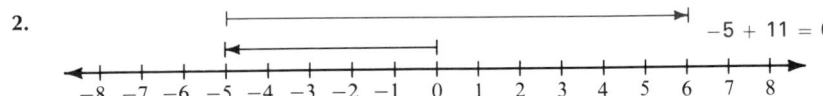 $-5 + 11 = 6$

3.   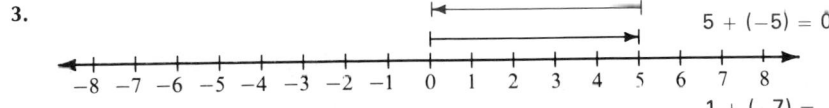 $5 + (-5) = 0$

$-1 + (-7) = -8$

4.

Suggested Assignments

Minimum
First day: 70/1–15
Second day: 70/16–36
         S 60/28–36 even
Average
   70/7–53 odd
S 60/38–48 even
Maximum
   70/7–53 odd, 55–60
S 60/50–56 even

Additional A Exercises

Simplify.
1. $(-4 + 5) + 7$ 8
2. $12 + (-10 + 5)$ 7
3. $-14 + [(-5) + (-14)]$
   -33
4. $-5.6 + (-5.4)$ -11
5. $[-12 + (-14)] + (-9)$
   -35
6. $-\frac{1}{7} + \left[-\frac{2}{7} + \left(-\frac{3}{7}\right)\right]$ $-\frac{6}{7}$

Mixed Review

Simplify.
1. $6|1 + 8| - 4|2 - 8|$ 30
Solve.
2. $-m = 14$ $\{-14\}$
3. $14 = |x|$ $\{14, -14\}$
4. $-|-y| = -6$ $\{6, -6\}$
5. $-n + 5 = -6$ $\{11\}$
6. $|p| + |-4| = 16$
   $\{12, -12\}$

Simplify. (Think of displacements along a number line.)

5. $-5 + 0$ -5
6. $0 + (-7)$ -7
7. $4 + (-4)$ 0
8. $-9 + 9$ 0
9. $-3 + (-6)$ -9
10. $-8 + (-2)$ -10
11. $9 + (-5)$ 4
12. $-3 + 4$ 1
13. $-\frac{2}{5} + \left(-\frac{1}{5}\right)$ $-\frac{3}{5}$
14. $\frac{6}{7} + \left(-\frac{1}{7}\right)$ $\frac{5}{7}$
15. $\frac{1}{9} + \left(-\frac{5}{9}\right)$ $-\frac{4}{9}$
16. $\frac{9}{10} + \left(-\frac{9}{10}\right)$
17. $3.5 + (-1)$ 2.5
18. $-2.9 + 1$ -1.9
19. $4.5 + (-3.5)$ 1
20. $-8.9 + 1.9$
                16. 0   20. -7

Written Exercises

Simplify.

                                                                          9. -25

A
1. $(-5 + 6) + (-9)$ -8
2. $-10 + (-4 + 7)$ -7
3. $12 + [7 + (-9)]$ 10
4. $[5 + (-8)] + 14$ 11
5. $[-4 + (-7)] + 8$ -3
6. $15 + [-6 + (-3)]$ 6
7. $-10 + [-14 + 6]$ -18
8. $[7 + (-16)] + (-5)$ -14
9. $[-12 + (-8)] + (-5)$
10. $-9 + [-11 + (-13)]$ -33
11. $\left[\frac{4}{7} + \left(-\frac{1}{7}\right)\right] + \frac{2}{7}$ $\frac{5}{7}$
12. $-\frac{1}{5} + \left[\frac{2}{5} + \left(-\frac{4}{5}\right)\right]$ $-\frac{3}{5}$

13. $2\frac{1}{9} + \left[\frac{4}{9} + \left(-\frac{5}{9}\right)\right]$ 2
14. $\left[1\frac{6}{7} + \left(-\frac{1}{7}\right)\right] + \left(-\frac{4}{7}\right)$ $1\frac{1}{7}$
15. $(-16 + 7) + (-11 + 2)$ -18
16. $[13 + (-6)] + [17 + (-9)]$ 15
17. $[7.5 + (-4)] + [8.5 + (-6)]$ 6
18. $(-9.5 + 7) + (-6.5 + 1)$ -8
19. $[6.3 + (-1.3)] + (-9.3 + 1.3)$ -3
20. $[4.7 + (-9.7)] + [3.4 + (-10.4)]$
21. $[-5 + (-14 + 19)] + (-23)$ -23
22. $-28 + [(-6 + 13) + 21]$ 0 -12
23. $35 + [(-21 + 4) + (-18)]$ 0
24. $[14 + (-27 + 13)] + (-41)$ -41

Solve.

25. $-2 + a = -6$ $\{-4\}$
26. $-6 + b = -11$ $\{-5\}$
27. $3 + r = -5$ $\{-8\}$
28. $2 + s = -7$ $\{-9\}$
29. $-4 + x = 7$ $\{11\}$
30. $-5 + y = 9$ $\{14\}$
31. $4 + j = 3$ $\{-1\}$
32. $9 + k = 1$ $\{-8\}$
33. $-9 + c = -9$ $\{0\}$
34. $-6 + d = 0$ $\{6\}$
35. $-10 + u = 10$ $\{20\}$
36. $8 + v = -8$ $\{-16\}$

B
37. $-9 = 8 + s$ $\{-17\}$
38. $6 = -7 + t$ $\{13\}$
39. $a + (-6) = -15$ $\{-9\}$
40. $b + 9 = -3$ $\{-12\}$
41. $c + (-6) = 8$ $\{14\}$
42. $d + 8 = 2$ $\{-6\}$
43. $-8 + j = -3 + (-5)$ $\{0\}$
44. $-5 + 5 = k + (-7)$ $\{7\}$
45. $x + (-5) = -3 + 7$
46. $1 + (-6) = -3 + d$ $\{-2\}$
47. $-7 + (-3) + v = 6$ $\{16\}$
48. $-12 = w + (-6) + 4$
49. $m + m = -8$ $\{-4\}$
50. $-18 = n + n$ $\{-9\}$
51. $p + p = -5$ $-2\frac{1}{2}$
52. $q + q = 0$ $\{0\}$
53. $x + x + x = -12$ $\{-4\}$
54. $y + y + y + y = -28$
                                   45. $\{9\}$ 48. $\{-10\}$ 54. $\{-7\}$

Evaluate each expression when $a = -2$, $b = 3$, $c = 4$, and $d = -8$.

C
55. $(-a + b) + (-c + d)$ -7
56. $-(a + b) + [-(c + d)]$ 3
57. $-a + [(-b + c) + d]$ -5
58. $a + [-b + (c + d)]$ -9
59. $-[a + (-b + c) + (-d)]$ -7
60. $-[-(-a + b) + (-c + d)]$ 17

2–6 Rules for Addition

The expression $-(4 + 5)$ represents the opposite of the sum of 4 and 5. Since $4 + 5 = 9$, it follows that

$$-(4 + 5) = -9.$$

The expression $-4 + (-5)$ represents the sum of the opposite of 4 and the opposite of 5. Using a number line, you can show that

$$-4 + (-5) = -9.$$

Since $-(4 + 5) = -9$ and $-4 + (-5) = -9$, it follows that

$$-(4 + 5) = -4 + (-5).$$

Using similar reasoning, you can also show the following facts.

$$-[-4 + (-5)] = 4 + 5$$
$$-[4 + (-5)] = -4 + 5$$
$$-(-4 + 5) = 4 + (-5)$$

These examples suggest the following property of addition.

Property of the Opposite of a Sum

The opposite of a sum of real numbers is equal to the sum of the opposites of the numbers. That is, for all real numbers a and b,

$$-(a + b) = -a + (-b).$$

By using the property of the opposite of a sum along with axioms you have learned and the familiar addition facts for positive numbers, you can compute sums of *any* real numbers without using a number line.

EXAMPLE 1 Simplify.

  a. $-11 + (-5)$ **b.** $11 + (-5)$ **c.** $5 + (-11)$

SOLUTION **a.** $-11 + (-5) = -(11 + 5)$
  $$= -16$$

  b. $11 + (-5) = (6 + 5) + (-5)$
  $$= 6 + [5 + (-5)]$$
  $$= 6 + 0 = 6$$

  c. $5 + (-11) = 5 + [-(5 + 6)]$
  $$= 5 + [-5 + (-6)]$$
  $$= [5 + (-5)] + (-6)$$
  $$= 0 + (-6) = -6$$

Teaching Suggestions
p. T74

Key Ideas

Add real numbers.
Solve word problems involving addition.

Chalkboard Examples

1. Simplify $13 + (-4) + 16 + (-5)$.
  Solution 1
  Add the numbers in order from left to right.
  $13 + (-4) = 9$;
  $9 + 16 = 25$;
  $25 + (-5) = 20$
  Solution 2
  Add the positive numbers:
  $13 + 16 = 29$
  Add the negative numbers:
  $-4 + (-5) = -9$
  Add the sums:
  $29 + (-9) = 20$

2. Simplify $5m + (-15m)$.
  $= [5 + (-15)]m$
  $= -10m$

3. Simplify
  $4y^3 + (-4y) + 3y + 6y^3$.
  $= [4y^3 + 6y^3] +$
  $\qquad\qquad [(-4y) + 3y]$
  $= [4 + 6]y^3 + [-4 + 3]y$
  $= 10y^3 + (-1)y$

It is helpful to most students to have specific referrals for general terms, definitions, rules, and the like. Have students read pages 71–72 silently to themselves, looking for the main ideas. Then have them go back and for each numerical example or illustration in the text, have them supply a different one. Then encourage students to ask each other for examples from their classmates as they cite the five rules on page 72. Have one student put the *Property of the Opposite of a Sum* on the chalkboard. Then have each student write an example to illustrate it. Discuss the helpfulness of this technique, and remark that many mathematicians read "with a pencil."

Common Errors

When adding negative integers, students tend to make the following type of mistake: $-5 + (-3) = -2$. Remind students to think of $-5 + (-3)$ as $-(5 + 3)$, and to visualize the number line when they add integers. Both techniques will help students avoid this kind of error.

After computing many sums by using either a number line or the methods of Example 1, you would probably discover the short-cut methods that are permitted by the following rules.

Rules for Addition of Positive and Negative Numbers

1. If a and b are both positive, then $a + b = |a| + |b|$.
   Example: $9 + 5 = 14$

2. If a and b are both negative, then $a + b = -(|a| + |b|)$.
   Example: $-8 + (-3) = -(8 + 3) = -11$

3. If a is positive and b is negative and a has the greater absolute value, then $a + b = |a| - |b|$.
   Example: $7 + (-6) = 7 - 6 = 1$

4. If a is positive and b is negative and b has the greater absolute value, then $a + b = -(|b| - |a|)$.
   Example: $2 + (-9) = -(9 - 2) = -7$

5. If a and b are opposites, then $a + b = 0$.
   Example: $6 + (-6) = 0$

The following example illustrates how to use these rules when adding more than two real numbers.

EXAMPLE 2   Simplify $6 + (-9) + 7 + (-8)$.

SOLUTION 1   Add the numbers in order from left to right.
$$6 + (-9) = -3; \ -3 + 7 = 4; \ 4 + (-8) = -4$$

SOLUTION 2   Add the positive numbers: $6 + 7 = 13$
Add the negative numbers: $-9 + (-8) = -17$
Add the sums: $13 + (-17) = -4$

The rules for addition of positive and negative numbers also apply when simplifying variable expressions.

EXAMPLE 3   Simplify.
   a. $(-10)m + 2m$ **b.** $2n^2 + (-3)n + 7n + (-8)n^2$

SOLUTION   **a.** $(-10)m + 2m = (-10 + 2)m$
$$= (-8)m$$

   b. $2n^2 + (-3)n + 7n + (-8)n^2 = [2n^2 + (-8)n^2] + [(-3)n + 7n]$
$$= [2 + (-8)]n^2 + [(-3) + 7]n$$
$$= (-6)n^2 + 4n$$

Oral Exercises

Add.

1. 12	**2.** −12	**3.** −14	**4.** 14	**5.** −25	**6.** −25
18	−18	20	−20	25	−25
$\overline{30}$	$\overline{-30}$	$\overline{6}$	$\overline{-6}$	$\overline{0}$	$\overline{-50}$

7. −1.2	**8.** 1.2	**9.** −1.2	**10.** $-\frac{3}{7}$	**11.** $\frac{3}{7}$	**12.** $-\frac{3}{7}$
−2.4	−2.4	2.4	$-\frac{2}{7}$	$-\frac{2}{7}$	$\frac{2}{7}$
$\overline{-3.6}$	$\overline{-1.2}$	$\overline{1.2}$	$\overline{-\frac{5}{7}}$	$\overline{\frac{1}{7}}$	$\overline{-\frac{1}{7}}$

Replace each __?__ with one of the words _always_, _sometimes_, or _never_ to make a true statement that has the widest application.

13. The sum of two negative numbers is __?__ a negative number. always
14. The sum of a positive number and a negative number is __?__ a negative number. sometimes
15. The sum of two positive numbers is __?__ zero. never
16. The sum of a positive number and zero is __?__ a negative number. never
17. The sum of a positive number and a negative number is __?__ zero. sometimes
18. The sum of zero and a negative number is __?__ a negative number. always

Tell whether each statement is true or false.

19. $-(3 + 5) = -3 + 5$ false
20. $-[-7 + (-9)] = 7 + 9$ true
21. $-(-10 + 4) = -10 + (-4)$ false
22. $-[14 + (-8)] = -14 + 8$ true
23. $4 + 7 = |4| + |7|$ true
24. $-17 + (-8) = |-17| - |-8|$ false
25. $-16 + 9 = |-16| + |9|$ false
26. $5 + (-13) = |-13| - |5|$ false
27. $-12 + 3 = -(|-12| - |3|)$ true
28. $11 + (-3) = -(|11| - |-3|)$ false

Written Exercises

Add.

A 1. 11	**2.** −21	**3.** −35	**4.** −47	**5.** 146	**6.** −223
19	7	−24	26	−97	351
−14	19	18	−19	−128	−269
−10	−12	−31	−38	119	−144
$\overline{6}$	$\overline{-7}$	$\overline{72}$	$\overline{-78}$	$\overline{40}$	$\overline{-285}$

Simplify.

7. $19 + (-11) + (-3) + 4$ 9
8. $-17 + 5 + (-21) + 14$ −19
9. $209 + (-401) + (-20) + (-103)$ −315
10. $-821 + 579 + (-37) + (-163)$ −442
11. $4.2 + (-1.3) + (-0.8) + 0.9$ 3
12. $-8.6 + (-2.2) + 9.4 + (-3.9)$ −5.3

Working with Real Numbers **73**

Suggested Assignments

Minimum
First day: 73/1–12
              75/*P*: 1–5
Second day: 74/13–28
              75/*P*: 6–10

Average
First day: 73/5–29 odd
              75/*P*: 5–11
Second day: 74/30–36
              76/*P*: 12–16

Maximum
First day: 73/5–33 odd
              75/*P*: 7–13
Second day: 74/34–40
              76/*P*: 14–17

Additional A Exercises

Add.

1. 15		**2.** 21
−24		7
−19		−8
18		−40
$\overline{-10}$		$\overline{-20}$

3. $(-11) + (-17) + 13 + 43$ 28
4. $-1.3 + (-2.7) + 9.1 + (-3.0)$ 2.1
5. $-\frac{1}{12} + \frac{5}{12} + \left(-\frac{7}{12}\right) + \frac{11}{12}$ $\frac{8}{12}$ or $\frac{2}{3}$
6. Jackie's expenses for one month were $15.00, $35.00, $10.00, $125.00 and $80.00. Her income was $400.00 for that month. How much money would Jackie have left after paying her expenses? $135.00

73

Simplify.

13. $-\dfrac{4}{9} + \left(-\dfrac{1}{9}\right) + \left(-\dfrac{2}{9}\right) + \dfrac{8}{9}$ $\dfrac{1}{9}$

14. $-\dfrac{3}{5} + \dfrac{1}{5} + \dfrac{2}{5} + \left(-\dfrac{4}{5}\right)$ $-\dfrac{4}{5}$

15. $-(-5 + 8) + (-4) + 3$ -4

16. $-[14 + (-9)] + 17 + (-2)$ 10

17. $-(-12 + 5) + [-14 + (-8)]$ -15

18. $-(-8 + 3) + -[16 + (-7)]$ -4

19. $6x + (-14)x$ $(-8)x$ $(-17)m + (-9)n$

20. $(-19)y^2 + 3y^2$ $(-16)y^2$

21. $(-5)m + (-9)n + (-12)m$

22. $6a + (-13)b + (-9)b$ $6a + (-22)b$

23. $3r^2 + (-5)r + (-7)r^2 + 13r$ $(-4)r^2 + 8r$

24. $(-6)s + 18s^2 + (-4)s^2 + 6s$ $14s^2$

B

25. $(-9)a + 6b + (-16)c + 9b + (-12)a + 21c$ $(-21)a + 15b + 5c$

26. $4z + (-10)y + (-8)x + (-12)y + 17x + (-13)z$ $9x + (-22)y + (-9)z$

27. $10j^2 + (-1)j^3 + 12j + (-19)j^2 + (-14)j^3 + (-8)j$ $(-15)j^3 + (-9)j^2 + 4j$

28. $(-5)k^4 + 7k^2 + (-13) + (-9)k^2 + 8 + (-10)k^4$ $(-15)k^4 + (-2)k^2 + (-5)$

29. $17p^2 + (-15)p^2 + (-4)p + (-17) + 7p + (-2)p^2$ $3p + (-17)$

30. $(-12)q + 19q^2 + (-6)q^3 + (-1)q^2 + (-14)q + 2q$ $(-6)q^3 + 18q^2 + (-24)q$

Evaluate each expression when r and s have the given values.

a. $-(r + s)$ **b.** $-[r + (-s)]$ **c.** $-|r + s|$ **d.** $-(|r| + |s|)$

31. $r = -1;\ s = 2$

32. $r = 3;\ s = 0$

33. $r = -4;\ s = 4$

34. $r = -5;\ s = -5$

35. $r = -0.7;\ s = 0.3$

36. $r = \dfrac{1}{2};\ s = -\dfrac{1}{4}$

Replace each __?__ with one of the symbols $=, <, >, \le,$ or $\ge$ to make a true statement that has the widest application.

C

37. For all real numbers $a \ge 0$ and $b \ge 0$, $|a + b|$ __?__ $|a| + |b|$. $=$

38. For all real numbers $a \le 0$ and $b \le 0$, $|a + b|$ __?__ $|a| + |b|$. $=$

39. For all real numbers $a < 0$ and $b > 0$, $|a + b|$ __?__ $|a| + |b|$. $<$

40. For all real numbers a and b, $|a + b|$ __?__ $|a| + |b|$. $\le$

Problems

a. Name a positive or a negative number to represent each measurement in the given problem.

b. Compute the sum of the numbers.

c. Answer the question.

EXAMPLE An elevator started at the 38th floor of an office building. It then went down 15 floors, up 6 floors, and down 18 floors. At what floor was the elevator then located?

SOLUTION **a.** 38, −15, 6, −18
b. 38 + (−15) + 6 + (−18) = 11
c. The elevator was located at the 11th floor.

A 1. A helicopter flew 43 km directly north from its base at Eagle Point and then flew 51 km directly south. Where was the helicopter then located relative to its base?

2. A salesperson drove 97 km directly east from the sales office to see a customer and then drove 123 km directly west to see a second customer. Where was the salesperson then located relative to the sales office?

3. A glider that was flying at an altitude of 3200 m dropped 340 m, rose 75 m, and then dropped 800 m. What was its new altitude?

4. A diving bell that was at a depth of 285 m below the surface of the ocean ascended 105 m, descended 220 m, and then ascended 175 m. What was its new depth?

5. Andrea bought four silver coins at prices of $12.05, $13.15, $13.45, and $12.05. She later sold these coins for $12.25, $12.85, $13.75, and $12.95, respectively. How much money did she make or lose in selling these coins?

6. A stock that opened on Monday morning at $45 per share gained $2.50 per share on each of Monday and Tuesday. It then dropped $1.75 per share on each of Wednesday, Thursday, and Friday. What was the closing price of this stock on Friday afternoon?

7. The temperature at midnight of one day was 8°C. The temperature fell 15°C during the next six hours, rose 9°C during the next twelve hours, and then fell 2°C during the next six hours. What was the temperature at midnight of the second day?

8. Robert's normal pulse rate is 72 beats per minute. After he jogged for 10 min, his pulse rate rose by 31 beats. It dropped by 29 beats after a 5 min rest and then rose by 34 beats after another 10 min of jogging. What was his pulse rate after this 25 min period?

9. Using her new revolving charge account, Alice Thornton purchased clothing worth $38.50 and stereo equipment worth $220.95. She then made two monthly payments of $84 each to her account. The finance charge on her account for this two-month period was $6.55. How much did Alice still owe on her account?

10. Earl Washington had a balance of $453.80 in his checking account at the start of the month. During the month, he wrote checks for $33.90, $14.62, $119.65, and $164.80, and he made a deposit of $360 from his paycheck. On the last day of the month, a $3.00 service charge and a $6.50 check-printing fee were deducted from his account. What was the starting balance of his account for the next month?

Working with Real Numbers **75**

b. −285 + 105 + (−220) + 175 = −225
c. The diving bell's new depth was 225 m below the surface of the ocean.

5. **a.** −12.05, −13.15, −13.45, −12.05, 12.25, 12.85, 13.75, 12.95
b. −12.05 + (−13.15) + (−13.45) + (−12.05) + 12.25 + 12.85 + 13.75 + 12.95 = 1.10
c. She made $1.10 in selling the coins.

6. **a.** 45, 2.50, 2.50, −1.75, −1.75, −1.75
b. 45 + 2.50 + 2.50 + (−1.75) + (−1.75) + (−1.75) = 44.75
c. The closing price of the stock was $44.75 on Friday afternoon.

7. **a.** 8, −15, 9, −2
b. 8 + (−15) + 9 + (−2) = 0
c. The temperature was 0°C at midnight of the second day.

8. **a.** 72, 31, −29, 34
b. 72 + 31 + (−29) + 34 = 108
c. His pulse rate was 108 beats per minute after the 25 min period.

9. **a.** −38.50, −220.95, 84, 84, −6.55
b. −38.50 + (−220.95) + 84 + 84 + (−6.55) = −98
c. Alice still owed $98 on her account.

10. **a.** 453.80, −33.90, −14.62, −119.65, −164.80, 360, −3.00, −6.50
b. 453.80 + (−33.90) + (−14.62) + (−119.65) + (−164.80) + 360 + (−3.00) + (−6.50) = 471.33
c. The starting balance was $471.33 for the next month.

Use what you know about addition of positive and negative numbers to answer each question.

B 11. A submarine fired a rocket that rose a total of 138 m and that reached an altitude of 125 m above the surface of the ocean. What was the position of the submarine relative to the surface? 13 m below

12. A passenger on a moving train walked toward the back of the train at a rate of 7 km/h. As a result, the passenger's rate of travel relative to the ground was 143 km/h. At what rate was the train traveling? 150 km/h

13. In one week, Diane deposited $150 in her savings account and then withdrew $65. If the amount of money in her account at the end of the week was $340, how much was in her account at the beginning? $255

14. The population of Bay City increased by 15,000 in one decade, then decreased by 8500 in the next decade. If the population at the end of these two decades was 23,000, what was the population at the beginning? 16,500

15. In one 5 min period, an elevator in an office building went up 9 floors, down 12 floors, up 5 floors, down 7 floors, and up 10 floors. If the elevator was then at the 11th floor, where was it located at the beginning of this 5 min period? at the 6th floor

16. At the four bus stops on Walnut Street, the 10:00 bus discharged 5 passengers and picked up 7; discharged 10 and picked up 3; discharged 8 and picked up 2; and discharged 6 and picked up 15. If there were 26 passengers on the bus after the fourth stop, how many were on the bus before the first stop? 28 passengers

C 17. Luis walked the following route as he was sightseeing in the city: 6 blocks north and 4 blocks east; 2 blocks south and 6 blocks east; 8 blocks west and 3 blocks north; and 7 blocks south and 2 blocks west. Where was Luis then located relative to his starting point?
at his starting point

Mixed Review

Simplify.

1. $10 + [6 + (-14)]$ 2

Solve.

2. $8 + m = 2$ $\{-6\}$

3. $-5 + (-y) = -7$ $\{2\}$

4. $7 + w = -7$ $\{-14\}$

5. $m + m + 1 = -5$ $\{-3\}$

PROGRAMMING IN BASIC

The computer program on page 34 can be expanded to provide graphs of additional open sentences over different domains. For example, in line 70 of the program that follows,

$$ABS(X) = |X|.$$

In line 60, the domain of the variable is changed to $\{-10, -9, \ldots, 9, 10\}$. Lines 150–220 locate and mark the zero point of the graph.

```
10  ⎫
 ⋮  ⎬  From program on page 34
50  ⎭
60   FOR X = -10 TO 10
70   IF ABS(X) = 2 THEN 120
```

```
80  ⎫
 ⋮  ⎬  From program on page 34
140 ⎭
150  FOR I = −10 TO 10
160  IF I <> 0 THEN 210
170  REM *ELSE
180  PRINT "0"
190  GOTO 230
200  REM *THEN            ⎧ Type a single space
210  PRINT " ";  ←———     ⎨ between the
220  NEXT I                ⎩ quotation marks.
230  END
```

Exercises

1. Type in and RUN the revised program. Compare the result with the graph shown in part (a) of Example 5 on page 64.

Change line 70 as indicated, then RUN the revised program.

2. 70   IF ABS(X) < 1 THEN 120

3. 70   IF ABS(X) >= 3 THEN 120

4. 70   IF 0 <= ABS(X) AND ABS(X) <= 2 THEN 120

5. 70   IF 3 > ABS(X) AND ABS(X) > 0 THEN 120

6. 70   IF 7 >= ABS(X) AND ABS(X) >= 3 THEN 120

The following change in this graphing program uses

$$INT(X) = \text{the } greatest \ integer \text{ less than or equal to X}$$

to mark every fifth unit on the number line in the graphs.

```
 72   REM *ELSE
 74   IF X/5 = INT(X/5) THEN 104
102   REM *THEN
104   PRINT "+";
106   GOTO 130
```

7. Type in these new lines and RUN the revised program.

8–12. Change line 70 as indicated in Exercises 2–6 and RUN the program again.

13. What changes do you need to make in the program if you wish to obtain graphs of open sentences over the domain {−20, −19, . . . , 1, 2}?

14. What changes do you need to make in the program if you wish to mark *every other unit* on the number line in the graphs?

Working with Real Numbers   **77**

2–7 Proving Statements: Theorems About Addition

As you have seen, axioms for the real numbers are statements that are *assumed* to be true. Other statements about the real numbers can be *proved* to be true, and these statements are called **theorems.**

A theorem consists of two parts, a *hypothesis* and a *conclusion*. The **hypothesis** states what is assumed to be true, and the **conclusion** states something that follows logically from this assumption. When you start with the hypothesis of a theorem and arrive at its conclusion through a logical chain of statements, you give a **direct proof** of the theorem. The reason that justifies each statement in a direct proof may be the hypothesis itself, or it may be an axiom, a definition, or a previously proved theorem.

For example, consider the following equations.

$$(8 + 5) + (-5) = 8$$
$$[2 + (-9)] + 9 = 2$$
$$[-4.5 + (-7)] + 7 = -4.5$$

These equations are true sentences that suggest the following theorem.

Theorem. For all real numbers a and b,

$$(a + b) + (-b) = a.$$

The hypothesis of this theorem is the statement that a and b are real numbers. In order to prove the theorem, you must reason from this hypothesis to the conclusion that $(a + b) + (-b) = a$. The following is a direct proof of this theorem.

PROOF

Statements	*Reasons*
1. a and b are real numbers.	Hypothesis
2. $-b$ is a real number.	Axiom of additive inverses
3. $(a + b) + (-b) = a + [b + (-b)]$	Associative axiom for addition
4. $b + (-b) = 0$	Axiom of additive inverses
5. $(a + b) + (-b) = a + 0$	Substitution principle
6. $a + 0 = a$	Identity axiom for addition
7. $\therefore (a + b) + (-b) = a$	Transitive axiom of equality

78 Chapter 2

The three dots, ∴, in statement (7) of the preceding proof are read as "Therefore."

To shorten the writing of a proof, simple statements involving closure, substitution, and basic properties of equality are frequently omitted. Also, the left side of an equation generally is not written when it is the same as the right side of the equation in the previous statement. For example, the proof just given may be shortened to the following form.

PROOF

Statements	*Reasons*
1. a and b are real numbers.	Hypothesis
2. $-b$ is a real number.	Axiom of additive inverses
3. $(a + b) + (-b) = a + [b + (-b)]$	Associative axiom for addition
4. $\quad\quad\quad\quad = a + 0$	Axiom of additive inverses
5. $\quad\quad\quad\quad = a$	Identity axiom for addition
6. $\therefore (a + b) + (-b) = a$	Transitive axiom of equality

Some familiar properties of the real numbers are actually theorems that can be proved using the fundamental axioms for the real numbers. The following example outlines a proof of the *property of the opposite of a sum*, which was discussed in Section 2-6.

EXAMPLE   Prove: For all real numbers a and b, $-(a + b) = -a + (-b)$.

SOLUTION   *Plan*: By the axiom of additive inverses, the additive inverse of the real number $(a + b)$ is a *unique* real number $-(a + b)$. If it can be shown that $(a + b) + [-a + (-b)] = 0$, then by the axiom of additive inverses $[-a + (-b)]$ is the additive inverse of $(a + b)$ and is therefore equal to $-(a + b)$.

PROOF

Statements	*Reasons*
1. a and b are real numbers.	Hypothesis
2. $-a$ and $-b$ are real numbers.	Axiom of additive inverses
3. $(a + b) + [-a + (-b)]$ $\quad = [a + (-a)] + [b + (-b)]$	Commutative and associative axioms for addition
4. $\quad = 0 + 0$	Axiom of additive inverses
5. $\quad = 0$	Identity axiom for addition
6. $(a + b) + [-a + (-b)] = 0$	Transitive axiom of equality
7. $-a + (-b) = -(a + b)$	Axiom of additive inverses
8. $\therefore -(a + b) = -a + (-b)$	Symmetric axiom of equality

Working with Real Numbers   **79**

Suggested Assignments
Minimum
First day: 81/1–13
Second day: 81/14–18
Average
  81/1–11 odd, 13–16,
    17–23 odd
Maximum
  81/5–11 odd, 13–28

Additional A Exercises
Simplify.
1. $[83 + (-23)] + 23$ 83
2. $-(22 + 33) + 22$ -33
3. $(13 + 4) + (-13)$ 4
4. $-15 + [-(-17 + 15)]$ 17
5. $(-25 + 32) + 25$ 32
6. $70 + [5 + (-70)]$ 5

Mixed Review
Simplify.
1. $-12 + 6 + (-31) - 16$
   -53
2. $-\dfrac{5}{11} + \left(-\dfrac{3}{11}\right) + \left(-\dfrac{1}{11}\right) +$
   $\dfrac{10}{11}\quad\dfrac{1}{11}$
3. $-6p^3 + 7p^2 - (-4p) + 7p^3$
   $p^3 + 7p^2 + 4p$
Evaluate each expression when $x = -1$ and $y = 2$.
4. $-(x + y)$ -1
5. $-|x + y|$ -1
6. $-|x^2 + y^2|$ -5

13. 1. Hypothesis
    2. Ax. of add. inverses
    3. Comm. ax. for add.
    4. Assoc. ax. for add.
    5. Ax. of add. inverses
    6. Identity ax. for add.
    7. Trans. ax. of equality
14. 1. Hypothesis
    2. Ax. of add. inverses
    3. Comm. ax. for add.
    4. Assoc. ax. for add.
    5. Ax. of add. inverses
    6. Identity ax. for add.
    7. Trans. ax. of equality
15. 1. Hypothesis
    2. Prop. of the opposite
       of a sum
    3. Comm. ax. for add.
    4. Assoc. ax. for add.
    5. Ax. of add. inverses
    6. Identity ax. for add.
    7. Trans. ax. of equality
16. 1. Hypothesis
    2. Ax. of add. inverses
    3. Prop. of the opposite
       of a sum
    4. Cancellation prop. of
       opposites
    5. Assoc. ax. for add.
    6. Ax. of add. inverses
    7. Identity ax. for add.
    8. Trans. ax. of equality
17. 1. a and b are real
       numbers.
       Hypothesis
    2. $-a$ is a real number.
       Ax. of add. inv.
    3. $-a + (a + b)$
       $= (-a + a) + b$
       Assoc. ax. for add.
    4. $= 0 + b$
       Ax. of add. inv.
    5. $= b$
       Iden. ax. for add.
    6. $\therefore -a + (a + b) = b$
       Trans. ax. of $=$

Another theorem for the real numbers is the *cancellation property of opposites*, which was discussed in Section 2-4. Exercise 14 below illustrates how the axioms can be used to prove this theorem.

Note that any theorem which has been proved can thereafter be used as a reason in other proofs.

Oral Exercises

Simplify.

1. $(9 + 4) + (-4)$ 9
2. $(-7 + 6) + 7$ 6
3. $8 + [5 + (-8)]$ 5
4. $-3 + [3 + (-9)]$ -9
5. $-(9 + 5) + 5$ -9
6. $-(10 + 2) + 10$ -2
7. $11 + [-(7 + 11)]$ -7
8. $5 + [-(5 + 12)]$ -12
9. $-(-3 + 8) + 8$ 3
10. $-(-9 + 7) + (-9)$ -7
11. $-10 + [-(-10 + 3)]$ -3
12. $4 + [-(-16 + 4)]$ 16

Replace each __?__ with the reason that justifies the statement to its left.

13. Prove: For all real numbers a and b, $(a + b) + (-a) = b$.

PROOF

Statements	*Reasons*
1. a and b are real numbers.	Hypothesis
2. $-a$ is a real number.	__?__ Axiom of additive inverses
3. $(a + b) + (-a) = (b + a) + (-a)$	Commutative axiom for addition
4. $\qquad\qquad = b + [a + (-a)]$	__?__ Associative axiom for addition
5. $\qquad\qquad = b + 0$	__?__ Axiom of additive inverses
6. $\qquad\qquad = b$	__?__ Identity axiom for addition
7. $\therefore (a + b) + (-a) = b$	Transitive axiom of equality

14. Prove: For all real numbers a, $-(-a) = a$.

PROOF

Statements	*Reasons*
1. a is a real number.	__?__ Hypothesis
2. $-a$ and $-(-a)$ are real numbers.	__?__ Axiom of additive inverses
3. $-(-a) = -(-a) + 0$	Identity axiom for addition
4. $\qquad = -(-a) + (-a + a)$	Axiom of additive inverses
5. $\qquad = [-(-a) + (-a)] + a$	__?__ Associative axiom for addition
6. $\qquad = 0 + a$	__?__ Axiom of additive inverses
7. $\qquad = a$	__?__ Identity axiom for addition
8. $\therefore -(-a) = a$	__?__ Transitive axiom of equality

80   Chapter 2

Written Exercises

Simplify.

4. −141   7. −49

A
1. $(82 + 65) + (-65)$ 82   2. $(-56 + 90) + 56$ 90   3. $38 + [104 + (-38)]$ 104
4. $-22 + [22 + (-141)]$ 5. $-(93 + 52) + 93$ −52   6. $17 + [-(47 + 17)]$ −47
7. $-(-18 + 49) + (-18)$ 8. $61 + [-(-14 + 61)]$ 14   9. $-[53 + (-16)] + (-16)$
10. $-85 + [-(-85 + 91)]$ 11. $-[24 + (-81)] + 24$ 12. $-[-43 + (-78)] + (-78)$
        −91   81   9. −53   12. 43

Give the reason that justifies each statement in the given proof.

13. Prove: For all real numbers
    a and b, $-b + (a + b) = a$.

    PROOF
    1. a and b are real numbers.
    2. $-b$ is a real number.
    3. $-b + (a + b) = -b + (b + a)$
    4. $= (-b + b) + a$
    5. $= 0 + a$
    6. $= a$
    7. $\therefore -b + (a + b) = a$

14. Prove: For all real numbers
    a and b, $(-a + b) + a = b$.

    PROOF
    1. a and b are real numbers.
    2. $-a$ is a real number.
    3. $(-a + b) + a = [b + (-a)] + a$
    4. $= b + (-a + a)$
    5. $= b + 0$
    6. $= b$
    7. $\therefore (-a + b) + a = b$

15. Prove: For all real numbers
    a and b, $-(a + b) + a = -b$.

    PROOF
    1. a and b are real numbers.
    2. $-(a + b) + a = [-a + (-b)] + a$
    3. $= [-b + (-a)] + a$
    4. $= -b + (-a + a)$
    5. $= -b + 0$
    6. $= -b$
    7. $\therefore -(a + b) + a = -b$

16. Prove: For all real numbers
    a and b, $-a + [-(-a + b)] = -b$.

    PROOF
    1. a and b are real numbers.
    2. $-a$ is a real number.
    3. $-a + [-(-a + b)]$
            $= -a + [-(-a) + (-b)]$
    4. $= -a + [a + (-b)]$
    5. $= (-a + a) + (-b)$
    6. $= 0 + (-b)$
    7. $= -b$
    8. $\therefore -a + [-(-a + b)] = -b$

Write a direct proof of each theorem.

B
17. For all real numbers a and b, $-a + (a + b) = b$.
18. For all real numbers a and b, $[a + (-b)] + b = a$.
19. For all real numbers a and b, $[a + (-b)] + (-a) = -b$.
20. For all real numbers a and b, $-b + (-a + b) = -a$.

18. 1. a and b are real
       numbers.
       Hypothesis
    2. $-b$ is a real number.
       Ax. of add. inv.
    3. $[a + (-b)] + b$
       $= a + [(-b) + b]$
       Assoc. ax. for add.
    4. $= a + 0$
       Ax. of add. inv.
    5. $= a$
       Iden. ax. for add.
    6. $\therefore [a + (-b)] + b = a$
       Trans. ax. of $=$

19. 1. a and b are real
       numbers.
       Hypothesis
    2. $-a$ and $-b$ are real
       numbers.
       Ax. of add. inv.
    3. $[a + (-b)] + (-a)$
       $= (-b + a) + (-a)$
       Comm. ax. for add.
    4. $= -b + [a + (-a)]$
       Assoc. ax. for add.
    5. $= -b + 0$
       Ax. of add. inv.
    6. $= -b$
       Iden. ax. for add.
    7. $\therefore [a + (-b)] + (-a)$
       $= -b$
       Trans. ax. of $=$

20. 1. a and b are real
       numbers.
       Hypothesis
    2. $-a$ and $-b$ are real
       numbers.
       Ax. of add. inv.
    3. $-b + (-a + b)$
       $= -b + [b + (-a)]$
       Comm. ax. for add.
    4. $= [(-b) + b] + (-a)$
       Assoc. ax. for add.
    5. $= 0 + (-a)$
       Ax. of add. inv.
    6. $= -a$
       Iden. ax. for add.
    7. $\therefore -b + (-a + b) = -a$
       Trans. ax. of $=$

(continued)

Write a direct proof of each theorem.

21. For all real numbers a and b, $-(a + b) + b = -a$.
22. For all real numbers a and b, $a + [-(a + b)] = -b$.
23. For all real numbers a and b, $-[a + (-b)] + (-b) = -a$.
24. For all real numbers a and b, $b + [-(-a + b)] = a$.

C 25. For all real numbers a and b, $(a + b) + [-a + (-b)] = 0$.
26. For all real numbers a and b, $-[a + (-b)] + [-(-a + b)] = 0$.
27. For all real numbers a, b, and c,
    $-[(a + b) + c] = [-a + (-b)] + (-c)$.
28. For all real numbers a, b, and c, $-[a + (b + c)] = -(a + b) + (-c)$.

Teaching Suggestions
p. T75

2–8 Subtracting Real Numbers

The first column below lists a few examples of subtracting 3. The second
column lists related examples of adding -3.

Supplementary Material

Test 6

Subtracting 3	Adding -3
$7 - 3 = 4$	$7 + (-3) = 4$
$6 - 3 = 3$	$6 + (-3) = 3$
$5 - 3 = 2$	$5 + (-3) = 2$
$4 - 3 = 1$	$4 + (-3) = 1$

Key Ideas

Use the definition of subtraction to subtract real numbers.

Comparing the entries in the two columns shows that *subtracting* 3 gives
the same result as *adding the opposite* of 3. This relationship between
addition and subtraction suggests the following definition.

Definition of Subtraction

For all real numbers a and b, the **difference** denoted as $a - b$ is
defined by:

$$a - b = a + (-b)$$

That is, to subtract b from a, add the opposite of b to a.

Chalkboard Examples

Simplify.
1. $19 - 23$
   $= 19 + (-23)$
   $= -4$
2. $-2r - (3 - 5r)$
   $= -2r - 3 + 5r$
   $= -2r + 5r - 3$
   $= 3r - 3$
3. $(7m - 5) - (3m + 8)$
   $= (7m - 5) + [-(3m + 8)]$
   $= (7m - 5) + [-3m + (-8)]$
   $= (7m + (-3m)) + [-5 + (-8)]$
   $= 4m + (-13)$
   $= 4m - 13$

EXAMPLE 1 Simplify.

a. $4 - 15$ b. $6 - (-21)$ c. $-9 - 17$ d. $-3 - (-18)$

SOLUTION a. $4 - 15 = 4 + (-15) = -11$
b. $6 - (-21) = 6 + 21 = 27$
c. $-9 - 17 = -9 + (-17) = -26$
d. $-3 - (-18) = -3 + 18 = 15$

82 *Chapter 2*

EXAMPLE 2   Simplify $14 - 29 + 16 - 43$.

SOLUTION 1   Add or subtract in order from left to right.
$$14 - 29 = -15; \; -15 + 16 = 1; \; 1 - 43 = -42$$

SOLUTION 2   Group positive and negative numbers.
$$14 - 29 + 16 - 43 = (14 + 16) - (29 + 43) = 30 - 72 = -42$$

Subtraction of real numbers is *not* commutative. For example:
$$7 - 2 = 5 \quad \text{but} \quad 2 - 7 = -5$$

Nor is subtraction of real numbers associative. For example:
$$(11 - 4) - 1 = 7 - 1 = 6 \quad \text{but} \quad 11 - (4 - 1) = 11 - 3 = 8$$

On the other hand, note that
$$7(9 - 4) = 7 \times 9 - 7 \times 4,$$

since $7(9 - 4) = 7(5) = 35$ and $7 \times 9 - 7 \times 4 = 63 - 28 = 35$. This example illustrates the fact that multiplication is *distributive with respect to subtraction*, which will be proved as a theorem in Exercise 62 on page 92.

For all real numbers a, b, and c,
$$a(b - c) = ab - ac \quad \text{and} \quad (b - c)a = ba - ca.$$

Certain sums are usually replaced by differences. For example:

$5n + (-7)$ is usually written $5n - 7$.
$10y^2 + (-3)y$ is usually written $10y^2 - 3y$.

EXAMPLE 3   Simplify $12 - 3a - 21 + 14a$.

SOLUTION
$$
\begin{aligned}
12 - 3a - 21 + 14a &= (-3a + 14a) + (12 - 21) \\
&= 11a + (-9) \\
&= 11a - 9
\end{aligned}
$$

When you subtract a sum within grouping symbols, you use the property of the opposite of a sum.

EXAMPLE 4   Simplify $(10b - 7) - (2b + 5)$.

SOLUTION
$$
\begin{aligned}
(10b - 7) - (2b + 5) &= (10b - 7) + [-(2b + 5)] \\
&= (10b - 7) + [-2b + (-5)] \\
&= [10b + (-2b)] + [-7 + (-5)] \\
&= 8b + (-12) \\
&= 8b - 12
\end{aligned}
$$

Working with Real Numbers   **83**

Mixed Review

Simplify.
1. $(4 + 5) + (-5)$ 4
2. $-(46 + 13) + 13$ -46
3. $18 + [81 + (-18)]$ 81
4. $-16 + [-(-16 + 10)]$
   -10

What is the opposite of the *difference* of two real numbers a and b? If the difference is denoted $a - b$, then

$$-(a - b) = -a + b.$$

This fact will be proved as a theorem in Exercise 48 on page 85.

Oral Exercises

Simplify.

1. $15 - 8$ 7
2. $15 - (-8)$ 23
3. $-15 - 8$ -23
4. $-15 - (-8)$ -7
5. $8 - 15$ -7
6. $8 - (-15)$ 23
7. $-8 - 15$ -23
8. $-8 - (-15)$ 7
9. $10 - 0$ 10
10. $0 - 10$ -10
11. $0 - (-10)$ 10
12. $-10 - 0$ -10
13. $(9 - 3) - 5$ 1
14. $9 - (3 - 5)$ 11
15. $(3 - 5) - 9$ -11
16. $3 - (5 - 9)$ 7
17. $6(8 - 3)$ 30
18. $6 \times 8 - 6 \times 3$ 30
19. $6 \times 8 - 3 \times 8$ 24
20. $(6 - 3)8$ 24
21. $2a - (2a + 7)$ -7
22. $2a - (7 + 2a)$ -7
23. $2a - (2a - 7)$ 7
24. $2a - (7 - 2a)$ $4a - 7$
25. $(3x + 20) - (3x + 12)$ 8
26. $(3x + 20) - (3x - 12)$ 32
27. $(3x - 20) - (3x + 12)$ -32
28. $(3x - 20) - (3x - 12)$ -8
29. $(20 - 3x) - (12 + 3x)$ $-6x + 8$
30. $(20 - 3x) - (12 - 3x)$ 8

Replace each _?_ with the reason that justifies the statement to its left.

31. Prove: For all real numbers a and b, $a - (-b) = a + b$.

PROOF

Statements	Reasons
1. a and b are real numbers.	Hypothesis
2. $-b$ is a real number.	_?_ Axiom of additive inverses
3. $a - (-b) = a + [-(-b)]$	Definition of subtraction
4. $\quad\quad = a + b$	_?_ Cancellation property of opposites
5. $\therefore a - (-b) = a + b$	_?_ Transitive axiom of equality

Written Exercises

Simplify.

A
1. $25 - 48$ -23
2. $39 - (-81)$ 120
3. $-16 - 37$ -53
4. $-34 - (-56)$ 22
5. $61 - 17$ 44
6. $-42 - (-35)$ -7
7. $78 - (-22)$ 100
8. $-97 - 32$ -129
9. $-6.9 - 3.7$ -10.6
10. $3.6 - 8.5$ -4.9
11. $-1.4 - (-5.8)$ 4.4
12. $-9.2 - (-4.9)$ -4.3
13. $0 - \left(-\frac{5}{8}\right)$ $\frac{5}{8}$
14. $-\frac{4}{9} - \frac{7}{9}$ $-\frac{11}{9}$
15. $8\frac{2}{7} - 1\frac{5}{7}$ $6\frac{4}{7}$
16. $1\frac{2}{3} - \left(-\frac{2}{3}\right)$ $2\frac{1}{3}$

84 *Chapter 2*

17. $18 - 54 - 33$ -69
18. $-97 - (-43) - 24$ -78
19. $74 - 92 - 16 - (-58)$ 24
20. $-26 - 83 + 65 - 49$ -93
21. $-47 + 50 + 66 - 12 - 34$ 23
22. $36 - 52 + 71 - 38 + 13$ 30
23. $93 - (23 + 45)$ 25
24. $67 - (-49 + 27)$ 89
25. $46 - (-63 - 24)$ 133
26. $-82 - (-12 - 93)$ 23
27. $(43 - 71) - (-71 - 52)$ 95
28. $(63 - 84) - (-19 + 63)$ -65
29. $-(97 - 38) - (38 + 43)$ -140
30. $-(-81 - 59) - (59 - 15)$ 96
31. $-12x + 7 - 4x - 10$ $-16x - 3$
32. $3y - 4 - 11 - 7y$ $-4y - 15$
33. $14 - 9z^2 - 22 - 12z^2$ $-21z^2 - 8$
34. $-15w^2 - 12w + 6w - 18w^2$ $-33w^2 - 6w$
35. $-13m^2 - 9m + 8 - 10m^2 - 14$ $-23m^2 - 9m - 6$
36. $3n - 5n^3 - 7n + 7n^2 - 4n^3$ $-9n^3 + 7n^2 - 4n$

B
37. $(-11a - 7) - (-1 + 4a)$ $-15a - 6$
38. $(13b - 9) - (-7b - 15)$ $20b + 6$
39. $-(16p + 9) - (5p - 6)$ $-21p - 3$
40. $-(6q - 14) - (8q + 5)$ $-14q + 9$
41. $(8m - 5n) - (-18n + 5m)$ $3m + 13n$
42. $(-7h + 8g) - (12g + 11h)$ $-18h - 4g$
43. $-(7x^3 + 4x) - (4x^3 - 11x)$ $-11x^3 + 7x$
44. $(3y^2 - 10y) - (-5y - 10y^2)$ $13y^2 - 5y$
45. $-9z^2 - (13z^2 + 2z) - (-6z - 7)$ $-22z^2 + 4z + 7$
46. $-(7w + 2w^2) - (19w^2 - 5w^3) - 9w^3$ $-4w^3 - 21w^2 - 7w$

Give the reason that justifies each statement in the given proof.

47. Prove: For all real numbers
a and b, $(a + b) - b = a$.

PROOF

1. a and b are real numbers.
2. $(a + b) - b = (a + b) + (-b)$
3. $\qquad = a + [b + (-b)]$
4. $\qquad = a + 0$
5. $\qquad = a$
6. $\therefore (a + b) - b = a$

48. Prove: For all real numbers
a and b, $-(a - b) = -a + b$.

PROOF

1. a and b are real numbers.
2. $-(a - b) = -[a + (-b)]$
3. $\qquad = -a + [-(-b)]$
4. $\qquad = -a + b$
5. $\therefore -(a - b) = -a + b$

Write a direct proof of each theorem.

C
49. For all real numbers a and b, $(a - b) + b = a$.
50. For all real numbers a and b, $-(-a - b) = a + b$.
51. For all real numbers a, b, and c, $-[(a + b) + c] = (-a - b) - c$.
52. For all real numbers a, b, and c, $a - (b - c) = (a - b) + c$.

A set of numbers is said to be *closed under subtraction* if all possible differences between members of the set are also members of the set. Tell whether each set is closed or not closed under subtraction.

53. $\{0\}$ closed
54. $\{1\}$ not closed
55. $\{0, 1\}$ not closed
56. $\{-1, 0, 1\}$ not closed
57. $\{$the natural numbers$\}$ not closed
58. $\{$the integers$\}$ closed
59. $\{$the even integers$\}$ closed
60. $\{$the odd integers$\}$ not closed
61. $\{$the rational numbers$\}$ closed
62. $\{$the real numbers$\}$ closed

Working with Real Numbers **85**

47. 1. Hypothesis
2. Def. of subt.
3. Assoc. ax. for add.
4. Ax. of add. inverses
5. Identity ax. for add.
6. Trans. ax. of equality

48. 1. Hypothesis
2. Def. of subt.
3. Prop. of the opp. of a sum
4. Cancellation prop. of opposites
5. Trans. ax. of equality

49. 1. a and b are real numbers. Hypothesis
2. $(a - b) + b$ $= [a + (-b)] + b$ Def. of subt.
3. $= a + (-b + b)$ Assoc. ax. for add.
4. $= a + 0$ Ax. of add. inv.
5. $= a$ Iden. ax. for add.
6. $\therefore (a - b) + b = a$ Trans. ax. of =

50. 1. a and b are real numbers. Hypothesis
2. $-(-a - b)$ $= -[-a + (-b)]$ Def. of subt.
3. $= -(-a) + [-(-b)]$ Prop. of opp. of a sum
4. $= a + b$ Canc. prop. of opp.
5. $\therefore -(-a - b) = a + b$ Trans. ax. of =

(continued)

Problems

a. **Express the answer to each question as the difference between two real numbers and compute this difference.**

b. **Interpret the sign of the difference (positive or negative) and answer the question.**

EXAMPLE What is the difference in altitude between Mount McKinley, Alaska, which is 6194 m above sea level, and Death Valley, California, which is 86 m below sea level?

SOLUTION 1 **a.** $6194 - (-86) = 6194 + 86 = 6280$
b. Mount McKinley is 6280 m higher than Death Valley.

SOLUTION 2 **a.** $-86 - 6194 = -86 + (-6194) = -6280$
b. Death Valley is 6280 m lower than Mount McKinley.

A 1. Carlos rode the elevator from the 29th floor to the parking garage 2 floors below street level. How many floors did he ride the elevator?

2. Tania rode her bicycle from her home 55 blocks south of Main Street to a store 36 blocks north of Main Street. How many blocks did she ride?

3. The German mathematician Emmy Noether was born in 1882 and died after her birthday in 1935. What was her age in years when she died?

4. If the Greek mathematician Pythagoras was born in 572 B.C. and died after his birthday in 497 B.C., what was his age in years when he died?

5. The highest temperature recorded in the world is 58°C, while the lowest temperature recorded is −88.3°C. What is the difference between these record high and low temperatures?

6. Mercury melts at a temperature of −38.86°C, but it does not boil until it is at a temperature of 356.86°C. What is the difference between the melting and boiling points of mercury?

7. When the wind is blowing at 72 km/h, an actual temperature of 2°C can feel like a temperature of −17°C. What is the difference between these temperatures?

8. An actual temperature of 32°C can feel like a temperature of 50°C when the relative humidity is 90%. What is the difference between these temperatures?

Use what you know about subtraction of positive and negative numbers to answer the question.

B 9. Mount Everest rises 8848 m above sea level, which is 19,763 m higher than the floor of the Marianas Trench in the Pacific Ocean. What is the depth of the Marianas Trench? 10,915 m below sea level

10. The melting point of nitrogen is 14.21°C lower than its boiling point. If its boiling point is −195.79°C, what is its melting point? −210°C

11. There is a difference of approximately 52° latitude between Point Barrow, Alaska and Ka Lae, Hawaii. If Ka Lae is located at 19° north latitude, what is the latitude of Point Barrow? 71° north latitude

Computer Exercises For students with computer experience

1. Write a program that will print in chart form the values of the following expressions when you input values for a and b.

   $|a| - |b|$ $\qquad$ $|b| - |a|$ $\qquad$ $|a - b|$ $\qquad$ $|b - a|$

   RUN the program for several test values of a and b. Which two expressions have equal values for all real values of a and b? $|a - b|$ and $|b - a|$

2. Write a program that will compute the distance between any two points on a number line when you input their coordinates.

Self-Test 2

VOCABULARY identity element for addition (p. 68)
additive inverse (p. 68)
theorem (p. 78)
hypothesis (p. 78)

conclusion (p. 78)
direct proof (p. 78)
subtraction (p. 82)
difference (p. 82)

Simplify.

1. $-7 + 2$ −5 $\qquad$ 2. $-3 + (-6)$ −9 $\qquad$ 3. $5 + (-5)$ 0 $\qquad$ *Obj. 1, p. 67*

4. $-9 + 17 + 35 + (-4)$ 39 $\qquad$ 5. $3a + (-7)b + (-8)a$ $\qquad$ *Obj. 2, p. 67*
$\qquad\qquad\qquad\qquad\qquad\qquad\qquad$ $(-5)a + (-7)b$

Give the reason that justifies each statement.

6. $x + 0 = x$ $\qquad$ 7. $x + (-x) = 0$ $\qquad$ 8. $-(-x) = x$ $\qquad$ *Obj. 3, p. 67*
Identity axiom for addition $\qquad$ Axiom of additive inverses $\qquad$ Cancellation property of

Simplify. $\qquad\qquad\qquad\qquad\qquad\qquad\qquad\qquad\qquad\qquad\qquad$ opposites

9. $-90 - (-36)$ −54 $\qquad$ 10. $(4n + 3) - (-2n - 7)$ $\qquad$ *Obj. 4, p. 67*
$\qquad\qquad\qquad\qquad\qquad\qquad\qquad\qquad$ $6n + 10$

Check your answers with those at the back of the book.

Working with Real Numbers **87**

Multiplication and Division

OBJECTIVES for Sections 2-9 through 2-11:
1. *To multiply real numbers and to simplify variable expressions involving multiplication.*
2. *To use multiplicative inverses and to simplify variable expressions involving multiplicative inverses.*
3. *To divide real numbers and to simplify variable expressions involving division.*

Teaching Suggestions
p. T75

Key Ideas

Use the rules for multiplication in multiplying real numbers.
Simplify variable expressions involving multiplication.

Chalkboard Examples

1. Simplify.
   a. $(-a)(-a)$
      $= [(-1)a][(-1)a]$
      $= [(-1)(-1)][a \cdot a]$
      $= 1 \cdot a^2 = a^2$
   b. $(-1)(-5)$
      $= (-1)[(-1)(5)]$
      $= [(-1)(-1)](5)$
      $= 1 \cdot 5 = 5$
2. Tell whether the expression names a positive number, a negative number, or zero. Then simplify the expression.
   a. $(-2)^4(5)^2(0)^4$ zero; 0
   b. $(-1)^9(-1)^5$ positive; 1
3. Simplify $-2(-3t + 5u)$.
   $= -2[(-3)t + (5)u]$
   $= (-2)(-3)t + (-2)(5)u$
   $= 6t + (-10)u$
   $= 6t - 10u$

2–9 Rules for Multiplication

When you multiply any given real number by 1, the product is identical to the given number. For example,

$$5 \times 1 = 5 \quad \text{and} \quad 1 \times 5 = 5.$$

Thus 1 is called the **identity element for multiplication.**

Identity Axiom for Multiplication
There is a unique real number 1, $1 \neq 0$, such that, for every real number a,

$$a \cdot 1 = a \quad \text{and} \quad 1 \cdot a = a.$$

The equations

$$5 \times 0 = 0 \quad \text{and} \quad 0 \times 5 = 0$$

illustrate the *multiplicative property of zero*: When one of the factors of a product is zero, the product itself is zero. The following is a formal statement of this property.

Multiplicative Property of Zero
For every real number a,

$$a \cdot 0 = 0 \quad \text{and} \quad 0 \cdot a = 0.$$

This property is a theorem that can be proved as follows.

Plan: By the identity axiom for addition, there is a *unique* real number 0 such that $a \cdot 0 + 0 = a \cdot 0$. If it can be shown that $a \cdot 0 + a \cdot 0 = a \cdot 0$, then by the identity axiom for addition $a \cdot 0$ must itself be equal to 0.

88   Chapter 2

PROOF

Statements	Reasons
1. a is a real number.	Hypothesis
2. $0 = 0 + 0$	Identity axiom for addition
3. $a(0 + 0) = a \cdot 0$	Substitution principle
4. $a \cdot 0 + a \cdot 0 = a \cdot 0$	Distributive axiom
5. $\therefore a \cdot 0 = 0$	Identity axiom for addition
6. $a \cdot 0 = 0 \cdot a$	Commutative axiom for multiplication
7. $\therefore 0 \cdot a = 0$	Transitive axiom of equality

Would you guess that $a(-1) = -a$? You might arrive at this guess by noticing that

$$2 \times (-1) = (-1) + (-1) = -2,$$
$$3 \times (-1) = (-1) + (-1) + (-1) = -3,$$
$$4 \times (-1) = (-1) + (-1) + (-1) + (-1) = -4,$$

and so on. These examples suggest the *multiplicative property of* -1: multiplying any real number by -1 produces the opposite of the number. You can state this property formally as follows.

Multiplicative Property of -1

For every real number a,

$$a(-1) = -a \quad \text{and} \quad (-1)a = -a.$$

This property will be proved as a theorem in Exercise 28 on page 91. Note that a special case of this property occurs when the value of a is -1:

$$(-1)(-1) = -(-1) = 1$$

The multiplicative property of -1 and the identity axiom for multiplication are used frequently in simplifying variable expressions.

EXAMPLE 1   Simplify.

a. $x(-x)$ b. $(-y)(-y)$ c. $-2z^3 + z^3$

SOLUTION   a. $x(-x) = x[(-1)x] = (-1)(x \cdot x) = (-1)x^2 = -x^2$

b. $(-y)(-y) = [(-1)y][(-1)y] = [(-1)(-1)](y \cdot y) = 1 \cdot y^2 = y^2$

c. $-2z^3 + z^3 = -2 \cdot z^3 + 1 \cdot z^3 = (-2 + 1)z^3 = (-1)z^3 = -z^3$

You can compute the product of *any* two real numbers by using the multiplicative property of -1 along with the familiar multiplication facts for positive numbers and with the commutative and associative axioms.

Working with Real Numbers   **89**

Suggested Assignments

Minimum
   91/1–25, 27–39 odd
Average
   91/7–57 odd, 58–63
Maximum
   91/7–59 odd, 61–68

Additional A Exercises

Simplify.

1. $(-1)(-5)(-3)$ -15

2. $(-1)^4(-2)^3$ -8

3. $-36(2) - (-2)$ -70

4. $(-14)(-7)(0)$ 0

5. $(-2n + 4m)3$ $-6n + 12m$

6. $-3(2b + 5r)$ $-6b - 15r$

Mixed Review

Simplify.

1. $16 - 49$ -33

2. $2\frac{7}{12} - \left(-\frac{5}{12}\right)$ 3

3. $(y^2 - y - 2) + (y - 2)$
$y^2 - 4$

4. $(a^2 + 2a - 1) -$
$(a + 1)$ $a^2 + a - 2$

5. $(-c - 2d) -$
$(-4c - 2d)$ $3c$

Additional Answers
Written Exercises (p. 92)

61. 1. Hypothesis
    2. Ax. of add. inverses
    3. Mult. prop. of -1
    4. Assoc. ax. for mult.
    5. Mult. prop. of -1
    6. Trans. ax. of equality

62. 1. Hypothesis
    2. Def. of subt.
    3. Distrib. axiom
    4. Prop. of opposites in products
    5. Def. of subt.
    6. Trans. ax. of equality

Consider the following examples.

$$4(3) = 12$$
$$(-4)3 = [(-1)4]3 = (-1)[(4)(3)] = (-1)12 = -12$$
$$4(-3) = 4[3(-1)] = [(4)(3)](-1) = 12(-1) = -12$$
$$(-4)(-3) = [(-1)4][(-1)3] = [(-1)(-1)][(4)(3)] = (1)12 = 12$$

These examples suggest the following theorem, which will be proved in Exercises 61, 63, and 64 on pages 92–93.

Property of Opposites in Products

For all real numbers a and b:

$$(-a)b = -ab \qquad a(-b) = -ab \qquad (-a)(-b) = ab$$

Practice in computing products of real numbers may lead you to discover the following rules.

Rules for Multiplication of Positive and Negative Numbers

1. The product of a positive number and a negative number is a negative number.

2. The product of two positive numbers or of two negative numbers is a positive number.

3. The absolute value of the product of two real numbers is equal to the product of the absolute values of the numbers. That is, for all real numbers a and b, $|ab| = |a| \cdot |b|$.

When a product has more than two factors, pairing the negative numbers should lead you to find that

the product of an *even* number of negative numbers is *positive*;

the product of an *odd* number of negative numbers is *negative*.

EXAMPLE 2   Tell whether the expression names a positive number, a negative number, or zero. Then simplify the expression.

    a. $-7(-5)(-2)$ **b.** $5^3(-2)^2(0)$ **c.** $(-1)^9(-2)^3(7)$

SOLUTION   **a.** negative; -70 **b.** zero; 0 **c.** positive; 56

EXAMPLE 3   Simplify.   **a.** $(-4a)(-6a)$ **b.** $-7(b - 2c)$

SOLUTION   **a.** $(-4a)(-6a) = [(-4)a][(-6)a] = [(-4)(-6)](a \cdot a) = 24a^2$

               b. $-7(b - 2c) = -7[b + (-2)c] = -7b + (-7)(-2)c = -7b + 14c$

Oral Exercises

Simplify.

1. $(-9)4$ -36
2. $(6)(-5)$ -30
3. $(-3)(-7)$ 21
4. $(-4)(-8)$ 32
5. $(-1)(-9)(5)$ 45
6. $(-7)(-1)(-4)$ -28
7. $(-3)(-4)(-2)$ -24
8. $(-5)(2)(-8)$ 80
9. $(19)(0)(41)$ 0
10. $(18)(-11)(0)$ 0
11. $6(-a)$ $-6a$
12. $(-13)(-z)$ 13z
13. $2(-5x)$ $-10x$
14. $(-6)(-9y)$ 54y
15. $(-8a)(-7b)$ 56ab
16. $(-6p)(3q)(2r)$ $-36pqr$
17. $(-10m)(7m)$ $-70m^2$
18. $(-4n)(5n)(-2n)$ $40n^3$
19. $-12g + g$ $-11g$
20. $-10h - h$ $-11h$
21. $-6j + 7j$ j
22. $-11k + 10k$ $-k$
23. $4v - 3v$ v
24. $5w - 6w$ $-w$
25. $(-a)(-a)$ a^2
26. $-b + (-b)$ $-2b$
27. $-c - (-c)$ 0

Replace each ? with the reason that justifies the statement to its left.

28. **Prove:** For all real numbers a, $a(-1) = -a$ and $(-1)a = -a$.

Plan: By the axiom of additive inverses, the additive inverse of the real number a is the *unique* real number $-a$. If it can be shown that $a + a(-1) = 0$, then by the axiom of additive inverses $a(-1)$ is the additive inverse of a and is therefore equal to $-a$.

PROOF

Statements	*Reasons*
1. a is a real number.	_?_ Hypothesis
2. $a + a(-1) = a \cdot 1 + a(-1)$	Identity axiom for multiplication
3. $\qquad = a[1 + (-1)]$	_?_ Distributive axiom
4. $\qquad = a \cdot 0$	_?_ Axiom of additive inverses
5. $\qquad = 0$	Multiplicative property of zero
6. $a + a(-1) = 0$	_?_ Transitive axiom of equality
7. $\therefore a(-1) = -a$	_?_ Axiom of additive inverses
8. $a(-1) = (-1)a$	_?_ Commutative axiom for mult.
9. $\therefore (-1)a = -a$	_?_ Transitive axiom of equality

Written Exercises

Simplify.

A

1. $(17)(-35)(-1)$ 595
2. $(-22)(-1)(-29)$ -638
3. $(-36)(0)(-14)$ 0
4. $(-2)(-16)(31)$ 992
5. $(-6)(-2)(15)(-3)$ -540
6. $(-8)(19)(-7)(0)$ 0
7. $(-2)(-3)(-5)^2$ 150
8. $(-2)^3(-3)(-5)$ -120
9. $(-2)^3(-3)^2(11)$ -792
10. $(-2)(-3)^3(-7)^2$ 2646
11. $(-1)^6(-2)(-5)^3$ 250
12. $(-1)^7(-2)^4(7)$ -112

Working with Real Numbers **91**

63. 1. a and b are real numbers.
Hypothesis
2. $-a$ is a real number.
Ax. of add. inv.
3. $(-a)b = [(-1)a]b$
Mult. prop. of -1
4. $\quad = (-1)(ab)$
Assoc. ax. for mult.
5. $\quad = -ab$
Mult. prop. of -1
6. $\therefore (-a)b = -ab$
Trans. ax. of $=$

64. 1. a and b are real numbers.
Hypothesis
2. $-a$ and $-b$ are real numbers.
Ax. of add. inv.
3. $(-a)(-b)$
$= [a(-1)](-b)$
Mult. prop. of -1
4. $\quad = a[-1(-b)]$
Assoc. ax. for mult.
5. $\quad = a[-(-b)]$
Mult. prop. of -1
6. $\quad = ab$
Canc. prop. of opp.
7. $\therefore (-a)(-b) = ab$
Trans. ax. of $=$

65. 1. a, b, and c are real numbers.
Hypothesis
2. $-a$ is a real number.
Ax. of add. inv.
3. $-a(b + c)$
$= (-a)b + (-a)c$
Dist. ax.
4. $\quad = -ab + (-ac)$
Prop. of opp. in prod.
5. $\quad = -ab - ac$
Def. of subt.
6. $\therefore -a(b + c)$
$= -ab - ac$
Trans. ax. of $=$

(continued)

(continued)

66. 1. a, b, and c are real
   numbers.
   Hypothesis
2. $-a$ is a real number.
   Ax. of add. inv.
3. $-a(b - c)$
   $= -a(b) - (-a)(c)$
   Exercise 62, page 92
4. $= -ab - (-ac)$
   Prop. of opp. in prod.
5. $= -ab + [-(-ac)]$
   Def. of subt.
6. $= -ab + ac$
   Canc. prop. of opp.
7. $= ac + (-ab)$
   Comm. ax. for add.
8. $= ac - ab$
   Def. of subt.
9. $\therefore -a(b - c)$
   $= ac - ab$
   Trans. ax. of $=$

67. 1. a, b, c, and d are real
   numbers.
   Hypothesis
2. $a[(b - c) - d]$
   $= a(b - c) - ad$
   Exercise 62, page 92
3. $= (ab - ac) - ad$
   Exercise 62, page 92
4. $= [ab + (-ac)] +$
   $(-ad)$
   Def. of subt.
5. $= ab +$
   $[-ac + (-ad)]$
   Assoc. ax. for add.
6. $= ab +$
   $[(-1)(ac) + (-1)(ad)]$
   Mult. prop. of -1
7. $= ab +$
   $[(-1)(ac + ad)]$
   Dist. ax.
8. $= ab + [-(ac + ad)]$
   Mult. prop. of -1
9. $= ab - (ac + ad)$
   Def. of subt.
10. $\therefore a[(b - c) - d]$
    $= ab - (ac + ad)$
    Trans. ax. of $=$

Simplify.

13. $27(-4) + 27(-6)$ -270
14. $-98(3) - 2(3)$ -300
15. $-5(-33) + 5(-33)$ **15.** 0
16. $-12(-8) - 12(8)$ 0
17. $-1(-56) - 9(-56)$ 560
18. $-83(11) - 83(-1)$ **18.** -830
19. $-49(7) + (-7)$ -350
20. $-12 + 12(-39)$ -480
21. $-22 - 29(22)$ -660
22. $-15(-21) - 15$ 300
23. $-89(11) + 89$ -890
24. $-16(41) + 16$ -640
25. $(-x)(-x)(-x)$ $-x^3$
26. $(-y)(-y)(-y)(-y)$ y^4
27. $(-9a)(4a)(-2a)$ **27.** $72a^3$
28. $(-6b)(-8b)(-2)$ $-96b^2$
29. $(-10p)(-8q)(-r)$ $-80pqr$
30. $(-x)(6y)(-14z)$
31. $-2(5m + 3)$ $-10m - 6$
32. $(-6n + 13)(-3)$ $18n - 39$
33. $(-10j - 4k)7$
34. $-8(9g - 4h)$ $-72g + 32h$
35. $-3(-x + 2y)$ $3x - 6y$
36. $(-9u - v)(-11)$
30. $84xyz$ **33.** $-70j - 28k$ **36.** $99u + 11v$

B
37. $2w - 3x - w + x$ $w - 2x$
38. $-9y - z + 8y - 2z$ $-y - 3z$
39. $-12j^2 + 7j - 8j + j^2$ $-11j^2 - j$
40. $-6k + 9k^3 + k^3 + 5k$ $10k^3 - k$
41. $-a^2 + 9a - 2a^2 - 9 - 10a$ $-3a^2 - a - 9$
42. $-b^2 - b + 7b + 2b^2 - 6b$ b^2
43. $m + 5n - (3m + 4n)$ $-2m + n$
44. $7p + q - (7q - 6p)$ $13p - 6q$
45. $7r + 3(s - 2r)$ $r + 3s$
46. $-2(5u - v) - v$ $-10u + v$
47. $-5(-c + d) + 6c - 4d$ $11c - 9d$
48. $-11g + h - 4(-3g - 2h)$ $g + 9h$
49. $4(-a + 2b) + 3(-2a - 3b)$ $-10a - b$
50. $7(-c - 2d) + 3(2c - 5d)$ $-c - 29d$
51. $5(3x - 4y) - 8(2x + y)$ $-x - 28y$
52. $-2(2v - 11w) - 3(v + 7w)$ $-7v + w$
53. $-7(p - 3q) - 5(4q - 3p)$ $8p + q$
54. $4(-7r - 2s) - 3(-3s - 4r)$ $-16r + s$

Evaluate each expression when $a = -2$ and $b = -5$.

55. $2[-3(a - 4b) - 9b] + 7a - 5b$ -7
56. $[2(-3a + b) + 8a]3 - 5a - 7b$ 3
57. $-2[2(2a - 5b) + b] + 2(5a - 9b)$ -4
58. $[-2(3a - b) + 2a]3 - 2(-6a + 5b)$ 20
59. $2[a - 3(2a - b)] - 3[-4b - 3(a - 2b)]$ 2
60. $-3[2a - 2(5a - 2b)] + 2[-8a - 2(2a - 3b)]$ 0

Give the reason that justifies each statement in the given proof.

61. Prove: For all real numbers
   a and b, $a(-b) = -ab$.

PROOF

1. a and b are real numbers.
2. $-b$ is a real number.
3. $a(-b) = a[b(-1)]$
4. $= (ab)(-1)$
5. $= -ab$
6. $\therefore a(-b) = -ab$

62. Prove: For all real numbers
   a, b, and c, $a(b - c) = ab - ac$.

PROOF

1. a, b, and c are real numbers.
2. $a(b - c) = a[b + (-c)]$
3. $= ab + [a(-c)]$
4. $= ab + (-ac)$
5. $= ab - ac$
6. $\therefore a(b - c) = ab - ac$

Write a direct proof of each theorem.

C 63. For all real numbers a and b, $(-a)b = -ab$.
64. For all real numbers a and b, $(-a)(-b) = ab$.
65. For all real numbers a, b, and c, $-a(b + c) = -ab - ac$.
66. For all real numbers a, b, and c, $-a(b - c) = ac - ab$.
67. For all real numbers a, b, c, and d, $a[(b - c) - d] = ab - (ac + ad)$.
68. For all real numbers a, b, c, and d, $a[b - (c - d)] = (ab - ac) + ad$.

Computer Exercises   For students with computer experience

Write a program that will allow you to input any pair of real numbers and will tell you, *without adding*, whether their sum is positive, negative, or zero. RUN the program for each of the following pairs of numbers.

1. 10, 2 POSITIVE
2. -10, -2 NEGATIVE
3. -10, 2 NEGATIVE
4. 10, -2 POSITIVE
5. 10, -10 ZERO
6. 0, -10 NEGATIVE
7. 2, 0 POSITIVE
8. 0, 0 ZERO

Write a program that will allow you to input any pair of real numbers and will tell you, *without multiplying*, whether their product is positive, negative, or zero. RUN the program for each of the following pairs of numbers.

9. 5, 7 POSITIVE
10. -5, 7 NEGATIVE
11. -5, -7 POSITIVE
12. -5, 0 ZERO

2–10 The Multiplicative Inverse of a Real Number

Two numbers whose product is 1 are called **multiplicative inverses,** or **reciprocals,** of each other. For example:

4 and $\frac{1}{4}$ are multiplicative inverses because $4 \times \frac{1}{4} = 1$.

$-\frac{3}{5}$ and $-\frac{5}{3}$ are multiplicative inverses because $(-\frac{3}{5})(-\frac{5}{3}) = 1$.

5 and 0.2 are multiplicative inverses because $5(0.2) = 1$.

1 is its own multiplicative inverse because $1 \times 1 = 1$.

(-1) is its own multiplicative inverse because $(-1)(-1) = 1$.

0 has no multiplicative inverse because the product of 0 and any real number is 0, *not* 1.

The symbol for the multiplicative inverse of a *nonzero* real number a is $\frac{1}{a}$. It is assumed that every real number except 0 has a multiplicative inverse. This assumption is stated formally as the following axiom.

Working with Real Numbers   **93**

68. 1. a, b, c, and d are real numbers.
      Hypothesis
   2. $a[b - (c - d)]$
      $= ab - a(c - d)$
      Exercise 62, page 92
   3. $= ab - [ac - ad]$
      Exercise 62, page 92
   4. $= ab - [ac + (-ad)]$
      Def. of subt.
   5. $= ab + \{-[ac + (-ad)]\}$
      Def. of subt.
   6. $= ab + \{(-ac) + [-(-ad)]\}$
      Prop. of opp. of a sum
   7. $= ab + [(-ac) + ad]$
      Prop. of opp. in prod.
   8. $= [ab + (-ac)] + ad$
      Assoc. ax. for add.
   9. $= (ab - ac) + ad$
      Def. of subt.
  10. $\therefore a[b - (c - d)]$
      $= (ab - ac) + ad$
      Trans. ax. of $=$

Teaching Suggestions
p. T75

Related Activities p. T76

Key Ideas
Use multiplicative inverses. Simplify variable expressions involving multiplicative inverses.

Simplify.

1. $(6 \cdot 5)\dfrac{1}{5}$

$$= 6\left(5 \cdot \dfrac{1}{5}\right)$$
$$= 6 \cdot 1$$
$$= 6$$

2. $-\dfrac{1}{2}(10f^2 - 6f)$

$$= -\dfrac{1}{2}(10f^2) - \left(-\dfrac{1}{2}\right)(6f)$$
$$= -\dfrac{1}{2}(2 \cdot 5f^2) - \left(-\dfrac{1}{2}\right)(2 \cdot 3f)$$
$$= \left(-\dfrac{1}{2} \cdot 2\right)5f^2 - \left(-\dfrac{1}{2} \cdot 2\right)3f$$
$$= (-1)5f^2 - (-1)3f$$
$$= -5f^2 + 3f$$

3. $-16y\left(\dfrac{1}{2y}\right)$

$$= [(-8)2y]\left(\dfrac{1}{2} \cdot \dfrac{1}{y}\right)$$
$$= (-8)\left(2 \cdot \dfrac{1}{2}\right)\left(y \cdot \dfrac{1}{y}\right)$$
$$= (-8)(1)(1)$$
$$= -8$$

Axiom of Multiplicative Inverses

For every nonzero real number a, there is a unique real number $\dfrac{1}{a}$ such that

$$a \cdot \dfrac{1}{a} = 1 \quad \text{and} \quad \dfrac{1}{a} \cdot a = 1.$$

EXAMPLE 1 Simplify.

    a. $(6 \cdot 7)\dfrac{1}{7}$ **b.** $(-24)\dfrac{1}{8}$

SOLUTION **a.** $(6 \cdot 7)\dfrac{1}{7} = 6\left(7 \cdot \dfrac{1}{7}\right) = 6 \cdot 1 = 6$

    b. $(-24)\dfrac{1}{8} = (-3 \cdot 8)\dfrac{1}{8} = (-3)\left(8 \cdot \dfrac{1}{8}\right) = (-3)(1) = -3$

Example 1 suggests the following theorem, which will be proved in Exercise 51 on page 98.

Theorem. For all real numbers a and all nonzero real numbers b,

$$(ab)\dfrac{1}{b} = a.$$

This theorem can be useful in simplifying variable expressions.

EXAMPLE 2 Simplify.

    a. $(-7m^3)\left(-\dfrac{1}{7}\right)$ **b.** $\dfrac{1}{5}(45a - 70b)$

SOLUTION **a.** $(-7m^3)\left(-\dfrac{1}{7}\right) = \left[(-7)\left(-\dfrac{1}{7}\right)\right]m^3 = 1 \cdot m^3 = m^3$

    b. $\dfrac{1}{5}(45a - 70b) = \dfrac{1}{5}(45a) - \dfrac{1}{5}(70b)$

$$= \dfrac{1}{5}(5 \cdot 9a) - \dfrac{1}{5}(5 \cdot 14b)$$
$$= \left(\dfrac{1}{5} \cdot 5\right)9a - \left(\dfrac{1}{5} \cdot 5\right)14b$$
$$= (1)9a - (1)14b$$
$$= 9a - 14b$$

You can use the axiom of multiplicative inverses to show that, for every nonzero real number a, $(-a)\left(-\dfrac{1}{a}\right) = 1$. Therefore, $-a$ and $-\dfrac{1}{a}$ are multiplicative inverses. This fact can be stated formally as follows.

> **Theorem.** For all nonzero real numbers a,
> $$\frac{1}{-a} = -\frac{1}{a}.$$

A proof of this theorem is outlined in Exercise 17 on page 96.
Now consider the following example.

EXAMPLE 3 Simplify.

  a. $(4 \cdot 5)\left(\frac{1}{4} \cdot \frac{1}{5}\right)$ **b.** $(ab)\left(\frac{1}{a} \cdot \frac{1}{b}\right)$, $a \neq 0$, $b \neq 0$

SOLUTION **a.** $(4 \cdot 5)\left(\frac{1}{4} \cdot \frac{1}{5}\right) = \left(4 \cdot \frac{1}{4}\right)\left(5 \cdot \frac{1}{5}\right) = (1)(1) = 1$

  b. $(ab)\left(\frac{1}{a} \cdot \frac{1}{b}\right) = \left(a \cdot \frac{1}{a}\right)\left(b \cdot \frac{1}{b}\right) = (1)(1) = 1$

Part (b) of Example 3 suggests that, for nonzero real numbers a and b, the product of ab and $\frac{1}{a} \cdot \frac{1}{b}$ is 1. Therefore, $\frac{1}{a} \cdot \frac{1}{b}$ is the multiplicative inverse, or reciprocal, of ab. This example suggests the following property, which will be proved as a theorem in Exercise 18 on pages 96–97.

> ### Property of the Reciprocal of a Product
>
> The reciprocal of a product of nonzero real numbers is equal to the product of the reciprocals of the numbers. That is, for all nonzero real numbers a and b,
> $$\frac{1}{ab} = \frac{1}{a} \cdot \frac{1}{b}.$$

EXAMPLE 4 Simplify.

  a. $\frac{1}{4} \cdot \frac{1}{-5}$ **b.** $-48v\left(\frac{1}{4v}\right)$, $v \neq 0$

SOLUTION **a.** $\frac{1}{4} \cdot \frac{1}{-5} = \frac{1}{4(-5)} = \frac{1}{-20} = -\frac{1}{20}$

  b. $-48v\left(\frac{1}{4v}\right) = [(-12)4v]\left(\frac{1}{4} \cdot \frac{1}{v}\right)$

  $= (-12)\left(4 \cdot \frac{1}{4}\right)\left(v \cdot \frac{1}{v}\right)$

  $= (-12)(1)(1)$

  $= -12$

Working with Real Numbers **95**

Suggested Assignments

Minimum
First day: 97/1–22
            S 87/9, 10, 11
Second day: 97/23–30
Average
First day: 97/13–39 odd
            S 85/26–34 even
Second day: 98/40–50
Maximum
First day: 97/13–49 odd
            S 85/38–46 even
Second day: 98/51–56

Additional A Exercises

1. $-\frac{1}{2}(46)$ -23

2. $(-32)\left(-\frac{1}{16}\right)$ 2

3. $51\left(-\frac{1}{17}\right)\left(\frac{1}{3}\right)$ -1

4. $-\frac{1}{6}(6x^2)$ $-x^2$

5. $(-25)\left(\frac{1}{5}\right)\left(-\frac{1}{5}\right)$ 1

6. $\frac{1}{2}(-10x + 28)$ $-5x + 14$

Oral Exercises

State the multiplicative inverse of each number.

1. $7\frac{1}{7}$

2. -7 $-\frac{1}{7}$

3. -1 -1

4. 11

5. $\frac{1}{8}$ 8

6. $-\frac{3}{8}$ $-\frac{8}{3}$

7. $-\frac{9}{8}$ $-\frac{8}{9}$

8. $1\frac{5}{8}$ $\frac{8}{13}$

9. -0.3 $-\frac{10}{3}$

10. 1.7 $\frac{10}{17}$

11. $s, s \neq 0$ $\frac{1}{s}$

12. $-t, t \neq 0$ $\frac{1}{-t}$

13. $-\frac{1}{x}, x \neq 0$ $-x$

14. $\frac{3}{y}, y \neq 0$ $\frac{y}{3}$

15. $\frac{z}{2}, z \neq 0$ $\frac{2}{z}$

16. $-\frac{u}{v}, u \neq 0, v \neq 0$ $-\frac{v}{u}$

Replace each __?__ with the reason that justifies the statement to its left.

17. Prove: For all nonzero real numbers a, $\frac{1}{-a} = -\frac{1}{a}$.

Plan: If it can be shown that $(-a)\left(-\frac{1}{a}\right) = 1$, then by the axiom of multiplicative inverses $-\frac{1}{a}$ is the multiplicative inverse of $-a$ and is therefore equal to $\frac{1}{-a}$.

PROOF

Statements	*Reasons*
1. a is a nonzero real number.	__?__ Hypothesis
2. $\frac{1}{a}$ is a real number.	__?__ Axiom of multiplicative inverses
3. $-a$ and $-\frac{1}{a}$ are real numbers.	__?__ Axiom of additive inverses
4. $(-a)\left(-\frac{1}{a}\right) = a \cdot \frac{1}{a}$	Property of opposites in products
5. $= 1$	__?__ Axiom of multiplicative inverses
6. $(-a)\left(-\frac{1}{a}\right) = 1$	__?__ Transitive axiom of equality
7. $-\frac{1}{a} = \frac{1}{-a}$	Axiom of multiplicative inverses
8. $\therefore \frac{1}{-a} = -\frac{1}{a}$	__?__ Symmetric axiom of equality

18. Prove: For all nonzero real numbers a and b, $\frac{1}{ab} = \frac{1}{a} \cdot \frac{1}{b}$.

Plan: If it can be shown that $(ab)\left(\frac{1}{a} \cdot \frac{1}{b}\right) = 1$, then by the axiom of multiplicative inverses $\left(\frac{1}{a} \cdot \frac{1}{b}\right)$ is the multiplicative inverse of ab and is therefore equal to $\frac{1}{ab}$.

Statements	Reasons
1. a and b are nonzero real numbers.	__?__ Hypothesis
2. $\frac{1}{a}$ and $\frac{1}{b}$ are real numbers.	Axiom of multiplicative inverses
3. $(ab)\left(\frac{1}{a}\cdot\frac{1}{b}\right)=\left(a\cdot\frac{1}{a}\right)\left(b\cdot\frac{1}{b}\right)$	__?__ Commutative and associative axioms for multiplication
4. $= 1\cdot 1$	__?__ Axiom of multiplicative inverses
5. $= 1$	Identity axiom for multiplication
6. $(ab)\left(\frac{1}{a}\cdot\frac{1}{b}\right)=1$	Transitive axiom of equality
7. $\frac{1}{a}\cdot\frac{1}{b}=\frac{1}{ab}$	__?__ Axiom of multiplicative inverses
8. $\therefore\ \frac{1}{ab}=\frac{1}{a}\cdot\frac{1}{b}$	__?__ Symmetric axiom of equality

Written Exercises

Simplify.

A

1. $\frac{1}{2}\cdot\frac{1}{9}$ $\frac{1}{18}$

2. $\frac{1}{3}\cdot\frac{1}{-5}$ $-\frac{1}{15}$

3. $\frac{1}{-7}\cdot\frac{1}{4}$ $-\frac{1}{28}$

4. $\frac{1}{-8}\cdot\frac{1}{-3}$ $\frac{1}{24}$

5. $\frac{1}{5}(-75)$ -15

6. $-\frac{1}{4}(92)$ -23

7. $-72\left(-\frac{1}{3}\right)$ 24

8. $-100\left(\frac{1}{25}\right)$ -4

9. $\frac{1}{12}(-9)(-4)3$

10. $-5\left(-\frac{1}{10}\right)(-16)$ -8

11. $-56\left(-\frac{1}{7}\right)\left(-\frac{1}{8}\right)$ -1

12. $\frac{1}{5}(-70)\left(-\frac{1}{14}\right)1$

13. $-\frac{1}{9}(-54)\left(-\frac{1}{2}\right)$ -3

14. $-\frac{1}{3}\left(-\frac{1}{4}\right)(96)8$

15. $-\frac{1}{8}(-8jk)\,jk$

16. $7gh\left(-\frac{1}{7}\right)$ $-gh$

17. $-\frac{1}{3}(81a^3)$ $-27a^3$

18. $-\frac{1}{4}(-56c^2)14c^2$

19. $\frac{1}{n}(-18mn),\ n\neq 0$ $-18m$

20. $-\frac{1}{y}(-24xyz),\ y\neq 0$ $24xz$

21. $\frac{1}{3p}(-12pq),\ p\neq 0$ $-4q$

22. $28ro^3\left(\frac{1}{2r}\right),\ r\neq 0$ $-14s^3$

23. $\frac{1}{4g}(\ 60gh^2)\left(\frac{1}{3}\right),\ g\neq 0$ $5h^2$

24. $\frac{1}{2}(\ 84cd)\left(\frac{1}{6c}\right),\ c\neq 0$ $-7d$

25. $\frac{1}{3}(-15a+33)$ $-5a+11$

26. $\frac{1}{4}(-36b-52)-9b$ -13

27. $-\frac{1}{6}(24c-36d)-4c+6d$

28. $(-48g+88h)\left(-\frac{1}{8}\right)$ $6g-11h$

29. $-35\left(\frac{1}{5}j+\frac{1}{7}k\right)$ $-7j-5k$

30. $\left(-\frac{1}{2}m-\frac{1}{3}n\right)(-42)$ $21m+14n$

---

53. 1. a and b are real numbers such that $b\neq 0$. Hypothesis

2. $\frac{1}{b}$ is a real number. Ax. of mult. inv.

3. $-\frac{1}{b}(ab)=-\left[\frac{1}{b}(ab)\right]$ Prop. of opp. in prod.

4. $=-\left[\left(\frac{1}{b}\cdot b\right)a\right]$ Comm. and assoc. ax. for mult.

5. $=-(1\cdot a)$ Ax. of mult. inv.

6. $=-a$ Iden. ax. for mult.

7. $\therefore\ -\frac{1}{b}(ab)=-a$ Trans. ax. of $=$

54. 1. a and b are real numbers such that $a\neq 0$. Hypothesis

2. $\frac{1}{a}$ is a real number. Ax. of mult. inv.

3. $-\frac{1}{a}(-ab)$
$=-\frac{1}{a}[(-a)\cdot b]$ Prop. of opp. in prod.

4. $=\left[-\frac{1}{a}(-a)\right]b$ Assoc. ax. of mult.

5. $=\left[\frac{1}{a}\cdot a\right]b$ Prop. of opp. in prod.

6. $=1\cdot b$ Ax. of mult. inv.

7. $=b$ Iden. ax. tor mult.

8. $\therefore\ -\frac{1}{a}(-ab)=b$ Trans. ax. of $=$

(continued)

55. 1. a and b are real numbers such that $a, b \neq 0$. Hypothesis

2. $\frac{1}{a}$ and $\frac{1}{b}$ are real numbers. Ax. of mult. inv.

3. $-\frac{1}{a}\left(-\frac{1}{b}\right) = \frac{1}{a} \cdot \frac{1}{b}$ Prop. of opp. in prod.

4. $= \frac{1}{ab}$ Prop. of recip. of a product

5. $\therefore -\frac{1}{a}\left(-\frac{1}{b}\right) = \frac{1}{ab}$ Trans. ax. of $=$

56. 1. a and b are real numbers such that $a, b \neq 0$. Hypothesis

2. $\frac{1}{a}$ and $\frac{1}{b}$ are real numbers. Ax. of mult. inv.

3. $\frac{1}{a}\left(-\frac{1}{b}\right) = -\left[\frac{1}{a} \cdot \frac{1}{b}\right]$ Prop. of opp. in prod.

4. $= -\frac{1}{ab}$ Exercise 18, page 96

5. $\therefore \frac{1}{a}\left(-\frac{1}{b}\right) = -\frac{1}{ab}$ Trans. ax. of $=$

Mixed Review

Simplify.
1. $(-2)(-7)(-6)$ -84
2. $(-1)^5(-5)(-2)$ -10
3. $-4(-2m + n)$ $8m - 4n$
Evaluate each expression when $e = -2$, $f = -3$, and $g = -1$.
4. $e(f + g)$ 8
5. $ef + fg$ 9
6. $(ef)^2$ 36

Simplify.

B **31.** $\frac{1}{9}(-45w - 9z) + \frac{1}{4}(44w - 32z)$ $6w - 9z$

32. $\frac{1}{4}(20x - 8y) + \frac{1}{3}(-27x - 30y)$ $-4x - 12y$

33. $12\left(\frac{1}{2}s - \frac{1}{3}t\right) - 14\left(-\frac{1}{7}s + \frac{1}{2}t\right)$ $8s - 11t$

34. $-3\left(\frac{1}{3}c + \frac{1}{3}d\right) - 10\left(\frac{1}{2}c - \frac{1}{5}d\right)$ $-6c + d$

35. $-\frac{1}{5}(5m^2 + 15m - 40) + \frac{1}{2}(-8m^2 - 6m - 30)$ $-5m^2 - 6m - 7$

36. $\frac{1}{6}(12n^2 - 6n + 18) - \frac{1}{3}(-12 + 6n - 6n^2)$ $4n^2 - 3n + 7$

37. $-\frac{1}{4}(16w^2 - 12w) + \frac{1}{3}(6w^2 - 3w) - \frac{1}{2}(8w^2 + 4w)$ $-6w^2$

38. $\frac{1}{2}(-4z^3 - 2z^2) - \frac{1}{6}(6z^3 - 12z^2) + \frac{1}{3}(9z^3 - 15z^2)$ $-4z^2$

39. $-\frac{1}{2}[3(2r - 4s)] - \frac{1}{3}[2(6r + 9s)] + \frac{1}{4}[5(4r - 8s)]$ $-2r - 10s$

40. $\frac{1}{4}[2(-2j + 6k) + 6(2j - 4k) - 3(4j + 12k)]$ $-j - 12k$

41. $18\left[\frac{1}{3}(5a - 2b)\right] - 12\left[\frac{1}{2}(3a + 7b)\right] - 16\left[\frac{1}{4}(2a - b)\right]$ $4a - 50b$

42. $-6\left[\frac{1}{3}(p - 2q) + \frac{1}{2}(5p + 3q) - \frac{1}{6}(7p - q)\right]$ $-10p - 6q$

Solve.

43. $\frac{1}{n} = -\frac{1}{4}$ $\{-4\}$ **44.** $\frac{1}{n} = -5$ $\{-\frac{1}{5}\}$ **45.** $\frac{1}{n} = 2\frac{1}{2}$ $\{\frac{2}{5}\}$ **46.** $\frac{1}{n} = 1.75$ $\{\frac{4}{7}\}$

47. $\frac{1}{n} = 0$ $\varnothing$ **48.** $\frac{1}{n} = n$ $\{1, -1\}$ **49.** $\frac{1}{n} = -n$ $\varnothing$ **50.** $\frac{1}{n} = n^2$ $\{1\}$

Write a direct proof of each theorem.

C **51.** For all real numbers a and all nonzero real numbers b, $(ab)\left(\frac{1}{b}\right) = a$.

52. For all real numbers b and all nonzero real numbers a, $(ab)\left(\frac{1}{a}\right) = b$.

53. For all real numbers a and all nonzero real numbers b, $-\frac{1}{b}(ab) = -a$.

54. For all real numbers b and all nonzero real numbers a, $-\frac{1}{a}(-ab) = b$.

55. For all nonzero real numbers a and b, $-\frac{1}{a}\left(-\frac{1}{b}\right) = \frac{1}{ab}$.

56. For all nonzero real numbers a and b, $\frac{1}{a}\left(-\frac{1}{b}\right) = -\frac{1}{ab}$.

2–11 Dividing Real Numbers

The first column below lists a few examples of dividing by 3.

The second column lists related examples of multiplying by $\frac{1}{3}$.

Dividing by 3	Multiplying by $\frac{1}{3}$
$3 \div 3 = 1$	$3 \times \frac{1}{3} = 1$
$6 \div 3 = 2$	$6 \times \frac{1}{3} = 2$
$9 \div 3 = 3$	$9 \times \frac{1}{3} = 3$
$12 \div 3 = 4$	$12 \times \frac{1}{3} = 4$

Comparing the entries in the two columns shows that *dividing* by 3 gives the same result as *multiplying by the reciprocal* of 3. This relationship between multiplication and division suggests the following definition.

Definition of Division

For all real numbers a and all *nonzero* real numbers b, the **quotient** denoted as $a \div b$ is defined by:

$$a \div b = a \cdot \frac{1}{b}$$

That is, to divide a by b, multiply a by the reciprocal of b.

Note that a quotient is often represented as a fraction.

$$a \div b = \frac{a}{b}$$

You can use the definition of division to replace any quotient with a product. For example:

$$\frac{56}{7} = 56 \times \frac{1}{7} = 8$$

$$\frac{56}{-7} = 56\left(-\frac{1}{7}\right) = -8$$

$$\frac{-56}{7} = -56 \times \frac{1}{7} = -8$$

$$\frac{-56}{-7} = -56\left(-\frac{1}{7}\right) = 8$$

Teaching Suggestions
p. T76

Supplementary Material
Test 7

Key Ideas

Use the definition of division to divide real numbers. Simplify variable expressions involving division.

Chalkboard Examples

Simplify.

1. $-30 \div \frac{1}{6}$

$= -30 \cdot 6$

$= -180$

2. $\frac{-28b}{-4}$

$= -28b\left(-\frac{1}{4}\right)$

$= -28\left(-\frac{1}{4}\right)b$

$= 7b$

3. $9a^2 \div \left(-\frac{3}{4}\right)$

$= 9a^2\left(-\frac{4}{3}\right)$

$= \left[9\left(-\frac{4}{3}\right)\right]a^2$

$= -12a^2$

Suggested Assignments

Minimum
  101/1–28
  R 103/Self-Test 3
Average
  101/11–41 odd
  R 103/Self-Test 3
Maximum
  101/11–57 odd
  R 103/Self-Test 3

Additional A Exercises

Simplify.

1. $144 \div (-12)$ -12

2. $\frac{1}{3} \div \left(-\frac{3}{4}\right)$ $-\frac{4}{9}$

3. $\frac{74xy}{-37y}$, $y \ne 0$ $-2x$

4. $-\frac{8}{9}s \div \left(-\frac{1}{9}\right)$ $8s$

5. The average of a set of numbers is the quotient when the sum of the numbers is divided by the number of members in the set. Find the average of the set of numbers {12, 18, 19, 10, 16}.   15

6. Evaluate each expression when $x = -3$, $y = -1$, and $z = 2$.

  a. $\frac{-z}{xy}$ $-\frac{2}{3}$

  b. $\frac{x - z}{y}$ 5

Division of real numbers is *not* commutative. For example:

$$12 \div 3 = 4 \qquad \text{but} \qquad 3 \div 12 = 0.25$$

Nor is division of real numbers associative. For example:

$$(36 \div 6) \div 2 = 6 \div 2 = 3 \qquad \text{but} \qquad 36 \div (6 \div 2) = 36 \div 3 = 12$$

On the other hand, note the following.

$$\frac{6 + 15}{3} = \frac{21}{3} = 7 \qquad \text{and} \qquad \frac{6}{3} + \frac{15}{3} = 2 + 5 = 7$$

$$\frac{20 - 8}{2} = \frac{12}{2} = 6 \qquad \text{and} \qquad \frac{20}{2} - \frac{8}{2} = 10 - 4 = 6$$

These examples illustrate the fact that *division is distributive with respect to both addition and subtraction*, which will be proved in Exercises 42 and 44 on page 102.

For all real numbers a, b, and c such that $c \ne 0$,

$$\frac{a + b}{c} = \frac{a}{c} + \frac{b}{c} \quad \text{and} \quad \frac{a - b}{c} = \frac{a}{c} - \frac{b}{c}.$$

Why can you never divide by zero? Using the relationship between division and multiplication, dividing by 0 would mean multiplying by the reciprocal of 0. As you have seen, 0 has no reciprocal. Therefore, *division by zero has no meaning in the set of real numbers.*

You cannot divide zero by zero, but can you divide zero by any other number? Consider these examples:

$$\frac{0}{7} = 0 \times \frac{1}{7} = 0 \qquad\qquad \frac{0}{-4} = 0\left(-\frac{1}{4}\right) = 0$$

When zero is divided by any nonzero number, the quotient is zero. That is, for any nonzero real number a,

$$0 \div a = 0 \cdot \frac{1}{a} = 0.$$

EXAMPLE   Simplify.

  a. $-40 \div \frac{1}{8}$ **b.** $\frac{45a}{-3}$ **c.** $8z^2 \div \left(-\frac{2}{3}\right)$

SOLUTION   **a.** $-40 \div \frac{1}{8} = -40 \times 8 = -320$

  b. $\frac{45a}{-3} = 45a\left(-\frac{1}{3}\right) = \left[45\left(-\frac{1}{3}\right)\right]a = -15a$

  c. $8z^2 \div \left(-\frac{2}{3}\right) = 8z^2\left(-\frac{3}{2}\right) = \left[8\left(-\frac{3}{2}\right)\right]z^2 = -12z^2$

100   *Chapter 2*

Oral Exercises

Simplify.

1. $\frac{20}{4}$ 5

2. $\frac{20}{-4}$ −5

3. $\frac{-4}{20}$ −$\frac{1}{5}$

4. $\frac{-4}{-20}$ $\frac{1}{5}$

5. $\frac{0}{12}$ 0

6. $\frac{0}{-12}$ 0

7. $\frac{12}{-12}$ −1

8. $\frac{-12}{-12}$ 1

9. $\frac{x}{-1}$ −x

10. $\frac{-x}{-1}$ x

11. $\frac{-x}{-x}$, $x \neq 0$ 1

12. $\frac{x}{-x}$, $x \neq 0$ −1

Read each quotient as a product. Then simplify.

13. $8 \div \frac{1}{4}$
$8 \cdot 4 = 32$

14. $16 \div \left(-\frac{1}{4}\right)$
$16 \cdot (-4)$
$= -64$

15. $-12 \div \frac{3}{4}$
$-12 \cdot \frac{4}{3}$
$= -16$

16. $-1 \div \left(-\frac{1}{10}\right)$
$-1 \cdot (-10) = 10$

17. $-9a \div \frac{1}{3}$
$-9a \cdot 3 = -27a$

18. $-12b \div \left(-\frac{2}{3}\right)$
$-12b \cdot \left(-\frac{3}{2}\right) = 18b$

19. $6 \div \frac{1}{c}$, $c \neq 0$
$6 \cdot c = 6c$

20. $7d \div (-d)$, $d \neq 0$
$7d \cdot \left(-\frac{1}{d}\right) = -7$

Replace each __?__ with one of the words *positive*, *negative*, or *zero* to make a true statement.

21. The quotient when a positive number is divided by a negative number is always __?__ . negative

22. The quotient when a negative number is divided by a negative number is always __?__ . positive

23. The quotient when zero is divided by a negative number is always __?__ . zero

24. In a quotient, the divisor can never be __?__ . zero

Written Exercises

Simplify.

A

1. $160 \div (-5)$ −32

2. $-124 \div (-4)$ 31

3. $-600 \div 24$ −25

4. $0 \div (-18)$ 0

5. $-15 \div \left(-\frac{1}{5}\right)$ 75

6. $18 \div \left(-\frac{1}{3}\right)$ −54

7. $\frac{1}{7} \div \left(-\frac{3}{7}\right)$ −$\frac{1}{3}$

8. $-\frac{3}{8} \div \left(-\frac{2}{5}\right)$ $\frac{15}{16}$

9. $\frac{87jk}{-3}$ −29jk

10. $\frac{-95m^2}{-5}$ 19m²

11. $\frac{-108n}{9n}$, $n \neq 0$ −12

12. $\frac{156pq}{-13p}$, $p \neq 0$ −12q

13. $-68w^3 \div \frac{1}{4}$
$-272w^3$

14. $-22uv \div \left(-\frac{1}{11}\right)$
242uv

15. $-\frac{7}{8}r \div \left(-\frac{1}{16}\right)$
14r

16. $\frac{4}{5}st \div \left(-\frac{1}{10}\right)$
−8st

The *average* of a set of numbers is the quotient when the sum of the numbers is divided by the number of numbers in the set. Find the average of each set of numbers.

17. $\{42, 17, 51, 37, 48\}$ 39

18. $\{-29, 14, 3, -11, -19, 24\}$ −3

19. $\{32, -47, -16, 19, 50, -38\}$ 0

20. $\{5, -1, 3, -4, 1, 1, -7, -2\}$ −$\frac{1}{2}$

Working with Real Numbers **101**

101

(continued on p. 104)

Evaluate each expression when $x = -3$, $y = -1$, $z = 2$, and $w = 6$.

21. a. $\dfrac{xy}{w}$ $\tfrac{1}{2}$ b. $\dfrac{-xy}{-w}$ $\tfrac{1}{2}$
22. a. $\dfrac{-x}{wz}$ $\tfrac{1}{4}$ b. $\dfrac{x}{-wz}$ $\tfrac{1}{4}$

23. a. $\dfrac{x+z}{y}$ 1 b. $\dfrac{-x-z}{-y}$ 1
24. a. $\dfrac{x-w}{y}$ 9 b. $\dfrac{w-x}{y}$ -9

25. a. $\dfrac{y+w}{x-z}$ -1 b. $\dfrac{w+y}{z-x}$ 1
26. a. $\dfrac{x-y}{z-w}$ $\tfrac{1}{2}$ b. $\dfrac{y-x}{w-z}$ $\tfrac{1}{2}$

27. a. $\dfrac{-x-w}{z-y}$ -1 b. $\dfrac{x+w}{y-z}$ -1
28. a. $\dfrac{-w-x}{y-z}$ 1 b. $-\dfrac{x+w}{z-y}$ -1

Evaluate each expression when $a = -4$, $b = -2$, $c = -1$, and $d = 6$.

B 29. $\dfrac{a^2}{-c}$ 16
30. $\dfrac{d^2}{a}$ -9
31. $\dfrac{b^3}{-a}$ -2
32. $\dfrac{c^3}{b}$ $\tfrac{1}{2}$

33. $\dfrac{(b+c)^2}{-d}$ $-\tfrac{3}{2}$
34. $\dfrac{(b-c)^2}{-ac}$ $-\tfrac{1}{4}$
35. $\dfrac{a^2+b^2}{-a-c}$ 4
36. $\dfrac{c^2-d^2}{a+b+c}$ 5

37. $\dfrac{d^2-6c}{a-2}$ -7
38. $\dfrac{5a-b^2}{-c-4}$ 8
39. $\dfrac{3a-7b}{c^2+3}$ $\tfrac{1}{2}$
40. $\dfrac{-4b-9c}{b^2+1}$ $\tfrac{17}{5}$

Give the reason that justifies each statement in the given proof.

41. Prove: For all real numbers a and b
such that $b \neq 0$, $(ab) \div b = a$.

PROOF

1. a and b are real numbers such
that $b \neq 0$.

2. $(ab) \div b = (ab) \cdot \dfrac{1}{b}$

3. $\qquad = a\left(b \cdot \dfrac{1}{b}\right)$

4. $\qquad = a \cdot 1$

5. $\qquad = a$

6. $\therefore (ab) \div b = a$

42. Prove: For all real numbers a, b, and
c such that $c \neq 0$, $\dfrac{a+b}{c} = \dfrac{a}{c} + \dfrac{b}{c}$.

PROOF

1. a, b, and c are real numbers
such that $c \neq 0$.

2. $\dfrac{a+b}{c} = (a+b)\dfrac{1}{c}$

3. $\qquad = a \cdot \dfrac{1}{c} + b \cdot \dfrac{1}{c}$

4. $\qquad = \dfrac{a}{c} + \dfrac{b}{c}$

5. $\therefore \dfrac{a+b}{c} = \dfrac{a}{c} + \dfrac{b}{c}$

Write a direct proof of each theorem.

C 43. For all real numbers a and b such that $b \neq 0$, $(a \div b)b = a$.

44. For all real numbers a, b, and c such that $c \neq 0$, $\dfrac{a-b}{c} = \dfrac{a}{c} - \dfrac{b}{c}$.

45. For all real numbers a such that $a \neq 0$, $\dfrac{a}{a} = 1$.

46. For all real numbers a such that $a \neq 0$, $\dfrac{-a}{a} = -1$.

102 *Chapter 2*

47. For all real numbers a and b such that $b \neq 0$, $\dfrac{a+b}{b} = 1 + \dfrac{a}{b}$.

48. For all real numbers a and b such that $a \neq 0$, $\dfrac{a-b}{a} = 1 - \dfrac{b}{a}$.

A set of numbers is said to be *closed under division* if all possible quotients when a member of the set is divided by a nonzero member of the set are also members of the set. Tell whether each set is closed or not closed under division.

$\qquad\qquad\qquad\qquad\qquad\qquad\qquad\qquad\qquad\qquad\qquad$ not closed

49. $\{1\}$ closed   50. $\{-1\}$ not closed   51. $\{-1, 0, 1\}$ closed   52. $\{\frac{1}{2}, 1, 2\}$

53. {the natural numbers} not closed   54. {the integers} not closed

55. {the positive real numbers} closed   56. {the negative real numbers}

57. {the rational numbers} closed   58. {the real numbers} closed

$\qquad\qquad\qquad\qquad\qquad\qquad\qquad\qquad\qquad\qquad$ **56.** not closed

Computer Exercises For students with computer experience

1. Write a program that will print the average of any two real numbers that you input.

2. Modify the program that you wrote for Exercise 1 to allow you to input two *or more* real numbers.

3. Write a program that will allow you to input the coordinates of any two points on a number line and will print the coordinate of the point that is located exactly halfway between the given points.

■ Self-Test 3

VOCABULARY   identity element for   division (p. 99)
 multiplication (p. 88)   quotient (p. 99)
 multiplicative inverse or
 reciprocal (p. 93)

Simplify.

1. $(-2)^2(-11)$ -44 2. $(-7x)(-9x)$ $63x^2$ *Obj. 1, p. 88*

3. $7(-y + 1) - 5y$ $-12y + 7$ 4. $4z^3 - (9z + 3z^3)$ $z^3 - 9z$

5. $\dfrac{1}{-7} \cdot \dfrac{1}{-3}$ $\frac{1}{21}$ 6. $\dfrac{1}{6}(-90w)$ $-15w$ 7. $-\dfrac{1}{2}(14p - 2q)$ *Obj. 2, p. 88*

$\qquad\qquad\qquad\qquad\qquad\qquad\qquad\qquad\qquad\qquad$ $-7p + q$

8. $65 \div (-5)$ -13 9. $-\dfrac{4}{5} \div \left(-\dfrac{1}{2}\right)$ $\frac{8}{5}$ 10. $\dfrac{-32n}{8}$ $-4n$ *Obj. 3, p. 88*

Check your answers with those at the back of the book.

Working with Real Numbers   **103**

Mixed Review

Simplify.
1. $210 \div (-7)$ -30

2. $\dfrac{-42r}{6r}$, $r \neq 0$ -7

3. $\dfrac{1}{3}x \div \left(-\dfrac{1}{12}\right)$ $-4x$

4. Find the average of the following numbers.
   $32, 10, -6, -20$ 4

Evaluate each expression when $a = -4$, $b = 2$, and $c = -1$.

5. $\dfrac{a - b}{-c}$ -6

6. $-\dfrac{-ac}{ab}$ $-\dfrac{1}{2}$

Quick Quiz

1. $(-1)^5(-12)$ 12
2. $(-2x)(-3x)(4x)$ $24x^3$
3. $8(m - 1) - 9m$ $-m - 8$
4. $16 - (7y^2 - 18)$
   $34 - 7y^2$
5. $\dfrac{1}{4}\left(-\dfrac{2}{3}\right)\left(-\dfrac{6}{7}\right)$ $\dfrac{1}{7}$
6. $(-55n)\left(\dfrac{1}{5}\right)$ $-11n$
7. $-\dfrac{1}{3}(-12y + 9)$ $4y - 3$
8. $\dfrac{35n}{-5}$ $-7n$
9. $-\dfrac{6}{7} \div \dfrac{3}{4}$ $-\dfrac{8}{7}$
10. $\dfrac{-48x}{12}$ $-4x$

(continued from p. 102)

47. 1. a and b are real numbers such that $b \neq 0$.
Hypothesis

2. $\dfrac{a + b}{b} = (a + b)\dfrac{1}{b}$

Def. of div.

3. $\quad = a \cdot \dfrac{1}{b} + b \cdot \dfrac{1}{b}$

Dist. ax.

4. $\quad = a \cdot \dfrac{1}{b} + 1$

Ax. of mult. inv.

5. $\quad = \dfrac{a}{b} + 1$

Def. of div.

6. $\quad = 1 + \dfrac{a}{b}$

Comm. ax. for add.

7. $\therefore \dfrac{a + b}{b} = 1 + \dfrac{a}{b}$

Trans. ax. of $=$

48. 1. a and b are real numbers such that $a \neq 0$.
Hypothesis

2. $\dfrac{a - b}{a} = (a - b)\dfrac{1}{a}$

Def. of div.

3. $\quad = a \cdot \dfrac{1}{a} - b \cdot \dfrac{1}{a}$

Exercise 62, page 92

4. $\quad = 1 - b \cdot \dfrac{1}{a}$

Ax. of mult. inv.

5. $\quad = 1 - \dfrac{b}{a}$

Def. of div.

6. $\therefore \dfrac{a - b}{a} = 1 - \dfrac{b}{a}$

Trans. ax. of $=$

Chapter Summary

1. The *additive inverse*, or the *opposite*, of the real number a is denoted by $-a$. The positive number of any pair of opposite nonzero real numbers is called the *absolute value* of each number in the pair. The absolute value of the real number a is denoted by $|a|$.

2. The *multiplicative inverse* of the real number a is denoted by $\dfrac{1}{a}$.

3. *Axioms*, or *postulates*, are statements about the real numbers that are assumed to be true. The following axioms are assumed to be true for all real values of each variable except as noted.

Axioms of equality	Reflexive axiom: $a = a$ Symmetric axiom: If $a = b$, then $b = a$. Transitive axiom: If $a = b$ and $b = c$, then $a = c$.	
	Addition	Multiplication
Axioms of closure	$a + b$ is a unique real number.	ab is a unique real number.
Commutative axioms	$a + b = b + a$	$ab = ba$
Associative axioms	$(a + b) + c = a + (b + c)$	$(ab)c = a(bc)$
Identity axioms	$a + 0 = a$ and $0 + a = a$	$a \cdot 1 = a$ and $1 \cdot a = a$
Axioms of inverses	$a + (-a) = 0$ and $-a + a = 0$	$a \cdot \dfrac{1}{a} = 1$ and $\dfrac{1}{a} \cdot a = 1$, if $a \neq 0$
Distributive axiom	$a(b + c) = ab + ac$ and $(b + c)a = ba + ca$	

4. *Theorems* are statements about the real numbers that can be proved to be true. The *hypothesis* of a theorem states what is assumed to be true, and the *conclusion* states something that follows logically from this assumption. By reasoning from the hypothesis to the conclusion, you can give a *direct proof* of a theorem. The following properties of the real numbers are theorems that can be proved to be true for all real values of each variable except as noted.

Cancellation property of opposites: $-(-a) = a$

Property of the opposite of a sum: $-(a + b) = -a + (-b)$

104   Chapter 2

Multiplicative property of zero: $a \cdot 0 = 0$ and $0 \cdot a = 0$

Multiplicative property of -1: $a(-1) = -a$ and $(-1)a = -a$

Property of opposites in products: $-a(b) = -ab; \; a(-b) = -ab; \; -a(-b) = ab$

Property of the reciprocal of a product: $\dfrac{1}{ab} = \dfrac{1}{a} \cdot \dfrac{1}{b},$ if $a \ne 0,\, b \ne 0$

5. A number line can be used to find the sum of two real numbers.

6. Subtraction and division are defined as follows:

$$a - b = a + (-b) \qquad a \div b = a \cdot \frac{1}{b}, \text{ if } b \ne 0$$

Chapter Review

Write the letter of the correct answer.

1. Which of the following statements is false?　　　　　　　　　　　*2-1*
   a. For each whole number n, $n^2 + 1 > 0$.
   b. There exists an integer k such that $k < 0$.
   c. For every natural number z, $z^2 > 1$.
   d. For some real number c, $c + 3 < 4$.

2. Name the axiom of equality that is illustrated by the following:　　*2-2*
   "If $a + 5 = b$ and $b = 2a$, then $a + 5 = 2a$."
   a. reflexive
   b. symmetric
   c. transitive
   d. It illustrates none of the axioms; the conclusion is false.

Simplify.

3. $(3k + 8) + (4m + 7)$
   a. $7km + 15$　　b. $12km + 15$　　c. $11k + 11m$　　d. $3k + 4m + 15$

4. $5p + 9q + 7p$　　　　　　　　　　　　　　　　　　　　　　　　*2-3*
   a. $21pq$　　b. $35p + 9q$　　c. $12p + 9q$　　d. $12p^2 + 9q$

5. $5y + 3(y + 2) + 10$
   a. $8y + 16$　　b. $8y + 12$　　c. $10y + 10$　　d. $16y^2 + 10$

6. $-(117 - 49)$　　　　　　　　　　　　　　　　　　　　　　　　*2-4*
   a. 166　　b. -166　　c. 68　　d. -68

7. $9|-8| - 2|-7|$
   a. -58　　b. 58　　c. -86　　d. 86

8. $-4 + [3 + (-5)]$　　　　　　　　　　　　　　　　　　　　　　*2-5*
   a. -6　　b. -12　　c. 2　　d. 6

Additional Answers
Written Exercises (p. 82)

(continued from p. 82)

21. 1. a and b are real numbers.
Hypothesis
2. $-a$ and $-b$ are real numbers.
Ax. of add. inv.
3. $-(a + b) + b$
$= [-a + (-b)] + b$
Prop. of opp. of a sum
4. $= -a + (-b + b)$
Assoc. ax. for add.
5. $= -a + 0$
Ax. of add. inv.
6. $= -a$
Iden. ax. for add.
7. $\therefore -(a + b) + b = -a$
Trans. ax. of $=$

22. 1. a and b are real numbers.
Hypothesis
2. $-a$ and $-b$ are real numbers.
Ax. of add. inv.
3. $a + [-(a + b)]$
$= a + [(-a) + (-b)]$
Prop. of opp. of a sum
4. $= [a + (-a)] + (-b)$
Assoc. ax. for add.
5. $= 0 + (-b)$
Ax. of add. inv.
6. $= -b$
Iden. ax. of add.
7. $\therefore a + [-(a + b)] = -b$
Trans. ax. of $=$

(continued)

9. Solve $-2 + u = 11$.

   a. $\{-9\}$ **b.** $\{9\}$ **c.** $\{13\}$ **d.** $\{-13\}$

10. Simplify $-17 + 28 + (-56) + 19$.   *2-6*

   a. -48 **b.** -26 **c.** 86 **d.** 26

11. The temperature at the end of one school day was 6°C. The temperature fell 10°C overnight, rose 15°C by noon, then fell 10°C by the end of the school day. What was the temperature at the end of the second school day?

   a. 15°C   **b.** 11°C   **c.** 1°C   **d.** −11°C

12. Give the reason that justifies the following statement: *2-7*
   $-(x + y) + x = -(y + x) + x$

   a. property of the opposite of a sum

   b. cancellation property of opposites

   c. associative axiom for addition

   d. commutative axiom for addition

Simplify.

13. $-38 - (-14)$ *2-8*

   a. -52 **b.** 52 **c.** 24 **d.** -24

14. $(14 - 2a) - (5 - 6a)$

   a. $9 - 8a$ **b.** $9 - 4a$ **c.** $9 + 4a$ **d.** $19 + 4a$

15. $(-3v)(2v)(-5v)$ *2-9*

   a. $-6v$ **b.** $-6v^3$ **c.** $-30v^3$ **d.** $30v^3$

16. $-4m + n - (5m + 3n)$

   a. $-9m + 4n$ **b.** $-9m^2 + 4n^2$ **c.** $m - 2n$ **d.** $-9m - 2n$

17. $\frac{1}{16}(-12)(-20)$ *2-10*

   a. 15 **b.** 2 **c.** -2 **d.** -15

18. $\left(\frac{1}{4}a - \frac{1}{3}b\right)(-48)$

   a. $12a - 16b$ **b.** $-12a + 16b$ **c.** $4ab$ **d.** $-4ab$

19. $\frac{-36d}{-4}$ *2-11*

   a. $-9d$ **b.** -9 **c.** $9d$ **d.** 9

20. $-\frac{4}{5}h \div \left(-\frac{3}{10}\right)$

   a. $\frac{8}{3}h$ **b.** $-\frac{8}{3}h$ **c.** $\frac{3}{8}h$ **d.** $-\frac{3}{8}h$

106   *Chapter 2*

Chapter Test

1. Find a value of the variable that makes this statement true: *2-1*
 "There exists a natural number r such that $r(r - 5) = 0$." $r = 5$

2. Name the axiom that is illustrated by the following: *2-2*
 $(g + 8) + 11 = g + (8 + 11)$ Associative axiom for addition

Simplify.

3. $12 + 46 + 18 + 94$ 170   4. $(2a)(16b)(5c)$ 160*abc*

5. $5c^2 + 6c + 8c^2 + 9 + 12c$ 6. $4(2d + 1) + 3(5d + 2)$ 23*d* + 10   *2-3*
      $13c^2 + 18c + 9$

7. $-[-(74 - 58)]$ 16   8. $|-11| - |7 - 3|$ 7   *2-4*

9. $-14 + 3$ −11   10. $[5 + (-8)] + (-7)$ −10   *2-5*

11. Solve $12 + t = 4$. {−8}

12. Simplify $-[47 + (-10)] + (-35) + 66$. −6   *2-6*

13. In the first four months of this year, Tami's baseball card collection increased by 12 cards, decreased by 19 cards, increased by 3 cards, and then decreased by 8 cards. If she had 185 cards at the end of that time, how many cards did she have at the beginning of the year?
      197 cards

14. Give the reason that justifies the following statement: *2-7*
 $(x + y) + (-x) = (y + x) + (-x)$ Commutative axiom for addition

Simplify.

15. $33 - (-9 + 7)$ 35   16. $8w - (5w - 3) - (2 - 7w)$ 10*w* + 1   *2-8*

17. $(-4t)(3t)(-6t)$ 72t^3 18. $(-5r - s)(-9) - 10s$ 45*r* − *s*   *2-9*

19. $-\frac{1}{2}\left(-\frac{2}{3}\right)(-30)$ −10   20. $-\frac{1}{3}(-6p - 15q)$ 2*p* + 5*q*   *2-10*

21. $\frac{-48xy}{16x}$ −3*y*   22. $\frac{3}{8}ab \div \left(-\frac{6}{5}\right)$ $-\frac{5}{16}ab$ *2-11*

Mixed Review

Simplify.

1. $-43 + 25$ −18   2. $18 - (-56)$ 74   3. $(-28)(-5)$ 140   4. $-26 \div 4$ $-6\frac{1}{2}$

5. $-15a - a$ −16*a*   6. $-2b + b$ −*b*   7. $-18c \div \left(-\frac{1}{2}\right)$ 36*c*   8. $(6d)(-9d)$ −54d^2

9. $3.26 + 4.3 + 2.74 + 6.7$ 17   10. $16 - 28 + 42 - 64 + 87$ 53

11. $(11 - 3)9 \div 3(2) - 2$ 10   12. $(8^2 - 2^3)2 - 2(6 - 2)^2 + 12$ 92

---

25. 1. a and b are real numbers.
 Hypothesis
 2. $-a$ and $-b$ are real numbers.
 Ax. of add. inv.
 3. $(a + b) +$
      $[-a + (-b)]$
    $= (a + b) +$
      $[-(a + b)]$
 Prop. of opp. of a sum
 4. $= 0$
 Ax. of add. inv.
 5. $\therefore (a + b) +$
      $[-a + (-b)] = 0$
 Trans. ax. of =

26. 1. a and b are real numbers.
 Hypothesis
 2. $-a$ and $-b$ are real numbers.
 Ax. of add. inv.
 3. $-[a + (-b)] +$
      $[-(-a + b)]$
    $= [(-a) + (-(-b))] +$
      $[-(-a) + (-b)]$
 Prop. of opp. of a sum
 4. $= [(-a) + b] +$
      $[a + (-b)]$
 Canc. prop. of opp.
 5. $= [b + (-a)] +$
      $[a + (-b)]$
 Comm. ax. for add.
 6. $= b + [(-a) + a] +$
      $(-b)$
 Assoc. ax. for add.
 7. $= b + 0 + (-b)$
 Ax. of add. inv.
 8. $= [b + (-b)] + 0$
 Comm. and assoc. ax. for add.
 9. $= 0 + 0$
 Ax. of add. inv.
 10. $= 0$
 Iden. ax. for add.
 11. $\therefore -[a + (-b)] +$
      $[-(-a + b)] = 0$
 Trans. ax. of =

(continued on p. 109)

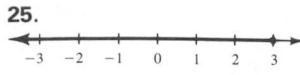
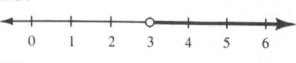
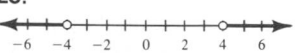

Solve if $n \in \{-4, -3, -2, -1, 0, 1, 2, 3, 4\}$.

13. $9 - n = 16$ Ø   **14.** $2n + 7 = 15$ {4}   **15.** $3n = 4$ Ø   **16.** $2 = \frac{n}{2}$ {4}

17. $n < -2$ {-4, -3}   **18.** $n + 1 > 5$ Ø   **19.** $-n \geq 0$ {-4, -3, -2, -1, 0}   **20.** $|n| < 2$ {-1, 0, 1}

Evaluate each expression when $a = -2$, $b = 3$, and $c = 10$.

21. $-a(b - c)$ -14   **22.** $\frac{a - b}{c}$ $-\frac{1}{2}$ **23.** $a^2 - 2bc$ -56   **24.** $\frac{5a - b^2}{-c - 9}$ 1

Graph each inequality on a number line.

25. $x \leq 3$ **26.** $x + 2 > 5$ **27.** $-1 < x < 4$ **28.** $|x| > 4$

Tell whether each statement is true or false.

29. {the natural numbers} = {the whole numbers greater than 0} True
30. {the irrational numbers} = {the real numbers that are not rational} True
31. {the rational numbers} $\not\subset$ {the integers} True
32. {the integers} $\subset$ {the natural numbers} False

33. Write a variable expression for the following word phrase: the total value in cents of n nickels, q quarters, and x dollars. $5n + 25q + 100x$
34. Write a mathematical sentence for the following word sentence: The product when the sum of a number x and two is multiplied by six is not equal to the square of x. $(x + 2)6 \neq x^2$
35. An airplane on a test flight was flying at an altitude of 9200 m. The airplane dropped 800 m, leveled off, rose 1100 m, and then dropped 500 m. What was its new altitude? 9000 m
36. The temperature at 6:00 one morning was $-3°C$, which was $9°C$ lower than the temperature at midnight the night before. What had the temperature been at midnight? 6°C

ON THE CALCULATOR

Use a calculator to find the reciprocal of each number. You may wish to use the reciprocal key if the calculator has one.

1. 16 0.0625   **2.** -1250 -0.0008 **3.** -0.0625 -16 **4.** 0.00032 3125

Use a calculator to evaluate each expression when $a = 8$, $b = -12$, $c = 5$, and $d = -4$.

5. $\frac{1}{a} \cdot \frac{1}{c}$ 0.025   **6.** $\frac{1}{a} + \frac{1}{d}$ -0.125 **7.** $\frac{1}{a + d}$ 0.25   **8.** $\frac{1}{bc + d}$ -0.015625

108   *Chapter 2*

PREPARING FOR
COLLEGE ENTRANCE EXAMS

Strategy for Success: When you are taking a college entrance exam, it is important to work quickly, but not so quickly that you lose accuracy. In particular, be sure to read the directions, the questions, and the answer choices very carefully. You may wish to underline important words such as *not, exactly, false, never,* and *except.* Cross out answer choices that are clearly incorrect. Mark the answer sheet carefully, and check the numbering after every few questions to avoid misplaced markings.

Decide which is the best of the choices given and write the corresponding letter on your answer sheet.

1. Name the coordinate of the point that is halfway between the points whose coordinates are -3 and 4 on a number line. E

   (A) -1 **(B)** 1 **(C)** $-3\frac{1}{2}$ **(D)** $-\frac{1}{2}$ **(E)** $\frac{1}{2}$

2. Which of the following statements must be true? D
   I. $\emptyset = \{0\}$ II. $5 \notin \{2, 4, 6\}$ III. $\{1, 3, 5, 7\} \subset \mathcal{R}$
   (A) I only   **(B)** II only   **(C)** III only   **(D)** II and III only   **(E)** I, II, and III

3. Which of the following statements must be true? D
   I. $2^5 > 5^2$ II. $10^3 = 30$ III. $3^4 = 81$
   (A) I only   **(B)** II only   **(C)** III only   **(D)** I and III only   **(E)** I, II, and III

4. Which word sentence is represented by $5n + 10d = 90$? C
   (A) The total value of five nickels and ten dimes is ninety cents.
   (B) The total number of five nickels and ten dimes is ninety coins.
   (C) The total value of n nickels and d dimes is ninety cents.
   (D) The total value of n nickels and d dimes is ninety dollars.
   (E) The total number of n nickels and d dimes is ninety coins.

5. Solve $-16 = 8 + 8 + a$. C
   (A) $\{0\}$ **(B)** $\{32\}$ **(C)** $\{-32\}$ **(D)** $\{-16\}$ **(E)** $\mathcal{R}$

6. Simplify $-(-3|-5|) + (-|-7 + 2|) + |-20|$. A
   (A) 30 **(B)** 0 **(C)** 40 **(D)** 44 **(E)** -30

7. For which of the following is $3m - m + 7 = 7 + 2m$ true? C
   (A) exactly one real number m
   (B) exactly two real numbers m
   (C) all real numbers m
   (D) some, but not all, real numbers m
   (E) no real numbers m

8. An operation $*$ is defined as $a * b = 2a + b + 1$ for all real numbers a and b. Which of the following are properties of $*$? A
   I. closure   II. commutativity   III. associativity
   (A) I only   **(B)** II only   **(C)** III only   **(D)** I and II only   **(E)** I, II, and III

Working with Real Numbers **109**

APPLICATION

Temperature Scales

How hot or cold will it be today? Weather forecasters give you this information in terms of temperature readings. *Temperature* is a measure of how hot or cold something is.

Temperature can be measured with a thermometer. A common type of thermometer is one that contains a liquid, usually mercury or alcohol, in a narrow glass tube. The volumes of these liquids increase when the temperature increases and decrease when the temperature decreases. Therefore, when there is a change in temperature there is also a change in the height of the liquid in the thermometer. In order to determine temperature by observing the height of a liquid in a tube, a thermometer must have a *temperature scale* marked on it. A temperature scale can be considered as a type of number line. Different types of thermometers have different scales.

Anders Celsius, a Swedish scientist who lived from 1701 to 1744, devised a temperature scale that today bears his name. Celsius called the temperature at which water freezes 0 degrees and the temperature at which water boils 100 degrees. He divided the space between these two points into 100 equal parts, and each part represented one *degree*. One of these degrees is now referred to as one degree Celsius (1°C). Celsius extended the scale above 100°C and below 0°C, keeping the size of the degree the same. Temperatures below 0°C are indicated by negative numbers.

A few years before the development of the Celsius thermometer, a German instrument maker, Gabriel Daniel Fahrenheit, designed a thermometer using a different scale. He chose the zero point to be the lowest temperature obtainable with a mixture of ice and common salt. The second point he chose was the body temperature of a healthy person. At first he divided the space between these two points into twelve equal divisions. Soon, however, he wanted to be able to take more accurate readings with his thermometer. Therefore, he divided each of the twelve parts into eight equal parts, resulting in 96 divisions between the two fixed points. Each of these divisions represented one degree, which is now referred to as one degree Fahrenheit (1°F). On the Fahrenheit scale, 32°F corresponds to the freezing point of water and 212°F corresponds to the boiling point of water.

Another important temperature scale is the *absolute*, or *Kelvin*, scale. It was named for the 19th century British scientist, Lord Kelvin, and used the concept that there is a lower limit to how cold any substance can be. This "lowest possible temperature" is referred to as *absolute zero*. Both theory and experiment indicate that absolute zero is approximately −273.15°C. On the Kelvin scale, absolute zero is the zero point. The divisions are the same size as on the Celsius scale, but they are called *kelvins* (K) rather than degrees Celsius. On this scale, water freezes at 273.15 K and boils at 373.15 K. This scale is often used in scientific work.

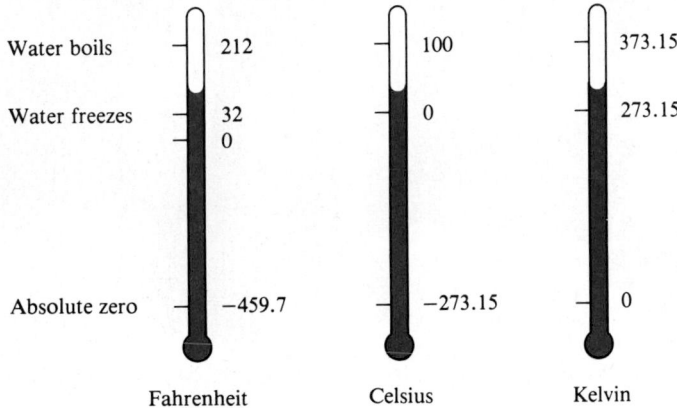

	Fahrenheit	Celsius	Kelvin
Water boils	212	100	373.15
Water freezes	32 / 0	0	273.15
Absolute zero	−459.7	−273.15	0

Exercises

1. A person reads an outside thermometer early one evening and finds the temperature to be 14°C. Early the next morning the same thermometer reads −11°C. How much did the temperature drop overnight? 25°C

2. A certain antifreeze solution for an automobile's cooling system freezes at a temperature that is 45°C below the temperature at which water freezes. What is the freezing point of this antifreeze solution in degrees Celsius? in Kelvins? −45°C; 228.15 K

3. Mercury freezes at a temperature of approximately 234 K and boils at approximately 629 K. What is the temperature range, expressed in degrees Celsius, to which a mercury thermometer is limited? −39.15°C to 355.85°C

4. A student made a thermometer and devised a scale for it. The scale was divided into 200 equal parts between the freezing point and the boiling point of water. The freezing point was then assigned the value of 50 degrees, and the boiling point was assigned 250 degrees. What temperature on the Celsius scale corresponds to a reading of 150 degrees on the student's scale? *Hint*: You may want to draw and compare simple diagrams of the two scales. 50°C

An astronaut aboard the space shuttle Challenger rides the remote manipulator system (RMS) arm during a space walk. The arm is controlled by another astronaut from within the cabin and may be used to retrieve and to release satellites and to shift equipment within the cargo bay.

Chapter 3

Solving Equations and Problems

Transforming Equations

OBJECTIVES for Sections 3-1 through 3-4:
1. *To use the addition, subtraction, multiplication, and division properties of equality as reasons for statements in proving theorems.*
2. *To solve equations using an addition, subtraction, multiplication, or division transformation.*
3. *To solve equations using several transformations.*

3–1 Properties of Equality

Pictured below is a device that is called a beam balance. When a 5 g mass is placed in each pan of the balance, the pans balance each other as shown.

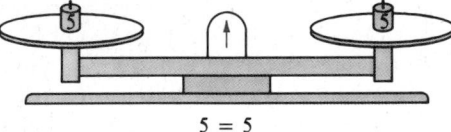

$$5 = 5$$

If a 10 g mass is now added to each pan, the pans remain in balance.

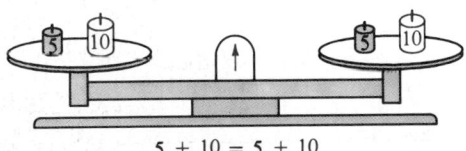

$$5 + 10 = 5 + 10$$

Word Problem Plan

A five-step *Plan for Solving a Word Problem,* which will be used throughout the course, is introduced in Section 3–5.

Using Charts

The use of a chart to organize the data given in a word problem is discussed in Section 3–7.

Using Formulas

The Problems in Section 3–8 involve a variety of formulas. Students will often need to transform a formula to solve for a given variable.

Teaching Suggestions
p. T76

Key Ideas

Use the properties of equality for addition, subtraction, multiplication, and division as reasons in proofs.

Name the axiom of equality
that is illustrated.

1. If $a - 6 = -7$, then
   $(a - 6) + 6 = -7 + 6$.
   Add. prop. of eq.
2. If $b + 5 = d$, then
   $(b + 5) - 5 = d - 5$.
   Subt. prop. of eq.
3. If $-9z = 108$, then
   $$\frac{-9z}{-9} = \frac{108}{-9}.$$
   Division prop. of eq.
4. If $\frac{y}{c} = h$, and $c \neq 0$, then
   $$c\left(\frac{y}{c}\right) = ch.$$
   Mult. prop. of eq.

Suggested Assignments

Minimum
First day: 116/1–7
Second day: 117/8–12
Average
  116/1–12, 13–19 odd
Maximum
  117/3–15, 16–22 even

Mixed Review

Name the property or axiom
illustrated in these equations.

1. $m + (-m) = 0$
   Ax. of add. inverses
2. $5 + (-3) = (-3) + 5$
   Commut. ax. of add.
3. $\frac{1}{4}(4y) = \left(\frac{1}{4} \cdot 4\right)y$

   Assoc. ax. of mult.
4. $a \cdot \frac{1}{a} = 1$

   Ax. of mult. inverses
5. $y + 0 = y$
   Identity ax. for add.
6. $m \cdot 1 = m$
   Identity ax. for mult.

The example on the preceding page illustrates the **addition property of equality:** *If the same number is added to equal numbers, the sums are equal.*

> ### Addition Property of Equality
> For all real numbers a, b, and c, if $a = b$, then
> $$a + c = b + c \quad \text{and} \quad c + a = c + b.$$

The addition property of equality is a theorem that can be proved as follows.

PROOF

Statements	Reasons
1. a, b, and c are real numbers.	Hypothesis
2. $a + c = a + c$	Reflexive axiom of equality
3. $a = b$	Hypothesis
4. $\therefore a + c = b + c$	Substitution principle
5. $a + c = c + a$ $b + c = c + b$	Commutative axiom for addition
6. $\therefore c + a = c + b$	Substitution principle

You can use similar reasoning to prove the **multiplication property of equality:** *If equal numbers are multiplied by the same number, the products are equal.* (See Exercise 9 on page 116.)

> ### Multiplication Property of Equality
> For all real numbers a, b, and c, if $a = b$, then
> $$ac = bc \quad \text{and} \quad ca = cb.$$

Since $a - c = a + (-c)$ and $b - c = b + (-c)$, you can write the **subtraction property of equality** as a special case of the addition property of equality: *If the same number is subtracted from equal numbers, the differences are equal.* (See Exercise 7 on page 117.)

> ### Subtraction Property of Equality
> For all real numbers a, b, and c, if $a = b$, then
> $$a - c = b - c.$$

114   *Chapter 3*

Similarly, since $\frac{a}{c} = a \cdot \frac{1}{c}$ and $\frac{b}{c} = b \cdot \frac{1}{c}$, you can write the **division property of equality** as a special case of the multiplication property of equality: *If equal numbers are divided by the same nonzero number, the quotients are equal.* (See Exercise 8 on page 117.)

Division Property of Equality
For all real numbers a and b and all nonzero real numbers c, if $a = b$, then

$$\frac{a}{c} = \frac{b}{c}.$$

EXAMPLE   Name the property of equality that is illustrated.

    a. If $m - 4 = -7$, then $(m - 4) + 4 = -7 + 4$.

    b. If $-6n = 78$, then $\frac{-6n}{-6} = \frac{78}{-6}$.

    c. If $\frac{x}{a} = b$, and $a \neq 0$, then $a\left(\frac{x}{a}\right) = ab$.

    d. If $\frac{1}{2}y + d = c$, then $\left(\frac{1}{2}y + d\right) - d = c - d$.

SOLUTION   **a.** addition property of equality

    b. division property of equality

    c. multiplication property of equality

    d. subtraction property of equality

Oral Exercises

Replace each ? with the number that makes a true statement.

1. If $a = b$, then $a + 8 = b +$ _?_ . 8

2. If $x - 8 = -10$, then $(x - 8) + 8 =$ _?_ $+ 8$. -10

3. If $a = b$, then $a -$ _?_ $= b - 12$. 12

4. If $5y + 12 = 7$, then $(5y + 12) - 12 =$ _?_ $- 12$. 7

5. If $a = b$, then $-5a = ($ _?_ $)b$. -5

6. If $\frac{z}{-5} =$ _?_ , then $-5\left(\frac{z}{-5}\right) = -5(10)$. 10

7. If $a = b$, then $\frac{a}{?} = \frac{b}{4}$. 4

8. If $4w = -68$, then $\frac{4w}{4} = \frac{?}{4}$. -68

Additional A Exercises

Additional A Exercises

Name the axiom, theorem, or definition that justifies each lettered step.

1. $m - 5 = 8$

$(m - 5) + (5) = 8 + (5)$ **(a)**

$m + [-5 + (5)] =$
                       $8 + (5)$ **(b)**

$m + 0 = 8 + (5)$ **(c)**

$m = 8 + (5)$ **(d)**

$m = 13$

    (a) Add. prop. of eq.

    (b) Assoc. ax. of add.

    (c) Ax. of add. inverses

    (d) Identity ax. for add.

2. $4x = -36$

    $\frac{1}{4}(4x) = \frac{1}{4}(-36)$ **(a)**

    $\left(\frac{1}{4} \cdot 4\right)x = \frac{1}{4}(-36)$ **(b)**

    $(1)x = \frac{1}{4}(-36)$ **(c)**

    $x = \frac{1}{4}(-36)$ **(d)**

    $x = -9$

    (a) Mult. prop. of eq.

    (b) Assoc. ax. of mult.

    (c) Ax. of mult. inverses

    (d) Identity ax. for mult.

3. $n + 6 = 7$

$(n + 6) + (-6) = 7 + (-6)$ **(a)**

$n + [6 + (-6)] = 7 + (-6)$ **(b)**

$n + 0 = 7 + (-6)$ **(c)**

$n = 7 + (-6)$ **(d)**

$n = 1$

    (a) Add. prop. of eq.

    (b) Assoc. ax. of add.

    (c) Ax. of add. inverses

    (d) Identity ax. for add.

(continued)

Additional A Exercises
(continued)

4. $\frac{1}{3}q = 13$

$$3\left(\frac{1}{3}q\right) = 3(13) \quad \textbf{(a)}$$

$$\left(3 \cdot \frac{1}{3}\right)q = 3(13) \quad \textbf{(b)}$$

$$(1)q = 3(13) \quad \textbf{(c)}$$
$$q = 3(13) \quad \textbf{(d)}$$
$$q = 39$$

(a) Mult. prop. of eq.
(b) Assoc. ax. of mult.
(c) Ax. of mult. inverses
(d) Identity ax. for mult.

Additional Answers
Written Exercises

1. (a) Add. prop. of $=$
   (b) Assoc. ax. for add.
   (c) Ax. of add. inverses
   (d) Identity ax. for add.
2. (a) Add. prop. of $=$
   (b) Assoc. ax. for add.
   (c) Ax. of add. inverses
   (d) Identity ax. for add.
3. (a) Mult. prop. of $=$
   (b) Assoc. ax. for mult.
   (c) Ax. of mult. inverses
   (d) Identity ax. for mult.
4. (a) Mult. prop. of $=$
   (b) Assoc. ax. for mult.
   (c) Ax. of mult. inverses
   (d) Identity ax. for mult.
5. (a) Def. of subt.
   (b) Add. prop. of $=$
   (c) Assoc. ax. for add.
   (d) Ax. of add. inverses
   (e) Identity ax. for add.
6. (a) Def. of division
   (b) Mult. prop. of $=$
   (c) Assoc. ax. for mult.
   (d) Ax. of mult. inverses
   (e) Identity ax. for mult.
7. 1. Hypothesis
   2. Ax. of add. inverses
   3. Hypothesis
   4. Add. prop. of $=$
   5. Def. of subt.

Replace each __?__ with the reason that justifies the statement to its left.

9. Prove: For all real numbers a, b, and c, if $a = b$, then $ac = bc$ and $ca = cb$.

PROOF

Statements	Reasons
1. a, b, and c are real numbers.	__?__ Hypothesis
2. $ac = ac$	__?__ Reflex. axiom of $=$
3. $a = b$	__?__ Hypothesis
4. $\therefore ac = bc$	Substitution principle
5. $ac = ca$ $\quad bc = cb$	Commutative axiom for multiplication
6. $\therefore ca = cb$	__?__ Substitution principle

10. Prove: For all real numbers a and b, if $a = b$, then $-a = -b$.

PROOF

Statements	Reasons
1. a and b are real numbers.	__?__ Hypothesis
2. $a + (-a) = 0$ $\quad b + (-b) = 0$	__?__ Axiom of additive inverses
3. $a + (-a) = b + (-b)$	Transitive axiom of equality
4. $a = b$	__?__ Hypothesis
5. $a + (-a) = a + (-b)$	Substitution principle
6. $-a$ is a real number.	__?__ Axiom of add. inv.
7. $-a + [a + (-a)] = -a + [a + (-b)]$	Addition property of equality
8. $-a + [a + (-a)] = (-a + a) + (-b)$	__?__
9. $\quad -a + 0 = 0 + (-b)$	__?__
10. $\quad \therefore -a = -b$	__?__

8. Associative axiom for addition **9.** Axiom of additive inverses **10.** Identity axiom for addition

Written Exercises

Name the axiom, theorem, or definition that justifies each lettered step.

A 1.
$$m + 3 = -5$$
$$(m + 3) + (-3) = -5 + (-3) \quad \textbf{(a)}$$
$$m + [3 + (-3)] = -5 + (-3) \quad \textbf{(b)}$$
$$m + 0 = -5 + (-3) \quad \textbf{(c)}$$
$$m = -5 + (-3) \quad \textbf{(d)}$$
$$m = -8$$

2.
$$-3 = -7 + n$$
$$7 + (-3) = 7 + (-7 + n) \quad \textbf{(a)}$$
$$7 + (-3) = [7 + (-7)] + n \quad \textbf{(b)}$$
$$7 + (-3) = 0 + n \quad \textbf{(c)}$$
$$7 + (-3) = n \quad \textbf{(d)}$$
$$4 = n$$

116 Chapter 3

3.

$$3x = -21$$

$$\tfrac{1}{3}(3x) = \tfrac{1}{3}(-21) \qquad \textbf{(a)}$$

$$\left(\tfrac{1}{3}\cdot 3\right)x = \tfrac{1}{3}(-21) \qquad \textbf{(b)}$$

$$(1)x = \tfrac{1}{3}(-21) \qquad \textbf{(c)}$$

$$x = \tfrac{1}{3}(-21) \qquad \textbf{(d)}$$

$$x = -7$$

4.

$$9 = -\tfrac{3}{4}y$$

$$-\tfrac{4}{3}(9) = -\tfrac{4}{3}\left(-\tfrac{3}{4}y\right) \qquad \textbf{(a)}$$

$$-\tfrac{4}{3}(9) = \left[-\tfrac{4}{3}\left(-\tfrac{3}{4}\right)\right]y \qquad \textbf{(b)}$$

$$-\tfrac{4}{3}(9) = (1)y \qquad \textbf{(c)}$$

$$-\tfrac{4}{3}(9) = y \qquad \textbf{(d)}$$

$$-12 = y$$

5.

$$a - 9 = -2$$

$$a + (-9) = -2 \qquad \textbf{(a)}$$

$$[a + (-9)] + 9 = -2 + 9 \qquad \textbf{(b)}$$

$$a + (-9 + 9) = -2 + 9 \qquad \textbf{(c)}$$

$$a + 0 = -2 + 9 \qquad \textbf{(d)}$$

$$a = -2 + 9 \qquad \textbf{(e)}$$

$$a = 7$$

6.

$$\tfrac{b}{4} = -8$$

$$b \cdot \tfrac{1}{4} = -8 \qquad \textbf{(a)}$$

$$\left(b \cdot \tfrac{1}{4}\right)4 = (-8)4 \qquad \textbf{(b)}$$

$$b\left(\tfrac{1}{4} \cdot 4\right) = (-8)4 \qquad \textbf{(c)}$$

$$b(1) = (-8)4 \qquad \textbf{(d)}$$

$$b = (-8)4 \qquad \textbf{(e)}$$

$$b = -32$$

Give the reason that justifies each statement in the given proof.

7. Prove: For all real numbers a, b, and c, if $a = b$, then $a - c = b - c$.

PROOF

1. a, b, and c are real numbers.
2. $-c$ is a real number.
3. $a = b$
4. $a + (-c) = b + (-c)$
5. $\therefore a - c = b - c$

8. Prove: For all real numbers a, b, and c such that $c \neq 0$, if $a = b$, then $\tfrac{a}{c} = \tfrac{b}{c}$.

PROOF

1. a, b, and c are real numbers such that $c \neq 0$.
2. $\tfrac{1}{c}$ is a real number.
3. $a = b$
4. $a \cdot \tfrac{1}{c} = b \cdot \tfrac{1}{c}$
5. $\therefore \tfrac{a}{c} = \tfrac{b}{c}$

8.
  1. Hypothesis
  2. Ax. of mult. inverses
  3. Hypothesis
  4. Mult. prop. of =
  5. Def. of division

9.
  1. a, b, and c are real numbers.
     Hypothesis
  2. $-c$ is a real number.
     Ax. of add. inv.
  3. $a + c = b + c$
     Hypothesis
  4. $(a + c) + (-c)$
     $= (b + c) + (-c)$
     Add. prop. of =
  5. $a + [c + (-c)]$
     $= b + [c + (-c)]$
     Assoc. ax. for add.
  6. $a + 0 = b + 0$
     Ax. of add. inv.
  7. $\therefore a = b$
     Iden. ax. for add.

10.
  1. a, b, and c are real numbers such that $c \neq 0$.
     Hypothesis
  2. $\tfrac{1}{c}$ is a real number.
     Ax. of mult. inv.
  3. $ac = bc$
     Hypothesis
  4. $(ac)\left(\tfrac{1}{c}\right) = (bc)\left(\tfrac{1}{c}\right)$
     Mult. prop. of =
  5. $a\left[c\left(\tfrac{1}{c}\right)\right] = b\left[c\left(\tfrac{1}{c}\right)\right]$
     Assoc. ax. for mult.
  6. $a \cdot 1 = b \cdot 1$
     Ax. of mult. inv.
  7. $\therefore a = b$
     Iden. ax. for mult.

(continued on p. 142)

Write a direct proof of each theorem.

B

9. For all real numbers a, b, and c, if $a + c = b + c$, then $a = b$.

10. For all real numbers a, b, and c such that $c \neq 0$, if $ac = bc$, then $a = b$.

11. For all real numbers a, b, and c, if $a - c = b - c$, then $a = b$.

12. For all real numbers a, b, and c such that $c \neq 0$, if $\tfrac{a}{c} = \tfrac{b}{c}$, then $a = b$.

Solving Equations and Problems **117**

Additional Answers
Written Exercises
(continued on p. 142)

Teaching Suggestions
p. T76

Key Ideas

Use addition and subtraction transformations to solve equations.

Chalkboard Examples

Solve.

1. $-4 + y = 10$
$-4 + y + 4 = 10 + 4$
$y = 14$
∴ the solution set is $\{14\}$.

2. $5 - (2 - x) = -9$
$5 - 2 + x = -9$
$3 + x = -9$
$3 + x - 3 = -9 - 3$
$x = -12$
∴ the solution set is $\{-12\}$.

3. $-1.5 + (x - 2.7) = -3.8$
$-1.5 + x - 2.7 = -3.8$
$x - 4.2 = -3.8$
$x - 4.2 + 4.2 =$
$\qquad -3.8 + 4.2$
$x = 0.4$
∴ the solution set is $\{0.4\}$.

Write a direct proof of each theorem. Assume that each variable represents a real number.

13. If $a + b = 0$, then $a = -b$.

14. If $ab = 1$ and $b \neq 0$, then $a = \frac{1}{b}$.

15. If $a - b = x$, then $b + x = a$.

16. If $a \div b = x$ and $b \neq 0$, then $bx = a$.

C **17.** If $a = b$ and $c = d$, then $a + c = b + d$.

18. If $a + c = b + d$ and $b = c$, then $a = d$.

19. If $a = b$ and $c = d$, then $ac = bd$.

20. If $ac = bd$, $a = d$, and $a \neq 0$, then $b = c$.

21. If $a = b$ and $c = d$, then $a - c = b - d$.

22. If $a = b$, $c = d$, $c \neq 0$, and $d \neq 0$, then $\frac{a}{c} = \frac{b}{d}$.

3–2 Transforming Equations: Addition and Subtraction

You can use the addition and subtraction properties of equality when solving certain equations. For example, study the following set of equations:

$$
\begin{aligned}
(1) \qquad & y - 7 = -3 \\
(2) \qquad & y - 7 + 7 = -3 + 7 \\
(3) \qquad & y = 4
\end{aligned}
$$

When you add 7 to each side of equation (1) and simplify the result, you obtain equation (3). On the other hand, when you subtract 7 from (or add -7 to) each side of equation (3), you obtain equation (1):

$$
\begin{aligned}
(3) \qquad & y = 4 \\
(2) \qquad & y - 7 = 4 - 7 \\
(1) \qquad & y - 7 = -3
\end{aligned}
$$

The addition and subtraction properties of equality guarantee that any root of equation (1) is also a root of equation (3) and that any root of equation (3) is also a root of equation (1). As a result, the equations $y - 7 = -3$ and $y = 4$ have the same solution set, namely $\{4\}$.

Equations that have the same solution set over a given domain are called **equivalent equations** over that domain. To solve an equation, you usually try to change, or **transform,** it into a simple equivalent equation whose solution or solutions can be seen at a glance, or by *inspection.* The

118   *Chapter 3*

properties of real numbers guarantee that each of the following *transformations* of a given equation will produce an equivalent equation.

Suggested Assignments
Minimum
  120/1–24, 25–39 odd
Average
  120/11–49 odd, 50–58
Maximum
  120/11–57 odd, 58–66

> ### *Transformations that Produce an Equivalent Equation*
> **Transformation by Substitution:** Substituting for any expression in a given equation an equivalent expression.
>
> **Transformation by Addition:** Adding the same real number to each side of a given equation.
>
> **Transformation by Subtraction:** Subtracting the same real number from each side of a given equation.

EXAMPLE 1   Solve $-8 + n = 11$.

SOLUTION
$$-8 + n = 11$$
$$-8 + n + 8 = 11 + 8$$
$$n = 19$$

To obtain n alone as the left side, add the opposite of -8, or 8, to each side.

Because errors may occur in transforming equations, always check your work by showing that each root of the transformed equation satisfies the *original equation*.

Check: $-8 + n = 11$ ⟵ original equation
$$-8 + 19 \overset{?}{=} 11$$
$$11 = 11 \quad \checkmark$$

∴ the solution set is {19}.

When solving an equation, sometimes you first need to simplify one or both sides using the properties of real numbers.

EXAMPLE 2   Solve $15 - (9 + z) = 20$.

SOLUTION
$$15 - (9 + z) = 20$$
$$15 - 9 - z = 20$$
$$6 - z = 20$$
$$6 - z - 6 = 20 - 6$$
$$-z = 14$$
$$z = -14$$

Use the property of the opposite of a sum to help simplify the left side of the equation.

Check: $15 - (9 + z) = 20$
$$15 - [9 + (-14)] \overset{?}{=} 20$$
$$15 - (-5) \overset{?}{=} 20$$
$$20 = 20 \quad \checkmark$$

∴ the solution set is {−14}.

Solving Equations and Problems   **119**

Oral Exercises

Tell what transformation can be used to obtain an equivalent equation with the variable alone as one side.

1. $m + 5 = 2$
2. $9 + g = 3$
3. $x - 3 = 6$
4. $r - 8 = 7$
5. $a + (-2) = -5$
6. $b + (-4) = -3$
7. $-4 + n = 4$
8. $-10 + s = 10$
9. $b - (-3) = -12$
10. $p - (-2) = 0$
11. $-7 = 7 + x$
12. $-5 = n + 5$
13. $0 = r - 6$
14. $-11 = m - 4$
15. $-0.6 + y = 1.5$
16. $-1\frac{1}{2} + b = -\frac{3}{4}$
17. $5 + x = -\frac{2}{3}$
18. $-4.6 = -7 + a$

Written Exercises

Solve. **18.** {−2} **21.** {0.22} **29.** {−0.72} **33.** {1, −1} **37.** {−11} **38.** {−16}

A
1. $t + 7 = 5$ {−2}
2. $p + 18 = 15$ {−3}
3. $a - 2 = 10$ {12}
4. $r - 7 = 6$ {13}
5. $-4 + k = -3$ {1}
6. $-9 + a = -2$ {7}
7. $b + (-11) = 32$ {43}
8. $(-3) + t = 5$ {8}
9. $s - (-4) = 9$ {5}
10. $w - (-1) = 8$ {7}
11. $-15 = 6 + j$ {−21}
12. $-13 = 12 + q$ {−25}
13. $10 = v - 10$ {20}
14. $x - 9 = 9$ {18}
15. $-4 + f = -4$ {0}
16. $-23 = -23 + b$ {0}
17. $-13 + 2 = x - 8$ {−3}
18. $f - 12 = -17 + 3$
19. $h - \frac{1}{2} = -\frac{2}{3}$ $\{-\frac{1}{6}\}$
20. $-\frac{4}{5} + e = -\frac{1}{4}$ $\{\frac{11}{20}\}$
21. $w + (-0.61) = -0.39$
22. $a - 2.7 = -4.32$ {−1.62}
23. $-t + 2 = -5$ {7}
24. $3 - p = 8$ {−5}
25. $3 = -8 - k$ {−11}
26. $-5 = -v + 7$ {12}
27. $-a + |-3| = 4$ {−1}
28. $k - |5| = -6$ {−1}
29. $-1.08 = -1.8 - t$
30. $0.1 = 0.01 - j$ {−0.09}
31. $|g| + 4 = 18$ {14, −14}
32. $|p| + 9 = 24$ {15, −15}
33. $-3 + |h| = -2$
34. $|t| - 15 = -7$ {8, −8}
35. $|k| - 4 = -6$ Ø
36. $-5 + |r| = -12$ Ø

B
37. $11 + (-1 + n) = -1$
38. $8 + (-2 + c) = -10$
39. $5 = (p - |-2|) - 16$
40. $-8 = 12 + (|-5| + r)$ {−25}
41. $2 - (3 - a) = -4$ {−3}
42. $7 - (12 - p) = -9$
43. $12 = 3 - (5 + b)$ {−14}
44. $-2 = -5 + (-x + 7)$ {4}
45. $1\frac{1}{2} - \left(\frac{2}{3} + x\right) = \frac{7}{2}$
46. $\left(\frac{1}{4} - y\right) - \frac{3}{8} = -\frac{5}{6}$ $\{\frac{17}{24}\}$
47. $-3 = \frac{1}{6} - \left(\frac{4}{3} - a\right)$ $\{-\frac{11}{6}\}$
48. $-3\frac{1}{2} = \frac{3}{2} - \left(\frac{3}{4} - k\right)$
49. $0.3 - (p - 0.5) = 0.7$
50. $2.5 - (7.8 + r) = 1.9$
51. $3 - (5 - |t|) = 6$
52. $4 - (-2 - |w|) = 10$
53. $-(|a| - 7) = 10$
54. $-(|s| - 7) = 5$
    {4, −4}   Ø   {2, −2}

Solve each equation for the value of r when $s = -2$ and $t = 6$.

55. $-3s + 2t = 6 - (st + r)$ {0}
56. $s - (3 - t) = 2s - (t - r)$ {11}
57. $r - s = -t - (1 + s^2)$ {−13}
58. $r - (s + t) = -3(t^2 - s^2)$ {−92}

39. {23} **42.** {−4} **45.** $\{-\frac{8}{3}\}$ **48.** $\{-\frac{17}{4}\}$ **49.** {0.1} **50.** {−7.2} **51.** {8, −8}

Solve each equation for the value of x when $a = \frac{1}{2}$ and $b = -\frac{3}{4}$.

C **59.** $-a(a^2 - 2b + ab) = -\frac{2}{3}a - x + \frac{5}{16}$ $\{\frac{2}{3}\}$

60. $x - 2 = \dfrac{-6a^3 + (2b)^2}{a - (2b - 3a)}$ $\{2\frac{3}{7}\}$

For which real values of t is the solution set of each equation empty?

EXAMPLE $|x| = t$

SOLUTION Since $|x| \geq 0$ for all real values of x, the solution set is $\emptyset$ if $t < 0$.

61. $|x| = -t$ $t > 0$ **62.** $|-x| = -t$ $t > 0$ **63.** $|x| = t + 1$ $t < -1$

64. $|x| = t - 1$ $t < 1$ **65.** $|x| + 2 = t$ $t < 2$ **66.** $|x| - 2 = t$

$ t < -2$

3-3 Transforming Equations: Multiplication and Division

Teaching Suggestions p. T77

Related Activities p. T77

Key Ideas
Use multiplication and division transformations to solve equations in one variable.

As you have just seen, certain equations can be solved using the addition and subtraction properties of equality. You will now see how certain other equations can be solved using the multiplication and division properties of equality. For example, consider the following set of equations:

$$(1) \qquad \frac{z}{-9} = -4$$

$$(2) \qquad \frac{z}{-9}(-9) = -4(-9)$$

$$(3) \qquad z = 36$$

When you multiply each side of equation (1) by -9 and simplify the result, you obtain equation (3). On the other hand, when you divide each side of equation (3) by -9, you obtain equation (1):

$$(3) \qquad z = 36$$

$$(2) \qquad \frac{z}{-9} = \frac{36}{-9}$$

$$(1) \qquad \frac{z}{-9} = -4$$

The multiplication and division properties of equality guarantee that any root of equation (1) is also a root of equation (3) and that any root of equation (3) is also a root of equation (1). Therefore the equations $\frac{z}{-9} = -4$ and $z = 36$ have the same solution set, $\{36\}$, and are equivalent equations. From this example, you see that the multiplication and division properties of equality provide two more ways to transform an equation into an equivalent equation.

Solving Equations and Problems **121**

Solve.

1. $4y = 16$

$$\frac{4y}{4} = \frac{16}{4}$$

$$y = 4$$

$\therefore$ the solution set is $\{4\}$.

2. $\dfrac{x}{-2} = -6$

$$(-2)\left(\frac{x}{-2}\right) = (-2)(-6)$$

$$x = 12$$

$\therefore$ the solution set is $\{12\}$.

3. $\dfrac{2}{5}p = -8$

$$\left(\frac{5}{2}\right)\frac{2}{5}p = \left(\frac{5}{2}\right)(-8)$$

$$p = -20$$

$\therefore$ the solution set is $\{-20\}$.

4. Solve the equation when $a = 4$ and $b = -2$.

$$\frac{b^2}{2}x = |4ab| + 6$$

$$\frac{4}{2}x = |4(4)(-2)| + 6$$

$$2x = |-32| + 6$$

$$2x = 38$$

$$\frac{2x}{2} = \frac{38}{2}$$

$$x = 19$$

$\therefore$ the solution set is $\{19\}$.

Suggested Assignments

Minimum
  123/1–32
Average
  123/5–37 odd, 39–46
Maximum
  123/5–41 odd, 43–50

> **Transformation by Multiplication:** Multiplying each side of a given equation by the same *nonzero* real number.
> **Transformation by Division:** Dividing each side of a given equation by the same *nonzero* real number.

EXAMPLE 1 Solve $3a = -87$.

SOLUTION

$$3a = -87$$

$$\frac{3a}{3} = \frac{-87}{3} \longleftarrow \begin{cases} \text{To obtain } a \text{ alone as the left} \\ \text{side, divide each side by 3 (or} \\ \text{multiply by } \tfrac{1}{3}, \text{ the reciprocal of 3).} \end{cases}$$

$$a = -29$$

Check: $3a = -87$

$$3(-29) \stackrel{?}{=} -87$$

$$-87 = -87 \quad \checkmark$$

$\therefore$ the solution set is $\{-29\}$.

EXAMPLE 2 Solve $12 = -\dfrac{2}{3}m$.

SOLUTION

$$12 = -\frac{2}{3}m$$

$$-\frac{3}{2}(12) = -\frac{3}{2}\left(-\frac{2}{3}m\right) \longleftarrow \begin{cases} \text{To obtain } m \text{ alone as the right} \\ \text{side, multiply each side by } -\tfrac{3}{2}, \\ \text{the reciprocal of } -\tfrac{2}{3}. \end{cases}$$

$$-18 = m$$

Check: $12 = -\dfrac{2}{3}m$

$$12 \stackrel{?}{=} -\frac{2}{3}(-18)$$

$$12 = 12 \quad \checkmark$$

$\therefore$ the solution set is $\{-18\}$.

You know that you can never divide by zero, and so zero is certainly not allowed as a divisor in a division transformation. Do you know why zero is not allowed as a *multiplier* in transforming an equation? Look at the following set of equations:

$$(1) \qquad 7x = 28$$

$$(2) \qquad 0 \cdot 7x = 0 \cdot 28$$

$$(3) \qquad (0 \cdot 7)x = 0$$

$$(4) \qquad 0 \cdot x = 0$$

Equation (1) has just one root, namely 4, but equation (4) is satisfied by *any* real number. Therefore, equations (1) and (4) are *not* equivalent. *In transforming an equation, never multiply by zero.*

122 *Chapter 3*

Oral Exercises

Tell what transformation can be used to obtain an equivalent equation with the variable alone as one side.

1. $7a = 35$
2. $9x = -207$
3. $-4b = -24$
4. $-3p = -54$
5. $\frac{1}{3}d = 15$
6. $\frac{1}{5}t = 5$
7. $-\frac{3}{10}f = 12$
8. $-\frac{3}{4}a = 6$
9. $-5 = -5c$
10. $0 = -6r$
11. $1.5j = 9$
12. $-5 = 0.1v$
13. $\frac{p}{3} = -4$
14. $\frac{8}{-5} = -3$
15. $x \div 5 = -3$
16. $6 = y \div (-3)$
17. $\frac{c}{-9} = 1$
18. $-1 = \frac{a}{21}$
19. $-\frac{3}{8} = \frac{3}{8}k$
20. $-\frac{4}{5} = -\frac{5}{4}h$

Written Exercises

Solve. 30. $\{4, -4\}$ **33.** $\{\frac{9}{8}, -\frac{9}{8}\}$ **39.** $\{400, -400\}$ **42.** $\{\frac{25}{6}, -\frac{25}{6}\}$

A

1. $13r = -143$ $\{-11\}$
2. $15t = -105$ $\{-7\}$
3. $\frac{1}{7}p = -3$ $\{-21\}$
4. $\frac{1}{6}x = -4$ $\{-24\}$
5. $-8t = -112$ $\{14\}$
6. $-144 = -9e$ $\{16\}$
7. $5 = \frac{v}{12}$ $\{60\}$
8. $7 = \frac{p}{-6}$ $\{-42\}$
9. $\frac{3}{20}n = 0$ $\{0\}$
10. $0 = -\frac{5}{3}t$ $\{0\}$
11. $-60 = -\frac{2}{3}w$ $\{90\}$
12. $48 = -\frac{3}{5}t$ $\{-80\}$
13. $\frac{3}{5}x = -\frac{7}{10}$ $\{-\frac{7}{6}\}$
14. $-\frac{8}{9} = -\frac{4}{15}x$ $\{\frac{10}{3}\}$
15. $\frac{1}{3}t = 3\frac{2}{3}$ $\{11\}$
16. $\frac{1}{2}k = 2\frac{5}{6}$ $\{5\frac{2}{3}\}$
17. $\frac{7}{2}t = -\frac{2}{7}$ $\{-\frac{4}{49}\}$
18. $-\frac{3}{10}y = \frac{3}{10}$ $\{-1\}$
19. $-\frac{3}{11}r = -\frac{3}{11}$ $\{1\}$
20. $\frac{5}{8}c = -\frac{8}{5}$ $\{-\frac{64}{25}\}$
21. $0.01 = -0.1p$ $\{-0.1\}$
22. $-0.15 = -0.003t$ $\{50\}$
23. $-\frac{1}{5}t = -0.3$ $\{1.5\}$
24. $-0.4 = \frac{1}{7}k$ $\{-2.8\}$
25. $\frac{m}{0.2} = -4$ $\{-0.8\}$
26. $\frac{h}{0.5} = -0.5$ $\{-0.25\}$
27. $\frac{-4}{3} = \frac{t}{-2}$ $\{\frac{8}{3}\}$
28. $\frac{-b}{7} = \frac{-2}{3}$ $\{\frac{14}{3}\}$
29. $5|x| = 10$ $\{2, -2\}$
30. $-3|r| = -12$

B

31. $\frac{3}{5}|d| = 9$ $\{15, -15\}$
32. $\frac{7}{12}|x| = 14$ $\{24, -24\}$
33. $-\frac{2}{3}|c| = \frac{3}{4}$
34. $\frac{5}{4}|s| = -\frac{3}{2}$ $\{\frac{6}{5}, -\frac{6}{5}\}$
35. $-\frac{1}{3}|t| = 15$ $\emptyset$
36. $\frac{1}{4}|t| = -6$ $\emptyset$
37. $0 = -\frac{2}{5}|y|$ $\{0\}$
38. $-\frac{3}{5}|z| = 0$ $\{0\}$
39. $-0.012|p| = -4.8$
40. $2.5|x| = -4.5$ $\emptyset$
41. $7 = -1.001|t|$ $\emptyset$
42. $-5 = -1.2|k|$

Solving Equations and Problems **123**

Mixed Review

Find the missing numbers.

1. $0.5 \div 0.1 = \underline{\ ?\ }$ 5

2. $-54\left(-\dfrac{2}{3}\right) = \underline{\ ?\ }$ 36

3. $\dfrac{7}{10}(\underline{\ ?\ }) = 1$ $\dfrac{10}{7}$

4. $-2(\underline{\ ?\ }) = 1$ $-\dfrac{1}{2}$

Solve each equation for the value of x when $a = -3$ and $b = 4$.

43. $-\dfrac{2b}{a}x = |a - 2b| + 5$ $\{6\}$

44. $|a|x = -1 - |b^2 - 3a^2|$ $\{-4\}$

45. $\dfrac{a^2b}{a + b}x = -a^2 + (-a)^2b$ $\{\frac{3}{4}\}$

46. $\dfrac{2a + 3b}{a^2 - b^2}x = a^2 + 2ab + b^2$ $\{-\frac{7}{6}\}$

Solve each equation for the value of r when $s = -\frac{1}{4}$ and $t = \frac{2}{5}$.

C 47. $\dfrac{1}{10}(s + t)|r| = s^2 + 2st + t^2$ $\{\frac{3}{2}, -\frac{3}{2}\}$

48. $s^2 - t^2 = 13s(s + t)|r|$ $\{\frac{1}{5}, -\frac{1}{5}\}$

49. $4r + 9st = 4\left(\dfrac{t}{s} + \dfrac{s}{t}\right)$ $\{-2\}$

50. $\left(\dfrac{7}{s^2 - 4t^2}\right)r = \dfrac{5}{s + 2t}$ $\{-\frac{3}{4}\}$

Teaching Suggestions
p. T77

Related Activities p. T78

Supplementary Material
Test 9

Key Idea

Solve more difficult equations using more than one transformation.

Chalkboard Examples

Solve.

1. $5x + 6 = 31$
   $5x + 6 - 6 = 31 - 6$
   $\quad\quad 5x = 25$
   $\quad\quad \dfrac{5x}{5} = \dfrac{25}{5}$
   $\quad\quad\quad x = 5$
   Check: $5x + 6 = 31$
   $\quad\quad 5(5) + 6 = 31$
   $\quad\quad 25 + 6 = 31$
   $\quad\quad\quad 31 = 31$ ✓
   ∴ the solution set is $\{5\}$.

3–4 Using Several Transformations

From your work in Section 2-7, recall the theorem which states that, for all real numbers a and b, $(a + b) + (-b) = a$. Using this theorem together with the definition of subtraction, the following related theorem has already been proved. (See Exercises 47 and 49 on page 85.)

> For all real numbers a and b,
>
> $$(a + b) - b = a \quad \text{and} \quad (a - b) + b = a.$$

This theorem can be stated informally as follows:

> *To "undo" addition of a number, you subtract that number.*
> *To "undo" subtraction of a number, you add that number.*

Operations that "undo" each other, such as addition and subtraction, are called **inverse operations**.

In Section 2-10, the theorem was proved which states that, for all real numbers a and all nonzero real numbers b, $(ab)\dfrac{1}{b} = a$. Using this theorem together with the definition of division, this related theorem has already been proved. (See Exercises 41 and 43 on page 102.)

> For all real numbers a and all nonzero real numbers b,
>
> $$(ab) \div b = a \quad \text{and} \quad (a \div b)b = a.$$

124   Chapter 3

This theorem can be stated informally as follows:

To "undo" multiplication by a nonzero number, you divide by that number.
To "undo" division by a nonzero number, you multiply by that number.

That is, multiplication and division are also inverse operations.

When deciding which transformations to use in solving an equation, you may find it helpful to think of inverse operations.

EXAMPLE 1   Solve $7x + 29 = 15$.

SOLUTION
$$7x + 29 = 15$$
$$7x + 29 - 29 = 15 - 29 \longleftarrow \begin{cases} \text{To undo the addition of 29 to } 7x, \\ \text{subtract 29 from each side.} \end{cases}$$
$$7x = -14$$
$$\frac{7x}{7} = \frac{-14}{7} \longleftarrow \begin{cases} \text{To undo the multiplication of } x \text{ by 7,} \\ \text{divide each side by 7.} \end{cases}$$
$$x = -2$$

Check:
$$7x + 29 = 15$$
$$7(-2) + 29 \stackrel{?}{=} 15$$
$$-14 + 29 \stackrel{?}{=} 15$$
$$15 = 15 \quad \checkmark$$

∴ the solution set is $\{-2\}$.

The following plan summarizes the steps that are usually helpful in solving an equation in which all the variables are on the same side.

---

1. Simplify each side of the equation.
2. If there are still indicated additions or subtractions, use the inverse operations to undo them.
3. If there are indicated multiplications or divisions involving the variable, use the inverse operations to undo them.

EXAMPLE 2   Solve $5 - 3(2n - 3) = 44$.

SOLUTION
$$5 - 3(2n - 3) = 44$$
$$5 - 6n + 9 = 44 \longleftarrow \begin{cases} \text{Use the distributive axiom to} \\ \text{help simplify the left side.} \end{cases}$$
$$14 - 6n = 44$$
$$14 - 6n - 14 = 44 - 14 \longleftarrow \{\text{Subtract 14 from each side.}$$
$$-6n = 30$$
$$\frac{-6n}{-6} = \frac{30}{-6} \longleftarrow \{\text{Divide each side by } -6.$$
$$n = -5$$

Solution continued on next page.

Solving Equations and Problems   **125**

2. $9 - 2(3n - 1) = 83$
$$9 - 6n + 2 = 83$$
$$11 - 6n = 83$$
$$-6n = 72$$
$$n = -12$$
Check:
$$9 - 2(3n - 1) = 83$$
$$9 - 2[3(-12) - 1] = 83$$
$$9 - 2(-37) = 83$$
$$9 + 74 = 83$$
$$83 = 83 \quad \checkmark$$
∴ the solution set is $\{-12\}$.

3. $\frac{3}{4}m + \frac{1}{4} = 4$

Solution 1

$$\frac{3}{4}m + \frac{1}{4} = 4$$

First subtract $\frac{1}{4}$ from each side of the equation, then multiply each side by 4.

$$\frac{3}{4}m = \frac{15}{4}$$
$$3m = 15$$
$$m = 5$$

Solution 2

$$\frac{3}{4}m + \frac{1}{4} = 4$$

First multiply each side by 4.

$$3m + 1 = 16$$
$$3m = 15$$
$$m = 5$$

Check: $\frac{3}{4}m + \frac{1}{4} = 4$

$$\frac{3}{4}(5) + \frac{1}{4} = 4$$
$$\frac{15}{4} + \frac{1}{4} = 4$$
$$4 = 4 \quad \checkmark$$

∴ the solution set is $\{5\}$.

Reading Algebra

Have the students read each equation in words. Encourage various translations.

1. $5n - 6 = 17$

2. $2(n + 1) = 6$

3. $\dfrac{n + 3}{5} = 0$

4. $2(n^2 + 1) = 71$

Next have them make up expressions like: "Three times a number cubed is the same as the number increased by 2." Ask them to write equations for the expressions others make up. Encourage different translations as you did before.

CONDENSED SOLUTION

$$5 - 3(2n - 3) = 44$$
$$5 - 6n + 9 = 44$$
$$14 - 6n = 44$$
$$-6n = 30$$
$$n = -5$$

$\therefore$ the solution set is $\{-5\}$.

Check:

$$5 - 3(2n - 3) = 44$$
$$5 - 3[2(-5) - 3] \stackrel{?}{=} 44$$
$$5 - 3(-13) \stackrel{?}{=} 44$$
$$5 + 39 \stackrel{?}{=} 44$$
$$44 = 44 \quad \checkmark$$

Often there is more than one way to solve a given equation.

EXAMPLE 3 Solve $\frac{1}{2}x + \frac{3}{2} = 4$.

SOLUTION 1
$$\frac{1}{2}x + \frac{3}{2} = 4 \longleftarrow \begin{cases} \text{First subtract } \frac{3}{2} \text{ from each side,} \\ \text{then multiply each side by 2.} \end{cases}$$
$$\frac{1}{2}x = \frac{5}{2}$$
$$x = 5$$

SOLUTION 2
$$\frac{1}{2}x + \frac{3}{2} = 4 \longleftarrow \begin{cases} \text{First multiply each side by 2,} \\ \text{then subtract 3 from each side.} \end{cases}$$
$$2\left(\frac{1}{2}x + \frac{3}{2}\right) = 2(4)$$
$$x + 3 = 8$$
$$x = 5$$

Check:
$$\frac{1}{2}x + \frac{3}{2} = 4$$
$$\frac{1}{2}(5) + \frac{3}{2} \stackrel{?}{=} 4$$
$$\frac{5}{2} + \frac{3}{2} \stackrel{?}{=} 4$$
$$\frac{8}{2} \stackrel{?}{=} 4$$
$$4 = 4 \quad \checkmark$$

$\therefore$ the solution set is $\{5\}$.

Oral Exercises

Tell what transformation was used to transform the first equation into the second and the second equation into the third.

1. $3x - 1 = -5$; $3x = -4$; $x = \dfrac{-4}{3}$ Add 1 to each side; divide each side by 3.

2. $-2x + 3 = 1$; $-2x = -2$; $x = 1$ Subtract 3 from each side; divide each side by -2.

3. $\frac{1}{2}b - 3 = -14$; $\frac{1}{2}b = -11$; $b = -22$ Add 3 to each side; multiply each side by 2.

4. $-4 + \frac{1}{3}c = 5$; $\frac{1}{3}c = 9$; $c = 27$ Add 4 to each side; multiply each side by 3.

5. $-8 = -\frac{2}{3}p + 6$; $-14 = -\frac{2}{3}p$; $21 = p$ Subtract 6 from each side; multiply each side by $-\frac{3}{2}$.

6. $5 - \frac{3}{4}x = -7$; $-\frac{3}{4}x = -12$; $x = 16$ Subtract 5 from each side; multiply each side by $-\frac{4}{3}$.

7. $\frac{1}{3}t + 5 = -3$; $t + 15 = -9$; $t = -24$ Multiply each side by 3; subtract 15 from each side.

8. $2 = 3 + \frac{k}{5}$; $10 = 15 + k$; $-5 = k$ Multiply each side by 5; subtract 15 from each side.

9. $\frac{c-1}{2} = -7$; $c - 1 = -14$; $c = -13$ Multiply each side by 2; add 1 to each side.

10. $\frac{-3+b}{4} = 8$; $-3 + b = 32$; $b = 35$ Multiply each side by 4; add 3 to each side.

11. $\frac{1}{2}(p - 1) = 10$; $p - 1 = 20$; $p = 21$ Multiply each side by 2; add 1 to each side.

12. $-3(x - 5) = 12$; $x - 5 = -4$; $x = 1$ Divide each side by -3; add 5 to each side.

Tell which of the equations (a), (b), or (c) are equivalent to the given equation. For each equivalent equation, tell what transformation was used to obtain it from the given equation.

13. Given $\frac{1}{2}x - 3 = 7$ **c.** Multiply each side by 2.

  a. $\frac{1}{2}x = 4$ **b.** $x - 3 = 14$ **c.** $x - 6 = 14$

14. Given $-3(x - 4) = 15$

  a. $x - 4 = -5$ **b.** $-3x + 12 = 15$ **c.** $-3x + 12 = -45$

15. Given $-\frac{1}{3}x + \frac{2}{3} = 3$ **a.** Multiply each side by 3.
                               c. Subtract $\frac{2}{3}$ from each side.

  a. $-x + 2 = 9$ **b.** $-x + 2 = 3$ **c.** $-\frac{1}{3}x = \frac{7}{3}$

                        a. Multiply each side by 4.

16. Given $\frac{3}{4}x + 6 = 9$ **b.** Multiply each side by $\frac{4}{3}$.
                       c. Subtract 6 from each side.

  a. $3x + 24 = 36$ **b.** $x + 8 = 12$ **c.** $\frac{3}{4}x = 3$

14. a. Divide each side by -3.
   b. Substitute $-3x + 12$ for $-3(x - 4)$ using the distributive axiom.

Written Exercises

Solve.

                                                           $\{2.4\}$

A **1.** $4c + 3 = 7$ $\{1\}$ **2.** $-4 + 2t = -1$ $\{\frac{3}{2}\}$ **3.** $0.25 = -0.5t + 1.45$

    4. $-\frac{3}{8} = \frac{5}{4} - 2x$ $\{\frac{13}{16}\}$ **5.** $-2 = \frac{b}{3} - 4$ $\{6\}$ **6.** $2 - \frac{1}{4}t = 16$ $\{-56\}$

    7. $5 - \frac{3}{4}x = 11$ $\{-8\}$ **8.** $5 = \frac{3}{5}p + 14$ $\{-15\}$ **9.** $\frac{2s - 3}{4} = -5$ $\{-\frac{17}{2}\}$

Additional A Exercises

Solve.
1. $3p + 2 = 8$ $\{2\}$
2. $2k - 1 = 7$ $\{4\}$
3. $5.25 = -0.5m + 0.25$
  $\{-10\}$

4. $4 = \frac{2}{3}n - 2$ $\{9\}$

5. $1 - 2x - 3x = 16$ $\{-3\}$
6. $-(7x + 1) = 6$ $\{-1\}$

Mixed Review

Solve.

1. $\frac{1}{3}x = -6$ {−18}

2. $-0.1x = -0.10$ {1}

Solve each equation for the value of x when $a = -4$ and $b = 0$.

3. $|x| = -a$ {±4}

4. $|b - a|x = -a$ {1}

Solve. **12.** {−2} **13.** {9} **14.** $\{\frac{1}{2}\}$ **18.** {14} **21.** {0} **24.** {−1}

10. $\frac{5 - 3g}{2} = 0$ $\{\frac{5}{3}\}$

11. $9t + 4 - 3t = 36$ $\{\frac{16}{3}\}$

12. $-7s + 6 - 8s = 36$

13. $315 - 22y - 13y = 0$

14. $-13 = -16b + (-10b)$

15. $3(a - 4) = -5$ $\{\frac{7}{3}\}$

16. $6(4 - 3r) = 24$ {0}

17. $\frac{1}{3}(2 - 5p) = -13$ $\{\frac{41}{5}\}$

18. $-\frac{4}{5}(x - 4) = -8$

19. $\frac{1}{2}x + \frac{1}{3} = \frac{3}{4}$ $\{\frac{5}{6}\}$

20. $-\frac{1}{4}g + \frac{1}{6} = \frac{5}{12}$ {−1}

21. $2(3 - c) - 7 = -1$

22. $-3(2 - y) = 0$ {2}

23. $-8 = -3(y - 3) - 8$ {3}

24. $5(2 - 3b) - 13 = 12$

25. $15g - (13g + 14) - 5g = 0$ $\{-\frac{14}{3}\}$

26. $10k - 8k - (3 - k) = -6$ {−1}

27. $(3r - 7) - (4r - 15) = 6$ {2}

28. $(21 - 2p) - (15 + 6p) = 0$ $\{\frac{3}{4}\}$

29. $2v - 8 - \frac{4}{3}v = 0$ {12}

30. $5 = u + 3 - \frac{8}{9}u$ {18}

31. $3(d - 3) - 4d = 5$ {−14}

32. $7y + 4(y + 1) = 15$ {1}

33. $6e - (2e - 5) = 17$ {3}

34. $7z - (2z - 8) = -2$ {−2}

35. $-2(3 - g) + 2g - 9 = 1$ {4}

36. $-2b - 5(b - 1) - 3 = 16$ {−2}

Solve each equation using two different methods. The first steps of the methods should be different from each other.

B **37.** $\frac{1}{3}a + 4 = 2$

38. $\frac{2}{5}x + 3 = \frac{1}{5}$

39. $3(p - 2) = 12$

40. $4(k + 1) = 7$

41. $\frac{1}{4}x + \frac{3}{4} = 5$

42. $\frac{3}{2}t - \frac{1}{2} = 2$

43. $\frac{1}{3}(t - 3) = 4$

44. $\frac{1}{2}(6x - 1) = 4$

45. $\frac{4}{5}x - 3 = \frac{2}{5}$

46. $\frac{2}{3}y - 6 = 8$

47. $\frac{1}{5}a - \frac{1}{4} = \frac{1}{2}$

48. $\frac{1}{4}t + \frac{1}{6} = \frac{2}{3}$

Solve.

49. $5 = \frac{2}{3}(10 - i) + 1$ {4}

50. $8 - \frac{3}{4}(9 - d) = 5$ {5}

51. $3(k - 2) - 2(2 - 3k) = -10$ {0}

52. $5(6f - 1) - 4(3 - 4f) = 6$ $\{\frac{1}{2}\}$

53. $\frac{2}{5}(j - 4) + 2(j + 3) = -10$ {−6}

54. $\frac{1}{2}(e + 5) + \frac{3}{4}(3 - e) = 5$ {−1}

C **55.** $6(1 - c) - 5(c + 1) + 2[(5 - 3c) + 3] = -34$ {3}

56. $3(x - 1) - 2(1 - 3x) - 3[(6 + x) - 1] = -15$ $\{\frac{5}{6}\}$

57. $3[5(4j - 3) - (5 - 8j)] + 2[2(5 - j) + 7] = -6$ $\{\frac{1}{4}\}$

58. $4[2(q - 10) + 3q - 12] - 3[3(q - 5) - (4 - q)] = 17$ {11}

59. $\frac{1}{3}(1 + p) - \frac{2}{3}[(2p + 7) - 3(p - 1)] = -7$ $\{-\frac{2}{3}\}$

60. $\frac{1}{2}[5 - 2(x - 3)] + \frac{1}{4}[(2 - x) - (3 - 2x)] = 5$ $\{\frac{1}{3}\}$

Computer Exercises For students with computer experience

Write a program that will solve an equation of the form $ax + b = c$ when you input values for a, b, and c. What value of a must be excluded? RUN the program to solve each of the following equations.

1. $4x + 9 = 1$ $\{-2\}$ **2.** $7y - 8 = 20$ $\{4\}$ **3.** $\frac{3}{4}z + 2 = -7$ $\{-12\}$ **4.** $\frac{w}{2} - 6 = 15$ $\{42\}$

5. $m - 19 = 7$ $\{26\}$ **6.** $28 - n = -6$ $\{34\}$ **7.** $-3 = -5p + 2$ $\{1\}$ **8.** $-6q = 18$ $\{-3\}$

Modify the program that you wrote for Exercises 1–8 to solve an equation of the form $a|x| + b = c$. RUN the program to solve each of the following.

9. $|x| - 5 = 9$ **10.** $|y| + 8 = 6$ **11.** $3|s| - 5 = 16$ **12.** $\frac{|t|}{2} = -7$
    $\{14, -14\}$ Ø $\{7, -7\}$ Ø

▪ Self-Test 1

VOCABULARY   addition property of equality   transformation by substitution
                     (p. 114)   (p. 119)
                  multiplication property of   transformation by addition
                     equality (p. 114)   (p. 119)
                  subtraction property of   transformation by subtraction
                     equality (p. 114)   (p. 119)
                  division property of equality   transformation by
                     (p. 115)   multiplication (p. 122)
                  equivalent equations (p. 118)   transformation by division
                  transform an equation (p. 118)   (p. 122)
                                                         inverse operations (p. 124)

Give the reason that justifies each statement, if x is a real number.

1. If $3x + 5 = 7$, then $(3x + 5) + (-5) = 7 + (-5)$.   *Obj. 1, p. 113*

2. If $\frac{x}{-2} = -8$, then $-2\left(\frac{x}{-2}\right) = -2(-8)$.   **1.** Addition property of equality

2. Multiplication property of equality
Solve.

3. $-17 = 9 + a$ $\{-26\}$ **4.** $b - |-4| = 12$ $\{16\}$ *Obj. 2, p. 113*

5. $-\frac{2}{5}c = 30$ $\{-75\}$ **6.** $-54 = -6|d|$ $\{9, -9\}$

7. $2j - 6 = 4$ $\{5\}$ **8.** $-\frac{1}{3}k + 9 = 5$ $\{12\}$ *Obj. 3, p. 113*

9. $3(s - 4) = -15$ $\{-1\}$ **10.** $1 = 9t - (2t - 8)$ $\{-1\}$

Check your answers with those at the back of the book.

Solving Equations and Problems **129**

Equations and Problem Solving

OBJECTIVES for Sections 3-5 through 3-8:
1. *To use equations to solve word problems.*
2. *To solve equations having the variable in both sides.*
3. *To arrange the facts of a word problem in a chart.*
4. *To transform formulas so that one variable is expressed in terms of the other variables.*

Teaching Suggestions
p. T78

Related Activities p. T78

Key Idea

Use equations to solve word problems.

Chalkboard Examples

1. One number is 20 greater than another. If the lesser number is subtracted from three times the greater number, the difference is 84. Find the numbers.
S.1 The problem asks to find the numbers.
S.2 Let x = the greater number. Then $x - 20$ = the lesser number.
S.3 $3x - (x - 20) = 84$
S.4 $3x - (x - 20) = 84$
$ 2x + 20 = 84$
$ 2x = 64$
$ x = 32$
Greater number = $x = 32$
Lesser number =
$x - 20 = 12$
∴ the two numbers are 32 and 12.

3–5 Using Equations to Solve Problems

Often the information that you are given in a word problem can be translated into relationships among numbers. If you can represent these relationships by an equation, then you can solve the word problem by solving the equation.

EXAMPLE 1 In planning his campaign for class president, Sean has ordered three types of prints of his photograph: regular prints, wallet-size prints, and enlargements. The number of regular prints he has ordered is ten more than twice the number of enlargements. The number of wallet-size prints equals the number of the other two types combined. Sean has ordered 200 prints altogether. How many of each type of print has he ordered?

SOLUTION

Step 1   Read the problem carefully a few times. What numbers are asked for? What information is given?

The problem asks for the number of each type of print.

There are ten more regular prints than twice the number of enlargements.

There are as many wallet-size prints as there are regular prints and enlargements together.

There are 200 prints altogether.

Step 2   Choose a variable. Use it with the given facts to represent the numbers described in the problem.

Let e = the number of enlargements. Then:

Number of regular prints = $2e + 10$

Number of wallet-size prints = $e + (2e + 10) = 3e + 10$

Step 3   Write an equation that represents the relationships among the numbers in the problem.

$$\underbrace{\text{The total number of prints}}_{e + (2e + 10) + (3e + 10)} \; \overset{\downarrow}{\text{is}} \; \overset{\downarrow}{200} $$
$$e + (2e + 10) + (3e + 10) = 200$$

130   *Chapter 3*

Step 4   Solve the equation. Then find the required numbers.

$$e + (2e + 10) + (3e + 10) = 200$$
$$6e + 20 = 200$$
$$6e = 180$$
$$e = 30$$

Number of enlargements = e = 30

Number of regular prints = $2e + 10 = 2(30) + 10 = 70$

Number of wallet-size prints = $3e + 10 = 3(30) + 10 = 100$

Step 5   Check your results with the words of the problem. Give the answer.
Are there 200 prints altogether?

$$30 + 70 + 100 \stackrel{?}{=} 200$$
$$200 = 200 \quad \sqrt{}$$

∴. Sean has ordered 30 enlargements, 70 regular prints, and 100 wallet-size prints.

In solving some problems, drawing a sketch may help you to identify the relationships among the numbers.

EXAMPLE 2   The length of a rectangular field is 10 m greater than three times its width. A roll of fencing that is 180 m long and 1 m high will enclose the field with no fencing left over. What are the dimensions of the field?

SOLUTION

Step 1   The problem asks for the number of meters in the length and width of a rectangle.

The length of the rectangle is 10 m greater than three times the width.

The perimeter of the rectangle is 180 m.

Step 2   Let w = the width in meters.

Then $3w + 10$ = the length in meters.

Step 3 $w + (3w + 10) + w + (3w + 10) = 180$

Step 4 $8w + 20 = 180$
$$8w = 160$$
$$w = 20$$

Width = 20 m

Length = $3(20) + 10 = 70$ m

Step 5   Is the perimeter 180 m?

$$20 + 70 + 20 + 70 \stackrel{?}{=} 180$$
$$180 = 180 \quad \sqrt{}$$

∴. the field is 20 m wide and 70 m long.

Solving Equations and Problems   **131**

2. The length of a certain rectangle is 15 cm more than three times its width. If the perimeter of the rectangle is 94 cm, what is the area?

S.1 The problem asks to find the area of the rectangle.

S.2 Let w = width. Then $3w + 15$ = length.

S.3 $2w + 2(3w + 15) = 94$

S.4 $2w + 2(3w + 15) = 94$
$$8w + 30 = 94$$
$$8w = 64$$
$$w = 8$$

Width = w = 8 cm
Length = $3w + 15$ = 39 cm
Area = lw = 8 cm × 39 cm = 312 cm²
∴ the area is 312 cm².

3. Erik ordered three different types of plants for his vegetable garden: beans, potatoes, and carrots. He ordered a dozen more potato plants than twice the number of bean plants, and the same number of carrot plants as bean plants. He ordered 72 plants altogether. How many of each plant did Erik order?

S.1 The problem asks for the number of each type of plant ordered: bean, carrot, potato.

S.2 Let b = number of bean plants. Then b = number of carrot plants; $2b + 12$ = number of potato plants.

S.3 $b + b + 2b + 12 = 72$

S.4 $b + b + 2b + 12 = 72$
$$4b + 12 = 72$$
$$4b = 60$$
$$b = 15$$

Bean plants = b = 15
Carrot plants = b = 15
Potato plants = $2b + 12$ = 42
∴ there are 15 bean plants, 15 carrot plants, and 42 potato plants ordered.

Suggested Assignments

Minimum
First day: 132/*P*: 1–10
Second day: 133/*P*: 11–18
               S 120/26–34 even
Average
   132/*P*: 1–11 odd, 13–22
S 120/40–48 even
Maximum
   132/*P*: 5–15 odd, 17–22
S 120/48–56 even

Additional A Exercises

1. Five more than three times a number is 14. Find the number.
   number = 3

2. The sum of two numbers is 78. If five times the lesser number is subtracted from four times the greater number, the difference is 69. Find the numbers.
   greater number = 51
   lesser number = 27

3. The number of ski trails where Carol skis is $\frac{3}{5}$ the number of trails in Joe's favorite ski place. If there are a total of 56 ski trails in both places, how many ski trails are there where Carol skis?   21 ski trails

4. Sarah's age is 10 years more than twice Tanya's age. The sum of their ages is 64. How old is Sarah?   Sarah is 46 years old.

5. Six is five less than the quotient obtained when a number is divided by two. Find the number.
   number = 22

6. The length of a rectangle is 2 m more than its width. The perimeter is 60 m. What is its width?
   width = 14 m

Notice that in Example 2 you did not use the fact that the height of the fencing is 1 m. Sometimes a word problem contains unnecessary information, and you must select only the facts needed for the solution.

The five steps used to solve the problems in Examples 1 and 2 form a plan that often helps in solving word problems.

Plan for Solving a Word Problem

Step 1   Read the problem carefully a few times. Decide what numbers are asked for and what information is given. Making a sketch may be helpful.

Step 2   Choose a variable and use it with the given facts to represent the number(s) described in the problem.

Step 3   Reread the problem. Then write an open sentence that represents the relationships among the numbers in the problem.

Step 4   Solve the open sentence and find the required numbers.

Step 5   Check your results with the words of the problem. Give the answer.

Problems

Solve.   **5.** Darcy's: 1.2 m; Damon's: 1.6 m   **7.** width: 1 m; length: 9 m

A   1. One number is six greater than another. If the sum of the two numbers is 44, find the numbers. 19, 25

   2. Angela has twelve fewer records than Craig. Together they have 86 records. How many records does Angela have? 37

   3. It took Lo Kwan 10 min more to walk home from school than it took him to walk to school from home. If his total walking time from home to school and back again was 54 min, how long did it take him to walk to school? 22 min

   4. A shirt cost $2.56 less than a pair of shorts. Together the two items cost $22.38. How much did the shirt cost? $9.91

   5. Darcy's walking stick is three fourths as long as Damon's. Placed end-to-end, the two sticks measure 2.8 m. How long is each stick?

   6. The number of boys in the Sharon Sports Club is four fifths the number of girls. If there are 360 boys and girls altogether in the club, how many girls are there? 200 girls

   7. The length of a certain rectangle is 4 m greater than five times its width. Find the dimensions of the rectangle if its perimeter is 20 m.

   8. Tanya's age is ten years greater than half Aaron's age. If the sum of their ages is 55, how old is Tanya? 25 years

9. A certain molecule contains twice as many atoms of hydrogen as oxygen and one more atom of carbon than hydrogen. If there are 21 atoms altogether in the molecule, how many atoms of carbon are there? 9 atoms

10. Last month Amy sold three times as many solar calculators as printing calculators and five fewer scientific calculators than solar calculators. If Amy sold 58 calculators altogether, how many of each type did she sell? solar: 27; printing: 9; scientific: 22

11. One number is 35 greater than a second number. If the lesser number is subtracted from twice the greater number, the difference is 87. Find the numbers. 52, 17

12. The difference of two numbers is 33. If four times the lesser number is subtracted from three times the greater number, the difference is 62. Find the numbers. 37, 70

B 13. Wade has three more quarters than dimes. The total value of the dimes and quarters together is $13. How many quarters does he have? 38 quarters

14. Lee withdrew $200 from his account. He asked the bank clerk to give the money to him only in $5 and $10 bills. If he received two more $10 bills than $5 bills, how many bills did he receive in all? 26 bills

15. The cost of an Aardvark personal computer is $40 less than twice the cost of an Afghan computer, while an Armadillo computer costs $70 more than an Afghan. If an order of two Afghans, one Aardvark, and three Armadillos totals $6820, what is the cost of each type of computer? Aardvark: $1860; Afghan: $950; Armadillo: $1020

16. In a mail-order catalogue, a cassette recorder costs $10 less than a clock radio, and a portable stereo costs twice as much as a clock radio. Three cassette recorders and two portable stereos cost $250 in all. What is the cost of a clock radio? $40

17. There are eight fewer students in the Debating Club than on the Math Team, and there are only half as many students on the Ecology Force as in the Debating Club. How many students are in each group if there are 68 students in all and each student participates in only one of the activities? Debating: 24; Math: 32; Ecology: 12

18. The length of a certain rectangle is 3 cm less than twice its width. If each dimension were reduced by 5 cm, the resulting figure would have a perimeter of 16 cm. What are the dimensions of the original rectangle? width: 7 cm; length: 11 cm

19. A collection of nickels and quarters has a total value of $6.40. If there are forty coins in the collection, how many are there of each kind? 18 nickels, 22 quarters

20. Jason babysat for sixty hours during July. He charged $1.50 for each hour that he worked before midnight and $3.00 for each hour after midnight. If he earned $112.50 in all, how many hours did he work after midnight? 15 hours

Solving Equations and Problems **133**

Mixed Review

Write an algebraic expression for each:

1. twice a number decreased by nine $2n - 9$

2. five less than the quotient obtained when a number is divided by three $\frac{n}{3} - 5$

Write the formula for each:

3. the perimeter of a rectangle $P = 2l + 2w$ or $P = 2(l + w)$

4. the area of a rectangle $A = lw$

Solve.

5. $\frac{5 - 2h}{3} = -10$ $\{17\frac{1}{2}\}$

6. $\frac{1}{4}y - 5 = -\frac{3}{4}$ $\{17\}$

Additional Answers
Problems

25. boxer: $200; Dalmatian:
$220; Airedale: $250;
miniature schnauzer:
$200

21. The length of a certain rectangle is 5 m less than twice its width. If the perimeter of the rectangle is 68 m, find its area. 273 m²

22. In 1985 Maria was three times as old as Sue. Five years earlier, the sum of their ages was eighteen. In what year was Maria born? 1964

C 23. The length of a rectangular field is 10 m less than four times its width. It cost $1584 to put a fence around the field. If the fencing cost $4.80 per meter, what is the area of the field? 4550 m²

24. At the time when Jay's son was born, Jay's age was twenty years less than twice his wife's age, and the sum of their ages was 64. If Jay was born in 1948, how old will his son be in 2001? 17 years

25. In Pat's Pet Palace there are two boxer pups, three Dalmatians, one Airedale, and one miniature schnauzer. The boxers are each priced $20 less than a Dalmatian, the price of the Airedale is $150 less than two times the price of a boxer, and the price of the miniature schnauzer is four fifths the price of the Airedale. What is the price of each type of dog if the total cost of all the dogs in the store is $1510?

26. The three sides of a triangle are labeled r, s, and t. The length of side s is 8 m longer than two thirds the length of side r. The length of side t is 2 m shorter than one fourth the sum of the lengths of the other two sides. If the perimeter of the triangle is 33 m, determine the length of each of the three sides. r: 12 m; s: 16 m; t: 5 m

3–6 Equations Having the Variable in Both Sides

Note that the variable n appears in both sides of the equation

$$6n = 2n - 56.$$

Are you permitted to transform this equation by subtracting $2n$ from each side? The answer is yes. Because a variable represents a real number, a variable expression such as $2n$ also represents a real number. Thus the addition and subtraction properties of equality permit you to transform an equation by adding a variable expression to each side or subtracting a variable expression from each side.

EXAMPLE 1   Solve $6n = 2n - 56$.

SOLUTION
$$6n = 2n - 56$$
$$6n - 2n = 2n - 56 - 2n$$
$$4n = -56$$
$$\frac{4n}{4} = \frac{-56}{4}$$
$$n = -14$$

Chalkboard Examples

Solve.

1. $5m - 3 = 2m$
$5m - 3 - 5m = 2m - 5m$
$-3 = -3m$
$1 = m$
∴ the solution set is $\{1\}$.

2. $4(a - 1) = -3(a + 4)$
$4a - 4 = -3a - 12$
$4a - 4 + 3a =$
$-3a - 12 + 3a$
$7a - 4 = -12$
$7a = -8$
$a = -\dfrac{8}{7}$

∴ the solution set is $\left\{-\dfrac{8}{7}\right\}$.

134   *Chapter 3*

CONDENSED SOLUTION

$6n = 2n - 56$
$4n = -56$
$n = -14$

Check: $6n = 2n - 56$
$6(-14) \overset{?}{=} 2(-14) - 56$
$-84 \overset{?}{=} -28 - 56$
$-84 = -84 \quad \checkmark$

∴ the solution set is $\{-14\}$.

When solving equations, be aware of the fact that not all equations have exactly one root. It is possible that an equation is true for *no* values of the variable. It is also possible that an equation is true for *all* values of the variable. Consider the following two examples.

EXAMPLE 2 Solve $3(c + 1) - c = 21 + 2(1 + c)$.

SOLUTION

$3(c + 1) - c = 21 + 2(1 + c)$
$3c + 3 - c = 21 + 2 + 2c$
$2c + 3 = 23 + 2c$
$3 = 23$

Since the given equation is equivalent to the false statement $3 = 23$, the equation has no root.   ∴ the solution set is ∅.

EXAMPLE 3 Solve $3(d + 1) + 8 = 2(d + 3) + d + 5$.

SOLUTION

$3(d + 1) + 8 = 2(d + 3) + d + 5$
$3d + 3 + 8 = 2d + 6 + d + 5$
$3d + 11 = 3d + 11$
$11 = 11$

Since the given equation is equivalent to the true statement $11 = 11$, the equation is satisfied by every real number.   ∴ the solution set is $\mathcal{R}$.

An equation that is true for all values of the variable(s) is called an **identity**. Thus, in Example 3, $3(d + 1) + 8 = 2(d + 3) + d + 5$ is an identity.

Oral Exercises

Match each equation at the left with its solution set at the right.

1. $6 - x = x$ e
2. $4x + 3 = 4x - 2$ b
3. $-10x + 2 = 7 - 9x$ f
4. $3 - 3x = 4 - 4x$ d
5. $3 - 8x = 8x + 3$ c
6. $2x - 5 = -x - (5 - 3x)$ a

a. $\mathcal{R}$
b. ∅
c. $\{0\}$
d. $\{1\}$
e. $\{3\}$
f. $\{-5\}$

Solving Equations and Problems **135**

---

3. $-3(a - 2) - 2a = -5a + 9$
   $-3a + 6 - 2a = -5a + 9$
   $-5a + 6 = -5a + 9$
   $6 = 9$
   Since the equation is equivalent to the false statement $6 = 9$, the equation has no root.
   ∴ the solution set is ∅.

Reading Algebra

Direct attention to the first three Written Exercises on page 136. Ask the students to describe these equations in words.

1. $10a = 56 + 2a$
   Say: "Ten times a certain number a is the same as 56 increased by twice the same number."

2. $12m = 9m - 6$
   Say: "Twelve times a certain number m is equal to nine times the number decreased by 6."

3. $b = 55 - 4b$
   Say: "A certain number b is equivalent to fifty-five decreased by four times the number."

Suggested Assignments

Minimum
First day: 136/1–15
Second day: 136/16–32
Average
First day: 136/1–29 odd, 31–34
Second day: 137/35–50
Maximum
First day: 136/11–41 odd
Second day: 137/42–56

Additional Answers
Oral Exercises

7. a. Mult. using the distrib.
ax. on each side.
   c. Divide each side by 4.
   d. Divide each side by 2.

8. a. Mult. using the distrib.
ax. on each side.
   b. Divide each side by 2.
   d. Divide each side by 8.

9. b. Mult. using the distrib.
ax. on each side.
   c. Mult. each side by 2.

   d. Mult. each side by $\frac{2}{3}$.

10. a. Mult. each side by $-\frac{3}{2}$.

   b. Mult. using the distrib.
ax. on each side.
   c. Mult. each side by -3.
   d. Mult. each side by 3.

Additional A Exercises

Solve.

1. $4b - 12 = 2b$ {6}
2. $5m = 6m - 1$ {1}
3. $2m + 3m - 7 = 4m$ {7}
4. $7y - 6 = 6 + 5y$ {6}
5. $4 - 9y = -5y$ {1}
6. $4m - 8 = 4m$ $\emptyset$

Mixed Review

Solve.

1. $2m + 3 = 11$ {4}
2. $2(c - 1) = 56$ {29}

Simplify.

3. $6\left(\frac{a}{2} - 1\right)$ $3a - 6$

4. $-m - 3m - 2m$ $-6m$

Tell which of the equations (a), (b), (c), or (d) are equivalent to the given equation. For each equivalent equation, tell what transformation was used to obtain it from the given equation.

7. Given: $4(3a - 8) = 8(a - 3) - 1$
   a. $12a - 32 = 8a - 24 - 1$
   b. $3a - 8 = 2(a - 3) - 1$
   c. $3a - 8 = 2(a - 3) - \frac{1}{4}$
   d. $2(3a - 8) = 4(a - 3) - \frac{1}{2}$

8. Given: $8(x + 1) = 4(x - 2) + 2$
   a. $8x + 8 = 4x - 8 + 2$
   b. $4(x + 1) = 2(x - 2) + 1$
   c. $2(x + 1) = (x - 2) + 2$
   d. $x + 1 = \frac{1}{2}(x - 2) + \frac{1}{4}$

9. Given: $\frac{1}{2}(6a - 4) = \frac{3}{2}(2a - 5) + 6$

   a. $6a - 4 = 3(2a - 5) + 6$
   b. $3a - 2 = 3a - \frac{15}{2} + 6$
   c. $6a - 4 = 3(2a - 5) + 12$
   d. $\frac{1}{3}(6a - 4) = (2a - 5) + 4$

10. Given: $-\frac{2}{3}(6x - 1) = 3(x - 2) + 1$

   a. $6x - 1 = -\frac{9}{2}(x - 2) - \frac{3}{2}$
   b. $-4x + \frac{2}{3} = 3x - 6 + 1$
   c. $2(6x - 1) = -9(x - 2) - 3$
   d. $-2(6x - 1) = 9(x - 2) + 3$

Written Exercises

Solve. **9.** $\{\frac{18}{25}\}$ **14.** $\mathcal{R}$ **15.** $\{\frac{3}{10}\}$ **16.** $\{\frac{2}{5}\}$ **17.** {3}   **18.** $\{-\frac{1}{13}\}$

A
1. $10a = 56 + 2a$ {7}   **2.** $12m = 9m - 6$ {−2}   **3.** $b = 55 - 4b$ {11}
4. $11n - 8 = 3n$ {1}   **5.** $-13q = 4 - 13q$ $\emptyset$ **6.** $2 - 3p = 7p$ $\{\frac{1}{5}\}$
7. $10g - 7 = 3g - 70$ {−9}   **8.** $5e - 24 = 8e - 24$ {0}   **9.** $23t - 10 = 8 - 2t$
10. $63x - 9 = 16 + 63x$ $\emptyset$ **11.** $6 + 7i = 7i + 6$ $\mathcal{R}$ **12.** $70 - 2y = 6 - 10y$ {−8}
13. $w + 10 = 10 - 3w$ {0}   **14.** $2x + (-1) = -1 + 2x$ **15.** $19z - 10 + 4z = 3z - 4$
16. $5n - 4 + 2n = 2 - 8n$ **17.** $4(2b - 1) = 5(b + 1)$ **18.** $-2(x - 5) = 3(3 - 5x)$

19. $\frac{2}{3}(t - 6) = \frac{3}{4}(2 - 3t)$ $\{\frac{66}{35}\}$ **20.** $\frac{1}{5}(a - 4) = \frac{2}{5}(3 - 2a)$ {2}

21. $10(0.1d - 0.3) = 0.2(d + 21)$ {9}   **22.** $0.8(1.25t - 5) = 0.1(4 - t)$ {4}

23. $\frac{1}{5}(3s - 5) = 2s + 3$ $\{-\frac{20}{7}\}$ **24.** $2 - 3t = -\frac{1}{3}(5 - t) + 1$ $\{\frac{4}{5}\}$

25. $7(2p - 1) = 10 - 3p$ {1}   **26.** $-2a - 4 = 6(1 - a) - 1$ $\{\frac{9}{4}\}$

27. $5(x + 3) = -\frac{1}{2}(x - 1)$ $\{-\frac{29}{11}\}$ **28.** $-\frac{2}{3}(3h - 1) = -3(1 - h)$ $\{\frac{11}{15}\}$

29. $4(5h - 9) - 4(1 - h) = 9(h - 11) - 1$ {−4}   **30.** $7(v + 1) - 4(5 - 2v) = -5(v - 9)$ $\{\frac{29}{10}\}$

136   *Chapter 3*

Solve each equation using two different methods. The first steps of the methods should be different from each other.

B 31. $\frac{1}{2}(x + 3) = \frac{1}{2}(5x - 4)$

32. $\frac{1}{4}(3x - 2) = \frac{1}{4}(8x + 1)$

33. $\frac{1}{3}(6t + 5) = \frac{1}{4}(2t + 10)$

34. $\frac{1}{5}(3b - 4) = \frac{1}{4}(b - 2)$

35. $6(c - 4) = -6(3c + 2)$

36. $-4(2 + t) = 8(3 - t)$

37. $9(2d - 5) = -6(5d - 4)$

38. $-10(4 - 7h) = -15(2 - 3h)$

39. $0.1(25t - 6) = 0.4(6t + 7)$

40. $-0.04(5b - 8) = 0.08(10b - 1)$

41. $\frac{1}{3}(2x - 4) + 5 = -\frac{2}{3}(x + 1)$

42. $\frac{3}{5}(3a - 10) = -\frac{1}{5}(2a - 9)$

Solve.

43. $10(y - 1) - 5 = -20(3 + y)$ $\{-\frac{3}{2}\}$

44. $-6(2f + 1) - 3 = 7(f - 4)$ $\{1\}$

45. $12\left(\frac{1}{3}x - \frac{1}{2}\right) = 8\left(\frac{1}{2}x - 1\right)$ Ø

46. $\frac{1}{5}\left(\frac{1}{2}t + 1\right) = \frac{1}{3}\left(\frac{3}{5}t - 3\right)$ $\{12\}$

47. $\frac{1}{5}(5 - h) - \frac{1}{2}(3 - 2h) = \frac{1}{2}h + 1$ $\{5\}$

48. $\frac{3}{4}(x - 1) - \frac{1}{2} = 2(1 - 3x)$ $\{\frac{13}{27}\}$

49. $0.3(2.5t - 0.5) = 0.6(1.5t + 0.3)$ $\{-2.2\}$ 50. $\frac{1}{2}(3.6p - 2) = 3(4.1 - 3p) - 2.5$ $\{1\}$

C 51. $5 + 6[5(3i + 20) + 2(5i + 11)] = 5 - 2[4(i + 2) + 3(7i + 42)]$ $\{-5\}$

52. $3[2(j - 2) - 7(2j - 3)] - 5[4(2j - 5) - (3j - 2)] = 6(j + 1) + 1$ $\{2\}$

Replace each ___?___ with a numerical or a variable expression so that the following conditions are satisfied.

a. The solution set is {2}. b. The solution set is {0}. c. The solution set is $\mathcal{R}$.

d. The solution set is Ø. e. The equation is an identity.

53. $2p + 3 = $ ___?___

54. $4 - (t - 2) = $ ___?___

55. $2k + $ ___?___ $ = 5k + 1$

56. $-7x + $ ___?___ $ = 3x - 7$

Computer Exercises For students with computer experience

Write a program to solve an equation of the form $ax + b = cx + d$ when you input values for a, b, c, and d. Be sure that the program works correctly for an equation whose solution is the empty set and for an identity. RUN the program to solve each of the following equations.

1. $3x + 5 = 7x + 1$ $\{1\}$

2. $2y - 9 = 6 - 3y$ $\{3\}$

3. $10 - z = z + 4$ $\{3\}$

4. $\frac{v}{2} + 12 = 8 + \frac{v}{4}$ $\{-16\}$

5. $\frac{m}{5} - 6 = 6 + \frac{1}{5}m$ Ø

6. $-2n + 1 = 1 - 2n$
The equation is an identity.

Solving Equations and Problems **137**

34. Mult. by 20; mult. using

    distrib. ax.; $\left\{\dfrac{6}{7}\right\}$

35. Divide by 6; mult. using

    distrib. ax.; $\left\{\dfrac{1}{2}\right\}$

36. Divide by 4 or −4; mult. using distrib. ax.; {8}

37. Divide by 3; mult. using

    distrib. ax.; $\left\{\dfrac{23}{16}\right\}$

38. Divide by −5; mult.

    using distrib. ax.; $\left\{\dfrac{2}{5}\right\}$

39. Mult. by 10; mult. using distrib. ax.; {34}

40. Mult. by 100; mult. using distrib. ax.; {0.4}

41. Mult. by 3; mult. using

    distrib. ax.; $\left\{-\dfrac{13}{4}\right\}$

42. Mult. by 5; mult. using

    distrib. ax.; $\left\{\dfrac{39}{11}\right\}$

53. a. 7
    b. 3
    c. 2p + 3
    d. Example: 2p + 2
    e. 2p + 3

54. a. 4
    b. 6
    c. 4 − (t − 2)
    d. Example: 1 − t
    e. 4 − (t − 2)

55. a. 7
    b. 1
    c. 3k + 1
    d. Example: 3k + 2
    e. 3k + 1

56. a. 13
    b. −7
    c. 10x − 7
    d. Example: 10x + 1
    e. 10x − 7

PROGRAMMING IN BASIC

The programs in this section use the asterisk, ∗, to produce geometric designs with a computer. For example, when you RUN the following program, you obtain the rectangular design shown at the right.

```
10    PRINT "TO CREATE GEOMETRIC"
20    PRINT "DESIGNS WITH ASTERISKS:"
30    PRINT
40    FOR I = 1 TO 10
50    FOR J = 1 TO 10
60    PRINT "∗";
70    NEXT J
80    PRINT
90    NEXT I
100   END
```

```
RUN
TO CREATE GEOMETRIC
DESIGNS WITH ASTERISKS:

**********
**********
**********
**********
**********
**********
**********
**********
**********
**********
```

Notice that this program contains *nested loops.* That is, the J-loop is said to be *nested* entirely within the I-loop.

Exercises

1. Type in and RUN the program as given.

2. What is the purpose of line 80?

3. Change line 50 as follows: 50   FOR J = 1 TO I
   LIST and RUN the revised program.

4. Change line 50 as follows: 50   FOR J = 1 TO 11 − I
   LIST and RUN the revised program.

5. Change line 50 as follows: 50   FOR J = 1 TO I
   Then insert these lines: 42   FOR J = 1 TO 11 − I
                               44   PRINT " ";
                               46   NEXT J

   LIST and RUN the revised program. If possible, PRINT a LISTing of the program and mark the loops.

6. Change line 50 as follows: 50   FOR J = 1 TO 2 ∗ I − 1
   LIST and RUN the revised program.

7. Change lines 40, 42, and 50 as follows: 40   FOR I = −9 TO 9
                                              42   FOR J = 1 TO ABS(I) + 1
                                              50   FOR J = 1 TO 18 − 2 ∗ ABS(I) + 1

   LIST and RUN the revised program.

If possible, save this program for later use.

138   *Chapter 3*

3–7 Using Charts in Solving Problems

Using a chart to organize the facts of a word problem can be a helpful strategy in solving the problem.

EXAMPLE Kathy is six years older than her cat. Next year she will be twice as old as her cat will be. How old is Kathy now?

SOLUTION

Step 1   The problem asks for Kathy's age now.

Step 2   Let x = the cat's age now. Make a chart of the given facts.

	Age now	Age next year
Cat	x	$x + 1$
Kathy	$x + 6$	$(x + 6) + 1$

Step 3   The only fact not recorded on the chart is that next year Kathy will be *twice as old as* her cat will be. Use this fact to write an equation:
$$(x + 6) + 1 = 2(x + 1)$$

Step 4
$$(x + 6) + 1 = 2(x + 1)$$
$$x + 7 = 2x + 2$$
$$7 = x + 2$$
$$5 = x$$

$x = 5$ (cat's age)

$x + 6 = 5 + 6 = 11$ (Kathy's age)

Step 5   Checking the results is left to you.

∴ Kathy is 11 years old.

Problems

Solve. You may find a chart helpful.

A 1. When twelve is added to a certain number, the result is the number's additive inverse. What is the number? −6

2. Find the number that is sixteen less than its additive inverse. −8

3. Five more than a certain number is six less than twice the number. Find the number. 11

4. Three times a certain number is 32 less than five times the number. Find the number. 16

Teaching Suggestions
p. T79

Key Idea

Arrange the facts of a word problem in a chart.

Chalkboard Examples

1. Two more than a certain number is 15 less than twice the number. Find the number.
$$n + 2 = 2n - 15$$
$$17 = n$$
∴ the number is 17.

2. Antoine is 8 years older than his sister Michelle. Next year he will be three times as old as his sister. How old is Antoine now?

S.1 The problem asks for Antoine's age now.

S.2 Let x = Michelle's age.

	Age now	Age next year
Antoine	$x + 8$	$x + 8 + 1$
Michelle	x	$x + 1$

S.3 Next year, Antoine will be three times as old as Michelle.
$$x + 8 + 1 = 3(x + 1)$$

S.4
$$x + 9 = 3x + 3$$
$$6 = 2x$$
$$3 = x$$

∴ Antoine is 11 years old now.

Suggested Assignments

Minimum
First day: 139/*P*: 1–8
Second day: 140/*P*: 9–15
           S 133/19, 20
Average
First day: 139/*P*: 1–12
Second day: 140/*P*: 13–20
           S 132/4–12 even
Maximum
First day: 139/*P*: 1–15
Second day: 141/*P*: 16–24
           S 132/6–14 even

Additional A Exercises

1. Find a number which is 12 less than its additive inverse. −6

2. Six times a certain number is 12 more than 4 times the number. Find the number. 6

3. One fourth a number is equal to the number decreased by fifteen. Find the number. 20

4. Erika is now eight years older than her sister Alexis. In six years, Erika will be twice Alexis's present age. Find the age of each girl now.

    Erika is 22 years old.
    Alexis is 14 years old.

5. The sum of 5 more than a certain number and 10 more than twice the number is equal to the product of 2 and the number increased by eight. Find the number.

$$(n + 5) + (2n + 10) = 2(n + 8)$$
$$n = 1$$

5. The sum of two numbers is 16. The greater of the two numbers is one more than four times the lesser number. What are the numbers? 3, 13

6. When two numbers are added together, the result is 45. Twice the greater number is six more than five times the lesser number. What are the numbers? 12, 33

7. Michelle is eight years older than Adam. Three years ago Michelle was twice as old as Adam was then. How old is Michelle? 19 years

8. Maurice is now twice as old as Roberto. Maurice's present age is ten years greater than Roberto's age one year ago. How old is Maurice? 18 years

9. In 1985, Barry was 13 years old and his father was 43. In what year will Barry's age be two fifths his father's age? 1992

10. Sheila's age is two years more than twice Nicole's age. The sum of their ages is the same as Steve's age. If Steve's age is ten years less than five times Nicole's age, find the age of each. Nicole: 6 years; Sheila: 14 years; Steve: 20 years

11. The Rhine River is 160 km shorter than the St. Lawrence River. The Amazon River is 800 km longer than five times the length of the Rhine River. The Zambezi River is twice as long as the St. Lawrence River. How long is the Amazon if it is 1440 km longer than the sum of the lengths of the other three rivers?

12. The population of Concord, New Hampshire in 1980 was 400 greater than half the population of Euclid, Ohio. The sum of the two populations was 29,600 less than twice the population of Euclid. What was the population of each city in 1980?

13. The Panama Canal is 2 km shorter than twice the length of the Suez Canal. The sum of the lengths of the two canals is 121 km greater than three fourths the length of the Panama Canal. Find the length of each canal. Panama Canal: 160 km; Suez Canal: 81 km

14. The width of a certain rectangle is 2 m greater than half its length. Four times its length is 26 m greater than its perimeter. What are the dimensions of the rectangle? width: 17 m; length: 30 m

B **15.** Last year Kristen read one fourth as many biographies as mysteries and eighteen more science fiction books than biographies. The number of nature books that she read was one less than the difference between the number of mysteries and the number of biographies. If Kristen read four times as many science fiction books as nature books, how many books did she read in all? 35 books

16. Taurus is the zodiac constellation with the greatest number of stars visible to the naked eye, while Aries has the least number of visible stars. Twenty more stars are visible in Cancer than in Aries. Sagittarius has twice as many visible stars as Cancer, and Taurus has twelve more visible stars than Sagittarius. If the number of visible stars in Aries is ten greater than one third the number of visible stars in Taurus, how many stars are visible in Cancer? 102 stars

17. The length of a certain rectangle is 20 m greater than its width. If the width were reduced by 20 m and the length increased by 100 m, the perimeter of the new rectangle would be twice the perimeter of the original rectangle. What are the dimensions of the original rectangle? width: 30 m; length: 50 m

18. The length of a certain rectangle is 10 m greater than twice its width. If the length were doubled and the width were halved, the perimeter would be increased by 80 m. Find the original dimensions of the rectangle. width: 20 m; length: 50 m

19. Penny collected some change in preparation for a garage sale. She collected two more nickels than twice the number of dimes and eight fewer quarters than twice the number of nickels. If the value of the quarters was $1.60 greater than four times the value of the nickels and dimes together, what was the total value of the change that Penny collected? $17.10

20. Jani earns $15 per hour as a music teacher and $11 per hour as a swimming instructor. During the summer she worked part-time at each job. In one week she worked a total of thirty-seven hours and earned $459. How many hours did she work at each job that week?

C 21. The East Street School opened fifteen years after the Brown School opened. In 1978, the Brown School had been open twice as long as the East Street School. In what year will the East Street School celebrate its fiftieth anniversary? 2013

22. Dennis earns $7.25 per hour and Tony earns $8.50 per hour. In the first week of July, they worked a combined total of seventy-five hours and earned a total of $585. Who earned more money that week? How much more did he earn? Dennis; $24

23. In 1982 Tim was twice as old as Sue. In 1977 the sum of their ages was 32.

   a. In what year was Tim born? 1954

   b. In what year will Sue's age be three fourths Tim's age? 2010

24. Ken and Roberta each sold fifty tickets for the school play. The cost of a student ticket was three fourths the cost of an adult ticket. Roberta sold ten more student tickets than adult tickets and collected a total of $212.50.

   a. How many student tickets did Roberta sell? 30 tickets

   b. What was the cost of a student ticket? $3.75

   c. If Ken collected $200 in all, how many student tickets did he sell? 40 tickets

20. music teacher: 13 hours; swimming instructor: 24 hours

Mixed Review

Solve.

1. $m = 16 - 5m$ $\left\{2\frac{2}{3}\right\}$

2. $-y - 2y + 6 = 18$ $\{-4\}$

3. $15 - 3y = 10 + 2y$ $\{1\}$

4. $3(x - 4) = -2(x + 6)$ $\{0\}$

5. $7(t - 3) + 4 = 33 + 3(t - 2)$ $\{11\}$

11. 1. a, b, and c are real numbers;
    $a - c = b - c$
    Hypothesis
  2. $a + (-c) = b + (-c)$
    Def. of subt.
  3. $[a + (-c)] + c$
    $= [b + (-c)] + c$
    Add. prop. of $=$
  4. $a + [(-c) + c]$
    $= b + [(-c) + c]$
    Assoc. ax. for add.
  5. $a + 0 = b + 0$
    Ax. of add. inv.
  6. $\therefore a = b$
    Iden. ax. for add.

12. 1. a, b, and c are real numbers such that
    $c \neq 0; \dfrac{a}{c} = \dfrac{b}{c}$
    Hypothesis
  2. $a \cdot \left(\dfrac{1}{c}\right) = b \cdot \left(\dfrac{1}{c}\right)$
    Def. of div.
  3. $\left[a \cdot \left(\dfrac{1}{c}\right)\right]c = \left[b \cdot \left(\dfrac{1}{c}\right)\right]c$
    Mult. prop. of $=$
  4. $a\left[\left(\dfrac{1}{c}\right) \cdot c\right] = b\left[\left(\dfrac{1}{c}\right) \cdot c\right]$
    Assoc. ax. for mult.
  5. $a \cdot 1 = b \cdot 1$
    Ax. of mult. inv.
  6. $\therefore a = b$
    Iden. ax. for mult.

13. 1. a and b are real nos.
    Hypothesis
  2. $-b$ is a real number.
    Ax. of add. inv.
  3. $a + b = 0$
    Hypothesis
  4. $(a + b) + (-b)$
    $= 0 + (-b)$
    Add. prop. of $=$
  5. $a + [b + (-b)]$
    $= 0 + (-b)$
    Assoc. ax. for add.
  6. $a + 0 = 0 + (-b)$
    Ax. of add. inv.
  7. $\therefore a = -b$
    Iden. ax. for add.

Careers

Accounting

Accountants prepare and analyze the financial records of individuals, businesses, and government. Because financial transactions can be so complex and diverse, many accountants choose to specialize in a particular area such as budget, tax, cost, or auditing.

For example, some accountants prepare detailed cost studies or report on the profits, losses, and inventory of a company. Other accountants work for the government and audit financial records in order to verify compliance with government regulations, while still others make credit approval decisions for banks.

EXAMPLE One of the major expenses of a business is the cost of purchasing and maintaining equipment. Each year, equipment decreases in value, or *depreciates*, as it becomes older and technically inferior to more recently produced models.

The Richland Supply Company purchases a cash register for $920. The company also pays a 5% sales tax and a $25 delivery fee. The estimated useful life of the cash register is four years. At the end of the four years, the cash register is valued at $200 and is considered to be fully (100%) depreciated. Find the amount of the *annual* depreciation of the cash register.

SOLUTION One formula that can be used to find the annual depreciation, d, is
$$d = (c - s)r,$$
where c is the original cost of the item, s is the value of the item at the end of its useful life, and r is the percent of depreciation that occurs *in one year*.

Use the given information to determine the values of c, s, and r.

c = price + tax + delivery fee = \$920 + 0.05(\$920) + \$25 = \$991

s = \$200

$r = \dfrac{\text{total depreciation}}{\text{number of years}} = \dfrac{100\%}{4} = 25\% = 0.25$

Thus, $d = (c - s)r = (\$991 - \$200)0.25 = \$197.75$

$\therefore$ the annual depreciation of the cash register is \$197.75.

142   *Chapter 3*

READING ALGEBRA Problem Solving

Reading word problems requires a different type of strategy than other kinds of reading. When you read a word problem, read it once all the way through to find out what it is about. Then read it a second time, more slowly and carefully, concentrating on the major facts that are given and on the question or questions that you are asked. If any words are not familiar to you, look them up in a dictionary or, if they are mathematical terms, look them up in the glossary at the back of the book.

After you have read the problem carefully, try to answer any questions that are asked. If you write an equation, check to be sure that it represents *all* the conditions of the problem. Then check your answers with those printed at the back of the book, or check with your teacher. If your answer is wrong, try reading the problem again to see if you can discover the source of your error. Remember that good problem solvers learn from their mistakes as well as from their successes.

Exercises

The exercises below refer to the following word problem.

A collection of nickels, dimes, and quarters contains three times as many nickels as dimes and four more quarters than nickels. The total value of the collection is $9. How many quarters are there?

Let d represent the number of dimes in the collection. Write an expression for each of the following in terms of d.

1. the number of nickels $3d$
2. the number of quarters $3d + 4$
3. the total number of coins in the collection $d + 3d + (3d + 4)$
4. the total value in cents of the coins in the collection
$$10d + 5(3d) + 25(3d + 4)$$

Let d represent the number of dimes in the collection. Tell whether or not each equation represents the conditions of the problem.

5. $d + 3d + (3d + 4) = 900$ No
6. $10d + 5(3d) + 25(3d + 4) = 9$ No
7. $10d + 5(3d) + 25(3d + 4) = 900$ Yes
8. $0.1d + 0.5(3d) + 0.25(3d + 4) = 9$ No

9. Tell why each of the following is *not* an answer to the question.
   a. 8   **b.** $7.00   **a.** 8 is the number of *dimes*. **b.** $7.00 is the *value* of the quarters.

10. What is the correct answer to the question? 28 quarters

Solving Equations and Problems   **143**

---

14. 1. a and b are real numbers such that $c \neq 0$.
       Hypothesis
    2. $\frac{1}{b}$ is a real number.
       Ax. of mult. inv.
    3. $ab = 1$
       Hypothesis
    4. $ab\left(\frac{1}{b}\right) = 1 \cdot \frac{1}{b}$
       Mult. prop. of $=$
    5. $a\left[b\left(\frac{1}{b}\right)\right] = 1 \cdot \frac{1}{b}$
       Assoc. ax. for mult.
    6. $a \cdot 1 = 1 \cdot \frac{1}{b}$
       Ax. of mult. inv.
    7. $\therefore a = \frac{1}{b}$
       Iden. ax. for mult.

15. 1. a, b, and x are real numbers; $a - b = x$
       Hypothesis
    2. $x = a - b$
       Symm. ax. of $=$
    3. $x = a + (-b)$
       Def. of subt.
    4. $x + b = [a + (-b)] + b$
       Add. prop. of $=$
    5. $x + b = a + [(-b) + b]$
       Assoc. ax. for add.
    6. $x + b = a + 0$
       Ax. of add. inv.
    7. $x + b = a$
       Iden. ax. for add.
    8. $\therefore b + x = a$
       Comm. ax. for add.

(continued on p. 154)

Key Ideas

Transform formulas so that one variable is expressed in terms of the other variable.

Chalkboard Examples

Solve each equation for the variable in color. (Assume that the variables represent real numbers that do not result in division by zero.)

1. $a = b(c - z) - d$

$a = bc - bz - d$

$a - bc + d = -bz$

$\dfrac{a - bc + d}{-b} = z$

2. $mn - mc = 2a$

$m(n - c) = 2a$

$m = \dfrac{2a}{n - c}$

3. The formula $P = 2l + 2w$ gives the perimeter P of a rectangle in terms of its length l and width w.

a. Express the length in terms of P and w.

$l = \dfrac{P - 2w}{2}$

b. If a rectangle has a perimeter of 46 cm and a width of 5 cm, what is the length?

$l = \dfrac{P - 2w}{2}$

$l = \dfrac{(46) - 2(5)}{2}$

$l = \dfrac{36}{2}$

∴ the length is 18 cm.

3–8 Transforming Formulas

A **formula** is an equation that states a relationship among quantities represented by variables. For example, the following are some formulas that you may recognize.

$A = s^2$ Area of a square = square of the length of a side

$D = rt$ Distance traveled = rate × time traveled

$I = prt$ Simple interest = principal × rate of interest × time

When you are working with a formula, it is often helpful to *solve for* one of the variables. To solve for a given variable, you transform the formula until you obtain that variable alone as one side of the equation. This variable is then said to be *expressed in terms of* the other variables.

EXAMPLE   The formula

$$x = \tfrac{1}{2}at^2$$

gives the distance x in meters that is traveled in t seconds by a freely falling object near the surface of a planet where the acceleration due to gravity is a m/s² (meters per second squared).

a. Express a in terms of x and t.

b. If it takes 10 s for a heat shield to fall 445 m to the surface of Venus, what is the acceleration due to gravity on Venus?

SOLUTION   **a.** $x = \dfrac{1}{2}at^2$

$2x = at^2$

$\dfrac{2x}{t^2} = a$

∴ $a = \dfrac{2x}{t^2}$, provided $t \neq 0$

b. Substitute the values $x = 445$ and $t = 10$ into the formula as it was transformed in part (a).

$a = \dfrac{2x}{t^2}$

$a = \dfrac{2(445)}{10^2} = \dfrac{890}{100} = 8.9$

∴ the acceleration due to gravity on Venus is 8.9 m/s².

Notice that the formula obtained for the acceleration a in part (a) of the Example is not valid when $t = 0$ because division by zero has no meaning.

144   *Chapter 3*

Oral Exercises

In these exercises, assume that the variables represent real numbers that do not result in division by zero.

Tell what steps are needed to transform the first formula into the second.

1. a. $2rs = t$

$r = \dfrac{t}{2s}$

b. $2rs = t$

$s = \dfrac{t}{2r}$

c. $2rs = t$

$t = 2rs$

2. a. $4xyz = -t$

$x = \dfrac{-t}{4yz}$

b. $4xyz = -t$

$y = \dfrac{-t}{4xz}$

c. $4xyz = -t$

$t = -4xyz$

3. a. $A = 2(B + C)$

$A = 2B + 2C$

b. $A = 2(B + C)$

$B = \dfrac{A - 2C}{2}$

c. $A = 2(B + C)$

$C = \dfrac{A - 2B}{2}$

4. a. $P = 3(x - y)$

$P = 3x - 3y$

b. $P = 3(x - y)$

$x = \dfrac{P + 3y}{3}$

c. $P = 3(x - y)$

$y = \dfrac{P - 3x}{-3}$

5. a. $h = w(x - y)$

$w = \dfrac{h}{x - y}$

b. $h = w(x - y)$

$h = wx - wy$

c. $h = w(x - y)$

$y = \dfrac{h - wx}{-w}$

6. a. $d = -t(r - s)$

$t = \dfrac{-d}{r - s}$

b. $d = -t(r - s)$

$d = -tr + ts$

c. $d = -t(r - s)$

$r = \dfrac{d - ts}{-t}$

Each of the following equations is solved for the variable in color. Name the property of the real numbers that justifies each lettered step.

7. $3P + 2Prt = 6$

$P(3 + 2rt) = 6$ **(a)**

$P = \dfrac{6}{3 + 2rt}$ **(b)**

8. $6w - 5wx = -3$

$w(6 - 5x) = -3$ **(a)**

$w = \dfrac{-3}{6 - 5x}$ **(b)**

9. $2t^2 = \dfrac{r}{s}$

$2st^2 = r$ **(a)**

$s = \dfrac{r}{2t^2}$ **(b)**

10. $5a = \dfrac{3h^2}{-k}$

$-5ak = 3h^2$ **(a)**

$k = \dfrac{3h^2}{-5a}$ **(b)**

11. $I = \dfrac{E}{A + B}$

$I(A + B) = E$ **(a)**

$IA + IB = E$ **(b)**

$IB = E - IA$ **(c)**

$B = \dfrac{E - IA}{I}$ **(d)**

12. $T = \dfrac{4 + b}{2 - a}$

$(2 - a)T = 4 + b$ **(a)**

$2T - aT = 4 + b$ **(b)**

$-aT = 4 + b - 2T$ **(c)**

$a = \dfrac{4 + b - 2T}{-T}$ **(d)**

Solving Equations and Problems **145**

Additional Answers
Oral Exercises
(continued)

c. Following step (b),
subtract wx from each
side, divide each side
by $-w$, and use the
symmetric axiom
of $=$.

6. a. Divide each side by
$r - s$, multiply each
side by -1, and use
the symmetric axiom
of $=$.
b. Use the distributive
axiom to substitute
$-tr + ts$ for $-t(r - s)$.
c. Following step (b),
subtract ts from each
side, divide each side
by $-t$, and use the
symmetric axiom
of $=$.

7. (a) Distrib. ax.
(b) Div. prop. of $=$
8. (a) Distrib. ax.
(b) Div. prop. of $=$
9. (a) Mult. prop. of $=$
(b) Div. prop. of $=$
10. (a) Mult. prop. of $=$
(b) Div. prop. of $=$
11. (a) Mult. prop. of $=$
(b) Distrib. ax.
(c) Subt. prop. of $=$
(d) Div. prop. of $=$
12. (a) Mult. prop. of $=$
(b) Distrib. ax.
(c) Subt. prop. of $=$
(d) Div. prop. of $=$

Suggested Assignments

Minimum
First day: 146/1–24
Second day: 146/25–36
                 147/P: 1, 2
Third day: 146/37, 38
                 147/P: 3–6
                 R 148/Self-Test 2

(continued)

Written Exercises

In these exercises assume that the variables represent real numbers that do not result in division by zero.

Solve each equation for the variable in color.

A
1. $V = IR$ $I = \frac{V}{R}$
2. $-3xy = z$ $y = \frac{z}{-3x}$
3. $-4xy = 3vw$ $v = \frac{-4xy}{3w}$
4. $-5pq = -2st$
5. $a^2b = 3cx$
6. $-2tm = 4t^2s$
7. $-2xy = 6g^2h$ $h = \frac{-xy}{3g^2}$
8. $E = I^2R$ $R = \frac{E}{I^2}$
9. $b - 2c = -2ax$
10. $2pq = 4r - 3t$
11. $b - 2c = -2ax$
12. $2pq = 4r - 3t$
13. $a(b - c) = -4d$ $a = \frac{-4d}{b - c}$
14. $s(t - v) = 2w$ $s = \frac{2w}{t - v}$
15. $a(b - c) = -4d$
16. $s(t - v) = 2w$
17. $a(b - c) = -4d$
18. $s(t - v) = 2w$
19. $-3a^2(x - 3) = -4c$
20. $st^2(p - 2q) = 5$
21. $A = \frac{1}{2}bh$ $b = \frac{2A}{h}$
22. $T = -\frac{3}{4}ws$ $w = -\frac{4T}{3s}$
23. $E = \frac{1}{2}mc^2$ $m = \frac{2E}{c^2}$
24. $V = \frac{22}{7}r^2h$ $h = \frac{7V}{22r^2}$
25. $\frac{t + c}{3} = n$ $c = 3n - t$
26. $\frac{t - a + b}{2} = 3c$ $b = 6c - t + a$
27. $-1.5st^2 = 0.09k$
28. $0.6x^2y = -0.02z$
29. $A = \frac{1}{2}h(a + b)$ $h = \frac{2A}{a + b}$
30. $V = \frac{1}{3}a(x + y)$
31. $A = wt + \frac{1}{2}at^2$ $a = \frac{2(A - wt)}{t^2}$
32. $T = -3p^2r + \frac{1}{2}p$ $r = \frac{p - 2T}{6p^2}$
33. $C = \frac{2}{3}(a + b)$
34. $t = \frac{3}{2}(x - y)$ $x = \frac{2}{3}t + y$
35. $A = \frac{5}{2}h(3x - y)$
36. $V = \frac{3}{4}t(2a - 3b)$

B
37. $A = P + Prt$ $P = \frac{A}{1 + rt}$
38. $T = 2A - 5AB$ $A = \frac{T}{2 - 5B}$
39. $C = \frac{kA}{4\pi d}$ $d = \frac{kA}{4\pi C}$
40. $F = \frac{km_1m_2}{d_1d_2}$ $d_1 = \frac{km_1m_2}{Fd_2}$
41. $C = \frac{c_1 - c_2}{2t}$ $t = \frac{c_1 - c_2}{2C}$
42. $a = \frac{w - v}{t}$ $t = \frac{w - v}{a}$
43. $C = \frac{c_1 - c_2}{2t}$ $c_2 = c_1 - 2Ct$
44. $a = \frac{w - v}{t}$ $v = w - at$
45. $I = \frac{E}{R + r}$
46. $\frac{y_1 - y_2}{x_1 - x_2} = m$ $y_1 = m(x_1 - x_2) + y_2$
47. $I = \frac{E}{R + r}$
48. $\frac{y_1 - y_2}{x_1 - x_2} = m$
49. $at + 3 = bt - 4b$ $b = \frac{at + 3}{t - 4}$
50. $2rs - 4r = 4s - 5$
51. $at + 3 = bt - 4b$
52. $2rs - 4r = 4s - 5$ $s = \frac{4r - 5}{2r - 4}$
53. $F = \frac{f_1f_2}{f_1 + f_2}$
54. $T = \frac{d_1d_2}{d_1 - d_2}$ $d_2 = \frac{Td_1}{d_1 + T}$

Solve each equation for the value of t when $r = -\frac{2}{3}$ and $s = \frac{3}{4}$.

C
55. $2rs = \frac{t + r}{t - r}$ $\{0\}$
56. $\frac{-2t + 3r}{t - r} = -3s$ $\{2\}$
57. $r(t - s) - 2(3t - s) = -5(t + 2s) + rs$ $\{6\}$
58. $-\frac{1}{2}r(s - t) + \frac{1}{3}s(r - t) = \frac{1}{2}rs$ $\{\frac{4}{7}\}$

Problems

Solve.

A **1.** The formula $V = lwh$ gives the volume V of a rectangular box in terms of the length l, width w, and height h.
   a. Express h in terms of V, l, and w. $h = \frac{V}{lw}, \ l, w \neq 0$
   b. If a rectangular box with length 11 cm and width 5 cm has a volume of 770 cm³, what is the height of the box? 14 cm

2. The formula $C = 2\pi r$ gives the circumference C of a circle in terms of the radius r.
   a. Express r in terms of C. $r = \frac{C}{2\pi}$
   b. If the circumference of a circle is 157 m, find the radius. Use $\pi = 3.14$. 25 m

3. The formula $SA = 2\pi r^2 + 2\pi rh$ gives the surface area SA of a cylinder in terms of the height h and the radius of the base r.
   a. Express h in terms of the other variables. $h = \frac{SA}{2\pi r} - r, \ r \neq 0$
   b. Find the height of a cylinder whose base has a radius of 14 cm if its surface area is 2200 cm². Use $\pi = \frac{22}{7}$. 11 cm

4. The formula $C = \frac{5}{9}(F - 32)$ gives the temperature C in degrees Celsius in terms of the temperature F in degrees Fahrenheit.
   a. Express F in terms of C. $F = \frac{9}{5}C + 32$
   b. What Fahrenheit temperature would be equivalent to a temperature of 30 degrees Celsius? 86°F

B **5.** The formula $V = \frac{1}{3} lwh$ gives the volume V of a pyramid having a rectangular base in terms of the length of the base l, the width of the base w, and the height of the pyramid h. Find the height of a pyramid with a base of length 21 m and width 17.5 m if the volume is 2940 m³. 24 m

6. The formula $A = \frac{1}{2}h(a + b)$ gives the area A of a trapezoid in terms of the height h and the bases a and b. Find the length of base a if the length of base b is 14 cm, the height is 26 cm, and the area is 377 cm². 15 cm

7. The formula $V = \pi r^2 h$ gives the volume V of a cylinder in terms of the radius of the base r and the height h. Find the height of a cylindrical tank with a base of radius $3\frac{1}{2}$ ft if it takes 1848 ft³ of water to fill it completely. Use $\pi = \frac{22}{7}$. 40 ft

8. Two cylinders have the same volume. One of the cylinders has a base of radius 12 cm and a height of 25 cm, while the second cylinder has a base of radius 10 cm. What is the height of the second cylinder? (The formula for the volume of a cylinder is given in Exercise 7.) 36 cm

Solving Equations and Problems **147**

Average
First day: 146/1–25
Second day: 146/26–40
                  147/P: 1–3
Third day: 146/41–55
                  147/P: 4–8
                  R 148/*Self-Test 2*
Maximum
First day: 146/1–30
Second day: 146/31–49
                  147/P: 1–3
Third day: 146/50–58
                  147/P: 4–8
                  R 148/*Self-Test 2*

Additional A Exercises

Given that $a(m - r) = s - t$, solve for the following variables. Assume that the variables represent real numbers that do not result in division by zero.

1. s $s = am - ar + t$

2. m $m = \frac{s - t + ar}{a}$

3. a $a = \frac{s - t}{m - r}$

4. r $r = \frac{s - t - am}{-a}$

5. t $t = -am + ar + s$

Additional Answers
Written Exercises (p. 146)
(See p. 152.)

Mixed Review

Simplify.
1. $y(5 - a)$ $5y - ay$
2. $n(a + 7)$ $an + 7n$
Substitute and simplify.
3. $A = 0.5b + c$
$b = 50; c = 25$
$A = \underline{\ ?\ }$ 50
4. $B = r^2 + y^2$
$r = 0.5; y = 0.2$
$B = \underline{\ ?\ }$ 0.29

Write an equation that represents relationships among the numbers in the problem. Then solve the equation and answer the question.

1. Thirty-six more than 4 times a number is −56. What is the number?

$4n + 36 = -56$
$4n = -92$
$n = -23$

∴ the number is −23.

2. Find the length and width of a rectangle whose perimeter is 36 cm, if the length is 4 cm more than its width.

Let w = width in cm. Then $w + 4$ = length in cm.
$2w + 2(w + 4) = 36$
$4w + 8 = 36$
$w = 7$

Width = w = 7 cm
Length = $w + 4$ = 11 cm

∴ the length of the rectangle is 11 cm and the width is 7 cm.

Solve.

3. $3(m - 4) = \frac{1}{2}(4m + 8)$

$3m - 12 = 2m + 4$
$m = 16$

∴ the solution set is {16}.

4. $5(y + 2) = -(8 - 3y)$

$5y + 10 = -8 + 3y$
$2y = -18$
$y = -9$

∴ the solution set is {−9}.

Make a chart in Question 5 to organize the facts of the problem. Then solve the problem.

5. John has a collection of pennies, nickels, and dimes that have a total value of $4.42. He has 4 more dimes than pennies, and 6 more nickels than dimes. How many coins of each type does he have?

Self-Test 2

VOCABULARY   identity (p. 135)   formula (p. 144)

Write an equation that represents relationships among the numbers in the problem. Then solve the equation and answer the question.

1. One number is six less than another. If the sum of the two numbers is 42, what are the numbers? $x + (x + 6) = 42$; {18}; 18, 24   *Obj. 1, p. 130*

2. Ella has five more dimes than nickels. If the total value of her dimes and nickels is $1.10, how many dimes does she have?
$10d + 5(d - 5) = 110$; {9}; 9 dimes

Solve.

3. $5x - 4 = 12 - 3x$ {2}   4. $\frac{1}{3}(y - 2) = -4$ {−10}   *Obj. 2, p. 130*

Make a chart to organize the facts of the problem. Then solve the problem.

5. Ann is six years older than Mark. Five years ago she was twice as old as Mark was then. How old is Ann? 17 years   *Obj. 3, p. 130*

6. The length of a certain rectangle is 3 cm greater than its width. If the width were doubled and the length were decreased by 1 cm, the new rectangle would have the same perimeter as the original. Find the dimensions of the original rectangle. width: 1 cm; length: 4 cm

Solve for b.

7. $A = \frac{1}{2}ab$ $b = \frac{2A}{a}, a \neq 0$ 8. $A = \frac{1}{2}h(a + b)$ *Obj. 4, p. 130*
$b = \frac{2A}{h} - a, h \neq 0$

Check your answers with those at the back of the book.

EXTRA

Indirect Proof

The proofs that you have studied so far in this book are each made up of a sequence of true statements that lead *directly* from the hypothesis of a theorem to its conclusion. Such proofs are called *direct proofs*. However, it is also possible to prove a theorem by a process of *indirect* reasoning. As an example of this type of reasoning, consider the following.

Suppose that a detective who is investigating a theft narrows the list of suspects to three people: Abel, Bolton, and Carter. Further investigation reveals that, at the time of the theft, Abel was out of town and Bolton was in the hospital. On the basis of these facts, the detective knows that Abel and Bolton are innocent, since neither could have been in two places at once. The detective has eliminated all possible suspects except one, and that one must be the prime suspect. Carter is arrested.

In mathematics, this method of reasoning appears in an **indirect proof.** To write an indirect proof of a theorem, you begin by assuming that the conclusion is *false*, even though the hypothesis is accepted as true. You then show that a logical chain of statements leads you to contradict an accepted fact. That fact might be the hypothesis of the theorem, or it might be an axiom, a definition, or a previously proved theorem. As a result of this contradiction, you know your assumption must be incorrect and the conclusion of the theorem must be *true*.

The following example illustrates how indirect reasoning might be used in proving a theorem for the real numbers.

EXAMPLE 1   Prove: For all real numbers a and b such that $a \neq 0$, if $ab = 0$, then $b = 0$.

SOLUTION   *Plan*: Assume $b \neq 0$.

PROOF

Statements	Reasons
1. $a \neq 0$ $ab = 0$	Hypothesis
2. $\frac{1}{b}$ is a real number.	Axiom of multiplicative inverses
3. $(ab) \cdot \frac{1}{b} = 0 \cdot \frac{1}{b}$	Multiplication property of equality
4. $a = 0 \cdot \frac{1}{b}$	Theorem on page 94
5. $a = 0$	Multiplicative property of zero

However, statement (5) contradicts that part of the hypothesis which states $a \neq 0$. Therefore, the assumption $b \neq 0$ must be incorrect and the conclusion $b = 0$ is true.

EXAMPLE 2   Prove $0 \neq -1$.

SOLUTION   *Plan*: Assume $0 = -1$.

PROOF

Statements	Reasons
1. $0 + 1 = 1 + 1$	Addition property of equality
2. $0 + 1 = 0$	Axiom of additive inverses
3. $1 = 0$	Identity axiom for addition

However, statement (3) contradicts that part of the identity axiom for multiplication which states $1 \neq 0$. Therefore, the assumption $0 = -1$ must be incorrect and the conclusion $0 \neq -1$ is true.

Solving Equations and Problems   **149**

Let p = number of pennies.

	Number of Coins	Value of Coins
Pennies	p	p
Dimes	$p + 4$	$10(p + 4)$
Nickels	$(p + 4) + 6$	$5[(p + 4) + 6]$

$p + 10(p + 4) +$
$\qquad 5[(p + 4) + 6] = 442$
$\qquad\qquad 16p + 90 = 442$
$\qquad\qquad\qquad p = 22$
Number of pennies =
$p = 22$
Number of dimes =
$p + 4 = 26$
Number of nickels =
$(p + 4) + 6 = 32$
∴ John has 80 coins in all.

6. Karen's age is 4 years less than 3 times Sandra's age. Half of Karen's age increased by Sandra's age is 2 years more than twice Sandra's age. Find the age of each girl.
Let x = Sandra's age.
Then $3x - 4$ = Karen's age.

$\frac{1}{2}(3x - 4) + x = 2x + 2$

$\quad 3x - 4 + 2x = 4x + 4$
$\qquad\qquad\qquad x = 8$
Sandra's age = x =
8 years old
Karen's age = $3x - 4$ =
20 years old
∴ Sandra is 8 years old, and Karen is 20.

7. Solve for x.
$k = \pi + 4x$
$k - \pi = 4x$

$\dfrac{k - \pi}{4} = x$

8. Solve for y.
$\dfrac{a + y - b}{4} = c$

$a + y - b = 4c$
$\qquad\quad y = 4c + b - a$

1. *Plan:* Assume $a = b$.
   1. $a + c = b + c$
      Add. prop. of $=$
   2. $a + c \neq b + c$
      Hypothesis
      Statement (1) contradicts the hypothesis
      $a + c \neq b + c$.
      $\therefore$ the assumption $a = b$ is incorrect and the conclusion $a \neq b$ is true.

2. *Plan:* Assume $a = b$.
   1. $a - c = b - c$
      Subt. prop. of $=$
   2. $a - c \neq b - c$
      Hypothesis
      Statement (1) contradicts the hypothesis
      $a - c \neq b - c$.
      $\therefore$ the assumption $a = b$ is incorrect and the conclusion $a \neq b$ is true.

3. *Plan:* Assume $a = b$.
   1. $ac = bc$
      Mult. prop. of $=$
   2. $ac \neq bc$
      Hypothesis
      Statement (1) contradicts the hypothesis $ac \neq bc$.
      $\therefore$ the assumption $a = b$ is incorrect and the conclusion $a \neq b$ is true.

4. *Plan:* Assume $a = b$.
   1. $c \neq 0$
      Hypothesis
   2. $a \div c = b \div c$
      Div. prop. of $=$
   3. $a \div c \neq b \div c$
      Hypothesis
      Statement (2) contradicts the hypothesis
      $a \div c \neq b \div c$.
      $\therefore$ the assumption $a = b$ is incorrect and the conclusion $a \neq b$ is true.

The following is a summary of the steps used in proving a theorem by the process of indirect reasoning.

To Write an Indirect Proof of a Theorem

1. Assume that the conclusion of the theorem is false.
2. Reason from this assumption until you obtain a statement that contradicts an accepted fact.
3. Point out that the assumption must be incorrect and that the conclusion of the theorem must therefore be true.

Exercises

Write an indirect proof of each theorem.

1. For all real numbers a, b, and c, if $a + c \neq b + c$, then $a \neq b$.
2. For all real numbers a, b, and c, if $a - c \neq b - c$, then $a \neq b$.
3. For all real numbers a, b, and c, if $ac \neq bc$, then $a \neq b$.
4. For all real numbers a, b, and c such that $c \neq 0$, if $a \div c \neq b \div c$, then $a \neq b$.
5. For all real numbers a and b, if $a \neq b$, then $-a \neq -b$.
6. For all real numbers a and b such that $a \neq 0$ and $b \neq 0$, if $a \neq b$, then $\frac{1}{a} \neq \frac{1}{b}$.
7. $1 \neq 2$
8. $-1 \neq -2$

Chapter Summary

1. The following *properties of equality* are theorems that can be proved to be true for all real values of each variable except as noted.

   Addition: If $a = b$, then $a + c = b + c$ and $c + a = c + b$.

   Multiplication: If $a = b$, then $ac = bc$ and $ca = cb$.

   Subtraction: If $a = b$, then $a - c = b - c$.

   Division: If $a = b$ and $c \neq 0$, then $\frac{a}{c} = \frac{b}{c}$.

2. Equations that have the same solution set over a given domain are called *equivalent equations* over that domain. Each of the following

transformations of a given equation will produce an equivalent equation.

1. Substituting for any expression in the given equation an equivalent expression.
2. Adding the same real number to, or subtracting the same real number from, each side of the given equation.
3. Multiplying or dividing each side of the given equation by the same *nonzero* real number.

3. Operations that "undo" each other are called *inverse operations*. Addition and subtraction are a pair of inverse operations, as are multiplication and division. Inverse operations can be used in solving equations.

4. Using the plan outlined on page 132, a word problem can often be solved by first writing an open sentence that represents relationships among the numbers in the problem and then solving the open sentence. Organizing the facts of a word problem in a chart is often helpful.

5. A *formula* is an equation that states a relationship among quantities represented by variables. Transformations are often used to *solve for* one variable in the formula. This variable is then said to be *expressed in terms of* the other variables.

Chapter Review

Write the letter of the correct answer.

1. Give the reason that justifies the following statement: $\qquad$ *3-1*
   "If $r + 5 = 8$, then $(r + 5) - 5 = 8 - 5$."
   a. axiom of additive inverses
   b. subtraction property of equality
   c. symmetric axiom of equality
   d. substitution principle

Solve.

2. $c - 14 = -5$ $\qquad$ *3-2*
   a. $\{9\}$ b. $\{-19\}$ c. $\{-9\}$ d. $\{19\}$

3. $1 = 7 + r$
   a. $\{8\}$ b. $\{-8\}$ c. $\{-6\}$ d. $\{6\}$

4. $-x - 3 = -4$
   a. $\{1\}$ b. $\{-1\}$ c. $\{-1, 1\}$ d. $\emptyset$

5. *Plan:* Assume $-a = -b$.
   1. $-1(-a) = -1(-b)$
      Mult. prop. of $=$
   2. $1 \cdot a = 1 \cdot b$
      Prop. of opp. in prod.
   3. $a = b$
      Iden. ax. for mult.
   4. $a \neq b$
      Hypothesis
   Statement (3) contradicts the hypothesis $a \neq b$.
   $\therefore$ the assumption $-a = -b$ is incorrect and the conclusion $-a \neq -b$ is true.

6. *Plan:* Assume $\frac{1}{a} = \frac{1}{b}$.
   1. $a \neq 0$ and $b \neq 0$
      Hypothesis
   2. $a\left(\frac{1}{a}\right) = 1, b\left(\frac{1}{b}\right) = 1$
      Ax. of mult. inv.
   3. $a\left(\frac{1}{a}\right) = b\left(\frac{1}{b}\right)$
      Subs. prin.
   4. $a\left(\frac{1}{a}\right) = b\left(\frac{1}{a}\right)$
      Subs. prin.
   5. $\left[a\left(\frac{1}{a}\right)\right]a = \left[b\left(\frac{1}{a}\right)\right]a$
      Mult. prop. of $=$
   6. $a\left[\left(\frac{1}{a}\right)a\right] = b\left[\left(\frac{1}{a}\right)a\right]$
      Assoc. ax. for mult.
   7. $a \cdot 1 = b \cdot 1$
      Ax. of mult. inv.
   8. $a = b$
      Iden. ax. for mult.
   9. $a \neq b$
      Hypothesis
   Statement (8) contradicts the hypothesis $a \neq b$.
   $\therefore$ the assumption $\frac{1}{a} = \frac{1}{b}$ is incorrect and the conclusion $\frac{1}{a} \neq \frac{1}{b}$ is true.

(continued)

7. *Plan*: Assume $1 = 2$.
   1. $1 + (-1) = 2 + (-1)$
      Add. prop. of $=$
   2. $0 = 2 + (-1)$
      Ax. of add. inv.
   3. $0 = 1$
      Subs. prin.
   Statement (3) contradicts
   the iden. ax. for mult.
   which states $1 \neq 0$.
   $\therefore$ the assumption $1 = 2$
   is incorrect and the conclusion $1 \neq 2$ is true.

8. *Plan*: Assume $-1 = -2$.
   1. $(-1) + 2 = (-2) + 2$
      Add. prop. of $=$
   2. $(-1) + 2 = 0$
      Ax. of add. inv.
   3. $1 = 0$
      Subs. prin.
   Statement (3) contradicts
   the iden. ax. for mult.
   which states $1 \neq 0$.
   $\therefore$ the assumption
   $-1 = -2$ is incorrect and
   the conclusion $-1 \neq -2$
   is true.

4. $t = \dfrac{5pq}{2s}$

5. $c = \dfrac{a^2 b}{3x}$

6. $m = -2ts$

9. $x = \dfrac{b - 2c}{-2a}$

Solve.

5. $-2y = -10$ 3-3
   a. $\{-5\}$ **b.** $\{5\}$ c. $\{-8\}$ d. $\{20\}$

6. $-72 = \frac{2}{3}u$
   a. $\{-48\}$ b. $\{48\}$ **c.** $\{-108\}$ d. $\{108\}$

7. $-5|d| = 30$
   a. $\{-6, 6\}$ b. $\{-6\}$ c. $\{6\}$ **d.** $\emptyset$

8. $8a + 5 = 21$ 3-4
   a. $\left\{\frac{13}{4}\right\}$ **b.** $\{2\}$ c. $\left\{\frac{21}{40}\right\}$ d. $\{16\}$

9. $2(5 - p) - 3 = -5$
   a. $\{-6\}$ b. $\{12\}$ c. $\{-1\}$ **d.** $\{6\}$

10. One number is five greater than another. If the sum of the lesser   3-5
    number and twice the greater is 61, what is the greater number?
    a. 28   b. 33   c. 17   **d.** 22

11. Jerry has eight more nickels than quarters. The total value of his
    nickels and quarters together is $10. How many nickels does he
    have?
    a. 32   **b.** 40   c. 20   d. 28

12. $8k + 6 = 5k$ 3-6
    a. $\left\{-\frac{6}{13}\right\}$ b. $\left\{-\frac{13}{6}\right\}$ **c.** $\{-2\}$ d. $\{2\}$

13. $-3(t - 7) = 4(9 - 2t)$
    a. $\{3\}$ b. $\{-43\}$ c. $\left\{\frac{57}{6}\right\}$ d. $\{-15\}$

14. Kevin is five years older than Lisa. Four years ago Kevin was twice   3-7
    as old as Lisa was then. How old is Kevin now?
    a. 9 years   b. 12 years   **c.** 14 years   d. 15 years

15. The length of a certain rectangle is 4 cm less than twice its width.
    Five times its width is 6 cm less than its perimeter. What is the
    length of the rectangle?
    a. 14 cm   **b.** 24 cm   c. 48 cm   d. 76 cm

16. Solve for h if $V = \frac{1}{3}bh$ and $V, b, h \neq 0$.   3-8
    a. $h = \dfrac{V}{3b}$ **b.** $h = \dfrac{3V}{b}$ c. $h = \dfrac{b}{3V}$ d. $h = V - \dfrac{b}{3}$

17. Solve for h if $SA = 2\pi r(r + h)$.

a. $h = SA - 2\pi r^2$

b. $h = \dfrac{SA}{2\pi r^2}, r \neq 0$

(c.) $h = \dfrac{SA}{2\pi r} - r, r \neq 0$

d. $h = \dfrac{SA}{2\pi r^2} + r, r \neq 0$

18. Use the formula $C = \dfrac{5}{9}(F - 32)$ to find the Fahrenheit temperature equivalent to a temperature of 25 degrees Celsius.

(a.) $77°F$

b. $13°F$

c. $45°F$

d. $-13°F$

Chapter Test

3-1

1. Name the property that is illustrated by the following:
"If $8x = 96$, then $\dfrac{1}{8} \cdot 8x = \dfrac{1}{8} \cdot 96$." Multiplication property of equality

Solve.

2. $k + 9 = -11$ {-20} 3. $5 - t = 3$ {2} 4. $|-9| = b + 2$ {7} 3-2

5. $-17d = 306$ {-18} 6. $-108 = -\dfrac{1}{9}z$ {972} 7. $-4|j| = 48$ Ø 3-3

8. $3n - 7 = 8$ {5} 9. $\dfrac{3h + 2}{5} = -2$ {-4} 10. $7c - 3(c - 4) = 8$ {-1} 3-4

11. One number is two thirds another number. If the sum of the numbers is forty, what is the lesser number? 16 3-5

12. The length of a certain rectangle is 2 cm less than three times its width. If the perimeter of the rectangle is 44 cm, what is its length? 16 cm

13. $6 - 4w = 13 - 3w$ {-7} 14. $8(c + 11) = 5(c + 1)$ {$-\frac{83}{3}$} 3-6

15. Find the number that is four greater than three times its additive inverse. 1 3-7

16. Jackie is five years older than her brother. In four years, Jackie will be twice as old as her brother will be. How old is Jackie now? 6 years

17. Solve for x if $S = \dfrac{n(a - x)}{2}$. $x = \dfrac{na - 2S}{n}$, or $a - \dfrac{2S}{n}$, $n \neq 0$ 3-8

18. Use the formula $V = lwh$ to find the width of a rectangular box having length 12 cm, height 4 cm, and volume 384 cm³. 8 cm

Solving Equations and Problems **153**

10. $p = \dfrac{4r - 3t}{2q}$

11. $c = \dfrac{-2ax - b}{-2}$, or $\dfrac{2ax + b}{2}$

12. $t = \dfrac{2pq - 4r}{-3}$, or $\dfrac{4r - 2pq}{3}$

15. $b = \dfrac{ac - 4d}{a}$, or $c - \dfrac{4d}{a}$

16. $t = \dfrac{2w + sv}{s}$, or $\dfrac{2w}{s} + v$

17. $c = \dfrac{ab + 4d}{a}$, or $b + \dfrac{4d}{a}$

18. $v = \dfrac{st - 2w}{s}$, or $t - \dfrac{2w}{s}$

19. $x = \dfrac{9a^2 + 4c}{3a^2}$, or $3 + \dfrac{4c}{3a^2}$

20. $q = \dfrac{st^2p - 5}{2st^2}$, or $\dfrac{p}{2} - \dfrac{5}{2st^2}$

27. $s = \dfrac{-0.06k}{t^2}$

28. $y = \dfrac{-0.02z}{0.6x^2} = \dfrac{-z}{30x^2}$

30. $a = \dfrac{3V}{x + y}$

33. $b = \dfrac{3}{2}C - a$

35. $y = 3x - \dfrac{2A}{5h}$, or

$\dfrac{15hx - 2A}{5h}$

36. $b = \dfrac{2a}{3} - \dfrac{4V}{9t}$, or

$\dfrac{6at - 4V}{9t}$

45. $E = I(R + r)$

47. $r = \dfrac{E - IR}{I}$, or $\dfrac{E}{I} - R$

48. $x_2 = x_1 - \dfrac{y_1 - y_2}{m}$, or

$\dfrac{mx_1 - y_1 + y_2}{m}$

50. $r = \dfrac{4s - 5}{2s - 4}$

51. $t = \dfrac{-4b - 3}{a - b}$

53. $f_1 = \dfrac{Ff_2}{f_2 - F}$, or $\dfrac{-Ff_2}{F - f_2}$

Additional Answers
Written Exercises (p. 118)
(continued from p. 143)

16. 1. a, b, and x are real
numbers and $b \neq 0$;
$a \div b = x$
Hypothesis

2. $x = a \div b$
Symm. ax. of =

3. $x = a \cdot \dfrac{1}{b}$
Def. of div.

4. $xb = \left(a \cdot \dfrac{1}{b}\right)b$
Mult. prop. of =

5. $xb = a\left(\dfrac{1}{b} \cdot b\right)$
Assoc. ax. for mult.

6. $xb = a \cdot 1$
Ax. of mult. inv.

7. $xb = a$
Iden. ax. for mult.

8. $\therefore bx = a$
Comm. ax. for mult.

17. 1. a, b, c, and d are real
numbers; $a = b$
Hypothesis

2. $a + c = b + c$
Add. prop. of =

3. $c = d$
Hypothesis

4. $\therefore a + c = b + d$
Subs. prin.

18. 1. a, b, c, and d are real
numbers; $a + c =$
$b + d$; $b = c$
Hypothesis

2. $-c$ is a real number.
Ax. of add. inv.

3. $a + c = c + d$
Subs. prin.

4. $a + c = d + c$
Comm. ax. for add.

5. $(a + c) + (-c)$
$= (d + c) + (-c)$
Add. prop. of =

6. $a + [c + (-c)]$
$= d + [c + (-c)]$
Assoc. ax. for add.

7. $a + 0 = d + 0$
Ax. of add. inv.

8. $\therefore a = d$
Iden. ax. for add.

Cumulative Review

Chapter 1

Tell whether each statement is true or false.

1. $\emptyset \subset$ {the natural numbers} True
2. $0 \notin$ {the whole numbers} False
3. {the integers} $\subset$ {the real numbers} True
4. {the irrational numbers} $\subset$ {the rational numbers} False

Simplify.

5. $5\dfrac{7}{8} + 1\dfrac{2}{3}$ $7\dfrac{13}{24}$

6. $72 \div \dfrac{1}{3}$ 216

7. $44 - 6(10 - 7)$ 26

8. $\dfrac{36 - 16}{12 - 2}$ 2

9. $18 + 2^3$ 26

10. $28 - 6 \div 2 + (9 - 7)^2$ 29

Evaluate each expression when $a = 2$, $b = 3$, and $c = \frac{1}{2}$.

11. $\dfrac{ab}{c}$ 12

12. $4b - 3ac$ 9

13. $3b - a(a + 2c)$ 3

14. $\dfrac{4abc}{b + 4c}$ $2\dfrac{2}{5}$

15. $(a^2 + b^2)c$ $\dfrac{13}{2}$

16. $\dfrac{a^2 + 2b - 6c}{a}$ $3\dfrac{1}{2}$

Chapter 2

Simplify. 23. $-4a^2 + 7$

17. $-56 + (-27)$ -83

18. $-49 - (-31)$ -18

19. $(-23)(-11)$ 253

20. $-36 \div (-2)$ 18

21. $-[-(42 - 70)]$ -28

22. $-|-24 + 5|$ -19

23. $-5a^2 + 7 + a^2$

24. $6 - 9b^3 - 23$ $-9b^3 - 17$

25. $(-3c)(5c)(-7c)$ $105c^3$

26. $-76xy \div 4$ $-19xy$

27. $-3(-7u + 4v)$
$21u - 12v$

28. $9m - 2(6n - 5m)$
$19m - 12n$

Give the reason that justifies each statement, assuming that m and n are real numbers. 29. Ax. of add. inv.

29. $n + (-n) = 0$

30. $n + 0 = n$ Iden. ax. for add.

31. $-(m + n) = -m + -n$ Prop. of opp. of a sum

32. $-(-n) = n$
Canc. prop. of opp.

33. $(-m)(-n) = mn$
Prop. of opp. in prod.

34. $n \cdot \dfrac{1}{n} = 1$, $n \neq 0$
Ax. of mult. inv.

Chapter 3

Solve.

35. $-4 + p = 16$ {20}

36. $26 = q - 13$ {39}

37. $-12s = 6$ $\{-\frac{1}{2}\}$

38. $\dfrac{t}{-2} = -14$ {28}

39. $17 - x = 28$ {-11}

40. $-2 = -\dfrac{1}{8}|y|$ {16, -16}

154 *Chapter 3*

41. $9m - 2 = -20$ $\{-2\}$ **42.** $-9 = \frac{n}{4} + 11$ $\{-80\}$ **43.** $-15 = -3(g - 2)$ $\{7\}$

44. $\frac{h - 8}{3} = -10$ $\{-22\}$ **45.** $-2(j + 6) = j$ $\{-4\}$ **46.** $6k - (1 - k) = 7k - 1$
    $\mathcal{R}$

47. The length of a certain rectangle is 4 cm longer than its width. If the width were increased by 6 cm and the length were doubled, the new perimeter would be twice the perimeter of the original rectangle. Find the dimensions of the original rectangle. width: 6 cm; length: 10 cm

48. The formula $C = \pi d$ gives the circumference of a circle C in terms of the diameter d. If the circumference of a certain circle is 176 cm, find its diameter. Use $\pi = \frac{22}{7}$. 56 cm

Contest Problems

1. If $\frac{8c - 7d}{d} = 2$, find the value of $\frac{5c - 6d}{2c}$. $-\frac{1}{6}$

2. Given that $t \in \{1, 2, \ldots, 9\}$ and $u \in \{0, 1, \ldots, 9\}$, when a two-digit number of the form $10t + u$ is divided by the sum of its digits, the quotient is 5.5. What must be the relationship between t and u? $t = u$

3. If $4 = \frac{3}{y}$, $5 = \frac{2}{x}$, and $6 = \frac{5}{z}$, find the value of $\frac{15x - 10y}{9z}$. $-\frac{1}{5}$

4. Given that r, s, t, and u are positive integers and that $9t = 2u$, $5r = 3s$, and $10u = 9s$, arrange r, s, t, and u in order from least to greatest. t, r, u, s

5. An operation $*$ is defined for real numbers u and v as
$$u * v = \tfrac{1}{2}(u + v - |u - v|).$$

a. Is $*$ a commutative operation? Yes

b. If u is an integer and v is the next greater integer, what is the value of $u * v$? u

19. 1. a, b, c, and d are real numbers; $a = b$
Hypothesis
2. $ac = bc$
Mult. prop. of $=$
3. $c = d$
Hypothesis
4. $\therefore ac = bd$
Subs. prin.

20. 1. a, b, c, and d are real numbers such that $a \neq 0$; $ac = bd$
Hypothesis
2. $bd = ac$
Symm. ax. of $=$
3. $a = d$
Hypothesis
4. $bd = dc$
Subs. prin.
5. $bd = cd$
Comm. ax. for mult.
6. $(bd)\frac{1}{d} = (cd)\frac{1}{d}$
Mult. prop. of $=$
7. $b\left(d \cdot \frac{1}{d}\right) = c\left(d \cdot \frac{1}{d}\right)$
Assoc. ax. for mult.
8. $b \cdot 1 = c \cdot 1$
Ax. of mult. inv.
9. $\therefore b = c$
Iden. ax. for mult.

21. 1. a, b, c, and d are real numbers; $a = b$
Hypothesis
2. $a - c = b - c$
Subt. prop. of $=$
3. $c = d$
Hypothesis
4. $\therefore a - c = b - d$
Subs. prin.

22. 1. a, b, c, and d are real numbers such that $c \neq 0$ and $d \neq 0$; $a = b$
Hypothesis
2. $\frac{a}{c} = \frac{b}{c}$
Div. prop. of $=$
3. $c = d$
Hypothesis
4. $\therefore \frac{a}{c} = \frac{b}{d}$
Subs. prin.

Quick acceleration is important in motorcycle racing. The Application on page 212 gives a formula relating acceleration to velocity and time.

Chapter 4

Solving Inequalities and Problems

Solving Inequalities

OBJECTIVES for Sections 4-1 through 4-5:
1. *To use the properties of order as reasons for statements in proving theorems.*
2. *To solve inequalities and graph their solution sets.*
3. *To find the intersection and the union of two sets.*
4. *To solve conjunctions and disjunctions.*
5. *To solve open sentences involving absolute value.*

4–1 Properties of Order

One and only one of the following statements is true.

$$2 < -3 \qquad 2 = -3 \qquad -3 < 2$$

The true statement is $-3 < 2$. This situation illustrates the following fact that is used in comparing real numbers.

Axiom of Comparison

For all real numbers a and b, one and only one of the following statements is true:

$$a < b \qquad a = b \qquad b < a$$

Solving Inequalities and Problems **157**

Problem Solving Strategies

Word Problem Plan
The *Plan for Solving a Word Problem* introduced on page 132 is used in Sections 4-6 through 4-10.
Recognizing a Problem Type
Students should realize that similar word problems are often set up in the same manner. For example, the consecutive integer problems in Section 4-6 all use the pattern x, $x + 1$, $x + 2$, . . . to represent the integers, the measures of complementary angles can always be represented by $m°$ and $(90 - m)°$ in the problems in Section 4-7, and the equations used to solve the uniform motion problems in Section 4-8 all involve the formula $r \cdot t = d$.
Drawing a Diagram
Diagrams may be useful to students in picturing the motions described in Section 4-8.
Using a Chart
Charts are used in Sections 4-8 and 4-9 to help organize the information given in the uniform motion and mixture problems.
Recognizing No Solution
In Section 4-10, students learn to identify problems without solutions.

Teaching Suggestions
p. T80

Key Ideas

Use the properties of order to justify steps in proofs.

1. What happens when 2 is added to both sides of this inequality?

$$8 > 4$$
$$8 + 2 > 4 + 2$$
$$10 > 6$$

The order of inequality remains the same.

2. What happens when each side of this inequality is multiplied by 4?

$$4 < 5$$
$$4(4) < 5(4)$$
$$16 < 20$$

The order of inequality remains the same.

3. What happens when each side of this inequality is multiplied by -3?

$$6 < 8$$
$$6(-3) > 8(-3)$$
$$-18 > -24$$

The order of inequality is reversed.

Suggested Assignments

Minimum
   161/1–24
Average
   161/9–33 odd, 34–38
Maximum
   161/9–37 odd, 38–46

Recall from Section 1-1 that the statement $a < b$ gives the same information as the statement $b > a$.

Another fact about order in the set of real numbers can be easily shown on a number line. Suppose you know that a, b, and c are real numbers such that $a < b$ and $b < c$. What is the relationship between a and c? If the graphs of a, b, and c are on a horizontal number line that is marked with positive numbers to the right, the graph of a is to the left of the graph of b, and the graph of b is to the left of the graph of c. Thus you can see that the graph of a is to the left of the graph of c and, therefore, $a < c$.

This example suggests the following property of the real numbers.

Transitive Property of Order

For all real numbers a, b, and c:

1. If $a < b$ and $b < c$, then $a < c$.

2. If $a > b$ and $b > c$, then $a > c$.

What happens when the same number is added to each side of an inequality such as $-4 < 3$?

Add 5: Is it true that $-4 + 5 < 3 + 5$?
            Yes, $1 < 8$.

Add -5: Is it true that $-4 + (-5) < 3 + (-5)$?
            Yes, $-9 < -2$.

These examples suggest another property of order for the real numbers.

Addition Property of Order

For all real numbers a, b, and c:

1. If $a < b$, then $a + c < b + c$ and $c + a < c + b$.

2. If $a > b$, then $a + c > b + c$ and $c + a > c + b$.

Since subtraction is defined in terms of addition, the addition property of order applies when you subtract a real number from each side of an inequality. For example,

$$\text{if } a < b,$$
$$\text{then } a - 2 < b - 2$$

since $a + (-2) < b + (-2)$.

158 *Chapter 4*

What happens when each side of the inequality $-4 < 3$ is *multiplied* by a nonzero real number?

Multiply by 5: Is it true that $-4(5) < 3(5)$?

Yes, $-20 < 15.$

Multiply by -5: Is it true that $-4(-5) < 3(-5)$?

No, $20 > -15.$

These examples suggest the following rules.

1. Multiplying each side of an inequality by a positive number *preserves* the order of the inequality.
2. Multiplying each side of an inequality by a negative number *reverses* the order of the inequality.

These rules are stated formally as the following property of order.

Multiplication Property of Order

For all real numbers a, b, and c:

1. If $a < b$ and $c > 0$, then $ac < bc$ and $ca < cb$.
   If $a > b$ and $c > 0$, then $ac > bc$ and $ca > cb$.

2. If $a < b$ and $c < 0$, then $ac > bc$ and $ca > cb$.
   If $a > b$ and $c < 0$, then $ac < bc$ and $ca < cb$.

Note that multiplying both sides of an inequality by zero does not produce an inequality; the result is the identity $0 = 0$.

Since division is defined in terms of multiplication, the multiplication property of order applies when each side of an inequality is divided by a nonzero real number. For example,

if $a > b$,

then $\dfrac{a}{-2} < \dfrac{b}{-2}$

since $(-\frac{1}{2})(a) < (-\frac{1}{2})(b)$.

The following assertion, which was accepted without proof in Section 2-4, can now be proved using the properties of order.

Theorem.   If a is a real number and $a > 0$, then $-a < 0$.
Similarly, if $a < 0$, then $-a > 0$.

The proof of the first part of the above theorem is given on the following page.

Solving Inequalities and Problems **159**

Replace each ? with one of
the symbols >, =, < so that
the statement is true for all
real values of the variables.

1. If $q < r$, then $q + 2$? $r + 2$.   <

2. If $n > m$, then $n - 5$? $m - 5$.   >

3. If $c > d$, then $2c$? $2d$.   >

4. If $a < b$, then $\frac{1}{4}a$? $\frac{1}{4}b$.   <

5. If $g < h$, then $\frac{g}{-2}$? $\frac{h}{-2}$.   >

6. If $r + 1 > s + 1$, then $r - 1$? $s - 1$.   >

**Additional Answers
Written Exercises** (p. 162)

21. 1. Hypothesis
2. Add. prop. of order
3. Hypothesis
4. Add. prop. of order
5. Trans. prop. of order

22. 1. Hypothesis
2. Mult. prop. of order
3. Hypothesis
4. Mult. prop. of order
5. Trans. prop. of order

23. 1. Hypothesis
2. Add. prop. of order
3. Def. of subt.
4. Ax. of add. inverses

24. 1. Hypothesis
2. Def. of subt.
3. Add. prop. of order
4. Assoc. ax. for add.
5. Ax. of add. inverses
6. Identity ax. for add.

33. 1. a is a real number;
$a < 0$
Hypothesis
2. $-1(a) > -1(0)$
Mult. prop. of order
3. $-1(a) = -a$
Mult. prop. of -1
4. $-1(0) = 0$
Mult. prop. of 0
5. $\therefore -a > 0$
Subs. prin.

Prove: If a is a real number and $a > 0$, then $-a < 0$.

PROOF

Statements	Reasons
1. a is a real number.	Hypothesis
2. $-a$ is a real number.	Axiom of additive inverses
3. $a > 0$	Hypothesis
4. $a + (-a) > 0 + (-a)$	Addition property of order
5. $a + (-a) = 0$	Axiom of additive inverses
6. $0 + (-a) = -a$	Identity axiom for addition
7. $\therefore 0 > -a$ (or $-a < 0$)	Substitution principle

The second part of this theorem will be proved in Exercise 33 on page 162.

You can use the properties of order to prove additional theorems about the real numbers. For example, you have probably observed that the square of any nonzero real number is a positive real number. You can state this fact as the following theorem.

Theorem.   If a is a real number and $a \neq 0$, then $a^2 > 0$.

This theorem can be proved as follows.

Plan: Since $a \neq 0$, the axiom of comparison tells you that there are two cases to consider. Case 1 is $a < 0$, and Case 2 is $a > 0$. The proof of Case 1 is given below. Case 2 will be proved in Exercise 34 on page 162.

PROOF

Statements	Reasons
1. $a < 0$	Hypothesis
2. $a \cdot a > 0 \cdot a$	Multiplication property of order
3. $a \cdot a = a^2$	Definition of a^2
4. $0 \cdot a = 0$	Multiplicative property of zero
5. $\therefore a^2 > 0$	Substitution principle

Since $1 \neq 0$ and $1 = 1^2$, the preceding theorem shows that $1 > 0$. Because of this fact and the theorem on page 159, you know that the familiar statement $-1 < 0$ is also true. Indeed, all of the order facts that you have learned in studying arithmetic are consequences of the properties of the real numbers discussed in this section.

160   *Chapter 4*

Oral Exercises

Name the property that justifies each statement. **3.** Axiom of comparison

1. If $d > 7$, then $d + 2 > 7 + 2$. Addition property of order
2. If $s < -1$, then $s + 8 < -1 + 8$. Addition property of order
3. Of two different real numbers, one must be greater than the other.
4. Any real number that is neither zero nor a positive number must be a negative number. Axiom of comparison
5. If $5c < 15$, then $\frac{5c}{5} < \frac{15}{5}$. Multiplication property of order
6. If $z < 4$, then $(-1)z > (-1)4$. Multiplication property of order
7. If $t > 5$, then $t - 1 > 5 - 1$. Addition property of order
8. If $x < -6$, then $x \div 2 < -6 \div 2$. Multiplication property of order
9. If j is not less than k and k is not less than j, then $j = k$. Axiom of comparison
10. Either the statement $a > b$ or the statement $b > a$ is false. Axiom of comparison
11. If $a < 0$, then $9a < 7a$. Multiplication property of order
12. If $c > 0$, then $-7c > -12c$. Multiplication property of order

Written Exercises

Replace each __?__ with one of the symbols >, =, or < so that the statement is true for all real values of the variables.

A
1. If $m > n$, then n __?__ m. <
2. If $p < q$, then q __?__ p. >
3. If $a < b$, then $a + 1$ __?__ $b + 1$. <
4. If $r > s$, then $r + 5$ __?__ $s + 5$. >
5. If $x > y$, then $x - 3$ __?__ $y - 3$. >
6. If $m < n$, then $m - 8$ __?__ $n - 8$. <
7. If $c < d$, then $2c$ __?__ $2d$. <
8. If $x > z$, then $\frac{1}{3}x$ __?__ $\frac{1}{3}z$. >
9. If $e > f$, then $-4e$ __?__ $-4f$. <
10. If $b < c$, then $-c$ __?__ $-b$. <

11. If $s < t$ and $u = 0$, then su __?__ tu. =
12. If $v > w$ and $t = 0$, then vt __?__ wt. =
13. If $k < 0$ and $m < 0$, then km __?__ 0. >
14. If $a > 0$ and $b < 0$, then ab __?__ 0. <
15. If $x > y$ and $w < 0$, then $x + w$ __?__ $y + w$. >
16. If $a < b$ and $c > 0$, then $a - c$ __?__ $b - c$. <
17. If $r - 1 < s - 1$, then $r - 5$ __?__ $s - 5$. <
18. If $p + 3 > q + 3$, then $p - 6$ __?__ $q - 6$. >
19. If $-3g > -18$, then g __?__ 6. <
20. If $-4t < 20$, then t __?__ -5. >

34.
1. a is a real number; $a > 0$
   Hypothesis
2. $a \cdot a > 0 \cdot a$
   Mult. prop. of order
3. $a \cdot a = a^2$
   Def. of a^2
4. $0 \cdot a = 0$
   Mult. prop. of 0
5. $\therefore a^2 > 0$
   Subs. prin.

35.
1. x is a real number; $x > 0$
   Hypothesis
2. $x^2 > 0$
   Exercise 34
3. $x^2 \cdot x > 0 \cdot x$
   Mult. prop. of order
4. $x^2 \cdot x = x \cdot x \cdot x$
   Def. of x^2
5. $x \cdot x \cdot x = x^3$
   Def. of x^3
6. $x^2 \cdot x = x^3$
   Trans. ax. of =
7. $0 \cdot x = 0$
   Mult. prop. of 0
8. $\therefore x^3 > 0$
   Subs. prin.

36.
1. x is a real number; $x < 0$
   Hypothesis
2. $x^2 > 0$
   Theorem, page 160
3. $x^2 \cdot x < 0 \cdot x$
   Mult. prop. of order
4. $x^2 \cdot x = x \cdot x \cdot x$
   Def. of x^2
5. $x \cdot x \cdot x = x^3$
   Def. of x^3
6. $x^2 \cdot x = x^3$
   Trans. ax. of =
7. $0 \cdot x = 0$
   Mult. prop. of 0
8. $\therefore x^3 < 0$
   Subs. prin.

(continued)

Solving Inequalities and Problems **161**

37. 1. x and y are real numbers; $x < 0$ and $y < 0$
Hypothesis
2. $x \cdot y > 0 \cdot y$
Mult. prop. of order
3. $\therefore xy > 0$
Mult. prop. of 0

38. 1. x and y are real numbers; $x < 0$ and $y > 0$
Hypothesis
2. $x \cdot y < 0 \cdot y$
Mult. prop. of order
3. $\therefore xy < 0$
Mult. prop. of 0

39. 1. m and n are real numbers; $m > 0$ and $n > 0$; $m < n$
Hypothesis
2. $m \cdot m < m \cdot n$
Mult. prop. of order
3. $m \cdot n < n \cdot n$
Mult. prop. of order
4. $m \cdot m < n \cdot n$
Trans. prop. of order
5. $m \cdot m = m^2$
Def. of m^2
6. $n \cdot n = n^2$
Def. of n^2
7. $\therefore m^2 < n^2$
Subs. prin.

40. 1. m and n are real numbers; $m < 0$ and $n < 0$; $m < n$
Hypothesis
2. $m \cdot m > n \cdot m$
Mult. prop. of order
3. $n \cdot m > n \cdot n$
Mult. prop. of order
4. $m \cdot m > n \cdot n$
Trans. prop. of order
5. $m \cdot m = m^2$
Def. of m^2
6. $n \cdot n = n^2$
Def. of n^2
7. $\therefore m^2 > n^2$
Subs. prin.

Give the reason that justifies each statement in the given proof.

21. Prove: For all real numbers a, b, c, and d, if $a > b$ and $c > d$, then $a + c > b + d$.

PROOF

1. a, b, c, and d are real numbers; $a > b$
2. $a + c > b + c$
3. $c > d$
4. $b + c > b + d$
5. $\therefore a + c > b + d$

22. Prove: For all positive real numbers a, b, c, and d, if $a > b$ and $c > d$, then $ac > bd$.

PROOF

1. a, b, c, and d are positive real numbers; $a > b$
2. $ac > bc$
3. $c > d$
4. $bc > bd$
5. $\therefore ac > bd$

23. Prove: For all real numbers a and b, if $a < b$, then $a - b < 0$.

PROOF

1. a and b are real numbers; $a < b$
2. $a - b < b - b$
3. $a - b < b + (-b)$
4. $\therefore a - b < 0$

24. Prove: For all real numbers a and b, if $a - b < 0$, then $a < b$.

PROOF

1. a and b are real numbers; $a - b < 0$
2. $a + (-b) < 0$
3. $[a + (-b)] + b < 0 + b$
4. $a + [(-b) + b] < 0 + b$
5. $a + 0 < 0 + b$
6. $\therefore a < b$

If $r > s$, specify the set of all values of t for which each open sentence is a true statement.

B **25.** $rt = st$ **26.** $rt > st$ **27.** $rt^2 > st^2$ **28.** $rt^2 = st^2$ **29.** $\dfrac{r}{t} < \dfrac{s}{t}$ **30.** $\dfrac{r}{t^2} < \dfrac{s}{t^2}$

$\{0\}$ $\{t: t > 0\}$ $\{t: t \ne 0\}$ $\{0\}$ $\{t: t < 0\}$ $\varnothing$

31. If $x < y$, specify the set of all values of x and y such that $x^2 < y^2$.

32. If $x < y$, specify the set of all values of x and y such that $x^2 > y^2$.

31. $\{x, y: |x| < |y|\}$ **32.** $\{x, y: |x| > |y|\}$

Write a direct proof of each theorem.

33. If a is a real number and $a < 0$, then $-a > 0$.

34. If a is a real number and $a > 0$, then $a^2 > 0$.

35. If x is a positive real number, then $x^3 > 0$.

36. If x is a negative real number, then $x^3 < 0$.

37. For all real numbers x and y, if $x < 0$ and $y < 0$, then $xy > 0$.

38. For all real numbers x and y, if $x < 0$ and $y > 0$, then $xy < 0$.

39. For all positive real numbers m and n, if $m < n$, then $m^2 < n^2$.

40. For all negative real numbers m and n, if $m < n$, then $m^2 > n^2$.

162   *Chapter 4*

C **41.** For all real numbers x and y, if $x < y$, then $x < \dfrac{x+y}{2}$.

42. For all real numbers x and y, if $x < y$, then $\dfrac{x+y}{2} < y$.

Tell whether each statement is true or false for all values of a and b such that $a > b$. If it is false, give an example to justify your answer.

43. $a^2 - b > 0$ **44.** $a^2 - b^2 > 0$ **45.** $ab - a > 0$ **46.** $ab - b^2 > 0$

False; $a = \frac{1}{2}$, $b = \frac{1}{3}$ True   False; $a = \frac{1}{2}$ False; $a = -1$

                                                              $b = \frac{1}{3}$ $b = -2$

Computer Exercises For students with computer experience

1. Write a program that will display the order of any two numbers that you input. The output should be in the form of a mathematical sentence that uses one of the symbols $<$, $>$, or $=$.

2. Write a program that will allow you to input any two *different* real numbers and will display the lesser number first, then the greater number. The output should be in the form of an inequality that uses the symbol $<$.

3. Write a program that will allow you to input any *three* different real numbers and will display the numbers in order from least to greatest. The output should be in the form of an inequality that uses the symbol $<$.

4–2 Equivalent Inequalities

Inequalities that have the same solution set over a given domain are called **equivalent inequalities** over that domain. The properties that have been stated in Section 4-1 guarantee that the following transformations of a given inequality always produce an equivalent inequality.

> ### *Transformations that Produce an Equivalent Inequality*
> 1. Substituting for either side of the inequality an expression equivalent to that side.
> 2. Adding to (or subtracting from) each side the same real number.
> 3. Multiplying (or dividing) each side by the same positive number.
> 4. Multiplying (or dividing) each side by the same negative number and *reversing* the order of the inequality.

Solving Inequalities and Problems **163**

Solve each inequality and graph its solution set.

1. $y - 4 > 2$
   $y - 4 + 4 > 2 + 4$
   $y > 6$
   ∴ the solution set is
   $\{y: y > 6\}$.

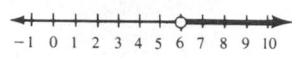

2. $2a + 3 < -3$
   $2a < -6$
   $a < -3$
   ∴ the solution set is
   $\{a: a < -3\}$.

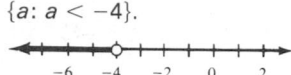

3. $3(a - 1) + 1 > (4a + 2)$
   $3a - 3 + 1 > 4a + 2$
   $3a - 2 > 4a + 2$
   $3a > 4a + 4$
   $-a > 4$
   $a < -4$
   ∴ the solution set is
   $\{a: a < -4\}$.

Reading Algebra

Solution sets of equations and inequalities may be expressed in terms of set-builder notation to enhance precision and compactness of answers. Students must be able to read set-builder notation properly to be able to state the correct solution set. Have them read $\{x: x > 6\}$ as "the set of all x such that x is greater than six." Also have them read $\{y: 3 \leq y \leq 9\}$ as "the set of all y such that y is greater than or equal to three and less than or equal to nine." Have students make cards with solution sets in set-builder notation and then pass them around for other students to read the sets aloud.

To solve an inequality, you usually try to transform it into a simple equivalent inequality whose solution set can be found by inspection.

EXAMPLE 1 Solve $7(c - 2) + 2 > 2(5c + 9)$ and graph its solution set.

SOLUTION

$7(c - 2) + 2 > 2(5c + 9)$

$7c - 14 + 2 > 10c + 18$ ⟵ { Use the distributive axiom to help simplify each side.

$7c - 12 > 10c + 18$

$7c - 12 + 12 > 10c + 18 + 12$ ⟵ { Add 12 to each side.

$7c > 10c + 30$

$7c - 10c > 10c + 30 - 10c$ ⟵ { Subtract $10c$ from each side.

$-3c > 30$

$\dfrac{-3c}{-3} < \dfrac{30}{-3}$ ⟵ { Divide each side by -3 and reverse the direction of the inequality.

$c < -10$

CONDENSED SOLUTION

$7(c - 2) + 2 > 2(5c + 9)$

$7c - 12 > 10c + 18$

$7c > 10c + 30$

$-3c > 30$

$c < -10$

∴ the solution set is $\{c: c < -10\}$.

One method used to check that an equation has been solved correctly is to test each member of the solution set to see if it satisfies the original equation. This method of checking can be used for equations and inequalities if the solution set is finite, but it cannot be used if the solution set is infinite.

The solution set in the previous example, $\{c: c < -10\}$, is an infinite set. You cannot test every number in the solution set, but errors can usually be found by testing one value in each region of the graph.

Try $c = -11$. Since -11 is in the solution set, the original inequality should be true when $c = -11$.

$$7(c - 2) + 2 > 2(5c + 9)$$
$$7(-11 - 2) + 2 > 2[5(-11) + 9]$$
$$-89 > -92 \qquad \text{True}$$

Try $c = -9$. Since -9 is not in the solution set, the original inequality should be false when $c = -9$.

$$7(c - 2) + 2 > 2(5c + 9)$$
$$7(-9 - 2) + 2 > 2[5(-9) + 9]$$
$$-75 > -72 \qquad \text{False}$$

164 *Chapter 4*

To solve an inequality, you take the same steps used to solve equations.

---

1. Simplify each side of the inequality.
2. Use the inverse operations to undo any indicated additions or subtractions.
3. Use the inverse operations to undo any indicated multiplications or divisions.

Certain inequalities are true for all real numbers, and others have no solution.

EXAMPLE 2 Solve.   **a.** $2x > 2(x + 1)$ **b.** $3y < 3(y + 2)$

SOLUTION   **a.** $2x > 2(x + 1)$
$$2x > 2x + 2$$
$$0 > 2$$

Since the given inequality is equivalent to the false statement $0 > 2$, the inequality has no solution.

∴ the solution set is $\emptyset$.

b. $3y < 3(y + 2)$
$$3y < 3y + 6$$
$$0 < 6$$

Since the given inequality is equivalent to the true statement $0 < 6$, the inequality is satisfied by every real number.

∴ the solution set is $\mathcal{R}$.

Oral Exercises

State the transformation used to transform the first inequality into the second.

1. $b + 5 < 8$
    $b < 3$

2. $z - 6 > 5$
    $z > 11$

3. $w - 7 > -13$
    $w > -6$

4. $p + 12 < -7$
    $p < -19$

5. $8t > 32$
    $t > 4$

6. $3w < -18$
    $w < -6$

7. $\frac{e}{3} < -4$
    $e < -12$

8. $\frac{h}{4} > 2$
    $h > 8$

9. $-5t > 30$
    $t < -6$

10. $-3d < 12$
    $d > -4$

11. $-8x < -4$
    $x > \frac{1}{2}$

12. $-7a > -14$
    $a < 2$

Mixed Review

Solve.
1. $2n + 6 = 20$ $\{7\}$

2. $\dfrac{2a + 6}{10} = -4$ $\{-23\}$

3. $|-6| = a + 1$ $\{5\}$

Replace each ? with one of the symbols $>$, $=$, $<$ so that the statement is true for all real values of the variables.

4. If $a < b$ and $b < c$, then a ? c. $<$

5. If $a < b$, then $a + c$? $b + c$. $<$

Suggested Assignments

Minimum
    166/1–24
Average
    166/1–29 odd, 30–36
Maximum
    166/1–33 odd, 34–40

Additional Answers
Oral Exercises

1. Subt. 5 from each side.
2. Add 6 to each side.
3. Add 7 to each side.
4. Subt. 12 from each side.
5. Divide each side by 8.
6. Divide each side by 3.
7. Mult. each side by 3.
8. Mult. each side by 4.
9. Divide each side by -5 and reverse the order of the inequality.
10. Divide each side by -3 and reverse the order of the inequality.
11. Divide each side by -8 and reverse the order of the inequality.
12. Divide each side by -7 and reverse the order of the inequality.

1. $c - 4 > -20$ $\{c: c > -16\}$

2. $5 < a + 9$ $\{a: a > -4\}$

3. $9 > \frac{a}{4}$ $\{a: a < 36\}$

4. $\frac{2x}{3} < 0$ $\{x: x < 0\}$

5. $\frac{y}{8} + 4 < 10$ $\{y: y < 48\}$

6. $6 > \frac{2 - a}{4}$ $\{a: a > -22\}$

Additional Answers Written Exercises

1. $\{a: a > -1\}$

2. $\{n: n < 6\}$

3. $\{c: c > -6\}$

4. $\{t: t < -6\}$

5. $\{b: b < -11\}$

6. $\{p: p > 7\}$

Written Exercises

Solve each inequality and graph its solution set.

A

1. $a - 13 > -14$ **2.** $n + 11 < 17$ **3.** $2 < c + 8$ **4.** $-15 > t - 9$

5. $11b < -121$ **6.** $15p > 105$ **7.** $4 > \frac{m}{3}$ **8.** $6 < \frac{n}{2}$

9. $\frac{2}{3}t > -4$ **10.** $\frac{3}{5}x < 0$ **11.** $0 > -3w$ **12.** $-5p > -20$

13. $29 < 5c - 6$ **14.** $132 > 7q + 6$ **15.** $3 - d > 16$ **16.** $0 > -4 - r$

17. $23 < 5 - 3e$ **18.** $5 - 6s > 71$ **19.** $-2 < 2 + \frac{f}{3}$ **20.** $\frac{t}{11} + 6 > 6$

21. $\frac{4}{5}u - 3 > 9$ **22.** $16 < 10 - \frac{2}{3}g$ **23.** $8 > \frac{h - 3}{2}$ **24.** $\frac{v + 1}{-3} > -1$

Solve. **28.** $\{x: x > -6\}$ **30.** $\{y: y < 1\}$ **32.** $\{z: z > -12\}$ **34.** $\{m: m < -\frac{5}{4}\}$

B **25.** $5i - 9 < 3 - i$ $\{i: i < 2\}$ **26.** $11 - 2w < 3w - 9$ $\{w: w > 4\}$

27. $3j + 25 > 6j - 23 + j$ $\{j: j < 12\}$ **28.** $12x - 11 - 10x < 9 + 7x + 10$

29. $3k - 7(k + 5) - 5 < 0$ $\{k: k > -10\}$ **30.** $11 + 6(y - 2) - 5y < 0$

31. $-4(t - 4) < 5(t + 3)$ $\{t: t > \frac{1}{9}\}$ **32.** $8(z + 5) > -(20 - 3z)$

33. $7(a + 4) - 13 < 13(3 + a) + 12$ $\{a: a > -6\}$ **34.** $7 - 2(m - 4) < 5(1 - 2m)$

35. $3(1 + 3t) - 7 > 4(3t + 2) - 3t$ $\emptyset$ **36.** $3(d - 8) < 5(d - 1) + 2(10 - d)$

$\mathcal{R}$

C **37.** $5[3(s - 6) - 2(3s - 5)] + 3[3(s + 5) + 8s + 7] < 2(4s + 3)$ $\{s: s < -2\}$

38. $5[3(2 - 3f) - 2(5 - f)] - 6[5(f - 2) - 2(4f - 3)] < 3f + 19$ $\{f: f > -\frac{3}{4}\}$

39. $\frac{2}{3}\left[2(b - 9) - \frac{1}{2}(b - 4)\right] < -\frac{5}{6}\left[3(b + 2) - 2(b + 1)\right]$ $\{b: b < 4\}$

40. $\frac{1}{2}\left[\frac{1}{5}(a - 2) - \frac{2}{5}(a - 3)\right] > \frac{3}{4}\left[2a - \frac{1}{3}(4 + 5a)\right]$ $\{a: a < 4\}$

Computer Exercises   For students with computer experience

Write a program that will solve an inequality of the form $ax + b < c$, $a \neq 0$, when you input values for a, b, and c. Be careful to account for negative values of a. RUN the program to solve each of the following.

1. $4x + 3 < 15$ **2.** $3y - 2 < 16$ **3.** $-2z + 5 < 17$ **4.** $-5w - 2 < 18$
   $\{x: x < 3\}$ $\{y: y < 6\}$ $\{z: z > -6\}$ $\{w: w > -4\}$

Modify the program that you wrote for Exercises 1–4 to solve an inequality of either form $ax + b < c$ or $ax + b > c$, $a \neq 0$. RUN the program to solve each of the following.

5. $4a + 5 < 21$ **6.** $4a + 5 > 21$ **7.** $-2b + 9 < 15$ **8.** $-5c - 7 > 3$
   $\{a: a < 4\}$ $\{a: a > 4\}$ $\{b: b > -3\}$ $\{c: c < -2\}$

166   *Chapter 4*

Modify the program that you wrote for Exercises 5–8 to solve any inequality of the form $ax + b < cx + d$ or $ax + b > cx + d$. Be careful to account for inequalities whose solution set is empty and for inequalities whose solution set is $\mathcal{R}$. RUN the program to solve each of the following.

9. $5m + 1 < 3m + 11$ 10. $2n - 4 > 7n - 9$ 11. $3 + 4r > 9 + 2r$
{n: n < 1}   {r: r > 3}

12. $8 + s < 4 - s$ {m: m < 5}   13. $-4p + 7 < 6 - 4p$ 14. $2q - 9 > 2(q - 5)$
{s: s < -2}   ∅ $\mathcal{R}$

(continued on p. 207)

4–3 Intersection and Union of Sets

Just as you can perform operations on real numbers, you can also perform operations on sets. *Intersection* and *union* are two examples of set operations.

The **intersection** of any two sets S and T is the set consisting of the members belonging to *both* S and T. For example, if

$$S = \{0, 2, 4, 6\} \quad \text{and} \quad T = \{0, 4, 8, 12\},$$

the intersection of S and T is the set $\{0, 4\}$. To indicate the intersection of these sets, you use the symbol ∩ and write

$$S \cap T = \{0, 4\}.$$

The **union** of any two sets S and T is the set consisting of the members belonging to *at least one* of the sets S and T. The union of the sets S and T just given is the set $\{0, 2, 4, 6, 8, 12\}$. To indicate the union of these sets, you use the symbol ∪ and write

$$S \cup T = \{0, 2, 4, 6, 8, 12\}.$$

Recall from Section 1-2 that you can use Venn diagrams to show how sets are related. In the Venn diagrams that follow, shading is used as shown to represent the intersection and union of the sets S and T.

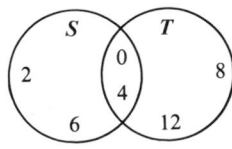

$S \cap T = \{0, 4\}$ $S \cup T = \{0, 2, 4, 6, 8, 12\}$

Sets that have no members in common are called **disjoint sets**. For example, if

$$R = \{1, 3, 5, 7, 9\} \quad \text{and} \quad S = \{0, 2, 4, 6\},$$

R and S are disjoint sets. Note that the intersection of two disjoint sets is the empty set. Thus you can write

$$R \cap S = \emptyset.$$

Solving Inequalities and Problems **167**

---

7. $\{m: m < 12\}$

8. $\{n: n > 12\}$

(continued on p. 207)

Teaching Suggestions p. T80

Related Activities p. T80

Key Ideas

Find the union and intersection of sets.

Chalkboard Examples

1. Let $A = \{-4, -2, 0, 2, 4\}$, $B = \{0\}$, and $C = \{2, 4, 6, 8\}$. Specify the following by roster: $(A \cup B) \cap C$.
$A \cup B =$
   $\{-4, -2, 0, 2, 4\} \cup \{0\}$
   $= \{-4, -2, 0, 2, 4\}$
$(A \cup B) \cap C =$
   $\{-4, -2, 0, 2, 4\} \cap$
   $\{2, 4, 6, 8\} = \{2, 4\}$

2. Let $P = \{$the natural numbers between 1 and 6$\}$ and $Q = \{$the integers between -2 and 4$\}$. Graph each of the following sets.
   a. P

   b. Q

   c. $P \cap Q$

   d. $P \cup Q$

167

When there is more than one set operation in an expression, parentheses are used to indicate which operations are to be performed first.

EXAMPLE 1 Let $A = \{-2, -1, 1, 2\}$, $B = \{0, 1, 3\}$, and $C = \{1, 2, 3, 4\}$. Specify by roster the set $A \cup (B \cap C)$.

SOLUTION First specify $B \cap C$.

$$B \cap C = \{0, 1, 3\} \cap \{1, 2, 3, 4\}$$
$$= \{1, 3\}$$

Then specify $A \cup (B \cap C)$.

$$A \cup (B \cap C) = \{-2, -1, 1, 2\} \cup \{1, 3\}$$
$$= \{-2, -1, 1, 2, 3\}$$

The operations of intersection and union can be performed on infinite sets as well as on finite sets. In the next example, a number line is used to represent the intersection and the union of infinite sets of numbers.

EXAMPLE 2 Let $R = \{$the real numbers between -3 and $3\}$ and $S = \{$the positive real numbers$\}$. Graph each of the following sets.
  a. R **b.** S **c.** $R \cap S$ **d.** $R \cup S$

SOLUTION   **a.**

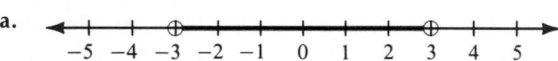

  b.

  c.

  d.

Oral Exercises

Exercises 1–12 refer to the Venn diagram at the right. Specify each set by roster.

1. $A \cap B$ **2.** $B \cap C$ **3.** $A \cup B$
4. $B \cup C$ **5.** $C \cap A$ **6.** $C \cup A$
7. $(A \cap B) \cap C$ **8.** $A \cup (B \cup C)$
9. $(A \cap B) \cup C$ **10.** $A \cap (B \cup C)$
11. $(A \cap C) \cap (B \cap C)$ **12.** $(A \cup B) \cap (B \cap C)$

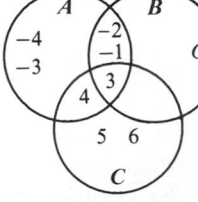

Exs. 1–12

168 *Chapter 4*

Written Exercises

Specify the intersection and the union of the given sets.

A **1.** {-3, -1, 1, 3}, {-1, 0, 1}
  3. {-6, -4, 1}, {2, 4, 6}
  5. {-2, 1, 8, 10}, {-2, 1, 8, 10}
  7. {1, 2, 3, 4}, Ø Ø; {1, 2, 3, 4}

  2. {-10, -6, -2}, {-2, 2, 6}
  4. {2, 3, 4}, {2, 3, 4}
  6. {-1, -3, 5}, {1, 3, -5}
  8. {0}, Ø Ø; {0}

  9. {the even whole numbers}, {2, 4}; {1, 3, the even whole numbers}
    {1, 2, 3, 4}

  10. {the even integers}, Ø; {the integers}
    {the odd integers}

  11. {the integers less than 4}, Ø; {the integers less than 4 and the
    {the integers greater than 6}   integers greater than 6}

  12. {the integers greater than -2}, {the integers greater than -2};
    {the integers greater than -5} {the integers greater than -5}

  13. {the natural numbers less than 5}, {2, 3, 4}; {1, 2, 3, 4, 5}
    {the natural numbers between 1 and 6}

  14. {the whole numbers less than or equal to 3}, {1, 2, 3}; {0, 1, 2, 3, 4, 5}
    {5 and the natural numbers less than 5}

  15. {the whole numbers less than 8}, {0, 1, 2, 3, 4, 5, 6}; {0, 1, 2, 3, 4, 5, 6, 7}
    {the whole numbers less than or equal to 6}

  16. {the negative integers greater than -5}, {-1}; {-4, -3, -2, -1, 0, 1}
    {the integers between -2 and 2}

Graph each of the following when R and S are the given sets.

  a. $R \cup S$ **b.** $R \cap S$

B **17.** R = {the real numbers between -2 and 6},
    S = {the positive real numbers}

  18. R = { the real numbers between -5 and 5},
    S = {the negative real numbers}

  19. R = {the real numbers greater than $-\frac{1}{2}$},
    S = {the real numbers less than $2\frac{1}{2}$}

  20. R = {the real numbers greater than or equal to 4.5},
    S = {the real numbers less than or equal to 5.5}

  21. R = {the real numbers greater than or equal to 4},
    S = {the real numbers less than -2}

  22. R = {the real numbers greater than -5},
    S = {the real numbers less than -5}

  23. R = {the real numbers greater than -3},
    S = {the real numbers greater than -6}

  24. R = {the real numbers less than 3},
    S = {the real numbers less than -1}

Solving Inequalities and Problems **169**

6. Ø; {-5, -3, -1, 1, 3, 5}

17. a.
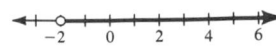
  b.
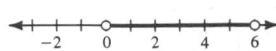

18. a.
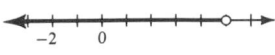
  b.
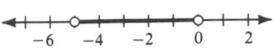

19. a.
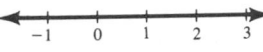
  b.
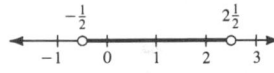

20. a.
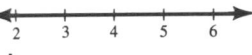
  b.
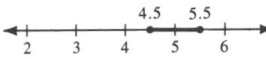

21. a.
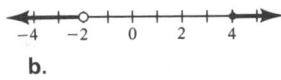
  b.

22. a.
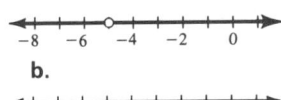
  b.

23. a.
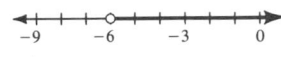
  b.
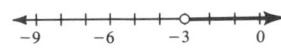

24. a.
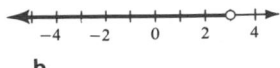
  b.
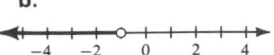

Additional Answers
Written Exercises
(continued on p. 208)

Mixed Review

1. Specify by roster {the months having only 30 days}.
   {April, June, September, November}
2. Specify by rule {0, 1, 2, 3}.
   {the whole numbers less than 4}
3. Draw a Venn diagram to illustrate $B \subset A$.

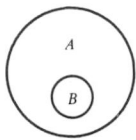

4. Draw a Venn diagram to illustrate $A \subset B \subset C$.

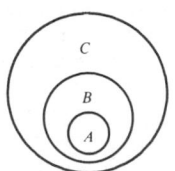

Teaching Suggestions
p. T80

Key Ideas

Solve conjunctions and disjunctions.

In each exercise, copy the Venn diagram shown at the right. Then shade the region that represents the given set.

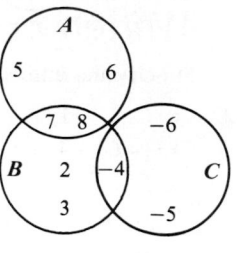

Exs. 25–34

25. $A \cap (B \cup C)$
26. $(A \cap B) \cup C$
27. $(A \cup C) \cap B$
28. $A \cup (C \cap B)$
29. $A \cup (B \cap C)$
30. $C \cap (A \cup B)$
31. $(A \cup B) \cap (A \cup C)$
32. $(C \cap A) \cup (C \cap B)$
33. $(A \cap B) \cup (A \cap C)$
34. $(A \cup B) \cap (A \cap C)$

Replace each __?__ with one of the symbols R, S, T, or $\emptyset$ to make a true statement.

35. If $R \subset S$, then $R \cap S = $ __?__ . R
36. If $T \subset S$, then $S \cup T = $ __?__ . S
37. If R and S are disjoint sets and $T \subset S$, then $R \cap T = $ __?__ . $\emptyset$
38. If $R \subset S$ and $S \subset T$, then $(R \cup S) \cup T = $ __?__ . T

Let A, B, and C be subsets of $U = \{0, 1, 2, 3, 4, 5, 6\}$. Given the following information about sets A, B, and C, specify B by roster.

C   39. $A = \{1, 2\}$, $A \cap B = \{1\}$, $A \cup B = \{1, 2, 3, 6\}$ {1, 3, 6}
40. $A = \{1, 2, 3, 6\}$, $A \cap B = \{1, 6\}$, $A \cup B = \{1, 2, 3, 5, 6\}$ {1, 5, 6}
41. $A \cap A = \emptyset$, $C \cup A = \{6\}$, $B \cup C = U$, $B \cap C = \emptyset$ {0, 1, 2, 3, 4, 5}
42. $A = \{3, 5\}$, $B \cap C = \{3\}$, $B \cup C = \{1, 2, 3, 4, 6\}$,
    $A \cup C = \{1, 3, 4, 5\}$ {2, 3, 6}
43. $A \cap C = \{0, 4\}$, $A \cap B = \{0, 3\}$, $B \cap C = \{0, 6\}$,
    $A \cup C = \{0, 1, 2, 3, 4, 6\}$, $B \cup C = \{0, 2, 3, 4, 5, 6\}$ {0, 3, 5, 6}
44. $A \cup B = \{1, 2, 3, 4, 5\}$, $A \cup C = \{1, 3, 4, 5\}$,
    $A \cap C = A$, $A \cap B = \emptyset$, $B \cap C = \{3, 5\}$ {2, 3, 5}

4–4 Combined Inequalities

A sentence formed by joining two sentences with the word *and* is called a conjunction. An example of a conjunction is

$$-3 < x \qquad \text{and} \qquad x < 4.$$

For a conjunction to be true, *both* of the joined sentences must be true. Therefore, the solution set of the conjunction $-3 < x$ and $x < 4$ is the *intersection* of the solution set of $-3 < x$ and the solution set of $x < 4$. That is,

$$\{x: -3 < x\} \cap \{x: x < 4\} = \{x: -3 < x < 4\},$$

as shown by the diagram on the following page.

170   *Chapter 4*

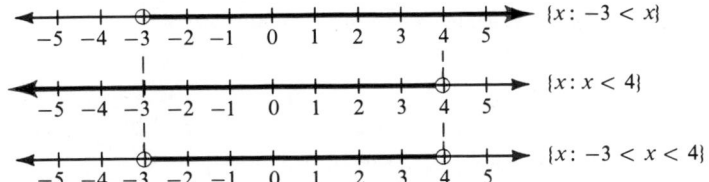

A sentence formed by joining two sentences with the word *or* is called a **disjunction**. An example of a disjunction is

$$y > -2 \qquad \text{or} \qquad y > 3.$$

For a disjunction to be true, *at least one* of the joined sentences must be true. Therefore, the solution set of the disjunction $y > -2$ or $y > 3$ is the *union* of the solution set of $y > -2$ and the solution set of $y > 3$. That is,

$$\{y: y > -2\} \cup \{y: y > 3\} = \{y: y > -2\},$$

as shown by the following diagram.

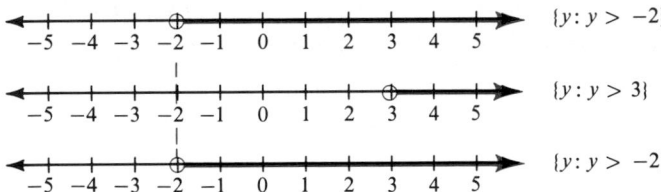

EXAMPLE 1   Solve.

        a. $x < -2$ and $x \geq 5$ **b.** $y < 3$ or $y \geq -1$

SOLUTION   **a.** Since the combined inequality is a *conjunction*, the solution set consists of all those numbers that satisfy *both* the joined sentences. However, there is no real number that satisfies both $x < -2$ and $x \geq 5$. ∴ the solution set is ∅.

               b. Since the combined inequality is a *disjunction*, the solution set consists of all those numbers that satisfy *at least one* of the joined sentences. However, every real number satisfies either $y < 3$ or $y \geq -1$. ∴ the solution set is $\mathcal{R}$.

Can you describe the solution set of the open sentence

$$-2 \leq x + 3 < 7?$$

This open sentence is equivalent to the conjunction

$$-2 \leq x + 3 \qquad \text{and} \qquad x + 3 < 7.$$

Therefore, to solve the given open sentence you must find the values of x that satisfy both of the inequalities $-2 \leq x + 3$ and $x + 3 < 7$. Example 2 shows two methods of solving this open sentence.

Solving Inequalities and Problems   **171**

1. Solve $2 < x - 4 < 5$ and graph its solution set.
$2 < x - 4$ and $x - 4 < 5$
$6 < x$ and $x < 9$
∴ the solution set is $\{x: 6 < x < 9\}$.

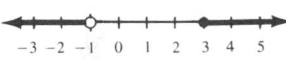

2. Solve the open sentence $x + 1 \geq 4$ or $2x - 10 < -12$ and graph its solution set.
$x + 1 \geq 4$ or $2x - 10 < -12$
  $x \geq 3$ $2x < -2$
         or $x < -1$
∴ the solution set is $\{x: x \geq 3$ or $x < -1\}$.

Reading Algebra

Be sure that the students read conjunctions and disjunctions properly. Have them read the exercises aloud to show that they grasp the true meaning of the statements. For $-9 < 2d - 1 \leq 15$ the students should say, "two times d minus one is greater than negative nine and less than or equal to fifteen." They could also read this open sentence as the conjunction: $-9 < 2d - 1$ and $2d - 1 \leq 15$, that is, "two times d minus one is greater than negative nine *and* two times d minus one is less than or equal to fifteen."

Suggested Assignments

Minimum
  173/1–24
S 166/25–27
Average
  173/1–31 odd, 32–38
S 166/26, 28, 37
Maximum
  173/1–37 odd, 39–46
S 166/26–32 even

Additional A Exercises

Solve each open sentence and graph its solution set.

1. $a > 0$ and $a \leq 4$
$\{a: 0 < a \leq 4\}$

2. $b < 2$ or $b > 2$
$\{b: b < 2 \text{ or } b > 2\}$

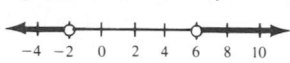

3. $4t < -8$ or $t > 6$
$\{t: t < -2 \text{ or } t > 6\}$

4. $2m > 4$ and $m - 2 > 2$
$\{m: m > 4\}$

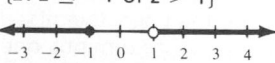

5. $-2 \leq x + 3 \leq 8$
$\{x: -5 \leq x \leq 5\}$

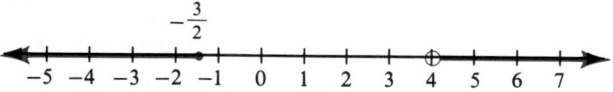

6. $2z + 1 > 3$ or $3z - 1 \leq -4$
$\{z: z \leq -1 \text{ or } z > 1\}$

EXAMPLE 2 Solve $-2 \leq x + 3 < 7$ and graph its solution set.

SOLUTION 1

$$
\begin{array}{ccc}
-2 \leq x + 3 & \text{and} & x + 3 < 7 \\
-2 - 3 \leq x + 3 - 3 & \big| & x + 3 - 3 < 7 - 3 \\
-5 \leq x & \text{and} & x < 4
\end{array}
$$

$\therefore$ the solution set is $\{x: -5 \leq x < 4\}$.

SOLUTION 2 It is possible to solve $-2 \leq x + 3 < 7$ more compactly as follows.

$$-2 \leq x + 3 < 7$$
$$-2 - 3 \leq x + 3 - 3 < 7 - 3$$
$$-5 \leq x < 4$$

Again, the solution set is $\{x: -5 \leq x < 4\}$.

EXAMPLE 3 Solve the open sentence $2z - 3 \leq -6$ or $3z - 12 > 0$ and graph its solution set.

SOLUTION

$$
\begin{array}{ccc}
2z - 3 \leq -6 & \text{or} & 3z - 12 > 0 \\
2z \leq -3 & \big| & 3z > 12 \\
z \leq -\dfrac{3}{2} & \text{or} & z > 4
\end{array}
$$

$\therefore$ the solution set is $\left\{z: z \leq -\dfrac{3}{2} \text{ or } z > 4\right\}$.

Note that the familiar open sentence $y \geq 3$ is an example of a disjunction since it means $y > 3$ or $y = 3$. Similarly, $y \leq 5$ is a disjunction since it means $y < 5$ or $y = 5$.

Oral Exercises

Replace each ? with the word or phrase that makes a true statement.

1. In a conjunction, sentences are joined by the word ? . and
2. In a disjunction, sentences are joined by the word ? . or
3. A conjunction of two sentences is true provided ? of the joined sentences (is/are) true. both
4. A disjunction of two sentences is true provided ? of the joined sentences (is/are) true. at least one

172 *Chapter 4*

Tell whether each statement is true or false.

5. $5 < 9$ or $0 > -1$ True
6. $4 < 10$ and $6 < 11$ True
7. $-5 > -4$ and $3 < 2$ False
8. $5 > 6$ or $-2 = 2$ False
9. $-2 \le -1$ or $0 \le -1$ True
10. $0 < -5$ and $6 > 5$ False
11. $5.6 > 0.3$ and $0.33 < 0.3$ False
12. $3.4 < 0$ or $1.6 > 0$ True

Match each open sentence with the graph of its solution set.

13. $x \ge -1$ or $x < 3$ c

a.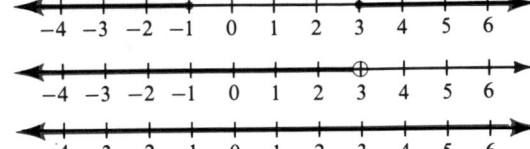

14. $x \ge -1$ and $x < 3$ f

b.

15. $x \le -1$ or $x \ge 3$ a

c.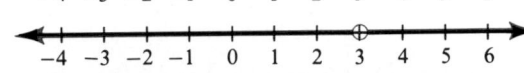

16. $x \le -1$ and $x \ge 3$ e

d.

17. $x < 3$ or $x \le -1$ b

e.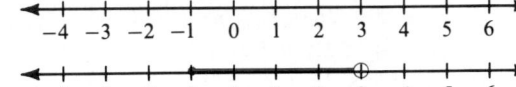

18. $x < 3$ and $x \le -1$ h

f.

19. $x < 3$ or $x > 3$ d

g.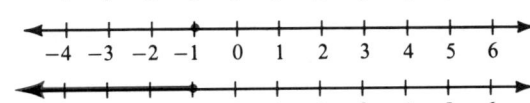

20. $x \le -1$ and $x \ge -1$ g

h.

Written Exercises

Solve each open sentence and graph each solution set that is not empty.

A
1. $a < -3$ or $a > 1$
2. $m < 5$ and $m > -3$
3. $e \ge 2$ and $e < -1$
4. $n > -2$ or $n < 6$
5. $t > -4$ or $t < 1$
6. $r > 4$ and $r < 0$
7. $z > -1$ and $z < 2$
8. $t < 3$ or $t > 5$
9. $w > -3$ and $w > 0$
10. $p > -2$ or $p > 1$
11. $-7 \le x + 5 < 2$
12. $3 < c - 3 < 5$
13. $z + 1 > 2$ or $z + 1 < -6$
14. $5x < -20$ and $3x > -18$
15. $2a > -6$ and $3a \le 15$
16. $20 \ge 4c \ge -2$
17. $-10 \le 2t \le 12$
18. $3a + 1 > -8$ and $a - 4 \le 2$
19. $2x - 1 < 5$ and $3x + 2 > -4$
20. $-21 \le 4n - 5 < -1$
21. $15 \ge 2d - 1 > -9$
22. $3z - 5 < -1$ or $3z - 5 > -4$
23. $-6 < \frac{1}{2}r - 4 < -3$
24. $-3 \le \frac{2}{3}t - 5 < -1$

Solving Inequalities and Problems **173**

Additional Answers
Written Exercises

1. $\{a: a < -3 \text{ or } a > 1\}$

2. $\{m: -3 < m < 5\}$

3. ∅

4. $\mathcal{R}$

5. $\mathcal{R}$

6. ∅

7. $\{z: -1 < z < 2\}$

8. $\{t: t < 3 \text{ or } t > 5\}$

9. $\{w: w > 0\}$

10. $\{p: p > -2\}$

11. $\{x: -12 \le x < -3\}$

12. $\{c: 6 < c < 8\}$

13. $\{z: z > 1 \text{ or } z < -7\}$

14. $\{x: -6 < x < -4\}$

(continued on p. 209)

Mixed Review

1. Solve $2x - 8 \leq 20$.
   $\{x: x \leq 14\}$
2. Solve $8 - 3x < -4$.
   $\{x: x > 4\}$
3. Specify $M \cup N$ if $M = \{0, 3, 5\}$ and $N = \{1, 7, 9\}$.
   $M \cup N = \{0, 1, 3, 5, 7, 9\}$
4. Specify $P \cap Q$ if $P = \{$the natural numbers less than 5$\}$ and $Q = \{$the integers greater than $-4\}$.
   $P \cap Q = \{1, 2, 3, 4\}$

Solve. **27.** $\{h: 2 < h < 6\}$ **29.** $\{y: y < \frac{5}{2}\}$ **30.** $\{x: x \geq \frac{4}{3}$ or $x < -2\}$

B **25.** $-2 < -3k \leq 3$ $\{k: -1 \leq k < \frac{2}{3}\}$ **26.** $-4 \leq -2t \leq -1$ $\{t: \frac{1}{2} \leq t \leq 2\}$
   27. $5 - 2h < 1$ and $5 - 2h > -7$ **28.** $5 - r < -6$ and $5 - r \geq -2$ $\emptyset$
   29. $3 - 2y > -2$ or $3y + 6 < 8$ **30.** $6 - 3x \leq 2$ or $2 - x > 4$
   31. $-8 < 2 - 3g < 10$ $\{g: -\frac{8}{3} < g < \frac{10}{3}\}$ **32.** $5 \geq -2t + 1 \geq -3$ $\{t: -2 \leq t \leq 2\}$
   33. $6 \geq 3 - \frac{1}{2}x \geq -2$ $\{x: -6 \leq x \leq 10\}$ **34.** $7 \leq 1 - \frac{3}{5}z \leq 8$ $\{z: -\frac{35}{3} \leq z \leq -10\}$

   35. $7d > 2d + 10$ and $2d > 5d - 15$ $\{d: 2 < d < 5\}$
   36. $4t - 3 > t - 9$ and $6 - 5t > 1 - 3t$ $\{t: -2 < t < \frac{5}{2}\}$
   37. $4j + 3 > 13 - j$ or $16 - 3j \geq -8$ $\mathcal{R}$
   38. $26 - 3v > v - 14$ or $1 - 8v < 31 - 5v$ $\mathcal{R}$

C **39.** $2e - 3 < 3e + 1 < 2e + 2$ $\{e: -4 < e < 1\}$
   40. $8 - 4v \geq 7 - 5v \geq 5 - 4v$ $\{v: -1 \leq v \leq 2\}$
   41. $6 - 3x \geq 4 - 2x > 5 - 4x$ $\{x: \frac{1}{2} < x \leq 2\}$
   42. $3t - 5 \leq 7t - 1 \leq 6t + 5$ $\{t: -1 \leq t \leq 6\}$
   43. $2d - 4 < 2d + 5$ and $2d + 1 > 2d - 4$ $\mathcal{R}$
   44. $3p - 5 > 3p + 6$ or $3p - 5 < 3p - 6$ $\emptyset$

Let $A = \{x: x \geq -2\}$, $B = \{x: x < 5\}$, and $C = \{x: -3 \leq x < 3\}$. Specify each of the following sets.

   45. $A \cap (B \cap C)$ $\{x: -2 \leq x < 3\}$ **46.** $A \cup (B \cup C)$ $\mathcal{R}$
   47. $(A \cup B) \cap (A \cup C)$ $\{x: x \geq -3\}$ **48.** $(B \cap C) \cup (B \cap A)$ $\{x: -3 \leq x < 5\}$

4–5 Absolute Values in Open Sentences

In Section 2-4, you saw that the expression $|x|$ could be interpreted as the distance between the graph of x and the origin on a number line. Similarly, the expression $|a - b|$ can be interpreted as the distance between the graphs of a and b. Since the distance from the graph of a to the graph of b is the same as the distance from the graph of b to the graph of a, you can write

$$|a - b| = |b - a|.$$

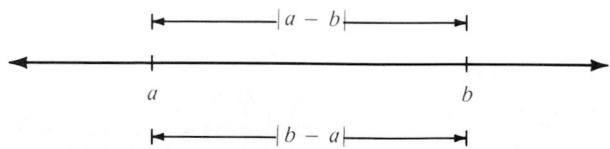

Note that, for simplicity, you usually speak of the distance between a

174   *Chapter 4*

Teaching Suggestions
p. T81

Supplementary Material
Test 13

Key Ideas

Solve open sentences involving absolute value.

and b instead of the distance between the graphs of a and b.

The following example shows how the relationship between absolute value and distance can be used in solving open sentences involving absolute value.

EXAMPLE 1   Solve.

a. $|x - 3| = 4$ **b.** $|x - 3| > 4$ **c.** $|x - 3| < 4$ **d.** $|x + 3| \geq 4$

SOLUTION   **a.** $|x - 3| = 4$

The distance between x and 3 is 4 units.

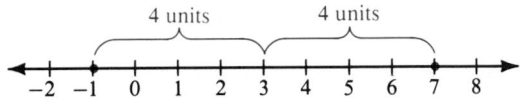

$\therefore$ the solution set is $\{-1, 7\}$.

b. $|x - 3| > 4$

The distance between x and 3 is more than 4 units.

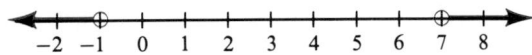

$\therefore$ the solution set is $\{x: x < -1 \text{ or } x > 7\}$

c. $|x - 3| < 4$

The distance between x and 3 is less than 4 units.

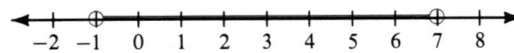

$\therefore$ the solution set is $\{x: -1 < x < 7\}$.

d. $|x + 3| \geq 4$ is equivalent to $|x - (-3)| \geq 4$.

The distance between x and -3 is 4 units or more than 4 units.

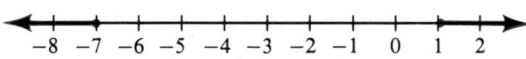

$\therefore$ the solution set is $\{x: x \geq 1 \text{ or } x \leq -7\}$.

Some open sentences involving absolute value can be expressed as conjunctions; others can be expressed as disjunctions. For example:

$|2x + 7| = 9$ is equivalent to $2x + 7 = 9$ or $2x + 7 = -9$.
$|2x + 7| > 9$ is equivalent to $2x + 7 > 9$ or $2x + 7 < -9$.
$|2x + 7| < 9$ is equivalent to $-9 < 2x + 7 < 9$.

Often you use an equivalent conjunction or disjunction in solving an open sentence involving absolute value.

Solving Inequalities and Problems   **175**

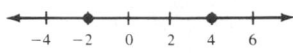
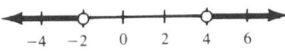

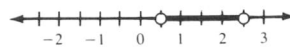

3. Solve $6 + 2|4a - 3| > 20$
and graph the solution
set.

$$6 + 2|4a - 3| > 20$$
$$2|4a - 3| > 14$$
$$|4a - 3| > 7$$

$4a - 3 > 7$ or $4a - 3 < -7$
$\quad 4a > 10 \quad | \quad\quad 4a < -4$

$\quad a > 2\frac{1}{2}$ or $a < -1$

∴ the solution set is

$\left\{ a: a < -1 \text{ or } a > 2\frac{1}{2} \right\}.$

Suggested Assignments

Minimum
    177/1–10, 12–30 even
R 178/Self-Test 1
Average
    177/1–55 odd
R 178/Self-Test 1
Maximum
    177/5–65 odd
R 178/Self-Test 1

Additional A Exercises

Write an equation or inequality involving absolute value to describe the set of all real numbers x that are at the given location on a number line.

1. 4 units from 6
$\quad |x - 6| = 4$

2. at least 8 units from 1
$\quad |x - 1| \geq 8$

3. 6 units from -7
$\quad |x + 7| = 6$

EXAMPLE 2   Solve $|1 - 3y| < 4$ and graph the solution set.

SOLUTION $|1 - 3y| < 4$ is equivalent to $-4 < 1 - 3y < 4$.

$$-4 < 1 - 3y < 4$$
$$-4 - 1 < 1 - 3y - 1 < 4 - 1$$
$$-5 < -3y < 3$$
$$\frac{-5}{-3} > \frac{-3y}{-3} > \frac{3}{-3}$$
$$\frac{5}{3} > y > -1$$

∴ the solution set is $\left\{ y: \frac{5}{3} > y > -1 \right\}$.

Another way of expressing this solution is $\left\{ y: -1 < y < \frac{5}{3} \right\}$.

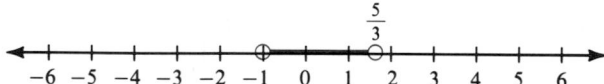

Since $|1 - 3y| = |3y - 1|$, the inequality in Example 2 can also be solved by finding the solution set of the equivalent inequality

$$|3y - 1| < 4.$$

You may find it easier to solve $|3y - 1| < 4$ than the original inequality.

EXAMPLE 3   Solve $2 + 3|2z - 1| > 8$ and graph the solution set.

SOLUTION   First transform the inequality into an equivalent inequality with the absolute value term alone as one side.

$$2 + 3|2z - 1| > 8$$
$$3|2z - 1| > 6$$
$$|2z - 1| > 2$$

Then solve the equivalent inequality.

$2z - 1 > 2$ or $2z - 1 < -2$
$\quad 2z > 3 \quad\quad | \quad\quad 2z < -1$

$\quad z > \frac{3}{2}$ or $z < -\frac{1}{2}$

∴ the solution set is $\left\{ z: z > \frac{3}{2} \text{ or } z < -\frac{1}{2} \right\}.$

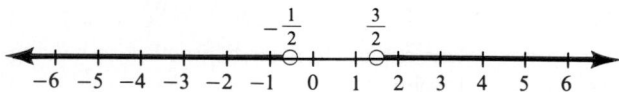

Oral Exercises

Replace each __?__ with one of the words *and* or *or* to make a true statement.

1. $|d| = 4$ is equivalent to $d = -4$ __?__ $d = 4$. or
2. $|m| > 5$ is equivalent to $m < -5$ __?__ $m > 5$. or
3. $|w| \leq 3$ is equivalent to $-3 \leq w$ __?__ $w \leq 3$. and   and
4. $|c| - 2 < 1$ is equivalent to $|c| < 3$, which is equivalent to $-3 < c$ __?__ $c < 3$.

Match each open sentence with its solution set.

5. $|x + 4| \leq 6$ d   a. $\emptyset$
6. $|x - 4| \leq 6$ e   b. $\mathcal{R}$
7. $|4 - x| > 6$ f   c. $\{x: -2 < x < 10\}$
8. $|4 - x| < 6$ c   d. $\{x: -10 \leq x \leq 2\}$
9. $|x + 4| \geq -6$ b   e. $\{x: -2 \leq x \leq 10\}$
10. $|x + 4| \leq -6$ a   f. $\{x: x > 10 \text{ or } x < -2\}$

State a conjunction or a disjunction equivalent to each open sentence.

$x - 3 = 5 \text{ or } x - 3 = -5$ $m + 1 = 8 \text{ or } m + 1 = -8$
11. $|x - 3| = 5$ 12. $|m + 1| = 8$ 13. $|y| > 3$
14. $|n| \leq 2$ $-2 \leq n \leq 2$ 15. $|u - 4| < 2$ $-2 < u - 4 < 2$ 16. $|r - 1| \geq 3$
                                   13. $y > 3 \text{ or } y < -3$ 16. $r - 1 \geq 3 \text{ or } r - 1 \leq -3$

Written Exercises

Write an equation or inequality involving absolute value to describe the set of all real numbers x that are at the given location on a number line.

A
1. 2 units from 5 $|x - 5| = 2$ 2. 3 units from -2 $|x + 2| = 3$
3. less than 3 units from -1 $|x + 1| < 3$ 4. more than 4 units from 7 $|x - 7| > 4$
5. no more than 6 units from 1 $|x - 1| \leq 6$ 6. no less than 2 units from -4 $|x + 4| \geq 2$
7. at least 5 units from 10 $|x - 10| \geq 5$ 8. 6 units from a $|x - a| = 6$
9. more than b units from 5 $|x - 5| > b$ 10. less than or equal to a units from b
                                                            $|x - b| \leq a$

Solve each open sentence and graph each solution set that is not empty.

11. $|t - 4| = 6$ 12. $|x - 5| = 3$ 13. $|a + 1| = 2$ 14. $|w + 5| = 6$
15. $|v - 6| \leq 6$ 16. $|p - 2| \geq 4$ 17. $|c - 8| > -2$ 18. $|t - 4| < -4$
19. $|10 - v| \leq 2$ 20. $|1 - r| \geq 3$ 21. $|1 - x| > 7$ 22. $|-p + 4| < 10$
23. $|5v| \geq 15$ 24. $\left|\frac{1}{2}t\right| \leq 1$ 25. $\left|-\frac{1}{3}w\right| \leq 2$ 26. $|-6x| > -3$
27. $|2w - 7| \geq 3$ 28. $|3q - 1| < 5$ 29. $|6 - 3c| \geq 2.1$ 30. $|4 - 2p| \leq -6.4$
31. $|5 - 2a| \geq -5$ 32. $|1 - 3x| < 10$ 33. $\left|\frac{1}{2}x + 3\right| < \frac{5}{2}$ 34. $\left|\frac{2}{3}t - 4\right| > \frac{1}{6}$

Solving Inequalities and Problems **177**

Solve.
4. $|m - 3| = 7$ $\{-4, 10\}$
5. $|u + 1| < 3$
   $\{u: -4 < u < 2\}$
6. $|3 - 2v| \geq 7$
   $\{v: v \leq -2 \text{ or } v \geq 5\}$

Additional Answers
Written Exercises

11. $\{-2, 10\}$
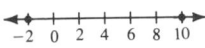

12. $\{2, 8\}$
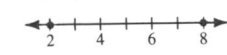

13. $\{-3, 1\}$
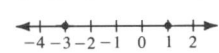

14. $\{-11, 1\}$
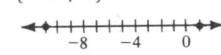

15. $\{v: 0 \leq v \leq 12\}$
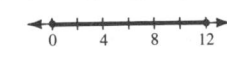

16. $\{p: p \leq -2 \text{ or } p \geq 6\}$
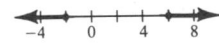

17. $\mathcal{R}$
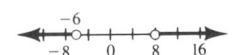

18. $\emptyset$

19. $\{v: 8 \leq v \leq 12\}$

20. $\{r: r \leq -2 \text{ or } r \geq 4\}$

21. $\{x: x < -6 \text{ or } x > 8\}$

22. $\{p: -6 < p < 14\}$

(continued on p. 211)

Additional Answers
Written Exercises
(continued on p. 212)

Mixed Review

1. If $x = 4$, what is the meaning of $|x|$ on a number line? The distance between 0 and 4, or 4 units.

2. If $y = -3$, what is the meaning of $|y|$ on a number line? The distance between 0 and -3, or 3 units.

3. Solve $\dfrac{b+1}{2} > -3$.
$\{b: b > -7\}$

4. Solve $-8 \le -2a \le 4$.
$\{a: -2 \le a \le 4\}$

Additional Answers
Self-Test 1
(See p. 213.)

Quick Quiz

Give the reason that justifies each statement.

1. If $2 - 3x > 9$, then $(2 - 3x) - 6 > 9 - 6$.
Addition property of order

2. If $-2x > 18$, then $x < -9$.
Multiplication property of order

Solve.

3. $4a + 6 > -3$
$\left\{a: a > -2\dfrac{1}{4}\right\}$

4. $-2b \le 6$
$\{b: b \ge -3\}$

178

Solve.

B 35. $5 + |3 - c| \ge 8$ $\{c: c \le 0 \text{ or } c \ge 6\}$ 36. $3 + |c - 1| \le 12$ $\{c: -8 \le c \le 10\}$
37. $13 + |4 - z| \le 9$ $\emptyset$ 38. $12 + |2 - k| > 8$ $\mathcal{R}$
39. $6 - |2c + 3| > -4$ $\{c: -\frac{13}{2} < c < \frac{7}{2}\}$ 40. $5 - |4k - 1| \le -7$
41. $3|x - 6| - 4 \le 11$ $\{x: 1 \le x \le 11\}$ 42. $2|3a - 5| + 1 > 7$ $\{a: a < \frac{2}{3} \text{ or } a > \frac{8}{3}\}$
43. $|8 - (3 - d)| - 5 > -1$ 44. $|5 - (v - 3)| + 8 < 15$
40. $\{k: k \le -\frac{11}{4} \text{ or } k \ge \frac{13}{4}\}$ 43. $\{d: d < -9 \text{ or } d > -1\}$ 44. $\{v: 1 < v < 15\}$

Write an equation or inequality involving absolute value whose solution set is the set given.

45. $\{-2, 2\}$ 46. $\{-6, 6\}$ 47. $\emptyset$ 48. $\mathcal{R}$
49. $\{5, 9\}$ 50. $\{6, 12\}$ 51. $\{-8, -12\}$ 52. $\{-8, 2\}$

53. $\{x: -4 < x < 4\}$ 54. $\{x: x > 2 \text{ or } x < -2\}$
55. $\{x: x > 5 \text{ or } x < -1\}$ 56. $\{x: -3 < x < 11\}$

Specify all the real numbers x for which each statement is true.

C 57. a. $|x + 3| < x + 3$ b. $|x + 3| = x + 3$ c. $|x + 3| > x + 3$
58. a. $|x - 2| < x - 2$ b. $|x - 2| = x - 2$ c. $|x - 2| > x - 2$
59. a. $|x^2| = x^2$ b. $|-x^2| = x^2$ c. $|-x^3| = -x^3$
60. a. $|-3x| = -3x$ b. $|-3x| = -3|x|$ c. $-|3x| = -3x$

61. a. $|-2x + 3| < |-2x| + 3$ 62. a. $|-2x - 3| < |-2x| - 3$
b. $|-2x + 3| = |-2x| + 3$ b. $|-2x - 3| = |-2x| - 3$
c. $|-2x + 3| > |-2x| + 3$ c. $|-2x - 3| > |-2x| - 3$

Solve. *Hint*: Using a number line may be helpful.

63. $3 < |d| < 5$ 64. $1 < |c| < 6$ 65. $-5 < |a| < -1$ 66. $-4 < |t| < 4$

Self-Test 1

VOCABULARY   axiom of comparison (p. 157)   equivalent inequalities (p. 163)
transitive property of order   intersection (p. 167)
(p. 158)   union (p. 167)
addition property of order   disjoint sets (p. 167)
(p. 158)   conjunction (p. 170)
multiplication property of   disjunction (p. 171)
order (p. 159)

Give the reason that justifies each statement, if c is a real number.

Addition property of order
1. If $3 - 2c < 17$, then $(3 - 2c) - 3 < 17 - 3$.   *Obj. 1, p. 157*
2. If $-2c < 14$, then $c > -7$. Multiplication property of order

178 *Chapter 4*

Solve.

3. $2x + 3 > -5$
{x: x > -4}

4. $5 - 4y \le 9$
{y: y \ge -1}

Obj. 2, p. 157

Let A = {the integers between −5 and 5} and B = {the whole numbers less than or equal to 6}. Specify each set.

5. $A \cap B$
{the whole numbers less than 5}

6. $A \cup B$
{the integers greater than −5 and less than or equal to 6}

Obj. 3, p. 157

Solve each open sentence and graph its solution set.

7. $-7 \le t + 3 < 5$

8. $m - 2 > 3$ or $m + 5 < -6$

Obj. 4, p. 157

9. $|z - 1| = 6$

10. $|r + 4| = 2$

Obj. 5, p. 157

11. $|p - 5| > 3$

12. $|2x - 3| \le 6$

Check your answers with those at the back of the book.

Inequalities and Equations in Solving Problems

OBJECTIVES for Sections 4-6 through 4-10:
1. To solve word problems involving consecutive integers, consecutive multiples, angle relationships, uniform motion, and mixtures.
2. To solve word problems involving inequalities.

4–6 Problems about Integers

If you count by ones from any given integer, you obtain **consecutive integers**. The following are examples of consecutive integers.

five consecutive integers	3, 4, 5, 6, 7
four consecutive integers	−2, −1, 0, 1
three consecutive integers when x is the least integer	$x, x + 1, x + 2$
three consecutive integers when x is the middle integer	$x - 1, x, x + 1$
three consecutive integers when x is the greatest integer	$x - 2, x - 1, x$

Sometimes an inequality is needed to translate a word problem into the language of algebra. Consider, for instance, the following problem about consecutive integers.

Solving Inequalities and Problems **179**

Let A = {the integers between −2 and 4} and B = {the whole numbers less than 7}. Specify each set.

5. $A \cup B$
{−1, 0, 1, 2, 3, 4, 5, 6}

6. $A \cap B$
{0, 1, 2, 3}

Solve each open sentence and graph its solution set.

7. $-4 \le x + 4 < 12$
{x: −8 ≤ x < 8}

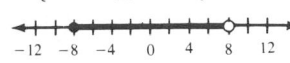

8. $v - 3 < 7$ or $v + 5 > 17$
{v: v < 10 or v > 12}

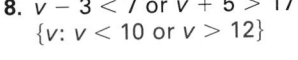

9. $|c + 2| = 3$ {1, −5}

10. $|q - 4| > 4$
{q: q < 0 or q > 8}

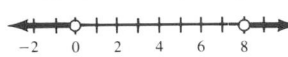

11. $|2x - 1| \le 7$
{x: −3 ≤ x ≤ 4}

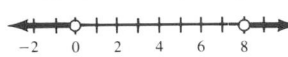

Teaching Suggestions
p. T81

Related Activities p. T81

Key Ideas
Find consecutive integers.

1. Find three consecutive odd integers such that the sum of the first and third integers is 42.
Let x = the first of the three odd integers. Then $x + 2$ = the second integer; $x + 4$ = the third integer.

$$x + (x + 4) = 42$$
$$2x + 4 = 42$$
$$2x = 38$$
$$x = 19$$

Thus $x = 19$, $x + 2 = 21$, and $x + 4 = 23$.
∴ the solution set is {19, 21, 23}.

2. Find all sets of five consecutive positive integers such that the greatest integer in the set is more than four times the least integer in the set.
Let x = the least of the five positive integers. Then $x + 1$ = the next greater integer; $x + 2$ = the next integer; $x + 3$ = the next integer; $x + 4$ = the greatest integer.

$$x + 4 > 4x$$
$$4 > 3x$$
$$1\frac{1}{3} > x$$

Since x must be a positive integer, the only choice for x is 1. Thus $x = 1$, $x + 1 = 2$, $x + 2 = 3$, $x + 3 = 4$, and $x + 4 = 5$.
∴ the required set is {1, 2, 3, 4, 5}.

EXAMPLE Find all sets of four consecutive positive integers such that the greatest integer in the set is more than twice the least integer in the set.

SOLUTION

Step 1   The problem asks for sets of four consecutive integers that are positive. The greatest integer in the set is more than twice the least. All such sets are to be found.

Step 2   Let x = the least of the four positive integers. Then:
$x + 1$ = the next greater integer,
$x + 2$ = the next integer,
$x + 3$ = the greatest integer

Step 3   The greatest integer is more than twice the least integer.

$$x + 3 \qquad > \qquad 2x$$

Step 4 $x + 3 > 2x$
$3 > x$

The problem states that the integers must be positive. Since $x < 3$, the only possible choices for x are 1 and 2.
Thus, the only possible sets of four consecutive positive integers are $A = \{1, 2, 3, 4\}$ and $B = \{2, 3, 4, 5\}$.

Step 5   In each set, is the greatest integer more than twice the least?

In A: $4 \overset{?}{>} 2(1)$ In B: $5 \overset{?}{>} 2(2)$
$\quad\;\; 4 > 2 \quad \checkmark$ $\quad\;\; 5 > 4 \quad \checkmark$

Are these two sets *all* the solutions?
Check the next possible set of four consecutive integers, {3, 4, 5, 6}, to see if the condition is satisfied.

$$6 \overset{?}{>} 2(3)$$
$$6 > 6 \qquad \text{False}$$

No other set will satisfy the condition.

∴ the required sets are {1, 2, 3, 4} and {2, 3, 4, 5}.

The product of any real number and an integer is called a **multiple** of that real number. If you multiply a given number n by consecutive integers, you obtain **consecutive multiples** of n. For example, if x is an integer, then

$$\ldots, 8(x - 2), 8(x - 1), 8x, 8(x + 1), 8(x + 2), \ldots$$

are consecutive multiples of 8.
The multiples of 2 are the *even integers*:

$$\ldots, -6, -4, -2, 0, 2, 4, 6, \ldots$$

180 *Chapter 4*

The integers that are not even are the *odd integers*:

$$\ldots, -5, -3, -1, 1, 3, 5, \ldots$$

If you count by twos from any given even integer, you obtain consecutive even integers. If you count by twos from any given odd integer, you obtain **consecutive odd integers**. In general,

$$\ldots, a - 4, a - 2, a, a + 2, a + 4, \ldots$$

will represent *consecutive even integers* if a is an even integer and will represent *consecutive odd integers* if a is an odd integer.

Suggested Assignments

Minimum
  182/*P*: 1–14
Average
  182/*P*: 1–13 odd, 15–22
Maximum
  182/*P*: 1–13 odd, 15–24

Oral Exercises

Let c = 8. Express each of the given numbers in terms of c.

1. 8, 9, 10 $c, c + 1, c + 2$ 2. 7, 8, 9 $c - 1, c, c + 1$ 3. 10, 12, 14 $c + 2, c + 4, c + 6$

4. 2, 4, 6 $c - 6, c - 4, c - 2$ 5. 3, 5, 7 $c - 5, c - 3, c - 1$ 6. 7, 9, 11 $c - 1, c + 1, c + 3$

Let e = −5. Express each of the given numbers in terms of e.

7. −5, −4, −3 $e, e + 1, e + 2$ 8. −6, −5, −4 $e - 1, e, e + 1$ 9. −5, −3, −1 $e, e + 2, e + 4$

10. −7, −5, −3 $e - 2, e, e + 2$ 11. −10, −9, −8 $e - 5, e - 4, e - 3$ 12. −6, −4, −2 $e - 1, e + 1, e + 3$

13. Let z represent an even integer.
    a. What are the next two greater even integers? $z + 2, z + 4$
    b. What is the preceding even integer? $z - 2$

14. Let $w - 3$ represent an integer.
    a. What are the next two greater integers? $w - 2, w - 1$
    b. What are the two preceding integers? $w - 5, w - 4$

15. Let $7r$ represent a multiple of 7.
    a. What are the next three greater multiples of 7? $7(r + 1), 7(r + 2), 7(r + 3)$
    b. What is the preceding multiple of 7? $7(r - 1)$

16. Let k represent an odd integer.
    a. What is the next greater odd integer? $k + 2$
    b. What is the next greater integer? $k + 1$
    c. What is the preceding odd integer? $k - 2$
    d. What is the preceding integer? $k - 1$

17. Let n represent an integer.
    a. Tell whether $2n$ is an even integer or an odd integer. even integer
    b. Tell whether $2n + 1$ is an even integer or an odd integer. odd integer

18. Let d represent an integer. Can you tell whether $d + 1$ is an even integer or an odd integer? Explain. No. If d is even, $d + 1$ is odd, but if d is odd, $d + 1$ is even.

Solving Inequalities and Problems **181**

Translate each word sentence into a mathematical sentence.

19. The sum of two consecutive integers is 71.
20. The sum of two consecutive even integers is 98.
21. The sum of three consecutive integers is less than 14.
22. The sum of three consecutive odd integers is no less than 15.
23. The sum of three consecutive multiples of 4 is greater than 16.
24. The sum of three consecutive even integers is not more than 42.
25. The sum of two consecutive integers is at least 25.
26. The sum of two consecutive multiples of 5 is at most 45.

Problems

Solve.

A 1. Find two consecutive integers whose sum is 383. 191, 192
2. Find three consecutive integers whose sum is 66. 21, 22, 23
3. Find two consecutive odd integers whose sum is -160. $-81, -79$
4. Find three consecutive even integers whose sum is -186. $-64, -62, -60$
5. Find three consecutive multiples of 7 whose sum is 84. 21, 28, 35
6. Find two consecutive multiples of 9 whose sum is -153. $-81, -72$
7. Find three consecutive even integers such that the sum of the first and third integers is 192. 94, 96, 98
8. Find four consecutive odd integers such that the sum of the least integer and the greatest integer is 164. 79, 81, 83, 85
9. Find the least two consecutive odd integers whose sum is more than 46. 23, 25
10. Find the greatest two consecutive even integers whose sum is less than 180. 88, 90
11. Three times an odd integer is eleven less than four times the next greater even integer. What is the odd integer? 7
12. The sum of an even integer and twice the next greater even integer is eight more than four times the greater integer. Find the lesser integer. -12
13. The lengths in meters of two sides of a rectangle are consecutive even integers. If ten more than half the length of the shorter side is added to twice the length of the longer side, the result is 59 m. Find the area of the rectangle. 360 m²
14. The lengths in centimeters of two sides of a rectangle are consecutive odd integers. The perimeter is 120 cm. What is the area of the rectangle? 899 cm²

B 15. The present ages in years of four cousins are consecutive multiples of 3. Five years ago the sum of their ages was 46. Find their ages now. 12 years, 15 years, 18 years, 21 years

16. The dimensions in meters of a rectangular field are consecutive multiples of 7. If each dimension were increased by 10 m, the perimeter of the field would be 306 m. Find the original dimensions of the field. 63 m × 70 m

17. Find all sets of three consecutive integers whose sum is less than 24 if the sum of the least integer and twice the next greater integer is more than twice the greatest integer.

18. Find all sets of two consecutive even integers whose sum is no greater than 30 if five times the lesser of the two consecutive even integers is more than four times the greater.

19. Find all sets of three consecutive multiples of 3 whose sum is between −134 and −109.

20. Find all sets of three consecutive multiples of 6 whose sum is between −6 and 50.

21. The sum of an even integer and twice the next greater even integer is the same as the difference between five times the greater even integer and 14. Find the integers. *Hint:* There are two solutions.

22. Find all sets of three consecutive multiples of 11 for which the sum of the two lesser numbers is greater than 100 and the sum of the two greater numbers is less than 200.

C 23. Show that the sum of any two consecutive odd integers is even.
24. Show that the sum of any three consecutive odd integers is odd.

Computer Exercises For students with computer experience

1. Write a program that will allow you to input any integer and will display whether the integer is even or odd.

2. Write a program that will compute the sum of ten consecutive integers when you input the least of the integers.

3. Modify the program that you wrote for Exercise 2 to compute the sum of ten consecutive *even* or *odd* integers when you input the least of the integers.

4. Modify the program that you wrote for Exercise 3 to compute the sum of *any number* of consecutive even or odd integers when you input the least integer and the number of integers to be added.

5. Write a program that will allow you to input a sum, S, and a number of consecutive integers, n, and will output n consecutive integers whose sum is S. Be careful to account for values that you may input for which there is no solution to the problem.

Mixed Review
Solve.

1. $x + 3(x + 2) = 26$ {5}
2. $4n + 4(n + 1) + 4(n + 2) > 60$ {$n: n > 4$}
3. $a + 3(a + 4) \le 30$
   $\left\{a: a \le 4\frac{1}{2}\right\}$
4. Find the area of a rectangle if the longer sides measure 9 m each and the shorter sides measure 4 m each. 36 m²

Chalkboard Examples

1. The measure of $\angle X$ is 45° and the measure of $\angle Y$ is 35°.
   a. Find the measure of the complement of $\angle X$.
      90 − 45 = 45
      ∴ the measure of the complement of $\angle X$ is 45°.
   b. Find the measure of the supplement of $\angle Y$.
      180 − 35 = 145
      ∴ the measure of the supplement of $\angle Y$ is 145°.
   c. If $\angle X$ and $\angle Y$ are two of the angles of a triangle, find the measure of the third angle.
      $z + 45 + 35 = 180$
      $z + 80 = 180$
      $z = 100$
      ∴ the measure of the third angle is 100°.

2. The measure of $\angle A$ is 30°.
   a. If $\angle B$ is the complement of $\angle A$, what is its measure?
      90 − 30 = 60
      ∴ the measure of $\angle B$ is 60°.
   b. If $\angle A$ and $\angle B$ are two angles of a triangle, find the measure of the third angle.
      $x + 30 + 60 = 180$
      $x + 90 = 180$
      $x = 90$
      ∴ the measure of the third angle is 90°.

4–7 Problems about Angles

The facts shown in the chart that follows may be familiar to you. These facts can be used in solving certain problems about angles.

Right angle

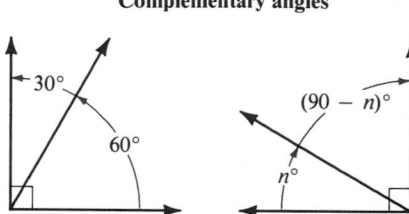

The measure of a right angle is 90°.

Complementary angles

The sum of the measures of complementary angles is 90°.

Two adjacent right angles

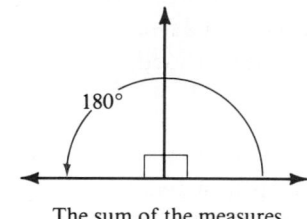

The sum of the measures of two right angles is 180°.

Supplementary angles

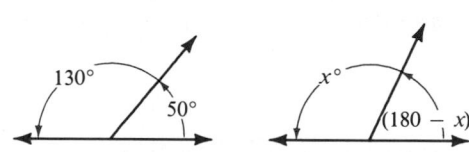

The sum of the measures of supplementary angles is 180°.

Triangle

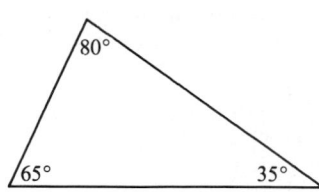

The sum of the measures of the angles of a triangle is 180°.

Parallelogram

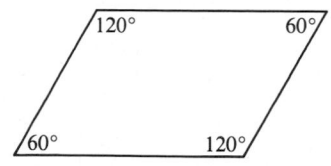

The opposite angles of a parallelogram are equal in measure.

EXAMPLE The measure of $\angle A$ (read "angle A") is $75°$ and the measure of $\angle B$ is $60°$.

    a. Find the measure of the complement of $\angle A$.

    b. Find the measure of the supplement of $\angle B$.

    c. If $\angle A$ and $\angle B$ are two of the angles of a triangle, find the measure of the third angle.

SOLUTION   **a.** The measure of $\angle A$ is $75°$.

$$90 - 75 = 15$$

    $\therefore$ the measure of the complement of $\angle A$ is $15°$.

    b. The measure of $\angle B$ is $60°$.

$$180 - 60 = 120$$

    $\therefore$ the measure of the supplement of $\angle B$ is $120°$.

    c. The sum of the measures of the three angles of a triangle is $180°$.

    Let $x° =$ the measure of the third angle.

$$x + 75 + 60 = 180$$
$$x + 135 = 180$$
$$x = 45$$

    $\therefore$ the measure of the third angle is $45°$.

Oral Exercises

State the measure of the angle that is the complement of the angle with the given measure.

1. $14°$ $76°$ **2.** $71°$ $19°$ **3.** $33°$ $57°$ **4.** $52°$ $38°$

5. $m°$ $(90 - m)°$ **6.** $5a°$ $(90 - 5a)°$ **7.** $(b + 2)°$ $(88 - b)°$ **8.** $(c - t)°$ $(90 - c + t)°$

State the measure of the angle that is the supplement of the angle with the given measure.

9. $125°$ $55°$ **10.** $50°$ $130°$ **11.** $10°$ $170°$ **12.** $150°$ $30°$

13. $n°$ $(180 - n)°$ **14.** $3r°$ $(180 - 3r)°$ **15.** $(e + 10)°$ $(170 - e)°$ **16.** $(2d - 40)°$ $(220 - 2d)°$

The measures of two angles of a triangle are given. State the measure of the third angle.

17. $20°$, $30°$ $130°$ **18.** $100°$, $15°$ $65°$ **19.** $15°$, $70°$ $95°$ **20.** $83°$, $96°$ $1°$

21. $x°$, $5x°$ **22.** $c°$, $(17 - c)°$ **23.** $3b°$, $(25 - b)°$ **24.** $(5t + 47)°$, $(19 - 5t)°$
    $(180 - 6x)°$ $163°$ $(155 - 2b)°$ $114°$

State an equation that represents the given information.

25. The measure of a certain angle is $40°$ less than the measure of its supplement.

26. The measure of a certain angle is $10°$ more than the measure of its complement.

Solving Inequalities and Problems   **185**

Suggested Assignments

Minimum
First day: 186/*P*: 1–8
Second day: 186/*P*: 9–16
Average
    186/*P*: 1–13 odd, 15–22
Maximum
    186/*P*: 1–13 odd, 15–24
S 178/58–66 even

Additional A Exercises

Solve.

1. The measure of the complement of an angle is 25°. Find the measure of the angle.   65°

2. The measure of an angle is 20° more than the measure of its complement. Find the measure of its supplement.   125°

3. Find the measure of an angle if the sum of the measure of its supplement and twice the measure of its complement is 207°.   51°

4. Find the measure of two supplementary angles if the measure of one is 39° less than two times the measure of the other.   73°, 107°

5. The measures of two angles of a triangle are equal. The measure of the third angle is 40° less than the sum of the measures of the first two angles. Find the measure of each angle.   55°, 55°, 70°

6. The sum of the measures of the four angles of a parallelogram is 360°. If the measure of one angle is 60°, find the measure of each of the other angles.   60°, 120°, 120°

Problems

Solve.

A

1. The measure of an angle is 58° more than the measure of its supplement. Find the measure of its supplement.   61°

2. The measure of an angle is 26° less than the measure of its complement. Find the measure of the angle.   32°

3. Find the measure of two complementary angles if the measure of one is 42° more than twice the measure of the other.   74°, 16°

4. Find the measure of two supplementary angles if the measure of one is 24° more than three times the measure of the other.   141°, 39°

5. Find the measure of an angle if the measure of its supplement is 4° more than twice the measure of its complement.   4°

6. Find the measure of an angle if the sum of the measures of its complement and supplement is 162°.   54°

7. The measure of one angle of a triangle is three times the measure of a second angle, and the measure of the third angle is 12° less than the sum of the measures of the other two. Find the measure of each angle.   72°, 24°, 84°

8. The measure of one angle of a triangle is 20° more than the measure of the second angle. The measure of the third angle is 30° more than twice the sum of the measures of the first two angles. Find the measure of each angle in the triangle.   35°, 15°, 130°

9. The sum of the measures of the four angles of a parallelogram is 360°. If the measure of one angle is 50°, find the measure of each of the other angles.   50°, 130°, 130°

10. The sum of the measures of the four angles of a parallelogram is 360°. If the measure of one angle is 30° more than the measure of a second angle, find the measure of each of the four angles.   75°, 105°, 75°, 105°

11. A triangle in which the three angles have equal measure is called an *equiangular triangle*. What is the degree measure of the complement of each angle of an equiangular triangle?   30°

12. A triangle in which two angles, called base angles, have equal measure is called an *isosceles triangle*. In a certain isosceles triangle, the sum of the measures of the base angles is 88° less than the measure of the remaining angle. Find the measure of each angle.   23°, 23° 134°

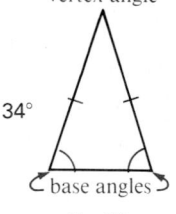
vertex angle

base angles

Ex. 12

13. The degree measures of two angles of a triangle are consecutive even integers. If the measure of the third angle is 22° more than the measure of the least angle of the triangle, what is the measure of each angle?   52°, 54°, 74°

14. The degree measures of two angles of a triangle are consecutive multiples of 10. The measure of the third angle is 50° more than twice the measure of the least angle. Find the measure of each angle.   30°, 40°, 110°

186   *Chapter 4*

Exercises 15–18 refer to the Law of Reflection, $i = r$. That is, when a light ray strikes a reflecting surface, the measure, i, of the *angle of incidence* is equal to the measure, r, of the *angle of reflection*.

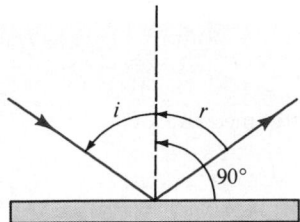 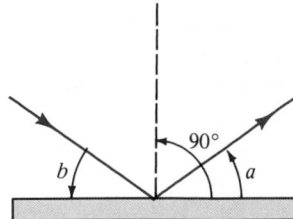

B **15.** $i = (3x - 10)°$
$r = (50 - x)°$
Find x. 15

16. $a = (5t + 5)°$
$b = (3t + 75)°$
Find t. 35

17. $i = (5m - 14)°$
$r = (7m - 86)°$
Find m. 36

18. $a = (18 - 4d)°$
$b = (113 - 9d)°$
Find d. 19

19. The measure of $\angle W$ is twice the measure of $\angle X$ and is 1° less than the measure of $\angle Y$. If the sum of the measures of the three angles is at least 56°, what is the least possible measure of $\angle W$? 22°

20. The measure of $\angle R$ is three times the measure of $\angle S$. If the measure of $\angle R$ is at least 80° more than the measure of $\angle S$, what is the least possible measure of $\angle S$? 40°

21. In $\triangle ABC$ (read "triangle A, B, C"), $\angle C$ is a right angle, and the measure of $\angle B$ is at least five times the measure of $\angle A$.

  a. Find the least possible degree measure of $\angle B$. 75°

  b. Find the greatest possible degree measure of $\angle B$ if the measure of $\angle B$ is an integer. 89°

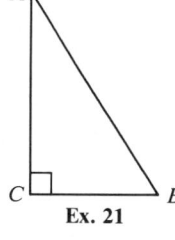

22. The measure of $\angle P$ is 40° less than the measure of $\angle D$ and is one half the measure of $\angle Q$. If the sum of the measures of the three angles is at most 140°, what is the greatest possible measure of $\angle P$? 25°

Ex. 21

23. measure of $\angle A$ is 80°, measure of $\angle B$ is 60°, measure of $\angle C$ is 40°

C **23.** In $\triangle ABC$, the measure of $\angle B$ is three fifths the measure of the supplement of $\angle A$. The measure of $\angle C$ is four thirds the measure of the complement of $\angle B$. Find the degree measures of the three angles of the triangle.

24. The measure of an angle is at least one fourth the measure of its supplement and at most four times the measure of its complement. Find the possible degree measures of the angle. $36° \le x \le 72°$

Solving Inequalities and Problems **187**

Solve.

1. $3x - 11 < 28$ $\{x: x < 13\}$

2. $|x - 2| = 26$ $\{-24, 28\}$

3. $x + (180 - 2x) = 120$ $\{60\}$

4. $x + 3(x - 2) + (x - 4) = 370$ $\{76\}$

5. The degree measures of the three angles of a triangle are consecutive integers. Find the measure of each angle. 59°, 60°, 61°

6. a. If the sum of the measures of two angles is 90° and the measure of one of the angles is $a°$, what is the measure of the other angle? $(90 - a)°$

  b. If the sum of the measures of two angles is 180° and the measure of one of the angles is $a°$, what is the measure of the other angle? $(180 - a)°$

Careers

Broadcasting

Broadcasting is one of the major modes of public communication. Through radio and television the public is informed with newscasts and documentaries and is entertained with music, movies, comedies, dramas, and sports.

People interested in the field of broadcasting have a wide variety of career opportunities available to them. For example, *broadcast technicians* are the people who operate and maintain technical equipment such as cameras and videotape machines. On the other hand, *program planners* are the people who actually determine what types of programs are to be aired and decide where commercials are to be placed.

EXAMPLE — It is going to cost television network CTN $400,000 to broadcast a popular movie. The network will receive an average of $48 in revenue from sponsors for each thousand television sets that is tuned in to the movie. How many sets must be tuned in to make a profit of at least $140,000 for CTN?

SOLUTION — Let x = the number in thousands of television sets tuned in. Then:

$48x$ = the network's revenue in dollars

CTN's profit will be equal to the difference when the cost of broadcasting the movie is subtracted from the revenue received for broadcasting the commercials. That is:

$$\text{Profit} = \underbrace{\text{Revenue}} - \underbrace{\text{Cost}}$$
$$= \quad 48x \quad - 400,000$$

Since the network needs to make a profit of *at least* $140,000, solve the following inequality:

$$\underbrace{\text{Profit}} \geq \underbrace{140,000}$$
$$48x - 400,000 \geq 140,000$$
$$48x \geq 540,000$$
$$x \geq 11,250$$

∴ at least 11,250 thousand, or 11.25 million, television sets must be tuned in to the movie.

4–8 Problems of Uniform Motion

An object that moves at a constant speed, or *rate*, is said to be in uniform motion. A formula that is often used in solving problems involving uniform motion is:

$$\text{rate} \cdot \text{time} = \text{distance}$$
$$r \cdot t = d$$

This formula may also be used for motion that is not uniform if the *average rate* is considered. To apply the formula correctly, the units used for the time and distance measurements must be the same as those used for the rate.

Three types of problems that can be solved using the formula $r \cdot t = d$ are illustrated in the examples that follow. The following standard symbols are used: s for seconds; min for minutes; h for hours.

EXAMPLE 1   (Motion in the same direction)

Two delivery trucks heading for Windsor along the same route leave a warehouse at the same time. Their average speeds differ by 15 km/h. The faster truck reaches Windsor after 5 h of travel time. The slower truck takes an hour longer. Find the average speed of each truck.

SOLUTION

Step 1   The problem asks for the average speed of each truck.

Step 2   Let x = rate of slower truck in km/h.

Then $x + 15$ = rate of faster truck in km/h.

	rate (km/h)	time (h)	distance (km)
Faster truck	$x + 15$	5	$5(x + 15)$
Slower truck	x	6	$6x$

Step 3 $\underline{\text{Distance that faster truck went}}$ = $\underline{\text{Distance that slower truck went}}$

$5(x + 15)$ $\qquad = \qquad$ $6x$

Step 4
$$5(x + 15) = 6x$$
$$5x + 75 = 6x$$
$$75 = x$$

Thus $x = 75$ and $x + 15 = 90$.

Step 5   Checking the results is left to you.

∴ the average speed of the slow truck is 75 km/h, and the average speed of the fast truck is 90 km/h.

Teaching Suggestions p. T82

Related Activities p. T82

Key Ideas

Solve uniform rate of motion problems.

Chalkboard Examples

1. (Motion in the same direction)

Two speedboats leave a dock in Los Angeles at the same time. The faster boat arrives in San Diego in 6 h; the slower boat arrives in 10 h. The slower boat travels at an average speed that is 14 km/h slower than the faster boat. What is the average speed of each boat?

Let x = the rate of the fast boat in km/h. Then $x - 14$ = the rate of the slow boat in km/h.

	rate (km/h)	time (h)	distance (km)
Fast boat	x	6	$6x$
Slow boat	$x - 14$	10	$10(x - 14)$

Distance of fast boat = Distance of slow boat.
$$6x = 10(x - 14)$$
$$6x = 10x - 140$$
$$-4x = -140$$
$$x = 35$$

Thus $x = 35$ and $x - 14 = 21$.

∴ the average speed of the fast boat is 35 km/h, and the average speed of the slow boat is 21 km/h.

(continued)

2. (Motion in opposite directions)
A partially disabled tank is headed for an emergency depot at a base 141 km away. The tank travels at an average speed of 24 km/h. An emergency van from the depot heads out to meet the tank at an average speed of 70 km/h. How long will it take for the tank and van to meet? Let t = the number of hours it will take for the tank and the van to meet.

	rate (km/h)	time (h)	distance (km)
Tank	24	t	$24t$
Van	70	t	$70t$

Distance traveled by tank + Distance traveled by van = Total distance.
$24t + 70t = 141$
$94t = 141$
$t = 1.5$
∴ the tank and the van would meet in 1.5 h.

3. (Round trip)
Each afternoon after school Fred rides his bicycle to the park and back home again. On the way to the park he rides at an average rate of 25 km/h and on the return trip at an average rate of 20 km/h. He takes 2.25 h for the trip. How far is the park from Fred's home?

EXAMPLE 2 (Motion in opposite directions)
An oil tanker with engine trouble radios the mainland for a seagoing tugboat. At the time that the tugboat leaves the dock, the tanker is 125 km away and heading directly toward the dock. If the average speed of the tugboat is 20 km/h and that of the tanker is 5 km/h, how long will it take the two vessels to meet?

SOLUTION

Step 1 The problem asks for the number of hours the tugboat and tanker will travel before they meet.

Step 2 Let t = the number of hours it will take for the tugboat and tanker to meet.

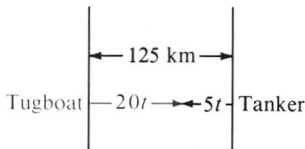

	rate (km/h)	time (h)	distance (km)
Tugboat	20	t	$20t$
Tanker	5	t	$5t$

Step 3

Distance traveled by tugboat	+	Distance traveled by tanker	=	Total distance traveled by tugboat and tanker
$20t$	+	$5t$	=	125

Step 4
$20t + 5t = 125$
$25t = 125$
$t = 5$

Step 5 Checking the result is left to you.
∴ the tugboat and tanker will meet in 5 h.

EXAMPLE 3 (Round trip)
Each day Kim jogs from her home to the lake and then walks back again along the same route. Her average speed is 5 m/s while jogging and 2 m/s while walking. If the whole trip takes 35 min, how far is it from Kim's home to the lake?

SOLUTION

Step 1 The problem asks for the distance from Kim's home to the lake.

Step 2 If you know the time it takes to go in one direction, you can find the distance. Since the rate is given in meters per second, the time will be found in seconds. Thus, the total time of 35 min must be changed into seconds.

Let t = time in seconds that it takes to jog to the lake.

Then $2100 - t$ = time in seconds that it takes to walk back.

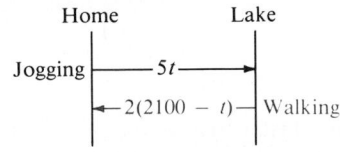

	rate (m/s)	time (s)	distance (m)
Jogging	5	t	$5t$
Walking	2	$2100 - t$	$2(2100 - t)$

Step 3 $\underbrace{\text{Distance to lake}}_{5t}$ = $\underbrace{\text{Distance back again}}_{2(2100 - t)}$

Step 4

$$5t = 2(2100 - t)$$
$$5t = 4200 - 2t$$
$$7t = 4200$$
$$t = 600$$

Since you know that it takes 600 s for Kim to jog to the lake at 5 m/s, you can find the distance from Kim's home to the lake.

$$d = 5 \times 600$$
$$= 3000$$

Step 5   Checking the result is left to you.

∴ the distance from Kim's home to the lake is 3000 m, or 3 km.

Oral Exercises

Let r = average rate, t = time, and d = distance. Find the quantity indicated using the given information. Give the correct units in each case.

1. $r = 12$ m/s, $t = 8$ s; d 96 m
2. $r = 15$ km/h, $d = 120$ km; t 8 h
3. $d = 300$ m, $t = 50$ min; r 6 m/min
4. $r = 90$ km/h, $t = 40$ min; d 60 km
5. $r = 100$ m/min, $d = 2$ km; t 20 min
6. $d = 80$ mi, $t = 2$ h 30 min; r 32 mi/h

Solve.

7. Tom traveled 180 km in 3 h. What was his average speed?   60 km/h
8. Jane traveled at an average speed of 100 km/h. How far did she go in 90 min?   150 km
9. Alice drove a distance of 220 km at an average speed of 80 km/h. How long did the trip take?   2 h 45 min
10. Mike's home is 35 km from his office. On Tuesday, Mike leaves home at 8:00 A.M. At what time will he reach his office if he travels at an average speed of 50 km/h?   8:42 A.M.

Solving Inequalities and Problems   **191**

Let t = the time in hours that it takes to ride to the park. Then $2.25 - t$ = the time it takes to ride back from the park.

	rate (km/h)	time (h)	distance (km)
To park	25	t	$25t$
To home	20	$2.25 - t$	$20(2.25 - t)$

Distance to park = Distance back home.
$$25t = 20(2.25 - t)$$
$$25t = 45 - 20t$$
$$45t = 45$$
$$t = 1$$
Thus $25t = 25$.
∴ the distance from Fred's home to the park is 25 km.

Common Errors

Students often forget to change units in word problems to a single unit of comparison. First, they must determine if it is necessary to change to like units. If it is, then they must change each measurement accordingly. For example, if the time is given in minutes and the rate is given in m/s, then the minutes could be converted into seconds.

Suggested Assignments

Minimum
First day: 192/*P*: 1–9
Second day: 193/*P*: 10–20
    S 177/31–34
Average
First day: 192/*P*: 1–10
Second day: 193/*P*: 11–22
    S 178/42–50 even
Maximum
    192/*P*: 1–13 odd, 20–25

Additional A Exercises

1. Two boats left an island at the same time to return to their home port. Their average speeds differed by 4 km/h. The faster boat reached port after 3 h. The slower boat took 30 min longer. Find the average speed of each boat. 28 km/h and 24 km/h.

2. John and Sam live 24 km apart. They agree to meet at a movie theater located directly between their homes. John needs 12 min and Sam needs 36 min to get to the movie. If they both travel at the same average speed, how far do John and Sam live from the theater? John lives 6 km and Sam 18 km away.

3. Joshua delivers newspapers every afternoon. The trip delivering them takes 45 min. The return trip home over the same route takes 30 min. If his average speed going is 6 km/h slower than returning home, how long is his paper route? 9 km

Problems

Solve.

A
1. The average speed of a moving van is 24 km/h faster than that of a delivery truck. If the van travels the same distance in 5 h as the delivery truck travels in 7 h, find the average speed of each vehicle. van, 84 km/h; truck, 60 km/h

2. Two airplanes start toward each other at the same time from airports located 1950 km apart. One plane flies at an average speed of 360 km/h. What should the average speed of the other plane be if they are to meet in 3 h? 290 km/h

3. Two campers leave Square Lake at the same time, one traveling west at an average rate of 75 km/h and the other traveling east at an average rate of 65 km/h. After how many hours will the campers be 490 km apart? $3\frac{1}{2}$ h

4. Lionel left Reedsville at 9:00 A.M. one day and drove to Fitzwilliam at an average rate of 80 km/h. At 10:00 A.M. the same day, Nancy left Reedsville and followed the same route. If both Lionel and Nancy arrived in Fitzwilliam at 3:00 P.M., at what average rate had Nancy traveled? How far is Reedsville from Fitzwilliam? 96 km/h; 480 km

5. Reggie rode his tractor from his home to the agricultural exhibit and then walked back home along the same route. His average speed was 8 km/h while riding and 5 km/h while walking. How far is it from his home to the exhibit if his traveling time totaled 78 min? 4 km

6. Gina flew from Toronto to Chicago and back again. The average speed of the plane was 332 km/h to Chicago and 415 km/h back to Toronto. If the actual flying time for the round trip was 4.5 h, how far is it from Toronto to Chicago? 830 km

7. Becky and Rita are 648 m apart. Becky walks toward Rita at the rate of 2.25 m/s and Rita runs toward Becky. What is Rita's average speed if she reaches Becky in 1.6 min? 4.5 m/s

8. A passenger train traveling due west at 40 km/h passes a freight train traveling due east at 90 km/h. How long after they pass each other will the trains be 325 km apart? 2.5 h

9. Riding mostly downhill, Joan rode her ten-speed bicycle to the state fair in 1.5 h. It took 4.2 h to make the return trip, since her average speed was 36 km/h less than her average speed on the way to the fair. Find her average speed in each direction. 56 km/h; 20 km/h

10. A commercial jet and a private airplane leave the same airfield at the same time and travel in opposite directions at average speeds of 880 km/h and 450 km/h, respectively. In how many minutes will they be 1463 km apart?   66 min

11. A helicopter traveling at an average speed of 225 km/h left San Jose 1 h after a train that had departed at 7:00 A.M. If the helicopter overtook the train in 0.8 h, find the average speed of the train.   100 km/h

12. Madeline made a certain trip by moped in 2 h. It took her 2.5 h to make the return trip, since her average speed returning was 12 km/h less than her speed going. Find Madeline's average speed in each direction.   60 km/h going; 48 km/h returning

13. Maria drove for two hours to get to the Fort William airport. She then flew from Fort William to Green Island. During the four hour flight to Green Island, the average speed of the plane was three times the average speed that Maria traveled on the drive to the airport. If Maria traveled a total distance of 1120 km, what was the average speed on each part of her trip?   80 km/h; 240 km/h

14. An armored truck left the Federal Bank at noon. When a shipping error was discovered, an agent was sent by helicopter to overtake the truck. The helicopter left at 1:48 P.M. and flew at an average speed of 170 km/h. If the truck was overtaken at 3:24 P.M., what was the average speed of the truck?   80 km/h

B 15. A commercial plane had been flying for three hours when a change in the wind decreased the plane's average speed by 30 km/h. If the entire trip of 2540 km took 5 h, how far did the plane travel before the wind changed?   1560 km

16. A private airplane had been flying for 1 h when a change of wind direction doubled the plane's average speed. If the entire trip of 384 km took 2.5 h, how far did the plane travel in the first hour?   96 km

17. A car traveled 462 km in 7 h. For the last 3 h of the trip its average speed was 20 km/h less than twice its average speed for the first 4 h. Find the two speeds at which the car traveled. 52.2 km/h; 84.4 km/h

18. Calvin drove from Greenfield to Munsonville at an average speed of 64 km/h. By traveling at an average speed of 78 km/h, he could have arrived 10 min earlier. How far is it from Greenfield to Munsonville? Give your answer to the nearest tenth of a kilometer.   59.4 km

19. A cargo ship must travel at an average speed of 25 km/h to make its 14 h run on schedule. During the first 4 h, bad weather forced the captain to reduce speed to 20 km/h. What should the average speed of the ship be for the rest of the trip to keep on schedule?   27 km/h

20. A Great Lakes ore carrier must travel at an average speed of 24 km/h for 9 h to reach its destination on schedule. Fog forced the captain to reduce the speed to 21 km/h for the first 6 h. To arrive on time, what should be the ship's average speed for the rest of the trip?   30 km/h

4. Two spaceships separate from their space station at the same time, traveling in opposite directions. The first ship moves at 82,000 km/h and the second at 43,000 km/h. How many hours will it take them to be 1 million kilometers apart?   8 h

5. Eric is a member of the cross country team at his school. On Tuesday he ran from school to the monument in the town square and back again. On the way to the monument he ran at a rate of 10 km/h and on the return trip at a rate of 8 km/h. He took 2 h 15 min to run the entire distance. How far is the monument from the school?   10 km

6. The average speed of a passenger train is 32 km/h faster than that of a freight train. In 6 h the passenger train travels the same distance that the freight train travels in 9 h. Find the average rate of speed of each train.
freight train: 64 km/h
passenger train: 96 km/h

Mixed Review

Solve.

1. If $r = 55$ and $t = 6$, find d if $r \cdot t = d$.   330

2. If $r = 30$ and $d = 135$, find t if $r \cdot t = d$.   4.5

3. $35x = 25(x + 3)$ 7.5

4. $7t + 12.5t = 64.35$ 3.3

5. $4.4m + 3.5m \leq 49.77$
$\{m: m \leq 6.3\}$

6. $|q - 4| > 6$
$\{q: q < -2 \text{ or } q > 10\}$

21. In still water Meg can row 6 km/h. However, on the Branch River it took her 4 h to row upstream to her friend's cabin and only 2 h to return. Find the speed at which the river was flowing. 2 km/h

22. To go fishing, Art must row upstream for 3 h 20 min. The return trip only takes 1 h 40 min. If Art can row 4.5 km/h in still water, what is the speed at which the river is flowing? How far is it to the fishing location? 1.5 km/h; 10 km

C 23. Ann lived in Bristol and Amy lived in Fairvale. At 2:00 P.M. they left their respective towns and walked in opposite directions, with Amy walking twice as fast as Ann. By 4:00 P.M., they were 66 km apart. If, instead, they had walked from the towns toward each other, they would have been 12 km apart at 3:00 P.M. How far apart are the towns? 30 km

24. Three jets depart from the airport at the same time. The average speed of the westbound plane is 50 km/h less than that of the southbound plane. The average speed of the eastbound plane is 60 km/h more than that of the southbound plane. The eastbound and westbound planes are 3150 km apart after 3 h. How far from the airport is the southbound plane at that time? 1560 km

25. Two robots, T4-2 and B4-U, are programmed to roll toward each other, bump, return to their original positions, and then repeat the moves. T4-2 moves at 1.8 m/s and B4-U moves at 1.2 m/s. If eight seconds elapse between their first and second bumps, how far apart are their original positions? 12 m

26. Gerry and Tony live 11 km apart. Gerry leaves his home at 2:00 P.M. and bikes toward Tony's home at 25 km/h. At the same time, Tony leaves his home and bikes toward Gerry's home at 30 km/h. As soon as they reach each other's homes, they turn around and start back along the same route. They stop when they meet each other on the ride back to their own homes. How far are they from Gerry's home when they stop? 7 km

27. A freight train traveling 100 km/h took 48 s to pass a motorcyclist going in the same direction at 40 km/h. Assuming the length of the motorcycle is negligible, find the length of the train. If the train and the motorcycle were traveling in opposite directions, find to the nearest second how long they would have taken to pass each other completely. 0.8 km; 21 s

28. The Midland Flyer, traveling at 90 km/h, overtakes another train that is traveling at 60 km/h in the same direction on a parallel track. The trains pass each other completely in 1.2 min. If the Midland Flyer is twice as long as the other train, what is the length of each train? If the trains were traveling in opposite directions, find to the nearest second how long they would have taken to pass each other completely. 0.6 km; 0.3 km; 22 s

194 *Chapter 4*

PROGRAMMING IN BASIC

Often a computer program can be made more flexible by using INPUT statements. For example, the following program is a modification of the program on page 138. Here the INPUT statements allow you to change the dimensions of the figure without changing the lines of the program.

```
10 PRINT "TO CREATE GEOMETRIC"
20 PRINT "DESIGNS WITH ASTERISKS:"
30 PRINT
32 PRINT "GIVE A VALUE FOR M";
34 INPUT M
36 PRINT "GIVE A VALUE FOR N";
38 INPUT N
40 FOR I = 1 TO M
50 FOR J = 1 TO N
60 PRINT "*";
70 NEXT J
80 PRINT
90 NEXT I
100 END
```

Exercises

1. Type in the program as given. RUN the program for several values of M and N. M × N rectangle

2. Delete lines 36 and 38.
   Change line 50 as follows: 50 FOR J = 1 TO I
   LIST and RUN the revised program for several values of M. triangle

3. Change line 50 as follows: 50 FOR J = 1 TO M + 1 − I
   LIST and RUN the revised program for several values of M. triangle

4. Change line 50 as follows: 50 FOR J = 1 TO 2 * I − 1
   Then insert these lines: 42 FOR J = 1 TO M + 1 − I
                               44 PRINT " ";
                               46 NEXT J
   LIST and RUN the revised program for several values of M. triangle

Sometimes you can use the TAB function in a PRINT statement instead of using a loop that prints blank spaces. Note how TAB is used in the following exercises.

5. Delete lines 42, 44, and 46.
   Then insert this line: 45 PRINT TAB(M + 1 − I);
   LIST and RUN the revised program for several values of M. triangle

6. Change lines 40, 45, and 50 as follows:

40 FOR I = −M TO M
45 PRINT TAB(ABS(I) + 1);
50 FOR J = 1 TO 2 * M − 2 * ABS(I) + 1

LIST and RUN the revised program for several values of M. diamond

7. To explore further how TAB works, type in and RUN this program.

NEW
10 FOR I = 1 TO 6
20 PRINT " ";
30 NEXT I
40 PRINT "6"
50 PRINT TAB(6);"6"
60 END

If the "6" printed in line 50 aligns with the "6" printed in line 40, the TAB on the computer you are using *skips over* 6 spaces. If the numbers do *not* align, consult the computer manual for a description of its TAB function.

4–9 Mixture Problems

A merchant often mixes goods of two or more kinds in order to sell a blend at a given price. Similarly, a chemist mixes solutions of different strengths of a chemical to obtain a solution of a desired strength. Problems related to these situations are often called *mixture problems*.

EXAMPLE 1 Find the number of kilograms of nuts worth $4.00 per kilogram and the number of kilograms of raisins worth $5.20 per kilogram that should be mixed to produce 3 kg of a snack worth $4.52 per kilogram.

SOLUTION

Step 1 The problem asks for the number of kilograms of nuts and the number of kilograms of raisins to use in 3 kg of a mixture.

Step 2 Let n = the number of kilograms of nuts.
Then $3 − n$ = the number of kilograms of raisins.

	Number of kg	Value per kg	Total value
Nuts	n	4	$4n$
Raisins	$3 − n$	5.2	$5.2(3 − n)$
Mixture	3	4.52	$4.52(3)$

196 Chapter 4

Teaching Suggestions
p. T82

Key Ideas

Solve mixture problems.

Chalkboard Examples

1. Kirk bought some 18¢ stamps and some 22¢ stamps. He bought 35 stamps in all and paid $7.10 for them. How many stamps of each kind did he buy?
Let s = the number of 18¢ stamps. Then $35 − s$ = the number of 22¢ stamps.

	Number of stamps	Value per stamp	Total value
18¢ stamps	s	0.18	$0.18s$
22¢ stamps	$35 − s$	0.22	$0.22(35 − s)$
Mixture	35		7.10

Step 3   The sum of the values of the original ingredients must equal the value of the mixture.

$$\underbrace{\substack{\text{Value} \\ \text{of nuts}}}_{4n} \quad + \quad \underbrace{\substack{\text{Value} \\ \text{of raisins}}}_{5.2(3-n)} \quad = \quad \underbrace{\substack{\text{Value} \\ \text{of mixture}}}_{4.52(3)}$$

Step 4
$$4n + 5.2(3 - n) = 4.52(3)$$
$$4n + 15.6 - 5.2n = 13.56$$
$$-1.2n = -2.04$$
$$n = 1.7$$
$$3 - n = 1.3$$

Step 5   Checking the results is left to you.
∴ 1.7 kg of nuts and 1.3 kg of raisins are needed.

EXAMPLE 2   A scientist has 200 mL of a solution that is 40% acid. How many milliliters of water should be added to make a solution that is 25% acid?

SOLUTION
Step 1   The problem asks for the number of milliliters of water that should be added to make a solution that is 25% acid.

Step 2   Let x = number of milliliters of water to be added.

	Volume of solution	% acid	Volume of acid
Original solution	200	40	0.40(200)
Added water	x	0	0
New solution	200 + x	25	0.25(200 + x)

Step 3
$$\underbrace{\substack{\text{Original volume} \\ \text{of acid}}}_{0.40(200)} \quad + \quad \underbrace{\substack{\text{Added volume} \\ \text{of acid}}}_{0} \quad = \quad \underbrace{\substack{\text{Total volume} \\ \text{of acid}}}_{0.25(200 + x)}$$

Step 4
$$0.40(200) + 0 = 0.25(200 + x)$$
$$80 = 50 + 0.25x$$
$$30 = 0.25x$$
$$x = 120$$

Step 5   Checking the results is left to you.
∴ 120 mL of water should be added.

Cost of 18¢ stamps + Cost of 22¢ stamps = Total cost.
$$0.18s + 0.22(35 - s) = 7.10$$
$$0.18s + 7.7 - 0.22s = 7.10$$
$$-0.04s = -0.60$$
$$s = 15$$
$$35 - s = 20$$
∴ Kirk bought fifteen 18¢ stamps and twenty 22¢ stamps.

2. The students in a chemistry class have 75 mL of pure acid which must be diluted to a solution that is 10% acid. How many milliliters of water should they add?
Let w = the number of milliliters of water to be added.

	Volume of solution	% acid	Volume of acid
Original solution	75	100	75(1.00)
Added water	w	0	0
New solution	(75 + w)	10	0.1(75 + w)

Original volume of acid + Added volume of acid = Total volume of acid.
$$75(1.00) + 0 = 0.1(75 + w)$$
$$75 = 7.5 + 0.1w$$
$$67.5 = 0.1w$$
$$675 = w$$
∴ 675 mL of water must be added.

Suggested Assignments

Minimum
First day: 199/*P*: 1–5
Second day: 199/*P*: 6–14
Average
   199/*P*: 1–14
Maximum
   199/*P*: 1–9 odd, 11–18

Mixed Review

Solve.

1. $4(c - 3) + 9 > 2(3c - 4)$
   $\{c: c < 2\frac{1}{2}\}$

2. $0.07(4x) + 0.5(x) = 11.7$
   $\{15\}$

3. $4n - 3 \leq 9$ or $3n - 12 > 9$
   $\{n: n \leq 3 \text{ or } n > 7\}$

4. $|a| \leq 6$ $\{a: -6 \leq a \leq 6\}$

5. $|2 - 4y| \leq 6$
   $\{y: -1 \leq y \leq 2\}$

Many problems that appear to be unrelated to each other can be thought of as "mixture" problems. If you can identify a problem as a mixture problem, then you can use the techniques of this section in solving it. All the problems in the following exercises can be treated as mixture problems.

Oral Exercises

1. Eric wishes to mix together some common stone worth $.10 per kilogram and some granite chips worth $.15 per kilogram to produce a decorative garden mixture. How much of each kind of stone should be mixed to obtain 300 kg of a mixture that is worth $.13 per kilogram?

   a. Copy and complete the chart.

	Number of kg	Value per kg	Total value	
Common stone	x	_?_	_?_	0.10, 0.10 x
Granite chips	_?_	_?_	_?_	300 − x, 0.15, 0.15(300 − x)
Mixture	_?_	_?_	_?_	300, 0.13, 0.13(300)

   $0.10x + 0.15(300 - x) = 0.13(300)$

   b. State an equation that can be used to solve the problem.

   c. Solve the problem. stone, 120 kg; granite, 180 kg

2. Lucy has a collection of dimes and quarters. If she has 80 coins in all and their total value is $12.20, how many coins are there of each type?

   a. Copy and complete the chart. Note that one space is left blank since it is not necessary to find the "average value" of each coin.

	Number of coins	Value per coin	Total value	
Dimes	x	_?_	_?_	0.10, 0.10x
Quarters	_?_	_?_	_?_	80 − x, 0.25, 0.25(80 − x)
Mixture of Coins	_?_		_?_	80, 12.20

   $0.10x + 0.25(80 - x) = 12.20$

   b. State an equation that can be used to solve the problem.

   c. Solve the problem. 52 dimes, 28 quarters

Problems

Use a chart in solving each of the following "mixture" problems.

A   1. How much gold valued at $12 per gram must be mixed with silver valued at $.80 per gram to obtain 20 g of an alloy worth $5 per gram?   7.5 g

2. How many liters of orange juice worth $1.80 per liter should be mixed with 2.4 L of lemonade worth $1.15 per liter to produce a punch worth $1.54 per liter?   3.6 L

3. A 400 g solution is 25% salt. How much salt must be added to produce a solution that is 40% salt?   100 g

4. A 30 L solution is 80% antifreeze. How much water must be added to produce a solution that is 60% antifreeze?   10 L

5. Matt has twice as many quarters as dimes in his coin collection. If the dimes and quarters together total $9.00, how many of each kind of coin does he have?   15 dimes, 30 quarters

6. Cecelia bought some 22¢ and some 17¢ stamps for the school office. Altogether she purchased 37 stamps for $7.54. How many stamps of each kind did she buy?   22¢: 25; 17¢: 12

7. Bus fares from Amherst to Mount Mohawk are $26 for adults and $18 for students. How many students are on the bus if a total of $1668 was collected from the 70 passengers?   19 students

8. The registration fee for a whale watching trip was $20 for nonmembers and $15 for members. If $15,950 was collected from 830 people, how many of the people who registered were members?   130 members

9. Jane took three hours to drive to her friend's home on the weekend. Her average speed was 85 km/h for the first two hours. What was her average speed for the last hour of the trip if her average speed was 75 km/h for the whole trip?   55 km/h

10. For the first hour of a four hour trip, the average speed of a train was 40 km/h. If the average speed for the whole trip was 70 km/h, what was the average speed for the last three hours of the trip?
   80 km/h

Solve.

B   11. Shirley has a quantity of yellow tulip bulbs worth 10¢ each, pink tulip bulbs worth 20¢ each, and red tulip bulbs worth 25¢ each. In her total collection of bulbs, there are five more pink bulbs than yellow bulbs and twice as many red bulbs as the sum of the other two kinds. If the bulbs altogether are worth $133.50, how many of each kind does Shirley have?   100 yellow, 105 pink, 410 red

12. Kevin has $6.45 in coins in his cash box. The number of quarters is one less than twice the number of dimes. The number of nickels is one less than twice the number of quarters. The value of the pennies is the same as the value of the nickels. How many of each type of coin does he have?   7 dimes, 13 quarters, 25 nickels, 125 pennies

1. Ed has $7.80 in nickels and half dollars. If he has three times as many nickels as half dollars, how many of each does he have?
   12 half dollars, 36 nickels

2. A 20 L solution is 15% alcohol. How much water must be added to make it an 8% alcohol solution?
   17.5 L

3. On a test each question in Part A was worth 4 points and each question in Part B was worth 6 points. George answered 15 questions correctly and had a point total of 76. How many questions did he answer correctly in each part of the test?
   Part A: 7 questions
   Part B: 8 questions

4. The average speed of a plane flying from Boston to Los Angeles is 830 km/h. If the plane flew the first and last hours of the 5 h flight at a speed of 500 km/h, what is the average cruising speed for the middle hours of the flight?   1050 km/h

5. Tickets to the school play cost $3 in advance and $4 at the door. In all, 684 tickets were sold for a total of $2457. How many tickets were sold at the door?
   405 tickets

6. How many kilograms of dried apricots worth $7.50 per kilogram must be added to 10 kg of raisins worth $5.50 per kilogram to form a mixture worth $6.50 per kilogram?
   10 kg

13. The average mark on a test in an algebra class is 80. If the two lowest scores of 34 and 48 are not counted, the remaining scores would average 83. How many students are in the algebra class? 28 students

14. At a family reunion, the average age of all those present was 45 years. If the two oldest people, aged 86 and 84 years, had not been present, the average age would have been 41 years. How many people were at the reunion? 22 people

C 15. Dominique flew her light plane from Buffalo to Cleveland at an average speed of 150 km/h. She waited at Cleveland one half the time that she had been in the air on her trip to Cleveland. Returning home, she took a route that was 60 km longer and flew at an average speed of 180 km/h. If the total trip took five hours, how long did she stay in Cleveland? 1 h

16. Stan wants to fill a 204 L drum with a mixture of oil and gasoline so that the ratio of oil to gasoline is 1 to 16. He bought oil at $.85 per liter and gasoline at $.34 per liter. What is the minimum he can charge per liter if he wants to make at least 20% profit? Give the answer in a whole number of cents. $.44

17. A grass seed mixture is 30% rye seed and 10% bluegrass seed. How many kilograms of seed that contains neither bluegrass nor rye seed should be added to 80 kg of the mix to produce a blend that is 20% rye seed? What percent of the new blend will be bluegrass seed? 40 kg; $6\frac{2}{3}$%

18. A 22 kg solution of salt and water is 24% salt. Water is evaporated from this solution to produce a solution that is 32% salt. If 2.2 kg of salt is then added to the solution that is 32% salt, what percent of this new solution will be salt? 40%

Computer Exercises For students with computer experience

1. Write a program that will allow you to input values for $x, y, z,$ and w and will compute the percent acid of the solution that results when x mL of a solution that is y% acid is added to z mL of a solution that is w% acid.

2. Write a program that will allow you to input values for $x, y,$ and z and will compute the percent acid of the solution that results when x mL of water is added to y mL of a solution that is z% acid.

3. Write a program that will allow you to input values for $x, y,$ and z and will compute the number of mL of water that must be added to x mL of a solution that is y% acid to obtain a solution that is z% acid.

4. Modify the program that you wrote for Exercise 3 so that it will allow you to input values for $x, y, z,$ and w and will compute the number of mL of a solution that is w% acid that must be added to x mL of a solution that is y% acid to obtain a solution that is z% acid.

200 *Chapter 4*

4–10 Problems without Solutions

Not every problem has a solution. Consider this example.

EXAMPLE 1   Find three consecutive odd integers whose sum is 144.

SOLUTION

Step 1   The problem asks for three consecutive odd integers whose sum is 144.

Step 2   Let n = first odd integer. Then:

$n + 2$ = next greater odd integer,

$n + 4$ = greatest consecutive odd integer

Step 3   The sum of the three consecutive odd integers is 144.

$$n + (n + 2) + (n + 4) = 144$$

Step 4
$$n + (n + 2) + (n + 4) = 144$$
$$3n + 6 = 144$$
$$3n = 138$$
$$n = 46$$

Step 5   The numbers obtained are 46, 48, and 50. However, these are consecutive even integers, not consecutive odd integers. Therefore they do not satisfy the conditions in the problem.

This example is an illustration of a problem in which the conditions given are *inconsistent*. This means that all the conditions in the problem cannot be true at the same time. In fact, you can prove that the sum of three consecutive odd integers is always odd. (See Exercise 24 on page 183.) Therefore, such a sum can never equal 144.

In reading problems, you should be on the lookout for inconsistent conditions. You should also be able to recognize problems in which not enough facts are given for you to obtain a definite answer.

EXAMPLE 2   The sum of two numbers is 10. Find the numbers.

SOLUTION   There is an infinite set of pairs of numbers whose sum is 10. For example:

$$1 + 9 = 10, \qquad 2 + 8 = 10, \qquad 3\frac{1}{2} + 6\frac{1}{2} = 10, \qquad 5.25 + 4.75 = 10,$$

and so on. In order to determine the required numbers, additional information is needed.

∴ the solution cannot be determined.

Some of the problems in the following set can be solved. Others fail to have solutions either because their conditions are inconsistent or because too few facts are given.

Solving Inequalities and Problems **201**

Teaching Suggestions
p. T83

Supplementary Material
Test 14

Key Ideas
Recognize problems without solutions.

Chalkboard Examples

1. Find five consecutive positive integers whose sum is 8.

   Let n = the first integer. Then $n + 1$, $n + 2$, $n + 3$, and $n + 4$, are the next four integers.
   $$n + (n + 1) + (n + 2) + (n + 3) + (n + 4) = 8$$
   $$5n + 10 = 8$$
   $$5n = -2$$
   $$n = -\frac{2}{5}$$

   ∴ there is no solution because $-\frac{2}{5}$ is neither positive nor an integer. (Inconsistent conditions)

2. Find three consecutive even integers whose sum is 131.

   Let n = the least even integer. Then $n + 2$ = the next even integer; $n + 4$ = the greatest even integer.
   $$n + (n + 2) + (n + 4) = 131$$
   $$3n + 6 = 131$$
   $$3n = 125$$
   $$n = 41\frac{2}{3}$$

   ∴ there is no solution because n is neither even nor an integer. (Inconsistent conditions)

Suggested Assignments

Minimum
  202/P: 1–12
  R 203/Self-Test 2
Average
  202/P: 1–16
  R 203/Self-Test 2
Maximum
  202/P: 1–18
  R 203/Self-Test 2

Additional Answers
Problems
(See p. 204.)

Additional A Exercises

Solve. If a problem has no solution, explain why.

1. Find three consecutive positive integers whose sum is 5. Inconsistent conditions

2. ∠A and ∠B are supplementary angles, each measuring between 0° and 180°. ∠B is twice ∠A. Find the complement of ∠B. Inconsistent conditions

3. Jon and Susan ride to school along the same path, a journey of 4 miles. If Susan rides faster than Jon and Jon needs 15 min to ride to school, how long does it take Susan to ride to school? Insufficient information

4. Lois has some coins whose total value is $11.25. She has 3 nickels, 6 dimes, and some half dollars. How many half dollars does she have? 21

Problems

See Additional Answers for explanations.

Solve. If a problem has no solution, explain why.

A 1. Find two consecutive integers whose sum is 102. No solution

2. Find the least two consecutive integers whose difference is 54. No solution

3. The lengths in meters of the sides of a triangle are three consecutive even integers. If the perimeter of the triangle is 132 m, find the length of each side. 42 m, 44 m, 46 m

4. The lengths in meters of two adjacent sides of a rectangle are consecutive multiples of 5. If the perimeter of the rectangle is 125 m, find the dimensions. No solution

5. In △ABC, the measure of ∠B is three more than four times the measure of ∠A. If ∠B and ∠C are complementary angles, find the measure of each angle. No solution

6. ∠A and ∠B are complementary angles, each with measure between 0° and 90°. The sum of twice the measure of ∠A and three times the measure of ∠B is 180°. Find the measures of ∠A and ∠B. No solution

7. A clerk took a twenty-dollar bill to the bank to get change. The clerk asked for twice as many dimes as nickels and three times as many quarters as nickels. Was the teller able to fulfill the request?

8. The difference between a two-digit number and the number obtained by reversing its digits is 9. If the digits are consecutive integers, find the number. No solution

9. Tom has 18 coins. Some are dimes and the rest are nickels. What is the greatest number of dimes he can have if the value of the nickels is greater than that of the dimes? 5 dimes

10. A total of $458 was collected for admission from 132 people who attended a show. If adults paid $4.00 each and children paid $2.50 each, how many children attended the show? No solution

11. On a trip of 210 km, Gary traveled by train for 3 h and by bus for the rest of the trip. The average speed of the train was 15 km/h more than that of the bus. Find the average speed of the bus. No solution

12. Susan walks at an average speed of 2.5 m/s to her school, which is 3 km from her home. Fifteen minutes after Susan left for school, her brother discovered she had forgotten her lunch. How fast must he ride his bicycle in order to overtake her before she reaches school?

B 13. A freight train takes 16 h to travel from Moncton to Parkroyal. A passenger train makes the trip in 12 h. If the average speed of the passenger train is 30 km/h more than the freight train, how far is it from Moncton to Parkroyal? 1440 km

14. Two cars leave a rest stop at the same time and travel south on the same road. Five hours later the slower car passes a service plaza that the faster car had passed an hour earlier. If their average speeds differ by 17 km/h, how far is the service plaza from the rest stop?
340 km

15. Eighteen-carat gold contains 18 parts by mass of gold to 6 parts of other metals. Fourteen-carat gold contains 14 parts of gold and 10 parts of other metals. How many kilograms of eighteen-carat gold should be mixed with fourteen-carat gold to obtain 60 kg of an alloy containing 17 parts gold and 7 parts of other metals? 45 kg

16. Terry's average grade on the first six tests in her history class was 75%. She got 100% on each of the next two tests. If the highest score possible on any test is 100%, what grades must Terry get on the final two tests in order to have an average of 90% for the course? No solution

C 17. Lisa travels at 30 km/h on the way to visit a friend who lives 90 km away. How fast must she go on the return trip if she wants her average speed for the whole trip to be 40 km/h? 50 km/h? 60 km/h?

18. Find the least two positive integers whose sum is an even integer and whose difference is an odd integer. No solution

■ Self-Test 2

VOCABULARY   consecutive integers (p. 179)   right angle (p. 184)
multiple (p. 180)   complementary angles (p. 184)
consecutive multiples (p. 180)   supplementary angles (p. 184)
consecutive even integers (p. 181)   uniform motion (p. 189)
consecutive odd integers (p. 181)

Solve.

1. Find three consecutive even integers whose sum is 154 more than the greatest of these three integers. 76, 78, 80   *Obj. 1, p. 179*

2. In △ABC, the measure of ∠A is 12° less than the measure of ∠B and the measure of ∠C is 18° more than the measure of ∠A. Find the measure of each angle.   measure of ∠A is 50°, measure of ∠B is 62°, measure of ∠C is 68°

3. Ed's average speed is 20 km/h when he is jogging. He leaves his school at 10:00 A.M. and starts jogging to the next town. His coach leaves at 10:15 A.M. in his car and follows the same route. If the coach's average speed is 60 km/h, how far from the school will the coach overtake Ed? 7.5 km

4. How many kilograms of dried apricots worth $8.30 per kilogram should be mixed with dried apples worth $6.50 per kilogram to produce 12 kg of a mixture worth $6.95 per kilogram? 3 kg

5. Sally has twenty coins in her purse. Some of them are dimes and the remainder are quarters. If she does not have enough money to buy a book for $3.98, what is the most number of quarters she can have? 13 quarters   *Obj. 2, p. 179*

Check your answers with those at the back of the book.

Solving Inequalities and Problems   **203**

Solve.

1. If ∠A and ∠B are supplementary angles and the measure of ∠A = 35°, find the measure of ∠B. 145°

2. If ∠A and ∠B are complementary angles and the measure of ∠B = 60°, find the measure of ∠A. 30°

3. Find three consecutive odd integers whose sum is 75. 23, 25, 27

4. Two planes leave Cincinnati at the same time. One flies due east at 992 km/h and the other flies due west at 896 km/h. In how many hours will they be 7552 km apart? 4 h

Quick Quiz

Solve.

1. Find four consecutive odd integers whose sum is 213 more than the greatest of these integers. 69, 71, 73, 75

2. In △ABC the measure of ∠A is 20° more than the measure of ∠B and the measure of ∠C is 3 times the measure of ∠B. Find the measure of each angle. 52°, 32°, 96°

3. The captain of a luxury liner radioed for a naval station helicopter to pick up an injured passenger. When the helicopter left its base the liner was 460 km away and heading toward the naval station. If the average speed of the liner was 20 km/h and that of the helicopter was 180 km/h, how long did it take the helicopter to reach the liner? 2.3 h

(continued)

4. A grocer mixed some dried fruit selling for $6.00 per kilogram with 60 kg of nuts worth $7.50 per kilogram. The resulting mixture sold for $6.60 per kilogram. How many kilograms of dried fruit did the grocer use?   90 kg

5. Anne has 35 coins in her purse. Some of them are nickels and some are dimes. She would like to buy a magazine for $2.70 but she does not have enough money. What is the greatest number of dimes she can have?   18

**Additional Answers
Problems** (p. 202)

1. No solution; the sum of two consecutive integers must be odd because the sum of an odd integer and an even integer is odd.

2. No solution; the difference of two consecutive integers is always 1 and can never be 54.

4. No solution; the lengths of the sides must be integers since they are multiples of 5, so the perimeter must be a multiple of 2 since $P = 2(l + w)$ and cannot be odd.

5. No solution; $\angle B$ and $\angle C$ are complementary angles so they are acute angles whose measures total 90°. The measure of $\angle A$ is 90°, and so the measure of $\angle B$ cannot be greater than the measure of $\angle A$.

EXTRA

Symbolic Logic: Boolean Algebra

Most people think of algebra as the study of operations with numbers and variables. Another kind of algebra, which is used in the design of electronic digital computers, involves operations with logical statements. This "algebra of logic" is called Boolean algebra. (Boolean algebra is named after George Boole, a nineteenth-century British mathematician who developed the fundamental principles on which this algebra is based.)

In Boolean algebra you use letters such as p, q, r, s, and so on, to represent statements. For example, you might let p represent the statement "3 is an odd integer," and q, the statement "5 is less than 3." In this case, the statement p has the truth value T (the statement is true), whereas q has the truth value F (the statement is false).

The following table shows the operations that are used in Boolean algebra to produce compound statements from any given statements p and q.

Operation	How Read	Symbols
conjunction	p and q	$p \land q$
disjunction	p or q	$p \lor q$
conditional	If p, then q	$p \to q$
equivalence	p if and only if q	$p \leftrightarrow q$
negation	not p	$\sim p$

The rules for assigning truth values to compound statements are shown in the five *truth tables* that follow.

Conjunction

p	q	$p \land q$
T	T	T
T	F	F
F	T	F
F	F	F

Disjunction

p	q	$p \lor q$
T	T	T
T	F	T
F	T	T
F	F	F

Notice that the conjunction $p \land q$ is true when *both* p and q are true; otherwise, $p \land q$ is false. On the other hand, the disjunction $p \lor q$ has the value T provided *at least one* of the statements p, q is true.

204   *Chapter 4*

Conditional

p	q	$p \to q$
T	T	T
T	F	F
F	T	T
F	F	T

Equivalence

p	q	$p \leftrightarrow q$
T	T	T
T	F	F
F	T	F
F	F	T

When is the conditional $p \to q$ false? Only when p is true and q is false.

The equivalence $p \leftrightarrow q$ is a brief way of stating the conjunction

$$(p \to q) \land (q \to p).$$

The equivalence is true when p and q are both true or both false.

Negation

p	$\sim p$
T	F
F	T

The negation $\sim p$ is the denial of p. Therefore, it is reasonable to agree that $\sim p$ is false when p is true, and true when p is false.

EXAMPLE 1 Let r represent "$1 > 2$," and s, "$2 < 4$." Read each of the following statements. Then, referring to the preceding truth tables, give the truth value of the statement and a reason for the answer.

a. $r \land s$ **b.** $r \lor s$ **c.** $r \to s$ **d.** $r \leftrightarrow s$ **e.** $\sim r$

SOLUTION

a. $r \land s$: $1 > 2$ *and* $2 < 4$.
F, because the truth value of r is F.

b. $r \lor s$: $1 > 2$ *or* $2 < 4$.
T, because the truth value of s is T.

c. $r \to s$: *If* $1 > 2$, *then* $2 < 4$.
T, because the truth value of r is F, and that of s is T.

d. $r \leftrightarrow s$: $1 > 2$ *if and only if* $2 < 4$.
F, because r and s have different truth values.

e. $\sim r$: *Not* $(1 > 2)$, that is, $1 \le 2$.
T, because the truth value of r is F.

EXAMPLE 2 Show that for all truth values of r, p, and q, the truth value of

$$r \lor (p \land q) \leftrightarrow (r \lor p) \land (r \lor q)$$

is T.

SOLUTION Construct a truth table containing the eight combinations of truth values of r, p, and q.

Can you tell how the table on the following page was constructed? Entries in column 4 were obtained by "*and*-ing" the entries in columns 2 and 3. The entries in column 5 result from "*or*-ing" the entries in columns 1 and 4. The rest of the table is obtained similarly.

6. No solution; the measure of $\angle A$ would have to be 90°, but the measure of $\angle A$ is given as *between* 0° and 90°.

7. Yes; the clerk received 20 nickels, 40 dimes, and 60 quarters.

8. No unique solution; if the first two-digit number is greater than the second it will be 9 greater in every case.

10. No solution; the solution of the equation $2.5x + 4(132 - x) = 458$ is not a whole number. The number of children's tickets must be a whole number.

11. No solution; not enough information is given; the time traveled by bus is needed.

12. He must ride faster than 10 m/s.

16. No solution; Terry would have to average 125% on the final two tests to average 90%, but no score may be greater than 100%.

17. 60 km/h; 150 km/h; no solution because the return trip would have to be made in zero hours.

18. No solution; if the sum of two positive integers is even, either they are both even or they are both odd. In either case their difference is also even.

r	p	q	$p \wedge q$	$r \vee (p \wedge q)$	$r \vee p$	$r \vee q$	$(r \vee p) \wedge (r \vee q)$
T	T	T	T	T	T	T	T
T	T	F	F	T	T	T	T
T	F	T	F	T	T	T	T
T	F	F	F	T	T	T	T
F	T	T	T	T	T	T	T
F	T	F	F	F	T	F	F
F	F	T	F	F	F	T	F
F	F	F	F	F	F	F	F

When you compare the fifth column of this table to the last column, you can see that $r \vee (p \wedge q)$ and $(r \vee p) \wedge (r \vee q)$ have the same array of truth values. Therefore,

$$r \vee (p \wedge q) \leftrightarrow (r \vee p) \wedge (r \vee q)$$

is a true statement no matter what truth values are assigned to r, p, and q.

A compound statement that is true for all truth values of its component statements is called a **tautology**.

Exercises

Assume that r and p are true statements and that q is a false statement. Determine the truth value of each statement.

1. $q \rightarrow r$ T

2. $\sim r \wedge p$ F

3. $p \vee \sim q$ T

4. $r \wedge (p \vee q)$ T

5. $p \vee (q \wedge r)$ T

6. $(p \vee r) \rightarrow q$ F

7. $\sim p \rightarrow (q \vee \sim r)$ T

8. $r \rightarrow (q \rightarrow r)$ T

9. $(p \rightarrow q) \leftrightarrow (\sim p \vee q)$ T

Given that $p \rightarrow q$ is a false statement, show that each statement is true.

10. $q \rightarrow p$

11. $p \wedge \sim q$

12. $(p \vee q) \wedge p$

Construct a truth table for each of the following statements. Then tell whether or not each statement is a tautology. Truth tables are not shown.

13. $(p \vee q) \rightarrow p$ No

14. $(q \wedge \sim q) \rightarrow p$ Yes

15. $(p \vee q) \leftrightarrow (q \vee p)$ Yes

16. $(p \rightarrow q) \leftrightarrow (\sim p \vee q)$ Yes

17. $r \wedge (p \vee q) \rightarrow (r \wedge p) \vee (r \wedge q)$ Yes

18. $[(p \rightarrow q) \wedge (q \rightarrow r)] \rightarrow (p \rightarrow r)$ Yes

Chapter Summary

Statements 1–4 are true for all values of each variable except as noted.

1. *Axiom of comparison* One and only one of the following statements is true: $a < b$, $a = b$, or $b < a$

2. *Transitive property of order*
   a. If $a < b$ and $b < c$, then $a < c$.
   b. If $a > b$ and $b > c$, then $a > c$.

3. *Addition property of order*
   a. If $a < b$, then $a + c < b + c$ and $c + a < c + b$.
   b. If $a > b$, then $a + c > b + c$ and $c + a > c + b$.

4. *Multiplication property of order*
   a. If $a < b$ and $c > 0$, then $ac < bc$ and $ca < cb$.
   b. If $a > b$ and $c > 0$, then $ac > bc$ and $ca > cb$.
   c. If $a < b$ and $c < 0$, then $ac > bc$ and $ca > cb$.
   d. If $a > b$ and $c < 0$, then $ac < bc$ and $ca < cb$.

5. Inequalities that have the same solution set over a given domain are called *equivalent inequalities*. Each of the following transformations of a given inequality will produce an equivalent inequality.
   1. Substituting for either side of the inequality an expression equivalent to that side.
   2. Adding to (or subtracting from) each side the same real number.
   3. Multiplying (or dividing) each side by the same positive number.
   4. Multiplying (or dividing) each side by the same negative number and reversing the order of the inequality.

6. If S and T are any sets, the set consisting of the members belonging to both S and T is called the *intersection* of S and T and is denoted by $S \cap T$. The set consisting of the members belonging to *at least one* of the sets S and T is the *union* of S and T and is denoted by $S \cup T$.

7. Sets having no members in common are *disjoint sets*.

8. A sentence that is formed by joining two sentences with the word *and* is called a *conjunction*. For a conjunction to be true, *both* of the joined sentences must be true. A sentence that is formed by joining two sentences with the word *or* is a *disjunction*. For a disjunction to be true, *at least one* of the joined sentences must be true.

9. Some open sentences involving absolute value can be expressed as conjunctions while others can be expressed as disjunctions.

10. Problems involving integers, angles, uniform motion, and mixtures can often be solved by following the plan outlined on page 132.

11. Some problems cannot be solved either because of insufficient information or because the information given is inconsistent.

Solving Inequalities and Problems **207**

Additional Answers
Written Exercises
(continued from p. 167)

9. $\{t: t > -6\}$

10. $\{x: x < 0\}$

11. $\{w: w > 0\}$

12. $\{p: p < 4\}$

13. $\{c: c > 7\}$

14. $\{q: q < 18\}$

15. $\{d: d < -13\}$

16. $\{r: r > -4\}$

17. $\{e: e < -6\}$

18. $\{s: s < -11\}$

19. $\{f: f > -12\}$

20. $\{t: t > 0\}$

21. $\{u: u > 15\}$

22. $\{g: g < -9\}$

23. $\{h: h < 19\}$

24. $\{v: v < 2\}$

25.

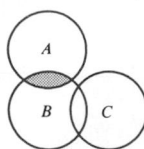

26.

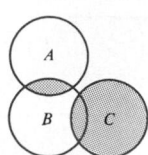

27.

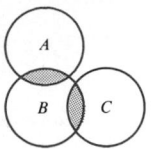

28.

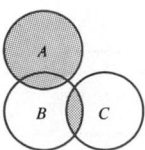

29.

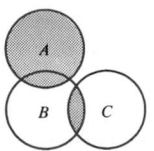

30.

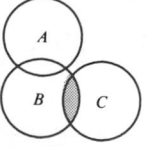

31.

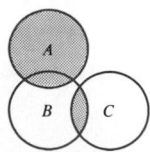

32.

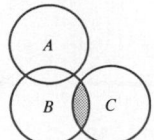

Chapter Review

Write the letter of the correct answer.

1. If $b < c$, which of the following statements is false? 4-1
   a. $b - 5 < c - 5$
   b. $5 + b < 5 + c$
   c. $\dfrac{b}{-5} < \dfrac{c}{-5}$
   d. $5b < 5c$

2. Solve $10a + 4 > 36 + 2a$. 4-2
   a. $\{a: a > 5\}$
   b. $\{a: a < 4\}$
   c. $\{a: a < 5\}$
   d. $\{a: a > 4\}$

3. Solve $5z + 3 - 7z \le 11$.
   a. $\{z: z \ge -4\}$
   b. $\{z: z \le -4\}$
   c. $\{z: z \le -7\}$
   d. $\{z: z \ge -7\}$

4. Specify the union of $\{1, 2, 3\}$ and $\{2, 3, 4\}$. 4-3
   a. $\{2, 3\}$
   b. $\{1, 4\}$
   c. $\{1, 2, 3, 4\}$
   d. $\emptyset$

5. Specify the intersection of {the real numbers between -6 and 2} and {the real numbers greater than -1}.
   a. {the real numbers between -6 and -1}
   b. {the real numbers greater than -6}
   c. {the real numbers between -1 and 2}
   d. $\emptyset$

Solve.

6. $-6 < x - 2 < 4$ 4-4
   a. $\{x: -8 < x < 2\}$
   b. $\{x: -4 < x < 2\}$
   c. $\{x: -4 < x < 6\}$
   d. $\{x: -8 < x < 6\}$

7. $2y + 3 \le -2$ or $3y - 5 > 4$
   a. $\{y: y \le -\frac{5}{2}$ or $y > 3\}$
   b. $\{y: -\frac{5}{2} \le y < 3\}$
   c. $\{y: y \ge -\frac{5}{2}$ and $y < 3\}$
   d. $\{y: y < -3$ or $y \ge \frac{5}{2}\}$

8. $5 + |2m - 9| \le 6$ 4-5
   a. $\{m: m \le 5\}$
   b. $\{m: m \le 4\}$
   c. $\{m: 4 \le m \le 5\}$
   d. $\{m: -5 \le m \le -4\}$

9. $|1 - 3n| > 7$
   a. $\{n: -\frac{8}{3} < n < 2\}$
   b. $\{n: n < -2$ or $n > \frac{8}{3}\}$
   c. $\{n: -2 < n < \frac{8}{3}\}$
   d. $\{n: n < -\frac{8}{3}$ or $n > 2\}$

10. Find the least three consecutive integers whose sum is greater than -87. 4-6
    a. $-29, -28, -27$
    b. $-31, -32, -33$
    c. $31, 32, 33$
    d. $-29, -30, -31$

208 Chapter 4

11. The measure of one angle of a triangle is 1° less than twice the measure of the second angle, and the measure of the third angle is 21° more than twice the sum of the measures of the other two. Find the measure of the third angle.   *4-7*

    a. 18°   (b.) 127°   c. 35°   d. 106°

12. A Coast Guard helicopter flew toward a ship that was heading for shore with an injured sailor. The average speed of the ship was 40 km/h and the average speed of the helicopter was 300 km/h. The helicopter started from the shore when the ship was 510 km away. How long did it take them to meet?   *4-8*

    a. 0.5 h   b. 1 h   c. 1.75 h   (d.) 1.5 h

13. A 60 L solution is 60% acid. How many liters of water must be added to produce a solution that is 45% acid?   *4-9*

    a. 18   (b.) 20   c. 24   d. 32

14. Tell why this problem has no solution: "Find four consecutive odd integers whose sum is −60."   *4-10*

    (a.) The sum of four odd integers cannot be an even integer.

    b. The sum of four odd integers cannot be negative.

    c. There are not enough facts given to write an equation.

    d. The only solution of an equation that repesents the relationships in the problem is an even integer.

Chapter Test

1. Name the property that is illustrated by the following: *4-1*
   "If $-3m > 15$, then $m < -5$." Multiplication property of order

Solve.

2. $7 - 2p > 15$ 3. $2q + 50 < 7q + 2 + 3q$ *4-2*
   $\{p: p < -4\}$ $\{q: q > 6\}$

Let A = {the natural numbers less than 5} and B = {the integers greater than −3}. Specify the following.

4. $A \cup B$ 5. $A \cap B$ {the natural numbers less than 5}   *4-3*
   {the integers greater than −3}

Solve.

6. $-8 \leq 1 - 3x < 10$ 7. $3y + 5 < 3$ or $2y - 5 > 3$ *4-4*
   $\{x: -3 < x \leq 3\}$ $\{y: y < -\frac{2}{3}$ or $y > 4\}$

8. $|3 - 2m| \geq 5$ 9. $2|4 + n| - 5 < 9$ $\{n: -11 < n < 3\}$ *4-5*
   $\{m: m \leq -1$ or $m \geq 4\}$

10. Find the greatest two consecutive odd integers whose sum is less than 85. 41 and 43   *4-6*

Solving Inequalities and Problems **209**

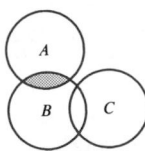
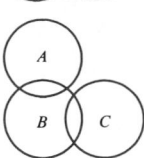
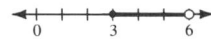

14. The measure of the third angle must be an even integer since the sum of the measures of the first and second angles is even, and the sum of the measures of the three angles is 180°. But the measure of the third angle is odd, since it is the sum of 37° and an even integer.

11. The measure of an angle is 16° more than the measure of its supplement. Find the measure of the angle. 98°   *4-7*

12. Jim's average speed is 8 km/h while walking and 12 km/h while jogging. It takes him 0.5 h longer to walk from the school to the pool than it takes him to jog the same distance. How far is it from the school to the pool? 12 km   *4-8*

13. How many liters of water must be added to 90 L of a solution that is 80% antifreeze to obtain a solution that is 60% antifreeze? 30 L   *4-9*

14. Explain why the following problem has no solution: The degree measures of two angles of a triangle are consecutive even integers. The measure of the third angle is 37° more than the measure of the smallest angle. Find the measure of each angle.   *4-10*

Mixed Review

Simplify.   **4.** $-\dfrac{3}{7}$ **8.** $-5t^2 + 2t$

1. $\dfrac{16 - 4(2)}{12 - 3(3)}$ $\dfrac{8}{3}$

2. $\dfrac{3^2 + 5^2}{(5 - 3)^2}$ $\dfrac{17}{2}$

3. $-\dfrac{2}{3} \div \dfrac{4}{9}$ $-\dfrac{3}{2}$

4. $-\dfrac{2}{7} - \dfrac{5}{7} + \dfrac{4}{7}$

5. $-4a^2 + a^2$ $-3a^2$ **6.** $(-5b)(-2b)$ $10b^2$ **7.** $-12s^2 \div (-3)$ $4s^2$ **8.** $3t - 5t^2 - t$

9. $-(-5 + 3) + (-8)$ -6 **10.** $24 \div 3(7 - 3) + 4$ 6 **11.** $15 + 2(3^2 - 2)^2$ 113

12. $9m^2 - 4(m^2 + 2)$ **13.** $3(2n - 1) - 2(3n + 1)$ **14.** $(4x^2 - 2x) - (x^2 - 5)$
       $5m^2 - 8$ -5 $3x^2 - 2x + 5$

Graph the solution set of each open sentence on a number line.

15. $0 \le x < 8$ **16.** $y < 2$ or $y \ge 6$ **17.** $|z| < 2$

18. $|a| > -3$ **19.** $b > -2$ and $b < 3$ **20.** $2 < |c| < 5$

Solve.   **24.** $\{y: y < -1\}$ **25.** $\{m: m < 3\}$ **26.** $\{-6\}$ **31.** $\{j: -4 < j < \frac{3}{2}\}$

21. $\dfrac{1}{a} = -2$ $\{-\frac{1}{2}\}$ **22.** $-\dfrac{b}{3} = \dfrac{1}{6}$ $\{-\frac{1}{2}\}$ **23.** $4 - 2x = 6$ $\{-1\}$

24. $4y > 5y + 1$ **25.** $5 - (2 - m) < 6$ **26.** $3n - 4(n - 1) = 10$

27. $|p| < -2$ ∅   **28.** $-q > 7$ $\{q: q < -7\}$ **29.** $|x - 5| = 2$ $\{3, 7\}$

30. $|y + 1| + |-7| = 4$ ∅   **31.** $-9 < 2j - 1 < 2$ **32.** $-2k < 8$ or $-3k > 9$
                                                                   $\mathcal{R}$

33. Jane has six more nickels than quarters and two fewer dimes than nickels. If she has a total of $2.70, how many of each type of coin does she have? 11 nickels, 9 dimes, 5 quarters

34. Find three consecutive odd integers whose sum is -162. no solution

35. Three years from now, Bob will be three times as old as he was three years ago. How old is Bob now? 6 years

36. How many liters of water must be added to 20 L of a solution that is 45% alcohol to obtain a solution that is 30% alcohol? 10 L

210   *Chapter 4*

15.

16.

17.

18.

19.

20.

PREPARING FOR COLLEGE ENTRANCE EXAMS

Strategy for Success: Before you actually answer any questions, you may find it helpful to skim an entire section of an exam in order to get an *overview* of the questions. You may wish to answer the easiest questions first, then proceed to the harder ones. Do not take time to double-check your answers unless you finish all the questions before the deadline and have extra time.

Decide which is the best of the choices given and write the corresponding letter on your answer sheet.

1. Under which conditions is $x + y = |x| - |y|$ a true statement? A
   I. $x > 0$ and $y \le 0$ II. $x < 0$ and $y \ge 0$ III. $x > 0$ and $y \ge 0$
   (A) I only   **(B)** II only   **(C)** III only   **(D)** I and II only   **(E)** I, II, and III

2. Solve $x + 3 \le -4$ or $-2x < 14$. B
   (A) $\emptyset$ **(B)** $\mathcal{R}$ **(C)** $\{x: x \ge 7\}$ **(D)** $\{x: x \le -7\}$ **(E)** $\{x: -7 < x \le 7\}$

3. Find the least of four consecutive even integers such that the third integer is equal to the sum of the first integer and twice the fourth integer. B
   (A) 2   **(B)** -4 **(C)** -2 **(D)** 18   **(E)** -12

4. Which of the following sets are closed under division, excluding division by zero? C
   I. $\{-3, 0, 3\}$ II. $\{-3, -1, 1, 3\}$ III. {the positive real numbers}
   (A) I only   **(B)** II only   **(C)** III only   **(D)** I and III only   **(E)** II and III only

5. Solve $\frac{1}{5}|-x| = 10$. D
   (A) $\{-2\}$ **(B)** $\{-50\}$ **(C)** $\{2, -2\}$ **(D)** $\{50, -50\}$ **(E)** $\emptyset$

6. Solve for m in the equation $b = \frac{1}{2}mr^2$. B
   (A) $m = \frac{b}{2r^2}$ **(B)** $m = \frac{2b}{r^2}$ **(C)** $m = \frac{r^2}{2b}$ **(D)** $m = \frac{2r^2}{b}$ **(E)** $m = \frac{2m}{r^2}$

7. Carla's monthly salary for part of the year was $1015. After she received a raise, her monthly salary for the remainder of the year was $1145. Her total earnings for the year, including a bonus of $1500, were $14,720. For how many months did she work at the lesser salary? A
   (A) 4   **(B)** 6   **(C)** 8   **(D)** 9   **(E)** 10

8. Which of the following is equivalent to $\frac{1}{2}x + \frac{1}{3} = 4$? E
   I. $\frac{1}{2}x = \frac{11}{3}$ II. $x + \frac{2}{3} = 8$ III. $3x + 2 = 24$
   (A) I only   **(B)** II only   **(C)** III only   **(D)** I and II only   **(E)** I, II, and III

Solving Inequalities and Problems   **211**

Additional Answers
Written Exercises
(continued from p. 177)

23. $\{v: v \le -3 \text{ or } v \ge 3\}$
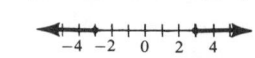

24. $\{t: -2 \le t \le 2\}$
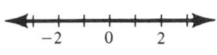

25. $\{w: -6 \le w \le 6\}$

26. $\mathcal{R}$

27. $\{w: w \le 2 \text{ or } w \ge 5\}$

28. $\left\{q: -\frac{4}{3} < q < 2\right\}$
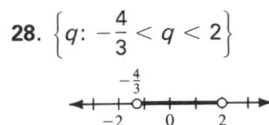

29. $\{c: c \le 1.3 \text{ or } c \ge 2.7\}$

30. $\emptyset$

31. $\mathcal{R}$

32. $\left\{x: -3 < x < \frac{11}{3}\right\}$

33. $\{x: -11 < x < -1\}$
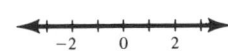

34. $\left\{t: t < \frac{23}{4} \text{ or } t > \frac{25}{4}\right\}$

APPLICATION

Acceleration

It is important for an automobile driver to be able to change the rate at which an automobile is moving. When the rate of motion, or speed, of a car increases, the car is said to *accelerate*.

In physics, acceleration is defined in terms of *velocity* rather than speed. A velocity specifies the direction of travel as well as the speed. For example, it is more informative to know that a car is traveling at a velocity of 40 km/h *due north* than to know only that its speed is 40 km/h.

The acceleration of an object can be defined as the rate at which its velocity changes over a given period of time. That is:

$$\text{acceleration} = \frac{\text{change in velocity}}{\text{time}}$$

Since velocity involves both direction and speed, a change in direction produces a change in acceleration whether or not a change in speed is also involved. For example, when you drive along a curved road at a constant speed, the direction of motion, and thus the acceleration, is changing.

The simplest case of acceleration to analyze involves an object that is moving in a straight line at a constant acceleration. The acceleration in this case can be expressed mathematically as

$$a = \frac{v_f - v_0}{t},$$

where a = constant acceleration, v_0 = initial velocity, v_f = final velocity, and t = time.

Suppose that a car is traveling along a straight road at a constant velocity of 25 km/h. The car then begins to accelerate at a constant rate. After 5 s of constant acceleration, the car is traveling at a velocity of 50 km/h. What is the acceleration of the car over the 5 s period?

Since $v_0 = 25$ km/h, $v_f = 50$ km/h, and $t = 5$ s,

$$a = \frac{v_f - v_0}{t} = \frac{50 \text{ km/h} - 25 \text{ km/h}}{5 \text{ s}} = 5 \text{ km/h/s}.$$

This acceleration is read as "five kilometers per hour per second" and means that the velocity of the car increases by 5 km/h each second.

Now suppose that a car starts from rest, that is, $v_0 = 0$. It then accelerates at a constant rate of 4 km/h/s along a straight road. At what velocity will it be traveling at the end of 15 s?

Since $a = \dfrac{v_f - v_0}{t}$,

$$v_f = v_0 + at = 0 + (4 \text{ km/h/s})(15 \text{ s}) = 60 \text{ km/h}.$$

If you want to find the *average velocity*, v_{ave}, of the car over the 15 s time period, you can use the following formula.

$$v_{ave} = \frac{v_0 + v_f}{2} = \frac{0 + 60 \text{ km/h}}{2} = 30 \text{ km/h}$$

To find the *distance* traveled in the 15 s time period, you can use the formula

$$d = v_{ave} \cdot t.$$

To apply this formula correctly, the units must be compatible. Since the average velocity is given in kilometers per hour,

$$d = v_{ave} \cdot t = 30 \text{ km/h} \cdot \frac{15}{3600} \text{ h} = 0.125 \text{ km}.$$

Exercises

Solve. Approximate answers to the nearest hundredth.

1. A racing car starts from rest and accelerates at a constant rate along a straight road. After 5 s of constant acceleration, the car is traveling at a velocity of 85 km/h.
   a. What is the acceleration of the car over the 5 s time period? 17 km/h/s
   b. What is the average velocity of the car over the 5 s time period? 42.5 km/h
   c. Find the distance traveled by the car over the 5 s time period. approx. 0.06 km

2. Starting from rest, a family car accelerates at a constant rate of 5 km/h/s along a straight road.
   a. At what velocity will it be traveling after 10 s of constant acceleration? 50 km/h
   b. Find the distance traveled by the car over the 10 s time period. approx. 0.07 km

3. Sometimes it is important to know how far an automobile must travel to attain a certain velocity. The acceleration lanes that run roughly parallel to major highways allow cars to accelerate to highway speeds by the time they enter the highway. How long must an acceleration lane be to enable an automobile to accelerate at a constant rate from 36 km/h to 80 km/h in 10 s? approx. 0.16 km

Solving Inequalities and Problems **213**

The photo shows archaeological workers at the site of a subway excavation in Boston. Archaeologists often use string grids such as those shown in order to keep track of the location where objects are uncovered.

Chapter 5

Graphs and Functions

Ordered Pairs and Functions

OBJECTIVES for Sections 5-1 through 5-3:
1. *To graph ordered pairs of numbers on a coordinate plane and to find the coordinates of any point on the plane.*
2. *To identify the domain and range of a relation specified by a set of ordered pairs and to draw a mapping diagram for the relation.*
3. *To determine whether a relation with a given finite domain is a function.*
4. *To graph relations and functions with given finite domains.*

5–1 The Coordinate Plane

Teaching Suggestions
p. T83

Key Ideas

Construct a coordinate plane.
Find the coordinates of points on the coordinate plane.
Graph points on the coordinate plane, given their coordinates.

In Section 1-1 you learned how to graph real numbers as points on a number line. In this section, you will learn how to graph *ordered pairs* of real numbers as points on a *number plane*. The following steps outline a general procedure for constructing a number plane.

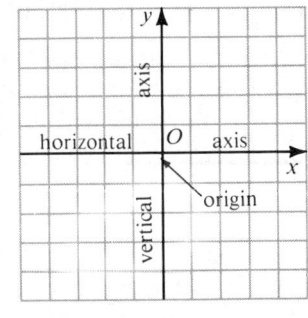

1. Draw two perpendicular number lines that intersect at the origin of each. These two number lines are called **axes.** As shown in Figure 1, one axis is usually horizontal and the other, vertical. The point of intersection of these axes is called the **origin** of the number plane and is labeled *O*.

2. Indicate the positive direction on each axis by an arrowhead. The positive direction is usually to the right on the horizontal axis and upward on the vertical axis, as shown in Figure 1.

Figure 1

Graphs and Functions   **215**

1. Name the coordinates of
   the given point in the
   form (abscissa, ordinate).

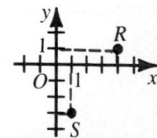

a. *R*
   To find the coordinates,
   draw vertical and horizontal lines from point
   R to both axes. The *x*-
   coordinate is 4 and the
   y-coordinate is 1; write
   R(4, 1).

b. *S*
   To find the coordinates,
   draw vertical and horizontal lines from point
   S to both axes. The *x*-
   coordinate is 1 and the
   y-coordinate is −3;
   write *S*(1, −3).

2. Graph the ordered pairs of
   numbers on a coordinate
   plane: (2, 3) and (−4, −2).

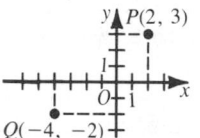

Common Errors

Students may confuse or-
dered pair notation, (*x*, *y*),
with set notation, {*x*, *y*} or
{(*x*, *y*)}. Explain that {*x*, *y*}
names a set containing two
elements, *x* and *y*, and that
{(*x*, *y*)} names a set containing only one element, the ordered pair (*x*, *y*).

Emphasize that (*x*, *y*) is
an *ordered* pair. Reinforce
this by showing that (5, 2)
does not name the same
point as (2, 5).

On a number plane, the horizontal axis is usually labeled with an x and is referred to as the x-**axis.** Similarly, the vertical axis is labeled with a y and is referred to as the y-**axis.** The x- and y-axes together are called **coordinate axes,** and the number plane is called a **coordinate plane.** The coordinate axes separate a coordinate plane into four **quadrants,** which are numbered as shown in Figure 2. Points on the coordinate axes are not considered to be in any quadrant.

	y	
second quadrant (II)		first quadrant (I)
	O	x
third quadrant (III)		fourth quadrant (IV)

Figure 2

Each point on a coordinate plane can be assigned to a pair of real numbers. These numbers are called the **coordinates** of the point. You can find the coordinates of a given point P as follows.

1. Draw a *vertical line* from P to the x-axis. The coordinate of the point at which this line intersects the x-axis is called the x-**coordinate,** or **abscissa,** of P. In Figure 3 below, the abscissa of P is 3.

2. Next, draw a *horizontal line* from P to the y-axis. The coordinate of the point at which this line intersects the y-axis is called the y-**coordinate,** or **ordinate,** of P. In Figure 3, the ordinate of P is 2.

3. Name the coordinates of P as an **ordered pair** of real numbers in the form (*abscissa, ordinate*). In Figure 3, you refer to P as the point (3, 2) and write P(3, 2).

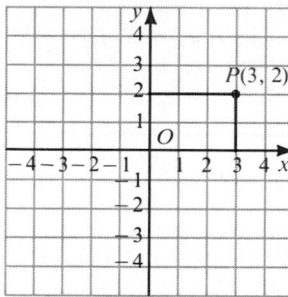

Figure 3

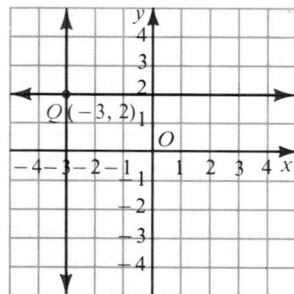

Figure 4

By reversing the process just described, each ordered pair of real numbers can be assigned to a point on a coordinate plane. When you find the point assigned to an ordered pair, you **graph the ordered pair,** or **plot the point** that corresponds to the ordered pair. For example, referring to Figure 4 above, the ordered pair (−3, 2) is graphed as follows.

1. Draw a *vertical line* through the graph of −3 on the x-axis.
2. Next, draw a *horizontal line* through the graph of 2 on the y-axis.
3. Locate the point of intersection of the vertical and horizontal lines. In Figure 4, the graph of (−3, 2) is the point Q.

216 *Chapter 5*

In working with a coordinate plane, you take the following facts for granted.

1. Each ordered pair of real numbers corresponds to exactly one point on a coordinate plane.
2. Each point on a coordinate plane corresponds to exactly one ordered pair of real numbers.

This one-to-one correspondence between the set of all points on a coordinate plane and the set of all ordered pairs of real numbers is called a **plane rectangular,** or **Cartesian, coordinate system.** (The Cartesian coordinate system is named after René Descartes, a seventeenth century French mathematician and philosopher who developed the early ideas about the system.)

Oral Exercises

Exercises 1–16 refer to the coordinate plane at the right. Name the coordinates of each point.

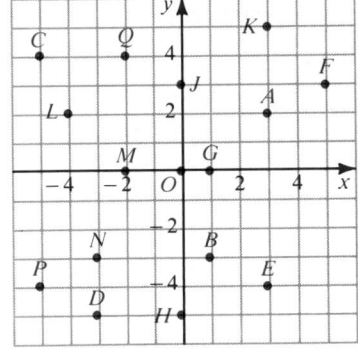

Exs. 1-16

1. A (3, 2)
2. B (1, −3)
3. C (−5, 4)
4. D (−3, −5)
5. G (1, 0)
6. H (0, −5)
7. N (−3, −3)
8. O (0, 0)

Name the point that has the given coordinates.

9. (3, 5) K
10. (5, 3) F
11. (−2, 4) Q
12. (−4, 2) L
13. (3, −4) E
14. (−5, −4) P
15. (−2, 0) M
16. (0, 3) J

Name all the quadrants, if any, that contain points satisfying the given requirements.

17. The ordinate is 2. I, II
18. The ordinate is −3. III, IV
19. The abscissa is −2. II, III
20. The abscissa is 3. I, IV
21. The ordinate is positive. I, II
22. The ordinate is negative. III, IV
23. The abscissa is positive. I, IV
24. The abscissa is negative. II, III
25. The ordinate is zero. none
26. The abscissa is zero. none
27. The ordinate equals the abscissa. I, III
28. The ordinate equals the opposite of the abscissa. II, IV
29. Name the ordinate of every point on the x-axis. 0
30. Name the abscissa of every point on the y-axis. 0

Graphs and Functions **217**

Additional A Exercises
Plot the given points on a set of coordinate axes.
1. P(3, 1)
2. Q(1, 2)
3. R(−2, −2)

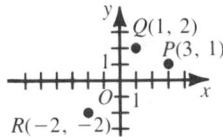

Name the quadrant or axis that contains each point.
4. (−3, 3) Quadrant II
5. (4, −5) Quadrant IV
6. Name the quadrants or axes that contain all the ordered pairs (x, y) where $x < 0$ and $y > 0$.
Quadrant II

Mixed Review
Plot the following points on a number line.
1. $P = 4$
2. $Q = 5$
3. $R = -1$
4. $S = -3$

Plot the following sets of points on a number line.
5. $\{x: x > 2\}$

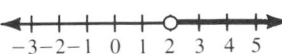

6. $\{x: x \le 0\}$

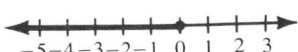

1-12.

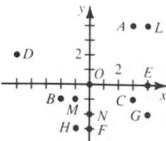

26.

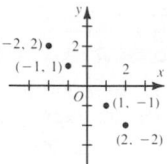

28.

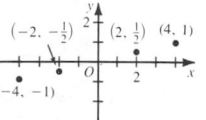

30.

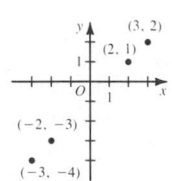

32.

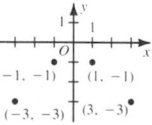

34.

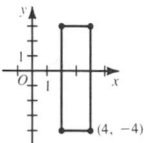

36.

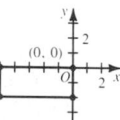

38.

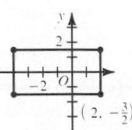

Written Exercises

Plot the given points on a coordinate plane.

A
1. $A(3, 4)$
2. $B(-2, -1)$
3. $C(3, -1)$
4. $D(-5, 2)$
5. $E(4, 0)$
6. $F(0, -3)$
7. $G(4, -2)$
8. $H(-1, -3)$
9. $L(4, 4)$
10. $M(-1, -1)$
11. $N(0, -2)$
12. $O(0, 0)$

Given a point with coordinates (x, y), name all the quadrants or axes in which the point might lie under the specified conditions.

13. $x > 0$ and $y > 0$ I
14. $x < 0$ and $y < 0$ III
15. $x > 0$ and $y < 0$ IV
16. $x < 0$ and $y > 0$ II
17. $x = 0$ y-axis
18. $y = 0$ x-axis
19. $xy > 0$ I, III
20. $xy < 0$ II, IV
21. $xy = 0$ x-axis, y-axis
22. $\frac{x}{y} = 0$ y-axis
23. $x - y = 0$ I, III, x-axis, y-axis
24. $x + y = 0$ II, IV, x-axis, y-axis

Plot four points that are in at least two different quadrants and whose coordinates are integers that satisfy the given requirements. Construct a different set of axes for each exercise. Examples are given.

B
25. $y = x$
26. $y = -x$
27. $y = 4x$
28. $y = \frac{1}{4}x$
29. $y = x + 1$
30. $y = x - 1$
31. $y = |x|$
32. $y = -|x|$

In Exercises 33–38, each set of ordered pairs lists the coordinates of three vertices of a rectangle on a coordinate plane. Graph the ordered pairs and sketch the rectangle. Then determine the coordinates of the fourth vertex.

33. $(-1, 1)$, $(5, -3)$, $(-1, -3)$
34. $(2, 3)$, $(4, 3)$, $(2, -4)$
35. $(1, 2)$, $(-4, 2)$, $(1, -5)$
36. $(-5, 0)$, $(0, -2)$, $(-5, -2)$
37. $(\frac{1}{2}, 1)$, $(3, 1)$, $(3, 6)$
38. $(-4, -\frac{3}{2})$, $(2, \frac{3}{2})$, $(-4, \frac{3}{2})$

For any three given points not all on one line, there are three possible ways of choosing a fourth point so that the four points are the vertices of a parallelogram. In each of Exercises 39 and 40, the coordinates of three vertices of a parallelogram are given. What are the coordinates of the three possible fourth vertices?

$(-4, 3)$, $(2, 5)$, $(0, -5)$

C
39. $(2, -3)$, $(6, -2)$, $(5, 1)$ $(1, 0)$, $(9, 2)$, $(3, -6)$
40. $(-2, -1)$, $(1, 0)$, $(-1, 4)$

41. Three vertices of an isosceles trapezoid are the points with coordinates $(0, 0)$, $(4, 0)$, and $(6, -2)$. Find the coordinates of two possible points for the fourth vertex. $(-2, -2)$, $(6, -6)$

42. The base of an isosceles triangle has endpoints $(-6, 0)$ and $(0, -4)$. Using only integers, name the coordinates of six possible points for the third vertex of the triangle.

43. What is the length of the segment that joins $P(a, b)$ and $Q(a, c)$? $|b - c|$

44. What is the length of the segment that joins $R(a, c)$ and $S(b, c)$? $|c - b|$

42. Answers will vary. For example: $(-9, -11)$, $(-7, -8)$, $(-5, -5)$, $(-1, 1)$, $(1, 4)$, $(3, 7)$

218 *Chapter 5*

218

5–2 Relations and Functions

In mathematics, any set of ordered pairs is called a **relation**. For example, the set

$$\{(0, 2), (-4, 3), (-3, -2), (2, -1)\}$$

is a relation. The set D of all the *first coordinates* of the ordered pairs is called the **domain** of the relation, while the set R of all the *second coordinates* is called the **range** of the relation. Thus, for the relation just specified,

$$D = \{0, -4, -3, 2\} \quad \text{and} \quad R = \{2, 3, -2, -1\}.$$

There are various ways to picture a relation. One of these is to use a table such as that shown in Figure 5, which pictures the relation in the preceding example. Another way is to use a *mapping diagram* such as that shown in Figure 6, which pictures this same relation.

D	R
0	2
-4	3
-3	-2
2	-1

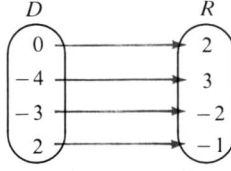

Figure 5   **Figure 6**

If a relation is such that no two ordered pairs have the same first coordinate, the relation is called a **function**.

EXAMPLE 1   Draw a mapping diagram of each relation. Then tell whether or not the relation is a function.

   a. $\{(0, 2), (0, -3), (1, 5), (-2, 4)\}$ **b.** $\{(-2, -1), (1, 2), (5, 2), (-1, 0)\}$

SOLUTION   **a.**   **b.**

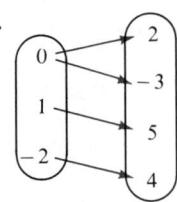   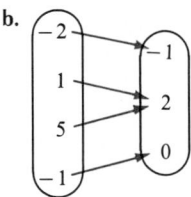

The mapping diagram clearly shows that two of the ordered pairs, (0, 2) and (0, -3), have the same first coordinate, 0.

∴ the relation is not a function.

The mapping diagram clearly shows that no two of the ordered pairs have the same first coordinate.

∴ the relation is a function.

Graphs and Functions   **219**

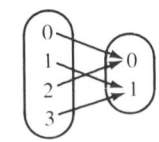

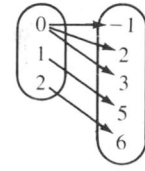

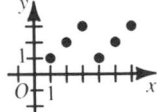

Teaching Suggestions
p. T83

Key Ideas

Identify the domains and ranges of given relations.
Draw mapping diagrams of relations.
Graph relations with finite domains.
Identify relations that are functions.

Chalkboard Examples

1. Draw a mapping diagram for each relation. Then tell whether or not the relation is a function.
   a. $\{(0, 0), (1, 1), (2, 0), (3, 1)\}$

   It is a function.
   b. $\{(0, 2), (1, 5), (0, 3), (0, -1), (2, 6)\}$

   It is not a function.
2. Graph each relation. Then use the vertical-line test to determine whether or not the relation is a function.
   a. $\{(1, 1), (2, 2), (3, 3), (4, 1), (5, 2), (6, 3)\}$

   It is a function.

(continued)

219

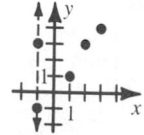
Notice in part (b) of Example 1 that a relation may be a function even if two ordered pairs of the relation have the same *second* coordinate.

Since a relation is a set of ordered pairs, another way to picture a relation is to graph the ordered pairs on a coordinate plane. For any given relation, the set of points on a coordinate plane that correspond to the ordered pairs is called the **graph of the relation**. For example, the relation

$$\{(-4, 1), (-1, 4), (3, 2), (3, -2)\}$$

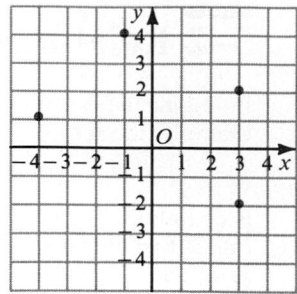

Figure 7

can be graphed as shown in Figure 7. Notice that numbers in the *domain* of the relation are associated with the *x*-axis, while numbers in the *range* are associated with the *y*-axis.

You know that the relation graphed in Figure 7 is *not* a function because two of the ordered pairs, (3, 2) and (3, −2), have the same first coordinate. On the graph, the points that correspond to these two ordered pairs lie directly above one another. This suggests the following "vertical-line test" to determine whether a given graph represents a function: *A relation is a function if and only if no vertical line can be drawn that intersects the graph of the relation in more than one point.*

EXAMPLE 2 Graph each relation. Then use the vertical-line test to determine whether or not the relation is a function.

    a. {(0, 0), (1, 1), (1, −1), (2, 2), (2, −2), (3, 3), (3, −3)}
    b. {(0, 0), (1, 1), (−1, 1), (2, 2), (−2, 2), (3, 3), (−3, 3)}

SOLUTION **a.** 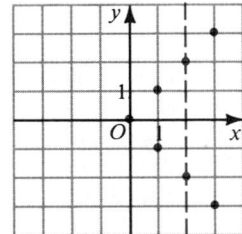   **b.** 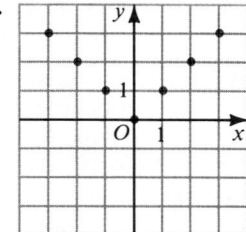

The vertical line drawn through two points indicates that two ordered pairs have the same first coordinate.

∴ the relation is not a function.

It is not possible to draw a vertical line that intersects two points of the graph, and so no two ordered pairs have the same first coordinate.

∴ the relation is a function.

In general, then, a relation is a pairing between two sets, the domain and the range. A function is a special type of relation in which no member of the domain is paired with more than one member of the range. Thus, the following is an alternative definition of a function.

220 *Chapter 5*

A **function** is a pairing that assigns to each member of one set, called the domain, *exactly one* member of a second set, called the range. Each member of the range is assigned to *at least one* member of the domain.

Oral Exercises

State the domain and range of each relation. Then tell whether or not the relation is a function.

1. {(1, 1), (2, 2), (3, 3)}
2. {(−4, 4), (−2, 2), (−1, 1)}
3. {(1, 1), (1, 2), (1, 3)}
4. {(1, 1), (2, 1), (3, 1)}
5. {(−2, 4), (−1, 1), (1, 1), (2, 4)}
6. {(4, −2), (1, −1), (1, 1), (4, 2)}

Tell whether or not each mapping diagram represents a function.

7. Yes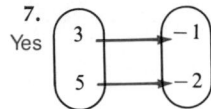
8. Yes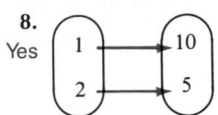
9. No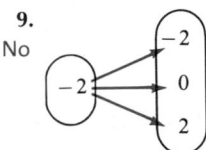
10. Yes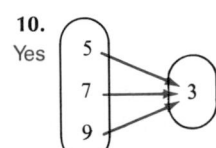

11. Yes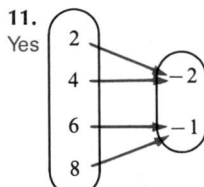
12. No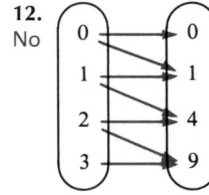
13. Yes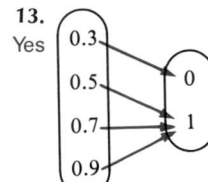
14. No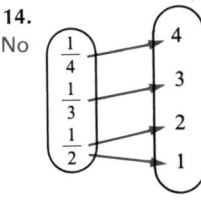

Tell whether or not each graph represents a function.

15. Yes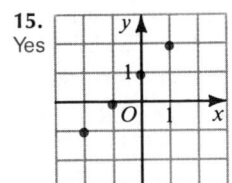
16. Yes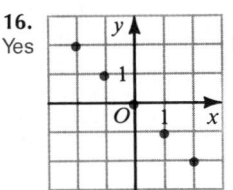
17. No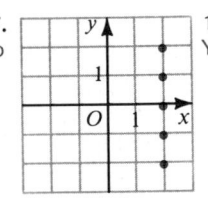
18. Yes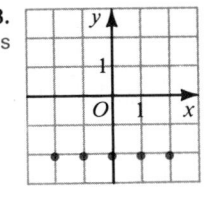

19. Yes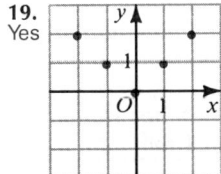
20. No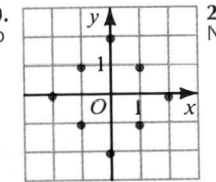
21. No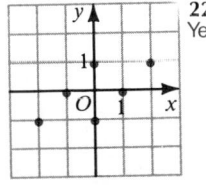
22. Yes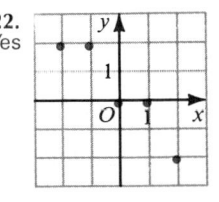

Graphs and Functions **221**

 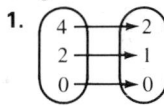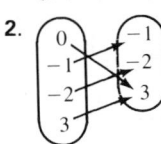
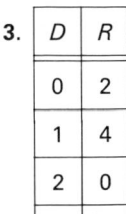 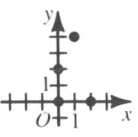

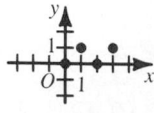
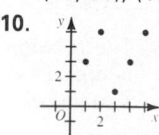
Written Exercises

Write the set of ordered pairs in the relation that is represented by each mapping diagram.

A

1.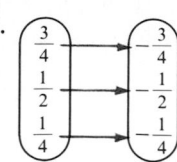

2.

3.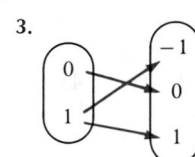

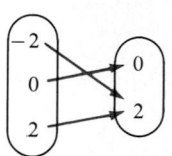

5.

6.

7.

8.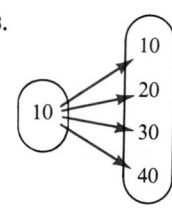

In Exercises 9–16:

a. Graph the relation whose ordered pairs are shown in the given table.

b. Tell whether or not the relation is a function.

9.

D	R
1	3
2	5
3	1
4	4
5	2

Yes

10.

D	R
1	3
2	5
3	1
4	3
5	5

Yes

11.

D	R
-1	1
-1	2
0	-3
3	4
5	-2

No

12.

D	R
-3	-1
-2	4
0	-1
1	5
4	-1

Yes

13.

D	R
0	0
1	2
2	4
3	6
4	8

Yes

14.

D	R
-6	-2
-3	-1
0	0
3	1
6	2

Yes

15.

D	R
4	-4
2	-2
0	0
2	2
4	4

No

16.

D	R
-4	4
-2	2
0	0
2	2
4	4

Yes

Draw a mapping diagram that represents each relation. Then tell whether or not the relation is a function.

B 17. {(0, 5), (1, 6), (2, 7), (3, 8)} Yes

18. {(1, 3), (2, 5), (3, 5), (4, 7)}

19. {(−10, −1), (0, −1), (1, 3), (5, 3)} Yes

20. {(1, 1), (1, −1), (3, 5), (3, −5)}

21. {(2, −7), (2, −3), (2, 0), (2, 4)} No

22. {(−6, 1), (−5, 1), (3, 1), (7, 1)}

23. {(−2, 4), (−1, 1), (1, 1), (2, 4)} Yes

24. {(9, −3), (1, −1), (1, 1), (9, 3)}

18. Yes   20. No   22. Yes   24. No

In each of Exercises 25 and 26 there is a mapping diagram of a relation. Write the ordered pairs in the relation and tell whether or not it is a function. Then state a rule that can be used to calculate the number in the range that is associated with any given ordered pair in the domain. (*Note:* The first coordinate of an ordered pair may itself be an ordered pair.)

25.
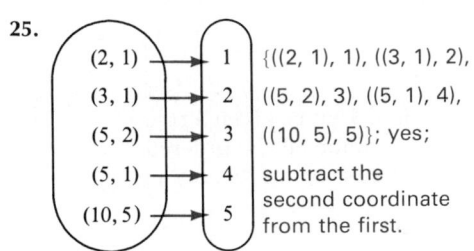

{((2, 1), 1), ((3, 1), 2), ((5, 2), 3), ((5, 1), 4), ((10, 5), 5)}; yes; subtract the second coordinate from the first.

26.
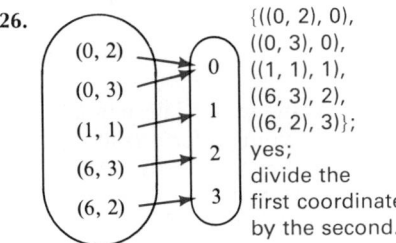

{((0, 2), 0), ((0, 3), 0), ((1, 1), 1), ((6, 3), 2), ((6, 2), 3)}; yes; divide the first coordinate by the second.

Exercises 27–30 refer to the mapping diagrams in Written Exercises 1–8. Tell which of these mapping diagrams represents a *function* that might be described by the given "rule."

C 27. reciprocal 2   28. additive inverse 1   29. square 5   30. absolute value 4

31. A certain mapping diagram shows a domain of three members being mapped onto a range of four members. Can this relation be a function? Why or why not? No, at least one element of the domain must be mapped onto at least two elements of the range. ∴ Two ordered pairs will have the same first coordinate.

Computer Exercises   For students with computer experience

1. Write a program that will allow you to input the coordinates of any point and will display the name of the quadrant or axis (or axes) in which the point is located. RUN the program for the points given in Exercises 1–12 on page 218.

2. Write a program that will determine whether a relation that consists of *four* ordered pairs is a function. RUN the program for the relations given in Exercises 17–24 above.

3. Modify the program that you wrote for Exercise 2 so that it will determine whether a relation that consists of *any number* of ordered pairs is a function.

Graphs and Functions   **223**

12.
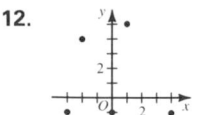

14.
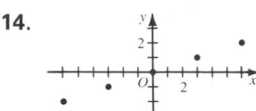

16.
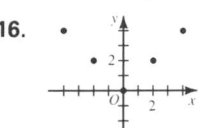

18.
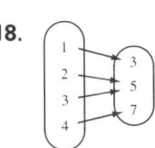

(continued on p. 241)

Mixed Review

Plot the given points on a set of coordinate axes.

1. (1, 3)   2. (4, 2)
3. (−2, −1)   4. (1, −4)

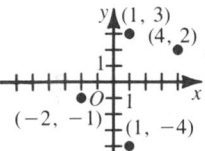

Plot 3 points whose coordinates are integers that satisfy the given requirements.

5. $y = 2x$

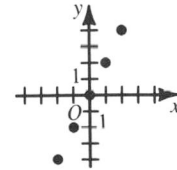

Any 3 of the points indicated above answer the exercise correctly.

Key Ideas

Define relations and functions.
Use arrow notation for
defining functions.
Use the symbol $f(x)$ to specify values of functions.

Common Errors

Students may confuse the
functional notation $f(x)$ with
the notation for multiplication, $f \cdot x$. To aid students,
note that $f(x) = x + 4$ means
$f: x \rightarrow x + 4$.

Chalkboard Examples

1. Determine the range R of
the relation defined by the
rule $y = \frac{1}{2}x - 1$, if the
domain $D = \{2, 4, 6\}$.

x	$\frac{1}{2}x - 1$	y
2	$\frac{1}{2}(2) - 1 = 1 - 1$	0
4	$\frac{1}{2}(4) - 1 = 2 - 1$	1
6	$\frac{1}{2}(6) - 1 = 3 - 1$	2

$\therefore R = \{0, 1, 2\}$

5-3 Defining Relations and Functions

Recall that in Section 1-2 you learned how to specify sets both by roster
and by rule. Thus far, each of the relations that you have studied has
been specified by a *roster* of its ordered pairs. Often, however, a relation
is specified by a *rule* that describes exactly how the members of the
domain and the range are related. Generally this rule is an open
sentence. For example, consider the relation

$$\{(3, 1), (6, 2), (9, 3), (12, 4), (15, 5)\},$$

in which each member of the domain is the triple of the related member
of the range. If x represents a member of the domain and y represents a
member of the range, a rule for this relation is

$$x = 3y, \quad \text{or} \quad y = \frac{1}{3}x.$$

A rule for a relation is said to *define* the relation. That is, given a rule
and a domain for a relation, it is possible to determine all the ordered
pairs that form the relation. Therefore, it is also possible to determine
the range of the relation.

EXAMPLE 1   Determine the range R of the relation defined by the rule $y = 2x + 6$, if
the domain $D = \{-5, -3, 0, 4\}$.

SOLUTION   Replace x in the expression $2x + 6$ with each member of D and find the
corresponding member of R. A table may be helpful.

x	$2x + 6$	
-5	$2(-5) + 6 = -10 + 6 = -4$	
-3	$2(-3) + 6 = -6 + 6 = 0$	
0	$2(0) + 6 = 0 + 6 = 6$	
4	$2(4) + 6 = 8 + 6 = 14$	

$\therefore R = \{-4, 0, 6, 14\}$

Notice that the relation specified in Example 1 is a function. Often a
function is named by a single letter, such as f, F, or g. For example, the
function defined by the rule $y = 2x + 6$ may be called f, and it may then
be specified using *arrow notation*, as follows.

$$f: x \rightarrow 2x + 6$$

This expression is read "the function f that pairs a number x with the
number $2x + 6$."

224   *Chapter 5*

EXAMPLE 2   Graph $f: x \rightarrow 2x - 1$ if $D = \{-2, -1, 0, 1, 2\}$.

SOLUTION   Replace x in $2x - 1$ with each member of D to find the corresponding value of y. Then graph the ordered pairs (x, y) on a coordinate plane.

x	$2x - 1 = y$	(x, y)
-2	$2(-2) - 1 = -5$	$(-2, -5)$
-1	$2(-1) - 1 = -3$	$(-1, -3)$
0	$2(0) - 1 = -1$	$(0, -1)$
1	$2(1) - 1 = 1$	$(1, 1)$
2	$2(2) - 1 = 3$	$(2, 3)$

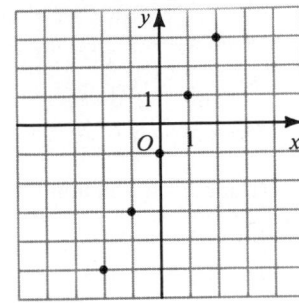

Members of the range of a function are called the **values** of the function. Thus, the values of the function specified in Example 2 are -5, -3, -1, 1, and 3. The symbol $f(x)$, read "f of x," is used to denote the specific value of the function f that is paired with the number x. Therefore, for the function just specified:

$$f(-2) = -5 \qquad f(-1) = -3 \qquad f(0) = -1 \qquad f(1) = 1 \qquad f(2) = 3$$

Note that the symbol $f(x)$ does *not* denote the product of f and x.

Oral Exercises

State the domain and range of each relation. Then give a rule for the relation, letting x represent a member of the domain and y represent a member of the range.

1. $\{(2, -2), (1, -1), (1, 1), (2, 2)\}$
2. $\{(-5, 5), (-3, 3), (3, 3), (5, 5)\}$
3. $\{(-2, 4), (-1, 1), (1, 1), (2, 4)\}$
4. $\{(9, -3), (1, -1), (1, 1), (9, 3)\}$
5. $\{(0, 0), (1, 5), (2, 10), (3, 15)\}$
6. $\{(-6, -3), (-4, -2), (-2, -1), (0, 0)\}$
7. $\{(-1, 0), (0, 1), (1, 2), (2, 3)\}$
8. $\{(0, 2), (2, 4), (4, 6), (6, 8)\}$

Let $D = \{-2, -1, 0, 1, 2\}$. Determine the range R of the relation defined by each rule.

9. $y = x$ $\{-2, -1, 0, 1, 2\}$
10. $y = -x$ $\{2, 1, 0, -1, -2\}$
11. $y = |x|$ $\{0, 1, 2\}$
12. $y = -|x|$ $\{-2, -1, 0\}$
13. $y = x + 1$ $\{-1, 0, 1, 2, 3\}$
14. $y = 5x$ $\{-10, -5, 0, 5, 10\}$
15. $y = x^2$ $\{0, 1, 4\}$
16. $y = x^3$ $\{-8, -1, 0, 1, 8\}$

Given the function $f: x \rightarrow 2 - 3x$, find the following values of f.

17. $f(0)$ 2
18. $f(1)$ -1
19. $f(-1)$ 5
20. $f(-2)$ 8
21. $f(2)$ -4
22. $f\left(\frac{1}{3}\right)$ 1
23. $f\left(\frac{2}{3}\right)$ 0
24. $f\left(-\frac{2}{3}\right)$ 4

Graphs and Functions   **225**

2. Graph $g: x \rightarrow x + 3$ if $D = \{-3, -1, 1, 2\}$.

x	$x + 3 = y$	(x, y)
-3	$-3 + 3 = 0$	$(-3, 0)$
-1	$-1 + 3 = 2$	$(-1, 2)$
1	$1 + 3 = 4$	$(1, 4)$
2	$2 + 3 = 5$	$(2, 5)$

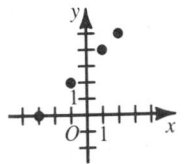

Additional Answers
Oral Exercises
(See p. 247.)

Additional A Exercises

Find the range of the function when $D = \{0, 1, 2, 3\}$.

1. $g: x \rightarrow 2x + 1$

x	$2x + 1$	$g(x)$
0	$2(0) + 1 = 0 + 1$	1
1	$2(1) + 1 = 2 + 1$	3
2	$2(2) + 1 = 4 + 1$	5
3	$2(3) + 1 = 6 + 1$	7

$R = \{1, 3, 5, 7\}$

2. Graph $F: x \rightarrow |x|$; $D = \{-2, -1, 0, 1, 2\}$.

x	-2	-1	0	1	2
$|x|$	2	1	0	1	2

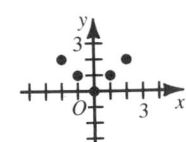

Suggested Assignments

Minimum
226/1–24
R 227/Self-Test 1
Average
226/13–31 odd, 33–45
R 227/Self-Test 1
Maximum
226/13–55 odd
R 227/Self-Test 1

Additional Answers
Written Exercises
(See p. 256.)

Mixed Review

Draw a mapping diagram of
the relation. Then tell
whether or not the relation is
a function.

1. $\{(-3, 3), (2, 2), (-1, 1), (1, -1), (2, -2)\}$

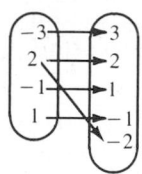

The relation is not a function.

2. Graph the function de-
scribed in the following
table.

D	3	2	0	-1	-2
R	-4	-3	-1	0	1

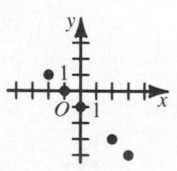

Written Exercises

6. $\{1, 3, 5, 7, 9\}$ 9. $\{4, 1, 0\}$

Find the range of each function when $D = \{-2, -1, 0, 1, 2\}$.

A   1. $f: x \longrightarrow x + 3$ $\{1, 2, 3, 4, 5\}$ 2. $g: x \longrightarrow x - 1$ $\{-3, -2, -1, 0, 1\}$ 3. $j: x \longrightarrow 2x$ $\{-4, -2, 0, 2, 4\}$

4. $k: x \longrightarrow -\frac{1}{2}x$ $\{1, \frac{1}{2}, 0, -\frac{1}{2}, -1\}$ 5. $F: x \longrightarrow 3x - 1$ $\{-7, -4, -1, 2, 5\}$ 6. $G: x \longrightarrow 5 + 2x$

7. $h: x \longrightarrow |x|$ $\{2, 1, 0\}$ 8. $H: x \longrightarrow -|x|$ $\{-2, -1, 0\}$ 9. $r: x \longrightarrow x^2$

10. $s: x \longrightarrow x^3$ $\{-8, -1, 0, 1, 8\}$ 11. $p: x \longrightarrow (x - 1)^2$ $\{9, 4, 1, 0\}$ 12. $q: x \longrightarrow x(x + 1)$ $\{2, 0, 6\}$

Graph each function.

13. $g: x \longrightarrow x$; $D = \{-3, -1, 1, 3\}$ 14. $h: x \longrightarrow -x$; $D = \{-3, -1, 1, 3\}$

15. $J: x \longrightarrow 3x$; $D = \{-1, 0, 1, 2\}$ 16. $K: x \longrightarrow \frac{x}{2}$; $D = \{-4, 0, 2, 4\}$

17. $f: x \longrightarrow |x| - 1$; $D = \{-1, 0, 1, 2\}$ 18. $F: x \longrightarrow |x| + 3$; $D = \{-2, -1, 0, 1\}$

19. $r: x \longrightarrow x^2 - 3$; $D = \{-3, -2, 0, 2\}$ 20. $s: x \longrightarrow 5 - x^2$; $D = \{-2, 0, 2, 3\}$

21. $M: x \longrightarrow \frac{x - 1}{x + 1}$; $D = \{-3, -2, 0, 1\}$ 22. $N: x \longrightarrow \frac{3 - x}{1 - x}$; $D = \{-1, 0, 2, 3\}$

23. $G: x \longrightarrow \frac{x^2 + 1}{2x + 1}$; $D = \{-3, -1, 0, 2\}$ 24. $H: x \longrightarrow \frac{1 - x^2}{3x + 1}$; $D = \{-3, -1, 0, 1\}$

In Exercises 25–36, the function f is defined as given.
a. Find $f(0)$.
b. Find all values of x such that $f(x) = 0$.

27. 0; 0
30. 3; 1
33. −9; 3, −3

B   25. $f: x \longrightarrow x + 5$ 5; −5   26. $f: x \longrightarrow x - 3$ −3; 3   27. $f: x \longrightarrow -5x$

28. $f: x \longrightarrow \frac{x}{7}$ 0; 0   29. $f: x \longrightarrow 2x + 6$ 6; −3   30. $f: x \longrightarrow 3 - 3x$

31. $f: x \longrightarrow |x| - 1$ −1; 1, −1   32. $f: x \longrightarrow 2 - |x|$ 2; 2, −2   33. $f: x \longrightarrow x^2 - 9$

34. $f: x \longrightarrow 1 - x^2$ 1; 1, −1   35. $f: x \longrightarrow x^3 + x$ 0; 0   36. $f: x \longrightarrow x^2 + x$ 0; 0, −1

In Exercises 37–44, let $f(x) = x^2$ and $g(x) = 3x - 1$. Find each of the
following.

37. $f(2) + g(2)$ 9   38. $f(2) \cdot g(2)$ 20   39. $4f(3)$ 36   40. $2g(5)$ 28

41. $2f(3) + g(0)$ 17   42. $f(2) - 2g(1)$ 0   43. $\frac{5f(2)}{2g(1)}$ 5   44. $\frac{f(3) + 1}{g(0)}$ −10

In Exercises 45–52, let $f(x) = x^2$ and $g(x) = x - 1$. Find each of the
following. (Hint: To find $g[f(x)]$, first find $f(x)$.)

C   45. $g[f(1)]$ 0   46. $f[g(1)]$ 0   47. $g[f(-1)]$ 0   48. $f[g(-1)]$ 4

49. $g[f(2)]$ 3   50. $f[g(2)]$ 1   51. $g[f(0)]$ −1   52. $f[g(0)]$ 1

53. If $f(x) = x - 2$ and $g[f(x)] = x$, what is $g(x)$? $x + 2$

54. If $g(x) = x - 2$ and $g[f(x)] = x$, what is $f(x)$? $x + 2$

226   *Chapter 5*

The *greatest integer function* is denoted $f: x \rightarrow [x]$, where the symbol $[x]$ is used to represent the greatest integer that is *less than or equal to* the real number x.

55. Simplify.

    a. $[3]$ 3   **b.** $[3.5]$ 3   **c.** $[-3]$ −3   **d.** $[-3.5]$ −4

56. Graph $f: x \rightarrow [x]$ if $D = \{x: -4 \le x \le 4\}$.

■ Self-Test 1

VOCABULARY   axes (p. 215)
                        origin (p. 215)
                        x-axis (p. 216)
                        y-axis (p. 216)
                        coordinate axes (p. 216)
                        coordinate plane (p. 216)
                        quadrants (p. 216)
                        coordinates (p. 216)
                        x-coordinate (p. 216)
                        abscissa (p. 216)
                        y-coordinate (p. 216)
                        ordinate (p. 216)

ordered pair (p. 216)
graph of an ordered pair
  (p. 216)
plot a point (p. 216)
plane rectangular or Cartesian
  coordinate system (p. 217)
relation (p. 219)
domain (p. 219)
range (p. 219)
function (pp. 219, 221)
graph of a relation (p. 220)
value of a function (p. 225)

Plot the given points on a coordinate plane.

  1. $A(2, -4)$ **2.** $B(-3, -1)$ **3.** $C(5, 0)$ *Obj. 1, p. 215*

State the domain and range of each relation. Then draw a mapping diagram that represents the relation.

                                               $D = \{0, -1, 5, -3\}$
  4. $\{(0, 2), (-1, 3), (5, -2), (-3, 6)\}$ $R = \{2, 3, -2, 6\}$ *Obj. 2, p. 215*
  5. $\{(2, 1), (-1, 0), (0, 0), (5, 4)\}$
     $D = \{2, -1, 0, 5\}$, $R = \{1, 0, 4\}$

Tell whether or not each relation is a function.

  6. $\{(-3, 3), (-1, 1), (0, 0), (1, 1), (3, 3)\}$ Yes   *Obj. 3, p. 215*
  7. $\{(4, 2), (1, 1), (0, 0), (1, -1), (4, -2)\}$ No

Graph each of the following.

  8. $\{(-3, 2), (0, 0), (4, -1), (-2, -3)\}$ *Obj. 4, p. 215*
  9. $f: x \rightarrow 3x - 1$; $D = \{-1, 0, 1, 2\}$
  10. $g: x \rightarrow |x| + 2$; $D = \{-4, -2, 0, 2, 4\}$

Check your answers with those at the back of the book.

Graphs and Functions **227**

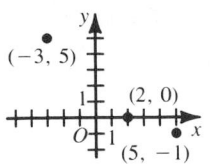
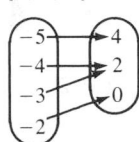
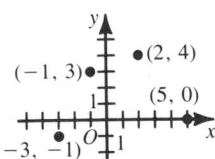
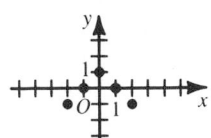

Key Ideas

Solve open sentences in two variables over given replacement sets of the variables. Graph the solution set of open sentences in two variables.

Chalkboard Examples

1. Solve $2x + y = 10$ if the values of x and y are whole numbers.

$$2x + y = 10$$
$$y = 10 - 2x$$

x	$10 - 2x$	y
0	$10 - 2(0)$	10
1	$10 - 2(1)$	8
2	$10 - 2(2)$	6
3	$10 - 2(3)$	4
4	$10 - 2(4)$	2
5	$10 - 2(5)$	0

∴ the solution set is $\{(0, 10), (1, 8), (2, 6), (3, 4), (4, 2), (5, 0)\}$.

2. Solve $y = 3 - 2x$ if $x \in \{-2, -1, 0, 1, 2\}$ and $y \in \{$the whole numbers$\}$. Graph the solution set.

x	$3 - 2x$	y
-2	$3 - 2(-2) = 3 + 4$	7
-1	$3 - 2(-1) = 3 + 2$	5
0	$3 - 2(0) = 3 - 0$	3
1	$3 - 2(1) = 3 - 2$	1
2	$3 - 2(2) = 3 - 4$	-1

Open Sentences in Two Variables

OBJECTIVES *for Sections 5-4 through 5-6:*
1. To solve open sentences in two variables over given replacement sets of the variables.
2. To graph linear equations and inequalities in two variables on a coordinate plane.

5–4 Solving Open Sentences in Two Variables

In Section 1-7 you learned that a solution of an open sentence in *one* variable is any value of the variable for which the open sentence is a true statement. For example, the equation $7x - 5 = 9$ has just one solution, namely 2.

A **solution of an open sentence in two variables**, x and y, is any ordered pair of numbers, (x, y), that *together* make the sentence a true statement. For example, consider the open sentence

$$2x + 3y = 11.$$

The ordered pair $(4, 1)$ is a solution of this open sentence because

$$2(4) + 3(1) = 11,$$

but $(5, 1)$ is *not* a solution because

$$2(5) + 3(1) \neq 11.$$

The set of *all* solutions of an open sentence in two variables is called the **solution set** of the open sentence. Any member of the solution set is said to **satisfy** the open sentence. You **solve** an open sentence in two variables when you determine its solution set.

EXAMPLE 1   Solve $3x + 2y = 14$ if the replacement set for x and y is the set of whole numbers.

SOLUTION   1. Solve the given equation for y.
$$3x + 2y = 14$$
$$2y = 14 - 3x$$
$$y = 7 - \tfrac{3}{2}x$$

2. Starting with 0, replace x in the expression $7 - \tfrac{3}{2}x$ with consecutive whole numbers in order to find the corresponding values of y. A table may be helpful.

x	$7 - \tfrac{3}{2}x$	y
0	$7 - \tfrac{3}{2}(0)$	7
1	$7 - \tfrac{3}{2}(1)$	$\tfrac{11}{2}$
2	$7 - \tfrac{3}{2}(2)$	4
3	$7 - \tfrac{3}{2}(3)$	$\tfrac{5}{2}$
4	$7 - \tfrac{3}{2}(4)$	1
5	$7 - \tfrac{3}{2}(5)$	$-\tfrac{1}{2}$
6	$7 - \tfrac{3}{2}(6)$	-2

228   *Chapter 5*

3. If the value of y found in Step 2 is a whole number, the ordered pair (x, y) satisfies the open sentence.

As the table on the previous page indicates, values of x greater than 4 yield negative values of y, and so x cannot be greater than 4.

∴ the solution set is $\{(0, 7), (2, 4), (4, 1)\}$.

Notice that the solution set of the equation in Example 1 is a function with domain $\{0, 2, 4\}$ and range $\{7, 4, 1\}$. You can specify this function as

$$f: x \rightarrow 7 - \tfrac{3}{2}x, \qquad x \in \{0, 2, 4\}.$$

The **graph of an open sentence in two variables** is the graph of its solution set.

EXAMPLE 2   Solve $y + 2x < 1$ if $x \in \{-1, 0, 1\}$ and $y \in \{-2, -1, 0, 1, 2\}$. Graph the solution set.

SOLUTION   Solve the given inequality for y: $y + 2x < 1$
$$y < 1 - 2x$$

x	$1 - 2x$	$y < 1 - 2x$	y
-1	$1 - 2(-1)$	$y < 3$	$-2, -1, 0, 1, 2$
0	$1 - 2(0)$	$y < 1$	$-2, -1, 0$
1	$1 - 2(1)$	$y < -1$	-2

∴ the solution set is $\{(-1, -2), (-1, -1), (-1, 0), (-1, 1), (-1, 2),$
$(0, -2), (0, -1), (0, 0), (1, -2)\}$.

The graph is shown below.

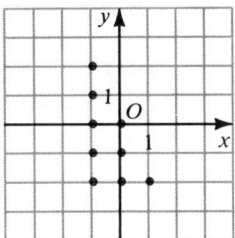

Although the solution set of the inequality in Example 2 is *not* a function, it is a relation. In general, the solution set of *any* open sentence in two variables is a relation whose domain is the set of first coordinates and whose range is the set of second coordinates of the ordered pairs that satisfy the sentence.

$(2, -1)$ is not a member of the solution set since $-1 \notin$ {the whole numbers}.

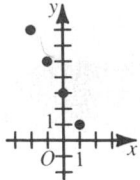

Suggested Assignments

Minimum
   230/1–24
S 226/25–27
Average
   230/1–31 odd, 33–42
S 226/26–32 even
Maximum
   230/5–41 odd, 43–48
S 226/50–56 even

Additional Answers
Written Exercises

13. $\{(-2, -2), (-1, -1),$
   $(0, 0), (1, 1), (2, 2)\}$

14. $\{(-2, 2), (-1, 1), (0, 0),$
   $(1, -1), (2, -2)\}$

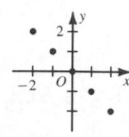

15. $\{(-2, -5), (-1, -2),$
   $(0, 1), (1, 4), (2, 7)\}$

16. $\{(-2, 10), (-1, 8), (0, 6),$
   $(1, 4), (2, 2)\}$

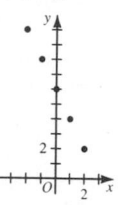

17. $\{(-2, 9), (-1, 7), (0, 5),$
   $(1, 3), (2, 1)\}$

18. $\{(-2, -8), (-1, -5),$
   $(0, -2), (1, 1), (2, 4)\}$

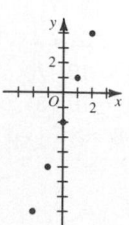

Oral Exercises

Tell whether or not the given ordered pair is a solution of the open sentence.

1. $x - y = -4$; $(1, -5)$ no
2. $x + 5y = -2$; $(3, -1)$ yes
3. $y + 2x = 2$; $(4, -6)$ yes
4. $2y - 3x = 1$; $(-2, 2)$ no
5. $x < 3y - 4$; $(2, 2)$ no
6. $y \geq -4x + 1$; $(0, 2)$ yes
7. $x - y^2 = 12$; $(4, 4)$ no
8. $x^2 + xy = 6$; $(-6, 5)$ yes

Solve each open sentence for y.

9. $y + 2x = 3$ $y = 3 - 2x$
10. $3x - y = 7$ $y = 3x - 7$
11. $2y = 4x - 6$ $y = 2x - 3$
12. $12 = 4x + 3y$ $y = 4 - \frac{4}{3}x$
13. $x + y \leq 3$ $y \leq 3 - x$
14. $2x - y > 7$ $y < 2x - 7$
15. $8x - 10 > 2y$ $y < 4x - 5$
16. $9x - 3y \leq 3$ $y \geq 3x - 1$

Solve each equation if the replacement set for x and y is the set of whole numbers.

18. $\{(0, 9), (1, 7), (2, 5), (3, 3), (4, 1)\}$ $\{(0, 1), (1, 0)\}$

17. $x + y = 3$
18. $2x + y = 9$
19. $2xy = 16$
20. $x^2 + y^2 = 1$
$\{(0, 3), (1, 2), (2, 1), (3, 0)\}$ $\{(1, 8), (2, 4), (4, 2), (8, 1)\}$

Written Exercises

Complete each ordered pair to form a solution of the given equation.

A
  1. $y = -x$; $(5, \underline{}), (0, \underline{}), (-3, \underline{})$ $-5, 0, 3$
  2. $y = 4x$; $(4, \underline{}), (0, \underline{}), (-3, \underline{})$ $16, 0, -12$
  3. $y = x - 5$; $(9, \underline{}), (5, \underline{}), (-7, \underline{})$ $4, 0, -12$
  4. $y = 2x + 3$; $(6, \underline{}), (0, \underline{}), (-4, \underline{})$ $15, 3, -5$
  5. $y = 2x - 1$; $(\underline{}, 1), (\underline{}, -1), (\underline{}, -7)$ $1, 0, -3$
  6. $y = 5 + 3x$; $(\underline{}, 8), (\underline{}, 5), (\underline{}, -7)$ $1, 0, -4$
  7. $4x + y = 9$; $(2, \underline{}), (0, \underline{}), (-5, \underline{})$ $1, 9, 29$
  8. $x - 2y = 10$; $(\underline{}, 10), (\underline{}, -1), (\underline{}, -10)$ $30, 8, -10$
  9. $2x + 3y = 11$; $(\underline{}, 3), (\underline{}, 0), (\underline{}, -5)$ $1, \frac{11}{2}, 13$
  10. $5x - 2y = 7$; $(3, \underline{}), (0, \underline{}), (-1, \underline{})$ $4, -\frac{7}{2}, -6$
  11. $x^2 - y = 5$; $(3, \underline{}), (2, \underline{}), (-2, \underline{})$ $4, -1, -1$
  12. $2x + y^2 = 18$; $(\underline{}, 4), (\underline{}, 0), (\underline{}, -6)$ $1, 9, -9$

Solve each equation if $x \in \{-2, -1, 0, 1, 2\}$ and $y \in \{$the integers$\}$. Graph the solution set.

13. $y = x$
14. $y = -x$
15. $y = 3x + 1$
16. $y = 6 - 2x$
17. $y + 2x = 5$
18. $3x - y = 2$
19. $6x - 2y = 4$
20. $9x + 3y = 9$
21. $3x - 2y = 1$
22. $2y - 5x = 3$
23. $6x = 1 - 2y$
24. $4x = 2y - 3$

Solve each inequality if $x \in \{-1, 0, 1\}$ and $y \in \{-2, -1, 0, 1, 2\}$. Graph the solution set.

B 25. $y \geq x$ 26. $y > -x$ 27. $y > x - 1$ 28. $y \leq 2x$
29. $x + y \leq 0$ 30. $y - x > 0$ 31. $y + 1 \leq 2x$ 32. $2x > y - 1$
33. $2x + y \geq 2y$ 34. $3y < y - 2x$ 35. $x - y \leq x + y$ 36. $x + y < y - x$

Solve each open sentence using the given replacement sets for the variables.

37. $x + y \leq 4$; $x \in \{2, 3, 4\}$, $y \in \{\text{the positive integers}\}$
38. $x - y > 2$; $x \in \{3, 4, 5\}$, $y \in \{\text{the positive integers}\}$
39. $y > 2x$; $x \in \{-2, -1, 0\}$, $y \in \{\text{the negative integers}\}$
40. $y \leq -\dfrac{x}{2}$; $x \in \{-4, -2, 0\}$, $y \in \{\text{the whole numbers}\}$
41. $x + 2y \leq 5$; $x \in \{3, 4, 5\}$, $y \in \{\text{the whole numbers}\}$
42. $3x + y > 0$; $x \in \{0, 1, 2\}$ $y \in \{\text{the negative integers}\}$

C 43. $|x| + y = 5$; $x, y \in \{-3, -2, -1, 0, 1, 2, 3\}$
44. $|x| + |y| = 6$; $x, y \in \{-4, -2, -1, 0, 1, 2, 4\}$
45. $|x| - |y| = 2$; $x, y \in \{-5, -3, -1, 0, 1, 3, 5\}$
46. $|x - y| = 3$ $x, y \in \{-4, -3, -2, -1, 0, 1, 2, 3, 4\}$
47. $x^2 - y = 1$ $x, y \in \{-2, -1, 0, 1, 2\}$
48. $2x + y^2 = 4$ $x, y \in \{-4, -2, -1, 0, 1, 2, 4\}$

Computer Exercises   For students with computer experience

Write a program that will allow you to input values for a, b, c, x, and y and will determine whether the ordered pair (x, y) is a solution of the equation $ax + by = c$. RUN the program to determine whether $(2, -3)$ is a solution of each of the following.

1. $5x + 2y = 4$ Yes   2. $2x - 4y = 10$ No   3. $3x - y = 6$ No   4. $x + 4y = -10$ Yes
5. $-4x = 3y + 1$ Yes   6. $x = 2$ Yes   7. $y = -3$ Yes   8. $3x = -2y$ Yes

9. Modify the program that you wrote for Exercises 1–8 so that, if (x, y) is *not* a solution of $ax + by = c$, the program will determine whether it is a solution of $ax + by < c$ or of $ax + by > c$.
10. $\{(3, 6), (6, 4), (9, 2)\}$

Write a program that will solve the equation $2x + 3y = 24$ when you input two integral intervals as replacement sets for x and y, respectively. RUN the program for each of the following replacement sets.

10. $x, y \in \{1, 2, 3, \ldots, 12\}$ 11. $x, y \in \{-12, -11, -10, \ldots, 12\}$

12. $x \in \{-12, -11, -10, \ldots, 0\}; y \in \{0, 1, 2, \ldots, 12\}$ $\{(-6, 12), (-3, 10), (0, 8)\}$
11. $\{(-6, 12), (-3, 10), (0, 8), (3, 6), (6, 4), (9, 2), (12, 0)\}$

Graphs and Functions   **231**

19. $\{(-2, -8), (-1, -5),$
    $(0, -2), (1, 1), (2, 4)\}$
20. $\{(-2, 9), (-1, 6), (0, 3),$
    $(1, 0), (2, -3)\}$

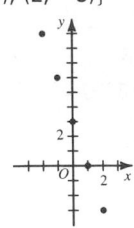

21. $\{(-1, -2), (1, 1)\}$
22. $\{(-1, -1), (1, 4)\}$

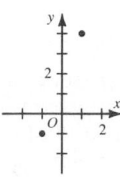

23. $\varnothing$ 24. $\varnothing$
25. $\{(-1, -1), (-1, 0),$
    $(-1, 1), (-1, 2), (0, 0),$
    $(0, 1), (0, 2), (1, 1), (1, 2)\}$
26. $\{(-1, 2), (0, 1), (0, 2),$
    $(1, 0), (1, 1), (1, 2)\}$

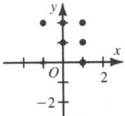

27. $\{(-1, -1), (-1, 0),$
    $(-1, 1), (-1, 2), (0, 0),$
    $(0, 1), (0, 2), (1, 1), (1, 2)\}$
28. $\{(-1, -2), (0, -2),$
    $(0, -1), (0, 0), (1, -2),$
    $(1, -1), (1, 0), (1, 1), (1, 2)\}$

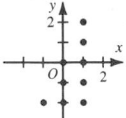

(continued on p. 260)

5–5 The Graph of a Linear Equation in Two Variables

Each of the following three figures shows a graph associated with the equation

$$x + y = 2.$$

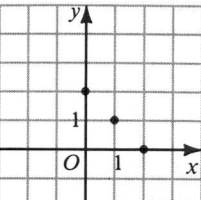 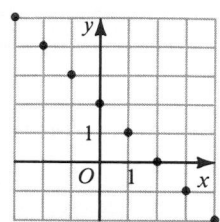 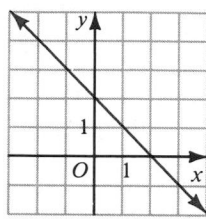

     Figure 8   **Figure 9**   **Figure 10**

In Figure 8, the replacement set of the variables x and y is the set of whole numbers, W, and so the graph consists of just three points: (0, 2), (1, 1), and (2, 0). In Figure 9 you see only a partial graph; the replacement set of each variable is the set of integers, J, and the resulting graph is an infinite set of isolated points. In Figure 10, the replacement set of each variable is the set of real numbers, $\mathcal{R}$, and the graph is an infinite set of points that together form a line.

In fact, if the replacement set of each of the variables x and y is $\mathcal{R}$, each solution of the equation $x + y = 2$ gives the coordinates of a point on the line shown in Figure 10. Similarly, the coordinates of each point on this line satisfy the equation. Thus, the line shown in Figure 10 is the set of *all those points* and *only those points* whose coordinates satisfy the equation $x + y = 2$. This line is called the **graph of the equation** on the coordinate plane, and the equation is called an **equation of the line**.

In general, the graph of any equation of the form

$$ax + by = c,$$

where a, b, and c are real numbers and a and b are not both zero, is a line. The number a is called the *coefficient of x*, and b is called the *coefficient of y*. Any equation of this form is called a **linear equation in two variables,** x and y. Thus,

$$6x - 5y = 7 \quad \text{and} \quad 2y = 15$$

are linear equations in two variables, but

$$x^2 + y = 5, \quad xy = 6, \quad \text{and} \quad \frac{1}{x} + y = 3$$

are not.

The following facts about linear equations are taken for granted, provided that the variables represent real numbers.

1. The graph of each linear equation in two variables is a line on a coordinate plane.
2. Each line on a coordinate plane is the graph of a linear equation in two variables.

Throughout the rest of this book, the replacement set of each variable in a linear equation is $\mathscr{R}$ unless otherwise specified.

Since two points determine a line, you need to find only two solutions of a linear equation in order to graph it. However, it is a good practice to find at least a third solution as a check on your work. Often the most convenient solutions to find are those where the line intersects the y-axis ($x = 0$) and where the line intersects the x-axis ($y = 0$).

EXAMPLE 1 Graph $3x - 2y = 6$ on a coordinate plane.

SOLUTION

Let $x = 0$.
$3(0) - 2y = 6$
$-2y = 6$
$y = -3$
Solution: $(0, -3)$

Let $y = 0$.
$3x - 2(0) = 6$
$3x = 6$
$x = 2$
Solution: $(2, 0)$

Any third solution, such as $(4, 3)$, can be used as a check. The graph is shown at the right.

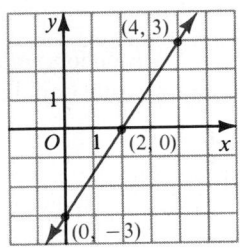

EXAMPLE 2 Graph each equation on a coordinate plane.
a. $x = -2$ b. $y = 3$

SOLUTION

a. The equation places no restriction on y, and so all points with abscissa -2 are graphs of solutions. Therefore, the graph of $x = -2$ is a vertical line two units to the left of the y-axis, as shown at the right.

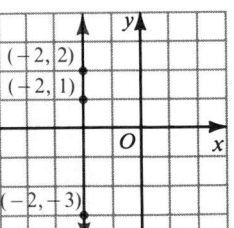

b. The equation places no restriction on x, and so all points with ordinate 3 are graphs of solutions. Therefore, the graph of $y = 3$ is a horizontal line three units above the x-axis, as shown at the right.

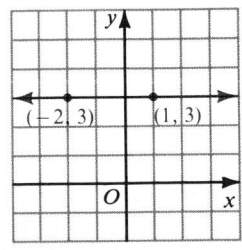

Graphs and Functions **233**

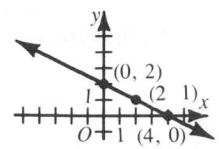

2. Graph each equation on a coordinate plane.
a. $x = 4$

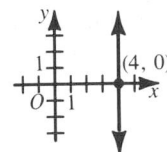

b. $y = -3$

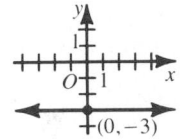

Additional A Exercises

Transform each equation into an equivalent equation of the form $ax + by = c$, where a, b, and c are integers.

1. $y - 1 = -x$ $x + y = 1$
2. $\frac{1}{2}x + y = 2$ $x + 2y = 4$
3. $-y + 2 = 3x$ $3x + y = 2$
4. $4y + 2 = -3x$
   $3x + 4y = -2$

Graph each equation on a coordinate plane.

5. $x + y = 0$

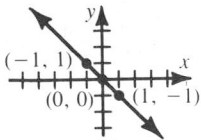

6. $y = x + 1$

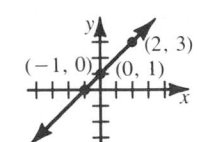

Suggested Assignments

Minimum
    235/1–22
Average
    235/1–35 odd, 36
Maximum
    235/5–37 odd, 38–42

**Additional Answers
Written Exercises**

8.
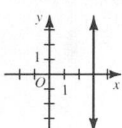

10.
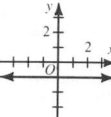

12.
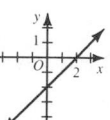

14.
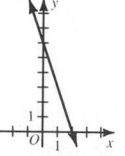

16.
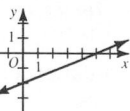

18.
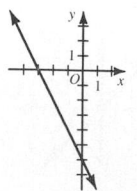

Note that graphing on a number line and on a coordinate plane are different. For example, Figure 11 shows $x = 3$ considered as an equation in one variable. Its solution set is graphed as a single point on a number line. Figure 12 shows $x = 3$ considered as an equation in two variables, $x + 0y = 3$. Its solution set consists of all ordered pairs $(3, y)$ where the replacement set for y is $\mathcal{R}$, and its graph is a line on a coordinate plane.

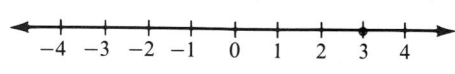

Figure 11

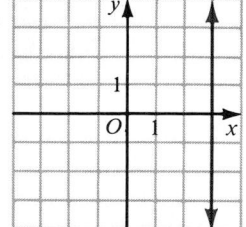

Figure 12

A function whose ordered pairs satisfy a linear equation is called a **linear function**. For example,

$$f: x \longrightarrow 2 - x, \quad x \in \mathcal{R}$$

is a linear function. Its graph is the line that is graphed on the coordinate plane in Figure 10 (page 232).

Oral Exercises

Tell whether or not each equation is a linear equation.

1. $x + y = 6$ Yes   2. $xy = 6$ No   3. $y = x$ Yes

4. $y = |x|$ No   5. $y = 2x$ Yes   6. $y = x^2$ No

7. $y = -2$ Yes   8. $x = 5$ Yes   9. $\frac{y}{3} = x$ Yes

10. $y = \frac{x}{3}$ Yes   11. $\frac{1}{3}x + \frac{1}{2}y = 7$ Yes   12. $\frac{3}{x} + \frac{2}{y} = 7$ No

13. **a.** vert. line 1 unit to right of y-axis

Describe the graph of each equation.   13. **b.** vert. line 2 units to left of y-axis

13. **a.** $x = 1$ **b.** $x = -2$ **c.** $x = 0$ y-axis

14. **a.** $y = 5$ **b.** $y = -4$ **c.** $y = 0$ x-axis

a. horiz. line 5 units above x-axis   **b.** horiz. line 4 units below x-axis

Give the coordinates of the points where the graph of each equation intersects the x-axis and the y-axis.

                                                                (5, 0); (0, 2)

15. $x + 3y = 6$ (6, 0); (0, 2)   16. $2x - y = 4$ (2, 0); (0, −4)   17. $2x + 5y = 10$

18. $2x - 3y = 12$ (6, 0); (0, −4)   19. $2x = 8$ (4, 0); does not   20. $y - 3 = 7$
                                                 intersect the y-axis   does not intersect the
                                                                       x-axis; (0, 10)

234   *Chapter 5*

State the relationship between the ordinate and the abscissa of each point on the graph of the given equation. 22. Ordinate is seven less than the abscissa.
23. Ordinate is two less than five times the abscissa.
24. Ordinate is one more than half the abscissa.

EXAMPLE $y = 3x + 1$

SOLUTION The ordinate is one more than three times the abscissa.
21. Ordinate is four times the abscissa.

21. $y = 4x$ 22. $y = x - 7$ 23. $y = 5x - 2$

24. $y = \frac{1}{2}x + 1$ 25. $y = |x| + 2$ 26. $y = x^2 - 1$

25. Ordinate is two more than the absolute value of the abscissa.
26. Ordinate is one less than the square of the abscissa.

Written Exercises

Transform each equation into an equivalent equation of the form $ax + by = c$, where a, b, and c are integers.

A   1. $3x = 4y - 5$ $3x - 4y = -5$ 2. $y = 2x + 7$ $-2x + y = 7$ 3. $5x - 8 = 3y$ $5x - 3y = 8$

4. $4y + 9 = 5x$ $-5x + 4y = -9$ 5. $\frac{1}{2}x + y = 3$ $x + 2y = 6$ 6. $2x + \frac{y}{3} = 5$ $6x + y = 15$

Graph each equation on a coordinate plane.

7. $x = -4$ 8. $x = 3$ 9. $y = 5$ 10. $y = -1$

11. $x + y = 5$ 12. $x - y = 2$ 13. $x + 2y = -4$ 14. $3x + y = 6$

15. $2x + 3y = 12$ 16. $2x - 5y = 10$ 17. $\frac{x}{3} + y = 2$ 18. $x + \frac{y}{2} = -3$

19. $5y - 3x = 15$ 20. $7y - 2x = 14$ 21. $2y = 6 - 3x$ 22. $4x = 20 - 5y$

Graph each function over $\mathcal{R}$.

B   23. $f: x \longrightarrow x - 3$ 24. $g: x \longrightarrow 3x + 1$ 25. $j: x \longrightarrow \frac{x}{2} + 1$

26. $k: x \longrightarrow \frac{x + 1}{3}$ 27. $R: x \longrightarrow \frac{1}{2}(x - 3)$ 28. $S: x \longrightarrow -3(2 - x)$

In each of Exercises 29–32, graph the given equations on the same coordinate plane. Label each graph.

29. a. $y = x$ b. $y = x + 1$ c. $y = x + 2$
30. a. $y = x$ b. $y = x - 2$ c. $y = x - 3$
31. a. $y = x$ b. $y = 2x$ c. $y = 3x$
32. a. $y = x$ b. $y = \frac{1}{2}x$ c. $y = \frac{1}{3}x$

33. For any real number k, describe how the graph of $y = x + k$ is related to the graph of $y = x$.
34. For any real number k, describe how the graph of $y = kx$ is related to the graph of $y = x$.

Graphs and Functions   **235**

20.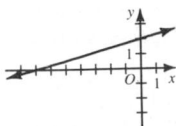

22.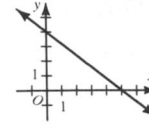

24.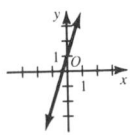

26.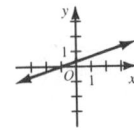

28.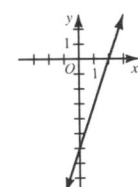

30.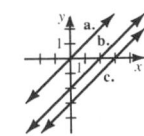

32.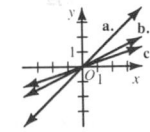

33. The graph of $y = x + k$ is parallel to the graph of $y = x$, which intersects the y-axis at the origin, but the graph of $y = x + k$ intersects the y-axis at $(0, k)$.

(continued on p. 258)

Graph the solution sets of the following open sentences in two variables, where $x \in \{-2, 0, 2\}$ and $y \in$ {the integers}.

1. $y = x$

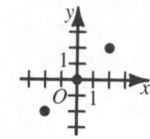

2. $y = 2x$

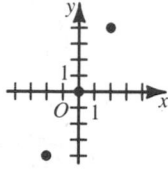

3. $y = x + 3$

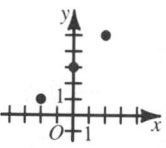

In each of Exercises 35–38, graph the given equations on the same coordinate plane. Find the coordinates of the point at which the two graphs intersect, then determine if this ordered pair satisfies *both* equations.

C **35.** $y = x + 1$; $y = -x + 1$
(0, 1); satisfies both equations

36. $y = x - 2$; $y = \frac{1}{3}x$
(3, 1); satisfies both equations

37. $2y - x = 4$; $2y + 3x = 12$
(2, 3); satisfies both equations

38. $4y + 5x = 8$; $4y + x = -8$
(4, -3); satisfies both equations

Graph each equation on a coordinate plane. Then tell whether or not the equation is a linear equation.

39. $y = |x|$ No

40. $y = -|x|$ No

41. $y = |x| + 2$ No

42. $y = |x| - 2$ No

43. $y = |x + 1|$ No

44. $y = |x - 1|$ No

45. $y = -|x| + 3$ No

46. $y = -|x + 3|$ No

47. For any real number k, describe how the graph of $y = |x| + k$ is related to the graph of $y = |x|$.

48. For any real number k, describe how the graph of $y = |x + k|$ is related to the graph of $y = |x|$.

5–6 The Graph of a Linear Inequality in Two Variables

The graph of the linear equation

$$y = x + 1$$

separates a coordinate plane into two regions that are called **open half-planes**. One of these regions is *above* the line, as shown by the colored shading in Figure 13, and the other region is *below* the line, as shown by the gray shading. The line itself is called the **boundary** of each half-plane.

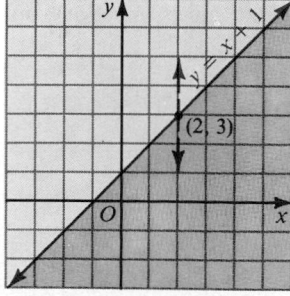

Figure 13

236 *Chapter 5*

If you start at any point of this line, say (2, 3), and move vertically upward, the y-coordinates of the points on the plane increase. Thus the open half-plane above the line is the graph of

$$y > x + 1,$$

as indicated in Figure 13. If you move vertically *downward* from (2, 3), the y-coordinates of the points *decrease*, and so the open half-plane below the line is the graph of

$$y < x + 1.$$

Figure 14 shows the graph of four inequalities.

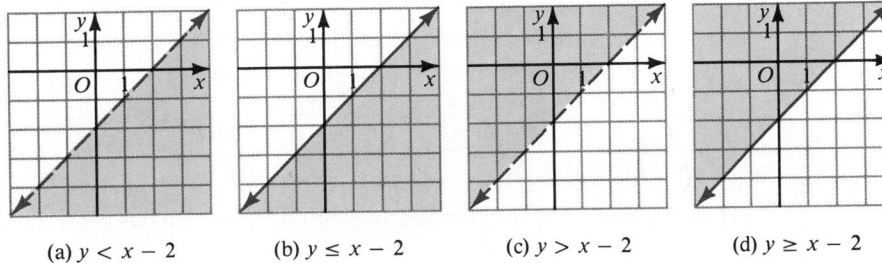

(a) $y < x - 2$　　(b) $y \le x - 2$　　(c) $y > x - 2$　　(d) $y \ge x - 2$

Figure 14

Each of these graphs has as its boundary the line with the equation

$$y = x - 2.$$

Notice that the boundary is drawn as a *dashed* line in parts (a) and (c) to show that the boundary is not part of the graph. That is, the dashed line indicates that the graph is an *open half-plane*. In parts (b) and (d), however, the boundary must be included as part of the graph, and so it is drawn as a *solid* line. In such cases, the graph is the *union* of the open half-plane and its boundary and is called a **closed half-plane**.

The equation of the boundary of the graph of an inequality is called the **associated equation** of the inequality. Thus, $y = x - 2$ is the associated equation of each inequality graphed in Figure 14. An inequality whose associated equation is a linear equation in two variables is called a **linear inequality in two variables**.

In general, any linear equation in two variables,

$$ax + by = c,$$

is the associated equation of four linear inequalities in two variables:

$$ax + by < c \qquad ax + by > c$$
$$ax + by \le c \qquad ax + by \ge c$$

On a coordinate plane, the *graph of a linear inequality in two variables* is either an open or closed half-plane.

Graphs and Functions　**237**

Chalkboard Examples

1. Graph $y - 2x < 1$ on a coordinate plane.
   $y - 2x < 1$
   　　$y < 2x + 1$
   Graph $y = 2x + 1$ as a dashed line. Shade the open half-plane below the line.

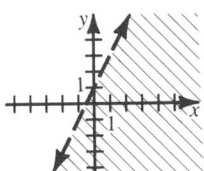

2. Graph each inequality on a coordinate plane.
   a. $y \ge -2$
   　Graph $y = -2$ as a solid line.

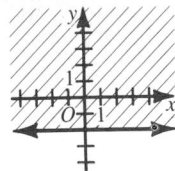

   b. $x < -1$
   　Graph $x = -1$ as a dashed line.

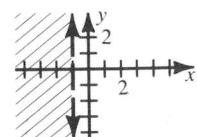

Graph each inequality on a coordinate plane.

1. $x \geq 3$

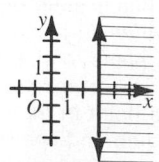

2. $x < -3$

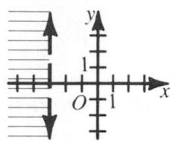

3. $y > -1$

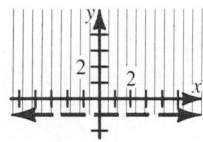

4. $y < 1$

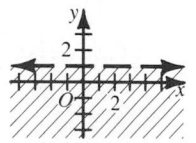

5. $y \geq x + 1$

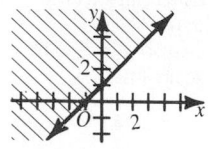

6. $y < 2x$

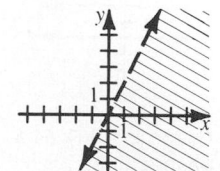

EXAMPLE 1 Graph $3x - y > 1$ on a coordinate plane.

SOLUTION

1. Solve the given inequality for y.
$$3x - y > 1$$
$$-y > -3x + 1$$
$$y < 3x - 1$$

2. Graph the associated equation
$$y = 3x - 1$$
and show it as a *dashed* line.

3. Shade the open half-plane *below* the line.

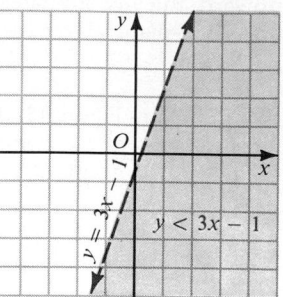

Check: Select any point in the shaded region and determine whether its coordinates satisfy the original inequality.

$$(2, 2): \qquad 3x - y > 1$$
$$3(2) - 2 \overset{?}{>} 1$$
$$4 > 1 \quad \checkmark$$

Thus, $(2, 2)$ is in the solution set, and the correct region has been shaded.

EXAMPLE 2 Graph each inequality on a coordinate plane.

a. $y \geq 3$ **b.** $x < -2$

SOLUTION

a. Graph the associated equation $y = 3$ as a solid line. Then shade the region *above* the line to graph all the points with ordinate *greater than* 3.

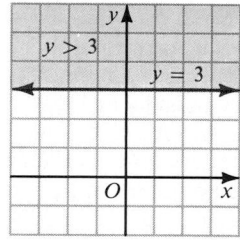

b. Graph the associated equation $x = -2$ as a dashed line. Then shade the region to the *left* of the line to graph all the points with abscissa *less than* -2.

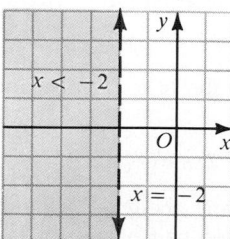

Oral Exercises

Tell whether the point $(-1, 0)$ lies above, on, or below the graph of the given equation on a coordinate plane.

1. $y = x + 1$ On
2. $y = x - 3$ Above   Below **3.** $y = x + 5$
4. $y = -2x$ Below
5. $y = 3x + 1$ Above   On **6.** $y = 2 + 2x$

Tell which of the points with given coordinates belong to the graph of the inequality.

7. $(1, -2), (-1, 3)$; $x + y < 0$ $(1, -2)$ **8.** $(-2, -3), (-3, 5)$; $x - y \geq 0$ $(-2, -3)$
9. $(1, 2), (3, -1)$; $3x + y \leq 5$ $(1, 2)$ **10.** $(5, 4), (3, 1)$; $x - 2y > 0$ $(3, 1)$

Solve each inequality for y. **13.** $y < -3x + 2$ **16.** $y \geq 2x$ **19.** $y < \frac{1}{3}x$

11. $x + y > 7$ $y > -x + 7$
12. $y + x \leq 0$ $y \leq -x$
13. $3x + y < 2$
14. $-3x + y \geq 4$ $y \geq 3x + 4$
15. $x - y > 2$ $y < x - 2$
16. $2x - y \leq 0$
17. $4x + 2y > 6$ $y > -2x + 3$
18. $12x + 4y < 4$ $y < -3x + 1$
19. $3y < x$
20. $-4y \geq x$ $y \leq -\frac{1}{4}x$
21. $x + 3y < 6$ $y < -\frac{1}{3}x + 2$
22. $x - 2y \geq 3$ $y \leq \frac{1}{2}x - \frac{3}{2}$

Written Exercises

Graph each inequality on a coordinate plane.

A
1. $y < 3$
2. $y \geq -3$
3. $x \leq -1$
4. $x > 4$
5. $x > 0$
6. $y \leq 0$
7. $y > x$
8. $y \leq x$
9. $y > -x$
10. $y < -2x$
11. $y \leq 2x - 1$
12. $y > 2 - 3x$
13. $x + y \geq 4$
14. $x + y < 5$
15. $x - y < 3$
16. $x - y \geq 1$
17. $3x + y \leq 2$
18. $4x - y > 1$

B
19. $4x + 2y > 4$
20. $6x + 2y < 8$
21. $3x - 2y \leq 6$
22. $2x - 3y > 9$
23. $2x - 5y > 5$
24. $3x + 4y \leq 2$

In each of Exercises 25–30, graph the given inequalities on the same coordinate plane.

25. $y > -1; y \leq 2$
26. $y > 1; y < x$
27. $y > -3; x < 1$
28. $y \leq 5; x \geq -2$
29. $x + y > -1; 2x + y > 3$
30. $4x + y < -1; 2x - y < 5$

Graph each inequality on a coordinate plane.

C
31. $|x| > 2$
32. $|x| < 5$
33. $|y| > 3$
34. $|y| < 1$
35. $y > |x|$
36. $y < -|x|$

Graphs and Functions   **239**

Suggested Assignments
Minimum
  239/1–18
R 240/*Self-Test 2*
Average
  239/5–23 odd, 26–30
R 240/*Self-Test 2*
Maximum
  239/9–29 odd, 30–36
R 240/*Self-Test 2*

Additional Answers
Written Exercises
(See p. 264.)

Mixed Review

Graph the following linear equations in two variables.

1. $y = x - 2$

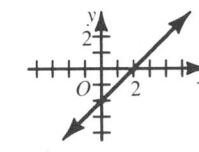

2. $y = \frac{1}{2}x - 2$

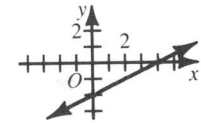

3. $y = -2x + 1$

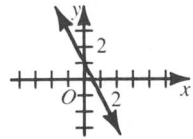

4. $y = \frac{2}{3}x - 1$

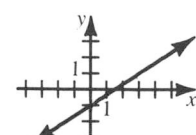

Let $x \in \{-1, 0, 1\}$ and $y \in \{-2, -1, 0, 1, 2\}$. Solve each open sentence.

1. $2y - 3x = 1$
$\{(-1, -1), (1, 2)\}$

2. $y < |x| - 1$
$\{(-1, -2), (-1, -1), (0, -2), (1, -2), (1, -1)\}$

Graph each open sentence on a coordinate plane.

3. $2x - y = 3$

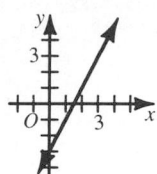

4. $3y = 5x - 15$

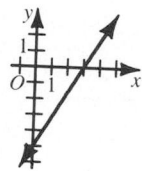

5. $x + y > -2$

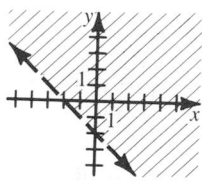

6. $3x + y < 2$

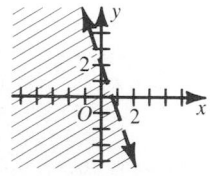

7. $y < 3$

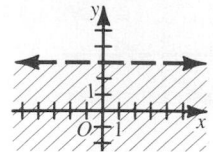

Self-Test 2

VOCABULARY

solution of an open sentence in two variables (p. 228)
graph of an open sentence in two variables (p. 229)
graph of an equation (p. 232)
equation of a line (p. 232)
linear equation in two variables (p. 232)

linear function (p. 234)
open half-plane (p. 236)
boundary of a half-plane (p. 236)
closed half-plane (p. 237)
associated equation (p. 237)
linear inequality in two variables (p. 237)

Let $x \in \{-2, 0, 2\}$ and $y \in \{-4, -2, 4\}$. Solve each open sentence.

1. $y = -x$
$\{(2, -2)\}$

2. $3x + 2y = -2$
$\{(2, -4)\}$

3. $y < x + 2$ *Obj. 1, p. 228*
$\{(-2, -4), (-2, -2), (0, -4),$

Graph each open sentence on a coordinate plane. $(0, -2), (2, -4), (2, -2)\}$

4. $y = x$

5. $2x - 5y = 10$

6. $x = -3$ *Obj. 2, p. 228*

7. $y \geq -2$

8. $x + y < 6$

9. $2x - y > 1$

Check your answers with those at the back of the book.

Ada Lovelace
1815–1852

Ada Lovelace was born Augusta Ada Byron, daughter of Lord Byron, the famous English poet, and Annabella Milbanke. Encouraged by her mother and by friends, she studied mathematics enthusiastically. In 1835 she married William, Lord King, who soon became the Earl of Lovelace.

During the 1830's, the English mathematician Charles Babbage was designing his "Analytical Engine," a machine that could make calculations, store data, and print out results. Lady Lovelace first wrote to him in her search for a mathematics tutor, but she soon became involved in the project of translating and annotating a paper describing his Analytical Engine. Her detailed notes describing the process of communicating with the machine are considered to be the first description of computer programming.

In 1979 a newly developed programming language was named *Ada* in recognition of Lady Lovelace's contribution to computer science.

READING ALGEBRA Independent Study

When you begin a new chapter or section of your textbook, you will want to have some goals in mind. The title of each main chapter division and the objectives that follow it will tell you what you will be reading about. Do not skip over them. Like the headlines in a newspaper story, they will highlight what is important.

After you know what you will be reading about, read through each section at a moderate speed. Look at each diagram and read through the examples and their solutions. If you encounter an unfamiliar word, look it up in the Glossary at the back of the book or in a dictionary. When you are finished, try to say aloud in your own words what you have read.

Usually you will need to go back and read a section a second time. When you do this, have a pencil, your notebook, and a piece of scrap paper at hand. Read slowly and carefully this time, making sure that you understand each word. Pay particular attention to mathematical symbols, translating each one into words. For example, on page 69, the mathematical sentence "$a + (-a) = 0$" translates into "a plus the opposite of a equals zero."

When you come to an example in the text, cover the solution with your scrap paper and try to work it out yourself. Then compare your solution with the one given in the book. If you did not work the example correctly, copy the given solution, one step at a time, understanding each step before going on to the next. Then cover the solution and try again to do the example by yourself.

Exercises

Name the objective for the given section.

1. Section 5–4
2. Section 5–5
3. Section 5–6

Tell which section you would review if you had trouble with the following exercises in Self-Test 2 on page 240.

4. Exercise 3
   Section 5-4
5. Exercise 5
   Section 5-5
6. Exercise 9
   Section 5-6

Translate into words.

7. $-(a + b) = -a + (-b)$

8. $f: x \longrightarrow 7 - \frac{3}{2}x,\ x \in \{0, 2, 4\}$

9. Write out a solution to the following example.

   EXAMPLE Given $f: x \longrightarrow 5 - |x|$, find all values of x such that $f(x) = 0$.

Graphs and Functions **241**

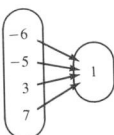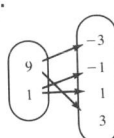

Lines on a Coordinate Plane

OBJECTIVES for Sections 5-7 through 5-9:
1. *To find the slope of a line on a coordinate plane.*
2. *To use the slope-intercept form of a linear equation.*
3. *To determine an equation of a line given its slope and y-intercept.*
4. *To determine an equation of a line given the slope of the line and the coordinates of one point through which the line passes, or given the coordinates of two points on the line.*

5–7 The Slope of a Line

There are many everyday terms that have a special meaning in mathematics. Sometimes the everyday use of such terms helps in understanding their use in mathematics. As an example, consider the word "slope."

To describe the steepness, or slope, of a hill, you may estimate the amount of vertical *rise* of the hill that corresponds to a certain amount of horizontal *run*, then calculate the ratio of the rise to the run. For example, Figure 15 represents a hill that rises 10 m for every 50 m of horizontal run. Its slope is the ratio

$$\frac{\text{rise}}{\text{run}} = \frac{10}{50} = 0.2.$$

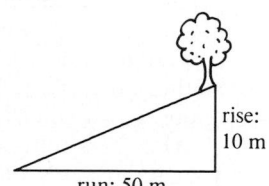

rise: 10 m

run: 50 m

Figure 15

Similarly, to describe the steepness, or slope, of a line on a coordinate plane, you may choose any two points on the line, compute the units in the rise and the run from one point to the other, and calculate the ratio of the rise to the run. For example, consider the lines that are graphed in Figures 16 and 17.

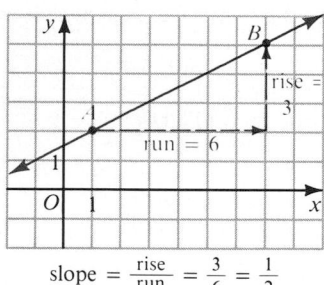

$$\text{slope} = \frac{\text{rise}}{\text{run}} = \frac{3}{6} = \frac{1}{2}$$

Figure 16

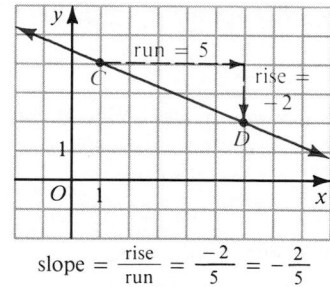

$$\text{slope} = \frac{\text{rise}}{\text{run}} = \frac{-2}{5} = -\frac{2}{5}$$

Figure 17

In Figure 16, the line is shown passing through the points $A(1, 2)$ and $B(7, 5)$. Notice that the rise, or *vertical change*, in moving from A to B is equal to the difference between the ordinates of these points: $5 - 2 = 3$.

242 *Chapter 5*

The run, or *horizontal change*, in moving from A to B is equal to the difference between the abscissas: $7 - 1 = 6$. Similarly, in Figure 17, the rise in moving from $C(1, 4)$ to $D(6, 2)$ is equal to the difference between the ordinates, $2 - 4 = -2$, and the run is equal to the difference between the abscissas, $6 - 1 = 5$. Thus, the slope of a line on a coordinate plane may be defined as follows.

$$\textbf{slope} = \frac{\text{rise}}{\text{run}} = \frac{\text{vertical change}}{\text{horizontal change}} = \frac{\text{difference between ordinates}}{\text{difference between abscissas}}$$

More formally, if (x_1, y_1), read "x sub one, y sub one," and (x_2, y_2), read "x sub two, y sub two," are any two different points on a line,

$$\textbf{slope} = \frac{y_2 - y_1}{x_2 - x_1} \qquad (x_1 \neq x_2).$$

Notice that the differences between the ordinates and the abscissas must be taken *in the same order*.

EXAMPLE 1 Find the slope of the line that passes through $(-5, 1)$ and $(7, -3)$.

SOLUTION Let $x_1 = -5$, $y_1 = 1$, $x_2 = 7$, and $y_2 = -3$.

$$\text{slope} = \frac{y_2 - y_1}{x_2 - x_1} = \frac{-3 - 1}{7 - (-5)} = \frac{-4}{12} = -\frac{1}{3}$$

$\therefore$ the slope of the line is $-\frac{1}{3}$.

EXAMPLE 2 Graph the line that passes through the point $(-6, 7)$ and has slope $-\frac{3}{2}$.

SOLUTION
1. Plot $(-6, 7)$ on a coordinate plane.
2. Since $-\frac{3}{2} = \frac{-3}{2}$, count 3 units *down* and 2 units to the *right* to obtain a second point on the line.

$$(-6 + 2, \quad 7 - 3) = (-4, 4)$$

3. Draw a line through $(-6, 7)$ and $(-4, 4)$.

Check: Since $-\frac{3}{2} = -\frac{6}{4}$ a third point on the line should be

$$(-6 + 4, \quad 7 - 6) = (-2, 1).$$

The line is graphed on the coordinate plane above.

Notice in Example 2 that $-\frac{3}{2} = \frac{3}{-2}$ is also true. Thus, in Step 2 you may obtain a second point on the line by counting 3 units *up* and 2 units to the *left* from the point $(-6, 7)$.

A basic property of a line is that its slope is constant.

Graphs and Functions **243**

1. Find the slope of the line that passes through $(-4, -2)$ and $(4, 4)$.
   Let $x_1 = -4$, $y_1 = -2$, $x_2 = 4$, and $y_2 = 4$.
   $$\text{slope} = \frac{y_2 - y_1}{x_2 - x_1} = \frac{4 - (-2)}{4 - (-4)}$$
   $$= \frac{6}{8} = \frac{3}{4}$$
   $\therefore$ the slope of the line is $\frac{3}{4}$.

2. Graph the line that passes through the point $(-3, -1)$ and has slope $\frac{2}{3}$.
   Plot $(-3, -1)$.
   Count 2 units up and 3 units to the right to $(0, 1)$.
   Draw a line through $(-3, -1)$ and $(0, 1)$.

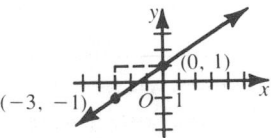

3. Determine whether or not the points $(-4, 3)$, $(-2, -1)$, $(1, -7)$, and $(4, -11)$ lie on the same line.

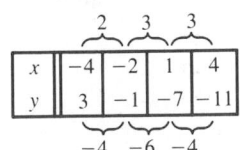

   The ratio,
   $$\frac{\text{difference of ordinates}}{\text{difference of abscissas}},$$
   is not constant.
   $\therefore$ the points are not collinear.

Determine the slope of the line that passes through the given points.

1. (0, 0), (2, 4) 2
2. (2, 1), (3, 2) 1
3. (−3, 4), (2, −1) −1
4. (−1, −5), (3, 3) 2

Graph the line that passes through the given point and that has the given slope.

5. (−4, 3); slope = $-\frac{1}{2}$

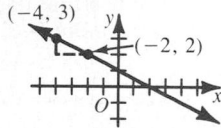

6. (−1, −4); slope = 1

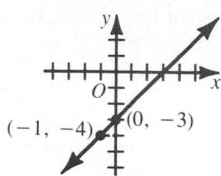

EXAMPLE 3   Determine whether or not the points (0, −2), (2, 1), (4, 4), and (8, 10) lie on the same line.

SOLUTION   Arrange the ordered pairs in a table in increasing order of abscissas. Then compute the differences between the abscissas and ordinates from point to point.

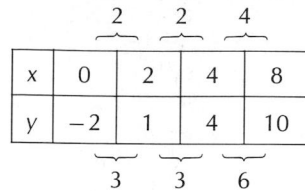

The ratio $\dfrac{\text{difference between ordinates}}{\text{difference between abscissas}}$ is constant: $\dfrac{3}{2} = \dfrac{3}{2} = \dfrac{6}{4}$

∴ the points lie on the same line.

Points that lie on the same line are called **collinear points**.

When a line rises from left to right on a coordinate plane, as in Figure 16 (page 242), the slope of the line is *positive*. When a line "falls" from left to right, as in Figure 17, its slope is *negative*.

Now consider the following figures.

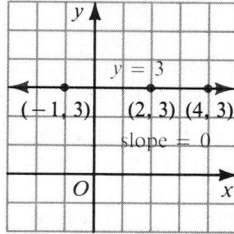

Figure 18

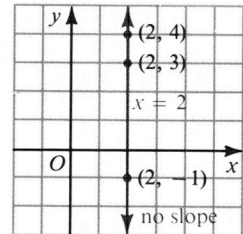

Figure 19

What is the slope of the horizontal line $y = 3$ that is shown in Figure 18? Because the ordinate of every point on this line is 3, the *difference* between the ordinates of any two points will always be 0. Thus, the numerator of the ratio

$$\frac{y_2 - y_1}{x_2 - x_1}$$

will always be 0, and so the slope of the line is 0. In fact, *the slope of any horizontal line is 0.*

On the other hand, every point on the vertical line $x = 2$, shown in Figure 19, has the same abscissa. Thus, the *denominator* of the above ratio will always be 0. Since division by 0 has no meaning, the line has no slope. In fact, *vertical lines have no slope.*

244   *Chapter 5*

Oral Exercises

Determine the slope of each line.

1. 3
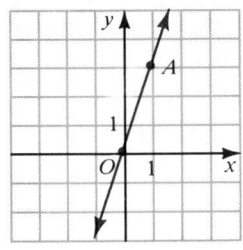

2. −2
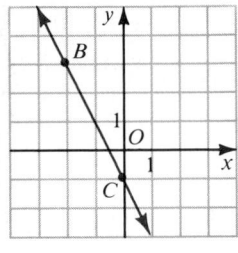

3. $\frac{3}{4}$
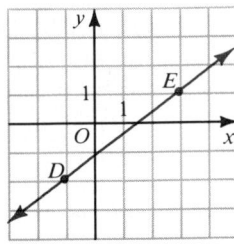

4. $-\frac{6}{5}$
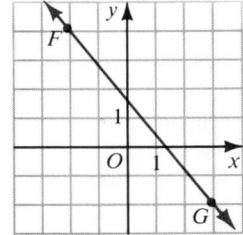

5. 0
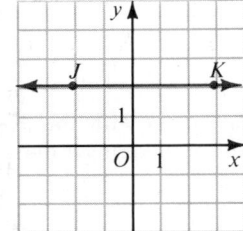

6.
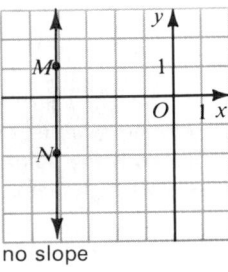
no slope

Tell whether or not the points that have the coordinates listed in each table are collinear. If the points are collinear, determine the slope of the line that passes through them.

7.

x	0	1	2	3
y	5	7	9	11

collinear; slope = 2

8.

x	2	3	4	5
y	9	6	3	0

collinear; slope = −3

9.

x	−1	0	1	2
y	4	4	4	4

collinear; slope = 0

10.

x	−2	0	2	4
y	−2	1	4	10

not collinear

11.

x	0	2	6	8
y	11	6	−4	−9

collinear; slope = $-\frac{5}{2}$

12.

x	1	3	7	11
y	−5	−1	7	15

collinear; slope = 2

13.

x	−4	−1	5	14
y	−1	−4	−10	−19

collinear; slope = −1

14.

x	−1	−3	−11	−17
y	−10	−10	−10	−10

collinear; slope = 0

Graphs and Functions **245**

Graph the following linear equations on the same set of coordinate axes.

1. $y = x + 2$

2. $y = 2x + 2$

3. $y = \frac{1}{2}x + 2$

4. $y = -x + 2$

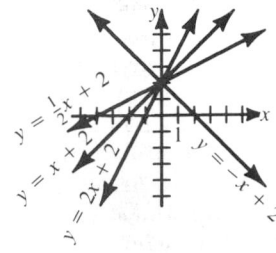

Graph the given linear inequalities on the same set of coordinate axes.

5. $y > 0;\ x > 0$

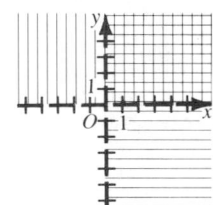

6. $y > x;\ x < -1$

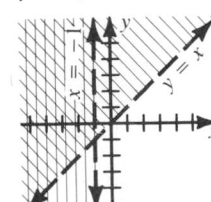

Suggested Assignments

Minimum
  246/1–18
Average
  246/1–23 odd, 24–32
Maximum
  246/5–31 odd, 32–34

Additional Answers
Written Exercises

14.

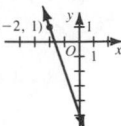

16.

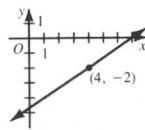

18.

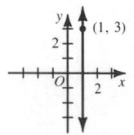

33. The slope of $\overline{BP}$ is $\frac{1}{2}$ and

the slope of $\overline{PC}$ is $\frac{1}{2}$. $\therefore$ B,

C, and P are collinear, so

P lies on side $\overline{BC}$.

34. The slope of $\overline{AQ}$ is 7 and
the slope of $\overline{QC}$ is 7. $\therefore$ A,
Q, and C are collinear.

The slope of $\overline{BQ}$ is $-\frac{1}{7}$

and the slope of $\overline{QD}$ is

$-\frac{1}{7}$; $\therefore$ B, Q, and D are

collinear. $\therefore$ Q lies on
both diagonals.

Written Exercises

Determine the slope of the line that passes through the given points.

A   **1.** $(2, 3)$, $(1, 1)$ 2   **2.** $(3, 5)$, $(4, 2)$ -3
      3. $(5, -4)$, $(-3, -4)$ 0   **4.** $(-2, 1)$, $(-2, 5)$ no slope
      5. $(2, -5)$, $(4, -7)$ -1 **6.** $(-3, 2)$, $(0, 8)$ 2
      7. $(4, -6)$, $(-1, 4)$ -2 **8.** $(3, 2)$, $(-2, -1)$ $\frac{3}{5}$
      9. $(-1, -7)$, $(1, 0)$ $\frac{7}{2}$ **10.** $(8, 1)$, $(-1, 1)$ 0
      11. $(-7, -2)$, $(-7, -3)$ no slope   **12.** $(-4, -1)$, $(0, 2)$ $\frac{3}{4}$

Graph the line that passes through the given point and has the given slope.

      13. $(1, 4)$; slope $= 2$ **14.** $(-2, 1)$; slope $= -3$
      15. $(-3, -2)$; slope $= -\frac{1}{2}$ **16.** $(4, -2)$; slope $= \frac{2}{3}$
      17. $(2, -5)$; slope $= 0$ **18.** $(1, 3)$; no slope

Determine whether or not the points with the given coordinates are collinear. If the points are collinear, determine the slope of the line that passes through them.   **20.** collinear; -3 **22.** not collinear

B   **19.** $(0, 3)$, $(1, 5)$, $(2, 7)$, $(3, 9)$ collinear; 2   **20.** $(4, 5)$, $(5, 2)$, $(6, -1)$, $(7, -4)$
      21. $(-4, 0)$, $(-1, -1)$, $(5, -3)$, $(8, -4)$ coll.; $-\frac{1}{3}$ **22.** $(1, -5)$, $(3, -2)$, $(7, 1)$, $(9, 4)$
      23. $(-6, -1)$, $(-4, -1)$, $(-3, -1)$, $(-1, -1)$ **24.** $(-3, 5)$, $(-3, 3)$, $(-3, 1)$, $(-3, -1)$
        collinear; 0   collinear; no slope

Determine the value of t so that the slope of the line through each pair of points has the given value.

      25. $(6, 3)$, $(-4, t)$; slope $= 2$ -17 **26.** $(t, 5)$, $(5, 1)$; slope $= -2$ 3
      27. $(5, 2t)$, $(1, 3t)$; slope $= -\frac{1}{2}$ 2   **28.** $(-4, 5t)$, $(5, 2t)$; slope $= \frac{1}{3}$ -1
      29. $(7, -4t)$, $(-8, 6t)$; slope $= -\frac{1}{3}$ $\frac{1}{2}$ **30.** $(3t, 5)$, $(-9t, -5)$; slope $= \frac{5}{4}$ $\frac{2}{3}$

      31. The vertices of a triangle are $A(-3, 2)$, $B(-1, -2)$, and $C(3, 0)$.
          Determine the slope of each side of the triangle. $\overline{AB}$: -2; $\overline{AC}$: $-\frac{1}{3}$; $\overline{BC}$: $\frac{1}{2}$
      32. The vertices of a parallelogram are $A(-3, -3)$, $B(2, -5)$, $C(5, -1)$,
          and $D(0, 1)$. Determine the slope of each side of the parallelogram
          and the slope of the two diagonals that connect the opposite
          vertices. $\overline{AB}$, $\overline{CD}$: $-\frac{2}{5}$; $\overline{AD}$, $\overline{BC}$: $\frac{4}{3}$; $\overline{AC}$: $\frac{1}{4}$; $\overline{BD}$: -3

C   **33.** The vertices of a right triangle are $A(-3, 2)$, $B(-3, -1)$, and
          $C(3, 2)$. Show algebraically that the point $P(1, 1)$ lies on one of the
          sides of the triangle.
      34. The vertices of a square are $A(-1, -3)$, $B(7, 3)$, $C(1, 11)$, and
          $D(-7, 5)$. Show algebraically that the point $Q(0, 4)$ lies both on the
          diagonal that connects A and C and on the diagonal that connects
          B and D.

PROGRAMMING IN BASIC

The following program will print a graph of an open sentence in two variables, X and Y, when you input the same replacement set for both X and Y. Note that, in line 110, the program uses a negative STEP in the loop for values of Y. Thus, if you input M = 9, the values of Y will run from 9 down to −9, and in line 120, the values of X will run from −9 to 9.

```
10    PRINT "TO GRAPH AN OPEN SENTENCE"
20    PRINT "IN TWO VARIABLES"
30    PRINT "(SENTENCE IS IN LINE 130):"
40    PRINT
50    PRINT "INPUT EXTENT OF GRAPH, −M TO M"
60    PRINT "(INPUT M> 0):";
70    INPUT M
80    REM *PRINT Y
90    PRINT TAB(M);"Y"
100   PRINT TAB(M);"!"
110   FOR Y = M TO −M STEP −1
120   FOR X = −M TO M
130   IF Y = X + 2 THEN 250
140   REM *PRINT Y-AXIS
150   IF X <> 0 THEN 190
160   PRINT "!";
170   GOTO 260
180   REM *PRINT X-AXIS
190   IF Y <> 0 THEN 220
200   PRINT "-";
210   GOTO 260
220   PRINT " ";
230   GOTO 260
240   REM *PRINT GRAPH
250   PRINT "*";
260   NEXT X
270   REM *PRINT X
280   IF Y = 0 THEN 310
290   PRINT
300   GOTO 320
310   PRINT "-X"
320   NEXT Y
330   PRINT TAB(M);"!"
340   END
```

Be aware that, in lines 90, 100, and 330, you may need to adjust the PRINT TAB statements in order to accommodate the specifics of the TAB function on the computer that you are using. (See page 196.)

Graphs and Functions **247**

Exercises

1. Type in and RUN the program as given. INPUT 9 for the extent of the graph.

Change line 130 as necessary to RUN the program for each of the following open sentences. INPUT 9 for the extent of the graph.

2. $y > x + 2$ 3. $y \le x + 2$ 4. $x + y = 8$ 5. $x - y = 6$

6. $x + y > 5$ 7. $x - y > 3$ 8. $2x + y \le -6$ 9. $x - 2y \ge 6$

If possible, save this program for later use.

Teaching Suggestions
p. T86

Related Activities p. T86

Key Ideas

State an equation of a line given its slope and its *y*-intercept.
Determine the slope and the *y*-intercept of a line given its linear equation.
Recognize that lines having the same slope but different *y*-intercepts are parallel.

5–8 The Slope-Intercept Form of a Linear Equation

The graph of the equation

$$y = 2x$$

is shown on the coordinate plane in Figure 20. As you can see, whenever the ordinates of two points on this line differ by 2, their abscissas differ by 1. Therefore, the slope of this line is $\frac{2}{1}$, or 2. Notice that the line passes through the origin.

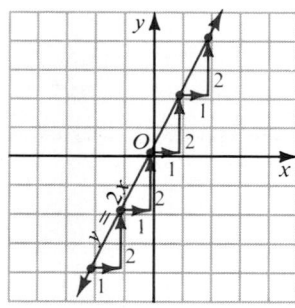

Figure 20

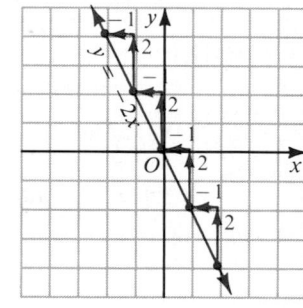

Figure 21

In Figure 21 you see the graph of the equation

$$y = -2x.$$

In this case, whenever the ordinates of two points on the line differ by 2, the abscissas differ by -1. Thus the slope of this line is $\frac{2}{-1}$, or -2. Notice that this line, too, passes through the origin.

248 *Chapter 5*

The graphs in Figures 20 and 21 illustrate the following fact.

For every real number m, the graph on a coordinate plane of the equation

$$y = mx$$

is the line that has slope m and passes through the origin.

Now consider the following equations, which are graphed on the coordinate plane in Figure 22.

$$y = 2x \qquad \text{and} \qquad y = 2x + 3$$

Notice that the lines have equal slopes, but they intersect the y-axis at different points. The ordinate of the point at which a line intersects the y-axis is called the y-intercept of the line. Since the abscissa of any point on the y-axis is 0, you can determine the y-intercept of a line by replacing the variable x with 0 in the equation of the line. Thus, for the lines graphed in Figure 22:

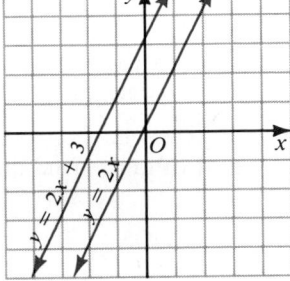

Figure 22

$y = 2x$	$y = 2x + 3$
$y = 2(0)$	$y = 2(0) + 3$
$y = 0$	$y = 3$
The y-intercept is 0.	The y-intercept is 3.

Notice that the lines graphed in Figure 22 are *parallel*. **Parallel lines** are lines in a plane that do not intersect. On a coordinate plane, nonvertical lines that have the same slope, but different y-intercepts, are parallel. Also, all vertical lines are parallel to each other.

Since the equations

$$y = 2x \qquad \text{and} \qquad y = 2x + 0,$$

are equivalent, the lines graphed in Figure 22 suggest the following fact about linear equations.

For all real numbers m and b, the graph on a coordinate plane of

$$y = mx + b$$

is a line whose slope is m and whose y-intercept is b.

When a linear equation is written in the form $y = mx + b$, it is said to be written in **slope-intercept form**.

Graphs and Functions **249**

 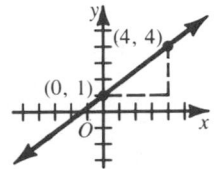

Determine the slope and y-intercept of the line whose equation is given. Graph each equation on a coordinate plane.

1. $y = x + 1$
The slope is 1, and the y-intercept is 1.

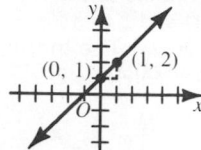

2. $y = 2x + 2$
The slope is 2 and the y-intercept is 2.

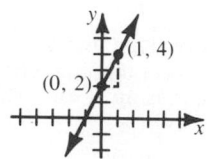

3. $y + 3x = -1$
The slope is -3 and the y-intercept is -1.

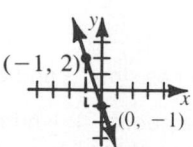

4. $2y = 4x + 2$
The slope is 2 and the y-intercept is 1.

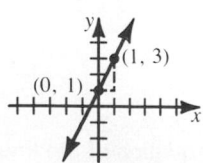

Write a linear equation for each line with the given slope m and y-intercept b.

5. $m = 3, b = 2$
$y = 3x + 2$

6. $m = 4, b = -5$
$y = 4x - 5$

EXAMPLE 1   Graph $y = \frac{5}{3}x - 4$ on a coordinate plane.

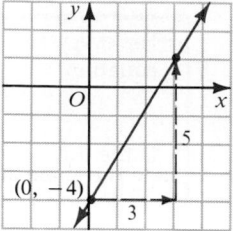

SOLUTION
1. Since the y-intercept is -4, plot $(0, -4)$.

2. The slope of the line is $\frac{5}{3}$. Count 5 units up from $(0, -4)$ and 3 units to the right to obtain a second point on the line.

3. Draw a line through the two points.

The equation is graphed on the coordinate plane at the right.

Often you are given a linear equation in a different form, and you need to determine the slope-intercept form.

EXAMPLE 2   Find the slope and the y-intercept of the line whose equation is $5x - 2y = 6$.

SOLUTION   Solve the given equation for y to obtain the slope-intercept form.

$$5x - 2y = 6$$
$$-2y = -5x + 6$$
$$y = \frac{5}{2}x - 3$$

$\therefore$ the slope is $\frac{5}{2}$ and the y-intercept is -3.

Sometimes you will need to find an equation of a line in some form other than slope-intercept form.

EXAMPLE 3   Find an equation of the line with slope $-\frac{2}{3}$ and y-intercept 5. The equation should be in the form $ax + by = c$, where a, b, and c are integers.

SOLUTION   Substitute the given slope and y-intercept into the slope-intercept form for a linear equation. Then transform the slope-intercept form into an equation of the required form.

$$y = mx + b$$
$$y = -\frac{2}{3}x + 5$$
$$3y = -2x + 15$$
$$2x + 3y = 15$$

$\therefore$ an equation of the line is $2x + 3y = 15$.

Oral Exercises

State the slope-intercept form of the equation of the line that has the given slope *m* and *y*-intercept *b*.

1. $m = 2$; $b = 3$ $y = 2x + 3$ 2. $m = -5$; $b = 4$ $y = -5x + 4$ 3. $m = -3$; $b = -1$ $y = -3x - 1$

4. $m = -1$; $b = 6$ $y = -x + 6$ 5. $m = 4$; $b = 0$ $y = 4x$ 6. $m = 0$; $b = -2$ $y = -2$

State the slope and the *y*-intercept, if any, of the line whose equation is given.

7. $y = 4x + 3$ $4; 3$ 8. $y = 5x - 1$ $5; -1$ 9. $2x + y = 5$ $-2; 5$ 10. $y - 8x = 5$ $8; 5$

11. $x + y = 7$ $-1; 7$ 12. $x - y = 2$ $1; -2$ 13. $2y = 4x - 12$ 14. $6x + 3y = 3$

15. $2x + 3y = 9$ 16. $4x - 2y = 7$ 17. $y = 3$ $0; 3$ 18. $x = -1$ no slope; no *y*-int.

Tell whether or not the lines whose equations are given are parallel.

19. $y = 3x + 1$ 20. $y = 2x + 9$ 21. $x + y = 4$ 22. $x + y = 3$

     $y = 3x - 2$ $y = -7x + 9$ $x - y = 4$ $x + y = -5$

     Yes   No   No   Yes

Written Exercises

Determine the slope and the *y*-intercept of the line whose equation is given. Then use the slope and *y*-intercept to graph the equation.   4. $-2; -3$ 6. $-1; -1$ 8. $2; 2$

A   1. $y = 2x + 1$ $2; 1$ 2. $y = 3x - 4$ $3; -4$ 3. $y = -3x + 2$ $-3; 2$ 4. $y = -2x - 3$

     5. $x - y = 4$ $1; -4$ 6. $x + y = -1$ 7. $3x - y = -5$ $3; 5$ 8. $y - 2x = 2$

     9. $2y - x = 6$ $\frac{1}{2}; 3$ 10. $x - 4y = 4$ $\frac{1}{4}; -1$ 11. $6x + 2y = 1$ $-3; \frac{1}{2}$ 12. $5x - 3y = 3$ $\frac{5}{3}; -1$

Determine an equation of the line with the given slope *m* and *y*-intercept *b*. Use the form $ax + by = c$, where *a*, *b*, and *c* are integers.

13. $m = 5$; $b = -1$ 14. $m = 6$; $b = 0$ 15. $m = -2$; $b = 0$

16. $m = -3$; $b = -5$ 17. $m = 0$; $b = 2$ 18. $m = 0$; $b = -6$

19. $m = \frac{1}{2}$; $b = -4$ 20. $m = -\frac{1}{3}$; $b = -2$ 21. $m = \frac{3}{2}$; $b = 2$

22. $m = \frac{4}{3}$; $b = 4$ 23. $m = -\frac{3}{5}$; $b = \frac{1}{2}$ 24. $m = -\frac{5}{2}$; $b = -\frac{1}{3}$

Determine an equation of the line that satisfies the given requirements. Use the form $ax + by = c$, where *a*, *b*, and *c* are integers.

B   25. *y*-intercept 3; parallel to the graph of $y = -5x + 7$ $5x + y = 3$

     26. *y*-intercept -1; parallel to the graph of $3x + y = 8$ $3x + y = -1$

     27. slope $= 3$; *y*-intercept the same as the graph of $y = 4x - 11$ $3x - y = 11$

     28. slope $= -1$; *y*-intercept the same as the graph of $y - 2x = 9$ $x + y = 9$

     29. *y*-intercept -7; parallel to the graph of $y = x$ $x - y = 7$

     30. *y*-intercept 2; parallel to the graph of $y = -3$ $y = 2$

     31. no slope; passes through the point $(-3, 5)$ $x = -3$

     32. no slope; intersects the *x*-axis at the same point as the graph of $y = x$ $x = 0$

Graphs and Functions   **251**

 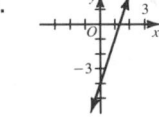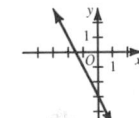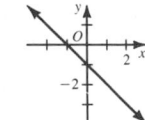
(continued on p. 259)

251

Graph on the same set of co-ordinate axes the lines that pass through $(-1, 0)$ and have the given slopes.

1. slope $= 1$ **2.** slope $= -2$
3. slope $= 3$ **4.** slope $= -4$

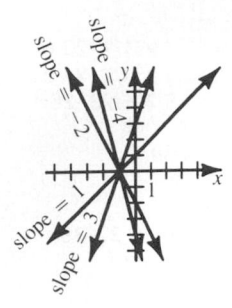

Teaching Suggestions
p. T87

Supplementary Material
Test 18

Key Ideas

Determine an equation of a line, given its slope and the coordinates of one point lying on the line.
Determine an equation of a line, given the coordinates of any two points lying on the line.

Determine the value of r so that the graph of the equation has the given slope.

33. $y = 2rx - 4$; slope $= 3\frac{3}{2}$

34. $y = \frac{2x}{r}$; slope $= 4\frac{1}{2}$

35. $ry = 8x - 1$; slope $= -2$ $\quad -4$

36. $2ry = 5x + 9$; slope $= 1\frac{5}{2}$

C **37.** Rewrite the linear equation $ax + by = c$ in slope-intercept form. What restrictions, if any, must be placed on the values of a, b, and c?

38. If two points on the nonvertical line whose equation is $ax + by = c$ have coordinates (p, r) and (q, s), show that

$$\frac{a}{b} = \frac{s - r}{p - q}.$$

39. Show that, if k is the y-intercept of the graph of $ax + by = c$, k is a root of the equation $by - c = 0$.

Computer Exercises For students with computer experience

1. Write a program that will allow you to input the coordinates of any two points on a coordinate plane and will determine the slope of the line that passes through the two points. Be careful to account for lines that have no slope.

2. Write a program that will allow you to input the coordinates of any *three* points on a coordinate plane and will determine whether the three points lie on the same line.

3. Write a program that will determine the coordinates of four other points on the line when you input the coordinates of a point and the slope of a line that passes through the point.

4. Write a program that will allow you to input values for a, b, and c and will compute the slope, the y-intercept, and the x-intercept of the line whose equation is $ax + by = c$. (The x-intercept is the abscissa of the point at which the line intersects the x-axis.) Be careful to account for situations in which the line has no slope, no y-intercept, or no x-intercept.

5–9 Determining an Equation of a Line

One way to describe the position of a nonvertical line on a coordinate plane is to state its slope and the coordinates of a point through which it passes. A second way is to give the coordinates of any two points through which the line passes. When a nonvertical line is described in either of these two ways, it is possible to determine an equation of the line by using the slope-intercept form of a linear equation.

EXAMPLE 1   Determine an equation of the line that passes through the point (6, 1) and has slope $-\frac{2}{3}$.

SOLUTION

1. The slope-intercept form of the equation of the line with slope $-\frac{2}{3}$ is

$$y = -\tfrac{2}{3}x + b.$$

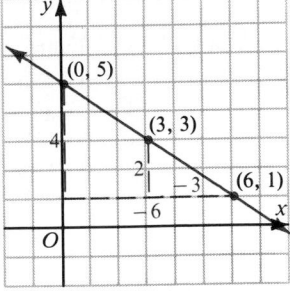

2. Since the point (6, 1) lies on the line, its coordinates must satisfy this equation.

$$y = -\tfrac{2}{3}x + b$$
$$1 = -\tfrac{2}{3}(6) + b$$
$$1 = -4 + b$$
$$5 = b$$

∴ an equation of the line is $y = -\frac{2}{3}x + 5$.

An equivalent equation for this line is $2x + 3y = 15$.

EXAMPLE 2   Determine an equation of the line that passes through the points (3, 2) and (−3, −6).

SOLUTION

1. Use the coordinates of the two points to compute m, the slope of the line.

$$m = \frac{2 - (-6)}{3 - (-3)} = \frac{8}{6} = \frac{4}{3}$$

2. The slope-intercept form of the equation of the line with slope $\frac{4}{3}$ is

$$y = \tfrac{4}{3}x + b.$$

3. Since the points (3, 2) and (−3, −6) both lie on the line, the coordinates of either point, say (3, 2), may be used to find the value of b.

$$y = \tfrac{4}{3}x + b$$
$$2 = \tfrac{4}{3}(3) + b$$
$$2 = 4 + b$$
$$-2 = b$$

Thus $y = \frac{4}{3}x - 2$ is an equation of the line.

4. To check, show that the coordinates of the other point, (−3, −6), satisfy the equation.

$$y = \tfrac{4}{3}x - 2$$
$$-6 \overset{?}{=} \tfrac{4}{3}(-3) - 2$$
$$-6 = -6 \quad \checkmark$$

∴ an equation of the line is $y = \frac{4}{3}x - 2$.

An equivalent equation for this line is $-4x + 3y = -6$.

Graphs and Functions   **253**

Chalkboard Examples

1. Determine an equation of a line that passes through (−3, −4) and has slope $\frac{3}{2}$.

$$-4 = \tfrac{3}{2}(-3) + b, \ \tfrac{1}{2} = b$$

An equation is $y = \frac{3}{2}x + \frac{1}{2}$, or $3x - 2y = -1$.

2. Determine an equation of a line that passes through (−5, 4) and (3, −2).

$$m = \frac{-2 - 4}{3 - (-5)} = \frac{-6}{8} = -\frac{3}{4}$$

$$-2 = -\tfrac{3}{4}(3) + b, \ \tfrac{1}{4} = b$$

$$y = -\tfrac{3}{4}x + \tfrac{1}{4}$$

Check using (−5, 4).

An equation is $y = -\frac{3}{4}x + \frac{1}{4}$, or $3x + 4y = 1$.

3. Determine an equation of the vertical line through the point (4, −1). $x = 4$

Additional A Exercises

Determine an equation of the line that has slope m and passes through the given point. The equation should be in the form $ax + by = c$, where a, b, and c are integers.

1. (2, 1); $m = 1$ $x - y = 1$
2. (2, 3); $m = -1$ $x + y = 5$

Determine an equation of the line that passes through the given points. The equation should be in the form $ax + by = c$, where a, b, and c are integers.

3. (1, 2), (3, 4) $x - y = -1$
4. (−1, 4), (5, −2) $x + y = 3$
5. (−4, −4), (4, 2)
   $3x - 4y = 4$

Suggested Assignments

Minimum
First day: 254/1–12
Second day: 254/13–24
      R 255/*Self-Test 3*
Average
First day:
      254/1–23 odd
Second day:
      254/25–38
  R 255/*Self-Test 3*
Maximum
      254/7–33 odd, 34–40
  R 255/*Self-Test 3*

Mixed Review

Determine the slope and the
y-intercept of the line whose
equation is given. Then use
the slope and y-intercept to
graph the equation on a coordinate plane.

1. $y = 2x - 1$ 2; -1

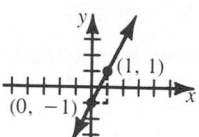

2. $y = \frac{1}{2}x + 2$ $\frac{1}{2}$; 2

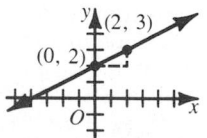

3. $y = \frac{2}{3}x + 3$ $\frac{2}{3}$; 3

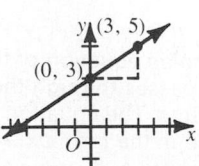

Since vertical lines have no slope, you cannot use the slope-intercept
form of a linear equation to determine the equation of a vertical line.

EXAMPLE 3   Determine an equation of the vertical line through the point $(-2, 3)$.

SOLUTION   Every point on the vertical line through the
point $(-2, 3)$ has -2 as its abscissa.

∴ an equation of this line is $x = -2$.

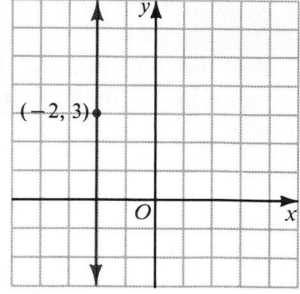

Written Exercises

6. $x + 3y = -3$ **7.** $x + 5y = -13$
8. $3x - 4y = -17$ **9.** $y = 5$

Determine an equation of the line that has slope m (if any) and passes
through the given point. The equation should be in the form $ax + by = c$,
where a, b, and c are integers.

A   **1.** $m = 2$; (1, 1) $2x - y = 1$ **2.** $m = 5$; $(-1, 3)$ $5x - y = -8$ $2x + y = 4$
3. $m = -2$; $(3, -2)$

4. $m = -1$; $(-1, -2)$ **5.** $m = \frac{1}{2}$; (4, 5)   **6.** $m = -\frac{1}{3}$; $(3, -2)$
  $x + y = -3$ $x - 2y = -6$

7. $m = -\frac{1}{5}$; $(2, -3)$ **8.** $m = \frac{3}{4}$; $(-3, 2)$ **9.** $m = 0$; (2, 5)

10. $m = 0$; $(-1, -4)$ $y = -4$ **11.** no slope; $(-2, 7)$ $x = -2$ **12.** no slope; $(-3, 3)$
  $x = -3$

Determine an equation of the line that passes through the given points.
The equation should be in the form $ax + by = c$, where a, b, and c are
integers.

13. (1, 5), (2, 4) $x + y = 6$ **14.** (3, 2), (2, 1) $x - y = 1$
15. $(2, -6)$, (0, 0) $3x + y = 0$ **16.** (0, 0), $(-1, -4)$ $4x - y = 0$
17. (4, 4), $(5, -1)$ $5x + y = 24$ **18.** (3, 1), (5, 5) $2x - y = 5$
19. $(5, -1)$, $(-1, -3)$ $x - 3y = 8$ **20.** $(-1, -4)$, $(5, -2)$ $x - 3y = 11$
21. $(-1, 7)$, (6, 7) $y = 7$ **22.** (7, 2), $(7, -1)$ $x = 7$
23. $(-3, 3)$, $(-3, 2)$ $x = -3$ **24.** $(5, -4)$, $(-1, -4)$ $y = -4$

Determine the value of k so that the graph of the equation passes through
the given point.

B   **25.** $3x + ky = 15$; (2, 3) 3   **26.** $kx + 4y = 2$; $(-1, 2)$ 6
27. $kx - 5y = 14$; $(2, -4)$ -3 **28.** $3x - ky = 1$; (3, 4) 2
29. $3x + ky = 9$; (3, 5) 0   **30.** $kx - 2y = -10$; $(-2, 5)$ 0

Find the linear function f with domain $\mathcal{R}$ that satisfies the given requirements.

31. $f(4) = -1; f(-2) = 5$ $f\colon x \to -x + 3$ **32.** $f(0) = 3; f(-2) = 7$ $f\colon x \to -2x + 3$

33. $f(1) = 3; f(3) = 8$ $f\colon x \to \frac{5}{2}x + \frac{1}{2}$ **34.** $f(-1) = 5; f(2) = -3$ $f\colon x \to -\frac{8}{3}x + \frac{7}{3}$

35. $f(-5) = 0; f(1) = 8$ $f\colon x \to \frac{4}{3}x + \frac{20}{3}$ **36.** $f(4) = -6; f(-4) = 0$ $f\colon x \to -\frac{3}{4}x - 3$

Find the value of r and the value of s for which the graphs of the given equations are the same line.

$r = -9, s = 4$

C **37.** $sx - y = -3; 12x - 3y = r$ **38.** $sx = 5(y - 2); ry = 2(3x + 10)$ $r = 10, s = 3$

39. Show that the y-intercept of the line with slope m that passes through the point (x_1, y_1) is $y_1 - mx_1$.

40. Use the result of Exercise 39 to show that an equation of the line with slope m that passes through the point (x_1, y_1) is $y - y_1 = m(x - x_1)$.

Self-Test 3

VOCABULARY   slope (p. 243)
collinear points (p. 244)
y-intercept (p. 249)
parallel lines (p. 249)
slope-intercept form of a linear equation (p. 249)

Determine the slope of the line that passes through the given points.

1. $(2, 5), (3, 8)$ 3   **2.** $(-5, 1), (-3, -2)$ $-\frac{3}{2}$ *Obj. 1, p. 242*

Determine the slope and the y-intercept of the line with the given equation. slope = 3, y-intercept = -5 slope = $-\frac{3}{2}$, y − intercept = $\frac{7}{2}$

3. $y = 3x - 5$ **4.** $3x + 2y = 7$ *Obj. 2, p. 242*

Determine an equation of the line that satisfies the given requirements. The equation should be in the form $ax + by = c$, where a, b, and c are integers.

5. slope = 4; y-intercept = -1 $4x - y = 1$ *Obj. 3, p. 242*

6. slope = $-\frac{2}{5}$; y-intercept = 3 $2x + 5y = 15$

7. passes through the point $(-2, -6)$ and has slope 3 $3x - y = 0$ *Obj. 4, p. 242*

8. passes through the points $(0, -2)$ and $(3, 2)$ $4x - 3y = 6$

Check your answers with those at the back of the book.

Additional Answers
Written Exercises

39. The equation of the line with slope m and y-intercept b is $y = mx + b$. Since point (x_1, y_1) is on the line, the coordinates satisfy the equation. So $y_1 = mx_1 + b$. $\therefore b = y_1 - mx_1$

40. From Ex. 39, $b = y_1 - mx_1$ so $y = mx + y_1 - mx_1$ is the equation of the line that passes through (x_1, y_1). This equation is equivalent to $y - y_1 = mx - mx_1$, or $y - y_1 = m(x - x_1)$.

Quick Quiz

Determine the slope of the line passing through the given points.

1. $(-2, -5)$ and $(3, 4)$ $\frac{9}{5}$

2. $(3, -2)$ and $(-1, -4)$ $\frac{1}{2}$

Determine the slope and y-intercept of the line with the given equation.

3. $y - 3x = 5$ 3; 5

4. $4x - 3y = -5$ $\frac{4}{3}; \frac{5}{3}$

Determine an equation of the line that satisfies the given requirements. The equation should be in the form $ax + by = c$, where a, b, and c are integers.

5. slope = 2; y-int. = -3
$-2x + y = -3$

6. passes through the point $(-4, -1)$ and has slope $\frac{3}{4}$
$-3x + 4y = 8$

7. passes through the points $(3, -1)$ and $(4, 3)$
$-4x + y = -13$

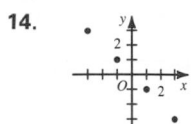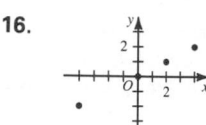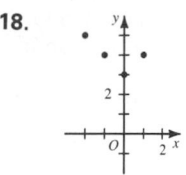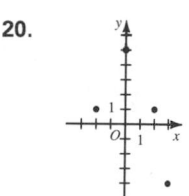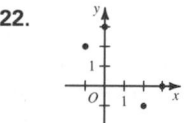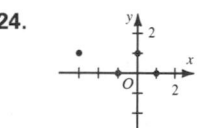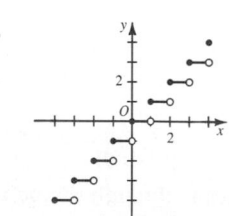
EXTRA

Transformations of the Plane: Translations

Figure 23 shows coordinate axes on a plane, together with a square whose vertices are at the points (1, 1), (2, 1), (2, 2), and (1, 2).

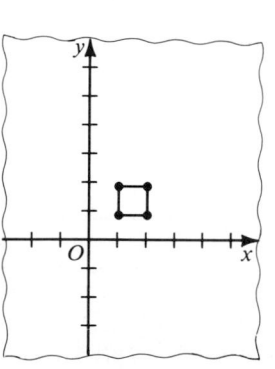

Figure 23   **Figure 24**

Now, imagine that the axes remain fixed while the plane, like a sheet of glass, slides rigidly into a new position, such as the one shown in Figure 24. The new coordinates of the vertices of the square after this "sliding" are (3, 4), (4, 4), (4, 5), and (3, 5).

In fact, *each* point on the plane now has a new pair of coordinates, (x', y'), where x' is 2 greater than the original x-coordinate and y' is 3 greater than the original y-coordinate. Thus for each point (x', y'),

$$x' = x + 2 \quad \text{and} \quad y' = y + 3.$$

Such a "sliding" of the plane is an example of a *transformation*, or *mapping*, of the plane called a **translation**. In a translation of the plane, if the point originally coinciding with the origin slides to a new location with coordinates (h, k), then every point on the plane has new coordinates (x', y') that are related to the original coordinates (x, y) by the *equations of translation*

$$x' = x + h \quad \text{and} \quad y' = y + k.$$

EXAMPLE   A translation maps the point with coordinates (5, 4) onto the point with coordinates (8, −2).

a. Find the equations of translation.

b. Find the new coordinates of the point with old coordinates (−4, 0).

256   *Chapter 5*

SOLUTION   **a.** Let $(x, y) = (5, 4)$ and $(x', y') = (8, -2)$, and substitute in the general translation equations:

$$x' = x + h \qquad \text{and} \qquad y' = y + k$$
$$8 = 5 + h \qquad\qquad -2 = 4 + k$$
$$h = 3 \qquad\qquad k = -6$$

$\therefore$ the equations of translations are $x' = x + 3$ and $y' = y - 6$.

b. Use the equations $x' = x + 3$ and $y' = y - 6$, and substitute -4 for x and 0 for y.

$$x' = -4 + 3 \qquad\qquad y' = 0 - 6$$
$$x' = -1 \qquad\qquad y' = -6$$

$\therefore$ the new coordinates of $(-4, 0)$ are $(-1, -6)$.

The following result is proved in more advanced courses.

Under a translation:

1. Every line is mapped (translated) onto itself or onto a line parallel to the original line.
2. Every line segment is mapped (translated) onto a line segment of equal length.
3. Every angle is mapped (translated) onto an angle of equal measure.

Exercises

In Exercises 1–8:

a. Find equations of translation that map the first point onto the second.

b. Find the new coordinates under this translation of the point whose original coordinates are $(-7, 3)$.

1. $(5, 5), (6, 2)$
2. $(3, 8), (5, 1)$
3. $(-7, 2), (6, -1)$
4. $(3, -2), (8, 1)$
5. $(-7, 0), (0, 2)$
6. $(0, -3), (5, 0)$
7. $(-5, -2), (4, -7)$
8. $(-6, -3), (-7, 1)$

9. Under what kind of translation of the plane will a vertical line be mapped onto itself? a horizontal line be mapped onto itself?

10. Use the slope formula to show that the slope of the line through the points (a, b) and (c, d), $a \neq c$, is equal to the slope of the line through the points $(a + h, b + k)$ and $(c + h, d + k)$, and thus show that under a translation a nonvertical line is mapped onto a parallel line.

Graphs and Functions   **257**

34. If $k \neq 0$, the graphs of $y = x$ and $y = kx$ intersect at the origin. If $|k| > 1$, the graph of $y = kx$ is steeper than the graph of $y = x$. If $0 < |k| < 1$, the graph of $y = kx$ is less steep than the graph of $y = x$.

36.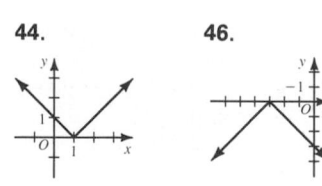

38.

40.   **42.**

44.   **46.**

47. The graph of $y = |x| + k$ is the graph of $y = |x|$ translated (slid) vertically k units (up if $k > 0$ and down if $k < 0$).

48. The graph of $y = |x + k|$ is the graph of $y = |x|$ translated horizontally k units (to the left if $k > 0$ and to the right if $k < 0$).

Chapter Summary

1. A *coordinate plane* can be set up using two perpendicular number lines that intersect at the origin of each. One line, called the *x-axis*, is usually horizontal and the other, called the *y-axis*, is usually vertical.

2. Each point on a coordinate plane is assigned a unique pair of *coordinates* that together form an *ordered pair* of real numbers. The first coordinate of an ordered pair is called the *x-coordinate*, or *abscissa*, and the second coordinate is called the *y-coordinate*, or *ordinate*. Conversely, each ordered pair of real numbers is assigned a unique point on a coordinate plane.

3. A *relation* is any set of ordered pairs. The set of all the first coordinates of the ordered pairs is called the *domain* of the relation, and the set of all the second coordinates is called the *range*.

4. A *function* is a relation in which no two ordered pairs have the same first coordinate.

5. Members of the range of a function are called the *values* of the function. The notation $f(x)$ is used to denote the specific value of the function f that is paired with the number x from the domain.

6. A *solution of an open sentence in two variables* is an ordered pair of values of the variables that together make the sentence a true statement. The set of all such ordered pairs is called the *solution set* of the open sentence. The *graph of an open sentence in two variables* is the graph of its solution set.

7. A *linear equation in two variables*, x and y, is any equation that can be written equivalently in the form $ax + by = c$, where a, b, and c are real numbers and a and b are not both zero. If the replacement set of each variable is $\mathcal{R}$, the graph of such an equation on a coordinate plane is a straight line.

8. A function whose ordered pairs satisfy a linear equation is called a *linear function*.

9. On a coordinate plane, the graph of a *linear inequality in two variables* is a *half-plane*. The equation of the line that forms the *boundary* of the half-plane is called the *associated equation* of the inequality.

10. Given that (x_1, y_1) and (x_2, y_2) are any two different points on a line on a coordinate plane and $x_1 \neq x_2$, the *slope* of the line is defined as the ratio

$$\frac{y_2 - y_1}{x_2 - x_1}.$$

If $x_1 = x_2$, the line has no slope.

11. The point at which a line intersects the y-axis is called the *y-intercept* of the line.

12. A line with slope m and y-intercept b is the graph of the equation $y = mx + b$, which is called the *slope-intercept* form of the equation.

13. On a coordinate plane, nonvertical lines with the same slope but different y-intercepts are *parallel*. Also, all vertical lines are parallel.

14. An equation of a line can be found given: (a) the slope and the y-intercept of the line; (b) the slope and any point on the line; (c) any two points on the line.

Chapter Review

Write the letter of the correct answer.

1. Which of the following statements is true?　　　　　　　　　5-1
   a. The y-coordinate of every point on the y-axis is 0.
   b. The x-coordinate of every point in the second quadrant is positive.
   (c.) The x-coordinate of every point in the fourth quadrant is greater than the y-coordinate of the point.
   d. None of the above statements is true.

2. Given the point $P(-4, 1)$, which of the following terms describes the number -4?
   a. slope 　　　　　　b. ordinate
   (c.) abscissa 　　　　d. y-intercept

3. Give the coordinates of the point plotted on the coordinate plane at the right.
   a. $(3, -2)$ 　　　　b. $(-3, 2)$
   c. $(2, -3)$ 　　　　(d.) $(-2, 3)$

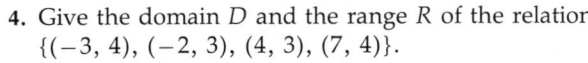

4. Give the domain D and the range R of the relation $\{(-3, 4), (-2, 3), (4, 3), (7, 4)\}$.　　　　　　　5-2
   a. $D = \{-3, -2, 4, 7\}$; 　$R = \{-4, -3, 3, 4\}$
   (b.) $D = \{-3, -2, 4, 7\}$; 　$R = \{3, 4\}$
   c. $D = \{3, 4\}$; 　$R = \{-3, -2, 4, 7\}$
   d. $D = \{-4, -3, 3, 4\}$; 　$R = \{-3, -2, 4, 7\}$

5. Which of the following relations is pictured in the mapping diagram at the right?
   a. $\{(1, -1), (2, -2), (3, -3)\}$
   (b.) $\{(1, -2), (2, -3), (3, -1)\}$
   c. $\{(-1, 1), (-2, 2), (-3, 3)\}$
   d. $\{(-1, 3), (-2, 2), (-3, 1)\}$

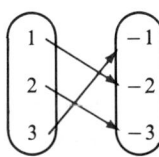

Graphs and Functions **259**

Additional Answers
Written Exercises
(continued from p. 251)

10. 　　　　12.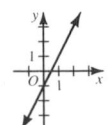

13. $5x - y = 1$

14. $6x - y = 0$

15. $2x + y = 0$

16. $3x + y = -5$

17. $y = 2$

18. $y = -6$

19. $x - 2y = 8$

20. $x + 3y = -6$

21. $3x - 2y = -4$

22. $4x - 3y = -12$

23. $6x + 10y = 5$

24. $15x + 6y = -2$

37. $y = -\dfrac{a}{b}x + \dfrac{c}{b}, b \neq 0$

38. $ap + br = c$
$aq + bs = c$
$ap + br = aq + bs$
$ap - aq = bs - br$
$a(p - q) = b(s - r)$
$\dfrac{a(p - q)}{b(p - q)} = \dfrac{b(s - r)}{b(p - q)}$
$\dfrac{a}{b} = \dfrac{s - r}{p - q}$

39. If $by - c = 0$, $y = \dfrac{c}{b}$.

From Exercise 37, the y-intercept of the graph of $ax + by = c$ is $\dfrac{c}{b}$. Replacing y with $\dfrac{c}{b}$ in the equation $by - c = 0$, you obtain $b\left(\dfrac{c}{b}\right) - c = 0$, or $0 = 0$. Thus, $\dfrac{c}{b}$ is a root of $by - c = 0$.

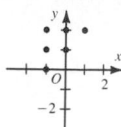
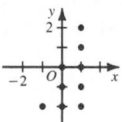
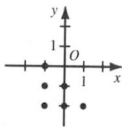
6. Which of the following relations is *not* a function?
  (a.) $\{(16, 2), (1, 1), (0, 0), (1, -1), (16, -2)\}$
  b. $\{(2, 16), (1, 1), (0, 0), (-1, 1), (-2, 16)\}$
  c. $\{(2, 16), (1, 16), (0, 16), (-1, 16), (-2, 16)\}$
  d. $\{(2, 2), (1, 1), (0, 0), (-1, 1), (-2, 2)\}$

7. Determine the function that is specified by $f: x \longrightarrow x - 5$ if the   5-3
    domain of the function is $\{0, -2, -3\}$.
  a. $\{(0, 5), (-2, 3), (-3, 2)\}$
  b. $\{(0, 5), (-2, -3), (-3, 2)\}$
  c. $\{(0, -5), (-2, -3), (-3, -2)\}$
  (d.) $\{(0, -5), (-2, -7), (-3, -8)\}$

8. Given the function $g: x \longrightarrow 3 - x^2$, which of the following statements
    is *not* true?
  a. $g(-1) = g(1)$ **b.** $g(-2) \le g(-1)$
  (c.) $g(0) \le g(1)$ **d.** $g(0) + 3g(2) = 0$

9. Which of the following ordered pairs is *not* a solution of   5-4
    $3x - 2y \ge 5$?
  (a.) $(-1, -3)$ **b.** $(2, -1)$ **c.** $(-2, -6)$ **d.** $(1, -2)$

10. Solve $x - 6y \le 3$ for y.
  a. $y \le -\dfrac{1}{6}x + \dfrac{1}{2}$ **b.** $y \le -\dfrac{1}{6}x + 2$
  (c.) $y \ge \dfrac{1}{6}x - \dfrac{1}{2}$ **d.** $y \ge -\dfrac{1}{6}x + \dfrac{1}{2}$

11. Solve $y - x < 1$ if $x \in \{-1, 0, 1\}$ and $y \in \{-2, 0, 2\}$.
  a. $\{(-2, -1), (-2, 0), (-2, 1), (0, -1), (0, 0)\}$
  b. $\{(0, -1), (0, 0), (2, -1), (2, 0), (2, 1)\}$
  c. $\{(-1, -2), (-1, 0), (0, -2), (0, 0), (1, 0)\}$
  (d.) $\{(-1, -2), (0, -2), (0, 0), (1, -2), (1, 0)\}$

12. Which of the following equations is graphed   5-5
    on the coordinate plane at the right?
  a. $2x + y = 1$ **(b.)** $2x - y = 1$
  c. $x + 2y = 1$ **d.** $x - 2y = 1$

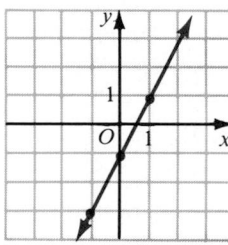

13. Which of the following describes the graph
    of $y = 5$ on a coordinate plane?
  a. parallel to the y-axis
  b. passes through quadrants I and IV
  (c.) passes through the point $(-9, 5)$
  d. none of the above

260   *Chapter 5*

14. Which of the following inequalities is graphed on the coordinate plane at the right?

    a. $x - y < -2$ (b.) $x - y > -2$

    c. $x - y \leq -2$ d. $x - y \geq -2$

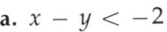

15. Determine the slope of the line that passes through the points $(-7, -1)$ and $(5, -6)$.

    a. $\frac{5}{12}$ b. $-\frac{12}{5}$ (c.) $-\frac{5}{12}$ d. $\frac{12}{5}$

16. Through which of the following points does a line pass if it passes through $(-4, 1)$ and has slope $-\frac{2}{3}$?

    a. $(0, 0)$ b. $(-6, 4)$ (c.) $(-1, -1)$ d. $(-6, 2)$

17. Determine the y-intercept of the line whose equation is $6x - 5y = 10$.

    a. 2   (b.) -2 c. 10   d. -10

18. Determine an equation of the line that has slope $-\frac{1}{3}$ and y-intercept 2.

    a. $x - 3y = 2$ b. $3x - y = 2$ c. $x + 2y = 6$ (d.) $x + 3y = 6$

19. Determine an equation of the line that passes through the point $(-2, -6)$ and has slope $\frac{4}{3}$.

    a. $4x + 3y = -10$ (b.) $-4x + 3y = -10$

    c. $3x - 4y = 10$ d. $3x + 4y = -10$

20. Determine an equation of the line that passes through the points $(-3, 2)$ and $(5, -2)$.

    (a.) $x + 2y = 1$ b. $-x + 2y = 1$ c. $2x + y = 1$ d. $2x - y = -1$

5-6

5-7

5-8

5-9

36. $\{(-1, -2), (-1, -1), (-1, 0), (-1, 1), (-1, 2)\}$

37. $\{(2, 1), (2, 2), (3, 1)\}$
38. $\{(4, 1), (5, 1), (5, 2)\}$
39. $\{(-2, -3), (-2, -2), (-2, -1), (-1, -1)\}$
40. $\{(-4, 0), (-4, 1), (-4, 2), (-2, 0), (-2, 1), (0, 0)\}$
41. $\{(3, 0), (3, 1), (4, 0), (5, 0)\}$
42. $\{(1, -2), (1, -1), (2, -5), (2, -4), (2, -3), (2, -2), (2, -1)\}$
43. $\{(-3, 2), (-2, 3), (2, 3), (3, 2)\}$
44. $\{(-4, -2), (-4, 2), (-2, -4), (-2, 4), (2, -4), (2, 4), (4, -2), (4, 2)\}$
45. $\{(-5, -3), (-5, 3), (-3, -1), (-3, 1), (3, -1), (3, 1), (5, -3), (5, 3)\}$
46. $\{(-4, -1), (-3, 0), (-2, 1), (-1, -4), (-1, 2), (0, -3), (0, 3), (1, -2), (1, 4), (2, -1), (3, 0), (4, 1)\}$
47. $\{(-1, 0), (0, -1), (1, 0)\}$
48. $\{(0, -2), (0, 2), (2, 0)\}$

Chapter Test

Name the coordinates of each point plotted on the coordinate plane at the right.

1. A (0, 2)   2. B (1, -3)   3. C (2, 2)   4. D (-3, 0)

Plot each point on a coordinate plane.

5. $N(1, 0)$ 6. $O(0, 0)$

7. $P(3, -1)$ 8. $Q(-2, -3)$

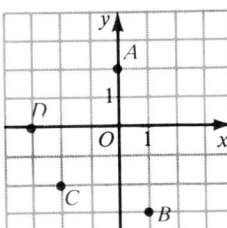

5-1

Additional Answers Chapter Test

5-8.

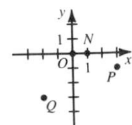

9.

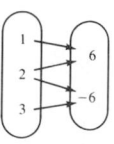

11.

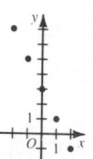

15.

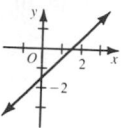

16.

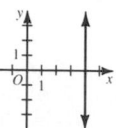

17.

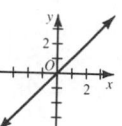

18.

19.

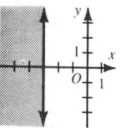

20.

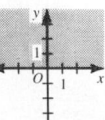

21.

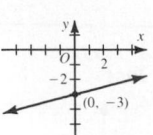

Given the relation $\{(1, 6), (2, -6), (2, 6), (3, -6)\}$:

9. Draw a mapping diagram that represents the relation.   5-2

10. Determine whether or not the relation is a function. not a function

Given $f: x \longrightarrow 3 - 2x$, with domain $D = \{-2, -1, 0, 1, 2\}$:

11. Graph the function.   12. Compute $f(-2) + 3f(1)$. 10   5-3

Solve when $x \in \{-2, 0, 2\}$ and $y \in \{-3, -1, 0, 3\}$.

13. $y = 1 - x^2$ $\{(-2, -3), (2, -3)\}$ 14. $y < -x$ $\{(-2, -3), (-2, -1), (-2, 0),$ 5-4
$(0, -3), (0, -1), (2, -3)\}$

Graph on a coordinate plane.

15. $2x - 2y = 3$ 16. $x = 4$ 17. $y = x$ 5-5

18. $x - y < 1$ 19. $x \le -3$ 20. $y \ge 0$ 5-6

21. Graph the line that passes through the point $(0, -3)$ and has slope $\frac{1}{4}$.   5-7

22. Determine the slope of the line that passes through the points $(-1, 2)$ and $(4, 6)$. $\frac{4}{5}$

Determine an equation of the line that satisfies the given requirements. Use the form $ax + by = c$, where a, b, and c are integers.

23. has slope $\frac{1}{2}$ and y-intercept -3 $x - 2y = 6$ 5-8

24. has no slope and passes through $(-3, 7)$ $x = -3$

25. passes through $(-4, -5)$ and has slope $-\frac{1}{2}$ $x + 2y = -14$ 5-9

26. passes through the points $(-3, -5)$ and $(2, -1)$ $4x - 5y = 13$

Cumulative Review

Chapter 1

Graph each set of numbers on a number line.

1. $\left\{-3, -\frac{7}{3}, -2, -\frac{5}{3}, \frac{2}{3}\right\}$ 2. $\{-1.75, -0.25, 0, 0.5, 2.75\}$

3. {the integers greater than -4 and less than 2}

4. {the natural numbers greater than 6}

5. {the real numbers between -4 and 3}

6. $\left\{\text{the positive real numbers less than } 4\frac{1}{4}\right\}$

262   *Chapter 5*

Simplify.

7. $3^2 + 5(3)^2$ 54

8. $[5(4) - 4] + [7 + 6(4)]$ 47

9. $\dfrac{(3)7 + 5}{(8)4 - 19}$ 2

10. $2^2(6) - 2(3^2) + 2(2) + 1$ 11

11. $2^6 \div 2^4 \div 2^2 \div 2$ $\frac{1}{2}$

12. $\frac{1}{3}[18 \div 2(6)] \div [(12)2 \div 6]$ $\frac{1}{8}$

Chapter 3

Name the property that justifies each statement.

13. If $a + 9 = 14$, $(a + 9) - 9 = 14 - 9$. Subtraction property of equality

14. If $7x = 21$, $\frac{1}{7} \cdot 7x = \frac{1}{7} \cdot 21$. Multiplication property of equality

Solve.

15. $m + (-17) = 22$ {39}

16. $-39 - n = -56$ {17}

17. $\frac{1}{8}a = -12$ {−96}

18. $18 = -4b + 18$ {0}

19. $3y + 4y + 7 = 49$ {6}

20. $2z + z + 5 = 3z + 5$ $\mathcal{R}$

21. $2|s| = -3$ $\emptyset$

22. $|t| - 9 = -1$ {8, −8}

23. $(1 - 2j) - (3 - 3j) = 0$ {2}

24. $[2(k - 4) + 6] + k = -117$ $\{-\frac{115}{3}\}$

25. Larry is twice as old as Amy. In five years, Larry will be three times as old as Amy was seven years ago. How old is Larry now? 52 years

26. Seven less than a certain number is nine more than three times the number. Find the number. −8

Chapter 4

Solve each inequality and graph its solution set.

31. $\{p\colon p < -8 \text{ or } p > 4\}$
32. $\{q\colon q < -1\}$

27. $-3v \geq 24$ {$v\colon v \leq -8$}

28. $-4w + 9 < w - 21$ {$w\colon w > 6$}

29. $6(a - 2) < 7(a - 3)$ {$a\colon a < 9$}

30. $-4(b + 5) + 8b \geq 4(b - 5)$ $\mathcal{R}$

31. $p + 2 > 6$ or $p + 2 < -6$

32. $10 - 2q > 12$ and $5q < 2q + 9$

33. $8 > 5 - 3g > -13$ {$g\colon -1 < g < 6$}

34. $7 \geq -2h + 3 \geq -1$ {$h\colon -2 \leq h \leq 2$}

35. $|x - 4| \geq 3$ {$x\colon x \leq 1 \text{ or } x \geq 7$}

36. $|2y + 9| - 9 < 2$ {$y\colon -10 < y < 1$}

41. $\mathcal{R}$; {the real numbers between $-\frac{1}{3}$ and $2\frac{2}{3}$}

Specify the union and the intersection of the given sets.

37. $\{-2, -1, 2, 3\}$, $\{-1, 0, 1\}$ {−2, −1, 0, 1, 2, 3}; {−1}

38. $\{1, 3, 5\}$, $\emptyset$ {1, 3, 5}; $\emptyset$

39. {the odd whole numbers}, $\{1, 2, 3, 4\}$ {2, 4, the odd whole numbers}; {1, 3}

40. {the integers greater than −2}, {the integers less than 1} {the integers}; {−1, 0}

41. {the real numbers greater than $-\frac{1}{3}$}, {the real numbers less than $2\frac{2}{3}$}

42. {the positive real numbers}, {the negative real numbers} {all the real numbers except 0}; $\emptyset$

Graphs and Functions **263**

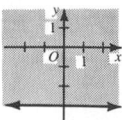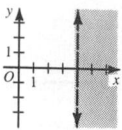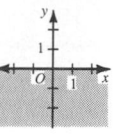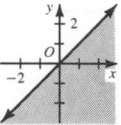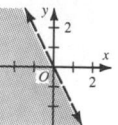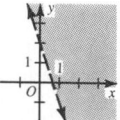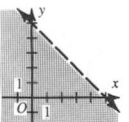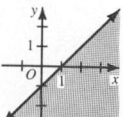
APPLICATION

Days of the Week

On what day of the week did astronauts first land on the moon? On what day of the week will New Year's Day fall in the year 2001? If you ever wonder about questions like these, you do not necessarily need to look for a calendar for the year in question. Using a specially devised formula, you can arithmetically calculate the day of the week for a given date, provided that the date occurs later than 1752. In that year, England and its colonies adopted the *Gregorian calendar*, which is the calendar still in use today. (The Gregorian calendar was created in order to correct errors that had been made in past calculations of leap years.)

The formula is as follows.

$$w = d + 2m + \left[\frac{3(m+1)}{5}\right] + y + \left[\frac{y}{4}\right] - \left[\frac{y}{100}\right] + \left[\frac{y}{400}\right] + 2$$

In this formula, d represents the day of the month, y represents the number of the year, and m represents the number of the month. The months March through December are numbered 3 through 12, as you might expect. However, January and February must be considered as the 13th and 14th months of the *previous* year, and so they are numbered 13 and 14, respectively. Correspondingly, when using the formula for a date in January or February, the value of y must be the number of the previous year.

The square brackets are symbols for the *greatest integer function*, meaning that any expression within the brackets is assigned its greatest integral value. For example:

$$\left[\frac{3(4+1)}{5}\right] = [3] = 3$$

$$\left[\frac{3(5+1)}{5}\right] = [3.6] = 3$$

Once you have used the formula to find the value of w, you divide w by 7, the number of days in a week. The *remainder* of this division represents the day of the week. Sunday through Friday are represented by remainders of 1 through 6, in that order, and Saturday is represented by 0. Thus, a remainder of 3 indicates that the day of the week was Tuesday.

264 *Chapter 5*

264

EXAMPLE 1 Astronauts first landed on the moon on July 16, 1969. What day of the week was this?

SOLUTION Given the date July 16, 1969, $d = 16$, $m = 7$, and $y = 1969$.

Substitute these values into the formula:

$$w = 16 + 2(7) + \left[\frac{3(8)}{5}\right] + 1969 + \left[\frac{1969}{4}\right] - \left[\frac{1969}{100}\right] + \left[\frac{1969}{400}\right] + 2$$

$$= 16 + 14 + 4 + 1969 + 492 - 19 + 4 + 2$$

$$= 2482$$

Divide the value of w by 7: $2482 \div 7 = 354$ R4

∴ the day of the week was the 4th day, or Wednesday.

EXAMPLE 2 On what day of the week will New Year's Day fall in the year 2001?

SOLUTION Given the date January 1, 2001, $d = 1$, $m = 13$, and $y = 2000$.

Substitute these values into the formula:

$$w = 1 + 2(13) + \left[\frac{3(14)}{5}\right] + 2000 + \left[\frac{2000}{4}\right] - \left[\frac{2000}{100}\right] + \left[\frac{2000}{400}\right] + 2$$

$$= 1 + 26 + 8 + 2000 + 500 - 20 + 5 + 2$$

$$= 2522$$

Divide the value of w by 7: $2522 \div 7 = 360$ R2

∴ the day of the week will be the 2nd day, or Monday.

Exercises

Determine the day of the week for each event.

1. the first telephone call: June 3, 1875 Thursday
2. the first scheduled passenger service using airplanes:
   August 25, 1919 Monday
3. the first commercial television broadcast: July 1, 1941 Tuesday
4. the day you were born Answers will vary.

18.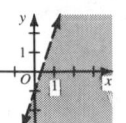

20.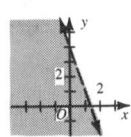

22.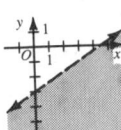

24.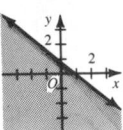

26.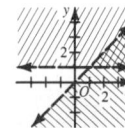

28.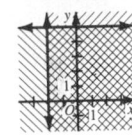

30.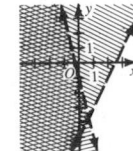

32.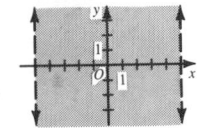

34.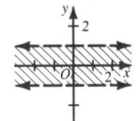

36.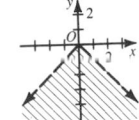

These wind turbines, located in Tehachapi Pass, California, are using wind currents to generate electricity. Using natural sources of power helps in the conservation of energy.

Chapter 6

Systems of Open Sentences

Solving Systems of Linear Equations in Two Variables

OBJECTIVES for Sections 6-1 through 6-4:
1. *To determine whether a system of two linear equations in two variables is consistent or inconsistent.*
2. *To solve a system of two linear equations in two variables by using graphs, addition or subtraction, linear combinations, or substitution.*

6–1 Using Graphs

In Chapter 5, you learned how to solve linear equations. In this chapter you will learn methods for solving *systems of linear equations*.

Two or more linear equations in the same variables, such as

$$x + y = 3 \quad \text{and} \quad x - 2y = 0,$$

together form a **system of linear equations** in two variables, or a set of **simultaneous linear equations**. To *solve* a system of linear equations *in two variables*, you find all ordered pairs of numbers that satisfy each equation in the system. Each such ordered pair is called a **solution of the system;** the set of all solutions is called the **solution set of the system.**

One method for solving a system of linear equations in two variables is to graph the equations on the same coordinate plane and find the points that are common to the graphs. Recall that the graph of a linear equation on a coordinate plane is a line. Thus, when two linear equations in two variables are graphed on the same coordinate plane, their graphs will be related in *exactly one* of the following three ways.

Systems of Open Sentences **267**

Problem Solving Strategies

Types of Word Problems
Sections 6–5 through 6–7 provide examples of the use of two equations in two variables to solve many types of word problems, such as digit or wind and water current problems. In Section 6–9, students graph systems of linear inequalities to solve linear programming problems. Students use three equations in three variables to solve a variety of problems in Section 6–10.

Teaching Suggestions
p. T87

Related Activities p. T88

Key Ideas

Solve systems of linear equations graphically.
Determine whether systems are consistent or inconsistent.
Determine the number of solutions of each system: none, one, or infinite.

Reading Algebra

Reading graphs of systems of equations is a helpful algebraic skill. Students should be able to tell by inspection if the solution set contains no members, one member, or an infinite number of members.

1. Solve each system of linear equations using graphs.

   a. $3x + y = -4$
   $3x + y = 2$

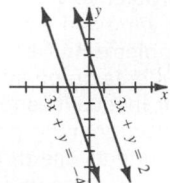

The graphs are parallel lines which have no points in common.
∴ the solution set is Ø.

   b. $3y = 3x - 6$
   $y = x - 2$

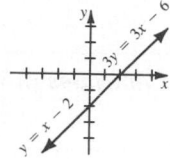

The graphs coincide. Thus the two equations represent the same line and have all their points in common.
∴ the solution set is the infinite set, $\{(x, y): y = x - 2\}$.

   c. $y + 2x = 8$
   $y - 2x = -4$

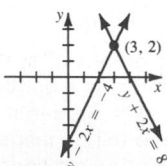

The graphs intersect in exactly one point, (3, 2).
Check: $2 + 2(3) \overset{?}{=} 8$
$8 = 8$ √
$2 - 2(3) \overset{?}{=} -4$
$-4 = -4$ √
∴ the solution set is $\{(3, 2)\}$.

---

1. The graphs will be *parallel lines*.
2. The graphs will be *coincident lines*. (Two lines that have *all* their points in common are called **coincident lines**.)
3. The graphs will intersect in exactly *one* point. (The one point that is common to both lines is called their **point of intersection**.)

EXAMPLE 1   Solve each system of linear equations using graphs.

   a. $y = -x + 3$
   $y = -x - 2$

   b. $x + y = 3$
   $2x + 2y = 6$

   c. $x + y = 3$
   $x - 2y = 0$

SOLUTION   Graph each equation in the system on the same coordinate plane.

a.

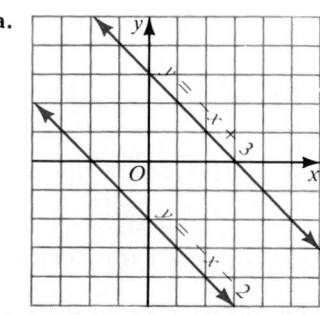

The graphs of these linear equations are parallel lines. The graphs have no points in common, and so the system has no solution.

∴ the solution set is the empty set, Ø.

b.

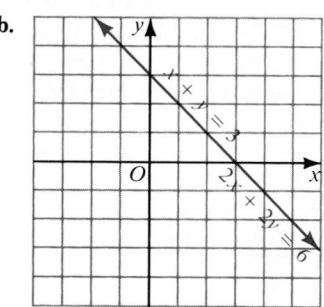

The graphs of these linear equations coincide. Thus, the two equations represent the same line. The graphs have all their points in common, and so all the points on the line are solutions of the system.

∴ the solution set is the infinite set $\{(x, y): x + y = 3\}$.

c.

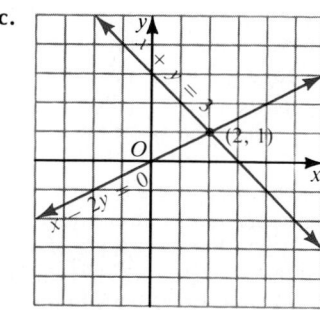

The graphs of these linear equations intersect in exactly one point, (2, 1). The graphs have one point in common, and so the ordered pair (2, 1) is the only solution of the system.

Check: Does (2, 1) satisfy both of the original equations?

$x + y = 3$ $x - 2y = 0$
$2 + 1 \overset{?}{=} 3$ $2 - 2(1) \overset{?}{=} 0$
$3 = 3$ √ $0 = 0$ √

∴ the solution set is $\{(2, 1)\}$.

Notice in part (b) of Example 1 that the solution set is written in set-builder notation and is read as, "the set of all ordered pairs x, y such that x plus y equals 3."

In part (c) of Example 1, the proposed solution $(2, 1)$ was checked in *both* of the given equations. This is essential. Because any linear equation has an infinite number of solutions, you cannot conclude that you have found the solution of the *system* of equations until you have shown that the proposed solution satisfies *each* equation in the system.

The graphing method for solving a system of linear equations in two variables can be summarized as follows.

The Graphing Method

To solve a system of linear equations in two variables graphically:
1. Graph the equations on the same coordinate plane.
2. Determine the coordinates of all points common to the graphs.

A system of equations that has at least one solution is called a **consistent** system. A system of equations whose solution set is the empty set is called an **inconsistent** system. You can tell whether a system of linear equations in two variables representing *nonvertical* lines is consistent or inconsistent by comparing the slope-intercept forms of the equations of the system.

EXAMPLE 2   Determine whether each system of equations in Example 1 is consistent or inconsistent.

SOLUTION   **a.** $\left. \begin{aligned} y &= -x + 3 \\ y &= -x - 2 \end{aligned} \right\}$ same slope but different y-intercepts

The graphs are parallel lines.
The solution set is the empty set.

∴ the system is inconsistent.

b. $\begin{aligned} x + y &= 3 \\ 2x + 2y &= 6 \end{aligned}$ $\longrightarrow$ $\left. \begin{aligned} y &= -x + 3 \\ y &= -x + 3 \end{aligned} \right\}$ same slope and same y-intercept

The graphs are coincident lines.
The solution set is an infinite set.

∴ the system is consistent.

c. $\begin{aligned} x + y &= 3 \\ x - 2y &= 0 \end{aligned}$ $\longrightarrow$ $\left. \begin{aligned} y &= -x + 3 \\ y &= \tfrac{1}{2}x \end{aligned} \right\}$ different slopes

The graphs are lines that intersect in exactly one point.
The solution set has exactly one member.

∴ the system is consistent.

Systems of Open Sentences   **269**

2. Determine whether each system in Example 1 is consistent or inconsistent.
   a. $3x + y = -4$
   $3x + y = 2$

   $y = -3x - 4$
   $y = -3x + 2$
   They have the same slopes but different y-intercepts. The graphs are parallel lines. The solution set is $\emptyset$.
   ∴ the system is inconsistent.

   b. $3y = 3x - 6$
   $y = x - 2$

   $y = x - 2$
   $y = x - 2$
   They have the same slope and the same y-intercept. The graphs are coincident lines. The solution set is an infinite set.
   ∴ the system is consistent.

   c. $y + 2x = 8$
   $y - 2x = -4$

   $y = -2x + 8$
   $y = 2x - 4$
   They have different slopes. The graphs are lines that intersect in exactly one point. The solution set has one member.
   ∴ the system is consistent.

3. Find the point on the graph of $x + y = 3$ where the ordinate is twice the abscissa.
   $x + y = 3$
   $y = 2x$

   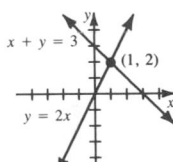

   The required point is $(1, 2)$.

Mixed Review

1. Write an equation of a line that is parallel to the line $y = -3x + 4$.
   Answers may vary.
   Example: $y = -3x + 2$

2. A line is described by the equation $3x + 2y = 4$. Write another equation that describes the same line. Answers may vary.
   Example: $6x + 4y = 8$

3. Graph $5x - y = 3$ on a coordinate plane.

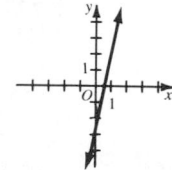

4. Name a point, where possible, in each of the four quadrants on a coordinate plane such that the ordinate of the point is twice the abscissa.
   Quad. I: (3, 6); Quad. III (−3, −6); not possible in Quad. II and IV.

Suggested Assignments

Minimum
  272/1–23 odd
Average
  272/7–27, odd, 28–32
Maximum
  272/7–31 odd, 32–38

Additional A Exercises

Solve each system of equations using graphs.

1. $y - x = 4$
   $3y + x = 0$ {(−3, 1)}

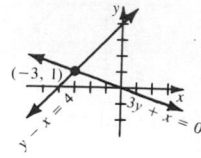

EXAMPLE 3   Find the point on the graph of $3x + 2y = 6$ where the ordinate is 3 more than the abscissa.

SOLUTION   The ordinate is 3 more than the abscissa for every point on the line $y = x + 3$. Thus, you can find the required point by solving the system

$$3x + 2y = 6 \quad \text{and} \quad y = x + 3.$$

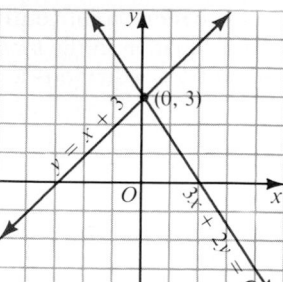

The graphs intersect at (0, 3).

Check:

Is (0, 3) on the graph of $3x + 2y = 6$?

$$3x + 2y = 6$$
$$3(0) + 2(3) \overset{?}{=} 6$$
$$6 = 6 \ \checkmark$$

Is the ordinate 3 more than the abscissa?

$$y = x + 3$$
$$3 \overset{?}{=} 0 + 3$$
$$3 = 3 \ \checkmark$$

∴ the solution is the point (0, 3).

Oral Exercises

Each graph represents a system of linear equations. Determine the solution of each system.

1. (1, 0)

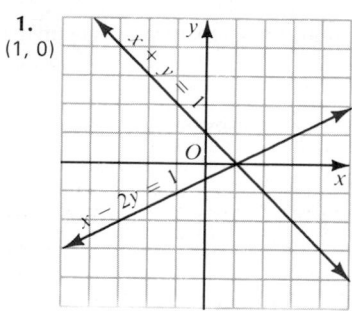

2. (1, −1)

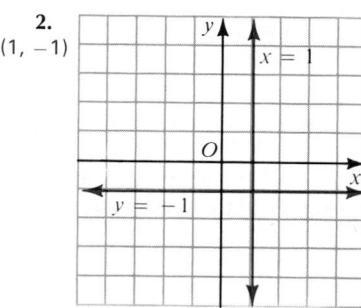

3. (0, −1)

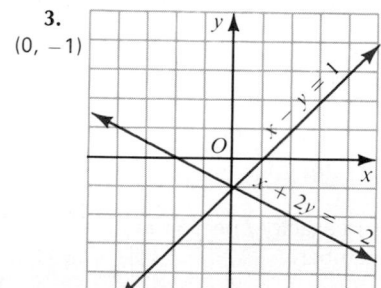
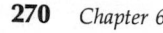

4. (−1, 1)

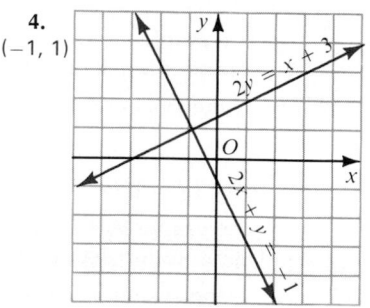

Tell whether or not (−3, −5) is a solution of the given system.

5. $x - y = 2$ Yes
   $2x + y = -11$

6. $x - 2y = 7$ No
   $2x - 3y = 8$

7. $x = 3y + 12$ Yes
   $2y = 5x + 5$

8. $x + 2y = -13$ Yes
   $y = -5$

9. $3x + y = -14$ Yes
   $x = -3$

10. $x + 3 = 0$ Yes
    $y + 5 = 0$

The following sentences refer to lines on the same coordinate plane. Replace each __?__ with one of the words or phrases in (a), (b), (c), or (d) to make a true statement.

11. Lines that have the same slope are either coincident or __?__ .
    a. intersecting   b. vertical   (c.) parallel   d. none of these

12. Nonvertical lines that intersect in exactly one point have __?__ .
    a. the same slope   b. no slope   (c.) different slopes   d. none of these

13. Parallel lines either have the same slope or are __?__ .
    a. coincident lines   (b.) vertical lines   c. horizontal lines   d. none of these

14. If a vertical line intersects another line in exactly one point, the second line cannot be __?__ .
    (a.) vertical   b. horizontal   c. of negative slope   d. of positive slope

15. If the solution set of a system of two linear equations in x and y is the empty set, the graphs of the equations are __?__ .
    a. coincident   (b.) parallel   c. intersecting   d. none of these

16. If a system of two linear equations in x and y has at least two solutions, the graphs of the equations are __?__ .
    a. parallel lines   (b.) coincident lines   c. intersecting lines.   d. none of these

17. In the figure at the right you see the graphs of the equations of the system:

    $$x = -1$$
    $$y = 1$$
    $$y = x + 2$$
    $$2y + x = 1$$

    What is the solution set of the system? $\{(-1, 1)\}$

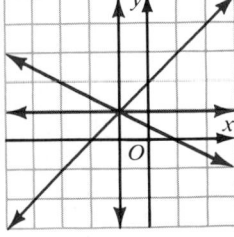

18. In the figure at the right you see the graphs of the equations of the system:

    $$y = x + 1$$
    $$3x + 2y = 2$$
    $$y = -2$$

    What is the solution set of the system? $\emptyset$

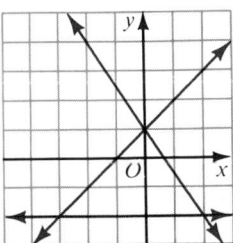

2. $4x + 3y = -3$
   $y = 3$ $\{(-3, 3)\}$

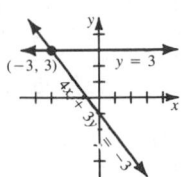

3. $2y = 6 - x$
   $x = 2 - 2y$ $\emptyset$

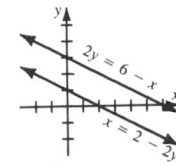

Determine whether the solution set of the given system is the empty set, an infinite set, or a set with exactly one member. Then tell whether the system is consistent or inconsistent.

4. $3x = -3y - 9$
   $-4y = 4x + 12$
   infinite set; consistent

5. $y - 3x = 3$
   $4y - x = 1$
   exactly one member; consistent

6. $y + 3x = 10$
   $8y + 3x = -4$
   exactly one member; consistent

1.
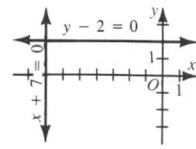

2.
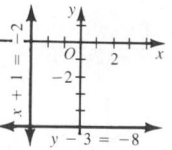

3.
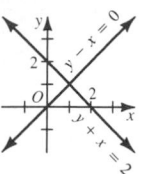

4.
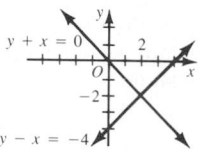

5.
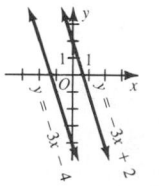

6.
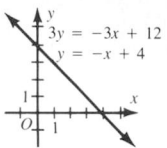

7.
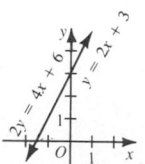

Written Exercises

Solve each system of equations using graphs. See Additional Answers for graphs.

A 1. $x + 7 = 0$
  $y - 2 = 0$ {(−7, 2)}

2. $y - 3 = -8$
  $x + 1 = -2$ {(−3, −5)}

3. $y - x = 0$
  $y + x = 2$ {(1, 1)}

4. $y + x = 0$
  $y - x = -4$ {(2, −2)}

5. $y = -3x + 2$
  $y = -3x - 4$ Ø

6. $y = -x + 4$
  $3y = -3x + 12$

7. $y = 2x + 3$
  $2y = 4x + 6$

8. $y = 5x - 2$
  $y = 5x + 1$ Ø

9. $y = 2x + 2$
  {(−3, −4)} $3y = -x - 15$

10. $x = 2 - 2y$
  $2x = -2 - y$ {(−2, 2)}

11. $x = 4y - 12$
  $2x = 8y - 24$

12. $2x = 3y + 4$
  $6x = 9y - 1$ Ø

6. {(x, y): y = −x + 4}   7. {(x, y): y = 2x + 3}   11. {(x, y): x = 4y − 12}

Determine whether the solution set of the given system is the empty set, an infinite set, or a set with exactly one member. Then tell whether the system is consistent or inconsistent.

13. $y = 3x + 1$
  $y = 2x + 2$

14. $y = -3x + 2$
  $y = -3x - 1$

15. $y + 5x = -1$
  $2y + 10x = -2$

16. $y - 4x = 5$
  $y + 4x = -3$

17. $3y - 2x = -7$
  $3y + 2x = 1$

18. $6y - 9x = -3$
  $2y - 3x = -1$

19. $2y - x = 2$
  $3x + 6 = 6y$

20. $2x + y = -5$
  $4x + 1 = y$

21. $x + 3 = 0$
  $x - 5 = 0$

22. $y - 7 = 4$
  $y + 1 = -3$

23. $3y + 2x - 6 = 0$
  $3 + 2x + 3y = 0$

24. $3y - 2x - 3 = 0$
  $3x - 2y - 8 = 0$

Solve each system of equations using graphs. Estimate the coordinates of the point of intersection to the nearest half unit.

B 25. $x + y = 5$
  $x - y = 2$ {(3½, 1½)}

26. $x - 3y = 5$
  $x + 3y = 5$ {(5, 0)}

27. $4y + 2 = -3x$
  $4y - 28 = 7x$ {(−3, 2)}

28. $-16y + 4 = 4x$
  $4y + 9 = 3x$ {(2½, −½)}

Solve.

29. Find the point on the graph of $x + 2y = 6$ where the ordinate is equal to the abscissa. (2, 2)

30. Find the point on the graph of $3y - x = 15$ where the ordinate is twice the abscissa. (3, 6)

31. Find the point on the graph of $3x + 5y = -10$ where the abscissa is the opposite of the ordinate. (5, −5)

32. Find the point on the graph of $2x - 5y = -13$ where the ordinate is one more than two thirds the abscissa. (6, 5)

In the following diagrams, a base b and the corresponding height h are indicated on a triangle and on a parallelogram. In Exercises 33–38 use the given formulas for area.

Area $= \frac{1}{2} \times$ base $\times$ height

$A = \frac{1}{2}bh$

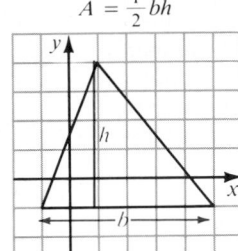

Area $=$ base $\times$ height

$A = bh$

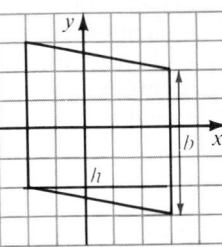

The points of intersection of the graphs of the given equations are the vertices of a triangle.

a. Find the vertices of the triangle. **b.** Find the area of the triangle.

C **33.** $y - x = 1$
$y + x = -3$
$x = 3$

34. $3x + 2y = 8$
$2x - 3y = 1$
$y = -5$

35. $x + 4y = 29$
$y = 4x + 3$
$x = -3$

The points of intersection of the graphs of the given equations are the vertices of a parallelogram.

a. Find the vertices of the parallelogram. **b.** Find the area of the parallelogram.

36. $y = 3x + 3$
$y = 6$
$y = 3x - 3$
$y = 3$

37. $3y - x = 6$
$x = 6$
$3y - x = -18$
$x = -3$

38. $2x + y = 7$
$y = -1$
$2y + 8 = -4x$
$y = -7$

Computer Exercises For students with computer experience

Write a program that will determine if an ordered pair (x, y) is a solution of a system of the form

$$ax + by = c$$
$$dx + ey = f$$

when you input values for x and y as well as a, b, c, d, e, and f. RUN the program to determine if $(3, -1)$ is a solution of each of the following.

1. $2x + 5y = 1$
$2x - 4y = 1$
No

2. $x - y = 4$
$x + y = 3$
No

3. $x + 5y = -2$
$y = -1$
Yes

4. $2x = 2y + 8$
$6y = -2x$
Yes

Systems of Open Sentences **273**

8.

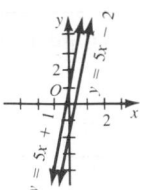

9.

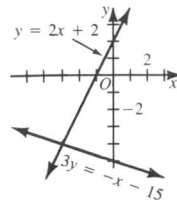

$y = 2x + 2$

$3y = -x - 15$

10.

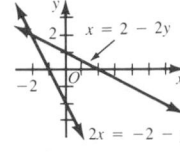

$x = 2 - 2y$

$2x = -2 - y$

11.

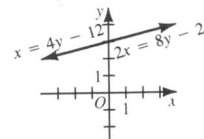

$x = 4y - 12$

$2x = 8y - 24$

12.

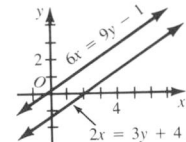

$6x = 9y - 1$

$2x = 3y + 4$

13. exactly one member; consistent
14. ∅; inconsistent
15. infinite set; consistent
16. exactly one member; consistent
17. exactly one member; consistent
18. infinite set; consistent
19. infinite set; consistent
20. exactly one member; consistent
21. ∅; inconsistent
22. ∅; inconsistent
23. ∅; inconsistent
24. exactly one member; consistent

(continued on p. 312)

273

Chalkboard Examples

Use addition or subtraction to solve each system of equations.

1. $u - v = 5$
   $u + 4v = -10$
   $-5v = 15$
   $v = -3$
   $u - (-3) = 5$
   $u = 2$
   Check:
   $(2) - (-3) \overset{?}{=} 5$
   $5 = 5 \ \checkmark$
   $(2) + 4(-3) \overset{?}{=} -10$
   $-10 = -10 \ \checkmark$
   $\therefore$ the solution is $\{(2, -3)\}$.

2. $c - d = 7$
   $4c + d = -2$
   $5c = 5$
   $c = 1$
   $(1) - d = 7$
   $d = -6$
   Check:
   $(1) - (-6) \overset{?}{=} 7$
   $7 = 7 \ \checkmark$
   $4(1) + (-6) \overset{?}{=} -2$
   $-2 = -2 \ \checkmark$
   $\therefore$ the solution is $\{(1, -6)\}$.

Common Errors

Remind students to transform equations so that equal signs and like terms line up vertically before solving by addition or subtraction. For instance the system:
$$u + 2v = 6$$
$$-u = 3v + 3$$

6–2 Using Addition or Subtraction

Often it is difficult to determine the exact solution of a system of linear equations by the graphing method. Therefore, it is important to learn algebraic methods that will enable you to find the exact solution set of any system of linear equations.

To solve a system of linear equations algebraically, you use number properties to find an *equivalent system* that can be solved by inspection. **Equivalent systems** of equations over a given domain are systems that have the same solution set over that domain. For example, consider the following systems of equations.

$$x = -3 \qquad\qquad\qquad -x + y = 5$$
$$y = 2 \qquad\qquad\qquad 4x + y = -10$$

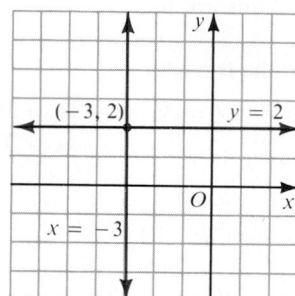

The solution set is $\{(-3, 2)\}$.   The solution set is $\{(-3, 2)\}$.

Since the two systems have the same solution set, they are equivalent systems of equations.

The following two examples show how you can use addition or subtraction to solve a system of equations in two variables in which a coefficient in one equation has the same absolute value as the corresponding coefficient in the other equation.

EXAMPLE 1   Solve the system: $2x - 3y = -6$
$2x - y = 2$

SOLUTION   1. Subtract corresponding terms of the two equations.

$$\begin{aligned} 2x - 3y &= -6 \\ 2x - y &= 2 \\ \hline -2y &= -8 \end{aligned}$$

2. Solve for y. $y = 4$

3. Replace one of the original equations with the equation $y = 4$. This system of equations is equivalent to the original system.

$$2x - 3y = -6$$
$$y = 4$$

4. Substitute 4 for y in the first equation of Step 3 and solve for x.

$$2x - 3y = -6$$
$$2x - 3(4) = -6$$
$$2x = 6$$
$$x = 3$$

5. The resulting system is equivalent to the original system and can be solved by inspection.

$$x = 3$$
$$y = 4$$

6. Check that (3, 4) satisfies both of the *original* equations.

$$2x - 3y = -6 \qquad\qquad 2x - y = 2$$
$$2(3) - 3(4) \overset{?}{=} -6 \qquad 2(3) - 4 \overset{?}{=} 2$$
$$-6 = -6 \;\checkmark \qquad\qquad 2 = 2 \;\checkmark$$

∴ the solution set of the given system is {(3, 4)}.

In the first step of Example 1, each term of the second equation was subtracted from the corresponding term of the first equation. Since the coefficients of the variable x were identical, this produced an equation in which the coefficient of x was zero. You can say that the variable x was *eliminated* from the system.

In solving a system of equations similar to the one in Example 1, it is usual to omit some of the steps.

EXAMPLE 2 Solve the system: $3s - 2t = -9$
$$5s + 2t = 1$$

SOLUTION 1. Add corresponding terms of the two equations to eliminate the variable t.

$$3s - 2t = -9$$
$$5s + 2t = 1$$
$$8s \qquad = -8$$

2. Solve for s.

$$s = -1$$

3. Substitute -1 for s in either of the original equations.

$$3s - 2t = -9$$
$$3(-1) - 2t = -9$$
$$-2t = -6$$
$$t = 3$$

4. Check that $(-1, 3)$ satisfies both of the original equations.

$$3s - 2t = -9 \qquad\qquad 5s + 2t = 1$$
$$3(-1) - 2(3) \overset{?}{=} -9 \qquad 5(-1) + 2(3) \overset{?}{=} 1$$
$$-9 = -9 \;\checkmark \qquad\qquad 1 = 1 \;\checkmark$$

∴ the solution set of the given system is {(−1, 3)}.

In working with equations in two variables, the order of the numbers in the ordered pair representing the solution should be the same as the alphabetical order of the corresponding variables. Thus the solution in the preceding example is given as $(-1, 3)$ to correspond to (s, t).

Systems of Open Sentences **275**

should be transformed so that it can be solved by the addition method with less chance of error. Rewrite as:
$$u + 2v = 6$$
$$-u - 3v = 3$$
Now the students may add the two equations together to enable them to solve the system.

Mixed Review

Solve each system of equations using graphs.

1. $3x - 4y = -2$
   $x + 2y = 6$ {(2, 2)}

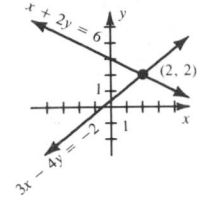

2. $8x - y + 3 = 0$
   $4x + y - 3 = 0$ {(0, 3)}

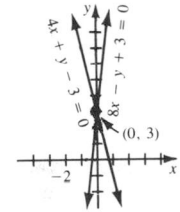

Solve.

3. On which point on the graph of $x + 3y - 6 = 0$ is the abscissa three times the ordinate?   (3, 1)

4. The points of intersection of the equations given below are the vertices of a triangle. What are the coordinates of the vertices and what is the area of the triangle?
   $$y = 0$$
   $$6 - x = 3y$$
   $$2x = 2 + 4y$$
   vertices: (1, 0), (6, 0), and
   (3, 1); area: $2\frac{1}{2}$ sq. units

Suggested Assignments

Minimum
276/1–12
Average
276/1–23, odd
Maximum
276/5–23 odd, 25–28

Additional A Exercises

Use addition or subtraction to solve each system of equations.

1. $a + 2b = 14$
$a - 3b = -11$ $\{(4, 5)\}$

2. $3r - s = 18$
$3r + s = 60$ $\{(13, 21)\}$

3. $2e + 5f = -10$
$4e - 5f = 10$ $\{(0, -2)\}$

4. $3p = -19 + 5q$
$3p = -5 - 2q$ $\{(-3, 2)\}$

5. $-5k = -2j + 12$
$-3k = -2j + 10$
$\left\{\left(3\frac{1}{2}, -1\right)\right\}$

6. $6d = 8 - 5e$
$-6d = 4 + 5e$ $\emptyset$

**Additional Answers
Written Exercises**

6. $\{(0, 0)\}$

9. $\left\{\left(-4, \frac{3}{4}\right)\right\}$

12. $\{(x, y): -7x + 3y = -4\}$

15. $\left\{\left(-\frac{11}{2}, -\frac{3}{2}\right)\right\}$

18. $\{(-5, 7)\}$

21. $\{(-5, 7)\}$

Oral Exercises

Add or subtract the equations in each system to eliminate one of the variables. Then solve for the other variable. Answers may vary.

1. $x + 2y = 12$ $8x = 16$;
$7x - 2y = 4$ $x = 2$

2. $5a + b = -2$ $3a = -3$;
$2a + b = 1$ $a = -1$

3. $5r + 2s = 19$ $3s = 2; s = \frac{2}{3}$
$5r - s = 17$

4. $5m - 4n = -6$ $2m = -16$;
$-3m + 4n = -10$ $m = -2$

5. $4e = 6 - 3f$ $0 = 8 + 4f$;
$4e = -2 - 7f$ $f = -2$

6. $9w = 6 + 4z$
$-5w = 2 - 4z$
$4w = 8; w = 2$

Written Exercises

Solve each system of equations using addition or subtraction.

A

1. $3x + y = 7$
$4x + y = 13$ $\{(6, -11)\}$

2. $4a - 3b = -5$
$a + 3b = -5$ $\{(-2, -1)\}$

3. $3s - 2t = -9$
$-3s + 5t = 21$ $\{(-\frac{1}{3}, 4)\}$

4. $4r + 3s = 11$
$4r - s = 23$ $\{(5, -3)\}$

5. $4u + v = 0$
$4u - 3v = 8$ $\{\frac{1}{2}, -2)\}$

6. $8c + 5d = 0$
$11c - 5d = 0$

7. $6w = 1 - 3x$
$6w = 7 + 6x$ $\{(\frac{1}{2}, -\frac{2}{3})\}$

8. $3d = 4e - 16$
$3d = 12e - 18$ $\{(-5, \frac{1}{4})\}$

9. $8c = 3b + 18$
$4c = -3b - 9$

10. $10k = -19 - 8j$
$-20k = 18 + 8j$ $\{(-\frac{5}{2}, \frac{1}{10})\}$

11. $2a - 3b = 8$
$-2a + 3b = 8$ $\emptyset$

12. $-7x + 3y = -4$
$7x - 3y = 4$

B

13. $6p - 5t = -4$
$5t + 9p = -6$ $\{(-\frac{2}{3}, 0)\}$

14. $4t - 3u = -1$
$9u + 4t = -5$ $\{(-\frac{1}{2}, -\frac{1}{3})\}$

15. $7f = -38 - 5e$
$5e = 3f - 23$

16. $8v = 5u - 22$
$10u = -10 - 8v$ $\{(\frac{4}{5}, -\frac{9}{4})\}$

17. $6m = 8 + 10n$
$11 = 6m - 5n$ $\{(\frac{7}{3}, \frac{3}{5})\}$

18. $13 = 3c + 4d$
$4d = 7c + 63$

19. $2w + 2x = -2 - 4w$
$7w + 27 = 3x + w$ $\{(-2, 5)\}$

20. $5c - 3f = 31 + f$
$2f - 8 = -2f - 8c$ $\{(3, -4)\}$

21. $3(c + d) = 6$
$3(c - d) = -36$

22. $4(r + s) = -20$
$4(r - s) = -12$ $\{(-4, -1)\}$

23. $3(b - 3) - 2c = 0$
$2(b - c) = -b - 3$ $\emptyset$

24. $3(j - 1) - 4k = 0$
$4(j - k) = j + 3$
$\{(j, k): 3(j - 1) - 4k = 0\}$

Solve each system of equations for a and b in terms of c.

C

25. $4a + 3b + c = 18$
$3a + 3b + c = 17$ $\{(1, \frac{14 - c}{3})\}$

26. $3a - 2b + 4c = -28$
$4a + 2b - 4c = -7$ $\{(-5, \frac{13 + 4c}{2})\}$

27. If $a \neq 0$ and $b \neq 0$, under what conditions will the system of equations at the right have no solution? $a = -b$
$ax + by = c$
$bx + ay = c$

28. If $a \neq 0$ and $b \neq 0$, solve the system of equations at the right for x and y in terms of a, b, c, and d. $\{(\frac{c + d}{2a}, \frac{c - d}{2b})\}$
$ax + by = c$
$ax - by = d$

276 Chapter 6

Write a program that will solve a system of the form

$$ax + by = c$$
$$dx + ey = f$$

when you input values for *a, b, c, d, e,* and *f* as well as two integral replacement sets for *x* and *y*. RUN the program to solve the following systems of linear equations.

1. $x - 3y = 1$
  $x + y = 13$ $x, y \in \{1, 2, \ldots, 10\}$ $\{(10, 3)\}$

2. $2x - 4y = 8$
  $2y - x = -4$ $x, y \in \{-5, -4, \ldots, 5\}$
              $\{(-4, -4), (-2, -3), (0, -2), (2, -1), (4, 0)\}$

3. $2x + 4y = 6$
  $3x + 2y = 3$ $x \in \{0, 1, \ldots, 8\}; \quad y \in \{-8, -7, \ldots, 0\}$ ∅

6–3 Using Linear Combinations

You can use the multiplicative property of equality to transform one or both of the given equations in a system of equations. This can be done to any given system of equations to produce an equivalent system in which a coefficient in one equation has the same absolute value as the corresponding coefficient in the other. You can then use the addition and subtraction method discussed in the preceding section to find the solution set of the given system.

EXAMPLE 1   Solve the system: $5x - 2y = -25$
                  $3x + y = -4$

SOLUTION   1. To obtain equations in which the coefficients of the *y*-terms have the same absolute value, multiply both sides of the second equation by 2.

$$5x - 2y = -25$$
$$\underline{6x + 2y = -8}$$

  2. Add the equations. $11x \qquad = -33$

  3. Solve for *x*. $x = -3$

  4. Substitute -3 for *x* in one of the given equations and solve for *y*.

$$3x + y = -4$$
$$3(-3) + y = -4$$
$$y = 5$$

  5. Checking that $(-3, 5)$ satisfies both of the original equations is left to you.
    ∴ the solution set of the given system is $\{(-3, 5)\}$.

Systems of Open Sentences   **277**

2. $5e - 4f = -4$
$4e - 7f = 12$

$\quad 20e - 16f = -16$
$\quad \underline{20e - 35f = 60}$
$\qquad\qquad 19f = -76$
$\qquad\qquad\quad f = -4$
$5e - 4(-4) = -4$
$\qquad\quad 5e = -20$
$\qquad\qquad e = -4$

Check:
$5(-4) - 4(-4) \overset{?}{=} -4$
$\qquad\qquad -4 = -4 \;\checkmark$
$4(-4) - 7(-4) \overset{?}{=} 12$
$\qquad\qquad 12 = 12 \;\checkmark$
∴ the solution is
$\{(-4, -4)\}$.

Mixed Review

Simplify.
1. $3x + y - (2x + 3y)$
$\quad x - 2y$
2. $-2x + 5y - 3(x - y)$
$\quad -5x + 8y$
Use addition or subtraction
to solve each system of
equations.
3. $2x - 2y = 2$
$\quad x - 2y = -2 \quad \{(4, 3)\}$
4. $3y - 17 = x - 3y$
$\quad 7 - y = x + y \quad \{(1, 3)\}$

Suggested Assignments

Minimum
$\quad$ 279/1–12
S 272/12, 29, 30
Average
$\quad$ 279/1–15
Maximum
$\quad$ 279/6–20

The equation that you obtain by multiplying one equation of a system by a nonzero constant and another equation of the system by another nonzero constant and adding or subtracting the two resulting equations is called a **linear combination** of the given equations. Thus, the equation in Step 2 of Example 1,

$$11x = -33,$$

is a linear combination of the equations in the given system of equations. The method for solving a system of linear equations illustrated in Example 1 is often referred to as the *linear-combination method*.

EXAMPLE 2 Solve the system: $\quad 6a + 5b = 9$
$\qquad\qquad\qquad\qquad\qquad\quad 5a + 2b = 14$

SOLUTION

1. Multiply both sides of the first equation by 2. Multiply both sides of the second equation by 5.

$\qquad 12a + 10b = 18$
$\qquad \underline{25a + 10b = 70}$

2. Subtract the resulting equations.

$\qquad -13a \qquad\quad = -52$

3. Solve for a.

$\qquad\qquad\quad a = 4$

4. Substitute 4 for a in one of the given equations and solve for b.

$\qquad\quad 6a + 5b = 9$
$\qquad\quad 6(4) + 5b = 9$
$\qquad\qquad\quad 5b = -15$
$\qquad\qquad\quad\; b = -3$

5. Checking that $(4, -3)$ satisfies both of the original equations is left to you.

$\qquad$ ∴ the solution set of the given system is $\{(4, -3)\}$.

There are several ways to transform each equation in a given system of equations to produce an equivalent system in which a coefficient in one equation has the same absolute value as the corresponding coefficient in the other. For example, in the first step of the preceding solution, you might choose to multiply both sides of the first equation by 2 and to multiply both sides of the second equation by -5.

$$2(6a + 5b) = 2(9) \quad\longrightarrow\quad 12a + 10b = 18$$
$$-5(5a + 2b) = -5(14) \quad\longrightarrow\quad -25a - 10b = -70$$

You would then add the resulting equations to eliminate the variable b.

You could also choose to multiply both sides of the first equation by 5 and to multiply both sides of the second equation by 6.

$$5(6a + 5b) = 5(9) \quad\longrightarrow\quad 30a + 25b = 45$$
$$6(5a + 2b) = 6(14) \quad\longrightarrow\quad 30a + 12b = 84$$

In this case you would then subtract the resulting equations to eliminate the variable a.

Oral Exercises

In solving each of the following systems of equations by the linear-combination method, explain how you can eliminate one of the variables by answering the following questions.

a. Which variable would you eliminate?
b. Which equation(s) would you transform by multiplication?
c. By what number(s) would you multiply?
d. Would you then add or subtract the equations?

1. $r - 3s = -1$
   $2r - 2s = 3$

2. $3y - 4z = 5$
   $2y + z = 1$

3. $a + b = 8$
   $5a + 8b = -1$

4. $4c + 2d = -3$
   $12c - 5d = -2$

5. $9s - 6t = -7$
   $18s + 12t = 1$

6. $8w - 20x = 4$
   $2w - 5x = 7$

7. $11d - 3e = 4$
   $6d + 2e = -5$

8. $4p + 7q = -5$
   $5p - 9q = 0$

9. $9f - 2g = -3$
   $12f + 7g = 6$

10. $4j - 5k = 7$
    $3j + 9k = -8$

11. $5r - 5s = 14$
    $8r + 12s = -3$

12. $4c - 9d = 5$
    $9c + 4d = -7$

Written Exercises   6. $\{(8, 7)\}$ 15. $\{(\frac{k}{3}, -\frac{k}{2})\}$

Solve each system of equations using the linear-combination method.

A
1. $2d - 5e = -3$
   $3d + 2e = 5$ $\{(1, 1)\}$

2. $2x + 3y = 8$
   $6x + 5y = 0$ $\{(-5, 6)\}$

3. $3p - 4q = 12$
   $5p + 8q = 20$ $\{(4, 0)\}$

4. $2a + 5b = 3$
   $3a - 2b = -5$ $\{(-1, 1)\}$

5. $3j - 5k = 15$
   $4j - 7k = 21$ $\{(0, -3)\}$

6. $3w - 4x = -4$
   $5w - 7x = -9$

7. $3c + 6d = -7$
   $11c + 9d = -30$ $\{(-3, \frac{1}{3})\}$

8. $8r + 6s = 0$
   $6r + 9s = -3$ $\{(\frac{1}{2}, -\frac{2}{3})\}$

9. $7x - 2y = 0$
   $-14x + 4y = 3$ Ø

10. $6a + 15b = -3$
    $2a - 5b = -11$ $\{(-3, 1)\}$

11. $4d - 5e = -10$
    $6d + 8e = 16$ $\{(0, 2)\}$

12. $8s - 3t = -5$
    $-16s + 6t = 10$
    $\{(s, t): 8s - 3t = -5\}$

Solve each system of equations for x and y in terms of the other variable(s).

B
13. $5x - n = 2y$
    $y = 3n - x$ $\{(n, 2n)\}$

14. $2x = 5y - 7n$
    $-3x + 4y = 7n$ $\{(-n, n)\}$

15. $9x + k + 8y = 0$
    $12y + k = -15x$

16. $8x - 3cy = 9c$
    $12x - 7cy = -39c$ $\{(9c, 21)\}$

17. $3ax + 2by = -5ab$
    $4ax - 5by = 24ab$ $\{(b, -4a)\}$

18. $8bx + 12ab = -7ay$
    $4ay + 9ab = -5bx$
    $\{(-5a, 4b)\}$

C
19. If a, b, and c are positive integers such that $a = b$ and $c = 3a$, find the vertices of the triangle formed by the x-axis and the graphs of the lines
    $$ax + by = c \quad \text{and} \quad by - ax = c.$$ (0, 3), (−3, 0), (3, 0)

Systems of Open Sentences **279**

20. a. Solve the system: $9x - 8y = 1$
                             $6x + 12y = 5$ $\{(\frac{1}{3}, \frac{1}{4})\}$

b. Show that for all nonzero real numbers e and f, the solution of the system in part (a) satisfies the equation

$$e(9x - 8y - 1) + f(6x + 12y - 5) = 0.$$

(This shows that the solution of the system in part (a) is a solution of every linear combination of the equations in that system.)

$$e[9(\tfrac{1}{3}) - 8(\tfrac{1}{4}) - 1] + f[6(\tfrac{1}{3}) + 12(\tfrac{1}{4}) - 5] = e(0) + f(0) = 0$$

Key Ideas

Solve systems of linear equations by the substitution method.

Chalkboard Examples

1. Solve the system:
$$u = 2v$$
$$3v + u = 5$$
$$3v + (2v) = 5$$
$$5v = 5$$
$$v = 1$$
$$u = 2(1)$$
$$u = 2$$
Check: $(2) \overset{?}{=} 2(1)$
            $2 = 2$ √
        $3(1) + (2) \overset{?}{=} 5$
            $5 = 5$ √
∴ the solution is $\{(2, 1)\}$.

6–4 Using Substitution

A method that is sometimes easier to use than the linear-combination method in solving a system of linear equations is the *substitution method*. Given two equations in x and y, you can transform one of them to express one variable in terms of the other. You can then use the substitution principle to replace one of the given equations by a third equation involving only one variable. The following example illustrates the substitution method.

EXAMPLE   Solve the system: $3x + 7y = -6$
                                $x - 2y = 11$

SOLUTION

1. Solve the second equation for x in terms of y.

$$x - 2y = 11$$
$$x = 2y + 11$$

2. Substitute this expression for x in the first equation, and solve for y.

$$3x + 7y = -6$$
$$3(2y + 11) + 7y = -6$$
$$6y + 33 + 7y = -6$$
$$13y = -39$$
$$y = -3$$

3. Substitute this value of y in the second equation in Step 1, and solve for x.

$$x = 2y + 11$$
$$x = 2(-3) + 11$$
$$x = 5$$

4. Checking that $(5, -3)$ satisfies both of the original equations is left to you.

∴ the solution of the given system of equations is $\{(5, -3)\}$.

In the first step of the preceding example, you could have solved for either variable in either equation. However, it is usually most convenient to solve for a variable that has a coefficient of one.

The following is a summary of the substitution method for solving a system of linear equations in two variables.

280   *Chapter 6*

The Substitution Method

To solve a system of linear equations in two variables:
1. Solve one equation for one of the variables.
2. Substitute this expression in the other equation and solve.
3. Find the corresponding value of the other variable.
4. Check the solution in both of the original equations.

The transformations used in solving systems of linear equations algebraically are summarized below.

Transformations That Produce an Equivalent System of Linear Equations

1. Replacing any equation of the system with an equivalent equation in the same variables.
2. Replacing any equation with a linear combination of itself and another equation of the system.
3. Substituting for one variable in any equation either
   (a) the actual value of the variable, or
   (b) an equivalent expression for that variable obtained from another equation of the system.

Oral Exercises

To use the substitution method, first select one equation. Using that equation, solve for one variable in terms of the other. In Exercises 1–6:

a. Which equation would you select?
b. Which variable would you solve for?

1. $2a + b = 7$
$5a - 3b = 1$

2. $2u - 3v = -2$
$2u - v = 15$

3. $4d - 3e = 1$
$d - 7e = -4$

4. $w + 8x = -3$
$4w - 3x = 7$

5. $2m - n = 5$
$2m + 5n = -2$

6. $3 - r = 2s$
$5 + 7r = 5s$

Solve each system of equations using the substitution method.

7. $c = 3d$
$c + d = 16$ $\{(12, 4)\}$

8. $s = 5r$
$r + s = -6$ $\{(-1, -5)\}$

9. $v = u - 3$
$u + v = 13$ $\{(8, 5)\}$

10. $f = e + 7$
$f - 2e = 4$ $\{(3, 10)\}$

11. $m - 2n = 7$
$m = n + 2$ $\{(-3, -5)\}$

12. $8 - p = q$
$q = 2p - 10$ $\{(6, 2)\}$

Systems of Open Sentences **281**

---

2. Solve the system:
$5a - 3b = -1$
$a + b = 3$
$a = 3 - b$
$5(3 - b) - 3b = -1$
$15 - 8b = -1$
$-8b = -16$
$b = 2$
$a + 2 = 3$
$a = 1$
Check:
$5(1) - 3(2) \overset{?}{=} -1$
$-1 = -1$ ✓
$(1) + (2) \overset{?}{=} 3$
$3 = 3$ ✓
∴ the solution is $\{(1, 2)\}$.

Suggested Assignments

Minimum
282/1–23 odd
R 283/Self-Test 1
Average
282/1–25 odd, 26–30
R 283/Self-Test 1
Maximum
282/7–29 odd, 31-36
R 283/Self-Test 1

Additional A Exercises

Solve each system of equations using the substitution method.

1. $r + 5s = 2$
   $r = -3s$ $\{(-3, 1)\}$

2. $x = y + 1$
   $y = -x - 5$ $\{(-2, -3)\}$

3. $c + 2d = -2$
   $c = 6d + 30$ $\{(6, -4)\}$

4. $3x = 5y + 8$
   $x + 2y = -1$ $\{(1, -1)\}$

5. $4p - 15 = 3q$
   $p - 2q = 0$ $\{(6, 3)\}$

6. $3a = 5 + 4b$
   $a - 10 = -7b$ $\{(3, 1)\}$

Additional Answers
Written Exercises

35. Assume that (x_1, y_1) is a solution of the system. Then $m_1x_1 + b_1 = m_2x_1 + b_2$. By hypothesis, $m_1 = m_2$, so $m_1x_1 + b_1 = m_1x_1 + b_2$. Then $b_1 = b_2$, but this contradicts the hypothesis $b_1 \neq b_2$. $\therefore$ the system has no solution.

36. Let (r, s) be a solution of the system. Then $ar + bs + c = 0$ and $s = mr + k$. $ar + b(mr + k) + c = 0$ Thus (r, s) is also a solution of $ax + b(mx + k) + c = 0$.

Written Exercises
10. $\{(v, w): v + 3w = 2\}$

Solve each system of equations using the substitution method.

A

1. $e + 2f = 8$
   $2e - 3f = -19$ $\{(-2, 5)\}$

2. $v + 4w = 5$
   $5v - 7w = -2$ $\{(1, 1)\}$

3. $p - 6q = 2$
   $2p - 7q = 9$ $\{(8, 1)\}$

4. $x - 8y = 4$
   $3x + 5y = 12$ $\{(4, 0)\}$

5. $2j - 15k = 7$
   $j - 6k = 4$ $\{(6, \frac{1}{3})\}$

6. $6d - 2e = 2$
   $e + 7d = 3$ $\{(\frac{2}{5}, \frac{1}{5})\}$

7. $2m - n = 13$
   $3m - 8n = 13$ $\{(7, 1)\}$

8. $2a - b = 7$
   $4a - 3b = 9$ $\{(6, 5)\}$

9. $8c - 7d = 6$ $\{(-1, -2)\}$
   $5c - d = -3$

10. $3v + 9w = 6$
    $v + 3w = 2$

11. $8b + 2c = 4$
    $4b + c = 1$ $\emptyset$

12. $2s + 3t = 15$ $\{(0, 5)\}$
    $9s - t = -5$

13. $8w = x - 1$
    $2x = 10w + 5$ $\{(\frac{1}{2}, 5)\}$

14. $3f = g - 11$
    $3g = 2f + 19$ $\{(-2, 5)\}$

15. $n = \frac{2}{3}m$ $\{(\frac{3}{5}, \frac{2}{5})\}$
    $2m + 7n = 4$

16. $s = \frac{3}{4}r$
    $9r - 4s = -2$ $\{(-\frac{1}{3}, -\frac{1}{4})\}$

17. $2x + 6y = 12$
    $3x + 4y = 13$ $\{(3, 1)\}$

18. $4a + 12b = -8$
    $3a - 5b = 8$
    $\{(1, -1)\}$

Solve each system of equations by any method. (You may transform any equation in a system of equations into an equivalent equation before you begin to solve the system.)

B

19. $3x + 5y = -9$
    $5x - 2y = 16$ $\{(2, -3)\}$

20. $6p - 7q = 11$
    $8p + 3q = -47$ $\{(-4, -5)\}$

21. $4(a + b) = 6(b - 1)$ $\{(-\frac{2}{3}, \frac{5}{3})\}$
    $8(a + 1) = b + 1$

22. $4(e + f) = 8(f - 4)$
    $2(e - 1) = f - 15$ $\{(-5, 3)\}$

23. $7(t + u) = 5 + 6t$
    $8(t - 1) = 2(1 + 3u)$

24. $2x - 3y = 8$
    $2(2x - 8) = 6y$

25. $5a + 2b = -5$
    $4(a + b) = 6(2 - a)$ $\emptyset$

26. $5(p + q) = 9q - 1$
    $2(2q + 3) = 5(p + 1)$ $\emptyset$

27. $\frac{y}{7} + z = \frac{-11}{7}$
    $y - \frac{5z}{3} = 2$ $\{(-\frac{1}{2}, -\frac{3}{2})\}$

28. $\frac{h}{2} + k = \frac{1}{2}$
    $\frac{h}{7} + 1 = \frac{4k}{7}$ $\{(-\frac{5}{3}, \frac{4}{3})\}$

29. $\frac{x}{2} + \frac{y}{3} = 6$
    $3x - 2y = 12$ $\{(8, 6)\}$

30. $\frac{s}{5} - \frac{t}{3} = -2$
    $5s - 3t = 14$
    $\{(10, 12)\}$

In Exercises 31 and 32 two equivalent systems of linear equations are given. Find the values of a and b. $a = \frac{1}{2}, b = -1$ $a = 5, b = 3$

C

31. $-2x + 3y = -13$ $-2ax + by = 1$
    $4x + y = 5$ $ax + by = 4$

32. $3x - 4y = -11$ $ax + 4by = 19$
    $5x + 2y = -1$ $ax + 2by = 7$

33. The graphs of the equations $2ax + by = -20$ and $-ax + 2by = -10$ intersect at $(-2, 2)$. Find a and b. $a = 3, b = -4$

34. The graphs of the equations $-3ax + 5by = 14$ and $2ax - 4by = -4$ intersect at $(\frac{1}{3}, -1)$. Find a and b. $a = -54, b = 8$

282 *Chapter 6*   23. $\{(\frac{50}{31}, \frac{15}{31})\}$ 24. $\{(x, y): 2x - 3y = 8\}$

35. Show that if $m_1 = m_2$, but $b_1 \neq b_2$, then the following system has no solution.

$$y = m_1 x + b_1$$
$$y = m_2 x + b_2$$

36. Show that any solution of the system

$$ax + by + c = 0$$
$$y = mx + k$$

is also a solution of the equation $ax + b(mx + k) + c = 0$.

Self-Test 1

VOCABULARY system of linear equations, or simultaneous linear equations (p. 267)
solution of a system (p. 267)
coincident lines (p. 268)
point of intersection (p. 268)

consistent system of equations (p. 269)
inconsistent system of equations (p. 269)
equivalent systems (p. 274)
linear combination (p. 278)
substitution method (p. 281)

1. Tell whether the given system is consistent or inconsistent.   *Obj. 1, p. 267*

$$3y + 2x = 6$$
$$6y - 6 = -4x$$ inconsistent

2. Solve the given system of equations using graphs.   *Obj. 2, p. 267*

$$3y = 2x + 5$$
$$3x + y = 9 \ \{(2, 3)\}$$

3. Solve the given system of equations using addition or subtraction.

$$3m - 3n = 9$$
$$3n + 2m = 1 \ \{(2, -1)\}$$

4. Solve the given system of equations using the linear-combination method.

$$5a + 3b = -3$$
$$4a + 5b = 8 \ \{(-3, 4)\}$$

5. Solve the given system of equations using the substitution method.

$$c + 4d = 1$$
$$2c + 7d = 3 \ \{(5, -1)\}$$

Check your answers with those at the back of the book.

Systems of Open Sentences **283**

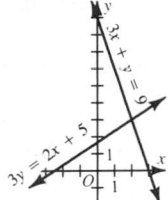
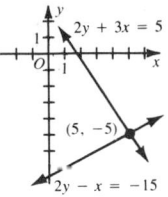

Key Ideas

Solve word problems using two equations in two variables.

Chalkboard Examples

Solve:

1. Terry has 32 coins worth a total of $6.80. Some of the coins are dimes and the rest are quarters. How many of each type of coin does Terry have?
Solution 1
(Using two variables)
Let d = the number of dimes.
Let q = the number of quarters.

Number of coins	Value of coins	Total Value
d	0.10	$0.10d$
q	0.25	$0.25q$

$$d + q = 32$$
$$0.10d + 0.25q = 6.80$$
$$10(32 - q) + 25q = 680$$
$$320 - 10q + 25q = 680$$
$$15q = 360$$
$$q = 24$$
$$d + (24) = 32$$
$$d = 8$$

∴ there are 8 dimes and 24 quarters.
Solution 2:
(Using one variable)
Let q = the number of quarters. Then:
$32 - q$ = the number of dimes.

Solving Problems

OBJECTIVES for Sections 6-5 through 6-7:
1. To use two variables to solve problems.
2. To solve wind and water current problems.
3. To solve digit problems.

6–5 Using Two Variables to Solve Problems

In Chapter 4 you used equations in one variable to solve problems. Now that you know how to solve a system of linear equations, you may find it more convenient to use two variables when solving certain problems. When you use two variables, you ordinarily need to form two equations. The next example illustrates how a problem may be solved using either one variable or two variables.

EXAMPLE A realtor has two homes for sale. Ten years ago the older home was four times as old as was the newer home. Thirty years from now, the older home will be only twice as old as the newer home will be. How old are the homes now?

SOLUTION 1 (Using two variables)

Step 1 The problem asks for the ages of the homes now.

Step 2 Let x = the present age in years of the older home.
Let y = the present age in years of the newer home.

	Age in years 10 years ago	Age in years now	Age in years 30 years from now
Older home	$x - 10$	x	$x + 30$
Newer home	$y - 10$	y	$y + 30$

Step 3 Ten years ago, older home was four times as old as newer home.
$$x - 10 = 4(y - 10)$$
Thirty years from now, older home will be twice as old as newer home.
$$x + 30 = 2(y + 30)$$

284 Chapter 6

Step 4 You can find x and y by solving the following system of equations.

$$x - 10 = 4(y - 10) \longrightarrow x - 4y = -30$$
$$x + 30 = 2(y + 30) \longrightarrow \underline{x - 2y = \ \ \ 30}$$
$$-2y = -60$$
$$y = \ \ \ 30$$

$$x - 2y = 30$$
$$x - 2(30) = 30$$
$$x = 90$$

Step 5 Ten years ago, was the older home four times as old as the newer one?
$$90 - 10 \overset{?}{=} 4(30 - 10)$$
$$80 = 80 \ \ \checkmark$$

Thirty years from now, will the older home be twice as old as the newer one?
$$90 + 30 \overset{?}{=} 2(30 + 30)$$
$$120 = 120 \ \ \checkmark$$

∴ the older home is 90 years old, and the newer home is 30 years old.

SOLUTION 2 (Using one variable)

Step 1 The problem asks for the ages of the homes now.

Step 2 Let x = the age in years of the newer home 10 years ago.
Then $4x$ = the age in years of the older home 10 years ago.

	Age in years 10 years ago	Age in years now	Age in years 30 years from now
Older home	$4x$	$4x + 10$	$4x + 40$
Newer home	x	$x + 10$	$x + 40$

Step 3 Thirty years from now, older home will be twice as old as newer home.
$$\underbrace{4x + 40} \ \ \underset{\downarrow}{=} \ \ \underbrace{2(x + 40)}$$

Step 4
$$4x + 40 = 2(x + 40)$$
$$4x + 40 = 2x + 80$$
$$2x = 40$$
$$x = 20$$
present age of older home = $4x + 10 = 4(20) + 10 = 90$ years
present age of newer home = $x + 10 = 20 + 10 = 30$ years

Step 5 The check is left to you.
∴ the older home is 90 years old, and the newer home is 30 years old.

Systems of Open Sentences **285**

Number of coins	Value of coins	Total Value
$(32 - q)$	0.10	$0.10(32 - q)$
q	0.25	$0.25q$

$$0.10(32 - q) + 0.25q = 6.80$$
$$3.2 - 0.10q + 0.25q = 6.80$$
$$0.15q = 3.60$$
$$q = 24$$
$$32 - q = 8$$
∴ there are 8 dimes and 24 quarters.

2. The sum of two numbers is 60. Twice the larger number is 10 less than eight times the smaller. What are the two numbers?

Solution 1:
(Using two variables)
Let x = the larger number.
Let y = the smaller number.
$$x + y = 60$$
$$\underline{2x = 8y - 10}$$

$$2x + 2y = 120$$
$$\underline{2x - 8y = -10}$$
$$10y = 130$$
$$y = 13$$
$$x + 13 = 60$$
$$x = 47$$

∴ the larger number is 47 and the smaller number is 13.

Solution 2:
(Using one variable)
Let x = the smaller number. Then:
$60 - x$ = the larger number.
$$2(60 - x) = 8x - 10$$
$$120 - 2x = 8x - 10$$
$$130 = 10x$$
$$13 = x$$
$$60 - x = 47$$
∴ the smaller number is 13 and the larger number is 47.

Suggested Assignments

Minimum
First day: 286/P: 1–7
Second day: 287/P: 8–16
         S 276/13–17
Average
First day: 286/P: 1–8
Second day: 287/P: 9–20
         S 276/16–24 even
Maximum
   286/P: 5–15 odd, 17–22
S 276/16–24 even

Additional A Exercises

Solve each problem using two equations in two variables.

1. The cost of five animal figures and three batteries is $8.10. The cost of three animal figures and four batteries is $6.40. If each animal figure costs the same amount and each battery costs the same amount, what does one animal figure cost? $1.20

2. Two angles are supplementary. The sum of the measures of the larger angle and twice the smaller angle is 265°. Find the measure of the larger angle.   95°

3. A washer and dryer together cost $350. The dryer costs $20 more than two times the washer. What is the price of each appliance?
   washer: $110
   dryer: $240

4. A rectangle has a width 7 inches less than the length. The perimeter is 46 inches. What are the dimensions of the rectangle?
   width: 8 in
   length: 15 in

Oral Exercises

Solve each problem by using the following:

a. **two equations in two variables**

b. **one equation in one variable**

1. The sum of two numbers is 24. One number is three times the other. What are the two numbers? 6, 18; $x + y = 24$, $x = 3y$; $x + 3x = 24$

2. One number is six less than another. The sum of the numbers is 17. What are the two numbers? $5\frac{1}{2}$, $11\frac{1}{2}$; $x = y - 6$, $x + y = 17$; $x + (x + 6) = 17$

3. The sum of two numbers is 61. Their difference is five. What are the two numbers? 28, 33; $x + y = 61$, $x - y = 5$; $x + (x + 5) = 61$

4. A wire that is 16 m long is cut into two pieces. One piece is three times the length of the other. What is the length of the shorter piece of wire? 4 m; $x + y = 16$, $x = 3y$; $x + 3x = 16$

5. Les has 20 coins in his pocket. Some of them are nickels and the rest are dimes. The nickels and dimes altogether are worth $1.40. How many dimes does Les have?

6. The perimeter of a rectangle is 108 cm. If the length of the rectangle is five times the width, what are the dimensions of the rectangle? 45 cm by 9 cm; $2x + 2y = 108$, $x = 5y$; $2x + 2(5x) = 108$

   5. 8 dimes; $n + d = 20$, $5n + 10d = 140$; $5n + 10(20 - n) = 140$

Problems

Solve each problem using two equations in two variables.

A
1. Susan has seven more fish than Tammy. They have 43 fish altogether. How many fish does each have? Susan: 25 fish; Tammy: 18 fish

2. Bob is three years older than his brother. The sum of their ages is 33. How old is Bob? 18 years

3. Two angles are supplementary. The measure of one angle is 30° more than the measure of the other. What is the measure of the larger angle? 105°

4. Two angles are complementary. The sum of the measures of the larger angle and three times the measure of the smaller angle is 114°. What is the measure of each angle? 78°, 12°

5. Two games and five puzzles cost $33.25 altogether. Three games and two puzzles cost $32.00. If every game costs the same amount, and every puzzle costs the same amount, what is the cost of each puzzle? $3.25

6. The difference between three times one number and a lesser one is 37. The sum of the greater number and twice the lesser number is 38. Find the numbers. 11, 16

7. The length of a rectangular garden is three times the width. If the perimeter is 32 m, what are the dimensions of the garden?
   12 m by 4 m

286 *Chapter 6*

8. Two sides of a triangle have the same length. The remaining side is one third as long as each of the other sides. If the perimeter of the triangle is 315 cm, what is the length of each side? 135 cm, 135 cm, 45 cm

9. During the year, Charles read five more books than twice the number of books read by Frank. If together they read a total of 29 books, how many books did each read? Charles: 21 books; Frank: 8 books

10. Two angles are supplementary. The measure of one of these angles is 12° less than one third the measure of the other. What is the measure of each angle? 36°, 144°

11. Kim has 40 coins worth a total of $8.80. Some of the coins are nickels and the rest are quarters. How many of each kind of coin does Kim have? 6 nickels, 34 quarters

12. Lou has $8.60 in dimes and quarters. If he has twelve more quarters than dimes, how many dimes does he have? 16 dimes

13. Jessica purchased some 20¢ stamps and some 25¢ stamps at the post office. If she paid $7.75 for 35 stamps, how many of each kind did she purchase? 20 20¢ stamps 15 25¢ stamps

14. The charge for admission to the zoo is $3.25 for each adult and $1.50 for each child. On a day when 500 people paid to visit the zoo, the receipts totalled $1275. Find the number of adult tickets purchased that day. 300 adult tickets

15. The ages of Lee's mother and aunt total 83 years. If her mother's age were doubled, then her mother would be 58 years older than her aunt. How old are Lee's mother and aunt? mother: 47 years; aunt: 36 years

16. In four years Cathy's cat Byte will be three fourths as old as Cathy will be. Four years ago, Byte was only half as old as Cathy was. How old are Cathy and her cat now? Cathy: 12 years; Byte: 8 years

Solve.

B 17. Bath towels sell for $13.25 each, while hand towels sell for $4.50 each. Theresa buys some of each type of towel for a total of $62.25. If she spends $17.25 more on bath towels than she spends on hand towels, how many of each type does she buy? 3 bath towels, 5 hand towels

18. It costs $5.40 to ship a radial saw and $6.50 to ship a table saw to Ohio. On Monday an order consisting of table saws and radial saws was sent to Ohio for a total shipping cost of $140.70. If nine more table saws had been sent, the number of table saws in the order would have been three times the number of radial saws. How many saws of each type were sent in the order? 8 radial saws, 15 table saws

Systems of Open Sentences **287**

5. The sum of a number and three times a lesser one is 1. Twenty more than three times the lesser number is twice the greater number. What are the two numbers?
   greater number: 7
   lesser number: −2

6. Cindy has 17 coins, all dimes and nickels. The total amount is $1.15. How many of each type of coin does she have? 6 dimes and 11 nickels

---

Reading Algebra

Have each student make a chart similar to the one below supplying their own information. Then have them take turns writing word problems which fit the given information. They can make similar charts showing distance or age relationships. Although it is possible to make charts with unrealistic values, encourage them to keep their chart values within reason.

	Wt. in kg	Value per kg	Total Value
Dried Fruit	x	$3.45	3.45x
Raisins	y	$2.10	2.10y
Mixture	10.5	$2.64	2.64(10.5)

Solve the following systems of equations using the substitution method.

1. $2r + 3s = 4$
$\quad 4r + s = -2$ $\{(-1, 2)\}$

2. $\quad 6p - q = -10$
$\quad 10p - 5q = 50$
$\quad \{(-5, -20)\}$

Solve each system of equations by any method.

3. $\dfrac{x}{4} - \dfrac{y}{3} = \dfrac{-5}{12}$

$\quad \dfrac{x}{10} + \dfrac{y}{5} = \dfrac{1}{2}$ $\{(1, 2)\}$

4. $\dfrac{x}{2} = y + 3$

$\quad x = 6 + \dfrac{y}{4}$ $\{(6, 0)\}$

19. Sarah purchased two picture frames. The perimeter of the larger frame is 240 cm, and the perimeter of the smaller frame is 140 cm. The height of the smaller frame is the same as the width of the larger frame, and the width of the smaller frame is 10 cm less than the height of the smaller frame. Find the dimensions of the larger frame. 40 cm by 80 cm

20. A video disk rental company charges a fixed amount for the first two days of a rental and an additional charge for each day thereafter. Juan paid $19.50 for a disk that he kept six days, and Tom paid $27.00 for a disk that he kept for nine days. How much would it cost to rent a video disk for ten days from the same company? $29.50

C 21. Erik has 50 coins, all nickels, dimes, and quarters. They are worth $3.60 altogether. There are two more dimes than quarters, and four times as many nickels as the combined number of dimes and quarters. How many of each kind of coin does Erik have? 40 nickels, 6 dimes, 4 quarters

22. The sum of the present ages of Pam, Sue, and Jim is 63 years. The age of Pam three years ago is the same as Jim's age two years from now. The sum of Pam's present age and the ages of Sue and Jim three years ago is one more than twice the age that Pam will be in two years. How old are Pam, Sue, and Jim now?
Pam: 26 years; Sue: 16 years; Jim: 21 years

ON THE CALCULATOR

Using either the substitution or linear-combination method, you can verify that the solution of the system

$$ax + by = c$$
$$dx + ey = f$$

is

$$\left(\frac{ce - bf}{ae - bd}, \frac{af - cd}{ae - bd} \right),$$

provided that $ae - bd \neq 0$. Sometimes using these expressions can simplify the process of solving a system of equations. In particular, when the values of a, b, c, d, e, and f are decimals, it may be easier to solve the system by using a calculator to evaluate these expressions.

Solve each system using a calculator.

1. $2.3x + 4.2y = 0.24$
$\quad 1.2x + 1.4y = 0.6$ $\{(1.2, -0.6)\}$

2. $3.4x + 5.6y = 39$
$\quad 0.7x + 0.93y = 5.8$ $\{(-5, 10)\}$

3. $4.9x - 8.3y = 0.4$
$\quad 3.2x + 5.6y = 21.2$
$\quad \{(3.3, 1.9)\}$

4. $4.3x + 3.8y = 9.8$
$\quad 1.8x - 7.2y = 25.2$ $\{(4.4, -2.4)\}$

6–6 Wind and Water Current Problems

Teaching Suggestions
p. T90

Related Activities p. T90

In solving motion problems about airplanes flying with or against the wind, you may need to know the meaning of the following terms.

air speed   the speed of the airplane in still air

wind speed   the speed of the wind

tail wind   a wind blowing in the same direction as the one in which the airplane is heading

head wind   a wind blowing in the direction opposite to the one in which the airplane is heading

ground speed   the speed of the airplane relative to the ground

     with a tail wind: ground speed = air speed + wind speed

     with a head wind: ground speed = air speed − wind speed

Common Errors

Remind students that it may be necessary to convert some of the given information in a problem so the measurement units are consistent throughout.

EXAMPLE   With a tail wind, a light plane can fly 720 km in 2 h. Going against the wind, the plane can fly the same distance in 3 h. What are the wind speed and the air speed of the plane?

Key Ideas

Solve motion problems involving wind speed and water currents.

SOLUTION

Step 1   The problem asks for the wind speed and the air speed of the plane.

Step 2   Let x = the air speed of the plane in kilometers per hour.

     Let y = the wind speed in kilometers per hour.

	ground speed r (km/h)	time t (h)	distance $d = rt$ (km)
With a tail wind	$x + y$	2	$2(x + y)$
With a head wind	$x - y$	3	$3(x - y)$

Step 3   The plane can fly 720 km in 2 h with a tail wind.
$$2(x + y) = 720$$

The plane can fly 720 km in 3 h with a head wind.
$$3(x - y) = 720$$

Step 4   You can find x and y by solving the following system of equations.
$$2(x + y) = 720 \longrightarrow x + y = 360$$
$$3(x - y) = 720 \longrightarrow x - y = 240$$

Completing Step 4 and checking your results (Step 5) are left to you. You should find that the air speed of the plane is 300 km/h and that the wind speed is 60 km/h.

Chalkboard Examples

Solve.

1. Traveling downstream, Tom can go 6 km in 45 min. Returning, the trip takes 1.5 h. What is the boat's speed in still water and the rate of the current?

Let x = the speed of the boat in still water.

Let y = the rate of the current.

$$0.75(x + y) = 6$$
$$\underline{1.5(x - y) = 6}$$
$$x + y = 8$$
$$\underline{x - y = 4}$$
$$2x = 12$$
$$x = 6$$
$$0.75(6 + y) = 6$$
$$6 + y = 8$$
$$y = 2$$

∴ the boat's speed in still water is 6 km/h, and the rate of the current is 2 km/h.

(continued)

2. Flying against a head-wind, Steve took 5.5 h to fly 2750 km to Montreal. With no change in the wind, the return flight took 5 h. What is the wind speed and the speed of the plane?
Let x = the air speed of the plane.
Let y = the wind speed.

$$5(x + y) = 2750$$
$$5.5(x - y) = 2750$$
$$\overline{}$$
$$x + y = 550$$
$$x - y = 500$$
$$\overline{}$$
$$2x = 1050$$
$$x = 525$$
$$5(525 + y) = 2750$$
$$y = 25$$

∴ the wind speed is 25 km/h and the plane's air speed is 525 km/h.

Suggested Assignments

Minimum
First day: 290/P: 1–5
Second day: 291/P: 6–10
            S 279/13–15
Average
First day: 290/P: 1–6
Second day: 291/P: 7–12
            S 282/20, 22, 24
Maximum
   290/P: 1–7 odd, 9–15

Additional A Exercises

Solve.

1. Flying with a tail wind, a plane took 30 min to complete a 280 km flight. With no change in the wind conditions, the return trip took 10 min longer. What was the wind speed and the air speed of the plane? wind speed: 70 km/h; plane's air speed: 490 km/h

The method of solution used in the example on the preceding page can also be used to solve problems involving boats moving in a current.

Oral Exercises

In Exercises 1–4, Sharon rows at the rate 5 km/h in still water, and the rate of the current is 3 km/h.

1. How fast does Sharon move rowing upstream? 2 km/h
2. How fast does Sharon move rowing downstream? 8 km/h
3. How long would it take her to row 8 km upstream and 8 km back? 5 h
4. What would happen if she tried to row upstream in a current flowing at 5.1 km/h?
   She would go downstream at 0.1 km/h.

In Exercises 5–8, the plane's air speed is 500 km/h.

5. What is the plane's ground speed on a windless day? 500 km/h
6. What is its ground speed if it has a 25 km/h tail wind? 525 km/h
7. What is its ground speed when it encounters head winds of 50 km/h? 450 km/h
8. a. How long will the plane take to fly 1200 km with a 100 km/h tail wind? 2 h
   b. How long will the plane take to fly 1200 km with a 100 km/h head wind? 3 h

Problems

Solve.

wind speed: 50 km/h; air speed: 550 km/h

A 1. Flying against a head wind, a plane could fly 3000 km in 6 h. The plane would require only 5 h for the return trip with no change in the wind. Find the wind speed and the air speed of the plane.

2. The air speed of a plane was 132 km/h. Flying with the wind, the plane traveled twice the distance in 5 h as it traveled against the wind in 3 h. What was the wind speed? 12 km/h

3. With a tail wind, a plane flew 240 km in 45 min. With no change in the wind, the return trip took 48 min. Find the wind speed and the air speed of the plane. air speed: 310 km/h; wind speed: 10 km/h

4. A boat travels 4 km in 20 min with the current. The return trip takes 24 min. Find the speed of the current and the speed of the boat in still water. speed of boat: 11 km/h; speed of current: 1 km/h

5. Rose took a half hour to row 3 km with the current. When she returned, she took 90 min. Find her rowing rate and the speed of the current. rowing rate: 4 km/h; speed of current: 2 km/h

290   Chapter 6

6. A sea lion swims 18 km from one feeding ground to another in 27 min. The return trip against the ocean current takes 36 min. How fast does the sea lion swim in still water? 35 km/h

7. Steve flew his experimental plane 56.25 km with the wind in 45 min. The return trip took 75 min with no change in the wind. What was the wind speed? 15 km/h

8. Walking down a long moving escalator, Phil covered the 75 m distance in 25 s. Walking back up against the motion of the escalator, the distance was covered in 75 s. What was the speed of the escalator? 1 m/s

B 9. When Anthony went bass fishing, he rowed 4 km against the current in 2 h. On his return trip, he still had 1 km to go to his starting point after he had been traveling for 0.5 h. Find Anthony's rate of rowing in still water and the speed of the current. rate in still water: 4 km/h; rate of current: 2 km/h

10. Jennifer can paddle a certain distance with the current in 2.5 h. To cover the same distance against the current, she takes 5 h. How many times faster is her rate of paddling in still water than the speed of the current? 3 times as fast

11. A sightseeing cruiser takes the same time to sail a certain distance up a river as it takes to sail three times that distance down the river. If the speed of the cruiser is s and that of the current is c, find the relationship between s and c. $s = 2c$

12. Jack's plane will normally cover 52 km in 12 min in still air. On one trip, he flew with the wind for 2.5 h. Returning, he still had 100 km to go after 3.5 h. What was the wind speed? 60 km/h

C 13. Ann regularly swam 0.4 km in 20 min at the school pool. Swimming in a river against the current, she swam 0.25 km in the same time that she swam 0.75 km with the current. Find the speed of the current and the time it took Ann to swim 0.75 km downstream. 0.6 km/h; 25 min

14. Walking at 4 km/h, Bruce can make the round trip between his campsite and Lookout Point in 2.5 h. Rowing on Crooked River, he can row upstream from the campsite to Lookout Point in 1 h and can row back again in 40 min. Find Bruce's rate of rowing in still water and the speed of the current in Crooked River. 6.25 km/h; 1.25 km/h

15. Flying against the wind, Lisa made the flight to Reno in 3 h with a steady air speed of 400 km/h. Returning later with a tail wind that had doubled in magnitude, she landed 225 km beyond the starting point in 3 h. What was the original wind speed? How far away from Reno did she begin the trip? original wind speed: 25 km/h; distance from Reno: 1125 km

Systems of Open Sentences **291**

2. Chris completed the 24 km downstream leg of a canoe race in 3 h. If the return trip took 4 h, how fast can she paddle in still water? 7 km/h

3. Swimming 32 km upstream takes a salmon 8 h. Returning the same distance takes only 2 h. What is the speed of the current and the fish's speed in still water? speed of the current: 6 km/h speed of the fish: 10 km/h

4. A plane takes $2\frac{1}{2}$ hours for a 1000 km flight against a strong head wind. A return trip with no change in the wind takes 2 h. What is the plane's air speed and the wind speed? plane's air speed: 450 km/h; wind speed: 50 km/h

---

Mixed Review

Let r = average rate, t = time, and d = distance. Find the quantity indicated using the given information.

1. $r = 32$ km/h, $t = 4$ h, $d =$ _?_ 128 km

2. $r =$ _?_, $t = 2.5$ min, $d = 40$ m 16 m/min

3. $r = 90$ km/h, $t =$ _?_, $d = 585$ km 6.5 h

4. $r = 50$ km/h, $t = 45$ min, $d =$ _?_ 37.5 km

5. Solve $1.5x = 2.3 - 3.1x$. {0.5}

6. Solve the system:
$3x + 2y = 14$
$4x - 2y = 14$ {(4, 1)}

Supplementary Material

Test 21

Key Ideas

Solve digit problems using two variables.

Chalkboard Examples

Solve.

1. The sum of the digits of a two-digit number is 11. The value of the number is 46 times the units' digit. What is the number?
Let t = the tens' digit. Let u = the units' digit.

$$t + u = 11$$
$$10t + u = 46u$$
$$t = 11 - u$$
$$10(11 - u) + u = 46u$$
$$110 - 10u + u = 46u$$
$$110 = 55u$$
$$u = 2$$
$$t + (2) = 11$$
$$t = 9$$

∴ the number is 92.

2. The sum of the digits of a two digit number is 9. The number with the digits reversed is 45 more than the original number. What is the original number?
Let t = the tens' digit. Let u = the units' digit.

$$t + u = 9$$
$$10u + t = 45 + 10t + u$$
$$u = 9 - t$$
$$9u = 45 + 9t$$
$$9(9 - t) = 45 + 9t$$
$$t = 2$$
$$u = 7$$

∴ the number is 27.

6–7 Digit Problems

Using a system of two equations in two variables is often a good way to solve problems involving two-digit numbers.

You can write a two-digit decimal number like 83 in expanded form as follows.

$$\overset{\text{tens' digit}}{\underset{\downarrow}{10}} \cdot \overset{}{8} + \overset{\text{units' digit}}{\underset{\downarrow}{3}}$$

In fact, any two-digit decimal number can be written in the expanded form

$$\overset{\text{tens' digit}}{\underset{\downarrow}{10t}} + \overset{\text{units' digit}}{\underset{\downarrow}{u}},$$

where $t \in \{1, 2, 3, 4, 5, 6, 7, 8, 9\}$ and $u \in \{0, 1, 2, 3, 4, 5, 6, 7, 8, 9\}$.

To represent a number with the same digits in reverse order, you write

$$10u + t.$$

In each case, the sum of the values of the digits is represented by $t + u$.

EXAMPLE 1 A catalog clerk mistakenly reversed the two digits in the price of a radio fuse and overcharged the customer 36¢. If the sum of the digits was 14, what was the correct price?

SOLUTION

Step 1   The problem asks for the correct price of the fuse.

Step 2   Let t = the tens' digit of the correct price.
Let u = the units' digit of the correct price. Then:
$10t + u$ = the correct price;
$10u + t$ = the price mistakenly charged

Step 3   The price charged was 36¢ more than the correct price.

$$10u + t = 10t + u + 36$$
$$9u - 9t = 36$$
$$u - t = 4$$

$\left\{\begin{array}{l}\text{Transform the first equation} \\ \text{into an equivalent equation in} \\ \text{the same variables.}\end{array}\right.$

The sum of the digits was 14.

$$u + t = 14$$

Step 4   You can find u and t by solving the following system of equations.

$$\begin{array}{l} u - t = 4 \\ \underline{u + t = 14} \\ 2u = 18 \\ u = 9 \end{array}$$

Then: $u + t = 14$
$9 + t = 14$
$t = 5$

Step 5   When you reverse the digits of 59, is the result 36 more than 59?

$$95 \stackrel{?}{=} 59 + 36$$
$$95 = 95 \ \checkmark$$

∴ the correct price of the fuse was 59¢.

In Section 6-5, you saw that the same problem could be solved using either two variables and two equations or one variable and one equation. In the following example, you will see a problem that can be solved using either three variables and three equations or two variables and two equations.

EXAMPLE 2 The sum of the digits of a three-digit number is 15. The units' digit is one more than four times the hundreds' digit. The tens' digit is two times the hundreds' digit. Find the number.

SOLUTION 1 (Using three variables)

Step 1   The problem asks for a three-digit number the sum of whose digits is 15.

Step 2   Let h = the hundreds' digit.
       Let t = the tens' digit.
       Let u = the units' digit.

Step 3   The sum of the digits is 15.

$$h + t + u = 15$$

The units' digit is one more than four times the hundreds' digit.

$$u = 1 + 4h$$

The tens' digit is two times the hundreds' digit.

$$t = 2h$$

Step 4   Solve the system: $h + t + u = 15$
             $u = 1 + 4h$
             $t = 2h$

In this case, the expressions for t and u can be substituted into the first equation.

$h + t + u = 15$
$h + (2h) + (1 + 4h) = 15$ Then: $u = 1 + 4h$ and $t = 2h$
      $7h + 1 = 15$ $u = 1 + 4(2)$ $t = 2(2)$
        $h - 2$ $u - 9$ $l = 4$

Step 5   Checking that 249 is the solution is left to you.
       ∴ the number is 249.

On the following page, the same problem is solved using two variables and two equations.

3. The sum of the digits of a three-digit number is 12. The tens' digit is two times the hundreds' digit. The sum of the units' and tens' digits is three times the hundreds' digit. Find the number.
Let u = the units' digit.
Let h = the hundreds' digit. Then:
$2h$ = the tens' digit.
$h + 2h + u = 12$
$\underline{u + 2h = 3h}$
   $u = h$
   $h = 3$
   $u = 3$
   $2h = 6$
∴ the number is 363.

SOLUTION 2 (Using two variables)

Step 1   The problem asks for a three-digit number the sum of whose digits is 15.

Step 2   Let h = the hundreds' digit.

Let t = the tens' digit.

Then $15 - (h + t)$ = the units' digit.

Step 3   The units' digit is one more than four times the hundreds' digit.

$$15 - (h + t) \overset{?}{=} 1 + 4h$$

The tens' digit is two times the hundreds' digit.

$$t \overset{?}{=} 2h$$

Step 4   Solve the system:

$$15 - (h + t) = 1 + 4h \longrightarrow \quad 5h + t = 14$$
$$t = 2h \longrightarrow \quad -2h + t = 0$$
$$\overline{\quad 7h \quad\quad = 14}$$
$$h = 2$$
$$-2h + t = 0$$
$$-2(2) + t = 0$$
$$t = 4$$

The units' digit is $15 - (h + t) = 15 - (2 + 4) = 9$.

Step 5   Checking that 249 is the solution is left to you.

$\therefore$ the number is 249.

You will learn to solve more complicated systems of equations in three variables in Section 6-10. In general, the problems in this section should be attempted using two variables.

The following table lists some *divisibility tests* that you may have learned in an earlier course.

Divisibility by	Test
2	Units' digit of number must be 0, 2, 4, 6, or 8.
3	Sum of digits must be divisible by 3.
4	Number formed by last two digits must be divisible by 4.
5	Units' digit of number must be 0 or 5.
6	Number must be divisible by both 2 and 3.
8	Number formed by last three digits must be divisible by 8.
9	Sum of digits must be divisible by 9.

These divisibility tests have their basis in the structure of the decimal system of numeration. They can be demonstrated by using some basic properties of the real numbers, such as the commutative, distributive, and associative axioms.

EXAMPLE 3 Show that a three-digit number is divisible by 3 if and only if the sum of its digits is divisible by 3.

SOLUTION Let h = the hundreds' digit.

Let t = the tens' digit.

Let u = the units' digit.

Then $100h + 10t + u$ represents a three-digit number.

Notice that the expression $100h + 10t + u$ can be rewritten as follows.

$$100h + 10t + u = (99h + h) + (9t + t) + u$$
$$= (99h + 9t) + (h + t + u)$$
$$= 3(33h + 3t) + (h + t + u).$$

Since $3(33h + 3t)$ is divisible by 3, the entire right side is divisible by 3 if and only if the sum of the digits, $h + t + u$, is divisible by 3.

∴ a three-digit number is divisible by 3 if and only if the sum of its digits is divisible by 3.

Oral Exercises

A two-digit number has tens' digit t and units' digit u. Express the following in terms of t and u.

1. the sum of the digits $t + u$
2. the value of the two-digit number $10t + u$
3. the value of the two-digit number obtained by reversing the digits $10u + t$

A three-digit number has hundreds' digit h, tens' digit t, and units' digit u. Express the following in terms of h, t, and u.

4. the sum of the digits $h + t + u$
5. the value of the three-digit number $100h + 10t + u$
6. the value of the three-digit number obtained by reversing the digits $100u + 10t + h$

Use the divisibility tests given on page 294 to answer the following.

7. Is 4270 divisible by 2? Yes
8. Is 3130 divisible by 3? No
9. Is 39,495 divisible by 3? Yes
10. Is 587,646 divisible by 4? No
11. Is 38,750 divisible by 5? Yes
12. Is 45,678 divisible by 6? Yes
13. Is 469,152 divisible by 8? Yes
14. Is 363,927 divisible by 9? No

Systems of Open Sentences **295**

Suggested Assignments

Minimum
  296/*P*: 1–10
R 297/*Self-Test 2*
Average
  296/*P*: 1–14
R 297/*Self-Test 2*
Maximum
  296/*P*: 9–20
R 297/*Self-Test 2*

Mixed Review

Solve.
1. $|x| - 4 > -1$
   $\{x: x > 3 \text{ or } x < -3\}$
2. $|a - 2| + 3 = 6$ $\{-1, 5\}$
Solve each system of equations using any method.
3. $x - 3y = 7$
   $2x - y = 14$ $\{(7, 0)\}$
4. $3x + y = 7$
   $2x - 5y = -1$ $\{(2, 1)\}$

1. The units' digit of a two digit number is four times the tens' digit. The sum of the digits is 5. What is the number? 14

2. The sum of the digits of a two-digit number is 8. The number with its digits reversed is 7 times the original tens' digit. What is the original number? 53

3. The sum of the digits of a two-digit number is 10. The tens' digit is 30 less than 7 times the units' digit. What is the number? 55

4. Sixteen more than five times the units' digit of a two-digit number is twice the tens' digit. The sum of the digits is 8. What is the number? 80

5. The units' digit of a three-digit number is three times the hundreds' digit. The tens' digit is 1. If the sum of the digits is 9, what is the number? 216

6. If the digits of a three-digit number are reversed, the new number is 198 less than the original number. If the hundreds' digit is 3, and three less than the tens' digit is five times the units' digit, what is the original number? 381

___

Additional Answers
Problems

13. $100h + 10t + u$ may be rewritten $99h + h + 9t + t + u$, or $9(11h + t) + h + t + u$. Since $9(11h + t)$ is divisible by 9, the entire number is divisible by 9 if and only if $h + t + u$ is divisible by 9.

Problems

Solve.

A

1. The sum of the digits of a two-digit number is 10. The value of the number is 16 times the units' digit. Find the number. 64

2. The sum of the digits of a two-digit number is 12. The number with the digits reversed is 15 times the original tens' digit. Find the original number. 57

3. The sum of the digits of a two-digit number is 4. If the order of the digits is reversed, the result is a number that exceeds the original number by 18. Find the original number. 13

4. The sum of the digits of a two-digit number is 11. The number obtained by reversing the order of the digits is 27 less than the original number. Find the original number. 74

5. The tens' digit of a two-digit number exceeds the units' digit by 2. The sum of the tens' digit and twice the units' digit is 17. Find the number. 75

6. The units' digit of a two-digit number exceeds three times the tens' digit by 3. If the tens' digit is subtracted from the units' digit, the difference is 7. Find the number. 29

7. If the tens' digit of a two-digit number is subtracted from the units' digit, the difference is 8. The number with the digits reversed is 10 more than nine times the units' digit of the original number. Find the number. 19

8. The sum of the digits of a two-digit number is 15. The number with the digits reversed is 30 more than six times the tens' digit of the original number. Find the number. 87

9. The sum of the digits of a three-digit number is 17. The units' digit is 6. If the order of the digits is reversed, the result is a number that is 297 more than the original number. What is the original number? 386

10. The tens' digit of a three-digit number is 5. The units' digit is twice the hundreds' digit. If the order of the digits is reversed, the new number is 396 more than the original number. What is the original number? 458

B

11. The sum of the digits of a three-digit number is 14. The hundreds' digit is four times the tens' digit and twice the units' digit. Find the number. 824

12. The sum of the digits of a three-digit number is 13. The hundreds' digit is two more than the tens' digit, and the tens' digit is four more than the units' digit. Find the number. 751

13. Show that a three-digit number is divisible by 9 if and only if the sum of the digits is divisible by 9.

14. Show that a four-digit number is divisible by 4 if and only if the number formed by the last two digits is divisible by 4.

15. Show that the sum of a two-digit number and the number with the order of the digits reversed is always divisible by 11.

16. Show that the difference between a three-digit number and the number with the order of the digits reversed is always divisible by 99.

C 17. The hundreds' digit of a three-digit number is 2 more than twice the tens' digit. The tens' digit is 1 less than twice the units' digit. If the order of the digits is reversed, the number obtained is 594 less than the original number. Find the original number. 832

18. The hundreds' digit of a three-digit number is one more than twice the units' digit. The units' digit is three more than the tens' digit. If the order of the digits is reversed, the number obtained is 396 less than the original number. Find the original number. 703

19. Show that the difference between a two-digit number and the number obtained when the tens' digit is decreased by 1 and the units' digit is increased by 1 is always 9.

20. Find all three-digit numbers, if any, that satisfy *all* of the requirements (a) through (c). 863
    a. The units' digit is one half the tens' digit.
    b. The hundreds' digit is 1 less than 3 times the units' digit.
    c. The difference between the number and the number obtained when the order of its digits is reversed is 495.

■ Self-Test 2

Solve.

1. Five years ago, Jake was five times as old as Jenny was at that time. Four years from now, Jake will be twice as old as Jenny will be. Find Jenny's age now. 8 years   *Obj. 1, p. 284*

2. Bridget needed 1 h to row 4 km upstream, but only 30 min to row the same distance back downstream. Find Bridget's rowing rate in still water and the speed of the current. rowing rate: 6 km/h; rate of current: 2 km/h   *Obj. 2, p. 284*

3. The sum of the digits of a two-digit number is 9. When the digits are reversed, the new number is 63 less than the original number. Find the original number. 81   *Obj. 3, p. 284*

Check your answers with those at the back of the book.

Systems of Open Sentences **297**

14. Let s = the thousands' digit. Then a four-digit number, $1000s + 100h + 10t + u$, may be rewritten $4(250s + 25h) + (10t + u)$. Since $4(250s + 25h)$ is divisible by 4, the entire number is divisible by 4 if and only if $10t + u$ is divisible by 4.

15. $10t + u + 10u + t = 11t + 11u = 11(t + u)$, which is divisible by 11.

16. $100h + 10t + u - (100u + 10t + h) = 99h - 99u = 99(h - u)$, which is divisible by 99.

19. $10t + u - [10(t - 1) + (u + 1)] = 10t + u - (10t - 10 + u + 1) = 9$

Quick Quiz

Solve.

1. Seven years ago, Stan was five times Pam's age at that time. In 8 years, Stan will be twice Pam's age at that time. How old is each of them now? Stan: 32 yr; Pam: 12 yr

2. A plane flew 750 km to Miami in 3 h with the wind. Returning against the wind, the trip took 5 hours. What was the plane's air speed and the wind speed? plane's air speed: 200 km/h; wind speed: 50 km/h

3. The sum of the digits of a two-digit number is 9. The number formed with the digits reversed is six times the tens' digit of the original number. What is the original number? 63

Chalkboard Examples

Graph the solution set of each system of inequalities.

1. $y - x \leq -1$
   $y + x \leq 1$
   $y \leq x - 1$
   $y \leq -x + 1$

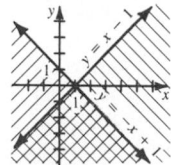

Check with $(1, -1)$.
$(-1) - (1) \stackrel{?}{\leq} -1$
$\qquad -2 \leq -1$ ✓
$(-1) + (1) \stackrel{?}{\leq} 1$
$\qquad 0 \leq 1$ ✓

2. $2x + y \leq 1$
   $x - 2y \leq 2$
   $\qquad x \geq -2$
   $y \leq -2x + 1$

   $y \geq \dfrac{1}{2}x - 1$

   $x \geq -2$

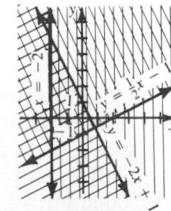

Check with $(0, 0)$.
$\qquad 0 \geq -2$ ✓
$2(0) + (0) \stackrel{?}{\leq} 1$
$\qquad 0 \leq 1$ ✓
$(0) - 2(0) \stackrel{?}{\leq} 2$
$\qquad 0 \leq 2$ ✓

Systems of Inequalities; Linear Equations in Three Variables

OBJECTIVES for Sections 6-8 through 6-10:
1. To graph the solution set of a system of linear inequalities in two variables.
2. To solve linear programming problems that involve two variables.
3. To solve a system of three linear equations in three variables.

6–8 Graphs of Systems of Linear Inequalities

You can use graphs to determine the solution set of a *system of linear inequalities* in two variables.

EXAMPLE 1 Graph the solution set of the following system of inequalities.

$$3x - 2y \leq 4$$
$$x + 2y < 2$$

SOLUTION

1. Transform each inequality into an equivalent inequality with y alone as one side.

   $$3x - 2y \leq 4 \longrightarrow y \geq \tfrac{3}{2}x - 2$$
   $$x + 2y < 2 \longrightarrow y < -\tfrac{1}{2}x + 1$$

2. Draw the graph of $y = \tfrac{3}{2}x - 2$.
   The graph of $y \geq \tfrac{3}{2}x - 2$ is the *closed* half-plane *on* and *above* the line $y = \tfrac{3}{2}x - 2$.

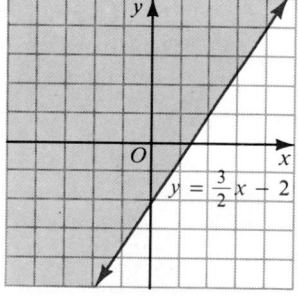

3. Draw the graph of $y = -\tfrac{1}{2}x + 1$.
   The graph of $y < -\tfrac{1}{2}x + 1$ is the *open* half-plane *below* the line $y = -\tfrac{1}{2}x + 1$.

4. The intersection of the half-planes found in Steps 2 and 3 (double shading) is the graph of the given system:

   $$3x - 2y \leq 4$$
   $$x + 2y < 2$$

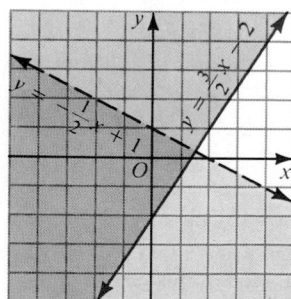

5. Check your work by selecting any point within the double-shaded region. A convenient point to use is $(0, 0)$. This ordered pair should satisfy each of the original inequalities.

298 *Chapter 6*

$$3x - 2y \le 4 \qquad\qquad x + 2y < 2$$
$$3(0) - 2(0) \overset{?}{\le} 4 \qquad 0 + 2(0) \overset{?}{<} 2$$
$$0 \le 4 \ \checkmark \qquad\qquad 0 < 2 \ \checkmark$$

EXAMPLE 2 Graph the solution set of the following system of inequalities.

$$y - x \le 1$$
$$y + x \ge -1$$
$$x - 4 \le 0$$

SOLUTION

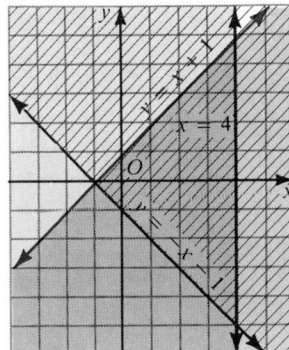

1. The graph of $y - x \le 1$ consists of the points on and below the line $y = x + 1$.
2. The graph of $y + x \ge -1$ consists of the points on and above the line $y = -x - 1$.
3. The graph of $x - 4 \le 0$ consists of the points on and to the left of the vertical line $x = 4$.
4. The graph of the solution set of the given system is the intersection of the three shaded regions. This intersection is represented by all the points on the sides and in the interior of the triangle formed by the three lines.
5. *Check:* Since $(0, 0)$ is in the region representing the solution set, the ordered pair $(0, 0)$ should satisfy each of the original inequalities.

$$y - x \overset{?}{\le} 1 \qquad\qquad y + x \overset{?}{\ge} -1 \qquad\qquad x - 4 \overset{?}{\le} 0$$
$$0 - 0 \overset{?}{\le} 1 \qquad\qquad 0 + 0 \overset{?}{\ge} -1 \qquad\qquad 0 - 4 \overset{?}{\le} 0$$
$$0 \le 1 \ \checkmark \qquad\qquad 0 \ge -1 \ \checkmark \qquad\qquad -4 \le 0 \ \checkmark$$

Oral Exercises

Name a system of two linear inequalities whose solution set is shown by the shaded region in each graph.

1.

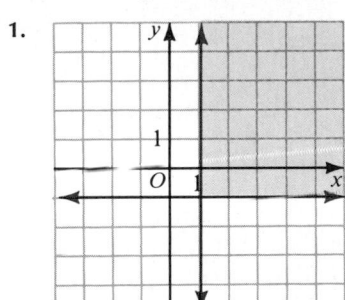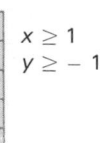

$$x \ge 1$$
$$y \ge -1$$

2.

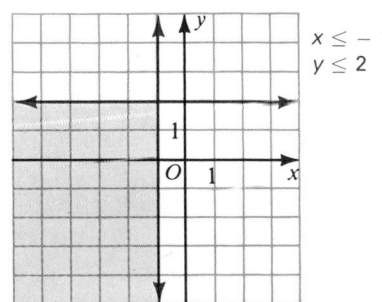

$$x \le -1$$
$$y \le 2$$

Graph the solution set of each system of inequalities.

1. $y \ge x$
$\quad y \ge 2 - x$

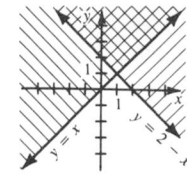

2. $y - 2x > 3$
$\quad y + 2x > 1$

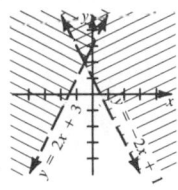

3. $x + y < 3$
$\quad x - y \le 2$

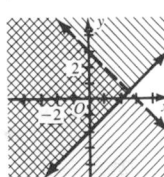

4. $y - 3x \le -2$
$\quad y + 2x > 1$

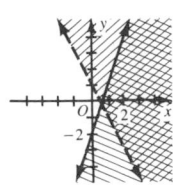

5. $y - 2x \ge 3$
$\quad y + 5x \le -2$

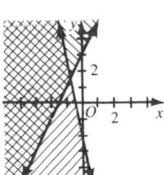

Suggested Assignments

Minimum
   300/1–15 odd
Average
   300/5–15 odd, 16–20
Maximum
   300/1–21 odd
S 290/P: 2–8 even

Mixed Review

Solve for y.

1. $2y + 4 > -5$
   $\{y: y > -4.5\}$
2. $8 - 2y \leq 6$
   $\{y: y \geq 1\}$
3. $3x + 2y \geq 6$
   $\{y: y \geq -\frac{3}{2}x + 3\}$
4. $2x - 3y < -9$
   $\{y: y > \frac{2}{3}x + 3\}$

Graph each inequality on a coordinate plane.

5. $2x - y > 3$

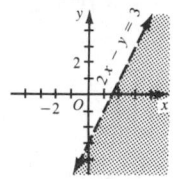

6. $x - 3 \leq y + 2x$

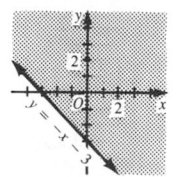

Additional Answers
Written Exercises

1.

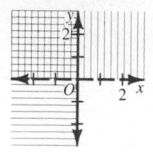

Name a system of two linear inequalities whose solution set is shown by the shaded region in each graph.

3.

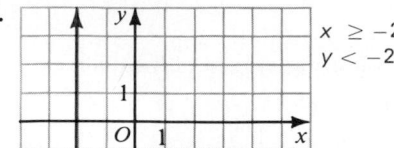

$x \geq -2$
$y < -2$

4.

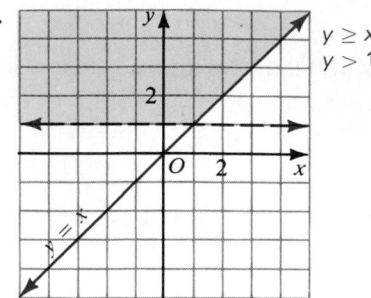

$y \geq x$
$y > 1$

Tell which of the systems in Exercises 1–4, if any, have solution sets that contain the given ordered pair.

                           2, 4   None   1, 4   2, 3

5. $(3, 1)$ 1   6. $(-1, 3)$ 4   7. $(-3, 2)$ 8. $(1, -2)$ 9. $(2, 4)$ 10. $(-2, -3)$

Tell whether or not each point belongs to the graph of the system.

$$x + 2 > 0 \quad \text{and} \quad y \geq x - 2$$

11. $(4, 1)$ 12. $(0, -2)$ 13. $(-3, 0)$ 14. $(-2, 2)$ 15. $(3, 1)$ 16. $(-1, -1)$
   No   Yes   No   No   Yes   Yes

Written Exercises

Graph the solution set of each system of inequalities.

A

1. $y > 0$
   $x \leq 0$
2. $y \leq 0$
   $x > 0$
3. $y < 1$
   $x \geq -1$
4. $y \geq -3$
   $x \leq 2$

5. $y < x$
   $x \geq -1$
6. $y > x$
   $y \leq 3$
7. $y \geq x + 1$
   $y \leq -x + 2$
8. $y \leq x - 2$
   $y > -x - 1$

9. $y + 1 \leq 2x$
   $y + x > -2$
10. $y - 2 > 2x$
    $y - x > 3$
11. $2y - x > -2$
    $-2x - y > -1$
12. $2y + x > 2$
    $3x - y \geq -3$

B

13. $y > -1$
    $x \leq 3$
    $y < x$
14. $x \leq 5$
    $y < 4$
    $y \geq -x$
15. $y + x \geq 0$
    $y - x \leq 1$
    $x - 2 \leq 0$
16. $y + x \leq 1$
    $y - x \leq -1$
    $y + 2 \geq 0$

17. $y - 1 < 2x$
    $y + x > -1$
    $x < 0$
18. $y + 3 > 3x$
    $y + x > -2$
    $y < 3$
19. $3 - y > 0$
    $y + x \geq -2$
    $y + 3 \geq 3x$
20. $y + 2 > 0$
    $x - y \geq -3$
    $y - 2 \leq -5x$

C

21. $y - 2 < 0$
    $x + 3 > 0$
    $2y + x < 2$
    $3y + 3 > 2x$
22. $2y - 6 \leq x$
    $2y + 3x \geq -6$
    $5y + 15 \geq 2x$
    $3y + 5x \leq 15$

PROGRAMMING IN BASIC

Using the program listed on page 247, you were able to print a graph of an open sentence in two variables. By making the following changes in that program, you will now be able to print a graph of *two* open sentences in two variables. In this program, the capital letter "O" is used to identify the point or region that is the graph of the solution set of the system formed by the two open sentences.

```
10    PRINT "TO GRAPH TWO OPEN SENTENCES"
20    PRINT "IN TWO VARIABLES"
30    PRINT "(SENTENCES ARE IN LINES 125-135):"
40  ⎫
 ⋮  ⎬  from program on page 247
120 ⎭
125   IF Y = X + 2 AND Y = 4 − X THEN 258
130   IF Y = X + 2 THEN 250
135   IF Y = 4 − X THEN 254
140 ⎫
 ⋮  ⎬  from program on page 247
250 ⎭
252   GOTO 260
254   PRINT ".";
256   GOTO 260
258   PRINT "O";
260 ⎫
 ⋮  ⎬  from program on page 247
340 ⎭
```

Exercises

1. Type in and RUN the revised program.

Change lines 125–135 as necessary to RUN the program for each of the following systems of open sentences. In each case, INPUT 9 for the extent of the graph.

2. $x = -3$
$y = 4$

3. $y = 4$
$2x + 3y = 6$

4. $x = -3$
$2x - 3y = -18$

5. $2x + 3y = 6$
$2x - 3y = -18$

6. $x > -3$
$y < 4$

7. $y > x + 2$
$y < 4 - x$

8. $x > -3$
$2x - 3y < -18$

9. $2x + 3y > 6$
$2x - 3y < -18$

Systems of Open Sentences **301**

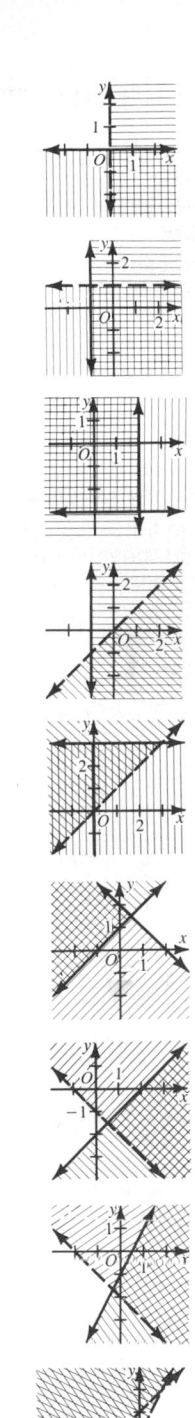

2.

3.

4.

5.

6.

7.

8.

9.

10.

(continued on p. 314)

Chalkboard Examples

Graph the following systems; find the minimum value of the expression $3x + 4y$ over the region.

1. $x \geq 0$
   $y + x \leq 4$
   $2y + x \leq 6$
   $y \geq 0$
   $y \leq -x + 4$
   $y \leq -\frac{1}{2}x + 3$

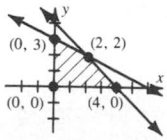

Evaluate $3x + 4y$ at the corner points $(0, 0)$, $(0, 3)$, $(2, 2)$, and $(4, 0)$.

$(0, 0)$: $3(0) + 4(0) = 0$
$(0, 3)$: $3(0) + 4(3) = 12$
$(2, 2)$: $3(2) + 4(2) = 14$
$(4, 0)$: $3(4) + 4(0) = 12$

∴ the minimum value occurs at $(0, 0)$, and is 0.

2. $x \geq 0$
   $y \geq 1$
   $y + 2x \leq 8$
   $y + x \geq 3$
   $y \leq -2x + 8$
   $y \geq -x + 3$

6–9 Linear Programming

Many practical problems in business, science, and industry can be solved using the techniques of *linear programming*. **Linear programming** is a method of solving problems in which a quantity represented by a linear equation, often profit or cost, is to be maximized or minimized subject to conditions expressed by a system of linear inequalities. The following example illustrates this method.

During an illness Gregory supplements his daily diet with vitamin pills. Each day he needs at least 4 mg of thiamine, 6 mg of riboflavin, and 80 mg of niacin. To meet these needs, he can buy either Brand X pills at 3¢ apiece or Brand Y pills at 4¢ apiece. The following table shows the amount of vitamins in one pill of each brand.

	Brand X	Brand Y
Thiamine	1 mg	4 mg
Riboflavin	3 mg	5 mg
Niacin	30 mg	40 mg

What combination of pills will provide his minimum daily needs for the three vitamins at the lowest cost?

Let x = the number of Brand X pills used daily.
Let y = the number of Brand Y pills used daily.
Let C = the daily cost in cents for the pills.

Then: $C = 3x + 4y$

Gregory wants to minimize the cost C subject to the restrictions, or *constraints*, described by the following inequalities.

$\left. \begin{array}{l} x + 4y \geq 4 \\ 3x + 5y \geq 6 \\ 30x + 40y \geq 80 \end{array} \right\}$ The total daily amount of each vitamin must equal at least the daily need.

$\left. \begin{array}{l} x \geq 0 \\ y \geq 0 \end{array} \right\}$ He cannot use a negative amount of either pill.

The graph of the solution set of this system of inequalities is indicated by shading in the diagram. Notice that $3x + 5y \geq 6$ has no effect on the result.

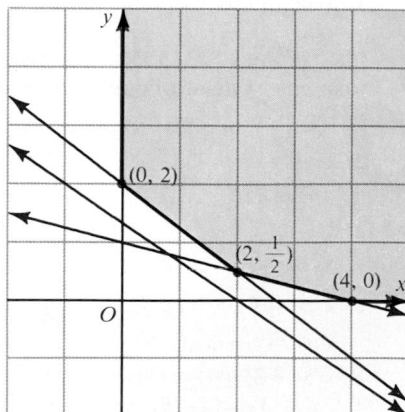

The shaded region of the graph on the preceding page is called the *feasible region*. A feasible region has the following characteristics.

1. Where it is bounded, the boundary is determined by straight lines. The points of the region where such boundary lines intersect are called **corner points.** In this case the corner points are $(0, 2)$, $(2, \frac{1}{2})$, and $(4, 0)$.
2. Every point on the boundary is a part of the region.
3. The region is *convex*.

A region is said to be **convex** if, whenever you choose any two points in the region and draw the segment joining them, the segment is contained in the region. For example, in the following diagram the shaded region at the left is convex, but the shaded region at the right is not.

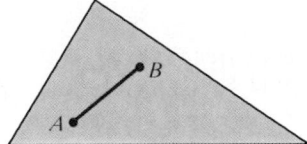 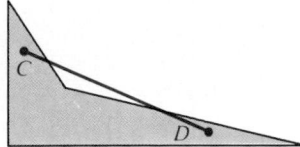

Any plane region that is the intersection of a finite number of closed half-planes has the three characteristics listed above and is called a **convex polygonal region.**

A remarkable result, not proved here, is the following:

> Over a convex polygonal region, any maximum or minimum values of a *linear expression* $ax + by$, where a and b are real numbers, occur for the coordinates of a corner point of the region.

It is possible that maximum or minimum values may also occur at other points besides the corner points. What is important is that by testing values at the corner points, you can, in a finite number of steps, find the greatest and least values, if they exist, of any linear expression over any closed polygonal region.

Thus the minimum value of C can be found by evaluating $3x + 4y$ at each corner point.

$$(0, 2): \quad 3(0) + 4(2) = 8$$
$$(2, \tfrac{1}{2}): \quad 3(2) + 4(\tfrac{1}{2}) = 8$$
$$(4, 0): \quad 3(4) + 4(0) = 12$$

In this case both $(0, 2)$ and $(2, \frac{1}{2})$ yield the minimum value, 8; however, it is inconvenient to take half a pill. Therefore the most economical and practical choice for Gregory is 2 Brand Y pills and no Brand X pills.

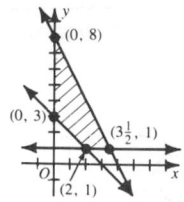

Evaluate $3x + 4y$ at the corner points $(0, 3)$, $(0, 8)$, $\left(3\frac{1}{2}, 1\right)$, and $(2, 1)$.

$$(0, 3): \ 3(0) + 4(3) \ = 12$$
$$(0, 8): \ 3(0) + 4(8) \ = 32$$
$$\left(3\tfrac{1}{2}, 1\right): 3\left(3\tfrac{1}{2}\right) + 4(1) = 14\tfrac{1}{2}$$
$$(2, 1): \ 3(2) + 4(1) \ = 10$$

∴ the minimum value occurs at $(2, 1)$, and is 10.

Additional A Exercises

1. Graph the solution set of the system of inequalities:

$$x \leq 5$$
$$x \geq 0$$
$$y \geq x - 3$$
$$x + y \geq 2$$
$$y \geq 0$$

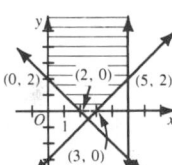

2. Does the system in Exercise 1 form a convex polygonal region? If so, give the coordinates of its corner points. Yes; $(0, 2)$, $(5, 2)$, $(2, 0)$, and $(3, 0)$

3. Find the minimum value of $x + y$ over the region graphed in Exercise 1.
   $(0, 2)$: $1(0) + 1(2) = 2$
   $(5, 2)$: $1(5) + 1(2) = 7$
   $(2, 0)$: $1(2) + 1(0) = 2$
   $(3, 0)$: $1(3) + 1(0) = 3$
   ∴ the minimum value occurs at two points, $(0, 2)$ and $(2, 0)$, and is 2.

Suggested Assignments

Minimum
   304/1–8
Average
   305/1–6, 7, 9, 11–14
Maximum
   304/1–6, 9–16

Mixed Review

Graph the solution set of the following systems of inequalities.

1. $x > -y$
   $y \le 3$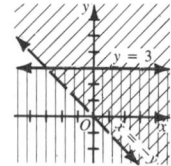

2. $x - y \ge 0$
   $x > 0$
   $y > 2$

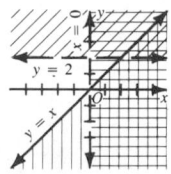

3. $y - 3 \ge x$
      $x \le -1$
   $2y + x \ge 4$

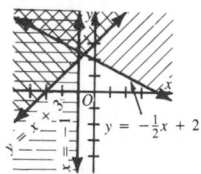

Additional Answers
Written Exercises

1.

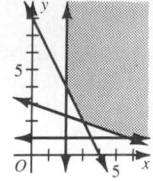

Oral Exercises

Tell whether or not each shaded region is convex.

1.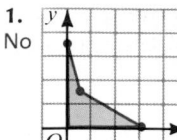
No

2.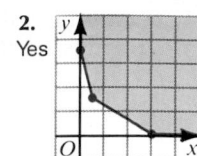
Yes

3.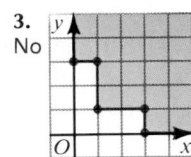
No

4.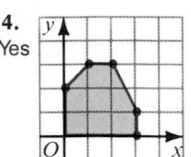
Yes

5. Evaluate the linear expression $3x + 2y$ at each corner point of the convex polygonal region shown in the graph below.

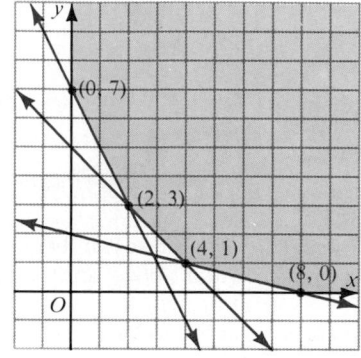

at (0, 7): 14
at (2, 3): 12
at (4, 1): 14
at (8, 0): 24

6. What is the minimum value of $3x + 2y$ over the region shown in the graph above? 12

Written Exercises

Exercises 1–3 refer to the following system of inequalities.

$$y \ge 1$$
$$x \ge 2$$
$$3y + x \ge 9$$
$$y + 2x \ge 8$$

A **1.** Graph the solution set of the system.

2. Determine the corner points of the region that represents the solution set. (2, 4), (3, 2), (6, 1)

3. Find the minimum value of each of the following linear expressions over the region graphed in Exercise 1 by evaluating the expression at each of the corner points.

   a. $x + 5y$ 11   **b.** $5x + y$ 14   **c.** $x + y$ 5

304   *Chapter 6*

Exercises 4–6 refer to the following system of inequalities.

$$x \geq 1$$
$$y \leq 6$$
$$y \geq -x + 4$$
$$2y \geq x - 1$$
$$3y \leq -x + 21$$

4. Graph the solution set of the system and label the corner points with their coordinates.

5. Find the minimum value of the expression $x + 5y$ over the region graphed in Exercise 4. 8

6. Find the maximum value of the expression $4x - 3y$ over the region graphed in Exercise 4. 24

In Exercises 7–10:

a. **Graph the solution set of the system of inequalities and label the corner points with their coordinates.**

b. **Find the minimum value of the expression $2x + 3y$ over the region graphed.**

7. $x \geq 0$ b. 8
    $y \geq -3x + 5$
    $y \geq -\frac{2}{3}x + \frac{8}{3}$
    $y \geq 0$

8. $3y + 4x \geq 18$ b. 12
    $3y + x \geq 9$
    $x - 1 \geq 0$
    $y \geq 0$

9. $y \geq 0$ b. 9
      $x \geq 0$
    $y + 4x \geq 7$
    $y + x \geq 4$
    $3y + x \geq 6$

10. $y \geq 0$ b. 18
      $x \geq 0$
    $y + 5x \geq 13$
    $y + 2x \geq 10$
    $3y + 2x \geq 18$

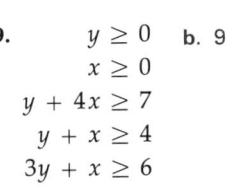

In Exercises 11–14, use the following information: The owners of a pet store plan to purchase some hamsters and some rabbits. They decide to spend at least $100 and purchase at least eight animals. Each hamster costs $5 and each rabbit costs $25.

B 11. Introduce variables for the number of hamsters and the number of rabbits to be purchased. Write four inequalities that express the conditions of the problem. (Remember that the number of each kind of animal must be at least zero.)

12. Graph the solution set of the system of inequalities. Determine the coordinates of the corner points.

13. Suppose that it costs 13¢ per day to feed each hamster and 30¢ per day to feed each rabbit. If daily feeding costs are to be held to a minimum, how many hamsters and rabbits should be purchased?
5 hamsters and 3 rabbits

Systems of Open Sentences **305**

4.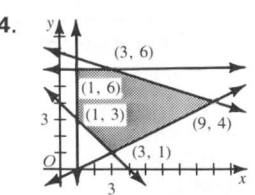

7. a.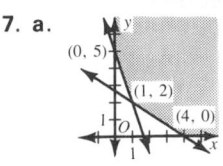

8. a.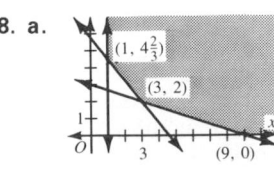

9. a.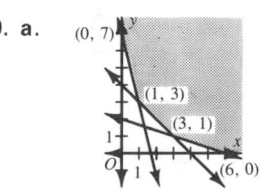

10. a.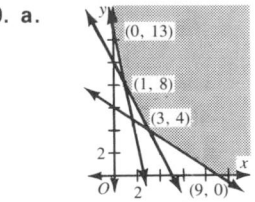

11. Let x = number of hamsters.
Let y = number of rabbits.
$$x \geq 0$$
$$y \geq 0$$
$$x + y \geq 8$$
$$5x + 25y \geq 100$$

12.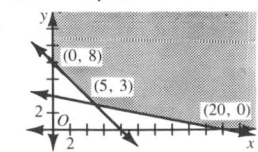

(continued)

14. 5 hamsters, 3 rabbits;
     10 hamsters, 2 rabbits;
     15 hamsters, 1 rabbit;
     20 hamsters, 0 rabbits
15. Shop *A*: 5 hours;
     Shop *B*: 6 hours
16. **a.** The minimum ex-
     pense, $98.55, occurs
     when the old machine is
     operated for 6 h and the
     new machine is operated
     for 3 h.
     The use of both
     machines would save
     approximately $6.97
     over the use of the old
     machine alone.
     b. Yes, since the min-
     imum expense, $101.55,
     would still occur when
     the old machine is op-
     erated for 6 h and the
     new machine for 3 h.

Teaching Suggestions
p. T91

Related Activities p. T91

Supplementary Material
Test 22

Key Ideas
Solve systems of three linear
equations in three variables.

14. Suppose that the cost of feeding each animal changes before the owners have purchased the animals. It now costs 14¢ per day to feed each hamster and 70¢ per day to feed each rabbit. The owners are surprised to find there are now several ways to make their selection while still meeting the original conditions and still keeping the daily feeding costs to a minimum. Give four possible choices that the owners have in making their selection.

C 15. A manufacturer has two plastic molding shops and must produce at least 1500 model cars and 900 model planes each day. Shop A can make 120 cars and 60 planes per hour. Shop B can make 150 cars and 100 planes per hour. The cost of running the machine in Shop A is $2.55 per hour, and the cost of running the machine in Shop B is $3.20 per hour. How many hours should each machine be used every day to minimize the costs?

16. Each day Central Post Office handles at least 52,800 letters and 16,800 advertising flyers. The cancelling machine now in use can handle 5500 letters and 1600 flyers per hour at an hourly cost of $10.05. The manufacturer of the machine is about to release an improved model which can do 6600 letters and 2400 flyers each hour at a cost of $12.75 per hour. Neither machine can be operated for more than 12 hours due to internal heating.

   a. How would the expense of handling the mail be affected if the post office could use both machines?

   b. Would the use of both machines be practical if the post office increased the cost of operating the old machine by 50¢ an hour?

6–10 Systems of Linear Equations in Three Variables

You have seen that a solution of an equation in *one* variable, such as

$$2x + 1 = 0,$$

is a *single* real number, and that a solution of an equation in *two* variables, such as

$$3x - 5y = 3,$$

is an *ordered pair* of real numbers.

   A **solution** of an equation in *three* variables, such as

$$x + 2y - 4z = 9,$$

is an **ordered triple** (x, y, z) of real-number values of the variables for which the equation is a true statement. The set of all such ordered triples is the **solution set** of the equation.

306   *Chapter 6*

EXAMPLE 1 Which of the following ordered triples are solutions of the equation $2x - y + 3z = 11$?

    a. $(3, 1, 2)$ **b.** $(2, 2, 3)$ **c.** $(1, 3, 2)$

SOLUTION Check each ordered triple in the original equation.

    a. $2(3) - 1 + 3(2) \stackrel{?}{=} 11$
                  $11 = 11$ ✓

    b. $2(2) - 2 + 3(3) \stackrel{?}{=} 11$
                  $11 = 11$ ✓

    c. $2(1) - 3 + 3(2) \stackrel{?}{=} 11$
                  $5 \neq 11$

$\therefore$ $(3, 1, 2)$ and $(2, 2, 3)$ are solutions.

Any equation of the form

$$ax + by + cz = d,$$

where a, b, c, and d are real numbers and a, b, and c are not all zero, is called a **linear equation in three variables.** (The graph of such an equation is a plane in space.) You can find as many solutions as you wish for such equations by choosing values for two of the variables and determining the corresponding value of the third variable.

EXAMPLE 2 Find three solutions of the equation $3x - 2y + z = 10$.

SOLUTION 1. Transform the given equation into an equivalent equation with z alone as one side.

$$3x - 2y + z = 10 \longrightarrow z = 10 - 3x + 2y$$

2. Substitute any values for x and y into the equation. Then solve the equation for z.

x	y	$z = 10 - 3x + 2y$	z
0	0	$z = 10 - 3(0) + 2(0)$	10
1	0	$z = 10 - 3(1) + 2(0)$	7
0	1	$z = 10 - 3(0) + 2(1)$	12

$\therefore$ three possible solutions are $(0, 0, 10)$, $(1, 0, 7)$, and $(0, 1, 12)$.

In Sections 6-2 through 6-4 you learned to solve systems of two linear equations in two variables using algebraic methods. Similar methods can be used to solve systems of three linear equations in three variables, since the transformations listed on page 281 are also valid for these systems. A **solution of a system** of three equations in three variables is an ordered triple that satisfies each equation in the system. The **solution set of the system** is the set of all such triples.

Systems of Open Sentences   **307**

EXAMPLE 3 Solve the system:
$$\begin{aligned} 2x - y + z &= -8 \quad (1) \\ 2x + 3y - 2z &= 7 \quad (2) \\ x + 2y + 3z &= 1 \quad (3) \end{aligned}$$

SOLUTION 1. Subtract equation (2) from equation (1) to eliminate x.

$$\begin{aligned} 2x - y + z &= -8 \\ 2x + 3y - 2z &= 7 \\ \hline -4y + 3z &= -15 \quad (4) \end{aligned}$$

2. Multiply equation (3) by 2 and subtract the result from equation (1) to eliminate x. Simplify the new equation when possible.

$$\begin{array}{ll} 2x - y + z = -8 \longrightarrow & 2x - y + z = -8 \\ x + 2y + 3z = 1 \longrightarrow & 2x + 4y + 6z = 2 \\ & \hline -5y - 5z = -10 \\ & y + z = 2 \quad (5) \end{array}$$

3. Equation (4) and equation (5) form a system of equations in two variables that can be solved for y and z.

$$\begin{array}{ll} -4y + 3z = -15 \longrightarrow & -4y + 3z = -15 \\ y + z = 2 \longrightarrow & 3y + 3z = 6 \\ & \hline -7y = -21 \\ & y = 3 \end{array}$$

$$\begin{aligned} y + z &= 2 \\ 3 + z &= 2 \\ z &= -1 \end{aligned}$$

4. Substitute 3 for y and -1 for z in one of the original equations. Then solve for x.

$$\begin{aligned} 2x - y + z &= -8 \\ 2x - 3 + (-1) &= -8 \\ 2x &= -4 \\ x &= -2 \end{aligned}$$

5. Checking that $(-2, 3, -1)$ satisfies each of the three original equations is left to you.

∴ the solution set is $\{(-2, 3, -1)\}$.

The method for solving a system of three equations in three variables illustrated in Example 3 involves working with pairs of equations. There are three ways to combine the original three equations into pairs: (1) and (2), (2) and (3), and (1) and (3). The same variable is eliminated from any two of these pairs. In Step 1 of Example 3, x is eliminated from equations (1) and (2). In Step 2, x is eliminated from equations (1) and (3). This results in two new equations, (4) and (5), in two variables that are solved for y and z.

308 *Chapter 6*

Oral Exercises

Tell whether or not each ordered triple is a solution of $x - 2y - 5z = 3$.

1. $(-5, 2, -3)$ No   2. $(4, -2, 1)$ Yes   3. $(3, 0, 0)$ Yes   4. $(-4, -1, -1)$ Yes

Tell whether or not each ordered triple is a solution of $4x - y + 2z = 7$.

5. $(\frac{3}{4}, 6, -1)$ No   6. $(0, 1, 4)$ Yes   7. $(0, -5, \frac{3}{2})$ No   8. $(\frac{1}{2}, 3, 4)$ Yes

Find three solutions of each equation. Examples are given.

9. $2x - y - z = 2$ (1, 0, 0), (0, -2, 0), (0, 0, -2)

10. $x - 3y + 2z = 5$ (5, 0, 0), (0, -1, 1), (8, 1, 0)

11. $3x + y - 2z = 8$
   (0, 8, 0), (0, 6, -1), (2, 0, -1)

12. $3x + 2y - z = -3$
   (-1, 0, 0), (0, 0, 3), (1, 0, 6)

Written Exercises

Solve each system of equations.

A

1. $x + y + z = 2$ {(1, -1, 2)}
   $x + 2y - z = -3$
   $2x - 2y - z = 2$

2. $x + y + z = 4$ {(6, 2, -4)}
   $x - y + z = 0$
   $2x - 3y + z = 2$

3. $x + y + z = 6$ {(2, 3, 1)}
   $x + y - z = 4$
   $2x - y = 1$

4. $2x + y - z = 1$ {(2, 0, 3)}
   $x + 2y + z = 5$
   $y - z = -3$

5. $x + y - 2z = 3$ {(2, -3, -2)}
   $3x + 2y - z = 2$
   $2x + 3y + z = -7$

6. $4x + y + 3z = 2$ {(0, 5, -1)}
   $5x - y - 2z = -3$
   $3x + 2y + 5z = 5$

B

7. $3(2a - b) = 4 + 5c$ {($\frac{1}{2}$, 3, -2)}
   $4a = 5b + 3c - 7$
   $2(a - 3b) = 7c - 3$

8. $3(r - 3q) = -4(1 + p)$ {($\frac{1}{2}$, $\frac{1}{3}$, -1)}
   $2(p - 3q + 4r) = -9$
   $3(2p + q) + 5r = -1$

9. $a + b - c = 1$ {($\frac{104}{19}$, $\frac{35}{19}$, $\frac{120}{19}$)}
   $\frac{1}{2}a - b + \frac{1}{3}c = 3$
   $\frac{1}{3}a + \frac{2}{3}b - \frac{1}{6}c = 2$

10. $r + 2s - t = 12$ {($-\frac{32}{5}$, $\frac{99}{5}$, $\frac{106}{5}$)}
   $\frac{3}{2}r + \frac{2}{3}s - \frac{1}{2}t = -7$
   $\frac{1}{4}r + \frac{2}{3}s - \frac{1}{2}t = 1$

11. $5x - 2y - 5z = 1$
   $-x + 2y - 5z = -9$
   $-10x + 4y + 10z = -2$
   $\{(x, y, z): 4x - 10z = -8\}$

12. $5x - 2y + 3z = 6$ Ø
   $3x + 2y - z = 5$
   $-10x + 4y - 6z = -6$

Find a, b, and c so that the following ordered triples will be solutions of the given equation.

C

13. $ax + by + cz = 8$; $(0, -2, 2)$, $(1, 1, 7)$, $(2, -1, -3)$ $a = 4, b = -3, c = 1$

14. $ax + 2y - bz = c$; $(1, 5, 3)$, $(3, 3, 5)$, $(0, -3, -4)$ $a = 5, b = 3, c = 6$

Systems of Open Sentences   **309**

Suggested Assignments

Minimum
 309/1–8
R 311/*Self-Test 3*
   Exercises 1, 2, 4
Average
 309/1–8
 310/*P*: 1–4
R 311/*Self-Test 3*
Maximum
 309/5–17 odd
 310/*P*: 1–6
R 311/*Self-Test 3*

Mixed Review

Solve the system.

1. $3x - 4y = -1$
   $-7x + 5y = -15$ {(5, 4)}

2. $3x - 2y = 6$
   $9x - 6y = 12$ Ø

Solve for x and y in terms of a.

3. $4x + 3y = 2a$
   $x + y = a$ {(-a, 2a)}

Solve.

4. $\frac{1}{3}x + 2 = \frac{3}{4}$ $\left\{-3\frac{3}{4}\right\}$

5. $\frac{3}{5}\left(x + \frac{5}{2}\right) - 2 =$
   $4 - \left(x + \frac{1}{2}\right)$ $\left\{2\frac{1}{2}\right\}$

Solve each system of equations.

1. $x + 2y + z = 5$
$3x + y + 4z = 1$
$-2x + y - z = -4$
$\{(5, 2, -4)\}$

2. $x + y + z = 2$
$2x - y + z = -1$
$x - y - z = 0$
$\{(1, 2, -1)\}$

3. $x + y + 2z = 0$
$2x - 2y + z = 8$
$3x + 2y + z = 2$
$\{(2, -2, 0)\}$

Solve.

4. Jackie has $6.25 in nickels, dimes, and quarters. She has 85 coins, with three times as many nickels as dimes. How many of each kind of coin has she? 60 nickels, 20 dimes, and 5 quarters

5. The sum of three numbers is 20. The sum of the first two is 4 less than twice the third. Three times the third number is twice the second. What are the three numbers?
first number: 0
second number: 12
third number: 8

6. Twice the length of a box is 1 cm more than the sum of its width and height. The sum of its height and length is 2 cm more than its width. The sum of the dimensions of the box is 14 cm. What are the dimensions of the box?
length: 5 cm
width: 6 cm
height: 3 cm

Find a, b, and c so that the following ordered pairs will be solutions of the given equation.

15. $y = ax^2 + bx + c$; $(0, 4)$, $(1, 5)$, $(-1, 9)$ $a = 3, b = -2, c = 4$
16. $y = ax^2 + bx + c$; $(0, -3)$, $(-1, 4)$, $(2, -5)$ $a = 2, b = -5, c = -3$
17. $y = ax^2 + bx + c$; $(0, 3)$, $(1, 3)$, $(-1, 5)$ $a = 1, b = -1, c = 3$
18. $y = ax^2 + bx + c$; $(-1, -1)$, $(0, -2)$, $(-2, 0)$ $a = 0, b = -1, c = -2$

Problems

Solve.

A 1. The sum of the length, width, and height of a rectangular box is 17 cm. The length is one third the height. The sum of the length and height exceeds twice the width by 2 cm. Find the length, width, and height of the box. length: 3 cm; width: 5 cm; height: 9 cm

2. The sum of three numbers is 12. One of the numbers is twice the sum of the other two numbers. This same number is also equal to the difference between one of the other numbers and triple the third number. What are the three numbers? 8, 5, −1

3. Marie has $1.95 in her purse, consisting entirely of nickels, dimes, and quarters. The number of quarters is one more than the sum of the number of nickels and the number of dimes. The number of dimes is two more than twice the number of nickels. How many of each kind of coin does Marie have? 1 nickel, 4 dimes, 6 quarters

4. Tino has one-dollar, five-dollar, and ten-dollar bills that total $153. He has as many five-dollar bills as one-dollar and ten-dollar bills combined. The sum of four times the number of one-dollar bills and twice the number of ten-dollar bills is one more than three times the number of five-dollar bills. How many of each kind of bill does he have? 8 one-dollar bills, 15 five-dollar bills, 7 ten-dollar bills

Find the three-digit number that satisfies *all* the conditions (a) through (c).

B 5. a. The sum of the three digits is 15.
564 b. The hundreds' digit is half the sum of the tens' digit and the units' digit.
c. If the hundreds' digit is subtracted from the tens' digit, the difference is one fourth as great as the units' digit.

6. a. The sum of the hundreds' digit and the tens' digit is 9.
273 b. The sum of the tens' digit and the units' digit is five times as great as the hundreds' digit.
c. The tens' digit is three less than twice the sum of the hundreds' digit and the units' digit.

Self-Test 3

VOCABULARY linear programming (p. 302)
corner point (p. 303)
convex region (p. 303)
convex polygonal region (p. 303)
solution of an equation in three variables (p. 306)

ordered triple (p. 306)
linear equation in three variables (p. 307)
solution of a system of three equations in three variables (p. 307)

1. Graph the solution set of the following system of inequalities.   *Obj. 1, p. 298*

$$y - 2x \leq 1$$
$$2y + x > 4$$
$$x - 5 \leq 0$$

2. Find the minimum value of the linear expression $50x + 80y$ over the   *Obj. 2, p. 298*
convex polygonal region that is shaded in the diagram below.

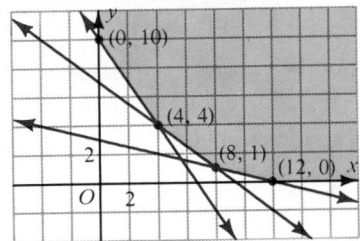

480

3. Calgary Shipping Company needs at least 500 barrels of diesel fuel, 600 barrels of gasoline, and 42 barrels of oil to keep its trucks in operation. Each order placed with Provincial Oil will provide 250 barrels of diesel fuel, 75 barrels of gasoline, and 7 barrels of oil for a total cost of $16,700. Each order placed with Rocky Mountain Oil will provide 50 barrels of diesel fuel, 150 barrels of gasoline, and 7 barrels of oil for a total cost of $10,900. How many orders should the Calgary Shipping Company place with each of the two oil companies if the shipping company wishes to keep all its trucks in operation at the minimum cost? 1 order from Provincial and 5 orders from Rocky Mountain

4. Solve the following system of equations.   *Obj. 3, p. 298*

$$2x + 3y - 2z = -6 \quad \{(5, -4, 2)\}$$
$$2x + 5y + 6z = 2$$
$$6x + 7y - 2z = -2$$

Check your answers with those at the back of the book.

Systems of Open Sentences   **311**

1. Graph the solution set of the system:
$$y - 1 \geq 0$$
$$x - 6 \leq 0$$
$$x - 2 \geq 0$$
$$2y + x \leq 12$$

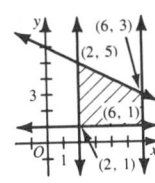

2. Find the minimum value of the linear expression $5y - 2x$ over the convex polygonal region determined in Exercise 1. -7

3. A children's zoo is given a sum of money to buy and maintain some animals. The zoo officials want to spend at least $200 for rabbits and (or) monkeys. They intend to buy at least 14 animals. Each rabbit costs $4 and each monkey costs $40. If it costs 10 cents per day to feed each rabbit and 50 cents each day to feed each monkey and the zoo officials want to keep their feeding costs at a minimum, how many of each animal should they buy? 10 rabbits, 4 monkeys

4. Solve the following system:
$$x + y - 3z = 3$$
$$x - y + 2z = 2$$
$$3x + y - z = 5$$
$$\{(2, -2, -1)\}$$

Additional Answers Self-Test 3

1.

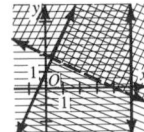

311

25.

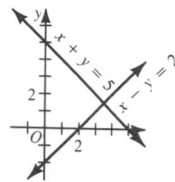

26.

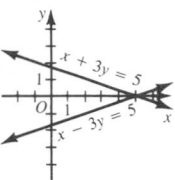

27.

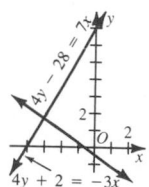

28.

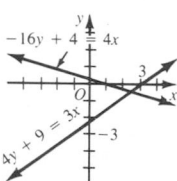

33. a. $(-2, -1)$, $(3, 4)$, $(3, -6)$
    b. 25

34. a. $(2, 1)$, $(6, -5)$, $(-7, -5)$
    b. 39

35. a. $(1, 7)$, $(-3, 8)$, $(-3, -9)$
    b. 34

36. a. $(1, 6)$, $(3, 6)$, $(2, 3)$, $(0, 3)$
    b. 6

37. a. $(6, 4)$, $(6, -4)$, $(-3, -7)$, $(-3, 1)$
    b. 72

38. a. $(4, -1)$, $\left(-\frac{3}{2}, -1\right)$, $\left(\frac{3}{2}, -7\right)$, $(7, -7)$
    b. 33

Careers

Aerospace Engineering

Aerospace engineers develop rockets, space exploration vehicles, satellites, and various aircraft. They are involved in projects from the earliest stages of research through the final stages of testing the finished product.

The early development stage of a project involves research, cost analysis, and testing. During this stage, a model of the proposed craft is built, and the design is tested and modified using this model. In the next stage, production, engineers supervise the manufacture of the craft in order to assure that each part conforms to the specifications developed in working with the model. In the final, or testing, stage, the performance of the manufactured craft is tested in actual flight.

EXAMPLE Two rockets are launched vertically at the same time. The acceleration of the first rocket is three times that of the second, and after 4 s the first rocket is 96 m higher than the second. The formula $d = \frac{1}{2}at^2$ gives the distance d in meters that is traveled in t seconds by an object whose acceleration is a m/s² (meters per second squared). Determine the acceleration of the first rocket.

SOLUTION Let h = height in meters of the second rocket, and
        a = acceleration in m/s² of the second rocket.

Then: $h + 96$ = height in meters of the first rocket, and
           $3a$ = acceleration in m/s² of the first rocket.

Substitute these expressions into the given formula.

Second rocket:	First rocket:
$d = \frac{1}{2}at^2$	$d = \frac{1}{2}at^2$
$h = \frac{1}{2}a(4)^2$	$h + 96 = \frac{1}{2}(3a)(4)^2$
$h = 8a$	$h + 96 = 24a$

The equations $h = 8a$ and $h + 96 = 24a$ form a system of two equations in two variables. Solving the system, the value of a is found to be 6, and so the value of $3a$ is $3(6)$, or 18.

∴ the acceleration of the first rocket is 18 m/s².

312   *Chapter 6*

EXTRA

Matrices

You can find the solution of a system such as

$$x + 4y = 9$$
$$2x + y = 4$$

by working only with the coefficients of x and y and the constant terms. To do this, you represent the coefficients and constants by means of an ordered array of numbers called a **matrix** (plural: *matrices*). Parentheses or brackets are used to group the *elements* of a matrix. Thus

$$\begin{bmatrix} 1 & 4 \\ 2 & 1 \end{bmatrix}$$

is the *coefficient matrix* of the given system, and

$$\begin{bmatrix} 1 & 4 & 9 \\ 2 & 1 & 4 \end{bmatrix}$$

is the *augmented matrix* of the system.

Now, compare the sequence of steps used to solve the given system by linear combinations, as shown at the left, with the corresponding sequence of matrices shown at the right.

$$\begin{array}{cc} x + 4y = 9 & \begin{bmatrix} 1 & 4 & 9 \\ 2 & 1 & 4 \end{bmatrix} \\ 2x + y = 4 & \end{array}$$

Multiplying each member of the first equation by -2 and adding the result to the second equation produces the equivalent system:

$$\begin{array}{cc} x + 4y = 9 & \begin{bmatrix} 1 & 4 & 9 \\ 0 & -7 & -14 \end{bmatrix} \\ 0x + (-7)y = -14 & \end{array}$$

Dividing the second equation by -7, or multiplying by $-\frac{1}{7}$, produces the equivalent system:

$$\begin{array}{cc} x + 4y = 9 & \begin{bmatrix} 1 & 4 & 9 \\ 0 & 1 & 2 \end{bmatrix} \\ 0x + y = 2 & \end{array}$$

Adding -4 times the second equation to the first equation yields the equivalent system:

$$\begin{array}{cc} x + 0y = 1 & \begin{bmatrix} 1 & 0 & 1 \\ 0 & 1 & 2 \end{bmatrix} \\ 0x + y = 2 & \end{array}$$

The only solution of the system, $(1, 2)$, is evident from this last set of equations and from the right-hand column of the corresponding matrix. At each step, the matrix shown is obtained from the preceding matrix in the sequence in exactly the same way as the corresponding system of equations is obtained from its preceding system.

Systems of Open Sentences **313**

11.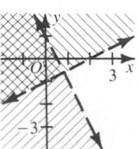

12.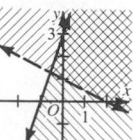

13.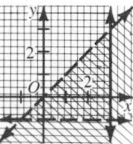

14.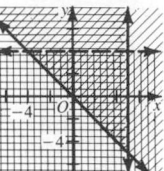

15.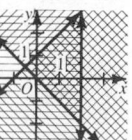

16.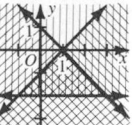

17.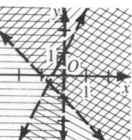

Each of the matrices in the example on the preceding page is said to be *row-equivalent* to each of the other matrices in the example. In general:

Two matrices are **row-equivalent** if one can be obtained from the other by means of one or more of the following *row transformations*:

1. Interchanging two rows.
2. Multiplying each entry in a row by the same nonzero real number.
3. Multiplying each entry in a row by a nonzero real number and adding the resulting product to the corresponding entry in another row.

As illustrated by the example, you can use the concept of equivalent matrices to solve a system of linear equations in two variables. The following is a summary of this method.

Steps in Solving Systems by Matrices

1. Represent the system by its augmented matrix.
2. Use row transformations to obtain an equivalent matrix of the form
$$\begin{bmatrix} 1 & 0 & p \\ 0 & 1 & q \end{bmatrix}.$$
3. Read the solution, (p, q), from this matrix.

EXAMPLE Solve the system: $3x - 4y = 5$
$$2x + 5y = -12$$

SOLUTION $\begin{bmatrix} 3 & -4 & 5 \\ 2 & 5 & -12 \end{bmatrix}$

$\begin{bmatrix} 1 & -\frac{4}{3} & \frac{5}{3} \\ 2 & 5 & -12 \end{bmatrix}$ ⟵ $\frac{1}{3} \times$ row 1

$\begin{bmatrix} 1 & -\frac{4}{3} & \frac{5}{3} \\ 0 & \frac{23}{3} & -\frac{46}{3} \end{bmatrix}$ ⟵ row 2 + [(−2) × row 1]

$\begin{bmatrix} 1 & -\frac{4}{3} & \frac{5}{3} \\ 0 & \frac{1}{3} & -\frac{2}{3} \end{bmatrix}$ ⟵ $\frac{1}{23} \times$ row 2

$\begin{bmatrix} 1 & 0 & -1 \\ 0 & \frac{1}{3} & -\frac{2}{3} \end{bmatrix}$ ⟵ row 1 + (4 × row 2)

$\begin{bmatrix} 1 & 0 & -1 \\ 0 & 1 & -2 \end{bmatrix}$ ⟵ 3 × row 2

∴ the solution set is $\{(-1, -2)\}$.

If, in the solution process, you can obtain a matrix of the form:

$$\begin{bmatrix} a & b & c \\ 0 & 0 & 0 \end{bmatrix} \text{ or } \begin{bmatrix} 0 & 0 & 0 \\ a & b & c \end{bmatrix},$$ the two equations in the system are equivalent;

$$\begin{bmatrix} a & b & c \\ 0 & 0 & d \end{bmatrix} \text{ or } \begin{bmatrix} 0 & 0 & d \\ a & b & c \end{bmatrix},$$ $d \neq 0$, the two equations in the system are inconsistent.

Exercises

Solve each system using matrices.

1. $2x - y = 0$ $\{(1, 2)\}$
    $x + 3y = 7$

2. $2x - 3y = -5$ $\{(-1, 1)\}$
    $5x - 8y = -13$

3. $-2x + y = -6$ $\{(4, 2)\}$
    $x - 3y = -2$

4. $4x - 3y = -15$ $\{(-3, 1)\}$
    $x + 2y = -1$

5. $3x - 5y = 17$ $\{(4, -1)\}$
    $-5x + y = -21$

6. $x - 3y = 5$ $\emptyset$
    $2x - 6y = 7$

Chapter Summary

1. The *solution set* of a *system of linear equations in two variables* is the set of all ordered pairs of numbers that satisfy each equation in the system.

2. The graphs of two linear equations in two variables on the same coordinate plane are either *parallel lines, coincident lines,* or *lines that intersect in exactly one point.*

3. A system of equations having one or more solutions is called a *consistent system.* A system whose solution set is the empty set is called *inconsistent.*

4. To solve a system of linear equations in two variables graphically, graph the equations on the same coordinate plane and determine the coordinates of all points common to the graphs.

5. The *linear-combination method* and the *substitution method* are algebraic methods that may be used to solve a system of linear equations.

6. A *system of linear inequalities* may be solved graphically by finding the intersection of the half-planes that are the graphs of the inequalities.

7. *Linear programming* is a method that may be used to solve problems in which a quantity that can be represented by a linear equation is to be maximized or minimized subject to conditions that can be represented by a system of linear inequalities.

8. The *solution set* of a *system of linear equations in three variables* is the set of all *ordered triples* of numbers that satisfy each equation in the system. Systems of three or more equations in *three or more variables* can be solved using the algebraic methods developed for systems of two equations in two variables.

Systems of Open Sentences   **315**

18.

19.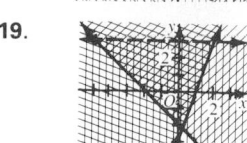

20.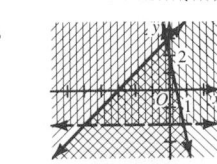

21.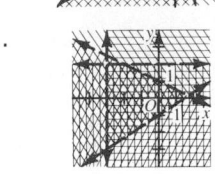

22.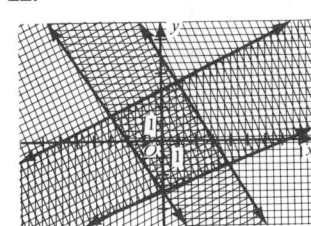

Chapter Review

Write the letter of the correct answer.

1. If the system $\begin{aligned} 2x - 3y &= 4 \\ nx - y &= 4 \end{aligned}$ is consistent, which of the following statements must be true?

   a. $n = 2$ **b.** $n = \frac{2}{3}$ **c.** $n \neq 0$ **(d.)** $n \neq \frac{2}{3}$ *6-1*

2. Given the system $\begin{aligned} y - 3x &= -2 \\ 3x - y &= 2 \end{aligned}$, how are the graphs of the two equations in the system related?

   (a.) The graphs coincide.

   b. The graphs are parallel lines.

   c. The graphs intersect at exactly one point, (2, 4).

   d. None of the above statements is true.

3. The solution set of the system $\begin{aligned} 3x - y &= 10 \\ x - 3y &= -2 \end{aligned}$ is $\{(4, 2)\}$. Which of the following is *not* an equivalent system?   *6-2*

   a. $\begin{aligned} x &= 4 \\ y &= 2 \end{aligned}$ **b.** $\begin{aligned} x + y &= 6 \\ x - y &= 2 \end{aligned}$

   c. $\begin{aligned} x &= y + 2 \\ y &= \frac{1}{2}x \end{aligned}$ **(d.)** $\begin{aligned} x + y &= 6 \\ 2x + 2y &= 12 \end{aligned}$

4. Use addition or subtraction to solve the system: $\begin{aligned} 4a &= 5b - 9 \\ -2b &= 4a - 5 \end{aligned}$

   a. $\{(2, \frac{1}{4})\}$ **b.** $\{(4, 2)\}$ **(c.)** $\{(\frac{1}{4}, 2)\}$ **d.** $\varnothing$

5. To solve the system $\begin{aligned} 5x - 3y &= 7 \\ 3x + 2y &= 9 \end{aligned}$ using linear combinations, the first equation is multiplied by a and the second equation is multiplied by b. Which values of a and b will eliminate one of the variables when the resulting equations are added?   *6-3*

   a. $a = 3, b = 5$ **b.** $a = -2, b = 3$

   (c.) $a = -3, b = 5$ **d.** $a = 2, b = -3$

6. Use linear combinations to solve the system: $\begin{aligned} 3x + 2y &= 5 \\ 2x - 3y &= 12 \end{aligned}$

   (a.) $\{(3, -2)\}$ **b.** $\{(-2, 3)\}$ **c.** $\{(3, 2)\}$ **d.** $\{(3, -\frac{1}{2})\}$

7. To solve the system $\begin{aligned} 2x - y &= 1 \\ 2x + 3y &= 7 \end{aligned}$ by substitution, which expression can be substituted for y in the *second* equation?   *6-4*

   a. $2x - y$ **b.** $1 - 2x$ **(c.)** $2x - 1$ **d.** $\frac{7}{3} - \frac{2}{3}x$

316   *Chapter 6*

8. Use substitution to solve the system: $\begin{array}{l} 2x + 3y = 7 \\ x - 4y = -13 \end{array}$

   a. $\{(2, 1)\}$ b. $\{(-5, 2)\}$ (c.) $\{(-1, 3)\}$ d. $\{(\frac{9}{5}, \frac{19}{5})\}$

9. A collection of coins contains a total of fourteen dimes and quarters and is worth $1.85 in all. How many dimes are in the collection?

   a. 8   (b.) 11   c. 3   d. 6

   6-5

10. With a tail wind, a plane travels 300 km in 1 h. With no change in the wind, the return trip takes $1\frac{1}{2}$ h. Find the speed of the wind.

    a. 100 km/h   b. 250 km/h   (c.) 50 km/h   d. 30 km/h

    6-6

11. The units' digit of a two-digit number is twice the tens' digit. If the digits of the number were reversed, the resulting number would be six less than twice the original number. Find the original number.

    a. 12   (b.) 24   c. 36   d. 48

    6-7

12. Which point belongs to the graph of the following system?

    6-8

    $$\begin{array}{l} x + y \geq 1 \\ x - y \leq 3 \\ x - 2 \leq 0 \end{array}$$

    a. $(0, 0)$ (b.) $(1, 2)$ c. $(-1, 0)$ d. $(2, -3)$

13. Find the coordinates of the corner points of the solution set of the following system.

    6-9

    $$\begin{array}{l} x \geq 0 \\ x \leq 2 \\ y \geq 0 \\ x + y \leq 4 \end{array}$$

    a. $(0, 0), (4, 0), (2, 2), (0, 4)$
    b. $(0, 0), (0, 2), (2, 0), (2, 2)$
    c. $(0, 0), (2, 0), (0, 2), (4, 0)$
    (d.) $(0, 0), (2, 0), (2, 2), (0, 4)$

14. Find the minimum value of the expression $4x + 3y$ over the region that has corner points $(1, 0)$, $(3, 0)$, $(3, 2)$, and $(1, 4)$.

    a. 18   b. 12   (c.) 4   d. 16

15. Solve the system: $\begin{array}{l} 2x + y - z = -5 \\ -5x - 3y + 2x = 7 \\ x + 4y - 3z = 0 \end{array}$

    6-10

    (a.) $\{(-2, 5, 6)\}$ b. $\{(3, -4, -7)\}$ c. $\{(4, 3, -2)\}$ d. $\{(-1, 0, 5)\}$

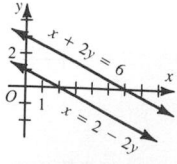
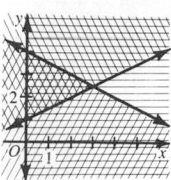
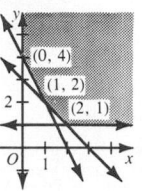
Chapter Test

1. Solve the system $\begin{aligned} x + 2y &= 6 \\ x &= 2 - 2y \end{aligned}$ using graphs.   6-1

2. Is the system $\begin{aligned} 3x + 3y &= -9 \\ -4x - 4y &= 12 \end{aligned}$ consistent or inconsistent? consistent

Solve each system using addition or subtraction.

3. $\begin{aligned} 32 &= 5a - 3b \\ -8 &= 5a + 7b \end{aligned}$ $\{(4, -4)\}$

4. $\begin{aligned} 3r + 3s &= -5 - 2r \\ 6r + 3 &= 7s + r \end{aligned}$ $\{(-\frac{22}{25}, -\frac{1}{5})\}$ 6-2

Solve each system using the linear-combination method.

5. $\begin{aligned} 7c + 5d &= 2 \\ 8c - 9d &= 17 \end{aligned}$ $\{(1, -1)\}$

6. $\begin{aligned} 5x + 4y &= 11 \\ -7x + 3y &= 19 \end{aligned}$ $\{(-1, 4)\}$ 6-3

Solve each system using the substitution method.

7. $\begin{aligned} 3m - 4n &= 5 \\ m + 7n &= 10 \end{aligned}$ $\{(3, 1)\}$

8. $\begin{aligned} 2(x + y) &= 4(x + 1) \\ 4(x + 2) &= y - 3 \end{aligned}$ $\{(-3, -1)\}$ 6-4

Solve.

9. Four years ago Sylvia was two thirds as old as Alex was then. Four   6-5
years from now, she will be four fifths as old as Alex will be. How
old are Sylvia and Alex now? Sylvia: 12 years; Alex: 16 years

10. Donna took 3 h to row 18 km upstream. Her return trip downstream   6-6
took 1 h less. Find her rate of rowing in still water and the rate of the
current. rate in still water: 7.5 km/h; rate of current: 1.5 km/h

11. The hundreds' digit of a three-digit number is two less than the tens'   6-7
digit, and the tens' digit is one more than the units' digit. The sum of
the digits is 12. Find the number. 354

12. Graph the solution set of the system: $2y - 2 \geq x$ 6-8
$2y + x \leq 8$
$x \geq 0$

13. Graph the solution set of the system at the right $2x + y \geq 4$ 6-9
and label the corner points with their coordinates. $x + y \geq 3$
$x \geq 0$
$y \geq 1$

14. Find the minimum value of the expression $2x + 3y$ over the region
graphed in Exercise 13. 10

15. Solve the system: $\begin{aligned} x + y + z &= 3 \\ x + 2y - z &= -1 \\ -2x - y + 3z &= 11 \end{aligned}$ $\{(-2, 2, 3)\}$ 6-10

Mixed Review

Simplify.

1. $-(-3 + 2) + 7$ 8
2. $2|3 - 6| - |-5 + 8|$ 3
3. $(3.2 + 6.7) - (12.7 + 4.2)$ −7
4. $-25\frac{1}{3} + 8\frac{2}{3} - 10\frac{1}{3}$ −27   18
5. $\frac{1}{2}[24 \div 2(6)] \div [12(3) \div 6]$ $\frac{1}{6}$
6. $[(11 - 3)(7 - 4)] - [18 \div (6 \div 2)]$
7. $(28 \div 2 + 20 \div 5) \div 9(2)^2$ $\frac{1}{2}$
8. $(2 + 3)^2 - [9 \div (8 - 5)^2]$ 24
9. $2x + 5y + 2x + 5y$ $4x + 10y$
10. $3q^2 - 2q - 13q^2 + 8q$ $-10q^2 + 6q$
11. $(6 + 4a) - (7 - 3a)$ $7a - 1$
12. $-3[x - (6 - x)]$ $-6x + 18$
13. $2(4m + 5) + 3(m + 8)$ $11m + 34$
14. $9(r + 3r^2) + 7(r^2 + 4r)$ $34r^2 + 37r$
15. $3(c - 2d) - 6(3c + d)$ $-15c - 12d$
16. $4(-x + 2y) - 3(x - 5y)$ $-7x + 23y$
17. $-4p(-3q)(5r)$ $60pqr$
18. $x(-x)(14x)$ $-14x^3$

Graph on a number line.

19. {the natural numbers between −2 and 5}
20. {the positive even integers less than or equal to 6}
21. {the real numbers between −3 and 4}
22. {the positive real numbers}
23. $\{x: |x| \geq 2\}$
24. $\{y: |y| < 4\}$

Graph on a coordinate plane.

25. $f: x \rightarrow x + 3; D = \{-2, -1, 0, 1, 2\}$
26. $g: x \rightarrow -3; D = \{-4, -2, 0, 2, 4\}$

27. $x = -1$
28. $y = 3$
29. $2x - y = 4$
30. $2x + 4y = 3$
31. $x - y \geq 3$
32. $6x + 3y > 2$

Evaluate each expression when $r = 4$, $s = -2$, and $t = \frac{1}{2}$.

33. $rt + s$ 0
34. $r^2 - st$ 17
35. $\dfrac{r - s}{-t}$ −12
36. $4(r - s) + 2t$ 25

Write an equation of the line that satisfies the given requirements. The equation should be in the form $ax + by = c$, where a, b, and c are integers.

37. has slope $-\frac{4}{5}$ and y-intercept $\frac{7}{4}$ $16x + 20y = 35$
38. has slope $\frac{2}{3}$ and passes through the point $(1, 2)$ $2x - 3y = -4$
39. is parallel to the x-axis and passes through the point $(-5, 6)$ $y = 6$
40. passes through the points $(4, -1)$ and $(-2, 3)$ $2x + 3y = 5$

Systems of Open Sentences **319**

25.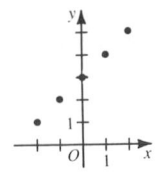
26.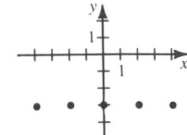
27.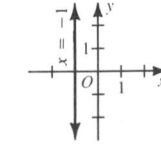
28.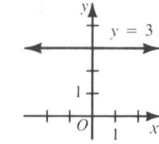
29.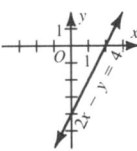
30.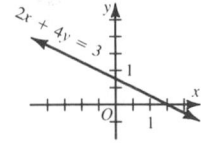
31.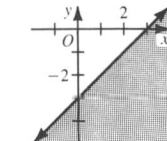
32.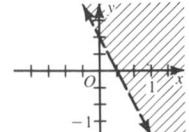

Strategy for Success: If you cannot solve a certain problem within a short period of time, leave that problem and go on to others that may be easier for you. Working on the easier problems may give you an idea that will help in solving the harder problem. Then, if time permits, you can go back and attempt the harder problem again.

Decide which is the best of the choices and write the corresponding letter on your answer sheet.

1. Determine an equation of the line that intersects the y-axis at the same point as the line containing $(3, 2)$ and $(-1, -2)$ and that is parallel to the line containing $(4, 5)$ and $(10, 1)$. E
   (A) $3y - 2x = 3$ **(B)** $2x - 3y = 3$ **(C)** $2x - 3y = 1$
   (D) $2x + 3y = 1$ **(E)** $2x + 3y = -3$

2. Which of the following points lies on the line that has slope $-\frac{1}{2}$ and that passes through the point $(-2, 3)$? A
   I. $(-1, \frac{5}{2})$ II. $(-3, \frac{1}{4})$ III. $(2, 3)$
   (A) I only **(B)** II only **(C)** III only **(D)** I and II only **(E)** I and III only

3. The domain of the function $f: x \rightarrow \dfrac{1}{|x|}$ is the set of all real numbers except 0. Which of the following statements is true? D
   (A) There are two values of x such that $f(x) = 0$.
   (B) The range of f is the set of all nonnegative real numbers.
   (C) The graph of f on a coordinate plane is a line.
   (D) The ordered pair $(-\frac{1}{2}, 2)$ satisfies the function.
   (E) None of the above statements is true.

4. Solve the system: $4x - \dfrac{y}{4} = -\dfrac{3}{2}$ and $4x + 7y = -4$ E
   (A) $\{(-8, 4)\}$ **(B)** $\{(4, -8)\}$ **(C)** $\emptyset$
   (D) $\{(x, y): x = -\frac{7}{4}y - 1\}$ **(E)** none of these

5. The sum of the digits of a three-digit number is 15. The units' digit is half the tens' digit, and the hundreds' digit is one less than the units' digit. Find the number. D
   (A) 438 **(B)** 843 **(C)** 348 **(D)** 384 **(E)** 483

6. Identify the inequality that is graphed on the coordinate plane at the right. D
   (A) $4x - y \geq 8$ **(B)** $x - 4y \leq -8$
   (C) $x - 4y \leq 8$ **(D)** $x - 4y \geq -8$
   (E) $x - 4y \geq 8$

Contest Problems

1. The coordinates of two opposite vertices of a rectangle are $(-1, -2)$ and $(2, 3)$. If the sides of the rectangle are parallel to the coordinate axes, what is the area of the rectangle in square units? 15 square units

2. The sum of the ages of a family of six persons is 160. If their ages range from six to fifty, what was the sum of their ages four years ago? 136 years

3. If $9x = 5x - 8c + 6$, what is the value of $6x + 5c$ in terms of c? $-7c + 9$

4. A straight line passes through the points $(2k + 1, -2), (-1 - 3k, 8)$, and $(1, 3)$. Through which of the following points does the line also pass?
   a. $(-7, -6)$ b. $(-7, -8)$ ⓒ $(9, 13)$ d. $(9, 15)$

5. How many solid cubes that measure 2 cm on each edge are needed to completely fill an empty cube that measures 6 cm on each edge? 27 cubes

6. Determine the least number of steps necessary to measure exactly 6 L of water if you have only a 4 L pail and a 9 L pail. If a represents the contents of the 4 L pail and b represents the contents of the 9 L pail, use an ordered pair (a, b) to describe the contents of each pail at each step.
   eight steps: (0,9), (4, 5), (0, 5), (4, 1), (0, 1), (1, 0), (1, 9), (4, 6)

Engineers design concert halls to ensure acoustic balance. The photo shows a performance at the Philharmonic Hall in West Berlin, where the audience surrounds a central stage.

Chapter 7

Polynomials and Factoring

Addition, Subtraction, and Multiplication of Polynomials

OBJECTIVES for Sections 7-1 through 7-4:
1. *To add and subtract polynomials.*
2. *To multiply monomials.*
3. *To find powers of a monomial.*
4. *To multiply a polynomial by a monomial.*
5. *To multiply polynomials.*

7–1 Adding and Subtracting Polynomials

A numeral, a variable, or an indicated product of a numeral and one or more variables is called a **monomial**. For example,

$$-3, \quad x, \quad 5y^2, \quad 12a^2b, \quad \text{and} \quad -\frac{1}{4}m^5n^2$$

are monomials. A monomial that contains no variable, such as -3, is called a **constant monomial,** or a **constant**.

The number of times that a variable occurs as a factor in a monomial is called the **degree of the variable** in that monomial. For example, in the monomial

$$-4x^2yz^3 \quad \begin{cases} x \text{ has degree 2,} \\ y \text{ has degree 1, and} \\ z \text{ has degree 3.} \end{cases}$$

Polynomials and Factoring **323**

1. Simplify
$7x^2 - 5x + 12x - 7$.
$= 7x^2 + (-5 + 12)x - 7$
$= 7x^2 + 7x - 7$

2. Add $3p^2 + 10p$ and
$p^2 - 8p + 3$.
Solution 1
$(3p^2 + 10p) +$
$\qquad (p^2 - 8p + 3)$
$= (3p^2 + p^2) +$
$\qquad [10p + (-8p)] + 3$
$= 4p^2 + 2p + 3$
Solution 2
$3p^2 + 10p$
$\underline{p^2 - 8p + 3}$
$4p^2 + 2p + 3$

3. Subtract $r^2 + 8t - 6$ from
$6r^2 - 2t$.
Solution 1
$(6r^2 - 2t) - (r^2 + 8t - 6)$
$= 6r^2 - 2t + (-r^2) +$
$\qquad (-8t) + 6$
$= (6r^2 - r^2) +$
$\qquad (-2t - 8t) + 6$
$= 5r^2 - 10t + 6$
Solution 2
$6r^2 - 2t$
$r^2 + 8t - 6$
$\begin{bmatrix} \text{change to the} \\ \text{opposite and add} \end{bmatrix}$
$6r^2 - 2t$
$\underline{-r^2 - 8t + 6}$
$5r^2 - 10t + 6$

The **degree of a monomial** is the total number of times that its variables occur as factors. Thus, the degree of $-4x^2yz^3$ is $2 + 1 + 3$, or 6. The degree of any nonzero constant monomial is 0. The monomial 0 has *no* degree.

The numerical factor of a monomial is called the **coefficient,** or **numerical coefficient,** of the monomial. Monomials that are identical or that differ only in their coefficients are called **similar,** or **like.** Thus,

$$-2xy^3, \qquad xy^3, \qquad -xy^3, \qquad \text{and} \qquad 7y^3x$$

are similar, but

$$5xy^3 \qquad \text{and} \qquad 5x^3y$$

are not similar.

A monomial or a sum of monomials is called a **polynomial.** Each of the monomials is called a **term** of the polynomial, and the coefficients of the terms are called the **coefficients** of the polynomial. The coefficient of a nonzero constant term is defined to be the constant itself. Thus, the terms of the polynomial

$$8x^3 + (-3x^2) + 0x + (-4)$$

are $8x^3$, $-3x^2$, $0x$, and -4; the coefficients are 8, -3, 0, and -4. Note that this polynomial is more commonly written

$$8x^3 - 3x^2 - 4,$$

where the term with coefficient 0 is omitted and the connecting + signs are taken to be understood.

It is sometimes convenient to say that a monomial is a polynomial of *one* term. A polynomial of *two* terms is called a **binomial,** and a polynomial of *three* terms is called a **trinomial.** For example, $x^2 + y^2$ is a binomial, and $x^2 + 2xy + y^2$ is a trinomial.

A polynomial is said to be simplified, or in **simplest form,** when no two of its terms are similar. For example,

$$8y^2 - 4y + 1$$

is in simplest form, but

$$9z^2 + 5z - 16z - 3$$

is not in simplest form. You simplify a polynomial by using the distributive axiom to add similar terms.

EXAMPLE 1   Simplify $9z^2 + 5z - 16z - 3$.

SOLUTION $9z^2 + 5z - 16z - 3 = 9z^2 + [5 + (-16)]z - 3$
$\qquad\qquad\qquad\qquad\qquad\qquad = 9z^2 - 11z - 3$

The **degree of a polynomial** is the greatest of the degrees of its terms after it has been simplified. Thus, $9z^2 - 11z - 3$ is of degree two.

To *add* two or more polynomials, you write the sum and simplify by adding similar terms.

EXAMPLE 2 Add $9n^2 + 5n$ and $n^2 - 6n + 12$.

SOLUTION 1 $(9n^2 + 5n) + (n^2 - 6n + 12) = (9n^2 + n^2) + [5n + (-6n)] + 12$
$$= 10n^2 - n + 12$$

SOLUTION 2 You can also line up the similar terms vertically and add.

$$
\begin{array}{r}
9n^2 + 5n \\
n^2 - 6n + 12 \\
\hline
10n^2 - n + 12
\end{array}
$$

To *subtract* one polynomial from another, you add the opposite of each term of the polynomial that you are subtracting. Then simplify the sum.

EXAMPLE 3 Subtract $t^2 - 8t + 2$ from $4t^2 + 7t$.

SOLUTION 1 $(4t^2 + 7t) - (t^2 - 8t + 2) = 4t^2 + 7t + (-t^2) + 8t + (-2)$
$$= (4t^2 - t^2) + (7t + 8t) - 2$$
$$= 3t^2 + 15t - 2$$

SOLUTION 2 $\begin{array}{r} 4t^2 + 7t \\ t^2 - 8t + 2 \\ \hline \end{array}$ $\longrightarrow$ $\begin{bmatrix} \text{change to the} \\ \text{opposite and add} \end{bmatrix}$ $\longrightarrow$ $\begin{array}{r} 4t^2 + 7t \\ -t^2 + 8t - 2 \\ \hline 3t^2 + 15t - 2 \end{array}$

Note that the terms of a simplified polynomial are usually arranged in order of *decreasing* or *increasing* degree of one of the variables. For example:

$7x^4 - 2x^3 - x + 5$ is in order of decreasing degree in x.

$5 - x - 2x^3 + 7x^4$ is in order of increasing degree in x.

$-4a^3 + a^2b + 6ab^2 + 5b^3$ is in order of $\begin{cases} \text{decreasing degree in } a, \\ \text{increasing degree in } b. \end{cases}$

Oral Exercises

Tell whether or not the given expression is a monomial. If the expression is a monomial, state its degree and coefficient.

Yes; 1; −1

1. $3a$ Yes; 1; 3
2. $-3bc$ Yes; 2; −3
3. x^4 Yes; 4; 1
4. $-y$

5. 12 Yes; 0; 12
6. 0 Yes; no degree; 0
7. $\frac{1}{2}m^2$ Yes; 2; $\frac{1}{2}$
8. $\frac{n^3}{7}$ Yes; 3; $\frac{1}{7}$

9. $-\frac{2}{z}$ No
10. $\frac{pq^2}{r}$ No
11. $4rs^2$ Yes; 3; 4
12. $5u + v^2$ No

Polynomials and Factoring **325**

Additional A Exercises

Replace each _?_ with the number or variable that makes a pair of similar monomials.

1. $9c^3$ _?_ ^{5}; $-2c$ _?_ d^5 d; 3
2. $-14t^5v^9$; 3 _?_ ^5v _?_ t; 9

Copy the given polynomial and underline similar terms in the same way. Then simplify the polynomial.

3. $8bp + pn + 9bp - 3pn - 4$
$$\underline{8bp} + \underline{pn} + \underline{9bp} - \underline{3pn} - 4$$
$$= 17bp - 2pn - 4$$

4. $4t^2m - 2m^2t - 6t^2m - 9m^2$
$$\underline{4t^2m} - 2m^2t - \underline{6t^2m} - 9m^2$$
$$= -2t^2m - 2m^2t - 9m^2$$

5. Add: $\begin{array}{r} 6m - 4t - 8 \\ 9m - 8t + 6 \\ \hline 15m - 12t - 2 \end{array}$

6. Subtract the second polynomial from the first:
$$\begin{array}{r} 6m - 4t - 8 \\ 9m - 8t + 6 \\ \hline -3m + 4t - 14 \end{array}$$

325

Name the similar monomials.

13. $3b$, $3b^2$, $-2b$, ab, $-2b^2$, $3ab$

14. $4xy$, $-2yz$, $9yx$, $4xz$, $-zy$, $2zx$

15. $-m^2n^2$, $3mn^2$, $-4m^2n$, $2m^2n^2$, $8mn$, $-2m^2n$, $3mn$, $-2mn^2$

16. rs^3, $-5r^3s^2$, $7r^3s$, $-2r^2s^3$, $-2r^3s$, $3r^2s^3$, $-8rs^3$, $2r^3s^2$

Tell whether the given polynomial is a monomial, binomial, trinomial, or none of these. Then state the degree and the coefficients of the polynomial.

17. $5c^2d$

18. $5c^2 + d$

19. $8p^2 - 2p + 4$

20. $3q + 7q^3 - 5q^2$

21. $6abc - 5a^2b^2c^2$

22. $-6a^3 + 2a^2b - ab^2 + 9b^3$

Compute the following for each pair of polynomials.
a. the sum when the second is added to the first
b. the difference when the second is subtracted from the first

25b. $4r + 6s + t$
26b. $-u + 7w$
27b. $5x^2 - 5x - 5$
28b. $3y^2 + 3$

23. $\begin{array}{r} 4a - 3 \\ a + 1 \\ \hline \end{array}$
 a. $5a - 2$ **b.** $3a - 4$

24. $\begin{array}{r} 2b + 7 \\ 3b - 2 \\ \hline \end{array}$
 a. $5b + 5$ **b.** $-b + 9$

25. $\begin{array}{r} 5r + 3s - t \\ r - 3s - 2t \\ \hline \end{array}$
 a. $6r - 3t$

26. $\begin{array}{r} u - 4v + 6w \\ 2u - 4v - w \\ \hline \end{array}$
 a. $3u - 8v + 5w$

27. $\begin{array}{r} 2x^2 - 3x - 5 \\ -3x^2 + 2x \\ \hline \end{array}$
 a. $-x^2 - x - 5$

28. $\begin{array}{r} 5y^4 + 2y^2 \\ 5y^4 - y^2 - 3 \\ \hline \end{array}$
 a. $10y^4 + y^2 - 3$

Arrange the given polynomial in order of *decreasing* degree in x.

29. $3x + x^2 - 7$ $x^2 + 3x - 7$

30. $9x^2 - 2x^4 + 8 - 5x^3$ $-2x^4 - 5x^3 + 9x^2 + 8$

31. $x^2y - 3y^3 + 4x^3 - 2xy^2$ $4x^3 + x^2y - 2xy^2 - 3y^3$

32. $3x^2y^2 - 8xy^3 + x^4 - y^4 + 2x^3y$ $x^4 + 2x^3y + 3x^2y^2 - 8xy^3 - y^4$

Written Exercises

Replace each __?__ with the number or variable that makes a pair of similar monomials.

A **1.** $8c^3d^5$; $-2c^?d^?$ 3; 5

2. $-3u^2v^4$; $7\underline{}^2\underline{}^4$ u; v

3. $3j^5k^?$; $-2\underline{}^5k^3$ j

4. $10m^6\underline{}^2$; $-3\underline{}^6n^2$ n; m

5. $8x^2$; $8x^?$ 2

6. $-5y^3$; $\underline{}y^3$ any nonzero real number

Copy the given polynomial and underline similar terms in the same way. Then simplify the polynomial.

EXAMPLE $5ac^2 + 3ac - 7ac^2 - 4ac + 8$

SOLUTION $\underline{5ac^2} + \underline{\underline{3ac}} - \underline{7ac^2} - \underline{\underline{4ac}} + 8 = (5ac^2 - 7ac^2) + (3ac - 4ac) + 8$
$= -2ac^2 - ac + 8$

7. $\underline{3rs} + \underline{\underline{2}} - \underline{2rs} - \underline{\underline{3}}$ $rs - 1$

8. $\underline{-5bc} + \underline{\underline{d}} - \underline{bc} - \underline{\underline{4d}}$ $-6bc - 3d$

9. $\underline{b^2} - \underline{\underline{7b}} - \underline{6b^2} + \underline{\underline{b}}$ $-5b^2 - 6b$

10. $\underline{4m^2} + \underline{\underline{2m^3}} - \underline{2m^2} - \underline{\underline{m^3}}$ $m^3 + 2m^2$

11. $-3t + t^3 - 5t - t^3 - 1$

12. $2w^2 - w^4 - 3w^4 + 4w^2 + w$

13. $9ab + bc + 2ab + bc + 7$

14. $7xy - 3yz - 6xy + 4yz - xz$

15. $4p^3q^2 - 5p^2q^3 - 9pq + p^2q^3 - 2p^3q^2$

16. $8m^2 + 6m^2n - 3mn^2 - 2m^2n + 5mn^2$

17. $9rs - 3st + 6rt - 5st - 6rt - rs$

18. $12yz - yz^2 + 7y^2z - 7yz - 2y^2z - 5yz$

Add.

19. $2a + 5b - c$
$5a + 3b + 2c$
——————————
$7a + 8b + c$

20. $4r - 3s + 5t$
$-3r - 2s + 2t$
——————————
$r - 5s + 7t$

21. $9s^2 + 4s - 5$
$3s^2 + 3$
——————————
$12s^2 + 4s - 2$

22. $8t^2 - 2t + 1$
$ 2t - 3$
——————————
$8t^2 - 2$

23. $6m^4 - 5m^2$
$-2m^4 + m^2 - 1$
——————————
$4m^4 - 4m^2 - 1$

24. $4n^5 + 5n$
$-5n^5 - 3n^3 + 2n$
——————————
$-n^5 - 3n^3 + 7n$

25–30. In Exercises 19–24, subtract the second polynomial from the first.

Add or subtract the polynomials as indicated.

B
31. $(5p - 2q + 7r) + (3q - r - s)$

32. $(3a^2 - 2a + 7) + (a^3 - 2a^2 + a)$

33. $(2x^2 - 5x - 4) - (3x^2 + x + 1)$

34. $(3y^2 + 2y + 1) - (4y^2 - 3y - 1)$

35. $(5ab - 7b + 9a) + (3a + 4b - 4ab)$ $ab - 3b + 12a$

36. $(-2v^2w^2 + 3v - w) + (-4v + 2w + 3v^2w^2)$ $v^2w^2 - v + w$

37. $(2m + 9) + (-3m - 2) - (4m - 1)$ $-5m + 8$

38. $(4n - 7) - (n + 3) + (3n + 8)$ $6n - 2$

39. $(2bc - 4bc^2) - (2b^2c - bc) + (5b^2c + bc^2)$ $3b^2c + 3bc - 3bc^2$

40. $(2x^2y^2 - 5x^2y) - (2xy^2 - x^2y^2) - (x^2y - xy^2)$ $3x^2y^2 - 6x^2y - xy^2$

41. $(6x^3 - 5x + 2x^2 - 1) - (3 - 7x - x^2 - x^3)$ $7x^3 + 3x^2 + 2x - 4$

42. $(5y^4 - y^6 + 6 - y^2) + (5y^6 - 1 + y^2 - y^4)$ $4y^6 + 4y^4 + 5$

Solve.

43. $(3x - 1) - (5x + 1) = -6$ {2}

44. $(2y + 5) - (3 - y) = -7$ {−3}

45. $(4 - 2a) - (3a + 1) = 6 - 6a$ {3}

46. $(3b - 5) - (8 - 3b) = 8b - 9$ {−2}

47. $(j^2 + j + 3) - (j^2 + 2j - 1) = 8$ {−4}

48. $(k^3 + 8k - 3) - (k^3 - 6k + 5) = -1$ {$\frac{1}{2}$}

C
49. Subtract the sum of $2c - 3cd - d^2$ and $c + 5cd + 3d^2$ from $8cd - c^2d^2$. $-3c + 6cd - c^2d^2 - 2d^2$

50. Subtract the sum of $3v - 2w$ and $2v^2 + w$ from the sum of $5w - w^2$ and $4v - w$. $-2v^2 + v - w^2 + 5w$

51. What polynomial must be added to $3r^2 - 5rs - s^2$ to obtain the polynomial $r^2 + 2rs + 3r^2s^2 - 4s^2$? $7rs - 2r^2 + 3r^2s^2 - 3s^2$

52. What polynomial must be subtracted from $t^2 - 3t + 2$ to obtain the polynomial $6t + 2t^2 - 5t^3$? $5t^3 - t^2 - 9t + 2$

53. If two polynomials of degree three are added, must their sum be a polynomial of degree three? Explain.

54. If the difference between two polynomials is a polynomial of degree two, must one or both of the polynomials be of degree two?

25. $-3a + 2b - 3c$

26. $7r - s + 3t$

27. $6s^2 + 4s - 8$

28. $8t^2 - 4t + 4$

29. $8m^4 - 6m^2 + 1$

30. $9n^5 + 3n^3 + 3n$

31. $5p + q + 6r - s$

32. $a^3 + a^2 - a + 7$

33. $-x^2 - 6x - 5$

34. $-y^2 + 5y + 2$

53. No, a term of one polynomial that is the additive inverse of a term of the other may change the degree of the sum.

54. No, a term of one polynomial that is the same as a term of the other may change the degree of the difference.

___

Mixed Review

Simplify.

1. $(3 - 2^3) + (6 \cdot 2^2) + (4 \cdot 2^3)$ 80

2. $\left(7 \cdot \dfrac{1}{2}\right) + \left(12 \cdot \dfrac{1}{4}\right) + \left(9 \cdot \dfrac{1}{3}\right) + \dfrac{1}{2}$ 10

3. $4(7 - 2) - 12 \div 2(3)$ 18

Evaluate each expression when the variables have the given values.

4. $5xy - 3xy + 6$; $x = 2$, $y = 5$ 26

5. $3a^2b + 4ab^2$; $a = -3$, $b = 2$ 6

6. $2s - 3t + 4s + 5t$; $s = 4$, $t = -1$ 22

Key Ideas

Find the product of powers
of the same number.
Find a power of a power of a
real number.
Find a power of a product of
real numbers.

Chalkboard Examples

1. Simplify $(9b^2y^4)(-3y^3)$.
   $= [9 \cdot (-3)] \cdot (b^2)(y^4 \cdot y^3)$
   $= -27b^2(y^{4+3})$
   $= -27b^2y^7$

2. Simplify $(-b^3)^4$.
   $= (-b)^{3 \cdot 4} = (-b)^{12}$
   $= b^{12}$

3. Simplify $(-9pt)^3$.
   $= (-9)^3 \cdot p^3 \cdot t^3$
   $= -729p^3t^3$

4. Simplify $(xy^3)(-2x^2y)^2$.
   $= (xy^3)[(-2)^2(x^2)^2y^2]$
   $= (xy^3)(4x^4y^2)$
   $= 4(x \cdot x^4)(y^3 \cdot y^2)$
   $= 4x^5y^5$

Common Errors

Students will be tempted to
rewrite $(2^3)^2$ as 2^{3+2} or 2^5.
Rewrite $(2^3)^2$ as $(2^3) \cdot (2^3)$ so
they can easily see the cor-
rect equivalent, 2^6.

7–2 Multiplying Monomials

In Section 1-6 you learned that, if n is a positive integer, the nth *power* of
a real number a is defined as follows.

$$a^n = \underbrace{a \cdot a \cdot \ldots \cdot a}_{n \text{ factors}}$$

If you now apply this definition of power to the product

$$a^2 \cdot a^3,$$

you obtain the following.

$$a^2 \cdot a^3 = \overbrace{(a \cdot a)}^{2 \text{ factors}} \cdot \overbrace{(a \cdot a \cdot a)}^{3 \text{ factors}} = a^5$$
$$\underbrace{}_{2 + 3 = 5 \text{ factors}}$$

In general, for all real numbers a, if m and n are positive integers:

$$a^m \cdot a^n = \overbrace{(a \cdot a \cdot \ldots \cdot a)}^{m \text{ factors}}\overbrace{(a \cdot a \cdot \ldots \cdot a)}^{n \text{ factors}} = a^{m+n}$$
$$\underbrace{}_{m + n \text{ factors}}$$

Thus, to find the *product of powers* of the same number, you add the
exponents. You can use this fact together with the commutative and
associative axioms for multiplication to simplify many products.

EXAMPLE 1   Simplify $(6x^2y)(-4x^5y^3)$.

SOLUTION $(6x^2y)(-4x^5y^3) = [6 \cdot (-4)](x^2 \cdot x^5)(y \cdot y^3)$
$$= -24(x^{2+5})(y^{1+3})$$
$$= -24x^7y^4$$

Now consider the expression

$$(a^5)^3.$$

If you apply the definition of power and the preceding result to this
expression, you obtain the following.

$$(a^5)^3 = a^5 \cdot a^5 \cdot a^5 = a^{5+5+5} = a^{15} = a^{5 \cdot 3}$$

In general, for all real numbers a, if m and n are positive integers:

$$(a^m)^n = \overbrace{(a^m)(a^m) \ldots (a^m)}^{a^m \text{ is a factor } n \text{ times}} = a^{\overbrace{m+m+\ldots+m}^{n \text{ terms}}} = a^{mn}$$

Thus, to find a *power of a power* of a real number, you multiply the
exponents.

328   *Chapter 7*

EXAMPLE 2 Simplify.   **a.** $(x^4)^3$ **b.** $(-y^2)^5$

SOLUTION   **a.** $(x^4)^3 = x^{4 \cdot 3} = x^{12}$

             b. $(-y^2)^5 = (-y)^{2 \cdot 5} = (-y)^{10} = y^{10}$

To simplify the expression

$$(ab)^3,$$

you can apply the definition of power and the commutative and associative axioms for multiplication.

$$(ab)^3 = (ab)(ab)(ab) = (a \cdot a \cdot a)(b \cdot b \cdot b) = a^3 b^3$$

In general, for all real numbers a and b, if m is a positive integer:

$$
(ab)^m = \overbrace{(ab)(ab) \cdot \ldots \cdot (ab)}^{ab \text{ is a factor } m \text{ times}} = \overbrace{(a \cdot \ldots \cdot a)}^{m \text{ factors}}\overbrace{(b \cdot \ldots \cdot b)}^{m \text{ factors}} = a^m b^m
$$

That is, to find a *power of a product* of real numbers, you find the power of each factor and then multiply.

EXAMPLE 3 Simplify.   **a.** $(5x)^3$ **b.** $(-2rs)^4$

SOLUTION   **a.** $(5x)^3 = 5^3 \cdot x^3 = 125x^3$

             b. $(-2rs)^4 = (-2)^4 \cdot r^4 \cdot s^4 = 16r^4 s^4$

The rules that have been used to simplify expressions in the three preceding examples can be summarized as the following *laws of exponents* for multiplication.

Laws of Exponents for Multiplication

For all real numbers a and b, if m and n are positive integers:

    1. $a^m \cdot a^n = a^{m+n}$ **2.** $(a^m)^n = a^{mn}$ **3.** $(ab)^m = a^m b^m$

Sometimes you need to use more than one of these laws of exponents in simplifying a single expression.

EXAMPLE 4 Simplify.   **a.** $(c^2 d)(2cd)^3$ **b.** $(-2x^2 y^3)(3xy^2)^3$

SOLUTION   **a.** $(c^2 d)(2cd)^3 = (c^2 d)(2^3 c^3 d^3)$

                          $= 2^3(c^2 \cdot c^3)(d \cdot d^3)$

                          $= 8c^5 d^4$

         b. $(-2x^2 y^3)(3xy^2)^3 = (-2x^2 y^3)[3^3 x^3 (y^2)^3]$

                                $= (-2x^2 y^3)(27x^3 y^6)$

                                $= (-2 \cdot 27)(x^2 \cdot x^3)(y^3 \cdot y^6)$

                                $= -54x^5 y^9$

Oral Exercises

Simplify. Assume that any variable used as an exponent represents a positive integer.

1. $z^5 \cdot z^2$ z^7
2. $c \cdot c^4$ c^5
3. $2r^2 \cdot r$ $2r^3$
4. $b^3 \cdot 5b^2$ $5b^5$
5. $3w^4 \cdot 2w$ $6w^5$
6. $4d^3 \cdot 3d^5$ $12d^8$
7. $a^2 \cdot a^3 \cdot a^5$ a^{10}
8. $t \cdot t^4 \cdot t^2$ t^7
9. $b^2 \cdot b^x$ b^{2+x}
10. $j^r \cdot j^s$ j^{r+s}
11. $m \cdot m^a \cdot m^b$ m^{1+a+b}
12. $u^y \cdot u \cdot u^y$ u^{2y+1}
13. $(cd)^3$ c^3d^3
14. $(rs)^5$ r^5s^5
15. $(5pq)^2$ $25p^2q^2$
16. $(-2uv)^3$ $-8u^3v^3$
17. $(a^3)^2$ a^6
18. $(z^2)^5$ z^{10}
19. $(2s^3)^2$ $4s^6$
20. $(5b^2)^3$ $125b^6$
21. $(a^3)^x$ a^{3x}
22. $(x^j)^k$ x^{jk}
23. $(3c)^d$ 3^dc^d
24. $(4c^2)^r$ 4^rc^{2r}

State the simplest form of the square of each monomial.

25. c^3 c^6
26. $-5m^4$ $25m^8$
27. $3ab$ $9a^2b^2$
28. $4u^3v^5$ $16u^6v^{10}$

State the simplest form of the cube of each monomial.

29. t^6 t^{18}
30. $-2n^2$ $-8n^6$
31. $-4st$ $-64s^3t^3$
32. $5r^3s^4$ $125r^9s^{12}$

Written Exercises

Simplify.

A

1. $(2b^2)(-5b)$ $-10b^3$
2. $(3z^3)(2z^2)$ $6z^5$
3. $(6mn)(-2m)^5$ $-192m^6n$
4. $(5cd)(-3d^2)$ $-15cd^3$
5. $(3a^2z^3)(3az)$ $9a^3z^4$
6. $(-2pq^6)(3p^2q)$ $-6p^3q^7$
7. $(4y)(-y^3)(2y^3)$ $-8y^7$
8. $(5c^3)(-2c^2)(-c^5)$ $10c^{10}$
9. $(-3uv^2)(5u^2v^2)$ $-15u^3v^4$
10. $(4a^3b)(-3a^2b)$ $-12a^5b^2$
11. $(2c^2)^2(c^5)$ $4c^9$
12. $(-3r^3)^2(r^4)$ $9r^{10}$
13. $(3g^2)^2(2g)^3$ $72g^7$
14. $(5f^3)^2(4f^2)^2$ $400f^{10}$
15. $(a^2b)^3(ab^2)^4$ $a^{10}b^{11}$
16. $(u^3v^2)^2(uv^4)^3$ u^9v^{16}
17. $(-2a^2b)^2(-a^4b^3)^3$ $-4a^{16}b^{11}$
18. $(3r^4s)^3(-5r^3s)^2$ $675r^{18}s^5$
19. $(x^2y)(xy^2)(x^2y^2)$ x^5y^5
20. $(a^3b^3)(a^2b)(a^2b^3)$ a^7b^7
21. $(-2mn)(3m^3n^2)(2m^4n)$ $-12m^8n^4$
22. $(-3jk^2)(7j^3k)(-j^4k^2)$ $21j^8k^5$
23. $(r^2s^2t)(r^3s)(s^4t^3)$ $r^5s^7t^4$
24. $(-p^2q^3)(-qr^4)(-p^5r^2)$ $-p^7q^4r^6$

Find a monomial that is equivalent to the given expression.

25. $(2z^3)(3z) + 5z^4$ $11z^4$
26. $8m^5 - 2m^2(3m^3)$ $2m^5$
27. $(5b^3c)(2bc^3) - (2bc)^4$ $-6b^4c^4$
28. $(-3r^2s^2)^2 + (3rs^2)(2r^3s^2)$ $15r^4s^4$
29. $(2xy^2)^2(5x^5y) - (6xy^2)(x^2y)^3$ $14x^7y^5$
30. $(3cd)(5c^3d^2)^3 - (4c^2d)^3(2c^2d^2)^2$ $119c^{10}d^7$

Additional A Exercises

Simplify.

1. $(4p^2)(5p^3)$ $20p^5$
2. $(4q^2)^3(q^4)$ $64q^{10}$
3. $(3s^2t^2)^3(-2st^3)^2$ $108s^8t^{12}$

Find a monomial that is equivalent to the given expression.

4. $(z^4)(2z) + 3z^5$ $5z^5$
5. $5p^6 - p^3(16p^3)$ $-11p^6$
6. $(b^2n)(3bn^2) - (3bn)^3$
   $-24b^3n^3$

Find a simplified polynomial that is equivalent to the given expression.

B **31.** $(3y^4)(-y)^5 - (4y)(2y^2)^3$ $-3y^9 - 32y^7$ **32.** $(5m^3)(3m)^4 + (-2m^2)(3m^2)^3$ $405m^7 - 54m^8$

33. $b^3(b^2)^2 + (2b)^4 - b(3b)^3$ $b^7 - 11b^4$ **34.** $q(2q^2)^3 - (3q^3)^2 - q(q^3)^2$ $7q^7 - 9q^6$

35. $(3c)^2(cd)^2 - (2c^2d)^2 + (5cd^2)^3$ **36.** $(5u)^2(2v^2)^3 + (3u^2v^2)^3 - (4uv^3)^2$

37. $-3(-2r^2st)^2(-rs^3)^3(rt^2)^4 + (-3s^4t^2)^2(-rs)^3(r^4t^3)^2$ $3r^{11}s^{11}t^{10}$

38. $(-x^4yz^2)^3(xy^2)^5(5yz^3)^2 + (-x^3z)^2(3xy^5)^3(-y^2z)^4$ $-25x^{17}y^{15}z^{12} + 27x^9y^{23}z^6$

35. $5c^4d^2 + 125c^3d^6$ **36.** $184u^2v^6 + 27u^6v^6$

Simplify. Assume that each expression that is an exponent represents a positive integer.

C **39.** $x^n \cdot x^n \cdot x^n$ x^{3n} **40.** $y^{n+1} \cdot y^{n-1}$ y^{2n} **41.** $(z^n)^2$ z^{2n}

42. $(a^{n+1})^2$ a^{2n+2} **43.** $(r^{n+1})^2 \cdot r^{n-2}$ r^{3n} **44.** $(s^2)^{n+1} \cdot s^{1-n}$ s^{n+3}

Solve for x.

45. $2^x \cdot 2^2 = 2^8$ $\{6\}$ **46.** $3^x \cdot 3^x = 3^{16}$ $\{8\}$ **47.** $(3^x)^5 = 3^{20}$ $\{4\}$

48. $(5^x)^x = 5^{16}$ $\{4\}$ **49.** $2 \cdot 2^{x-1} = 2^7$ $\{7\}$ **50.** $5^x \cdot 5 = 5^{11}$ $\{10\}$

Computer Exercises   For students with computer experience

1. Without using the computer's exponentiation operation, write a program that will compute the value of a^n when you input a value for a and a positive integral value for n.

2. Modify the program that you wrote for Exercise 1 so that it will allow you to input *two* numbers, a and b, and it will compute the values of a^n, b^n, and $(ab)^n$.

3. Write a program that will evaluate a polynomial of the form $ax^3 + bx^2 + cx + d$ when you input values for a, b, c, and d as well as a value for x. (*Note:* Use the computer's exponentiation operation, but be aware that it will often produce only *approximations* of the correct value because of the method that the computer uses to calculate the result.)

4. Write a program that will perform the same task as the program that you wrote for Exercise 3, but this time do *not* use the computer's exponentiation operation. The method that you use should be based on the fact that

$$ax^3 + bx^2 + cx + d = [(ax + b)x + c]x + d.$$

5. Modify the program that you wrote for Exercise 4 so that it will evaluate any polynomial of degree n in the variable x when you input a value for n. (*Hint:* Write the program so that the values of x and n are input first. Then have the program evaluate the polynomial as the coefficients are input, adding each coefficient to the running total and multiplying this total by x.)

Polynomials and Factoring **331**

Mixed Review

Simplify.

1. $(2m^2 - 4) + (m^2 + m + 4)$
$3m^2 + m$

2. $(3mn^2 + 2m^2n) + (5mn^2 - 2m^2n)$ $8mn^2$

Evaluate each expression when $a = 2$ and $b = -1$.

3. a. $(a^3)^2$ 64 **b.** $(a^2)^3$ 64
 c. a^5 32 **d.** a^6 64

4. a. $(a^2b)(ab^2)$ -8
 b. a^3b^3 -8

Key Ideas

Multiply a polynomial by a monomial.
Multiply a polynomial by a polynomial.

Chalkboard Examples

1. Simplify $a^2(3a - ab - b)$.
   $= a^2(3a) + a^2(-ab) +$
   $\qquad\qquad\qquad a^2(-b)$
   $= 3a^3 - a^3b - a^2b$

2. Simplify
   $4t(6t + 1) - 2t^2(t - 2)$.
   $= [4t(6t) + 4t(1)] -$
   $\qquad [2t^2(t) + 2t^2(-2)]$
   $= (24t^2 + 4t) - (2t^3 - 4t^2)$
   $= 24t^2 + 4t - 2t^3 + 4t^2$
   $= -2t^3 + 28t^2 + 4t$

3. Simplify
   $(k + 3)(k^3 + 4k^2 - k)$.
   $$\begin{array}{r} k^3 + 4k^2 - k \\ \underline{k \qquad\quad + 3} \\ k^4 + 4k^3 - \quad k^2 \\ \underline{\quad 3k^3 + 12k^2 - 3k} \\ k^4 + 7k^3 + 11k^2 - 3k \end{array}$$

4. Solve $9p(p + 3) =$
   $9(p^2 + 6)$.
   $9p(p + 3) = 9(p^2 + 6)$
   $9p^2 + 27p = 9p^2 + 54$
   $\qquad\qquad p = 2$
   $\therefore$ the solution is $\{2\}$.

7–3 Multiplying Polynomials

To multiply a polynomial *by a monomial*, you use the distributive axiom together with the laws of exponents for multiplication. For example:

$$2x(3x + 7) = 2x(3x) + 2x(7)$$
$$= 6x^2 + 14x$$

The expression $6x^2 + 14x$ is called the *product* of the monomial $2x$ and the polynomial $3x + 7$.

EXAMPLE 1 Simplify $-3a^3(2a^2 - 5a + 2)$.

SOLUTION $-3a^3(2a^2 - 5a + 2) = -3a^3(2a^2) + (-3a^3)(-5a) + (-3a^3)(2)$
$$= -6a^5 + 15a^4 - 6a^3$$

EXAMPLE 2 Simplify $2m^2(3m + 1) - 5m(m + 4)$.

SOLUTION Follow the order of operations and first perform the indicated multiplications. Then simplify the resulting polynomial by adding the similar terms.

$$2m^2(3m + 1) - 5m(m + 4)$$
$$= [2m^2(3m) + 2m^2(1)] - [5m(m) + 5m(4)]$$
$$= (6m^3 + 2m^2) - (5m^2 + 20m)$$
$$= 6m^3 + 2m^2 + (-5m^2) + (-20m)$$
$$= 6m^3 - 3m^2 - 20m$$

To multiply a polynomial *by a polynomial*, you multiply each term of one of the polynomials by each term of the other, then add similar terms as shown in Example 2. For example:

$$(x + 4)(2x + 5) = x(2x + 5) + 4(2x + 5)$$
$$= x(2x) + x(5) + 4(2x) + 4(5)$$
$$= 2x^2 + 5x + 8x + 20$$
$$= 2x^2 + 13x + 20$$

If you wish, you can use a vertical arrangement when you multiply two polynomials, as shown in the following example.

EXAMPLE 3 Simplify $(y + 2)(y^3 - 5y^2 + 2)$.

SOLUTION
$$\begin{array}{l} y^3 - 5y^2 \qquad + 2 \\ \underline{y \ + 2} \\ y^4 - 5y^3 \qquad + 2y \qquad \longleftarrow \quad \{ \ y(y^3 - 5y^2 + 2) \\ \underline{\quad 2y^3 - 10y^2 \qquad + 4} \quad \longleftarrow \quad \{ \ 2(y^3 - 5y^2 + 2) \\ y^4 - 3y^3 - 10y^2 + 2y + 4 \quad \longleftarrow \quad \{(y + 2)(y^3 - 5y^2 + 2) \end{array}$$

332 *Chapter 7*

Notice in the first line of the solution to Example 3 that a space was left for the "missing term" $0y$.

Sometimes you will need to multiply polynomials in the process of solving an equation.

EXAMPLE 4   Solve $4n(n + 2) = 4(n^2 - 4)$.

SOLUTION

$$4n(n + 2) = 4(n^2 - 4)$$
$$4n^2 + 8n = 4n^2 - 16$$
$$4n^2 + 8n - 4n^2 = 4n^2 - 16 - 4n^2$$
$$8n = -16$$
$$n = -2$$

Check:

$$4n(n + 2) = 4(n^2 - 4)$$
$$4(-2)(-2 + 2) \overset{?}{=} 4[(-2)^2 - 4]$$
$$4(-2)(0) \overset{?}{=} 4(4 - 4)$$
$$0 = 0 \;\checkmark$$

∴ the solution set is $\{-2\}$.

Oral Exercises

Simplify.

1. $3(w - 2)$ $3w - 6$
2. $-2(a + 5)$ $-2a - 10$
3. $-12 + 4c$
3. $-4(3 - c)$
4. $(5 - z)5$ $25 - 5z$
5. $(3a + 2)(-2)$ $-6a - 4$
6. $-12d + 3$
6. $-3(4d - 1)$
7. $f(2f - 3)$ $2f^2 - 3f$
8. $g(5 - 4g)$ $5g - 4g^2$
9. $u^2(2u + 7)$
9. $2u^3 + 7u^2$
10. $-p^2(3p - 4)$ $-3p^3 + 4p^2$
11. $3c(4 - 5c^2)$ $12c - 15c^3$
12. $5v(3v^2 + 8)$
12. $15v^3 + 40v$

Replace each __?__ with the number or expression that makes a true statement.

13. $(3b + 2)(2b + 3) = \underline{\;?\;}(2b + 3) + \underline{\;?\;}(2b + 3)$ $3b; 2$
14. $(7a + 1)(2a - 3) = (7a + 1)\underline{\;?\;} - (7a + 1)\underline{\;?\;}$ $2a; 3$
15. $(5w - 3)(3w - 2) = 5w(\underline{\;?\;}) - 3(\underline{\;?\;})$ $3w - 2; 3w - 2$
16. $(8v - 9)(v + 5) = (8v - 9)\underline{\;?\;} + (8v - 9)\underline{\;?\;}$ $v; 5$
17. $(2f - 5)(f^2 + 3f - 1) = \underline{\;?\;}(f^2 + 3f - 1) - \underline{\;?\;}(f^2 + 3f - 1)$ $2f; 5$
18. $(4u + 3)(2u^2 - u + 2) = 4u(\underline{\;?\;}) + 3(\underline{\;?\;})$ $2u^2 - u + 2; 2u^2 - u + 2$

19. $(2m + 3)(m + 4) = 2m(m + 4) + 3(m + 4)$
     $= \underline{\;?\;}m^2 + \underline{\;?\;}m + \underline{\;?\;}$ $2; 11; 12$

20. $(3d - 2)(2d + 5) = 3d(2d + 5) - 2(2d + 5)$
     $= \underline{\;?\;}d^2 + \underline{\;?\;}d + \underline{\;?\;}$ $6; 11; -10$

Polynomials and Factoring   **333**

Suggested Assignments

Minimum
334/1–30
Average
334/1–43 odd
Maximum
334/7–57 odd

**Additional Answers
Written Exercises**

8. $-6t^6 - 15t^5 + 9t^4 + 3t^3$

9. $2p^4q - 2p^3q^2 + 2p^2q^3 - 2pq^4$

10. $6c^2d^4 - 15c^3d^3 + 3c^4d^2 - 12c^5d$

11. $-10u^2v^7 + 4u^4v^5 - 6u^6v^3$

12. $15j^7k^4 - 5j^6k^5 + 5j^4k^7 - 35j^3k^8$

13. $2x^3 + 3x^2 - 10x$

14. $3y^4 - 3y^3 - 2y^2$

32. $3s^3 - 4s^2 - 7s + 6$

34. $3n^3 + 17n^2 + 22n + 8$

48. $12n^3 + 13n^2 - 20n + 4$

57. The degree of the product is the degree of the product of the terms of greatest degree of the two polynomials, which is the sum of the degrees of the polynomials.

Mixed Review

Fill in each _?_ to make a true statement.

1. $8(3 + 9) = 8(3) + 8(\underline{?})$ 9

2. $(a + b)c = ac + \underline{?}$ bc

Evaluate when $a = -3$.

3. a. $2a(a^2 - 3)$ -36
 b. $2a^3 - 6a$ -36

4. a. $(2a + 3)(a - 2)$ 15
 b. $2a^2 - 4a + 3a - 6$ 15

Written Exercises

Simplify.

A

1. $2m(m^2 + 5m - 9)$ $2m^3 + 10m^2 - 18m$
2. $(4n^2 - n + 7)(-3n)$ $-12n^3 + 3n^2 - 21n$
3. $z^2(z^2 - 5z + 2)$ $z^4 - 5z^3 + 2z^2$
4. $y^4(5 - 2y - y^2)$ $5y^4 - 2y^5 - y^6$
5. $-a^3(3 - 7a - 4a^2)$ $-3a^3 + 7a^4 + 4a^5$
6. $(c^4 + 3c^2 - 5)(-c^5)$ $-c^9 - 3c^7 + 5c^5$
7. $4s^2(3s^3 - 2s^2 + s - 5)$ $12s^5 - 8s^4 + 4s^3 - 20s^2$
8. $-3t^3(2t^3 + 5t^2 - 3t - 1)$
9. $2pq(p^3 - p^2q + pq^2 - q^3)$
10. $3c^2d(2d^3 - 5cd^2 + c^2d - 4c^3)$
11. $-2u^2v^3(5v^4 - 2u^2v^2 + 3u^4)$
12. $5j^3k^4(3j^4 - j^3k + jk^3 - 7k^4)$
13. $x^2(2x + 1) + 2x(x - 5)$
14. $y^3(4y - 3) - y^2(y^2 + 2)$
15. $2a^2(4a - 3) - a(6a^2 - a)$ $2a^3 - 5a^2$
16. $b^3(3b - 1) - 3b^2(b^2 - 7b)$ $20b^3$
17. $2j(4j^2 + 3j) - (j + 1)j^2$ $7j^3 + 5j^2$
18. $3k^2(k - 2) - (4k - 1)k^2$ $-k^3 - 5k^2$
19. $(a + 3)(a + 1)$ $a^2 + 4a + 3$
20. $(z - 1)(z - 4)$ $z^2 - 5z + 4$
21. $(m + 2)(m - 4)$ $m^2 - 2m - 8$
22. $(c + 5)(c - 2)$ $c^2 + 3c - 10$
23. $(2b + 1)(b + 3)$ $2b^2 + 7b + 3$
24. $(3w + 2)(w - 4)$ $3w^2 - 10w - 8$
25. $(4t - 3)(2t + 3)$ $8t^2 + 6t - 9$
26. $(5c + 3)(5c - 3)$ $25c^2 - 9$
27. $(3z - 7)(3z + 7)$ $9z^2 - 49$
28. $(2x - 5)(5x + 6)$ $10x^2 - 13x - 30$
29. $(2c + 5)^2$ $4c^2 + 20c + 25$
30. $(3d - 2)^2$ $9d^2 - 12d + 4$

33. $2c^3 - 9c^2 + 10c - 3$

B

31. $(r + 3)(r^2 - 2r - 2)$ $r^3 + r^2 - 8r - 6$
32. $(s - 2)(3s^2 + 2s - 3)$
33. $(2c^2 - 3c + 1)(c - 3)$
34. $(3n^2 + 5n + 2)(n + 4)$
35. $(u - v)(u^2 + uv - v^2)$ $u^3 - 2uv^2 + v^3$
36. $(x + y)(x^2 - xy + y^2)$ $x^3 + y^3$
37. $(2c + d)(3c^2 - 2cd - 5d^2)$
$6c^3 - c^2d - 12cd^2 - 5d^3$
38. $(3a - b)(5a^2 + 3ab - 2b^2)$
$15a^3 + 4a^2b - 9ab^2 + 2b^3$

Solve.

39. $3a(a + 1) = 3(a^2 + 5)$ {5}
40. $2(b^2 - 8) = 2b(b + 4)$ {−2}
41. $2x(x + 3) = 2x(x + 1) - 20$ {−5}
42. $4y(y + 1) = y(4y + 3) - 7$ {−7}
43. $r(3r + 8) - 15 = 3r(r + 1)$ {3}
44. $2t(2t + 3) + 14 = t(4t - 1)$ {−2}

Simplify.

47. $6m^3 + 11m^2 - 3m - 2$

C

45. $(x + 1)(x - 2)(x + 3)$ $x^3 + 2x^2 - 5x - 6$
46. $(y - 2)(y + 3)(y - 5)$ $y^3 - 4y^2 - 11y + 30$
47. $(m + 2)(2m - 1)(3m + 1)$
48. $(4n - 1)(n + 2)(3n - 2)$
49. $(r + 2)^3$ $r^3 + 6r^2 + 12r + 8$
50. $(2s + 1)^3$ $8s^3 + 12s^2 + 6s + 1$
51. $(x^2 - 2x + 3)(x^2 + 2x + 1)$
 $x^4 + 4x + 3$
52. $(3a^2 + 2ab + b^2)(a^2 - ab - b^2)$
 $3a^4 - a^3b - 4a^2b^2 - 3ab^3 - b^4$

Simplify. Assume that n is a positive integer.

53. $x^n(x^n + 1)$ $x^{2n} + x^n$
54. $x^n(x^n + x)$ $x^{2n} + x^{n-1}$
55. $x^{n+1}(x^n + 1)$ $x^{2n+1} + x^{n-1}$
56. $x^{n+1}(x^n + x)$ $x^{2n+1} + x^{n-2}$

57. Explain why the degree of the product of two nonzero polynomials is equal to the sum of the degrees of the polynomials.

334 *Chapter 7*

7–4 Multiplying Binomials Mentally

When you multiply two binomials, you usually can save time if you learn to perform the multiplication mentally. For example, the multiplication at the right shows the steps used to find the product of the binomials $2x + 1$ and $3x - 2$. The procedure for obtaining each term of the product is outlined below.

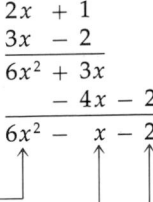

$$
\begin{array}{r}
2x + 1 \\
3x - 2 \\
\hline
6x^2 + 3x \\
-4x - 2 \\
\hline
6x^2 - x - 2
\end{array}
$$

1. Multiply the *first terms* of the binomials.
2. Multiply the *first term* of each binomial by the *second term* of the other; add the products if possible.
3. Multiply the *second terms* of the binomials.

When the binomials are arranged horizontally, this procedure can be shown as follows.

$$
(2x + 1)(3x - 2) = 6x^2 - x - 2
$$

Special patterns emerge when you multiply certain pairs of binomial factors. For example, the multiplication at the right shows the steps used to *square the binomial* $a + b$. Again, notice how each term of the product is obtained.

$$
\begin{array}{r}
a + b \\
a + b \\
\hline
a^2 + ab \\
ab + b^2 \\
\hline
a^2 + 2ab + b^2
\end{array}
$$

1. Square the first term of the binomial.
2. Double the product of the two terms.
3. Square the second term of the binomial.

Thus,

$$(a + b)^2 = a^2 + 2ab + b^2.$$

Similarly, when you square the binomial $a - b$, you obtain the result

$$(a - b)^2 = a^2 - 2ab + b^2.$$

Because each of the expressions $a^2 + 2ab + b^2$ and $a^2 - 2ab + b^2$ is a trinomial that can be obtained by squaring a binomial, a polynomial that can be written in one of these forms is called a **trinomial square**. For example, since

$$m^2 + 6m + 9 = m^2 + 2(3)m + 3^2 = (m + 3)^2,$$

the trinomial $m^2 + 6m + 9$ is a trinomial square.

Polynomials and Factoring **335**

Supplementary Material
Test 26

Key Ideas
Find the product of two binomials mentally.
Solve word problems by multiplying polynomials.

1. Simplify $(2m - 5)^2$.
 $= (2m - 5)(2m - 5)$
 $= 4m^2 - 2(2m)(5) + 5^2$
 $= 4m^2 - 20m + 25$

2. Simplify $(2c + 3)(2c - 3)$.
 $= 4c^2 + 6c - 6c - 9$
 $= 4c^2 - 9$

3. Simplify $31 \cdot 29$.
 $31 = 30 + 1$
 $29 = 30 - 1$
 $31 \cdot 29 = (30 + 1)(30 - 1)$
    $= 30^2 - 1^2$
    $= 899$

Solve.

4. A certain rectangle is 2 cm longer than it is wide. A second rectangle is 3 cm longer and 2 cm narrower than the first rectangle. Find the dimensions of the first rectangle if both rectangles have the same area.
Let x = the width of the first rectangle. Then $x + 2$ = the length of the first rectangle; $x - 2$ = the width of the second rectangle; $x + 5$ = the length of the second rectangle. Area of first rectangle equals area of second rectangle.
$x(x + 2) = (x - 2)(x + 5)$
$x^2 + 2x = x^2 + 3x - 10$
   $x = 10$
  $x + 2 = 12$
$\therefore$ the first rectangle is 10 cm wide and 12 cm long.

EXAMPLE 1 Simplify.

  a. $(a + 5)^2$ **b.** $(3b - 2)^2$ **c.** $(2x - 5y)^2$ **d.** $(4z^3 + 1)^2$

SOLUTION **a.** $(a + 5)^2$ $= a^2 + 2(a)(5) + 5^2$
          $= a^2 + 10a + 25$

     b. $(3b - 2)^2$ $= (3b)^2 - 2(3b)(2) + 2^2$
           $= 9b^2 - 12b + 4$

     c. $(2x - 5y)^2 = (2x)^2 - 2(2x)(5y) + (5y)^2$
           $= 4x^2 - 20xy + 25y^2$

     d. $(4z^3 + 1)^2 = (4z^3)^2 + 2(4z^3)(1) + 1^2$
           $= 16z^6 + 8z^3 + 1$

You can extend the procedure for finding the square of a binomial to finding higher powers. For example, to find the *cube* of the binomial $a + b$, observe the following.

$$(a + b)^3 = (a + b)(a + b)^2$$
$$= (a + b)(a^2 + 2ab + b^2)$$

To complete the multiplication, you can use a vertical arrangement as shown at the right. Thus,

$$\begin{array}{r} a^2 + 2ab + b^2 \\ a + b \\ \hline a^3 + 2a^2b + ab^2 \\ a^2b + 2ab^2 + b^3 \\ \hline a^3 + 3a^2b + 3ab^2 + b^3 \end{array}$$

$$(a + b)^3 = a^3 + 3a^2b + 3ab^2 + b^3.$$

Similarly, the cube of the binomial $a - b$ is

$$(a - b)^3 = a^3 - 3a^2b + 3ab^2 - b^3.$$

Another special pattern occurs when you multiply two binomials in which *the first terms are the same*, but *the second terms are opposites of each other*. For example, the multiplication at the right shows the steps used to find the product of the binomials $a + b$ and $a - b$. Notice that two terms of the product, ab and $-ab$, are opposites, and so their sum is zero. Thus, you obtain the result

$$\begin{array}{r} a + b \\ a - b \\ \hline a^2 + ab \\ - ab - b^2 \\ \hline a^2 \qquad - b^2 \end{array}$$

$$(a + b)(a - b) = a^2 - b^2.$$

A polynomial that can be written in the form $a^2 - b^2$ is called a **difference of squares**.

EXAMPLE 2 Simplify.

  a. $(s + 4)(s - 4)$

  b. $(3t - 1)(3t + 1)$

  c. $(m^2 + 7n)(m^2 - 7n)$

SOLUTION **a.** $(s + 4)(s - 4) = s^2 - 4^2 = s^2 - 16$

     b. $(3t - 1)(3t + 1) = (3t)^2 - 1^2 = 9t^2 - 1$

     c. $(m^2 + 7n)(m^2 - 7n) = (m^2)^2 - (7n)^2 = m^4 - 49n^2$

336 *Chapter 7*

Sometimes you can mentally find the product of two real numbers if you recognize that one factor is the *sum* of two particular numbers and the other factor is the *difference* of these same numbers. You can compute such a product by finding the difference of the squares of these numbers.

EXAMPLE 3 Simplify $51 \cdot 49$.

SOLUTION $51 = 50 + 1$

$49 = 50 - 1$

$51 \cdot 49 = (50 + 1)(50 - 1) = 50^2 - 1^2 = 2500 - 1 = 2499$

$\therefore 51 \cdot 49 = 2499$

In solving a word problem, you sometimes need to multiply polynomials.

EXAMPLE 4 A certain rectangle is three times as long as it is wide. A second rectangle is 4 cm longer and 1 cm narrower than the first rectangle, but the areas of the two rectangles are equal. Find the dimensions of the first rectangle.

SOLUTION

Step 1   The problem asks for the dimensions of the first rectangle.

Step 2   Let w = the width of the first rectangle. Then:

$3w$ = the length of the first rectangle

$w - 1$ = the width of the second rectangle

$3w + 4$ = the length of the second rectangle

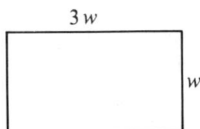

Step 3 $\underbrace{\text{Area of first rectangle}}_{3w(w)}$ equals $\underset{\overset{\downarrow}{=}}{}$ $\underbrace{\text{area of second rectangle}}_{(3w + 4)(w - 1)}$

Step 4 $3w(w) = (3w + 4)(w - 1)$

$3w^2 = 3w^2 + w - 4$

$3w^2 - 3w^2 = 3w^2 + w - 4 - 3w^2$

$0 = w - 4$

$4 = w$

width $= 4$

length $= 3(4) = 12$

Step 5   Checking the results is left to you.

$\therefore$ the first rectangle is 4 cm wide and 12 cm long.

Polynomials and Factoring   **337**

Oral Exercises

Name the coefficient of x in each simplified product.

1. $(x + 3)(x + 1)$ 4   **2.** $(x - 3)(x - 1)$ −4   **3.** $(x + 3)(x - 1)$ 2   **4.** $(x - 3)(x + 1)$ −2

5. $(x + 3)(x - 3)$ 0   **6.** $(x - 3)(x + 3)$ 0   **7.** $(x + 3)(x + 3)$ 6   **8.** $(x - 3)(x - 3)$ −6

Complete.

9. $(a + 1)(a + 4) = a^2 + \underline{?}a + 4$ 5

10. $(b - 2)(b - 3) = b^2 + \underline{?}b + 6$ −5

11. $(m - 5)(m + 2) = m^2 + \underline{?}m - 10$ −3

12. $(n + 4)(n - 3) = n^2 + n + \underline{?}$ −12

13. $(3x + 2)(x + 1) = 3x^2 + \underline{?}x + 2$ 5

14. $(y - 4)(2y + 5) = 2y^2 + \underline{?}y - 20$ −3

15. $(r - 5)(r - 5) = r^2 - 10r + \underline{?}$ 25

16. $(s - 3)(s + 3) = s^2 + \underline{?}s - 9$ 0

17. $(v + 7)^2 = v^2 + \underline{?}v + 49$ 14

18. $(3w - 1)^2 = 9w^2 + \underline{?}w + 1$ −6

19. $(2c + 3)^2 = 4c^2 + \underline{?}c + 9$ 12

20. $(4d - 1)(3d + 2) = 12d^2 + \underline{?}d - 2$ 5

Written Exercises

Simplify.

A

1. $(j + 5)(j - 3)$ $j^2 + 2j - 15$ **2.** $(k - 2)(k + 5)$ $k^2 + 3k - 10$ **3.** $(3 + u)(7 + u)$ $21 + 10u + u^2$

4. $(7 - v)(1 - v)$ $7 - 8v + v^2$ **5.** $(s + 11)(s - 11)$ $s^2 - 121$ **6.** $(t - 9)(t + 9)$

7. $(m - 7)^2$ $m^2 - 14m + 49$ **8.** $(n + 10)^2$ $n^2 + 20n + 100$ **9.** $(3y - 1)(y + 2)$

10. $(2z - 5)(z + 3)$ **11.** $(2c + 3)^2$ $4c^2 + 12c + 9$ **12.** $(3s - 7)^2$

13. $(2p + 1)(3p - 2)$ **14.** $(5q - 3)(5q + 3)$ $25q^2 - 9$ **15.** $(4a - 7)(4a + 7)$

16. $(4b - 3)(2b + 3)$ **17.** $(4r - 3s)(3r + 2s)$ **18.** $(3j + 4k)(2j - 3k)$

19. $(6c - 5d)(2c - 3d)$ **20.** $(3m + 5n)(4m + 3n)$ **21.** $(5b + 2c)^2$

22. $(4x - 3y)^2$ $16x^2 - 24xy + 9y^2$ **23.** $(3a^2 + 1)(3a^2 - 1)$ $9a^4 - 1$ **24.** $(2z^3 - 5)(2z^3 + 5)$

25. $(4m^2 + 3)^2$ $16m^4 + 24m^2 + 9$ **26.** $(5n^3 - 1)^2$ $25n^6 - 10n^3 + 1$ **27.** $(j^2 - k^2)^2$

28. $(s^3 + t^3)^2$ $s^6 + 2s^3t^3 + t^6$ **29.** $(5c^3 - d^3)^2$ $25c^6 - 10c^3d^3 + d^6$ **30.** $(p^2 + 3q^2)^2$

B

31. $(x + 2)^3$ **32.** $(y - 1)^3$ **33.** $(3m - 1)^3$

34. $(2n + 3)^3$ **35.** $(a^2 + 1)^3$ **36.** $(b^2 - 2)^3$

$8n^3 + 36n^2 + 54n + 27$ $a^6 + 3a^4 + 3a^2 + 1$ $b^6 - 6b^4 + 12b^2 - 8$

Compute each product by first rewriting it as a difference of squares.

37. $99 \cdot 101$ **38.** $38 \cdot 42$ **39.** $28 \cdot 32$ **40.** $81 \cdot 79$

41. $87 \cdot 93$ **42.** $16 \cdot 24$ **43.** $8.9 \cdot 9.1$ **44.** $48\frac{1}{2} \cdot 51\frac{1}{2}$

$90^2 - 3^2 = 8091$ $20^2 - 4^2 = 384$ $9^2 - 0.1^2 = 80.99$ $50^2 - (\frac{3}{2})^2 = 2497\frac{3}{4}$

Solve.

45. $(4x + 1)(x - 2) = (2x - 3)(2x + 3)$ {1}   **46.** $(4y - 1)(4y + 1) = (8y - 3)(2y + 1)$ {1}

47. $(8a - 3)(a + 1) = (4a + 5)(2a - 1)$ {2}   **48.** $(6b - 5)(2b + 1) = (4b - 3)(3b + 1)$

49. $(2m - 1)^2 = (4m + 1)(m - 2)$ {−1}   **50.** $(n - 2)(4n + 7) = (2n + 1)^2$ {−3} {2}

51. $(x + 1)^2 - x^2 = 3(x - 2)$ {7}   **52.** $y^2 - (y - 1)^2 = 3(y + 1)$ {−4}

53. $(r - 1)^2 - (r + 1)^2 = 2(r + 4)$ {−$\frac{4}{3}$}   **54.** $(s + 1)^2 - (s - 1)^2 = 2(s - 3)$ {−3}

338   *Chapter 7*

Suggested Assignments

Minimum
  338/1–25
  339/P: 1–4
R 340/ Self-Test 1
Average
  338/5–33 odd
  339/P: 1–8
R 340/Self-Test 1
Maximum
  338/25–53 odd, 55–60
  339/P: 5–12
R 340/Self-Test 1

Additional Answers
Written Exercises

6. $t^2 - 81$

9. $3y^2 + 5y - 2$

10. $2z^2 + z - 15$

12. $9s^2 - 42s + 49$

13. $6p^2 - p - 2$

15. $16a^2 - 49$

16. $8b^2 + 6b - 9$

17. $12r^2 - rs - 6s^2$

18. $6j^2 - jk - 12k^2$

19. $12c^2 - 28cd + 15d^2$

20. $12m^2 + 29mn + 15n^2$

21. $25b^2 + 20bc + 4c^2$

24. $4z^6 - 25$

27. $j^4 - 2j^2k^2 + k^4$

30. $p^4 + 6p^2q^2 + 9q^4$

31. $x^3 + 6x^2 + 12x + 8$

32. $y^3 - 3y^2 + 3y - 1$

33. $27m^3 - 27m^2 + 9m - 1$

37. $100^2 - 1^2 = 9999$

38. $40^2 - 2^2 = 1596$

39. $30^2 - 2^2 = 896$

40. $80^2 - 1^2 = 6399$

Simplify. Assume that n represents a positive integer.

C **55.** $(x^n + y^n)^2$ **56.** $(x^{n+1} - y)^2$

57. $(x^n - y^n)^3$ **58.** $(x^{2n} + y)(x^{2n} - y)$

59. Show that the absolute value of the difference of the squares of two consecutive integers is equal to the absolute value of the sum of the integers.

60. Show that the absolute value of the difference of the squares of two consecutive even integers is equal to twice the absolute value of the sum of the integers.

Problems

Solve.

A **1.** The side of one square is 3 cm longer than the side of a second square, and the area of the second square is 51 cm^2 less than the area of the first square. Find the length of a side of the second square. 7 cm

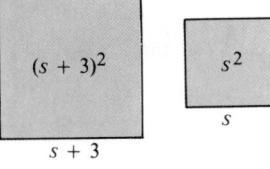

$(s + 3)^2$ s^2

s

$s + 3$

Ex. 1

2. A rectangle is 2 cm wider and 5 cm longer than a certain square. The area of the rectangle is 38 cm^2 greater than that of the square. What are the dimensions of the rectangle? 6 cm by 9 cm

3. The difference of the squares of two consecutive positive even integers is 164. Find the integers. 40 and 42

4. The difference of the squares of two consecutive negative integers is 9. Find the integers. -5 and -4

B **5.** The product of two consecutive positive odd integers is 38 less than the square of the greater integer. Find the integers. 17 and 19

6. The product of three consecutive integers is 22 less than the cube of the middle integer. Find the integers. 21, 22, 23

7. The height of a rectangular box is 2 cm less than the width. The length of the box is 5 cm greater than the width. The sum of the areas of the top and bottom of the box is 22 cm less than the sum of the areas of the sides of the box. Find the dimensions of the box. $h = 5$ cm; $w = 7$ cm; $l = 12$ cm

8. A rectangular box is 4 cm longer and 3 cm narrower than a certain cube. The rectangular box and the cube have equal heights and equal surface areas. Find the length and width of the rectangular box. 10 cm; 3 cm

C **9.** A rectangular picture is 3 cm longer than it is wide. The picture is surrounded by a frame that is 2 cm wide. The area of the picture and the frame together is 100 cm^2 greater than the area of the picture. Find the width of the picture. 9 cm

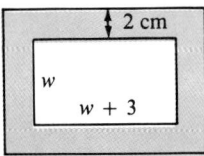

2 cm

w

$w + 3$

Ex. 9

Polynomials and Factoring   **339**

---

55. $x^{2n} + 2x^n y^n + y^{2n}$

56. $x^{2n+2} - 2x^{n+1}y + y^2$

57. $x^{3n} - 3x^{2n}y^n + 3x^n y^{2n} - y^{3n}$

58. $x^{4n} - y^2$

59. $|(x + 1)^2 - x^2| = |x^2 + 2x + 1 - x^2| = |2x + 1| = |x + x + 1|$

60. $|(x + 2)^2 - x^2| = |x^2 + 4x + 4 - x^2| = |4x + 4| = |2(2x + 2)| = 2|(x + x + 2)|$

10. A certain castle, twice as long as it is wide, is surrounded by a moat 6 m wide. The area of the castle floor (the area of the rectangle inside the moat) is 864 m^2 less than the area of the region bounded by the outer edge of the moat. Find the dimensions of the castle. 40 m by 20 m

11. Find three consecutive integers such that the product of the first and third is one less than the square of the second.

12. Kevin has three boxes, each the shape of a cube. An edge of the largest box is 2 cm longer than an edge of the middle box, which in turn is 2 cm longer than an edge of the smallest box. The surface area of the largest box is 240 cm^2 greater than the surface area of the smallest box. Find the measure of an edge of each box. 7 cm, 5 cm, 3 cm
   11. Any three consecutive integers have this property.

Quick Quiz

Add or subtract the polynomials as indicated.

1. $(2p^2 - 4p + 9) + (p^2 + 2p + 2)$
   $3p^2 - 2p + 11$

2. $(9t^3 - 3t^2 - 3) - (4t^3 - 2t^2 - 4)$
   $5t^3 - t^2 + 1$

3. $(4c^2 + 3c) - (2c^3 - 3c^2 + 2c - 6)$
   $-2c^3 + 7c^2 + c + 6$

4. $(6m^2n - mn^2) + (5m^2 - 6m^2n + 2n^2)$
   $5m^2 - mn^2 + 2n^2$

Simplify.

5. $x^3 \cdot x^4$ x^7

6. $(3k^2)(-3k^4)$ $-9k^6$

7. $(2x^2z)(6xz^4)$ $12x^3z^5$

8. $(p^3)^4$ p^{12}

9. $(-5de)^3$ $-125d^3e^3$

10. $(3e^4f^2)^2$ $9e^8f^4$

11. $2g(g^2 - 3g)$ $2g^3 - 6g^2$

12. $h^2i(2h^2 - 4hi - i^2)$
   $2h^4i - 4h^3i^2 - h^2i^3$

13. $(j - 2)(j + 5)$
   $j^2 + 3j - 10$

14. $(3k - 2)(k + 3)$
   $3k^2 + 7k - 6$

15. $(2q - 3)^2$ $4q^2 - 12q + 9$

16. $(3r - 4s)(3r + 4s)$
   $9r^2 - 16s^2$

■ Self-Test 1

VOCABULARY

monomial (p. 323)
constant monomial or constant (p. 323)
degree of a variable in a monomial (p. 323)
degree of a monomial (p. 324)
coefficient or numerical coefficient of a monomial (p. 324)
similar or like monomials (p. 324)
polynomial (p. 324)

term of a polynomial (p. 324)
coefficients of a polynomial (p. 324)
binomial (p. 324)
trinomial (p. 324)
simplest form of a polynomial (p. 324)
degree of a polynomial (p. 324)
laws of exponents for multiplication (p. 329)
trinomial square (p. 335)
difference of squares (p. 336)

Add or subtract the polynomials as indicated.

1. $(3r^2 - 2r + 6) + (r^2 + r - 2)$ $4r^2 - r + 4$ *Obj. 1, p. 323*

2. $(5c^2 + 2c) - (4c^3 + 2c^2 + c - 1)$ $-4c^3 + 3c^2 + c + 1$

3. $(8u^2v - 3uv^2) + (2u^2 - 3u^2v + v^2)$ $2u^2 + 5u^2v - 3uv^2 + v^2$

Simplify.

4. $p^2 \cdot p^5$ p^7 **5.** $(4a^3)(-2a^2)$ $-8a^5$ **6.** $(3x^2y)(xy^3)$ $3x^3y^4$ *Obj. 2, p. 323*

7. $(m^2)^5$ m^{10} **8.** $(-3st)^2$ $9s^2t^2$ **9.** $(2u^2v^3)^3$ $8u^6v^9$ *Obj. 3, p. 323*

10. $5y(y^2 - 2y)$ $5y^3 - 10y^2$ **11.** $a^2b(3a^2 - 2ab + b^2)$ *Obj. 4, p. 323*

12. $(c - 2)(c + 3)$ $c^2 + c - 6$ **13.** $(2d - 5)(d + 4)$ $2d^2 + 3d - 20$ *Obj. 5, p. 323*

14. $(3x - 2)^2$ $9x^2 - 12x + 4$ **15.** $(2m - 5n)(2m + 5n)$ $4m^2 - 25n^2$
                                                    11. $3a^4b - 2a^3b^2 + a^2b^3$

Check your answers with those at the back of the book.

Factoring

OBJECTIVES for Sections 7-5 through 7-9:
1. *To factor integers over the set of prime numbers.*
2. *To find the greatest common factor and the least common multiple of two or more monomials.*
3. *To find the greatest monomial factor of a polynomial.*
4. *To factor quadratic polynomials completely.*

7–5 Factoring Monomials

Teaching Suggestions
p. T94

Key Ideas

Factor integers over the set of prime numbers.
Find the greatest common factor and least common multiple of two or more integers.
Find the greatest common factor and least common multiple of two or more monomials.

As you have learned, two or more numbers that are multiplied to form a product are called the *factors* of the product. Thus, when a number is expressed as the product of two or more members of a given set, the number is said to be **factored** over that set. The set of numbers from which the factors are selected is called the **factor set.** For example, the number 6 can be factored over the set of *integers* as follows.

$$6 = 1 \cdot 6 \qquad 6 = (-1)(-6) \qquad 6 = 2 \cdot 3 \qquad 6 = (-2)(-3)$$

Although it is also true that

$$6 = \tfrac{1}{3} \cdot 18 \qquad \text{and} \qquad 6 = (-0.5)(-12),$$

in these cases 6 is *not* factored over the set of integers because $\frac{1}{3}$ and -0.5 are not integers. *In this book, integers will be factored over the set of integers unless some other factor set is specified.* When a number is factored over the set of integers, the factors are called *integral factors* of the number.

A set of numbers that is often used as a factor set is the set of *prime numbers.* A **prime number,** or **prime,** is an integer *greater than 1* that has no positive integral factors other than itself and 1. For example, the first ten prime numbers are

$$2, 3, 5, 7, 11, 13, 17, 19, 23, \text{ and } 29.$$

Over the set of primes, then, the only factors of 6 are 2 and 3. Thus, 2 and 3 are called the *prime factors* of 6. The expression of a positive integer as the product of prime factors is called the **prime factorization** of the integer.

To find the prime factorization of a positive integer, you can sometimes proceed in more than one way. Consider the following example.

$$
\begin{aligned}
108 &= 2 \cdot 54 \\
&= 2 \cdot 2 \cdot 27 \\
&= 2 \cdot 2 \cdot 3 \cdot 9 \\
&= 2 \cdot 2 \cdot 3 \cdot 3 \cdot 3
\end{aligned}
\qquad
\begin{aligned}
108 &= 9 \cdot 12 \\
&= 3 \cdot 3 \cdot 12 \\
&= 3 \cdot 3 \cdot 4 \cdot 3 \\
&= 3 \cdot 3 \cdot 2 \cdot 2 \cdot 3
\end{aligned}
$$

Either way, you find that $108 = 2^2 \cdot 3^3$.

1. Find the GCF and the LCM
   of 50 and 70.
   $50 = 5^2 \cdot 2$
   $70 = 7 \cdot 5 \cdot 2$
   $GCF = 5 \cdot 2 = 10$
   $LCM = 7 \cdot 5^2 \cdot 2 = 350$

2. a. Find the GCF of $36e^5f^3$
      and $30e^3f^2$.
      $36 = 2^2 \cdot 3^2$
      $30 = 2 \cdot 3 \cdot 5$
      $GCF = 2 \cdot 3 = 6$
      e^5 and e^3; choose e^3
      f^3 and f^2; choose f^2
      $GCF = 6e^3f^2$
   b. Find the LCM of $36e^5f^3$
      and $30e^3f^2$.
      $36 = 2^2 \cdot 3^2$
      $30 = 2 \cdot 3 \cdot 5$
      $LCM = 2^2 \cdot 3^2 \cdot 5 = 180$
      e^5 and e^3; choose e^5
      f^3 and f^2; choose f^3
      $LCM = 180e^5f^3$

In fact, no matter how you proceed, you will obtain the prime factorization $2^2 \cdot 3^3$; it is essentially the *one and only one*, or *unique*, factorization of 108 over the set of prime numbers. Other prime factorizations merely vary the order in which the prime factors 2 and 3 appear.

Once you have found the prime factorization of a number, you can use it to find all the positive integral factors of the number.

EXAMPLE 1   List all the positive integral factors of 108.

SOLUTION   The prime factorization of 108 is $2^2 \cdot 3^3$.

The positive integral factors of 108 can be listed as follows.

1	$2^2 = 4$	$2 \cdot 3 = 6$	$2^2 \cdot 3 = 12$
2	$3^2 = 9$	$2 \cdot 3^2 = 18$	$2^2 \cdot 3^2 = 36$
3	$3^3 = 27$	$2 \cdot 3^3 = 54$	$2^2 \cdot 3^3 = 108$

∴ the positive integral factors of 108 are
   1, 2, 3, 4, 9, 27, 6, 18, 54, 12, 36, and 108.

You can also use prime factorization to find the *greatest common factor* and the *least common multiple* of two or more integers. The greatest integer that is a factor of each of the given integers is called their **greatest common factor (GCF)**. The least positive integer that is a multiple of each of the given integers is called their **least common multiple (LCM)**.

EXAMPLE 2   Find the GCF and the LCM of 48 and 60.

SOLUTION   First find the prime factorization of each integer.
$$48 = 2^4 \cdot 3$$
$$60 = 2^2 \cdot 3 \cdot 5$$

The GCF is the product of the lesser powers of each *common* prime factor.
$$GCF = 2^2 \cdot 3 = 12$$

The LCM is the product of the greater powers of *each* prime factor.
$$LCM = 2^4 \cdot 3 \cdot 5 = 240$$

∴ the GCF is 12 and the LCM is 240.

The factoring of monomials is similar to the factoring of integers. For example, since

$$-2x^2 = -2x \cdot x,$$

$-2x^2$ is a *multiple of* $-2x$ and of x, and $-2x$ and x are *factors* of $-2x^2$. In fact, if the factor set is specified as the set of all monomials with integral coefficients, you can list all the factors of $-2x^2$ as follows:

$$1, -1, 2, -2, x, -x, 2x, -2x, x^2, -x^2, 2x^2, -2x^2$$

342   *Chapter 7*

In this book, monomials with integral coefficients will be factored over the set of all monomials with integral coefficients unless some other factor set is specified.

Using methods similar to those used with integers, you can also find the *greatest common factor* and the *least common multiple* of two or more monomials. The monomial with the greatest degree and the greatest numerical coefficient that is a factor of each of the given monomials is called their **greatest common factor (GCF).** The monomial with the least degree and the least positive numerical coefficient that is a multiple of each of the given monomials is called their **least common multiple (LCM).**

The GCF of $9ab^2$ and $-6a^2b$ is $3ab$.

The LCM of $9ab^2$ and $-6a^2b$ is $18a^2b^2$.

Note that, although numerical coefficients of monomials may be negative, the numerical coefficients of the GCF and LCM of monomials are always positive.

EXAMPLE 3   **a.** Find the GCF of $27x^2y^2$ and $18x^3yz$.

b. Find the LCM of $27x^2y^2$ and $18x^3yz$.

SOLUTION   First find the prime factorizations of the numerical coefficients.

$$27 = 3^3 \qquad 18 = 2 \cdot 3^2$$

a. Find the GCF of the numerical coefficients.

$$\text{GCF} = 3^2 = 9$$

Compare the powers of each variable that occurs in *both* monomials, and choose the power with the *lesser* exponent.

Compare x^2 and x^3; choose x^2.

Compare y^2 and y; choose y.

The GCF of the two monomials is the product of the GCF of the numerical coefficients and the lesser powers of the common variables.

$$\text{GCF} = 9x^2y$$

b. Find the LCM of the numerical coefficients.

$$\text{LCM} = 2 \cdot 3^3 = 54$$

Compare the powers of each variable that occurs in *either* monomial, and choose the power with the *greater* exponent.

Compare x^2 and x^3; choose x^3.

Compare y^2 and y; choose y^2.

Choose z.

The LCM of the two monomials is the product of the LCM of the numerical coefficients and the greater powers of the variables.

$$\text{LCM} = 54x^3y^2z$$

To make it easier to read and understand polynomials, it is customary to write the terms of the polynomial in either increasing or decreasing order of the powers of one of the variables. For example,

$$x^2 + x^4 + 4 + 2x + x^3$$

is easier to read and understand if it is written as

$$x^4 + x^3 + x^2 + 2x + 4.$$

When there is a second variable, if one variable is written in decreasing order of powers, the other variable will often be in increasing order. For example,

$$3xy^2 + 3x^2y + y^3 + x^3$$

would best be written as

$$x^3 + 3x^2y + 3xy^2 + y^3.$$

Have students write polynomials with the powers in any order they prefer. Then other students should put them into a more convenient form if possible and tell why they chose the order they did.

Polynomials and Factoring   **343**

Oral Exercises

Tell whether the given statement is true or false.

1. 2 is a factor of 10 over the set of integers. T
2. 2 is a factor of 21 over the set of integers. F
3. 0 is a factor of 5 over the set of integers. F
4. 5 is a factor of 0 over the set of integers. T

5. 2 is a factor of every even integer. T
6. 1 is a factor of every odd integer. T
7. 1 is a prime number. F
8. 7 has no prime factors. F
9. 3 and 7 are the prime factors of 21. T
10. 1 and 3 are the prime factors of 3. F
11. Every integer is a factor of itself. T
12. 1 is a factor of every integer. T

Give the prime factorization of each integer.

13. 15 $3 \cdot 5$
14. 24 $2^3 \cdot 3$
15. 32 2^5
16. 60 $2^2 \cdot 3 \cdot 5$

Name the GCF and the LCM of each pair of integers.

17. 3 and 7 1; 21
18. 6 and 14 2; 42
19. 7 and 49 7; 49
20. 14 and 49 7; 98

Name the GCF and the LCM of each pair of monomials.

21. x and y 1; xy
22. xy and xz x; xyz
23. x and x^2 x; x^2
24. xy and y^2 y; xy^2

Written Exercises

Factor each integer over the set of prime numbers.

A
1. 75 $3 \cdot 5^2$
2. 98 $2 \cdot 7^2$
3. 154 $2 \cdot 7 \cdot 11$
4. 195 $3 \cdot 5 \cdot 13$
5. 144 $2^4 \cdot 3^2$
6. 200 $2^3 \cdot 5^2$
7. 1296 $2^4 \cdot 3^4$
8. 1024 2^{10}

List all the positive integral factors of each number.

9. 18 1, 2, 3, 6, 9, 18
10. 36 1, 2, 3, 4, 6, 9, 12, 18, 36
11. 54 1, 2, 3, 6, 9, 18, 27, 54
12. 59 1, 59
13. 101 1, 101
14. 105 1, 3, 5, 7, 15, 21, 35, 105
15. 96 1, 2, 3, 4, 6, 8, 12, 16, 24, 32, 48, 96
16. 216

For each monomial, list all the factors that are monomials with integral coefficients.

17. $-8a$
18. $5b^2$
19. $3rs^2$
20. $2u^2v^2$

Find the GCF and the LCM of each pair of integers.

21. 22, 33 11; 66
22. 14, 35 7; 70
23. 45, 75 15; 225
24. 28, 98 14; 196
25. 6, 35 1; 210
26. 15, 22 1; 330
27. 90, 135 45; 270
28. 68, 119 7; 476

**Additional Answers
Written Exercises**

16. 1, 2, 3, 4, 6, 8, 9, 12, 18, 24, 27, 36, 54, 72, 108, 216

17. 1, −1, 2, −2, 4, −4, 8, −8, a, −a, 2a, −2a, 4a, −4a, 8a, −8a

18. 1, −1, 5, −5, b, −b, b^2, −b^2, 5b, −5b, 5b^2, −5b^2

19. 1, −1, 3, −3, r, −r, rs, −rs, rs^2, −rs^2, s, −s, s^2, −s^2, 3r, −3r, 3rs, −3rs, 3rs^2, −3rs^2, 3s, −3s, 3s^2, −3s^2

20. 1, −1, 2, −2, u, −u, u^2, −u^2, uv, −uv, u^2v, −u^2v, uv^2, −uv^2, u^2v^2, −u^2v^2, v, −v, v^2, −v^2, 2u, −2u, 2u^2, −2u^2, 2uv, −2uv, 2u^2v, −2u^2v, 2uv^2, −2uv^2, 2u^2v^2, −2u^2v^2, 2v, −2v, 2v^2, −2v^2

Find the GCF and the LCM of each pair of monomials.

29. $6ab^2$, $45a^3b$ $3ab$; $90a^3b^2$

30. $48m^2n^2$, $80mn^4$ $16mn^2$; $240m^2n^4$

31. $-48r^5s^2$, $144qr^3s^2$ $48r^3s^2$; $144qr^5s^2$

32. $34u^2v$, $-85u^3v^3w$ $17u^2v$; $170u^3v^3w$

33. $-110ab^2c^5$, $-154a^2bc^3$ $22abc^3$; $770a^2b^2c^5$

34. $-105r^2s^2t^3$, $-175rst^4$ $35rst^3$; $525r^2s^2t^4$

Name the monomial factor, if any, by which the first monomial can be multiplied so that the product is the second monomial. In Exercises 47–52, assume that n is a positive integer.

B

35. $5j$; $15jk$ $3k$

36. $-6p$; $-54pq$ $9q$

37. $9rs^2$; $-45r^3s^3$ $-5r^2s$

38. $-8b^2c$; $96b^2c^2$ $-12c$

39. $-10s^3t^5$; $-130s^6t^6$ $13s^3t$

40. $-3x^4y$; $-57x^6y^8$ $19x^2y^7$

41. $-12uv^2w$; $60u^2v^3w^2$

42. $-8p^2qr$; $56p^3q^3r^3$ $-7pq^2r^2$

43. $9a^2c$; $54a^2b^3c^4$ $6b^3c^3$

44. $-14f^3g^2$; $-42f^4g^4h^4$

45. $-16r^2s^2t^3$; 0 0

46. 0; $-4x^2y^4z^3$

C

47. $5x^n$; $15x^{n+1}$ $3x$

48. $-6y^n$; $-48y^{n+2}$ $8y^2$

49. $2a^n$; $-18a^{2n}$ $-9a^n$

50. $-7z^n$; $42z^{3n}$ $-6z^{2n}$

51. $-3j^n$; $-18j^{2n+1}$ $6j^{n+1}$

52. $-4k^n$; $-32k^{5n+3}$ $8k^{4n+3}$

53. The GCF of two monomials is $3a^3b^4$ and their LCM is $15a^7b^6$. If one of the monomials is $15a^3b^6$, find the other monomial. $3a^7b^4$

41. $-5uvw$ 44. $3fg^2h^4$ 46. no monomial factor

Computer Exercises For students with computer experience

1. Write a program that will determine whether or not one positive integer is a factor of another by determining whether any integral multiple of the first integer is equal to the second.

2. Write a program that will determine *all* the positive integral factors of a given positive integer.

3. Modify the program that you wrote for Exercise 2 so that it will determine all the *common* factors of two given positive integers.

4. Write a program that will determine the LCM of two given positive integers. (*Hint*: Start with either of the two integers, say a, and determine whether or not it is a multiple of the other. If it is, print it. If it is not, try $2a$, $3a$, and so on until you find a multiple of a that is also a multiple of the other integer.)

5. Write a program that will determine whether or not a given positive integer is a prime number.

6. Modify the program that you wrote for Exercise 2 so that it will determine all the *prime* factors of a given positive integer. Be sure that repeated factors are listed the number of times that they occur in the unique prime factorization of the integer. (*Hint*: Try each prime number starting with 2 and determine whether or not it is a factor of the given integer. If it is, print it, then try it as a factor again. If it is not, go on to the next greater prime number.)

Polynomials and Factoring **345**

Key Ideas

Factor monomials from polynomials.
Recognize common binomial factors of a polynomial.

Chalkboard Examples

1. Factor $21y^3 - 7y^2 - 7y$.
The GCF of all terms is $7y$.
$\therefore 21y^3 - 7y^2 - 7y =$
$7y(3y^2 - y - 1)$

2. Factor $j(h^2 + 1) -$
$\qquad\qquad 2(h^2 + 1)$.
The common binomial factor is $(h^2 + 1)$.
$\therefore j(h^2 + 1) - 2(h^2 + 1) =$
$(h^2 + 1)(j - 2)$

3. Factor $p(h - 1) -$
$\qquad\qquad 4(1 - h)$.
Notice that $h - 1$ and $1 - h$ are opposites.
$p(h - 1) - 4(1 - h)$
$= p(h - 1) - 4[-(h - 1)]$
$= p(h - 1) + 4(h - 1)$
$= (h - 1)(p + 4)$

7–6 Factoring Monomials from Polynomials

To factor a polynomial, you express it as a product of polynomials that are members of a specified factor set. *In this book, polynomials with integral coefficients will be factored over the set of all polynomials with integral coefficients unless some other factor set is specified.*

The first step in factoring a polynomial that is in simplest form is to determine its *greatest monomial factor*. The **greatest monomial factor** of a polynomial is the greatest common factor of its terms. If there is a greatest monomial factor other than 1, you use the distributive axiom to rewrite the given polynomial as the product of this greatest monomial factor and a polynomial whose greatest monomial factor is 1.

EXAMPLE 1 Factor. **a.** $6x^4 - 15x^3 + 3x^2$ **b.** $4m^3n - 7m^2n^2$

SOLUTION **a.** The GCF of all the terms is $3x^2$.
$\qquad \therefore 6x^4 - 15x^3 + 3x^2 = 3x^2(2x^2 - 5x + 1)$

b. The GCF of both terms is m^2n.
$\qquad \therefore 4m^3n - 7m^2n^2 = m^2n(4m - 7n)$

Sometimes you can use the distributive axiom to factor a polynomial that is *not* in simplest form by recognizing a common *binomial* factor.

EXAMPLE 2 Factor. **a.** $y(y - 3) + 7(y - 3)$ **b.** $a(z^2 + 5) - (z^2 + 5)$

SOLUTION **a.** The common binomial factor is $(y - 3)$.
$\qquad \therefore y(y - 3) + 7(y - 3) = (y - 3)(y + 7)$

b. The common binomial factor is $(z^2 + 5)$.
$\qquad a(z^2 + 5) - (z^2 + 5) = a(z^2 + 5) - 1(z^2 + 5)$
$\qquad\qquad\qquad\qquad\qquad = (z^2 + 5)(a - 1)$
$\qquad \therefore a(z^2 + 5) - (z^2 + 5) = (z^2 + 5)(a - 1)$

In working with common binomial factors, you should learn to recognize factors that are opposites of each other. For example:

$$x - y = x + (-y) = -y + x = -(y - x)$$

EXAMPLE 3 Factor $n(n - 3) - 7(3 - n)$.

SOLUTION Notice that $n - 3$ and $3 - n$ are opposites.
$\qquad n(n - 3) - 7(3 - n) = n(n - 3) - 7[-(n - 3)]$
$\qquad\qquad\qquad\qquad\qquad = n(n - 3) + 7(n - 3)$
$\qquad\qquad\qquad\qquad\qquad = (n - 3)(n + 7)$

346 *Chapter 7*

Oral Exercises

Name the greatest monomial factor of each polynomial.

1. $3x - 12$ 3
2. $90 + 15y$ 15
3. $a^2 + 5a$ a
4. $3b - 4b^2$ b
5. $7c^2 - 21c$ $7c$
6. $25d + 30d^3$ $5d$
7. $7t^2 + 15$ 1
8. $9m^2 + n^2$ 1
9. $18rs - 24r$ $6r$
10. $30g + 54gh$ $6g$
11. $24uv + 40u^2v^2$ $8uv$
12. $27jk^2 - 18j^2k$ $9jk$
13. $28c^2d^2 - 21c^2d$ $7c^2d$
14. $36y^3z + 48y^2z^2$ $12y^2z$
15. $6p + 8p^2 - 4p^3$ $2p$
16. $12a^3 - 15a^2 - 7a$ a
17. $4v^4 - 13v^3 + 8v^2$ v^2
18. $6w + 10w^3 + 14w^5$ $2w$

Name a binomial factor of each polynomial.

19. $a(a - 6) + 2(a - 6)$ $a - 6$ (or $a + 2$)
20. $5(2 + b) - b(2 + b)$ $2 + b$ (or $5 - b$)
21. $x(x + 8) + (x + 8)$ $x + 8$ (or $x + 1$)
22. $y(5 - y) - (5 - y)$ $5 - y$ (or $y - 1$)
23. $m(m - 2) + 7(2 - m)$ $m - 2$ (or $m - 7$)
24. $n(7 - n) - 10(n - 7)$ $7 - n$ (or $n + 10$)

Written Exercises

Write each polynomial as the product of its greatest monomial factor and another polynomial.

A
1. $7r + 14s$ $7(r + 2s)$
2. $3w - 3y^2$ $3(w - y^2)$
3. $5m^2 - 6m$ $m(5m - 6)$
4. $9c + 4c^2$ $c(9 + 4c)$
5. $15b^2 + 6b$ $3b(5b + 2)$
6. $12u - 20uv$ $4u(3 - 5v)$
7. $10pq - 12p^2q$ $2pq(5 - 6p)$
8. $6f^2g + 9fg$ $3fg(2f + 3)$
9. $36r^3s^2 - 60r^2s^3$ $12r^2s^2(3r - 5s)$
10. $27u^4v^5 - 63u^3v^6$ $9u^3v^5(3u - 7v)$
11. $16x^4y^5 + 16x^2y^2$ $16x^2y^2(x^2y^3 + 1)$
12. $10a^2b - 10a^5b^2$ $10a^2b(1 - a^3b)$
13. $5b^3 - 35b^2 + 10b$ $5b(b^2 - 7b + 2)$
14. $12r - 30r^2 + 6r^3$ $6r(2 - 5r + r^2)$
15. $21z^5 - 77z^3 - 49z$ $7z(3z^4 - 11z^2 - 7)$
16. $20p^4 - 28p^5 + 44p^6$
17. $6b^2c - 4bc^2 + 4bc$ $2bc(3b - 2c + 2)$
18. $8r^3s^2 - 12rs^3 + 4r^2s$
19. $15y^4z + 10y^3z^2 - 20y^3z^3$
20. $14a^5b^2 + 28a^2b^4 - 21a^2b^3$
21. $45a^2b^4 + 18a^4b^3 - 81a^3b^4$
22. $40u^4v^6 - 48u^3v^5 - 16u^5v^3$
23. $42r^2s^5t^3 + 54r^3s^2t$ $6r^2s^2t(7s^3t^2 + 9r)$
24. $84xy^4z^3 - 60x^4yz^2$ $12xyz^2(7y^3z - 5x^3)$

Write each polynomial as the product of two binomials.

25. $z(z - 1) + 2(z - 1)$ $(z - 1)(z + 2)$
26. $6(3 + r) + r(3 + r)$ $(3 + r)(6 + r)$
27. $2m(m + 5) - 3(m + 5)$ $(m + 5)(2m - 3)$
28. $5t(t - 4) - 6(t - 4)$ $(t - 4)(5t - 6)$
29. $d(d - 5) + 7(5 - d)$ $(d - 5)(d - 7)$
30. $p(p - 2) - 4(2 - p)$ $(p - 2)(p + 4)$
31. $9(1 - q) - q(q - 1)$ $(1 - q)(9 + q)$
32. $4(m + 9) - m(9 + m)$ $(m + 9)(4 - m)$

Polynomials and Factoring **347**

Mixed Review

Find the GCF and LCM of each pair of monomials.

1. $12m^2n$, $15\,mn$ $3mn$; $60m^2n$

2. $-10stu$, $-60s^2t^2u^2$ $10stu$; $60s^2t^2u^2$

Simplify.

3. $a(b + c)$ $ab + ac$

4. $3(4x - 2y)$ $12x - 6y$

5. $-3x^2(xy - 4)$ $-3x^3y + 12x^2$

6. $5a^2b^3(3a - 2b)$ $15a^3b^3 - 10a^2b^4$

Write each polynomial as the product of two binomials.

EXAMPLE $(z - 2)^2 + 9(z - 2)$

SOLUTION $(z - 2)^2 + 9(z - 2) = (z - 2)\,[(z - 2) + 9]$
$$= (z - 2)(z + 7)$$

B

33. $(x + 7)^2 + 5(x + 7)$ $(x + 7)(x + 12)$

34. $(y - 2)^2 - 3(y - 2)$ $(y - 2)(y - 5)$

35. $(a - 6)^2 - 4(6 - a)$ $(a - 6)(a - 2)$

36. $9(b - 5) + (5 - b)^2$ $(b - 5)(b + 4)$

37. $(m - 8)^2 + (m - 8)$ $(m - 8)(m - 7)$

38. $(n - 10)^2 + (10 - n)$ $(n - 10)(n - 11)$

39. $4(z + 1)^2 + 3(z + 1)$ $(z + 1)(4z + 7)$

40. $7(w - 3)^2 + 5(w - 3)$ $(w - 3)(7w - 16)$

41. $2(p - 3) + 5(p - 3)^2$ $(p - 3)(5p - 13)$

42. $8(q + 1) - 2(q + 1)^2$ $(q + 1)(6 - 2q)$

43. $3(a - 1)^2 + 2(1 - a)$ $(a - 1)(3a - 5)$

44. $5(z - 4)^2 - (4 - z)$ $(z - 4)(5z - 19)$

Replace each ? with the binomial that will make the sentence a true statement. Assume that n is a positive integer. **52.** $c^4 - d^2$

C

45. $x^n + x^{n+1} = x^n(\underline{\ ?\ })$ $1 + x$

46. $y^{n+3} - y^{n+2} = y^n(\underline{\ ?\ })$ $y^3 - y^2$

47. $j^n - j^{2n} = j^n(\underline{\ ?\ })$ $1 - j^n$

48. $k^{4n} - k^{5n} = k^{4n}(\underline{\ ?\ })$ $1 - k^n$

49. $r^{9n}s^2 - r^{5n}s^4 = r^{5n}s^2(\underline{\ ?\ })$ $r^{4n}s - s^3$

50. $a^3b^{3n} - ab^{2n} = ab^{2n}(\underline{\ ?\ })$ $a^2b^n - 1$

51. $v^nw^{n+3} + v^{n+1}w^n = v^nw^n(\underline{\ ?\ })$ $w^3 + v$

52. $c^{n+5}d^{n+2} - c^{n+1}d^{n+4} = c^{n+1}d^{n+2}(\underline{\ ?\ })$

53. $a^{n+2}b^{3n+1} - a^{n+1}b^{3n+4} = a^nb^{3n}(\underline{\ ?\ })$ $a^2b - ab^4$

54. $x^{4n}y^{2n+1} + x^{2n}y^{2n+5} = x^{2n}y^{2n}(\underline{\ ?\ })$ $x^{2n}y + y^5$

François Viète

1540–1603

François Viète (or Vieta) was born in Poitou, France. A lawyer by profession, he served as councilor to the parliaments at Rennes and Tours and as royal privy councilor. His hobby was algebra. During the war with Spain, Viète aided Henry IV by decoding Spanish messages. To Viète, code breaking was simply a matter of solving algebraic equations.

Banished from court by his political enemies from 1584 to 1589, Viète turned his full attention to mathematics. He introduced the use of letters to express known numbers, such as constants and coefficients, as well as unknown numbers. He used vowels for the unknown numbers and consonants for the known numbers. Viète also used signs of operation to indicate addition, subtraction, multiplication, and division. Viète simplified the subject of algebra by this use of signs and symbols to replace words.

READING ALGEBRA Prefixes

Many of the words that are used in algebra have *prefixes*. A prefix is a letter or a group of letters that is placed before a *base word* or a *root* to make another word. Usually the prefix changes or modifies the meaning of the base word or root. For example, the prefix *in*, meaning "not," is placed before the base word *consistent* to form the word *inconsistent*, meaning "not consistent."

In your study of algebra, knowing the meaning of the prefix of a word may help you to better understand and remember the word. The following chart lists some of the common prefixes and their meanings, as well as examples of how these prefixes are used in algebra.

Prefix	Meaning	Example
bi	two	binomial
co	together	coefficient
equi	the same	equivalent
in	not	inequality
mono	one	monomial
poly	many	polynomial
quad	four	quadrant
tri	three	trinomial

Exercises

Identify each term whose definition is given.

1. a mathematical sentence which states that two expressions name the same number an equation
2. an operation that pairs any two real numbers with a third real number binary operation
3. two lines that have all their points in common coincident lines
4. a set that is not finite an infinite set
5. the number that corresponds to a point on a number line a coordinate

6. Why is a trinomial of the form $ax^2 + bx + c$ called a *quadratic* trinomial?

Polynomials and Factoring **349**

Key Ideas

Factor special polynomials:
    difference of squares
    trinomial squares
    sum of cubes
    difference of cubes
Factor polynomials by grouping.

Chalkboard Examples

1. Factor $t^6 - 121$.
= $(t^3)^2 - 11^2$
= $(t^3 - 11)(t^3 + 11)$

2. Factor $125 + g^3$.
= $g^3 + 5^3$
= $(g + 5)[g^2 - (5)(g) + 5^2]$
= $(g + 5)(g^2 - 5g + 25)$

3. Factor $3a^3 + 24a^2 + 48a$.
= $3a(a^2 + 8a + 16)$
= $3a(a + 4)^2$

4. Factor $p^2 + 4p + 4 - 4n^2$.
= $(p^2 + 4p + 4) - 4n^2$
= $(p + 2)^2 - (2n)^2$
= $[(p + 2) - 2n][(p + 2) + 2n]$
= $(p - 2n + 2)(p + 2n + 2)$

7–7 Factoring Special Polynomials

When the greatest monomial factor of a polynomial in simplest form is 1, you may still be able to factor the polynomial if you recognize a special *factor pattern*. In particular, the special products of binomials that you learned in Section 7-4 are used frequently in the factoring of polynomials. That is:

$$\left.\begin{array}{l} a^2 + 2ab + b^2 = (a + b)^2 \\ a^2 - 2ab + b^2 = (a - b)^2 \end{array}\right\} \longrightarrow \textit{trinomial squares}$$

$$a^2 - b^2 = (a + b)(a - b) \ \} \longrightarrow \textit{difference of squares}$$

EXAMPLE 1   Factor.

   a. $x^2 + 6x + 9$ **b.** $9y^2 - 12y + 4$ **c.** $z^4 - 25$

SOLUTION   **a.** $x^2 + 6x + 9 = x^2 + 2(x)(3) + 3^2$ ⟵ trinomial square
                 $= (x + 3)^2$

   b. $9y^2 - 12y + 4 = (3y)^2 - 2(3y)(2) + 2^2$ ⟵ trinomial square
                 $= (3y - 2)^2$

   c. $z^4 - 25 = (z^2)^2 - 5^2$ ⟵ difference of squares
                 $= (z^2 + 5)(z^2 - 5)$

Two other special factor patterns occur when a polynomial is a *sum of cubes* or a *difference of cubes*. A **sum of cubes** is a polynomial that can be written in the form $a^3 + b^3$. Similarly, a **difference of cubes** is a polynomial that can be written in the form $a^3 - b^3$. To factor sums and differences of cubes, you use the following patterns.

$$a^3 + b^3 = (a + b)(a^2 - ab + b^2)$$
$$a^3 - b^3 = (a - b)(a^2 + ab + b^2)$$

EXAMPLE 2   Factor.

   a. $r^3 + 27$ **b.** $8s^3 - t^3$ **c.** $q^6 + 1$

SOLUTION   **a.** $r^3 + 27 = r^3 + 3^3$
                 $= (r + 3)[r^2 - (r)(3) + 3^2]$
                 $= (r + 3)(r^2 - 3r + 9)$

   b. $8s^3 - t^3 = (2s)^3 - t^3$
                 $= (2s - t)[(2s)^2 + (2s)(t) + t^2]$
                 $= (2s - t)(4s^2 + 2st + t^2)$

   c. $q^6 + 1 = (q^2)^3 + 1^3$
                 $= (q^2 + 1)[(q^2)^2 - (q^2)(1) + 1^2]$
                 $= (q^2 + 1)(q^4 - q^2 + 1)$

350   *Chapter 7*

Sometimes you will be able to use more than one method in factoring a given polynomial. For example, the terms of the polynomial

$$3x^5 - 75x^3$$

have a greatest monomial factor, $3x^3$, and so you can write

$$3x^5 - 75x^3 = 3x^3(x^2 - 25).$$

In this factorization, however, notice that the binomial factor, $x^2 - 25$, is itself a difference of squares. Therefore, you can further factor the polynomial as follows.

$$3x^5 - 75x^3 = 3x^3(x + 5)(x - 5)$$

EXAMPLE 3 Factor.

  a. $6a^2 + 12a + 6$ **b.** $15b^4 + 15b$ **c.** $c^4 - 81$

SOLUTION **a.** $6a^2 + 12a + 6 = 6(a^2 + 2a + 1)$ ← trinomial square
  $= 6(a + 1)^2$

  b. $15b^4 + 15b = 15b(b^3 + 1)$ ←——— sum of cubes
  $= 15b(b + 1)(b^2 - b + 1)$

  c. $c^4 - 81 = (c^2 + 9)(c^2 - 9)$ ←——— difference of squares
  $= (c^2 + 9)(c + 3)(c - 3)$

When you are unable to factor a polynomial either by using a greatest monomial factor or a special factor pattern, you may still be able to *factor by grouping* the terms of the polynomial. In the next example, for instance, notice that the first three terms of the polynomial, when grouped, form a trinomial square. When you factor this trinomial, the original polynomial can then be factored as a difference of squares.

EXAMPLE 4 Factor $m^2 - 6m + 9 - n^2$.

SOLUTION $m^2 - 6m + 9 - n^2 = (m^2 - 6m + 9) - n^2$
  $= (m - 3)^2 - n^2$
  $= [(m - 3) + n][(m - 3) - n]$
  $= (m + n - 3)(m - n - 3)$

You may also find it helpful to rearrange terms in a polynomial before trying to factor it. In the following example, the arrows indicate an appropriate grouping of the terms.

EXAMPLE 5 Factor $3xy - 20zw - 15xz + 4yw$.

SOLUTION $3xy - 20zw - 15xz + 4yw = (3xy - 15xz) + (4yw - 20zw)$
  $= 3x(y - 5z) + 4w(y - 5z)$
  $= (y - 5z)(3x + 4w)$

Polynomials and Factoring **351**

Suggested Assignments

Minimum
First day: 352/1–15
Second day: 352/16–30
Average
First day:
   352/1–29 odd
Second day:
   352/31–48
Maximum
   352/1–47 odd, 49–54

Additional Answers
Written Exercises

4. $(12 + 7d^3)(12 - 7d^3)$
5. $(8u + 11v)(8u - 11v)$
6. $(4j^3 + 9k^4)(4j^3 - 9k^4)$
15. $(2g - 5)^2$
16. $(4 + 3h)^2$
17. $(5y - 4z)^2$
18. $(3m - 7n)^2$
19. $(r + 2)(r^2 - 2r + 4)$
20. $(z - 3)(z^2 + 3z + 9)$
21. $(4u - 1)(16u^2 + 4u + 1)$
22. $(1 + 5v)(1 - 5v + 25v^2)$
23. $(3j - k)(9j^2 + 3jk + k^2)$
24. $(a^3 + 2b^2)(a^6 - 2a^3b^2 + 4b^4)$
25. $3(c^2 - 25)$
    $= 3(c + 5)(c - 5)$
26. $2(9 - s^2)$
    $= 2(3 + s)(3 - s)$
27. $5(16 + 8x + x^2)$
    $= 5(4 + x)^2$
28. $7(y^2 - 2y + 1)$
    $= 7(y - 1)^2$
29. $5(9v^2 - 24v + 16)$
    $= 5(3v - 4)^2$
30. $4(25w^2 - 20w + 4)$
    $= 4(5w - 2)^2$
31. $4m(4m^2 - 9)$
    $= 4m(2m + 3)(2m - 3)$
32. $5ab(4 - 9b^2)$
    $= 5ab(2 + 3b)(2 - 3b)$
33. $3x(x^2 + 2x + 1)$
    $= 3x(x + 1)^2$

Oral Exercises

8. Yes; $(c^3 + 10)(c^3 - 10)$

Tell whether or not each polynomial is a difference of squares. If it is, give the factored form of the polynomial.

Yes; $(x + 6)(x - 6)$ Yes; $(3 + a)(3 - a)$ No
1. $x^2 - 36$ 2. $y^2 + 25$ No   3. $9 - a^2$ 4. $8 - b^2$
5. $16m^2 - 49$ 6. $1 - 64n^2$ 7. $25g^2 - 4h^2$ 8. $c^6 - 100$
   Yes; $(4m + 7)(4m - 7)$ Yes; $(1 + 8n)(1 - 8n)$ Yes; $(5g + 2h)(5g - 2h)$

Tell whether or not the given polynomial is a trinomial square. If it is, give the factored form of the polynomial.

Yes; $(x + 5)^2$ Yes; $(y - 4)^2$
9. $x^2 + 10x + 25$ 10. $y^2 - 8y + 16$ 11. $z^2 - 7z + 49$ No
12. $a^2 + 2a + 1$ 13. $b^2 + 4b + 4$ 14. $x^2 - 2xy + y^2$
    Yes; $(a + 1)^2$ Yes; $(b + 2)^2$ Yes; $(x - y)^2$

State the value or values of k, if any, for which the given polynomial is a trinomial square.

20 or -20
15. $m^2 + 6m + k$ 9   16. $n^2 - 16n + k$ 64   17. $p^2 + kp + 100$
18. $q^2 + kq - 81$ 19. $s^2 + kst + t^2$ 2 or -2 20. $u^2 + kuv - v^2$
    not a trinomial square   not a trinomial square

Written Exercises

Factor each polynomial.

$(10 + 3m)(10 - 3m)$ $(11c^2 + 1)(11c^2 - 1)$
A   1. $64z^2 - 25$ $(8z + 5)(8z - 5)$ 2. $100 - 9m^2$ 3. $121c^4 - 1$
    4. $144 - 49d^6$ 5. $64u^2 - 121v^2$ 6. $16j^6 - 81k^8$
    7. $a^2 + 14a + 49$ $(a + 7)^2$ 8. $b^2 - 16b + 64$ $(b - 8)^2$ 9. $36 - 12r + r^2$ $(6 - r)^2$
    10. $81 + 18s + s^2$ $(9 + s)^2$ 11. $m^2 + 2mn + n^2$ $(m + n)^2$ 12. $p^2 - 2pq + q^2$ $(p - q)^2$
    13. $4c^2 + 4c + 1$ $(2c + 1)^2$ 14. $9d^2 - 6d + 1$ $(3d - 1)^2$ 15. $4g^2 - 20g + 25$
    16. $16 + 24h + 9h^2$ 17. $25y^2 - 40yz + 16z^2$ 18. $9m^2 - 42mn + 49n^2$
    19. $r^3 + 8$ 20. $z^3 - 27$ 21. $64u^3 - 1$
    22. $1 + 125v^3$ 23. $27j^3 - k^3$ 24. $a^9 + 8b^6$

Factor each polynomial in two steps. The first step should be to write the polynomial as the product of its greatest monomial factor and another polynomial.

25. $3c^2 - 75$ 26. $18 - 2s^2$ 27. $80 + 40x + 5x^2$
28. $7y^2 - 14y + 7$ 29. $45v^2 - 120v + 80$ 30. $100w^2 - 80w + 16$
B   31. $16m^3 - 36m$ 32. $20ab - 45ab^3$ 33. $3x^3 + 6x^2 + 3x$
    34. $20y^2 - 20y^3 + 5y^4$ 35. $50j^3 + 40j^2 + 8j$ 36. $27m^2n - 36mn + 12n$

Factor each polynomial by grouping.

$(c - 2d + 5)(c - 2d - 5)$
37. $a^2 + 12a + 36 - b^2$ $(a + 6 + b)(a + 6 - b)$ 38. $c^2 - 4cd + 4d^2 - 25$
39. $25x^2 - 10x + 1 - y^2$ $(5x - 1 + y)(5x - 1 - y)$ 40. $4 + 28s + 49s^2 - t^2$
                                                             $(2 + 7s + t)(2 + 7s - t)$

41. $3a^2 - 3ab + 2a - 2b$ $(a - b)(3a + 2)$ 42. $c^2 - c + 3cd - 3d$ $(c - 1)(c + 3d)$
43. $5r + r^2 - 5s - rs$ $(5 + r)(r - s)$ 44. $p^2 - 2p + pq - 2q$ $(p - 2)(p + q)$
45. $10ac + 6bd + 15bc + 4ad$ 46. $2xz - 21wy - 14yz + 3wx$
47. $2p^2 + 15qr + 10pq + 3pr$ 48. $6a^2 - 5bc + 2ab - 15ac$
45. $(2a + 3b)(5c + 2d)$ **46.** $(x - 7y)(2z + 3w)$ **47.** $(2p + 3r)(p + 5q)$ **48.** $(3a + b)(2a - 5c)$
Factor each polynomial. Assume that n represents a positive integer.

C   49. $x^{2n} - 1$ 50. $x^{2n} - y^2$ $(x^n + y)(x^n - y)$ 51. $x^{2n} - 2x^n + 1$
     52. $x^{4n} - 6x^{2n} + 9$ $(x^{2n} - 3)^2$ 53. $x^{3n} - y^3$ 54. $x^{4n} - y^2$
                                            $(x^n - y)(x^{2n} + x^ny + y^2)$ $(x^{2n} + y)(x^{2n} - y)$

Computer Exercises   For students with computer experience

1. Write a program that will compute the product of two binomials of the form $px + q$ and $rx + s$ when you input values for p, q, r, and s. A sample output would be

    $$5X \uparrow 2 + 7X + 2.$$

    (*Note:* Some computers will display a different exponent symbol.)

2. Write a program that will allow you to input values for a, b, and c and will determine whether or not a trinomial of the form $ax^2 + bx + c$ is a trinomial square. If it is, the program should display the binomial factors of the trinomial. A sample output would be

    $$(4X + 1) \uparrow 2.$$

    If it is *not* a trinomial square, the output should so state.

3. Modify the program that you wrote for Exercise 1 so that it will compute the product of a binomial of the form $px + q$ and a trinomial of the form $rx^2 + sx + t$ when you input values for p, q, r, s, and t. A sample output would be

    $$6X \uparrow 3 + 7X \uparrow 2 + 4X + 1.$$

7–8 Factoring Quadratic Trinomials of the Form $x^2 + bx + c$

A polynomial that can be expressed in the form

$$ax^2 + bx + c, a \neq 0,$$

is called a **quadratic polynomial**. For example,

$$x^2 - 5x + 4, \qquad 3y^2 + 9y, \qquad \text{and} \qquad -5z^2$$

are quadratic polynomials. In a quadratic polynomial that is expressed in simplest form, the term of *degree two* is called the **quadratic term**; the term of *degree one* is called the **linear term**; and the numerical term is called the **constant term**.

Polynomials and Factoring   **353**

Mixed Review
Name the greatest monomial factor of each polynomial.
1. $7p^3 + 14p^4 - 91p^5$ $7p^3$
2. $6a^2b^2 - 3ab + 6$ 3
Simplify.
3. $(2t + 1)^2$ $4t^2 + 4t + 1$
4. $(2t - 1)^2$ $4t^2 - 4t + 1$
Write each of the following monomials as the square of a monomial.
5. $49x^8$ $(7x^4)^2$
6. $4x^2y^2$ $(2xy)^2$

Teaching Suggestions
p. T95

Key Ideas
Factor quadratic trinomials of the form $x^2 + bx + c$. Distinguish reducible from irreducible quadratic trinomials.

Thus, in the quadratic polynomial $ax^2 + bx + c$:

ax^2 is the quadratic term,

bx is the linear term, and

c is the constant term.

A quadratic polynomial in which $b \neq 0$ and $c \neq 0$ is called a **quadratic trinomial.** In Section 7-7 you learned to factor a special type of quadratic trinomial, the *trinomial square.* In this section you will learn to factor other quadratic trinomials in which the coefficient of the quadratic term is 1. That is, you will learn a method of factoring quadratic trinomials of the general form $x^2 + bx + c$.

If a quadratic trinomial of the form $x^2 + bx + c$ can be factored, its factors will be binomials of the form $x + p$ and $x + q$. That is,

$$x^2 + bx + c = (x + p)(x + q).$$

Applying the distributive axiom to the right side of this equation, you obtain:

$$\begin{aligned} x^2 + bx + c &= (x + p)x + (x + p)q \\ &= x^2 + px + qx + pq \\ &= x^2 + (p + q)x + pq \end{aligned}$$

Therefore,

$$x^2 + bx + c = x^2 + (p + q)x + pq,$$

and the following relationships must hold true:

$$b = p + q$$
$$c = pq$$

These relationships suggest the following technique for factoring a quadratic trinomial of the form $x^2 + bx + c$.

---

1. List the pairs of factors of the constant term, c, that have a product equal to c.
2. Find the pair of factors in the list that have a *sum* equal to the coefficient of the linear term, b.
3. Write the factorization as $(x + p)(x + q)$, using the factors in the chosen pair as the values of p and q.

Note that, once you have written a factorization, you should check that it is correct by multiplying the binomial factors to determine if their product is the original trinomial.

EXAMPLE 1   Factor.

    a. $x^2 - 5x + 6$ **b.** $y^2 + y - 12$

SOLUTION **a.** 1. The constant term is 6.

Thus the possible factor pairs are:

$$1, 6 \quad -1, -6 \quad 2, 3 \quad -2, -3$$

2. The coefficient of the linear term is -5.

Choose the factor pair whose sum is -5:

$$-2 + (-3) = -5$$

3. Write the factorization as $(x - 2)(x - 3)$.

Check: $(x - 2)(x - 3) = x^2 - 2x - 3x + 6 = x^2 - 5x + 6$

$\therefore x^2 - 5x + 6 = (x - 2)(x - 3)$

b. 1. The constant term is -12.

Thus the possible factor pairs are:

$$1, -12 \quad -1, 12 \quad 2, -6 \quad -2, 6 \quad 3, -4 \quad -3, 4$$

2. The coefficient of the linear term is 1.

Choose the factor pair whose sum is 1:

$$-3 + 4 = 1$$

3. Write the factorization as $(y - 3)(y + 4)$.

Check: $(y - 3)(y + 4) = y^2 - 3y + 4y - 12 = y^2 + y - 12$

$\therefore y^2 + y - 12 = (y - 3)(y + 4)$

As you gain experience in factoring, you will probably learn to review the factors of the constant term mentally instead of writing them down.

When the coefficient of the quadratic term is -1, it is usually helpful to begin by factoring -1 from each term of the trinomial.

EXAMPLE 2 Factor $8 + 2w - w^2$.

SOLUTION First, rearrange the terms of the trinomial in order of decreasing degree.

$$8 + 2w - w^2 = -w^2 + 2w + 8$$

Then factor -1 from each term.

$$-w^2 + 2w + 8 = -(w^2 - 2w - 8)$$

Now factor the trinomial within parentheses.

$$w^2 - 2w - 8 = (w + 2)(w - 4)$$

Finally, return the factor -1 to this product.

$$-(w^2 - 2w - 8) = -(w + 2)(w - 4)$$

$\therefore 8 + 2w - w^2 = -(w + 2)(w - 4)$

Each of the quadratic trinomials in Examples 1 and 2 is said to be *reducible* over the set of polynomials with integral coefficients. A polynomial is **reducible** over a given factor set if it can be expressed as the product of two or more polynomials *of lower positive degree* taken from that set. A polynomial that is *not* reducible over a given factor set is said to be **irreducible** over that set.

Polynomials and Factoring **355**

Minimum
First day: 357/1–18
Second day: 357/19–30
        S 339/P: 5–8
Average
First day:
     357/1–29 odd
Second day:
     357/31–46
  S 348/34–44 even
Maximum
First day:
     357/1–41 odd
Second day:
     357/43–52
  S 348/46–54 even

Additional A Exercises

Factor. If the trinomial is irreducible, so state.

1. $y^2 - 3y - 28$
    $(y - 7)(y + 4)$
2. $p^2 + 12p + 35$
    $(p + 7)(p + 5)$
3. $y^2 - 2y + 3$ Irreducible
4. $3k^3 + 6k^2 + 3k$ $3k(k + 1)^2$
5. $5m^2 - 15m + 10$
    $5(m - 2)(m - 1)$
6. $3n^2 - 27n - 108$
    $3(n - 12)(n + 3)$

Mixed Review

Factor.

1. $p^2 + 14p + 49$ $(p + 7)^2$
2. $16t^2 - 72t + 81$ $(4t - 9)^2$

Use a common binomial factor to write each polynomial as the product of binomials.

3. $8(n - 3) + p(3 - n)$
    $(8 - p)(n - 3)$
4. $a(b - c) - 2d(b - c)$
    $(a - 2d)(b - c)$

Simplify.

5. $(x - 2)(x + 3)$ $x^2 + x - 6$
6. $(x - 2)(x - 3)$ $x^2 - 5x + 6$

EXAMPLE 3   Factor $t^2 - 6t - 5$.

SOLUTION   The constant term is -5, and so the only possible factor pairs to consider are $1, -5$ and $-1, 5$.

The coefficient of the linear term is -6, but neither pair of factors has a sum of -6.

$\therefore t^2 - 6t - 5$ is irreducible.

Sometimes a common monomial factor "conceals" a quadratic trinomial in which the coefficient of the quadratic term is 1. Therefore, it is generally a good practice to begin any factorization of a polynomial by determining the greatest monomial factor of its terms.

EXAMPLE 4   Factor.

    a. $2a^3 + 10a^2 - 28a$ **b.** $12b + 15 - 3b^2$ **c.** $3cd^2 + 6cd + 6c$

SOLUTION   **a.** $2a^3 + 10a^2 - 28a = 2a(a^2 + 5a - 14)$
                           $= 2a(a + 7)(a - 2)$

    b. $12b + 15 - 3b^2 = -3b^2 + 12b + 15$
                      $= -3(b^2 - 4b - 5)$
                      $= -3(b - 5)(b + 1)$

    c. $3cd^2 + 6cd + 6c = 3c(d^2 + 2d + 2)$

In part (c) of Example 4, notice that there is no further factorization of $3c(d^2 + 2d + 2)$ because the trinomial factor, $d^2 + 2d + 2$, is irreducible.

Oral Exercises

For each of the following quadratic trinomials, name the quadratic term, the linear term, and the constant term.

1. $x^2 + 5x + 4$ x^2; $5x$; 4 **2.** $y^2 - 5y - 14$ y^2; $-5y$; -14 **3.** $10 + 3z - z^2$ $-z^2$; $3z$; 10
4. $8 + 2w^2 - 7w$ $2w^2$; $-7w$; 8 **5.** $9u^2 + 7u$ $9u^2$; $7u$; 0 **6.** $4 + v^2$ v^2; 0; 4

For each of the following quadratic trinomials, find a pair of integers whose product is the constant term and whose sum is the coefficient of the linear term.

7. $a^2 + 9a + 18$ $3, 6$ **8.** $b^2 - 7b + 12$ $-3, -4$ **9.** $c^2 - 5c - 6$ $-6, 1$
10. $d^2 + 3d - 10$ $5, -2$ **11.** $f^2 + f - 2$ $2, -1$ **12.** $g^2 - g - 20$ $-5, 4$

Tell whether each of the following quadratic trinomials is reducible or irreducible.

13. $m^2 + 6m + 8$ reducible   **14.** $n^2 - 6n + 8$ reducible   **15.** $p^2 - 7p + 8$ irreducible
16. $q^2 + 7q - 8$ reducible   **17.** $x^2 - 9x - 8$ irreducible   **18.** $y^2 - 6y - 8$ irreducible

Written Exercises

Factor. If the trinomial is irreducible, so state.

A 1. $a^2 + 7a + 12$ $(a + 3)(a + 4)$
    3. $r^2 + 4r - 5$ $(r + 5)(r - 1)$
    5. $j^2 - j + 2$ irreducible
    7. $g^2 - 5g - 24$ $(g - 8)(g + 3)$
    9. $m^2 - 9m + 20$ $(m - 5)(m - 4)$
  11. $-x^2 - 11x - 24$ $-(x + 8)(x + 3)$
  13. $6 + 7x + x^2$ $(6 + x)(1 + x)$
  15. $36 + 9j - j^2$ $(3 + j)(12 - j)$
  17. $m^2 + 3mn + 2n^2$ $(m + 2n)(m + n)$

  2. $b^2 - 7b + 10$ $(b - 5)(b - 2)$
  4. $s^2 - s - 12$ $(s - 4)(s + 3)$
  6. $k^2 - 5k - 6$ $(k - 6)(k + 1)$
  8. $h^2 - 7h + 18$ irreducible
10. $n^2 - 2n - 24$ $(n - 6)(n + 4)$
12. $-y^2 + 12y - 32$ $-(y - 8)(y - 4)$
14. $21 - 10y + y^2$ $(7 - y)(3 - y)$
16. $48 - 2k - k^2$ $(8 + k)(6 - k)$
18. $a^2 - 2ab - 3b^2$ $(a - 3b)(a + b)$

Factor each trinomial in two steps. The first step should be to write the trinomial as the product of its greatest monomial factor and another trinomial.

  19. $4c^2 - 44c + 120$
  21. $k^3 - 8k^2 + 15k$
  23. $m^4 - 7m^3 - 18m^2$
  25. $3v^3 - 3v^2 - 60v$
  27. $5b^4 - 15b^3 + 10b^2$
  29. $-6x^5 + 24x^4 - 18x^3$

B 31. $2x^2y^2 - 8x^2y - 90x^2$
  33. $15st + 8s^2t + s^3t$
  35. $40u^3 + 16u^3v - 2u^3v^2$
  37. $s^3 - s^2t - 12st^2$
  39. $m^4 - 5m^3n + 6m^2n^2$
  41. $2j^2 - 10j^2k + 8j^2k^2$

  20. $3d^2 - 18d - 48$
  22. $a^3 + 13a^2 + 42a$
  24. $r^5 + 10r^4 + 24r^3$
  26. $4w^3 - 28w^2 - 120w$
  28. $3c^5 - 18c^4 - 48c^3$
  30. $-2y^3 + 22y^2 + 24y$
  32. $az^3 + 9az^2 - 22az$
  34. $9q^2 - 10pq^2 + p^2q^2$
  36. $30xy - 5x^2y - 5x^3y$
  38. $a^3 + 3a^2b - 28ab^2$
  40. $b^2c^3 + 8bc^4 + 12c^5$
  42. $3c^4 + 12c^3d - 36c^2d^2$

Determine all positive and negative integral values of k for which the given trinomial is reducible.

  43. a. $x^2 + kx + 2$ $3, -3$
      b. $x^2 + kx - 2$ $1, -1$
  45. a. $x^2 + kx + 12$ $7, -7, 8, -8, 13, -13$
      b. $x^2 + kx - 12$ $1, -1, 4, -4, 11, -11$
  46. a. $11, -11, 13, -13, 17, -17, 31, -31$

  44. a. $x^2 + kx + 6$ $5, -5, 7, -7$
      b. $x^2 + kx - 6$ $1, -1, 5, -5$
  46. a. $x^2 + kx + 30$
      b. $x^2 + kx - 30$
  b. $1, -1, 7, -7, 13, -13, 29, -29$

Factor. In Exercises 51 and 52, assume that n represents a positive integer.

C 47. $x^4 - x^2 - 2$ $(x^2 - 2)(x^2 + 1)$
  49. $x^8 + 5x^4 + 6$ $(x^4 + 3)(x^4 + 2)$
  51. $x^{2n} + 3x^n - 10$ $(x^n + 5)(x^n - 2)$

  48. $x^6 + 2x^3 - 15$ $(x^3 + 5)(x^3 - 3)$
  50. $x^{10} - x^5 - 30$ $(x^5 - 6)(x^5 + 5)$
  52. $x^{4n} + 3x^{2n} + 2$ $(x^{2n} + 2)(x^{2n} + 1)$

Polynomials and Factoring **357**

Key Ideas

Factor trinomials of the form
$ax^2 + bx + c$ in which
$a \neq \pm 1$

Chalkboard Examples

1. Factor $2c^2 + 9c + 9$.
   $= (2c + 3)(c + 3)$
2. Factor $6m^3 - 39m^2 + 45m$
   completely.
   $= 3m(2m^2 - 13m + 15)$
   $= 3m(2m - 3)(m - 5)$
3. Factor $4x^4 - 8x^2 - 32$
   completely.
   $= 4(x^4 - 2x^2 - 8)$
   $= 4(x^2 + 2)(x^2 - 4)$
   $= 4(x^2 + 2)(x - 2)(x + 2)$

7–9 Factoring Quadratic Trinomials of the Form $ax^2 + bx + c$

In Section 7-8 you learned to factor quadratic trinomials in which the coefficient of the quadratic term is 1. As you will see, the factoring technique developed in that section can be extended to the factoring of quadratic trinomials in which the coefficient of the quadratic term is an integer other than 1.

If a quadratic trinomial of the form $ax^2 + bx + c$ can be factored, its factors will be binomials of the form $px + r$ and $qx + s$. That is,

$$ax^2 + bx + c = (px + r)(qx + s).$$

Applying the distributive axiom to the right side of this equation, you obtain:

$$ax^2 + bx + c = (px + r)qx + (px + r)s$$
$$= pqx^2 + rqx + psx + rs$$
$$= pqx^2 + (rq + ps)x + rs$$

Therefore,

$$ax^2 + bx + c = pqx^2 + (rq + ps)x + rs,$$

and the following relationships must hold true:

$$a = pq$$
$$b = rq + ps$$
$$c = rs$$

These equations suggest the method for finding the values of p, q, r, and s that is illustrated in the following example.

EXAMPLE 1   Factor $5x^2 + x - 4$.

SOLUTION   The coefficient of the quadratic term is 5. The only *positive* factor pair that has a product of 5 is 5, 1. Thus the factorization of the trinomial must *begin* as:

$$(5x \quad)(x \quad)$$

The constant term is -4. The factor pairs that have a product of -4 are 1, -4; -1, 4; and 2, -2. Thus the factorization of the trinomial must *end* as:

$$(\quad 1)(\quad -4),$$
$$(\quad -1)(\quad 4),$$
$$\text{or} \quad (\quad 2)(\quad -2)$$

There are six *trial* factorizations to check to determine which gives a linear term of x in its product.

<table>
<thead>
<tr><th>Trial Factorization</th><th>Linear Term</th></tr>
</thead>
<tbody>
<tr><td>$(5x + 1)(x - 4)$</td><td>$-20x + 1x = -19x$</td></tr>
<tr><td>$(5x - 1)(x + 4)$</td><td>$20x + (-1x) = 19x$</td></tr>
<tr><td>$(5x + 4)(x - 1)$</td><td>$-5x + 4x = -1x$</td></tr>
<tr><td>$(5x - 4)(x + 1)$</td><td>$5x + (-4x) = 1x$</td></tr>
<tr><td>$(5x + 2)(x - 2)$</td><td>$-10x + 2x = -8x$</td></tr>
<tr><td>$(5x - 2)(x + 2)$</td><td>$10x + (-2x) = 8x$</td></tr>
</tbody>
</table>

The fourth factorization gives a linear term of $1x$, or x.

$$\therefore 5x^2 + x - 4 = (5x - 4)(x + 1)$$

The factorization of a polynomial is said to be a **complete factorization** when each factor is either a monomial or an irreducible polynomial whose greatest monomial factor is 1.

EXAMPLE 2 Factor $12y^3 - 14y^2 - 10y$ completely.

SOLUTION The greatest monomial factor is $2y$. Thus:

$$12y^3 - 14y^2 - 10y = 2y(6y^2 - 7y - 5)$$

Find the binomial factors, if any, of $6y^2 - 7y - 5$.

<table>
<thead>
<tr><th>Trial Factorization</th><th>Linear Term</th></tr>
</thead>
<tbody>
<tr><td>$(6y + 1)(y - 5)$</td><td>$-30y + 1y = -29y$</td></tr>
<tr><td>$(6y - 1)(y + 5)$</td><td>$30y + (-1y) = 29y$</td></tr>
<tr><td>$(6y + 5)(y - 1)$</td><td>$-6y + 5y = -1y$</td></tr>
<tr><td>$(6y - 5)(y + 1)$</td><td>$6y + (-5y) = 1y$</td></tr>
<tr><td>$(3y + 1)(2y - 5)$</td><td>$-15y + 2y = -13y$</td></tr>
<tr><td>$(3y - 1)(2y + 5)$</td><td>$15y + (-2y) = 13y$</td></tr>
<tr><td>$(3y + 5)(2y - 1)$</td><td>$-3y + 10y = 7y$</td></tr>
<tr><td>$(3y - 5)(2y + 1)$</td><td>$3y + (-10y) = -7y$</td></tr>
</tbody>
</table>

The last factorization gives a linear term of $-7y$, and so $6y^2 - 7y - 5 = (3y - 5)(2y + 1)$. The greatest monomial factor of each of these binomials is 1.

$\therefore$ factored completely, $12y^3 - 14y^2 - 10y = 2y(3y - 5)(2y + 1)$.

Except for trivial changes in the factors, such as changing their order or introducing -1 as a factor, the complete factorization of a polynomial is *unique*. Thus, each of the following is an equivalent form of the factorization given as an answer in Example 2.

$$2y(2y + 1)(3y - 5)$$
$$2y(-5 + 3y)(1 + 2y)$$
$$-2y(5 - 3y)(1 + 2y)$$

(continued on p. 363)

360

EXAMPLE 3   Factor $15x^4 + 3x^2 - 18$ completely.

SOLUTION $15x^4 + 3x^2 - 18 = 3(5x^4 + x^2 - 6)$
$$= 3(5x^2 + 6)(x^2 - 1)$$
$$= 3(5x^2 + 6)(x + 1)(x - 1)$$

Oral Exercises

Name the pairs of expressions that you would try as linear terms and as constant terms of the binomials when looking for binomial factors of the given polynomial.

EXAMPLE $6x^2 + 7x - 10$

SOLUTION   Linear terms: $6x, x$ and $3x, 2x$
Constant terms: 1, -10; -1, 10; 2, -5; and -2, 5

1. $2c^2 + 3c + 1$	**2.** $3d^2 - 5d + 2$	**3.** $2m^2 - m - 10$
4. $5n^2 + 3n - 14$	**5.** $5p^2 - 16p + 12$	**6.** $2q^2 + 13q + 18$
7. $6a^2 + a - 7$	**8.** $14b^2 - 3b - 2$	**9.** $9x^2 - 9x - 10$
10. $4y^2 + 4y - 15$	**11.** $10r^2 - r - 21$	**12.** $15s^2 - 2s - 8$

Written Exercises

Factor completely.

A
1. $5a^2 + 7a + 2$	**2.** $3b^2 - 7b + 2$	**3.** $2c^2 - 3c - 5$
4. $5d^2 + 2d - 3$	**5.** $3x^2 - 5x - 8$	**6.** $2y^2 + 7y + 6$
7. $6r^2 - r - 2$	**8.** $15s^2 + s - 2$	**9.** $4z^2 - 16z + 15$
10. $4w^2 + 5w - 21$	**11.** $8m^2 + 2m - 15$	**12.** $18n^2 - n - 4$
13. $10j^2 - 37j - 12$	**14.** $8k^2 - 26k + 21$	**15.** $12p^2 + 23p + 10$
16. $12q^2 + 19q - 10$	**17.** $20u^2 + 17u - 10$	**18.** $20v^2 - 37v - 6$
19. $4x^2 + 24x + 20$	**20.** $6y^2 - 12y - 18$	**21.** $4m^2 + 14m + 6$
22. $6n^2 - 22n + 12$	**23.** $12a^2 + 2a - 4$	**24.** $12b^2 - 24b - 15$
25. $12m^3 + 6m^2 - 6m$	**26.** $9n^3 - 3n^2 - 6n$	**27.** $18z^4 + 12z^3 + 2z^2$
28. $27d^5 - 18d^4 + 3d^3$	**29.** $16a^5 - 4a^3$	**30.** $20b^4 - 45c^2$

B
31. $6x^5 - 22x^3 - 8x$	**32.** $6y^5 + 3y^3 - 9y$
33. $6z^6 - 21z^4 - 12z^2$	**34.** $3w^6 - 9w^4 + 3w^2$
35. $-12b^5 - b^3 + b$	**36.** $-18c^7 - c^5 + 4c^3$
37. $6u^3 + 23u^2v - 18uv^2$	**38.** $10r^3 + 3r^2s - 18rs^2$
39. $12w^4 - 35w^3z + 8w^2z^2$	**40.** $20x^4 - 7x^3y - 6x^2y^2$
41. $6y^3z + 10y^2z^2 - 4yz^3$	**42.** $8a^3b^2 + 2a^2b^3 - 3ab^4$

360   *Chapter 7*

C **43.** $x^4 - y^4$ **44.** $x^8 - y^8$ **45.** $x^6 - y^6$ **46.** $x^{10} - y^{10}$

47. $y^4 + y^2 + 1$ (*Hint:* $y^4 + y^2 + 1 = y^4 + 2y^2 + 1 - y^2$)

48. $x^4 - 7x^2 + 9$ (*Hint:* $x^4 - 7x^2 + 9 = x^4 - \underline{\;\;?\;\;} + 9 - \underline{\;\;?\;\;}$)

49. Find integral values of a, b, and c such that $ax + b$ is a factor of both $2x^2 - 5x + c$ and $4x^2 + 4x + 1$. $a = 2$, $b = 1$, $c = -3$

50. Show that there is an infinite set of integral values of c for which the trinomial $x^2 + x + c$ can be factored over the set of polynomials with integral coefficients.

▎Self-Test 2

VOCABULARY
    factor a number (p. 341)
    factor set (p. 341)
    prime number, or prime
      (p. 341)
    prime factorization (p. 341)
    greatest common factor (GCF)
      of integers (p. 342)
    least common multiple (LCM)
      of integers (p. 342)
    greatest common factor (GCF)
      of monomials (p. 343)
    least common multiple (LCM)
      of monomials (p. 343)

    greatest monomial factor
      (p. 346)
    sum of cubes (p. 350)
    difference of cubes (p. 350)
    quadratic polynomial (p. 353)
    quadratic term (p. 353)
    linear term (p. 353)
    constant term (p. 353)
    quadratic trinomial (p. 354)
    reducible (p. 355)
    irreducible (p. 355)
    complete factorization (p. 359)

1. Factor 240 over the set of prime numbers. $2^4 \cdot 3 \cdot 5$ *Obj. 1, p. 341*

2. Find the greatest common factor and the least common   *Obj. 2, p. 341*
    multiple of $48c^3d^5$ and $32bc^2d^6$. $16c^2d^5$; $96bc^3d^6$

Write each polynomial as the product of its greatest monomial factor and another polynomial.

3. $25m^2 - 40m$ $5m(5m - 8)$ **4.** $6x^4y^2 - 10x^5y^3 + 14x^6y$ *Obj. 3, p. 341*
                                             $2x^4y(3y - 5xy^2 + 7x^2)$

Factor completely.

5. $s^2 + 6s + 9$ $(s + 3)^2$ **6.** $81m^2 - 49$ $(9m + 7)(9m - 7)$ *Obj. 4, p. 341*

7. $x^2 - 2xy + y^2$ $(x - y)^2$ **8.** $n^3 - 1$ $(n - 1)(n^2 + n + 1)$

9. $t^2 - 7t + 12$ $(t - 3)(t - 4)$ **10.** $2u^2 - 3u - 20$ $(u - 4)(2u + 5)$

11. $3g^3 - 27g$ $3g(g + 3)(g - 3)$ **12.** $6h^3 - 10h^2 + 8h$ $2h(3h^2 - 5h + 4)$

Check your answers with those at the back of the book.

PROGRAMMING IN BASIC

The following program can be used to factor a quadratic trinomial of the form $ax^2 + bx + c$ when you input values for a, b, and c. The program uses a *subroutine* in lines 300–350. That is, the line

200 GOSUB 300

takes the execution of the program to line 300, and the line

350 RETURN

takes the execution back to line 210.

```
10     PRINT "TO FIND FACTORS OF"
20     PRINT "A TRINOMIAL OF THE FORM"
30     PRINT "AX↑2 + BX + C"
40     PRINT "INPUT A(>0), B, C(<>0):";
50     INPUT A, B, C
60     IF A <= 0 THEN 40
70     IF C = 0 THEN 40
80     PRINT
90     REM *FIND FACTORS OF A
100      FOR A1 = 1 TO A
110      LET A2 = A/A1
120      IF A2 <> INT(A2) THEN 250
130      IF A2 < A1 THEN 380
140      PRINT "A1 = ";A1,"A2 = ";A2
150      REM *FIND FACTORS OF C
160      FOR C1 = 1 TO ABS(C)
170      LET C2 = C/C1
180      IF C2 <> INT(C2) THEN 240
190      LET C3 = C1
200      GOSUB 300
210      LET C3 = −C3
220      LET C2 = −C2
230      GOSUB 300
240      NEXT C1
250      NEXT A1
260      GOTO 380
270      REM *SUBROUTINE
280      REM *COMPUTE AND PRINT
290      REM *POSSIBLE MIDDLE TERMS
300      PRINT A1;" X ";C2;" + ";
310      PRINT A2;" X ";C3;" = ";
320      LET M = A1 * C2 + A2 * C3
330      PRINT M;" B =";B
340      IF M = B THEN 410
```

```
350   RETURN
360   REM *END OF SUBROUTINE
370   REM *OUTPUT
380   PRINT
390   PRINT "IRREDUCIBLE"
400   GOTO 440
410   PRINT
420   PRINT "(";A1;"X + (";C3;"))";
430   PRINT "(";A2;"X + (";C2;"))"
440   END
```

Exercises

Type in the program as given. Then RUN the program to factor the following quadratic trinomials.

IRREDUCIBLE
1. $x^2 + 3x + 4$

$(1x + (2))(1x + (2))$
2. $x^2 + 4x + 4$

$(1x + (-6))(1x + (-6))$
3. $x^2 - 12x + 36$

4. $2x^2 + 14x - 36$
$(1x + (-2))(2x + (18))$

5. $8x^2 + 5x - 6$
IRREDUCIBLE

6. $6x^2 + 29x + 35$
$(2x + (5))(3x + (7))$

Applications of Factoring

OBJECTIVES for Sections 7-10 and 7-11:
1. To solve polynomial equations by factoring.
2. To use quadratic equations to solve word problems.

7–10 Solving Polynomial Equations by Factoring

An equation that can be written equivalently in the form

$$ax^2 + bx + c = 0, \quad a \neq 0,$$

is called a **quadratic equation.** In this section you will learn to solve certain quadratic equations by using the factoring techniques that have been presented in this chapter. The basis of this method of solution is the *zero-product property* of real numbers.

To understand the zero-product property, first note that the multiplicative property of zero guarantees the following theorem to be true.

Theorem. For all real numbers a and b,

if $a = 0$ or $b = 0$, then $ab = 0$.

Additional Answers
Oral Exercises
(continued from p. 360)

8. Linear terms: 14*b*, *b*;
7*b*, 2*b*
Constant terms: 1, −2;
−1, 2

9. Linear terms: 9*x*, *x*;
3*x*, 3*x*
Constant terms: 1, −10;
−1, 10; 2, −5; −2, 5

10. Linear terms: 4*y*, *y*;
2*y*, 2*y*
Constant terms: 1, −15;
−1, 15; 3, −5; −3, 5

11. Linear terms: 10*r*, *r*;
5*r*, 2*r*
Constant terms: 1, −21;
−1, 21; 3, −7; −3, 7

12. Linear terms: 15*s*, *s*;
5*s*, 3*s*
Constant terms: 1, −8;
−1, 8; 2, −4; −2, 4

Teaching Suggestions
p. T96

Key Ideas

Use the zero-product property to solve polynomial equations.

1. Solve $(4x - 1)(x + 2) = 0$
$4x - 1 = 0$ or $x + 2 = 0$

$$x = \frac{1}{4} \text{ or } \quad x = -2$$

Check:

$$\left[4\left(\frac{1}{4}\right) - 1\right]\left(\frac{1}{4} + 2\right) \stackrel{?}{=} 0$$

$$(0)\left(\frac{9}{4}\right) \stackrel{?}{=} 0$$

$$0 = 0 \quad \checkmark$$

$$[4(-2) - 1](-2 + 2) \stackrel{?}{=} 0$$

$$(-9)(0) \stackrel{?}{=} 0$$

$$0 = 0 \quad \checkmark$$

∴ the solution set is

$$\left\{\frac{1}{4}, -2\right\}.$$

2. Solve $m^2 = -3m + 10$.
$m^2 + 3m - 10 = 0$
$(m + 5)(m - 2) = 0$
$m + 5 = 0$ or $m - 2 = 0$
$\quad m = -5 \qquad m = 2$

Check:
$(-5)^2 \stackrel{?}{=} -3(-5) + 10$
$\quad 25 = 25 \quad \checkmark$
$(2)^2 \stackrel{?}{=} -3(2) + 10$
$\quad 4 = 4 \quad \checkmark$

∴ the solution set is
$\{-5, 2\}$.

3. Solve $6y^3 - y^2 = 12y$.
$6y^3 - y^2 - 12y = 0$
$y(6y^2 - y - 12) = 0$
$y(2y - 3)(3y + 4) = 0$
$\quad y = 0 \quad$ or $\quad 2y - 3 = 0$

$$y = \frac{3}{2}$$

or $3y + 4 = 0$

$$y = -\frac{4}{3}$$

∴ the solution set is

$$\left\{0, \frac{3}{2}, -\frac{4}{3}\right\}.$$

Compare the theorem on the preceding page with the following one, which will be proved in Exercise 53 on page 367.

Theorem. For all real numbers a and b,

$$\text{if } ab = 0, \text{ then } a = 0 \text{ or } b = 0.$$

As you can see, the hypothesis of the first theorem, $a = 0$ or $b = 0$, is the conclusion of the second theorem. Moreover, the hypothesis of the second theorem, $ab = 0$, is the conclusion of the first theorem. Thus, each of these theorems is said to be the *converse* of the other. Two statements are **converses** when the hypothesis of each statement is the conclusion of the other statement.

A statement and its converse can be combined by using the phrase *if and only if*. Thus the following *zero-product property* is a combination of the two preceding theorems.

Zero-Product Property
For all real numbers a and b,

$$ab = 0 \text{ if and only if } a = 0 \text{ or } b = 0.$$

The zero-product property can be shown to be true for any number of factors. In general, then, the zero-product property states that *a product is equal to zero if and only if at least one of its factors is equal to zero.* This property is often used in solving quadratic equations.

EXAMPLE 1 Solve $(2x - 3)(x + 5) = 0$.

SOLUTION Using the zero-product property, write the equivalent disjunction.

$$2x - 3 = 0 \qquad \text{or} \qquad x + 5 = 0$$

$$x = \frac{3}{2} \qquad\qquad\qquad x = -5$$

Check *both* solutions in the original equation.

$$(2x - 3)(x + 5) = 0 \qquad\qquad (2x - 3)(x + 5) = 0$$

$$\left[2\left(\frac{3}{2}\right) - 3\right]\left(\frac{3}{2} + 5\right) \stackrel{?}{=} 0 \qquad [2(-5) - 3](-5 + 5) \stackrel{?}{=} 0$$

$$(0)\left(\frac{13}{2}\right) \stackrel{?}{=} 0 \qquad\qquad\qquad (-13)(0) \stackrel{?}{=} 0$$

$$0 = 0 \quad \checkmark \qquad\qquad\qquad\qquad 0 = 0 \quad \checkmark$$

∴ the solution set is $\left\{\frac{3}{2}, -5\right\}$.

EXAMPLE 2 Solve $y^2 = 5y + 36$.

SOLUTION 1. Transform the given equation into an equivalent equation that has 0 as one side. $\qquad y^2 - 5y - 36 = 0$

2. Factor the trinomial completely. $\qquad (y + 4)(y - 9) = 0$

3. Solve the equivalent disjunction.

$$y + 4 = 0 \qquad \text{or} \qquad y - 9 = 0$$
$$y = -4 \qquad\qquad\qquad y = 9$$

4. Check both solutions in the original equation.

$$y^2 = 5y + 36 \qquad\qquad y^2 = 5y + 36$$
$$(-4)^2 \overset{?}{=} 5(-4) + 36 \qquad 9^2 \overset{?}{=} 5(9) + 36$$
$$16 = 16 \;\checkmark \qquad\qquad 81 = 81 \;\checkmark$$

∴ the solution set is $\{-4, 9\}$.

In general, a **polynomial equation** is any equation that can be written equivalently with 0 as one side and a polynomial as the other side. Polynomial equations are often named by the term of highest degree. Thus, as you have learned, an equation of the form

$$ax + b = 0, \quad a \ne 0, \quad \text{is a } \textit{linear equation};$$
$$ax^2 + bx + c = 0, \quad a \ne 0, \quad \text{is a } \textit{quadratic equation}.$$

Similarly, an equation of the form

$$ax^3 + bx^2 + cx + d = 0, \quad a \ne 0, \quad \text{is a } \textbf{cubic equation}.$$

Since the zero-product property is true for any number of factors, it can also be used in solving cubic equations and other polynomial equations of higher degree.

EXAMPLE 3 Solve $10a^3 + a^2 = 3a$.

SOLUTION 1. Transform the given equation into an equivalent equation that has 0 as one side. $\qquad 10a^3 + a^2 - 3a = 0$

2. Factor the polynomial completely. $\qquad a(10a^2 + a - 3) = 0$
$$a(5a + 3)(2a - 1) = 0$$

3. Solve the equivalent disjunction.

$$a = 0 \quad \text{or} \quad 5a + 3 = 0 \quad \text{or} \quad 2a - 1 = 0$$
$$a = 0 \qquad\qquad a = -\frac{3}{5} \qquad\qquad a = \frac{1}{2}$$

4. Checking the results is left to you.

∴ the solution set is $\left\{0, -\frac{3}{5}, \frac{1}{2}\right\}$.

Polynomials and Factoring **365**

Additional Answers
Oral Exercises

1. $c = 0$ or $c - 5 = 0$
2. $4x = 0$ or $x + 6 = 0$
3. $m - 2 = 0$ or $m + 3 = 0$
4. $w + 1 = 0$ or $w - 1 = 0$
5. $2a + 1 = 0$ or $a + 3 = 0$
6. $n = 0$ or $3n - 2 = 0$ or $n + 1 = 0$
7. a. True
   b. If $a = b$, then $a + c = b + c$; true
   c. $a + c = b + c$ if and only if $a = b$.
8. a. True
   b. If $ac = bc$, then $a = b$; false
9. a. True
   b. If $x + y = 7$, then $x = 3$ and $y = 4$; false
10. a. True
    b. If $x^2 - 2x = 0$, then $x^2 = 2x$; true
    c. $x^2 = 2x$ if and only if $x^2 - 2x = 0$.
11. a. False
    b. If $r = 1$ and $s = 1$, then $rs = 1$; true
12. a. True
    b. If $a - c < b - c$, then $a < b$; true
    c. $a < b$ if and only if $a - c < b - c$.
13. a. True
    b. If $a < 0$, then $-a > 0$; true
    c. $-a > 0$ if and only if $a < 0$.
14. a. False
    b. If $ac > bc$, then $a > b$; false
15. For all real numbers a and b, if $a > b$, then $a - b > 0$. If $a - b > 0$, then $a > b$.

(continued on p. 376)

Oral Exercises

State a disjunction that is equivalent to the given open sentence.

1. $c(c - 5) = 0$
2. $4x(x + 6) = 0$
3. $(m - 2)(m + 3) = 0$
4. $(w + 1)(w - 1) = 0$
5. $(2a + 1)(a + 3) = 0$
6. $n(3n - 2)(n + 1) = 0$

For each statement in Exercises 7–14:

a. Tell whether the statement is true or false for all real values of the variables.

b. Give the converse of the statement and tell whether it is true or false.

c. If a statement and its converse are *both* true, reword them in combined form using the words "if and only if."

7. If $a + c = b + c$, then $a = b$.
8. If $a = b$, then $ac = bc$.
9. If $x = 3$ and $y = 4$, then $x + y = 7$.
10. If $x^2 = 2x$, then $x^2 - 2x = 0$.
11. If $rs = 1$, then $r = 1$ and $s = 1$.
12. If $a < b$, then $a - c < b - c$.
13. If $-a > 0$, then $a < 0$.
14. If $a > b$, then $ac > bc$.

Reword each theorem in the form of two theorems that are converses.

15. For all real numbers a and b, $a > b$ if and only if $a - b > 0$.

16. For all real numbers a and b, $ab > 0$ if and only if either $a > 0$ and $b > 0$ or $a < 0$ and $b < 0$.

17. For all positive real numbers a and b, $a < b$ if and only if $\dfrac{1}{a} > \dfrac{1}{b}$.

18. For all real numbers a and b and all positive real numbers c and d, $\dfrac{a}{c} < \dfrac{b}{d}$ if and only if $ad < bc$.

Written Exercises

Solve.

A
1. $(x - 4)(x + 1) = 0$ $\{4, -1\}$
2. $(y + 3)(y - 2) = 0$ $\{-3, 2\}$
3. $(5m - 1)(2m - 3) = 0$ $\{\frac{1}{5}, \frac{3}{2}\}$
4. $(3n + 4)(5n + 4) = 0$ $\{-\frac{4}{3}, -\frac{4}{5}\}$
5. $a(a - 2)(a + 5) = 0$ $\{0, 2, -5\}$
6. $b(2b - 1)(b + 3) = 0$ $\{0, \frac{1}{2}, -3\}$
7. $3p(p + 7)(5p - 3) = 0$ $\{0, -7, \frac{3}{5}\}$
8. $5q(3q + 5)(4q + 3) = 0$ $\{0, -\frac{5}{3}, -\frac{3}{4}\}$
9. $w^2 + 3w - 10 = 0$ $\{-5, 2\}$
10. $z^2 + 2z - 24 = 0$ $\{4, -6\}$
11. $s^2 - 9 = 0$ $\{3, -3\}$
12. $t^2 - 1 = 0$ $\{1, -1\}$
13. $3c^2 + 5c - 12 = 0$ $\{\frac{4}{3}, -3\}$
14. $8d^2 - 2d - 3 = 0$ $\{\frac{3}{4}, -\frac{1}{2}\}$
15. $6u^2 + 7u = 3$ $\{\frac{1}{3}, -\frac{3}{2}\}$
16. $3v^2 - v = 10$ $\{2, -\frac{5}{3}\}$
17. $20f^2 = -21f - 4$ $\{-\frac{1}{4}, -\frac{4}{5}\}$
18. $6g^2 = 11g - 5$ $\{1, \frac{5}{6}\}$

19. $29v = 10 + 10v^2$ $\{\frac{5}{2}, \frac{2}{5}\}$ **20.** $14w = -16 - 3w^2$ $\{-2, -\frac{8}{3}\}$

21. $4j^2 = 25$ $\{\frac{5}{2}, -\frac{5}{2}\}$ **22.** $9k^2 = 49$ $\{\frac{7}{3}, -\frac{7}{3}\}$

23. $3g^2 = 12g$ $\{0, 4\}$ **24.** $2h^2 = 5h$ $\{0, \frac{5}{2}\}$

B   **25.** $a^3 - 5a^2 - 36a = 0$ $\{0, 9, -4\}$ **26.** $b^3 - 7b^2 + 12b = 0$ $\{0, 4, 3\}$

27. $r^3 = 4r^2 - 3r$ $\{0, 3, 1\}$ **28.** $s^3 - 4s^2 = 32s$ $\{0, 8, -4\}$

29. $4x^3 + 2x^2 = 6x$ $\{0, 1, -\frac{3}{2}\}$ **30.** $9y^3 - 3y^2 = 12y$ $\{0, \frac{4}{3}, -1\}$

31. $9z^3 = 25z$ $\{0, \frac{5}{3}, -\frac{5}{3}\}$ **32.** $18w^3 = 8w$ $\{0, \frac{2}{3}, -\frac{2}{3}\}$

33. $x^4 - 8x^2 + 16 = 0$ $\{2, -2\}$ **34.** $y^5 - 2y^3 + y = 0$ $\{0, 1, -1\}$

35. $p^4 - 5p^2 = -4$ $\{1, -1, 2, -2\}$ **36.** $q^4 + 36 = 13q^2$ $\{2, -2, 3, -3\}$

37. $(2m - 1)(m - 3) = 18$ $\{5, -\frac{3}{2}\}$ **38.** $(4n - 3)(2n + 1) = 63$ $\{3, -\frac{11}{4}\}$

39. $(3w + 2)^2 = 9$ $\{\frac{1}{3}, -\frac{5}{3}\}$ **40.** $(5w - 2)^2 = 16$ $\{\frac{6}{5}, -\frac{2}{5}\}$

41. $(p + 3)^2 - p = 15$ $\{-6, 1\}$ **42.** $(4q - 1)^2 + 2 = 8q$ $\{\frac{3}{4}, \frac{1}{4}\}$

43. $(4a - 3)^2 + 11a = (a + 2)(a - 2) + 33$ **44.** $(3b - 5)^2 + 9b = (4 + b)(4 - b)$

                                                 $\{\frac{5}{3}, -\frac{4}{5}\}$ $\{\frac{3}{2}, \frac{3}{5}\}$

Write a quadratic equation that has the given solution set. The equation should be written in the form $ax^2 + bx + c = 0$, where a, b, and c are integers.

C   **45.** $\{3, 7\}$ **46.** $\{-5, -6\}$ **47.** $\{-2, 3\}$ **48.** $\{7, -2\}$

49. $\{\frac{1}{2}, \frac{3}{4}\}$ **50.** $\{-\frac{2}{5}, -\frac{2}{3}\}$ **51.** $\{-\frac{3}{2}, \frac{3}{5}\}$ **52.** $\{\frac{5}{3}, -\frac{3}{4}\}$

Give the reason that justifies each statement in the following proof.

53. Prove: For all real numbers a and b, if $ab = 0$, then $a = 0$ or $b = 0$.

PROOF

Case 1:   If $a = 0$, the conclusion is true whether or not $b = 0$.

Case 2:   If $a \neq 0$, reason as follows.

Statements	*Reasons*
1. $\frac{1}{a}$ is a real number.	_?_ Ax. of mult. inv.
2. $ab = 0$	_?_ Hypothesis
3. $\frac{1}{a}(ab) = \frac{1}{a}(0)$	_?_ Mult. prop. of $=$
4. $\frac{1}{a}(ab) = b$	_?_ Iden. ax. for mult.
5. $\frac{1}{a}(0) = 0$	_?_ Mult. prop. of zero
6. $\therefore b = 0$	Substitution principle

Therefore, given $ab = 0$, it follows that $a = 0$ or $b = 0$.

54. Prove: For all real numbers a and b, $ab > 0$ if and only if either $a > 0$ and $b > 0$ or $a < 0$ and $b < 0$.

Polynomials and Factoring   **367**

Mixed Review

Solve.

1. $2x + 7 = 0$ $\{-3\frac{1}{2}\}$

2. $4 - 3a = 0$ $\{1\frac{1}{3}\}$

Factor completely.

3. $2u - 2u^2$ $2u(1 - u)$

4. $a^2 - 3a - 18$
   $(a + 3)(a - 6)$

5. $2x^2 - 6x - 20$
   $2(x + 2)(x - 5)$

6. $15n^2 - 43n - 8$
   $(15n + 2)(n - 3)$

Key Ideas

Solve word problems using quadratic equations.

Chalkboard Examples

1. A picture frame of uniform width surrounds a 30 cm by 20 cm picture. The combined area of the picture and frame is 2000 cm². What is the width of the frame?

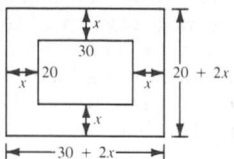

Let x = width of the frame in centimeters. Then $30 + 2x$ = total length in centimeters; $20 + 2x$ = total width in centimeters.
$(30 + 2x)(20 + 2x) = 2000$
$4x^2 + 100x - 1400 = 0$
$4(x^2 + 25x - 350) = 0$
$4(x + 35)(x - 10) = 0$
Then $x = -35$ or $x = 10$. Discard -35 as a possible solution.
∴ the width of the frame is 10 cm.

2. Todd uses a sling shot to throw a rock upward at a speed of 29.4 m/s. If you assume that the rock starts at ground level, when will it be 32.9 m above the ground? (Use the formula $h = rt - 4.9t^2$ where h = height in me-

7–11 Solving Problems by Factoring

The zero-product property of the real numbers is also used in solving problems that can be represented by quadratic equations. When you use a quadratic equation to solve a problem, you may find that the equation has more than one solution. Remember that each solution of the equation is only a *possible* solution of the problem. Each solution of the equation must be checked against the conditions of the problem to determine the *actual* solutions of the problem.

EXAMPLE 1 The flower garden in the Bay Area Park measures 24 m by 30 m. The garden is surrounded by a paved walk of uniform width. If the combined area of the paved walk and the garden is 1080 m², what is the width of the walk?

SOLUTION

Step 1   The problem asks for the width of the walk.

Step 2   Let x = width of the walk in meters. Then:
$30 + 2x$ = total length in meters;
$24 + 2x$ = total width in meters.

Step 3   The total area of the paved walk and the garden is 1080 m.
$$(30 + 2x)(24 + 2x) = 1080$$

Step 4 $720 + 48x + 60x + 4x^2 = 1080$
$4x^2 + 108x + 720 = 1080$
$4x^2 + 108x - 360 = 0$
$4(x^2 + 27x - 90) = 0$
$4(x + 30)(x - 3) = 0$

Since $4 \neq 0$, $x + 30 = 0$ or $x - 3 = 0$
$x = -30$ $x = 3$

Step 5   Since it is not meaningful to have a width of -30, that value is discarded as a possible solution. Check the value 3 in the original problem.
Is the total area 1080 m² when the walk is 3 m wide?

total width in meters = $24 + 2(3) = 30$
total length in meters = $30 + 2(3) = 36$
total area in square meters = $30 \cdot 36 = 1080$

∴ the width of the walk is 3 m.

In the following example, both solutions of the equation satisfy the conditions of the original problem.

368   *Chapter 7*

368

EXAMPLE 2   The height h in meters that an object will reach in t seconds when it is thrown upward from the ground with an initial speed of r meters per second is given by the formula

$$h = rt - 4.9t^2.$$

In how many seconds after it is thrown will an object that is thrown upward from the ground with an initial speed of 34.3 m/s be 49 m above the ground?

SOLUTION

Step 1   The problem asks for the number of seconds after being thrown upward when the object will be 49 m above the ground.

Step 2   Let t = required number of seconds. Then:
h = height in meters = 49
r = initial speed in m/s = 34.3

Step 3   The given formula is $h = rt - 4.9t^2$

Step 4
$$49 = 34.3t - 4.9t^2$$ Multiply
$$490 = 343t - 49t^2$$ both sides
$$49t^2 - 343t + 490 = 0$$ by 10
$$49(t^2 - 7t + 10) = 0$$
$$49(t - 2)(t - 5) = 0$$
$$t - 2 = 0 \quad \text{or} \quad t - 5 = 0$$
$$t = 2 \quad \quad t = 5$$

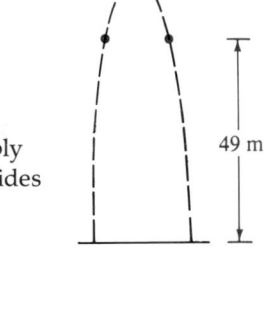
49 m

Step 5   Checking the results is left to you.

∴ the object will be 49 m above the ground both 2 s and 5 s after being thrown upward.

ters at the end of t seconds, r = initial speed in meters per second, and t = time in seconds.)
Let t = required number of seconds; h = 39.2; r = 29.4.
$$h = rt - 4.9t^2$$
$$39.2 = 29.4t - 4.9t^2$$
$$392 = 294t - 49t^2$$
$$49(t^2 - 6t + 8) = 0$$
$$49(t - 4)(t - 2) = 0$$
$$t = 4$$
$$t = 2$$
∴ the rock will be 39.2 m above the ground both 2 s and 4 s after being shot.

Problems

Solve.

A   1. Find two consecutive positive integers whose product is 132. 11 and 12
2. Find two negative integers that differ by 5 and whose product is 126. −14 and −9
3. The square of a positive integer exceeds four times the integer by 32. Find the integer. 8
4. The square of a negative integer is 28 greater than three times the integer. Find the integer. −4
5. Find two consecutive negative integers such that the sum of their squares is 113. −8 and −7
6. Find two integers that differ by 6 while their squares differ by 132.
   8 and 14; −8 and −14

Polynomials and Factoring   **369**

Solve.

1. Find two consecutive positive integers such that the sum of their squares is 244. 10, 12

2. Find two consecutive positive integers whose product is 210. 14, 15

3. Find two odd integers whose squares differ by 64. 15, 17

4. Find two consecutive negative integers such that the sum of their squares is 265. −11, −12

5. The sum of two integers is 15 and their product is 36. Find the numbers. 12, 3

6. A rectangle is 5 cm longer than it is wide. Find the length and width if the area of the rectangle is 66 cm². length: 11 cm; width: 6 cm

Mixed Review

Solve.

1. $k^2 - 196 = 0$ $\{14, -14\}$

2. $16a^2 = 25$ $\left\{\dfrac{15}{4}, -\dfrac{15}{4}\right\}$

3. $3m^2 = 14m$ $\left\{0, \dfrac{14}{3}\right\}$

4. $w^2 - w - 56 = 0$ $\{8, -7\}$

5. $rp^3 = 9p$ $\left\{0, \dfrac{3}{2}, -\dfrac{3}{2}\right\}$

7. A rectangular garden is 12 m long and 10 m wide. Surrounding the garden is a paved walk of uniform width. The combined area of the garden and the walk is 168 m². Find the width of the walk. 1 m

8. A rectangular picture is 4 cm longer than it is wide. It is surrounded by a mat that is 2 cm wide. The combined area of the picture and the mat is 140 cm². Find the dimensions of the picture. 6 m by 10 m

9. The width of a rectangle is 7 m less than twice the length. The area of the rectangle is 30 m². Find the length of the rectangle. 6 m

10. A certain rectangle is 3 cm longer than it is wide. The area of the rectangle is 550 cm². Find the dimensions of the rectangle. 22 cm by 25 cm

11. The sum of two integers is 20 and their product is 36. Find the integers. 2 and 18

12. Find two positive integers whose product is 240 and whose difference is 8. 12 and 20

For problems 13–16, use the formula $h = rt - 4.9t^2$.

13. A ball is thrown upward with an initial speed of 24.5 m/s. When is the ball 29.4 m high? after 3 s and after 2 s

14. A projectile is fired upward with an initial speed of 2940 m/s. After how many minutes does it hit the ground? 10 min

B 15. A signal flare is fired upward with an initial speed of 245 m/s. A helicopter pilot at a height of 1960 m sees the flare pass on its way upwards. Assuming that the helicopter remains at the same height, how long will it be before the flare passes the helicopter on its way down? 30 s

16. A ball is thrown upward from the top of a tower that is 98 m high with an initial speed of 39.2 m/s. When does it hit the ground? (*Hint:* If h is the height of the ball above the tower, then $h = -98$ when the ball hits the ground.) after 10 s

17. The length of one leg of a right triangle is 2 cm less than three times the length of the other leg. The area of the triangle is 48 cm². Find the length of each leg. 6 cm and 16 cm

18. The length of one side of a triangle is 2 cm less than twice the length of the altitude to that side. The area of the triangle is 30 cm². Find the length of the altitude. 6 cm

19. Marie made a rectangular pen for her dog using a side of the barn for one side and 26 m of fencing for the remaining three sides. If the area enclosed was 72 m², find the dimensions of the pen. 4 m by 18 m or 9 m by 8 m

20. A side of a house is in the shape of a triangle on top of a rectangle. The rectangle is four times as long as it is high, and the altitude of the triangular part is 1 m greater than the height of the rectangle. The total area of the side of the house is 60 m². Find the height of this side of the house. **3 m**

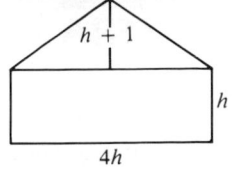

Ex. 20

21. Jackie is k years old. Her brother Joe is k^2 years old. In 7 years, Joe will be one year older than twice Jackie's age at that time. How old is each now? **4 years and 16 years**

22. Ken is z years old. His aunt is $(z - 1)^2$ years old. In two years, his aunt's age will be six times Ken's age then. How old is each now? **9 years, 64 years**

23. Dan owns a vacant lot that is 32 m wide and 40 m long. He makes a rectangular basketball court in the middle by subtracting equal amounts from the length and width. The area of the court is 560 m². How far from the edge of the lot is an edge of the court? **6 m**

24. An open rectangular box with length four times its width is made from a rectangular piece of metal by cutting a 2 cm square from each corner and turning up the sides. If the volume of the box is 144 cm³, find the dimensions of the original piece of metal.
20 cm long by 8 cm wide

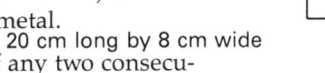
2 cm

Ex. 24

C 25. Show that the sum of the squares of any two consecutive integers is one greater than a multiple of four.

26. Show that the square of an odd integer is one greater than a multiple of eight.

Self-Test 3

VOCABULARY   quadratic equation (p. 363)   polynomial equation (p. 365)
converse (p. 364)   cubic equation (p. 365)
zero-product property (p. 364)

Solve.

1. $(m + 2)(m - 7) = 0$ $\{-2, 7\}$ 2. $n(3n + 5) = 0$ $\{0, -\frac{5}{3}\}$ *Obj. 1, p. 363*
3. $x^2 - 3x - 18 = 0$ $\{6, -3\}$ 4. $2y^3 - 5y^2 - 25y = 0$ $\{0, 5, -\frac{5}{2}\}$
5. $6g^2 + 5g = 6$ $\{\frac{2}{3}, -\frac{3}{2}\}$ 6. $2h^3 = h^2 + 6h$ $\{0, 2, -\frac{3}{2}\}$

7. The width of a certain rectangle is 13 m less than its length. The area   *Obj. 2, p. 363*
of the rectangle is 48 m². Find the dimensions of the rectangle. **3 m by 16 m**

8. Find two consecutive positive integers such that the sum of their squares is 145. **8 and 9**

Check your answers with those at the back of the book.

Polynomials and Factoring   **371**

Quick Quiz

Solve.
1. $(z - 5)(2z + 1) = 0$

$\left\{5, -\frac{1}{2}\right\}$

2. $h(3h - 1) = 0$ $\left\{0, \frac{1}{3}\right\}$

3. $r^2 + 3r - 18 = 0$ $\{3, -6\}$

4. $6t^2 - 19t = -15$ $\left\{1\frac{2}{3}, 1\frac{1}{2}\right\}$

5. $2m^2 - 9m = -10$

$\left\{2\frac{1}{2}, 2\right\}$

6. $3h^3 = -14h^2 + 5h$

$\left\{0, \frac{1}{3}, -5\right\}$

7. The length of a certain rectangle is 5 cm more than its width. The area of the rectangle is 456 cm². Find the dimensions of the rectangle.   width: 19 cm; length: 24 cm

8. Find two consecutive negative integers such that the sum of their squares is 421.   −14, −15

25. The sum of the squares of two consecutive integers, $x^2 + (x + 1)^2$, may be represented as $2(x^2 + x) + 1$. If x is even, x may be represented as $2y$ and the sum above as $2[(2y)^2 + 2y] + 1$, or $4(2y^2 + y) + 1$. If x is odd, x may be represented as $2y + 1$ and the sum as $2[(2y + 1)^2 + 2y + 1] + 1$, or $4(2y^2 + 3y + 1) + 1$. Both expressions are one greater than a multiple of four.

26. Every odd integer may be represented as $2x + 1$. The square of the odd integer, $(2x + 1)^2$, may be represented as $4(x^2 + x) + 1$. If x is even, x may be represented as $2y$ and the square above as $8(2y^2 + y) + 1$. If x is odd, x may be represented as $2y + 1$ and the square as $8(2y^2 + 3y + 1) + 1$. Both expressions are one greater than a multiple of 8.

EXTRA

Transformations of the Plane: Reflections

Recall from page 256 that a *translation* is a type of transformation, or mapping, of a plane. In this section you will learn about another type of mapping that is called a **reflection.** In a reflection, each point on a plane is transformed into its mirror image across an *axis of reflection.* For example, if the axis of reflection is the x-axis, as in Figure 1, then the square with vertices $(1, 1)$, $(2, 1)$, $(2, 2)$, and $(1, 2)$ is mapped onto the square with vertices $(1, -1)$, $(2, -1)$, $(2, -2)$, and $(1, -2)$.

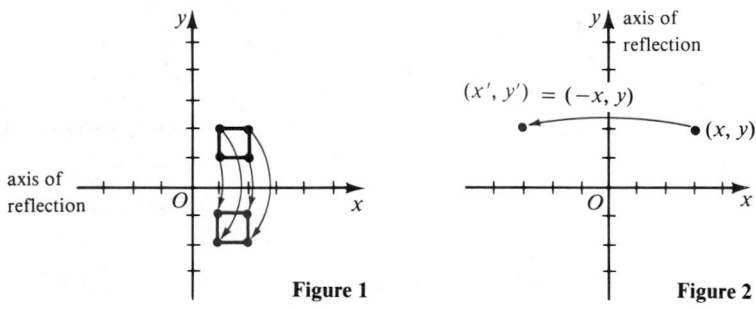

Figure 1   **Figure 2**

A reflection across the x-axis maps each point (x, y) on the plane onto a new point (x', y') whose coordinates are related to those of the point (x, y) by the *equations of reflection*

$$x' = x \quad \text{and} \quad y' = -y.$$

Similarly, as suggested by Figure 2, the equations for a reflection across the y-axis are

$$x' = -x \quad \text{and} \quad y' = y.$$

EXAMPLE   Sketch the line segment with endpoints $(-1, 3)$ and $(2, 1)$ and then sketch the reflections of this segment across the x-axis and the y-axis.

SOLUTION   A reflection across the x-axis maps the endpoints $(-1, 3)$ and $(2, 1)$ onto $(-1, -3)$ and $(2, -1)$, respectively. These points are the endpoints of the reflection of the segment across the x-axis. A reflection across the y-axis maps the endpoints $(-1, 3)$ and $(2, 1)$ onto $(1, 3)$ and $(-2, 1)$, respectively. These points are the endpoints of the reflection of the segment across the y-axis. The reflections of the segment are sketched on the coordinate plane at the right.

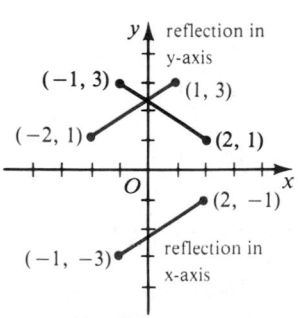

372   *Chapter 7*

The following result is proved in more advanced courses.

Under a reflection:
1. Every line in the plane is mapped (reflected) onto a line in the plane.
2. Every line segment is mapped (reflected) onto a line segment of equal length.
3. Every angle is mapped (reflected) onto an angle of equal measure.

Translations and reflections of the plane are called *rigid transformations* because they preserve the size and shape of geometric figures on the plane. Another rigid transformation of the plane that you will study in later courses is a *rotation* of the plane. In Figure 3, you see the effect of a 45° rotation of the plane on the square with vertices (1, 1), (2, 1), (2, 2), and (1, 2).

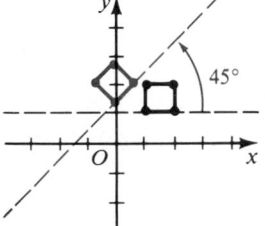

Figure 3

Exercises

In Exercises 1–8, the coordinates of the endpoints of a line segment are given.

a. Sketch the segment on a coordinate plane and label its endpoints with their coordinates.

b. On the same coordinate plane, sketch the reflections of the segment across the x-axis and the y-axis. Label the endpoints of each reflected segment with their coordinates.

1. (3, 1), (5, 0)
2. (0, 2), (3, 1)
3. (−2, 3), (1, 4)
4. (4, −2), (1, 1)
5. (−1, −1), (1, 1)
6. (−1, 1), (1, −1)
7. (1, −4), (3, 2)
8. (−3, 2), (−1, −2)

9. What must be true of a line other than an axis if it is transformed into itself by a reflection across the y-axis? the x-axis? It is perpendicular to the y-axis; it is perpendicular to the x-axis.
10. What must be true of a line other than an axis if it is transformed into a parallel line by a reflection across the y-axis? the x-axis? It is parallel to the y-axis; it is parallel to the x-axis.
11. What is the reflection of the line segment with endpoints (2, 1) and (5, 2) across the line $y = x$? the line segment with endpoints (1, 2) and (2, 5)
12. What is the reflection of the line segment with endpoints (2, 1) and (5, 2) across the line $y = -x$? the line segment with endpoints (−1, −2) and (−2, −5)

Polynomials and Factoring **373**

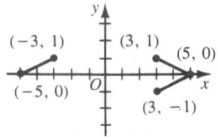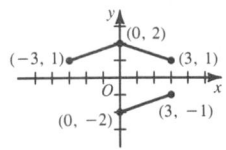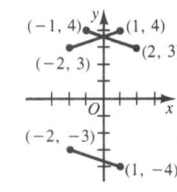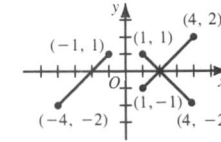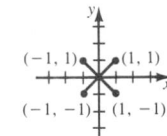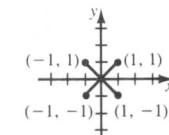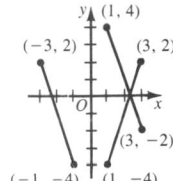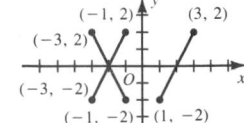
373

1. $(5a + 2)(a + 1)$
2. $(3b - 1)(b - 2)$
3. $(2c - 5)(c + 1)$
4. $(5d - 3)(d + 1)$
5. $(3x - 8)(x + 1)$
6. $(2y + 3)(y + 2)$
7. $(3r - 2)(2r + 1)$
8. $(3s - 1)(5s + 2)$
9. $(2z - 5)(2z - 3)$
10. $(4w - 7)(w + 3)$
11. $(4m - 5)(2m + 3)$
12. $(2n - 1)(9n + 4)$
13. $(j - 4)(10j + 3)$
14. $(4k - 7)(2k - 3)$
15. $(3p + 2)(4p + 5)$
16. $(12q - 5)(q + 2)$
17. $(5u - 2)(4u + 5)$
18. $(v - 2)(20v + 3)$
19. $4(x + 1)(x + 5)$
20. $6(y - 3)(y + 1)$
21. $2(2m + 1)(m + 3)$
22. $2(n - 3)(3n - 2)$
23. $2(2a - 1)(3a + 2)$
24. $3(2b - 5)(2b + 1)$
25. $6m(2m - 1)(m + 1)$
26. $3n(n - 1)(3n + 2)$
27. $2z^2(3z + 1)^2$
28. $3d^3(3d - 1)^2$
29. $4a^3(2a + 1)(2a - 1)$
30. $5(2b^2 + 3c)(2b^2 - 3c)$
31. $2x(x + 2)(x - 2)(3x^2 + 1)$
32. $3y(y + 1)(y - 1)(2y^2 + 3)$
33. $3z^2(z + 2)(z - 2)(2z^2 + 1)$
34. $3w^2(w^4 - 3w^2 + 1)$
35. $b(1 - 2b)(1 + 2b)(1 + 3b^2)$
36. $-c^3(3c + 2)(3c - 2)(2c^2 + 1)$
37. $u(3u - 2v)(2u + 9v)$
38. $r(5r - 6s)(2r + 3s)$
39. $w^2(3w - 8z)(4w - z)$
40. $x^2(4x - 3y)(5x + 2y)$
41. $2yz(3y - z)(y + 2z)$
42. $ab^2(2a - b)(4a + 3b)$

Chapter Summary

1. A *monomial* is a numeral, a variable, or an indicated product of a numeral and one or more variables. A sum of monomials is called a *polynomial*.

2. If a and b are real numbers and m and n are positive integers, the following *laws of exponents* for multiplication are true:
$$a^m a^n = a^{m+n} \qquad (a^m)^n = a^{mn} \qquad (ab)^n = a^n b^n$$

3. To find the product of two polynomials, multiply each term of one polynomial by each term of the other using the laws of exponents, then simplify the result by adding similar terms.

4. The *prime factorization* of a positive integer is the expression of the integer as the product of primes. Prime factorization can be used to find the *greatest common factor (GCF)* and the *least common multiple (LCM)* of two or more integers.

5. To factor a polynomial, you express it as the product of polynomials that are members of a specified factor set. A factorization of a polynomial is *complete* when each of the factors is either a monomial or a polynomial whose greatest monomial factor is 1.

6. The following three factor patterns occur frequently.

difference of squares: $\qquad a^2 - b^2 = (a + b)(a - b)$

trinomial squares: $\qquad a^2 + 2ab + b^2 = (a + b)^2$
$$a^2 - 2ab + b^2 = (a - b)^2$$

7. A *quadratic equation* is an equation that can be written equivalently in the form $ax^2 + bx + c = 0$, $a \neq 0$. Many quadratic equations can be solved by the use of factoring and the *zero-product property*.

Chapter Review

Write the letter of the correct answer.

1. What is the degree of the polynomial $7x^2y^3 - xy^3 + 9x^3y$?   7-1
   a. 2   b. 3   c. 4   (d.) 5

Simplify.

2. $(5m - 7mn - 10n) - (-6m + 8mn + 12n)$
   a. $-m + mn + 2n$ b. $-m + 15mn - 2n$
   (c.) $11m - 15mn - 22n$ d. $-11m + mn + 2n$

3. $(6m^3)(-2m^3)^2$ 7-2
   a. $4m^8$ b. $-12m^8$ c. $-12m^9$ (d.) $24m^9$

374   *Chapter 7*

4. $(5x^3)(3x^3) + (2x^3)^2$
   a. $17x^{12}$ **b.** $17x^6$ **c.** $19x^{14}$ **(d.)** $19x^6$

5. $-a^3(4 - 6a - 4a^3)$
   a. $4a^3 + 6a^4 - 4a^6$ **b.** $-4a^3 + 6a^4 + 4a^9$
   (c.) $-4a^3 + 6a^4 + 4a^6$ **d.** $2a^3 + 4a^9$

6. $(b + 4)(b^2 - 2b + 9)$
   a. $4b^3 - 8b^2 + 36b$ **b.** $b^2 + b + 13$
   c. $b^3 - 2b^2 + 9b + 36$ **(d.)** $b^3 + 2b^2 + b + 36$

7. $(6 - 5q)(8 + 3q)$
   a. $48 + 22q - 15q^2$ **b.** $14 - 58q - 15q^2$
   c. $14 - 22q - 15q^2$ **(d.)** $48 - 22q - 15q^2$

8. $(2r - 5)^3$
   a. $8r^3 - 125$ **(b.)** $8r^3 - 60r^2 + 150r - 125$
   c. $6r^3 - 15$ **d.** $4r^3 - 10r^2 + 50r - 50$

9. The difference of the squares of two consecutive positive odd integers is 64. Find the integers.
   a. 31, 33   **b.** 9, 11   **c.** 25, 27   **(d.)** 15, 17

10. Factor 126 over the set of prime numbers.
   a. $2 \cdot 63$ **b.** $3 \cdot 42$ **c.** $2 \cdot 7 \cdot 9$ **(d.)** $2 \cdot 3^2 \cdot 9$

11. What is the GCF of 160 and 240?
   a. 16   **b.** 24   **c.** 32   **(d.)** 80

12. What is the LCM of $51x^2z^3$ and $34xz^5$?
   a. $17xz^3$ **(b.)** $102x^2z^5$ **c.** $1734xz^3$ **d.** $1734x^2z^5$

Factor completely.

13. $18r^3s - 24rs^2 + 48r^2s^2$
   a. $6rs^2(3r^2 - 4 + 8r)$ **b.** $6r^2s(3r - 4 + 4s)$
   c. $6(3r^3s - 4rs^2 + 8r^2s^2)$ **(d.)** $6rs(3r^2 - 4s + 8rs)$

14. $t(t - 5) + 9(5 - t)$
   a. $(t + 9)(t - 5)$ **b.** $(t + 9)(5 - t)$
   c. $(t - 9)(t + 5)$ **(d.)** $(t - 9)(t - 5)$

15. $9w^2 - 24w + 16$
   a. $(3w - 4)(3w + 4)$ **b.** $(9w + 4)(w + 4)$
   (c.) $(3w - 4)^2$ **d.** $(3w + 4)^2$

16. $x^2 + zx - xy - zy$
   (a.) $(x - y)(x + z)$ **b.** $(x - z)(x + y)$
   c. $x(x + z - y) - zy$ **d.** $x^2 + x(z - y) - zy$

43. $(x^2 + y^2)(x + y)(x - y)$
44. $(x^4 + y^4)(x^2 + y^2)(x + y)(x - y)$
45. $(x + y)(x^2 - xy + y^2)(x - y)(x^2 + xy + y^2)$
46. $(x^5 - y^5)(x^5 + y^5) = (x - y)(x + y)(x^4 + x^3y + x^2y^2 + xy^3 + y^4)(x^4 - x^3y + x^2y^2 - xy^3 + y^4)$
47. $(y^2 + y + 1)(y^2 - y + 1)$
48. $(x^2 + x - 3)(x^2 - x - 3)$
50. For every nonnegative integer n, $c = -n(n + 1)$ is such that $x^2 + x + c$ can be factored over the set of polynomials with integral coefficients.

7-3

7-4

7-5

7-6

7-7

**Additional Answers
Oral Exercises**

(continued from p. 366)

16. For all real numbers a and b, if $ab > 0$, then either $a > 0$ and $b > 0$ or $a < 0$ and $b < 0$. If either $a > 0$ and $b > 0$ or $a < 0$ and $b < 0$, then $ab > 0$.

17. For all positive real numbers a and b, if $a < b$ then $\frac{1}{a} > \frac{1}{b}$. If $\frac{1}{a} > \frac{1}{b}$, then $a < b$.

18. For all real numbers a and b and all positive real numbers c and d, if $\frac{a}{c} < \frac{b}{d}$, then $ad < bc$. If $ad < bc$, then $\frac{a}{c} < \frac{b}{d}$.

**Additional Answers
Written Exercises**

(continued from p. 367)

54. Part I: If either $a > 0$ and $b > 0$ or $a < 0$ and $b < 0$, then $ab > 0$.
    Case 1: $a > 0$ and $b > 0$
    1. $b > 0$
       Hypothesis
    2. $ab > a \cdot 0$
       Mult. prop. of order
    3. $a \cdot 0 = 0$
       Mult. prop. of zero
    4. $\therefore ab > 0$
       Subs. prin.
    Case 2: $a < 0$ and $b < 0$
    1. $b < 0$
       Hypothesis
    2. $ab > a \cdot 0$
       Mult. prop. of order
    3. $a \cdot 0 = 0$
       Mult. prop. of zero
    4. $\therefore ab > 0$
       Subs. prin.

Factor completely.

17. $m^2 - 5m - 24$ 7-8
    a. $(m + 8)(m - 3)$ b. $(m - 6)(m + 4)$
    c. $(m - 12)(m + 2)$ ⓓ $(m - 8)(m + 3)$

18. $2x^4 - 10x^3 - 12x^2$
    ⓐ $2x^2(x - 6)(x + 1)$ b. $2x^2(x - 3)(x - 2)$
    c. $2(x^4 - 5x^3 - 6x^2)$ d. $2x^2(x^2 - 5x - 6)$

19. $5n^2 - 18n - 56$ 7-9
    a. $(5n - 7)(n + 8)$ ⓑ $(5n - 28)(n + 2)$
    c. $(5n - 8)(n + 7)$ d. $(5n + 28)(n - 2)$

20. $12p^6 - 27p^4$
    a. $(6p^5 + p^4)(2p - 3)$ b. $(6p^5 - 9p^4)(2p + 3)$
    ⓒ $3p^4(2p + 3)(2p - 3)$ d. $(4p^6 - 3)(3 + 9p^4)$

Solve.

21. $a(3a - 1)(a + 4) = 0$ 7-10
    a. $\{\frac{1}{3}, -4\}$ b. $\{-\frac{1}{3}, 4\}$ ⓒ $\{0, \frac{1}{3}, -4\}$ d. $\{0, -\frac{1}{3}, 4\}$

22. $13z - 63 = -6z^2$
    a. $\{-\frac{7}{3}, \frac{9}{2}\}$ b. $\{\frac{7}{6}, -9\}$ ⓒ $\{\frac{7}{3}, -\frac{9}{2}\}$ d. $\{-\frac{7}{6}, 9\}$

23. A rectangular garden measures 10 m by 8 m. Surrounding it is a brick walk of uniform width. The combined area of the garden and the walk is 120 m². Find the width of the walk.   7-11
    a. 10 m   b. 2 m   ⓒ 1 m   d. 8 m

Chapter Test

1. What is the degree of the polynomial $8p^5q - 5p^3q^2 + 7p^2q^3$? 6   7-1

Simplify.

2. $(-9ab + 3b - 7) + (2ab - 6a + 7)$ $-7ab + 3b - 6a$

3. $(-3g^2h)^2(-2g^3h^2)^2$ $36g^{10}h^6$ 4. $(2k)(4k^5) - (2k^2)^3$ 0   7-2

5. $-2v^2(v^2 + 3v - 1)$ $-2v^4 - 6v^3 + 2v^2$ 6. $(w + 3)(2w^2 - w - 2)$ 7-3

7. $(6c + 7d)(5c - 2d)$ $30c^2 + 23cd - 14d^2$ 8. $(9g + 11h)(9g - 11h)$ $81g^2 - 121h^2$ 7-4

9. A certain rectangle is 4 cm longer than it is wide. A second rectangle is 2 cm longer and 1 cm wider than the first, and its area is 12 cm² greater than the area of the first. Find the dimensions of the first rectangle.   6 cm by 2 cm   6. $2w^3 + 5w^2 - 5w - 6$

376   *Chapter 7*

10. Factor 216 over the set of prime numbers. $2^3 \cdot 3^3$ 7-5

11. Find the GCF and the LCM of $120a^3b^2c^2$ and $40a^2c^2$. $40a^2c^2$, $120a^3b^2c^2$

Factor completely.
$9m^2(3m^2n^3 + 5mn^2 - 2)$

12. $27m^4n^3 + 45m^3n^2 - 18m^2$ **13.** $q(q - 7) - 2(7 - q)$ $(q - 7)(q + 2)$ 7-6

14. $4x^2 - 28x + 49$ $(2x - 7)^2$ **15.** $8y^3 + 27$ $(2y + 3)(4y^2 - 6y + 9)$ 7-7

16. $j^2 + 11j + 18$ $(j + 2)(j + 9)$ **17.** $3k^3 - 21k^2 + 24k$ $3k(k^2 - 7k + 8)$ 7-8

18. $6a^2 - a - 40$ $(3a - 8)(2a + 5)$ **19.** $16b^3 - 28b^2 - 30b$ 7-9
$2b(2b - 5)(4b + 3)$

Solve.

20. $-4t(5t + 3)(3t - 5) = 0$ $\{0, -\frac{3}{5}, \frac{5}{3}\}$ **21.** $3s - 14 = 20s - 6s^2$ $\{\frac{7}{2}, -\frac{2}{3}\}$ 7-10

22. Find two consecutive negative integers such that the difference of   7-11
their squares is 63. -32 and -31

Cumulative Review

Chapter 2

Tell whether each statement is true or false.

1. For any natural number a, $a^3 > a$. False

2. There exists a whole number w such that $4w + 1 = 15$. False

3. The absolute value of any real number is greater than zero. False

4. The opposite of any real number is equal to the absolute value of
that number. False

Name the axiom that is illustrated by each statement.

5. $3(4a) = (3 \cdot 4)a$ Associative axiom for multiplication

6. $(3a - 2b)7 = 7(3a - 2b)$ Commutative axiom for multiplication

7. For any real number a, $a + 0 = a$. Identity axiom for addition

8. If $5 \cdot 3 = 15$ and $15 = 45 \div 3$, then $5 \cdot 3 = 45 \div 3$. Transitive axiom of equality

Simplify.

9. $\frac{2}{3} \times 25 \times 18 \times \frac{1}{5}$ 60

10. $[3.2 + (-9.7)] + (-10.2 + 4.7)$ -12

11. $(9x - 13) - (-7x - 15)$ $16x + 2$

12. $-(7y + 2y^2) - (16y^2 - 5y^3) - 6y^3$ $-y^3 - 18y^2 - 7y$

13. $9\left(\frac{1}{3}x + \frac{1}{3}y\right) - 10\left(\frac{1}{5}x - \frac{1}{2}y\right)$ $x + 8y$

14. $\frac{1}{3}(-45v - 9z) + \frac{1}{2}(44v$ $32z)$ $7v - 19z$

15. $\frac{3}{4}cd \div \left(-\frac{1}{8}\right)$ $-6cd$

16. $\frac{26ab}{-13a}$, $a \neq 0$ $-2b$

Polynomials and Factoring   **377**

Part II: If $ab > 0$, then
either $a > 0$ and $b > 0$ or
$a < 0$ and $b < 0$.
By the axiom of compari-
son, $a < 0$, $a = 0$, or $a > 0$.
Case 1: $a = 0$
Impossible; if $a = 0$, then
$ab = 0$.
Case 2: $a < 0$
1. $ab > 0$
   Hypothesis
2. $ab \cdot \frac{1}{a} < 0 \cdot \frac{1}{a}$
   Mult. prop. of order
3. $ab \cdot \frac{1}{a} = b$
   Exercise 52, p. 98
4. $0 \cdot \frac{1}{a} = 0$
   Mult. prop. of zero
5. $\therefore b < 0$
   Subs. prin.
Case 3: $a > 0$
1. $ab > 0$
   Hypothesis
2. $ab \cdot \frac{1}{a} > 0 \cdot \frac{1}{a}$
   Mult. prop. of order
3. $ab \cdot \frac{1}{a} = b$
   Exercise 52, p. 98
4. $0 \cdot \frac{1}{a} = 0$
   Mult. prop. of zero
5. $\therefore b > 0$
   Subs. prin.
Therefore, $ab > 0$ if and
only if either $a > 0$ and
$b > 0$ or $a < 0$ and $b < 0$.

25.

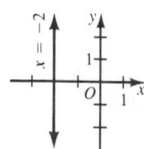

26.

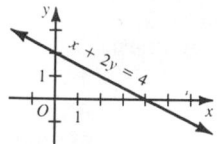

27.

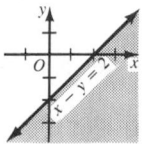

28.

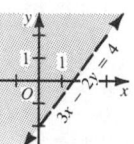

42.

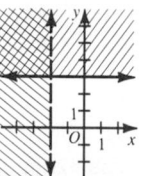

43.

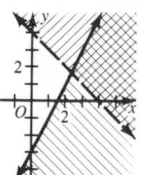

44.

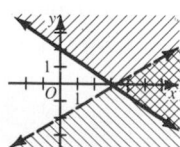

Chapter 5

Find the range of each relation. Then tell whether or not the relation is a function.

17. $\{(-2, 1), (-2, 3), (6, 1), (-4, 5)\}$ range: $\{1, 3, 5\}$; not a function

18. $\{(4, -2), (3, 4), (2, -2), (1, 0)\}$ range: $\{-2, 0, 4\}$; a function

19. $y = |x + 1|$; $D = \{-2, -1, 0, 1, 2\}$ range: $\{0, 1, 2, 3\}$; a function

20. $y = |x| + 1$; $D = \{-2, -1, 0, 1, 2\}$ range: $\{1, 2, 3\}$; a function

Given $f: x \longrightarrow 3x^2 - 2x - 1$. Compute the following.

21. $f(0)$ -1

22. $f(-1)$ 4

23. $f(1) - f(3)$ -20

24. $2f(2) - f(-2)$ -1

Graph on a coordinate plane.

25. $x = -2$

26. $x + 2y = 4$

27. $x - y \geq 2$

28. $3x - 2y < 4$

Determine an equation of the line that satisfies the given requirements.

29. has slope $\frac{1}{4}$ and y-intercept 1 $x - 4y = -4$

30. passes through the point $(-3, 4)$ and has slope $\frac{2}{3}$ $2x - 3y = -18$

31. passes through the point $(-2, -1)$ and is parallel to the y-axis $x = -2$

32. passes through the points $(-2, 4)$ and $(-5, 5)$ $x + 3y = 10$

Chapter 6

Solve each system by any method.

38. $\{(x, y): 9x = 6y - 12\}$

33. $2y = 21 - 5x$ $\{(5, -2)\}$
    $7x - 2y = 39$

34. $7y + 10z = 17$ $\{(1, 1)\}$
    $8y + 15z = 23$

35. $m + 3n = 2$ $\{(5, -1)\}$
    $2m + 3n = 7$

36. $4a + 3b = 14$ $\{(2, 2)\}$
    $9a - 14 = 2b$

37. $x + 2y = 6$ $\emptyset$
    $x = 2 - 2y$

38. $9x = 6y - 12$
    $y = \frac{3}{2}x + 2$

39. $2r - 3s = 3$ $\{(\frac{9}{4}, \frac{1}{2})\}$
    $r - \frac{s}{2} = 2$

40. $2x - 3y = -14$ $\{(\frac{41}{4}, \frac{23}{2})\}$
    $\frac{x}{2} - \frac{y}{4} = \frac{9}{4}$

41. $2(a - b) = -5(b - 1)$
    $3a - b = 9(a + 1)$
    $\{(-2, 3)\}$

Graph each system on a coordinate plane.

42. $x < -2$
    $y \geq 3$

43. $y > -x + 4$
    $y \leq 2x - 3$

44. $3x - 5y > 10$
    $3y + 2x \geq 6$

378   *Chapter 7*

Solve.   **46.** Wind speed: 50 km/h; Speed of plane: 250 km/h

45. Mike and Lisa live twenty blocks apart in opposite directions from their school. Mike lives one block less than twice as far from the school as Lisa does. How many blocks from the school does Mike live? 13 blocks

46. Flying against the wind, Francine flew her plane 1200 km in 6 h. With no change in the wind, she made the return trip in 4 h. Find the wind speed and the speed of the plane.

47. The sum of the digits of a three-digit number is nine. The units' digit is three times the hundreds' digit and is seven less than twice the tens' digit. Find the number. 153

48. Frank has 65 coins worth a total of $9.20. Some of the coins are dimes and the rest are quarters. How many of each type of coin does Frank have? 47 dimes, 18 quarters

Contest Problems

1. Factor $(x^2 - 5x - 1)^2 - 25$ into a product of binomials of the form $(x - r_1)(x - r_2) \ldots (x - r_n)$. What is the value of $r_1 \cdot r_2 \cdot \ldots \cdot r_n$? -24

2. Solve the following system: $|x - 2| < 3$
$$x - 1 \geq 0 \quad \{1\}$$
$$2x + 1 \leq 3$$

3. What is the units' digit of the simplified form of 1987^{1987}? 3

4. Simplify the following expression, given that n is a whole number.
$$[(-1)^n]^2 - [(+1)^n]^3 \quad 0$$

5. Central Junior High School has exactly 1000 students, and there are exactly 1000 lockers in its locker room. On a certain day, the lockers were opened and closed by the students in the following manner. The first student to enter the locker room opened all the lockers. The second student closed each even-numbered locker. The third student *changed* every *third* locker, opening each one that was closed and closing each one that was open. The fourth student changed every fourth locker, the fifth student changed every fifth locker, and so on. After all 1000 students had passed through the locker room, how many lockers were open? 31

Polynomials and Factoring   **379**

This great spiral galaxy, named M 31, has twice the mass of our galaxy, the Milky Way. The galaxy M 31 is located in the constellation Andromeda, which is 2.2×10^6 light-years from Earth.

Chapter 8

Polynomials and Rational Expressions

Division of Polynomials

OBJECTIVES for Sections 8-1 through 8-3:
1. *To divide monomials.*
2. *To divide a polynomial by a monomial.*
3. *To divide polynomials.*

8–1 Dividing Monomials

Numerical computations often reveal number properties that are useful in algebraic work. For example, notice that

$$\frac{12 \cdot 10}{3 \cdot 2} = \frac{120}{6} = 20 \quad \text{and} \quad \frac{12}{3} \cdot \frac{10}{2} = 4 \cdot 5 = 20.$$

Thus, by the transitive axiom of equality,

$$\frac{12 \cdot 10}{3 \cdot 2} = \frac{12}{3} \cdot \frac{10}{2}.$$

This result suggests the following *basic property of quotients*.

Basic Property of Quotients

For all real numbers r and s and all nonzero real numbers t and u,

$$\frac{rs}{tu} = \frac{r}{t} \cdot \frac{s}{u}.$$

Teaching Suggestions
p. T96

Key Ideas

Use the *Basic Property of Quotients*.
Use the *Laws of Exponents for Division* to simplify quotients of monomials.

The basic property of quotients is a theorem that can be proved as follows.

Statements	*Reasons*

1. $\dfrac{rs}{tu} = (rs) \cdot \left(\dfrac{1}{tu}\right)$ Definition of division

2. $\quad = (rs) \cdot \left(\dfrac{1}{t} \cdot \dfrac{1}{u}\right)$ Property of the reciprocal of a product

3. $\quad = \left(r \cdot \dfrac{1}{t}\right) \cdot \left(s \cdot \dfrac{1}{u}\right)$ Commutative and associative axioms for multiplication

4. $\quad = \dfrac{r}{t} \cdot \dfrac{s}{u}$ Definition of division

5. $\therefore \dfrac{rs}{tu} = \dfrac{r}{t} \cdot \dfrac{s}{u}$ Transitive axiom of equality

The following useful results are obtained by substituting 1 for t or 1 for r in the basic property of quotients just proved.

---

1. For all real numbers r and s and all nonzero real numbers u,

$$\frac{rs}{u} = r \cdot \frac{s}{u}.$$

2. For all real numbers s and all nonzero real numbers t and u,

$$\frac{s}{tu} = \frac{1}{t} \cdot \frac{s}{u}.$$

The basic property of quotients and the results just discussed can be used along with the laws of exponents for multiplication to simplify a quotient of monomials.

EXAMPLE 1   Simplify. Assume that $a \neq 0$.

     a. $\dfrac{a^3}{a^3}$ **b.** $\dfrac{a^7}{a^4}$ **c.** $\dfrac{a^2}{a^6}$

SOLUTION   **a.** $\dfrac{a^3}{a^3} = a^3 \cdot \dfrac{1}{a^3} = 1$

           b. $\dfrac{a^7}{a^4} = \dfrac{a^4 \cdot a^3}{a^4} = \dfrac{a^4}{a^4} \cdot a^3 = 1 \cdot a^3 = a^3$

           c. $\dfrac{a^2}{a^6} = \dfrac{a^2}{a^4 \cdot a^2} = \dfrac{1}{a^4} \cdot \dfrac{a^2}{a^2} = \dfrac{1}{a^4} \cdot 1 = \dfrac{1}{a^4}$

382   *Chapter 8*

The preceding example suggests a pattern for simplifying any quotient of powers of the form $\frac{a^m}{a^n}$, where a is a nonzero real number and m and n are positive integers.

If $m = n$, then $\dfrac{a^m}{a^n} = \dfrac{a^n}{a^n} = a^n \cdot \dfrac{1}{a^n} = 1.$

If $m > n$, then $\dfrac{a^m}{a^n} = \dfrac{a^{m-n} \cdot a^n}{a^n} = a^{m-n} \cdot \dfrac{a^n}{a^n} = a^{m-n} \cdot 1 = a^{m-n}.$

If $m < n$, then $\dfrac{a^m}{a^n} = \dfrac{a^m}{a^{n-m} \cdot a^m} = \dfrac{1}{a^{n-m}} \cdot \dfrac{a^m}{a^m} = \dfrac{1}{a^{n-m}} \cdot 1 = \dfrac{1}{a^{n-m}}.$

These results can be summarized as the following *laws of exponents* for division.

Laws of Exponents for Division

For all nonzero real numbers a and all positive integers m and n:

1. If $m = n$, then $\dfrac{a^m}{a^n} = 1.$

2. If $m > n$, then $\dfrac{a^m}{a^n} = a^{m-n}.$

3. If $m < n$, then $\dfrac{a^m}{a^n} = \dfrac{1}{a^{n-m}}.$

EXAMPLE 2 Simplify. Assume that no denominator equals zero.

a. $\dfrac{12x^4}{-3x^6}$ b. $\dfrac{27x^3y^5}{18x^4y^2}$ c. $\dfrac{-5xy^5}{-x^3y^5z}$

SOLUTION a. $\dfrac{12x^4}{-3x^6} = \dfrac{12}{-3} \cdot \dfrac{x^4}{x^6} = -4 \cdot \dfrac{1}{x^2} = -\dfrac{4}{x^2}$

b. $\dfrac{27x^3y^5}{18x^4y^2} = \dfrac{27}{18} \cdot \dfrac{x^3}{x^4} \cdot \dfrac{y^5}{y^2} = \dfrac{3}{2} \cdot \dfrac{1}{x} \cdot y^3 = \dfrac{3y^3}{2x}$

c. $\dfrac{-5xy^5}{-x^3y^5z} = \dfrac{-5}{-1} \cdot \dfrac{x}{x^3} \cdot \dfrac{y^5}{y^5} \cdot \dfrac{1}{z} = 5 \cdot \dfrac{1}{x^2} \cdot 1 \cdot \dfrac{1}{z} = \dfrac{5}{x^2z}$

CONDENSED SOLUTION a. $\dfrac{12x^4}{-3x^6} = -\dfrac{4}{x^2}$ b. $\dfrac{27x^3y^5}{18x^4y^2} = \dfrac{3y^3}{2x}$ c. $\dfrac{-5xy^5}{-x^3y^5z} = \dfrac{5}{x^2z}$

In simplifying a quotient, it is sometimes more convenient to first use the laws of exponents for multiplication to simplify either the numerator or the denominator, or both.

Polynomials and Rational Expressions **383**

Additional A Exercises

Simplify. Assume that no denominator equals zero.

1. $\dfrac{t^5}{t^2}$ t^3

2. $\dfrac{d^3}{d^4}$ $\dfrac{1}{d}$

3. $\dfrac{12w^4}{-6w}$ $-2w^3$

4. $\dfrac{3^4 \cdot x}{3^2 x^3}$ $\dfrac{9}{x^2}$

5. $\dfrac{(w^2)^3}{-w^3}$ $-w^3$

6. $\dfrac{b(a+5)^2}{b^3(a+5)^3}$ $\dfrac{1}{b^2(a+5)}$

Suggested Assignments

Minimum
  384/1–20
Average
  384/1–31 odd, 33–40
Maximum
  384/1–39 odd, 41–46

Mixed Review

Simplify.

1. $a^4 \cdot a^4$ a^8

2. $(a^4)^2$ a^8

3. $a^4 \cdot a^2$ a^6

4. $a^4 \cdot a^1$ a^5

5. $(a^2)^3(3a)^2$ $9a^8$

6. $(2a^3)(3a^2)$ $6a^5$

EXAMPLE 3   Simplify $\dfrac{(3x^2y)(2xy^5)^2}{(-4x^3y)(3y^6)}$, $x \neq 0$, $y \neq 0$.

SOLUTION $\dfrac{(3x^2y)(2xy^5)^2}{(-4x^3y)(3y^6)} = \dfrac{(3x^2y)(4x^2y^{10})}{(-4x^3y)(3y^6)} = \dfrac{12x^4y^{11}}{-12x^3y^7} = -xy^4$

Oral Exercises

Replace each __?__ with the factor that will make a true statement. Assume that no denominator equals zero.

1. $\dfrac{a^7}{a^2} = \dfrac{\text{?} \cdot a^2}{a^2}$ a^5

2. $\dfrac{b^6}{b^{10}} = \dfrac{b^6}{b^6 \cdot \text{?}}$ b^4

3. $\dfrac{x^9}{x^8} = \dfrac{\text{?} \cdot x^8}{x^8}$ x

4. $\dfrac{t}{t^{12}} = \dfrac{t}{t \cdot \text{?}}$ t^{11}

5. $\dfrac{10^5}{10^2} = \dfrac{10^2 \cdot \text{?}}{10^2}$ 10^3

6. $\dfrac{2^8}{2^{10}} = \dfrac{2^8}{2^8 \cdot \text{?}}$ 2^2

Simplify. Assume that $a \neq 0$.

7. $\dfrac{a^5}{a^2}$ a^3

8. $\dfrac{a^2}{a^5}$ $\dfrac{1}{a^3}$

9. $\dfrac{a^5}{a^5}$ 1

10. $\dfrac{a^5}{a^4}$ a

11. $\dfrac{5a^2}{-a^3}$ $-\dfrac{5}{a}$

12. $\dfrac{2a^2}{6a^2}$ $\dfrac{1}{3}$

Written Exercises

Simplify. Assume that no denominator equals zero.

A 1. $\dfrac{x^7}{x^3}$ x^4

2. $\dfrac{y^5}{y^8}$ $\dfrac{1}{y^3}$

3. $\dfrac{b}{b^5}$ $\dfrac{1}{b^4}$

4. $\dfrac{a^6}{a}$ a^5

5. $\dfrac{10c^3}{5c^2}$ $2c$

6. $\dfrac{3y^4}{4y^5}$ $\dfrac{3}{4y}$

7. $\dfrac{2n}{-22n^3}$ $-\dfrac{1}{11n^2}$

8. $\dfrac{-6r^2}{24r^8}$ $-\dfrac{1}{4r^6}$

9. $\dfrac{16e^5f^2}{2ef}$ $8e^4f$

10. $\dfrac{-8a^4b}{16ab^4}$ $-\dfrac{a^3}{2b^3}$

11. $\dfrac{5v^2w^5}{15v^3w^5}$ $\dfrac{1}{3v}$

12. $\dfrac{20c^3d}{5c^2d}$ $4c$

13. $\dfrac{3a^2b^3c}{-2a^2c^2}$ $-\dfrac{3b^3}{2c}$

14. $\dfrac{-24x^2y^{10}}{-12x^2z^{10}}$ $\dfrac{2y^{10}}{z^{10}}$

15. $\dfrac{2^6 \cdot p^6}{2^5 \cdot p^5}$ $2p$

16. $\dfrac{10^3 \cdot t^2}{10 \cdot t^2}$ 10^2

17. $\dfrac{0.8a^3b^2c}{0.2ab^4c^3}$ $\dfrac{4a^2}{b^2c^2}$

18. $\dfrac{5x^7y^4z^3}{0.2y^3z}$ $25x^7yz^2$

19. $\dfrac{10m^3n^2}{0.5mn^4}$ $\dfrac{20m^2}{n^2}$

20. $\dfrac{0.24r^5s^2t^2}{0.8r^3s^5t}$ $\dfrac{3r^2t}{10s^3}$

B 21. $\dfrac{(xy)^3}{x^2y}$ xy^2

22. $\dfrac{(a^3)^2}{(a^2)^3}$ 1

23. $\dfrac{(-a^4)^2}{-a^4}$ $-a^4$

24. $\dfrac{(-2x)^3}{-2x^3}$ 4

25. $\dfrac{(t^6)^2}{t^6 \cdot t^2}$ t^4

26. $\dfrac{(3s^2)^3 \cdot (-2s)^2}{-12(-s)^2}$ $-9s^6$

27. $\dfrac{(6cd^2)(2c^2d)}{(-3c^2d^3)(4cd)}$ $-\dfrac{1}{d}$

28. $\dfrac{(3s^2t)(-2st^2)^2}{(-4st)^3(-2st)}$ $-\dfrac{3t}{32}$

29. $\dfrac{(-3a^2)(-3a)^2}{-(3a^2)^2(3a^3)^2}$ $\dfrac{1}{3a^6}$

30. $\dfrac{(-5e)^2(5e)^2}{(10e)^2(-2e)^2}$ $\dfrac{25}{16}$

31. $\dfrac{s^2(3t-1)^5}{s^3(3t-1)^3}$ $\dfrac{(3t-1)^2}{s}$

32. $\dfrac{v^5(w-3)^5}{(-v)^7(w-3)^7}$ $-\dfrac{1}{v^2(w-3)^2}$

In Exercises 33–46, assume that variable expressions appearing as exponents denote positive integers and that no denominator equals zero.

Simplify. Express the answer so that all exponents are positive integers.

33. $\dfrac{d^{2n}}{d^n}$ d^n

34. $\dfrac{w^{2m}}{w^{3m}}$ $\dfrac{1}{w^m}$

35. $\dfrac{2p}{6p^{3n}}$ $\dfrac{1}{3p^{3n-1}}$

36. $\dfrac{2e^{5m-1}}{6e^{5m}}$ $\dfrac{1}{3e}$

37. $\dfrac{f^{n+1}}{f^{n+3}}$ $\dfrac{1}{f^2}$

38. $\dfrac{2g^{2n-1}}{6g^{2n-5}}$ $\dfrac{g^4}{3}$

39. $\dfrac{u^{2m+1}}{u^{m-2}}$ u^{m+3}

40. $\dfrac{k^{3m-1}}{k^{4m+1}}$ $\dfrac{1}{k^{m+2}}$

Given that all exponents are positive integers, specify the set of all possible values for m.

             {all integers greater than 5}

C **41. a.** $\dfrac{q^{m-3}}{q^2} = q^{m-5}$

       b. $\dfrac{q^{m-3}}{q^2} = \dfrac{1}{q^{5-m}}$ {4}

       c. $\dfrac{q^{m-3}}{q^2} = 1$ {5}

42. a. $\dfrac{t^5}{t^{8-m}} = t^{m-3}$

       b. $\dfrac{t^5}{t^{8-m}} = \dfrac{1}{t^{3-m}}$

       c. $\dfrac{t^5}{t^{8-m}} = 1$ {3}

     {4, 5, 6, 7}   {all integers less than 2}

Simplify. Express the answer so that all exponents are positive integers.

43. $\dfrac{x^{4-m}}{x^3}$

44. $\dfrac{w^6}{w^{n-2}}$

45. $\dfrac{a^{2-t}}{a^{t+3}}$

46. $\dfrac{b^{v+4}}{b^{8-v}}$

8–2 Dividing a Polynomial by a Monomial

In determining a method for dividing a polynomial by a monomial, it is again helpful to look at a numerical computation. For example,

$$\dfrac{56 + 14}{7} = \dfrac{70}{7} = 10 \quad \text{and} \quad \dfrac{56}{7} + \dfrac{14}{7} = 8 + 2 = 10.$$

Thus, by the transitive axiom of equality,

$$\dfrac{56 + 14}{7} = \dfrac{56}{7} + \dfrac{14}{7}.$$

This result illustrates the following theorem, discussed in Section 2-11.

> **Theorem.**   For all real numbers a, b, and c such that $c \neq 0$,
>
> $$\dfrac{a + b}{c} = \dfrac{a}{c} + \dfrac{b}{c} \quad \text{and} \quad \dfrac{a - b}{c} = \dfrac{a}{c} - \dfrac{b}{c}.$$

Teaching Suggestions
p. T96

Key Ideas

Use the *Laws of Exponents for Division* to simplify quotients of polynomials divided by monomials.

Express each quotient as a sum. Assume that no denominator equals zero.

1. $\dfrac{20r^2 - 12r}{4r}$ $\dfrac{20r^2}{4r} - \dfrac{12r}{4r} =$

$5r - 3$

2. $\dfrac{35t^5 + 21t}{7t^2}$ $\dfrac{35t^5}{7t^2} + \dfrac{21t}{7t^2} =$

$5t^3 + \dfrac{3}{t}$

3. $\dfrac{10a^4 - 6a^3 + 16a^2}{-2a}$

$\dfrac{10a^4}{-2a} - \dfrac{6a^3}{-2a} + \dfrac{16a^2}{-2a} =$

$-5a^3 + 3a^2 - 8a$

Reading Algebra

To help students understand why a polynomial like $ax^2 - 3ax + a$ is *not divisible* by ax, remind them that a polynomial is the sum of one or more monomials. The quotient when $ax^2 - 3ax + a$ is divided by ax is $x - 3 + \dfrac{a}{x}$.

This quotient is not a polynomial since the last term $\dfrac{a}{x}$ is not a monomial and cannot be written as ax^m where m is a positive integer. (In Section 8-5 the students will see that $\dfrac{a}{x}$ can be expressed as ax^{-1}.)

Suggested Assignments

Minimum
   386/1–20
Average
   386/1–23 odd, 24–28
Maximum
   386/1–29 odd, 30–34

EXAMPLE   Express each quotient as a sum. Assume that no denominator equals zero.

a. $\dfrac{8w^3 - 6w^2 + 2w}{2w}$ b. $\dfrac{ax^2 - 3ax + a^2}{ax}$

SOLUTION   a. $\dfrac{8w^3 - 6w^2 + 2w}{2w} = \dfrac{8w^3}{2w} - \dfrac{6w^2}{2w} + \dfrac{2w}{2w} = 4w^2 - 3w + 1$

b. $\dfrac{ax^2 - 3ax + a^2}{ax} = \dfrac{ax^2}{ax} - \dfrac{3ax}{ax} + \dfrac{a^2}{ax} = x - 3 + \dfrac{a}{x}$

The Example above illustrates the following rule.

To divide a polynomial by a monomial, divide each term of the polynomial by the monomial, then add all the quotients.

One polynomial is said to be **divisible** by another polynomial if the quotient is also a polynomial. The preceding Example shows that the polynomial $8w^3 - 6w^2 + 2w$ is divisible by $2w$ since the quotient, $4w^2 - 3w + 1$, is a polynomial. The polynomial $ax^2 - 3ax + a^2$ is *not* divisible by ax since the quotient, $x - 3 + \dfrac{a}{x}$, is not a polynomial.

Oral Exercises

Replace each __?__ with a monomial to make a true statement. Assume that no denominator equals zero.

1. $\dfrac{20 + 36}{4} = \dfrac{?}{4} + \dfrac{?}{4}$ 20; 36

2. $\dfrac{18 - 12 + 30}{6} = \dfrac{?}{6} - \dfrac{?}{6} + \dfrac{?}{6}$ 18; 12; 30

3. $\dfrac{5m^2 + 2n}{3n} = \dfrac{?}{3n} + \dfrac{?}{3n}$ $5m^2$; $2n$

4. $\dfrac{7b^3 - 9c^2}{2b} = \dfrac{?}{2b} - \dfrac{?}{2b}$ $7b^3$; $9c^2$

5. $\dfrac{a^3 + b^2 + c}{5a} = \dfrac{?}{5a} + \dfrac{?}{5a} + \dfrac{?}{5a}$ a^3; b^2; c

6. $\dfrac{3x^2y - 4xy - 5}{8xy} = \dfrac{?}{8xy} - \dfrac{?}{8xy} - \dfrac{?}{8xy}$
$3x^2y$; $4xy$; 5

Written Exercises   **4.** $2z^2 + 3$ **8.** $\dfrac{5x}{6y^2} + \dfrac{1}{x}$ **12.** $-\dfrac{3}{5s} + \dfrac{7s}{5}$

Express each quotient as a sum. Assume that no denominator equals zero.

A **1.** $\dfrac{6m + 9n}{3}$ $2m + 3n$ **2.** $\dfrac{8b - 12c}{2}$ $4b - 6c$ **3.** $\dfrac{15e^2 - 5e}{5e}$ $3e - 1$ **4.** $\dfrac{8z^3 + 12z}{4z}$

5. $\dfrac{7h + 3}{2h^3}$ $\dfrac{7}{2h^2} + \dfrac{3}{2h^3}$ **6.** $\dfrac{5a^2 - 6}{3a^5}$ $\dfrac{5}{3a^3} - \dfrac{2}{a^5}$ **7.** $\dfrac{2a^2 - 3b^3}{4a^3b^2}$ $\dfrac{1}{2ab^2} - \dfrac{3b}{4a^3}$ **8.** $\dfrac{5x^2 + 6y^2}{6xy^2}$

9. $\dfrac{3t^2 - 6t^5}{-12t^4}$ $-\dfrac{1}{4t^2} + \dfrac{1}{2}$ **10.** $\dfrac{8w^5 + 3w^2}{-6w^8}$ $-\dfrac{4}{3w^3} - \dfrac{1}{2w^6}$ **11.** $\dfrac{9v^4 + 27v^8}{-9v^4}$ $-1 - 3v^4$ **12.** $\dfrac{3s^3 - 7s^5}{-5s^4}$

13. $\dfrac{6r^3 + 8r^2 - 14r}{2r}$ $3r^2 + 4r - 7$

14. $\dfrac{12e^4 - 6e^2 + 4e}{-2e}$ $-6e^3 + 3e - 2$

15. $\dfrac{12z^4 - 6z^3 + 16z^2}{-2z^2}$ $-6z^2 + 3z - 8$

16. $\dfrac{24f^5 + 16f^3 - 12f}{4f}$ $6f^4 + 4f^2 - 3$

17. $\dfrac{15u^2v^3 + 6uv^2 - 12u^3v}{9uv}$ $\dfrac{5uv^2}{3} + \dfrac{2v}{3} - \dfrac{4u^2}{3}$

18. $\dfrac{12a^2b^2 + 16ab - 32b}{-16ab}$ $-\dfrac{3ab}{4} - 1 + \dfrac{2}{a}$

19. $\dfrac{20e^4 - 10e^2t^2 + 8t^4}{-4e^2t^2}$ $\dfrac{5e^2}{t^2} + \dfrac{5}{2} - \dfrac{2t^2}{e^2}$

20. $\dfrac{6a^4b^4 - 2b^2c^2 + 3ac}{12a^2b^2c^2}$ $\dfrac{a^2b^2}{2c^2} - \dfrac{1}{6a^2} + \dfrac{1}{4ab^2c}$

For the functions P and Q defined in Exercises 21–24, express $\dfrac{P(x)}{Q(x)}$ as a sum. Assume that $Q(x) \neq 0$.

B **21.** $P(x) = 12x^5 - 9x^3 + 15x;\quad Q(x) = 3x$ $4x^4 - 3x^2 + 5$

22. $P(x) = 40x^4 + 15x^3 - 20x^2 + 10;\quad Q(x) = -5x$ $-8x^3 - 3x^2 + 4x - \dfrac{2}{x}$

23. $P(x) = 32x^6 - 8x^4 + 24x^2 - 1;\quad Q(x) = 8x^2$ $4x^4 - x^2 + 3 - \dfrac{1}{8x^2}$

24. $P(x) = 36x^8 - 72x^6 - 12x^4 + 48x^2;\quad Q(x) = -12x^4$ $-3x^4 + 6x^2 + 1 - \dfrac{4}{x^2}$

Simplify. Assume that variable expressions appearing as exponents denote positive integers and that no denominator equals zero.

25. $\dfrac{b^{3n} - b^{2n} + b^{n+1}}{b^n}$ $b^{2n} - b^n + b$

26. $\dfrac{2k^{3n} - k^{2n+2} + 3k^{n+1} - 4k^2}{3k^{n-1}}$
$\dfrac{6k^{4n-1} - 3k^{3n+1} + 9k^{2n} - 12k^{n+1}}{3k^{n-1}}$

27. $\dfrac{a^{3n}b^{n+1} + 2a^{2n}b^2 - 3ab^n}{-a^nb}$ $-a^{2n}b^n - 2a^nb + \dfrac{3b^{n-1}}{a^{n-1}}$

28. $\dfrac{t^{5n+2}v^{2n} - (t^2v)^n - t(v)^{2n+1}}{(tv)^{2n}}$ $t^{3n+2} - \dfrac{1}{v^n} - \dfrac{v}{t^{2n-1}}$

C **29.** Evaluate $\dfrac{x}{y}$ when $\dfrac{x+y}{y} = 35$. 34

30. Evaluate $\dfrac{x}{y}$ when $\dfrac{x-y}{y} = 29$. 30

31. Find the positive value of $\dfrac{a}{x}$ that is a solution of $\dfrac{a^2 - x^2}{x^2} = -0.64$. 0.6

32. Solve $\dfrac{5r^2 + 20r}{5r} = r + 4$. {all real numbers except 0}

33. a. Are $\dfrac{2x^2 + 12x}{2x} = 8$ and $x + 6 = 8$ equivalent open sentences?

Explain. Yes. The solution set of each equation is {2}.

b. Are $\dfrac{2x^2 + 12x}{2x} = 6$ and $x + 6 = 6$ equivalent open sentences?

Explain. No. The solution set of the first equation is Ø. The solution set of the second equation is {0}.

34. a. Are $\dfrac{6m^2 - 20m}{2m} < 2$ and $3m - 10 < 2$ equivalent open sentences?

Explain. No. The solution set of the first inequality is {$m: m < 4$ and $m \neq 0$}. The solution set of the second inequality is {$m: m < 4$}.

b. Are $\dfrac{6m^2 - 20m}{2m} > 2$ and $3m - 10 > 2$ equivalent open sentences?

Explain. Yes. The solution set of each inequality is {$m: m > 4$}.

Additional A Exercises

Express each quotient as a sum. Assume that no denominator equals zero.

1. $\dfrac{4x + 8y}{2}$ $2x + 4y$

2. $\dfrac{-21c^3 + 14c^2}{7c}$ $-3c^2 + 2c$

3. $\dfrac{15 - t}{-3t}$ $-\dfrac{5}{t} + \dfrac{1}{3}$

4. $\dfrac{6x + 9}{3x}$ $2 + \dfrac{3}{x}$

5. $\dfrac{4y^2 - 10y}{2y^3}$ $\dfrac{2}{y} - \dfrac{5}{y^2}$

6. $\dfrac{20x^2 - 16xy^2 + 10x^2y^2}{-12xy}$
$-\dfrac{5x}{3y} + \dfrac{4y}{3} - \dfrac{5xy}{6}$

Mixed Review

Simplify. Express your result so that all exponents are positive integers.

1. $\dfrac{20a^5}{4a}$ $5a^4$

2. $\dfrac{6b^7}{15b^5}$ $\dfrac{2}{5}b^2$

3. $\dfrac{4c^4}{-7c^6}$ $-\dfrac{4}{7c^2}$

4. $\dfrac{14a^2b^3}{-7a^4b^2}$ $-\dfrac{2b}{a^2}$

Evaluate when $a = -1$ and $b = 3$.

5. a. $\dfrac{2a - b}{2ab}$ $\dfrac{5}{6}$

b. $\dfrac{1}{b} - \dfrac{1}{2a}$ $\dfrac{5}{6}$

6. a. $\dfrac{3a^2 + ab}{a^2}$ 0

b. $3 + \dfrac{b}{a}$ 0

Careers

Pharmacy

Pharmacists prepare and dispense the medicinal drugs that are prescribed by doctors and dentists. A pharmacist must know how to store these drugs properly and, in some cases, must know how to mix, or *compound*, the drugs that form a particular medicine. In dispensing medicines, a pharmacist may also need to inform the patient of the proper dosage and of possible side effects.

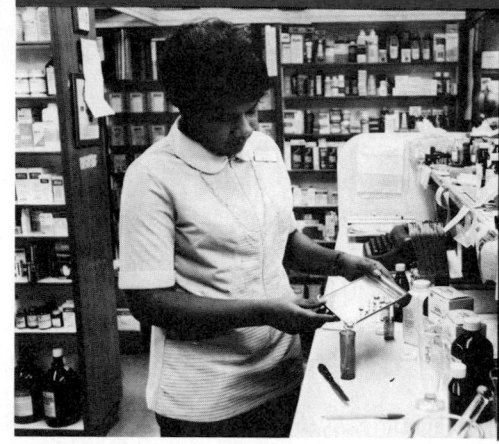

Although most pharmacists work in community pharmacies, some work in hospitals as consultants to the medical staff. Others work in private industry and may be involved in research, manufacturing, or sales. Still others teach at medical schools, schools of nursing, or schools of pharmacy.

In pursuing a degree in pharmacy, some people choose to concentrate on a particular field of study. For example, some pharmacists specialize in *pharmaceutics*, which is the study of the physical and chemical properties of drugs. Others choose to concentrate on *pharmacology*, which is the study of the effect of drugs on the human body, or to concentrate on *pharmacognosy*, which is the study of drugs derived from plant and animal sources.

EXAMPLE The prescribed adult dosage of a particular prescription medicine is twelve tablets per day. Determine the correct dosage of the same medicine for a three-year-old child.

SOLUTION One formula that is used to determine the child's dosage, c, is

$$c = \frac{a + 1}{24} \cdot d,$$

where a represents the child's age in years and d represents the amount of the adult dosage.

Substitute the given information into this formula.

$$c = \frac{3 + 1}{24} \cdot 12$$

$$= \frac{1}{6} \cdot 12 = 2$$

∴ the correct child's dosage is two tablets per day.

388 *Chapter 8*

8–3 Dividing Polynomials: Rational Expressions

In order to divide one polynomial by another, you can use a division algorithm very similar to the one used for integers. For example, first consider the following division of integers.

Step 1
$$\begin{array}{r} 1 \\ 28\overline{)\ 354} \\ \underline{28} \\ 74 \end{array}$$

Step 2
$$\begin{array}{r} 12 \\ 28\overline{)\ 354} \\ \underline{28} \\ 74 \\ \underline{56} \\ 18 \end{array}$$ ←—(partial quotient)

Check: $354 \stackrel{?}{=} 12 \cdot 28 + 18$
$354 \stackrel{?}{=} 336 + 18$
$354 = 354$ ✓

$\therefore \dfrac{354}{28} = 12\dfrac{18}{28}$, or $12\dfrac{9}{14}$

Now consider the following division of polynomials.

Step 1
$$\begin{array}{r} 3x \\ 2x-3\overline{)\ 6x^2 - 11x + 8} \\ \underline{6x^2 - 9x} \\ -2x + 8 \end{array}$$
←— {Multiply $2x - 3$ by $3x$
←— {Subtract

Step 2
$$\begin{array}{r} 3x \ - \ 1 \\ 2x-3\overline{)\ 6x^2 - 11x + 8} \\ \underline{6x^2 - 9x} \\ -2x + 8 \\ \underline{-2x + 3} \\ 5 \end{array}$$
←— (partial quotient)

←— {Multiply $2x - 3$ by -1
←— {Subtract

Check: $6x^2 - 11x + 8 \stackrel{?}{=} (3x - 1)(2x - 3) + 5$
$6x^2 - 11x + 8 \stackrel{?}{=} (6x^2 - 11x + 3) + 5$
$6x^2 - 11x + 8 = 6x^2 - 11x + 8$ ✓

$\therefore \dfrac{6x^2 - 11x + 8}{2x - 3} = 3x - 1 + \dfrac{5}{2x - 3}$

Notice that, in both divisions just shown, the answer was expressed in the form

$$\frac{\text{dividend}}{\text{divisor}} = \text{quotient} + \frac{\text{remainder}}{\text{divisor}}.$$

Unless the remainder is zero, the quotient in this equation is the *partial quotient* indicated in Step 2. By transforming this equation, you obtain the relationship that was used in checking each of the divisions:

$$\text{dividend} = \text{quotient} \times \text{divisor} + \text{remainder}$$

Polynomials and Rational Expressions **389**

Teaching Suggestions
p. T97

Related Activities p. T97

Supplementary Material
Test 30

Key Ideas

Use the division algorithm to simplify quotients of polynomials.

Chalkboard Examples

1. Divide $6a^2 + 10a + 4$ by $2a + 2$.
$$\begin{array}{r} 3a \ + \ 2 \\ 2a+2\overline{)\ 6a^2 + 10a + 4} \\ \underline{6a^2 + 6a} \\ 4a + 4 \\ \underline{4a + 4} \\ 0 \end{array}$$

2. Divide $4x^3 - 33x + 7$ by $x + 3$.
$$\begin{array}{r} 4x^2 - 12x \ + \ 3 \\ x+3\overline{)\ 4x^3 + 0x^2 - 33x + 7} \\ \underline{4x^3 + 12x^2} \\ -12x^2 - 33x \\ \underline{-12x^2 - 36x} \\ 3x + 7 \\ \underline{3x + 9} \\ -2 \end{array}$$

$4x^2 - 12x + 3 + \dfrac{-2}{x + 3}$

3. Is $2a^2 - 1$ a factor of $6a^4 + 8a^3 - 3a^2 - 4a$?
$$\begin{array}{r} 3a^2 + 4a \\ 2a^2-1\overline{)\ 6a^4 + 8a^3 - 3a^2 - 4a} \\ \underline{6a^4 \qquad - 3a^2} \\ 8a^3 \qquad - 4a \\ \underline{8a^3 \qquad - 4a} \\ 0 \end{array}$$

Since the remainder is 0, $2a^2 - 1$ is a factor of $6a^4 + 8a^3 - 3a^2 - 4a$.

A **rational expression** is any expression that can be written as the quotient of two polynomials, provided the denominator is not zero. Using division, a rational expression can also be written as the sum of a polynomial and a rational expression. When dividing one polynomial by another, it is important that the terms in both dividend and divisor are arranged in order of descending degree in one variable.

EXAMPLE 1   Divide $4y^2 + 15y^4 + 9y^3 + 1$ by $3y^2 - 1$.

SOLUTION   Arrange the terms of the dividend in order of descending degree. The dividend has no first-degree term. This "missing" term can be inserted as shown by using zero as its coefficient:

$$
\begin{array}{r}
5y^2 + 3y + 3 \\
3y^2 - 1 \overline{)\, 15y^4 + 9y^3 + 4y^2 + 0y + 1} \\
\underline{15y^4 - 5y^2 } \\
9y^3 + 9y^2 + 0y + 1 \\
\underline{9y^3 - 3y } \\
9y^2 + 3y + 1 \\
\underline{9y^2 - 3} \\
3y + 4
\end{array}
$$

Check: $15y^4 + 9y^3 + 4y^2 + 1 \overset{?}{=} (3y^2 - 1)(5y^2 + 3y + 3) + (3y + 4)$
$ \overset{?}{=} 15y^4 + 9y^3 + 4y^2 - 3y - 3 + 3y + 4$
$ = 15y^4 + 9y^3 + 4y^2 + 1 \;\checkmark$

$\therefore$ the quotient is $5y^2 + 3y + 3 + \dfrac{3y + 4}{3y^2 - 1}$, $3y^2 - 1 \neq 0$.

The division process ends when the remainder is either zero or a polynomial of lesser degree than that of the divisor.

You can use division to determine whether one polynomial is a factor of another polynomial.

EXAMPLE 2   **a.** Is $x^2 - x + 1$ a factor of $4x^4 - 4x^3 + 3x^2 + x - 1$?
   b. Factor $4x^4 - 4x^3 + 3x^2 + x - 1$ completely.

SOLUTION   **a.**

$$
\begin{array}{r}
4x^2 - 1 \\
x^2 - x + 1 \overline{)\, 4x^4 - 4x^3 + 3x^2 + x - 1} \\
\underline{4x^4 - 4x^3 + 4x^2 } \\
-x^2 + x - 1 \\
\underline{-x^2 + x - 1} \\
0
\end{array}
$$

Since the remainder is 0,

$$4x^4 - 4x^3 + 3x^2 + x - 1 = (4x^2 - 1)(x^2 - x + 1).$$

$\therefore x^2 - x + 1$ is a factor of $4x^4 - 4x^3 + 3x^2 + x - 1$.

b. $4x^4 - 4x^3 + 3x^2 + x - 1 = (4x^2 - 1)(x^2 - x + 1)$
$ = (2x + 1)(2x - 1)(x^2 - x + 1)$

Oral Exercises

In each exercise, tell how you would write the dividend before using the division algorithm.

1. $(9 + 6b + b^2) \div (b + 2)$ $b^2 + 6b + 9$
2. $(d^2 + 20 - 9d) \div (d - 5)$
3. $(y^2 - 9) \div (y + 3)$ $y^2 + 0y - 9$
4. $(z^3 - 8) \div (z - 2)$
5. $(2f^2 + 3f^3) \div (3f - 1)$ $3f^3 + 2f^2 + 0f + 0$
6. $(w^4 - 16) \div (w^2 + 4)$

      2. $d^2 - 9d + 20$ 4. $z^3 + 0z^2 + 0z - 8$ 6. $w^4 + 0w^2 + 16$

Written Exercises

Divide the first polynomial by the second. Assume that no divisor equals zero.

A

1. $m^2 - m - 12$; $m - 4$ $m + 3$
2. $a^2 + 9a + 20$; $a + 5$ $a + 4$
3. $18s^2 + 27s + 10$; $6s + 5$ $3s + 2$
4. $15b^2 + 4b - 3$; $3b - 1$ $5b + 3$
5. $35t^2 - 51t + 16$; $7t - 6$
6. $4u^2 + 28u + 14$; $2u + 7$
7. $3x^2 - 2x^3 + x - 1$; $x + 1$
8. $3y + 3y^3 - 2 - 2y^2$; $3y - 2$
9. $d^3 + 125$; $d - 2$
10. $p^4 + 2p + 1$; $p + 3$
11. $18x^4 + 9x^3 + 3x^2$; $3x^2 + 1$
12. $2b^5 - 8b^4 + 2b^3 + b^2$; $2b^3 + 1$
13. $t^6 + 1$; $t^2 - 1$
14. $2w^5 - 3$; $w^2 + 2$
15. $r^4 + 2r^2 - 1$; $r^2 + 2r - 3$
16. $h^5 - 1$; $h^2 - h + 2$

Determine whether or not the first polynomial is a factor of the second.

B

17. $3x + 2$; $6x^3 - 11x^2 - 7x + 2$ Yes
18. $a^2 + 3a + 1$; $a^3 + 2a^2 - 2a + 4$ No
19. $v^2 - 1$; $v^6 - 2v^4 + v^2 - 2$ No
20. $b^2 + 1$; $3b^7 + 3b^5 - b^2 - 1$ Yes

In Exercises 21–24, the first polynomial is a factor of the second polynomial. Factor the second polynomial completely.

21. $k - 3$; $k^3 - k^2 - 21k + 45$
22. $2r + 5$; $4r^3 + 28r^2 + 9r - 90$
23. $w^2 - 1$; $w^5 - w^3 + 8w^2 - 8$
24. $p^4 - 16$; $p^6 - 5p^5 + 4p^4 - 16p^2 + 80p - 64$

C

25. Determine p so that $4q + 3$ is a factor of $20q^3 + 23q^2 - 10q + p$. $p = -12$

26. Determine p so that $(6b^2 + pb + 36) \div (2b + 7) = 3b + 5 + \dfrac{1}{2b + 7}$. $p = 31$

27. **a.** Divide $a^{100} - 1$ by $a - 1$.
    b. Divide $a^n - 1$ by $a - 1$, where n is a positive integer.
28. **a.** Divide $a^{100} + 1$ by $a + 1$.
    b. Divide $a^{101} + 1$ by $a + 1$.
    c. Divide $a^n + 1$ by $a + 1$, where n is a positive integer.
29. Is the quotient of two polynomials always a polynomial? Explain.
30. Is the quotient of two polynomials, if the quotient exists, always a rational expression? Explain.

Polynomials and Rational Expressions **391**

Suggested Assignments

Minimum
  391/1–16
R 392/*Self-Test 1*
Average
  391/1–15 odd, 17–24
R 392/*Self-Test 1*
Maximum
  391/1-23 odd, 25–30
R 392/*Self-Test 1*

Additional Answers
Written Exercises
(See p. 425)

Additional A Exercises

Divide the first polynomial by the second. Assume that no divisor equals zero.

1. $a^2 + 4a + 3$; $a + 1$ $a + 3$
2. $2b^2 + 4b - 6$; $b + 3$
  $2b - 2$
3. $c^3 - 64$; $c - 4$
  $c^2 + 4c + 16$
4. $12d^2 + 13d + 7$; $3d + 1$
  $4d + 3 + \dfrac{4}{3d + 1}$
5. $x^4 - x^3 + x^2 - 5x + 6$;
  $x - 3$
  $x^3 + x - 2$
6. $x^4 - 3x^3 - 3x^2 - 3x - 4$;
  $x^2 + 1$ $x^2 - 3x - 4$

Mixed Review

Factor completely.
1. $b^2 - b - 6$ $(b + 2)(b - 3)$
2. $2c^2 + 7c + 6$
  $(2c + 3)(c + 2)$
3. $3x^2 - 11x - 4$
  $(3x + 1)(x - 4)$
Simplify.
4. $(2x + 1)(x^3 + x - 1)$
  $2x^4 + x^3 + 2x^2 - x - 1$
5. $(3x - 1)(3x + 1) - 4$
  $9x^2 - 5$

Quick Quiz

Simplify. Assume that no denominator equals zero.

1. $\dfrac{36r^5s^2}{15r^4s^3}$ $\dfrac{12r}{5s}$

2. $\dfrac{-56x^3t^2}{8x^2t^3}$ $\dfrac{-7x}{t}$

Express each quotient as a sum. Assume that no denominator equals zero.

3. $\dfrac{16u^4 - 20u^3 + 2u^2}{4u^2}$

$4u^2 - 5u + \dfrac{1}{2}$

4. $\dfrac{28a^6b + 24a^2b^5 - 12a^3}{8a^4b^2}$

$\dfrac{7a^2}{2b} + \dfrac{3b^3}{a^2} - \dfrac{3}{2ab^2}$

Divide as indicated. Assume that no divisor equals zero.

5. $(12p^2 - 32p - 35) \div$
   $(2p - 7)$ $6p + 5$

6. $x + 5\overline{)\,2x^2 + 7x - 16}$

$2x - 3 + \dfrac{-1}{x + 5}$

 **Self-Test 1**

VOCABULARY   basic property of quotients   divisible (p. 386)
                (p. 381)   rational expression (p. 390)
                laws of exponents
                  for division (p. 383)

Simplify. Assume that no denominator equals zero.

1. $\dfrac{28m^3t}{4mt}$ $7m^2$ 2. $\dfrac{-50r^3s^2}{-5r^3s^3}$ $\dfrac{10}{s}$ *Obj. 1, p. 381*

Express each quotient as a sum. Assume that no denominator equals zero.

3. $\dfrac{12b^3 - 16b^2 + 4b}{4b}$ $3b^2 - 4b + 1$ 4. $\dfrac{10e^4f^3 - 15e^3f^2 + 20f}{5ef^2}$ *Obj. 2, p. 381*

                                              $2e^3f - 3e^2 + \dfrac{4}{ef}$

Divide as indicated. Assume that no divisor equals zero.

5. $(c^2 + 2c - 35) \div (c - 5)$ $c + 7$ 6. $2r + 3\overline{)\,6r^2 - r - 10}$ *Obj. 3, p. 381*

Check your answers with those at the back of the book. 6. $3r - 5 + \dfrac{5}{2r + 3}$

Simplifying Rational Expressions

OBJECTIVES for Sections 8-4 and 8-5:
1. To simplify fractions.
2. To simplify expressions that contain zero or negative integral exponents.
3. To use scientific notation.

8–4 Simplifying Rational Expressions

Using the basic property of quotients given in Section 8-1, you can prove the following theorem, which is useful in simplifying rational expressions. (See Exercises 35 and 36 on page 396.)

> **Theorem.**   For all real numbers r and all nonzero real numbers s and t,
>
> $$\frac{r}{s} = \frac{r \cdot t}{s \cdot t} \quad \text{and} \quad \frac{r}{s} = \frac{r \div t}{s \div t}.$$

392   *Chapter 8*

The preceding theorem states that dividing or multiplying the numerator and denominator of a fraction by the same nonzero number produces a fraction equivalent to the given fraction. For example,

$$\frac{3}{4} = \frac{3 \times 5}{4 \times 5} = \frac{15}{20} \quad \text{and} \quad \frac{24}{30} = \frac{24 \div 6}{30 \div 6} = \frac{4}{5}.$$

Any fraction whose numerator and denominator are integers or polynomials with integral coefficients is in **simplest form** if the greatest common factor (GCF) of the numerator and denominator is 1. Thus, to simplify a fraction you divide the numerator and denominator by their GCF.

EXAMPLE 1   Simplify $\frac{18x^4}{24x}$.

SOLUTION   Divide the numerator and denominator by their GCF.

$$\frac{18x^4}{24x} = \frac{18x^4 \div 6x}{24x \div 6x} = \frac{3x^3}{4}, \; x \neq 0$$

In finding the simplest form of some rational expressions, you may first need to factor either the numerator or denominator, or both.

EXAMPLE 2   Simplify $\frac{5b - 15}{2b^2 - 18}$.

SOLUTION   Factor the numerator and the denominator. Then divide the numerator and denominator by their GCF.

$$\frac{5b - 15}{2b^2 - 18} = \frac{5(b - 3)}{2(b + 3)(b - 3)} = \frac{5}{2(b + 3)}, \; b \neq 3, \, -3$$

Another way to write the answer to Example 2 is $\frac{5}{2b + 6}$. However, *unless you are otherwise directed, you may leave an answer in its factored form,* as was done in Example 2.

Sometimes factors of the numerator and denominator are opposites of one another, as in the next example.

EXAMPLE 3   Simplify $\frac{6 - 3m}{m^2 + m - 6}$.

SOLUTION $\dfrac{6 - 3m}{m^2 + m - 6} = \dfrac{3(2 - m)}{(m + 3)(m - 2)}$

$$= \frac{3(-1)(m - 2)}{(m + 3)(m - 2)}$$

$$= \frac{-3}{m + 3}$$

$$= -\frac{3}{m + 3}, \; m \neq -3, 2$$

Polynomials and Rational Expressions   **393**

1. Simplify.

   a. $\dfrac{35a^5}{15a^2}$

   $\dfrac{35a^5 \div 5a^2}{15a^2 \div 5a^2} = \dfrac{7a^3}{3}, a \neq 0$

   b. $\dfrac{6y + 24}{2y^2 + 11y + 12}$

   $\dfrac{6(y + 4)}{(2y + 3)(y + 4)} =$

   $\dfrac{6}{2y + 3}, \, y \neq -4, \, -\dfrac{3}{2}$

2. Simplify $\dfrac{15 - 5x}{x^2 - x - 6}$.

   $\dfrac{5(3 - x)}{(x + 2)(x - 3)} =$

   $\dfrac{5(-1)(x - 3)}{(x + 2)(x - 3)} = \dfrac{-5}{x + 2}$

   $= -\dfrac{5}{x + 2}, x \neq 3, -2$

3. Solve the following equation for y. Restrict t so that no division by zero results.

   $yt - t^2 = -3y - 2t - 15$

   $yt + 3y = t^2 - 2t - 15$

   $y(t + 3) = t^2 - 2t - 15$

   If $t \neq -3$, then

   $y = \dfrac{t^2 - 2t - 15}{t + 3}$

   $= \dfrac{(t + 3)(t - 5)}{t + 3} = t - 5.$

   $\therefore$ if $t \neq -3$, then
   $y = t - 5.$

Notice that in Examples 1, 2, and 3, the *restrictions* on the values of the variables are stated. This is done since the original fraction and the fraction in lowest terms are equivalent only for those values of the variable for which neither denominator is zero.

Throughout the rest of this book, it will be assumed that the replacement sets of the variables in a fraction include no numbers for which the denominator is zero. If you use a rational expression in the process of solving an equation, however, you must be aware of restrictions on the values of the variables.

EXAMPLE 4 Solve the following equation for x. Restrict a so that no division by zero results.

$$ax - a^2 = 4x - 8a + 16$$

SOLUTION
$$ax - a^2 = 4x - 8a + 16$$
$$ax - 4x = a^2 - 8a + 16$$
$$x(a - 4) = a^2 - 8a + 16$$

If $a \neq 4$, then

$$x = \frac{a^2 - 8a + 16}{a - 4} = \frac{(a - 4)^2}{a - 4} = a - 4.$$

$\therefore$ if $a \neq 4$, then $x = a - 4$.

Notice in Example 4 that, if $a = 4$, the original equation is true for *all* real values of x.

Oral Exercises

State all restrictions on the values of the variables in the given rational expression.

1. $\dfrac{5}{x}$ $x \neq 0$

2. $\dfrac{4}{a - 2}$ $a \neq 2$

3. $\dfrac{2n - 8}{3}$ none

4. $\dfrac{y - 5}{y - 5}$ $y \neq 5$

5. $\dfrac{4}{x^2 + 1}$ none

6. $\dfrac{3}{(n + 1)(n + 2)}$ $n \neq -1, -2$

7. $\dfrac{a - 2}{a^2 + 3a}$ $a \neq 0, -3$

8. $\dfrac{x - 3}{(x - 2)(x + 4)}$ $x \neq 2, -4$

Name the greatest common factor of the numerator and denominator of each rational expression.

9. $\dfrac{15}{25}$ 5

10. $\dfrac{9b}{6b}$ $3b$

11. $\dfrac{2m^5}{2m^2}$ $2m^2$

12. $\dfrac{a + 5}{a - 5}$ 1

13. $\dfrac{7(3m - 5)}{8(3m - 5)}$ $3m - 5$

14. $\dfrac{r + 4}{(r + 3)(r + 4)}$ $r + 4$

15. $\dfrac{2c - 6}{8c - 24}$ $2(c - 3)$

16. $\dfrac{x + 2}{x^2 - 4}$ $x + 2$

17. Do the open sentences $\dfrac{x}{x} = 1$ and $x = x$ have the same solution set over $\mathcal{R}$? Explain. No. The solution set of the first equation is {all real numbers except 0}. The solution set of the second equation is $\mathcal{R}$.

394 Chapter 8

Written Exercises

Simplify.

A 1. $\dfrac{25a^2b^2}{35ab}$ $\dfrac{5ab}{7}$

2. $\dfrac{-9st^2}{15s^2t}$ $-\dfrac{3t}{5s}$

3. $\dfrac{a(b-4)^2}{3a(b-4)}$ $\dfrac{b-4}{3}$

4. $\dfrac{6(a-1)}{18(a-1)^2}$ $\dfrac{1}{3(a-1)}$

5. $\dfrac{(3w+2)(w-2)}{(w-2)(3w-2)}$ $\dfrac{3w+2}{3w-2}$

6. $\dfrac{(a-b)^2(2a+b)}{2(a+b)(a-b)}$

7. $\dfrac{6b^3-3b^2}{5-10b}$ $-\dfrac{3b^2}{5}$

8. $\dfrac{y^2-5y}{15-3y}$ $-\dfrac{y}{3}$

9. $\dfrac{2f+6}{f^2-2f-15}$

10. $\dfrac{6u-4}{3u^2-10u+8}$ $\dfrac{6u^2-4}{3u^2-10u+8}$

11. $\dfrac{c^2+c-6}{8-2c^2}$ $-\dfrac{c+3}{2(c+2)}$

12. $\dfrac{5w-4v}{16v^2-25w^2}$

13. $\dfrac{3d^2+13d+12}{9d^2+24d+16}$ $\dfrac{d+3}{3d+4}$

14. $\dfrac{8u^2+6u-9}{6-5u-4u^2}$ $-\dfrac{2u+3}{u+2}$

15. $\dfrac{a^2+ab-20b^2}{a^2+2ab-15b^2}$

16. $\dfrac{5w^2-21wq+4q^2}{5w^2-19wq-4q^2}$ $\dfrac{5w-q}{5w+q}$

17. $\dfrac{18e^2+21e-15}{12e^2+32e+20}$ $\dfrac{3(2e-1)}{4(e+1)}$

18. $\dfrac{15t^2-9t+18}{20t^2-12t+24}$

6. $\dfrac{(a-b)(2a+b)}{2(a+b)}$ 9. $\dfrac{2}{f-5}$ 12. $-\dfrac{1}{5w+4v}$ 15. $\dfrac{a-4b}{a-3b}$ 18. $\dfrac{3}{4}$

Solve each equation for x in terms of a. Restrict a so that no division by zero results.

B 19. $2ax-5a=2a^2+3x-12$

20. $ax+4a-4=a^2+2x$

21. $2a(x-a)=5(x-3)+a$

22. $4x-10=9(3a^3-1)-3x(a-1)$

Simplify.

23. $\dfrac{3a^3-24}{a^2-4a+4}$ $\dfrac{3(a^2+2a+4)}{a-2}$

24. $\dfrac{125-8x^3}{4x^2-20x+25}-\dfrac{4x^2+10x+25}{2x-5}$

25. $\dfrac{a^6+1}{a^4-1}\cdot\dfrac{a^4-a^2+1}{a^2-1}$

26. $\dfrac{b^{12}-1}{(b^4-1)(b^6-1)}\cdot\dfrac{b^4-b^2+1}{b^2-1}$

27. $\dfrac{3(x-2)^2+4(x-2)-32}{x^2-4}$ $\dfrac{3x-14}{x-2}$

28. $\dfrac{9a^2-4a-5}{3(3a+2)^2+5(3a+2)-2}$ $\dfrac{a-1}{3a+4}$

29. $\dfrac{4(2a^2+2b^2)+b(18a+b)}{(2a+3b)^2-4a^2}$ $\dfrac{2a+3b}{3b}$

30. $\dfrac{(2b-1)^2-4}{(2b-1)^3-8}$ $\dfrac{2b+1}{4b^2+3}$

Determine all values of x for which the given fraction has a value of zero.

C 31. $\dfrac{28x^2+5x-12}{16x^3-9x}$ $x=\dfrac{4}{7}$

32. $\dfrac{2x^2+9x}{16x^3+68x^2-18x}$ none

Solve the following equations for x in terms of a.
a. If $\mathcal{R}$ is the solution set, what is the value of a?
b. If $\emptyset$ is the solution set, what is the value of a?
c. Determine the solution set for all values of a except those found in (a) and (b).

33. $a^2(x-1)=4(x-1)+5(a+2)$

34. $2a^2(x-1)-3(x+1)-a(5x-7)=0$

Polynomials and Rational Expressions   **395**

Suggested Assignments

Minimum
  395/1–18
S 384/21–25
Average
  395/1–25 odd, 26–30
S 384/22–32 even
Maximum
  395/1–31 odd, 32–36
S 385/32–40 even

Additional Answers
Written Exercises

19. $x=a+4;\ a\neq\dfrac{3}{2}$

20. $x=a-2;\ a\neq2$

21. $x=a+3;\ a\neq\dfrac{5}{2}$

22. $x=9a^2-3a+1;\ a\neq-\dfrac{1}{3}$

33. a. $a=-2$
   b. $a=2$
   c. $x=\dfrac{a+3}{a-2}$

34. a. $a=3$
   b. $a=-\dfrac{1}{2}$
   c. $x=\dfrac{2a-1}{2a+1}$

Mixed Review

Simplify.

1. $\dfrac{4x^5}{5x^6}$ $\dfrac{4}{5x}$ 2. $\dfrac{4b^2c^4}{b^4c^3}$ $\dfrac{4c}{b^2}$

Evaluate when $x=-2$ and $y=-3$.

3. a. $\dfrac{x^2(x-y)}{y(y-x)}$ $\dfrac{4}{3}$

   b. $-\dfrac{x^2(x-y)}{y(x-y)}$ $\dfrac{4}{3}$

Factor completely.

4. a^2-4a+3
   $(a-3)(a-1)$

5. $4y^2+11y+6$
   $(4y+3)(y+2)$

Write a direct proof of each theorem.

35. For all real numbers r, and all nonzero real numbers s and t, $\frac{r}{s} = \frac{r \cdot t}{s \cdot t}$.

36. For all real numbers r, and all nonzero real numbers s and t, $\frac{r}{s} = \frac{r \div t}{s \div t}$.

Computer Exercises For students with computer experience

1. Write a program that will compute the greatest common factor (GCF) of any two integers that you input. (*Hint*: Start by using the absolute value of either integer as a "trial factor," and test to determine if it is a factor of *both* integers. If it is, then it is the GCF. If it is not, decrease the trial factor by one and test again. Continue this process down to the number one, if necessary, until you find the GCF.)

2. Incorporate the program that you wrote for Exercise 1 into a program that will compute the simplest form of any fraction that you input. The numerator and denominator of the fraction should be input separately, and the output should be in the form of a fraction, such as 7/9, not a decimal.

Julio Rey Pastor

1888–1962

A young poet turned scientist, Julio Rey Pastor published an article on number theory when he was only seventeen. At twenty-three he was named professor of mathematical analysis at the University of Oviedo in his native Spain. Five years later, he wrote a book that discussed the synthetic geometry of space in *n* dimensions, and he founded a mathematics laboratory in Madrid. In 1917 he published his classic work, *Elementos de análisis algebraico (Elements of Algebraic Analysis)*.

Also in 1917 Rey Pastor began dividing his professional life between two continents, teaching half the year in Spain and the other half in Buenos Aires, Argentina. It was during this time that he established the journal *Revista matemática hispanoamericana (Spanish American Mathematical Review)*. Julio Rey Pastor was also noted as an historian of Spanish cartography.

8–5 Zero and Negative Exponents

From your work in Section 1-6, you are already familiar with expressions that involve *positive* integral exponents, such as 3^5, 2^3, and $(-5)^2$. In this section, you will learn to assign a meaning to expressions that involve *zero* or *negative* integral exponents, such as 2^0 and 2^{-5}.

In extending the meaning of a power to include any integer as an exponent, it is convenient to make definitions so that operations with powers continue to obey the laws already established for positive integral exponents. For example, the first law of exponents for division in Section 8-1 implies that

$$\frac{2^7}{2^7} = 1.$$

If the second law were to hold for $m = n$, then this quotient could also be expressed as

$$\frac{2^7}{2^7} = 2^{7-7} = 2^0.$$

This suggests that it is appropriate to define 2^0 to be 1, and, in general, to make the following definition.

> For every nonzero real number a,
> $$a^0 = 1.$$

Notice that no meaning is assigned to the expression 0^0.

The third law of exponents for division implies that

$$\frac{2^3}{2^8} = \frac{1}{2^{8-3}} = \frac{1}{2^5}.$$

If the second law of exponents for division were to hold for $m < n$, then this quotient could be expressed as

$$\frac{2^3}{2^8} = 2^{3-8} = 2^{-5}.$$

This suggests that it is appropriate to define 2^{-5} to be $\frac{1}{2^5}$ and, in general, to make the following definition.

> For every nonzero real number a and every integer n,
> $$a^{-n} = \frac{1}{a^n}.$$

Teaching Suggestions
p. T98

Related Activities p. T98

Supplementary Material
Test 31

Key Ideas

Use the definitions of zero and negative exponents in simplifying expressions. Use scientific notation.

1. Simplify $x^2y^{-1} \cdot x^{-2}y$. Express the result using only positive exponents. Assume no variable has zero as a value.

$x^2 \cdot x^{-2} \cdot y^{-1} \cdot y^1 =$
$x^{2+(-2)} \cdot y^{1+(-1)} = x^0 \cdot y^0 =$
$1 \cdot 1 = 1$

2. Simplify the following expressions, using scientific notation.

a. $\dfrac{2250}{0.0075}$

$\dfrac{2.25 \times 10^3}{7.5 \times 10^{-3}} =$

$0.3 \times \dfrac{10^3}{10^{-3}} =$

$0.3 \times 10^6 = 300{,}000$

b. $(260{,}000)^2$

$(2.6 \times 10^5)^2 =$
$2.6^2 \times 10^{10} =$
$6.76 \times 10^{10} =$
$67{,}600{,}000{,}000$

Thus, for every *nonzero* real number a, a^{-n} is the reciprocal of a^n.

With these definitions, the laws of exponents that you have learned will apply to expressions involving negative or zero exponents if the *base* of the expression is not zero. For example, if $a \neq 0$, you can use the law of exponents for multiplication given in Section 7-2 to simplify the following expression.

$$a^{-4} \cdot a^{-3} = a^{-4+(-3)} = a^{-7}$$

You can justify this method by showing that the result is consistent with the result you would obtain using the laws of exponents for positive integral exponents together with the definition just given for negative exponents.

$$a^{-4} \cdot a^{-3} = \frac{1}{a^4} \cdot \frac{1}{a^3} = \frac{1 \cdot 1}{a^4 \cdot a^3} = \frac{1}{a^{4+3}} = \frac{1}{a^7} = a^{-7}$$

Similarly, you can justify the following statements for every nonzero value of each variable.

$$\frac{a^{-3}}{a^{-5}} = a^{-3-(-5)} = a^2$$

$$(3a^{-2})^4 = 3^4 \cdot a^{-2 \cdot 4} = 81a^{-8}$$

EXAMPLE 1   Simplify each expression. Express the answer using only positive exponents. Assume that no variable has zero as a value.

a. $a^2 \cdot a^{-2}$ b. $\dfrac{(a^0 b^{-2})^5}{2a^{-1}}$

SOLUTION   a. $a^2 \cdot a^{-2} = a^{2+(-2)} = a^0 = 1$

b. $\dfrac{(a^0 b^{-2})^5}{2a^{-1}} = \dfrac{b^{-10}}{2a^{-1}} = \dfrac{a}{2b^{10}}$

The next example illustrates some other useful results that can be obtained for expressions with zero and negative exponents.

EXAMPLE 2   Show that each of the following open sentences is true if $a \neq 0$ and $b \neq 0$.

a. $\left(\dfrac{a}{b}\right)^0 = 1$ b. $\left(\dfrac{1}{b}\right)^{-1} = b$ c. $\left(\dfrac{a}{b}\right)^{-1} = \dfrac{b}{a}$

SOLUTION   a. $\left(\dfrac{a}{b}\right)^0 = 1$ by definition since $\dfrac{a}{b}$ is a nonzero real number.

b. $\left(\dfrac{1}{b}\right)^{-1} = (1 \cdot b^{-1})^{-1} = b$

c. $\left(\dfrac{a}{b}\right)^{-1} = (a \cdot b^{-1})^{-1} = a^{-1} \cdot b = \dfrac{b}{a}$

398   *Chapter 8*

Using the laws of exponents, you can express any positive number in the form

$$k \times 10^n, \text{ where } 1 \le k < 10, \text{ and } n \text{ is an integer.}$$

For example:

$$52,000 = 5.2 \times 10,000 = 5.2 \times 10^4$$

$$0.006 = \frac{6}{1000} = \frac{6}{10^3} = 6 \times 10^{-3}$$

Numbers written in this form are said to be written in **scientific notation.** Scientific notation provides an efficient way to read and write the very large and very small numbers that frequently occur in scientific measurements. For example:

Earth is about 150,000,000 km from the sun.
This distance can be written as 1.5×10^8 km.

One of the nearest stars, Alpha Centauri, is about 39,000,000,000,000 km from Earth.
This distance can be written as 3.9×10^{13} km.

The diameter of a silver atom is about 0.00000000025 m.
This measurement can be written as 2.5×10^{-10} m.

Scientific notation is also useful in some numerical computations, since numbers written in scientific notation can be multiplied and divided easily using the laws of exponents.

EXAMPLE 3   Simplify.

   a. 220×4000 **b.** $\dfrac{0.072}{6000}$ **c.** $(15,000)^2$

SOLUTION   **a.** $220 \times 4000 = (2.2 \times 10^2)(4.0 \times 10^3)$
$$= (2.2 \times 4) \times (10^2 \times 10^3)$$
$$= 8.8 \times 10^5$$
$$= 880,000$$

   b. $\dfrac{0.072}{6000} = \dfrac{7.2 \times 10^{-2}}{6.0 \times 10^3}$
$$= \frac{7.2}{6} \times \frac{10^{-2}}{10^3}$$
$$= 1.2 \times 10^{-5}$$
$$= 0.000012$$

   c. $(15,000)^2 = (1.5 \times 10^4)^2$
$$= 1.5^2 \times 10^8$$
$$= 2.25 \times 10^8$$
$$= 225,000,000$$

Oral Exercises

Simplify.

1. 2^{-3} $\frac{1}{8}$
2. 7^0 1
3. $(-3)^0$ 1
4. $(-4)^{-1}$ $-\frac{1}{4}$

5. $\left(\frac{4}{5}\right)^{-1}$ $\frac{5}{4}$
6. $\left(-\frac{3}{4}\right)^0$ 1
7. $5^2 \cdot 5^{-3}$ $\frac{1}{5}$
8. $\frac{5^2}{5^{-3}}$ 3125

Simplify using only positive exponents. Assume that no variable has zero as a value.

9. a^{-3} $\frac{1}{a^3}$
10. $\frac{1}{b^{-2}}$ b^2
11. $m^{-3} \cdot m^5$ m^2
12. $b^{-8} \cdot b^0$ $\frac{1}{b^8}$

13. $\frac{d^{-2}}{d^{-6}}$ d^4
14. $\frac{r^{-3}}{r^2}$ $\frac{1}{r^5}$
15. $(x^{-2})^{-1}$ x^2
16. $\left(\frac{1}{y^{-2}}\right)^2$ y^4

17. $(e^0)^{-2}$ 1
18. $6(-c)^0$ 6
19. $(3w^{-2})^0$ 1
20. $\left(\frac{1}{4v}\right)^0$ 1

Express each number in scientific notation.

21. $41{,}600$ 4.16×10^4
22. $2{,}000{,}000$ 2×10^6
23. 0.12 1.2×10^{-1}
24. 0.00372 3.72×10^{-3}

Written Exercises

Throughout this set of exercises, assume that no variable has zero as a value.

Simplify using only positive exponents.

A 1. $\frac{5}{2^{-1}}$ 10
2. $\frac{3}{2^{-2}}$ 12
3. $\frac{2^{-3}}{-4}$ $-\frac{1}{32}$

4. $\frac{5^0}{-3}$ $-\frac{1}{3}$
5. $5z^{-3}$ $\frac{5}{z^3}$
6. $(4y)^{-2}$ $\frac{1}{16y^2}$

7. $6c^{-3}b^2$ $\frac{6b^2}{c^3}$
8. $-3j^{-2}k^{-1}$ $-\frac{3}{j^2k}$
9. $\frac{n^{-3}}{m^{-3}}$ $\frac{m^3}{n^3}$

10. $\frac{2t^{-3}}{x^{-1}}$ $\frac{2x}{t^3}$
11. $a^{-4} \cdot a^{-1}$ $\frac{1}{a^5}$
12. $2b^{-3} \cdot b$ $\frac{2}{b^2}$

13. $\frac{5r^3}{10r^{-1}}$ $\frac{r^4}{2}$
14. $\frac{6p^{-2}}{8p^2}$ $\frac{3}{4p^4}$
15. $(uv^2)^{-2}$ $\frac{1}{u^2v^4}$

16. $(cd^{-1})^{-1}$ $\frac{d}{c}$
17. $(g^{-3}y)^0$ 1
18. $(a^0b^2)^{-3}$ $\frac{1}{b^6}$

19. $(e+f)^{-1}$ $\frac{1}{e+f}$
20. $(-3x^2y^{-1})^0$ 1
21. $\left(\frac{2+3t^0}{s}\right)^{-1}$ $\frac{s}{5}$

22. $\left(\frac{3x^2 + 2x^2y^0}{10x^2}\right)^{-2}$ 4
23. $\frac{2^{-3}c^{-1}d^2}{4^{-1}cd^{-1}}$ $\frac{d^3}{2c^2}$
24. $\frac{r^{-3}s^2t^{-5}}{r^0s^{-1}t^{-7}}$ $\frac{s^3t^2}{r^3}$

400 Chapter 8

Suggested Assignments

Minimum
  400/1–32
R 402/Self-Test 2
Average
  400/1–31 odd, 33–44
R 402/Self-Test 2
Maximum
  400/1–43 odd, 45–50
R 402/Self-Test 2

Additional A Exercises

In all these exercises assume that no variable has zero as a value. Simplify each expression. Express your result using only positive exponents.

1. $\frac{1}{2^{-1}}$ 2

2. $\frac{3^0}{2^{-2}}$ 4

3. $3z^{-2}$ $\frac{3}{z^2}$

4. $(2y)^{-3}$ $\frac{1}{8y^3}$

Simplify each expression by first expressing each number in scientific notation.

5. $0.03 \times 30{,}000 \times 300{,}000$
   $(3 \times 10^{-2})(3 \times 10^4) \cdot$
   (3×10^5); 270,000,000

6. $4{,}000{,}000 \div 0.004$
   $(4 \times 10^6) \div (4 \times 10^{-3})$;
   1,000,000,000

400

In Exercises 25–32, first express each number in scientific notation. Then simplify.

25. $500,000 \times 0.005$

26. $200 \times 2,000,000 \times 0.002$

27. $0.00256 \div 160$

28. $81,000 \div 0.09$

29. $(0.00012)^2$

30. $(3000)^3$

31. $\dfrac{40,000,000 \times 0.006}{8000}$

32. $\dfrac{(0.00048)(0.0012)}{(3000)(0.000024)}$

Simplify using only positive exponents.

B 33. $\dfrac{(-2)^{-3}c^{-2}h}{(4c)^{-1}h^{-3}} \quad \dfrac{h^4}{2c}$

34. $\dfrac{a^0 b^{-3} c^2 d^{-1}}{(ac)^{-2} b^{-5} d^0} \quad \dfrac{a^2 b^2 c^4}{d}$

35. $\left[\dfrac{(-2)^3}{2^2}\right]^{-1} \quad -\dfrac{1}{2}$

36. $\left[\dfrac{3^{-1}}{(-2)^{-2}}\right]^{-2} \quad \dfrac{9}{16}$

37. $\left[\dfrac{5^{-1}r}{5s^{-1}}\right]^2 \quad \dfrac{r^2 s^2}{625}$

38. $\left[\dfrac{3^{-3}u^2 w^{-5}}{6^{-1}u^{-1}w^{-2}}\right]^{-2} \quad \dfrac{81w^6}{4u^6}$

Simplify each expression. Express the answer so that no denominator contains a variable.

39. $\dfrac{(3^2)^{-1}m^{-1}n^{-2}}{3^{-3}m^0 n^3} \quad 3m^{-1}n^{-5}$

40. $\dfrac{r^{-3}s^2 t^{-4}}{r^0 s^{-4} t^{-5}} \quad r^{-3}s^6 t$

41. $\dfrac{(2^{-3})^{-1}d^{-2}e^0 f^5}{(2^{-2})^{-1}d^{-1}e^6 f^{-2}} \quad 2d^{-1}e^{-6}f^7$

42. $\dfrac{10^{-1}k^{-3}t^{-5}m^2}{(2^{-1})^{-2}kt^{-3}m^{-1}} \quad \dfrac{k^{-4}t^{-2}m^3}{40}$

43. $\dfrac{(-5)^{-2}x^5(y^{-1}z^{-2})^2}{10^{-1}(x^{-1}y^2)^{-1}z^{-3}} \quad \dfrac{2x^4 z^{-1}}{5}$

44. $\dfrac{(2^{-1})^{-2}(a^{-1}b^2 c^{-3})^{-2}}{(6^{-1})^2(a^2 b^{-1}c^2)^{-3}} \quad 144a^8 b^{-7}c^{12}$

In Exercises 45–48, evaluate each expression when $r = -3$ and $s = 2$.

C 45. $\dfrac{5}{r^{-1} + s^{-1}} \quad 30$

46. $\dfrac{(r+s)^{-1}}{r^{-1} - s^{-1}} \quad \dfrac{6}{5}$

47. $\dfrac{r^{-2} - s^{-2}}{r^{-2} + s^{-2}} \quad -\dfrac{5}{13}$

48. $\left(\dfrac{r^{-1}}{s^{-1}} + \dfrac{s^{-2}}{r^0}\right)^{-1} \quad -\dfrac{12}{5}$

49. If $a = 2^m$ and $b = 2^{m+1}$, show that

$$\dfrac{8a^3}{b^2} = 2^{m+1}.$$

50. **a.** Express each of the following as a power of 2.

$8, \quad 8^x, \quad 16^{x+3} \quad 2^3,\ 2^{3x},\ 2^{4x+12}$

b. Simplify the following expression by first expressing each factor as a power of 2.

$$\dfrac{8^{2x-1} \cdot (2^x)^{-1} \cdot 16^{x+3}}{4^{3x+1}} \quad \dfrac{2^{6x+3} \cdot 2^{-x} \cdot 2^{4x+12}}{2^{6x+2}} = \dfrac{2^{9x+9}}{2^{6x+2}} = 2^{3x+7}$$

Polynomials and Rational Expressions **401**

Mixed Review
Simplify.

1. $\dfrac{x^3}{x^1} \quad x^2$

2. $\dfrac{x^3}{x^5} \quad \dfrac{1}{x^2}$

3. $\dfrac{x^3}{x^4} \quad \dfrac{1}{x}$

4. $\dfrac{x^3}{x^3} \quad 1$

5. $\dfrac{2a^3 - 6a}{2a} \quad a^2 - 3$

6. $\left(\dfrac{3x^2 - 4x^2 y}{x^2 y}\right)^2$
$\dfrac{9 - 24y + 16y^2}{y^2}$

Computer Exercises For students with computer experience

1. Without using the computer's exponentiation operation, write a program that will compute the value of a^n when you input any real value for a and any integral value—positive, negative, or zero—for n.

2. Write a program that will determine whether or not a given integer *greater than one* is a power of a second integer, the *base*. If it is, the program should also determine what power it is. (*Hint*: Test all positive integral powers of the base until one of them is either equal to or greater than the given integer. Do *not* use the computer's exponentiation operation, since this will give inexact results.)

3. Modify the program that you wrote for Exercise 2 so that it will determine whether any real number *greater than zero* and *less than one* is a power of a given integer.

4. Write a program that will express a given real number greater than one in scientific notation. (*Hint*: First find the greatest power of ten that is less than or equal to the given number.)

Quick Quiz

Simplify.

1. $\dfrac{-36r^2s}{60rs^3}$ $\dfrac{-3r}{5s^2}$

2. $\dfrac{4k + 10}{10k + 25}$ $\dfrac{2}{5}$

3. $\dfrac{x^2 - 2x - 3}{6 - 2x}$ $-\dfrac{x + 1}{2}$

Simplify using only positive exponents. Assume that no variable has zero as a value.

4. $a^3 \cdot a^{-4}$ $\dfrac{1}{a}$

5. $\dfrac{b^{-4}}{b^{-2}}$ $\dfrac{1}{b^2}$

6. $(7j^{-3}k^4)^0$ 1

7. $\dfrac{u^3v^{-4}}{u^{-2}}$ $\dfrac{u^5}{v^4}$

8. $\dfrac{4^0ab^{-3}}{a^{-1}b^{-2}}$ $\dfrac{a^2}{b}$

9. $(2x^{-3}y^5)^{-3}$ $\dfrac{x^9}{8y^{15}}$

Express each number in scientific notation.

10. $1{,}423{,}000$ 1.423×10^6

11. 0.00013 1.3×10^{-4}

Self-Test 2

VOCABULARY   simplest form of a fraction (p. 393)   scientific notation (p. 399)

Simplify.

1. $\dfrac{-16bc}{24cd}$ $-\dfrac{2b}{3d}$

2. $\dfrac{3s - 6t}{5s - 10t}$ $\dfrac{3}{5}$

3. $\dfrac{6a^2 - 7a - 5}{3a - 5}$ $2a + 1$ *Obj. 1, p. 392*

Simplify using only positive exponents. Assume that no variable has zero as a value.

4. $c^{-5} \cdot c^2$ $\dfrac{1}{c^3}$

5. $\dfrac{x^{-1}}{x^{-5}}$ x^4

6. $(3a^2b^{-5})^{-2}$ $\dfrac{b^{10}}{9a^4}$ *Obj. 2, p. 392*

7. $(5s^{-1}t^3)^0$ 1

8. $\left(\dfrac{2a^2}{3b}\right)^{-1}$ $\dfrac{3b}{2a^2}$

9. $\dfrac{r^4p^{-3}m^0}{r^{-1}p^{-4}m^{-1}}$ r^5pm

Express each number in scientific notation.

10. $270{,}000$ 2.7×10^5

11. 0.0027 2.7×10^{-3} *Obj. 3, p. 392*

Check your answers with those at the back of the book.

Operations with Rational Expressions

OBJECTIVES for Sections 8-6 through 8-9:
1. To simplify products and quotients of rational expressions.
2. To simplify sums and differences of rational expressions.
3. To simplify complex fractions.

8–6 Products and Quotients of Rational Expressions

By using the symmetric property of equality, you can rewrite the basic property of quotients given in Section 8-1 as follows.

> **Theorem.** For all real numbers r and s and all nonzero real numbers t and u,
>
> $$\frac{r}{t} \cdot \frac{s}{u} = \frac{rs}{tu}.$$

You will notice that this is the rule that is used in multiplying rational numbers. For example,

$$\frac{3}{8} \cdot \frac{5}{2} = \frac{3 \cdot 5}{8 \cdot 2} = \frac{15}{16}.$$

The same rule can be used in multiplying rational expressions.

EXAMPLE 1 Simplify.

    a. $\dfrac{a-2}{a} \cdot \dfrac{a+3}{a-3}$ **b.** $4t \cdot \dfrac{5t}{s}$ **c.** $\left(\dfrac{u}{v}\right)^3$

SOLUTION   **a.** $\dfrac{a-2}{a} \cdot \dfrac{a+3}{a-3} = \dfrac{(a-2)(a+3)}{a(a-3)} = \dfrac{a^2+a-6}{a^2-3a}$

    b. $4t \cdot \dfrac{5t}{s} = \dfrac{4t}{1} \cdot \dfrac{5t}{s} = \dfrac{20t^2}{s}$

    c. $\left(\dfrac{u}{v}\right)^3 = \left(\dfrac{u}{v}\right)\left(\dfrac{u}{v}\right)\left(\dfrac{u}{v}\right) = \dfrac{u \cdot u \cdot u}{v \cdot v \cdot v} = \dfrac{u^3}{v^3}$

    Part (c) of Example 1 illustrates the first part of the theorem on the following page.

Polynomials and Rational Expressions   **403**

Chalkboard Examples

1. Simplify
$$\frac{2x-1}{x+4} \cdot \frac{x+3}{x-3}.$$
$$\frac{(2x-1)(x+3)}{(x+4)(x-3)} =$$
$$\frac{2x^2+5x-3}{x^2+x-12}$$

2. Simplify
$$\left(\frac{2x^{-1}y^2z^{-2}}{3x^{-2}y^3z^{-1}}\right)^{-3}.$$
$$\frac{(2x^{-1}y^2z^{-2})^{-3}}{(3x^{-2}y^3z^{-1})^{-3}} =$$
$$\frac{2^{-3}x^3y^{-6}z^6}{3^{-3}x^6y^{-9}z^3} = \frac{3^3x^3y^9z^6}{2^3x^6y^6z^3} =$$
$$\frac{27y^3z^3}{8x^3}$$

3. Simplify
$$\frac{4a+20}{2a^2-a-6} \div \frac{12a+8}{a^2+3a-10}.$$
$$\frac{4a+20}{2a^2-a-6} \cdot \frac{a^2+3a-10}{12a+8}$$
$$= \frac{4(a+5)(a-2)(a+5)}{4(2a+3)(3a+2)(a-2)}$$
$$= \frac{(a+5)(a+5)}{(2a+3)(3a+2)}$$
$$= \frac{a^2+10a+25}{6a^2+13a+6}$$

Theorem.   Let a and b denote real numbers and let n denote an integer.

1.   If $b \neq 0$ and a and n are not both 0, then $\left(\dfrac{a}{b}\right)^n = \dfrac{a^n}{b^n}$.

2.   If $a \neq 0$ and $b \neq 0$, then $\left(\dfrac{a}{b}\right)^{-n} = \dfrac{a^{-n}}{b^{-n}} = \dfrac{b^n}{a^n}$.

3.   If $a \neq 0$ and $b \neq 0$, then $\left(\dfrac{a}{b}\right)^{0} = \dfrac{a^0}{b^0} = 1$.

EXAMPLE 2   Simplify $\left(\dfrac{3a^2b^{-3}}{2c^{-4}d}\right)^{-2}$.

SOLUTION $\left(\dfrac{3a^2b^{-3}}{2c^{-4}d}\right)^{-2} = \dfrac{(3a^2b^{-3})^{-2}}{(2c^{-4}d)^{-2}} = \dfrac{3^{-2}a^{-4}b^6}{2^{-2}c^8d^{-2}} = \dfrac{4b^6d^2}{9a^4c^8}$

You can often simplify products of rational expressions by factoring numerators and denominators.

EXAMPLE 3   Simplify $\dfrac{4c + 12}{5c} \cdot \dfrac{c^2 - 6c + 9}{c^2 - 9}$.

SOLUTION $\dfrac{4c + 12}{5c} \cdot \dfrac{c^2 - 6c + 9}{c^2 - 9} = \dfrac{4(c + 3)}{5c} \cdot \dfrac{(c - 3)(c - 3)}{(c + 3)(c - 3)}$

$$= \dfrac{4(c + 3)(c - 3)(c - 3)}{5c(c + 3)(c - 3)}$$

$$= \dfrac{4(c - 3)}{5c}$$

$$= \dfrac{4c - 12}{5c}$$

The definition of division given in Section 2-11, together with the fact that the reciprocal of $\dfrac{s}{u}$ is $\dfrac{u}{s}$ if $s \neq 0$ and $u \neq 0$, leads to the following theorem.

Theorem.   For all real numbers r and all nonzero real numbers s, t, and u,

$$\dfrac{r}{t} \div \dfrac{s}{u} = \dfrac{r}{t} \cdot \dfrac{u}{s}.$$

You will notice that this is the rule used in dividing rational numbers.

404   *Chapter 8*

For example,

$$\frac{5}{8} \div \frac{2}{3} = \frac{5}{8} \cdot \frac{3}{2} = \frac{5 \cdot 3}{8 \cdot 2} = \frac{15}{16}.$$

The same rule can be used in dividing rational expressions.

EXAMPLE 4 Simplify. Assume that no variable has a value that results in division by zero.

a. $\dfrac{3x}{4y} \div \dfrac{x}{2y}$ **b.** $\dfrac{4m^2 - 8m - 5}{6m + 3} \div \dfrac{2m^2 - 5m}{6m + 12}$

SOLUTION **a.** $\dfrac{3x}{4y} \div \dfrac{x}{2y} = \dfrac{3x}{4y} \cdot \dfrac{2y}{x} = \dfrac{6xy}{4xy} = \dfrac{3}{2}$

b. $\dfrac{4m^2 - 8m - 5}{6m + 3} \div \dfrac{2m^2 - 5m}{6m + 12} = \dfrac{4m^2 - 8m - 5}{6m + 3} \cdot \dfrac{6m + 12}{2m^2 - 5m}$

$$= \frac{(2m + 1)(2m - 5)}{3(2m + 1)} \cdot \frac{6(m + 2)}{m(2m - 5)}$$

$$= \frac{3 \cdot 2(2m + 1)(2m - 5)(m + 2)}{3m(2m + 1)(2m - 5)}$$

$$= \frac{2(m + 2)}{m}, \text{ or } \frac{2m + 4}{m}$$

Oral Exercises

Throughout this set of exercises, assume that no variable has a value that results in division by zero.

Simplify.

1. $\dfrac{a}{b} \cdot \dfrac{c}{d}$ $\dfrac{ac}{bd}$

2. $-\dfrac{3x}{2y} \cdot \dfrac{w}{z}$ $-\dfrac{3xw}{2yz}$

3. $\dfrac{a^2b}{c} \cdot \dfrac{c^2}{ab}$ ac

4. $\dfrac{y^2}{2y} \cdot \dfrac{8}{2y}$ 2

5. $\dfrac{a - 2}{a + 1} \cdot \dfrac{3}{a - 1}$ $\dfrac{3(a - 2)}{(a + 1)(a - 1)}$

6. $\dfrac{x + 5}{x - 2} \cdot \dfrac{x - 2}{x + 1}$ $\dfrac{x + 5}{x + 1}$

7. $\left(\dfrac{2x}{3y}\right)^0$ 1

8. $\left(\dfrac{-1}{b^2}\right)^{-1}$ $-b^2$

9. $\left(\dfrac{3w^2}{2v^3}\right)^{-1}$ $\dfrac{2v^3}{3w^2}$

10. $\left(\dfrac{a^2}{2b}\right)^4$ $\dfrac{a^8}{16b^4}$

11. $\left(\dfrac{x^3}{y^2}\right)^{-2}$ $\dfrac{y^4}{x^6}$

12. $\left(-\dfrac{3z}{2}\right)^{-3}$ $-\dfrac{8}{27z^3}$

Express each quotient as a product.

13. $\dfrac{b}{r} \div \dfrac{c}{q}$ $\dfrac{b}{r} \cdot \dfrac{q}{c}$

14. $\dfrac{x}{y} \div 2x$ $\dfrac{x}{y} \cdot \dfrac{1}{2x}$

15. $\dfrac{3k}{k - 2} \div \dfrac{2}{k}$ $\dfrac{3k}{k - 2} \cdot \dfrac{k}{2}$

16. $\dfrac{e}{e - 1} \div \dfrac{e - 1}{e + 1}$ $\dfrac{e}{e - 1} \cdot \dfrac{e + 1}{e - 1}$

17. $-\dfrac{r}{5s} \div \dfrac{s - 2}{3r}$ $-\dfrac{r}{5s} \cdot \dfrac{3r}{s - 2}$

18. $\dfrac{a}{b} \div \left(-\dfrac{a - 1}{b - 1}\right)$ $\dfrac{a}{b} \cdot \left(-\dfrac{b - 1}{a - 1}\right)$

Additional A Exercises

In these exercises assume that no variable has a value that results in division by zero.

1. $\dfrac{4a}{3} \cdot \dfrac{3}{2}$ $2a$

2. $\dfrac{16y^2}{3x} \div \dfrac{4y}{x^2}$ $\dfrac{4xy}{3}$

3. $\dfrac{3a + 6}{a^2b} \cdot \dfrac{a^3}{a^2 + 2a}$ $\dfrac{3}{b}$

4. $\dfrac{w^2 - 1}{10w + 15} \div \dfrac{5w + 5}{2w + 3}$ $\dfrac{w - 1}{25}$

Simplify. Express your results using only positive exponents.

5. $\left(\dfrac{s^{-1}t}{2t^2}\right)^{-2}$ $4s^2t^2$

6. $\left(\dfrac{a^{-2}b^3x^{-1}}{b^{-4}c^3}\right)^0$ 1

Suggested Assignments

Minimum
406/1–27
Average
406/1–29 odd, 31–42
Maximum
406/1–45 odd, 46–52

Additional Answers
Written Exercises

35. $\dfrac{(m + 2)(m + 3)}{(m - 2)(m + 1)}$

36. $\dfrac{v - 1}{v - 2}$

37. $\dfrac{x^2 + 2xy + 4y^2}{x - 2y}$

38. $-\dfrac{1}{(a - 1)(3a - 4b)^2}$

39. $\dfrac{2t + 3}{3t - 1}$ 40. $\dfrac{p - 3}{p + 6}$

41. $\dfrac{3x + 2}{x + 1}$ 42. $\dfrac{5}{5a + 1}$

49. Let $\dfrac{a}{b}$ and $\dfrac{c}{d}$ be rational

numbers. Then a, b, c, and d are integers, $b \neq 0$ and $d \neq 0$. By the basic property of quotients, $\dfrac{a}{b} \cdot \dfrac{c}{d} = \dfrac{ac}{bd}$. Since the set of integers is closed with respect to multiplication, ac and bd are integers. Since $b \neq 0$ and $d \neq 0$, $bd \neq 0$. $\therefore \dfrac{ac}{bd}$ is a rational number.

50. Let $\dfrac{a}{b}$ and $\dfrac{c}{d}$ be rational

numbers. Then a, b, c, and d are integers, $b \neq 0$ and $d \neq 0$. If $c \neq 0$, $\dfrac{a}{b} \div \dfrac{c}{d}$ $= \dfrac{a}{b} \cdot \dfrac{d}{c} = \dfrac{ad}{bc}$, which is a rational number. (See Exercise 49.)
(continued on p. 422)

Written Exercises

Throughout this set of exercises, assume that no variable has a value that results in division by zero.

Simplify. **12.** $\dfrac{g - h}{g}$ **15.** $\dfrac{t + 2}{3t - 1}$ **16.** $2(2d - 3)(d - 1)$ **18.** $c + d$ **21.** 15

A

1. $\dfrac{5m}{2} \cdot \dfrac{2}{5}$ m

2. $\dfrac{12}{7} \cdot \dfrac{7n}{4}$ $3n$

3. $\dfrac{5a^2}{b} \div \dfrac{2a}{3b}$ $\dfrac{15a}{2}$

4. $\dfrac{12x^2}{5y} \div \dfrac{3x}{2y}$ $\dfrac{8x}{5}$

5. $\dfrac{15k}{-4} \div 3k$ $-\dfrac{5}{4}$

6. $3k \div \dfrac{15k}{-4}$ $-\dfrac{4}{5}$

7. $\dfrac{8t^2}{3} \div \left(-\dfrac{4t}{9}\right)$ $-6t$

8. $\dfrac{-3r^2}{2s} \div \left(-\dfrac{2r}{s}\right)$ $\dfrac{3r}{4}$

9. $\dfrac{16pq}{25} \cdot \dfrac{5p}{8q^2}$ $\dfrac{2p^2}{5q}$

10. $\dfrac{-36j^2k^3}{7i} \cdot \dfrac{14i^2j^2}{9k^2}$ $-8ij^4k$

11. $\dfrac{rs}{4r - 4s} \cdot \dfrac{r^3 - rs^2}{r^2}$ $\dfrac{s(r + s)}{4}$

12. $\dfrac{2g - 2h}{g + h} \cdot \dfrac{2g + 2h}{4g}$

13. $\dfrac{5s + 10t}{10s - 10t} \div \dfrac{3s + 6t}{6s - 6t}$ 1

14. $\dfrac{3e - 9f}{2e + 4f} \div \dfrac{5e - 15f}{7e + 14f}$ $\dfrac{21}{10}$

15. $\dfrac{3t + 1}{t - 2} \cdot \dfrac{t^2 - 4}{9t^2 - 1}$

16. $\dfrac{4d^2 - 9}{d + 1} \cdot \dfrac{10d^2 - 10}{10d + 15}$

17. $\dfrac{3u - 6}{u^2 - 1} \div \dfrac{4u - 8}{2u^2 - 2}$ $\dfrac{3}{2}$

18. $\dfrac{c^2 - d^2}{2c + d} \div \dfrac{c^2 - cd}{2c^2 + cd}$

19. $\dfrac{4v - 12}{v^2} \div (2v^2 - 6v)$ $\dfrac{2}{v^3}$

20. $\dfrac{b^2 + 3b}{4b} \div (3b + 9)$ $\dfrac{1}{12}$

21. $(5w - 20) \div \dfrac{w^2 - 16}{3w + 12}$

22. $(4a - 24) \div \dfrac{2a^2 - 72}{3a}$ $\dfrac{6a}{a + 6}$ 23. $\dfrac{1}{3x + 9} \div \dfrac{10x}{5x + 15}$ $\dfrac{1}{6x}$ 24. $\dfrac{3t^2 - 27}{t - 2} \div \dfrac{1}{t^2 - 4}$

$3(t - 3)(t + 3)(t + 2)$

Simplify using only positive exponents.

25. $\left(\dfrac{3a^{-2}b^4}{2c^2d^{-2}}\right)^{-3}$ $\dfrac{8a^6c^6}{27b^{12}d^6}$

26. $\left(\dfrac{5x^0y^{-5}}{10xy^{-6}}\right)^{-1}$ $\dfrac{2x}{y}$

27. $\left(\dfrac{2a^4b^{-3}}{10a^{-2}b}\right)^0$ 1

28. $\dfrac{3}{4}\left(\dfrac{3x^4y^{-1}}{-2x^0y}\right)^{-2}$ $\dfrac{y^4}{3x^8}$

29. $\left(\dfrac{a^4b^{-1}}{6c^{-1}}\right)^{-2} \div \dfrac{ac^2}{-2b}$ $-\dfrac{72b^3}{a^9c^4}$ 30. $\left(\dfrac{x - 1}{3x}\right)^{-2} \cdot \left(\dfrac{1 - x}{2}\right)^3$

$\dfrac{9x^2(1 - x)}{8}$

Simplify.

B

31. $\dfrac{k^2 + 2k - 3}{k^2 + k - 2} \cdot \dfrac{3k + 6}{k + 3}$ 3

32. $\dfrac{z^2 - 2z}{z + 1} \cdot \dfrac{z^2 - 1}{z^2 - 2z + 1}$ $\dfrac{z(z - 2)}{z - 1}$

33. $\dfrac{t^2 + 4t + 4}{t^2 - 3t - 10} \cdot \dfrac{t^2 - 10t + 25}{t^2 - 4}$ $\dfrac{t - 5}{t - 2}$

34. $\dfrac{w^2 + 3w}{w^3 - 6w} \cdot \dfrac{w^2 - 6}{w^3 - 9w}$ $\dfrac{1}{w(w - 3)}$

35. $\dfrac{m^2 - 2m - 8}{m^2 - 5m + 6} \div \dfrac{m^2 - 3m - 4}{m^2 - 9}$

36. $\dfrac{v^2 - 2v - 24}{v^2 + 6v + 8} \div \dfrac{v^2 - 8v + 12}{v^2 + v - 2}$

37. $\left(\dfrac{x - 2y}{x + 2y}\right)^{-2} \cdot \dfrac{x^3 - 8y^3}{x^2 + 4xy + 4y^2}$

38. $\left(\dfrac{3a - 4b}{a - 1}\right)^3 \div \dfrac{(4b - 3a)^5}{a^2 - 2a + 1}$

39. $\dfrac{4t^2 + 4t - 3}{9t^2 - 1} \cdot \left(\dfrac{4t^2 - 8t + 3}{6t^2 - 7t - 3}\right)^{-1}$

40. $\dfrac{2p^2 - 5p - 3}{p^2 + 7p + 6} \div \left(\dfrac{3p^2 + 2p - 1}{6p^2 + p - 1}\right)^{-1}$

41. $\left(\dfrac{6x^2 + 5x - 6}{6x^2 - 4x - 6}\right)^0 \div \left(\dfrac{9x^2 - 4}{3x^2 + x - 2}\right)^{-1}$

42. $\left(\dfrac{5a^2 - 4a - 1}{5a - 5}\right)^{-1} \cdot \left(\dfrac{2a^2 + a - 3}{25a^2 + 10a + 1}\right)^0$

C 43. $\dfrac{3j + 9}{10j + 10} \cdot \dfrac{j - 2}{6j + 18} \div \dfrac{6 - 3j}{5j + 25}$ $-\dfrac{j + 5}{12(j + 1)}$

44. $\dfrac{z^2 - 1}{z^2 + 4} \cdot \dfrac{6z + 30}{z^2 + z} \div \left(\dfrac{z^3 + 4z}{3z - 3}\right)^{-1}$ $2(z + 5)$

45. $\dfrac{6h^2 - 7h - 20}{2 + 5h - 12h^2} \div \dfrac{15h^2 + 14h - 8}{3h^2 - 2h} \cdot \dfrac{20h^2 - 3h - 2}{2h^2 + 7h - 30}$ $-\dfrac{h}{h + 6}$

46. $\dfrac{y^2 - y - 12}{y^2 - 16} \div \dfrac{20 - y - y^2}{y^2 + 10y + 24} \div \dfrac{y^2 + 4y - 12}{y^2 - 8y + 16}$ $-\dfrac{(y + 3)(y - 4)}{(y + 5)(y - 2)}$

47. Is the product of two rational expressions always a rational expression? Yes

48. Is the quotient of two nonzero rational expressions always a rational expression? Yes

49. Explain why the set of rational numbers is closed with respect to multiplication.

50. Explain why the set of rational numbers is closed with respect to division, excluding division by zero.

Exercises 51–52 refer to the three parts of the theorem stated at the top of page 404.

51. Use the definition of a power (page 25) and the rule for multiplying rational expressions (page 403) to prove the first part of the theorem.

52. Use the first part of the theorem to prove the second and third parts of the theorem.

8–7 Sums and Differences of Rational Expressions with Equal Denominators

By using the symmetric axiom of equality, you can rewrite the theorem given in Section 8-2 as follows.

Theorem. For all real numbers a, b, and c such that $c \neq 0$,

$$\dfrac{a}{c} + \dfrac{b}{c} = \dfrac{a + b}{c} \quad \text{and} \quad \dfrac{a}{c} - \dfrac{b}{c} = \dfrac{a - b}{c}.$$

You will notice that this is the rule that is used in adding or subtracting rational numbers with equal denominators, as shown on the following page.

Polynomials and Rational Expressions **407**

1. Simplify.

a. $\dfrac{4}{b} + \dfrac{5}{b} - \dfrac{7}{b}$ $\dfrac{4 + 5 - 7}{b} =$

$\dfrac{2}{b}$

b. $\dfrac{2x}{3y} + \dfrac{4x}{3y} - \dfrac{x}{3y}$

$\dfrac{2x + 4x - x}{3y} = \dfrac{5x}{3y}$

2. Simplify.

a. $\dfrac{2n + 1}{n + 3} + \dfrac{n - 5}{n + 3}$

$\dfrac{2n + 1 + n - 5}{n + 3} =$

$\dfrac{3n - 4}{n + 3}$

b. $\dfrac{y^2}{y^2 - 4} - \dfrac{y + 6}{y^2 - 4}$

$= \dfrac{y^2 - y - 6}{y^2 - 4}$

$= \dfrac{(y + 2)(y - 3)}{(y + 2)(y - 2)} = \dfrac{y - 3}{y - 2}$

Common Errors

When simplifying differences of rational expressions, students may express

$\dfrac{x}{y} - \dfrac{2z - 2x}{y}$ as $\dfrac{x - 2z - 2x}{y}$

instead of $\dfrac{x - 2z + 2x}{y}$. Encourage students to include an extra step in these computations:

$\dfrac{x}{y} - \dfrac{2z - 2x}{y} = \dfrac{x - (2z - 2x)}{y}$

$= \dfrac{x - 2z + 2x}{y} = \dfrac{3x - 2z}{y}$

For example,

$$\frac{4}{15} + \frac{3}{15} = \frac{4 + 3}{15} = \frac{7}{15}$$

and

$$\frac{8}{11} - \frac{7}{11} = \frac{8 - 7}{11} = \frac{1}{11}.$$

The same rule can be used in adding and subtracting rational expressions that have equal denominators.

EXAMPLE   Simplify.

a. $\dfrac{2a}{5} + \dfrac{3a}{5}$ **b.** $\dfrac{a}{3b} - \dfrac{5a}{3b}$ **c.** $\dfrac{a^2}{a^2 - 9} + \dfrac{2a - 3}{a^2 - 9}$ **d.** $\dfrac{3a}{a - 1} - \dfrac{a + 1}{a - 1}$

SOLUTION   **a.** $\dfrac{2a}{5} + \dfrac{3a}{5} = \dfrac{2a + 3a}{5} = \dfrac{5a}{5} = a$

b. $\dfrac{a}{3b} - \dfrac{5a}{3b} = \dfrac{a - 5a}{3b} = \dfrac{-4a}{3b} = -\dfrac{4a}{3b}$

c. $\dfrac{a^2}{a^2 - 9} + \dfrac{2a - 3}{a^2 - 9} = \dfrac{a^2 + 2a - 3}{a^2 - 9}$

$= \dfrac{(a + 3)(a - 1)}{(a + 3)(a - 3)} = \dfrac{a - 1}{a - 3}$

d. $\dfrac{3a}{a - 1} - \dfrac{a + 1}{a - 1} = \dfrac{3a - (a + 1)}{a - 1}$

$= \dfrac{3a - a - 1}{a - 1} = \dfrac{2a - 1}{a - 1}$

Oral Exercises

Simplify.

1. $\dfrac{5}{13} + \dfrac{2}{13}$ $\frac{7}{13}$

2. $\dfrac{9}{16} - \dfrac{11}{16}$ $-\frac{1}{8}$

3. $\dfrac{3x}{7} - \dfrac{x}{7}$ $\frac{2x}{7}$

4. $\dfrac{10a}{7} + \dfrac{3b}{7}$ $\frac{10a + 3b}{7}$

5. $\dfrac{2}{5m} - \dfrac{n}{5m}$ $\frac{2 - n}{5m}$

6. $\dfrac{5b}{9a} - \dfrac{4b}{9a}$ $\frac{b}{9a}$

7. $\dfrac{7n}{3n + 1} + \dfrac{n + 2}{3n + 1}$ $\frac{8n + 2}{3n + 1}$

8. $\dfrac{5x}{x^2 + 1} + \dfrac{2x - 1}{x^2 + 1}$ $\frac{7x - 1}{x^2 + 1}$

9. $\dfrac{8b}{5} - \dfrac{2b + 1}{5}$ $\frac{6b - 1}{5}$

10. $\dfrac{4t}{9} - \dfrac{3t - 1}{9}$ $\frac{t + 1}{9}$

11. $\dfrac{2d - 1}{d - 4} - \dfrac{d - 1}{d - 4}$ $\frac{d}{d - 4}$

12. $\dfrac{3j + 4}{j - 7} - \dfrac{2j + 11}{j - 7}$ 1

Written Exercises

Simplify.

A 1. $\dfrac{7}{a} + \dfrac{5}{a} - \dfrac{3}{a}$ $\dfrac{9}{a}$

2. $\dfrac{3}{2b} - \dfrac{4}{2b} + \dfrac{7}{2b}$ $\dfrac{3}{b}$

3. $\dfrac{2n}{n+1} + \dfrac{n-1}{n+1}$ $\dfrac{3n-1}{n+1}$

4. $\dfrac{5c-2}{c-1} + \dfrac{3c-1}{c-1}$ $\dfrac{8c-3}{c-1}$

5. $\dfrac{3t}{4} - \dfrac{t-1}{4}$ $\dfrac{2t+1}{4}$

6. $\dfrac{5x}{3} - \dfrac{2x+1}{3}$ $\dfrac{3x-1}{3}$

7. $\dfrac{4a+1}{a-3} - \dfrac{3a+4}{a-3}$ 1

8. $\dfrac{3s-5}{s-2} - \dfrac{2s-3}{s-2}$ 1

9. $\dfrac{3a}{2a-6} - \dfrac{9}{2a-6}$ $\dfrac{3}{2}$

10. $\dfrac{3b}{2b+12} + \dfrac{18}{2b+12}$ $\dfrac{3}{2}$

11. $\dfrac{x^2}{x^2-9} + \dfrac{3x}{x^2-9}$ $\dfrac{x}{x-3}$

12. $\dfrac{3x}{x^2-1} - \dfrac{3}{x^2-1}$ $\dfrac{3}{x+1}$

13. $\dfrac{2h-i}{h-i} - \dfrac{3h-2i}{h-i}$ -1

14. $\dfrac{5t+4w}{2t-w} - \dfrac{9t+2w}{2t-w}$ -2

15. $\dfrac{3a^2}{a+1} - \dfrac{a^2+2}{a+1}$ $2(a-1)$

16. $\dfrac{x^2+3}{2x-3} - \dfrac{5x^2-6}{2x-3}$ $-2x-3$

17. $\dfrac{3d+4}{2d^2-8} - \dfrac{7d+12}{2d^2-8}$ $-\dfrac{2}{d-2}$

18. $\dfrac{8+5h}{3h^2-48} - \dfrac{6h+4}{3h^2-48}$ $-\dfrac{1}{3(h+4)}$

B 19. $\dfrac{5c-4}{2c^2-c-3} + \dfrac{1-3c}{2c^2-c-3}$ $\dfrac{1}{c+1}$

20. $\dfrac{4m+3}{9m^2-3m-2} - \dfrac{m+5}{9m^2-3m-2}$ $\dfrac{1}{3m+1}$

21. $\dfrac{3d+7}{9d^2+9d-4} - \dfrac{11-9d}{9d^2+9d-4}$ $\dfrac{4}{3d+4}$

22. $\dfrac{z^2+3z}{10z^2+11z-6} + \dfrac{5z^2+6z}{10z^2+11z-6}$ $\dfrac{3z}{5z-2}$

23. $\dfrac{7n-3}{8n^2+2n-3} + \dfrac{4-9n}{8n^2+2n-3}$ $-\dfrac{1}{4n+3}$

24. $\dfrac{3w^2-7}{4w^2-15w-4} - \dfrac{2w^2+9}{4w^2-15w-4}$ $\dfrac{w+4}{4w+1}$

25. $\dfrac{2e}{e-5} - \left(\dfrac{e-2}{e-5} + \dfrac{2}{e-5}\right)$ $\dfrac{e}{e-5}$

26. $\dfrac{3p}{(p+1)^2} - \left(\dfrac{5p+1}{(p+1)^2} - \dfrac{3p+2}{(p+1)^2}\right)$ $\dfrac{1}{p+1}$

27. $\dfrac{6v-1}{v^2-9} - \dfrac{2v+7}{v^2-9} + \dfrac{5-3v}{v^2-9}$ $\dfrac{1}{v+3}$

28. $\dfrac{7g+3f}{f^2-g^2} - \dfrac{5g+3f}{f^2-g^2} + \dfrac{f-g}{f^2-g^2}$ $\dfrac{1}{f-g}$

C 29. Find a rational expression which when added to $\dfrac{5r-1}{5-3r}$ gives a sum of $\dfrac{2r-3}{5-3r}$.

30. Find a rational expression which when added to $\dfrac{u^2-3v^2}{u^2-9v^2}$ gives a sum of $\dfrac{u+v}{u+3v}$.

31. Show that for all real numbers x and y such that $x \neq y$,

$$\dfrac{1}{x-y} + \dfrac{1}{y-x} = 0.$$

32. Simplify $\dfrac{5x}{2x-1} - \dfrac{1-x-6x^2}{1-4x^2}$. $\dfrac{2x+1}{2x-1}$

Polynomials and Rational Expressions   **409**

Suggested Assignments

Minimum
  409/1–18
Average
  409/1–23 odd, 24–28
Maximum
  409/1–27 odd, 29–32

Additional A Exercises

Simplify.

1. $\dfrac{6}{p} + \dfrac{7}{p} - \dfrac{2}{p}$ $\dfrac{11}{p}$

2. $\dfrac{11}{3s} - \dfrac{4}{3s} - \dfrac{1}{3s}$ $\dfrac{2}{s}$

3. $\dfrac{a}{3a+1} + \dfrac{2}{3a+1}$ $\dfrac{a+2}{3a+1}$

4. $\dfrac{2b}{4b-1} - \dfrac{3}{4b-1}$ $\dfrac{2b-3}{4b-1}$

5. $\dfrac{2x-1}{3x+2} + \dfrac{2x}{3x+2}$ $\dfrac{4x-1}{3x+2}$

Mixed Review

Express each quotient as a sum or difference.

1. $\dfrac{4a+2}{3b}$ $\dfrac{4a}{3b} + \dfrac{2}{3b}$

2. $\dfrac{4x+3x^2}{y^2}$ $\dfrac{4x}{y^2} + \dfrac{3x^2}{y^2}$

3. $\dfrac{3n^2-n-7}{n+9}$

   $\dfrac{3n^2}{n+9} - \dfrac{n}{n+9} - \dfrac{7}{n+9}$

Additional Answers
Written Exercises

29. $\dfrac{-3r-2}{5-3r}$, or $\dfrac{3r+2}{3r-5}$

30. $-\dfrac{2uv}{u^2-9v^2}$

31. $\dfrac{1}{x-y} + \dfrac{1}{y-x} = \dfrac{1}{x-y} +$

   $\dfrac{-1}{x-y} = \dfrac{1}{x-y} - \dfrac{1}{x-y} = 0$

Key Ideas

Simplify sums and differences of rational expressions with unequal denominators.

8–8 Sums and Differences of Rational Expressions with Unequal Denominators

The **least common denominator (LCD)** of two or more fractions is the *least common multiple* of their denominators. To add or subtract rational numbers having unequal denominators, first find the LCD of their denominators. Then express each fraction as an equivalent fraction with the LCD as denominator.

EXAMPLE 1   Simplify $\frac{3}{8} + \frac{5}{14}$.

SOLUTION   1. Find the LCM by factoring 8 and 14 over the set of primes.

$$8 = 2^3 \text{ and } 14 = 2 \cdot 7$$
$$\text{LCM} = 2^3 \cdot 7 = 56$$

Thus, the LCD is 56.

2. Express each fraction as an equivalent fraction with the LCD as denominator.

$$\frac{3}{8} = \frac{3 \cdot 7}{8 \cdot 7} = \frac{21}{56}; \qquad \frac{5}{14} = \frac{5 \cdot 4}{14 \cdot 4} = \frac{20}{56}$$

3. Simplify.

$$\frac{3}{8} + \frac{5}{14} = \frac{21}{56} + \frac{20}{56} = \frac{21 + 20}{56} = \frac{41}{56}$$

The method used in Example 1 to find the sum of two rational numbers can also be used to find the sum or difference of rational expressions that have unequal denominators. The **least common multiple (LCM)** of two or more polynomials is the polynomial of least degree and least positive constant factor that has each of the given polynomials as a factor. The least common denominator of two or more rational expressions is the LCM of their denominators.

To find the LCM of two or more polynomials you first factor each of the polynomials completely. The LCM will be the product of all the different factors, each factor occurring in the product the greatest number of times that it occurs in the complete factorization of any one of the given polynomials.

EXAMPLE 2   Find the LCM of $4x^2 - 36$ and $6x^2 - 36x + 54$.

SOLUTION $4x^2 - 36 = 2^2(x + 3)(x - 3)$
$6x^2 - 36x + 54 = 2 \cdot 3(x - 3)^2$
$\text{LCM} = 2^2 \cdot 3(x + 3)(x - 3)^2 = 12(x + 3)(x - 3)^2$

410   *Chapter 8*

EXAMPLE 3 Simplify $2 + \dfrac{5r}{3r + 12} - \dfrac{r - 1}{r^2 - 16}$.

SOLUTION 1. Find the LCM of the denominators.

$$3r + 12 = 3(r + 4); \qquad r^2 - 16 = (r + 4)(r - 4)$$
$$\text{LCM} = 3(r + 4)(r - 4)$$

Thus, the LCD $= 3(r + 4)(r - 4)$.

2. Express each fraction as an equivalent fraction with the LCD as denominator.

$$2 = \frac{2 \cdot 3(r + 4)(r - 4)}{1 \cdot 3(r + 4)(r - 4)} = \frac{6(r + 4)(r - 4)}{3(r + 4)(r - 4)}$$

$$\frac{5r}{3r + 12} = \frac{5r \cdot (r - 4)}{3(r + 4) \cdot (r - 4)} = \frac{5r(r - 4)}{3(r + 4)(r - 4)}$$

$$\frac{r - 1}{r^2 - 16} = \frac{(r - 1) \cdot 3}{(r + 4)(r - 4) \cdot 3} = \frac{3(r - 1)}{3(r + 4)(r - 4)}$$

3. Simplify.

$$2 + \frac{5r}{3r + 12} - \frac{r - 1}{r^2 - 16}$$

$$= \frac{6(r + 4)(r - 4)}{3(r + 4)(r - 4)} + \frac{5r(r - 4)}{3(r + 4)(r - 4)} - \frac{3(r - 1)}{3(r + 4)(r - 4)}$$

$$= \frac{6(r + 4)(r - 4) + 5r(r - 4) - 3(r - 1)}{3(r + 4)(r - 4)}$$

$$= \frac{6r^2 - 96 + 5r^2 - 20r - 3r + 3}{3(r + 4)(r - 4)}$$

$$= \frac{11r^2 - 23r - 93}{3(r + 4)(r - 4)}, \text{ or } \frac{11r^2 - 23r - 93}{3r^2 - 48}$$

Unless instructed otherwise, you need not number and describe steps as in Example 3. The way that you usually give the solution is shown in Example 4.

EXAMPLE 4 Simplify $\dfrac{3x}{x^2 - 2xy + y^2} - \dfrac{2y}{x^2 - y^2}$.

SOLUTION $\dfrac{3x}{x^2 - 2xy + y^2} - \dfrac{2y}{x^2 - y^2} = \dfrac{3x}{(x - y)^2} - \dfrac{2y}{(x + y)(x - y)}$

$$= \frac{3x(x + y)}{(x - y)^2(x + y)} - \frac{2y(x - y)}{(x + y)(x - y)(x - y)}$$

$$= \frac{3x^2 + 3xy - 2xy + 2y^2}{(x - y)^2(x + y)}$$

$$= \frac{3x^2 + xy + 2y^2}{(x - y)^2(x + y)}$$

1. Find the LCM of $6a^2 + 33a + 36$ and $12a^2 - 6a - 36$.

$$6a^2 + 33a + 36 = 3(2a + 3)(a + 4)$$
$$12a^2 - 6a - 36 = 3 \cdot 2(a - 2)(2a + 3)$$
$$\text{LCM} =$$
$$3 \cdot 2(2a + 3)(a + 4)(a - 2)$$
$$= 6(2a + 3)(a + 4)(a - 2)$$

2. Simplify

$$3 - \frac{p + 2}{p^2 - 3p + 2} + \frac{p - 3}{3p - 6}.$$

$$= \frac{9(p - 1)(p - 2)}{3(p - 1)(p - 2)} -$$

$$\frac{3(p + 2)}{3(p - 1)(p - 2)} +$$

$$\frac{(p - 3)(p - 1)}{3(p - 1)(p - 2)}$$

$$= \frac{10p^2 - 34p + 15}{3p^2 - 9p + 6}$$

3. Simplify

$$\frac{2x^2 - 1}{2x^2 + 3x + 1} + \frac{2x + 3}{3x + 3}.$$

$$= \frac{3(2x^2 - 1)}{3(2x + 1)(x + 1)} +$$

$$\frac{(2x + 3)(2x + 1)}{3(2x + 1)(x + 1)}$$

$$= \frac{6x^2 - 3 + 4x^2 + 8x + 3}{3(2x + 1)(x + 1)}$$

$$= \frac{10x^2 + 8x}{3(2x + 1)(x + 1)}$$

Suggested Assignments

Minimum
First day: 412/1–12
Second day: 413/13-24
       S   406/28–32

Average
First day:
      412/1–23 odd
Second day:
      413/25–36
S   406/22–30 even

Maximum
      412/1–37 odd, 38–44
S 406/38–44 even

Additional A Exercises

Simplify.

1. $\dfrac{3x}{5} + \dfrac{3x}{10} \quad \dfrac{9x}{10}$

2. $\dfrac{7a}{5} + \dfrac{5}{3} \quad \dfrac{21a + 25}{15}$

3. $\dfrac{3}{a} - \dfrac{2}{b} \quad \dfrac{3b - 2a}{ab}$

4. $\dfrac{3}{f} + \dfrac{1}{2f} \quad \dfrac{7}{2f}$

5. $\dfrac{m - n}{3m} + \dfrac{m + n}{2n}$
   $\dfrac{3m^2 + 5mn - 2n^2}{6mn}$

6. $\dfrac{r}{r - 1} - \dfrac{1}{2r - 2} \quad \dfrac{2r - 1}{2r - 2}$

Mixed Review

Simplify.

1. $3(x + y) - 2(2x - 3y)$
   $-x + 9y$

2. $4(a - b)^2 - a(4a - 3b)$
   $4b^2 - 5ab$

3. $\dfrac{5c + 3}{3c - 2} - \dfrac{2c + 4}{3c - 2} \quad \dfrac{3c - 1}{3c - 2}$

4. $\dfrac{6m - 5}{2m - 3} - \dfrac{2m - 6}{2m - 3} \quad \dfrac{4m + 1}{2m - 3}$

Factor completely.

5. $3a^2 - 27$
   $3(a + 3)(a - 3)$

6. $12x^3 + 2x^2 - 2x$
   $2x(3x - 1)(2x + 1)$

Note that, in adding and subtracting rational expressions, *any* common multiple of the denominators may be used as a common denominator. However, it is often most convenient to use the *least* common denominator.

Oral Exercises

Name the LCM of each of the following.

1. 16, 24, 30 240

2. 10, 44, 56 3080

3. $18x^2y, 15x^3y^3$ $90x^3y^3$

4. $10c, 25c^2d$ $50c^2d$

5. $4(x - 3)^2, (x + 3)^2(x - 3)$
   $4(x - 3)^2(x + 3)^2$

6. $5a^2(a - 2)^3, 6(a - 2)^2$
   $30a^2(a - 2)^3$

Name the LCD of the fractions in each expression.

7. $\dfrac{1}{4} + \dfrac{1}{6}$ 12

8. $\dfrac{3}{4a} - \dfrac{5}{2ab}$ 4ab

9. $\dfrac{a}{b^2c} + \dfrac{b}{a^2c}$ $a^2b^2c^2$

10. $3 + \dfrac{1}{a}$ a

11. $x + \dfrac{1}{x}$ x

12. $\dfrac{2}{x + 1} + \dfrac{3}{x - 1}$
    $(x + 1)(x - 1)$

13. $\dfrac{t}{2(t - 1)} - \dfrac{3t}{1 - t}$ 2(t − 1)

14. $\dfrac{1}{x} + \dfrac{1}{y} - \dfrac{1}{z}$
    xyz

15. $\dfrac{3}{(x + 1)(x - 4)} + \dfrac{5}{3(x - 4)}$ 3(x + 1)(x − 4)

16. $\dfrac{5s}{6(s - 1)^2(s + 2)^3} + \dfrac{5s}{8(s + 2)^5}$
    $6(s - 1)^2(s + 2)^5$

17. $\dfrac{3}{x^2y} - \dfrac{1}{y^2z} + \dfrac{5}{xyz}$ x^2y^2z

18. $4 - \dfrac{1}{x(x + 1)} + \dfrac{1}{x^2 - 1}$ x(x + 1)(x − 1)

Replace each ___?___ with the factor(s) that will make the fractions equivalent.

19. $\dfrac{2}{3a} = \dfrac{?}{15a^2b^2}$ 10ab²

20. $\dfrac{4}{c - 3} = \dfrac{?}{c(c - 3)}$ 4c

21. $\dfrac{3d}{d + 1} = \dfrac{?}{(d + 1)(d - 2)}$ 3d(d − 2)

22. $\dfrac{4a + 1}{(a - 1)^2} = \dfrac{?}{(a - 1)^3(a + 1)}$
    (4a + 1)(a − 1)(a + 1)

23. $\dfrac{5(x - 1)}{6(x + 1)^2(x - 2)} = \dfrac{?}{12(x + 1)^3(x - 2)^3}$
    $10(x - 1)(x + 1)(x - 2)^2$

24. $\dfrac{3t}{t^2 - 1} = \dfrac{?}{4(t + 1)^2(t - 1)}$ 12t(t + 1)

Written Exercises

Simplify.

A

1. $\dfrac{5x}{2} + \dfrac{5x}{6} \quad \dfrac{10x}{3}$

2. $\dfrac{5p}{7} - \dfrac{3}{11} \quad \dfrac{55p - 21}{77}$

3. $\dfrac{5}{c} + \dfrac{1}{d} \quad \dfrac{5d + c}{cd}$

4. $\dfrac{3e}{4} - \dfrac{5}{e} \quad \dfrac{3e^2 - 20}{4e}$

5. $3 + \dfrac{1}{t} \quad \dfrac{3t + 1}{t}$

6. $5s - \dfrac{1}{5s} \quad \dfrac{25s^2 - 1}{5s}$

7. $\dfrac{1}{2h} + \dfrac{1}{h^2} \quad \dfrac{h + 2}{2h^2}$

8. $\dfrac{3}{fg} - \dfrac{2}{f^2g^2} \quad \dfrac{3fg - 2}{f^2g^2}$

9. $\dfrac{u - v}{u} + \dfrac{u + v}{v} \quad \dfrac{u^2 + 2uv - v^2}{uv}$

10. $\dfrac{r + s}{2r} - \dfrac{r + s}{2s} \quad \dfrac{s^2 - r^2}{2rs}$

11. $\dfrac{4}{6 - 3a} - \dfrac{1}{a - 2} \quad \dfrac{7}{-3(a - 2)}$

12. $\dfrac{3}{2t - 6} + \dfrac{4}{3t - 9} \quad \dfrac{17}{6(t - 3)}$

13. $\dfrac{t}{t-3} + \dfrac{t-1}{t+3}$ $\dfrac{2t^2-t+3}{(t+3)(t-3)}$

14. $\dfrac{3b}{2b-4} - \dfrac{b+1}{b-2}$ $\dfrac{1}{2}$

15. $b - \dfrac{3}{b-2}$

16. $3x - \dfrac{1}{1-3x}$ $\dfrac{9x^2-3x+1}{3x-1}$

17. $\dfrac{3}{a(a-3)} - \dfrac{5}{a-3}$ $\dfrac{3-5a}{a(a-3)}$

18. $\dfrac{5t}{t-1} + \dfrac{3}{t(t-1)}$

19. $\dfrac{h}{g-3} - \dfrac{2h}{3-g}$ $\dfrac{3h}{g-3}$

20. $\dfrac{2}{a-b} + \dfrac{3}{2(b-a)}$ $\dfrac{1}{2(a-b)}$

21. $\dfrac{3}{w+2} - \dfrac{3w}{w^2-4}$

22. $\dfrac{2}{9q-6} + \dfrac{1+q}{6q^2-4q}$ $\dfrac{7q+3}{6q(3q-2)}$

23. $\dfrac{4}{3x-6} - \dfrac{x+1}{2x^2-4x}$ $\dfrac{5x-3}{6x(x-2)}$

24. $\dfrac{x-1}{x^2+5x} - \dfrac{x+1}{x^2-25}$

B 25. $\dfrac{5m}{m^2-2mn+n^2} - \dfrac{3}{m-n}$ $\dfrac{2m+3n}{(m-n)^2}$

26. $\dfrac{1}{u^2+u-6} + \dfrac{1}{u^2-9}$ $\dfrac{2u-5}{(u+3)(u-3)(u-2)}$

27. $\dfrac{b}{b^2-b-20} - \dfrac{b}{b^2+9b+20}$

28. $\dfrac{p+1}{p^2+4p+3} + \dfrac{1}{p^2-2p-15}$

29. $c - \dfrac{25c-5c^2}{c^2-10c+25}$

30. $\dfrac{2d-12}{d^2-5d-6} + \dfrac{1}{d-1}$

31. $\dfrac{a+5}{a^2-a-6} + \dfrac{a+3}{a^2+7a+10}$

32. $\dfrac{a+3b}{a^2-7ab+12b^2} - \dfrac{a-3b}{a^2-ab-12b^2}$

33. $\dfrac{p+1}{p} - \dfrac{p+q}{q} + \dfrac{q-1}{p}$ $\dfrac{q^2-p^2}{pq}$

34. $\dfrac{2}{3+c} - \dfrac{5}{3-c} - \dfrac{2c+1}{9-c^2}$ $\dfrac{9c+10}{(c+3)(c-3)}$

35. $s - \dfrac{s-1}{s-6} - \dfrac{s^2-6}{s^2-36}$ $\dfrac{s^3-2s^2-41s+12}{(s+6)(s-6)}$

36. $\dfrac{3}{e+2} - e + \dfrac{3e}{e^2-4}$ $\dfrac{-e^3+10e-6}{(e+2)(e-2)}$

C 37. $\left(\dfrac{1}{s}+\dfrac{1}{t}\right) \div \left(\dfrac{1}{s}-\dfrac{1}{t}\right)$ $\dfrac{t+s}{t-s}$

38. $\left(1+\dfrac{3}{x-2}\right)\left(2-\dfrac{6}{x+1}\right)$ 2

39. $\left(1+\dfrac{ab}{a^2-ab+b^2}\right)\left(1+\dfrac{ab}{a^2+b^2}\right)$

40. $\left(x-4-\dfrac{21}{x}\right) \div \left(x-8+\dfrac{7}{x}\right)$

41. $\left(\dfrac{1}{a+b}-\dfrac{1}{a-b}\right)^{-1}\left(1-\dfrac{a^2-3b^2}{2a^2-ab-3b^2}\right)$

42. $\left(\dfrac{1}{y^2-4}+\dfrac{1}{y^2+2y}\right)^{-1}\left(\dfrac{1}{y}-\dfrac{y-2}{y^2+y-2}\right)$

43. **a.** Is the sum of two rational expressions always a rational expression? Yes

    b. Is the difference of two rational expressions always a rational expression? Yes

44. Explain why the set of rational numbers is closed with respect to addition and subtraction.

Computer Exercises   For students with computer experience

1. Write a program that will compute the sum of two fractions with equal denominators. The output should be in the form of a fraction, not a decimal, and should be in simplest form.

2. Modify the program that you wrote for Exercise 1 so that it will compute the sum of *any* two fractions that you input, whether or not their denominators are equal.

PROGRAMMING IN BASIC

Given the program that follows, it is possible to use a computer to find the coefficients of the product of two polynomials in a single variable. The program uses an array of *subscripted variables* to store the coefficients of the polynomials and to execute the necessary multiplications and additions. Specifically, A(I) represents the coefficients of the first polynomial being multiplied, B(I) the coefficients of the second, and P(I) the coefficients of the product.

For example, to multiply the polynomials $3x + 1$ and $5x + 4$, you input the degree of the polynomials as 1 and their coefficients as:

$$A(1) = 3 \qquad B(1) = 5$$
$$A(2) = 1 \qquad B(2) = 4$$

$$
\begin{array}{r}
(3)x + (1) \\
(5)x + (4) \\
\hline
(5)(3)x^2 + (5)(1)x \\
(4)(3)x + (4)(1) \\
\hline
15x^2 + \quad 17x + \quad 4
\end{array}
$$

The coefficients of the product are then output as:

$$P(3) = 15 \qquad P(2) = 17 \qquad P(1) = 4$$

To understand how the program computes these coefficients, it may be helpful to study the multiplication as shown at the right above.

In using the program, notice that each of the polynomials being multiplied is considered as having the same degree, *n*, and the same number of terms, $n + 1$. Thus, coefficients of "missing terms" must be input as zero. Also note that, if the degree of the polynomials being multiplied is greater than 4, a DIMension statement may be needed.

```
10    PRINT "TO MULTIPLY TWO"
20    PRINT "POLYNOMIALS IN X:"
30    PRINT
40    PRINT "INPUT DEGREE N (N <= 4):";
50    INPUT N
60    REM *INPUT COEFFICIENTS
70    FOR I = N + 1 TO 1 STEP −1
80    PRINT "INPUT A(";I;"), B(";I;"):";
90    INPUT A(I), B(I)
100   NEXT I
110   REM *SET P(I) = 0
120   FOR I = 1 TO 2*N + 1
130   LET P(I) = 0
140   NEXT I
150   REM *COMPUTE COEFFICIENTS
160   REM *OF THE PRODUCT
170   FOR J = N + 1 TO 1 STEP −1
180   FOR I = N + 1 TO 1 STEP −1
190   LET M = I + J − 1
200   LET P(M) = P(M) + A(I) * B(J)
```

```
210   NEXT I
220   NEXT J
230   PRINT "COEFFICIENTS OF PRODUCT:"
240   FOR I = 2*N + 1 TO 1 STEP −1
250   PRINT P(I);" ";
260   NEXT I
270   END
```

Exercises

Type in the program as given. Then RUN it to find the product of each of the following pairs of polynomials.

1. $3x^2 + 2x + 1$
$3x^2 + 2x + 1$

2. $3x^3 - 4x^2 + 5x - 2$
$2x^3 + x^2 - x + 3$

3. $6x^4 + 5x^3 - 2x^2 + x - 1$
$x^4 + 3x^3 + x^2 - 3x + 1$

4. $x^3 - 2x^2 + 0x + 3$
$0x^3 + 0x^2 + x + 3$

5. $2x^2 + 3x - 1$
$x - 2$

6. $x^3 + x^2 + x + 1$
$x - 1$

COEFFICIENTS OF PRODUCT:
1. 9 12 10 4 1   **2.** 6 −5 3 14 −19 17 −6
3. 6 23 19 −18 −9 9 −6 4 −1   **4.** 0 0 1 1 −6 3 9
5. 0 2 −1 −7 2   **6.** 0 0 1 0 0 0 −1

8–9 Complex Fractions

If a fraction has a numerator or denominator that contains a fraction or a term with a negative exponent, the fraction is called a **complex fraction**. Complex fractions may be completely numerical expressions or may involve variables. The following expressions are examples of complex fractions.

$$\frac{\frac{1}{2} - \frac{1}{3}}{\frac{3}{5}}, \qquad \frac{\frac{a}{b^2}}{\frac{b}{c^2}}, \qquad \frac{5}{2 + \frac{1}{a}}, \qquad \frac{d^{-2} + 1}{d - 1}$$

Any complex fraction can be transformed into an expression that is free of negative exponents and fractions in the numerator and denominator. Such an expression is called a **simple fraction**. The complex fractions in the examples just given can be transformed, respectively, into the following simple fractions. (See Oral Exercises 6, 8, 9, and 11 on page 417.)

$$\frac{5}{18}, \qquad \frac{ac^2}{b^3}, \qquad \frac{5a}{2a + 1}, \qquad \frac{d^2 + 1}{d^3 - d^2}$$

Throughout this section the variables will be restricted to values for which each expression is defined. You can therefore assume that the complex fraction and its simplified form are equivalent expressions. Two common methods used to simplify complex fractions are illustrated in Example 1 on the following page.

Polynomials and Rational Expressions   **415**

Teaching Suggestions
p. T99

Related Activities p. T100

Supplementary Material
Test 32

Key Ideas
Simplify complex fractions.

Chalkboard Examples

1. Simplify

$$\dfrac{\dfrac{2x}{3y} - \dfrac{y}{2x}}{\dfrac{2}{x} + \dfrac{3}{y}}.$$

(Method 1)
First, simplify the numerator and denominator separately.

$$\dfrac{\dfrac{4x^2 - 3y^2}{6xy}}{\dfrac{2y + 3x}{xy}}$$

Then, rewrite the fraction as a quotient using the ÷ sign.

$$= \dfrac{4x^2 - 3y^2}{6xy} \div \dfrac{2y + 3x}{xy}$$

$$= \dfrac{4x^2 - 3y^2}{6xy} \cdot \dfrac{xy}{2y + 3x}$$

$$= \dfrac{4x^2 - 3y^2}{6(2y + 3x)}$$

(Method 2) Multiply the numerator and denominator by the LCD of all the

EXAMPLE 1 Simplify $\dfrac{\dfrac{r}{2s} - \dfrac{s}{2r}}{\dfrac{1}{r} + \dfrac{1}{s}}$.

SOLUTION (Method 1)

$$\dfrac{\dfrac{r}{2s} - \dfrac{s}{2r}}{\dfrac{1}{r} + \dfrac{1}{s}} = \dfrac{\dfrac{r^2 - s^2}{2rs}}{\dfrac{s + r}{rs}} \qquad \longleftarrow \left\{ \begin{array}{l}\text{Simplify the numerator and} \\ \text{denominator separately.}\end{array}\right.$$

$$= \dfrac{r^2 - s^2}{2rs} \div \dfrac{s + r}{rs} \qquad \longleftarrow \left\{ \begin{array}{l}\text{Rewrite the fraction as a} \\ \text{quotient using the ÷ sign.}\end{array}\right.$$

$$= \dfrac{r^2 - s^2}{2rs} \cdot \dfrac{rs}{s + r}$$

$$= \dfrac{(r + s)(r - s)}{2rs} \cdot \dfrac{rs}{s + r}$$

$$= \dfrac{r - s}{2}$$

(Method 2)

$$\dfrac{\dfrac{r}{2s} - \dfrac{s}{2r}}{\dfrac{1}{r} + \dfrac{1}{s}} = \dfrac{\left(\dfrac{r}{2s} - \dfrac{s}{2r}\right) \cdot 2rs}{\left(\dfrac{1}{r} + \dfrac{1}{s}\right) \cdot 2rs} \qquad \longleftarrow \left\{ \begin{array}{l}\text{Multiply the numerator and} \\ \text{denominator by the LCD of all} \\ \text{the simple fractions contained} \\ \text{in the complex fraction.}\end{array}\right.$$

$$= \dfrac{r^2 - s^2}{2s + 2r}$$

$$= \dfrac{(r + s)(r - s)}{2(s + r)}$$

$$= \dfrac{r - s}{2}$$

If the complex fraction contains terms with negative exponents, first rewrite the expression using positive exponents only.

EXAMPLE 2 Simplify $\dfrac{d^{-2} + 1}{d - 1}$.

SOLUTION

$$\dfrac{d^{-2} + 1}{d - 1} = \dfrac{\left(\dfrac{1}{d^2} + 1\right)}{(d - 1)}$$

$$= \dfrac{\left(\dfrac{1}{d^2} + 1\right) \cdot d^2}{(d - 1) \cdot d^2}$$

$$= \dfrac{1 + d^2}{d^3 - d^2}, \text{ or } \dfrac{d^2 + 1}{d^3 - d^2}$$

If the numerator or denominator of a complex fraction is itself a complex fraction, simplify the fraction in steps.

EXAMPLE 3 Simplify $\dfrac{\dfrac{5}{1-x}}{1+\dfrac{1}{1+\dfrac{1}{x}}}$.

SOLUTION First, simplify $\dfrac{1}{1+\dfrac{1}{x}}$: $\dfrac{1}{1+\dfrac{1}{x}} = \dfrac{1\cdot x}{\left(1+\dfrac{1}{x}\right)\cdot x} = \dfrac{x}{x+1}$

Then:
$$\frac{\dfrac{5}{1-x}}{1+\dfrac{1}{1+\dfrac{1}{x}}} = \frac{\left(\dfrac{5}{1-x}\right)\cdot(1-x)(1+x)}{\left(1+\dfrac{x}{x+1}\right)\cdot(1-x)(1+x)}$$

$$= \frac{5(1+x)}{(1-x)(1+x)+x(1-x)}$$

$$= \frac{5+5x}{1-x^2+x-x^2}$$

$$= \frac{5+5x}{1+x-2x^2}$$

Oral Exercises

a. Use Method 1 to simplify each complex fraction.
b. Use Method 2 to simplify each complex fraction.   9. $\frac{5a}{2a+1}$ 12. $\frac{x^2+1}{x^2(x^2-1)}$

1. $\dfrac{\frac{5}{9}}{\frac{7}{9}}$ $\frac{5}{7}$

2. $\dfrac{\frac{4}{7}}{\frac{2}{7}}$ 2

3. $\dfrac{\frac{5}{6}}{\frac{3}{2}}$ $\frac{5}{9}$

4. $\dfrac{\frac{15}{8}}{\frac{25}{12}}$ $\frac{9}{10}$

5. $\dfrac{1+\frac{5}{16}}{1-\frac{3}{4}}$ $\frac{21}{4}$

6. $\dfrac{\frac{1}{2}-\frac{1}{3}}{\frac{3}{5}}$ $\frac{5}{18}$

7. $\dfrac{\frac{a}{b^2}}{\frac{a}{b^3}}$ b

8. $\dfrac{\frac{a}{b^2}}{\frac{b}{c^2}}$ $\frac{ac^2}{b^3}$

9. $\dfrac{5}{2+\frac{1}{a}}$

10. $\dfrac{1+\frac{1}{a}}{3+\frac{1}{a}}$ $\frac{a+1}{3a+1}$

11. $\dfrac{d^{-2}+1}{d-1}$ $\frac{d^2+1}{d^3-d^2}$

12. $\dfrac{x^{-3}+\frac{1}{x}}{x-\frac{1}{x}}$

simple fractions contained in the complex fraction.

$$\frac{\dfrac{2x}{3y}-\dfrac{y}{2x}}{\dfrac{2}{x}+\dfrac{3}{y}}$$

$$=\frac{\left(\dfrac{2x}{3y}-\dfrac{y}{2x}\right)\cdot 6xy}{\left(\dfrac{2}{x}+\dfrac{3}{y}\right)\cdot 6xy}$$

$$=\frac{4x^2-3y^2}{12y+18x}=\frac{4x^2-3y^2}{6(2y+3x)}$$

2. Simplify $\dfrac{2a^{-3}+1}{3a^2-4}$.

First, rewrite the expression using only positive exponents.

$$\frac{\dfrac{2}{a^3}+1}{3a^2-4}$$

$$\frac{\left(\dfrac{2}{a^3}+1\right)\cdot a^3}{(3a^2-4)\cdot a^3}=\frac{2+a^3}{3a^5-4a^3}$$

3. Simplify $\dfrac{\dfrac{3}{2-3t}}{2+\dfrac{1}{2+\dfrac{1}{2t}}}$.

First, simplify $\dfrac{1}{2+\dfrac{1}{2t}}$.

$$\frac{1\cdot 2t}{\left(2+\dfrac{1}{2t}\right)\cdot 2t}=\frac{2t}{4t+1}$$

Then,

$$\frac{\dfrac{3}{2-3t}}{2+\dfrac{2t}{4t+1}}=\frac{\dfrac{3}{2-3t}}{\dfrac{8t+2+2t}{4t+1}}$$

$$=\frac{3}{2-3t}\div\frac{10t+2}{4t+1}$$

$$=\frac{3}{2-3t}\cdot\frac{4t+1}{10t+2}$$

$$=\frac{12t+3}{4+14t-30t^2}$$

Suggested Assignments

Minimum
First day: 418/1–8
Second day: 418/9–18
         R 419/Self-Test 3
Average
First day: 418/1–18
Second day: 418/19–28
         R 419/Self-Test 3
Maximum
First day: 418/1–22
Second day: 418/23–34
         R 419/Self-Test 3

Additional Answers
Written Exercises
(See p. 420)

Additional A Exercises

Simplify.

1. $\dfrac{\frac{2}{3}}{\frac{4}{5}}$ $\frac{5}{6}$

2. $\dfrac{\frac{3}{5}}{\frac{7}{25}}$ $\frac{15}{7}$

3. $\dfrac{\frac{a}{4}}{\frac{5a^2}{12}}$ $\frac{3}{5a}$

4. $\dfrac{\frac{a}{2} - \frac{b}{3}}{\frac{a}{3} + \frac{b}{2}}$ $\frac{3a - 2b}{2a + 3b}$

5. $\dfrac{3 + \frac{1}{k}}{1 - \frac{3}{k^2}}$ $\frac{3k^2 + k}{k^2 - 3}$

6. $\dfrac{4 - e^{-2}}{4 + e^{-2}}$ $\frac{4e^2 - 1}{4e^2 + 1}$

Written Exercises

Simplify.

A

1. $\dfrac{\frac{4}{5}}{\frac{8}{15}}$ $\frac{3}{2}$

2. $\dfrac{\frac{7}{9}}{\frac{15}{14}}$ $\frac{98}{135}$

3. $\dfrac{\frac{k}{3}}{\frac{2k}{9}}$ $\frac{3}{2}$

4. $\dfrac{\frac{2d}{5}}{\frac{3d}{10}}$ $\frac{4}{3}$

5. $\dfrac{\frac{c^2}{d}}{\frac{c}{d^3}}$ cd^2

6. $\dfrac{\frac{4r}{3}}{\frac{5r^2}{6}}$ $\frac{8}{5r}$

7. $\dfrac{\frac{e}{3} + \frac{f}{4}}{\frac{e}{2} - \frac{f}{3}}$ $\frac{4e + 3f}{6e - 4f}$

8. $\dfrac{\frac{2}{a} - \frac{1}{b}}{\frac{1}{a} - \frac{3}{b}}$

9. $\dfrac{\frac{5w}{6} - \frac{3z}{2}}{\frac{w}{3} + \frac{z}{6}}$ $\frac{5w - 9z}{2w + z}$

10. $\dfrac{\frac{2c}{5} + \frac{3d}{10}}{\frac{3c}{2} - \frac{7d}{10}}$ $\frac{4c + 3d}{15c - 7d}$

11. $\dfrac{\frac{u}{3v} - \frac{v}{3u}}{\frac{1}{3v} + \frac{1}{3u}}$ $u - v$

12. $\dfrac{1 - \frac{25}{k^2}}{1 - \frac{5}{k}}$

8. $\dfrac{2b - a}{b - 3a}$

12. $\dfrac{k + 5}{k}$

15. $\dfrac{x(x + 2y)}{y(y - 2x)}$

13. $\dfrac{d^{-1} + 3}{1 - d^{-1}}$ $\frac{3d + 1}{d - 1}$

14. $\dfrac{t^{-1}}{t^{-1} - (3t)^{-1}}$ $\frac{3}{2}$

15. $\dfrac{xy^{-1} + 2}{yx^{-1} - 2}$

16. $\dfrac{5 - r(2s)^{-1}}{4 + r(3s)^{-1}}$ $\frac{3(10s - r)}{2(12s + r)}$

17. $\dfrac{9 - c^{-2}}{3c^{-1} - c^{-2}}$ $3c + 1$

18. $\dfrac{2j^{-1} + 5j^{-2}}{4 - 25j^{-2}}$ $\frac{1}{2j - 5}$

B

19. $\dfrac{\frac{1}{x} + \frac{3}{x^2}}{1 + \frac{1}{x} - \frac{6}{x^2}}$ $\frac{1}{x - 2}$

20. $\dfrac{1 - \frac{4}{a^2}}{1 - \frac{5}{a} + \frac{6}{a^2}}$ $\frac{a + 2}{a - 3}$

21. $(r^{-1} - s^{-1})^{-1}$ $\frac{rs}{s - r}$

22. $(x^{-1} + y^{-1})^{-2}$ $\frac{x^2y^2}{x^2 + 2xy + y^2}$

23. $\dfrac{1}{(u^{-2} + v^{-2})^{-1}}$ $\frac{u^2 + v^2}{u^2v^2}$

24. $\dfrac{1}{(c^{-1} - d^{-1})^{-2}}$ $\frac{(c - d)^2}{c^2d^2}$

25. $\dfrac{\frac{c}{c - d} - \frac{d}{c + d}}{\frac{cd}{c^2 - d^2}}$ $\frac{c^2 + d^2}{cd}$

26. $\dfrac{\frac{r}{r - 2} - \frac{2}{r + 2} - \frac{8}{r^2 - 4}}{\frac{1}{r - 2}}$ $r - 2$

27. $[(e^2 - 1)^{-1} + 1][(e - 1)^{-1} + 1]^{-1}$

28. $[a(a - 1)^{-1} + 1][3(a^2 - 1)^{-1} + 4]^{-1}$ $\frac{a + 1}{2a + 1}$

C

29. $1 - \dfrac{1}{1 - \frac{1}{z - 2} - \frac{1}{z - 3}}$ $\frac{e}{e + 1}$

30. $2 + \dfrac{1}{1 + \dfrac{2}{m + \frac{1}{m}}}$ $\frac{3m^2 + 4m + 3}{m^2 + 2m + 1}$

31. $\dfrac{1}{1 - \dfrac{1}{2 - \frac{1}{3 - t}}}$ $\frac{5 - 2t}{2 - t}$

32. $1 - \dfrac{1}{1 - \dfrac{1}{1 - \frac{1}{1 - n}}}$ $1 - n$

33. If $r = \dfrac{1 - s}{1 + s}$ and $s = \dfrac{1 + t}{1 - t}$, show that $r + t = 0$.

34. Sketch the graphs of $f(x) = \dfrac{1}{2 + \frac{1}{x}}$ and $g(x) = \dfrac{x}{2x + 1}$. State how the two graphs are alike and how they are different.

Self-Test 3

VOCABULARY   least common denominator   complex fraction (p. 415)
(LCD) (p. 410)   simple fraction (p. 415)
least common multiple (LCM)
of polynomials (p. 410)

Simplify. Assume that no variable has a value that results in division by zero.

1. $\dfrac{3x+3}{2x} \cdot \dfrac{x-1}{x+1}$ $\dfrac{3(x-1)}{2x}$ 2. $\dfrac{c^2-16}{3c} \div \dfrac{c+4}{2c}$ $\dfrac{2(c-4)}{3}$ *Obj. 1, p. 403*

3. $\dfrac{5d-2}{4d^2+8d-5} - \dfrac{3d-7}{4d^2+8d-5}$ $\dfrac{1}{2d-1}$ 4. $\dfrac{3}{2e} + \dfrac{2}{3e^3}$ $\dfrac{9e^2+4}{6e^3}$ *Obj. 2, p. 403*

5. $\dfrac{3b+\dfrac{2}{b}}{b+\dfrac{1}{2}}$ $\dfrac{6b^2+4}{2b^2+b}$ 6. $\dfrac{a^{-2}+1}{1-a^{-1}}$ $\dfrac{a^2+1}{a^2-a}$ *Obj. 3, p. 403*

Check your answers with those at the back of the book.

Chapter Summary

1. The *basic property of quotients* asserts that, for all real numbers r and s and all nonzero real numbers t and u, $\dfrac{rs}{tu} = \dfrac{r}{t} \cdot \dfrac{s}{u}$.

2. If a is a nonzero real number and m and n are positive integers, the following *laws of exponents* for division are true:

   1. If $m = n$, then $\dfrac{a^m}{a^n} = 1$.

   2. If $m > n$, then $\dfrac{a^m}{a^n} = a^{m-n}$.

   3. If $m < n$, then $\dfrac{a^m}{a^n} = \dfrac{1}{a^{n-m}}$.

3. One polynomial can be divided by another using a division algorithm similar to the long division algorithm used for integers.

4. A *rational expression* is any expression that can be written as the quotient of two polynomials, provided that the denominator is not zero. Rational expressions can be added, subtracted, multiplied, or divided using methods similar to the methods used in numerical computations.

Polynomials and Rational Expressions   **419**

Mixed Review

Simplify. Express your result using only positive exponents. Assume that no variable has zero as a value.

1. $2t^{-1} + 3t^{-2}$ $\dfrac{2t+3}{t^2}$

2. $(4a-3)^{-1} - 1$ $\dfrac{4-4a}{4a-3}$

Simplify.

3. $t + \dfrac{3}{t-3}$ $\dfrac{t^2-3t+3}{t-3}$

4. $4x - \dfrac{7}{5-3x}$

$\dfrac{-12x^2+20x-7}{5-3x}$

Quick Quiz

Simplify. Assume that no variable has a value that results in division by zero.

1. $\dfrac{2x-1}{3x} \cdot \dfrac{x+2}{x-2}$

$\dfrac{2x^2+3x-2}{3x^2-6x}$

2. $\dfrac{a^2-9}{5} \div \dfrac{2a+6}{10}$ $a-3$

3. $\dfrac{3f-5}{6f^2-2f-4} -$

$\dfrac{f-3}{6f^2-2f-4}$ $\dfrac{1}{3f+2}$

4. $\dfrac{2}{3w} - \dfrac{w+1}{4w^2}$ $\dfrac{5w-3}{12w^2}$

5. $\dfrac{2x+\dfrac{5}{y}}{2x+\dfrac{1}{2}}$ $\dfrac{4xy+10}{4xy+y}$

6. $\dfrac{x^{-1}+1}{1-\dfrac{1}{x^2}}$ $\dfrac{x}{x-1}$

33. $r = \dfrac{1 - \dfrac{1 + t}{1 - t}}{1 + \dfrac{1 + t}{1 - t}} =$

$\dfrac{1 - t - (1 + t)}{1 - t + (1 + t)} = \dfrac{-2t}{2} = -t;$

$\therefore r + t = -t + t = 0$

34.

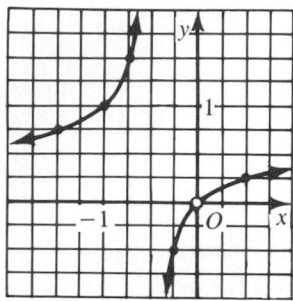

$$f(x) = \dfrac{1}{2 + \dfrac{1}{x}}$$

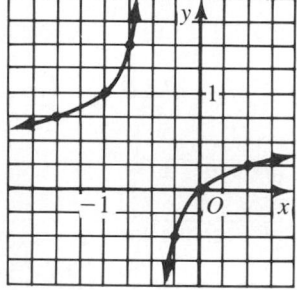

$$g(x) = \dfrac{x}{2x + 1}$$

The two graphs are alike in that neither graph contains a point whose x-coordinate is $-\dfrac{1}{2}$. They are different in that the graph of $f(x)$ also contains no point whose x-coordinate is 0.

5. Any fraction whose numerator and denominator are integers or polynomials can be simplified by dividing the numerator and denominator by their greatest common factor (GCF).

6. For every nonzero real number a, $a^0 = 1$.

7. For every nonzero real number a and every integer n, $a^{-n} = \dfrac{1}{a^n}$.

8. A number is said to be written in *scientific notation* if it is expressed in the form $k \times 10^n$, where $1 \le k < 10$ and n is an integer.

9. Let a and b denote real numbers and let n denote an integer.

1. If $b \ne 0$ and a and n are not both 0, then $\left(\dfrac{a}{b}\right)^n = \dfrac{a^n}{b^n}$.

2. If $a \ne 0$ and $b \ne 0$, then $\left(\dfrac{a}{b}\right)^{-n} = \dfrac{a^{-n}}{b^{-n}} = \dfrac{b^n}{a^n}$.

3. If $a \ne 0$ and $b \ne 0$, then $\left(\dfrac{a}{b}\right)^0 = \dfrac{a^0}{b^0} = 1$.

10. A *complex fraction* has a numerator or denominator that contains a fraction or a term with a negative exponent. Any complex fraction can be transformed into a *simple fraction* that is free of negative exponents and fractions in its numerator and denominator.

Chapter Review

Write the letter of the correct answer.

1. Simplify $\dfrac{26a^3b^7}{39ab^9}$ using only positive exponents.

   a. $\dfrac{2a^3}{3b^2}$ **(b.)** $\dfrac{2a^2}{3b^2}$ **c.** $\dfrac{2a^2b^2}{3}$ **d.** $\dfrac{13a^2}{b^2}$

   8-1

2. Simplify $\dfrac{(-2ax)^3(-3a^2x^3)^2}{(-9a^3x^4)(4ax^2)}$ using only positive exponents.

   (a.) $2a^3x^3$ **b.** a^3x^3 **c.** $\dfrac{a^3x^2}{2}$ **d.** $-\dfrac{a^3x^2}{2}$

3. Express $\dfrac{28u^7 - 16u^5 + 20u^2}{-4u^5}$ as a sum.

   8-2

   a. $-7u^2 - 4 + 5u^3$ **b.** $-7u^2 + 4 - 5u^3$

   (c.) $-7u^2 + 4 - \dfrac{5}{u^3}$ **d.** $-7u^2 - 4 + \dfrac{5}{u^3}$

4. Divide $x^2 + 6x + 5$ by $x + 1$, assuming $x + 1 \ne 0$.

   8-3

   (a.) $x + 5$ **b.** $x + 5 + \dfrac{10}{x + 1}$

   c. $x - 5$ **d.** $x + 5 - \dfrac{10}{x + 1}$

420   *Chapter 8*

5. Simplify $\dfrac{5x + 10}{15(x + 2)^2}$. 8-4

  a. $\dfrac{x}{(x + 2)^2}$ **b.** $\dfrac{3}{x + 2}$ **c.** $\dfrac{1}{3(x + 2)}$ **d.** $\dfrac{x + 10}{3(x + 2)^2}$

6. Simplify $\dfrac{3f^2 + 6f - 45}{27 - 3f^2}$.

  a. $\dfrac{f + 5}{f - 3}$ **b.** $-\dfrac{f + 5}{f + 3}$ **c.** $\dfrac{3f + 5}{f + 9}$ **d.** $-\dfrac{f - 5}{9 - 3f}$

7. Simplify $\dfrac{16a^{-2}}{8a^{-3}}$ using only positive exponents. 8-5

  a. $2a$ **b.** $\dfrac{2}{a^5}$ **c.** $\dfrac{a}{2}$ **d.** $\dfrac{2}{a}$

8. Simplify $(x^0 y^{-2})^{-1}$ using only positive exponents.

  a. y^2 **b.** $\dfrac{y^2}{x}$ **c.** $\dfrac{1}{y^2}$ **d.** $\dfrac{1}{xy^3}$

9. Express 0.000437 in scientific notation.
  a. 437×10^6 **b.** 437×10^{-6} **c.** 4.37×10^{-4} **d.** 0.437×10^3

10. Simplify $\dfrac{v^2 - 6v + 8}{v^2 + 4v} \cdot \dfrac{9v}{6 - 3v}$. 8-6

  a. -3 **b.** $3v^2 - 48$ **c.** $\dfrac{3(v - 4)}{v + 4}$ **d.** $\dfrac{3(4 - v)}{v + 4}$

11. Simplify $\dfrac{4m^2n^3}{3xw^2} \div \dfrac{20m^4n^2}{9x^3w}$, assuming m, n, x, and $w \neq 0$.

  a. $\dfrac{3x^2}{5w}$ **b.** $\dfrac{3nx^2}{5m^2w}$ **c.** $\dfrac{3m^2w}{5nx^2}$ **d.** $\dfrac{100m^6n^5}{27x^4w^3}$

12. Simplify $\dfrac{7m}{2} - \dfrac{5m - 1}{2}$. 8-7

  a. $\dfrac{2m - 1}{2}$ **b.** $\dfrac{2m + 1}{4}$ **c.** $\dfrac{2m + 1}{2}$ **d.** $\dfrac{m + 1}{2}$

13. Simplify $\dfrac{c - 6}{c^2 - 3c - 10} + \dfrac{1}{c^2 - 3c - 10}$.

  a. $c + 2$ **b.** $c - 2$ **c.** $\dfrac{1}{c - 2}$ **d.** $\dfrac{1}{c + 2}$

14. Simplify $\dfrac{2}{x - 1} - \dfrac{3}{x^2 - 1}$. 8-8

  a. $\dfrac{2x - 3}{x^2 - 1}$ **b.** $\dfrac{2x - 1}{x^2 - 1}$ **c.** $\dfrac{2x + 5}{x^2 - 1}$ **d.** $\dfrac{2}{x + 1}$

15. Simplify $\dfrac{2 - x^2}{x^2 - 4x} + \dfrac{2x - 1}{2x - 8}$.

  a. $-\dfrac{1}{2x}$ **b.** $\dfrac{-2}{x - 4}$ **c.** $\dfrac{1 + 2x - x^2}{2x^2 - 8x}$ **d.** $\dfrac{1 + 2x - x^2}{x^2 - 2x - 8}$

Polynomials and Rational Expressions   **421**

51. Case 1: $n > 0$
 1. a and b are real numbers.
  Hypothesis

 2. $\left(\dfrac{a}{b}\right)^n = \underbrace{\dfrac{a}{b} \cdot \dfrac{a}{b} \cdot \ldots \cdot \dfrac{a}{b}}_{n \text{ factors}}$
  Def. of a power

 3. $= \dfrac{a \cdot a \cdot \ldots \cdot a}{b \cdot b \cdot \ldots \cdot b}$
  Theorem, page 403

 4. $= \dfrac{a^n}{b^n}$
  Def. of a power

 5. $\therefore \left(\dfrac{a}{b}\right)^n = \dfrac{a^n}{b^n}$
  Trans. ax. of $=$

Case 2: $n = 0$
 1. a and b are real numbers; $a \neq 0$
  Hypothesis

 2. $\left(\dfrac{a}{b}\right)^n = 1$
  Def. of zero exponent

 3. $= \dfrac{1}{1}$
  Exercise 45, page 102

 4. $= \dfrac{a^n}{b^n}$
  Def. of zero exponent

 5. $\therefore \left(\dfrac{a}{b}\right)^n = \dfrac{a^n}{b^n}$
  Trans. ax. of $=$

Case 3: $n < 0$
 1. a and b are real numbers.
  Hypothesis

 2. $\left(\dfrac{a}{b}\right)^n = \left(\dfrac{1}{\frac{a}{b}}\right)^{-n}$
  Def. of neg. exponent

 3. $= \left(\dfrac{b}{a}\right)^{-n}$
  Ax. of mult. inverses

---

16. Simplify $\dfrac{\frac{r-s}{r}}{\frac{1}{2} + \frac{s}{2r}}$.

8-9

 a. $\dfrac{r+s}{r-s}$ **b.** $\dfrac{r-s}{r+s}$ **c.** $\dfrac{r+s}{2r-2s}$ **d.** $\dfrac{2r-2s}{r+s}$

Chapter Test

Simplify using only positive exponents.

1. $\dfrac{5x}{-60x^2}$ $-\dfrac{1}{12x}$

2. $\dfrac{-(v^2)^3}{v^4}$ $-v^2$

3. $\dfrac{21a^5bc^3}{0.3ab^2c^2}$ $\dfrac{70a^4c}{b}$

8-1

Express each quotient as a sum.

4. $\dfrac{6a^3 + 24a^5}{6a^4}$ $\dfrac{1}{a} + 4a$

5. $\dfrac{11c^3d - 33cd - 55c^2d^2}{-11cd}$ $-c^2 + 3 + 5cd$

8-2

Divide the first polynomial by the second. Assume that no divisor equals zero.

6. $6y^2 - 7y + 5$; $2y - 3$ $3y + 1 + \dfrac{8}{2y-3}$

7. $t^3 - 5t^2 + 10t - 12$; $t - 3$ $t^2 - 2t + 4$

8-3

Simplify using only positive exponents.

8. $\dfrac{4b^2 - 12b}{24 - 8b}$ $-\dfrac{b}{2}$

9. $\dfrac{2a^2 - 7a - 4}{6a^2 + a - 1}$ $\dfrac{a-4}{3a-1}$

8-4

10. $(-4g^3h^{-2})^0$ 1

11. $\dfrac{(-3)^0 a^4 b^{-1}}{a^{-1}bc^{-2}}$ $\dfrac{a^5c^2}{b^2}$

8-5

12. Express the product $500 \times 50{,}000 \times 0.0005$ in scientific notation. 1.25×10^4

Simplify using only positive exponents. Assume that no variable has a value that results in division by zero.

13. $\dfrac{-24m^3n^2}{14n^4} \cdot \dfrac{21n^3}{4m^5}$ $-\dfrac{9n}{m^2}$

14. $\dfrac{e^2 - 2e + 1}{e^2} \div (e - 1)$ $\dfrac{e-1}{e^2}$

8-6

15. $\dfrac{2x+3}{x-5} - \dfrac{3x-2}{x-5}$ -1

16. $\dfrac{3-x}{x^2-16} + \dfrac{2x-7}{x^2-16}$ $\dfrac{1}{x+4}$

8-7

17. $\dfrac{3}{g+2} + 8$ $\dfrac{g^2 + 2g + 3}{g+2}$

18. $\dfrac{w}{w^2-25} - \dfrac{1}{2w+10}$ $\dfrac{1}{2(w-5)}$

8-8

19. $\dfrac{\frac{m}{4n} - \frac{n}{4n}}{\frac{1}{n} - \frac{1}{m}}$ $\dfrac{m}{4}$

20. $\dfrac{(4x)^{-1} - xy^{-2}}{2x - y}$ $-\dfrac{2x+y}{4xy^2}$

8-9

Mixed Review

Simplify.

1. $(3c - 5)(2c + 3)$ $6c^2 - c - 15$

2. $-d^3 + 4d - 5 - (6d^3 + 6d - 5)$ $-7d^3 - 2d$

3. $(3xy)^2(2x^3y)(-2x)$ $-36x^6y^3$

4. $(4ab)(-3a^2b)^2 - (2ab)^3(5a^2)$ $-4a^5b^3$

5. $(2x - 3)^2$ $4x^2 - 12x + 9$

6. $(x + 3)^3$ $x^3 + 9x^2 + 27x + 27$

7. $(7x - 2y)(7x + 2y)$ $49x^2 - 4y^2$

8. $(r + 2s)(r^2 - 3rs + s^2)$
   $r^3 - r^2s - 5rs^2 + 2s^3$

Factor completely.

9. $32ax^3 - 18ax$ $2ax(4x + 3)(4x - 3)$

10. $3x^2 - 5x - 8$ $(3x - 8)(x + 1)$

11. $x^4 - 2x^3 - 24x^2$ $x^2(x - 6)(x + 4)$

12. $3m - 2mn - 4n^2 + 6n$

13. $18 - 21x - 4x^2$ $(6 + x)(3 - 4x)$

14. $9x^2 - 15x - 24$ $3(3x - 8)(x + 1)$

15. $y^4 - 16$ $(y^2 + 4)(y + 2)(y - 2)$

16. $84z^2 + 7z - 42$ $7(4z + 3)(3z - 2)$

17. $6a^3 + 9a^2 - 3a$ $3a(2a^2 + 3a - 1)$

18. $4b^3 - 48b^2 + 80b$ $4b(b - 10)(b - 2)$

19. $5r^3 - 40s^3$ $5(r - 2s)(r^2 + 2rs + 4s^2)$

20. $t^2(t - 1) + 4(1 - t)$ $(t - 1)(t + 2)(t - 2)$

12. $(m + 2n)(3 - 2n)$

Solve.

21. $4w - 3(1 - w) = -17$ $\{-2\}$

22. $3 - 5y \geq (2 - y)4 + 7$ $\{y: y \leq -12\}$

23. $\frac{3}{4}(4z - 8) - 4 < 5z + 8$ $\{z: z > -9\}$

24. $\frac{2}{3}(3y - 6) = \frac{1}{2}(10y + 4)$ $\{-2\}$

25. $|3u - 7| < 2$ $\{u: \frac{5}{3} < u < 3\}$

26. $3x > 4x - 3$ and $2x \leq 5x + 6$

27. $m^2 - 15 = 2m$ $\{-3, 5\}$

28. $(4n - 1)(3n + 2) = (6n + 5)(2n - 3)$

26. $\{x: -2 \leq x < 3\}$ 28. $\{-1\}$

Solve each system of equations.

29. $3x + 4y = -7$ $\{(-5, 2)\}$
    $-2x + 3y = 16$

30. $3a - 4b = 5$ $\{(3, 1)\}$
    $a + 7b = 10$

31. $n = \frac{m}{t + v}$

Solve for the variable in color.

31. $m = n(t + v)$

32. $x = 5(y + z)$

32. $z = \frac{x}{5} - y$

33. $T = 3a - 4b$

34. $P = q + qnt$

33. $b = \frac{3a - T}{4}$

35. $d = \frac{ef}{2bc}$ $c = \frac{ef}{2bd}$

36. $h = \frac{s + w}{u}$
    $s = hu - w$

34. $q = \frac{P}{1 + nt}$

Solve.

37. One train traveling at an average speed of 60 km/h left a station 2 h after another train left from the same station. Traveling in the same direction along a parallel track, the second train overtook the first in 1 h. What was the average speed of the first train?   20 km/h

38. The sum of the digits of a two-digit number is 9. When the digits are reversed, the new number is 27 more than the original number. Find the original number.   36

Polynomials and Rational Expressions   **423**

4. $= \underbrace{\dfrac{b}{a} \cdot \dfrac{b}{a} \cdot \ldots \cdot \dfrac{b}{a}}_{-n \text{ factors}}$

   Def. of a power

5. $= \dfrac{b \cdot b \cdot \ldots \cdot b}{a \cdot a \cdot \ldots \cdot a}$

   Theorem, page 403

6. $= \dfrac{b^{-n}}{a^{-n}}$

   Def. of a power

7. $= \dfrac{\dfrac{1}{b^n}}{\dfrac{1}{a^n}}$

   Def. of neg. exponent

8. $= \dfrac{1}{b^n} \cdot a^n = \dfrac{a^n}{b^n}$

   Def. of division

9. $\therefore \left(\dfrac{a}{b}\right)^n = \dfrac{a^n}{b^n}$

   Trans. ax. of $=$

52. Part 2:

1. a and b are real numbers; $a \neq 0$, $b \neq 0$; n is an integer. Hypothesis

2. $\left(\dfrac{a}{b}\right)^{-n} = \dfrac{1}{\left(\dfrac{a}{b}\right)^n}$

   Def. of neg. exponent

3. $= \dfrac{1}{\dfrac{a^n}{b^n}}$

   Part 1 of this theorem

4. $= \dfrac{1}{a^n \cdot \dfrac{1}{b^n}}$

   Def. of division

5. $= \dfrac{1}{a^n \cdot b^{-n}}$

   Def. of neg. exponent

6. $= \dfrac{1}{a^n} \cdot \dfrac{1}{b^{-n}}$

   Prop. of recip. of a prod.

7. $= a^{-n} \cdot \dfrac{1}{b^{-n}}$

   Def. of neg. exponent

(continued)

8. $= \dfrac{a^{-n}}{b^{-n}}$

Def. of division

9. $\therefore \left(\dfrac{a}{b}\right)^{-n} = \dfrac{a^{-n}}{b^{-n}}$

Trans. ax. of $=$

10. $= \dfrac{\dfrac{1}{a^n}}{\dfrac{1}{b^n}}$

Def. of neg. exponent

11. $= \dfrac{1}{a^n} \cdot b^n$

Def. of division

12. $= \dfrac{b^n}{a^n}$

Def. of division

13. $\therefore \left(\dfrac{a}{b}\right)^{-n} = \dfrac{a^{-n}}{b^{-n}}$

$= \dfrac{b^n}{a^n}$

Trans. ax. of $=$

Part 3:

1. a and b are real numbers; $a \neq 0$, $b \neq 0$; n is an integer.
Hypothesis

2. $\left(\dfrac{a}{b}\right)^0 = \dfrac{a^0}{b^0}$

Part 1 of this theorem

3. $= \dfrac{1}{1}$

Def. of zero exponent

4. $= 1$
Exercise 45, page 102

5. $\therefore \left(\dfrac{a}{b}\right)^0 = \dfrac{a^0}{b^0} = 1$

Trans. ax. of $=$

Strategy for Success: The method that is used in scoring a particular multiple-choice exam determines whether or not it is worthwhile to guess an answer. If you find that it *is* worthwhile, you may be able to guess by using your knowledge of algebra to eliminate one or more of the answer choices. For example, if you know that a certain answer must be a positive integer, you can eliminate any choice that is a negative number, zero, or a fraction or decimal.

Decide which is the best of the choices given and write the corresponding letter on your answer sheet.

1. How many integral values of k are there for which the polynomial $x^2 + kx + 36$ is factorable? B
   (A) 4   (B) 10   (C) 3   (D) 2   (E) 5

2. Which of the following polynomials is (are) irreducible? E
   I. $6x^2 + 55x + 51$ II. $8x^2 - 43x - 57$ III. $143a^3 + 297$
   (A) I only   (B) II only   (C) III only   (D) I and II only   (E) I and III only

3. For what value of n is the sentence $4^{3n-2} = 16^n$ true? B
   (A) 4   (B) 2   (C) 1   (D) 3   (E) 0

4. The product of two consecutive positive odd integers is 74 more than the square of the lesser integer. Find the greater integer. D
   (A) 51   (B) 37   (C) 43   (D) 39   (E) 45

5. Which of the following are factors of $x^3 - 9x - 7x^2 + 63$? E
   I. $x - 7$ II. $x + 3$ III. $x - 3$
   (A) I only   (B) II only   (C) III only   (D) I and II only   (E) I, II, and III

6. Find an integral value of c such that $3x^2 - 23x + c$ and $9x^2 + 6x + 1$ will have a common binomial factor. A
   (A) -8 (B) 8   (C) 5   (D) -5 (E) -7

7. The perimeter of a rectangle is 86 m and its area is 432 m^2. What is the length of the longer side of the rectangle? C
   (A) 16 m   (B) 32 m   (C) 27 m   (D) 21 m   (E) 14 m

8. Jill is x years old. Her brother Jim's age is the square of her age. Five years from now, Jim's age will be two years less than twice Jill's age at that time. How old is Jim now? D
   (A) 3 years   (B) 8 years   (C) 14 years   (D) 9 years   (E) 7 years

9. Express $\dfrac{2x - 3}{15 + 7x - 2x^2} \div \dfrac{(x - 5)^{-1}}{(2x - 3)^{-2}}$ in lowest terms. Assume that no variable has a value that results in division by zero. C
   (A) -1 (B) $\dfrac{3 - 2x}{3 + 2x}$ (C) $\dfrac{1}{9 - 4x^2}$ (D) $\dfrac{2x - 3}{5 - x}$ (E) $\dfrac{x - 5}{(5 - x)(3 + 2x)}$

APPLICATION

Metric Prefixes

You are probably familiar with such metric prefixes as kilo-, centi-, and milli-, but have you ever encountered the prefixes giga- (JIG uh), nano- (NAN oh), or pico- (PEEK oh)? These and other metric prefixes, together with their symbols, are given in Table 2 on page 681.

For example, the diameter of a galaxy, such as the Milky Way, might be expressed as one thousand exameters, or 1000 Em, and the time needed for the sun to revolve around the Milky Way as ten petaseconds, or 10 Ps. Astronomers are able to calculate these very large measurements with the aid of telescopes such as the one shown at the right.

Biologists, on the other hand, are able to calculate very *small* measurements with the aid of microscopes. For example, the diameter of a specimen that might be examined with a *light microscope* could range from one micrometer, or 1 μm, to one hundred micrometers, or 100 μm. A specimen that might be examined with the use of an *electron microscope* could range from one nanometer, or 1 nm, to one millimeter, or 1 mm.

Exercises

Solve. Use the table of metric prefixes on page 681.

1. The frequencies of radio waves range from one tenth megahertz, or 0.1 MHz, to one tenth gigahertz, or 0.1 GHz. Are radio frequencies greater or less than the frequencies of visible light, which range from 0.4 PHz to 0.75 PHz? less

2. Which of the following specimens (diameters given in parentheses) could be viewed by using an electron microscope but not by using a light microscope? b
   a. a bacterium (1 μm)
   b. a virus (100 nm)
   c. an amoeba (50 μm)

3. Express each of the following using the appropriate metric prefix.
   a. the average distance from the sun to the planet Pluto, 5.9×10^{12} m 5.9 Tm
   b. the average time needed for sunlight to travel to Pluto, 20×10^3 s 20 ks
   c. the mass of ten million plutonium atoms, 4.0×10^{-15} g 4 fg

Polynomials and Rational Expressions **425**

Robots are often used in manufacturing and industry to increase the speed and quality of production. The photo shows robots welding a car chassis in a St. Louis assembly plant.

Chapter 9

Rational Expressions in Open Sentences

Open Sentences Involving Fractions or Percents

OBJECTIVES for Sections 9-1 and 9-2:
1. To solve open sentences with whole-number denominators.
2. To solve problems involving percents.

9–1 Open Sentences Involving Whole-Number Denominators

Open sentences often contain one or more fractions with whole-number denominators. For example, consider the following open sentences.

$$\frac{1}{3}a + \frac{1}{6}a = 1 \qquad\qquad \frac{b}{2} - \frac{b-2}{4} \geq 1$$

$$\frac{3c+7}{12} - \frac{c-1}{3} = \frac{1}{2} \qquad\qquad \frac{3e+1}{15} + \frac{e-1}{12} < \frac{5}{6}$$

A convenient first step in solving such an open sentence is to transform the given sentence into an equivalent sentence that contains no fractions. If the open sentence is an equation, you can use the multiplication property of equality and multiply both sides of the equation by the LCD of all the fractions appearing in the given equation.

Rational Expressions in Open Sentences **427**

Solve.

1. $\dfrac{2t + 3}{15} - \dfrac{t - 2}{5} = \dfrac{2}{5}$

Multiply both sides by 15, the LCD of the fractions.

$15\left(\dfrac{2t + 3}{15} - \dfrac{t - 2}{5}\right) = 15\left(\dfrac{2}{5}\right)$

$2t + 3 - 3t + 6 = 6$

$t = 3$

∴ the solution set is {3}.

2. $\dfrac{k + 1}{2} > \dfrac{k - 1}{5} + \dfrac{1}{10}$

Multiply both sides by 10, the LCD of the fractions.

$10\left(\dfrac{k + 1}{2}\right) > 10\left(\dfrac{k - 1}{5} + \dfrac{1}{10}\right)$

$5(k + 1) > 2(k - 1) + 1$

$5k + 5 > 2k - 2 + 1$

$3k > -6$

$k > -2$

∴ the solution set is {k: k > −2}.

Common Errors

When students multiply each term of an open sentence by the LCD of the fractions in that sentence, be sure that they realize that the purpose of the multiplication is to simplify the open sentence. Some students will do the following.

$$\dfrac{3x}{5} - \dfrac{x}{2} = 3$$

$$\dfrac{10(3x)}{5} - \dfrac{10(x)}{2} = 10(3)$$

$$\dfrac{30x}{5} - \dfrac{10x}{2} = 30$$

The third equation above should not have fractions. The multiplication was performed so the fractions would be eliminated. The proper third line would be
$6x − 5x = 30$.

EXAMPLE 1   Solve $\dfrac{3c + 7}{12} - \dfrac{c - 1}{3} = \dfrac{1}{2}$.

SOLUTION

$$\dfrac{3c + 7}{12} - \dfrac{c - 1}{3} = \dfrac{1}{2}$$

$$12\left(\dfrac{3c + 7}{12} - \dfrac{c - 1}{3}\right) = 12\left(\dfrac{1}{2}\right) \longleftarrow \begin{cases} \text{Multiply both sides by 12,} \\ \text{the LCD of the fractions.} \end{cases}$$

$$3c + 7 - 4(c - 1) = 6$$

$$3c + 7 - 4c + 4 = 6$$

$$-c + 11 = 6$$

$$-c = -5$$

$$c = 5$$

Check:
$$\dfrac{3c + 7}{12} - \dfrac{c - 1}{3} = \dfrac{1}{2}$$

$$\dfrac{3(5) + 7}{12} - \dfrac{5 - 1}{3} \overset{?}{=} \dfrac{1}{2}$$

$$\dfrac{22}{12} - \dfrac{16}{12} \overset{?}{=} \dfrac{1}{2}$$

$$\dfrac{1}{2} = \dfrac{1}{2} \checkmark$$

∴ the solution set is {5}.

In Example 1 both sides of the original equation were multiplied by 12, the LCD of all the fractions in the equation. Notice that it was not necessary to use the LCD. Instead, both sides of the equation could have been multiplied by 24, 36, or any other common multiple of the denominators of all the fractions.

A similar method can be used to solve inequalities involving fractions. Using the multiplication property of order stated on page 159, multiply both sides of the inequality by the LCD of the fractions.

EXAMPLE 2   Solve $\dfrac{3e + 1}{15} + \dfrac{e - 1}{12} < \dfrac{5}{6}$.

SOLUTION

$$\dfrac{3e + 1}{15} + \dfrac{e - 1}{12} < \dfrac{5}{6}$$

$$60\left(\dfrac{3e + 1}{15} + \dfrac{e - 1}{12}\right) < 60\left(\dfrac{5}{6}\right) \longleftarrow \begin{cases} \text{Multiply both sides by 60,} \\ \text{the LCD of the fractions.} \end{cases}$$

$$4(3e + 1) + 5(e - 1) < 50$$

$$17e - 1 < 50$$

$$17e < 51$$

$$e < 3$$

∴ the solution set is {e: e < 3}.

Oral Exercises

State the LCD of the terms of each open sentence. Then give the equivalent sentence formed by multiplying both sides by the LCD.

1. $\frac{z}{4} + \frac{z}{8} = 3$ 8; $2z + z = 24$

2. $\frac{x}{3} + \frac{3x}{2} > \frac{11}{2}$ 6; $2x + 9x > 33$

3. $\frac{3w}{4} - 2 \geq \frac{1}{4}$ 4; $3w - 8 \geq 1$

4. $\frac{3t + 1}{4} = \frac{1}{2}$ 4; $3t + 1 = 2$

5. $\frac{x + 1}{4} + \frac{2x - 1}{6} = \frac{5}{4}$
   12; $3(x + 1) + 2(2x - 1) = 15$

6. $\frac{a}{6} - \frac{2a + 3}{8} \leq 0$
   24; $4a - 3(2a + 3) \leq 0$

Written Exercises

Solve.

A

1. $\frac{8u}{15} - \frac{u}{5} = 2$ {6}

2. $\frac{2m}{3} - \frac{m}{6} = -1$ {−2}

3. $\frac{a}{5} - 1 = \frac{a}{30}$ {6}

4. $2v - \frac{v}{5} = \frac{3}{2}$ $\{\frac{5}{6}\}$

5. $\frac{f}{4} \leq \frac{3}{2} - \frac{f}{5}$ $\{f: f \leq \frac{10}{3}\}$

6. $\frac{t}{3} > 3 - \frac{t}{6}$ {t: t > 6}

7. $\frac{5n - 4}{12} < 3$ {n: n < 8}

8. $\frac{4b + 5}{3} > -5$ {b: b > −5}

9. $\frac{q - 7}{7} = \frac{7 - q}{3}$ {7}

10. $\frac{d - 6}{3} = \frac{22 - d}{5}$ {12}

11. $\frac{p}{2} - 1 \leq \frac{4}{5} + \frac{3p}{10}$ {p: p ≤ 9}

12. $3c - 1 \geq \frac{c + 3}{2}$ {c: c ≥ 1}

13. $\frac{3s}{5} - \frac{s}{10} > 5$ {s: s > 10}

14. $\frac{3p}{8} - p < \frac{5}{6}$ $\{p: p > -\frac{4}{3}\}$

15. $\frac{2s + 1}{2} + \frac{11}{10} > \frac{s}{5}$ {s: s > −2}

16. $\frac{r - 6}{6} < \frac{1}{4} - \frac{r}{12}$ {r: r < 5}

17. $\frac{5t}{9} - \frac{2t - 3}{6} = 0$ $\{-\frac{9}{4}\}$

18. $\frac{3b}{14} - \frac{5 - b}{21} = \frac{1}{7}$ $\{\frac{16}{11}\}$

19. $\frac{45k + 43}{100} + \frac{k}{5} = 1 - \frac{5k + 6}{50}$ $\{\frac{3}{5}\}$

20. $\frac{2h + 1}{2} + \frac{3h - 9}{4} > \frac{1}{8} - 2h$ $\{h: h > \frac{1}{2}\}$

21. $\frac{3t - 4}{5} - \frac{2t + 1}{4} = -\frac{1}{2}$ $\{\frac{11}{2}\}$

22. $\frac{b - 1}{16} - \frac{1 - b}{12} = \frac{7b}{8}$ $\{-\frac{1}{5}\}$

23. $\frac{a - 2}{6} - \frac{2a + 1}{10} < \frac{1}{15}$ {a: a > −15}

24. $\frac{w + 8}{12} - \frac{3w - 5}{15} < \frac{w}{20}$ {w: w > 6}

B

25. $2.2(i - 3) - 3.5(i - 2) = 0.1(13 - 63i)$ {0.18}

26. $0.25(3x - 5) + 0.07(5x + 3) + 0.06(15x + 14) = 0$ {0.1}

27. $\frac{2}{3}\left(c - \frac{1}{4}\right) - \frac{1}{6}(c + 2) = \frac{3}{2}$ {4}

28. $\frac{3}{8}(2 - d) - \frac{5}{6}\left(3d - \frac{1}{4}\right) = 1 - 3d$ $\{\frac{1}{3}\}$

Rational Expressions in Open Sentences   **429**

Suggested Assignments

Minimum
  429/1–24
Average
  429/1–27 odd, 29–32
Maximum
  429/1–31 odd, 33–38

Additional A Exercises

Solve.

1. $\frac{w}{5} - \frac{2}{3} = \frac{w}{3}$ {−5}

2. $\frac{u}{3} - \frac{3u}{4} \leq \frac{5}{12}$ {u: u ≥ −1}

3. $\frac{3a}{5} - \frac{a}{15} > 8$ {a: a > 15}

4. $\frac{b - 2}{6} - \frac{b}{10} > 4$
   {b: b > 65}

5. $\frac{v - 2}{4} - \frac{v + 3}{6} = \frac{1}{2}$ {18}

6. $\frac{5e - 2}{3} - \frac{1}{2} = 3$ $\{2\frac{1}{2}\}$

Mixed Review

Solve.

1. $\frac{1}{3}x + \frac{1}{2} = -1$ $\{-4\frac{1}{2}\}$

2. $0.4t - 3.2 = -1.2t$ {2}

3. $5m - 4 \geq 1$ {m: m ≥ 1}

4. $3 - 2y < -5$ {y: y > 4}

Simplify.

5. $\frac{3r - 2}{8} + \frac{4 - r}{12}$ $\frac{7r + 2}{24}$

6. $\frac{x + 2}{3} - \frac{2x - 1}{4}$ $\frac{-2x + 11}{12}$

In Exercises 29–32 find the slope of the graph of the given equation by first expressing the equation in the form $y = mx + b$. $y = -\frac{4}{5}x + \frac{6}{5}$; $m = -\frac{4}{5}$

29. $\dfrac{2x - 1}{3} + \dfrac{2y + 1}{2} = \dfrac{17}{36}$ $y = -\frac{2}{3}x + \frac{11}{36}$; $m = -\frac{2}{3}$

30. $\dfrac{2x + 3}{5} + \dfrac{4y - 9}{8} = \dfrac{3}{40}$

31. $\dfrac{14x - 15}{4} - 2y = -\dfrac{45}{12}$ $y = \frac{7}{4}x$; $m = \frac{7}{4}$

32. $\dfrac{4x - 3}{4} - \dfrac{2y + 5}{5} = \dfrac{1}{4}$ $y = \frac{5}{2}x - 5$; $m = \frac{5}{2}$

Solve each system of equations.

C 33. $\dfrac{y}{2} = \dfrac{x}{3} + 2$

$\dfrac{y}{4} = x - \dfrac{23}{2}$ $\{(15, 14)\}$

34. $\dfrac{9y}{7} = \dfrac{10x}{7} - 11$

$\dfrac{y}{2} = \dfrac{5x}{6} - \dfrac{17}{3}$ $\{(5, -3)\}$

35. $\dfrac{8y + 5}{4} = \dfrac{4x - 3}{3} - \dfrac{23}{12}$

$\dfrac{y - 5}{10} = \dfrac{7x - 4}{15} - \dfrac{7}{12}$ $\{(\frac{17}{48}, -\frac{133}{72})\}$

36. $\dfrac{y - 2}{6} = \dfrac{3x + 2}{5} + \dfrac{7}{15}$

$\dfrac{5y - 9}{6} = \dfrac{7x - 8}{10} - \dfrac{2}{15}$ $\{(-\frac{163}{69}, -\frac{30}{23})\}$

Solve.

37. $\dfrac{2t^3 - t}{8} - \dfrac{t^2 - 4t}{6} = \dfrac{8t - 1}{12}$ $\{\frac{2}{3}, \frac{\sqrt{2}}{2}, -\frac{\sqrt{2}}{2}\}$

38. $\dfrac{8 - 2k}{7} - \dfrac{k^2 - 11k}{14} - \dfrac{7 - k}{2} = 0$ $\{3, 11\}$

Computer Exercises For students with computer experience

Write a program that will solve an equation of the form

$$\frac{x}{a} + \frac{x}{b} = \frac{c}{d}$$

when you input values of a, b, c, and d such that a, b, and d are nonzero and $a \neq -b$. RUN the program to solve each of the following.

1. $\dfrac{x}{2} + \dfrac{x}{3} = \dfrac{5}{6}$ $\{1\}$

2. $\dfrac{z}{4} - \dfrac{z}{8} = \dfrac{1}{4}$ $\{2\}$

3. $\dfrac{1}{3}y - \dfrac{1}{4}y = \dfrac{2}{3}$ $\{8\}$

4. $\dfrac{w}{4} + w = \dfrac{5}{2}$ $\{2\}$

Modify the program that you wrote for Exercises 1–4 to solve an equation of the form

$$\frac{ax}{b} + \frac{cx}{d} = \frac{e}{f}$$

when you input values for a, b, c, d, e, and f such that b, d, and f are nonzero and $ad \neq -bc$. RUN the program to solve each of the following.

5. $\dfrac{3x}{4} + \dfrac{2x}{3} = \dfrac{17}{4}$ $\{3\}$

6. $\dfrac{3v}{5} = \dfrac{2v}{3} - \dfrac{2}{15}$ $\{2\}$

7. $\dfrac{12}{5} = \dfrac{2}{5}m - \dfrac{3}{2}m$ $\{-\frac{24}{11}\}$

8. $\dfrac{3}{2}n + 1 = \dfrac{n}{6}$ $\{-\frac{3}{4}\}$

430 Chapter 9

9–2 Percent Problems

The word **percent** (often denoted by %) means "per 100" or "divided by 100." Thus, 7% is another way of writing $\frac{7}{100}$ or 0.07. Any percent can be expressed as a fraction or decimal. For example:

$$25\% = \frac{25}{100} = \frac{1}{4} \qquad\qquad \frac{2}{3}\% = \frac{2}{3} \div 100 = \frac{1}{150}$$

$$150\% = \frac{150}{100} = 1.5 \qquad\qquad 0.2\% = \frac{0.2}{100} = 0.002$$

Any rational number can be expressed as a percent. For example:

$$\frac{1}{2} = \frac{50}{100} = 50\% \qquad\qquad \frac{6}{5} = \frac{120}{100} = 120\%$$

$$3.4 = \frac{340}{100} = 340\% \qquad\qquad \frac{2}{3} = \frac{66\frac{2}{3}}{100} = 66\frac{2}{3}\%$$

When you multiply a number called the *base* (b) by a percent, or *rate* (r), the result is the *percentage* (p). This can be expressed as

$$p = br.$$

Generally, in percent problems you are asked to find one of the three quantities p, b, or r given in the formula $p = br$.

EXAMPLE 1 **a.** What is 9% of 52? **b.** What percent of 185 is 148?
c. 525 is 175% of what number?

SOLUTION

a. $p = br$
$p = 52\,(9\%)$
$\quad= 52(0.09)$
$\quad= 4.68$

b. $p = br$
$148 = 185r$
$r = \dfrac{148}{185}$
$\quad= 0.8$
$\quad= 80\%$

c. $p = br$
$525 = b(175\%)$
$525 = b\left(\dfrac{7}{4}\right)$
$4(525) = 4\left(\dfrac{7b}{4}\right)$
$2100 = 7b$
$300 = b$

Percents are often used to describe changes in quantities. For example:

A sweater was marked down 20%.

The population increased by 15%.

In each of these cases, the percent change is based on the original quantity or amount. The formula $p = br$ can be used to describe each situation where b is the original amount, r is the rate or percent change, and p is the change in the amount. This can be expressed as follows.

change in amount = original amount × % change

Rational Expressions in Open Sentences **431**

Supplementary Material

Test 34

Key Ideas

Solve percent problems.
Solve interest problems.
Solve mixture problems.

Common Errors

Students frequently make errors in the conversion of percents to decimals. Showing the following percents in order should help them to remember the correct placement of the decimal point.

$$200\% = 2.00$$
$$100\% = 1.00$$
$$50\% = 0.50$$
$$25\% = 0.25$$
$$10\% = 0.10$$
$$1\% = 0.10$$
$$0.5\% = 0.005$$

$$\frac{1}{2}\% = 0.005$$

1. **a.** What is 108% of 45?
$$p = br$$
$$p = 45(108\%)$$
$$p = 45(1.08) = 48.6$$
   b. What percent of 600 is 153?
$$p = br$$
$$153 = 600r$$
$$r = \frac{153}{600} = 0.255$$
$$= 25.5\%$$
   c. 15.6 is 65% of what number?
$$p = br$$
$$15.6 = b(65\%)$$
$$15.6 = b(0.65)$$
$$b = 24$$

2. Maria bought her groceries during a special sale at the supermarket. If everything in the store was reduced 5% and she paid $64.98 for her groceries, how much money did Maria save by buying her groceries on sale?

Let x = the original price. Then $x - 64.98$ = the amount of the discount.
$$x - 64.98 = x(5\%)$$
$$x - 64.98 = 0.05x$$
$$-64.98 = -0.95x$$
$$68.40 = x$$
$$68.40 - 64.98 = 3.42$$
$\therefore$ Maria saved $3.42.

EXAMPLE 2 Blair bought a sweater at a 20% discount. If she paid $36 for the sweater, what was the original price?

SOLUTION

Step 1   The problem asks for the original price of the sweater.

Step 2   Let x = the original price.
Then $x - 36$ = the amount of the discount.

Step 3

Amount of the discount	=	Original amount	×	% discount
$x - 36$	=	x	×	20%

Step 4
$$x - 36 = x(20\%)$$
$$x - 36 = x\left(\frac{1}{5}\right)$$
$$5(x - 36) = 5\left(\frac{x}{5}\right)$$
$$5x - 180 = x$$
$$4x = 180$$
$$x = 45$$

Step 5   Is $36 twenty percent less than $45?
Yes, since
$$20\% \text{ of } \$45 = \frac{1}{5} \times \$45 = \$9$$
and
$$\$45 - \$9 = \$36.$$
$\therefore$ the original price was $45.

When money is invested, the rate of interest is expressed as a percent of the amount of money invested. The *simple interest* earned annually can be found by multiplying the *principal* (the amount of money invested) by the *annual interest rate*. This can be expressed as follows.

simple annual interest = principal × annual interest rate

This formula can be written as

$$i = pr,$$

where i is the simple annual interest, p is the principal, and r is the annual rate of interest. Thus, the simple annual interest, or income, from $2000 invested at an annual rate of interest of 5% is

$$i = \$2000 \times 5\% = \$2000 \times 0.05 = \$100.$$

EXAMPLE 3   Murray invests part of $9000 in bank accounts that pay 6% simple annual interest and the rest in bonds that pay 11% simple annual interest. How much money is invested in each way if his total annual income from these investments is $890?

SOLUTION

Step 1   The problem asks for the amounts of money invested in bank accounts and in bonds.

Step 2   Let x = amount invested in bank accounts.
Then $9000 - x$ = amount invested in bonds.

Step 3

$$\underbrace{\text{Total interest}}_{890} = \underbrace{\text{Interest from bank accounts}}_{x(6\%)} + \underbrace{\text{Interest from bonds}}_{(9000 - x)(11\%)}$$

Step 4

$$890 = x(6\%) + (9000 - x)(11\%)$$
$$890 = x(0.06) + (9000 - x)(0.11)$$
$$890 = 0.06x + 990 - 0.11x$$
$$0.05x = 100$$
$$x = 2000$$

Thus, $x = 2000$ and $9000 - x = 7000$.

Step 5   Checking the results is left to you.

$\therefore$ Murray invested $2000 in bank accounts and $7000 in bonds.

Chemists, druggists, and others frequently are confronted with situations where they find it necessary to mix ingredients. As you saw in Section 4-9, percent is commonly used in such problems to describe the composition of the mixture. Additional problems of this type are found in the following set of problems.

Oral Exercises

Replace each ___?___ with a real number to make a true statement.

1. 16% of 150 = __?__   24
2. $10\frac{1}{2}$% of 400 = __?__   42
3. __?__ % of 64 = 16   25
4. __?__ % of 300 = 225   75
5. 25% of __?__ = 13   52
6. 120% of __?__ = 60   50
7. __?__ % of 175 = 350   200
8. 250% of __?__ = 40   16
9. 0.7% of 1000 = __?__   7
10. $133\frac{1}{3}$% of 150 = __?__   200
11. __?__ % of 90 = 60 $66\frac{2}{3}$
12. $\frac{2}{3}$% of __?__ = 30   4500

Problems

Solve.

A

1. Linda received 54% of the votes cast for president of the student council of Fisher High School. If Linda received 594 votes, how many votes were cast? 1100 votes

2. Del paid a 6% sales tax on his new car. What was the price of the car if the sales tax was $705? $11,750

---

3. The DeLucas have invested $8500 in two money market accounts. One account has an 8% rate and the other has a 9% rate. If their annual interest is $735, how much is invested at each rate?

Let x = amount invested at 8%. Then $8500 - x$ = amount invested at 9%.
$$735 = x(8\%) + (8500 - x)(9\%)$$
$$735 = x(0.08) + (8500 - x)(0.09)$$
$$735 = 0.08x + 765 - 0.09x$$
$$0.01x = 30$$
$$x = 3000$$
$$8500 - x = 5500$$
$\therefore$ the DeLucas have $3000 invested at 8% and $5500 invested at 9%.

Suggested Assignments

Minimum
  433/*P*: 1–10
R 436/*Self-Test 1*
Average
  433/*P*: 1–16
R 436/*Self-Test 1*
Maximum
  433/*P*: 1–20
R 436/*Self-Test 1*

1. On Monday, 13% of the students at Buckingham School were absent because of the flu. If 52 students were absent, how many students attended school that day? 348

2. If an ore contains 8% copper, how many kilograms of ore are needed to obtain 250 kg of copper? 3125 kg

3. The Weirs paid $2820 in closing costs when purchasing their new home. If this amount represents 3% of the purchase price, how much did they pay for their house? $94,000

4. Acme Electronics sells a certain VCR for $872. If the store paid $645.28 for each VCR, what percent of the selling price is the store's profit? 26%

5. How many liters of water must be added to 80 L of a 40% salt solution to obtain a 10% salt solution? 240 L

6. Manuel has to pay 12% of his salary to the government for taxes. If he paid $2082 last year, what was his base salary? $17,350

3. A survey found that 1950 people out of 3000 people polled read at least one newspaper per day. What percent of the people polled did not read at least one newspaper per day? 35%

4. A dress was on sale for $52. If the original selling price was $78, what was the percent discount? $33\frac{1}{3}$%

5. How many liters of pure acid must be added to 5 L of a solution that is 20% acid to make a solution that is 60% acid? 5 L

6. How many liters of water must be evaporated from 220 L of a solution that is 5% salt, to leave a solution that is $5\frac{1}{2}$% salt? 20 L

7. Computer Universe stock lost $12\frac{1}{2}$% of its value in one day's trading in the stock market. If the stock sold for $28 per share at the end of the day, what was its price at the start of the day? $32

8. A buyer for a computer store paid $194 each for some printers. What should be the selling price of each printer, if the markup is to be 25% of the selling price? $258.67

9. Liz invested $3000, part at an annual interest rate of 5% and the rest at an annual interest rate of 12%. How much did she invest at each rate if her total income on the investment for one year was $220? $2000 at 5%; $1000 at 12%

10. Andrew invested a total of $15,000 in two local businesses. He received 6% per year on one investment and 8% per year on the other. If the total income from these two investments for one year was $1090, how much did Andrew invest in each business? $5500 at 6%; $9500 at 8%

B 11. The bowling team of Franco's Bakery has a record of 8 wins and 12 losses. What is the least number of the remaining 35 games the team must win if they are to finish the season winning at least 60% of all the games played? 25 games

12. Aubrey has decided to invest a total of $10,000, some at an annual interest rate of $5\frac{3}{4}$% and the rest at an annual interest rate of $9\frac{1}{2}$%. What is the most he can invest at $5\frac{3}{4}$% if he wants to earn at least $920 interest on the investments during the year? $800

13. Tom bought several appliances and a new car. He paid a sales tax of $7\frac{1}{2}$% on the appliances and an excise tax of $6\frac{1}{2}$% on the car. Before these taxes, the appliances and car together cost $15,200. If he paid a total of $1015 in taxes, how much did the car cost? $12,500

14. When a rubber ball is dropped, it rebounds to a height that is 80% of that from which it is dropped. From what height was it dropped if it has traveled a total of 306.5 cm at the instant it strikes the ground for the fourth time? 62.5 cm

15. Michelle has some money invested at an annual interest rate of 9%. She has three fourths as much money invested at an annual interest rate of 8% as she does at 9%. She also has $300 less invested at an annual interest rate of $6\frac{1}{2}$% than she does at 8%. If Michelle's income for one year from these three investments is $378, how much is invested at each rate?
$2000 at 9%; $1500 at 8%; $1200 at $6\frac{1}{2}$%

16. Juanita purchased vacuum cleaners at a cost of $171 each to sell in her store. During a summer sale, she marked them down 5%. If this discount price gave her a 25% profit on the cost price, what was the selling price of each vacuum cleaner before the sale? $225

C 17. Donna invested $2400 in bonds that earn 8% per year. She also invested money in a bank account that earns $6\frac{1}{2}\%$ per year. Her yearly return on the two investments was the same as if both sums had been invested at $7\frac{1}{6}\%$ per year. Find the amount of money invested in the bank account. $3000

18. Jon invested part of a $200,000 fund at 12% per year and the rest at 8% per year. If he had invested twice as much of the fund at 12% and the rest at 8%, he would have increased his annual income from the fund by $2400. How much did he invest at 8%? $140,000

19. A chemist has a can full of paint thinner that is 70% alcohol. After replacing 7 L of the solution with 7 L of pure alcohol, the resulting solution is $87\frac{1}{2}\%$ alcohol. How many liters does the can hold? 12 L

20. A car radiator that can hold 20 L is full of a solution that is 16% antifreeze. How many liters of the solution should be replaced with pure antifreeze in order to have the radiator full of a solution that is 37% antifreeze? 5 L

Computer Exercises For students with computer experience

1. Write a program that will allow you to input the price of an item, a percent, and a code that indicates whether the percent is to be used for discount or markup. The program should then compute the new price of the item after the given percent discount or markup.

2. Write a program that will allow you to input two positive numbers and will compute the percent that the first number is of the second. A sample output would be

60 IS 25% OF 240.

3. Modify the program that you wrote for Exercise 2 so that you can input two positive numbers and the program will compute the percent increase or decrease of the second number from the first. A sample output would be

60 IS A 20% INCREASE FROM 50.

4. In baseball, a pitcher's *earned run average*, or *ERA*, is the average number of runs that the pitcher allows the opposing team per nine innings pitched. To find a pitcher's ERA, you divide the number of innings pitched by nine, then divide the number of earned runs that the pitcher has allowed by this result. Write a program that will compute a pitcher's ERA when you input a number of innings pitched and the number of earned runs allowed.

Mixed Review

Solve.

1. $\dfrac{f}{4} + \dfrac{f}{8} = -\dfrac{3}{4}$ $\{-2\}$

2. $\dfrac{1}{5} + \dfrac{1}{r} = \dfrac{1}{2}$ $\left\{\dfrac{10}{3}\right\}$

3. $\dfrac{y}{3} + \dfrac{5y}{12} \geq \dfrac{3}{4}$ $\{y: y \geq 1\}$

4. $0.5(2n - 3) = 0.75 + 0.125(n - 4)$ $\{2\}$

5. $\dfrac{x - 2}{6} + \dfrac{3x}{8} - \dfrac{3x + 8}{24} = 1$

   $\{4\}$

Rational Expressions in Open Sentences   **435**

Quick Quiz

Solve.

1. $\frac{m}{3} = 5 - \frac{m}{2}$ {6}

2. $\frac{c+1}{4} - \frac{c}{6} = 2$ {21}

3. $2 - \frac{5s}{3} \leq -3$ {s: s ≥ 3}

4. $\frac{2y-1}{2} + \frac{2y+1}{1} > \frac{1}{2}$
   {y: y > 0}

5. If the Smythe Department Store has a one day sale when everything is marked down 25%, what is the regular price of a coat that sells for $107.10 on sale day? $142.80

6. How much of $6500 must be invested at $9\frac{1}{2}$% interest and how much at 5% interest to yield an annual income of $550?

   $5000 at $9\frac{1}{2}$%; $1500

   at 5%

Self-Test 1

VOCABULARY   percent (p. 431)

Solve.

1. $\frac{a}{4} = 20 - \frac{a}{6}$ {48}   2. $\frac{t+3}{4} + \frac{t-1}{3} = 1$ {1}   *Obj. 1, p. 427*

3. $b - \frac{5b}{2} \leq 6$ {b: b ≥ −4}   4. $\frac{5x-1}{3} - \frac{3x+1}{2} > \frac{5}{6}$ {x: x > 10}

5. Bill purchased a sweater that had been marked down 20%. If he paid   *Obj. 2, p. 427*
   $35.92 for the sweater, what was the selling price of the sweater before it was marked down? $44.90

6. Doris invested $5000, part at 8% per year and the rest at $10\frac{1}{2}$% per year. How much did she invest at each rate if her total income from the investments was $467.50 for one year? $2300 at 8%; $2700 at $10\frac{1}{2}$%

Check your answers with those at the back of the book.

Charlotte Angas Scott

1858–1931

In 1876 Charlotte Angas Scott graduated from Cambridge University in England with high honors, but she did not receive a degree. At that time, although women were allowed to take courses and examinations at Cambridge, they were not degree candidates. The University of London, however, awarded her the bachelor of science degree in 1882 and the doctorate of science in 1885.

Dr. Scott then came to the United States to teach at Bryn Mawr, a newly founded women's college in Pennsylvania. She directed undergraduate and graduate mathematical studies at the college for forty years until her retirement in 1925. In addition, she served on the council of the American Mathematical Society and wrote many articles for mathematical journals. Her major theoretical contribution to the field of mathematics was in the study of algebraic geometry.

436   *Chapter 9*

Rational Expressions in Equations and Problems

OBJECTIVES for Sections 9-3 through 9-6:
1. To solve fractional equations.
2. To use fractional equations to solve number problems, work problems, and motion problems.

9–3 Fractional Equations

An equation in which a variable appears in the denominator of one or more terms is called a **fractional equation.** The method used to solve a fractional equation is similar to the one used in Section 9-1 to solve an equation in which the fractions have whole-number denominators. Note, however, that when you multiply each side of a fractional equation by the LCD of the terms, you may not obtain an equivalent equation.

EXAMPLE Solve $3 + \dfrac{16}{x^2 - 1} = \dfrac{8}{x - 1}$.

SOLUTION

$$3 + \frac{16}{x^2 - 1} = \frac{8}{x - 1}$$

$$(x^2 - 1)\left(3 + \frac{16}{x^2 - 1}\right) = (x^2 - 1)\left(\frac{8}{x - 1}\right) \longleftarrow \begin{cases} \text{Multiply each} \\ \text{side by } x^2 - 1, \\ \text{the LCD of the} \\ \text{fractions.} \end{cases}$$

$$3(x^2 - 1) + 16 = 8(x + 1)$$
$$3x^2 + 13 = 8x + 8$$
$$3x^2 - 8x + 5 = 0$$
$$(3x - 5)(x - 1) = 0$$
$$x = \tfrac{5}{3} \quad \text{or} \quad x = 1$$

Notice that $x \neq 1$ is a restriction on the variable in the original equation. If $x = 1$, the denominators $x - 1$ and $x^2 - 1$ would equal zero. Thus, 1 is not a solution of the original equation. Check $\tfrac{5}{3}$ in the original equation.

$$3 + \frac{16}{x^2 - 1} = \frac{8}{x - 1}$$

$$3 + \frac{16}{(\frac{5}{3})^2 - 1} \stackrel{?}{=} \frac{8}{\frac{5}{3} - 1}$$

$$3 + \frac{16}{\frac{16}{9}} \stackrel{?}{=} \frac{8}{\frac{2}{3}}$$

$$3 + 9 \stackrel{?}{=} 12$$

$$12 = 12 \quad \checkmark$$

∴ the solution set is $\{\tfrac{5}{3}\}$.

Rational Expressions in Open Sentences **437**

Teaching Suggestions
p. T101

Key Ideas

Solve fractional equations. Identify the "true" roots of an equation.

Chalkboard Examples

1. Solve

$$\frac{2w - 9}{w - 7} + \frac{w}{2} = \frac{5}{w - 7}.$$

Multiply each side by $2(w - 7)$, the LCD of the fractions.

$$2(w - 7)\left(\frac{2w - 9}{w - 7} + \frac{w}{2}\right)$$
$$= 2(w - 7)\left(\frac{5}{w - 7}\right)$$

$$2(2w - 9) + (w - 7)w = 2(5)$$
$$4w - 18 + w^2 - 7w = 10$$
$$w^2 - 3w - 28 = 0$$
$$(w + 4)(w - 7) = 0$$
$$w = -4 \quad \text{or} \quad w = 7$$

The denominator of the original equation is zero when $w = 7$. Thus 7 is not a solution of the original equation.
∴ the solution set is $\{-4\}$.

2. Solve $\dfrac{3}{w - 2} = \dfrac{w^2 + 2}{w^2 - 4}$.

$$(w^2 - 4)\left(\frac{3}{w - 2}\right) -$$
$$(w^2 - 4)\left(\frac{w^2 + 2}{w^2 - 4}\right)$$

$$3(w + 2) = w^2 + 2$$
$$w^2 - 3w - 4 = 0$$
$$(w - 4)(w + 1) = 0$$
$$w = 4 \quad \text{or} \quad w = -1$$

The solution set is $\{4, -1\}$.

437

The equation obtained by multiplying each side of the original equation by $x^2 - 1$ has the extra, or *extraneous*, root 1, a number for which the multiplier $x^2 - 1$ represents 0. Whenever you multiply an equation in a variable by a polynomial in that variable, the solution set of the resulting equation always contains all the roots of the original equation. But it sometimes also contains numbers that are *not* roots of the original equation. Therefore, *you must check each root of the resulting equation to see that it satisfies the original equation.*

Oral Exercises

For each equation:

a. State all restrictions on the value of the variable.

b. State the LCD of the fractions.

c. State the equation that results when both sides are multiplied by the LCD.

d. Solve.

1. $\dfrac{7}{m + 5} = \dfrac{3}{5}$

2. $\dfrac{6}{y + 2} - \dfrac{3}{y} = -1$

3. $\dfrac{4}{p^2} + \dfrac{1}{p - 3} = \dfrac{1}{p}$

4. $\dfrac{2x}{x - 1} = x + \dfrac{2}{x - 1}$

5. $\dfrac{e + 2}{e - 2} - \dfrac{e - 2}{e + 2} = 0$

6. $\dfrac{5}{t^2 - 9} = \dfrac{1}{t + 3}$

Written Exercises

Solve.

A

1. $\dfrac{b}{b + 2} = \dfrac{7}{8}$ $\{14\}$

2. $\dfrac{5}{m - 3} = 5$ $\{4\}$

3. $\dfrac{9}{z} - \dfrac{3}{z} = \dfrac{1}{2}$ $\{12\}$

4. $\dfrac{7}{c} = 2 + \dfrac{3}{c}$ $\{2\}$

5. $\dfrac{9}{2y} + \dfrac{3}{y} = \dfrac{8}{y} - \dfrac{1}{16}$ $\{8\}$

6. $\dfrac{6}{n} - \dfrac{1}{2} = \dfrac{9}{2n} - \dfrac{1}{n}$ $\{5\}$

7. $\dfrac{7}{3d} = \dfrac{5}{6d} + \dfrac{d + 1}{2d}$ $\{2\}$

8. $\dfrac{1}{2} + \dfrac{2x - 5}{2x} = \dfrac{14}{x} - \dfrac{1}{3}$ $\{9\}$

9. $\dfrac{3}{y - 3} + 9 = \dfrac{y}{y - 3}$ $\varnothing$

10. $1 + \dfrac{4}{q + 2} = \dfrac{q + 6}{q + 2}$ $\{q : q \neq -2\}$

11. $\dfrac{2}{e} - \dfrac{2e + 1}{6e} = \dfrac{2e - 1}{18}$ $\{-\frac{11}{2}, 3\}$

12. $\dfrac{5}{2} - \dfrac{3z - 6}{3z} = \dfrac{z - 8}{2z}$ $\{-6\}$

13. $f = \dfrac{f + 5}{f - 2} - 5$ $\{-5, 3\}$

14. $4x = \dfrac{14 - x}{x - 1} + 14$ $\{0, \frac{17}{4}\}$

15. $\dfrac{20}{p} - 3 = \dfrac{22}{3p - 3}$ $\{3, \frac{20}{9}\}$

16. $\dfrac{10 - z}{4z} = \dfrac{1}{z - 1}$ $\{5, 2\}$

17. $\dfrac{5b - 4}{3b + 1} = \dfrac{1}{14}(2b - 5) - \dfrac{1}{7}b$ $\{\frac{3}{5}\}$

18. $\dfrac{5c^2 + 4c - 7}{5c + 8} = \dfrac{1}{15}(15c - 13)$ $\{\frac{1}{5}\}$

19. $\frac{2}{r^2 - 2r} - \frac{1}{r} = \frac{1}{3}$ $\{-4, 3\}$

20. $\frac{4}{s - 1} = \frac{s^2 + 4}{s^2 - 1}$ $\{0, 4\}$

21. $\frac{2t}{t - 1} = \frac{t + 6}{1 - t}$ $\{-2\}$

22. $\frac{5 - x}{x - 3} - \frac{1}{3 - x} = 0$ $\{6\}$

23. $\frac{a}{a - 1} - \frac{a - 1}{a} = \frac{3}{2}$ $\{\frac{1}{3}, 2\}$

24. $\frac{2b^2 + 3b}{2b + 1} + \frac{1}{3b} = b + 1$ $\{1\}$

25. $\frac{9}{t + 5} - \frac{1}{t - 5} = \frac{3t}{t^2 - 25}$ $\{10\}$

26. $\frac{10w}{w + 2} - \frac{2w - 3}{w - 2} = \frac{2w^2 - 3}{w^2 - 4}$ $\{\frac{1}{2}, 3\}$

27. $\frac{4t - 36}{t^2 - 9} + \frac{11}{3 - t} = \frac{11}{t + 3}$ $\{-2\}$

28. $\frac{5}{3 - x} = \frac{10}{x + 3} - \frac{7x + 1}{x^2 - 9}$ $\{2\}$

Solve for x.

B 29. $y = \frac{x}{x + b}$ $\{\frac{by}{1 - y}\}$

30. $t = \frac{x - \pi r^2}{\pi x}$ $\{\frac{\pi r^2}{1 - \pi t}\}$

31. $P = \frac{n}{x} - \frac{m}{cx}$ $\{\frac{cn - m}{cP}\}$

32. $R = \frac{x}{a + \frac{x}{t}}$ $\{\frac{aRt}{t - R}\}$

Solve.

33. $\frac{5}{u + 1} + \frac{1}{u - 1} - \frac{7}{3u - 5} = 0$ $\{\frac{9}{11}, 3\}$

34. $\frac{1}{c - 3} - \frac{2}{2c - 1} = \frac{5}{3c + 3}$ $\{0, 5\}$

35. $\frac{2}{y - 2} - \frac{1}{y + 2} = \frac{4}{y^2 - 2y}$ $\{-4\}$

36. $\frac{b + 1}{b^2 - b} - \frac{b}{b^2 - 1} = \frac{b - 1}{b^2 + b}$ $\{4\}$

37. $\frac{5}{x - 2} - \frac{x}{x + 5} = \frac{x^2 - 4}{10 - 3x - x^2}$ $\{-3\}$

38. $\frac{a - 1}{3a + 2} + \frac{3a + 4}{1 - 2a} = \frac{3a^2 - 5}{6a^2 + a - 2}$ $\{-2, -\frac{1}{10}\}$

39. a. Are $\frac{x^2 - 1}{x^2 - 1} = 1$ and $x^2 - 1 = x^2 - 1$ equivalent equations over $\mathscr{R}$? Explain.

b. Are $\frac{x^2 + 1}{x^2 + 1} = 1$ and $x^2 + 1 = x^2 + 1$ equivalent equations over $\mathscr{R}$? Explain.

C 40. For what value of k will the solution set of $\frac{7x + 4}{k} = x - 13$ be $\{-6\}$? 2

41. For what value of k will the solution set of $\frac{4x - k}{x - 5} = 3$ be the empty set? 20

Solve. (*Hint*: If $\frac{a}{b} > 0$, then either (1) $a > 0$ and $b > 0$ or (2) $a < 0$ and $b < 0$.)

42. $\frac{x + 1}{x - 2} > 0$

$\{x: x < -1 \text{ or } x > 2\}$

43. $\frac{x - 4}{x - 1} < 0$

$\{x: 1 < x < 4\}$

44. $\frac{x + 3}{x - 1} < -1$

$\{x: -1 < x < 1\}$

45. $\frac{8 - x}{6 + x} > 1$

$\{x: -6 < x < 1\}$

Rational Expressions in Open Sentences **439**

Additional A Exercises

Solve.

1. $\frac{1}{2} + \frac{5}{t} = \frac{6}{t}$ $\{2\}$

2. $\frac{11}{b} - 2 = \frac{5}{b}$ $\{3\}$

3. $\frac{2}{2f} = \frac{5}{f} - 14$ $\{\frac{2}{7}\}$

4. $\frac{5}{t + 4} = \frac{3}{t - 2}$ $\{11\}$

5. $\frac{z}{z - 3} - 2 = \frac{3}{z - 3}$ $\varnothing$

6. $\frac{3v - 7}{v - 5} + \frac{v}{2} = \frac{8}{v - 5}$ $\{-6\}$

Mixed Review

Factor completely.

1. $2a^3 - 16$
   $2(a - 2)(a^2 + 2a + 4)$

2. $3x^2 + 9x - 30$
   $3(x + 5)(x - 2)$

Find the LCM of A and B.

3. $A = 4x^2y^3z$
   $B = 6xy^2z^4$
   LCM: $12x^2y^3z^4$

4. $A = a^2 - 16$
   $B = 3a - 12$
   LCM: $3a^2 - 48$

Solve.

5. $\frac{1}{2}(x + 3) = x - 2$ $\{7\}$

6. $\frac{1}{5}(2x + 1) + \frac{1}{2} = \frac{1}{10}x + 1$

$\{1\}$

Key Ideas
Solve number problems in-
volving fractional equations.

Chalkboard Examples

Solve.

1. The difference of the reciprocals of two positive

   numbers is $\frac{3}{16}$, and one of

   the numbers is 4 times
   the other. Find the numbers.
   Let x = the lesser number.
   Then $4x$ = the greater
   number. Since $4x > x$,

   then $\frac{1}{4x} < \frac{1}{x}$.

   $$\frac{1}{x} - \frac{1}{4x} = \frac{3}{16}$$

   $$16x\left(\frac{1}{x} - \frac{1}{4x}\right) = 16x\left(\frac{3}{16}\right)$$

   $$16 - 4 = 3x$$
   $$x = 4$$
   $$4x = 16$$

   ∴ the two numbers are 4
   and 16.

2. The difference of the reciprocals of two numbers

   is $\frac{3}{16}$. If one of the num-

   bers is 4 times the other
   number, find the two
   numbers.
   Let x = one of the numbers. Then $4x$ = the other
   number.

   $$\left|\frac{1}{x} - \frac{1}{4x}\right| = \frac{3}{16}$$

   $$\frac{1}{x} - \frac{1}{4x} = \frac{3}{16}$$

   $$x = 4$$
   $$4x = 16$$

   or

9–4 Number Problems

Fractional equations are often used in solving number problems.

EXAMPLE 1 The numerator of a fraction is 15 less than the denominator, and the fraction is equal to $\frac{4}{7}$. Find the fraction.

SOLUTION

Step 1 The problem asks for a fraction whose numerator is 15 less than the denominator.

Step 2 Let x = the denominator of the fraction.
        Then $x - 15$ = the numerator of the fraction.

Step 3 $$\frac{x - 15}{x} = \frac{4}{7}$$

Step 4 $$7x\left(\frac{x - 15}{x}\right) = 7x\left(\frac{4}{7}\right)$$
        $$7(x - 15) = 4x$$
        $$7x - 105 = 4x$$
        $$3x = 105$$
        $$x = 35, \text{ and } x - 15 = 20$$

Step 5 Checking the results is left to you.

        ∴ the fraction is $\frac{20}{35}$.

The following theorem is often helpful in solving problems involving the difference of two reciprocals. It is proved in Exercise 23 on page 443.

> **Theorem.** For all positive real numbers a and b,
>
> $$\text{if } a < b, \text{ then } \frac{1}{a} > \frac{1}{b}.$$

EXAMPLE 2 The difference of the reciprocals of two positive numbers is $\frac{1}{6}$, and one of the numbers is 3 times the other. Find the numbers.

SOLUTION

Step 1 The problem asks for two positive numbers, one of which is 3 times the other.

Step 2 Let x = the lesser number.
        Then $3x$ = the greater number.

        The reciprocals are $\frac{1}{x}$ and $\frac{1}{3x}$.

440 *Chapter 9*

Step 3   Since $x > 0$, $3x > x$. Thus, $\frac{1}{3x} < \frac{1}{x}$.

$$\underbrace{\text{Greater reciprocal}}_{\frac{1}{x}} \quad - \quad \underbrace{\text{Lesser reciprocal}}_{\frac{1}{3x}} \quad = \quad \underbrace{\text{Difference}}_{\frac{1}{6}}$$

Step 4
$$\frac{1}{x} - \frac{1}{3x} = \frac{1}{6}$$
$$6x\left(\frac{1}{x} - \frac{1}{3x}\right) = 6x\left(\frac{1}{6}\right)$$
$$6 - 2 = x$$
$$4 = x$$
Thus $x = 4$, and $3x = 12$.

Step 5   Checking the results is left to you.

$\therefore$ the numbers are 4 and 12.

In the next example, a problem that is almost identical to the problem solved in Example 2 is considered. The only difference is that you are not told that the two numbers are positive.

EXAMPLE 3   The difference of the reciprocals of two numbers is $\frac{1}{6}$, and one of the numbers is 3 times the other. Find the numbers.

SOLUTION

Step 1   The problem asks for two numbers, one of which is 3 times the other.

Step 2   Let x = one of the numbers.
Then $3x$ = the other number.

The reciprocals are $\frac{1}{x}$ and $\frac{1}{3x}$.

Step 3   If $\frac{1}{3x} < \frac{1}{x}$, then $\frac{1}{x} - \frac{1}{3x} = \frac{1}{6}$.

If $\frac{1}{3x} > \frac{1}{x}$, then $\frac{1}{3x} - \frac{1}{x} = \frac{1}{6}$. That is, $\frac{1}{x} - \frac{1}{3x} = -\frac{1}{6}$.

The disjunction

$$\frac{1}{x} - \frac{1}{3x} = \frac{1}{6} \quad \text{or} \quad \frac{1}{x} - \frac{1}{3x} = -\frac{1}{6}$$

can be represented by the following equation involving the absolute value of the difference.

$$\left|\frac{1}{x} - \frac{1}{3x}\right| = \frac{1}{6}$$

Steps 4   By solving the disjunction, you will find that there are two possible
and 5   solutions, 4 and 12 or -4 and -12.

Rational Expressions in Open Sentences   **441**

$$\frac{1}{x} - \frac{1}{4x} = -\frac{3}{16}$$
$$x = -4$$
$$4x = -16$$
$\therefore$ there are two possible solutions, 4 and 16 or -4 and -16.

Additional A Exercises

Solve.

1. One sixth of a number is 9 less than two thirds of the number. Find the number.   18

2. The difference of two integers is 24 and their quotient is $\frac{2}{5}$. Find the two numbers.   16 and 40 or -16 and -40

3. The sum of the reciprocals of 2 consecutive positive odd integers is $\frac{8}{15}$. Find the integers.   3 and 5

4. What number added to both the numerator and denominator of the fraction $\frac{2}{5}$ results in a fraction equal to $\frac{3}{4}$?   7

5. The sum of a number and its reciprocal is $\frac{85}{18}$. What is the number? $\frac{2}{9}$ or $\frac{9}{2}$

6. The denominator of a fraction is 6 more than four times the numerator, and the fraction equals $\frac{3}{13}$. Find the fraction. $\frac{18}{78}$

Suggested Assignments

Minimum
   442/*P*: 1–12
Average
   442/*P*: 1–11 odd, 13–20
Maximum
   442/*P*: 5–11 odd, 13–24

**Additional Answers
Problems**

23. 1. $a < b, 0 < a, 0 < b$
    Hypothesis
  2. $0 \cdot b < a \cdot b$
    Mult. prop. of order
  3. $0 < ab$
    Mult. prop. of zero
  4. $\dfrac{a}{ab} < \dfrac{b}{ab}$
    Mult. prop. of order
  5. $\dfrac{a}{a} \cdot \dfrac{1}{b} < \dfrac{1}{a} \cdot \dfrac{b}{b}$
    Basic prop. of quot.
  6. $1 \cdot \dfrac{1}{b} < \dfrac{1}{a} \cdot 1$
    Exercise 45, p. 102
  7. $\dfrac{1}{b} < \dfrac{1}{a} \left(\text{or} \dfrac{1}{a} > \dfrac{1}{b} \right)$
    Iden. ax. for mult.

24. a. 1. $a < b, b < 0$
    Hypothesis
  2. $\dfrac{a}{b} > \dfrac{b}{b}$
    Mult. prop. of order
  3. $\dfrac{a}{b} > 1$
    Subs. prin.
b. 1. $a < b, 0 < b$
    Hypothesis
  2. $\dfrac{a}{b} < \dfrac{b}{b}$
    Mult. prop. of order
  3. $\dfrac{a}{b} < 1$
    Subs. prin.

Problems

Solve.

A 1. The sum of two numbers is 30 and their quotient is $\frac{2}{3}$. Find the numbers. 12 and 18

2. The difference of two positive numbers is 8 and their quotient is $\frac{5}{7}$. Find the two numbers. 28 and 20

3. Five sixths of a number is 14 more than half of the number. Find the number. 42

4. What number added to both the numerator and denominator of the fraction $\frac{3}{8}$ results in a fraction equal to $\frac{2}{3}$? 7

5. The sum of the reciprocals of two consecutive positive integers is $\frac{17}{72}$. Find the integers. 8 and 9

6. Find two consecutive even integers such that 4 times the reciprocal of the lesser integer is equal to 5 times the reciprocal of the greater integer. 8 and 10

7. One number is 14 more than another number. Five ninths of the lesser number is equal to five sixteenths of the greater number. Find the numbers. 32 and 18

8. The denominator of a fraction is 2 greater than 3 times the numerator, and the fraction is equal to $\frac{5}{16}$. Find the fraction. $\frac{10}{32}$

9. The numerator of a fraction is 11 more than twice the denominator, and the fraction is equal to $\frac{7}{3}$. Find the fraction. $\frac{77}{33}$

10. When a number is added to the numerator of $\frac{5}{16}$ and twice the same number is subtracted from the denominator, the result is a fraction equal to 6. Find the number. 7

11. Find two numbers whose sum is 14 and the sum of whose reciprocals is $\frac{7}{24}$. 6 and 8

12. The sum of a number and its reciprocal is $\frac{34}{15}$. Find the two numbers. $\frac{5}{3}$ and $\frac{3}{5}$

B 13. The difference of the reciprocals of two numbers is $\frac{1}{4}$ and one of the numbers is 4 times the other. Find the numbers. 3 and 12 or −3 and −12

14. One number is 3 times another. If the difference of the reciprocals of the two numbers is 12, find the numbers. $\frac{1}{6}$ and $\frac{1}{18}$ or $-\frac{1}{6}$ and $-\frac{1}{18}$

15. The difference between two numbers is 4. When the reciprocal of the lesser number is subtracted from the reciprocal of the greater number, the result is $-\frac{1}{24}$. Find the numbers. 12 and 8 or −12 and −8

16. Find the least positive integer for which the sum of its reciprocal and $\frac{9}{10}$ is greater than 4 times the reciprocal. 4

17. The difference of a number and its reciprocal is $\frac{9}{20}$. Find the two numbers. $\frac{5}{4}$ and $\frac{4}{5}$ or $-\frac{4}{5}$ and $-\frac{5}{4}$

18. The difference of two numbers is 10 and their quotient is $\frac{3}{8}$. Find the two numbers. 16 and 6 or −6 and −16

19. One positive number is 4 times another. If 240 is divided by each number, the greater quotient exceeds the lesser by 15. Find the two numbers. 48 and 12

20. Find two consecutive integers such that twice the reciprocal of the lesser, increased by the reciprocal of the greater, is equal to 11 times the reciprocal of the product of the integers. 3 and 4

C 21. The sum of two numbers is 100. When the greater number is divided by the lesser, the partial quotient is 7 and the remainder is 4. Find the numbers. 88 and 12

22. An integer is to be subtracted from both the numerator and the denominator of $\frac{7}{12}$ to yield a fraction whose value is greater than $\frac{1}{3}$. Find the greatest value of such an integer less than 10 and the least value greater than 10. 4 and 13

Write a direct proof of each theorem.

23. If $0 < a < b$, then $\frac{1}{a} > \frac{1}{b}$.

24. **a.** If $a < b < 0$, then $\frac{a}{b} > 1$.   **b.** If $0 < a < b$, then $\frac{a}{b} < 1$.

Mixed Review
Solve.

1. $\frac{x}{3} - \frac{x}{4} = 1$ $\{12\}$

2. $\frac{3}{2x} - \frac{4}{3x} = -1$ $\left\{-\frac{1}{6}\right\}$

3. $|2x - 5| = 3$ $\{4, 1\}$

4. $\left|\frac{1}{3x} - \frac{1}{4x}\right| = \frac{1}{6}$ $\left\{\frac{1}{2}, -\frac{1}{2}\right\}$

5. $\frac{4}{g} = 3 + \frac{5}{2g + 3}$ $\{-2, 1\}$

6. $\frac{2}{y - 3} + \frac{2}{y} = 1$ $\{1, 6\}$

9–5 Work Problems

It is often useful to be able to solve problems that involve finding how long it takes to accomplish a task when a steady rate of work is assumed. If the *work rate* is the fraction of the whole job that can be done per unit of time, then

$$\text{work rate} \quad \times \quad \text{time} \quad = \quad \text{work done.}$$

EXAMPLE 1   Molly can paint a house in four days. Tess could paint the same house in six days. How long would it take them to paint the house if they worked together?

SOLUTION

Step 1   The problem asks for the number of days it would take to paint the house if they worked together.

Step 2   Let x = number of days required to do the job together.

Since Molly can do the whole job in 4 days, she can do $\frac{1}{4}$ of the job per day. In x days, she could do $\frac{1}{4} \cdot x$ of the job, or $\frac{x}{4}$ of the job. Similarly, in x days Tess could do $\frac{x}{6}$ of the job.

Solution continued on following page.

Rational Expressions in Open Sentences   **443**

Teaching Suggestions
p. T101

Key Ideas

Solve work problems involving fractional equations.

Chalkboard Examples

Solve.

1. Tom can mow the lawn in 5 h, but it takes his brother only 3 h. How long would it take them to mow the lawn if they work together using two mowers?

Let x = number of hours required to do the job together. In x hours Tom can do $\frac{1}{5} \cdot x$ or $\frac{x}{5}$ of the job and Tom's brother can do $\frac{x}{3}$ of the job.

$$\frac{x}{5} + \frac{x}{3} = 1$$
$$3x + 5x = 15$$
$$x = \frac{15}{8}, \text{ or } 1\frac{7}{8} \text{ hours}$$

∴ it would take $1\frac{7}{8}$ hours to mow the lawn.

2. Steve can seed his garden in 4 h. Working together, Steve and Robin do the same job in $1\frac{1}{2}$ h. How long would it take Robin to plant all the seed alone?

Let x = the time in hours for Robin to plant the seed.

	Work rate	Time	Work done
Steve	$\frac{1}{4}$	1.5	$\frac{1.5}{4}$
Robin	$\frac{1}{x}$	1.5	$\frac{1.5}{x}$

$$\frac{1.5}{4} + \frac{1.5}{x} = 1$$
$$1.5x + 6 = 4x$$
$$x = 2.4$$

∴ it would take Robin 2.4 h.

Step 3

Part of job done by Molly	+	Part of job done by Tess	=	Whole job
$\frac{x}{4}$	+	$\frac{x}{6}$	=	1

Step 4
$$\frac{x}{4} + \frac{x}{6} = 1$$
$$3x + 2x = 12$$
$$5x = 12$$
$$x = \frac{12}{5}, \text{ or } 2\frac{2}{5}$$

Step 5 In $2\frac{2}{5}$ days, Molly would paint $\frac{1}{4} \times \frac{12}{5} = \frac{3}{5}$ of the house and Tess would paint $\frac{1}{6} \times \frac{12}{5} = \frac{2}{5}$ of the house. Since $\frac{3}{5} + \frac{2}{5} = 1$, the whole house would be painted in $2\frac{2}{5}$ days.

∴ it would take $2\frac{2}{5}$ days to paint the house if they worked together.

EXAMPLE 2 There are two intake pipes to a large storage tank. Using the smaller pipe alone, it takes twice as long to fill the tank as it does using the larger pipe alone. The tank can be filled in 8 min if both pipes are used. How long would it take using only the smaller pipe?

SOLUTION

Step 1 The problem asks for the time it would take to fill the tank using the smaller pipe alone.

Step 2 Let x = time in minutes to fill tank using large pipe alone.
Then $2x$ = time in minutes to fill tank using smaller pipe alone.

	Work rate	Time	Work done
Larger pipe	$\frac{1}{x}$	8	$\frac{8}{x}$
Smaller pipe	$\frac{1}{2x}$	8	$\frac{8}{2x}$

Step 3

Part of tank filled by larger pipe	+	Part of tank filled by smaller pipe	=	Whole tank
$\frac{8}{x}$	+	$\frac{8}{2x}$	=	1

Step 4
$$\frac{8}{x} + \frac{8}{2x} = 1$$
$$16 + 8 = 2x$$
$$24 = 2x, \text{ and } x = 12$$

444 Chapter 9

Step 5   In 8 min, the larger pipe would fill $\frac{1}{12} \times 8 = \frac{2}{3}$ of the tank and the smaller

pipe would fill $\frac{1}{24} \times 8 = \frac{1}{3}$ of the tank. Since $\frac{2}{3} + \frac{1}{3} = 1$, the whole tank would be filled in 8 min.

∴ it would take the smaller pipe 24 min to fill the whole tank.

Problems

Solve.

A 1. Jack can enter the payroll data into the computer in 6 h; Amanda requires 8 h to complete the job. How long would it take both people to enter the data if they work together? $3\frac{3}{7}$ h

2. It takes one crew of cleaners 12 h to wash and wax the floors in the Euclid Office Building. Another crew does the same job in 15 h. How long would it take both crews working together to do the job? $6\frac{2}{3}$ h

3. One outlet of a grain elevator will empty the elevator in 1.5 h; a second outlet will empty the same elevator in 4 h. How long would it take both outlets working together to discharge the stored grain from the elevator? $1\frac{1}{11}$ h

4. A computer can process a company's invoices in 2 h. A newer computer can process the same number of invoices in 1.2 h. How long would it take to process the invoices using both computers? 0.75 h

5. To reduce the oil level in a fuel storage tank, two pumps are used to transfer the oil into fuel trucks. If one pump can lower the oil level 1 m in 2 h and the other pump can lower the oil level 1 m in 5 h, how long must both pumps operate together to lower the surface of the oil 14 m? 20 h

6. The Warren County Pool uses its own pump along with a pump on the local fire truck to fill the swimming pool. If one pump alone takes 8 h to fill the pool and the other pump alone takes 12 h to fill the pool, how long would it take both pumps working together to fill the pool? 4.8 h

7. One optical scanner can read and grade a set of standard answer sheets in $\frac{1}{3}$ the time it takes another scanner to do the same job. Together they can read and grade the sheets in 12 min. How long would it take each scanner alone to do the job? 16 min and 48 min

8. It takes Gary twice as long to mow his lawn as it takes his father to do the same job. If they work together they can complete the job in $1\frac{1}{2}$ h. How long would it take Gary to do the job alone? $4\frac{1}{2}$ h

B 9. Michele can type the school newspaper in $1\frac{1}{2}$ h. Working together with Paul, the job is completed in $1\frac{1}{8}$ h. How long would it take Paul alone to do the job? $4\frac{1}{2}$ h

Rational Expressions in Open Sentences   **445**

Suggested Assignments

Minimum
First day: 445/P: 1–6
Second day: 445/P: 7–11
              S 439/20–28 even
Average
First day: 445/P: 1–8
Second day: 445/P: 9–14
              S 439/30–40 even
Maximum
First day: 445/P: 1–10
Second day: 446/P: 11–16
              S 439/30–40 even

Additional A Exercises

Solve.

1. It takes the varsity football team 60 minutes to prepare the field before the first game. The junior varsity can do the same job in 90 minutes. How long would it take both teams to prepare the field working together?   36 min

2. Jeannine can clean the attic in 6 h. If John helps her, the attic can be cleaned in 4 h. How much time would it take if John worked alone?   12 h

3. Beth can erase a number of computer disks in 5 min while Tammy requires 6 min to do the same job. How long will the job take if the women work together on two computers? $2\frac{8}{11}$ min

4. Donna can lay floor tiles in a large room in 12 hours. If Kenneth helps her, the job is done in 7 hours. How many hours would it take Kenneth to do the job alone? $16\frac{4}{5}$ h

(continued)

Additional A Exercises
(continued)

5. Ray can mow a lawn in 75 min and Jorge can do it in $\frac{2}{3}$ the time. How quickly can they mow the lawn working together using two mowers?   30 min

6. Using a power roller, Mrs. Janus can paint 3 average sized rooms in 8 hours. Her son prefers to use a regular roller and needs 12 hours to do the same job. If they work together, how long will it take to paint six average sized rooms?   9.6 h

---

Mixed Review

Solve.

1. $\frac{5}{2c^2} = \frac{3}{c^2} - \frac{1}{2}$ $\{-1, 1\}$

2. $\frac{3}{j + 1} = \frac{6}{3j - 1}$ $\{3\}$

3. $x^2 - 6x + 9 = 0$ $\{3\}$

4. $|5 + x| \geq 3$
   $\{x: x \geq -2 \text{ or } x \leq -8\}$

Teaching Suggestions p. T102

Related Activities p. T102

Supplementary Materials

Test 35

Key Ideas

Solve uniform motion problems involving fractional equations.

10. Sandie can complete her paper route in 45 min. When her sister Christina helps her, it takes them 18 min to complete the route. How long would it take Christina alone to complete the route?   30 min

11. It takes 6 min to fill a certain pool and 18 min to drain the same pool when it is full. With the drain open and the pool empty, how long would it take to fill the pool?   9 min

12. An aquarium can be filled by two inlets in 0.4 h and 0.25 h, respectively, and emptied by an outlet in 0.5 h. How long would it take to fill the aquarium if the inlets and outlet were operating simultaneously? $\frac{2}{9}$ h

C 13. Diann can complete a roofing job in 4 h. Sue can complete the same job in 5 h. After working together on the job for 1.25 h, Diann leaves. How long will it take Sue to complete the work? $2\frac{3}{16}$ h

14. Tony can build a fence in 8 h, while his brother Sam can do it in 6 h. If Tony works alone for 5 h and then lets Sam finish the job, how long will it take Sam working alone to finish the fence? $2\frac{1}{4}$ h

15. Ruth and Terry planted tulip and crocus bulbs for 6 h. Ruth planted 5 bulbs in the time it took Terry to plant 3 bulbs. How long would it take Terry alone to complete the job?   16 h

16. In 2 h, Kelly laid new floor tile over $\frac{1}{5}$ of the room. When she was joined by Annette, the rest of the work was completed in 3 h. How long would it have taken Annette to do the entire job alone?   6 h

9–6 Motion Problems

In Sections 4-8 and 6-6 you solved problems involving uniform motion using the following formula.

$$\text{distance} = \text{rate} \times \text{time}$$
$$d = rt$$

Now that you know how to solve fractional equations, you may sometimes find it convenient to use the formula for uniform motion rewritten as

$$t = \frac{d}{r} \qquad \text{or} \qquad r = \frac{d}{t}.$$

EXAMPLE   Lorraine can ride 17 km on her bicycle in the same time that it takes her to walk 9 km. If her riding speed is 4 km/h faster than her walking speed, how fast does she walk?

SOLUTION

Step 1   The problem asks for Lorraine's speed walking.

Step 2   Let x = speed walking in km/h.
Then $x + 4$ = speed riding in km/h.

	rate (km/h)	distance (km)	time (h)
Riding	$x + 4$	17	$\dfrac{17}{x + 4}$
Walking	x	9	$\dfrac{9}{x}$

Step 3 $\underbrace{\text{Time walking}} = \underbrace{\text{Time riding}}$

$$\frac{9}{x} = \frac{17}{x + 4}$$

Step 4
$$\frac{9}{x} = \frac{17}{x + 4}$$
$$9(x + 4) = 17x$$
$$9x + 36 = 17x$$
$$36 = 8x$$
$$4.5 = x$$

Step 5   Can Lorraine ride 17 km in the same time that she can walk 9 km?
Yes, since
$$\frac{17}{x + 4} = \frac{17}{4.5 + 4} = \frac{17}{8.5} = 2 \quad \text{and} \quad \frac{9}{x} = \frac{9}{4.5} = 2.$$

∴ Lorraine walks at a speed of 4.5 km/h.

Problems

Solve.

A   1. A freight train travels 280 km in the same time an express train travels 420 km. If the speed of the express train is 40 km/h greater than that of the freight train, find the speed of each train. 80 km/h, 120 km/h

2. Joe drove 480 km in the same amount of time it took Marie, traveling 10 km/h faster, to travel 540 km. How fast was Joe traveling? 80 km/h

Rational Expressions in Open Sentences   **447**

Solve.

1. A plane can travel 2250 km in 1 hour longer than it takes a train to travel 300 km. If the plane's rate is six times that of the train, find the train's rate of travel.   75 km/h

2. In 3 h, Jack rode his bike 6 km from his house to the park, then walked home along the same route. If his cycling speed is 5 times his walking speed, how long did Jack walk? $2\frac{1}{2}$ h

3. A Canada goose can fly 40 km/h in still air. It can fly 104 km with a tail wind in the same time that it flies 56 km against the same wind. What is the speed of the wind? 12 km/h

4. Greta drove to the ocean at an average speed of 90 km/h. She returned home on the same road at an average speed of 60 km/h. If the round trip took 2 h, how far does Greta live from the ocean? 72 km

5. A truck travels 150 km in the same time that a slower truck travels 100 km. If the fast truck is 30 km/h faster than the slow truck, find the rate of each truck. 90 km/h and 60 km/h

Mixed Review

Solve.

1. $\dfrac{4c - 3}{c - 4} - \dfrac{2c}{3} = \dfrac{2c + 5}{c - 4}$ {3}

2. $|6 - a| < 3$ $\{a: 3 < a < 9\}$

3. $|2t + 3| > -4$ $\mathscr{R}$

4. $|2t + 3| < -4$ $\varnothing$

3. An airplane whose speed in still air is 370 km/h can travel 1000 km with the wind in the same time it travels 850 km against the wind. Find the speed of the wind. 30 km/h

4. Mike can row in still water at a speed twice that of the current in a certain river. It takes Mike 2 h more to row 10 km up the river than it takes him to row 15 km down the river. What is the speed of the current in the river? 2.5 km/h

5. A train required 2 h longer to travel 500 km than a plane required to travel 1200 km. Find the speed of the train, if the plane travels 4 times as fast as the train. 100 km/h

6. The Bedford family made a 43.25 km trip in 5.5 h. On the first part of the trip they crossed a lake by boat traveling at 12 km/h. On the rest of the trip they walked along a scenic trail. If their average walking speed was 5 km/h, how far did they walk? 16.25 km

7. Flying at a steady speed, a pilot flew 765 km due south. He then changed course and flew west at the same speed for 918 km. If the second part of the trip took 36 min more than the first part, what was the plane's air speed? 255 km/h

B   8. Lynn drove 225 km to visit a college campus. On the trip home, she averaged 15 km/h more than on the trip to the college. If her total travel time was $6\frac{3}{4}$ h, what was her average speed on the trip home? 75 km/h

9. In still water, Sarah's boat can travel 15 km/h. If it takes her a total of $4\frac{1}{2}$ h to travel 30 km up a river and then to return by the same route, what is the speed of the current in the river? 5 km/h

10. A bus trip of 200 km to the marching band competition would have taken five eighths as long if the average speed that the bus traveled had been increased by 30 km/h. Find the speed at which the bus traveled. 50 km/h

11. A racer on a motorcycle required 1 h longer to travel 180 km than a bicyclist required to travel 30 km. If the speed of the motorcycle was $2\frac{1}{2}$ times that of the bicycle, find both speeds. Motorcycle: 105 km/h; bicycle: 42 km/h

12. After driving at a steady speed for 30 km, Roberto reached the expressway, where he increased his speed by 45 km/h. If after 120 km of travel on the expressway his total travel time had been 2 h, what was his speed before reaching the expressway? 45 km/h

13. Maurice left his home at 1:30 P.M. and drove to the airport at an average speed of 45 km/h. After a 40 min wait, he took off on a flight with an average speed of 350 km/h. He reached his destination at 4:20 P.M. If the total distance that Maurice traveled by car and by plane was 555 km, how far was Maurice's home from the airport? 30 km

C   14. Tony and Jason are test driving two cars on a 30 km track. Just as Jason starts his car, Tony passes him traveling at 120 km/h. After 2 min, Jason is traveling at 130 km/h and has traveled 2.5 km. If Tony and Jason maintain these speeds, how far will Jason travel before he passes Tony? 22 km

448   *Chapter 9*

Self-Test 2

VOCABULARY   fractional equation (p. 437)

Solve.

1. $5 - \dfrac{3}{a} = 4$ {3}

2. $\dfrac{m-1}{m+1} = \dfrac{6}{7}$ {13}

Obj. 1, p. 437

3. $\dfrac{7}{z-3} = \dfrac{14}{z+1}$ {7}

4. $\dfrac{2}{b-1} - \dfrac{1}{b^2-1} = \dfrac{1}{b+1}$ {−2}

5. Find two numbers whose sum is 96 and whose quotient is $\frac{1}{5}$. 16 and 80

Obj. 2, p. 437

6. One pipe can fill a tank in 5 h. A second pipe can fill the same tank in 10 h. How long would it take both pipes together to fill the tank? $3\frac{1}{3}$ h

7. The Driscoll family went on a weekend trip to their cottage, which was 90 km from their home. Their average speed going to the cottage was 20 km/h less than their average speed returning home. If the traveling time for the round trip was $3\frac{3}{4}$ h, what was the family's average speed on the way to the cottage? 40 km/h

Check your answers with those at the back of the book.

Ratio, Proportion, and Variation

OBJECTIVES for Sections 9-7 through 9-10:
1. To use the concepts of ratio and proportion to solve problems.
2. To solve problems involving direct, inverse, joint, and combined variation.

9–7 Ratio and Proportion

Quotients of real numbers can be used to describe many situations. For example, in the packages of tulip bulbs that Larry sells, there are 2 red bulbs for every 3 yellow bulbs. You can describe this situation mathematically as follows.

$$\frac{\text{number of red tulip bulbs}}{\text{number of yellow tulip bulbs}} = \frac{2}{3}$$

You can describe this relationship by stating that the number of red bulbs and the number of yellow bulbs are in the *ratio* 2 to 3, or, in symbols, $2:3$. The **ratio** of two numbers is the quotient of one number divided by a second number, provided that the second number is not zero.

Solve.

1. In an accounting office $4200 in bonuses is divided between Mrs. Ryan and Mr. Boulos in the ratio 7 to 3. How much money does each receive?
Let $7x$ = Mrs. Ryan's bonus. Then $3x$ = Mr. Boulos' bonus.
$$7x + 3x = 4200$$
$$x = 420$$
Thus $7x = 2940$, and $3x = 1260$.
∴ Mrs. Ryan receives $2940, and Mr. Boulos receives $1260.

2. Solve.

a. $\dfrac{5}{12} = \dfrac{x}{18}$
$12x = 90$
$x = \dfrac{15}{2}; \quad \left\{\dfrac{15}{2}\right\}$

b. $\dfrac{5}{x+4} = \dfrac{3}{x-2}$
$5(x - 2) = 3(x + 4)$
$5x - 10 = 3x + 12$
$2x = 22$
$x = 11; \quad \{11\}$

3. Find the ratio of x to y given that $\dfrac{5x - 2y}{4x} = 6$.
$5x - 2y = 24x$
$-2y = 19x$
$\dfrac{x}{y} = \dfrac{-2}{19}$
∴ the ratio of x to y is -2 to 19.

If you know that two numbers are in the ratio 4 to 5, you can represent one number as $4x$ and the other as $5x$, since

$$\frac{4x}{5x} = \frac{4}{5}, \quad \text{or} \quad 4x:5x = 4:5.$$

This fact is useful when solving problems involving ratios.

EXAMPLE 1 If $2700 is divided between George and Mary in the ratio 4 to 5, respectively, how much does each receive?

SOLUTION

Step 1 The problem asks how much money George and Mary each receive.

Step 2 Let $4x$ = the amount of money in dollars that George receives.
Then $5x$ = the amount of money in dollars that Mary receives.

Step 3

Amount of money George receives	+	Amount of money Mary receives	=	Total amount of money
$4x$	+	$5x$	=	2700

Step 4 $4x + 5x = 2700$
$9x = 2700$
$x = 300$
Thus $4x = 1200$, and $5x = 1500$.

Step 5 Are 1200 and 1500 in the ratio $4:5$?
$$\frac{1200}{1500} = \frac{4}{5} \quad \checkmark$$

Does the money total $2700?
$$1200 + 1500 \overset{?}{=} 2700$$
$$2700 = 2700 \quad \checkmark$$

∴ George receives $1200 and Mary receives $1500.

Often the idea of ratio is extended to include three or more numbers. For instance, you say that a, b, and c are in the ratio $2:3:7$ if the ratio of a to b is $\frac{2}{3}$, the ratio of a to c is $\frac{2}{7}$, and the ratio of b to c is $\frac{3}{7}$. If this situation occurred in a problem, you could represent a by $2x$, b by $3x$, and c by $7x$.

An equation which states that two ratios are equal is called a **proportion**. An example of a proportion is

$$\frac{a}{b} = \frac{c}{d}.$$

This proportion can also be written as

$$a:b = c:d$$

and can be read "a is to b as c is to d." In this proportion, c and d are called the **extremes**, and b and c are called the **means**.

450 *Chapter 9*

The next theorem is proved in Exercise 39 on page 454.

> **Theorem.** For all real numbers a and c and all nonzero real numbers b and d,
>
> $$\text{if } \frac{a}{b} = \frac{c}{d}, \text{ then } ad = bc.$$

Thus, in any proportion, *the product of the extremes equals the product of the means.* In Example 2, this property is used in solving fractional equations.

EXAMPLE 2 Solve.

    a. $\dfrac{2}{x} = \dfrac{4}{9}$ **b.** $\dfrac{6x - 2}{7} = \dfrac{5x + 7}{8}$

SOLUTION

a. $\dfrac{2}{x} = \dfrac{4}{9}$

$18 = 4x$

$\dfrac{9}{2} = x$

Check: $\dfrac{2}{x} = \dfrac{4}{9}$

$\dfrac{2}{\frac{9}{2}} \overset{?}{=} \dfrac{4}{9}$

$\dfrac{4}{9} = \dfrac{4}{9}$ ✓

∴ the solution set is $\left\{\dfrac{9}{2}\right\}$.

b. $\dfrac{6x - 2}{7} = \dfrac{5x + 7}{8}$

$8(6x - 2) = 7(5x + 7)$

$48x - 16 = 35x + 49$

$13x = 65$

$x = 5$

Check: $\dfrac{6x - 2}{7} = \dfrac{5x + 7}{8}$

$\dfrac{6(5) - 2}{7} \overset{?}{=} \dfrac{5(5) + 7}{8}$

$\dfrac{28}{7} \overset{?}{=} \dfrac{32}{8}$

$4 = 4$ ✓

∴ the solution set is $\{5\}$.

EXAMPLE 3 Find the ratio of x to y, given that

$$\frac{3x + 10y}{y} = 5.$$

SOLUTION $\dfrac{3x + 10y}{y} = \dfrac{5}{1}$

$3x + 10y = 5y$

$3x = -5y$

$\dfrac{x}{y} = \dfrac{-5}{3}$

∴ the ratio of x to y is -5 to 3.

4. If a plane can fly 2475 km in 6 hours, how far will it travel in 8 hours if its average speed remains the same?

Let x = number of kilometers the plane travels in 8 hours.

Using $r = \dfrac{d}{t}$, $\dfrac{2475}{6} = \dfrac{x}{8}$.

$6x = 19{,}800$

$x = 3300$

∴ the plane will travel 3300 km in 8 h.

Rational Expressions in Open Sentences **451**

EXAMPLE 4   If a train can travel 1600 mi in 30 h, how long will the train take to travel 1800 mi if it maintains the same average speed?

SOLUTION

Step 1   The problem asks how many hours the train will take to travel 1800 mi.

Step 2   Let x = the number of hours.

Step 3   The train travels 1600 mi in 30 h and 1800 mi in x h.

Since the average speed $\left(r = \dfrac{d}{t}\right)$ is the same,

$$\frac{1600}{30} = \frac{1800}{x}.$$

Step 4
$$\frac{1600}{30} = \frac{1800}{x}$$
$$1600x = 54{,}000$$
$$x = 33\tfrac{3}{4}$$

Step 5   Will the train travel 1800 miles in $33\tfrac{3}{4}$ h?

The train's average speed is $\dfrac{1600}{30} = \dfrac{160}{3}$ mi/h.

The distance in miles the train will travel in $33\tfrac{3}{4}$ h is

$$d = r \cdot t = \frac{160}{3} \cdot 33\tfrac{3}{4} = \frac{160}{3} \cdot \frac{135}{4} = 1800. \quad \checkmark$$

$\therefore$ it takes the train $33\tfrac{3}{4}$ h to travel 1800 mi.

Oral Exercises

Simplify each ratio.

1. $\dfrac{6}{8}$ $\tfrac{3}{4}$

2. $16:24$ $2:3$

3. $\dfrac{3c}{7c}$ $\tfrac{3}{7}$

4. 10 km to 8 km
   $5:4$

Replace each ___?___ with the number that makes a true statement.

5. If the ratio of profit to selling price is $3:11$, then out of every 11 cents in sales, ___?___ cents are profit. 3

6. If the ratio of new homes to old ones in a city is $2:5$, then there are 4 new homes for every ___?___ old ones. 10

7. If the ratio of iron to oxygen, by mass, in ferric oxide is $7:3$, then 10 grams of this compound contains ___?___ grams of iron and ___?___ grams of oxygen. 7; 3

8. If 4 out of every 9 foundry workers are 40 years old or older, then the ratio of the number of workers 40 or more years old to the number under 40 is ___?___. $4:5$

State the equation that results when you equate the product of the extremes and the product of the means in the following proportions.

9. $\dfrac{-5}{6} = \dfrac{x}{4}$
$-20 = 6x$

10. $\dfrac{4}{3b} = \dfrac{-6}{5}$
$20 = -18b$

11. $\dfrac{a-2}{3} = 2a$
$a - 2 = 6a$

12. $\dfrac{5}{x+1} = \dfrac{3}{x-2}$
$5(x-2) = 3(x+1)$

Written Exercises

Throughout this set of exercises, assume that the variables represent real numbers that do not result in division by zero.

Simplify each ratio.

A
1. Men to women in a college with 2400 men and 2000 women. $6 : 5$, or $\dfrac{6}{5}$
2. Women to men in a college with 10,000 students, of whom 7500 are women. $3 : 1$, or $\dfrac{3}{1}$
3. Profit to selling price if the profit is $2 and the selling price $9. $2 : 9$, or $\dfrac{2}{9}$
4. Profit to selling price if the cost is $8 and the selling price is $10. $1 : 5$, or $\dfrac{1}{5}$
5. The perimeter of a rectangle that measures 6 cm by 8 cm to the perimeter of a rectangle that measures 5 cm by 12 cm. $14 : 17$, or $\dfrac{14}{17}$
6. The circumference of a circle with radius 3 cm to the circumference of a circle with radius 5 cm. $3 : 5$, or $\dfrac{3}{5}$
7. The area of a rectangle that measures 7 cm by 9 cm to the area of a rectangle that measures 6 cm by 7 cm. $3 : 2$, or $\dfrac{3}{2}$
8. The area of a circle with radius 5 cm to the area of a circle with radius 6 cm. $25 : 36$, or $\dfrac{25}{36}$

Solve.

9. $\dfrac{5}{x} = \dfrac{3}{2}$ $\{\tfrac{10}{3}\}$

10. $\dfrac{4}{a} = \dfrac{-2}{3}$ $\{-6\}$

11. $\dfrac{n}{-5} = \dfrac{-2n}{5}$ $\{0\}$

12. $-2 = \dfrac{1}{y}$ $\{-\tfrac{1}{2}\}$

13. $\dfrac{x-3}{x+5} = \dfrac{11}{15}$ $\{25\}$

14. $\dfrac{-3}{5} = \dfrac{5-4c}{3c+10}$ $\{5\}$

15. $\dfrac{a+9}{a+3} = 2$ $\{3\}$

16. $\dfrac{3b-1}{4} = -\dfrac{3}{2}$ $\{-\tfrac{5}{3}\}$

17. $\dfrac{-4}{2r-9} = \dfrac{-16}{3r+14}$ $\{10\}$

18. $\dfrac{2}{10+3e} = \dfrac{10}{28-7e}$ $\{-1\}$

19. $\dfrac{u-7}{-12} = -\dfrac{2u+6}{72}$ $\{12\}$

20. $-\dfrac{2+w}{8} = -\dfrac{w-4}{24}$ $\{-5\}$

Find the ratio of a to b.

21. $a = 6b$ $6 : 1$

22. $a = b$ $1 : 1$

23. $4a + 2b = 0$ $-1 : 2$

24. $15b - 10a = 0$ $3 : 2$

Rational Expressions in Open Sentences **453**

Additional A Exercises

In these exercises, assume that the variables represent real numbers that do not result in division by zero. State each ratio in lowest terms.

1. Profit to selling price, if the profit is $6 and the selling price is $50. $\dfrac{3}{25}$

2. The area of a garden 10 m by 12 m to the area of a garden 24 m by 17 m. $\dfrac{5}{17}$

Solve.

3. $\dfrac{7}{4} = \dfrac{2}{x}$ $\{\tfrac{8}{7}\}$

4. $\dfrac{13}{5} = \dfrac{2x+1}{2x-7}$ $\{6\}$

5. Four cups of strawberries make 9 jars of jam. How many jars of jam can be made with 7 cups of strawberries? $15\tfrac{3}{4}$ jars

6. Two numbers are in the ratio of $3 : 4\tfrac{1}{4}$. The lesser number is 108. What is the larger number? 153

39. 1. $\frac{a}{b} = \frac{c}{d}$; $b, d \neq 0$

Hypothesis

2. $\frac{a}{b}(bd) = \frac{c}{d}(bd)$

Mult. prop. of =

3. $\left(\frac{a}{b} \cdot b\right)d = b\left(\frac{c}{d} \cdot d\right)$

Comm. and assoc. ax. for mult.

4. $ad = bc$

Exercise 43, p. 102

40. 1. $\frac{a}{b} = \frac{c}{d}$; $b, c, d \neq 0$

Hypothesis

2. $\left(\frac{a}{b}\right)\left(\frac{b}{c}\right) = \left(\frac{c}{d}\right)\left(\frac{b}{c}\right)$

Mult. prop. of =

3. $\frac{ab}{bc} = \frac{cb}{dc}$

Theorem, p. 403

4. $\frac{ab}{cb} = \frac{cb}{cd}$

Comm. ax. of mult.

5. $\frac{a}{c} \cdot \frac{b}{b} = \frac{c}{c} \cdot \frac{b}{d}$

Basic prop. of quot.

6. $\frac{a}{c} \cdot 1 = 1 \cdot \frac{b}{d}$

Subs. prin.

7. $\frac{a}{c} = \frac{b}{d}$

Iden. ax. for mult.

41. 1. $\frac{a}{b} = \frac{c}{d}$; $b, d \neq 0$

Hypothesis

2. $\frac{a}{b} + 1 = \frac{c}{d} + 1$

Add. prop. of =

3. $\frac{a}{b} + \frac{b}{b} = \frac{c}{d} + \frac{d}{d}$

Subs. prin.

4. $\frac{a + b}{b} = \frac{c + d}{d}$

Theorem, p. 407

Find the ratio of a to b.

25. $\frac{3}{a} = \frac{4}{b}$ $3 : 4$

26. $\frac{b}{3} = \frac{a}{7}$ $7 : 3$

B **27.** $\frac{a}{c} = \frac{b}{d}$ $c : d$

28. $\frac{d}{b} = \frac{c}{a}$ $c : d$

29. $\frac{4a + 3b}{3b} = 7$ $9 : 2$

30. $\frac{5a - b}{5a} = -\frac{1}{5}$ $1 : 6$

31. $\frac{2a + b}{a - b} = \frac{2}{3}$ $-5 : 4$

32. $\frac{3a - 2b}{6} = \frac{a + b}{5}$ $16 : 9$

Solve for x.

33. $R = \frac{1}{x} + \frac{1}{R}$ $\left\{\frac{R}{R^2 - 1}\right\}$

34. $M = \frac{1}{M + \frac{1}{x}}$ $\left\{\frac{M}{1 - M^2}\right\}$

35. $\frac{x - 4y}{2y} = \frac{5y}{x - y}$ $\{6y, -y\}$

36. $\frac{x^2 - 5y^2}{y^2} = \frac{2(3x - 7y)}{y}$ $\{3y\}$

C **37.** Find b, given that

$$\frac{a + b}{b} = \frac{c + 18}{18} \quad \text{and} \quad \frac{a - 12}{12} = \frac{c - 36}{36}. \quad 6$$

38. Find z, given that

$$\frac{6y + z}{z} = \frac{2 - 8x}{2} \quad \text{and} \quad \frac{2x - y}{6} = \frac{y - x}{3}. \quad -2$$

Write a direct proof of each theorem.

39. If $\frac{a}{b} = \frac{c}{d}$, and if b and d are nonzero, then $ad = bc$.

40. If $\frac{a}{b} = \frac{c}{d}$, and if b, c, and d are nonzero, then $\frac{a}{c} = \frac{b}{d}$.

41. If $\frac{a}{b} = \frac{c}{d}$, and if b and d are nonzero, then $\frac{a + b}{b} = \frac{c + d}{d}$.

42. If $\frac{a}{b} = \frac{c}{d}$, and if a, b, c, and d are nonzero, then $\frac{b}{a} = \frac{d}{c}$.

Problems

Solve.

A **1.** If 540 acres of land are divided between two people in the ratio 4 to 5, how many acres does each receive? 240 acres and 300 acres

2. Two numbers are in the ratio $2 : 3$. The lesser number is 164. What is the other number? 246

454   *Chapter 9*

3. The measures of two complementary angles are in the ratio $7:8$. Find the measure of each angle in degrees. 42° and 48°

4. The measures of two supplementary angles are in the ratio $3:1$. Find the measure of each angle in degrees. 135° and 45°

5. The lengths of the sides of a triangle are in the ratio $3:3:5$. The perimeter of the triangle is 44 cm. Find the lengths of the sides. 12 cm, 12 cm, 20 cm

6. The lengths of the sides of a quadrilateral are in the ratio $2:4:5:9$. The perimeter of the quadrilateral is 460 cm. Find the lengths of the sides. 46 cm, 92 cm, 115 cm, 207 cm

7. A certain type of rain gutter has a mass of 4 kg for every 3 m of length. What is the mass of a 7.5 m length? 10 kg

8. On an average, fifteen loblolly pine trees provide enough wood pulp for 12,000 paper grocery bags. How many bags could be made from the pulp of 1250 trees? 1,000,000 bags

9. Each cube of chicken bouillon is used to make $1\frac{1}{2}$ c of broth. How many cubes are needed to make 12 c of broth? 8 cubes

10. A person walks 6 km in three fourths of an hour. How long would it take the person to walk 10 km at the same speed? $1\frac{1}{4}$ h

11. Ron earned $27.30 working 6.5 h. How much would he earn working 20 h? $84

12. Aquarium gravel can be purchased at $3.65 for 3 kg. What would be the cost of 9 kg? $10.95

13. A certain kind of carpet sells for $16 per square meter. What area can be covered for $312? 19.5 m²

14. A certain fabric costs $2.50 per yard. How many yards of fabric can be bought for $17? 6.8 yd

B 15. In a collection of dimes and quarters worth $9.00, the ratio of the number of dimes to the number of quarters is $5:7$. How many coins are there in the collection? 48 (20 dimes and 28 quarters)

16. Pam earns $10.50 per hour and Jane earns $11.25 per hour. In one week, the ratio of the number of hours Pam worked to the number of hours Jane worked was $6:5$. Together they earned $715.50. How much will Pam earn the following week if she works 4 h more than she did the week before? $420

17. Show that a profit to selling-price ratio of $\frac{1}{4}$ is equivalent to a profit to cost ratio of $\frac{1}{3}$.

18. Show that a profit to cost ratio of $\frac{1}{4}$ is equivalent to a profit to selling-price ratio of $\frac{1}{5}$.

19. A segment n units long is to be separated into two parts whose lengths have the ratio $a:b$. Find the length of each part in terms of a, b, and n. $\frac{an}{a+b}$, $\frac{bn}{a+b}$

20. A polygon whose area is n square units is to be divided into three parts whose areas are in the ratio $h:j:k$. Find the area of each part in terms of h, j, k, and n. $\frac{hn}{h+j+k}$, $\frac{jn}{h+j+k}$, $\frac{kn}{h+j+k}$

Rational Expressions in Open Sentences **455**

42. 1. $\frac{a}{b} = \frac{c}{d}$; $a, b, c, d \neq 0$
    Hypothesis
    2. $ad = bc$
    Theorem, p. 451
    3. $bc = ad$
    Symm. prop. of =
    4. $\frac{bc}{ac} = \frac{ad}{ac}$
    Div. prop. of =
    5. $\frac{b}{a} \cdot \frac{c}{c} = \frac{a}{a} \cdot \frac{d}{c}$
    Basic prop. of quot.
    6. $\frac{b}{a} \cdot 1 = 1 \cdot \frac{d}{c}$
    Subs. prin.
    7. $\frac{b}{a} = \frac{d}{c}$
    Iden. ax. for mult.

17. Selling Price (S)
    = Profit (P) + Cost (C)
    $\frac{P}{S} = \frac{1}{4}$
    $\frac{P}{P+C} = \frac{1}{4}$
    $4P = P + C$
    $3P = C$
    $\frac{3P}{3C} = \frac{C}{3C}$
    $\frac{P}{C} = \frac{1}{3}$

18. Cost (C)
    = Selling Price (S) −
    Profit (P)
    $\frac{P}{C} = \frac{1}{4}$
    $\frac{P}{S-P} = \frac{1}{4}$
    $4P = S - P$
    $5P = S$
    $\frac{5P}{5S} = \frac{S}{5S}$
    $\frac{P}{S} = \frac{1}{5}$

Solve.

1. $\dfrac{x+1}{4} - \dfrac{x-1}{3} = \dfrac{1}{2}$ {1}

2. $\dfrac{2y-7}{5} = \dfrac{3}{4}$ $\left\{5\dfrac{3}{8}\right\}$

3. $\dfrac{x^2 - 5x}{x^2 - 1} + \dfrac{3}{x+1} = 0$ {3}

Solve the system of equations.

4. $2x - 4y = 5$

   $8x + 3y = 1$ $\left\{\left(\dfrac{1}{2}, -1\right)\right\}$

C 21. The ratio of Jane's speed paddling upstream to her speed paddling downstream is $5:6$. She can paddle 12 km upstream and then return to her starting point in a total of 8.8 h. Paddling at the same speed, how long would it take her to travel 5.5 km in still water? 2 h

22. George decided to invest a certain amount of money, some at an annual interest rate of 5% and the remainder at an annual interest rate of 8%. The ratio of the amount of money invested at 5% to the amount invested at 8% was $3:5$. In one year, he earned $200 more from this investment than if he had invested all the money in a fund earning $6\frac{1}{4}\%$ per year. How much would his income from the investment have been if he had invested all the money in a fund earning 10% per year? $3200

23. In calcium sulfate the ratio, by mass, of calcium to sulfur is $5:4$ and of calcium to oxygen is $5:8$. How many grams of each are there in 34 g of calcium sulfate? 10 g of calcium, 8 g of sulfur, 16 g of oxygen

24. A man wills $18,000 to his three sons. He specifies that the first and second sons are to receive amounts in the ratio $4:3$, and that the first and third sons are to receive amounts in the ratio $2:1$. Find the amount each son gets. $8000, $6000, $4000

25. A metallurgist decides to experiment with a new alloy by using iron and nickel in the ratio $21:5$ by mass and nickel and copper in the ratio $4:3$ by mass. How many grams of each metal does the metallurgist use to make 238 g of the alloy? iron: 168 g; nickel: 40 g; copper: 30 g

26. A woman offers to give n dollars to three university departments. She specifies that the mathematics and German departments receive sums in the ratio $a:b$, and that the chemistry and German departments receive sums in the ratio $j:k$. How much does the mathematics department receive in terms of n, a, b, j, and k? $\dfrac{akn}{ak + bk + bj}$

Computer Exercises For students with computer experience

Write a program that will compute the fourth term of a proportion when you input nonzero values for the other three terms. The program should first display the general form

$$\text{A/B} = \text{C/D},$$

then ask which of the four terms is to be found. The values of the other three terms should then be input in order. RUN the program to compute the fourth term of each of the following proportions.

1. $\dfrac{a}{9} = \dfrac{7}{3}$ 21

2. $\dfrac{2}{3} = \dfrac{14}{d}$ 21

3. $\dfrac{9}{-4} = \dfrac{c}{5}$ -11.25

4. $\dfrac{3}{b} = \dfrac{-2}{9}$ -13.5

5. $\dfrac{4}{x} = \dfrac{18}{12}$ 2.66666667

6. $\dfrac{z}{-15} = \dfrac{-3}{50}$ 0.9

7. $\dfrac{2}{25} = \dfrac{-1}{y}$ -12.5

8. $\dfrac{-3}{8} = \dfrac{w}{-11}$ 4.125

Careers

Architecture

Architects design buildings and supervise their construction. An architect is responsible for making sure that the structure meets the client's requirements and is safe, attractive, and compatible in design with other buildings in the area.

There are many steps in the creation of a design for a structure. The architect makes preliminary drawings of the floor plan and the exterior and interior details of the building before meeting with the client to decide on the final design. This final design is used by consulting engineers to prepare detailed working drawings of the plumbing, electrical connections, and climate-control systems.

In addition to planning the design of a building, the architect advises and represents the client in dealing with the building contractor. Periodic visits to the construction site ensure that the design and specifications are being followed.

EXAMPLE An architect is designing a house that will be 15 m long and 8 m wide. A scale drawing of the floor plan is 45 cm long.

    a. How wide is the floor plan?

    b. If the living room is 18 cm by 13.5 cm on the floor plan, what will be the actual dimensions?

SOLUTION **a.** Let P = the width in centimeters of the floor plan.

    Since

$$\frac{\text{true length in m}}{\text{plan length in cm}} = \frac{15}{45},$$

$$\frac{\text{true width in m}}{\text{plan width in cm}} = \frac{8}{P} = \frac{15}{45}.$$

Solving, you find that the width is 24 cm.

b. Let l and w be the actual dimensions in meters of the living room.

$$\frac{l}{18} = \frac{15}{45} \qquad\qquad \frac{w}{13.5} = \frac{15}{45}$$

$$45l = 270 \qquad\qquad 45w = 202.5$$

$$l = 6 \qquad\qquad w = 4.5$$

$\therefore$ the living room will be 6 m by 4.5 m.

Rational Expressions in Open Sentences **457**

Key Ideas

Find the constant of variation.
Solve problems involving direct variation.

Chalkboard Examples

Solve.

1. If y varies directly as x, and if $y = 21$ when $x = 7$, determine the following.
   a. the constant of variation
   b. the value of y when $x = 10$
   Let $y = kx$, where k is the constant of variation.
   a. Since $y = 21$ when $x = 7$, then $k = \dfrac{y}{x} = \dfrac{21}{7} = 3$
   b. Since $k = 3$, and $x = 10$, then $y = kx = 3 \cdot 10 = 30$.

2. The distance traveled by a car with a constant velocity is directly proportional to the length of time traveled. If the car travels 120 km in 3 h, then how far will it have gone in 5 h?
   Solution 1
   Let $d = kt$ where d = distance in kilometers that the car travels, t = number of hours the car travels, and k = constant of variation.
   $$k = \frac{d}{t} = \frac{120}{3} = 40$$
   Since $k = 40$ and $t = 5$, $d = kt = 40 \cdot 5 = 200$.
   $\therefore$ the car travels 200 km in 5 h.

9–8 Direct Variation

The following table shows the distance, d, in kilometers traveled by a car in t h at a speed of 85 km/h. You can see that

$$\frac{d}{t} = 85, \quad \text{or} \quad d = 85t.$$

The graph of the linear equation $d = 85t$ is shown at the right of the table. Note that this equation defines a linear function.

t (h)	d (km)
1	85
2	170
3	255
4	340

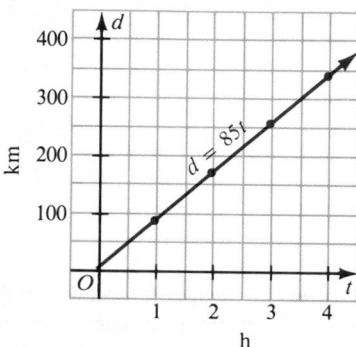

Notice from the table that, when the time is doubled, the distance is doubled; when the time is tripled, the distance is tripled; and so on. You say that the distance *varies directly* as the time. This function is an example of a *direct variation*.

Any function defined by an equation of the form

$$y = kx, \quad \text{where } k \text{ is a nonzero constant,}$$

is a **direct variation.** You say that y *varies directly as* x, or that y *is directly proportional to* x. The constant k is called the **constant of variation,** or the **constant of proportionality.**

When the domain is the set of real numbers, the graph of a direct variation defined by the equation $y = kx$ is a line with slope k that passes through the origin.

EXAMPLE 1 If r varies directly as s, and if $r = 2$ when $s = 10$, determine the following.
   a. the constant of variation   b. the value of r when $s = 30$

SOLUTION Let $r = ks$, where k is the constant of variation.
   a. Since $r = 2$ when $s = 10$,
   $$k = \frac{r}{s} = \frac{2}{10} = \frac{1}{5}.$$
   $\therefore$ the constant of variation is $\frac{1}{5}$.

   b. Since $k = \frac{1}{5}$ and $s = 30$,
   $$r = ks = \frac{1}{5} \cdot 30 = 6.$$
   $\therefore$ the value of r when $s = 30$ is 6.

458   *Chapter 9*

Consider any two ordered pairs (x_1, y_1) and (x_2, y_2) in the direct variation specified by $y = kx$. Then

$$y_1 = kx_1 \quad \text{and} \quad y_2 = kx_2.$$

Hence, if $x_1 \neq 0$ and $x_2 \neq 0$, then

$$\frac{y_1}{x_1} = k \quad \text{and} \quad \frac{y_2}{x_2} = k.$$

Therefore,

$$\frac{y_1}{x_1} = \frac{y_2}{x_2}.$$

Notice that this proportion does not involve the constant of variation k.

The next example illustrates two methods for solving a problem involving direct variation. In the first solution, the constant of variation is determined. In the second solution, a proportion is used and the constant of variation is not determined.

EXAMPLE 2   The distance that a spring is stretched is directly proportional to the mass of the object that is stretching the spring. If a mass of 15 g stretches a spring 6 cm, how far will an 8 g mass stretch the spring?

SOLUTION 1

Step 1   The problem asks how far an 8 g object will stretch the spring.

Step 2   Let d = the distance in centimeters that the spring is stretched,
 m = the mass in grams of the object stretching the spring,
and k = the constant of variation.

Step 3 $d = km$

Step 4   First find k.
Since $d = 6$ when $m = 15$,

$$k = \frac{d}{m} = \frac{6}{15} = 0.4.$$

Then find d when $m = 8$.
Since $k = 0.4$ and $m = 8$,

$$d = km = 0.4 \times 8 = 3.2.$$

Step 5   Checking the results is left to you.

$\therefore$ the spring is stretched 3.2 cm by an 8 g object.

Rational Expressions in Open Sentences   **459**

Solution 2

Let $\frac{d_1}{t_1} = \frac{d_2}{t_2}$, where $d_1 =$ 120 km, $t_1 = 3$ h, and $t_2 = 5$ h.

$$\frac{d_1}{t_1} = \frac{d_2}{t_2}$$
$$\frac{120}{3} = \frac{d_2}{5}$$
$$3d_2 = 600$$
$$d_2 = 200$$

$\therefore$ the car travels 200 km in 5 h.

Solve.

1. If u varies directly as v, and $u = 18$ when $v = 36$, find u when $v = 12$. 6

2. If p is directly proportional to q, and $p = 4$ when $q = 6$, find q when $p = 10$. 15

3. If c varies directly as d, and $c = -2$ when $d = \dfrac{1}{2}$, find c when $d = 12$. -48

4. The price of lumber varies directly with the number of board-feet purchased. If 80 board-feet cost $52.80, how many board-feet of lumber can be bought for $29.70? 45 board-feet

5. The interest earned on an investment varies directly with the interest rate. If a 9% rate yields $279, what interest rate yields $341? 11%

6. There are 42 serrations on a 3 cm segment of a knife blade. If the number of serrations is directly proportional to the length of the blade, how many serrations are on a 16 cm blade? 224 serrations

SOLUTION 2

Step 1   The problem asks how far the 8 g object will stretch the spring.

Step 2   Let d = the distance in centimeters that the spring is stretched, and m = the mass in grams of the object stretching the spring. Then $d_1 = 6$ cm, $m_1 = 15$ g, and $m_2 = 8$ g.

Step 3 $$\frac{d_1}{m_1} = \frac{d_2}{m_2}$$

Step 4 $$\frac{6}{15} = \frac{d_2}{8}$$
$$48 = 15d_2$$
$$d_2 = \frac{48}{15} = 3.2$$

Step 5   Does $\dfrac{d_1}{m_1} = \dfrac{d_2}{m_2}$?
$$\frac{6}{15} \overset{?}{=} \frac{3.2}{8}$$
$$6 \times 8 \overset{?}{=} 15 \times 3.2$$
$$48 = 48 \quad \checkmark$$

$\therefore$ the spring is stretched 3.2 cm by an 8 g object.

There are many situations in which one quantity *varies directly as the square* of another quantity. For example, the area of a circle varies directly as the square of the radius of the circle. In general, you can say that y *varies directly as the square of x* if

$$y = kx^2, \quad \text{where } k \text{ is a nonzero constant.}$$

You can also say that y *varies directly as*, or *is directly proportional to, the nth power of x* if

$$y = kx^n, \quad \text{where } n > 0, \text{ and } k \text{ is a nonzero constant.}$$

Oral Exercises

State whether the set of ordered pairs (x, y) satisfying the given equation, or specified by the given table, is a direct variation. For each direct variation, state the constant of variation.

1. $y = -8x$
   Yes; $k = -8$

2. $y = \frac{3}{4}x$
   Yes; $k = \frac{3}{4}$

3. $xy = 5$
   No

4. $y = \frac{3}{x}$
   No

5. $y + 4x = 0$
   Yes; $k = -4$

6. $y - 3x = 1$
   No

7.
x	1	2	3	4
y	-5	-10	-15	-20

Yes; $k = -5$

8.
x	2	4	6	8
y	-1	2	-3	4

No

9.

x	-2	0	2	4
y	-4	0	4	8

Yes; $k = 2$

10.

x	2	4	6	8
y	10	8	6	4

No

In Exercises 11 and 12, solve using the following two methods.

a. Determine the constant of variation. Use it to complete the solution.

b. Use a proportion. Do not determine the constant of variation.

11. If y varies directly as x, and if $y = 24$ when $x = 8$, find y when $x = 50$. **a.** $k = 3$; 150   **b.** 150

12. If p is directly proportional to t, and if $p = 2$ when $t = 10$, find p when $t = -1$. **a.** $k = \frac{1}{5}$; $-\frac{1}{5}$ **b.** $-\frac{1}{5}$

Problems

Solve.

A **1.** If d varies directly as t, and if $d = 4$ when $t = 9$, find d when $t = 21$. $\frac{28}{3}$

2. If v is directly proportional to w, and if $v = -6$ when $w = 15$, find w when $v = 3$. $-\frac{15}{2}$

3. If x is directly proportional to y, and if $x = a$ when $y = b$, find x when $y = 3b$. 3a

4. If a varies directly as b, and if $a = 5$ when $b = 6$, find a when $b - 1 = 11$. 10

5. A fish with a mass of 3 kg causes a fishing pole to bend 9 cm. If the amount of bending varies directly as the mass, how much will the pole bend for a 2 kg fish? 6 cm

6. The mass of a uniform copper bar varies directly as its length. If a bar 40 cm long has a mass of approximately 420 g, find the mass of a bar 136 cm long. 1428 g

7. A tropical fern 90 cm tall has fronds that are 3 cm long. A fossil of the same kind of fern was found. If the height of this type of fern varies directly as the length of its fronds, what must the height of the fossilized fern have been if one of its fronds was 4.8 cm long? 144 cm

8. The 220 km bus trip to the math league competition cost each of the students from Canton $26.40. Using the same bus company, the students from Hampton paid $42 each. If the bus fare is directly proportional to the number of kilometers traveled, how far was the Hampton students' trip? 350 km

B **9.** If c varies directly as $d - 3$, and if $c = 16$ when $d = 7$, find c when $d = 10$. 28

10. If r varies directly as $s + 5$, and if $r = 1$ when $s = -2$, find r when $s = -8$. -1

Rational Expressions in Open Sentences   **461**

Mixed Review

Graph each function on the same coordinate plane.

1. a. $f(x) = 2x$
   b. $f(x) = 2x - 3$
   c. $f(x) = -2x$

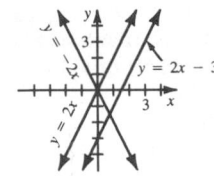

Simplify.

3. $\dfrac{12a^2 b^3}{15ab^4}$ $\dfrac{4a}{5b}$

4. $\dfrac{18n^4 p^5}{21n^3 p^7}$ $\dfrac{6n}{7p^2}$

Solve for t.

5. $\dfrac{2s + 3t}{s - t} = -\dfrac{4}{3}$ $\{-2s\}$

6. $\dfrac{w - wt}{6w} = \dfrac{2}{3}$ $\{-3\}$

Additional Answers
Problems

17. No; the graph of the linear function must go through the origin in order for the function to be a direct variation.
18. No; if the constant of proportionality is negative, then y decreases as x increases.
19. $r_1 = kt;$
    $r_2 = k(2t) = 2(kt);$
    $\therefore r_2 = 2r_1$
20. $b = k_1 c, c = k_2 d$
    $b = k_1(k_2 d) = (k_1 k_2)d$
    $\therefore b$ varies directly as d, with $k_1 k_2$ as the constant of proportionality.

Teaching Suggestions
p. T103

Related Activities p. T103

Key Ideas

Find the constant of variation in inverse variations.
Solve problems involving inverse variation.

---

11. If e varies directly as f^3, and if $e = 32$ when $f = \frac{2}{5}$, write a formula for e in terms of f. $e = 500f^3$

12. If s varies directly as v^4, and if $s = 8$ when $v = -\frac{2}{3}$, write a formula for s in terms of v. $s = \frac{81}{2}v^4$

13. The surface area of a cube varies directly as the square of the length of one edge. If the surface area of a cube is 150 cm² when the length of an edge is 5 cm, what is the surface area of a cube with an edge of length 8 cm? 384 cm²

14. The surface area of a sphere varies directly as the square of the radius. If the surface area of a sphere is 36π m² when the radius is 3 m, what is the surface area of a sphere with a radius of 6 m? 144π m²

15. A certain mineral's price varies directly as the square of its mass. If a sample with mass 4.2 g is worth $61.74, what will be the value of a sample with mass 100 g? $35,000

16. A plastic rod changes length in direct proportion to the change in temperature. At 22°C it is 500 mm long. At 42°C it has increased by 0.05 mm. What is the length if the temperature drops to 18°C? 499.99 mm

C 17. Is every linear function a direct variation? Explain.

18. If y varies directly as x, must y increase when x increases? Explain.

19. If r varies directly as t, show that r is doubled when t is doubled.

20. If b varies directly as c, and c varies directly as d, show that b varies directly as d.

21. If u varies directly as v, and u increases by 18 when v is increased by 3, express u in terms of v. $u = 6v$

22. If p varies directly as the square of q, and q varies directly as the square of r, what effect will doubling r have on the value of p?
It will be multiplied by a factor of 16.

9–9 Inverse Variation

The table at the right shows the time, t, that it takes a car traveling at a speed of r km/h to cover a distance of 90 km. You can see that

$$rt = 90 \quad \text{or} \quad t = \frac{90}{r}.$$

r (km/h)	t (h)
45	2
50	1.8
60	1.5
75	1.2
90	1

This relationship is an example of an *inverse variation*.
Any function defined by an equation of the form

$$xy = k, \quad \text{or} \quad y = \frac{k}{x},$$

where k is a nonzero constant, is an **inverse variation**. You say that y *is inversely proportional to* x, or that y *varies inversely as* x. As with direct variation, the nonzero constant k is called the *constant of variation*, or the *constant of proportionality*.

462   *Chapter 9*

The graph of an inverse variation is not a line, since an equation of the form

$$y = \frac{k}{x}, \quad \text{or} \quad xy = k,$$

is not linear. In a linear equation no term can be of degree greater than 1; the term xy is of degree 2.

Each of the following diagrams illustrates the graph of an inverse variation

$$y = \frac{k}{x}, \text{ where } k \text{ is a nonzero constant.}$$

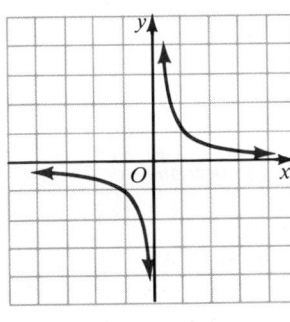

$y = \frac{k}{x}, k > 0$

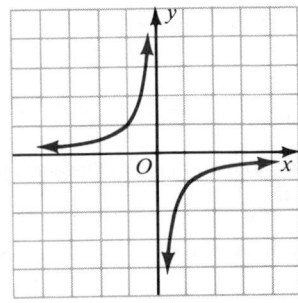

$y = \frac{k}{x}, k < 0$

In the diagram at the left, $k > 0$. In the diagram at the right, $k < 0$. Note that, in each case, the graph does not intersect the x-axis or the y-axis since neither x nor y can have the value 0.

The graph of an inverse variation is an example of a curve called a *hyperbola*. If the constant of variation is positive, the branches of the hyperbola will be in Quadrants I and III. If the constant of variation is negative, the branches of the hyperbola will be in Quadrants II and IV.

EXAMPLE If y varies inversely as x, and if $y = 4$ when $x = 72$, find the following.
  a. the constant of variation
  b. the value of y when $x = 9$

SOLUTION **a.** Let $xy = k$, where k is the constant of variation.
    Since $y = 4$ when $x = 72$,

$$k = xy = 72 \cdot 4 = 288.$$

    . . the constant of variation is 288.

  b. Since $k = 288$ and $x = 9$,

$$y = \frac{k}{x} = \frac{288}{9} = 32.$$

    $\therefore$ when $x = 9$, the value of y is 32.

Rational Expressions in Open Sentences **463**

Solve.

1. If r varies inversely as s, and $r = \frac{1}{2}$ when $s = 6$, find s when $r = 9$. $\frac{1}{3}$

2. If y varies inversely as z, and $y = -3$ when $z = 6$, find y when $z = \frac{1}{2}$. -36

3. If g varies inversely as h, and $g = 0.1$ when $h = 5$, find g when $h = 25$. 0.02

4. The time needed for George to jog is inversely proportional to his speed. If he jogs at 7 km/h for 2 h, how long will it take him to cover the same distance if he increases his speed to 10 km/h? 1.4 h

5. The resistance in an electric circuit varies inversely with the current. If the resistance is 6 ohms and the current is 20 amps, what is the current when the resistance is 2 ohms? 60 amps

6. A flower bed must be of a certain area to plant the 72 seedlings Mrs. Vaughn has bought. One that is 6 m wide and 10 m long is of the right size, but she prefers a bed 3 m wide. What should the length of the flower bed be? 20 m

Consider any two ordered pairs of nonzero real numbers (x_1, y_1) and (x_2, y_2) in the inverse variation specified by $y = \frac{k}{x}$, $k \neq 0$. Then

$$y_1 = \frac{k}{x_1} \quad \text{and} \quad y_2 = \frac{k}{x_2}.$$

Since $x_1 y_1 = k$ and $x_2 y_2 = k$, you have

$$x_1 y_1 = x_2 y_2.$$

Thus if three of these values are known, you can find the fourth value without first finding the constant of variation. For example, using the information given in the example on page 463, you can let $x_1 = 72$, $y_1 = 4$, $x_2 = 9$, and solve for y_2.

You say that y *varies inversely as the square of* x if

$$y = \frac{k}{x^2}, \quad \text{where } k \text{ is a nonzero constant.}$$

In general, you say that y *varies inversely as*, or *is inversely proportional to, the nth power of* x if

$$y = \frac{k}{x^n}, \quad \text{where } n > 0 \text{ and } k \text{ is a nonzero constant.}$$

Oral Exercises

State the relationship between the given variables as an equation, using k for the constant of variation.

1. The volume V of a gas at a fixed temperature varies inversely as the pressure P. $V = \frac{k}{P}$

2. The current I in an electrical circuit of fixed voltage varies inversely as the resistance R. $I = \frac{k}{R}$

3. The height h of a cylinder of fixed volume varies inversely as the area A of the base. $h = \frac{k}{A}$

4. The frequency f of an electromagnetic wave is inversely proportional to the length l of the wave. $f = \frac{k}{l}$

In Exercises 5 and 6, solve using the following two methods.

a. Determine the constant of variation. Use it to complete the solution.

b. Solve without first finding the constant of variation.

5. If y varies inversely as x, and if $y = 9$ when $x = 2$, find y when $x = 3$. **a.** $k = 18$; 6 **b.** 6

6. If u is inversely proportional to v, and if $u = 12$ when $v = 3$, find u when $v = 9$. **a.** $k = 36$; 4 **b.** 4

Problems

Solve.

A 1. If a varies inversely as b, and if $a = 4$ when $b = 25$, find a when $b = 5$. **20**

2. If r varies inversely as s, and if $r = 25$ when $s = 10$, find s when $r = 50$. **5**

3. If c is inversely proportional to d, and if $c = 18$ when $d = \frac{2}{3}$, find d when $c = \frac{6}{7}$. **14**

4. If m is inversely proportional to n, and if $m = 0.02$ when $n = 5$, find m when $n = 0.2$. **0.5**

5. The frequency of a radio wave is inversely proportional to the length of the wave. If a wave of length 500 m has a frequency of 600 kHz (kilohertz), find the frequency of a wave whose length is 750 m. **400 kHz**

6. The time needed to travel from one place to another is inversely proportional to the speed. A person traveling 72 km/h can go from New Glasgow to Pictou in 10 h. How fast must the person travel to make the trip in 9 h? **80 km/h**

7. The force needed to pry up a rock varies inversely as the length of the crowbar used. Using a crowbar 100 cm long, a force of 150 N (newtons) is needed to pry up a certain rock. How long a crowbar is needed to pry up the same rock when the force applied is 120 N? **125 cm**

8. The number of hours needed to clear the trees from some land is inversely proportional to the number of people who are working. If it would take 4 people 15 h to do the job, how many people would be needed to complete the job in 6 h? **10 people**

9. A pulley revolves at a speed that is inversely proportional to its diameter. A pulley with a diameter of 12 cm is belted to a pulley with a diameter of 8 cm. If the smaller pulley is revolving at a rate of 96 rpm (revolutions per minute), how fast is the larger pulley revolving? **64 rpm**

10. If the rotational speed of a gear wheel is inversely proportional to the number of teeth on the wheel, how fast is a gear wheel with 25 teeth revolving if it is meshed with a gear wheel with 40 teeth that is revolving at 150 rpm? **240 rpm**

B 11. If a varies inversely as $b + 2$, and if $a = 8$ when $b = 1.5$, find a when $b = 3$. **5.6**

12. If x varies inversely as $y - 4$, and if $x = -5$ when $y = \frac{1}{2}$, find x when $y = -1$. **$-\frac{7}{2}$**

13. If a varies inversely as t^2, and if $a = 2$ when $t = 0.3$, write a formula for a in terms of t. **$a = \frac{0.18}{t^2}$**

14. If f varies inversely as m^3, and if $f = 0.5$ when $m = 2$, write a formula for f in terms of m. **$f = \frac{4}{m^3}$**

Rational Expressions in Open Sentences   **465**

Mixed Review

Solve for the variable indicated.

1. $A = 2(B - 3C)$; C

$$\left\{ \frac{2B - A}{6} \right\}$$

2. $\dfrac{at - bt}{t^2} = \dfrac{4}{t}$; a

$$\{4 + b\}$$

Solve.

3. If r varies directly as s, and $r = -8$ when $s = -6$, find r when $s = 3$.   **4**

4. A length of elastic stretches in direct variation with the mass hung from it. If a 14 g mass stretches the elastic 6 cm, what mass stretches it 9 cm?   **21 g**

5. If m varies in direct proportion to n, and $m = -4$ when $n = -1$, find n when $m = 3$.   **$\frac{3}{4}$**

18. y is directly proportional to t because if $y = \dfrac{k_1}{x}$ and $x = \dfrac{k_2}{t}$, then $y = k_1 \div \left(\dfrac{k_2}{t}\right) = \left(\dfrac{k_1}{k_2}\right)t$. (The constant of proportionality is $\dfrac{k_1}{k_2}$.)

15. The interest rate required to yield a given income is inversely proportional to the amount of money invested. Chris receives income from \$16,000 that she has invested at an annual interest rate of 8%. How much money should she invest to receive the same income if the annual interest rate increases to 10%? $12,800

16. The height of a cylinder of fixed volume varies inversely as the area of the base. Ken uses a cylindrical vat that has a base area of 6 m² and a height of 1.5 m for mixing paint. He wants to replace this vat with a new one having the same volume but with a height of 2 m. What will be the base area of the new vat? 4.5 m²

C **17.** If y is inversely proportional to x, how does y change when x is doubled? y is halved.

18. If y is inversely proportional to x, and if x is inversely proportional to t, what is the relationship of y to t? Explain.

19. If a is inversely proportional to b, and if b is inversely proportional to c^2, what effect will doubling c have on a? a will be quadrupled.

20. Assume that a is inversely proportional to b, b is inversely proportional to c, and c is inversely proportional to d^2. What effect will doubling d have on a? a will be divided by 4.

PROGRAMMING IN BASIC

The program on page 247 graphs open sentences in two variables only for integral values of X and Y. In the following graphing program, the replacement set of X is restricted to {−10, −9, −8, . . . , 10}, but the corresponding values of Y may be fractional. In such cases, the TAB function *approximates* the value of Y by assigning to it the value of the greatest integer less than or equal to it. Since Y is associated with TAB, the program plots values of Y horizontally and values of X vertically.

```
10   PRINT "TO GRAPH THE FUNCTION"
20   PRINT "DEFINED IN LINE 70:"
30   PRINT "INPUT K (> 0)";
40   INPUT K
50   PRINT
60   FOR X = −10 TO 10
70   LET Y = K*X
80   IF ABS(Y) > 10 THEN 200
90   IF INT(Y) = 0 THEN 150
100  IF Y < 0 THEN 130
110  PRINT TAB(11);"!";TAB(Y + 11);"*";
120  GOTO 160
130  PRINT TAB(Y + 11);"*";TAB(11);"!";
140  GOTO 160
```

```
150   PRINT TAB(11);"*";
160   IF X = 0 THEN 190
170   PRINT
180   GOTO 200
190   PRINT TAB(22);"----Y"
200   NEXT X
210   PRINT TAB(11);"X"
220   END
```

Exercises

Type in the program as given. Note that the equation in line 70, $y = kx$, represents a *direct* variation. RUN the program for the following values of k.

1. 1   **2.** 2   **3.** 0.5

Change line 70 to represent a variation directly as the *square* of x, that is, $y = kx^2$. RUN the program for the following values of k.

4. 0.25   **5.** 0.5   **6.** 1

To graph that portion of an *inverse* variation that is in the first quadrant, first delete lines 80, 180, and 190, then type in these lines.

```
55    PRINT TAB(11);"!";TAB(K + 12);"----Y"
60    FOR X = 1 TO K
70    LET Y = K/X
160   GOTO 170
```

RUN the revised program for the following values of k.

7. 12   **8.** 15   **9.** 18

9–10 Joint and Combined Variation

If one variable varies directly as the product of two or more other variables, the resulting relationship is called a **joint variation**. For example, the distance traveled by a car starting from rest is directly proportional to the product of its acceleration, a, and the square of the time, t, it has been traveling. This joint variation is defined by the equation

$$d = \frac{1}{2}at^2,$$

where the constant of variation is $\frac{1}{2}$. You can say that the distance *varies jointly as* the acceleration and the square of the time.

Rational Expressions in Open Sentences   **467**

Teaching Suggestions
p. T103

Supplementary Material
Test 36

Key Ideas

Find the constant of variation in cases of joint and combined variation.
Solve problems involving joint variation.
Solve problems involving combined variation.

Solve.

1. If b varies jointly as t^2 and c, and $b = 90$ when $t = 3$ and $c = 2$, find b when $t = 2$ and $c = 6$.

Let $b = kct^2$ where k is the constant of variation.

$$k = \frac{b}{ct^2} = \frac{90}{2 \cdot 3^2} = 5$$

$b = kct^2 = 5 \cdot 6 \cdot 2^2 = 120$

$\therefore b = 120$ when $t = 2$ and $c = 6$.

2. The force required to spin a weight at the end of a string varies directly with the square of the velocity (v), and inversely with the length of the string (r). If the force is 5 newtons when the velocity is 5 m/s and the string is 1 m long, what is the force when the length of the string is 2 m and the velocity is 4 m/s?

Let $F = k\dfrac{v^2}{r}$ where k is a nonzero constant, and $r \neq 0$.

$$k = \frac{rF}{v^2} = \frac{1 \cdot 5}{5^2} = \frac{5}{25} = \frac{1}{5} = 0.2$$

$$F = k\frac{v^2}{r} = \frac{0.2(4)^2}{2} = \frac{0.2 \cdot 16}{2} = 1.6$$

$\therefore$ the required force is 1.6 newtons.

If you know that z varies jointly as x and y, you can express the relationship as

$$z = kxy, \text{ where } k \text{ is a nonzero constant.}$$

If (x_1, y_1, z_1) and (x_2, y_2, z_2) are two ordered triples that satisfy the equation $z = kxy$, and if x_1, x_2, y_1, and y_2 are nonzero, then

$$\frac{z_1}{x_1 y_1} = k \quad \text{and} \quad \frac{z_2}{x_2 y_2} = k,$$

and so

$$\frac{z_1}{x_1 y_1} = \frac{z_2}{x_2 y_2}.$$

EXAMPLE 1 If r varies jointly as s and the square of t, and if $r = 144$ when $s = 2$ and $t = 3$, find r when $s = 5$ and $t = 4$.

SOLUTION Let $r_1 = 144$, $s_1 = 2$, $t_1 = 3$, $s_2 = 5$, $t_2 = 4$, and use the proportion

$$\frac{r_1}{s_1(t_1)^2} = \frac{r_2}{s_2(t_2)^2}$$

$$\frac{144}{2(3)^2} = \frac{r_2}{5(4)^2}$$

$$\frac{144}{18} = \frac{r_2}{80}$$

$$144(80) = 18r_2$$

$$640 = r_2$$

Notice in Example 1 that the required value of r could also have been found by first determining the constant of variation.

If a variable *varies directly as* one variable (or a power of the variable) and *inversely as* another variable (or a power of the variable), the resulting relationship is called a **combined variation**. For example, if you know that z varies directly as x and inversely as y, you can express the relationship as

$$z = k\left(\frac{x}{y}\right), \text{ where } k \text{ is a nonzero constant.}$$

If (x_1, y_1, z_1) and (x_2, y_2, z_2) are two ordered triples that satisfy the equation $z = k\left(\frac{x}{y}\right)$, and if x_1 and x_2 are nonzero, then

$$\frac{z_1 y_1}{x_1} = k \quad \text{and} \quad \frac{z_2 y_2}{x_2} = k,$$

and so

$$\frac{z_1 y_1}{x_1} = \frac{z_2 y_2}{x_2}.$$

EXAMPLE 2   The pressure required to force water through a pipe varies directly as the square of the speed of the water and inversely as the diameter of the pipe. If it requires 80 N/m² (newtons per square meter) pressure to drive water at 40 km/h through a pipe with a 2 cm diameter, what would be the pressure required to drive water at 30 km/h through a pipe with a diameter of 1.5 cm?

SOLUTION

Step 1   The problem asks for the pressure needed to drive water at 30 km/h through a pipe with a 1.5 cm diameter.

Step 2   Let p = the pressure,
         s = the speed of the water,
         d = the diameter of the pipe,
         and k = the constant of variation.

Step 3 $p = k\left(\dfrac{s^2}{d}\right)$

Step 4   Since $p = 80$ when $s = 40$ and $d = 2$,

$$80 = k\left(\frac{40^2}{2}\right),$$

and

$$k = \frac{80 \cdot 2}{40^2} = \frac{1}{10}.$$

Since $k = \dfrac{1}{10}$, $s = 30$, and $d = 1.5$,

$$p = k\left(\frac{s^2}{d}\right) = \frac{1}{10}\left(\frac{30^2}{1.5}\right) = 60.$$

Step 5   Checking the results is left to you.

∴ the required pressure is 60 N/m².

Notice in Example 2 that the required pressure could also have been found by using the proportion

$$\frac{p_1 d_1}{(s_1)^2} = \frac{p_2 d_2}{(s_2)^2},$$

where $p_1 = 80$, $d_1 = 2$, $s_1 = 40$, $d_2 = 1.5$, and $s_2 = 30$.

Oral Exercises

Express the relationship in words, assuming that k is the constant of variation.

1. $c = kef$ **2.** $r = ks^2 t^2$ **3.** $e = \dfrac{kf^2}{g}$ **4.** $m = \dfrac{kp^3}{q^2}$

Rational Expressions in Open Sentences   **469**

Additional Answers
Oral Exercises

5. u varies jointly as the square of v and the square root of w.

6. g varies inversely as the product of h and i.

7. t varies directly as the square of c and inversely as the product of d and the cube of e.

8. h varies jointly as e_1 and the square of e_2 and inversely as the cube of e_3.

Express the relationship in words, assuming that k is the constant of variation.

5. $u = kv^2\sqrt{w}$ **6.** $g = \dfrac{k}{hi}$

7. $t = \dfrac{kc^2}{de^3}$ **8.** $h = \dfrac{ke_1e_2^2}{e_3^3}$

In Exercises 9 and 10, solve using the following two methods.

a. Determine the constant of variation. Use it to complete the solution.

b. Use a proportion. Do not determine the constant of variation.

9. If b varies jointly as g and h, and if $b = 12$ when $g = 10$ and $h = 6$, find b when $g = 15$ and $h = 3$. **a.** $k = \frac{1}{5}$; 9   **b.** 9

10. If n varies directly as s and inversely as r, and if $n = 1$ when $s = 8$ and $r = 12$, find n when $s = 10$ and $r = 15$. **a.** $k = \frac{3}{2}$; 1   **b.** 1

Problems

Solve.

A **1.** If d varies jointly as p and q, and if $d = 20$ when $p = 8$ and $q = 15$, find d when $p = 9$ and $q = 16.24$

    2. If q varies directly as b and inversely as c, and if $q = 0.2$ when $b = 16$ and $c = 8$, find q when $b = 2$ and $c = 5.0.04$

    3. If z varies directly as x^2 and inversely as y, and if $z = 8$ when $x = 4$ and $y = 6$, find z when $x = 0.1$ and $y = 0.2.0.15$

    4. If Q varies jointly as f^2 and g^3, and if $Q = 24$ when $f = 4$ and $g = 0.5$, find Q when $f = 0.25$ and $g = 10.750$

    5. If z varies jointly as u and v^2, and if $z = 12$ when $u = 6$ and $v = 4$, find u when $z = 63$ and $v = 6.14$

    6. If A varies jointly as m^2 and n, and if $A = 15$ when $m = 3$ and $n = 10$, find n when $A = 0.4$ and $m = 0.5.9.6$

    7. If T varies directly as s and inversely as v^2, and if $T = 0.1$ when $s = 72$ and $v = 6$, find s when $T = 5$ and $v = 0.2.4$

    8. If H varies directly as c^2 and inversely as d, and if $H = \frac{5}{3}$ when $c = 2$ and $d = 15$, find d when $H = \frac{1}{8}$ and $c = \frac{1}{2}$. $\frac{25}{2}$

In Exercises 9–12 assume that Q varies directly as x and inversely as y.

    9. If x is doubled and y is doubled, what happens to Q? No change

  10. If x is tripled and y is doubled, what happens to Q? Q is multiplied by $\frac{3}{2}$.

  11. If x is halved and y is doubled, what happens to Q? Q is divided by 4.

  12. If x is doubled and y is halved, what happens to Q? Q is multiplied by 4.

470 *Chapter 9*

Solve.

B 13. The volume of a cylinder varies jointly as its height and the square of its radius. When the height is 3 cm and the radius is 4 cm, the volume is 48π cm³. Find the volume in terms of π when the height is 3 cm and the radius is 5 cm. $75\,\pi$ cm³

14. If a wire carries an electric current for a given time, the heat developed varies jointly as the resistance and the square of the current. If a current of 4 A (amperes) produces 225 J (joules) of heat in a wire having a resistance of 15 Ω (ohms), find the heat produced by a current of 3 A in a wire having a resistance of 12 Ω. 101.25 J

15. The number of persons needed to do a job varies directly as the amount of work to be done and inversely as the time in which the job must be done. If 2 people can cut 6 trees into fireplace logs in 8 h, how long will it take 6 people to cut 12 trees into logs? $5\frac{1}{3}$ h

16. The heat loss through a window pane varies jointly as the difference of the inside and outside temperatures and the window area, and inversely as the thickness of the pane. In 1 h, 189 J (joules) are lost through a pane measuring 45 cm by 70 cm that is 1 cm thick, when the temperature difference is 5°C. How many joules are lost in 1 h through a pane measuring 30 cm by 60 cm that is 0.8 cm thick when the temperature difference is 12°C? 324 J

17. The volume of a pyramid varies jointly as its height and the area of its base. A pyramid whose base is a square 5 cm on each side and whose height is 12 cm has a volume of 100 cm³. Find the volume of a pyramid whose base is a square 6 cm on each side and whose height is 8 cm. 96 cm³

18. The centrifugal force of an object moving in a circle varies jointly with the object's mass and the radius of the circular path, and inversely with the square of the time it takes to complete one full circle. The centrifugal force of a 6 g object moving in a circle with radius 100 cm and completing each revolution in 2 s is 6000 dynes. What is the centrifugal force of an 18 g object moving in a circle with radius 200 cm and completing each revolution in 3 s? 16,000 dynes

C 19. In the formula $x = \dfrac{y^2zw}{4}$, z remains constant. If y is doubled and w is tripled, how is x changed? It is multiplied by 12.

20. In the formula $F = \dfrac{m\pi l^2}{p}$, m and π remain constant, l is halved, and p is doubled. How does F change? It is multiplied by $\frac{1}{8}$.

21. The heat generated by a stove element varies directly as the square of the voltage and inversely as the resistance. If the voltage is constant, what could be done to double the heat generated? Halve the resistance.

22. The power in an electric circuit varies jointly as the resistance and the square of the current. If the current is halved, what other change needs to be made if the power is to remain the same? Multiply the resistance by 4.

Rational Expressions in Open Sentences **471**

4. If e varies directly as f^3 and inversely as g, and $e = 1$ when $f = \dfrac{1}{2}$ and $g = \dfrac{1}{4}$, find e when $f = 2$ and $g = 4$. 4

5. If y varies directly as x and inversely as z^2, and $x = 2$ when $y = 2$ and $z = \dfrac{1}{2}$, find y when $x = \dfrac{2}{3}$ and $z = \dfrac{3}{4}$. $\dfrac{8}{27}$

6. If n varies directly as m and inversely as p^2, what happens to n when m is doubled and p is doubled? n is halved.

Mixed Review

Solve.

1. Find the ratio of polyester to cotton in a blouse that is 65% cotton and 35% polyester. $\dfrac{7}{13}$

2. Find the ratio of a to b if $\dfrac{-6a + 3b}{4b} = \dfrac{-1}{2}$. $\dfrac{5}{6}$

3. The circular area that a dog can range varies directly as the square of the length of chain to which it is tied. If the area is 100π m² when the chain is 10 m long, how long must the chain be when the area is to be 289π m²? 17 m

4. If a varies directly as b, and a increases by 12 when b is increased by 3, express a in terms of b. $a = 4b$

5. If m varies inversely as n, and $m = -24$ when $n = -3$, find m when $n = -8$. -9

Solve.

1. The $220 montly rent for a beach house is to be split by two families in the ratio $8:3$. How much must each family pay? $160, $60

2. If f varies directly with g, and if $f = 6$ when $g = 26$, find f when $g = 65$. 15

3. If z varies inversely as r, and $z = 12$ when $r = 2$, find z when $r = 3$. 8

4. If h varies jointly with i and j, and if $h = 1.25$ when $j = 2.5$ and $i = 5$, find h when $j = 4$ and $i = 2.5$. 1

5. If d varies directly as e and inversely as f, and $d = 3$ when $e = 2$ and $f = 4$, find d when $e = 5$ and $f = 10$. 3

Self-Test 3

VOCABULARY   ratio (p. 449)
proportion (p. 450)
extremes (p. 450)
means (p. 450)
direct variation (p. 458)
constant of variation (p. 458)

constant of proportionality (p. 458)
inverse variation (p. 462)
joint variation (p. 467)
combined variation (p. 468)

1. A rope that is 8.4 m long is to be cut into two pieces in the ratio of $5:7$. How long will each piece be? 3.5 m and 4.9 m   *Obj. 1, p. 449*

2. Property tax varies directly as assessed value. The tax on a certain house assessed at $60,000 is $1410. What is the tax on a house in the same community that is assessed at $100,000? $2350   *Obj. 2, p. 449*

3. If c varies inversely as d, and if $c = 4$ when $d = 9$, find c when $d = 6$. 6

4. The volume of a cone varies jointly as its height and the square of its radius. A cone whose height is 3 cm and whose radius is 7 cm has a volume of 154 cm³. What is the volume of a cone whose height is 7 cm and whose radius is 6 cm? 264 cm³

5. If r varies directly as s and inversely as t, and if $r = 4$ when $s = 16$ and $t = 2$, find r when $s = 36$ and $t = 9$. 2

Check your answers with those at the back of the book.

Chapter Summary

1. If an open sentence contains fractions with whole-number denominators, the fractions may be eliminated by multiplying both sides by their LCD.

2. Percent means "per 100." Percents are commonly used in problems involving rates and mixtures. When solving an equation that contains a percent, it is convenient to first express the percent as a fraction or as a decimal.

3. An equation with variables in the denominator of one or more terms is called a *fractional equation*. Fractions can be eliminated in a fractional equation by multiplying both sides by the LCD of the terms. This may introduce *extraneous roots*, and so each root of the new equation must be checked to determine whether it satisfies the original equation.

4. Fractional equations are often used in solving number problems, work problems, and motion problems.

472   *Chapter 9*

5. The *ratio* 3 to 4 can be written as $\frac{3}{4}$ or $3:4$. In solving problems, if the ratio of two numbers is given as $3:4$, the numbers can be represented as $3x$ and $4x$.

6. A *proportion* is an equation stating that two ratios are equal. In a proportion, the product of the *extremes* equals the product of the *means*. Thus,

$$\text{if } \frac{a}{b} = \frac{c}{d}, \quad \text{then } ad = bc.$$

7. Many relationships can be described in terms of *direct, inverse, joint,* or *combined variation*. The following is a summary of these types of variations. In each case, k is a nonzero constant called the *constant of variation*.

   a. A *direct variation* is a function defined by a linear equation of the form $y = kx$. You say that y varies directly as x, or that y is directly proportional to x. The graph of an equation of the form $y = kx$ is a line.

   b. An *inverse variation* is a function defined by an equation of the form $xy = k$, or $y = \frac{k}{x}$. You say that y varies inversely as x, or that y is inversely proportional to x. The graph of an inverse variation is a hyperbola.

   c. An equation of the form $z = kxy$ defines a *joint variation*. You say that z varies directly as the product xy, or that z varies jointly as x and y.

   d. An equation of the form $z = k\left(\frac{x}{y}\right)$ defines a *combined variation*. You say that z varies directly as x and inversely as y.

8. In solving problems involving variations, it is helpful to be familiar with the two methods illustrated in Example 2 on pages 459 and 460. In the first method, the constant of variation is determined. In the second method, the constant of variation is not determined.

Chapter Review

Write the letter of the correct answer.

1. Solve $\dfrac{x-4}{6} - \dfrac{20-2x}{8} = 1.$ $\qquad\qquad$ 9-1

   a. $\{12\}$ $\qquad$ **b.** $\{10\}$ $\qquad$ c. $\{-50\}$ $\qquad$ d. $\{8\}$

2. Solve $\dfrac{3a-14}{6} \le \dfrac{2a}{9} + \dfrac{a}{4}.$

   a. $\{a: a \le 84\}$ b. $\{a: a \ge 84\}$ c. $\{a: a \le -84\}$ d. $\{a: a \ge -84\}$

3. In 1980 the population of a certain town was 20,000. If the population of the same town in 1985 was 25,000, what was the percent increase from the 1980 population?

    a. 75%   **b.** 20%   (**c.**) 25%   **d.** 125%

4. Sharon invested $8000, part at an annual interest rate of 6% and the rest at an annual interest rate of 8%. How much did she invest at 8% if her total income for one year from these investments was $540?

    a. $2000   (**b.**) $3000   **c.** $5000   **d.** $6000

5. Solve $\dfrac{4}{3e} + \dfrac{3}{3e + 1} = -2$.

    a. $\left\{\dfrac{1}{6}, \dfrac{4}{3}\right\}$ **b.** $\left\{6, \dfrac{3}{4}\right\}$ (**c.**) $\left\{-\dfrac{4}{3}, -\dfrac{1}{6}\right\}$ **d.** $\left\{-\dfrac{1}{2}, -\dfrac{2}{3}\right\}$

6. Solve $\dfrac{30}{d^2 - 9} + 2 = \dfrac{5}{d - 3}$.

    a. $\left\{-\dfrac{1}{2}, 3\right\}$ (**b.**) $\left\{-\dfrac{1}{2}\right\}$ **c.** $\left\{-\dfrac{1}{2}, -3\right\}$ **d.** $\left\{\dfrac{1}{2}, 3\right\}$

7. One number is 40 less than another number. Two thirds of the lesser number is equal to one fourth of the greater number. Find the greater number.

    a. 24   **b.** 16   (**c.**) 64   **d.** 42

8. The denominator of a fraction is 8 less than twice the numerator, and the fraction is equal to $\dfrac{17}{32}$. Find the fraction.

    (**a.**) $\dfrac{68}{128}$ **b.** $\dfrac{34}{64}$ **c.** $\dfrac{51}{96}$ **d.** $\dfrac{85}{160}$

9. One pipe can fill a tank in 12 h and another pipe can fill the same tank in 8 h. How long will it take both pipes together to fill the tank?

    a. 4 h   (**b.**) $4\dfrac{4}{5}$ h   **c.** 5 h   **d.** $3\dfrac{1}{2}$ h

10. It takes John twice as long to clear out his garage as it would take Mark to do the same job. If they work together, they can complete the job in $1\frac{1}{3}$ h. How long would it take John to do the job alone?

    (**a.**) 4 h   **b.** 2 h   **c.** 1 h   **d.** 6 h

11. Sam drove his car during part of his trip at an average rate of 75 km/h. He then took a bus that averaged 90 km/h during the remainder of his trip. If he traveled by bus 55 km longer than he had traveled by car, and his total travel time was $5\frac{1}{2}$ h, how far did he travel by car?

    a. 130 km   **b.** 145 km   **c.** 255 km   (**d.**) 200 km

12. Jill travels 24 km up a river in the same amount of time it takes her to go 36 km down the river. If the current is flowing at a speed of 3 km/h, what is Jill's speed in still water?

   a. 10 km/h   **b.** 12 km/h   **c.** 15 km/h   **d.** 18 km/h

13. Solve $\dfrac{3w}{w + 2} = \dfrac{5}{2}$.   9-7

   a. {2}   **b.** {10}   **c.** {5}   **d.** $\{\frac{1}{5}\}$

14. The number of votes received by the winner and loser in a county election was in the ratio $8:5$. If the winner received 11,880 votes, how many votes did the loser receive?

   a. 8125   **b.** 7425   **c.** 7050   **d.** 6995

15. If w is directly proportional to z, and if $w = 8$ when $z = 24$, find w when $z = 12$.   9-8

   a. −4   **b.** $\frac{1}{3}$ **c.** 36   **d.** 4

16. The distance that a free falling body travels varies directly as the square of the time it has been falling. If an object falls 320 m in 8 s, how far will it fall in 3 s?

   a. 5 m   **b.** 15 m   **c.** 120 m   **d.** 45 m

17. If x varies inversely as y, and if $x = 18$ when $y = 52$, find x when $y = 234$.   9-9

   a. 4   **b.** 20   **c.** 14   **d.** 24

18. The base of a rectangle of fixed area varies inversely as its height. The height of a certain rectangle is 12 cm and its base is 42 cm. Find the base of another rectangle of equal area whose height is 14 cm.

   a. 15 cm   **b.** 30 cm   **c.** 36 cm   **d.** 12 cm

19. If e varies directly as g and inversely as h, and if $e = 6$ when $g = 18$ and $h = 9$, find g when $e = 12$ and $h = 3$.   9-10

   a. 12   **b.** 3   **c.** 4   **d.** 8

20. If A varies jointly as c and d^3, and if $A = 16$ when $c = 4$ and $d = 2$, find A when $c = 6$ and $d = 3$.

   a. 108   **b.** 18   **c.** 81   **d.** 36

Chapter Test

1. Solve $\dfrac{r + 1}{4} - \dfrac{3}{2} \le \dfrac{2r - 9}{10}$. $\{r: r \le 7\}$ 9-1

2. Solve $\dfrac{2x + 5}{8} - \dfrac{3x + 1}{7} = \dfrac{3 - 2x}{4}$. $\{\frac{5}{6}\}$

Rational Expressions in Open Sentences   **475**

3. A sports coat was on sale for $96. If the original selling price was $120, what was the percent discount? 20%

9-2

4. Carl has some money invested at an annual rate of 6%. He has twice as much money invested at an annual rate of 9%. How much has he invested at each rate if his total income for one year from these two investments is $168? $700 at 6%; $1400 at 9%

5. Solve $\dfrac{1}{k^2 - k} = \dfrac{3}{k} - 1$. {2}

9-3

6. Solve $\dfrac{12}{a^2 - 4} + 1 = \dfrac{2a - 1}{a - 2}$. {−5}

7. Three eighths of a number is 2 less than one half the number. Find the number. 16

9-4

8. The difference of two numbers is 3, and the sum of their reciprocals is $\dfrac{1}{2}$. Find the numbers. 3 and 6 or −2 and 1

9. Working alone, Tim can paint his house in 3 days. It takes his daughter 5 days to paint the house if she works alone. How long will it take them to paint the house if they work together? $1\frac{7}{8}$ days

9-5

10. Sylvia rows 4 km up a stream in twice the time it takes her to row 12 km down the stream. If the current in the river flows at a speed of 6 km/h, how fast does Sylvia row in still water? 8.4 km/h

9-6

11. Solve $\dfrac{9}{n} = \dfrac{7}{n - 4}$. {18}

9-7

12. Find the ratio of a to b when $\dfrac{2a - b}{a + 4b} = \dfrac{1}{2}$. 2 : 1

13. If m varies directly as w^2, and if $m = 1600$ when $w = 2$, find m when $w = \dfrac{3}{2}$. 900

9-8

14. If x is inversely proportional to y, and if $x = 60$ when $y = 80$, find y when $x = 240$. 20

9-9

15. The frequency of a radio wave is inversely proportional to the wave length. If the frequency is 60 Mc/s (megacycles per second) for a wave 5 m long, what is the frequency for a wave 3 m long? 100 Mc/s

16. If a varies jointly as b and c, and if $a = 6$ when $b = 0.5$ and $c = 0.3$, find a when $b = 0.75$ and $c = 8$. 240

9-10

17. The height of a square pyramid varies directly with the volume of the pyramid and inversely with the square of the length of one side of its base. A pyramid whose volume is 100 m³ and whose base is 5 m by 5 m has a height of 12 m. What is the height of a pyramid whose volume is 80 m³ and whose base is 4 m by 4 m? 15 m

Cumulative Review

Chapter 4   3. $\{a : a < -1 \text{ or } a > 5\}$ 4. $\{x : -1 \le x < 1\}$

Solve each open sentence and graph its solution set.

1. $-6 < 8 - 2x \le 4$ $\{x : 2 \le x < 7\}$

2. $4 - (3t + 1) < 5t + 2(t - 6)$ $\{t : t > \frac{3}{2}\}$

3. $5 + 2a < 3a$ or $2 - 4a > 6$

4. $x + 2 > 6x - 3$ and $7 - 2x \le 5x + 14$

5. $|6p + 12| \le 6$ $\{p : -3 \le p \le -1\}$

6. $15 + |2x - 1| > 20$ $\{x : x < -2 \text{ or } x > 3\}$

7. The measure of an angle is 20° less than three times the measure of its supplement. Find the measure of the angle. 130°

8. The lengths of two sides of a rectangle are consecutive even integers. The perimeter of the rectangle is 196 cm. Find the area. 2400 cm²

9. Margaret rode her bike to school one day, but walked home along the same route after school. If her average speed was 7 km/h while riding and 4 km/h while walking, and her total traveling time that day was 1.1 h, find the distance from Margaret's house to the school. 2.8 km

10. An 80 g solution is 16% salt. How much salt must be added to produce a solution that is 36% salt? 25 g

Chapter 5

Tell whether each statement is true or false.

11. The graph of the point $(-4, 0)$ lies in the third quadrant. False

12. The relation $\{(-4, 4), (-5, 4), (-6, 2)\}$ is a function. True

13. The ordered pair $(-2, -2)$ is a solution of $3x - y < -4$. False

14. The points $(1, 8), (2, 9),$ and $(3, 10)$ are collinear. True

18. $\{(-1, -2), (-1, -1), (0, -2), (0, -1), (0, 2), (1, -2), (1, -1), (1, 2)\}$

Solve each open sentence if $x \in \{-1, 0, 1\}$ **and** $y \in \{-2, -1, 2\}$.

15. $y = x$ 16. $3x - y = 3$ 17. $x + y > 0$ 18. $y - 3 \le 2x$

$\{(-1, -1)\}$ $\varnothing$ $\{(-1, 2), (0, 2), (1, 2)\}$

Given the function $f: x \rightarrow 2x^2 - 1$, **find the following values of** f.

19. $f(3)$ 17   20. $f(4)$ 31   21. $f(-2)$ 7   22. $f(-5)$ 49

Graph each of the following on a coordinate plane.

23. $y = 4x + 1$ 24. $3x - 2y = 10$ 25. $y < \frac{2}{3}x + 2$ 26. $5x - 4y \le 8$

Determine an equation of the line that satisfies the given requirements.

27. has slope $-\frac{3}{2}$ and y-intercept 1 $\quad y = -\frac{3}{2}x + 1$

28. has slope 0 and y-intercept 4 $\quad y = 4$

29. has slope -2 and passes through the point $(-3, 0)$ $\quad y = -2x - 6$

30. passes through the points $(5, 3)$ and $(9, 2)$ $\quad x + 4y = 17$

Rational Expressions in Open Sentences   **477**

 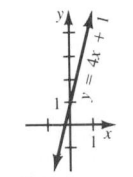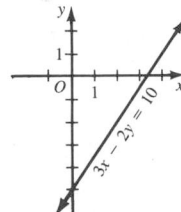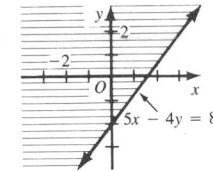

35. $-x^6y^2 + 4x^5y^4 - 2x^3y^5$

36. $2a^5 - 2a^3b - 8a^2b^3$

48. $(y + z)(x + 2)(x - 2)$

65. $\dfrac{(4x - y)(x + y)}{6(4x + y)}$

Chapter 7

Simplify.

31. $3x^4 + 5x^3 - 6 + (-2x^4 - 4x^3 + 3x)$ $x^4 + x^3 + 3x - 6$

32. $7rst^2 - 3r^2st - 2(rst^2 - rs^2t)$ $5rst^2 - 3r^2st + 2rs^2t$

33. $(3y^3)(-8y^2)(-y^4)$ $24y^9$

34. $(-4e^2f)^2(5ef^2)(-2ef)$ $-160e^6f^5$

35. $-x^3y^2(x^3 - 4x^2y^2 + 2y^3)$

36. $2a^3(a^2 - 3b) + 4a^2b(a - 2b^2)$

37. $(2x - 3)(x + 1)$ $2x^2 - x - 3$

38. $(10x - 2y)(10x + 2y)$ $100x^2 - 4y^2$

39. $(4c + 3)^2$ $16c^2 + 24c + 9$

40. $(x - 5y)^2$ $x^2 - 10xy + 25y^2$

41. $(2x + 1)^3$ $8x^3 + 12x^2 + 6x + 1$

42. $(3s - t)(s^2 - 5st + t^2)$

     $3s^3 - 16s^2t + 8st^2 - t^3$

Factor completely.

43. $16x^2 - 24x + 9$ $(4x - 3)^2$

44. $6m^2n^2 - 8m^3n^2 + 10m^2n$ $2m^2n(3n - 4mn + 5)$

45. $10n^3 + 10$ $10(n + 1)(n^2 - n + 1)$

46. $6x^2 - 11x - 10$ $(2x - 5)(3x + 2)$

47. $9x^2 - 25y^2$ $(3x + 5y)(3x - 5y)$

48. $x^2y - 4z - 4y + x^2z$

49. $12x^3 + 32x^2 - 12x$ $4x(3x - 1)(x + 3)$

50. $105 + 14x - 7x^2$ $7(5 - x)(3 + x)$

Solve.

51. $2x(x - 3)(3x + 1) = 0$ $\{0, 3, -\frac{1}{3}\}$

52. $(6x - 1)(4x + 3) = (3x - 4)(8x + 7)$ $\{-1\}$

53. $2a^2 + 5a = -3$ $\{-\frac{3}{2}, -1\}$

54. $4y^2 = 5 - 8y$ $\{-\frac{5}{2}, \frac{1}{2}\}$

55. Find two consecutive negative odd integers such that the square of the lesser integer is 40 more than the square of the greater integer. $-11, -9$

Chapter 8

Simplify.

56. $\dfrac{(-3x^3y)(3xy^2)^2}{y(6x^3)^2}$ $-\dfrac{3y^4}{4x}$

57. $\dfrac{e^{-2}h^4k^{-6}}{e^{-10}h^0k^{-4}}$ $\dfrac{e^8h^4}{k^2}$

58. $\dfrac{z^2 - 36}{2z - 12}$ $\dfrac{z + 6}{2}$

59. $\dfrac{25m^5 + 35m^3 - 45m}{-5m}$ $-5m^4 - 7m^2 + 9$

60. $\dfrac{5a^3 + 3a^2b^2 - 2ab}{ab^2}$ $\dfrac{5a^2 + 3ab^2 - 2b}{b^2}$

61. $\dfrac{12m^2 + 5m - 3}{6m^2 - 17m + 5}$ $\dfrac{4m + 3}{2m - 5}$

62. $\dfrac{t^2 - 3t - 4}{3t^2 + 10t - 8} \cdot \dfrac{6t^2 - t - 2}{2t^2 + 3t + 1}$ $\dfrac{t - 4}{t + 4}$

63. $\dfrac{7c - 2}{5c^2 + 2c - 3} - \dfrac{4c - 5}{5c^2 + 2c - 3}$ $\dfrac{3}{5c - 3}$

64. $\dfrac{a^2 + 3ab}{a^2 + 3ab + 2b^2} + \dfrac{a}{a + 2b}$ $\dfrac{2a}{a + b}$

65. $\dfrac{16x^2 - 8xy + y^2}{6x - 12y} \div \dfrac{16x^2 - y^2}{x^2 - xy - 2y^2}$

66. $\dfrac{1 + a^{-1}}{1 - a^{-2}}$ $\dfrac{a}{a - 1}$

67. $\dfrac{\dfrac{x}{2y} - \dfrac{y}{2x}}{\dfrac{2}{y} + \dfrac{2}{x}}$ $\dfrac{x - y}{4}$

Divide the first polynomial by the second.

68. $x^2 + 6x - 7;\ x - 1$ $x + 7$

69. $2a^2 + 11a - 18;\ 2a - 3$ $a + 7 + \dfrac{3}{2a - 3}$

478   *Chapter 9*

APPLICATION

Gravity

No matter how high a person throws a javelin, it always falls back to the ground. As the javelin falls, its velocity keeps increasing; that is, the javelin accelerates. A force is needed to cause an object to accelerate. The acceleration depends on the object's mass and the force acting on it. The force that pulls a javelin back to Earth is the **force of gravity,** or **gravitational force.** Gravitational force exists between any two objects in the universe. The equation for this force, F, is

$$F = G\frac{m_1 m_2}{d^2},$$

where G is the gravitational constant. The masses of the two objects are m_1 and m_2. The distance between their centers of mass is d. When F is in newtons, m_1 and m_2 are in kilograms, and d is in meters, G is equal to 0.667×10^{-10} m^3/(kg $\cdot$ s^2). (Newtons are expressed as kg $\cdot$ m/s^2).

Suppose that m_1 is the mass of an object on Earth's surface. The distance d between its center of mass and Earth's is Earth's radius. Since Earth's mass and radius are constant, $\frac{F}{m_1}$ is also constant.

$$\frac{F}{m_1} = G\frac{m_2}{d^2}$$

The ratio $\frac{F}{m_1}$ is the *acceleration due to gravity*. This acceleration, represented by the symbol g, does not depend on m_1. If air resistance is small, all objects fall to Earth at the same rate.

Exercises

What gravitational force, in newtons, acts between the two objects in each pair? (**Earth: mass = 5.98×10^{24} kg; radius = 6.38×10^6 m. Moon: mass = 7.34×10^{22} kg; radius = 1.74×10^6 m.**)

1. Earth and a 1-kg object at its surface 9.8 N

2. the moon and a 1-kg object at its surface 1.6 N

3. two 1-kg spheres, each with radius 0.333 m and center of mass at the center of the sphere, when the spheres are just touching 1.5×10^{-10} N

4. From your answer to Exercise 1, find g. 9.8 m/s^2

Rational Expressions in Open Sentences **479**

The photo shows a hot air balloon fiesta in Albuquerque, New Mexico. From these balloons, passengers are able to see great distances. The Application on page 540 gives a formula that can be used to calculate such distances.

Chapter 10

Irrational Numbers and Radicals

Problem Solving Strategies

Word Problem Plan
The *Plan for Solving a Word Problem* is used in conjunction with the Pythagorean Theorem to solve problems.
Drawing a Diagram
Students are encouraged to use diagrams as problem-solving aids in Section 10-5.

Rational and Irrational Numbers

OBJECTIVES for 10-1 through 10-4:
1. *To apply basic properties of the rational numbers.*
2. *To express rational numbers as fractions or decimals.*
3. *To find the square roots of expressions that have rational square roots.*
4. *To simplify radicals.*
5. *To find approximations of irrational square roots.*

10–1 Properties of Rational Numbers

Recall from Chapter 1 that a rational number is any number that can be expressed as the quotient of two integers, where the divisor is not zero. The following are examples of rational numbers.

$$\frac{5}{8} \qquad 3\frac{2}{9} = \frac{29}{9} \qquad 0 = \frac{0}{2} \qquad -4 = \frac{-4}{1} \qquad 0.43 = \frac{43}{100}$$

Any given rational number can be expressed as a quotient of integers in many different ways. For example,

$$\frac{5}{8} = \frac{10}{16} = \frac{15}{24} = \frac{-10}{-16},$$

and so on.

To decide which of a group of rational numbers is greatest, you can rewrite each fraction with the same positive denominator and then compare the numerators, as shown in Example 1 on the following page.

Irrational Numbers and Radicals   **481**

1. Compare $\frac{7}{3}$ and $\frac{5}{4}$.

$$\frac{7}{3} = \frac{7 \cdot 4}{3 \cdot 4} = \frac{28}{12}$$

$$\frac{5}{4} = \frac{5 \cdot 3}{4 \cdot 3} = \frac{15}{12}$$

Since $28 > 15$, $\frac{28}{12} > \frac{15}{12}$.

$$\therefore \frac{7}{3} > \frac{5}{4}.$$

2. Compare.

a. $\frac{1}{2}$ and $\frac{4}{7}$

$$1 \cdot 7 < 4 \cdot 2$$

$$\therefore \frac{1}{2} < \frac{4}{7}.$$

b. $-\frac{6}{7}$ and $-\frac{7}{8}$

Rewrite as $\frac{-6}{7}$ and $\frac{-7}{8}$.

$$(-6)(8) > (7)(-7)$$

$$\therefore -\frac{6}{7} > -\frac{7}{8}.$$

c. $\frac{65}{85}$ and $\frac{72}{90}$

$$\frac{65}{85} = \frac{13}{17} \text{ and } \frac{72}{90} = \frac{4}{5}$$

Compare $\frac{13}{17}$ and $\frac{4}{5}$.

$$13 \cdot 5 < 17 \cdot 4$$

$$\therefore \frac{13}{17} < \frac{4}{5}, \text{ or } \frac{65}{85} < \frac{72}{90}.$$

EXAMPLE 1 Compare $\frac{9}{5}$ and $\frac{11}{6}$.

SOLUTION Use the product of the denominators as a common denominator.

$$5 \times 6 = 30$$

Rewrite each fraction as an equivalent fraction with the common denominator.

$$\frac{9}{5} = \frac{9 \cdot 6}{5 \cdot 6} = \frac{54}{30} \qquad \frac{11}{6} = \frac{11 \cdot 5}{6 \cdot 5} = \frac{55}{30}$$

Compare the numerators. Since $54 < 55$, $\frac{54}{30} < \frac{55}{30}$.

$$\therefore \frac{9}{5} < \frac{11}{6}$$

The method used to compare rational numbers in Example 1 may be generalized as follows.

If a and b are integers and c and d are positive integers, then

$$\frac{a}{c} > \frac{b}{d} \quad \text{if and only if } ad > bc;$$

$$\frac{a}{c} < \frac{b}{d} \quad \text{if and only if } ad < bc.$$

EXAMPLE 2 Compare.

a. $\frac{2}{3}$ and $\frac{5}{6}$ **b.** $-\frac{3}{4}$ and $-\frac{4}{3}$ **c.** $\frac{80}{112}$ and $\frac{200}{125}$

SOLUTION **a.** $2 \cdot 6 < 5 \cdot 3$

$$\therefore \frac{2}{3} < \frac{5}{6}$$

b. Rewrite $-\frac{3}{4}$ and $-\frac{4}{3}$ as $\frac{-3}{4}$ and $\frac{-4}{3}$, respectively.

$$(-3)(3) > (-4)(4)$$

$$\therefore \frac{-3}{4} > \frac{-4}{3}, \quad \text{or} \quad -\frac{3}{4} > -\frac{4}{3}$$

c. First simplify each fraction.

$$\frac{80}{112} = \frac{5}{7} \qquad \frac{200}{125} = \frac{8}{5}$$

Then compare $\frac{5}{7}$ and $\frac{8}{5}$.

$$5 \cdot 5 < 8 \cdot 7$$

$$\therefore \frac{5}{7} < \frac{8}{5}, \quad \text{or} \quad \frac{80}{112} < \frac{200}{125}$$

482 *Chapter 10*

Rational numbers are different from integers in many ways. For example, on a number line there is not always another integer between any two given integers.

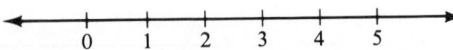

There are exactly four integers between 0 and 5: 1, 2, 3, and 4

There are *no* integers between 0 and 1.

Between any two rational numbers, however, it is always possible to find another rational number. For example, consider $\frac{1}{2}$ and $\frac{3}{4}$. You can find the rational number that is one half of the way from $\frac{1}{2}$ to $\frac{3}{4}$ by adding one half the difference between $\frac{1}{2}$ and $\frac{3}{4}$ to $\frac{1}{2}$:

$$\frac{1}{2} + \frac{1}{2}\left(\frac{3}{4} - \frac{1}{2}\right) = \frac{1}{2} + \frac{1}{2}\left(\frac{1}{4}\right) = \frac{1}{2} + \frac{1}{8} = \frac{5}{8}$$

You can then use this method to find the number that is one half of the way from $\frac{1}{2}$ to $\frac{5}{8}$:

$$\frac{1}{2} + \frac{1}{2}\left(\frac{5}{8} - \frac{1}{2}\right) = \frac{1}{2} + \frac{1}{2}\left(\frac{1}{8}\right) = \frac{1}{2} + \frac{1}{16} = \frac{9}{16}$$

Similarly, $\frac{17}{32}$ is one half of the way from $\frac{1}{2}$ to $\frac{9}{16}$, $\frac{33}{64}$ is one half of the way from $\frac{1}{2}$ to $\frac{17}{32}$, and so on.

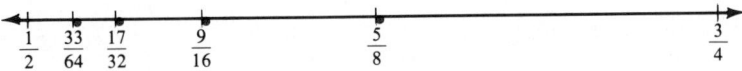

Therefore, there exist at least four other rational numbers between $\frac{1}{2}$ and $\frac{3}{4}$. In fact, continuing with this process, you could find an infinite number of rational numbers between $\frac{1}{2}$ and $\frac{3}{4}$.

In general, given two rational numbers a and b, $a < b$, the number that is one half of the way from a to b is

$$a + \frac{1}{2}(b - a);$$

the number one third of the way from a to b is

$$a + \frac{1}{3}(b - a);$$

and so on. These formulas suggest the *density property of rational numbers*, which is stated on the following page.

Irrational Numbers and Radicals   **483**

3. Find a rational number between $\frac{2}{3}$ and $\frac{9}{10}$.

Choose, for example, the number that is one half the distance between $\frac{2}{3}$ and $\frac{9}{10}$.

$$\frac{2}{3} + \frac{1}{2}\left(\frac{9}{10} - \frac{2}{3}\right) = \frac{2}{3} +$$
$$\frac{1}{2}\left(\frac{7}{30}\right) = \frac{2}{3} + \frac{7}{60} = \frac{47}{60}$$

Check: Is $\frac{47}{60}$ between $\frac{2}{3}$

and $\frac{9}{10}$?

$$\frac{2}{3} \stackrel{?}{<} \frac{47}{60} \stackrel{?}{<} \frac{9}{10}$$
$$(2)(60) \stackrel{?}{<} 3(47)$$
$$120 < 141 \quad \sqrt{}$$
and
$$(47)(10) \stackrel{?}{<} (9)(60)$$
$$470 < 540 \quad \sqrt{}$$
$\therefore \frac{47}{60}$ is between $\frac{2}{3}$ and $\frac{9}{10}$.

1. Write a decimal and a fractional equivalent for $\frac{1}{2}\%$. $0.005, \frac{1}{200}$

2. Solve
$$\frac{6n-2}{4} + \frac{n-13}{8} = 6. \quad 5$$

3. How much money must be invested for one year at 11% to earn $110 interest? $1000

4. A negative number is 49 times its reciprocal. Find the number. -7

5. If m varies jointly as n and the square of p, and if $m = 140$ when $n = 5$ and $p = 2$, find m when $n = 3$ and $p = 3$. 189

Density Property of Rational Numbers

Between any two different rational numbers, there is another rational number.

EXAMPLE 3 Find a rational number between $\frac{1}{5}$ and $\frac{7}{10}$.

SOLUTION Choose, for example, the number one fourth of the way from $\frac{1}{5}$ to $\frac{7}{10}$.

$$\frac{1}{5} + \frac{1}{4}\left(\frac{7}{10} - \frac{1}{5}\right) = \frac{1}{5} + \frac{1}{4}\left(\frac{1}{2}\right) = \frac{1}{5} + \frac{1}{8} = \frac{13}{40}$$

Check: Is $\frac{13}{40}$ between $\frac{1}{5}$ and $\frac{7}{10}$?

$$\frac{1}{5} \overset{?}{<} \frac{13}{40} \overset{?}{<} \frac{7}{10}$$

$$(1)(40) \overset{?}{<} (13)(5) \quad \text{and} \quad (13)(10) \overset{?}{<} (7)(40)$$
$$40 < 65 \quad \checkmark \qquad\qquad 130 < 280 \quad \checkmark$$

$\therefore$ one rational number between $\frac{1}{5}$ and $\frac{7}{10}$ is $\frac{13}{40}$.

Oral Exercises

Show that each number is a rational number by expressing it as a quotient of integers.

1. $5 \quad \frac{5}{1}$

2. $16\frac{1}{3} \quad \frac{49}{3}$

3. $0.05 \quad \frac{5}{100}$

4. $-2.6 \quad -\frac{26}{10}$

5. $75\% \quad \frac{75}{100}$

6. $36\% \quad \frac{36}{100}$

7. $0 \quad \frac{0}{1}$

8. $2 + \frac{1}{9} \quad \frac{19}{9}$

Suggested Assignments

Minimum
  484/1–24
Average
  484/1–23 odd, 25–32
Maximum
  485/7–29 odd, 30–34

Written Exercises

Replace each __?__ with one of the symbols <, =, or > to make a true statement.

A

1. $\frac{2}{5}$ __?__ $\frac{7}{12}$ $<$

2. $\frac{1}{8}$ __?__ $\frac{7}{63}$ $>$

3. $-\frac{5}{3}$ __?__ $-\frac{7}{5}$ $<$

4. $\frac{3}{7}$ __?__ $\frac{13}{19}$ $<$

5. $\frac{2}{9}$ __?__ $\frac{18}{81}$ $=$

6. $-\frac{16}{56}$ __?__ $-\frac{24}{84}$ $=$

484 *Chapter 10*

7. $\frac{32}{60}$ __?__ $\frac{325}{625}$ >

8. $\frac{96}{112}$ __?__ $\frac{140}{154}$ <

9. $-3\frac{7}{8}$ __?__ $-\frac{96}{27}$ <

10. $5\frac{4}{9}$ __?__ $\frac{138}{27}$ >

11. $\frac{147}{63}$ __?__ $\frac{315}{135}$ =

12. $\frac{90}{216}$ __?__ $\frac{108}{225}$ <

Write the given numbers in order from least to greatest.

13. $\frac{3}{16}, \frac{5}{14}, \frac{8}{29}$ $\frac{3}{16}, \frac{8}{29}, \frac{5}{14}$

14. $\frac{30}{18}, \frac{54}{36}, \frac{60}{75}$ $\frac{60}{75}, \frac{54}{36}, \frac{30}{18}$

15. $-\frac{15}{40}, -\frac{14}{63}, -\frac{27}{81}$ $-\frac{15}{40}, -\frac{27}{81}, -\frac{14}{63}$

16. $\frac{32}{5}, 6\frac{3}{7}, \frac{56}{9}$ $\frac{56}{9}, \frac{32}{5}, 6\frac{3}{7}$

17. $\frac{5}{16}, \frac{13}{65}, \frac{4}{24}, \frac{7}{43}$ $\frac{7}{43}, \frac{4}{24}, \frac{13}{65}, \frac{5}{16}$

18. $2\frac{3}{4}, \frac{84}{40}, \frac{46}{16}, \frac{49}{20}$ $\frac{84}{40}, \frac{49}{20}, 2\frac{3}{4}, \frac{46}{16}$

Find the rational number that is one half of the way from *a* to *b*.

19. $a = \frac{1}{5}, b = \frac{1}{4}$ $\frac{9}{40}$

20. $a = \frac{1}{8}, b = \frac{7}{12}$ $\frac{17}{48}$

21. $a = -\frac{5}{6}, b = -\frac{2}{5}$

22. $a = -\frac{4}{9}, b = \frac{11}{6}$ $\frac{25}{36}$

23. $a = 2\frac{1}{4}, b = 3\frac{1}{2}$ $2\frac{7}{8}$

24. $a = -4\frac{1}{3}, b = -4\frac{1}{6}$

B 25. Find the rational number that is one third of the way from $\frac{3}{5}$ to $\frac{8}{3}$. $\frac{58}{45}$ 21. $-\frac{37}{60}$

26. Find the rational number that is one fourth of the way from $-\frac{1}{12}$ to $\frac{4}{9}$. $\frac{7}{144}$ 24. $-4\frac{1}{4}$

27. Find the rational number that is one sixth of the way from $-\frac{5}{4}$ to $-\frac{1}{6}$. $-\frac{77}{72}$

28. Find the rational number that is one seventh of the way from $1\frac{3}{4}$ to $4\frac{3}{8}$. $2\frac{1}{8}$

29. Find the rational number that is three sevenths of the way from $5\frac{2}{3}$ to $13\frac{4}{9}$. 9

30. Find the rational number that is two fifths of the way from $-1\frac{2}{7}$ to $\frac{9}{2}$. $\frac{36}{35}$

31. Find an expression for the rational number that is one fifth of the way from $\frac{2a}{3}$ to $\frac{9a}{4}$, given that $a \neq 0$. $\frac{59a}{60}$

32. Find an expression for the rational number that is one half of the way from $-3b$ to $-\frac{5b}{2}$, given that $b \neq 0$. $-\frac{11b}{4}$

C 33. Find the rational number that is one half of the way from *a* to *b* if *a* and *b* are the numbers that are, respectively, one third and two thirds of the way from $\frac{3}{7}$ to $\frac{12}{5}$. $\frac{99}{70}$

34. Find two rational numbers *g* and *h* such that *g* is one fifth of the way from $\frac{5}{2}$ to *h* and *h* is one half of the way from *g* to $\frac{15}{4}$. $\frac{95}{36}, \frac{115}{36}$

Irrational Numbers and Radicals   **485**

Key Ideas

Express common fractions as terminating or repeating decimals.
Express terminating and repeating decimals as common fractions.

10–2 Decimals and Fractions

To find *decimal representations* of rational numbers, you can use the division process as follows.

$$\frac{5}{8} = 5 \div 8 \qquad\qquad \frac{7}{11} = 7 \div 11$$

$$
\begin{array}{r}
0.625 \\
8\overline{)5.000} \\
\underline{4\,8} \\
20 \\
\underline{16} \\
40 \\
\underline{40} \\
0
\end{array}
\qquad\qquad
\begin{array}{r}
0.6363 \\
11\overline{)7.0000} \\
\underline{6\,6} \\
40 \\
\underline{33} \\
70 \\
\underline{66} \\
40 \\
\underline{33} \\
7
\end{array}
$$

$$\therefore \frac{5}{8} = 0.625 \qquad\qquad \therefore \frac{7}{11} = 0.6363 \ldots$$

In the division of 5 by 8, the quotient 0.625 is called a **terminating decimal** because the division process *terminates* when a final remainder of 0 is reached.

In the division of 7 by 11, on the other hand, the quotient 0.6363 . . . is a *nonterminating* decimal. The division process never terminates because the remainders 4 and 7 (printed in red) keep appearing, and a remainder of 0 is never reached. The quotient is called a **repeating decimal** because it is a nonterminating decimal in which the same block of digits *repeats* without end. In a repeating decimal, a bar is often used to indicate the block of digits that repeats. Therefore,

$$0.6363 \ldots = 0.\overline{63}.$$

In general, when an integer p is divided by a positive integer q, the decimal quotient will either terminate or repeat. The remainder at each step in the division process will be $0, 1, 2, \ldots$, or $q - 1$. Therefore, after only zeros are left in the dividend, within at most q steps either 0 occurs as a remainder and the division process ends, or one of $1, 2, \ldots, q - 1$ recurs, producing a repeating sequence of remainders and hence a repeating block of digits in the quotient. Thus, if an integer is divided by 11, for example, the only remainders that can appear are the integers 0 through 10 and the repeating block of digits in the quotient will contain *at most* 10 digits.

Just as rational numbers can be represented by decimals, many decimals may be expressed as fractions. You already know that a *terminating* decimal is equivalent to a fraction that has a power of 10 as its denominator.

486   *Chapter 10*

Thus,

$$3.17 = \frac{317}{100}, \qquad 0.051 = \frac{51}{1000}, \qquad \text{and} \qquad 1.0009 = \frac{10{,}009}{10{,}000}.$$

To find a fraction equivalent to a *repeating* decimal, you can use the procedure shown in the following example.

EXAMPLE   Express each decimal as a fraction in simplest form.

      a. $0.\overline{14}$

      b. $3.2\overline{8}$

      c. $0.\overline{675}$

SOLUTION   **a.** Let $N = 0.\overline{14}$.

Multiply: $100N = 14.\overline{14}$

              $N = 0.\overline{14}$

Subtract: $99N = 14$

Divide: $N = \dfrac{14}{99}$

$\therefore\ 0.\overline{14} = \dfrac{14}{99}$

b. Let $N = 3.2\overline{8}$.

Multiply: $10N = 32.8\overline{8}$

           $N = 3.2\overline{8}$

Subtract: $9N = 29.6$

Divide: $N = \dfrac{29.6}{9} = \dfrac{296}{90} = \dfrac{148}{45}$

$\therefore\ 3.2\overline{8} = \dfrac{148}{45}$

c. Let $N = 0.\overline{675}$.

Multiply: $1000N = 675.\overline{675}$

            $N = 0.\overline{675}$

Subtract: $999N = 675$

Divide: $N = \dfrac{675}{999} = \dfrac{225}{333} = \dfrac{75}{111} = \dfrac{25}{37}$

$\therefore\ 0.\overline{675} = \dfrac{25}{37}$

Notice that in part (a) of Example 1 the multiplication of N by 100 shifts the repeating block 2 places to the left. The subtraction then produces a terminating decimal. A similar procedure is used in parts (b) and (c). In general, if a repeating block of digits contains n digits, you multiply by 10^n.

Irrational Numbers and Radicals   **487**

The discussion on the preceding pages suggests the following facts about rational numbers.

1. Each rational number can be named by a terminating or a repeating decimal.
2. Each repeating or terminating decimal names a unique rational number.

It is possible to construct a decimal numeral, however, that is nonterminating and nonrepeating. For example, the decimal numerals

$$0.3\,7\,3\,3\,7\,3\,3\,3\,7\,3\,3\,3\,3\,7\ldots$$

and

$$0.2\,4\,6\,8\,1\,0\,1\,2\,1\,4\,1\,6\,1\,8\ldots$$

both follow patterns and could be carried out to any number of decimal places, but no block of digits would keep repeating. Since every rational number can be expressed as a terminating or a repeating decimal, these numerals cannot be equivalent to rational numbers. These non-repeating, nonterminating decimals are called *irrational numbers*. Recall from Chapter 1 that an irrational number is any real number that is not a rational number.

Ordinarily, the decimal numerals for irrational numbers do not have systematic patterns for their digits as do the preceding examples. You may already know one such irrational number,

$$\pi = 3.14159\ldots.$$

Often, you may need to use π in a computation, such as finding the circumference or area of a circle. In a computation, you usually use a *rational approximation* for π, such as

$$\pi \approx 3.14 \text{ or } \pi \approx 3.1416,$$

where the symbol $\approx$ is read *equals approximately* (or *is approximately equal to*).

It is often convenient to *round* a lengthy or infinite decimal in order to use an approximation of the original decimal in a computation. For example, you may approximate $2\frac{6}{13}$, or $2.\overline{461538}$, as follows.

$$2.\overline{461538} \approx 2.4615 \text{ to the nearest ten-thousandth}$$
$$\approx 2.462 \quad \text{to the nearest thousandth}$$
$$\approx 2.46 \quad\, \text{to the nearest hundredth}$$
$$\approx 2.5 \quad\;\; \text{to the nearest tenth}$$
$$\approx 2 \quad\;\;\;\, \text{to the nearest unit}$$

In rounding you use the rules stated at the top of the following page.

1. If the first digit dropped is 5 or more, add 1 to the last digit retained.
2. If the first digit dropped is less than 5, leave the retained digits unchanged.

The decimals studied in this section have all represented rational or irrational numbers. Since these two sets of numbers make up the set of real numbers, the following property is true.

Property of Completeness
Each decimal represents a real number, and every real number can be represented as a decimal.

Oral Exercises

Replace each __?__ with one of the words *rational* or *irrational* to make a true statement.

1. Each terminating decimal numeral represents a(n) __?__ number. rational
2. 0.554555445555444 . . . names a(n) __?__ number. irrational
3. Nonterminating, nonrepeating decimal numerals represent __?__ numbers. irrational
4. Each repeating or terminating decimal names a unique __?__ number. rational
5. 0.979797 . . . names a(n) __?__ number. rational
6. Let $a = 2.232332333233332\ldots$ and $b = 0.101001000100001\ldots$.
   a. $a + b$ is a(n) __?__ number. rational
   b. $a - b$ is a(n) __?__ number. irrational

Replace each __?__ with the number that makes a true statement.

7. The decimal equivalent of the rational number $\frac{n}{13}$, where n is a nonzero integer, will start repeating or will terminate after at most __?__ digits beyond the decimal point. 12

8. The decimal equivalent of the rational number $\frac{n}{18}$, where n is a nonzero integer, will start repeating or will terminate after at most __?__ digits beyond the decimal point. 17

9. The repeating block of digits in the quotient $\frac{6}{29}$ will contain at most __?__ digits. 28

10. The repeating block of digits in the quotient $\frac{9}{47}$ will contain at most __?__ digits. 46

Round each decimal as indicated.

a. to the nearest thousandth   **b.** to the nearest hundredth   **c.** to the nearest tenth

11. 0.43792   **12.** $0.\overline{25}$ **13.** 12.1357   **14.** $-6.3\overline{84}$

Express each rational number as a terminating or repeating decimal.

15. $\frac{2}{3}$ $0.\overline{6}$ **16.** $\frac{8}{5}$ 1.6   **17.** $\frac{7}{2}$ 3.5   **18.** $\frac{5}{6}$ $0.8\overline{3}$ **19.** $\frac{1}{4}$ 0.25   **20.** $\frac{1}{12}$ $0.08\overline{3}$

21. a. Which denominators in Exercises 15–20 have prime factors other than 2 and 5? 3, 6, 12
  b. Which fractions in Exercises 15–20 are equivalent to terminating decimals? $\frac{8}{5}, \frac{7}{2}, \frac{1}{4}$

22. Construct a decimal numeral that names an irrational number.
  Answers may vary. Example: 0.1011011101111 . . .

Written Exercises

Express each rational number as a terminating or repeating decimal.

A   **1.** $\frac{1}{4}$ 0.25   **2.** $\frac{3}{8}$ 0.375   **3.** $\frac{5}{9}$ $0.\overline{5}$ **4.** $\frac{4}{9}$ $0.\overline{4}$ **5.** $\frac{5}{12}$ $0.41\overline{6}$ **6.** $\frac{7}{12}$ $0.58\overline{3}$

7. $-\frac{21}{40}$ -0.525 **8.** $\frac{35}{16}$ 2.1875   **9.** $-\frac{56}{9}$ $-6.\overline{2}$ **10.** $2\frac{4}{11}$ $2.\overline{36}$ **11.** $4\frac{3}{7}$ **12.** $-\frac{88}{7}$

11. $4.\overline{428571}$ **12.** $-12.\overline{571428}$

Express each rational number as a fraction in simplest form.

13. 0.84 $\frac{21}{25}$ **14.** 0.015 $\frac{3}{200}$ **15.** 52.54 $\frac{2627}{50}$ **16.** 129.6 $\frac{648}{5}$

17. $0.\overline{4}$ $\frac{4}{9}$ **18.** $0.\overline{7}$ $\frac{7}{9}$ **19.** $0.\overline{25}$ $\frac{25}{99}$ **20.** $0.\overline{43}$ $\frac{43}{99}$

21. $2.\overline{31}$ $\frac{229}{99}$ **22.** $12.\overline{5}$ $\frac{113}{9}$ **23.** $5.\overline{160}$ $\frac{5155}{999}$ **24.** $-7.0\overline{38}$

25. $-1.0\overline{3}$ $-\frac{31}{30}$ **26.** $2.7\overline{3}$ $\frac{41}{15}$ **27.** $-11.\overline{6655}$ $-\frac{10604}{909}$ **28.** $10.0\overline{27}$

24. $-\frac{7031}{999}$ **28.** $\frac{1103}{110}$

Multiply. Express the product as a fraction in simplest form. (Hint: First convert each decimal to fraction form.)

B   **29.** $(0.5)(0.\overline{4})$ $\frac{2}{9}$ **30.** $(0.\overline{6})(0.75)$ $\frac{1}{2}$ **31.** $(3.\overline{7})(0.6)$ $\frac{34}{15}$ **32.** $(1.5)(0.\overline{2})$ $\frac{1}{3}$

33. $(0.\overline{3})(0.\overline{5})$ $\frac{5}{27}$ **34.** $(0.\overline{18})(0.\overline{8})$ $\frac{16}{99}$ **35.** $(-0.\overline{45})(0.\overline{36})$ $-\frac{20}{121}$ **36.** $(0.\overline{81})(0.\overline{77})$ $\frac{7}{11}$

Add. Express the sum as a fraction in simplest form. (Hint: First convert each decimal to fraction form.)

37. $0.375 + \frac{3}{4}$ $\frac{9}{8}$ **38.** $0.625 + \frac{2}{3}$ $\frac{31}{24}$ **39.** $0.2 + 0.\overline{2}$ $\frac{19}{45}$ **40.** $0.4 + 0.\overline{4}$ $\frac{38}{45}$

41. $0.\overline{7} + \frac{1}{3}$ $\frac{10}{9}$ **42.** $0.2\overline{6} + \frac{3}{10}$ $\frac{17}{30}$ **43.** $0.\overline{25} + 0.\overline{62}$ $\frac{29}{33}$ **44.** $0.8\overline{3} + 0.\overline{54}$

45. $1.0\overline{4} + 2.0\overline{5}$ $\frac{31}{10}$ **46.** $-6.\overline{1} + 3.\overline{3}$ $-\frac{25}{9}$ **47.** $0.\overline{18} + 5.\overline{81}$ 6   **48.** $9.\overline{753} + 1.\overline{246}$

44. $\frac{91}{66}$ **48.** 11

C 49. a. Express $3.\overline{9}$, $7.\overline{9}$, and $15.\overline{9}$ in fraction form. $\frac{36}{9} = 4$; $\frac{72}{9} = 8$; $\frac{144}{9} = 16$

   b. Use the result of part (a) to determine the value of any repeating decimal of the form $\_.\overline{9}$. The value of the entire decimal is one greater than the whole-number part of the decimal.

50. Show that if a fraction in lowest terms can be represented by a terminating decimal, the only numbers that can be prime factors of the denominator are 2 and 5.

PROGRAMMING IN BASIC

When you use a computer's built-in division operation, the quotient that the computer outputs will contain a fixed number of digits. If you wish to have a greater number of digits in the quotient, however, the following program will instruct the computer to perform the division using the same steps that you might use. When you input any positive numbers N and D, N < D, the program will compute the quotient N/D digit-by-digit. If the remainder R is zero at any step, the division is complete and the word "TERMINATES" is printed. Otherwise the division is carried out for D + 3 steps.

```
10   PRINT "TO COMPUTE N/D"
20   PRINT "INPUT N, D (0 < N < D)";
30   INPUT N, D
40   PRINT N;"/";D;" = 0.";
50   LET R = N
60   FOR I = 1 TO D + 3
70   LET R = R*10
80   LET Q = INT(R/D)
90   PRINT Q;
100  LET R = R − Q*D
110  IF R = 0 THEN 150
120  NEXT I
130  PRINT " . . . "
140  GOTO 160
150  PRINT " TERMINATES "
160  END
```

Exercises

1. Type in the program as given. Then RUN the program to find a decimal equivalent for each of the following fractions:

$$\frac{1}{3}, \frac{1}{4}, \frac{1}{5}, \frac{1}{6}, \frac{1}{7}, \frac{1}{11}, \frac{1}{13}, \frac{1}{17}, \frac{1}{21}, \frac{1}{25}, \frac{1}{37}$$

2. If you can obtain a printout of the program, draw a red box around the loop in lines 60–120.

Irrational Numbers and Radicals   **491**

1. $1/3 = 0.333333\ldots$
   $1/4 = 0.25$ TERMINATES
   $1/5 = 0.2$ TERMINATES
   $1/6 = 0.1666666\ldots$
   $1/7 = 0.1428571428\ldots$
   $1/11 = 0.0909090909090$
   $9\ldots$
   $1/13 = 0.076923076923076$
   $9\ldots$
   $1/17 = 0.058823529411764$
   $70588\ldots$
   $1/21 = 0.047619047619047$
   $619047619\ldots$
   $1/25 = 0.04$ TERMINATES
   $1/37 = 0.027027027027027$
   027027027027027027
   $0270270\ldots$

3. $1/3 = 0.3\ldots$
   $1/4 = 0.25$ TERMINATES
   $1/5 = 0.2$ TERMINATES
   $1/6 = 0.16\ldots$
   $1/7 = 0.142857\ldots$
   $1/11 = 0.09\ldots$
   $1/13 = 0.076923\ldots$
   $1/17 = 0.058823529411764$
   $7\ldots$
   $1/21 = 0.047619\ldots$
   $1/25 = 0.04$ TERMINATES
   $1/37 = 0.027\ldots$

5. $1/3 = 0.3'3'\ldots$
   $1/4 = 0.25$ TERMINATES
   $1/5 = 0.2$ TERMINATES
   $1/6 = 0.16'6'\ldots$
   $1/7 = 0.142857'142857'\ldots$
   $1/11 = 0.09'09'\ldots$
   $1/13 = 0.076923'076923'\ldots$
   $1/17 = 0.058823529411764$
   $\qquad 7'058823529411764$
   $\qquad 7'\ldots$
   $1/21 = 0.047619'047619'\ldots$
   $1/25 = 0.04$ TERMINATES
   $1/37 = 0.027'027'\ldots$

In the program given on the preceding page, note that a nonterminating division is carried out for $D + 3$ steps regardless of the point at which the digits of the quotient begin to repeat. If you wish to stop the program when the first such repeat occurs, however, you can store the remainders in an array and compare each new remainder with those previously found. In order to do this, make the following changes in the program.

```
5    DIM R(200)          20   PRINT "INPUT N, D (0 < N < D < 100)";
```

Change the loop: Insert the following lines:

```
60   LET I = 1           112  REM *SEARCH FOR FIRST REPEAT
65   LET R(I) = R        113  FOR J = 1 to I
120  LET I = I + 1       114  IF R = R(J) THEN 130
125  GOTO 65             115  NEXT J
```

3. Type in the changes, then RUN the revised program to find a decimal equivalent of the same fractions listed in Exercise 1.

4. If you can obtain a printout of the program, draw a red box around the loop in lines 65–125 and a blue box around lines 112–115.

The following changes in the program will insert apostrophes to mark off the block of repeating digits in the quotient. The program uses a flag, F, and sets $F = 0$ until the first repeat is found. The program then sets $F = 1$ while it searches for the next repeat.

```
44   REM *SET FLAG TO ZERO
45   LET F = 0
111  IF F > 0 THEN 118
116  GOTO 120
117  REM *SEARCH FOR SECOND REPEAT
118  IF R = X THEN 138
123  IF F > 0 THEN 70
128  REM *FIRST REPEAT MARKED
130  PRINT "'";
131  REM *STORE REPEATING REMAINDER
132  LET X = R
133  REM *SET FLAG TO ONE
134  LET F = 1
135  GOTO 120
137  REM *SECOND REPEAT MARKED
138  PRINT "'..."
```

5. Type in the changes, then RUN the revised program to find a decimal equivalent of the same fractions listed in Exercise 1.

6. If you can obtain a printout of the program, draw a red box around the loop in lines 65–125. Draw a blue box around lines 112–115, and a second blue box around lines 117–118.

492 *Chapter 10*

10-3 Rational Square Roots

Just as the inverse of addition is subtraction and the inverse of multiplication is division, the inverse operation of squaring a number is finding a *square root*. In general, a number b is called a **square root** of a positive real number a if

$$b^2 = a.$$

Since $b^2 = (-b)^2$, a positive real number a has two square roots, a positive square root and a negative square root.

EXAMPLE 1 Solve $x^2 = 36$.

SOLUTION The problem asks for all numbers whose square is 36.
Since $(6)^2 = 36$ and $(-6)^2 = 36$, $x = 6$ or $x = -6$.
∴ the solution set is $\{6, -6\}$.

In equations of the type $x^2 = k$, it may be convenient to use the symbol $\pm$ (read "plus-or-minus") in the solution. Therefore, in Example 1,

$$x = 6 \text{ or } x = -6 \text{ may be written as } x = \pm 6.$$

The positive square root of a number a is called the **principal square root** and is represented by the symbol

$$\sqrt{a}.$$

The negative square root is therefore represented by the symbol

$$-\sqrt{a}.$$

Thus, if $a = 49$, then $\sqrt{a} = 7$ and $-\sqrt{a} = -7$. Of course, if $a = 0$, then both $\sqrt{a}$ and $-\sqrt{a}$ represent 0.

In an expression like $\sqrt{a}$, the symbol $\sqrt{}$ is called a **radical sign**. Any numeral or expression under a radical sign is called the **radicand**. When a mathematical expression appears under a radical sign, the entire expression is called a **radical**.

Because the square of each real number is a nonnegative real number, a negative real number has no real square root. Thus, expressions such as $\sqrt{-2}$, $\sqrt{-3}$, and so on do not name real numbers.

Since squaring a number and taking the square root of a number are inverse operations, you may see that if a number has a real square root, squaring that square root results in the number itself. Thus,

$$(\sqrt{16})^2 = 16 \quad \text{and} \quad (\sqrt{3})^2 = 3.$$

By the definition of principal square root, the expression $\sqrt{x^2}$ denotes a nonnegative number whether x represents a nonnegative or a negative number. Thus,

$$\sqrt{(5)^2} = 5 \quad \text{and} \quad \sqrt{(-5)^2} = 5.$$

Irrational Numbers and Radicals   **493**

Teaching Suggestions
p. T104

Key Ideas

Find rational square roots of expressions.

Chalkboard Examples

1. Solve $m^2 = 81$.
   The problem asks for a number whose square is 81. Since $(9)^2 = 81$ and $(-9)^2 = 81$, $m = 9$ or $m = -9$.
   ∴ the solution set is $\{9, -9\}$.

2. Simplify.
   a. $\sqrt{(m + 2)^2}$ $|m + 2|$
   b. $\sqrt{x^4 y^2}$ $|x^2 y|$
   c. $\sqrt{4a^8}$ $|2a^4| = 2a^4$

3. Simplify.
   a. $\sqrt{100}$ 10
   b. $\sqrt{\left(\frac{1}{4}\right)^2}$ $\frac{1}{4}$
   c. $\sqrt{(-2)^4}$ $|(-2)^2| = 4$

Common Errors

In an expression such as $\sqrt{x^9}$ or $\sqrt{x^{25}}$, a common error is to take the square root of the exponent. Thus, some students would mistakenly find that $\sqrt{x^9} = x^3$ or $\sqrt{x^{25}} = x^5$. Remind students that $x^3 \cdot x^3 = x^6$ and that $x^5 \cdot x^5 = x^{10}$. Have them rewrite expressions as products of squares when possible before trying to simplify. Therefore, $\sqrt{x^9} = \sqrt{x^4 \cdot x^4 \cdot x} = \sqrt{(x^4)^2 \cdot x} = x^4\sqrt{x}$. As a second example, $\sqrt{169x^{11}} = \sqrt{(13^2)(x^5)^2 x} = \sqrt{(13x^5)^2 x} = 13x^5\sqrt{x}$.

Have students use dictiona-
ries to look up the meanings
of the words "root" and
"radical." Be sure they pay
particular attention to the
fact that the word radical is
derived from the Latin word
radix which means root. See
if they can determine the relationship between the dictionary meanings of these
words and their meanings in
a mathematical sense. By re-
alizing that "root" means the
source or origin of some-
thing and "radical" means
going to the source of something, they should be able to
understand why 2 is the
square root of 4, or $\sqrt{4} = 2$.
The word root is also used to
signify those values which
satisfy a quadratic equation.
For example, 7 and -2 are
the roots of the equation
$x^2 - 5x - 14 = 0$.

Since $|a|$ is nonnegative for all real numbers a, the following is true.

> For any real number a,
> $$\sqrt{a^2} = |a|.$$

EXAMPLE 2   Simplify.

    a. $\sqrt{(n-2)^2}$ **b.** $\sqrt{m^2 n^2}$ **c.** $\sqrt{x^4}$

SOLUTION   **a.** $\sqrt{(n-2)^2} = |n-2|$

    b. $\sqrt{m^2 n^2} = |mn|$

    c. $\sqrt{x^4} = |x^2|$

    Since x^2 is nonnegative for any real number x,
    $|x^2| = x^2$.
    $\therefore \sqrt{x^4} = x^2$.

When a radicand does not involve a variable, however, it is not necessary to use absolute value notation.

EXAMPLE 3   Simplify.

    a. $\sqrt{64}$ **b.** $\sqrt{\left(\frac{2}{5}\right)^2}$ **c.** $\sqrt{(-4)^2}$

SOLUTION   **a.** $\sqrt{64} = 8$

    b. $\sqrt{\left(\frac{2}{5}\right)^2} = \frac{2}{5}$

    c. $\sqrt{(-4)^2} = 4$

Any number that can be expressed as the square of a rational number is called a **perfect square.** For example:

    64 is a perfect square because $64 = 8^2$.

    $\frac{9}{16}$ is a perfect square because $\frac{9}{16} = \left(\frac{3}{4}\right)^2$.

Equivalently, any number with a rational square root is a perfect square. The square roots of real numbers other than perfect squares are all irrational numbers. For example,

$$\sqrt{2}, \sqrt{15}, \text{ and } \sqrt{120}$$

all represent irrational numbers.

494   *Chapter 10*

Oral Exercises

Tell whether or not the given symbol represents a real number. If the number represented is real, tell whether it is rational or irrational.

1. $\sqrt{11}$ real, irrational   **2.** $\sqrt{-4}$ not real   **3.** $-\sqrt{49}$ real, rational   **4.** $\sqrt{(-3)^2}$
real, rational

5. $\sqrt{\dfrac{1}{5}}$ real, irrational   **6.** $\sqrt{\dfrac{16}{49}}$ real, rational   **7.** $\sqrt{0.81}$ real, rational   **8.** $\sqrt{-0.25}$
not real

9. $\sqrt{(-1)^3}$ not real   **10.** $-\sqrt{72}$ real, irrational   **11.** $\sqrt{0.1}$ real, irrational   **12.** $\dfrac{2}{\sqrt{3}}$
real, irrational

Simplify.

13. $\sqrt{36}$ 6   **14.** $-\sqrt{9}$ -3 **15.** $\sqrt{121}$ 11   **16.** $\sqrt{\dfrac{1}{4}}$ $\dfrac{1}{2}$

17. $(\sqrt{64})^2$ 64   **18.** $\sqrt{(-5)^2}$ 5   **19.** $-\sqrt{0.01}$ -0.1 **20.** $(\sqrt{81})^2$
81

21. $\sqrt{(-2)^4}$ 4   **22.** $\sqrt{10,000}$ 100   **23.** $\sqrt{\dfrac{4}{225}}$ $\dfrac{2}{15}$ **24.** $-\sqrt{1.44}$
-1.2

25. Is the sentence $\sqrt{a^2} = a$ true for every nonnegative real value of a? Yes
for every negative real value of a? No

26. Is the sentence $\sqrt{a^2} = |a|$ true for every nonnegative real value of a? Yes
for every negative real value of a? Yes

Written Exercises

Simplify.

A
1. $\sqrt{81}$ 9   **2.** $\sqrt{144}$ 12   **3.** $-\sqrt{9}$ -3 **4.** -0.2 **8.** 10
4. $-\sqrt{0.04}$

5. $\sqrt{(17)^2}$ 17   **6.** $-\sqrt{(21)^2}$ -21 **7.** $\sqrt{(-4)^2}$ 4   **8.** $\sqrt{(-10)^2}$

9. $(\sqrt{25})^2$ 25   **10.** $(\sqrt{49})^2$ 49   **11.** $-\sqrt{\dfrac{16}{81}}$ $-\dfrac{4}{9}$ **12.** $\sqrt{\dfrac{9}{169}}$ $\dfrac{3}{13}$

Solve.

13. $x^2 = 25$ {5, -5}   **14.** $y^2 = 64$ {8, -8}

15. $z^2 - 15 = 66$ {9, -9}   **16.** $x^2 + \dfrac{7}{12} = \dfrac{5}{6}$ {$\dfrac{1}{2}$, $-\dfrac{1}{2}$}

17. $100y^2 = 81$ {$\dfrac{9}{10}$, $-\dfrac{9}{10}$}   **18.** $3x^2 - 8 = 19$ {3, -3}

19. $5 - 4z^2 = -59$ {4, -4}   **20.** $2x^2 + 150 = 1400$ {25, -25}

Mixed Review

Perform the operation indicated. Express the results as a fraction in lowest terms.

1. $(0.\overline{4})(0.\overline{6})$ $\dfrac{8}{27}$

2. $3.2 + \dfrac{1}{6}$ $\dfrac{101}{30}$

Simplify. Assume that no variable equals zero. Express your result so that all exponents are positive integers.

3. $(-2x^2)^2 + (-x)^3$ $4x^4 - x^3$

4. $(t^{-4} \cdot t^{-1})^2$ $\dfrac{1}{t^{10}}$

5. $\left(\dfrac{3a^2b}{-4c}\right)^2$ $\dfrac{9a^4b^2}{16c^2}$

6. $\dfrac{2^{-1}c^{-2}d}{4c^2d^{-3}}$ $\dfrac{d^4}{8c^4}$

Suggested Assignments

Minimum
First day: 495/1–15
Second day: 495/16–32
Average
  495/1–33 odd, 34–44
Maximum
  495/1–45 odd, 47–56

Additional A Exercises

Simplify.
1. $\sqrt{169}$ 13
2. $\sqrt{(23)^2}$ 23
3. $-\sqrt{\dfrac{9}{36}}$ $-\dfrac{1}{2}$

Solve.
4. $m^2 = 121$ {11, -11}
5. $2x^2 + 7 = 105$ {7, -7}
6. $p^2 - \dfrac{1}{8} = \dfrac{1}{8}$ {$\dfrac{1}{2}$, $-\dfrac{1}{2}$}

55. $\sqrt{a^2} = \sqrt{(-a)^2} = |a|$. You must know if a is negative or nonnegative before you can decide if $|a| = a$ or $-a$.

56. Assume $b \neq c$. Then there are two cases to consider.
Case 1: $b > c$
1. a, b, and c are positive real numbers; $b > c$
   Hypothesis
2. $b \cdot b > b \cdot c$;
   $b \cdot c > c \cdot c$
   Mult. prop. of order
3. $b^2 > bc$; $bc > c^2$
   Def. of a power
4. $b^2 > c^2$
   Trans. prop. of order
5. $b^2 = a$; $c^2 = a$
   Hypothesis
6. $a > a$
   Subs. prin.
Statement (6) contradicts the reflexive axiom of equality. ∴ The assumption $b \neq c$ is incorrect and the conclusion $b = c$ is true.
Case 2: $b < c$
The proof of Case 2 is similar; it leads to the contradiction $a < a$.

Simplify.

B **21.** $\sqrt{64c^2}$ $8|c|$

22. $\sqrt{144m^2}$ $12|m|$

23. $\sqrt{9m^2n^2}$ $3|mn|$

24. $\sqrt{400x^2y^2}$ $20|xy|$

25. $\sqrt{16u^4}$ $4u^2$

26. $-\sqrt{4f^4}$ $-2f^2$

27. $\sqrt{\dfrac{c^2}{e^2}}$ $\left|\dfrac{c}{e}\right|$

28. $\sqrt{\dfrac{16z^2}{b^2}}$ $4\left|\dfrac{z}{b}\right|$

29. $\sqrt{25a^4b^2}$ $5a^2|b|$

30. $\sqrt{144c^6d^4}$ $12|c^3|d^2$

31. $\sqrt{\dfrac{x^4}{9y^2}}$ $\dfrac{x^2}{3|y|}$

32. $\sqrt{\dfrac{49m^2}{36n^6}}$ $\dfrac{7|m|}{6|n^3|}$

33. $-\sqrt{1.21g^4h^4}$ $-1.1g^2h^2$

34. $\sqrt{6.25a^4b^4}$ $2.5a^2b^2$

35. $\sqrt{(5ef)^2}$ $5|ef|$

36. $\sqrt{(-3jk)^2}$ $3|jk|$

37. $\sqrt{(r+s)^2}$ $|r+s|$

38. $\sqrt{(p+q)^4}$ $(p+q)^2$

39. $\sqrt{(f+g)^8}$ $(f+g)^4$

40. $\sqrt{(1-m)^6}$ $|1-m|^3$

41. $\sqrt{196x^4y^{16}}$ $14x^2y^8$

42. $\sqrt{2500r^{10}s^{18}}$ $50|r^5s^9|$

43. $\sqrt{\dfrac{169j^{30}}{64k^{20}}}$ $\dfrac{13j^{15}}{8k^{10}}$

44. $\sqrt{\dfrac{36d^{48}}{441m^{54}}}$ $\dfrac{2d^{24}}{7|m^{27}|}$

45. Show that $\sqrt{25a^2 + 30ab + 9b^2} = |5a + 3b|$ is true for all real values of a and b. $\sqrt{25a^2 + 30ab + 9b^2} = \sqrt{(5a + 3b)^2} = |5a + 3b|$

46. Show that $\sqrt{16x^2 - 72xy + 81y^2} = |4x - 9y|$ is true for all real values of x and y. $\sqrt{16x^2 - 72xy + 81y^2} = \sqrt{(4x - 9y)^2} = |4x - 9y|$

Simplify. Assume that variable expressions appearing as exponents denote positive integers.

C **47.** $\sqrt{c^{4m}}$ c^{2m} **48.** $\sqrt{(e^mf^{2n})^2}$ $|e^m|f^{2n}$ **49.** $\sqrt{(q^{3m})(q^{7m})}$ $|q^{5m}|$ **50.** $\sqrt{\dfrac{q^{2+m}}{q^{8-m}}}$ $|q^{m-3}|$

51. Does $\sqrt{a^2 - 1}$ represent a real number for all real values of a? No

52. Does $\sqrt{a^2 + 1}$ represent a real number for all real values of a? Yes

53. Does $f: x \rightarrow \sqrt{x}$ define a function? If so, what is the domain? the range? Yes; domain: $\{x: x \geq 0\}$; range: $\{y: y \geq 0\}$

54. Does $f: x \rightarrow -\sqrt{x}$ define a function? If so, what is the domain? the range? Yes; domain: $\{x: x \geq 0\}$; range: $\{y: y \leq 0\}$

55. Find the fallacy in the following argument that every real number is equal to its opposite.

   If a is a real number, then $a = \sqrt{a^2} = \sqrt{(-a)^2} = -a$.

56. Prove that each positive real number has at most one positive square root. That is, if a, b, and c are any positive real numbers such that $b^2 = a$ and $c^2 = a$, then $b = c$.

496 *Chapter 10*

10–4 Irrational Square Roots

Since the decimal representations of irrational numbers such as $\sqrt{2}$ or $\sqrt{15}$ are nonterminating and nonrepeating, you cannot name an exact decimal numeral for these numbers. However, you can find a rational approximation for an irrational number. To find a rational approximation for an irrational number, begin by locating the irrational number between two consecutive integers or between two rational numbers. For example, since

$$9 < 15 < 16,$$

you know that

$$\sqrt{9} < \sqrt{15} < \sqrt{16},$$

or

$$3 < \sqrt{15} < 4.$$

To obtain a more precise idea of the value of $\sqrt{15}$, find numbers whose squares are close to 15. Thus, since $(3.8)^2 = 14.44$, and since $(3.9)^2 = 15.21$, you know that

$$3.8 < \sqrt{15} < 3.9.$$

Continuing to find numbers whose squares approach 15, you may find that $(3.87)^2 = 14.9769$ and $(3.88)^2 = 15.0544$, and so

$$3.87 < \sqrt{15} < 3.88.$$

Similarly, you may find that

$$3.872 < \sqrt{15} < 3.873,$$

and

$$3.8729 < \sqrt{15} < 3.8730.$$

Therefore,

$$\sqrt{15} \approx 3.8730.$$

Rational approximations of all irrational square roots can be determined by the method outlined for $\sqrt{15}$. The *Table of Square Roots* on page 683 gives these approximations for the integers from 1 to 100.

EXAMPLE 1 Using the Table of Square Roots, give a rational approximation of each of the following.

        a. $\sqrt{51}$ **b.** $3 + \sqrt{5}$ **c.** $6\sqrt{13}$

SOLUTION **a.** $\sqrt{51} \approx 7.141$

        b. $3 + \sqrt{5} \approx 3 + 2.236 = 5.236$

        c. $6\sqrt{13} \approx 6(3.606) = 21.636$

Irrational Numbers and Radicals **497**

Teaching Suggestions
p. T105

Supplementary Material
Test 39

Key Ideas
Use the product property of square roots to simplify radicals.
Find approximations of irrational roots.

Chalkboard Examples

1. Using the Table of Square Roots, give a rational approximation of the following.
   a. $\sqrt{43} \approx 6.557$
   b. $7 + \sqrt{10}$
      $\approx 7 + 3.162 = 10.162$
   c. $4\sqrt{82} \approx 4(9.055) = 36.220$

2. Simplify.
   a. $\sqrt{243}$ $\sqrt{81 \cdot 3} = 9\sqrt{3}$
   b. $\sqrt{1296}$ $\sqrt{81 \cdot 16} = \sqrt{81} \cdot \sqrt{16} = 9 \cdot 4 = 36$

3. Solve $p^2 - 18 = 80$.
   $p^2 = 98$
   $p = \pm\sqrt{98}$
   $p = \pm\sqrt{49 \cdot 2}$
   $p = \pm 7\sqrt{2}$
   $\therefore$ the solution set is $\{7\sqrt{2}, -7\sqrt{2}\}$.

When a radicand is an integer greater than 100, its square root will not be listed in the table. To determine a method for finding these square roots, first consider the following.

$$\sqrt{16 \cdot 25} = \sqrt{400} = 20$$
$$\sqrt{16} \cdot \sqrt{25} = 4 \cdot 5 = 20$$

Then, by the transitive axiom of equality,

$$\sqrt{16 \cdot 25} = \sqrt{16} \cdot \sqrt{25}.$$

This relationship suggests the following fact about square roots.

Product Property of Square Roots

For any nonnegative real numbers a and b,

$$\sqrt{ab} = \sqrt{a} \cdot \sqrt{b}.$$

This property can be used to simplify irrational square roots as in the following example.

EXAMPLE 2 Simplify.　　**a.** $\sqrt{120}$　　**b.** $\sqrt{180}$

SOLUTION　　**a.** $\sqrt{120} = \sqrt{4 \cdot 30}$　　　　**b.** $\sqrt{180} = \sqrt{9 \cdot 20}$
$$= \sqrt{4} \cdot \sqrt{30} \qquad\qquad\qquad = \sqrt{9 \cdot 4 \cdot 5}$$
$$= 2\sqrt{30} \qquad\qquad\qquad\qquad = \sqrt{9} \cdot \sqrt{4} \cdot \sqrt{5}$$
$$\qquad\qquad\qquad\qquad\qquad\qquad = 3 \cdot 2 \cdot \sqrt{5}$$
$$\qquad\qquad\qquad\qquad\qquad\qquad = 6\sqrt{5}$$

Simplifying the radicals $\sqrt{120}$ and $\sqrt{180}$ now enables you to use the table to approximate these square roots. Thus,

$$\sqrt{120} = 2\sqrt{30} \approx 2(5.477) = 10.954$$

and

$$\sqrt{180} = 6\sqrt{5} \approx 6(2.236) = 13.416.$$

To approximate irrational square roots that cannot be found in the table or that cannot be simplified using the product property, you may wish to use the *divide-and-average method* discussed in Appendix A on page 685.

The product property of square roots is helpful in simplifying some expressions in which the radicand is a fraction. For example:

$$\sqrt{\frac{2}{25}} = \sqrt{\frac{1}{25} \cdot 2} = \sqrt{\frac{1}{25}} \cdot \sqrt{2} = \frac{1}{5}\sqrt{2}, \text{ or } \frac{\sqrt{2}}{5}$$

The product property of square roots may also be helpful in simplifying *rational* square roots.

EXAMPLE 3 Simplify $\sqrt{1024}$.

SOLUTION $\sqrt{1024} = \sqrt{4 \cdot 256}$
$= \sqrt{4 \cdot 4 \cdot 64}$
$= \sqrt{4} \cdot \sqrt{4} \cdot \sqrt{64}$
$= 2 \cdot 2 \cdot 8$
$= 32$

You may use the product property in solving equations with irrational roots.

EXAMPLE 4 Solve. Simplify irrational solutions.

a. $x^2 - 18 = 90$ **b.** $10x^2 - 2430 = 570$

SOLUTION

a. $x^2 - 18 = 90$
$x^2 = 108$
$x = \pm\sqrt{108}$
$= \pm\sqrt{4 \cdot 9 \cdot 3}$
$= \pm\sqrt{4} \cdot \sqrt{9} \cdot \sqrt{3}$
$= \pm 2 \cdot 3 \cdot \sqrt{3}$
$x = \pm 6\sqrt{3}$

$\therefore$ the solution set is
$\{6\sqrt{3}, -6\sqrt{3}\}$.

b. $10x^2 - 2430 = 570$
$10x^2 = 3000$
$x^2 = 300$
$x = \pm\sqrt{300}$
$= \pm\sqrt{100 \cdot 3}$
$= \pm\sqrt{100} \cdot \sqrt{3}$
$x = \pm 10\sqrt{3}$

$\therefore$ the solution set is
$\{10\sqrt{3}, -10\sqrt{3}\}$.

Oral Exercises

Name the two consecutive integers between which each square root lies.

1. $\sqrt{17}$ 4, 5 **2.** $\sqrt{50}$ 7, 8 **3.** $\sqrt{38.5}$ 6, 7 **4.** $\sqrt{92.8}$ 9, 10 **5.** $\sqrt{\frac{1}{2}}$ 0, 1 **6.** $\sqrt{\frac{32}{5}}$ 2, 3

7. $\sqrt{175}$
13, 14 **8.** $\sqrt{255}$
15, 16 **9.** $-\sqrt{402}$
$-21, -20$ **10.** $-\sqrt{628}$
$-26, -25$ **11.** $3\sqrt{15}$
11, 12 **12.** $4\sqrt{70}$
33, 34

Simplify.

13. $\sqrt{60}$
$2\sqrt{15}$ **14.** $\sqrt{84}$
$2\sqrt{21}$ **15.** $\sqrt{240}$
$4\sqrt{15}$ **16.** $\sqrt{726}$
$11\sqrt{6}$ **17.** $(\sqrt{5})^2$
5 **18.** $(\sqrt{48})^2$
48

Approximate each square root to the nearest thousandth. Use the Table of Square Roots as necessary. Approximations may vary.

19. $\sqrt{85}$
9.920 **20.** $\sqrt{56}$
7.483 **21.** $\sqrt{160}$
12.648 **22.** $\sqrt{384}$
19.596 **23.** $\sqrt{8200}$
90.550 **24.** $\sqrt{1728}$
41.568

Simplify.

1. $-\sqrt{\frac{24}{144}}$ $-\frac{\sqrt{6}}{6}$
2. $(\sqrt{9})^2$ 9
Solve.
3. $x^2 = 256$ $\{16, -16\}$
4. $m^2 - 196 = 0$ $\{14, -14\}$
Solve by factoring first and then using the zero-product property given in Section 7-10.
5. $x^2 = 256$ $\{16, -16\}$
6. $m^2 - 196 = 0$ $\{14, -14\}$

Suggested Assignments

Minimum
First day:
  500/1–27 odd
Second day:
  500/2–28 even, 29–36
  R 502/*Self-Test 1*
Average
First day: 500/1–24
          S 490/20–28 even
Second day: 500/25–47
          R 502/*Self-Test 1*
Maximum
First day: 500/1–43 odd
          S 490/26–40 even
Second day: 500/44–61
          R 502/*Self-Test 1*

Additional A Exercises

Simplify.
1. $\sqrt{882}$ $21\sqrt{2}$
2. $\sqrt{\dfrac{63}{100}}$ $\dfrac{3}{10}\sqrt{7}$
3. $\sqrt{1.28}$ $0.8\sqrt{2}$
Solve.
4. $m^2 - 12 = 0$
   $\{2\sqrt{3}, -2\sqrt{3}\}$
5. $2r^2 + 3 = 75$ $\{6, -6\}$
6. Simplify $6\sqrt{43}$. Approximate your answer to the nearest hundredth. Use the Table of Square Roots as necessary.   39.34

Written Exercises

Simplify.

A 1. $\sqrt{48}$ $4\sqrt{3}$ 2. $\sqrt{18}$ $3\sqrt{2}$ 3. $\sqrt{75}$ $5\sqrt{3}$ 4. $\sqrt{128}$ $8\sqrt{2}$

5. $\sqrt{320}$ $8\sqrt{5}$ 6. $\sqrt{1200}$ $20\sqrt{3}$ 7. $2\sqrt{96}$ $8\sqrt{6}$ 8. $3\sqrt{108}$

9. $\sqrt{\dfrac{5}{16}}$ $\dfrac{1}{4}\sqrt{5}$ 10. $\sqrt{\dfrac{27}{25}}$ $\dfrac{3}{5}\sqrt{3}$ 11. $\sqrt{0.72}$ $0.6\sqrt{2}$ 12. $\sqrt{0.32}$

    8. $18\sqrt{3}$ 12. $0.4\sqrt{2}$

Solve. Simplify irrational solutions.

13. $a^2 = 11$ $\{\sqrt{11}, -\sqrt{11}\}$ 14. $b^2 = 28$ $\{2\sqrt{7}, -2\sqrt{7}\}$
15. $c^2 - 7 = 0$ $\{\sqrt{7}, -\sqrt{7}\}$ 16. $z^2 - 10 = 0$ $\{\sqrt{10}, -\sqrt{10}\}$
17. $d^2 + 5 = 25$ $\{2\sqrt{5}, -2\sqrt{5}\}$ 18. $m^2 - 6 = 34$ $\{2\sqrt{10}, -2\sqrt{10}\}$
19. $3e^2 + 7 = 97$ $\{\sqrt{30}, -\sqrt{30}\}$ 20. $4x^2 - 12 = 48$ $\{\sqrt{15}, -\sqrt{15}\}$
21. $12x^2 - 50 = 670$ $\{2\sqrt{15}, -2\sqrt{15}\}$ 22. $10y^2 + 63 = 513$ $\{3\sqrt{5}, -3\sqrt{5}\}$
23. $8x^2 + 12 = 102$ $\{\frac{3}{2}\sqrt{5}, -\frac{3}{2}\sqrt{5}\}$ 24. $50z^2 - 26 = 190$ $\{\frac{6}{5}\sqrt{3}, -\frac{6}{5}\sqrt{3}\}$

Simplify. Approximate the answer to the nearest thousandth. Use the Table of Square Roots as necessary.   Approximations may vary.

25. $14 - \sqrt{91}$ 4.461   26. $\sqrt{20} - \sqrt{10}$ 1.310   27. $7\sqrt{67}$ 57.295   28. $4\sqrt{83}$

B 29. $5 + 2\sqrt{34}$ 16.662   30. $9\sqrt{24} - 16$ 28.091   31. $\dfrac{\sqrt{43} - 5}{2}$ 0.779   32. $\dfrac{2\sqrt{76} - 4}{3}$

33. $\sqrt{180} + \sqrt{250}$ 29.226   34. $\dfrac{6}{5}\sqrt{567}$ 28.577   35. $\dfrac{\sqrt{432} - 2}{3}$ 6.261   36. $\dfrac{8 - 3\sqrt{648}}{2}$

    28. 36.440   32. 4.479   36. −34.178

Determine which of the two numbers is greater.

37. $5 + \sqrt{10}$, $20 - \sqrt{85}$ $20-\sqrt{85}$ 38. $12 - 3\sqrt{5}$, $3 + 2\sqrt{6}$ $3 + 2\sqrt{6}$
39. $2 + \sqrt{8}$, $8 - \sqrt{11}$ $2 + \sqrt{8}$ 40. $\sqrt{75} - 5$, $\sqrt{5} + 3$ $\sqrt{5} + 3$
41. $1 + \sqrt{148}$, $\sqrt{204} - 1$ $\sqrt{204} - 1$ 42. $2\sqrt{260}$, $\sqrt{600} + 10$ $\sqrt{600} + 10$

43. Show that $(3 - \sqrt{5})^2 < 3 - \sqrt{5}$.

44. Show that $(8 - \sqrt{54})^2 < 8 - \sqrt{54}$.

45. Solve $2r + \sqrt{2} = 5$. Approximate the value of r to the nearest hundredth. $\{1.79\}$

46. Solve $\sqrt{12} - 3x = \sqrt{3}$. Approximate the value of x to the nearest hundredth. $\{0.58\}$

47. Find two nonzero rational numbers x and y such that $x\sqrt{5} + y\sqrt{5}$ is rational.   Any nonzero x, y such that $x = -y$. Example: $x = \frac{1}{2}$, $y = -\frac{1}{2}$

48. Find two nonzero rational numbers a and b such that $a\sqrt{34} + b\sqrt{34}$ is rational. Any nonzero a, b such that $a = -b$. Example: $a = 7$, $b = -7$

500   *Chapter 10*

Evaluate $(\sqrt{a} + \sqrt{b})(\sqrt{a} - \sqrt{b})$ for the given values of a and b.

C 49. $a = 4$, $b = 9$ -5 50. $a = 9$, $b = 4$ 5 51. $a = 16$, $b = 25$ -9

52. $a = 36$, $b = 4$ 32 53. $a = 64$, $b = 49$ 15 54. $a = 100$, $b = 100$ 0

55. Simplify $(\sqrt{a} + \sqrt{b})(\sqrt{a} - \sqrt{b})$ when a and b are nonnegative numbers. $a - b$

Let r and s be irrational numbers. Determine whether or not the expression may represent a rational number. Give examples to illustrate your hypothesis.

56. $2r$ 57. $r - 1$ 58. $r \cdot s$ 59. $r \div s$

60. If p is a rational number and $\sqrt{q}$ is an irrational number, show that $(p + \sqrt{q})$ is an irrational number.

61. On a number line, graph the solution set of $\sqrt{x} > x$.

Computer Exercises For students with computer experience

1. Write a program that will allow you to input a positive integer less than 1000 and will find the greatest square factor of the integer. A sample output would be

    THE GREATEST SQUARE FACTOR OF 162 IS 81.

    (*Hint*: Begin with $31^2 = 961$ as a trial factor and continue down to $1^2 = 1$. The first such square that is a factor of the given number is the greatest square factor.)

2. Incorporate the program that you wrote for Exercise 1 into a program that will simplify the square root of any integer less than 1000. A sample output would be

    SQR(162) = 9 * SQR(2).

3. Write a program that will *locate* the square root of a given positive integer *without using the computer's square-root function*. (*Hint*: Start with 1 as a *trial* square root, square it, then check whether the square is greater than or equal to the given integer. If it is, print it. If it is not, add 1 to the trial root and check again. Continue adding 1 if the square is *less than* the given integer, and print the first trial root whose square is *greater than or equal to* it.)

4. Modify the program that you wrote for Exercise 3 so that it *estimates* the square root in successive steps. That is, the program should first print out the greatest *integer* less than or equal to the square root, then the greatest *tenth*, the greatest *hundredth*, and finally the greatest *thousandth*. For comparison, the program should also print out the value obtained by using the computer's square-root function.

Irrational Numbers and Radicals **501**

501

Replace each ___?___ with one of the symbols <, =, or > to make a true statement.

1. $\frac{7}{8}$ _?_ $\frac{8}{9}$ <

2. $\frac{84}{23}$ _?_ $\frac{72}{35}$ >

3. Find the rational number that is one half the distance from $\frac{3}{4}$ to 7.

$\frac{31}{8}$, or $3\frac{7}{8}$

Find the decimal equivalent for each rational number.

4. $\frac{3}{8}$ 0.375

5. $\frac{16}{33}$ $0.\overline{48}$

Express as a fraction in lowest terms.

6. 0.214 $\frac{107}{500}$

7. $0.3\overline{91}$ $\frac{194}{495}$

Simplify.

8. $\sqrt{81}$ 9

9. $(\sqrt{24})^2$ 24

10. $\sqrt{(-16)^2}$ 16

11. $\sqrt{32}$ $4\sqrt{2}$

12. $3\sqrt{27}$ $9\sqrt{3}$

13. $\sqrt{450}$ $15\sqrt{2}$

Approximate each square root to the nearest hundredth. Use the Table of Square Roots as necessary.

14. $\sqrt{38}$ 6.16

15. $\sqrt{93}$ 9.64

16. $\sqrt{321}$ 17.92

Self-Test 1

VOCABULARY   density property of rational
  numbers (p. 484)
terminating decimal (p. 486)
repeating decimal (p. 486)
property of completeness
  (p. 489)

square root (p. 493)
principal square root (p. 493)
radical sign (p. 493)
radicand (p. 493)
radical (p. 493)
perfect square (p. 494)

Replace each ___?___ with one of the symbols <, =, or > to make a true statement.   *Obj. 1, p. 481*

1. $\frac{6}{5}$ ___?___ $\frac{7}{6}$ >

2. $\frac{96}{72}$ ___?___ $\frac{39}{26}$ <

3. Find the rational number that is one half of the way from $\frac{2}{3}$ to 5. $2\frac{5}{6}$

Express each rational number as a terminating or repeating decimal.

4. $\frac{5}{8}$ 0.625

5. $\frac{11}{24}$ $0.458\overline{3}$ *Obj. 2, p. 481*

Express each rational number as a fraction in simplest form.

6. 0.384 $\frac{48}{125}$

7. $0.4\overline{36}$ $\frac{24}{55}$

Simplify.

8. $\sqrt{64}$ 8

9. $(\sqrt{36})^2$ 36

10. $\sqrt{(-18)^2}$ 18   *Obj. 3, p. 481*

11. $\sqrt{24}$ $2\sqrt{6}$

12. $2\sqrt{84}$ $4\sqrt{21}$

13. $\sqrt{320}$ $8\sqrt{5}$ *Obj. 4, p. 481*

Approximate each square root to the nearest thousandth. Use the Table of Square Roots as necessary. Approximations may vary.

14. $\sqrt{88}$ 9.381

15. $\sqrt{69}$ 8.307

16. $\sqrt{245}$ 15.652   *Obj. 5, p. 481*

Check your answers with those at the back of the book.

ON THE CALCULATOR

Most calculators have a square root key that can be used to find the square root of a nonnegative number. Use a calculator to approximate each square root to the nearest hundredth.

1. $\sqrt{7}$ 2.65

2. $\sqrt{19}$ 4.36

3. $\sqrt{125}$ 11.18

4. $\sqrt{4000}$ 63.25

5. $\sqrt{2.5}$ 1.58

6. $\sqrt{3.15}$ 1.77

7. $\sqrt{0.625}$ 0.79

8. $\sqrt{0.004}$ 0.06

502   *Chapter 10*

Careers

Transportation Engineering

Transportation engineers are responsible for designing and developing surface transportation systems that are safe, efficient, and economical. Their work may involve the planning of new systems or the improvement of existing ones.

At the beginning of a project, transportation engineers draft plans that specify the details of the proposed construction or repair. They must then prepare cost estimates for required materials and labor, and they may also need to report on the effect of the proposed changes on the environment. Once a project is underway, they often supervise on-site construction and equipment maintenance.

Transportation engineers continue to be involved in a project even after the actual construction or repair is finished. For example, some engineers work to prepare accurate maps of the completed system, while others are involved in ongoing analysis of traffic flow.

EXAMPLE   Roads expand with a large increase in temperature. It is the job of the transportation engineer to design a road so that it can withstand such expansion without warping and breaking up. One formula that is used to determine the amount of expansion, I, is

$$I = \frac{9}{5}kl(T - t),$$

where T is the actual temperature in degrees Celsius, t is the temperature in degrees Celsius at which the road was built, l is the length of the road, and k is a constant. How many feet will a two-mile stretch of road expand at 38°C if $k = 0.000012$ and the road was built at 16°C?

SOLUTION   Substitute the given information into the formula.

$$I = \frac{9}{5}(0.000012)(2)(38 - 16)$$

$$\backsim 0.00095$$

Since the length of the road was given in *miles*, multiply this result by 5280 to determine the number of *feet* of expansion.

$$0.00095 \times 5280 = 5.016$$

∴ the road will expand *approximately* 5 ft.

Irrational Numbers and Radicals   **503**

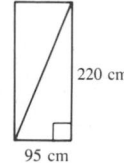
READING ALGEBRA Problem Solving

The first time that you read a problem, you may know how to solve it immediately. If you do, solve it carefully. As discussed on page 143, be sure that you have answered the question that was asked and that your answer is reasonable.

More often, however, you will not be able to solve a problem after just one reading. In such cases, reread the problem until you understand exactly what it is about. Questions like these may be helpful.

1. What am I asked to find?
2. What am I told?
3. Have I done problems like this before? How? Can my experience help in solving this problem?
4. Can I draw a diagram or make a chart using the given information?
5. What do I expect the answer to be?

Consider the following problem.

Suppose you plan to fly your airplane from Xavier to York. Looking at a road map, you see that York is 42 km south and 77 km east of Xavier. What is the distance if you fly there directly?

Your answers to the questions should be specific, like the following.

1. I am looking for the straight-line distance from Xavier to York.
2. York is 42 km south and 77 km east of Xavier.
3. I have done problems with distances in the same and opposite directions, by adding or subtracting. The distance should not be more than 42 + 77 = 119 km or less than 77 − 42 = 35 km.
4. I can draw a diagram like the one at the right and find the approximate distance by measuring the third side of the triangle.
5. I expect the answer to be a distance in kilometers.

Xavier
42 km
77 km York

Although you may not have seen a problem like this before, the answers to the questions show one way that you might approach it. In the following section you will learn a method of finding a more exact answer.

Exercises

Read the following problem and answer the five questions above.

A door is 95 cm wide and 220 cm high. You need to attach a diagonal brace to the door. How long must the brace be?

Using the Pythagorean Theorem

OBJECTIVES for Sections 10-5 and 10-6:
1. To use the Pythagorean Theorem and its converse.
2. To use the distance formula to find the distance between two points on a coordinate plane.

10–5 The Pythagorean Theorem

While the irrational number $\sqrt{2}$ cannot be represented as a terminating decimal, it is nevertheless possible to construct a picture of a line segment whose length is $\sqrt{2}$ units, as you will see later in this section.

First, look at the tile pattern shown in Figure 1. You can verify by counting the small triangles, each of which has the same area, that the area of the large square is equal to the sum of the areas of the two small red squares. This figure illustrates a special case of the *Pythagorean Theorem.* The development of this theorem is credited to Pythagoras, a famous Greek philosopher and mathematician who lived from about 580–500 B.C. It is believed that his method of proof was based on comparison of areas.

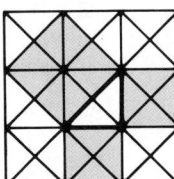

Figure 1

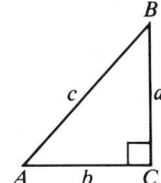

Figure 2

In stating the general theorem, it is convenient to use the customary labeling of a right triangle that is shown in Figure 2. The vertex of the right angle is labeled C. The length of the side opposite the right angle, called the **hypotenuse,** is represented by c. The lengths of the other two sides are represented by a and b.

Pythagorean Theorem

In any right triangle, the square of the length c of the hypotenuse is equal to the sum of the squares of the lengths a and b of the other two sides; that is:

$$c^2 = a^2 + b^2$$

Irrational Numbers and Radicals **505**

Key Ideas
Use the Pythagorean Theorem and its converse.

The diagrams in Figure 3 suggest a proof of this theorem. Each diagram shows a square, with sides of length $(a + b)$ units, separated into a series of smaller squares and triangles. Recall that the area of a square is the square of the length of one of its sides and use this fact to determine two expressions for the area of a square with side of $(a + b)$ units. If you equate these two expressions, you will obtain the equation stated in the theorem.

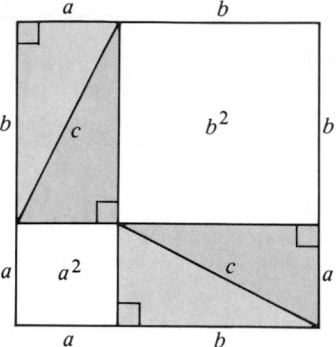
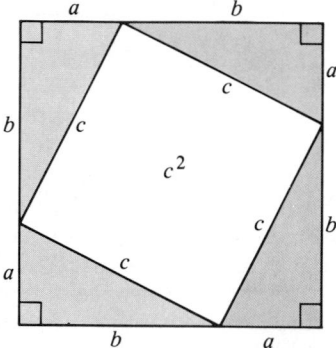

Figure 3

$$(a + b)^2 = a^2 + b^2 + 4\left(\tfrac{1}{2}ab\right) \qquad (a + b)^2 = c^2 + 4\left(\tfrac{1}{2}ab\right)$$

$$a^2 + b^2 + 4\left(\tfrac{1}{2}ab\right) = c^2 + 4\left(\tfrac{1}{2}ab\right)$$

$$a^2 + b^2 = c^2$$

Thus, by the symmetric axiom of equality,

$$c^2 = a^2 + b^2.$$

Using the Pythagorean Theorem, it is now possible to construct a line segment of length $\sqrt{2}$ units. First, draw a right triangle whose shorter sides are 1 unit long, as shown in Figure 4. Then the length of the hypotenuse, c, can be found by using the Pythagorean Theorem, with $a = 1$ and $b = 1$, as follows.

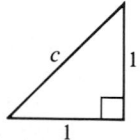

Figure 4

$$\begin{aligned} c^2 &= a^2 + b^2 \\ &= 1^2 + 1^2 \\ &= 1 + 1 \\ c^2 &= 2 \\ c &= \pm\sqrt{2} \end{aligned}$$

Therefore, the length of the hypotenuse is $\sqrt{2}$ units, as shown in Figure 5. (Notice that the negative root, $-\sqrt{2}$, is rejected because lengths cannot be negative.)

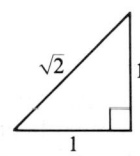

Figure 5

506   *Chapter 10*

The triangles in Figure 6 illustrate the fact that the Pythagorean Theorem may be used to construct line segments of length $\sqrt{3}$ units, $\sqrt{4}$ units, $\sqrt{5}$ units, $\sqrt{6}$ units, and so on.

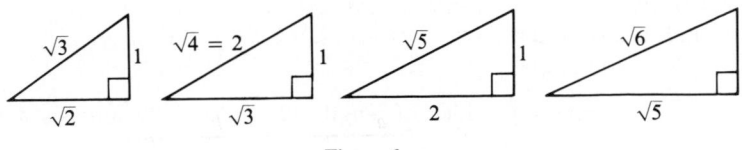

Figure 6

The triangles constructed in Figure 6 may now be used to locate the graphs of irrational square roots such as $\sqrt{2}$, $-\sqrt{2}$, and $\sqrt{3}$ on a number line, as shown in Figure 7. Here arcs are drawn to transfer the lengths of the hypotenuse of each triangle to a number line. Note that $-\sqrt{2}$ is located $\sqrt{2}$ units to the left of zero, $-\sqrt{3}$ is located $\sqrt{3}$ units to the left of zero, and so on.

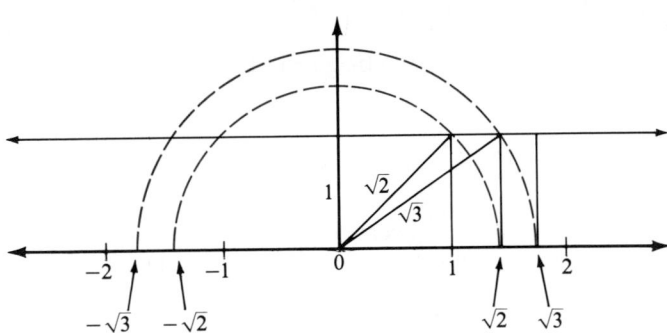

Figure 7

If you constructed a series of such triangles and arcs, you would be able to locate the graphs of $\sqrt{5}$, $\sqrt{6}$, and so on, on the number line. Note that the *converse* of the Pythagorean Theorem is also true.

Converse of the Pythagorean Theorem

If the lengths of the sides of a triangle are such that the sum of the squares of the lengths of the two shorter sides is equal to the square of the length of the longest side, the triangle is a right triangle. The right angle will be opposite the longest side.

Irrational Numbers and Radicals **507**

1. Determine whether or not each set of numbers could represent the lengths of the sides of a right triangle.

   a. $\{15, 36, 39\}$

   $15^2 + 36^2 \stackrel{?}{=} 39^2$

   $225 + 1296 \stackrel{?}{=} 1521$

   $1521 = 1521$ ✓

   ∴ a triangle with sides of length 15, 36, and 39 units is a right triangle.

   b. $\{12, 14, 16\}$

   $12^2 + 14^2 \stackrel{?}{=} 16^2$

   $144 + 196 \stackrel{?}{=} 256$

   $340 \neq 256$

   ∴ a triangle with sides of length 12, 14, and 16 units is not a right triangle.

2. Show that $\{18, 24, 30\}$ is a Pythagorean triple.

   $18^2 + 24^2 \stackrel{?}{=} 30^2$

   $324 + 576 \stackrel{?}{=} 900$

   $900 = 900$ ✓

3. A 13 m pole casts a shadow of 5 m. How far is it from the top of the pole to the end of the shadow? Assume that the pole forms a right angle with the ground.

   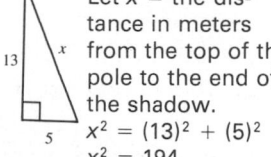

   Let x = the distance in meters from the top of the pole to the end of the shadow.

   $x^2 = (13)^2 + (5)^2$

   $x^2 = 194$

   $x = \pm\sqrt{194}$

   Select the positive square root.

   $x = \sqrt{194} \approx 13.93$ m

   ∴ the distance from the top of the pole to the end of the shadow is $\sqrt{194}$ m, or approximately 13.92 m.

EXAMPLE 1 Determine whether or not each set of numbers could represent the lengths of the sides of a right triangle.

a. $\{12, 16, 20\}$ b. $\{8, 10, 12\}$

SOLUTION   a. $12^2 + 16^2 \stackrel{?}{=} 20^2$

$144 + 256 \stackrel{?}{=} 400$

$400 = 400$ ✓

∴ a triangle with sides of length 12, 16, and 20 units is a right triangle.

b. $8^2 + 10^2 \stackrel{?}{=} 12^2$

$64 + 100 \stackrel{?}{=} 144$

$164 \neq 144$

∴ a triangle with sides of length 8, 10, and 12 units is not a right triangle.

Any set of positive integers that satisfies the equation

$$c^2 = a^2 + b^2$$

is called a set of **Pythagorean numbers,** or a **Pythagorean triple.**

EXAMPLE 2 Show that $\{15, 20, 25\}$ is a Pythagorean triple.

SOLUTION $15^2 + 20^2 \stackrel{?}{=} 25^2$

$225 + 400 \stackrel{?}{=} 625$

$625 = 625$ ✓

The Pythagorean Theorem and its converse can be used in solving problems.

EXAMPLE 3 A support wire 10 m long is attached to the top of a utility pole 7 m tall and is then stretched taut. How far from the base of the pole will the wire be attached to the ground?

SOLUTION

Step 1   The problem asks for the distance from the base of the pole to the point at which the wire touches the ground. Assume that the pole forms a right angle with the ground. Then the positions of the pole, the wire, and the ground can be represented by a right triangle.

Step 2   Let x = the distance in meters from the base of the pole to the ground end of the wire.

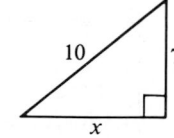

Step 3   Apply the Pythagorean Theorem, using $a = x$, $b = 7$, and $c = 10$.

$$a^2 + b^2 = c^2$$

$$x^2 + 7^2 = 10^2$$

508 *Chapter 10*

Step 4
$$x^2 + 7^2 = 10^2$$
$$x^2 + 49 = 100$$
$$x^2 = 51$$
$$x = \pm\sqrt{51}$$

Since a distance cannot be negative, the positive square root is selected. Thus,

$$x = \sqrt{51} \approx 7.14 \text{ m.}$$

Step 5   Checking the results is left to you.

∴ the distance from the base of the pole to the ground end of the wire is $\sqrt{51}$ m, or approximately 7.14 m.

Oral Exercises

State an equation that expresses the relationship between the lengths of the sides of each triangle.

1.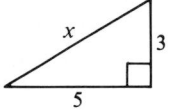

    $3^2 + 5^2 = x^2$

2.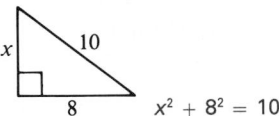

    $x^2 + 8^2 = 10^2$

3.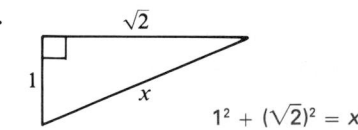

    $1^2 + (\sqrt{2})^2 = x^2$

4.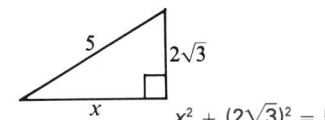

    $x^2 + (2\sqrt{3})^2 = 5^2$

5.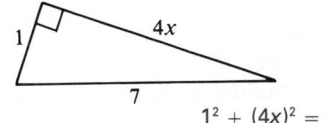

    $1^2 + (4x)^2 = 7^2$

6.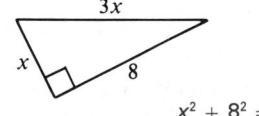

    $x^2 + 8^2 = (3x)^2$

Tell whether or not each set of numbers is a Pythagorean triple.

Yes

7. {15, 12, 9} Yes   8. {2, 3, 4} No   9. {5, 12, 13} Yes   10. {10, 24, 26}

11. $\left\{\frac{3}{4}, 1, \frac{5}{4}\right\}$ No   12. {7, 24, 25} Yes   13. {6, 10, 11} No   14. {2, $\sqrt{5}$, 3}

No

15. Explain how to find the length of a diagonal of a square whose sides are 4 cm long.

16. Explain how to find the length of a diagonal of a rectangle with dimensions 5 cm and 8 cm.

Irrational Numbers and Radicals **509**

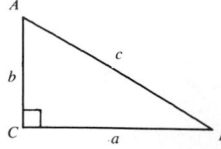
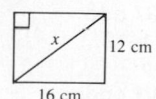
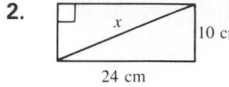
Written Exercises

Exercises 1–12 refer to the right triangle pictured below. Find the required length. Approximate your answer to the nearest hundredth, using a calculator or the Table of Square Roots as necessary. Approximations may vary.

A

1. $a = 2, b = 5, c = \underline{\quad?\quad}$ 5.39
2. $a = 3, b = 7, c = \underline{\quad?\quad}$ 7.62
3. $b = 5, c = 9, a = \underline{\quad?\quad}$ 7.48
4. $b = 11, c = 13, a = \underline{\quad?\quad}$ 6.93
5. $a = 10, c = 15, b = \underline{\quad?\quad}$ 11.18
6. $a = 12, c = 16, b = \underline{\quad?\quad}$ 10.58

7. $a = \sqrt{10}, b = 4, c = \underline{\quad?\quad}$ 5.10   8. $a = 6, b = \sqrt{23}, c = \underline{\quad?\quad}$ 7.68

9. $a = \sqrt{19}, c = \sqrt{87}, b = \underline{\quad?\quad}$ 8.25
5.66   10. $a = \sqrt{59}, c = \sqrt{95}, b = \underline{\quad?\quad}$ 6

11. $b = \sqrt{42}, c = \sqrt{74}, a = \underline{\quad?\quad}$ 12. $b = \sqrt{37}, c = \sqrt{109}, a = \underline{\quad?\quad}$ 8.49

Determine whether or not each set of numbers could represent the lengths of the sides of a right triangle.

16. Yes   20. No   24. No

13. {6, 8, 10}   Yes   14. {8, 15, 17} Yes   15. {2, 4, 6}   No   16. {9, 40, 41}
17. {7, 10, 11}   No   18. {6, 14, 15} No   19. {24, 25, 7} Yes   20. {20, 10, 15}
21. {9, 21, 30}   No   22. {14, 48, 50} Yes   23. {12, 12, 25} No   24. {35, 35, 35}

B

25. {5a, 12a, 13a}   Yes   26. {9q, 12q, 15q} Yes
27. {$\sqrt{29}$, 8, $\sqrt{93}$}   Yes   28. {$\sqrt{65}$, $\sqrt{15}$, 9} No
29. {$\sqrt{2}$, $\sqrt{2}$, 2}   Yes   30. {$\sqrt{3}$, $\sqrt{3}$, $\sqrt{6}$} Yes
31. {1.2, 1.6, 2}   Yes   32. {1.8, $\sqrt{7}$, 3.2} Yes

In Exercises 33–42, if c is the length of the hypotenuse and a and b the lengths of the other two sides of a right triangle, find the required length(s). Approximate your answer to the nearest hundredth, using a calculator or the Table of Square Roots as necessary.   Approximations may vary.

33. $a = \frac{3}{8}, b = \frac{1}{2}, c = \underline{\quad?\quad}$ 0.63   34. $a = \frac{1}{3}, b = \frac{4}{9}, c = \underline{\quad?\quad}$ 0.56

35. $a = \frac{1}{2}b, b = 8, c = \underline{\quad?\quad}$ 8.94   36. $a = \frac{2}{5}b, b = 25, c = \underline{\quad?\quad}$ 26.93

37. $a = 32, b = \frac{3}{4}a, c = \underline{\quad?\quad}$ 40   38. $a = 27, b = \frac{4}{3}a, c = \underline{\quad?\quad}$ 45
8.94   4.47   5.29   2.65

C

39. $a = 2b, c = 10, a = \underline{\quad?\quad}, b = \underline{\quad?\quad}$ 40. $b = \frac{1}{2}a, c = \sqrt{35}, a = \underline{\quad?\quad}, b = \underline{\quad?\quad}$
4.47   6.71   8.66   10.39

41. $a = \frac{2}{3}c, b = 5, a = \underline{\quad?\quad}, c = \underline{\quad?\quad}$ 42. $c = \frac{6}{5}a, b = \sqrt{33}, a = \underline{\quad?\quad}, c = \underline{\quad?\quad}$

510   *Chapter 10*

Problems

Make a sketch for each problem. Approximate each square root to the nearest hundredth. Use a calculator or the Table of Square Roots as necessary. Approximations may vary.
See Additional Answers for sketches.

A 1. Find the length of a diagonal of a rectangle whose dimensions are 12 cm by 16 cm. 20 cm

2. Find the length of a diagonal of a rectangle whose dimensions are 10 cm by 24 cm. 26 cm

3. Find the length of a diagonal of a square whose sides are each 10 m. 14.14 m

4. Find the length of a diagonal of a square whose sides are each $\sqrt{7}$ cm. 3.74 cm

5. Starting at the airfield, Ben flew his glider 8 km due east to point A and then 15 km due south to point B. If he flew directly back to the airfield from point B, how far did he travel in this third part of his trip? 17 km

6. A ship navigated a course due west for 18 km and then navigated a course due north for 80 km. At that point, what was the distance of the ship from its starting point? 82 km

7. A cable from the top of a circus tent pole is attached to the ground at a point 6 m from the base of the pole. If the cable is 44 m long, how high is the pole? 43.59 m

8. The foot of a 2.6 m ramp is 2.4 m from the base of a loading platform. Find the height of the platform. 1 m

B 9. A diagonal of a square is 12 cm long. Find the length of a side of the square. 8.49 cm

10. A diagonal of a square is 8 cm long. Find the length of a side of the square. 5.66 cm

11. The hypotenuse of a certain right triangle is twice as long as the shortest side, and the length of the third side is 15 m. Find the length of the shortest side. 8.66 m

12. Find the dimensions of a rectangle whose length is three times its width if one of its diagonals has length 30 cm. 9.49 cm by 28.46 cm

C 13. Two sides of a right triangle have lengths 5 m and 8 m. Find the two possible lengths for the third side. 9.43 m; 6.24 m

14. Two sides of a right triangle have lengths 6 cm and 10 cm. Find the two possible lengths for the third side. 11.66 cm; 8 cm

Irrational Numbers and Radicals **511**

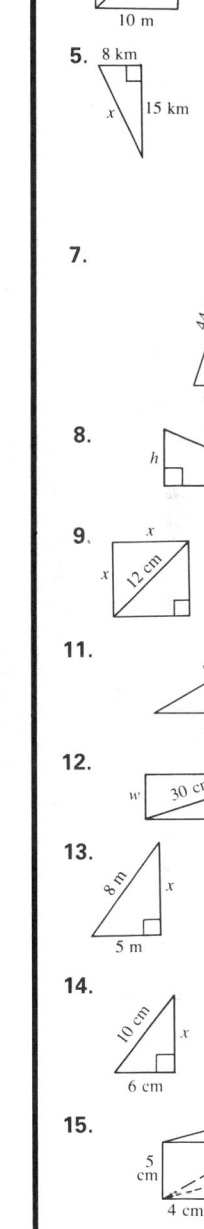

3. x, 10 m, 10 m

4. x, $\sqrt{7}$ cm, $\sqrt{7}$ cm

5. 8 km, x, 15 km

6. 80 km, x, 18 km

7. 44 m, h, 6 m

8. h, 2.6 m, 2.4 m

9. x, x, 12 cm

10. x, x, 8 cm

11. 2b, b, 15 m

12. w, 3w, 30 cm

13. 8 m, x, 5 m y, 5 m, 8 m

14. 10 cm, x, 6 cm x, 10 cm, 6 cm

15. 5 cm, d, 4 cm, 10 cm

(continued)

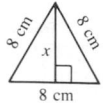
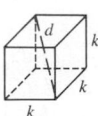
15. Find the length of a diagonal of a rectangular box of length 10 cm, width 4 cm, and height 5 cm. 11.87 cm

16. Find the altitude of an equilateral triangle in which each side has length 8 cm. 6.93 cm

17. Develop a formula for finding the length d of the diagonal of a cube whose edge has length k units. $d = k\sqrt{3}$

18. Show that if $\{r, s, t\}$ is a Pythagorean triple such that $r^2 + s^2 = t^2$, then for any positive integer k, $\{kr, ks, kt\}$ is also a Pythagorean triple.

Computer Exercises For students with computer experience

1. Write a program that will allow you to input three positive numbers *in any order* and will determine whether or not the numbers can represent the sides of a right triangle. (*Note*: Do *not* use the computer's exponentiation operation to square the numbers.)

2. If p and q are whole numbers such that $p > q$, the three numbers
$$a = p^2 - q^2, \qquad b = 2pq, \qquad \text{and} \qquad c = p^2 + q^2$$
always form a Pythagorean triple. Use this fact to write a program that lists all Pythagorean triples for which p and q are both less than 7. (*Hint*: Use two nested loops, one in which p ranges from 2 to 6 and the other in which q ranges from 1 to $p - 1$.)

10–6 The Distance Formula

Recall from Section 4-5 that the distance between two points on a number line is equal to the absolute value of the difference between their coordinates. To denote the distance between points P and Q, write PQ. Thus, in Figure 8,

$$AB = |1 - (-2)| = 3,$$
$$BC = |3 - 1| = 2, \text{ and}$$
$$CA = |-2 - 3| = 5.$$

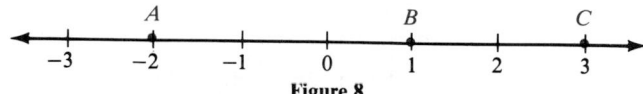

Figure 8

Since $|a - b| = |b - a|$, you can see that the order in which the coordinates are subtracted does not matter.

512   *Chapter 10*

You can use a similar method to find the distance between two points on a coordinate plane. In Figure 9, notice that $\overline{LM}$ (read "line segment L,M") is parallel to the y-axis and $\overline{NP}$ is parallel to the x-axis.

Since the x-coordinates of points L and M are both 2, the length of $\overline{LM}$ will be the difference of the y-coordinates of the points:

$$LM = |-1 - 4| = 5$$

Similarly, since the y-coordinates of points N and P are both -3, take the difference of the x-coordinates of these points to find the length of $\overline{NP}$:

$$NP = |1 - (-4)| = 5$$

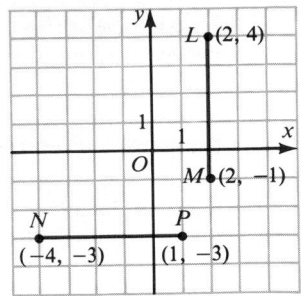

Figure 9

Suppose, now, that you wish to find the distance between two points that lie on a line which is *not* parallel to either axis, such as the points $A(3, 2)$ and $B(-2, 6)$ shown in Figure 10. By drawing the horizontal and vertical segments intersecting at $C(-2, 2)$, as shown, you can form a right triangle having $\overline{AB}$ as its hypotenuse. You can then find the length of each horizontal and vertical side and apply the Pythagorean Theorem.

$$AC = |-2 - 3| = 5$$
$$CB = |6 - 2| = 4$$
$$(AB)^2 = 5^2 + 4^2 = 25 + 16 = 41$$
$$AB = \sqrt{41}$$

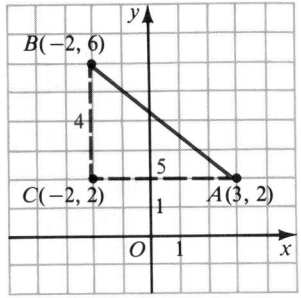

Figure 10

A formula for the distance between any two points $P_1(x_1, y_1)$ and $P_2(x_2, y_2)$ on a coordinate plane can be derived in a similar way. In Figure 11, notice that

$$P_1C = |x_2 - x_1| \quad \text{and} \quad CP_2 = |y_2 - y_1|.$$

Since triangle P_1P_2C is a right triangle:

$$(P_1P_2)^2 = (P_1C)^2 + (CP_2)^2$$
$$= |x_2 - x_1|^2 + |y_2 - y_1|^2$$

However,

$$|x_2 - x_1|^2 = (x_2 - x_1)^2$$

and

$$|y_2 - y_1|^2 = (y_2 - y_1)^2.$$

Thus,

$$(P_1P_2)^2 = (x_2 - x_1)^2 + (y_2 - y_1)^2,$$

and, therefore,

$$P_1P_2 = \sqrt{(x_2 - x_1)^2 + (y_2 - y_1)^2}.$$

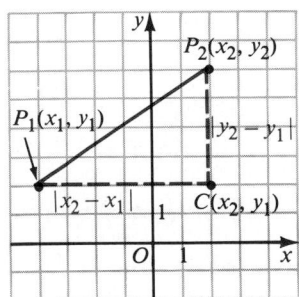

Figure 11

Irrational Numbers and Radicals **513**

Solve. Find lengths correct to the nearest hundredth.

1. If the hypotenuse of an isosceles right triangle has length 30, what is the length of the legs? 21.21

2. If the legs of a right triangle have lengths 9 and 15, what is the length of the hypotenuse? 17.49

3. State whether {0.75, 1, 1.25} is a Pythagorean triple. no

Simplify.

4. a. $|5 - 3| - |-3 - 2|$ −3
   b. $|-a + 2| + |a - 3|$, where $a > 3$ $2a - 5$

5. a. $[-8 - (-1)]^2 + [-4 - 3]^2$ 98
   b. $[a - (-2)]^2 + [-a - 3]^2$ $2a^2 + 10a + 13$

Minimum
   514/1–12
   R 516/Self-Test 2
Average
   514/1–25 odd, 27–30
   R 516/Self-Test 2
Maximum
   514/1–29 odd, 30–33
   R 516/Self-Test 2

Find the distance between the two points having the given coordinates. Simplify irrational distances.

1. (3, 4), (4, 3) $\sqrt{2}$
2. (−2, −5), (−5, −2) $3\sqrt{2}$
3. (7, 8), (−1, 8) 8
4. (4, 10), (5, 7) $\sqrt{10}$
5. (2, 0), (−2, 2) $2\sqrt{5}$
6. (−1, −2), (−3, −4) $2\sqrt{2}$

The method of finding the distance between two points that was discussed on the preceding page can be generalized as the following *distance formula*.

Distance Formula

Given any points $P_1(x_1, y_1)$ and $P_2(x_2, y_2)$:

$$P_1P_2 = \sqrt{(x_2 - x_1)^2 + (y_2 - y_1)^2}$$

EXAMPLE　Find the distance between $P(-2, -3)$ and $Q(4, -1)$.

SOLUTION 1　$PQ = \sqrt{[4 - (-2)]^2 + [-1 - (-3)]^2}$

$$= \sqrt{6^2 + 2^2} = \sqrt{40} = 2\sqrt{10}$$

SOLUTION 2　$QP = \sqrt{[-2 - 4]^2 + [-3 - (-1)]^2}$

$$= \sqrt{(-6)^2 + (-2)^2} = \sqrt{40} = 2\sqrt{10}$$

The two solutions in the Example show that it does not matter which point is considered (x_1, y_1) and which point (x_2, y_2).

Oral Exercises

Find the distance between the two points having the given coordinates.

1. (2, −1), (5, 3) 5
2. (1, −6), (6, 6) 13
3. (−3, 20), (5, 5) 17
4. (8, 10), (2, 2) 10
5. (6, −18), (−4, 6) 26
6. (10, 10), (1, −30) 41

Written Exercises

Find the distance between the two points having the given coordinates. Simplify irrational distances.

A
1. (1, 1), (3, 2) $\sqrt{5}$
2. (2, 2), (−2, 1) $\sqrt{17}$
3. (2, 1), (5, 5) 5
4. (−7, −3), (5, 2) 13
5. (3, −2), (−3, 1) $3\sqrt{5}$
6. (4, −1), (−1, 4) $5\sqrt{2}$
7. (5, −1), (2, 2) $3\sqrt{2}$
8. (−2, −1), (−1, 3) $\sqrt{17}$
9. (3, −6), (11, 9) 17
10. (8, 20), (−1, 8) 15
11. (24, −18), (6, 0) $18\sqrt{2}$
12. (−4, 40), (−20, 8) $16\sqrt{5}$

514　*Chapter 10*

B   13. (a, b), $(-a, -b)$ $2\sqrt{a^2 + b^2}$

14. $(4a, b)$, $(2a, -3b)$ $2\sqrt{a^2 + 4b^2}$

15. $(a + b, 2b)$, $(2b, a + b)$ $|a - b|\sqrt{2}$

16. $(a + b, a - b)$, $(-b, -a)$ $\sqrt{5a^2 + 5b^2}$

17. (a, b), (b, a) $|a - b|\sqrt{2}$

18. $(2a, b)$, $(-b, -2a)$ $|2a + b|\sqrt{2}$

Using the distance formula and the converse of the Pythagorean Theorem, determine whether or not the triangle with the given vertices is a right triangle.

19. $(1, 0)$, $(4, 4)$, $(4, 0)$ Yes

20. $(3, 1)$, $(6, 1)$, $(6, -5)$ Yes

21. $(2, -2)$, $(8, 0)$, $(4, -4)$ Yes

22. $(4, 2)$, $(6, -3)$, $(-1, 5)$ No

23. $(-1, 1)$, $(1, 5)$, $(6, -1)$ No

24. $(-4, -2)$, $(1, -3)$, $(-2, -5)$ Yes

25. $(-3, -2)$, $(0, 4)$, $(7, 3)$ No

26. $(10, 6)$, $(-1, 2)$, $(13, -1)$ No

27. Given that $(1, -2)$, $(0, 5)$, and $(-3, 1)$ are the vertices of a right triangle, determine an equation of the line that passes through the endpoints of the hypotenuse. $7x + y = 5$

28. Given that $(-1, -1)$, $(5, 3)$, and $(-3, 2)$ are the vertices of a right triangle, determine an equation of the line that passes through the endpoints of the shortest side. $3x + 2y = -5$

29. Show that the points $(18, 4)$, $(12, 12)$, and $(8, 4)$ are the vertices of an isosceles triangle.

30. Show that the points $(6, 2)$, $(2, -1)$, and $(-1, 3)$ are the vertices of an isosceles right triangle.

C   31. Show that the distance d between the origin and a point whose coordinates are (x, y) is given by the formula $d = \sqrt{x^2 + y^2}$.

32. Find two values of x such that the point $(x, 5)$ is 5 units from the point $(-1, 2)$. -5 and 3

33. Find two values of m such that the point $(m, -1)$ is 6.5 units from the point $(-4, -7)$. -6.5 and -1.5

Computer Exercises   For students with computer experience

1. Write a program that will allow you to input the coordinates of any two points on a coordinate plane, (x_1, y_1) and (x_2, y_2), and will compute the distance between them.

2. Modify the program that you wrote for Exercise 1 so that you can input the coordinates of a third point, (x_3, y_3), and the program will determine whether or not the distance between (x_1, y_1) and (x_2, y_2) is equal to the distance between (x_2, y_2) and (x_3, y_3).

3. Modify the program that you wrote for Exercise 2 so that it will allow you to input the coordinates of any three noncollinear points and it will determine whether the triangle that has these three points as vertices is equilateral, isosceles, or scalene.

*Additional Answers
Written Exercises*

29. Let $A = (18, 4)$, $B = (12, 12)$, and $C = (8, 4)$. Then $AB = 10$, $BC = 4\sqrt{5}$, and $AC = 10$. No two lengths have a sum equal to the third; thus, the points are not collinear and are the vertices of a triangle. Since $AB = AC = 10$, the triangle is isosceles.

30. Let $A = (6, 2)$, $B = (2, -1)$, and $C = (-1, 3)$. Then $AB = 5$, $BC = 5$, and $AC = 5\sqrt{2}$. No two lengths have a sum equal to the third; thus, the points are not collinear and are the vertices of a triangle. Since $AB = BC = 5$, the triangle is isosceles. Since $AB^2 + BC^2 = AC^2$, the triangle is a right triangle.

31. Since the coordinates of the origin are $(0, 0)$, compute the distance between (x, y) and $(0, 0)$.
$$d = \sqrt{(x - 0)^2 + (y - 0)^2}$$
$$= \sqrt{x^2 + y^2}$$

Determine whether or not each set of numbers could represent the lengths of the sides of a right triangle.

1. {10, 24, 26}   yes
2. {14, 18, 26}   no
3. {$\sqrt{9}$, 6, 3$\sqrt{5}$}   yes
4. {24, 81, 100}   no

If c is the length of the hypotenuse and a and b are the lengths of the other two sides of a right triangle, find the required length. Use a calculator or the Table of Square Roots as necessary and approximate each length to the nearest hundredth.

5. $a = 18$, $b = 80$, $c = $ _?_   82
6. $a = \sqrt{38}$, $c = 8$, $b = $ _?_   5.10
7. $b = \sqrt{83}$, $c = 18$, $a = $ _?_   15.52
8. $a = 14$, $c = \sqrt{276}$, $b = $ _?_   8.94

Find the distance between the two points having the given coordinates. Simplify irrational distances.

9. (−6, 4), (7, 0) $\sqrt{185}$
10. (−2, −4), (6, 1) $\sqrt{89}$
11. (17, 20), (6, 14) $\sqrt{157}$
12. (30, 3), (20, 2) $\sqrt{101}$

Teaching Suggestions
p. T106

Key Ideas

Use the product and quotient properties of square roots.

Self-Test 2

VOCABULARY   hypotenuse (p. 505)
Pythagorean Theorem (p. 505)
Pythagorean numbers or Pythagorean triple (p. 508)
distance formula (p. 514)

Determine whether or not each set of numbers could represent the lengths of the sides of a right triangle.

1. {16, 30, 34} Yes
2. {12, 16, 25} No
3. {$\sqrt{2}$, 5, 7} No
4. {21, 72, 75} Yes

Obj. 1, p. 505

If c is the length of the hypotenuse and a and b are the lengths of the other two sides of a right triangle, find the required length. Approximate your answer to the nearest hundredth, using a calculator or the Table of Square Roots as necessary. Approximations may vary.

5. $a = 18$, $b = 24$, $c = $ _?_ 30
6. $b = \sqrt{42}$, $c = 9$, $a = $ _?_ 6.24
7. $b = \sqrt{91}$, $c = 17$, $a = $ _?_ 14.07
8. $a = 13$, $c = \sqrt{281}$, $b = $ _?_ 10.58

Find the distance between the two points having the given coordinates. Simplify irrational distances.

9. (4, −2), (8, 3) $\sqrt{41}$
10. (2, −8), (−3, 4) 13
11. (20, 15), (8, 45) 6$\sqrt{29}$
12. (32, 4), (26, −12) 2$\sqrt{73}$

Obj. 2, p. 505

Check your answers with those at the back of the book.

Working with Radicals

OBJECTIVES for Sections 10-7 through 10-10:
1. To simplify expressions that contain radicals.
2. To simplify nth roots.
3. To solve equations involving radicals.

10–7 Products and Quotients of Square Roots

Recall that, in Section 10-4, you used the *product property of square roots* to simplify radicals whose radicands contain factors that are perfect squares. The following example shows how the same property is used to simplify *products* of square roots.

516   *Chapter 10*

EXAMPLE 1 Simplify. Assume that all variables denote positive real numbers.

a. $\sqrt{2} \cdot \sqrt{6}$ **b.** $\sqrt{15yz} \cdot \sqrt{3yz}$ **c.** $12\sqrt{10} \cdot 4\sqrt{2}$

SOLUTION   **a.** $\sqrt{2} \cdot \sqrt{6} = \sqrt{2 \cdot 6}$

$$= \sqrt{12}$$
$$= \sqrt{4} \cdot \sqrt{3}$$
$$= 2\sqrt{3}$$

b. $\sqrt{15yz} \cdot \sqrt{3yz} = \sqrt{45y^2z^2}$

$$= \sqrt{9y^2z^2} \cdot \sqrt{5}$$
$$= 3yz\sqrt{5}$$

c. $12\sqrt{10} \cdot 4\sqrt{2} = 48\sqrt{20}$

$$= 48 \cdot \sqrt{4} \cdot \sqrt{5}$$
$$= 48 \cdot 2 \cdot \sqrt{5}$$
$$= 96\sqrt{5}$$

To determine a property that you may use to simplify *quotients* of square roots or square roots of quotients, first consider the following. Since

$$\sqrt{\frac{36}{4}} = \sqrt{9} = 3$$

and

$$\frac{\sqrt{36}}{\sqrt{4}} = \frac{6}{2} = 3,$$

by the transitive axiom of equality,

$$\sqrt{\frac{36}{4}} = \frac{\sqrt{36}}{\sqrt{4}}.$$

This relationship suggests the following property of square roots.

Quotient Property of Square Roots

For any nonnegative real number a and positive real number b,

$$\sqrt{\frac{a}{b}} = \frac{\sqrt{a}}{\sqrt{b}}.$$

This property, together with the *basic property of quotients*, provides a means of expressing a fraction with an irrational denominator or a radical containing a fraction as an equivalent fraction with a rational denominator, as shown on the following page.

Irrational Numbers and Radicals **517**

EXAMPLE 2 Simplify. Assume that all variables represent positive real numbers.

a. $\sqrt{\dfrac{5}{6}}$ b. $\dfrac{3}{\sqrt{12}}$ c. $\sqrt{\dfrac{2}{5x}}$

SOLUTION

a. $\sqrt{\dfrac{5}{6}} = \dfrac{\sqrt{5}}{\sqrt{6}} = \dfrac{\sqrt{5}\cdot\sqrt{6}}{\sqrt{6}\cdot\sqrt{6}} = \dfrac{\sqrt{30}}{\sqrt{36}} = \dfrac{\sqrt{30}}{6}, \text{ or } \dfrac{1}{6}\sqrt{30}$

b. $\dfrac{3}{\sqrt{12}} = \dfrac{3\cdot\sqrt{3}}{\sqrt{12}\cdot\sqrt{3}} = \dfrac{3\sqrt{3}}{\sqrt{36}} = \dfrac{3\sqrt{3}}{6} = \dfrac{\sqrt{3}}{2}, \text{ or } \dfrac{1}{2}\sqrt{3}$

c. $\sqrt{\dfrac{2}{5x}} = \dfrac{\sqrt{2}}{\sqrt{5x}} = \dfrac{\sqrt{2}\cdot\sqrt{5x}}{\sqrt{5x}\cdot\sqrt{5x}} = \dfrac{\sqrt{10x}}{\sqrt{25x^2}} = \dfrac{\sqrt{10x}}{5x}, \text{ or } \dfrac{1}{5x}\sqrt{10x}$

The process of expressing a fraction with an irrational denominator as an equivalent fraction with a rational denominator is called **rationalizing the denominator.** Although sometimes it is helpful to use an expression with an irrational denominator, such as $\dfrac{1}{\sqrt{3}}$, in the process of a computation, it is customary to express a final answer in a form in which the denominator has been rationalized.

The product and quotient properties of square roots can be used to express radicals in *simplest form* as follows.

An expression that contains a radical is in **simplest form** when:

1. no integral radicand has a perfect square factor other than 1,
2. no fractions are under a radical, and
3. no radicals are in a denominator.

You may apply this rule to expressions containing radicals that have polynomials as radicands when such polynomials represent non-negative real numbers. Of course, if the polynomial is a denominator; the polynomial must represent a *positive* real number.

Rationalizing the denominator may simplify the calculations involved in finding an approximation of an expression containing a radical. For example, contrast the following computations.

$$\dfrac{1}{\sqrt{2}} \approx \dfrac{1}{1.414} \approx 0.707$$

and

$$\dfrac{1}{\sqrt{2}} = \dfrac{\sqrt{2}}{2} \approx \dfrac{1.414}{2} = 0.707$$

You would probably find that the second of the two computations is easier to complete. Therefore, simplify any expression containing radicals before evaluating the expression.

518 *Chapter 10*

Oral Exercises

Simplify. Assume that all variables represent positive real numbers.

1. $\sqrt{2} \cdot \sqrt{5}$ $\sqrt{10}$

2. $\sqrt{2} \cdot \sqrt{10}$ $2\sqrt{5}$

3. $\sqrt{8} \cdot \sqrt{6}$ $4\sqrt{3}$

4. $\sqrt{12} \cdot \sqrt{6}$ $6\sqrt{2}$

5. $\frac{1}{\sqrt{5}}$ $\frac{\sqrt{5}}{5}$

6. $\frac{1}{\sqrt{7}}$ $\frac{\sqrt{7}}{7}$

7. $\sqrt{\frac{1}{3}}$ $\frac{\sqrt{3}}{3}$

8. $\sqrt{\frac{2}{5}}$ $\frac{\sqrt{10}}{5}$

9. $6\sqrt{2} \cdot 4\sqrt{30}$ $48\sqrt{15}$

10. $5\sqrt{15} \cdot 8\sqrt{10}$ $200\sqrt{6}$

11. $\sqrt{2x} \cdot \sqrt{xy}$ $x\sqrt{2y}$

12. $\sqrt{3r} \cdot \sqrt{6r}$ $3r\sqrt{2}$

13. $\sqrt{\frac{1}{z}}$ $\frac{\sqrt{z}}{z}$

14. $\sqrt{\frac{5}{e}}$ $\frac{\sqrt{5e}}{e}$

15. $\frac{b}{\sqrt{b^3}}$ $\frac{\sqrt{b}}{b}$

16. $\frac{v^2}{\sqrt{v^5}}$ $\frac{\sqrt{v}}{v}$

Tell whether each statement is true or false for all positive real numbers a and b.

17. $5\sqrt{9a} = 8\sqrt{a}$ False

18. $\sqrt{a^{36}} = a^6$ False

19. $\frac{a}{\sqrt{b}} = \frac{a\sqrt{b}}{b}$ True

20. $\sqrt{3b} \cdot \sqrt{7b} = \sqrt{21b}$ False

21. $\frac{\sqrt{b}}{a} = \sqrt{\frac{b}{a}}$ False

22. $\sqrt{\frac{a}{b}} = \frac{1}{b}\sqrt{ab}$ True

23. $\sqrt{a} \cdot \sqrt{a + ab} = a\sqrt{1 + b}$ True

24. $(\sqrt{a^2b^3})^2 = a^2b^3$ True

Written Exercises

Simplify.

A

1. $\sqrt{5} \cdot \sqrt{10}$ $5\sqrt{2}$

2. $\sqrt{2} \cdot \sqrt{32}$ 8

3. $2\sqrt{3} \cdot \sqrt{6}$ $6\sqrt{2}$

4. $4\sqrt{5} \cdot \sqrt{18}$ $12\sqrt{10}$

5. $\sqrt{12} \cdot \sqrt{15}$ $6\sqrt{5}$

6. $\sqrt{24} \cdot \sqrt{10}$ $4\sqrt{15}$

7. $\frac{6}{\sqrt{2}}$ $3\sqrt{2}$

8. $\frac{8}{\sqrt{6}}$ $\frac{4\sqrt{6}}{3}$

9. $\sqrt{\frac{10}{3}}$ $\frac{\sqrt{30}}{3}$

10. $\sqrt{\frac{5}{12}}$ $\frac{\sqrt{15}}{6}$

11. $\frac{\sqrt{32}}{\sqrt{8}}$ 2

12. $\frac{\sqrt{75}}{\sqrt{10}}$ $\frac{\sqrt{30}}{2}$

13. $\sqrt{\frac{8}{3} \cdot \frac{3}{4}}$ $\sqrt{2}$

14. $\sqrt{\frac{2}{3}} \cdot \sqrt{\frac{15}{8}}$ $\frac{\sqrt{5}}{2}$

15. $\frac{2\sqrt{800}}{\sqrt{20}}$ $4\sqrt{10}$

16. $\frac{5\sqrt{600}}{\sqrt{75}}$ $10\sqrt{2}$

Simplify. Approximate your answer to the nearest hundredth. Use a calculator or the Table of Square Roots as necessary. Approximations may vary.

17. $\sqrt{2} \cdot \sqrt{10}$ 4.47

18. $\sqrt{18} \cdot \sqrt{12}$ 14.70

19. $\sqrt{48} \cdot \sqrt{6}$ 16.97

20. $\sqrt{32} \cdot \sqrt{24}$ 27.71

21. $\sqrt{\frac{1}{5}}$ 0.45

22. $\frac{1}{\sqrt{7}}$ 0.38

23. $\frac{2}{\sqrt{7}}$ 0.76

24. $\sqrt{\frac{5}{8}}$ 0.79

Irrational Numbers and Radicals **519**

Mixed Review

Simplify.

1. $\frac{6x - 4}{3x^2 + x - 2}$ $\frac{2}{x + 1}$

2. $\frac{(a - b)^2}{a^2 - b^2}$ $\frac{a - b}{a + b}$

3. Is the distance between (3, 4) and (4, 3) the same as that between (7, 6) and (6, 7)? If yes, tell that distance. yes, $\sqrt{2}$

4. Is the triangle whose coordinates are (3, 4), (7, 7), and (6, 0) a right triangle? yes

Suggested Assignments

Minimum
First day:
  519/1–19 odd
Second day:
  519/2–20 even, 21–32
Average
  519/1–35 odd, 36–40
Maximum
  519/1–43 odd, 45–48

Additional A Exercises

Simplify.

1. $\sqrt{3} \cdot \sqrt{21}$ $3\sqrt{7}$

2. $4\sqrt{5} \cdot \sqrt{35}$ $20\sqrt{7}$

3. $\sqrt{20} \cdot \sqrt{30}$ $10\sqrt{6}$

4. $\frac{9}{\sqrt{27}}$ $\sqrt{3}$

5. $\sqrt{\frac{3}{4}} \cdot \sqrt{\frac{7}{8}}$ $\frac{\sqrt{42}}{8}$

6. Simplify $\sqrt{38} \cdot \sqrt{3}$. Approximate your answer to the nearest hundredth. Use a calculator or the Table of Square Roots as necessary. 10.68

Simplify. Assume that all variables represent positive real numbers and that the value of any expression under a radical sign is a positive real number.

B 25. $\sqrt{2c} \cdot \sqrt{3c}$ $\;c\sqrt{6}$ 26. $\sqrt{20u} \cdot \sqrt{8u}$ $\;4u\sqrt{10}$ 27. $\sqrt{10e} \cdot \sqrt{10e^2}$ $\;10e\sqrt{e}$ 28. $\sqrt{3q^2} \cdot \sqrt{48q}$ $\;12q\sqrt{q}$

29. $\sqrt{\dfrac{3}{t}}$ $\;\dfrac{\sqrt{3t}}{t}$ 30. $\sqrt{\dfrac{5a}{b}}$ $\;\dfrac{\sqrt{5ab}}{b}$ 31. $\sqrt{8r^3} \cdot \sqrt{32r^5}$ $\;16r^4$ 32. $\sqrt{18j^6} \cdot \sqrt{12j^3}$ $\;6j^4\sqrt{6j}$

33. $\dfrac{5\sqrt{2e}}{\sqrt{e^3}}$ $\;\dfrac{5\sqrt{2}}{e}$ 34. $\dfrac{\sqrt{6h}}{\sqrt{2h^3}}$ $\;\dfrac{\sqrt{3}}{h}$ 35. $\sqrt{\dfrac{45}{8c}}$ $\;\dfrac{3\sqrt{10c}}{4c}$ 36. $\sqrt{\dfrac{75}{28c}}$ $\;\dfrac{5\sqrt{21c}}{14c}$

37. $n\sqrt{\dfrac{m^3}{mn}}$ $\;m\sqrt{n}$ 38. $\dfrac{2c\sqrt{2c}}{\sqrt{c^3}}$ $\;2\sqrt{2}$ 39. $\dfrac{k}{i}\sqrt{\dfrac{5i^2}{3k}}$ $\;\dfrac{\sqrt{15k}}{3}$ 40. $\dfrac{5i}{6t}\sqrt{\dfrac{9t^3}{10i}}$ $\;\dfrac{\sqrt{10it}}{4}$

41. $\sqrt{g^5} \cdot \sqrt{\dfrac{3}{g^2h}}$ $\;\dfrac{g\sqrt{3gh}}{h}$ 42. $\sqrt{8c^3} \cdot \sqrt{\dfrac{3d^2}{c^5d^3}}$ $\;\dfrac{2\sqrt{6d}}{cd}$ 43. $\dfrac{2t\sqrt{12t^5}}{\sqrt{6t^7}}$ $\;2\sqrt{2}$ 44. $\dfrac{e}{f}\sqrt{\dfrac{27f^3}{32e^3}}$ $\;\dfrac{3\sqrt{6ef}}{8e}$

C 45. $\dfrac{1}{\sqrt{a+b}}$ $\;\dfrac{\sqrt{a+b}}{a+b}$ 46. $\dfrac{m+n}{\sqrt{m+n}}$ $\;\sqrt{m+n}$ 47. $\sqrt{\dfrac{2x+1}{(2x+1)^2}}$ 48. $\sqrt{\dfrac{y+2}{y^2+4y+4}}$

49. $\sqrt{\dfrac{x-1}{x^2+2x-3}}$ $\;\dfrac{\sqrt{x+3}}{x+3}$ 50. $\sqrt{\dfrac{x-2y}{x^2+xy-6y^2}}$ $\;\dfrac{\sqrt{x+3y}}{x+3y}$ 51. $\dfrac{r^2-s^2}{\sqrt{(r-s)^3}}$ $\;\dfrac{(r+s)\sqrt{r-s}}{(r-s)}$ 52. $\dfrac{2a^2+8ab+8b^2}{\sqrt{(a+2b)^5}}$ $\;\dfrac{2\sqrt{a+2b}}{a+2b}$

10–8 Sums and Products of Expressions Containing Radicals

To simplify the sum

$$4\sqrt{5} + 9\sqrt{5},$$

note that the two terms have a common radical factor, $\sqrt{5}$, and apply the distributive axiom. Thus:

$$4\sqrt{5} + 9\sqrt{5} = (4+9)\sqrt{5} = 13\sqrt{5}$$

Terms that do *not* have a common radical factor cannot be combined. Thus:

$$2\sqrt{3} - 8\sqrt{2} + 12\sqrt{3} = 14\sqrt{3} - 8\sqrt{2}$$

To determine whether or not two or more terms have a common radical factor, write each radical in simplest form.

EXAMPLE 1 Simplify.

　　　　a. $\sqrt{20} + \sqrt{80} - \sqrt{45}$ 　　　**b.** $(5 + 2\sqrt{12}) - (3 + \sqrt{48})$

SOLUTION **a.** $\sqrt{20} + \sqrt{80} - \sqrt{45} = 2\sqrt{5} + 4\sqrt{5} - 3\sqrt{5} = 3\sqrt{5}$

　　　　b. $(5 + 2\sqrt{12}) - (3 + \sqrt{48}) = (5 + 4\sqrt{3}) - (3 + 4\sqrt{3}) = 2$

Key Ideas

Simplify radical expressions using the distributive axiom. Rationalize binomial denominators using conjugates.

Part (b) of Example 1 illustrates the fact that a *sum* of irrational numbers may be a rational number.

> To simplify a sum of radicals:
>
> 1. Express each radical in simplest form.
> 2. Use the distributive axiom to combine radical terms having the same radicand.

The distributive axiom also enables you to simplify products of expressions containing radicals.

EXAMPLE 2 Simplify.

a. $\sqrt{2}(3\sqrt{2} + \sqrt{3})$ b. $(2 + \sqrt{3})(3 - \sqrt{3})$

c. $(\sqrt{6} + \sqrt{5})(\sqrt{6} - \sqrt{5})$ d. $(x\sqrt{3} + \sqrt{2})(x\sqrt{3} - \sqrt{2})$

SOLUTION

a. $\sqrt{2}(3\sqrt{2} + \sqrt{3}) = (\sqrt{2})(3\sqrt{2}) + (\sqrt{2})(\sqrt{3})$

$$= 3 \cdot 2 + \sqrt{6}$$

$$= 6 + \sqrt{6}$$

b. $(2 + \sqrt{3})(3 - \sqrt{3}) = 6 - 2\sqrt{3} + 3\sqrt{3} - 3$

$$= 3 + \sqrt{3}$$

c. $(\sqrt{6} + \sqrt{5})(\sqrt{6} - \sqrt{5})$

$$= (\sqrt{6})(\sqrt{6}) - (\sqrt{6})(\sqrt{5}) + (\sqrt{5})(\sqrt{6}) - (\sqrt{5})(\sqrt{5})$$

$$= 6 - \sqrt{30} + \sqrt{30} - 5$$

$$= 6 - 5 = 1$$

d. $(x\sqrt{3} + \sqrt{2})(x\sqrt{3} - \sqrt{2}) = (x\sqrt{3})^2 - (\sqrt{2})^2$

$$= 3x^2 - 2$$

Part (c) of Example 2 illustrates the fact that a *product* of irrational numbers may be a rational number. In particular, the product of two irrational numbers, such as the expressions $\sqrt{6} + \sqrt{5}$ and $\sqrt{6} - \sqrt{5}$, may be an integer.

If b and d are nonnegative real numbers, then the binomials

$$a\sqrt{b} + c\sqrt{d} \text{ and } a\sqrt{b} - c\sqrt{d}$$

are called **conjugates.** If a and c are integers and b and d are nonnegative integers, the product

$$(a\sqrt{b} + c\sqrt{d})(a\sqrt{b} - c\sqrt{d})$$

will be an integer.

Irrational Numbers and Radicals   **521**

You can use conjugates to rationalize some binomial denominators that contain radicals.

EXAMPLE 3   Simplify.   **a.** $\dfrac{\sqrt{2}}{4 - \sqrt{3}}$ **b.** $\dfrac{\sqrt{6}}{\sqrt{2} + \sqrt{5}}$

SOLUTION   **a.** $\dfrac{\sqrt{2}}{4 - \sqrt{3}} = \dfrac{\sqrt{2}}{4 - \sqrt{3}} \cdot \dfrac{4 + \sqrt{3}}{4 + \sqrt{3}}$ **b.** $\dfrac{\sqrt{6}}{\sqrt{2} + \sqrt{5}} = \dfrac{\sqrt{6}}{\sqrt{2} + \sqrt{5}} \cdot \dfrac{\sqrt{2} - \sqrt{5}}{\sqrt{2} - \sqrt{5}}$

$$= \dfrac{4\sqrt{2} + \sqrt{6}}{16 - 3} \qquad\qquad = \dfrac{\sqrt{12} - \sqrt{30}}{2 - 5}$$

$$= \dfrac{4\sqrt{2} + \sqrt{6}}{13} \qquad\qquad = \dfrac{2\sqrt{3} - \sqrt{30}}{-3}$$

$$= \dfrac{\sqrt{30} - 2\sqrt{3}}{3}$$

In Chapter 7, you factored polynomials over the set of polynomials with integral coefficients. If the factor set is extended to include polynomials with irrational coefficients, you may now factor additional polynomials. For example, in part (d) of Example 2,

$$3x^2 - 2 = (x\sqrt{3} + \sqrt{2})(x\sqrt{3} - \sqrt{2}).$$

Oral Exercises

Simplify.

1. $2\sqrt{2} + 4\sqrt{2}$ $6\sqrt{2}$ 2. $12\sqrt{3} - 5\sqrt{3}$ $7\sqrt{3}$

3. $\sqrt{2}(5 + \sqrt{3})$ $5\sqrt{2} + \sqrt{6}$ 4. $\sqrt{3}(\sqrt{2} - \sqrt{5})$ $\sqrt{6} - \sqrt{15}$

5. $(3\sqrt{2})(2\sqrt{2})$ 12   6. $(2\sqrt{5})^2$ 20

7. $(4\sqrt{3})(3\sqrt{6})$ $36\sqrt{2}$ 8. $(3\sqrt{8})(2\sqrt{12})$ $24\sqrt{6}$

9. $\sqrt{2} + \sqrt{8}$ $3\sqrt{2}$ 10. $\sqrt{5} - \sqrt{20}$ $-\sqrt{5}$

11. $2\sqrt{3} - \sqrt{75}$ $-3\sqrt{3}$ 12. $7\sqrt{2} + \sqrt{128}$ $15\sqrt{2}$

13. $\sqrt{28} - \sqrt{63}$ $-\sqrt{7}$ 14. $\sqrt{72} + \sqrt{200}$ $16\sqrt{2}$

15. $(2 + \sqrt{3})(2 - \sqrt{3})$ 1   16. $(3 + \sqrt{5})(3 + \sqrt{5})$ $14 + 6\sqrt{5}$

17. $(\sqrt{2} + \sqrt{3})(\sqrt{2} + \sqrt{3})$ $5 + 2\sqrt{6}$ 18. $(\sqrt{10} - \sqrt{7})(\sqrt{10} - \sqrt{7})$ $17 - 2\sqrt{70}$

State the conjugate of each binomial.

19. $2 + \sqrt{3}$ $2 - \sqrt{3}$ 20. $4 - \sqrt{5}$ $4 + \sqrt{5}$

21. $\sqrt{2} + \sqrt{7}$ $\sqrt{2} - \sqrt{7}$ 22. $\sqrt{5} - \sqrt{15}$ $\sqrt{5} + \sqrt{15}$

23. $2\sqrt{2} - \sqrt{6}$ $2\sqrt{2} + \sqrt{6}$ 24. $4\sqrt{2} + 3\sqrt{3}$ $4\sqrt{2} - 3\sqrt{3}$

Mixed Review

Simplify. Assume that all variables denote positive real numbers.

1. $\sqrt{\dfrac{3}{5}}$ $\dfrac{\sqrt{15}}{5}$

2. $\dfrac{6}{\sqrt{3}}$ $2\sqrt{3}$

3. $\dfrac{6}{\sqrt{g}}$ $\dfrac{6\sqrt{g}}{g}$

4. $\sqrt{\dfrac{1}{2p}}$ $\dfrac{\sqrt{2p}}{2p}$

Simplify.

5. $(a + b)(a - b)$ $a^2 - b^2$

Suggested Assignments

Minimum
    523/1–10, 11–23 odd
Average
    523/1–49 odd
Maximum
    523/9–55 odd

Written Exercises

Simplify. Assume that all variables represent positive real numbers.

A
1. $5\sqrt{6} + \sqrt{6} - 2\sqrt{6}$ $4\sqrt{6}$
2. $8\sqrt{3} - 2\sqrt{3} + \sqrt{3}$ $7\sqrt{3}$
3. $7\sqrt{2} - \sqrt{18} + 3\sqrt{8}$ $10\sqrt{2}$
4. $\sqrt{75} - 2\sqrt{3} + 4\sqrt{27}$ $15\sqrt{3}$
5. $8\sqrt{2} - 4\sqrt{8} - 4\sqrt{50}$ $-20\sqrt{2}$
6. $6\sqrt{5} + \sqrt{500} - 2\sqrt{125}$ $6\sqrt{5}$
7. $\sqrt{16c} + \sqrt{c} - \sqrt{81c}$ $-4\sqrt{c}$
8. $t\sqrt{st^2} - t^2\sqrt{s} - \sqrt{49st^2}$ $-7t\sqrt{s}$
9. $(5\sqrt{5})(3\sqrt{6})$ $15\sqrt{30}$
10. $(6\sqrt{15})(2\sqrt{3})$ $36\sqrt{5}$
11. $\sqrt{3}(\sqrt{6} - \sqrt{12})$ $3\sqrt{2} - 6$
12. $\sqrt{2}(\sqrt{32} + \sqrt{10})$ $8 + 2\sqrt{5}$
13. $(\sqrt{3} + 1)(\sqrt{3} - 1)$ 2
14. $(3 + \sqrt{5})(3 - \sqrt{5})$ 4
15. $(2 + \sqrt{13})(4 - \sqrt{13})$ $-5 + 2\sqrt{13}$
16. $(\sqrt{6} + 1)(\sqrt{6} - 4)$ $2 - 3\sqrt{6}$
17. $(2\sqrt{2} + 3)(\sqrt{2} - 1)$ $1 + \sqrt{2}$
18. $(3\sqrt{3} - 2)(\sqrt{3} + 4)$ $1 + 10\sqrt{3}$
19. $(4\sqrt{3} + 5)(3\sqrt{3} - 2)$ $26 + 7\sqrt{3}$
20. $(8 - 2\sqrt{2})(3 - 4\sqrt{2})$ $40 - 38\sqrt{2}$
21. $(\sqrt{6} - \sqrt{11})(\sqrt{6} + \sqrt{11})$ -5
22. $(\sqrt{15} + \sqrt{8})(\sqrt{15} - \sqrt{8})$ 7
23. $(2\sqrt{5} - \sqrt{3})(2\sqrt{5} + \sqrt{3})$ 17
24. $(\sqrt{2} + 3\sqrt{6})(\sqrt{2} - 3\sqrt{6})$ -52

27. $9 + 6\sqrt{2}$ **30.** $9r - 12\sqrt{rs} + 4s$

B
25. $(\sqrt{a} - \sqrt{c})^2$ $a - 2\sqrt{ac} + c$
26. $(\sqrt{b} + \sqrt{d})^2$ $b + 2\sqrt{bd} + d$
27. $(\sqrt{3} + \sqrt{6})^2$
28. $(\sqrt{10} - \sqrt{2})^2$ $12 - 4\sqrt{5}$
29. $(2\sqrt{5} + \sqrt{3})^2$ $23 + 4\sqrt{15}$
30. $(3\sqrt{r} - 2\sqrt{s})^2$

31. $\dfrac{2}{\sqrt{3} - 1}$ $1 + \sqrt{3}$
32. $\dfrac{5}{\sqrt{2} + 3}$ $\dfrac{15 - 5\sqrt{2}}{7}$
33. $\dfrac{4\sqrt{6}}{2 + \sqrt{2}}$ $4\sqrt{6} - 4\sqrt{3}$

34. $\dfrac{5\sqrt{15}}{5 - \sqrt{3}}$ $\dfrac{25\sqrt{15} + 15\sqrt{5}}{22}$
35. $\dfrac{\sqrt{3} + 2}{\sqrt{3} - 1}$ $\dfrac{5 + 3\sqrt{3}}{2}$
36. $\dfrac{\sqrt{5} - 1}{\sqrt{2} + 5}$

37. $\dfrac{\sqrt{6} + 2\sqrt{2}}{\sqrt{2} - \sqrt{3}}$
38. $\dfrac{\sqrt{3} - \sqrt{5}}{3\sqrt{3} + \sqrt{5}}$
39. $\dfrac{\sqrt{2} + \sqrt{5}}{2\sqrt{5} - \sqrt{3}}$

40. $\dfrac{\sqrt{10} + \sqrt{3}}{\sqrt{2} + 3\sqrt{3}}$
41. $\dfrac{\sqrt{5} - 2\sqrt{3}}{\sqrt{5} + 3\sqrt{2}}$
42. $\dfrac{2\sqrt{6} - \sqrt{3}}{\sqrt{6} + \sqrt{5}}$

43. $15\sqrt{\dfrac{2}{3}} + \sqrt{150} - 8\sqrt{1\tfrac{1}{2}}$ $6\sqrt{6}$
44. $3\sqrt{2\tfrac{2}{5}} - \dfrac{2}{5}\sqrt{240} + \sqrt{540}$ $\dfrac{28\sqrt{15}}{5}$

45. $3j\sqrt{98k} - 5k\sqrt{\dfrac{242j^2}{k}}$ $-34j\sqrt{2k}$
46. $3\sqrt{a^5b} + a^2b\sqrt{\dfrac{25a}{b}}$ $8a^2\sqrt{ab}$

47. $\sqrt{\dfrac{35u^2}{72} + \dfrac{5u^2}{24}}$ $\dfrac{5u}{6}$
48. $\sqrt{\dfrac{5}{2c} - \dfrac{2}{9c} - \dfrac{1}{36c}}$ $\dfrac{3\sqrt{c}}{2c}$

49. Determine an expression in simplest form that represents the area of a square whose perimeter is $(16\sqrt{5} + 4\sqrt{2})$ m. $(82 + 8\sqrt{10})$ m²

50. Determine an expression in simplest form that represents the area of a square whose perimeter is $(12\sqrt{2} + 20\sqrt{3})$ m. $(93 + 30\sqrt{6})$ m²

Irrational Numbers and Radicals   **523**

Let $f(x) = 3x^2 - 3x + 1$. Compute each of the following.

C 51. $f(\sqrt{3})$ $10 - 3\sqrt{3}$ 52. $f(-\sqrt{2})$ $7 + 3\sqrt{2}$ 53. $f(\sqrt{2} - 1)$ $13 - 9\sqrt{2}$ 54. $f(\sqrt{3} + 2)$

55. Show that $(2 + \sqrt{5})$ and $(2 - \sqrt{5})$ are solutions of $x^2 - 4x - 1 = 0$. $16 + 9\sqrt{3}$

56. Show that $(-3 + 2\sqrt{5})$ and $(-3 - 2\sqrt{5})$ are solutions of $x^2 + 6x - 11 = 0$.

Teaching Suggestions
p. T107

Key Ideas

Simplify nth roots.

Chalkboard Examples

Simplify.

1. $\sqrt[3]{54}$
   $\sqrt[3]{27} \cdot \sqrt[3]{2} = 3\sqrt[3]{2}$

2. $\sqrt[4]{48} - \sqrt[4]{3} + \sqrt[3]{16}$
   $\sqrt[4]{16} \cdot \sqrt[4]{3} - 1 \cdot \sqrt[4]{3} + \sqrt[3]{8} \cdot \sqrt[3]{2} = 2\sqrt[4]{3} - 1\sqrt[4]{3} + 2\sqrt[3]{2} = \sqrt[4]{3} + 2\sqrt[3]{2}$

3. $\dfrac{7}{\sqrt[3]{9}}$
   $\dfrac{7}{\sqrt[3]{9}} \cdot \dfrac{\sqrt[3]{3}}{\sqrt[3]{3}} = \dfrac{7\sqrt[3]{3}}{\sqrt[3]{27}} = \dfrac{7\sqrt[3]{3}}{3}$

10–9 *n*th Roots

Recall from Section 10-3 that if $b^2 = a$, then b is called a *square root* of a. Similarly, if

$$b^n = a,$$

then b is called an *nth root* of a. In general, if n is a positive integer and a is a real number, any real number whose nth power equals a is called an *nth root* of a.

If $b^3 = a$, then b is called the **cube root** of a. There is only *one* real number, 4, whose cube is 64. Therefore, 4 is the cube root of 64. The number -4 is the cube root of -64, since $(-4)^3 = -64$. In general:

> *If a is a real number and n is a positive odd integer*, then there is only *one* real nth root of a, denoted by
>
> $$\sqrt[n]{a}.$$

For example, $\sqrt[3]{64} = 4$ and $\sqrt[3]{-64} = -4$.

On the other hand, if you want to find the sixth root of 64, there are *two* values, 2 and -2, since $2^6 = 64$ and $(-2)^6 = 64$. In general:

> *If a is a positive real number and n is a positive even integer*, then there are *two* real nth roots of a. The *positive* nth root of a is denoted by
>
> $$\sqrt[n]{a},$$
>
> and the *negative* nth root of a is denoted by
>
> $$-\sqrt[n]{a}.$$

The positive nth root is referred to as the **principal nth root**. For example, $\sqrt[6]{64} = 2$ and $-\sqrt[6]{64} = -2$. The *principal* sixth root of 64 is 2.

Although 64 has two real sixth roots, -64 has *no* real sixth roots, since there are no real numbers that satisfy $x^6 = -64$. In general:

> *If a is a negative real number and n is a positive even integer*, then there are *no* real nth roots of a.

When n is even and a is negative, the symbols $\sqrt[n]{a}$ and $-\sqrt[n]{a}$ do not represent real numbers.

If $a = 0$ and n is a positive integer, then 0 is the only nth root of a, no matter whether n is odd or even, since $0^n = 0$ for any positive integer n.

524 *Chapter 10*

Thus, for any positive integer n,

$$\sqrt[n]{0} = 0.$$

In the symbol $\sqrt[n]{a}$, the positive integer n is called the **index**. In a radical such as $\sqrt{10}$ the index is understood to be 2, but it is usually not written.

The properties of nth roots are similar to the properties of square roots.

Properties of nth Roots

If $\sqrt[n]{a}$ and $\sqrt[n]{b}$ are real numbers, then:

1. $\sqrt[n]{ab} = \sqrt[n]{a} \cdot \sqrt[n]{b}$ 2. If $b \neq 0$, $\sqrt[n]{\dfrac{a}{b}} = \dfrac{\sqrt[n]{a}}{\sqrt[n]{b}}$

If a number has a real nth root, raising the nth root to the nth power results in the number itself. Thus,

$$(\sqrt[3]{8})^3 = 8, \qquad (\sqrt[4]{81})^4 = 81, \qquad \text{and} \qquad (\sqrt[n]{a})^n = a.$$

If a real number is not the nth power of some rational number, any nth root of the number will be irrational. The following example shows how the properties of nth roots can be used to express these irrational nth roots in *simplest form*.

EXAMPLE Simplify.   **a.** $\sqrt[3]{16}$ **b.** $\sqrt[4]{162} + \sqrt[4]{32} - \sqrt[3]{54}$ **c.** $\dfrac{3}{\sqrt[3]{25}}$

SOLUTION   **a.** $\sqrt[3]{16} = \sqrt[3]{8} \cdot \sqrt[3]{2} = 2\sqrt[3]{2}$

b. $\sqrt[4]{162} + \sqrt[4]{32} - \sqrt[3]{54} = (\sqrt[4]{81})(\sqrt[4]{2}) + (\sqrt[4]{16})(\sqrt[4]{2}) - (\sqrt[3]{27})(\sqrt[3]{2})$
$$= 3\sqrt[4]{2} + 2\sqrt[4]{2} - 3\sqrt[3]{2}$$
$$= 5\sqrt[4]{2} - 3\sqrt[3]{2}$$

c. $\dfrac{3}{\sqrt[3]{25}} = \dfrac{3}{\sqrt[3]{25}} \cdot \dfrac{\sqrt[3]{5}}{\sqrt[3]{5}} = \dfrac{3\sqrt[3]{5}}{\sqrt[3]{125}} = \dfrac{3\sqrt[3]{5}}{5}$

Oral Exercises

Tell whether or not the given symbol represents a real number. If the number represented is real, tell whether it is rational or irrational.

1. $\sqrt[3]{8}$ real, rational

2. $\sqrt[4]{625}$ real, rational

3. $\sqrt[8]{-1}$ not real

4. $\sqrt[3]{0}$ real, rational

5. $\sqrt[n]{-\dfrac{3}{64}}$ not real

6. $-\sqrt[4]{-16}$ not real

7. $\sqrt{\dfrac{16}{125}}$ real, irrational

8. $\sqrt[4]{-\dfrac{1}{32}}$ not real

9. $\sqrt[3]{-\dfrac{81}{64}}$ real, irrational

10. $\sqrt[5]{\dfrac{12}{25}}$ real, irrational

11. $\sqrt[12]{-\dfrac{56}{75}}$ not real

12. $\sqrt[6]{\dfrac{64}{729}}$ real, rational

Irrational Numbers and Radicals **525**

Mixed Review

Simplify.

1. $3\sqrt{44} - 6\sqrt{99}$ $-12\sqrt{11}$

2. $(\sqrt{6} + \sqrt{2})(\sqrt{6} - \sqrt{2})$ 4

3. $(\sqrt{7} - \sqrt{2})^2$ $9 - 2\sqrt{14}$

4. $\dfrac{4}{\sqrt{2} + 1}$ $4\sqrt{2} - 4$

5. $\dfrac{\sqrt{\dfrac{t - 2}{(t - 1)(t^2 - t - 2)}}}{\dfrac{\sqrt{t^2 - 1}}{t^2 - 1}}$

Suggested Assignments

Minimum
  526/1–16
S 515/19–22
Average
  526/1–15 odd, 17–28
S 520/41–44
Maximum
  526/1–27 odd, 28–32
S 520/49–52

Additional A Exercises

Simplify.

1. $\sqrt[3]{64}$ 4

2. $\sqrt[4]{625}$ 5

3. $\sqrt[3]{-27}$ -3

4. $\sqrt[3]{40} - \sqrt[3]{320}$ $-2\sqrt[3]{5}$

5. $\dfrac{7}{\sqrt[5]{25}}$ $\dfrac{7\sqrt[5]{125}}{5}$

6. $\sqrt[3]{-\dfrac{1}{8}}$ $-\dfrac{1}{2}$

Written Exercises

Simplify.

A

1. $\sqrt[3]{27}$ 3

2. $\sqrt[3]{125}$ 5

3. $\sqrt[4]{81}$ 3

4. $\sqrt[5]{-32}$ -2

5. $(\sqrt[7]{-128})^7$ -128

6. $(\sqrt[6]{64})^6$ 64

7. $\sqrt[4]{80}$ $2\sqrt[4]{5}$

8. $\sqrt[3]{40}$ $2\sqrt[3]{5}$

9. $\sqrt[3]{24} + \sqrt[3]{81}$ $5\sqrt[3]{3}$

10. $\sqrt[4]{48} - \sqrt[4]{243}$ $-\sqrt[4]{3}$

11. $\dfrac{3}{\sqrt[3]{2}}$ $\dfrac{3\sqrt[3]{4}}{2}$

12. $\dfrac{9}{\sqrt[3]{3}}$ $3\sqrt[3]{27}$

13. $\dfrac{6}{\sqrt[4]{8}}$ $3\sqrt[4]{2}$

14. $\dfrac{12}{\sqrt[5]{4}}$ $6\sqrt[5]{8}$

15. $\sqrt[3]{\dfrac{5}{8}}$ $\dfrac{\sqrt[3]{5}}{2}$

16. $\sqrt[6]{\dfrac{7}{64}}$ $\dfrac{\sqrt[6]{7}}{2}$

B

17. $\sqrt[5]{\dfrac{9}{16}}$ $\dfrac{\sqrt[5]{18}}{2}$

18. $\sqrt[4]{\dfrac{16}{27}}$ $\dfrac{2\sqrt[4]{3}}{3}$

19. $\sqrt[3]{\dfrac{5}{32}}$ $\dfrac{\sqrt[3]{10}}{4}$

20. $\sqrt[3]{-\dfrac{1}{81}}$ $-\dfrac{\sqrt[3]{9}}{9}$

Simplify. Assume that all variables represent positive real numbers and that the value of any expression under a radical sign is positive.

21. $\sqrt[3]{x^9}$ x^3

22. $\sqrt[4]{y^{12}}$ y^3

23. $\sqrt[10]{x^{20}y^{30}z^{10}}$ x^2y^3z

24. $\sqrt[15]{a^{45}b^{30}c^{60}}$ $a^3b^2c^4$

25. $\sqrt[3]{x^7y^{11}z^8}$ $x^2y^3z^2\sqrt[3]{xy^2z^2}$

26. $\sqrt[6]{m^{13}p^{21}t^{48}}$ $m^2p^3t^8\sqrt[6]{mp^3}$

27. $\sqrt[4]{\dfrac{y^9}{x^2}}$ $\dfrac{y^2\sqrt[4]{x^2y}}{x}$

28. $\sqrt[5]{\dfrac{b^6}{a^4y^2}}$ $\dfrac{b\sqrt[5]{aby^3}}{ay}$

C

29. $\sqrt[3]{\dfrac{x}{x^2 - 2x + 1}}$ $\dfrac{\sqrt[3]{x(x - 1)}}{x - 1}$

30. $\sqrt[4]{\dfrac{y + 1}{(y - 3)^3}}$ $\dfrac{\sqrt[4]{(y + 1)(y - 3)}}{y - 3}$

31. $\sqrt[6]{\dfrac{x - 3}{x^2 - 2x - 3}}$ $\dfrac{\sqrt[6]{(x + 1)^5}}{x + 1}$

32. $\sqrt[5]{\dfrac{a + 4}{(a + 1)(a^2 + 5a + 4)}}$ $\dfrac{\sqrt[5]{(a + 1)^3}}{a + 1}$

ON THE CALCULATOR

The square root key of a calculator can be used to find the *n*th root of a nonnegative real number provided that *n* is a power of two. For example, since $2^2 = 4$, you can find the *fourth* root by pressing the square root key *twice*. Similarly, since $2^3 = 8$, you can find the *eighth* root by pressing the square root key *three times*. Following this pattern, you can use the square root key to find the *sixteenth* root, the *thirty-second* root, and so on.

Use a calculator to find the indicated roots. Approximate the answer to the nearest hundredth.

1. $\sqrt{3021}$ 54.96

2. $\sqrt[4]{72}$ 2.91

3. $\sqrt[32]{468}$ 1.21

4. $\sqrt{0.0142}$ 0.12

5. $\sqrt[16]{12}$ 1.17

6. $\sqrt[64]{100}$ 1.07

7. $\sqrt[8]{0.379}$ 0.89

8. $\sqrt[8]{\dfrac{13}{55}}$ 0.84

10–10 Equations Involving Radicals

Sometimes you will need to solve an equation in which a term has a variable in a radicand. In such cases, you transform the equation into an equivalent equation in which the term with the variable in the radicand is alone as one side of the equation. Then raise both sides of the equation to the power equal to the index of the radical and solve the resulting equation.

EXAMPLE 1   Solve.

    a. $\sqrt{d} - 3 = 4$

    b. $\sqrt[3]{x + 6} = 2$

    c. $2 + \sqrt{x - 5} = 3$

SOLUTION

a.
$$\sqrt{d} - 3 = 4$$
$$\sqrt{d} = 7$$
$$(\sqrt{d})^2 = 7^2$$
$$d = 49$$

Check:
$$\sqrt{d} - 3 = 4$$
$$\sqrt{49} - 3 \overset{?}{=} 4$$
$$7 - 3 \overset{?}{=} 4$$
$$4 = 4 \;\checkmark$$

∴ the solution set is {49}.

b.
$$\sqrt[3]{x + 6} = 2$$
$$(\sqrt[3]{x + 6})^3 = 2^3$$
$$x + 6 = 8$$
$$x = 2$$

Check:
$$\sqrt[3]{x + 6} = 2$$
$$\sqrt[3]{2 + 6} \overset{?}{=} 2$$
$$\sqrt[3]{8} \overset{?}{=} 2$$
$$2 = 2 \;\checkmark$$

∴ the solution set is {2}.

c.
$$2 + \sqrt{x - 5} = 3$$
$$\sqrt{x - 5} = 1$$
$$(\sqrt{x - 5})^2 = 1^2$$
$$x - 5 = 1$$
$$x = 6$$

Check:
$$2 + \sqrt{x - 5} = 3$$
$$2 + \sqrt{6 - 5} \overset{?}{=} 3$$
$$2 + \sqrt{1} \overset{?}{=} 3$$
$$2 + 1 \overset{?}{=} 3$$
$$3 = 3 \;\checkmark$$

∴ the solution set is {6}.

The equation you obtain by raising both sides of an equation to a power is not necessarily equivalent to the original equation. Therefore, the solutions of the new equation may not all be solutions of the original equation. However, every solution of the original equation will always be a solution of the new equation. It is important to check each solution of the new equation in the original equation to discover any *extraneous* solutions.

Irrational Numbers and Radicals   **527**

Teaching Suggestions
p. T107

Related Activities p. T107

Supplementary Material

Test 41

Key Ideas

Solve equations involving radicals.

Chalkboard Examples

Solve.

1. a.
$$\sqrt{n} + 7 = 9$$
$$\sqrt{n} = 2$$
$$(\sqrt{n})^2 = 2^2$$
$$n = 4$$
Check: $\sqrt{4} + 7 \overset{?}{=} 9$
$$2 + 7 \overset{?}{=} 9$$
$$9 = 9$$
The solution set is {4}.

b.
$$\sqrt[3]{a + 2} = 3$$
$$(\sqrt[3]{a + 2})^3 = 3^3$$
$$a + 2 = 27$$
$$a = 25$$
Check: $\sqrt[3]{25 + 2} \overset{?}{=} 3$
$$\sqrt[3]{27} \overset{?}{=} 3$$
$$3 = 3$$
The solution set is {25}.

c.
$$4 + \sqrt{n - 1} = 6$$
$$(\sqrt{n - 1})^2 = 2^2$$
$$n - 1 = 4$$
$$n = 5$$
Check: $4 + \sqrt{5 - 1} \overset{?}{=} 6$
$$4 + \sqrt{4} \overset{?}{=} 6$$
$$6 = 6$$
The solution set is {5}.

(continued)

2. $\sqrt{2x - 1} - x = -2$
$\sqrt{2x - 1} = x - 2$
$(\sqrt{2x - 1})^2 = (x - 2)^2$
$2x - 1 = x^2 - 4x + 4$
$0 = x^2 - 6x + 5$
$0 = (x - 5)(x - 1)$
$x = 5$ or $x = 1$
Check: $\sqrt{2(5) - 1} - 5 \overset{?}{=} -2$
$\sqrt{9} - 5 \overset{?}{=} -2$
$-2 = -2$
$\sqrt{2(1) - 1} - 1 \overset{?}{=} -2$
$1 - 1 \overset{?}{=} -2$
$0 \neq -2$
$\therefore$ the solution set is $\{5\}$.

EXAMPLE 2   Solve $\sqrt{5x - 4} - x = -2$

SOLUTION $\sqrt{5x - 4} - x = -2$
$\sqrt{5x - 4} = x - 2$
$(\sqrt{5x - 4})^2 = (x - 2)^2$
$5x - 4 = x^2 - 4x + 4$
$0 = x^2 - 9x + 8$
$0 = (x - 8)(x - 1)$
$x = 8$ or $x = 1$

Check: $\sqrt{5x - 4} - x = -2$ $\sqrt{5x - 4} - x = -2$
$\sqrt{5(8) - 4} - 8 \overset{?}{=} -2$ $\sqrt{5(1) - 4} - 1 \overset{?}{=} -2$
$\sqrt{40 - 4} - 8 \overset{?}{=} -2$ $\sqrt{5 - 4} - 1 \overset{?}{=} -2$
$\sqrt{36} - 8 \overset{?}{=} -2$ $\sqrt{1} - 1 \overset{?}{=} -2$
$6 - 8 \overset{?}{=} -2$ $1 - 1 \overset{?}{=} -2$
$-2 = -2$ ✓ $0 \neq -2$

$\therefore$ the solution set is $\{8\}$.

Not all equations involving radicals have a real-number solution. For example, consider the equation

$$\sqrt{2x - 3} + 5 = 2.$$

Transforming the equation, you will see that

$$\sqrt{2x - 3} = -3.$$

The equation $\sqrt{2x - 3} = -3$, however, will never have a real-number solution since a square root radical always denotes a non-negative number. Thus, the solution set of the equation is $\emptyset$.

Oral Exercises

Tell whether or not each equation has a real-number solution.

1. $\sqrt{x} = 5$ Yes   2. $\sqrt{y} = -2$ No

3. $-\sqrt{x} = -1$ Yes   4. $-\sqrt{y} = 3$ No

5. $\sqrt[4]{p} = -3$ No   6. $\sqrt[3]{q} = -5$ Yes

7. $\sqrt{z} - 4 = 0$ Yes   8. $\sqrt{x} + 6 - 9$ Yes

9. $\sqrt[3]{t} + 12 = 16$ Yes   10. $\sqrt[4]{m} + 2 = 4$ Yes

11. $\sqrt{z} - 4 = -12$ No   12. $6 - \sqrt{2x} = 4$ Yes

528   *Chapter 10*

Describe the first step you would take in solving each equation.

13. $\sqrt{4x - 3} = 3$ Square each side.

14. $\sqrt[3]{2m + 1} = 3$ Cube each side.

15. $\sqrt{2 - x} + 4 = 6$ Subtract 4 from each side.

16. $\sqrt{1 + 3x} - 10 = -3$ Add 10 to each side.

17. $\sqrt{5x} + 10 = x$ Subtract 10 from each side.

18. $4\sqrt{x} - 2x = 1$ Add 2x to each side.

Written Exercises

Solve.

A

1. $\sqrt{x} - 2 = 6$ {64}

2. $\sqrt{x} + 4 = 9$ {25}

3. $3\sqrt{y} + 2 = 8$ {4}

4. $4\sqrt{x} - 3 = 1$ {1}

5. $2\sqrt{n} = -12$ Ø

6. $-5\sqrt{z} = 10$ Ø

7. $3\sqrt{y} = 1$ $\{\frac{1}{9}\}$

8. $4\sqrt{y} = 3$ $\{\frac{9}{16}\}$

9. $\frac{\sqrt{m}}{2} + 2 = 7$ {100}

10. $2 - \frac{\sqrt{h}}{3} = 1$ {9}

11. $\sqrt{x - 4} = 3$ {13}

12. $\sqrt[3]{2 - x} = 2$ {-6}

13. $\sqrt[4]{2x + 4} = 1$ $\{-\frac{3}{2}\}$

14. $\sqrt{4 - 5y} = 8$ {-12}

15. $\sqrt[3]{\frac{4n}{3}} + 1 = 4$ $\{\frac{81}{4}\}$

16. $\sqrt[3]{\frac{2}{5}x} - 3 = 1$ {160}

17. $3\sqrt{n} - \frac{1}{5} = \frac{8}{5}$ $\{\frac{9}{25}\}$

18. $\frac{2}{3} - 4\sqrt{z} = 2$ Ø

B

19. $\sqrt{\frac{2x - 5}{2}} = 1$ $\{\frac{7}{2}\}$

20. $\sqrt{\frac{4 - 3x}{7}} = 2$ {-8}

21. $\sqrt{x} = 2\sqrt{3}$ {12}

22. $3\sqrt{y} = 6\sqrt{2}$ {8}

23. $\sqrt{x^2 - 2} = 4$ $\{3\sqrt{2}, -3\sqrt{2}\}$

24. $\sqrt{5 - y^2} = 1$ {2, -2}

25. $x + \sqrt{x} = 6$ {4}

26. $y - \sqrt{y} = 12$ {16}

27. $\sqrt{x + 2} - x = -10$ {14}

28. $\sqrt{3x} - x = -6$ {12}

29. $\sqrt{3x + 13} - 2x = 8$ {-3}

30. $\sqrt{33 - 4x} + x = 3$ {-4}

C

31. $\sqrt{a^2 + 5} = 2a - 1$ {2}

32. $\sqrt{2a^2 - 7a + 10} = 2a - 5$ {5}

33. $\sqrt[3]{a^3 + 3a^2} = a + 1$ $\{-\frac{1}{3}\}$

34. $\sqrt[4]{2b^2 - 1} = b$ {1}

35. $\sqrt{3x + 4} - \sqrt{x} = 2$ {0, 4}

36. $\sqrt{x - 5} - \sqrt{2x + 7} = -3$ {9, 21}

Solve each system of equations.

37. $3\sqrt{a} - 2\sqrt{b} = -2$ {(4, 16)}
$4\sqrt{a} + 5\sqrt{b} = 28$

38. $5\sqrt{a} + 6\sqrt{b} = 45$ {(9, 25)}
$15\sqrt{a} - 8\sqrt{b} = 5$

Irrational Numbers and Radicals   **529**

Suggested Assignments

Minimum
First day: 529/1–12
Second day: 520/13–25
      R 530/*Self-Test 3*

Average
First day: 529/1–18
Second day: 529/19–30
      R 530/*Self-Test 3*

Maximum
  529/9–37 odd, 38
R 530/*Self-Test 3*

Additional A Exercises

Solve.

1. $\sqrt{a} + 3 = 5$ {4}

2. $4\sqrt{p} - 9 = 7$ {16}

3. $\frac{\sqrt{h}}{3} + 1 = 2$ {9}

4. $\sqrt{n - 5} = 6$ {41}

5. $\sqrt[3]{\frac{1}{2}x} + 1 = 3$ {16}

6. $\frac{1}{3} + 2\sqrt{h} = 3$ $\{\frac{16}{9}\}$

Mixed Review

Simplify. Assume that all variables denote positive real numbers.

1. $\sqrt[3]{x^6}$ x^2

2. $\sqrt[4]{n^{16}y^3}$ $n^4\sqrt[4]{y^3}$

3. $\sqrt[5]{x^5y^{10}z^{15}}$ xy^2z^3

4. $\sqrt[4]{\frac{y^5}{x^2}}$ $\frac{y\sqrt[4]{x^2y}}{x}$

Solve each system of equations.

5. $2x + 3y = 7$
$x - 3y = -1$ {(2, 1)}

6. $3s - 2t = 3$
$5s + 3t = -14$ {(-1, -3)}

Quick Quiz

Simplify.

1. $\sqrt{12} \cdot 3\sqrt{20}$ $12\sqrt{15}$
2. $\sqrt{\dfrac{28}{54}}$ $\dfrac{\sqrt{42}}{9}$
3. $\sqrt{18} - \sqrt{50} + \sqrt{48}$
   $4\sqrt{3} - 2\sqrt{2}$
4. $\sqrt{5}(\sqrt{2} + \sqrt{5})$ $\sqrt{10} + 5$
5. $(7 + \sqrt{3})(3 - \sqrt{3})$
   $18 - 4\sqrt{3}$
6. $\dfrac{\sqrt{6} - 1}{\sqrt{8} + 2}$
   $\dfrac{2\sqrt{3} - \sqrt{2} - \sqrt{6} + 1}{2}$
7. $\sqrt[4]{256}$ 4
8. $\sqrt[4]{243} + \sqrt[4]{48}$ $5\sqrt{3}$
9. $\sqrt[4]{\dfrac{4}{27}}$ $\dfrac{\sqrt[4]{12}}{3}$

Solve.

10. $\sqrt{x + 2} + 4 = 7$ $\{7\}$
11. $\sqrt[3]{3x + 1} = 4$ $\{21\}$
12. $\sqrt{2x} + x = 12$ $\{8\}$

▮ Self-Test 3

VOCABULARY   rationalize a denominator (p. 518)
                 simplest form of a radical (p. 518)
                 conjugates (p. 521)

nth root (p. 524)
cube root (p. 524)
principal nth root (p. 524)
index (p. 525)

Simplify.

1. $\sqrt{18} \cdot 2\sqrt{27}$ $18\sqrt{6}$
2. $\sqrt{\dfrac{32}{45}}$ $\dfrac{4\sqrt{10}}{15}$
3. $\sqrt{48} - \sqrt{72} + \sqrt{75}$ $9\sqrt{3} - 6\sqrt{2}$ *Obj. 1, p. 516*

4. $\sqrt{6}(\sqrt{3} + \sqrt{2})$ $3\sqrt{2} + 2\sqrt{3}$
5. $\dfrac{\sqrt{5} - 2}{\sqrt{7} + 3}$
5. $\dfrac{\sqrt{35} - 2\sqrt{7} - 3\sqrt{5} + 6}{-2}$
6. $(4 - \sqrt{2})(5 + \sqrt{2})$ $18 - \sqrt{2}$

7. $\sqrt[5]{243}$ 3
8. $\sqrt[3]{\dfrac{5}{54}}$ $\dfrac{\sqrt[3]{20}}{6}$
9. $\sqrt[4]{32} + \sqrt[4]{1250}$ $7\sqrt[4]{2}$ *Obj. 2, p. 516*

Solve.

10. $\sqrt{x + 1} - 6 = 2$ $\{63\}$
11. $\sqrt[3]{2x + 5} = 3$ $\{11\}$
12. $\sqrt{3x} + x = 6$ $\{3\}$ *Obj. 3, p. 516*

Check your answers with those at the back of the book.

▬ EXTRA

Imaginary Numbers

In Sections 10-3 and 10-4, you learned to solve equations such as

$$x^2 = 16 \qquad \text{and} \qquad x^2 + 4 = 22.$$

You found, however, that you were unable to solve equations such as

$$x^2 = -4$$

over the set of real numbers because the square of a real number is always a nonnegative real number.

The solutions of this equation do exist, however, if the replacement set of the variable is extended to include *imaginary numbers*. **Imaginary numbers** involve the **imaginary unit,** i, which has the property that

$$i^2 + 1 = 0, \qquad \text{or} \qquad i^2 = -1.$$

Thus, i is called a "square root of -1," and you write

$$i = \sqrt{-1}.$$

530   *Chapter 10*

The imaginary unit can be used in solving equations as follows.

$$x^2 = -4$$
$$x = \pm\sqrt{-4}$$
$$= \pm\sqrt{-1} \cdot \sqrt{4}$$
$$= \pm i \cdot 2$$
$$x = \pm 2i$$

In fact, for every positive real number r,

$$\sqrt{-r} = i\sqrt{r} \quad \text{and} \quad -\sqrt{-r} = -i\sqrt{r}.$$

Thus,

$$(i\sqrt{r})^2 = -r \quad \text{and} \quad (-i\sqrt{r})^2 = -r.$$

These relationships can be used in simplifying expressions involving radicals whose radicands are negative numbers.

EXAMPLE 1   Simplify.

    a. $\sqrt{-48}$ **b.** $\sqrt{-20} + \sqrt{-45}$

    c. $\sqrt{-8} \cdot \sqrt{-18}$ **d.** $\sqrt{-\dfrac{5}{7}}$

SOLUTION   **a.** $\sqrt{-48} = \sqrt{-1} \cdot \sqrt{48} = i \cdot 4\sqrt{3} = 4i\sqrt{3}$

           b. $\sqrt{-20} + \sqrt{-45} = i\sqrt{20} + i\sqrt{45}$
$$= 2i\sqrt{5} + 3i\sqrt{5}$$
$$= 5i\sqrt{5}$$

           c. $\sqrt{-8} \cdot \sqrt{-18} = i\sqrt{8} \cdot i\sqrt{18} = 2i\sqrt{2} \cdot 3i\sqrt{2}$
$$= 6i^2 \cdot 2$$
$$= 6(-1) \cdot 2 = -12$$

           d. $\sqrt{-\dfrac{5}{7}} = \dfrac{\sqrt{-5}}{\sqrt{7}} = \dfrac{i\sqrt{5}}{\sqrt{7}} = \dfrac{i\sqrt{5}}{\sqrt{7}} \cdot \dfrac{\sqrt{7}}{\sqrt{7}} = \dfrac{i\sqrt{35}}{7}$

    Notice that in part (c) of Example 1 the product property of square roots could not be used until the radicals were expressed as real square roots.

When you simplify increasing powers of i, you find the values repeating in cycles of four according to the pattern i, -1, $-i$, 1.

$i^1 = i$ $i^5 = i^4 \cdot i = 1 \cdot i = i$

$i^2 = i \cdot i = -1$ $i^6 = i^5 \cdot i = i \cdot i = -1$

$i^3 = i^2 \cdot i = -1 \cdot i = -i$ $i^7 = i^6 \cdot i = -1 \cdot i = -i$

$i^4 = i^3 \cdot i = -i \cdot i = -i^2 = -(-1) = 1$ $i^8 = i^7 \cdot i = -i \cdot i = 1$

Thus, $i^9 = i$, $i^{10} = -1$, $i^{11} = -i$, $i^{12} = 1$, and so on.

Irrational Numbers and Radicals **531**

Since $i(-i) = (-i)(i) = -i^2 = -(-1) = 1$, i and $-i$ are reciprocals. That is,

$$\frac{1}{i} = -i \qquad \text{and} \qquad \frac{1}{-i} = i.$$

These facts about reciprocals are used in simplifying fractions whose denominators are imaginary numbers.

EXAMPLE 2 Simplify.

a. $\dfrac{18}{30i}$ b. $\dfrac{2}{\sqrt{-40}}$ c. $\dfrac{\sqrt{3}}{-\sqrt{-24}}$

SOLUTION a. $\dfrac{18}{30i} = \dfrac{18}{30} \cdot \dfrac{1}{i}$ b. $\dfrac{2}{\sqrt{-40}} = \dfrac{2}{2i\sqrt{10}}$ c. $\dfrac{\sqrt{3}}{-\sqrt{-24}} = \dfrac{\sqrt{3}}{-2i\sqrt{6}}$

$= \dfrac{3}{5}(-i)$ $\qquad = \dfrac{1}{\sqrt{10}} \cdot \dfrac{1}{i}$ $\qquad = \dfrac{1}{-i} \cdot \dfrac{\sqrt{3}}{2\sqrt{6}}$

$= -\dfrac{3}{5}i$ $\qquad = \dfrac{-i}{\sqrt{10}} \cdot \dfrac{\sqrt{10}}{\sqrt{10}}$ $\qquad = \dfrac{i\sqrt{3}}{2\sqrt{6}} \cdot \dfrac{\sqrt{6}}{\sqrt{6}}$

$\qquad\qquad = \dfrac{-i\sqrt{10}}{10}$ $\qquad = \dfrac{i\sqrt{18}}{2 \cdot 6}$

$\qquad\qquad\qquad\qquad\qquad = \dfrac{3i\sqrt{2}}{12} = \dfrac{i\sqrt{2}}{4}$

Exercises

Express in simplest form as i, -1, $-i$, or 1.

1. i^{16} 1 2. i^{29} i 3. i^{62} -1 4. i^{81} i 5. $\dfrac{1}{i^3}$ i 6. $\dfrac{1}{-i^4}$ -1

Simplify.

7. $\sqrt{-28}$ $2i\sqrt{7}$ 8. $\sqrt{-32}$ $4i\sqrt{2}$ 9. $\sqrt{-60}$ $2i\sqrt{15}$

10. $\sqrt{-72}$ $6i\sqrt{2}$ 11. $\sqrt{-36} + \sqrt{-81}$ $15i$ 12. $\sqrt{-54} - \sqrt{-150}$ $-2i\sqrt{6}$

13. $\sqrt{-12} \cdot \sqrt{-96}$ $-24\sqrt{2}$ 14. $\sqrt{-108} \cdot \sqrt{-75}$ -90 15. $\dfrac{16}{56i}$ $-\dfrac{2i}{7}$

16. $\dfrac{24}{-32i}$ $\dfrac{3i}{4}$ 17. $\dfrac{8}{\sqrt{-180}}$ $-\dfrac{4i\sqrt{5}}{15}$ 18. $\dfrac{\sqrt{6}}{2\sqrt{-240}}$ $-\dfrac{i\sqrt{10}}{40}$

Solve.

19. $x^2 + 49 = 0$ $\{7i, -7i\}$ 20. $x^2 + 13 = 1$ $\{2i\sqrt{3}, -2i\sqrt{3}\}$

21. $x^2 + 120 = 40$ $\{4i\sqrt{5}, -4i\sqrt{5}\}$ 22. $70 - x^2 = 95$ $\{5i, -5i\}$

23. $3x^2 + 50 = 2$ $\{4i, -4i\}$ 24. $2x^2 - 5 = -95$ $\{3i\sqrt{5}, -3i\sqrt{5}\}$

Chapter Summary

1. The *density property of rational numbers* asserts that between any two rational numbers there is always another rational number.

2. A number can be represented by a *terminating* or *repeating* decimal if and only if it is a *rational number*. Numbers that can only be expressed as nonterminating, nonrepeating decimals are *irrational numbers*.

3. If $b^2 = a$, then b is called a *square root* of a. The nonnegative, or *principal*, square root is denoted by the symbol $\sqrt{a}$. The negative square root is denoted by $-\sqrt{a}$.

4. For any real number a, $\sqrt{a^2} = |a|$.

5. If the square root of a given rational number is rational, then the given number is a *perfect square*. Square roots of numbers that are not perfect squares may be approximated using a table of square roots.

6. The *Pythagorean Theorem* is a statement of an important relationship among the sides of a right triangle: If the length of the hypotenuse is c and the lengths of the other sides are a and b, then

$$c^2 = a^2 + b^2.$$

7. The *distance formula* is a special application of the Pythagorean Theorem: The distance d between two points (x_1, y_1) and (x_2, y_2) is

$$d = \sqrt{(x_2 - x_1)^2 + (y_2 - y_1)^2}.$$

8. The following two properties are used to simplify radicals.
   a. *Product property of square roots*: For any nonnegative real numbers a and b,
   $$\sqrt{ab} = \sqrt{a} \cdot \sqrt{b}.$$
   b. *Quotient property of square roots*: For any nonnegative real number a and any positive real number b,
   $$\sqrt{\frac{a}{b}} = \frac{\sqrt{a}}{\sqrt{b}}.$$

9. For every positive integer n, a solution of $b^n = a$ is an *nth root* of a. The symbol $\sqrt[n]{a}$ denotes the *principal nth root* of a and $-\sqrt[n]{a}$ denotes the negative nth root of a. The positive integer n is called the *index*.

10. To solve an equation in which a term has a variable in a radicand, transform the equation into an equivalent equation in which the term with the variable in the radicand is alone as one side of the equation. Then raise both sides of the equation to the power equal to the index of the radical.

Chapter Review

Write the letter of the correct answer.

1. Which of the following statements is false?   *10-1*

   a. $\dfrac{126}{81} = \dfrac{84}{54}$ b. $\dfrac{28}{42} > \dfrac{32}{56}$ (c.) $4\dfrac{2}{9} > \dfrac{47}{11}$ d. $-\dfrac{8}{9} < -\dfrac{5}{12}$

2. Find the rational number that is one fifth of the way from $-\dfrac{3}{2}$ to 6.

   a. $\dfrac{3}{2}$ b. -1 (c.) 0 d. $\dfrac{9}{2}$

3. Express $2\dfrac{5}{18}$ as a decimal.   *10-2*

   a. $0.2\overline{7}$ b. $0.22\overline{7}$ c. $2.\overline{27}$ (d.) $2.2\overline{7}$

4. Which of the following is an irrational number?

   a. 2.01 b. $2.\overline{01}$ c. $2.\overline{01011}$ (d.) $2.010110111 \ldots$

5. Express $(0.\overline{4})(1.8)$ as a fraction in simplest form.

   a. $\dfrac{18}{25}$ (b.) $\dfrac{4}{5}$ c. $\dfrac{36}{5}$ d. $\dfrac{8}{25}$

6. Which statement is true for all real values of x?   *10-3*

   a. $\sqrt{x^2} = x$ b. $\sqrt{(x+1)^2} = x + 1$

   (c.) $\sqrt{x^2} = |x|$ d. $\sqrt{x}$ is a real number.

7. Simplify $-\sqrt{(-9)^2}$.

   a. 3 b. 9 (c.) -9 d. -3

8. Simplify $\sqrt{81a^2b^2}$.

   a. $9ab$ b. $-9ab$ (c.) $9|ab|$ d. $|9|ab$

9. Name the two consecutive integers between which $\sqrt{13.4}$ lies.   *10-4*

   (a.) 3 and 4   b. 4 and 5   c. 10 and 11   d. 11 and 12

10. Simplify $\sqrt{45}$.

    a. $9\sqrt{5}$ (b.) $3\sqrt{5}$ c. $5\sqrt{3}$ d. $5\sqrt{9}$

11. Solve $4y^2 + 3 = 6$.

    a. $\left\{\dfrac{3}{2}\right\}$ b. $\left\{-\dfrac{3}{2}, \dfrac{3}{2}\right\}$ (c.) $\left\{-\dfrac{\sqrt{3}}{2}, \dfrac{\sqrt{3}}{2}\right\}$ d. $\left\{-\dfrac{\sqrt{3}}{4}, \dfrac{\sqrt{3}}{4}\right\}$

12. In a right triangle, the shorter sides measure 10 cm and 24 cm. Find   *10-5*
    the length of the hypotenuse.

    (a.) 26 cm   b. 34 cm   c. $2\sqrt{119}$ cm   d. 20 cm

13. Which of the following sets of numbers is *not* a Pythagorean triple?

    a. $\{3, 4, 5\}$ **b.** $\{5, 12, 13\}$ **c.** $\{8, 15, 17\}$ **ⓓ** $\{1, 2, \sqrt{5}\}$

14. A ladder that is 10 m long leans against a wall. If the ladder meets the wall at a point that is 8 m above the ground, how far from the base of the wall is the foot of the ladder?

    a. 2 m   **ⓑ** 6 m   **c.** 18 m   **d.** $\sqrt{18}$ m

15. Find the distance between the points $A(2, -3)$ and $B(-1, 4)$.   *10-6*

    a. $\sqrt{10}$ **ⓑ** $\sqrt{58}$ **c.** $2\sqrt{119}$ **d.** 20

Simplify. Assume that all variables represent positive real numbers.

16. $\dfrac{2}{\sqrt{2}}$ *10-7*

    a. $\dfrac{\sqrt{2}}{2}$ **ⓑ** $\sqrt{2}$ **c.** $2\sqrt{2}$ **d.** 2

17. $\sqrt{6x^2} \cdot \sqrt{8x}$

    a. $16x\sqrt{3x}$ **b.** $3x\sqrt{4x}$ **ⓒ** $4x\sqrt{3x}$ **d.** $4x^2\sqrt{3x}$

18. $3\sqrt{50} + 2\sqrt{18}$ *10-8*

    a. $5\sqrt{68}$ **ⓑ** $21\sqrt{2}$ **c.** $13\sqrt{2}$ **d.** $6\sqrt{5} + 4\sqrt{3}$

19. $(\sqrt{5} - 1)(\sqrt{5} + 2)$

    a. 3   **b.** 23   **ⓒ** $\sqrt{5} + 3$ **d.** $\sqrt{5} + 23$

20. What is the conjugate of $2\sqrt{3} - 6\sqrt{5}$?

    a. $\sqrt{3} + \sqrt{5}$ **b.** $6\sqrt{5} - 2\sqrt{3}$

    c. $3\sqrt{2} - 5\sqrt{6}$ **ⓓ** $2\sqrt{3} + 6\sqrt{5}$

21. Simplify $\sqrt[3]{-27}$.   *10-9*

    a. -9 **ⓑ** -3 **c.** 9   **d.** none of these

22. Simplify $\sqrt[3]{\dfrac{3}{4}}$.

    a. $\sqrt[3]{3}$ **b.** $\dfrac{\sqrt[3]{3}}{2}$ **ⓒ** $\dfrac{\sqrt[3]{6}}{2}$ **d.** $\dfrac{\sqrt[3]{12}}{4}$

Solve.

23. $\sqrt[3]{4 + x} = 5$ *10-10*

    a. $\{1\}$ **b.** $\{11\}$ **c.** $\{21\}$ **ⓓ** $\{121\}$

24. $6 - \sqrt{x} = x$

    ⓐ $\{4\}$ **b.** $\{4, 9\}$ **c.** $\{9\}$ **d.** $\{3\}$

Irrational Numbers and Radicals   **535**

Chapter Test

1. Find the rational number that is one third of the way from $-\frac{3}{4}$ to $\frac{1}{3}$. $-\frac{7}{18}$ *10-1*

2. Write $\frac{36}{54}$, $\frac{60}{144}$, and $\frac{42}{96}$ in order from least to greatest. $\frac{60}{144}, \frac{42}{96}, \frac{36}{54}$

3. Express $\frac{14}{11}$ as a decimal. $1.\overline{27}$ *10-2*

4. Express $-0.41\overline{6}$ as a fraction in simplest form. $-\frac{5}{12}$

5. Express $0.8\overline{3} + \frac{1}{3}$ as a fraction in simplest form. $\frac{7}{6}$

Simplify.

6. $\sqrt{(-36)^2}$ 36   7. $-\sqrt{\frac{64}{81}}$ $-\frac{8}{9}$ 8. $\sqrt{(x-3)^2}$ $|x-3|$ *10-3*

9. $\left(\sqrt{24}\right)^2$ 24   10. $\sqrt{450}$ $15\sqrt{2}$ 11. $\sqrt{\frac{20}{9}}$ $\frac{2\sqrt{5}}{3}$ *10-4*

12. Solve $16z^2 + 7 = 25$. $\{\frac{3\sqrt{2}}{4}, -\frac{3\sqrt{2}}{4}\}$

13. Given that the length of the hypotenuse of a right triangle is $\sqrt{34}$ cm   *10-5*
    and that the length of a second side is 3 cm, find the length of the
    third side.   5 cm

14. Roger rode his bike from his home due east for 10 km and then due
    north for 6 km. How far was he from his home? Approximate the
    answer to the nearest hundredth.   11.66 km

15. Find the distance between the points $A(3, 2)$ and $B(9, 5)$. $3\sqrt{5}$ *10-6*

16. Determine whether or not the triangle with vertices $(-1, -1)$, $(1, 2)$,
    and $(5, -5)$ is a right triangle.   Yes

Simplify. Assume that all variables represent positive real numbers.

17. $\sqrt{32} \cdot \sqrt{6}$ $8\sqrt{3}$ 18. $\sqrt{\frac{2}{3}}$ $\frac{\sqrt{6}}{3}$ 19. $\sqrt{5y^3} \cdot \sqrt{20y^5}$ $10y^4$ *10-7*

20. $\sqrt{5} + \sqrt{12} - \sqrt{20}$ 21. $\sqrt{15}\left(\sqrt{5} - \sqrt{6}\right)$ 22. $\frac{3}{4 - \sqrt{2}}$ $\frac{12 + 3\sqrt{2}}{14}$ *10-8*
        $2\sqrt{3} - \sqrt{5}$ $5\sqrt{3} - 3\sqrt{10}$

23. $\sqrt[5]{-32}$ -2 24. $\sqrt[3]{270}$ $3\sqrt[3]{10}$ 25. $\sqrt[4]{\frac{7}{8}}$ $\frac{\sqrt[4]{14}}{2}$ *10-9*

Solve.

26. $\sqrt{x} = -3$ ∅   27. $\sqrt{y - 3} = 7$ {52}   28. $\sqrt{z} + 2 = z$ {4}   *10-10*

Mixed Review

Simplify.

2. $5w^3 + 7w^2 + 19$

6. $20a^5b + 12a^3b^2 - 8a^2b^3$

1. $(n)^2(2n) + (n)(n^2)$ $3n^3$

2. $(7w^3 - w^2 + 9) - (2w^3 - 10 - 8w^2)$

3. $(2x - 5)(2x + 5)$ $4x^2 - 25$

4. $(4t - 3)(3t - 2)$ $12t^2 - 17t + 6$

5. $(x - 5)^3$ $x^3 - 15x^2 + 75x - 125$

6. $4a^2b(5a^3 + 3ab - 2b^2)$

7. $\dfrac{2^{-2}x^2z^0}{4^{-1}x^{-2}z^{-3}}$ x^4z^3

8. $\dfrac{a^2 - ab - 12b^2}{a^2 - 3ab - 4b^2}$ $\dfrac{a + 3b}{a + b}$

9. $\dfrac{m^2 - 4}{2m - 6} \div \dfrac{2 - m}{m - 3}$ $-\dfrac{m + 2}{2}$

10. $\dfrac{4}{x^2 + 2xy + y^2} - \dfrac{3}{x + y}$ $\dfrac{4 - 3x - 3y}{x^2 + 2xy + y^2}$

11. $\dfrac{42a^3b^2 - 27a^2b^3 + 3a^2b}{-3a^2b}$ $-14ab + 9b^2 - 1$

12. $\dfrac{v^2 + 2v - 15}{v^2 + 3v - 10} \cdot \dfrac{v^2 - 4}{6 + v - v^2}$ -1

13. $\dfrac{d}{c + 3d} + \dfrac{c}{3c + d}$ $\dfrac{c^2 + 6cd + d^2}{3c^2 + 10cd + 3d^2}$

14. $\dfrac{a - 3}{a^2 - 5a + 4} - \dfrac{a - 1}{12 + a - a^2}$

15. $\dfrac{\dfrac{5x}{4} - \dfrac{5z}{24}}{\dfrac{z}{12} - \dfrac{x}{2}}$ $-\dfrac{5}{2}$

16. $\dfrac{1 - \dfrac{36}{w^2}}{1 + \dfrac{6}{w}}$ $\dfrac{w - 6}{w}$

$\dfrac{2a^2 - 2a - 8}{(a - 1)(a - 4)(a + 3)}$

Solve.

17. $2d^2 + d = 15$ $\{\frac{5}{2}, -3\}$

18. $y^3 = 7y^2 + 18y$ $\{0, 9, -2\}$

19. $(3 - 6a^2) - (5a - 6a^2) = -2$ $\{1\}$

20. $(2t + 5)(2t - 3) = (4t - 1)(t - 1)$ $\{\frac{16}{9}\}$

21. $\dfrac{3k}{8} - \dfrac{k}{4} < \dfrac{3}{2}$ $\{k: k < 12\}$

22. $\dfrac{5m - 2}{6} - \dfrac{2m + 1}{3} > \dfrac{1}{2}$ $\{m: m > 7\}$

23. $\dfrac{x - 16}{x} = \dfrac{3}{5}$ $\{40\}$

24. $\dfrac{-5}{2 - 3z} = \dfrac{2}{4 + z}$ $\{24\}$

25. $\dfrac{5}{2y + 3} = \dfrac{4}{y} - 3$ $\{1, -2\}$

26. $\dfrac{2t + 3}{t + 1} - \dfrac{4t + 2}{t^2 - 1} = \dfrac{t}{t - 1}$ $\{5\}$

27. $\dfrac{r^2 - 7}{r^2 - 4r + 3} + \dfrac{r + 2}{r - 3} = 2$ $\{\frac{5}{3}\}$

28. $\dfrac{x + 1}{x - 2} + \dfrac{x - 3}{x + 2} = \dfrac{7x + 4}{x^2 - 4}$ $\{\frac{1}{2}, 4\}$

Factor completely.

29. $x^2 - 3x - 18$ $(x - 6)(x + 3)$

30. $y(y - 5) + 2(5 - y)$ $(y - 5)(y - 2)$

31. $20z^2 + 3z - 2$ $(5z + 2)(4z - 1)$

32. $a^2 + 9$ irreducible

33. $3m^2 + 9m - 12$ $3(m + 4)(m - 1)$

34. $24n^2 + n - 10$ $(8n - 5)(3n + 2)$

35. $2ab^3 - 18ab$ $2ab(b - 3)(b + 3)$

36. $16c^2 - 81$ $(4c + 9)(4c - 9)$

Graph on a coordinate plane.

37. $3y - 2x \geq 9$

38. $4x + 2y = 3$

39. $5y - 3x = 10$

40. $15x - 5y > 20$

Irrational Numbers and Radicals **537**

37.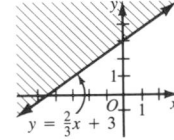

38.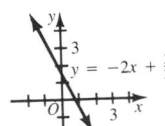

39.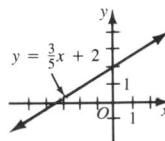

40.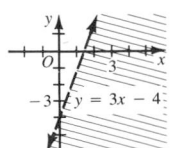

Determine an equation of the line that satisfies the given requirements.

41. passes through the point $(-1, -2)$ and is parallel to the y-axis $x = -1$
42. passes through the points $(-1, 0)$ and $(4, 10)$ $2x - y = -2$
43. has slope $\frac{2}{3}$ and y-intercept -5 $2x - 3y = 15$
44. passes through the point $(-2, 6)$ and has slope 3 $3x - y = -12$

Express in scientific notation.

45. 90,700,000,000,000 9.07×10^{13} 46. 0.0000000814 8.14×10^{-8}

Solve.

47. Nancy invested part of her $6500 savings at an annual interest rate of $6\frac{1}{2}\%$ and the rest at an annual interest rate of 8%. How much did she invest at each rate if her total income from these investments for one year is $490? $2000 at $6\frac{1}{2}\%$; $4500 at 8%

48. Find two consecutive positive integers such that the sum of their squares is 85.   6 and 7

49. The difference of two numbers is 8 and their quotient is $\frac{5}{3}$. Find the two numbers.   20 and 12 or -20 and -12

50. One pipe can fill a tank in 4 h. A second pipe can fill the same tank in 6 h. How long will it take both pipes working together to fill the tank? $2\frac{2}{5}$ h

51. A suitcase with a mass of 4 kg causes a spring to stretch 14 cm. If the amount that the spring stretches varies directly with the mass, how much will a 2.5 kg suitcase stretch the spring?   8.75 cm

52. The number of hours needed to enter a certain set of data into a computer is inversely proportional to the number of persons who are entering the data. If it would take 5 people 8 h to complete the job, how long would the job take if 12 people were entering the data? $3\frac{1}{3}$ h

53. The length of a certain rectangle is 12 m and the width is 5 m. If both the length and the width were increased by the same amount, the area would be increased by 60 m². Find the dimensions of the new rectangle.   15 m by 8 m

54. The length of a square field is increased by 4 m and the width is increased by 2 m, creating a rectangular field that has an area of 195 m². Find the dimensions of the original field.   11 m by 11 m

55. The volume of a pyramid varies jointly as the altitude of the pyramid and the area of its base. The volume of a pyramid with altitude 6 cm and base area 16 cm² is 32 cm³. Find the volume of a pyramid with altitude 8 cm and base area 9 cm².   24 cm³

56. The number of persons needed to do a job varies directly as the amount of work to be done and inversely as the time in which the job must be done. If two typists can type 210 pages of manuscript in three days, how many typists will be needed to type 700 pages in two days?   10 typists

PREPARING FOR COLLEGE ENTRANCE EXAMS

Strategy for Success: On a multiple-choice exam, you often are asked to determine the *best* answer to a question. For any given question, more than one answer may seem "right" to some degree, and you must be careful not to choose the first answer that seems reasonable. Be sure to evaluate all the possible choices before you determine which answer is "best."

Decide which is the best of the choices and write the corresponding letter on your answer sheet.

1. What is the range of the function $\{(x, y): y = -\sqrt{x}\}$? B
   (A) $\{y: y \geq 0\}$ (B) $\{y: y \leq 0\}$ (C) $\{y: y < 0\}$
   (D) all real numbers   (E) none of these

2. A rectangle is three times as long as it is wide. One of its diagonals measures 30 cm. Find the width of the rectangle to the nearest tenth of a centimeter. A
   (A) 9.5 cm   (B) 28.5 cm   (C) 15 cm   (D) 11.2 cm   (E) none of these

3. Simplify $\dfrac{4}{\sqrt[3]{3}}$. C
   (A) $\dfrac{3\sqrt[3]{9}}{4}$ (B) 4   (C) $\dfrac{4\sqrt[3]{9}}{3}$ (D) $\dfrac{4\sqrt[3]{27}}{3}$ (E) none of these

4. Which of the following are factors of $4x^3 + 8x^2 - 7x - 5$? D
   I. $2x + 1$ II. $2x - 1$ III. $2x + 5$
   (A) I only   (B) II only   (C) III only   (D) I and III only   (E) II and III only

5. For which value of p in the division $(2x^2 + 11x - p) \div (2x - 3)$ is the remainder 0? B
   (A) 18   (B) 21   (C) 3   (D) 5   (E) 12

6. Which of the following is true if $b > c$ and $\dfrac{ab - ac}{b - c} < 0$? D
   (A) $a > b$ (B) $a < b$ (C) $a = b$ (D) $a < 0$ (E) $b < 0$

7. The vertices of a triangle are at $(-4, 4)$, $(-2, -4)$, and $(6, -2)$. Which of the following statements is true? E
   (A) The perimeter of the triangle is $4\sqrt{17} + 2\sqrt{34}$.
   (B) The triangle is isosceles.
   (C) The triangle is a right triangle.
   (D) The area of the triangle is 34 square units.
   (E) All of the above statements are true.

8. A reservoir can be filled in 6 days by pipe A running alone, or in 4 days by pipe B alone. How many days would be needed to fill the reservoir if both pipes were running? C
   (A) 10   (B) 5   (C) $2\dfrac{2}{5}$ (D) $\dfrac{5}{12}$ (E) $3\dfrac{1}{2}$

Irrational Numbers and Radicals   **539**

APPLICATION

Raising Your Sights

Have you ever been up in a balloon? If so, you probably noticed that, the higher the balloon rose, the more of Earth's surface you could see. However, because the surface of Earth is curved and light travels in straight lines, from any given height you can see only as far as the *horizon*, where your line of sight just touches Earth's surface. Did you know that it is possible to calculate how far you can see?

The diagram at the right represents a balloon that is at height h above Earth's surface. The length of the line of sight from the balloon to the horizon is represented by d. Notice that, at the horizon, the line of sight forms a right angle with Earth's radius, r.

Now consider the line of sight d and the radius r as the two shorter sides of a right triangle. The distance from the balloon to the center of Earth, or $r + h$, is then the hypotenuse of the right triangle. Therefore, to find the length of the line of sight, you can use the Pythagorean Theorem as follows.

$$a^2 + b^2 = c^2$$
$$r^2 + d^2 = (r + h)^2$$
$$r^2 + d^2 = r^2 + 2rh + h^2$$
$$d^2 = 2rh + h^2$$
$$d = \sqrt{2rh + h^2}$$

Suppose that you are in a balloon that is 1 km above Earth's surface. Earth's radius, r, is about 6380 km. To calculate how far you could see from this height, substitute these values into the above equation.

$$d = \sqrt{2rh + h^2}$$
$$= \sqrt{2(6380)(1) + 1^2}$$
$$d = \sqrt{12{,}761} \approx 113$$

Therefore, from a height of 1 km you can see about 113 km to the horizon.

540 *Chapter 10*

Notice that in the equation

$$d = \sqrt{2rh + h^2},$$

the value of h^2 is frequently much less than the value of $2rh$. This happens because the height that an object rises above Earth's surface, h, is frequently very small when compared to the radius of Earth, r. In such cases, therefore, the value of h^2 may be omitted to obtain the alternate formula

$$d \approx \sqrt{2rh}.$$

Using this formula in the previous example, notice that you would obtain

$$d \approx \sqrt{12,760},$$

which again rounds to an approximate distance to the horizon of 113 km.

Exercises   All answers are approximations.

1. A balloon is at a height of 0.5 km. How far can the pilot see? 80 km

2. An airplane is flying at an altitude of 10 km. How far can the passengers see? 357 km

3. An airplane is flying at an altitude of 4000 m. How far can the passengers see? 226 km

4. An observer is in a fire tower that is 20 m high. How far can the observer see? 16 km

5. A lifeguard sitting in a tall chair can see 5 km out to sea. How tall is the chair? 2 m

6. The pilot of a balloon can see for 200 km. What is the altitude of the balloon? 3 km

7. A team of astronauts is 100,000 km above the surface of Earth.

  a. Use the formula $d = \sqrt{2rh + h^2}$ to calculate the distance they can see to the horizon. 106,189 km

  b. Use the formula $d \approx \sqrt{2rh}$ to calculate the distance they can see to the horizon. 35,721 km

  c. Explain why the answers to parts (a) and (b) are so different.
  Since h is large, the h^2 term is significant and must not be omitted in the calculation.

Irrational Numbers and Radicals   **541**

This suspension bridge in Portland, Maine, was completed in 1866, and its original iron cables are still in place. Each of the cables is shaped like part of a parabola.

Chapter 11

Quadratic Equations and Functions

Problem Solving Strategies

Word Problem Plan
The *Plan for Solving a Word Problem* is used with the methods learned in Sections 11-1 and 11-2 to solve word problems involving quadratic equations.

Quadratic Equations

OBJECTIVES for Sections 11-1 through 11-3:
1. *To solve quadratic equations by completing the square.*
2. *To solve quadratic equations by using the quadratic formula.*
3. *To use quadratic equations to solve problems.*

11–1 Completing the Square

Recall from Section 7-10 that any equation that can be written equivalently in the form

$$ax^2 + bx + c = 0, \quad a \neq 0,$$

is called a *quadratic equation*. In Chapter 7 you learned to solve some equations of this form by factoring. In this chapter you will learn methods that can be used to solve *any* quadratic equation.

As you learned in Chapter 10, you can solve a simple quadratic equation such as

$$x^2 = 9$$

by observing that it has two roots:

$$x = \sqrt{9} \quad \text{or} \quad x = -\sqrt{9}$$
$$x = 3 \quad \quad \quad x = -3$$

Thus the solution set is {3, −3}.

Quadratic Equations and Functions **543**

1. Solve $m^2 + 12m + 36 = 5$.
   $(m + 6)^2 = 5$
   $m + 6 = \sqrt{5}$ or
   $\qquad\qquad m + 6 = -\sqrt{5}$
   $m = -6 + \sqrt{5}$ or
   $\qquad\qquad m = -6 - \sqrt{5}$
   $\therefore$ the solution set is
   $\{-6 + \sqrt{5}, -6 - \sqrt{5}\}$.

2. Solve $y^2 - 20y + 1 = 0$ by completing the square.
   $$y^2 - 20y = -1$$
   $$y^2 - 20y + (-10)^2 =$$
   $$-1 + (-10)^2$$
   $$(y - 10)^2 = 99$$
   $y - 10 = \sqrt{99}$ or
   $\qquad\qquad y - 10 = -\sqrt{99}$
   $y = 10 + 3\sqrt{11}$ or
   $\qquad\qquad y = 10 - 3\sqrt{11}$
   $\therefore$ the solution set is
   $\{10 + 3\sqrt{11}, 10 - 3\sqrt{11}\}$.

3. Solve $a^2 + 3a = 9$ by completing the square. State the results as exact solutions and as approximations to the nearest hundredth.
   $$a^2 + 3a + \left(\frac{3}{2}\right)^2 = 9 + \left(\frac{3}{2}\right)^2$$
   $$\left(a + \frac{3}{2}\right)^2 = \frac{45}{4}$$
   $a + \frac{3}{2} = \sqrt{\frac{45}{4}}$ or
   $\qquad\qquad a + \frac{3}{2} = -\sqrt{\frac{45}{4}}$
   $a = -\frac{3}{2} + \frac{3\sqrt{5}}{2}$ or
   $\qquad\qquad a = -\frac{3}{2} - \frac{3\sqrt{5}}{2}$
   $a = \frac{-3 + 3\sqrt{5}}{2}$ or $\frac{-3 - 3\sqrt{5}}{2}$
   $\therefore$ the solution set is
   $\left\{\dfrac{-3 + 3\sqrt{5}}{2}, \dfrac{-3 - 3\sqrt{5}}{2}\right\}$,
   or $\{1.85, -4.85\}$.

Now consider the equation

$$(x - 2)^2 = 9.$$

You can solve this equation by noting that it also has two roots:

$$
\begin{array}{ccc}
x - 2 = \sqrt{9} & \text{or} & x - 2 = -\sqrt{9} \\
x - 2 = 3 & & x - 2 = -3 \\
x = 5 & \text{or} & x = -1
\end{array}
$$

Therefore, the solution set is $\{-1, 5\}$.

You can extend this method to solve quadratic equations such as those in the following examples.

EXAMPLE 1   Solve $2(y - 3)^2 = 50$.

SOLUTION
$$2(y - 3)^2 = 50$$
$$(y - 3)^2 = 25$$
$$
\begin{array}{ccc}
y - 3 = \sqrt{25} & \text{or} & y - 3 = -\sqrt{25} \\
y - 3 = 5 & & y - 3 = -5 \\
y = 8 & \text{or} & y = -2
\end{array}
$$

Check:
$$
\begin{array}{ll}
2(y - 3)^2 = 50 & \qquad 2(y - 3)^2 = 50 \\
2(8 - 3)^2 \overset{?}{=} 50 & \qquad 2(-2 - 3)^2 \overset{?}{=} 50 \\
\qquad 50 = 50 \;\checkmark & \qquad\qquad 50 = 50 \;\checkmark
\end{array}
$$

$\therefore$ the solution set is $\{8, -2\}$.

EXAMPLE 2   Solve $g^2 + 6g + 9 = 2$.

SOLUTION
$$g^2 + 6g + 9 = 2$$
$$(g + 3)^2 = 2$$
$$
\begin{array}{ccc}
g + 3 = \sqrt{2} & \text{or} & g + 3 = -\sqrt{2} \\
g = -3 + \sqrt{2} & & g = -3 - \sqrt{2}
\end{array}
$$

Check:
$$g^2 + 6g + 9 = 2$$
$$(-3 + \sqrt{2})^2 + 6(-3 + \sqrt{2}) + 9 \overset{?}{=} 2$$
$$9 - 6\sqrt{2} + 2 - 18 + 6\sqrt{2} + 9 \overset{?}{=} 2$$
$$2 = 2 \;\checkmark$$

$$g^2 + 6g + 9 = 2$$
$$(-3 - \sqrt{2})^2 + 6(-3 - \sqrt{2}) + 9 \overset{?}{=} 2$$
$$9 + 6\sqrt{2} + 2 - 18 - 6\sqrt{2} + 9 \overset{?}{=} 2$$
$$2 = 2 \;\checkmark$$

$\therefore$ the solution set is $\{-3 + \sqrt{2}, -3 - \sqrt{2}\}$.

Examples 1 and 2 suggest that any quadratic equation can be solved readily if it can be written in the form $(x + p)^2 = q$. Examples 3 and 4 illustrate how to take advantage of this fact by completing the trinomial

544   *Chapter 11*

square that is suggested by the coefficient of the linear term. This method of solution is called **completing the square.**

EXAMPLE 3 Solve $x^2 - 10x + 15 = 0$ by completing the square.

SOLUTION

$$x^2 - 10x + 15 = 0$$
$$x^2 - 10x = -15 \qquad \longleftarrow \text{\{Subtract 15 from both sides}$$
$$x^2 - 10x + (-5)^2 = -15 + (-5)^2 \qquad \longleftarrow \begin{cases} \text{Add the square of } half \text{ the} \\ \text{coefficient of } x \text{ to both sides} \end{cases}$$
$$= -15 + 25$$
$$(x - 5)^2 = 10 \qquad \longleftarrow \text{\{Factor the trinomial square}$$

$$x - 5 = \sqrt{10} \qquad \text{or} \qquad x - 5 = -\sqrt{10}$$
$$x = 5 + \sqrt{10} \qquad | \qquad x = 5 - \sqrt{10}$$

Checking the solutions is left to you.

∴ the solution set is $\{5 + \sqrt{10}, 5 - \sqrt{10}\}$.

EXAMPLE 4 Solve $d^2 + 5d = 5$ by completing the square.

SOLUTION

$$d^2 + 5d = 5$$
$$d^2 + 5d + \left(\frac{5}{2}\right)^2 = 5 + \left(\frac{5}{2}\right)^2 \qquad \longleftarrow \begin{cases} \text{Add the square of } half \text{ the} \\ \text{coefficient of } d \text{ to both sides} \end{cases}$$
$$= 5 + \frac{25}{4}$$
$$\left(d + \frac{5}{2}\right)^2 = \frac{45}{4} \qquad \longleftarrow \text{\{Factor the trinomial square}$$

$$d + \frac{5}{2} = \sqrt{\frac{45}{4}} \qquad \text{or} \qquad d + \frac{5}{2} = -\sqrt{\frac{45}{4}}$$
$$d + \frac{5}{2} = \frac{3\sqrt{5}}{2} \qquad \qquad d + \frac{5}{2} = -\frac{3\sqrt{5}}{2}$$
$$d = -\frac{5}{2} + \frac{3\sqrt{5}}{2} \qquad \qquad d = -\frac{5}{2} - \frac{3\sqrt{5}}{2}$$
$$d = \frac{-5 + 3\sqrt{5}}{2} \qquad \text{or} \qquad d = \frac{-5 - 3\sqrt{5}}{2}$$

Check:
$$d^2 + 5d = 5$$
$$\left(\frac{-5 + 3\sqrt{5}}{2}\right)^2 + 5\left(\frac{-5 + 3\sqrt{5}}{2}\right) \stackrel{?}{=} 5$$
$$\frac{25 - 30\sqrt{5} + 45}{4} + \frac{-25 + 15\sqrt{5}}{2} \stackrel{?}{=} 5$$
$$25 - 30\sqrt{5} + 45 + 2(-25 + 15\sqrt{5}) \stackrel{!}{=} 20$$
$$25 - 30\sqrt{5} + 45 - 50 + 30\sqrt{5} \stackrel{?}{=} 20$$
$$20 = 20 \quad \checkmark$$

Checking the second solution is left to you.

∴ the solution set is $\left\{\dfrac{-5 + 3\sqrt{5}}{2}, \dfrac{-5 - 3\sqrt{5}}{2}\right\}$.

Quadratic Equations and Functions **545**

Suggested Assignments

Minimum
  547/1–25 odd
Average
  547/1–35 odd
Maximum
  547/1–43 odd

546

If the coefficient of the quadratic term of a quadratic equation is a number other than 1, you must divide each term in the equation by this coefficient before completing the square, as shown in Example 5.

EXAMPLE 5   Solve $3x^2 - 8x + 2 = 0$ by completing the square.

SOLUTION

$$3x^2 - 8x + 2 = 0$$

$$x^2 - \frac{8}{3}x + \frac{2}{3} = 0 \quad \longleftarrow \quad \begin{cases} \text{First divide both sides by 3.} \\ \text{Then proceed to complete the square} \\ \text{as in the preceding examples.} \end{cases}$$

$$x^2 - \frac{8}{3}x = -\frac{2}{3}$$

$$x^2 - \frac{8}{3}x + \left(-\frac{4}{3}\right)^2 = -\frac{2}{3} + \left(-\frac{4}{3}\right)^2$$

$$\left(x - \frac{4}{3}\right)^2 = \frac{10}{9}$$

$$x - \frac{4}{3} = \sqrt{\frac{10}{9}} \quad \text{or} \quad x - \frac{4}{3} = -\sqrt{\frac{10}{9}}$$

$$x = \frac{4}{3} + \frac{\sqrt{10}}{3} \qquad\qquad x = \frac{4}{3} - \frac{\sqrt{10}}{3}$$

$$x = \frac{4 + \sqrt{10}}{3} \quad \text{or} \quad x = \frac{4 - \sqrt{10}}{3}$$

Checking the results is left to you.

$\therefore$ the solution set is $\left\{\dfrac{4 + \sqrt{10}}{3}, \dfrac{4 - \sqrt{10}}{3}\right\}$.

Notice that the solutions in Examples 2 through 5 are given as *exact solutions*, with all irrational solutions written in simplest form. If you use a calculator or the Table of Square Roots on page 683, you can also obtain *decimal approximations* of such solutions. For example, the solutions of the equation in Example 5 can be approximated as follows.

$$\frac{4 + \sqrt{10}}{3} \approx \frac{4 + 3.162}{3} = \frac{7.162}{3} = 2.387\overline{3} \approx 2.39$$

$$\frac{4 - \sqrt{10}}{3} \approx \frac{4 - 3.162}{3} = \frac{0.838}{3} = 0.279\overline{3} \approx 0.28$$

Throughout the rest of this chapter, irrational solutions should be approximated to the nearest hundredth unless otherwise specified.

Some quadratic equations have *no* solutions over the set of real numbers. For example, consider the equation

$$p^2 + 4p + 5 = 0.$$

By completing the square, you would obtain the equation

$$(p + 2)^2 = -1.$$

Since the square of a real number must be a *nonnegative* real number, the equation $p^2 + 4p + 5 = 0$ has no real-number solution.

Oral Exercises

Name the value(s) of q for which the expression is a trinomial square.

1. $x^2 + 10x + q$ 25
2. $d^2 - 12d + q$ 36
3. $b^2 + qb + 16$ 8
4. $r^2 - qr + 81$ 18
5. $a^2 - 5a + q$ $\frac{25}{4}$
6. $b^2 + 9b + q$ $\frac{81}{4}$
7. $y^2 + y + q$ $\frac{1}{4}$
8. $f^2 + \frac{2}{5}f + q$ $\frac{1}{25}$
9. $n^2 - qn + \frac{9}{16}$ $\frac{3}{2}$
10. $c^2 + qc + \frac{49}{64}$ $\frac{7}{4}$
11. $z^2 + mz + q$ $\frac{m^2}{4}$
12. $x^2 - \frac{r}{s}x + q$ $\frac{r^2}{4s^2}$

Written Exercises

Solve by completing the square. First give the exact solution, expressing any irrational roots in simplest form. Then approximate irrational roots to the nearest hundredth. Approximations may vary.

A
1. $c^2 - 2c - 3 = 0$ $\{-1, 3\}$
2. $z^2 + 8z - 9 = 0$ $\{-9, 1\}$
3. $p^2 + 6p = -8$ $\{-4, -2\}$
4. $q^2 = 12q - 27$ $\{3, 9\}$
5. $e^2 + 8e + 10 = 0$
6. $d^2 - 6d + 2 = 0$
7. $n^2 - 4n - 8 = 0$
8. $x^2 - 16x + 40 = 0$
9. $3m^2 - 12m + 9 = 0$
10. $5t^2 + 20t - 60 = 0$
11. $12x^2 + 84 = 96x$
12. $144 - 9y^2 = 54y$
13. $w^2 - w - 4 = 0$
14. $z^2 + z - 3 = 0$
15. $r^2 = 1 - 3r$
16. $p^2 - 5p = -5$
17. $z^2 - 9z + 10 = 0$
18. $b^2 + 17 = 11b$

Solve by completing the square. Give only the exact solutions, expressing any irrational roots in simplest form. If an equation has no real roots, so state.

B
19. $a^2 - \frac{1}{2}a - \frac{1}{2} = 0$ $\{-\frac{1}{2}, 1\}$
20. $b^2 - \frac{1}{8} = \frac{1}{4}b$ $\{-\frac{1}{4}, \frac{1}{2}\}$
21. $s^2 = \frac{2}{3}s + \frac{4}{3}$
22. $m^2 - \frac{2}{5}m + \frac{1}{5} = 0$
23. $5x^2 + 30x - 25 = 0$
24. $8x^2 - 32x + 16 = 0$
25. $2n^2 + 6n + 5 = 0$
26. $3t^2 + 15t + 12 = 0$
27. $5y^2 + y - 3 = 0$
28. $4m^2 - m - 6 = 0$
29. $25x^2 + 20x + 1 = 0$
30. $18z^2 - 48z - 30 = 0$
31. $49x^2 - 42x + 10 = 0$
32. $12a^2 + 54a + 45 = 0$
33. $2 - \frac{1}{p-1} = \frac{2}{p}$
34. $\frac{5}{r+4} - 3 = \frac{9}{x-4}$
35. $\frac{1}{u} + \frac{1}{u-2} = 3$
36. $\frac{3}{c+1} + \frac{4}{c-1} = 2$

C
37. $(2x - 1)^2 - (x + 3) = -1$
38. $(x + 6)^2 + (3x - 5)^2 = 65$
39. $x^2 + x\sqrt{3} - 3 = 0$
40. $2y^2 - y\sqrt{5} + 1 - 0$
41. $kx^2 - \frac{2}{5}x - \frac{1}{5k} = 0$
42. $\frac{x^2}{6} - \frac{1}{2}kx = \frac{2}{3}k^2$
43. $x^2 + bx + c = 0$
44. $ax^2 + bx + c = 0$

Quadratic Equations and Functions   **547**

---

21. $\left\{\dfrac{1 + \sqrt{13}}{3}, \dfrac{1 - \sqrt{13}}{3}\right\}$
22. no real roots
23. $\{-3 + \sqrt{14}, -3 - \sqrt{14}\}$
24. $\{2 + \sqrt{2}, 2 - \sqrt{2}\}$
25. no real roots
26. $\{-4, -1\}$
27. $\left\{\dfrac{-1 + \sqrt{61}}{10}, \dfrac{-1 - \sqrt{61}}{10}\right\}$
28. $\left\{\dfrac{1 + \sqrt{97}}{8}, \dfrac{1 - \sqrt{97}}{8}\right\}$
29. $\left\{\dfrac{-2 + \sqrt{3}}{5}, \dfrac{-2 - \sqrt{3}}{5}\right\}$
30. $\left\{\dfrac{4 + \sqrt{31}}{3}, \dfrac{4 - \sqrt{31}}{3}\right\}$
31. no real roots
32. $\left\{\dfrac{9 + \sqrt{21}}{4}, \dfrac{9 - \sqrt{21}}{4}\right\}$
33. $\left\{\dfrac{1}{2}, 2\right\}$
34. no real roots
35. $\left\{\dfrac{4 + \sqrt{10}}{3}, \dfrac{4 - \sqrt{10}}{3}\right\}$
36. $\left\{\dfrac{7 + \sqrt{73}}{4}, \dfrac{7 - \sqrt{73}}{4}\right\}$
37. $\left\{\dfrac{5 + \sqrt{41}}{8}, \dfrac{5 - \sqrt{41}}{8}\right\}$
38. $\left\{-\dfrac{1}{5}, 2\right\}$
39. $\left\{\dfrac{-\sqrt{3} + \sqrt{15}}{2}, \dfrac{-\sqrt{3} - \sqrt{15}}{2}\right\}$
40. no real roots
41. $\left\{\dfrac{1 + \sqrt{6}}{5k}, \dfrac{1 - \sqrt{6}}{5k}\right\}$, if $k \neq 0$
42. $\{-k, 4k\}$
43. $\left\{\dfrac{-b + \sqrt{b^2 - 4c}}{2}, \dfrac{-b - \sqrt{b^2 - 4c}}{2}\right\}$, if $b^2 - 4c \geq 0$
44. $\left\{\dfrac{-b + \sqrt{b^2 - 4ac}}{2a}, \dfrac{-b - \sqrt{b^2 - 4ac}}{2a}\right\}$, if $b^2 - 4ac \geq 0$

Computer Exercises For students with computer experience

1. Write a program that will use the process of completing the square to rewrite an equation of the form $x^2 + bx = c$ in the form $(x - d)^2 = e$ when you input values for b and c. A sample output would be

$$(X - 3) \uparrow 2 = 7.$$

2. Modify the program you wrote for Exercise 1 so that it will rewrite an equation of the form $ax^2 + bx = c$ in the form $a(x - d)^2 = e$ when you input values for a, b, and c, $a \neq 0$.

Teaching Suggestions
p. T108

Key Ideas

Use the quadratic formula to solve quadratic equations.

Chalkboard Examples

1. Use the quadratic formula to solve $3m^2 - 2m = 1$.
   Using the quadratic formula with $a = 3$, $b = -2$, $c = -1$, $m =$
   $$\frac{-(-2) \pm \sqrt{(-2)^2 - 4(3)(-1)}}{2(3)}$$
   $$= \frac{2 \pm \sqrt{16}}{6} = \frac{2 \pm 4}{6}$$
   $m = 1$ or $m = -\dfrac{1}{3}$

   the solution set is $\left\{1, -\dfrac{1}{3}\right\}$.

2. Use the discriminant to determine the number of real roots of the equation $x^2 + 3 = -6x$.
   $x^2 + 6x + 3 = 0$
   $b^2 - 4ac = (6)^2 - 4(1)(3)$
   $\qquad = 24$
   Since $24 > 0$, the equation has two real roots.

11–2 The Quadratic Formula

Quadratic equations arise so frequently in mathematics that it is useful to have a formula that you can use to obtain their solutions directly from the coefficients. Such a formula can be derived by applying the method of completing the square to the general form of a quadratic equation, that is, $ax^2 + bx + c = 0$, $a \neq 0$.

$$ax^2 + bx + c = 0$$

$$x^2 + \frac{b}{a}x + \frac{c}{a} = 0$$

$$x^2 + \frac{b}{a}x = -\frac{c}{a}$$

$$x^2 + \frac{b}{a}x + \left(\frac{b}{2a}\right)^2 = -\frac{c}{a} + \left(\frac{b}{2a}\right)^2 \quad \longleftarrow \quad \{\text{Complete the square}$$

$$\left(x + \frac{b}{2a}\right)^2 = -\frac{c}{a} + \frac{b^2}{4a^2}$$

$$\left(x + \frac{b}{2a}\right)^2 = \frac{b^2 - 4ac}{4a^2}$$

$$x + \frac{b}{2a} = \pm\sqrt{\frac{b^2 - 4ac}{4a^2}} \quad \longleftarrow \quad \{\text{ If } b^2 - 4ac \geq 0$$

$$x = -\frac{b}{2a} \pm \sqrt{\frac{b^2 - 4ac}{4a^2}}$$

$$x = -\frac{b}{2a} \pm \frac{\sqrt{b^2 - 4ac}}{2a}$$

$$x = \frac{-b \pm \sqrt{b^2 - 4ac}}{2a}$$

This open sentence is called the **quadratic formula.**

548 *Chapter 11*

The Quadratic Formula

If $ax^2 + bx + c = 0$, $a \neq 0$, and $b^2 - 4ac \geq 0$, then

$$x = \frac{-b \pm \sqrt{b^2 - 4ac}}{2a}.$$

EXAMPLE 1 Solve $2y^2 - 6y = -3$ using the quadratic formula.

SOLUTION First rewrite the given equation in the form $ax^2 + bx + c = 0$.

$$2y^2 - 6y = -3$$
$$2y^2 - 6y + 3 = 0$$

Then apply the formula $y = \frac{-b \pm \sqrt{b^2 - 4ac}}{2a}$, with $a = 2$, $b = -6$, $c = 3$.

$$y = \frac{-(-6) \pm \sqrt{(-6)^2 - 4(2)(3)}}{2(2)}$$

$$= \frac{6 \pm \sqrt{36 - 24}}{4}$$

$$= \frac{6 \pm \sqrt{12}}{4}$$

$$= \frac{6 \pm 2\sqrt{3}}{4}$$

$$= \frac{3 \pm \sqrt{3}}{2}$$

Thus, $y = \frac{3 + \sqrt{3}}{2}$ or $y = \frac{3 - \sqrt{3}}{2}$.

Checking the results is left to you.

∴ the solution set is $\left\{ \frac{3 + \sqrt{3}}{2}, \frac{3 - \sqrt{3}}{2} \right\}$.

Notice the role of the value of $b^2 - 4ac$ in the quadratic formula.

1. When $b^2 - 4ac > 0$, $\sqrt{b^2 - 4ac}$ is a positive real number and the quadratic formula gives *two different* real roots of the equation.

2. When $b^2 - 4ac = 0$, $\sqrt{b^2 - 4ac} = 0$ and the quadratic formula gives *only one* root, called a **double root.**
$$\frac{-b \pm 0}{2a} = \frac{-b}{2a}$$

3. When $b^2 - 4ac < 0$, there is *no* real value for $\sqrt{b^2 - 4ac}$ and, hence, no real root.

Because the value of $b^2 - 4ac$ distinguishes, or *discriminates*, the three possibilities, it is called the **discriminant** of the quadratic equation.

Quadratic Equations and Functions   **549**

Additional A Exercises

Use the quadratic formula to solve each equation. Express all radicals in simplified form. State the exact solution and then approximate each irrational root to the nearest hundredth. If the equation has no real roots, write "none."

1. $x^2 + 3x = 2$

$$\left\{ \frac{-3 + \sqrt{17}}{2}, \frac{-3 - \sqrt{17}}{2} \right\}$$
$$\{0.56, -3.56\}$$

2. $2x^2 + x - 6 = 0$

$$\left\{ \frac{3}{2}, -2 \right\}$$

3. $3x^2 = 5x - 2$

$$\left\{ \frac{2}{3}, 1 \right\}$$

Use the discriminant to determine the number of real roots of each equation.

4. $10x + 41 = x^2$ two real roots

5. $3x^2 - 4x - 2 = 0$ two real roots

6. $2x^2 + 3x + 5 = 0$ no real roots

EXAMPLE 2 Use the discriminant to determine the number of real roots of the equation $9x^2 = 12x - 5$.

EXAMPLE 2 Use the discriminant to determine the number of real roots of the equation $9x^2 = 12x - 5$.

SOLUTION First rewrite the given equation in the form $ax^2 + bx + c = 0$.

$$9x^2 = 12x - 5$$
$$9x^2 - 12x + 5 = 0$$

Then evaluate $b^2 - 4ac$, using $a = 9$, $b = -12$, and $c = 5$.

$$b^2 - 4ac = (-12)^2 - 4(9)(5)$$
$$= 144 - 180$$
$$= -36$$

Since $-36 < 0$, the equation has no real roots.

Oral Exercises Answers may vary. Likely choices are given.

Give the values that you would use for a, b, and c in the quadratic formula.

1. $2u^2 - 3u + 1 = 0$
2. $5r^2 + r - 4 = 0$ 5; 1; -4
3. $x^2 - x - 6 = 0$ 1; -1; -6
4. $3x^2 + 5x = 2$ 3; 5; -2
5. $s^2 + 5 = 3s$ 1; -3; 5
6. $y^2 = 7y - 4$ 1; -7; 4
7. $4m^2 - 7m = -1$
8. $x^2 + 13 = 4x$ 1; -4; 13
9. $2w^2 = 5$ 2; 0; -5
10. $6 - 3x^2 = 0$ -3; 0; 6
11. $3q^2 = -4q$ 3; 4; 0
12. $6z^2 = z$ 6; -1; 0
13. $2x^2 - \frac{1}{5}x - \frac{11}{5} = 0$
14. $\frac{2}{3}x^2 + 4x + 5 = 0$
15. $\frac{e^2}{5} + 6e = 1$
16. $\frac{3f^2}{5} + 4 = 2f$
17. $9x^2 - x\sqrt{2} + 3 = 0$
18. $4y^2 - y\sqrt{13} - 2 = 0$

$\frac{3}{5}$; -2; 4 or 3; -10; 20   9; $-\sqrt{2}$; 3   4; $-\sqrt{13}$; -2

Written Exercises

Solve each equation using the quadratic formula. First give the exact solution, expressing any irrational roots in simplest form. Then approximate irrational roots to the nearest hundredth. If the equation has no real roots, so state.

A
1. $r^2 - 4r - 5 = 0$
2. $2s^2 - 8s + 6 = 0$
3. $x^2 + 3x = 4$
4. $9y^2 + 2y = -4$
5. $3a^2 - 10a = 8$
6. $4b^2 + 4b - 3 = 0$
7. $j^2 + 2j = 1$
8. $k^2 + 6k + 4 = 0$
9. $7m^2 - 6m + 1 = 0$
10. $n^2 + 5n + 2 = 0$
11. $8p^2 - 2p = -1$
12. $2q^2 + 8q = -5$

Use the discriminant to determine the number of real roots of each equation.

13. $m^2 + 4m - 4 = 0$ 2
14. $2n^2 - 5n - 1 = 0$ 2
15. $3x^2 - 6x + 30 = 0$ 0
16. $x^2 - 4x + 5 = 0$ 0
17. $v^2 = 4v - 1$ 2
18. $4x^2 + 12x = -9$ 1

Suggested Assignments

Minimum
First day: 550/1–12
Second day: 550/13–24
Average
  550/1–35 odd
Maximum
  550/1–41 odd, 43–48

Additional Answers
Oral Exercises

Answers may vary.
Likely choices are given.

1. 2; −3; 1
7. 4; −7; 1
13. 2; $-\frac{1}{5}$; $-\frac{11}{5}$ or 10; −1; −11
14. $\frac{2}{3}$; 4; 5 or 2; 12; 15
15. $\frac{1}{5}$; 6; −1 or 1; 30; −5

Additional Answers
Written Exercises

Approximations may vary.

1. {5, −1}
2. {3, 1}
3. {−4, 1}
4. no real roots
5. $\{4, -\frac{2}{3}\}$
6. $\{-\frac{3}{2}, \frac{1}{2}\}$
7. $\{-1 + \sqrt{2}, -1 - \sqrt{2}\}$; {0.41, −2.41}
8. $\{-3 + \sqrt{5}, -3 - \sqrt{5}\}$; {−0.76, −5.24}
9. $\{\frac{3 + \sqrt{2}}{7}, \frac{3 - \sqrt{2}}{7}\}$; {0.63, 0.23}
10. $\{\frac{-5 + \sqrt{17}}{2}, \frac{-5 - \sqrt{17}}{2}\}$; {−0.44, −4.56}
11. no real roots

19. $16x^2 + 1 = 8x$ 1
20. $2t^2 = 5t - 3$ 2
21. $5u^2 + 6u + 2 = 0$ 0
22. $6x^2 - 4x = -1$ 0
23. $15 - 3y^2 = 0$ 2
24. $6p + 2 = -5p^2$ 0

Solve each equation using the quadratic formula. Give only the exact solutions, expressing any irrational roots in simplest form.

B 25. $x^2 + \dfrac{5}{2}x - \dfrac{3}{2} = 0$
26. $z^2 - \dfrac{1}{2}z - \dfrac{15}{16} = 0$
27. $\dfrac{m^2}{3} - 2m + \dfrac{5}{3} = 0$

28. $3y^2 - \dfrac{2}{5}y - \dfrac{1}{5} = 0$
29. $4x(x - 2) = 5(x - 1)$
30. $x(2x - 1) - 3(x + 4) = 0$

31. $\dfrac{3}{m^2} - \dfrac{8}{m} + 4 = 0$
32. $\dfrac{4z + 9}{z} + \dfrac{4}{z^2} = 0$
33. $\dfrac{3}{e - 2} + \dfrac{1}{e + 2} = 1$

34. $\dfrac{2}{n - 3} + \dfrac{5}{n + 3} = 1$
35. $\dfrac{6}{y + 1} = 1 - \dfrac{4}{y - 5}$
36. $\dfrac{5}{w - 2} - 2 = \dfrac{-1}{w + 3}$

C 37. $h^2 - 3 = h\sqrt{6}$
38. $r^2 + r\sqrt{3} = 3$

39. $v^2 - 2v\sqrt{5} = 3$
40. $x^2\sqrt{2} - x - 2\sqrt{2} = 0$

41. $2(c + 3)^2 = 5(c + 2)^2$
42. $(t + 1)^2 - 2(t - 2)^2 = 8$

43. Find a value of k so that the equation $kx^2 - 24x + 16 = 0$ has one real root. 9

44. Find a value of k so that the equation $25x^2 + kx + 4 = 0$ has one real root. Is there more than one such value of k? 20 or -20; Yes

Given the equation $ax^2 + bx + c = 0$ with $a \neq 0$ and $b^2 - 4ac > 0$, show that the following statements are true.

45. The sum of the roots of the equation is $-\dfrac{b}{a}$.

46. The product of the roots of the equation is $\dfrac{c}{a}$.

47. Determine a quadratic equation whose roots are $2 - \sqrt{5}$ and $2 + \sqrt{5}$. (*Hint:* Use the results of Exercises 45 and 46.) $x^2 - 4x - 1 = 0$

48. Determine a quadratic equation whose roots are $-3 \pm 2\sqrt{2}$. $x^2 + 6x + 1 = 0$

Computer Exercises For students with computer experience

1. Write a program that will allow you to input the coefficients a, b, and c of a quadratic equation and will determine whether the equation has two different real roots, one real root (a double root), or no real roots.

2. Modify the program that you wrote for Exercise 1 so that, if the equation has at least one real root, the program will output the values of the two different real roots or of the one double real root. If the equation has no real roots, the output should so state.

Quadratic Equations and Functions **551**

---

12. $\left\{\dfrac{-4 + \sqrt{6}}{2}, \dfrac{-4 - \sqrt{6}}{2}\right\}$; $\{-0.78, -3.22\}$

25. $\left\{-3, \dfrac{1}{2}\right\}$

26. $\left\{\dfrac{5}{4}, -\dfrac{3}{4}\right\}$

27. $\{1, 5\}$

28. $\left\{\dfrac{1}{3}, -\dfrac{1}{5}\right\}$

29. $\left\{\dfrac{13 + \sqrt{89}}{8}, \dfrac{13 - \sqrt{89}}{8}\right\}$

30. $\{1 + \sqrt{7}, 1 - \sqrt{7}\}$

31. $\left\{\dfrac{1}{2}, \dfrac{3}{2}\right\}$

32. $\left\{\dfrac{-9 + \sqrt{17}}{8}, \dfrac{-9 - \sqrt{17}}{8}\right\}$

33. $\{2 + 2\sqrt{3}, 2 - 2\sqrt{3}\}$

34. $\{0, 7\}$

35. $\{7 + 2\sqrt{7}, 7 - 2\sqrt{7}\}$

36. $\left\{\dfrac{2 + 3\sqrt{6}}{2}, \dfrac{2 - 3\sqrt{6}}{2}\right\}$

37. $\left\{\dfrac{\sqrt{6} + 3\sqrt{2}}{2}, \dfrac{\sqrt{6} - 3\sqrt{2}}{2}\right\}$

38. $\left\{\dfrac{-\sqrt{3} + \sqrt{15}}{2}, \dfrac{-\sqrt{3} - \sqrt{15}}{2}\right\}$

39. $\{\sqrt{5} + 2\sqrt{2}, \sqrt{5} - 2\sqrt{2}\}$

40. $\left\{\dfrac{\sqrt{2} + \sqrt{34}}{4}, \dfrac{\sqrt{2} - \sqrt{34}}{4}\right\}$

(continued on p. 571)

Mixed Review

Solve by completing the square.

1. $r^2 - 8r = 20$ $\{10, -2\}$
2. $y^2 - 10y = -9$ $\{1, 9\}$
3. $a^2 + 8a = 6$ $\{-4 + \sqrt{22}, -4 - \sqrt{22}\}$
4. $a^2 - a = \dfrac{3}{4}$ $\left\{1\dfrac{1}{2}, -\dfrac{1}{2}\right\}$

Solve.

5. $\dfrac{x}{3} - \dfrac{2x}{5} = -\dfrac{1}{10}$ $\left\{\dfrac{3}{2}\right\}$
6. $\dfrac{2}{x^2 - 1} - \dfrac{3}{2x + 2} = 0$ $\left\{\dfrac{7}{3}\right\}$

Supplementary Material
Test 43

Key Ideas

Solve word problems by using the quadratic formula.

Chalkboard Examples

Solve.

1. The sum of two numbers is 24 and their product is 135. What are the two numbers?
Let x = the first number. Then $24 - x$ = the second number.
$x(24 - x) = 135$
$24x - x^2 = 135$
$x^2 - 24x + (12)^2 =$
$-135 + (12)^2$
$(x - 12)^2 = 9$
$x - 12 = 3$ or $x - 12 = -3$
$x = 9$ or $x = 15$
$24 - x = 15$ or $24 - x = 9$
∴ the numbers are 15 and 9.

2. When a certain number is multiplied by itself, then increased by eight times itself, the result is negative five. What is the number?
Let y = the number.
$y^2 + 8y = -5$
$y^2 + 8y + 5 = 0$
$y = \dfrac{-(8) \pm \sqrt{(8)^2 - 4(1)(5)}}{2(1)} =$
$\dfrac{-8 \pm \sqrt{44}}{2} = \dfrac{-8 \pm 2\sqrt{11}}{2}$
$y = -4 \pm \sqrt{11}$
∴ the number is either $-4 + \sqrt{11}$ or $-4 - \sqrt{11}$.

11–3 Using Quadratic Equations to Solve Problems

Many problems involve solving quadratic equations.

EXAMPLE 1 One number is ten more than a second number. If the product of the two numbers is 144, find the numbers.

SOLUTION

Step 1   The problem asks for two numbers whose difference is 10 and whose product is 144.

Step 2   Let x = the first number. Then:
$x - 10$ = the second number

Step 3   The product of the numbers is 144.
$$x(x - 10) = 144$$

Step 4
$$x(x - 10) = 144$$
$$x^2 - 10x = 144$$
$$x^2 - 10x + (-5)^2 = 144 + (-5)^2$$
$$(x - 5)^2 = 169$$
$$x - 5 = \sqrt{169} \quad \text{or} \quad x - 5 = -\sqrt{169}$$
$$x - 5 = 13 \quad \mid \quad x - 5 = -13$$
$$x = 18 \quad \text{or} \quad x = -8$$
$$x - 10 = 8 \quad \mid \quad x - 10 = -18$$

Since the coefficient of x is an even number, completing the square is a convenient method to use in solving the equation.

Step 5   Checking the results is left to you.
∴ the numbers are either 18 and 8 or -8 and -18.

EXAMPLE 2 A landscaper designed a rectangular grass plot with length 3 m less than four times its width. If the landscaper purchased 51 m² of sod to use for the plot, what is the length of the plot?

SOLUTION

Step 1   The problem asks for the length of a rectangular grass plot.
The length is 3 m less than four times the width.
The area is 51 m².

Step 2   Let w = the width of the plot. Then:
$4w - 3$ = the length of the plot

Step 3   The area of the plot is 51 m².
$$w(4w - 3) = 51$$

552 *Chapter 11*

Step 4
$$w(4w - 3) = 51$$
$$4w^2 - 3w - 51 = 0$$

$$w = \frac{-(-3) \pm \sqrt{(-3)^2 - 4(4)(-51)}}{2(4)}$$

Since no other method of solution seems easier, use the quadratic formula.

$$= \frac{3 \pm \sqrt{9 + 816}}{8} = \frac{3 \pm \sqrt{825}}{8} = \frac{3 \pm 5\sqrt{33}}{8}$$

$$w = \frac{3 + 5\sqrt{33}}{8} \approx 3.97 \text{ or } w = \frac{3 - 5\sqrt{33}}{8} \approx -3.22$$

Since length cannot be negative, reject the second root.

$$w \approx 3.97$$
$$4w - 3 \approx 4(3.97) - 3 = 12.88$$

Step 5   *Check:* $(3.97)(12.88) \stackrel{?}{=} 51$

$$51.1336 \approx 51 \quad \checkmark$$

∴ the length of the grass plot is approximately 12.88 m.

EXAMPLE 3   The total cost of a skiing trip was to be $960, to be shared equally by each of the members of the Outdoor Club. At the last minute, ten people decided not to go on the trip, thus raising the cost to each of the other members by $16. How many members actually went on the trip?

SOLUTION

Step 1   The problem asks for the number of members who went on the ski trip.

Step 2   Let x = the number of members actually on the trip. Then:
$x + 10$ = the number of members in the club

Step 3   The actual cost to each member going on the trip is $16 more than the originally planned cost.

$$\frac{960}{x} = \frac{960}{x + 10} + 16$$

Step 4
$$\frac{960}{x} = \frac{960}{x + 10} + 16$$
$$960(x + 10) = 960x + 16x(x + 10)$$
$$960x + 9600 = 960x + 16x^2 + 160x$$
$$0 = 16x^2 + 160x - 9600$$
$$0 = x^2 + 10x - 600$$
$$0 = (x + 30)(x - 20)$$

Since the factors of $x^2 + 10x - 600$ can be seen readily, factoring is a convenient method of solving the equation.

$x + 30 = 0$ or $x - 20 = 0$
$x = -30$ $x = 20$

Since the number of people cannot be negative, reject the first root.

Step 5   Checking the results is left to you.

∴ twenty members of the club actually went on the ski trip.

3. A painting is 20 cm wider than it is high. Its area is 2400 cm². Find the height and width of the painting.
Let w = the width of the painting. Then $w - 20$ = the height of the painting.
$w(w - 20) = 2400$
$w^2 - 20w = 2400$
$w^2 - 20w - 2400 = 0$
$(w + 40)(w - 60) = 0$
$w = -40$ or $w = 60$
Since lengths cannot be negative, reject the first root.
$w = 60$
$w - 20 = 60 - 20 = 40$
∴ the width of the painting is 60 cm and the height is 40 cm.

Quadratic Equations and Functions   **553**

Suggested Assignments

Minimum
554/P: 1–10
R 556/*Self-Test 1*
Average
554/P: 1–14
R 556/*Self-Test 1*
Maximum
554/P: 8–17
R 556/*Self–Test 1*

Additional A Exercises

Solve. Approximate irrational roots to the nearest hundredth. Reject impossible roots.

1. The difference between a number and twice its reciprocal is $\frac{17}{10}$. What is the number?
$\frac{5}{2}$ or $-\frac{4}{5}$

2. The length and width of a rectangle are consecutive even integers. The area of the rectangle is 24 cm. Find the length and width. length: 6 cm; width: 4 cm

3. One number is 8 more than another number. Their product is 273. Find the numbers.
13 and 21, or −21 and −13

4. The square of a negative number plus six times the number is 27. What is the number? −9

5. The sum of a number and its reciprocal is $\frac{13}{6}$. What is the number? $\frac{3}{2}$ or $\frac{2}{3}$

Oral Exercises
a = factoring, b = completing the square, c = using the quadratic formula

Tell whether you would solve each equation by factoring, completing the square, or using the quadratic formula. Answers may vary.

1. $2x^2 - 6x = 0$ a
2. $x^2 - 5x + 5 = 0$ c
3. $x^2 + 12x - 8 = 0$ b
4. $32x^2 = 64$ a
5. $5x^2 - 7x + 2 = 0$ a
6. $x^2 - 2x + 15 = 0$ b
7. $9x^2 - 180 = 0$ a
8. $2x^2 - 3x - 2 = 0$ a
9. $7x^2 + 2x - 8 = 0$ c
10. $x^2 + 18x - 24 = 0$ b
11. $x^2 - 10x + 3 = 0$ b
12. $15x^2 - 7x + 1 = 0$ c

Problems

5. width: 2.21 m; length: 10.85 m
6. width: 4 m; length: 9 m

Solve. Approximate irrational roots to the nearest hundredth. Reject impossible roots. Approximations may vary.

A

1. A bulletin board is 30 cm longer than it is high. Its area is 2800 cm². Find the length and height of the bulletin board. 70 cm; 40 cm

2. A painting is 6 cm longer than it is high. Its area is 691 cm². Find the length and height of the painting. 29.46 cm; 23.46 cm

3. The sum of a number and its reciprocal is $\frac{25}{12}$. Find the number. $\frac{4}{3}$ or $\frac{3}{4}$

4. The difference between a number and twice its reciprocal is $\frac{1}{6}$. Find the number. $\frac{3}{2}$ or $-\frac{4}{3}$

5. If the area of the floor of an office is 24 m² and the length of the floor is 2 m longer than four times its width, find the dimensions of the floor.

6. The area of the Johnsons' dining room floor is 36 m². If the length of the floor is 1 m longer than twice its width, find the dimensions of the floor.

7. A tunnel through a mountain has a vertical ventilation shaft whose outside opening is 17 km up the slope of the mountain from the mouth of the tunnel. The distance from the base of the shaft to the mouth of the tunnel is 7 km more than the height of the ventilation shaft. Find the height of the shaft and its distance from the mouth of the tunnel. 8 km; 15 km

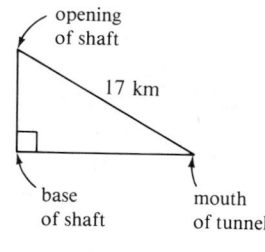

Ex. 7

8. Town A is directly north of town B and town C is directly east of town B. The distance between towns A and C is 100 km. If the distance from town A to town B is 20 km less than the distance from town C to town B, find the distance between towns A and B and the distance between towns C and B. 60 km; 80 km

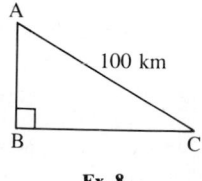

Ex. 8

554 *Chapter 11*

B 9. Kedra earned $192 from the sale of flower arrangements that she had made. If she had charged $8 more apiece, she could have sold two fewer arrangements and still have earned the same amount. How many arrangements did Kedra sell? 8

10. Pamela and Anita can complete a job in 6 h if they work together. Working alone, Pamela takes 4 h longer to complete the job than Anita does when she works alone. How long would it take each woman to complete the job working alone?

11. Jose rowed his fishing boat 4 km up a stream and then returned the same distance. The current in the stream was flowing at a speed of 3 km/h. If the round trip took Jose 1 h 20 min, find the speed of his boat in still water. 7.24 km/h **10.** Anita: 10.32 h; Pamela: 14.32 h

12. The members of the Historical Society sold a number of tickets to the Easton town picnic and raised $240 from the sale. If the Society had lowered its price by $2 a ticket, the members would have needed to sell 20 more tickets to earn the $240. How many tickets did the Society members actually sell? 40

13. Ron takes 2 h less than his brother Russ to cut enough firewood to fill his family's wood bin. Working together, they can fill the bin in $4\frac{4}{9}$ h. How long would it take each of them to cut the wood alone? Ron: 8 h; Russ: 10 h

14. The average speed in still air of a certain plane is 500 km/h. With a constant wind blowing, the plane takes $4\frac{1}{6}$ h to fly 1040 km into the wind and then return the same distance. What is the speed of the wind? 20 km/h

C 15. Lauren bought a certain number of shares of stock in the XYZ Company for $405. As soon as the price of a share in the company increased by $10, she sold all but 16 shares and received $483. How many shares did she buy? 40

16. An open box was formed from a rectangular sheet of metal by cutting a square of side 4 cm from each of the corners of the rectangular sheet and then folding up the edges. The length of the original sheet of metal was 3 cm less than twice its width, and the volume of the box was 532 cm³. Find the dimensions of the sheet of metal. width: 15 cm; length: 27 cm

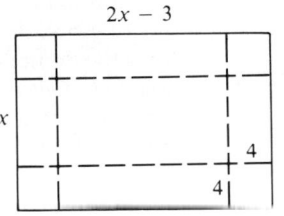

$2x - 3$

x

4

4

Ex. 16

17. A car starts traveling due east along a road. At the same time and from the same point, a second car starts traveling due north at a speed that is 20 km/h faster than that of the first car. After 2 h, the cars are 200 km apart. At what speeds are they traveling? 60 km/h; 80 km/h

6. Kris and Julie live 38 km apart. If Julie drives directly south and Kris drives directly west they will meet. If Julie drives 4 km more than Kris, how far would each drive?
Julie: 28.80 km
Kris: 24.80 km

Mixed Review

Solve, using the quadratic formula.

1. $x^2 - 5x - 3 = 0$
$\left\{ \dfrac{5 + \sqrt{37}}{2}, \dfrac{5 - \sqrt{37}}{2} \right\}$

2. $1 - 3x = x^2$
$\left\{ \dfrac{-3 + \sqrt{13}}{2}, \dfrac{-3 - \sqrt{13}}{2} \right\}$

3. $4x^2 - 6x - 3 = 0$
$\left\{ \dfrac{3 + \sqrt{21}}{4}, \dfrac{3 - \sqrt{21}}{4} \right\}$

4. $2x^2 - 3x = 5$ $\left\{ \dfrac{5}{2}, -1 \right\}$

Solve.

5. $\dfrac{3}{x} + \dfrac{3}{x-1} = 2$
$\left\{ \dfrac{4 + \sqrt{10}}{2}, \dfrac{4 - \sqrt{10}}{2} \right\}$

6. $\dfrac{1}{x^2 - 3x + 2} - \dfrac{5}{x^2 - 4} = \dfrac{1}{x + 2}$
$\left\{ \dfrac{-1 + \sqrt{21}}{2}, \dfrac{-1 - \sqrt{21}}{2} \right\}$

Quadratic Equations and Functions **555**

Solve by completing the square.

1. $n^2 - 16n = 36$ $\{18, -2\}$

2. $a^2 + 10a + 3 = 0$
$\{-5 + \sqrt{22}, -5 - \sqrt{22}\}$

Solve by using the quadratic formula.

3. $15x^2 = 8 + 14x$ $\left\{\dfrac{4}{3}, -\dfrac{2}{5}\right\}$

4. $3n^2 - 5n + 1 = 0$
$\left\{\dfrac{5 + \sqrt{13}}{6}, \dfrac{5 - \sqrt{13}}{6}\right\}$

5. The sum of the squares of two consecutive even integers is equal to 164. Find the integers.
8 and 10 or −8 and −10

Self-Test 1

VOCABULARY   completing the square (p. 545)
quadratic formula (p. 548)
double root (p. 549)
discriminant of a quadratic equation (p. 549)

Solve by completing the square.   **3.** $\left\{\dfrac{-5 + \sqrt{17}}{4}, \dfrac{-5 - \sqrt{17}}{4}\right\}$ **4.** $\left\{\dfrac{2 + \sqrt{10}}{3}, \dfrac{2 - \sqrt{10}}{3}\right\}$

1. $x^2 + 4x - 3 = 0$ **2.** $y^2 = 6y - 1$ *Obj. 1, p. 543*
    $\{-2 + \sqrt{7}, -2 - \sqrt{7}\}$ $\{3 + 2\sqrt{2}, 3 - 2\sqrt{2}\}$

Solve by using the quadratic formula.

3. $2z^2 + 5z + 1 = 0$ **4.** $3w^2 - 2 = 4w$ *Obj. 2, p. 543*

5. The perimeter of a rectangle is 40 cm. If the area of the rectangle is   *Obj. 3, p. 543*
96 cm², find the width and length of the rectangle. 8 cm; 12 cm

Check your answers with those at the back of the book.

Chu Shih-chieh
13th–14th centuries

Chu Shih-chieh was one of the greatest Chinese mathematicians, yet little is known of his personal life. He is thought to have lived in Yen-shan, near Beijing, and to have traveled throughout the country as a wandering teacher for twenty years.

Chu is the author of *Suan-hsüeh ch'i-meng (Introduction to Mathematical Studies)*, written in 1299. This work was used for centuries in East Asia as a textbook for beginners. Another of Chu's studies, *Ssu-yüan yü-chien (Precise Mirror of the Four Elements)*, written in 1303, marked the height of the development of algebra in China. In this text Chu demonstrated a method for representing up to four unknowns in one algebraic equation. Also in this work Chu developed a method for finding rational square roots. Chu Shih-chieh made substantial contributions as well to the theory of series and the method of finite differences.

Graphing Quadratic and Other Polynomial Functions

OBJECTIVES for Sections 11-4 through 11-6:
1. *To graph quadratic functions.*
2. *To graph simple polynomial functions.*
3. *To graph quadratic inequalities.*

11–4 Quadratic Functions

Teaching Suggestions
p. T109

Related Activities p. T109

Key Ideas

Graph quadratic functions.
Estimate zeros of quadratic
functions from their graphs.

Any function of the type

$$f: x \longrightarrow ax^2 + bx + c, \quad a \neq 0,$$

is called a **quadratic function.** If the domain of such a function is $\mathcal{R}$, then its graph on a coordinate plane is a smooth curve called a **parabola.**

The simplest quadratic function is $f: x \longrightarrow x^2$, whose graph is specified by the equation $y = x^2$. Figure 1 shows the curve, together with the table of values used to construct it.

x	y
-3	9
-2	4
-1	1
0	0
1	1
2	4
3	9

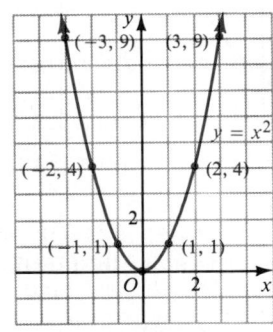

Figure 1

From Figure 1 you can see that the origin is the lowest point, or *minimum point*, on the graph of $y = x^2$. In general, a point on a curve is said to be a **minimum point** of the curve if its y-coordinate is less than or equal to the y-coordinate of every other point on the curve.

Recall from Chapter 5 that any member of the range of a function is called a *value* of the function. If the point (j, k) is a minimum point of the graph of a function, then k is called the **minimum value** of the function. The definitions for a **maximum point,** or highest point, and the **maximum value** are similar.

Quadratic Equations and Functions **557**

1. Graph $f: x \rightarrow x^2 + 4x - 5$ and give the zeros of the function, if any.
   Set $y = x^2 + 4x - 5$ and compare this equation with $y = ax^2 + bx + c$. Since $a = 1$ and $1 > 0$, the parabola opens upward.

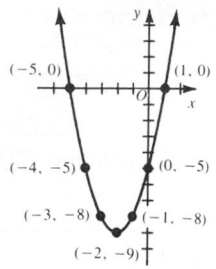

   The zeros are -5 and 1.

2. Graph $g: x \rightarrow -x^2 + 2x - 1$ and give the zeros of the function, if any.
   Set $y = -x^2 + 2x - 1$ and compare this equation with $y = ax^2 + bx + c$. Since $a = -1$ and $-1 < 0$, the parabola opens downward.

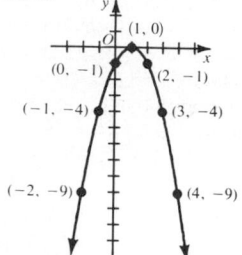

   The zero is 1.

3. a. Find the axis of symmetry for the parabola in Example 2.
   $$x = \frac{-2}{2(-1)} = 1$$
   ∴ the axis of symmetry is $x = 1$.

   b. Tell whether the graph in Example 2 has a maximum or a minimum point and name that point. max.; $(1, 0)$

It can be proved that each quadratic function with domain $\mathcal{R}$ has one maximum point or one minimum point, but not both. This maximum or minimum point is called the **vertex** of the parabola. Correspondingly, it can be proved that each quadratic function with domain $\mathcal{R}$ has one maximum value or one minimum value, but not both.

Now imagine that the graph of $y = x^2$ that is shown in Figure 1 could be folded along the y-axis. Do you see that the part of the curve which lies in the first quadrant would then *coincide* with the part of the curve that lies in the second quadrant? This illustrates the fact that, given any point (u, t) that lies on the parabola, the point $(-u, t)$ also lies on the parabola. This is confirmed by the fact that, if $t = u^2$, then $t = (-u)^2$ also.

Since it divides the curve $y = x^2$ into two matching parts, the y-axis is called the **axis of symmetry** of the curve, and the curve is said to be *symmetric with respect to the y-axis*. It can be proved that the graph of any quadratic function with domain $\mathcal{R}$ has an axis of symmetry.

Several other quadratic functions are graphed in Figure 2.

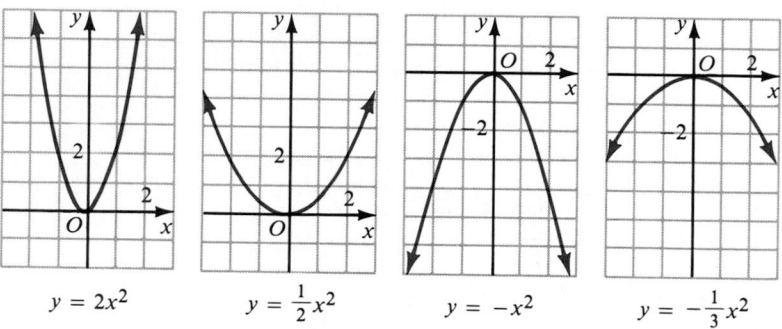

$$y = 2x^2 \qquad y = \frac{1}{2}x^2 \qquad y = -x^2 \qquad y = -\frac{1}{3}x^2$$

Figure 2

As you can see, the equations specifying the functions graphed in Figure 2 are of the form

$$y = ax^2,$$

where a is 2, $\frac{1}{2}$, -1, and $-\frac{1}{3}$, respectively. Notice that

> when $a > 0$, the parabola opens upward;
> when $a < 0$, the parabola opens downward.

Notice also that the less $|a|$ is, the broader the parabola appears to be when the same scale is used. These remarks about a also apply to the general quadratic function

$$f: x \rightarrow ax^2 + bx + c.$$

In Figure 2, the curve $y = -\frac{1}{3}x^2$ appears broader than the curve $y = 2x^2$. However, if the scale is changed suitably, all parabolas can be made to look alike, as suggested by Figure 3 on the following page.

558   *Chapter 11*

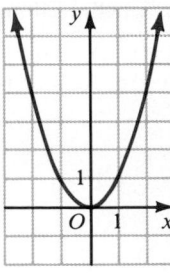 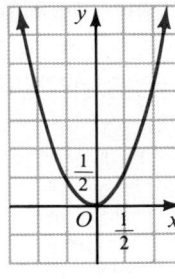 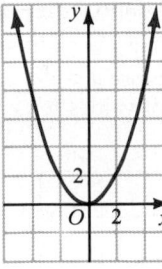

$$y = x^2 \qquad\qquad y = 2x^2 \qquad\qquad y = \tfrac{1}{2}x^2$$

Figure 3

Any member of the domain of a function for which the value of the function is 0 is called a **zero of the function**. For example, the zeros of the function

$$f: x \longrightarrow x^2 - 4x - 5$$

are 5 and −1 because, for these values of x,

$$x^2 - 4x - 5 = (x - 5)(x + 1) = 0.$$

EXAMPLE 1 Graph $f: x \longrightarrow x^2 + 2x - 3$ and give the zeros of the function, if any.

SOLUTION Set $y = x^2 + 2x - 3$ and compare this equation with $y = ax^2 + bx + c$. Since $a = 1$ and $1 > 0$, the parabola opens upward. Make a table of values for x and y, then sketch the graph.

x	y
−4	5
−3	0
−2	−3
−1	−4
0	−3
1	0
2	5

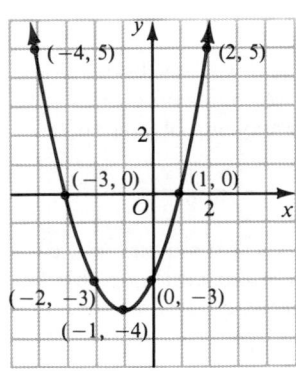

The graph intersects the x-axis at the points $(-3, 0)$ and $(1, 0)$.

∴ the function has two zeros, −3 and 1.

Notice that the zeros of a function are the x-coordinates of those points at which the graph of the function intersects the x-axis.

Quadratic Equations and Functions **559**

Reading Algebra

Have students read the beginning of Section 11-4 up to the first example and then write in their own words explanations of the following terms.

1. quadratic function
2. parabola
3. minimum point, minimum value
4. maximum point, maximum value
5. vertex
6. axis of symmetry
7. zero of the function

Students can then discuss their definitions with the rest of the group.

Common Errors

Many students make arithmetic errors in the substitution process when trying to determine points on the graph of a quadratic function. Students should be encouraged to use parentheses as a way to organize the information. For example, when substituting −1 in the function $f: x \rightarrow x^2 - 3x + 4$, write $y = (-1)^2 - 3(-1) + 4$.

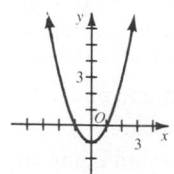
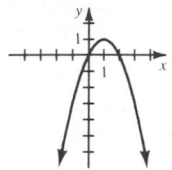
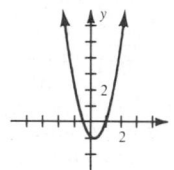
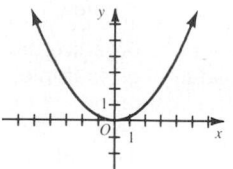
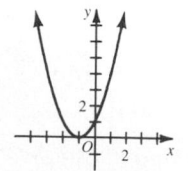
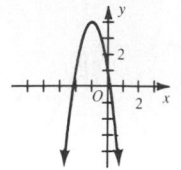
EXAMPLE 2 Graph $f: x \rightarrow 2x - x^2 + 3$ and give the zeros of the function, if any.

SOLUTION Set $y = 2x - x^2 + 3$ and compare this equation with $y = ax^2 + bx + c$. Since $a = -1$ and $-1 < 0$, the parabola opens downward. Make a table of values for x and y.

x	y
-2	-5
-1	0
0	3
1	4
2	3
3	0
4	-5

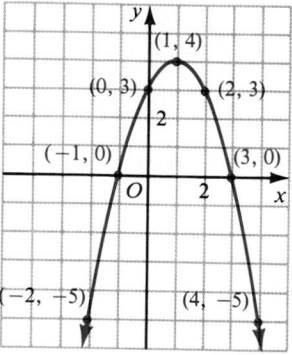

The graph intersects the x-axis at the points $(-1, 0)$ and $(3, 0)$.

$\therefore$ the zeros of the function are -1 and 3.

As you have seen, the zeros of a function are the values of x at the points where the graph of the function intersects the x-axis. Since $y = 0$ for every point on the x-axis, you can therefore find the zeros of the quadratic function specified by the equation

$$y = ax^2 + bx + c$$

by replacing y with 0 and solving the resulting quadratic equation.

Also, if you know two values of x, say x_1 and x_2, which are paired with the same value of y, then

$$x = \frac{x_1 + x_2}{2}$$

is the equation of the axis of symmetry, since the graphs of (x_1, y) and (x_2, y) are equidistant from this line. Thus, in Example 1 on the preceding page, note that $f(-2) = -3$ and $f(0) = -3$, and so the equation of the axis of symmetry is

$$x = \frac{-2 + 0}{2}, \quad \text{or} \quad x = -1,$$

as shown in Figure 4. The minimum point on this curve occurs then when $x = -1$. Thus, the vertex is at

$$(-1, -4).$$

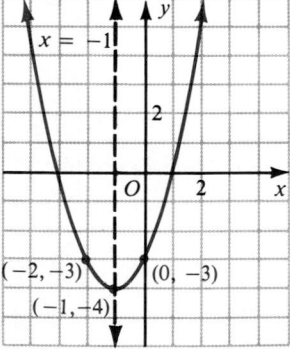

Figure 4

560 *Chapter 11*

Notice that the equation of the axis of symmetry was found by taking the average of the x-coordinates of two points on the curve. Remember that the zeros of the function occur when $ax^2 + bx + c = 0$, that is, when

$$x_1 = \frac{-b + \sqrt{b^2 - 4ac}}{2a} \quad \text{and} \quad x_2 = \frac{-b - \sqrt{b^2 - 4ac}}{2a}.$$

If you take the average of these x values, you obtain the equation

$$x = \frac{x_1 + x_2}{2} = -\frac{b}{2a}.$$

This formula gives the x-coordinate of the vertex of a parabola and the equation of the axis of symmetry of the parabola.

Given a parabola with equation $y = ax^2 + bx + c$, $a \neq 0$:

1. The x-coordinate of the vertex of the parabola is $-\dfrac{b}{2a}$.

2. The axis of symmetry of the parabola is the line $x = -\dfrac{b}{2a}$.

Once you have determined the x-coordinate of the vertex of a parabola, you can use this value to determine the maximum or minimum value of the quadratic function that is represented by the parabola.

EXAMPLE 3 Find the minimum value of $g: x \longrightarrow x^2 - 4x + 2$.

SOLUTION Find the x-coordinate of the vertex of the parabola that represents the function, using $a = 1$ and $b = -4$.

$$x = -\frac{b}{2a} = -\frac{-4}{2(1)} = -(-2) = 2$$

Substitute this value of x into the equation $y = x^2 - 4x + 2$ in order to find the y-coordinate of the vertex.

$$y = x^2 - 4x + 2 = (2)^2 - 4(2) + 2 = -2$$

$\therefore$ the minimum value of the function is -2.

Oral Exercises

u = upward, d = downward;
max. = maximum value, min. = minimum value

a. Tell whether the graph of each function opens upward or downward.

b. Tell whether the function has a maximum or a minimum value.

1. $f: x \longrightarrow 4x^2$ **a.** u   **b.** min.
2. $g: x \longrightarrow -\frac{1}{2}x$ **a.** d   **b.** max.
3. $h: x \longrightarrow 3x + x^2$ **a.** u   **b.** min.
4. $f: x \longrightarrow 5x - x^2$ **a.** d   **b.** max.
5. $g: x \longrightarrow -\frac{x^2}{4} + 2x$ **a.** d   **b.** max.
6. $f: x \longrightarrow 3x^2 + 6x - 1$ **a.** u   **b.** min.
7. $g: x \longrightarrow -(x - 1)^2$ **a.** d   **b.** max.
8. $h: x \longrightarrow -(2 - x)^2$ **a.** d   **b.** max.

Quadratic Equations and Functions   **561**

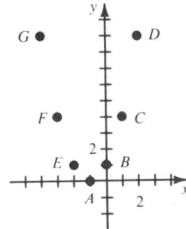
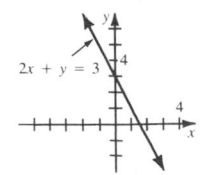

Additional Answers
Written Exercises

2.
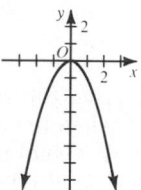

4.
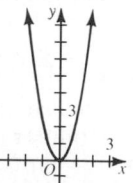

6.
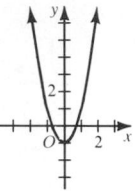

8.
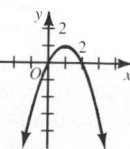

10.
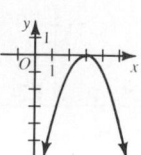

Estimate the zeros, if any, of the functions whose graphs are sketched.

9.
1
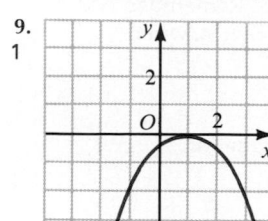

10.
1, 5
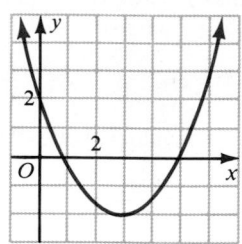

11.
none
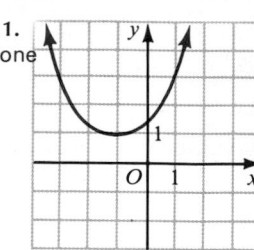

12.
$-2\frac{1}{2}$,
$2\frac{1}{2}$
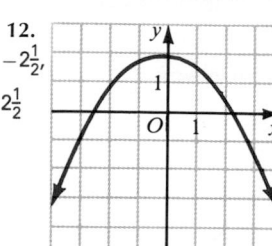

13.
none
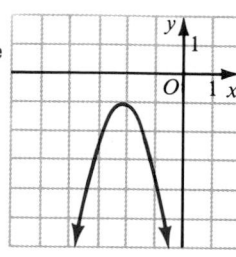

14.
$2\frac{1}{2}$
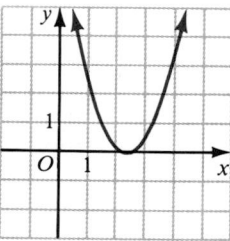

Written Exercises

Graph each function and determine its zeros, if any.   6. $\frac{\sqrt{2}}{2}, -\frac{\sqrt{2}}{2}$ 9. −1

A   1. $f: x \longrightarrow 2x^2$ 0

2. $g: x \longrightarrow -x^2$ 0

3. $h: x \longrightarrow -\frac{1}{2}x^2$ 0

4. $g: x \longrightarrow \frac{5}{2}x^2$ 0

5. $F: x \longrightarrow -x^2 + 3\sqrt{3}, -\sqrt{3}$

6. $G: x \longrightarrow 2x^2 - 1$

7. $g: x \longrightarrow x^2 + 4x$ 0, −4

8. $h: x \longrightarrow -x^2 + 2x$ 0, 2

9. $H: x \longrightarrow x^2 + 2x + 1$

10. $G: x \longrightarrow -x^2 + 6x - 9$ 3

11. $g: x \longrightarrow -x^2 - 6x - 7$
$-3 + \sqrt{2}, -3 - \sqrt{2}$

12. $f: x \longrightarrow x^2 - 2x - 3$
$-1, 3$

Find the minimum value of each function.

13. $f: x \longrightarrow x^2 + 3x$ $-\frac{9}{4}$

14. $g: x \longrightarrow x^2 + 4$ 4

15. $h: x \longrightarrow x^2 - x - 6$ $-\frac{25}{4}$

16. $H: x \longrightarrow 3x^2 - 6x + 4$ 1

17. $G: x \longrightarrow 4x^2 - 6$ −6

18. $F: x \longrightarrow \frac{1}{2}x^2$ 0

Find the maximum value of each function.

19. $f: x \longrightarrow -x^2 - 3x$ $\frac{9}{4}$

20. $F: x \longrightarrow -2x^2 + 4x$ 2

21. $g: x \longrightarrow -x^2 - 6x - 10$ −1

22. $G: x \longrightarrow -10x^2 + x + 5$ $\frac{201}{40}$

23. $h: x \longrightarrow -2x^2 + x$ $\frac{1}{8}$

24. $H: x \longrightarrow -\frac{1}{3}x^2$ 0

In each of Exercises 25-28, graph the given equations on the same set of coordinate axes.

B 25. **a.** $y = x^2$ **b.** $y = x^2 + 1$ **c.** $y = x^2 - 3$

26. **a.** $y = \frac{1}{2}x^2$ **b.** $y = \frac{1}{2}x^2 + 1$ **c.** $y = \frac{1}{2}x^2 - 3$

27. **a.** $y = x^2$ **b.** $y = (x + 1)^2$ **c.** $y = (x - 3)^2$

28. **a.** $y = -2x^2$ **b.** $y = -2(x + 1)^2$ **c.** $y = -2(x - 3)^2$

29. For any real numbers a and k, describe how the graph of $y = ax^2 + k$ is related to the graph of $y = ax^2$.

30. For any real numbers a and h, describe how the graph of $y = a(x + h)^2$ is related to the graph of $y = ax^2$.

31. Graph $A = \{(x, y): x = y^2\}$.
    a. Does the graph have a maximum point? a minimum point? No; No
    b. Is A a function? No

32. Graph $B = \{(x, y): x = -y^2\}$. See Additional Answers for graph.
    a. Does B have a maximum point? a minimum point? No; No
    b. Is B a function? No

33. Find and graph a linear function whose graph has no minimum points and no maximum points.

34. Find and graph a linear function whose graph has infinitely many minimum points and infinitely many maximum points. What is the minimum value of the function and what is the maximum value?

C 35. Find the coordinates of two points on the graph of the equation $y = x^2$ such that the x-coordinates, x_1 and x_2, differ by 1 and the y-coordinates, y_1 and y_2, differ by at least 1,000,000. What is the least possible positive integral value of x_1 that satisfies these conditions?

36. Find a value of k so that the line with equation $y = \dfrac{1}{1,000,000}x$ intersects the parabola with equation $y = kx^2$ at a point whose y-coordinate is 1. $k = \frac{1}{1,000,000,000,000}$

37. If the point $P(u, v)$ lies on the graph of $y = ax^2 + bx + c, a \neq 0$, show that the point $Q\left(-\dfrac{b}{a} - u, v\right)$ also lies on the graph.

38. **a.** Is it possible to find the equation of a parabola passing through any two points whose x-coordinates are given? How many parabolas are there that pass through points with the given x-coordinates? (*Hint*: Refer to Figure 2 on page 558 and consider the points where $x - 1$ and $x = -1$.) No; infinitely many
    b. It can be established that two points lie on exactly one line whose equation can be determined. What is the least number n of arbitrary points needed to determine the equation of a parabola?
    3 noncollinear points

Quadratic Equations and Functions **563**

12.

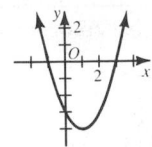

26.

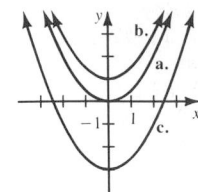

28.

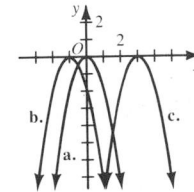

29. The graph of $y = ax^2 + k$ is the graph of $y = ax^2$ translated (slid) vertically k units (up if k is positive and down if k is negative).

30. The graph of $y = a(x + h)^2$ is the graph of $y = ax^2$ translated horizontally h units (to the left if h is positive and to the right if h is negative).

32.

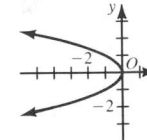

33. Answers may vary. (The graph is any nonhorizontal line.) Example: $y = x$.

34. Answers may vary. (The graph is any horizontal line.) Example: $y = 1$; the minimum value is 1, and the maximum value is 1.

(continued on p. 574)

Computer Exercises For students with computer experience

1. Write a program that will allow you to input values for a, b, and c and will determine an equation of the axis of symmetry of the parabola whose equation is $y = ax^2 + bx + c$. The output should also indicate whether the parabola opens upward or downward. Be sure to account for the fact that the value of a must be nonzero.

2. Modify the program that you wrote for Exercise 1 so that it will also determine the coordinates of the vertex of the parabola.

Teaching Suggestions
p. T109

Related Activities p. T110

Key Ideas

Graph simple polynomial functions.
Estimate zeros of polynomial functions from graphs.

Chalkboard Examples

1. Sketch the graph of the function $f: x \to x^3 - 1$ and state the zeros of the function, if any.

x	y
0	−1
1	0
−1	−2
2	7
−2	−9
3	26
−3	−28

∴ the zero of the function is 1.

11–5 Polynomial Functions

Any function of the form $x \to P(x)$, where $P(x)$ represents a polynomial, is called a **polynomial function.** For example, the following are polynomial functions:

$$f: x \to \frac{2}{3}x - 4 \qquad h: x \to 9x^3 \qquad g: x \to 5 - 3x - 2x^2$$

The graph of a polynomial function with domain $\mathcal{R}$ is a smooth curve, but many of these curves may be difficult to plot accurately. In more advanced courses you will learn many special plotting techniques. For the present, however, if you plot enough points, you will be able to make a rough sketch of the curve.

EXAMPLE 1   Sketch the graph of the function $f: x \to x^3 - 2x^2 - 5x + 6$ and give the zeros of the function, if any.

SOLUTION   Make a table of values for x and y.

x	y
−3	−24
−2	0
−1	8
0	6
1	0
2	−4
3	0
4	18

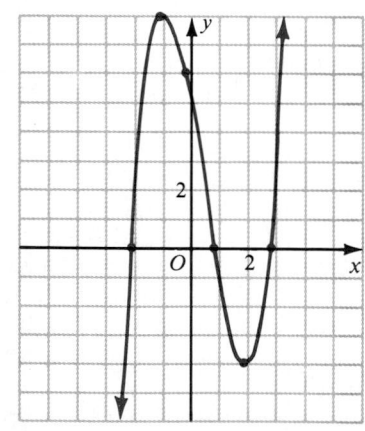

∴ the zeros of the function are −2, 1, and 3.

564   *Chapter 11*

EXAMPLE 2 Make a rough sketch of the graph of the function $f: x \longrightarrow -x^3 + 2$ and estimate the zeros, if any.

SOLUTION

x	y
-2	10
$-\frac{3}{2}$	$5\frac{3}{8}$
-1	3
$-\frac{1}{2}$	$2\frac{1}{8}$
0	2
$\frac{1}{2}$	$1\frac{7}{8}$
1	1
$\frac{3}{2}$	$-1\frac{3}{8}$
2	-6

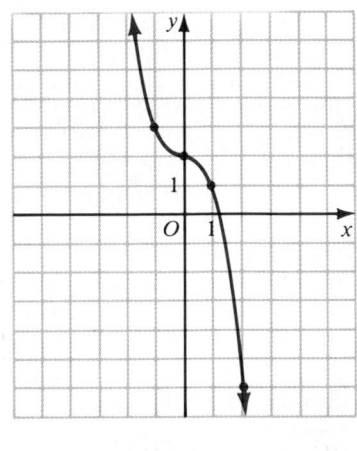

∴ there is one zero, and it lies between 1 and $1\frac{1}{2}$.
Estimate it from the graph as $1\frac{1}{4}$.

Oral Exercises

Name the zero(s) of each function.

1. $f: x \longrightarrow x^2 - 64$ $-8, 8$

2. $g: x \longrightarrow x^3 - 64$ 4

3. $h: x \longrightarrow x^3 + 64$ -4

4. $g: x \longrightarrow x(x - 2)(x + 1)$ $0, 2, -1$

5. $h: x \longrightarrow x(x^2 - 9)$ $0, 3, -3$

6. $f: x \longrightarrow x(x + 4)\left(x - \frac{1}{2}\right)$ $0, -4, \frac{1}{2}$

7. $f: x \longrightarrow (x - 1)(x + 2)\left(x + \frac{3}{8}\right)$ $1, -2, -\frac{3}{8}$ 8. $h: x \longrightarrow \left(x + \frac{1}{3}\right)\left(x - \frac{5}{8}\right)\left(x + \frac{1}{3}\right)$ $-\frac{1}{3}, \frac{5}{8}$

Written Exercises

Sketch the graph of the given function and estimate the zeros, if any.

A
1. $f: x \longrightarrow x^3$ 0

2. $g: x \longrightarrow -x^3$ 0

3. $h: x \longrightarrow x^3 - 2$

4. $f: x \longrightarrow 2 - x^3$

5. $g: x \longrightarrow -x^3 + 3x$

6. $h: x \longrightarrow -x^3 - 3x$ 0

7. $f: x \longrightarrow x^3 - 4x + 1$

8. $g: x \longrightarrow x^3 + 4x - 2$

9. $h: x \longrightarrow x^4 - 1$ $-1, 1$

10. $f: x \longrightarrow -x^4 + 1$ $-1, 1$

11. $g: x \longrightarrow x^4 + x^2$ 0

12. $h: x \longrightarrow x^4 - x^2$ $-1, 0, 1$

Quadratic Equations and Functions **565**

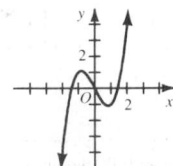
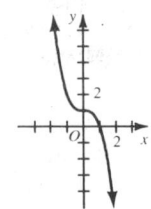
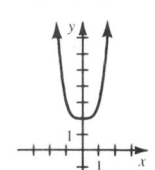

Estimate the zeros of the function whose graph is sketched below.

1.

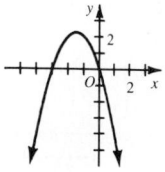

Zeros: −3 and 0

2. Tell whether the graph of $4x - x^2$ opens upward or downward.   downward

3. Tell whether the graph of $4x - x^2$ has a maximum or a minimum point. maximum

Find the required value of f when $f(x)$ is defined as follows.

a. $f(x) = x^3 - x^2$ b. $f(x) = -x^3 + x^2$

B 13. $f(-100)$ 14. $f(-1000)$ 15. $f(-10,000)$
16. $f(100)$ 17. $f(1000)$ 18. $f(10,000)$

19. a. What is true of the values of $f: x \longrightarrow x^3 - x^2$ as x takes on lesser and lesser values? greater and greater values?

b. What is true of the values of $f: x \longrightarrow -x^3 + x^2$ as x takes on lesser and lesser values? greater and greater values?

C 20. If P is a polynomial function such that $P(a) < 0$ and $P(b) > 0$ or such that $P(a) > 0$ and $P(b) < 0$, there is a theorem which states that P has *at least one* zero between a and b. Given $P: x \longrightarrow x^3 - 2x^2 + 12x - 10$, must P have at least one zero between -3 and 12? Yes

21. Given $Q: x \longrightarrow (x - 1)^4 - 1$, $Q(-1) = 15$ and $Q(4) = 80$. On the basis of the theorem cited in Exercise 20, can you conclude that Q has no zeros between -1 and 4? No. For example, $Q(0) = 0$.

PROGRAMMING IN BASIC

The graphing program given on page 247 can be expanded as follows to provide a variation designed specifically for quadratic equations.

```
10    PRINT "TO GRAPH A QUADRATIC"
15    PRINT "EQUATION, Y = AX↑2 + BX + C"
20    PRINT "(SENTENCE IS IN LINE 130):"
25    PRINT "INPUT A (<> 0), B, C";
30    INPUT A, B, C
40  ⎫
 ·  ⎬  from program on page 247
70  ⎭
73    LET D = B*B − 4*A*C
74    IF D < 0 THEN 79
75    LET D1 = SQR(D)/(2*A)
76    LET X1 = −B/(2*A)
77    PRINT "ZEROS ARE: "; X1 − D1;", ";X1 + D1
78    GOTO 80
79    PRINT "NO ZEROS"
80  ⎫
 ·  ⎬  from program on page 247
120 ⎭
130   IF Y = A*X*X + B*X + C THEN 250
140 ⎫
 ·  ⎬  from program on page 247
340 ⎭
```

566   *Chapter 11*

Exercises

1. Type in and RUN the program as given. INPUT 9 for the extent of the graph.

Change line 130 as necessary to RUN the program for each of the following quadratic functions. INPUT 9 for the extent of the graph.

2. $y = x^2$

3. $y = 0.25x^2$

4. $y = 0.25x^2 - 4$

5. $y = 0.25x^2 - 6$

6. $y = 0.25x^2 - 9$

7. $y = -0.25x^2$

8. $y = -0.25x^2 + 4$

9. $y = -0.25x^2 + 9$

10. $y = 0.25x^2 - x - 3$

11–6 Quadratic Inequalities

Recall from Sections 5-6 and 6-8 the methods used in graphing linear inequalities and systems of linear inequalities. Similar methods may be used in graphing *quadratic inequalities* as well.

A **quadratic inequality** in two variables is an inequality whose *associated equation* is a quadratic equation. For example, consider the quadratic equation

$$y = x^2 - 1.$$

The graph of this equation is a parabola that separates a coordinate plane into two regions. One of these regions is *above* the parabola, as shown by the colored shading in Figure 5, and the other region is *below* the parabola, as shown by the gray shading. The parabola itself is the *boundary* of the two regions.

If you start at any point of the parabola, say (2, 3), and move vertically *upward*, the y-coordinates of the points on the plane increase. Thus, the region above the parabola is the graph of the quadratic inequality

$$y > x^2 - 1.$$

On the other hand, if you move vertically *downward* from (2, 3), the y-coordinates of the points *decrease*, and so the region below the parabola is the graph of the quadratic inequality

$$y < x^2 - 1.$$

As in graphing linear inequalities, you use a *solid* line to indicate that the boundary is to be included as part of a graph and a *dashed* line to indicate that it is not.

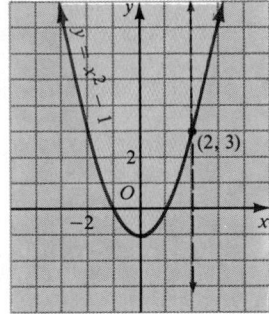

Figure 5

Quadratic Equations and Functions **567**

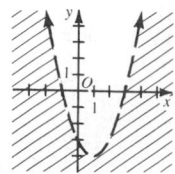
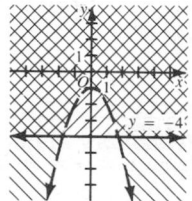
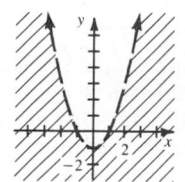
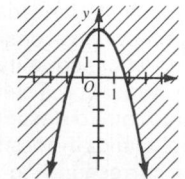
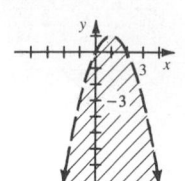
EXAMPLE 1 Graph $y < x^2 - 6x + 5$ on a coordinate plane.

SOLUTION 1. Graph the associated equation
$$y = x^2 - 6x + 5$$
as a *dashed* curve.

2. Shade the region *below* the curve.

Check: Select any point in the shaded region and determine whether its coordinates satisfy the original inequality.
$$(0, 0): y < x^2 - 6x + 5$$
$$0 \overset{?}{<} 0^2 - 6(0) + 5$$
$$0 < 5 \;\; \checkmark$$

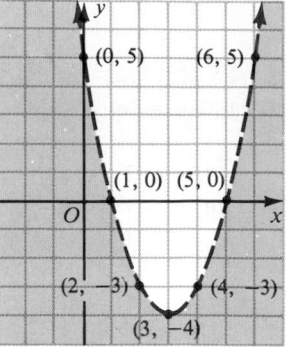

Thus $(0, 0)$ is in the solution set, and the correct region of the graph has been shaded.

The solution set of a *system of inequalities* includes all points on a coordinate plane that satisfy each inequality in the system. To solve a system of inequalities, you graph each inequality in the system on the same coordinate plane.

EXAMPLE 2 Graph the solution set of the following system of inequalities:
$$y > 1$$
$$y \geq x^2 - 2$$

SOLUTION 1. Graph $y = 1$ as a *dashed* line. The graph of $y > 1$ is the open half-plane above this line.

2. Graph $y = x^2 - 2$ as a *solid* curve. The graph of $y > x^2 - 2$ is the region above this curve.

3. The intersection of the two shaded regions is the graph of the given system (double shading).

4. Check your work by selecting any point within the double-shaded region, such as $(1, 2)$. This ordered pair should satisfy *each* of the original inequalities.

$$
\begin{array}{ll}
y > 1 & y \geq x^2 - 2 \\
2 > 1 \;\; \checkmark & 2 \overset{?}{\geq} (1)^2 - 2 \\
& 2 \geq -1 \;\; \checkmark
\end{array}
$$

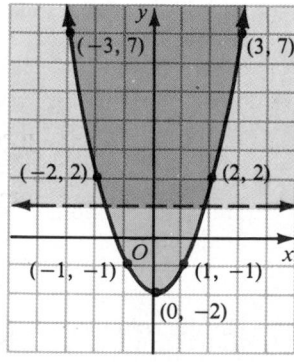

Thus $(1, 2)$ is in the solution set, and the correct region of the graph has been shaded.

Oral Exercises s = solid, d = dashed; a = above, b = below

a. Tell whether the graph of the inequality is a solid or dashed curve.
b. Tell whether the graph contains the region above or below the curve.

1. $y \le x^2 + 1$ **a.** s **b.** b   **2.** $y > x^2 - 2$ **a.** d **b.** a   **3.** $y > 4 - x^2$ **a.** d **b.** a

4. $y \le x^2 + 5$ **a.** s **b.** b   **5.** $y \ge x^2 - x$ **a.** s **b.** a   **6.** $y < x - x^2$ **a.** d **b.** b

7. $y < 3x - x^2$ **a.** d **b.** b   **8.** $y \le x^2 - 3x$ **a.** s **b.** b   **9.** $y < x^2 + 6x + 9$

10. $y \ge x^2 - x - 12$ **11.** $y \ge 5 - 4x - x^2$ **12.** $y > 8 + 2x - x^2$

    a. s **b.** a   **a.** s **b.** a   **a.** d **b.** a

Written Exercises

                                                             9. a. d   **b.** b

A   **1–12.** Graph each inequality in Oral Exercises 1–12.

Graph the solution set of each system of inequalities.

B   **13.** $y > x^2$ **14.** $y \ge x^2$ **15.** $y \le -\frac{1}{2}x^2$ **16.** $y < 3x^2$

        $y \le 5$ $x < 2$ $x > -1$ $y \ge 2$

   17. $y < x^2 - 2x$ **18.** $y > x^2 + 2x$ **19.** $y \ge x^2 + x - 6$ **20.** $y \le x^2 - 3x + 2$

        $y \ge -1$ $y < 2$ $x \le 1$ $x \ge 1$

   21. $y \ge x^2 - 1$ **22.** $y < x^2 - 2$ **23.** $y < x^2 + 2x - 3$ **24.** $y \ge 4 - 3x - x^2$

        $y \le x$ $y \ge -x$ $y > x - 2$ $y < x + 3$

C   **25.** $y \ge x^2 - 1$ **26.** $y \le x^2 - 2$ **27.** $y > x^2 - 2x - 8$ **28.** $y < 3 - 2x - x^2$

        $y \le x^2 + 1$ $y \ge 2 - x^2$ $y \le 8 - 2x - x^2$ $|y| \le 2$

◼ Self-Test 2

VOCABULARY   quadratic function (p. 557)   vertex of a parabola (p. 558)

                      parabola (p. 557)   axis of symmetry (p. 558)

                      minimum (or maximum) point   zero of a function (p. 559)

                         of a curve (p. 557)   polynomial function (p. 564)

                      minimum (or maximum) value   quadratic inequality (p. 567)

                         of a function (p. 557)

Graph each of the following on a coordinate plane.

1. $f: x \longrightarrow -x^2$ **2.** $g: x \longrightarrow x^2 + 3x - 5$ *Obj. 1, p. 557*

3. $F: x \longrightarrow -x^3$ **4.** $G: x \longrightarrow x^4 + 1$ *Obj. 2, p. 557*

5. $y \le x^2$ **6.** $y > x^2 + 2x$ *Obj. 3, p. 557*

Check your answers with those at the back of the book.

Suggested Assignments

Minimum
   569/1–12
R 569/*Self-Test 2*
Average
First day: 569/1–8, 13–16
Second day: 569/17–24
          R   569/*Self-Test 2*
Maximum
First day: 569/1–12
Second day: 569/13–28
          R   569/*Self-Test 2*

**Additional Answers
Written Exercises**
(See p. 576.)

Quick Quiz

1. Graph $g: x \longrightarrow x^2 + 2x + 3$.

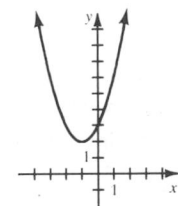

2. Graph $f: x \longrightarrow x - x^3$.

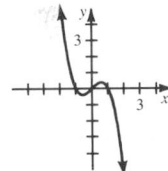

3. Graph the solution set of the system $y < x^2 - 3$
                     $y > 1$.

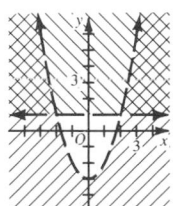

1. $y < 2x - 1$

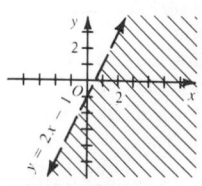

2. $y \le -x + 5$

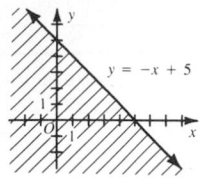

3. $f: x \to \dfrac{1}{3}x^2$

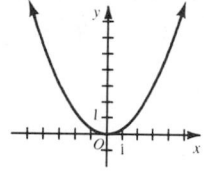

EXTRA

Rational Exponents

When you simplify $2^3 \cdot 2^2$ by writing $2^3 \cdot 2^2 = 2^{3+2} = 2^5$, you apply the law of exponents for products of powers, that is,

$$a^m \cdot a^n = a^{m+n}.$$

Can you determine what value n must have if this law is to be true for

$$2^n \cdot 2^n = 2?$$

Since

$$2^n \cdot 2^n = 2^{n+n} = 2^{2n},$$

you have

$$2^{2n} = 2.$$

But if powers of the same base are equal, the exponents of the powers must be equal provided the base is not -1, 0, or 1. This means, in this case, that $2n = 1$, or $n = \frac{1}{2}$. If

$$2^{\frac{1}{2}} \cdot 2^{\frac{1}{2}} = 2,$$

what meaning can be given to the symbol $2^{\frac{1}{2}}$? Because you know that

$$\sqrt{2} \cdot \sqrt{2} = 2 \quad \text{and} \quad (-\sqrt{2})(-\sqrt{2}) = 2,$$

it makes sense to *define* the symbol $2^{\frac{1}{2}}$ to represent one of the square roots of 2. Selecting the positive square root, you say that

$$2^{\frac{1}{2}} = \sqrt{2}.$$

This example suggests the following definition.

If a is a positive real number and n is a positive integer, or if a is a negative real number and n is a positive odd integer,

$$a^{\frac{1}{n}} = \sqrt[n]{a}.$$

Note that, if a is a negative real number and n is a positive even integer, $\sqrt[n]{a}$ does not name a real number, and hence neither does $a^{\frac{1}{n}}$.

EXAMPLE 1   Simplify.

     a. $49^{\frac{1}{2}}$ **b.** $81^{\frac{1}{4}}$ **c.** $(-27)^{\frac{1}{3}}$ **d.** $(-81)^{\frac{1}{4}}$

SOLUTION   **a.** $49^{\frac{1}{2}} = \sqrt{49} = 7$

           b. $81^{\frac{1}{4}} = \sqrt[4]{81} = 3$

           c. $(-27)^{\frac{1}{3}} = \sqrt[3]{-27} = -3$

           d. $(-81)^{\frac{1}{4}}$ does not name a real number.

570   *Chapter 11*

Now, consider the expression $(\sqrt{9})^3$. By the definition of a power with a positive integral exponent,

$$(\sqrt{9})^3 = \sqrt{9} \cdot \sqrt{9} \cdot \sqrt{9}$$
$$= 3 \cdot 3 \cdot 3 = 27.$$

Also, if the law of exponents for powers of powers, $(a^m)^n = a^{mn}$, is to be true for $(9^{\frac{3}{2}})^2$, you must have

$$(9^{\frac{3}{2}})^2 = 9^3 = 729 = 27^2, \quad \text{or} \quad 9^{\frac{3}{2}} = 27.$$

Thus, a reasonable definition of $9^{\frac{3}{2}}$ would be

$$9^{\frac{3}{2}} = (\sqrt{9})^3.$$

This example suggests the following definition.

If a is a positive real number, m is an integer, and n is a positive integer; or if a is a negative real number, m is an integer, and n is a positive odd integer; then

$$a^{\frac{m}{n}} = (a^{\frac{1}{n}})^m.$$

As before, if a is a negative real number and n is a positive *even* integer, $a^{\frac{m}{n}}$ does not name a real number.

EXAMPLE 2 Simplify.

      a. $4^{\frac{3}{2}}$ **b.** $(-27)^{-\frac{2}{3}}$ **c.** $(-9)^{\frac{5}{2}}$

SOLUTION **a.** $4^{\frac{3}{2}} = (4^{\frac{1}{2}})^3 = (2)^3 = 8$

             b. $(-27)^{-\frac{2}{3}} = [(-27)^{\frac{1}{3}}]^{-2} = (-3)^{-2} = \dfrac{1}{(-3)^2} = \dfrac{1}{9}$

             c. $(-9)^{\frac{5}{2}} = [(-9)^{\frac{1}{2}}]^5$ is not defined because $(-9)^{\frac{1}{2}}$ is not a real number.

It can be shown that under the definitions given here for $a^{\frac{1}{n}}$ and $a^{\frac{m}{n}}$, all the laws of exponents for positive integral exponents may be applied to powers with *rational exponents*, as can the definition of a negative exponent.

EXAMPLE 3 Simplify.

      a. $2^{\frac{1}{2}} \cdot 2^{\frac{1}{4}} \cdot 2^{\frac{5}{4}}$ **b.** $\dfrac{9^{\frac{3}{4}}}{9^{\frac{1}{4}}}$ **c.** $(3^{-\frac{2}{3}})^6$

SOLUTION **a.** $2^{\frac{1}{2}} \cdot 2^{\frac{1}{4}} \cdot 2^{\frac{5}{4}} = 2^{\frac{1}{2}+\frac{1}{4}+\frac{5}{4}} = 2^{\frac{8}{4}} = 2^2 = 4$

             b. $\dfrac{9^{\frac{3}{4}}}{9^{\frac{1}{4}}} = 9^{\frac{3}{4}-\frac{1}{4}} = 9^{\frac{1}{2}} = 3$

             c. $(3^{-\frac{2}{3}})^6 = 3^{(-\frac{2}{3})(6)} = 3^{-4} = \dfrac{1}{3^4} = \dfrac{1}{81}$

Quadratic Equations and Functions **571**

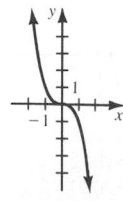
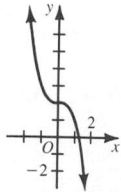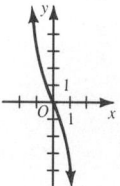
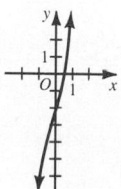
When the restriction $a > 0$ is placed on a, it can also be shown that

$$\left(a^{\frac{1}{n}}\right)^m = \left(a^m\right)^{\frac{1}{n}}$$

if m is an integer and n is a natural number. For example, $(16^2)^{\frac{1}{4}}$ and $(16^{\frac{1}{4}})^2$ name the same number, since

$$(16^2)^{\frac{1}{4}} = (256)^{\frac{1}{4}} = 4 \qquad \text{and} \qquad (16^{\frac{1}{4}})^2 = (2)^2 = 4.$$

Note that the form $(16^{\frac{1}{4}})^2$ is much easier to simplify.

Exercises

Simplify. If the expression does not represent a real number, so state. Assume that the domain of every variable is the set of positive real numbers.

1. $4^{\frac{1}{2}}$ 2

2. $16^{\frac{1}{2}}$ 4

3. $(-8)^{\frac{1}{3}}$ -2

4. $81^{\frac{1}{4}}$ 3

5. $(-4)^{\frac{1}{2}}$ not real

6. $(9)^{\frac{3}{2}}$ 27

7. $(16)^{\frac{3}{4}}$ 8

8. $(8)^{\frac{2}{3}}$ 4

9. $(36)^{-\frac{1}{2}}$ $\frac{1}{6}$

10. $(-8)^{-\frac{1}{3}}$ $-\frac{1}{2}$

11. $3^{\frac{1}{2}} \cdot 3^{\frac{3}{2}}$ 9

12. $5^{\frac{8}{3}} \div 5^{\frac{2}{3}}$ 25

13. $(2y^2)^{\frac{1}{2}}(2y^2)^{\frac{3}{2}}$ $4y^4$

14. $(27x)^{\frac{1}{3}} \div (27x)^{-\frac{2}{3}}$ $27x$

15. $[(2x)^{\frac{2}{3}}]^3$ $4x^2$

16. $[(81y)^{-\frac{1}{4}}]^2$ $\frac{\sqrt{y}}{9y}$

Chapter Summary

1. Any quadratic equation $ax^2 + bx + c = 0$, $a \neq 0$, can be solved by *completing the square*. It can also be solved by using the *quadratic formula*, producing the roots

$$x = \frac{-b + \sqrt{b^2 - 4ac}}{2a} \quad \text{and} \quad x = \frac{-b - \sqrt{b^2 - 4ac}}{2a}.$$

2. The *discriminant* $b^2 - 4ac$ can be used to determine the number of roots of a quadratic equation. If the discriminant is
   1. positive, there are two real roots;
   2. zero, there is one real root, called a *double root*;
   3. negative, there is no real root.

3. The graph of a quadratic function $f: x \longrightarrow ax^2 + bx + c$, where a, b, and c are real numbers and $a \neq 0$, is a *parabola*. Every parabola has a *vertex* (*maximum* or *minimum point*) and an *axis of symmetry*.

4. The graph of every *polynomial function* is a smooth curve.

5. A parabola separates a coordinate plane into two regions, the set of points above the parabola and the set of points below the parabola. These regions are used in graphing quadratic inequalities.

572   *Chapter 11*

Chapter Review

Write the letter of the correct answer.

1. Name the term that completes the square in the expression
   $x^2 - 8x + \underline{\ ?\ }$.

   a. 4   **(b.)** 16   **c.** 64   **d.** -64

   11-1

2. If $(n - 2)^2 = 6$, $n = \underline{\ ?\ }$.

   a. 38   **b.** -4 or 8
   (c.) $2 + \sqrt{6}$ or $2 - \sqrt{6}$ **d.** $-2 + \sqrt{6}$ or $-2 - \sqrt{6}$

3. Solve $4z^2 = 4z + 12$ by completing the square.

   a. $\left\{ \dfrac{\sqrt{13}}{2}, -\dfrac{\sqrt{13}}{2} \right\}$
   b. $\left\{ \dfrac{\sqrt{13}}{4}, -\dfrac{\sqrt{13}}{4} \right\}$

   c. $\left\{ 1 + \dfrac{\sqrt{13}}{2}, 1 - \dfrac{\sqrt{13}}{2} \right\}$
   (d.) $\left\{ \dfrac{1 + \sqrt{13}}{2}, \dfrac{1 - \sqrt{13}}{2} \right\}$

4. Solve $2x^2 - 2x - 1 = 0$ using the quadratic formula.

   (a.) $\left\{ \dfrac{1 + \sqrt{3}}{2}, \dfrac{1 - \sqrt{3}}{2} \right\}$
   b. $\left\{ \dfrac{-1 + \sqrt{3}}{2}, \dfrac{-1 - \sqrt{3}}{2} \right\}$

   c. $\left\{ -\dfrac{5}{4}, \dfrac{7}{4} \right\}$
   d. no real roots

   11-2

5. Solve $9x^2 = 12x - 2$ using the quadratic formula.

   a. $\left\{ \dfrac{2 + \sqrt{6}}{3}, \dfrac{2 - \sqrt{6}}{3} \right\}$
   b. $\left\{ \dfrac{-2 + \sqrt{6}}{3}, \dfrac{-2 - \sqrt{6}}{3} \right\}$

   (c.) $\left\{ \dfrac{2 + \sqrt{2}}{3}, \dfrac{2 - \sqrt{2}}{3} \right\}$
   d. no real roots

6. Use the discriminant to determine the number of real roots of the
   equation $5n^2 + 2n = 2$.

   a. 1   **(b.)** 2   **c.** 3   **d.** none

7. One of the two shorter sides of a right triangle is 7 cm longer than
   the other, and the hypotenuse is 3 cm longer than twice the measure
   of the shortest side. Find the length of the hypotenuse.

   a. 5 cm   **(b.)** 13 cm   **c.** 24 cm   **d.** 27 cm

   11-3

8. Which of the following functions has a
   graph that opens downward?

   a. $f: x \longrightarrow x^2 - 3x$ **(b.)** $f: x \longrightarrow 3x - x^2$
   c. $f: x \longrightarrow (3 - x)^2$ **d.** $f: x \longrightarrow -(3x - x^2)$

   11-4

   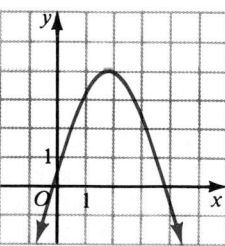

9. Which of the following functions is graphed
   on the coordinate plane at the right?

   (a.) $g: x \longrightarrow 4x - x^2$ **b.** $g: x \longrightarrow 4x + x^2$
   c. $g: x \longrightarrow x^2 - 2x$ **d.** $g: x \longrightarrow 4x^2 - x$

10.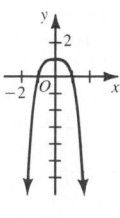

12.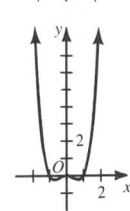

13. **a.** $-1,010,000$
    b. $1,010,000$
14. **a.** $-1,001,000,000$
    b. $1,001,000,000$
15. **a.** $-1,000,100,000,000$
    b. $1,000,100,000,000$
16. **a.** $990,000$
    b. $-990,000$
17. **a.** $999,000,000$
    b. $-999,000,000$
18. **a.** $999,900,000,000$
    b. $-999,900,000,000$
19. **a.** The values of f de-
       crease;
       the values of f in-
       crease.
    b. The values of f in-
       crease;
       the values of f de-
       crease.

35. Answers may vary.
Example: (500,000,
250,000,000,000) and
(500,001, 250,001,000,001).
The least possible
positive integral value of
x_1 is 500,000.

37. Let $x = -\dfrac{b}{a} - u$. Then:

$ax^2 + bx + c = a\left(-\dfrac{b}{a} -\right.$

$\left. u\right)^2 + b\left(-\dfrac{b}{a} - u\right) + c =$

$a\left(\dfrac{b^2}{a^2} + \dfrac{2ub}{a} + u^2\right) - \dfrac{b^2}{a} -$

$bu + c = \dfrac{b^2}{a} + 2ub +$

$au^2 - \dfrac{b^2}{a} - bu + c =$

$au^2 + bu + c$
Since $P(u, v)$ lies on the
graph, $au^2 + bu + c = v$.
Thus, $a\left(-\dfrac{b}{a} - u\right)^2 +$

$b\left(-\dfrac{b}{a} - u\right) + c = v$, and

so $Q\left(-\dfrac{b}{a} - u, v\right)$ lies on
the graph.

---

10. Find the minimum value of $h: x \longrightarrow x^2 - 2$.
- **(a.)** -2
- **b.** -1
- **c.** 0
- **d.** 1

11. Which of the following is *not* a zero of $f: x \longrightarrow x^3 - x$?   *11-5*
- **(a.)** -2
- **b.** -1
- **c.** 0
- **d.** 1

12. Find the zero(s) of $g: x \longrightarrow x^3 - 1$.
- **(a.)** 1
- **b.** $0, 1$
- **c.** $-1, 1$
- **d.** $-1, 0, 1$

13. Which of the following inequalities is graphed on the coordinate plane at the right?   *11-6*
- **a.** $y \le x^2 - 4x + 4$
- **b.** $y > x^2 - 4x + 4$
- **c.** $y < x^2 - 4x + 4$
- **(d.)** $y \ge x^2 - 4x + 4$

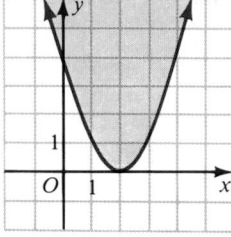

14. Describe the graph of the system $\begin{aligned} y &> x^2 - 2 \\ y &< 0 \end{aligned}$ on a coordinate plane.
- **a.** a region bounded by the graph of $y = x^2 - 2$ and the y-axis
- **(b.)** a region bounded by the graph of $y = x^2 - 2$ and the x-axis
- **c.** the entire coordinate plane
- **d.** none of the above

Chapter Test

1. Find a value of k so that $x^2 - 9x + k$ is a trinomial square. $\frac{81}{4}$ *11-1*

Solve by completing the square. Express irrational solutions in simplest form.

2. $a^2 - 8a + 4 = 0$
$\{4 + 2\sqrt{3}, 4 - 2\sqrt{3}\}$

3. $3z^2 = 18z - 24$ $\{2, 4\}$

Solve by using the quadratic formula. Express irrational solutions in simplest form. **4.** $\{\frac{-5 + \sqrt{33}}{2}, \frac{-5 - \sqrt{33}}{2}\}$

4. $j^2 + 5j - 2 = 0$

5. $4k^2 + 5 = 9k$ $\{\frac{5}{4}, 1\}$ *11-2*

6. Determine the number of real roots of the equation $4y^2 + 9 = 9y$. 0

Solve. Approximate irrational solutions to the nearest hundredth.

7. A rectangular swimming pool is 5 m wide and 10 m long. The pool is surrounded by a concrete walk of uniform width. The combined area of the pool and the walk is 50 m² more than the area of the swimming pool alone. Find the width of the walk. 1.40 m   *11-3*

8. The dimensions of a rectangle can be represented by consecutive odd integers, and its area is 255 cm². Find the dimensions. width: 15 cm; length: 17 cm

574 *Chapter 11*

Graph each function on a coordinate plane.

9. $f: x \longrightarrow x^2 + 1$ **10.** $g: x \longrightarrow -x^2 - 2x + 3$ *11-4*

For each function, determine its zeros, if any, and its maximum or minimum value.

11. $G: x \longrightarrow -x^2 + 4x$ **12.** $H: x \longrightarrow x^2 + 2x + 1$
      zeros: 0, 4; max. value: 4   zero: −1; min. value: 0

Sketch the graph of each function and estimate its zeros, if any.

13. $g: x \longrightarrow x^3 - x$ **14.** $h: x \longrightarrow 1 - x^3$ *11-5*

15. Graph $y < x^2 + 2$ on a coordinate plane.   *11-6*

16. Graph the solution set of the system $\begin{matrix} y \geq x^2 \\ x < 3 \end{matrix}$ on a coordinate plane.

Cumulative Review

Chapter 6

Solve each system of equations using any method.

1. $6x - y = 11 \{(2, 1)\}$
    $2x + y = 5$

2. $8x - 3y = 3 \ \{(3, 7)\}$
    $3x - 2y = -5$

3. $7r = 8 + 3s\ \emptyset$
    $21r - 9s = 42$

4. $5s + t = 12 \ \{(2, 2)\}$
        $s = t$

5. $3c + 5d = 10 \ \{(0, 2)\}$
      $c = d - 2$

6. $3a = 8 + 5b\ \emptyset$
    $12a - 20b = 24$

7. $2j - 3k = 4 \ \{(5, 2)\}$
    $3j + \ \ k = 17$

8. $u + v = 6 \ \{(8, -2)\}$
    $u - v = 10$

9. $9m + 5n = -6 \ \{(-4, 6)\}$
    $4m + 3n = 2$

10. $a + 3z = 10 \ \{(4, 2)\}$
     $a + 9z = 22$

Solve.

11. There are 28 students in a computer programming class. The number of sophomores is 24 less than three times the number of juniors. How many sophomores and how many juniors are in the class? 15 sophomores; 13 juniors

12. Flying with the wind, an airplane can travel 1500 km in 2.5 h, but flying against the wind, the airplane requires 0.5 h more to make the return trip. Find the speed of the wind. 50 km/h

Quadratic Equations and Functions   **575**

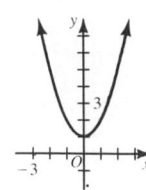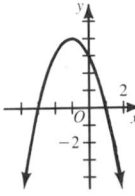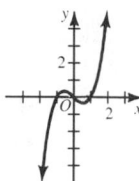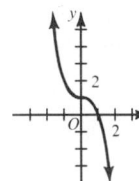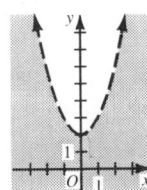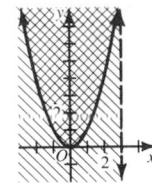

2.
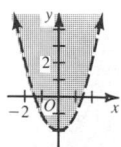

4.
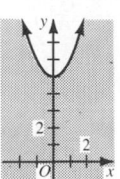

6.
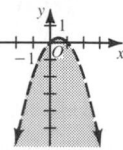

8.
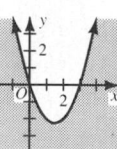

10.
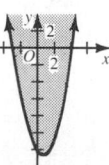

12.
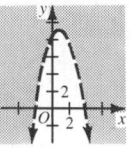

14.
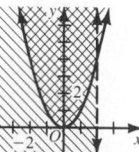

16.
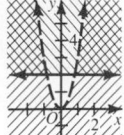

Chapter 7

Simplify.

13. $(u^2v - uv^2) - (u^2v + 2uv^2)$ $-3uv^2$

14. $-8m^2n^2 + 4m^2n + 3mn$ $-8m^2n^2 - 3mn + 4m^2n + 6mn$

15. $-rs(r^2 - rs + s^2)$ $-r^3s + r^2s^2 - rs^3$

16. $(-a)(a^2b)^3(ab^2)^2$ $-a^9b^7$

17. $(4x - 3)^2$ $16x^2 - 24x + 9$

18. $(x - 3)(x^2 + 3x + 9)$ $x^3 - 27$

Factor completely.

19. $-12cd^2e - 48c^3d^3e$ $-12cd^2e(1 + 4c^2d)$

20. $4m(2n + 9) - 3(2n + 9)$ $(2n + 9)(4m - 3)$

21. $1 + 12w + 36w^2$ $(1 + 6w)^2$

22. $3g^2 - 3gh + 2g - 2h$ $(g - h)(3g + 2)$

23. $16x^2 - 50x + 25$ $(2x - 5)(8x - 5)$

24. $24y^2 + 10y - 4$ $2(4y - 1)(3y + 2)$

Solve.

25. The sum of two integers is 36 and their product is 128. Find the integers. 32, 4

26. A rectangular plot of land has an area of 1050 m². Its length exceeds its width by 5 m. Find the length of the plot. 35 m

Chapter 9

Solve.

27. $\frac{b + 12}{3} = \frac{b}{5}$ $\{-30\}$

28. $\frac{2x + 3}{4} - \frac{4x + 1}{3} = \frac{5}{6}$ $\{-\frac{1}{2}\}$

29. $\frac{a}{14} - \frac{2}{7} \le \frac{a}{7}$ $\{a: a \ge -4\}$

30. $\frac{2x - 1}{9} > \frac{1}{3}$ $\{x: x > 2\}$

31. $\frac{w - 2}{w + 2} - 1 = \frac{12}{w^2 - 4}$ $\{-1\}$

32. $\frac{x - 2}{x} - \frac{x}{x - 1} = \frac{5}{2}$ $\{\frac{4}{5}, -1\}$

33. If y varies directly as x, and if $y = 42$ when $x = 14$, find y when $x = 8$. 24

34. If y varies inversely as x, and if $y = 18$ when $x = 24$, find y when $x = 12$. 36

35. If y varies jointly as x and z, and if $y = 4$ when $x = 6$ and $z = 2$, find y when $x = 12$ and $z = 10$. 40

36. If y varies directly as x and inversely as z, and if $y = 12$ when $x = 4$ and $z = 3$, find y when $x = 12$ and $z = 18$. 6

37. It takes Louise 6 h to do a certain job, and it takes Margaret 9 h to do the same job. How long would it take the two of them working together to complete the job? $3\frac{3}{5}$ h

38. When a car starts from rest and travels at a constant acceleration, the distance that it travels varies jointly as its acceleration and the square of the time that it has been traveling. If a car travels 320 m from rest in 4 s at an acceleration of 40 m/s², how far will the car travel from rest in 6 s at an acceleration of 25 m/s²? 450 m

576 *Chapter 11*

Chapter 10

Simplify.

39. $\sqrt{225}$ 15

40. $\sqrt{192}$ $8\sqrt{3}$

41. $(\sqrt{150})^2$ 150

42. $\sqrt{(-24)^2}$ 24

43. $4\sqrt{2} \cdot \sqrt{50}$ 40

44. $\sqrt[4]{80}$ $2\sqrt[4]{5}$

45. $\sqrt{\dfrac{7}{18}}$ $\dfrac{\sqrt{14}}{6}$

46. $\dfrac{\sqrt{96}}{\sqrt{8}}$ $2\sqrt{3}$

47. $\sqrt[3]{\dfrac{5}{9}}$ $\dfrac{\sqrt[3]{15}}{3}$

48. $(2\sqrt{3})(8\sqrt{21})$ $48\sqrt{7}$

49. $\dfrac{\sqrt{3}-2}{\sqrt{3}+4}$ $\dfrac{11-6\sqrt{3}}{-13}$

50. $\dfrac{\sqrt{5}+\sqrt{2}}{\sqrt{5}-\sqrt{2}}$ $\dfrac{7+2\sqrt{10}}{3}$

Solve. Approximate irrational roots to the nearest hundredth.

51. $2x^2 - 72 = 0$ $\{-6, 6\}$

52. $3z^2 + 35 = 107$ $\{4.90, -4.90\}$

53. $x - \sqrt{x} = 6$ $\{9\}$

54. $\sqrt{x^2 - 10} = 5$ $\{5.92, -5.92\}$

Contest Problems

1. Determine the total number of distinct squares that can be outlined on the four-by-four "checkerboard" that is shown at the right. 30

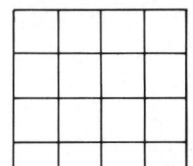

2. Simplify the following expression.

$$\frac{2x - 6}{2x - 10} \cdot \left(\frac{x^3 - 27}{3x^2 - 75}\right)^{-1} \div \frac{3x + 15}{2x^2 + 6x + 18}$$ 2

3. Find the least three-digit number that is itself a perfect square but the sum of whose digits is *not* a perfect square. 256

4. Solve $\sqrt{c^3 + c^3 + c^3 + c^3 + c^3} = 25$. $\{5\}$

5. If $x + y = 11$ and $y = \dfrac{15}{x}$, find the value of $x^2 + y^2$. 91

6. Solve $x = \dfrac{x^x}{\sqrt[x]{x}}$. $\{1, \dfrac{1+\sqrt{5}}{2}\}$

7. Complete this statement: If n is an integer, the average of the numbers 8, n, and $8n + 1$ is always divisible by 3 and $\underline{\quad?\quad}$. $n + 1$

8. Find all values of d for which the following expression is not defined.

$$\frac{\dfrac{d}{d - 2} - \dfrac{d}{d + 2}}{\dfrac{d}{d - 2} + \dfrac{d}{d + 2}}$$ 0, 2, −2

Quadratic Equations and Functions **577**

18.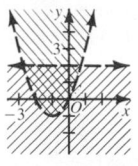

20.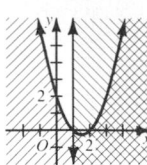

22.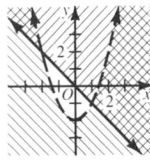

24.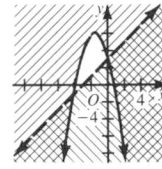

26.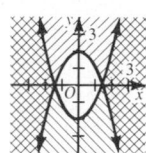

28.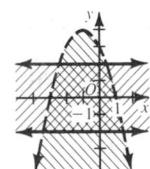

Modern roads and buildings, such as those pictured in this aerial view of Boston, are triumphs of civil engineering. This branch of engineering depends heavily on trigonometry.

Chapter 12

Trigonometry and Vectors

Problem Solving Strategies

Drawing a Diagram
Students are encouraged to use diagrams as problem solving aids in Sections 12-4 and 12-7.

Using a Formula
In Section 12-4 students use the Pythagorean Theorem and the formulas for the sine, the cosine, and the tangent of an angle to solve problems involving right triangles. When solving problems involving vectors in Section 12-7, students use those formulas to find the norm and the direction of a vector.

Trigonometry

OBJECTIVES for Sections 12-1 through 12-4:
1. To picture an angle as a rotation.
2. To determine the sine, cosine, and tangent of an angle in standard position.
3. To use trigonometric tables to find the sine, cosine, and tangent of an angle.
4. To solve right triangles.

12–1 Angles

When an airport radar antenna revolves, as shown in Figure 1, you can think of its beam as generating an *angle*. For example, since the measure of a right angle is 90°, when the antenna turns through one complete revolution the beam generates an angle of measure 360°. In two revolutions it generates an angle of measure 720°, and so on.

In geometry an **angle** is defined as the union of two rays that have the same endpoint. To develop the extended notion of an angle described above, you can call the ray of the angle that is at the starting position of the generating ray the **initial side** of the angle. The ray of the angle that is at the final position of the generating ray is then called the **terminal side** of the angle. Thus, a **directed angle** is defined as the union of two ordered rays with a common endpoint, called the **vertex** of the angle, *together with a rotation* from the initial side to the terminal side, as shown in Figure 2.

Figure 1

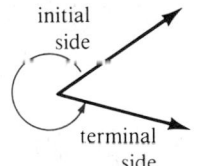

Figure 2

Teaching Suggestions
p. T110

Key Ideas

Determine the measure of an angle visualized as a rotation.

Trigonometry and Vectors **579**

1. For each figure, determine the measure of $\angle A$.

   a.

   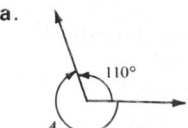

   $m \angle A$ is negative.
   $m \angle A = -[360° - 110°]$
   $= -250°$

   b.

   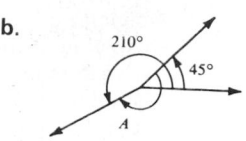

   $m \angle A$ is negative.
   $m \angle A =$
   $-[360° - (210° - 45°)]$
   $= -195°$

   c.

   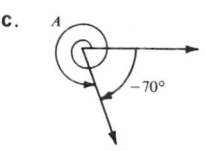

   $m \angle A$ is positive.
   $m \angle A = (360° \times 2) - 70°$
   $= 650°$

   d.

   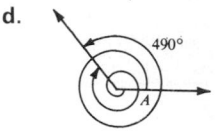

   An angle of 490° is coterminal with an angle of $(490 - 360)°$, or 130°.
   $m \angle A =$
   $-[(360° \times 2) - 130°]$
   $= -590°$

If two or more angles have the same initial side *and* the same terminal side, they are called **coterminal angles.** Figure 3 shows a pair of coterminal angles.

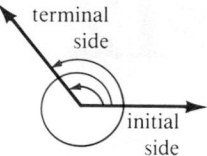

Figure 3

The number of degrees through which a ray rotates in turning from the initial side to the terminal side is the **measure** of the angle. Ordinarily, the measure is considered to be *positive* if the rotation is counterclockwise, and *negative* if the rotation is clockwise. In Figure 4, you can use a *protractor* to verify that $m \angle AOB$ (read "the measure of angle A, O, B") is approximately 145°, and that $m \angle COD$ is approximately −30°.

Figure 4

On a coordinate plane, an angle that has its vertex at the origin and has the positive x-axis as its initial side is said to be in **standard position.** In Figure 5, $\angle A$ and $\angle B$ are in standard position.

Recall from Section 5-1 that the x- and y-axes separate a coordinate plane into four quadrants, which are numbered as shown in Figure 5. The terminal side of an angle in standard position determines the quadrant in which the angle is said to *be*, or to *lie*. For example, $\angle A$ in Figure 5 lies in the fourth quadrant, or Quadrant IV. Similarly, an angle in standard position having a measure of 200° lies in the third quadrant, or Quadrant III.

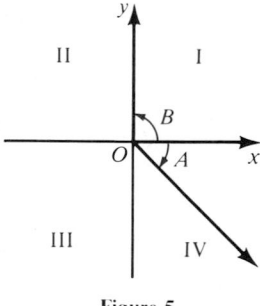

Figure 5

An angle in standard position whose terminal side coincides with an axis is called a **quadrantal angle.** For example, $\angle B$ in Figure 5 is a quadrantal angle; its terminal side lies on the positive y-axis. In fact, any angle in standard position having a measure of 0°, 90°, 180°, 270°, 360°, −90°, −180°, and so on, is a quadrantal angle. A quadrantal angle does *not* lie in any quadrant.

580 *Chapter 12*

EXAMPLE 1   Determine the measure of $\angle A$ in each figure.

a.

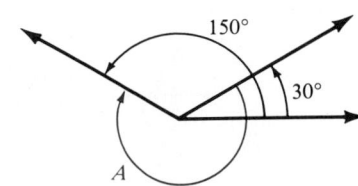

b.

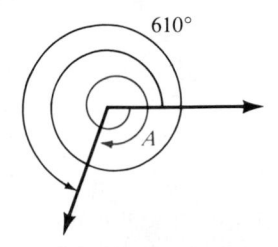

c.

d.

SOLUTION   The measures of coterminal angles differ by multiples of $360°$.

  a. $m\angle A$ is negative.

  $$m\angle A = -(360° - 75°)$$
  $$= -285°$$

  b. $m\angle A$ is negative.

  $$m\angle A = -[360° - (150° - 30°)]$$
  $$= -240°$$

  c. $m\angle A$ is positive.

  $$m\angle A = (360° \times 2) - 160°$$
  $$= 720° - 160°$$
  $$= 560°$$

  d. An angle of $610°$ is coterminal with an angle of $(610 - 360)°$, or $250°$.

  $$m\angle A = -[(360° \times 2) - 250°]$$
  $$= -470°$$

EXAMPLE 2   Assume that $\angle A$ is in standard position and that $0° \le m\angle A < 360°$. Find $m\angle A$ so that each set of conditions is satisfied.

  a. $\angle A$ is a quadrantal angle and $100° < m\angle A < 215°$.

  b. $\angle A$ is coterminal with an angle measuring $550°$.

  c. $\angle A$ is coterminal with an angle measuring $-220°$.

SOLUTION   **a.** The terminal side of $\angle A$ lies on an axis between Quadrant II and Quadrant III. Therefore, $m\angle A = 180°$.

  b. $m\angle A = (550 - 360)° = 190°$

  c. $m\angle A = (-220 + 360)° = 140°$

Trigonometry and Vectors   **581**

2. Assume that $\angle A$ is in standard position and that $0° \le m\angle A < 360°$. Find the measure of $\angle A$ given the following information.

  a. $\angle A$ is a quadrantal angle and $190° < m\angle A < 310°$. The terminal side of $\angle A$ lies on an axis between Quadrant III and Quadrant IV.

  $m\angle A = 270°$

  b. $\angle A$ is coterminal with an angle measuring $435°$.

  $m\angle A = (435 - 360)°$
  $= 75°$

  c. $\angle A$ is coterminal with an angle measuring $-165°$.

  $m\angle A = (-165 + 360)°$
  $= 195°$

Oral Exercises

Find $m\angle A$.

1. 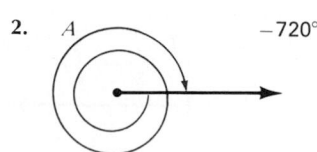 360°

2. 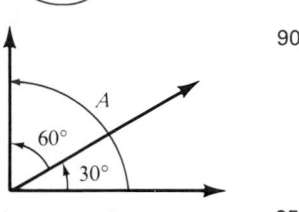 −720°

3. 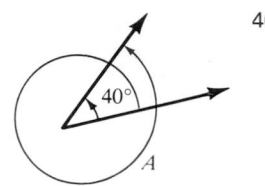 400°

4. 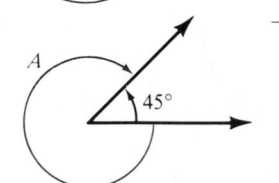 90°

5. 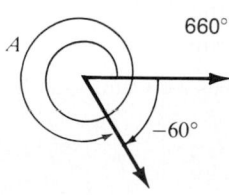 −315°

6. 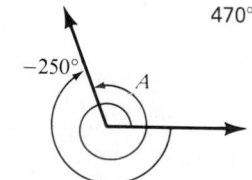 35°

Find $m\angle A$, given that $\angle A$ is coterminal with the angle whose measure is given and $0° \le m\angle A < 360°$.

7. 380° 20°
8. −40° 320°
9. −90° 270°
10. 560° 200°
11. −300° 60°
12. −10° 350°
13. −200° 160°
14. 720° 0°
15. 410° 50°
16. 800° 80°

17–26. In Exercises 7–16 above, name the quadrant, if any, in which an angle with the given measure would lie if it were in standard position. **17.** I **18.** IV **19.** none **20.** III **21.** I **22.** IV **23.** II **24.** none **25.** I **26.** I

Find the measure of the quadrantal angle A that satisfies the given conditions.

27. $45° < m\angle A < 135°$ 90°
28. $-200° < m\angle A < -100°$ −180°
29. $200° < m\angle A < 300°$ 270°

Written Exercises

Find $m\angle A$.

A 1. 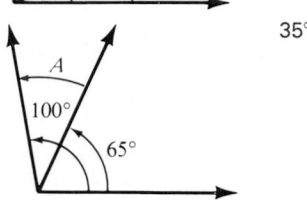 660° −60°

2. 470° −250°

3.

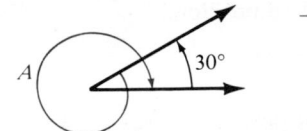

−390°

4.

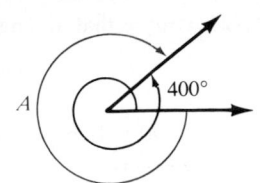

−320°

5.

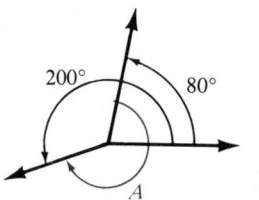

−240°

6.

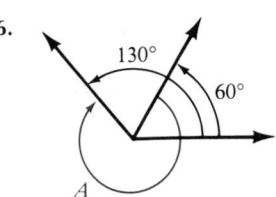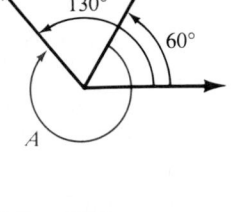

−290°

7.

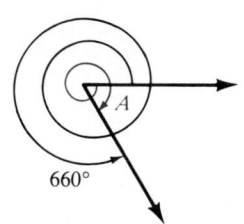

−420°

8.

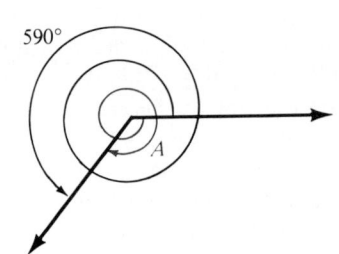

−490°

In Exercises 9–16 assume that ∠A is in standard position and that 0° ≤ m∠A < 360°.

9. If $m\angle A > 180°$ and $\angle A$ is not a quadrantal angle, then $\angle A$ is in either Quadrant __?__ or __?__.   III; IV

10. If $90° < m\angle A < 180°$, then $\angle A$ is in Quadrant __?__.   II

11. If $\angle A$ is in Quadrant I, then __?__ $< m\angle A <$ __?__.   0°; 90°

12. If $\angle A$ is in Quadrant IV, then __?__ $< m\angle A <$ __?__.   270°; 360°

13. If $\angle A$ is a quadrantal angle and $180° < m\angle A < 360°$, then $m\angle A =$ __?__.   270°

14. If $\angle A$ is a quadrantal angle and $50° < m\angle A < 200°$, then either $m\angle A =$ __?__ or $m\angle A =$ __?__.   90°; 180°

15. If $\angle A$ is coterminal with an angle measuring $-40°$, then $m\angle A =$ __?__.   320°

16. If $\angle A$ is coterminal with an angle measuring $550°$, then $m\angle A =$ __?__.   190°

B 17. If $\angle A$ is coterminal with an angle measuring $-520°$, and if $400° < m\angle A < 600°$, then $m\angle A =$ __?__.   560°

18. If $\angle A$ is coterminal with an angle measuring $1010°$, and if $-500° < m\angle A < -100°$, then $m\angle A =$ __?__.   −430°

Trigonometry and Vectors **583**

Additional A Exercises

In Exercises 1–4 determine $m \angle A$.

1.

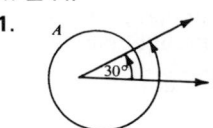

390°

2.

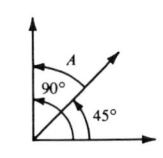

45°

3.

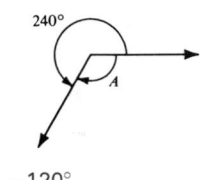

−120°

4.

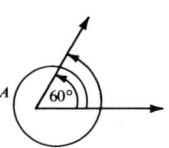

420°

In Exercises 5–6 assume that $\angle A$ is in standard position and that $0° \leq m \angle A < 360°$.

5. If $\angle A$ is in Quadrant III, then __?__ $< m \angle A <$ __?__.   180°, 270°

6. If $\angle A$ is a quadrantal angle and $0° < m \angle A < 100°$, then $m \angle A =$ __?__.   90°

Name the quadrants or axes that together contain all the ordered pairs (x, y) that satisfy the given requirements.

1. $x > 0$ and $y > 0$
Quadrant I

2. $x < 0$ and $y < 0$
Quadrant III

3. $x > 0$ and $y < 0$
Quadrant IV

4. $xy > 0$ Quadrants I and III

5. $xy < 0$ Quadrants II and IV

6. $x = 0$ x-axis, y-axis

In Exercises 19–24 assume that all angles are in standard position.

19. If $180° < |m\angle R| < 270°$, then $\angle R$ is in either Quadrant __?__ or __?__ . II; III

20. If $0° < |m\angle S| < 90°$, then $\angle S$ is in either Quadrant __?__ or __?__ . I; IV

21. If $|m\angle T| < 90°$ and $\angle T$ is not in Quadrant I or IV, then $m\angle T =$ __?__ . 0°

22. If $|90° - m\angle U| < 90°$ and $\angle U$ is not in Quadrant I or II, then $m\angle U =$ __?__ . 90°

23. If the terminal sides of $\angle G$ and $\angle H$ are collinear, $m\angle G = 495°$, and $|m\angle H| < 90°$, then $m\angle H =$ __?__ . −45°

24. If the terminal sides of $\angle J$ and $\angle K$ are collinear, $m\angle J = -340°$, and $90° < |m\angle K| < 180°$, then $m\angle K =$ __?__ . −160°

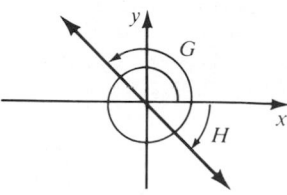

Ex. 23

C **25.** If $\angle R$ lies in Quadrant IV and $m\angle R = 3m\angle Q$, then $\angle Q$ lies in Quadrant __?__ if $0° < m\angle R < 360°$ and in Quadrant __?__ if $-360° < m\angle R < 0°$. II; IV

26. Given that $180° < 5m\angle E < 270°$ and $240° < 6m\angle E < 360°$, then __?__ $< m\angle E <$ __?__ . 40°; 54°

27. A wagon wheel makes 20 revolutions per minute. Through how many degrees does a spoke of the wheel turn in 1 s? 5 s? 2 min? 120°; 600°;

28. The beam of a radar antenna sweeps through an angle of 54° each 14,400° second. How many revolutions does it make in 5 min? 45

29. The arm of a garden sprinkler rotates counterclockwise at a constant rate. If the arm makes one rotation in 0.1 s, through how many degrees does it turn in 5 s? 18,000°

30. A carnival ride turns through 24° in 1.5 s. How long does it take the ride to complete 12 revolutions? 270 s

Computer Exercises   For students with computer experience

1. Write a program that will allow you to input a positive integer n and will compute a positive integer d, $0 \leq d \leq 360$, such that angles of measure $n°$ and $d°$ are coterminal.

2. Modify the program that you wrote for Exercise 1 so that it will allow you to input a negative integer. The output should still be an integer d such that $0 \leq d \leq 360$.

3. Write a program that will allow you to input the coordinates of a point and will determine in which quadrant or on which axis the angle lies whose terminal side in standard position passes through the given point. If the angle lies on an axis, the output should specify whether it lies on the positive or negative side of the axis.

584   *Chapter 12*

12–2 Trigonometric Functions

Teaching Suggestions
p. T111

A circle whose radius is one *unit* long and whose center is at the origin of a coordinate plane is called a **unit circle**. From the distance formula derived in Section 10-6, it follows that a point (x, y) is on the unit circle if and only if $\sqrt{x^2 + y^2} = 1$. Hence an equation of the unit circle is $x^2 + y^2 = 1$.

You can see from Figure 6 that the terminal side of each angle in standard position intersects the unit circle in exactly one point. For example, the terminal side of $\angle AOB$ intersects the circle in point B, and the terminal side of $\angle AOC$ intersects it in point C.

Key Ideas

Determine the sine, cosine, and tangent of an angle in standard position.

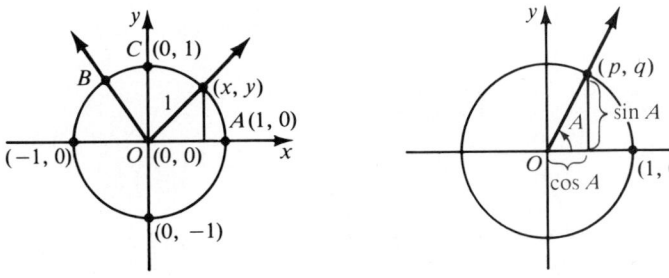

Figure 6   **Figure 7**

Special names are given to the ordinate and abscissa of the intersection point for each angle and also to the quotient of the ordinate and the abscissa. Namely, if (p, q) is the point of intersection of the unit circle and the terminal side of an angle A in standard position (Figure 7), then

$$\sin A = q \text{ (read ''the \textbf{sine} of } A \text{ is equal to } q\text{''),}$$
$$\cos A = p \text{ (read ''the \textbf{cosine} of } A \text{ is equal to } p\text{''),}$$

and, if $p \neq 0$,

$$\tan A = \frac{q}{p} \text{ (read ''the \textbf{tangent} of } A \text{ is equal to } q \text{ divided by } p\text{'').}$$

These definitions tell you what is meant by the sine, cosine, and tangent of an angle in standard position. However, any angle can be put into standard position by choosing the coordinate system so that the origin is the vertex of the angle and the positive x-axis is the initial side of the angle. Therefore, you can speak of the sine, cosine, and tangent of an angle whether or not it is given in standard position. You should realize that the terminal side of an angle is determined when its initial side and its measure are known.

The notation "tan 45°" means "the tangent of an angle of measure 45°." However, in the notation "tan A," the letter A does not denote a measure, but rather an angle, even though the symbol $\angle$ is omitted.

Common Errors

Students often have difficulty discerning between the domain and range of trigonometric functions. When finding sin 40°, they might write sin 40° ≈ sin 0.6428 rather than sin 40° ≈ 0.6428. Also, when finding the measure of angle A, they might write $m\angle A$ = sin 0.1908 ≈ 11° rather than sin A = 0.1908 and therefore $m\angle A$ ≈ 11°.

To help students avoid these errors, remind them that sine, cosine, and tangent are functions. The domain of these functions is always expressed in terms of degrees, and the range is expressed in terms of real numbers. At times, it may be useful to use more familiar function notation when discussing the trigonometric functions. (Example: Write $f(11°) ≈ 0.1908$ in place of sin 11° ≈ 0.1908.)

Trigonometry and Vectors **585**

Chalkboard Examples

1. Give the sine, cosine, and tangent of an angle A in standard position whose terminal side contains the given point on the unit circle.

a. $\left(\dfrac{1}{2}, -\dfrac{\sqrt{3}}{2}\right)$

$$\sin A = -\dfrac{\sqrt{3}}{2}$$

$$\cos A = \dfrac{1}{2}$$

$$\tan A = \dfrac{\sin A}{\cos A}$$

$$= \dfrac{-\dfrac{\sqrt{3}}{2}}{\dfrac{1}{2}}$$

$$= -\sqrt{3}$$

b. $\left(\dfrac{2\sqrt{2}}{3}, -\dfrac{1}{3}\right)$

$$\sin A = -\dfrac{1}{3}$$

$$\cos A = \dfrac{2\sqrt{2}}{3}$$

$$\tan A = \dfrac{\sin A}{\cos A}$$

$$= \dfrac{-\dfrac{1}{3}}{\dfrac{2\sqrt{2}}{3}} = \dfrac{-1}{2\sqrt{2}}$$

$$= -\dfrac{\sqrt{2}}{4}$$

The definitions of the numbers sin A, cos A, and tan A suggest the following three functions.

$$\text{sine:} \quad \angle A \longrightarrow \sin A$$
$$\text{cosine:} \quad \angle A \longrightarrow \cos A$$
$$\text{tangent:} \quad \angle A \longrightarrow \tan A$$

Since the numbers sin A and cos A are defined for every angle A, the sine and cosine functions have the same domain, namely the set of all angles. Further, since sin A and cos A are coordinates of points on the unit circle, these functions also have the same range, namely the set of all real numbers between -1 and 1, inclusive.

Notice that, for an angle A in standard position, tan A is not defined if the terminal side is on the y-axis because division by zero is not defined. Therefore, the domain of the tangent function is the set of all angles whose degree measure is *not* an odd multiple of 90. The range of the tangent function is the set of all real numbers.

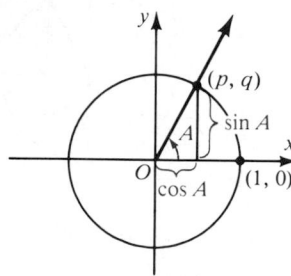

Figure 7

Sine, cosine, and tangent are examples of **trigonometric functions**. The word *trigonometric* comes from two Greek words: *trigonon* (triangle) and *metron* (measure).

The following are two useful trigonometric formulas:

1. For each point (p, q) on the unit circle,

$$q^2 + p^2 = 1.$$

But, as in Figure 7, $q = \sin A$ and $p = \cos A$; therefore

$$(\sin A)^2 + (\cos A)^2 = 1.$$

In place of the symbols $(\sin A)^2$ and $(\cos A)^2$, you ordinarily write $\sin^2 A$ and $\cos^2 A$, respectively. Thus,

$$\sin^2 A + \cos^2 A = 1.$$

2. If $\cos A \neq 0$, then

$$\tan A = \dfrac{\sin A}{\cos A},$$

because $\tan A = \dfrac{q}{p}$, and $q = \sin A$ and $p = \cos A$.

586 *Chapter 12*

EXAMPLE 1 Give the sine, cosine, and tangent of an angle A in standard position whose terminal side contains the given point on the unit circle.

a. $\left(\dfrac{\sqrt{5}}{5}, -\dfrac{2\sqrt{5}}{5}\right)$

b. $(-1, 0)$

c. $(0, 1)$

SOLUTION a. $\sin A = -\dfrac{2\sqrt{5}}{5};$ $\cos A = \dfrac{\sqrt{5}}{5};$ $\tan A = \dfrac{\sin A}{\cos A} = \dfrac{-\dfrac{2\sqrt{5}}{5}}{\dfrac{\sqrt{5}}{5}} = -2$

b. $\sin A = 0;$ $\cos A = -1;$ $\tan A = \dfrac{\sin A}{\cos A} = \dfrac{0}{-1} = 0$

c. $\sin A = 1;$ $\cos A = 0;$ $\tan A$ is undefined.

When the quadrant of an angle A is given, and also the value of $\sin A$, $\cos A$, or $\tan A$, you can find the remaining two values, using the formulas $\sin^2 A + \cos^2 A = 1$ and $\tan A = \dfrac{\sin A}{\cos A}$ as needed.

EXAMPLE 2 Find $\cos A$ and $\tan A$, given that angle A lies in Quadrant III, and that $\sin A = -\dfrac{\sqrt{2}}{2}.$

SOLUTION Substitute $-\dfrac{\sqrt{2}}{2}$ for $\sin A$ in the formula

$$\sin^2 A + \cos^2 A = 1.$$
$$\left(-\dfrac{\sqrt{2}}{2}\right)^2 + \cos^2 A = 1$$
$$\cos^2 A = 1 - \dfrac{2}{4} = \dfrac{1}{2}$$
$$\cos A = \dfrac{\sqrt{2}}{2} \quad \text{or} \quad \cos A = -\dfrac{\sqrt{2}}{2}$$

Since angle A is in Quadrant III, $\cos A$ is negative.

Thus, $\cos A = -\dfrac{\sqrt{2}}{2}.$

Then $\tan A = \dfrac{\sin A}{\cos A} = \dfrac{-\dfrac{\sqrt{2}}{2}}{-\dfrac{\sqrt{2}}{2}} = 1.$

$\therefore \cos A = -\dfrac{\sqrt{2}}{2}$ and $\tan A = 1.$

2. Find $\cos A$ and $\tan A$, given that angle A lies in Quadrant IV, and $\sin A = -\dfrac{\sqrt{10}}{10}.$

Substitute $-\dfrac{\sqrt{10}}{10}$ for $\sin A$ in $\sin^2 A + \cos^2 A = 1.$

$$\left(-\dfrac{\sqrt{10}}{10}\right)^2 + \cos^2 A = 1$$
$$\cos^2 A = 1 - \dfrac{1}{10} = \dfrac{9}{10}$$
$$\cos A = \dfrac{3\sqrt{10}}{10} \text{ or}$$
$$\cos A = -\dfrac{3\sqrt{10}}{10}$$

Since angle A is in Quadrant IV, $\cos A$ is positive.

$\therefore \cos A = \dfrac{3\sqrt{10}}{10}$

Then $\tan A = \dfrac{\sin A}{\cos A}$

$$= \dfrac{-\dfrac{\sqrt{10}}{10}}{\dfrac{3\sqrt{10}}{10}} = -\dfrac{1}{3}.$$

$\therefore \cos A = \dfrac{3\sqrt{10}}{10}$ and

$\tan A = -\dfrac{1}{3}$

___

Additional A Exercises

Give the sine, cosine, and tangent of an angle A in standard position whose terminal side contains the given point on the unit circle.

1. $\left(-\dfrac{\sqrt{10}}{10}, \dfrac{3\sqrt{10}}{10}\right)$

$\sin A = \dfrac{3\sqrt{10}}{10}$

$\cos A = -\dfrac{\sqrt{10}}{10}$

$\tan A = -3$

2. $\left(-\dfrac{3\sqrt{10}}{10}, -\dfrac{\sqrt{10}}{10}\right)$

$\sin A = -\dfrac{\sqrt{10}}{10}$

$\cos A = -\dfrac{3\sqrt{10}}{10}$

$\tan A = \dfrac{1}{3}$

3. If $\sin A = \dfrac{2\sqrt{2}}{3}$, and $\angle A$ lies in Quadrant I, find $\cos A$ and $\tan A$.

$\cos A = \dfrac{1}{3}$

$\tan A = 2\sqrt{2}$

4. If $\sin A = \dfrac{\sqrt{3}}{2}$, and $\angle A$ lies in Quadrant II, find $\cos A$ and $\tan A$.

$\cos A = -\dfrac{1}{2}$

$\tan A = -\sqrt{3}$

Oral Exercises

In the figure below, state the sine, cosine, and tangent (if it exists) of the angle in standard position whose terminal side contains the given point.

1. A 0; 1; 0
2. B $\frac{\sqrt{3}}{2}$; $\frac{1}{2}$; $\sqrt{3}$
3. C 1; 0; undefined
4. D $\frac{\sqrt{2}}{2}$; $-\frac{\sqrt{2}}{2}$; -1
5. E 0; -1; 0
6. F $-\frac{3}{5}$; $-\frac{4}{5}$; $\frac{3}{4}$
7. G -1; 0; undefined
8. H $-\frac{5}{13}$; $\frac{12}{13}$; $-\frac{5}{12}$

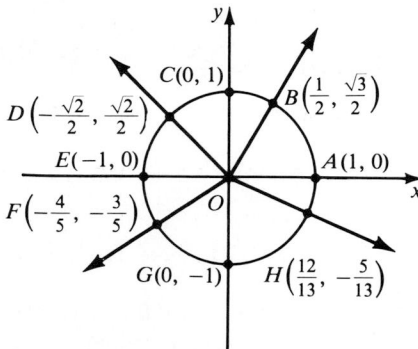

Use the figure above to state the value (if it exists) of each of the following.

9. $\sin 0°$ 0
10. $\cos 0°$ 1
11. $\sin 90°$ 1
12. $\cos 90°$ 0
13. $\tan 0°$ 0
14. $\tan 90°$ undefined
15. $\cos 180°$ -1
16. $\sin 270°$ -1
17. $\cos 360°$ 1
18. $\sin 180°$ 0
19. $\cos 450°$ 0
20. $\tan 180°$ 0
21. $\sin (-90°)$ -1
22. $\cos (-180°)$ -1
23. $\tan (-360°)$ 0
24. $\sin (-540°)$ 0

Explain why there is no angle A for which the given statement is true.

25. $\sin A = 2$
26. $\cos A = -1.5$

25, 26. The range of the sine and cosine functions is restricted to the set of real numbers between -1 and 1, inclusive.

Written Exercises

Give the sine, cosine, and tangent of an angle A in standard position whose terminal side contains the given point on the unit circle.

A
1. $\left(\dfrac{3}{5}, \dfrac{4}{5}\right)$
2. $\left(\dfrac{4}{5}, \dfrac{3}{5}\right)$
3. $\left(-\dfrac{4}{5}, \dfrac{3}{5}\right)$
4. $\left(\dfrac{3}{5}, -\dfrac{4}{5}\right)$

5. $\left(\dfrac{5}{13}, \dfrac{12}{13}\right)$
6. $\left(-\dfrac{5}{13}, \dfrac{12}{13}\right)$
7. $\left(\dfrac{\sqrt{2}}{2}, -\dfrac{\sqrt{2}}{2}\right)$
8. $\left(-\dfrac{\sqrt{2}}{2}, -\dfrac{\sqrt{2}}{2}\right)$

9. $\left(\dfrac{\sqrt{2}}{2}, \dfrac{\sqrt{2}}{2}\right)$
10. $\left(-\dfrac{\sqrt{3}}{2}, -\dfrac{1}{2}\right)$
11. $\left(\dfrac{\sqrt{3}}{2}, \dfrac{1}{2}\right)$
12. $\left(\dfrac{1}{2}, -\dfrac{\sqrt{3}}{2}\right)$

In Exercises 13–20 the quadrant of angle A is given, and also the value of sin A, cos A, or tan A. Find the remaining two values.

13. $\cos A = \frac{\sqrt{3}}{2}$; Quadrant I

14. $\sin A = \frac{3}{5}$; Quadrant I

15. $\sin A = -\frac{\sqrt{2}}{2}$; Quadrant IV

16. $\cos A = -\frac{4}{5}$; Quadrant II

17. $\tan A = 1$; Quadrant III

18. $\tan A = -\sqrt{3}$; Quadrant IV

19. $\cos A = -\frac{12}{13}$; Quadrant II

20. $\sin A = -\frac{5}{13}$; Quadrant III

21. Copy and complete the following chart.

p = positive; n = negative

	Quadrant I	Quadrant II	Quadrant III	Quadrant IV
sin A	positive	? p	? n	? n
cos A	? p	negative	? n	? p
tan A	? p	? n	? p	? n

Find sin A, cos A, and tan A such that the given conditions are satisfied.

B 22. $\sin A = \cos A$ and $\sin A < 0$
$\sin A = \cos A = -\frac{\sqrt{2}}{2}$; $\tan A = 1$

23. $\sin A = \tan A$ and $\cos A > 0$
$\sin A = \tan A = 0$; $\cos A = 1$

Using the trigonometric formulas given on page 586, show why there is no angle A for which the given statement is true.

24. $\sin A = 1$ and $\cos A = -1$

25. $\sin A = 0$ and $\cos A = 0$

26. $\sin A < 0$, $\cos A > 0$, and $\tan A > 0$

27. $\sin A < 0$, $\cos A < 0$, and $\tan A < 0$

28. $0 < \tan A < \sin A$

29. $\sin A < \tan A < 0$

In each of Exercises 30–33, k is a positive real number. Find the sine, cosine, and tangent of the angle A in standard position whose terminal side contains the given point. (*Hint:* Find the distance of the given point from the origin. Then solve for "p" and "q" by using the fact that lengths of corresponding sides of similar triangles have equal ratios.)

30. $(3k, 4k)$

31. $(5k, 12k)$

32. $(k, k\sqrt{3})$

33. $(k\sqrt{2}, k\sqrt{2})$

Find the value of k if the terminal side of angle A in standard position contains the point whose coordinates are given. (*Hint:* Use corresponding sides of similar triangles.)

34. $(k, 4)$, if $\tan A = 3$ $k = \frac{4}{3} = 1\frac{1}{3}$

35. $(3, k)$, if $\tan A = \frac{1}{2}$ $k = \frac{3}{2} = 1\frac{1}{2}$

36. $(6, 3k)$, if $\tan A = -1$ $k = -2$

37. $(6, -2k)$, if $\tan A = -\frac{5}{12}$ $k = \frac{5}{4} = 1\frac{1}{4}$

Trigonometry and Vectors **589**

(continued on p. 616)

C **38.** Draw a unit circle. Then use a protractor to find the intersections of the circle with the terminal sides of angles at 30° intervals: 30°, 60°, 90°, . . . , 360°. Accurately transfer to a function graph the lengths of line segments for sin A found from the unit circle diagram. To maintain a reasonable scale in the graph, make the length of the unit circle radius three times as long as whatever unit you choose for the 30° intervals on the function graph. Connect the plotted points with a smooth curve. Make a similar graph for cos A.

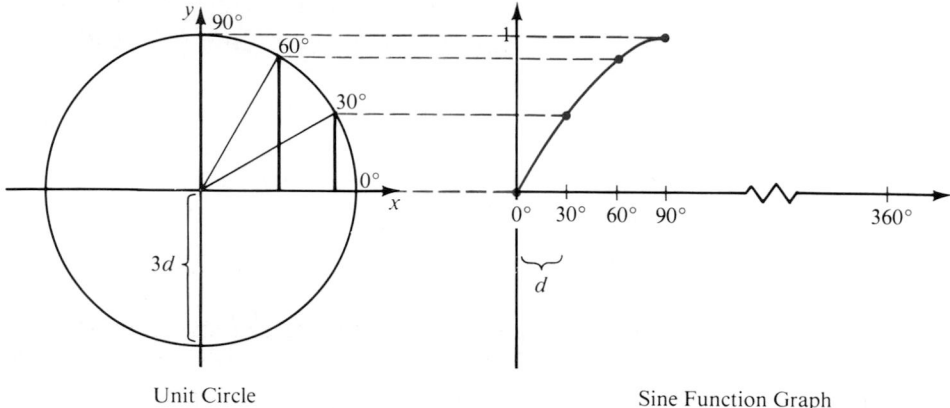

Unit Circle   Sine Function Graph

12–3 Trigonometric Tables

You can use the table on page 684 to find approximations for sin A, cos A, and tan A for angles *in the first quadrant* when you know the measure of angle A in degrees. A part of that table is shown below.

$m \angle A$	sin A	cos A	tan A
16°	0.2756	0.9613	0.2867
17°	0.2924	0.9563	0.3057
18°	0.3090	0.9511	0.3249

For example, to find an approximation for tan 17° from this table, you use the entry where the *row* containing 17° intersects the *column* headed "tan A." That entry is shown here in color and is 0.3057. The word *approximation* is used in describing the entries in this table because, except for sin 30°, tan 45°, cos 60°, sin 90°, and cos 90°, the entries represent rational-number approximations to irrational numbers.

If you are given a four-digit decimal as a value of sin A, cos A, or tan A, you can reverse the procedure just described to find an approximation for $m \angle A$. For example, if you are given cos A = 0.9511, you locate 0.9511 in the column headed "cos A" and read $m \angle A$ from the left-hand column in the same (horizontal) row. You find $m \angle A = 18°$.

If a given value for sin A, cos A, or tan A is not an entry in the table, you can use the nearest table entry.

EXAMPLE If sin A = 0.2806, find the measure of angle A to the nearest degree.

SOLUTION There is no entry for 0.2806 in the "sin A" column of the table. The nearest entries are 0.2756 and 0.2924.
Find which of the entries is closer to 0.2806.

$$0.2924 - 0.2806 = 0.0118$$
$$0.2806 - 0.2756 = 0.0050$$

The entry 0.2806 is closer to 0.2756.

$\therefore m \angle A = 16°$ to the nearest degree, or $m \angle A \approx 16°$.

Oral Exercises

Use the table on page 684 to find the required value. Then tell whether or not this value is exact. e = exact; n = not exact

1. sin 20° 0.3420; n
2. cos 30° 0.8660; n
3. tan 60° 1.7321; n
4. sin 30° 0.50̄; e
5. cos 45° 0.7071; n
6. tan 18° 0.3249; n
7. sin 90° 1; e
8. cos 60° 0.50̄; e

Use the table on page 684 to find the measure of angle A, $1° \le m \angle A \le 90°$, to the nearest degree.

9. sin A = 0.2079 12°
10. cos A = 0.4067 66°
11. tan A = 5.6713 80°
12. sin A = 0.9934 83°
13. cos A = 0.4142 66°
14. tan A = 42.9628 88°

For Exercises 15–23, refer to the table on page 684.

For which $m \angle A$ is each of the following statements true?

15. sin 40° = cos A 50°
16. cos 10° = sin A 80°
17. sin 25° = cos A 65°
18. cos 60° = sin A 30°
19. sin A = cos A 45°
20. sin A = cos 2A 30°
21. The greater the degree measure of $\angle A$, the ___?___ (greater/lesser) is the value of sin A. greater
22. The greater the degree measure of $\angle A$, the ___?___ (greater/lesser) is the value of cos A. lesser
23. The greater the degree measure of $\angle A$, the ___?___ (greater/lesser) is the value of tan A. greater

Trigonometry and Vectors **591**

2. Given that cos A = 0.9543, find the measure of angle A to the nearest degree. The entry 0.9543 is closer to 0.9563 than it is to 0.9511.
$\therefore m \angle A = 17°$ to the nearest degree, or $m \angle A \approx 17°$.

3. Given that sin A = 0.3002, find the measure of angle A to the nearest degree. The entry 0.3002 is closer to 0.2924 than it is to 0.3090.
$\therefore m \angle A = 17°$ to the nearest degree, or $m \angle A \approx 17°$.

Reading Algebra

To help students understand the generalized concept of a directed angle, point out some familiar examples: the winding of a watch, the turning of gears, and so on. Have students offer other examples.

To help students read and use the table on page 684, have them tell how they would find the approximate value for sin 59° or find the approximate angle whose cosine is 0.8975.

To assist in student understanding of the definitions of the trigonometric functions, point out that, given the point at which an angle in standard position intersects the unit circle, the sine of the angle Is the y-coordinate and the cosine is the x-coordinate. Consequently, the tangent of the angle is $\frac{y}{x}$.

Mixed Review

Solve.

1. $-5 < 3x + 4 \le 4$
$\{x: -3 < x \le 0\}$

2. $|2x - 1| > 4$
$\left\{x: x > \dfrac{5}{2} \text{ or } x < -\dfrac{3}{2}\right\}$

3. $|5 - x| = 1$ $\{4, 6\}$

4. $3|2 - 3x| < 0$ Ø

In Exercises 5–6, angle A is in Quadrant I, and the value of sin A, cos A, or tan A is given. Find the remaining two values.

5. $\sin A = \dfrac{1}{2}$

$\cos A = \dfrac{\sqrt{3}}{2}$

$\tan A = \dfrac{\sqrt{3}}{3}$

6. $\tan A = \dfrac{5}{12}$

$\sin A = \dfrac{5}{13}$

$\cos A = \dfrac{12}{13}$

Written Exercises

Use the table on page 684 to find the required value.

A

3.7321	0.6293	0.7071	0.9613	2.1445	0.5736
1. tan 75°	**2.** cos 51°	**3.** sin 45°	**4.** cos 16°	**5.** tan 65°	**6.** sin 35°
7. cos 55°	**8.** tan 89°	**9.** sin 81°	**10.** tan 6°	**11.** cos 73°	**12.** sin 60°
0.5736	57.2900	0.9877	0.1051	0.2924	0.8660

Use the table on page 684 to find the measure of angle A, $1° \le m \angle A \le 90°$, to the nearest degree.

13. sin $A = 0.2425$ 14°   **14.** cos $A = 0.2214$ 77°   **15.** cos $A = 0.1099$

16. tan $A = 0.3757$ 21°   **17.** cos $A = 0.5533$ 56°   **18.** sin $A = 0.6130$

19. tan $A = 2.8415$ 71°   **20.** cos $A = 0.8811$ 28°   **21.** sin $A = 0.9723$

22. tan $A = 16.8228$ 87°   **23.** tan $A = 0.5120$ 27°   **24.** sin $A = 0.3812$

15. 84°   **18.** 38°   **21.** 76°   **24.** 22°

Suppose that you have a table with sine values only for $1° \le m \angle A \le 90°$. What value would you look up to determine each of the following?

B

25. cos 45° = sin __?__   45°   **26.** cos 30° = sin __?__   60°   **27.** cos 70° = sin __?__   20°

28. cos 5° = sin __?__   85°   **29.** cos 60° = sin __?__   30°   **30.** cos R = sin __?__   90° − R

31. To the nearest degree, find $m \angle A$ if cos $A = 3$ sin A.   18°

Suppose that you have a table with sine values only for $1° \le m \angle A \le 90°$. How might you use this table to determine each of the following?

C

32. tan 40° $\tan 40° = \dfrac{\sin 40°}{\sin 50°}$ **33.** tan 10° $\tan 10° = \dfrac{\sin 10°}{\sin 80°}$ **34.** tan S

$$\tan S = \dfrac{\sin S}{\sin (90° - S)}$$

Computer Exercises   For students with computer experience

Note: **In the programs that you write for the following exercises, you will need to include a statement of the form**

LET A = A * 3.14159/180

before using A as the *argument* of any of the functions SIN, COS, or TAN. This is necessary because a computer is programmed to use a system of angle measurement that is different from degree measure.

1. Write a program that will print in chart form the sine, cosine, and tangent of any angle whose degree measure you input. Be sure to account for angles whose tangent is undefined.

2. Write a program that will print a table of the sines of angles from 0° through 90° in increments of 3°.

12–4 Solving Triangles

Teaching Suggestions
p. T111

Related Activities p. T112

Supplementary Material

Test 48

Key Ideas

Solve right triangles.

Suppose that you know the coordinates (x, y), $x \neq 0$, of some point B, other than the origin, on the terminal side of an angle A in standard position. Can you find $\sin A$, $\cos A$, and $\tan A$?

Figure 8 pictures an angle A in standard position. It also pictures points (p, q), $p \neq 0$, and $B(x, y)$, $x \neq 0$, on the terminal side of A; (p, q) is on the unit circle, and $B(x, y)$ is not on the unit circle.

You learned in Section 5-7 that the slope of a line is constant. The slope of line OB is given by $\frac{q}{p}$ and also by $\frac{y}{x}$. Therefore,

$$\frac{q}{p} = \frac{y}{x}.$$

But by definition of tangent, $\tan A = \frac{q}{p}$, and so

$$\tan A = \frac{y}{x}.$$

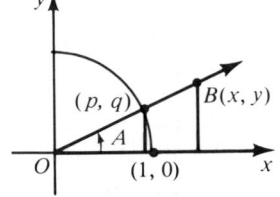

Figure 8

Since $\frac{q}{p} = \frac{y}{x}$, you have $q = \frac{y}{x}p$, and since $\sin A = q$ and $\cos A = p$, you have

$$\sin A = \frac{y}{x} \cos A.$$

Replacing $\sin A$ with $\frac{y}{x}\cos A$ in $\sin^2 A + \cos^2 A = 1$, you have:

$$\left(\frac{y}{x}\right)^2 \cos^2 A + \cos^2 A = 1$$

$$\left(\frac{y^2}{x^2} + 1\right) \cos^2 A = 1$$

$$\cos^2 A = \frac{1}{\dfrac{x^2 + y^2}{x^2}} = \frac{x^2}{x^2 + y^2}$$

Therefore:

$$\cos A = \frac{x}{\sqrt{x^2 + y^2}} \qquad \text{or} \qquad \cos A = \frac{-x}{\sqrt{x^2 + y^2}}$$

Since $\sqrt{x^2 + y^2} > 0$ and, for $\angle A$ in any quadrant, $\cos A$ and x are both positive or both negative, only the first of these equations applies.

$$\therefore \cos A = \frac{x}{\sqrt{x^2 + y^2}}$$

Since $\sin A = \frac{y}{x}\cos A$, you also have $\sin A = \frac{y}{x}\left(\dfrac{x}{\sqrt{x^2 + y^2}}\right)$, that is:

$$\sin A = \frac{y}{\sqrt{x^2 + y^2}}$$

1. Find sin A, cos A, and tan A for an angle A in standard position whose terminal side contains the point (3, 5).

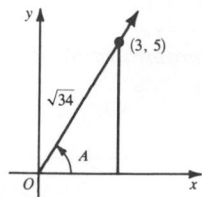

Since $x = 3$ and $y = 5$, you have:
$$\sqrt{x^2 + y^2} = \sqrt{3^2 + 5^2}$$
$$= \sqrt{34}$$

Then:

$$\sin A = \frac{y}{\sqrt{x^2 + y^2}} = \frac{5}{\sqrt{34}}$$

$$\cos A = \frac{x}{\sqrt{x^2 + y^2}} = \frac{3}{\sqrt{34}}$$

$$\tan A = \frac{y}{x} = \frac{5}{3}$$

2. Solve the right triangle ABC given that $m \angle A = 57°$ and $a = 8$. Give angle measures to the nearest degree and lengths to the nearest tenth of a unit. Make a sketch and label it.

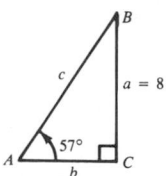

To find $m \angle B$, notice that $m \angle A + m \angle B = 90°$
Thus,
$$57° + m \angle B = 90°,$$
$$m \angle B = 33°.$$
To find b, use the fact that

$$\tan A = \frac{a}{b}, \text{ or}$$

$$b = \frac{a}{\tan A}.$$

For the case $x = 0$, you know that the terminal side of the angle is on the y-axis, and the formulas on the preceding page give sin $A = 1$ and cos $A = 0$ if $y > 0$, and sin $A = -1$ and cos $A = 0$ if $y < 0$.

For the case $y = 0$, the terminal side of the angle is on the x-axis, and the formulas give sin $A = 0$ and cos $A = 1$ if $x > 0$, and sin $A = 0$ and cos $A = -1$ if $x < 0$. These are also the values resulting from the definitions of sine and cosine on page 585.

Thus the formulas for sin A, cos A, and tan A shown on page 585 hold in every case, except that tan A is not defined if $x = 0$.

Notice that $\sqrt{x^2 + y^2}$ is the distance from the origin to point $B(x, y)$, because $\overline{OB}$ is the hypotenuse of a right triangle having remaining sides of lengths $|x|$ and $|y|$, as shown in Figure 9.

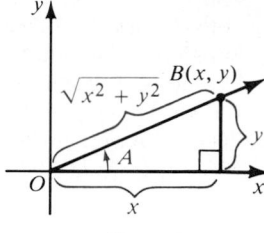

Figure 9

EXAMPLE 1 Find sin A, cos A, and tan A for an angle A in standard position whose terminal side contains the point (3, −2).

SOLUTION Since $x = 3$ and $y = -2$:
$$\sqrt{x^2 + y^2} = \sqrt{3^2 + (-2)^2} = \sqrt{13}$$
Then:

$$\sin A = \frac{y}{\sqrt{x^2 + y^2}} = \frac{-2}{\sqrt{13}}, \text{ or } -\frac{2}{\sqrt{13}}$$

$$\cos A = \frac{x}{\sqrt{x^2 + y^2}} = \frac{3}{\sqrt{13}}$$

$$\tan A = \frac{y}{x} = \frac{-2}{3}, \text{ or } -\frac{2}{3}$$

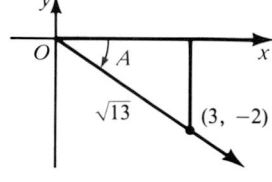

Any right triangle ACB in which $\angle C$ is the right angle can be placed so that $\angle A$ is in standard position, $\overline{AC}$ lies along the x-axis, and the point B is in the first quadrant, as shown in Figure 10. Then by applying the formulas on the preceding page, you arrive at the following.

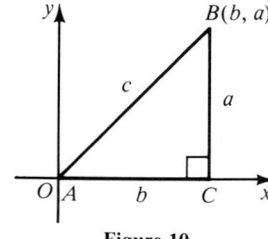

Figure 10

$$\tan A = \frac{a}{b} = \frac{\text{length of side opposite } \angle A}{\text{length of side adjacent to } \angle A}$$

$$\sin A = \frac{a}{\sqrt{a^2 + b^2}} = \frac{a}{c} = \frac{\text{length of side opposite } \angle A}{\text{length of hypotenuse}}$$

$$\cos A = \frac{b}{\sqrt{a^2 + b^2}} = \frac{b}{c} = \frac{\text{length of side adjacent to } \angle A}{\text{length of hypotenuse}}$$

594 *Chapter 12*

Finding measures, or approximations to measures, of angles or sides of a triangle when measures of other angles or sides are given is called **solving the triangle**. It is customary in labeling triangles to use capital letters for vertices and the corresponding lowercase letters for the lengths of the sides opposite the vertices, as shown in Figure 11. The capital letters are also used to denote the angles of the triangle. When A, B, and C and a, b, and c are used in this way, the right angle is ordinarily labeled C, and the length of the hypotenuse is represented by c.

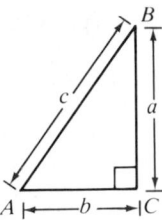

Figure 11

EXAMPLE 2 Solve the right triangle ABC given that $m\angle A = 42°$ and $b = 6$. Give angle measures to the nearest degree and lengths to the nearest tenth of a unit.

SOLUTION Make a sketch and label it. To find $m\angle B$, notice that

$$m\angle A + m\angle B = 90°.$$

Therefore:

$$42° + m\angle B = 90°$$
$$m\angle B = 48°$$

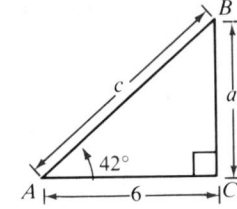

To find a, use the fact that $\tan A = \frac{a}{b}$, or $a = b \tan A$. Thus,

$$a = 6 \tan 42°.$$

From the table on page 684 you find that $\tan 42° \approx 0.9004$, and so

$$a \approx 6 \times 0.9004 \approx 5.4.$$

To find c, use

$$\frac{b}{c} = \sin B, \qquad \text{or} \qquad c = \frac{b}{\sin B} = \frac{6}{\sin 48°}.$$

From the table, $\sin 48° \approx 0.7431$, and so

$$c \approx \frac{6}{0.7431} \approx 8.1.$$

As a check, you might verify that the values of a, b, and c satisfy the Pythagorean Theorem. However, since the values of a and c were rounded, in this case you can only check that the value of $a^2 + b^2$ is *approximately* equal to the value of c^2.

$$a^2 + b^2 \approx c^2$$
$$5.4^2 + 6^2 \overset{?}{\approx} 8.1^2$$
$$29.16 + 36 \overset{?}{\approx} 65.61$$
$$65.16 \approx 65.61 \quad \checkmark$$

$\therefore m\angle B = 48°$, $a \approx 5.4$, $c \approx 8.1$

Then $b = \dfrac{8}{\tan 57°}$

and from the table on page 684 you find that $\tan 57° \approx 1.5399$, so

$$b \approx \frac{8}{1.5399} \approx 5.2.$$

To find c, you can use $\dfrac{a}{c} =$

$\sin A$ or $c = \dfrac{a}{\sin A} =$

$\dfrac{8}{\sin 57°}.$

From the table, $\sin 57° \approx$

0.8387, so $c \approx \dfrac{8}{0.8387} \approx$

$9.5.$

$\therefore m\angle B = 33°$, $b \approx 5.2$, $c \approx 9.5$.

Trigonometry and Vectors **595**

Oral Exercises

Exercises 1–10 refer to the three triangles below. Find the required value.

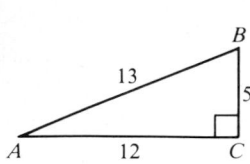
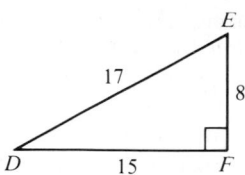
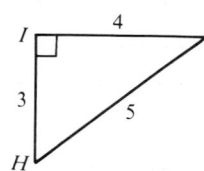

1. sin A $\frac{5}{13}$ **2.** cos B $\frac{5}{13}$ **3.** tan E $\frac{15}{8}$ **4.** cos D $\frac{15}{17}$ **5.** sin G $\frac{3}{5}$

6. cos H $\frac{3}{5}$ **7.** tan G $\frac{3}{4}$ **8.** sin D $\frac{8}{17}$ **9.** tan B $\frac{12}{5}$ **10.** sin H $\frac{4}{5}$

11. In triangle *MNP* below, would it be more convenient to find an approximation for *x* using tan 35° or using tan 55°? Why?

12. In triangle *XYZ* below, would it be more convenient to find an approximation for *x* using tan 20° or using tan 70°? Why?

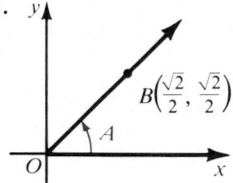

Ex. 11

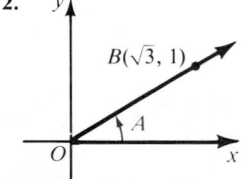

Ex. 12

Written Exercises

In Exercises 1–6 find the following.

a. the length of $\overline{OB}$ **b.** sin A **c.** cos A **d.** tan A

A

1.
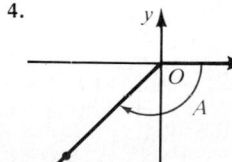

2.
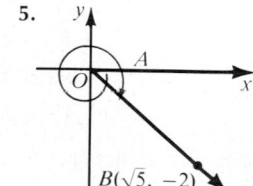

3.
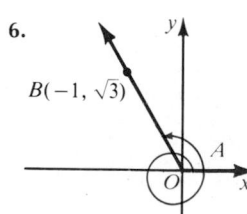

4.

5.

6.

Find sin A, cos A, and tan A for an angle A in standard position whose terminal side contains the given point.

7. (5, 5)
8. (1, √3)
9. (−√3, 1)
10. (−3, 4)
11. (−√3, √3)
12. (−5, −12)
13. (12, 5)
14. (0, −4)
15. (−3, 0)
16. (2, 2)
17. (−4, −3)
18. (√5, −√5)

Find an approximation for the length x or the measure of ∠A. Use the table on page 684 as necessary. Give angle measures to the nearest degree and lengths to the nearest tenth of a unit.

19.
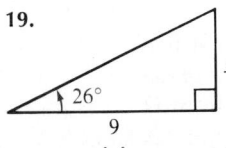
x = 4.4

20.
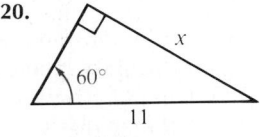
x = 9.5

21.
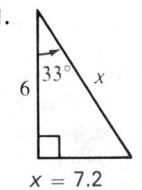
x = 7.2

22.
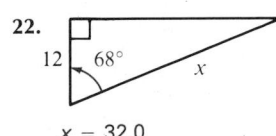
x = 32.0

23.
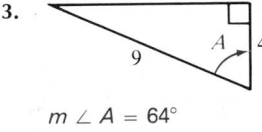
m ∠ A = 64°

24.
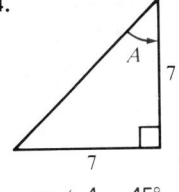
m ∠ A = 45°

25.
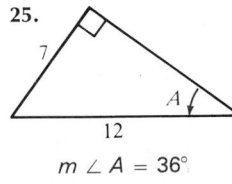
m ∠ A = 36°

26.
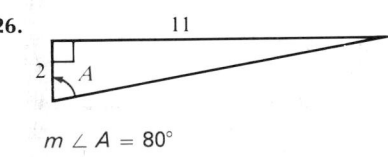
m ∠ A = 80°

In Exercises 27–34 solve the right triangle. Use the table on page 684 as necessary. Give angle measures to the nearest degree and lengths to the nearest tenth of a unit.

B

27.
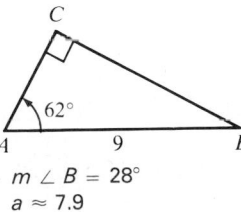
m ∠ B = 28°
a ≈ 7.9
b ≈ 4.2

28.
A ────── C
    42°
  40
m ∠ B = 48°
a ≈ 26.8
b ≈ 29.7

29.
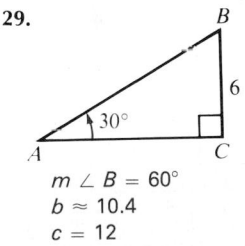
m ∠ B = 60°
b ≈ 10.4
c = 12

Trigonometry and Vectors **597**

---

4. a. √2   b. $-\dfrac{\sqrt{2}}{2}$

   c. $-\dfrac{\sqrt{2}}{2}$ d. 1

5. a. 3   b. $-\dfrac{2}{3}$

   c. $\dfrac{\sqrt{5}}{3}$ d. $-\dfrac{2\sqrt{5}}{5}$

6. a. 2   b. $\dfrac{\sqrt{3}}{2}$

   c. $-\dfrac{1}{2}$ d. $-\sqrt{3}$

7. $\sin A = \dfrac{\sqrt{2}}{2}$; $\cos A =$

   $\dfrac{\sqrt{2}}{2}$; $\tan A = 1$

8. $\sin A = \dfrac{\sqrt{3}}{2}$; $\cos A = \dfrac{1}{2}$;

   $\tan A = \sqrt{3}$

9. $\sin A = \dfrac{1}{2}$; $\cos A =$

   $-\dfrac{\sqrt{3}}{2}$; $\tan A = -\dfrac{\sqrt{3}}{3}$

10. $\sin A = \dfrac{4}{5}$; $\cos A = -\dfrac{3}{5}$;

    $\tan A = -\dfrac{4}{3}$

11. $\sin A = \dfrac{\sqrt{2}}{2}$; $\cos A =$

    $-\dfrac{\sqrt{2}}{2}$; $\tan A = -1$

12. $\sin A = -\dfrac{12}{13}$; $\cos A =$

    $-\dfrac{5}{13}$; $\tan A = \dfrac{12}{5}$

13. $\sin A = \dfrac{5}{13}$; $\cos A = \dfrac{12}{13}$;

    $\tan A = \dfrac{5}{12}$

14. sin A = −1; cos A = 0; tan A is undefined.

15. sin A = 0; cos A = −1; tan A = 0

(continued)

16. $\sin A = \dfrac{\sqrt{2}}{2}$; $\cos A =$

$\dfrac{\sqrt{2}}{2}$; $\tan A = 1$

17. $\sin A = -\dfrac{3}{5}$; $\cos A =$

$-\dfrac{4}{5}$; $\tan A = \dfrac{3}{4}$

18. $\sin A = -\dfrac{\sqrt{2}}{2}$; $\cos A =$

$\dfrac{\sqrt{2}}{2}$; $\tan A = -1$

36.
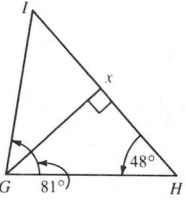

37.
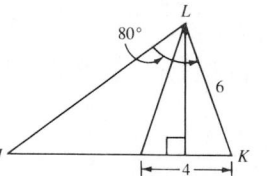

38.
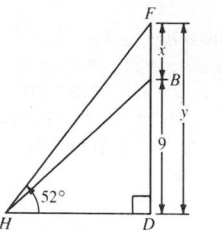

39.
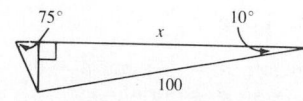

30. $m \angle B = 53°$
$b \approx 10.6$
$c \approx 13.3$
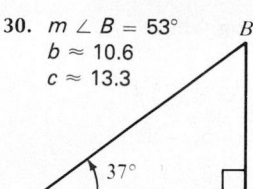

31.
$m \angle B = 36°$
$a \approx 40.5$
$b \approx 29.4$

32.
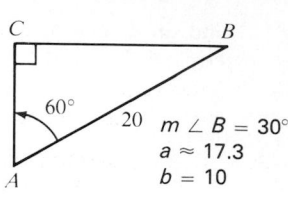
$m \angle B = 30°$
$a \approx 17.3$
$b = 10$

33.
$m \angle A \approx 66°$
$m \angle B \approx 24°$
$c \approx 9.8$

34.
$m \angle A \approx 53°$
$m \angle B \approx 37°$
$b = 9$

In Exercises 35–39 copy the figure, and by drawing an additional line segment introduce some right triangles to help you find a value for x to the nearest tenth of a unit. (*Hint for Exercise 35*: Draw a segment from *D* perpendicular to $\overline{AB}$. Find *z*, then *y*, and note that $x = y + z$.) See Additional Answers for sketches.

C 35. $x \approx 11.8$
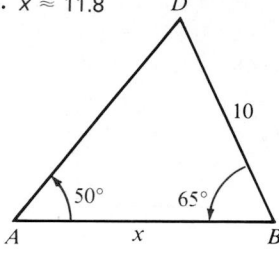

36. $x \approx 6.4$
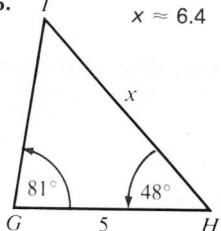

37. $x \approx 11.7$
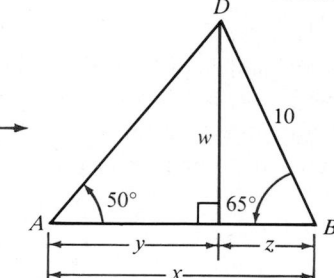

38. $x \approx 3.8$
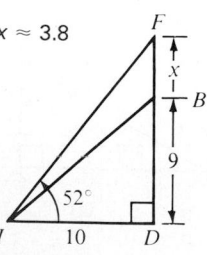

39.
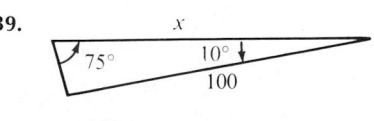

$x \approx 103.2$

Problems

Solve each problem using the table on page 684 as necessary. Give angle measures to the nearest degree and lengths to the nearest tenth of a unit.

A

1. After takeoff, an airplane maintained a flight angle of 8° with the ground as shown at the right. Find its elevation x after it had covered a ground distance of 1200 m. 168.6 m

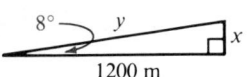

Exs. 1, 2

2. For the airplane in Problem 1, find the distance y it traveled in the air along the flight path while covering the ground distance of 1200 m. 1211.8 m

3. If an airplane attains an elevation of 210 m after takeoff while covering a ground distance of 1000 m, as shown below, what is the degree measure of its flight angle A with the ground? 12°

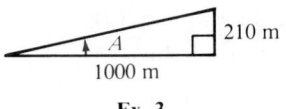

Ex. 3

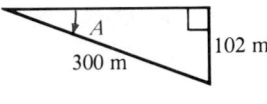

Ex. 4

4. Diving at a constant angle A, as shown above, a submarine descends 102 m while traveling 300 m. Find the degree measure of A. 20°

5. A tree is 20 m tall. When the angle of elevation of the sun is 34°, as shown below, how long is the shadow of the tree? 29.7 m

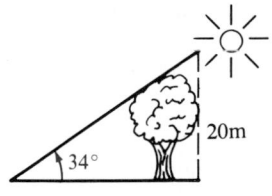

Ex. 5

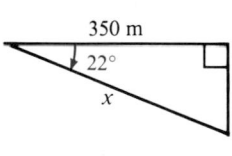

Ex. 6

6. A submarine maintains a diving angle of 22° as shown above. How far has it traveled when it is directly under a point 350 m along the surface from the point where it submerged? 377.5 m

B

7. At a point 80 m from a building, the angle of elevation of the top of the building is 32° and the angle of the top of a television antenna at the edge of the building is 34°, as shown at the right. What is the height x of the antenna? 4.0 m

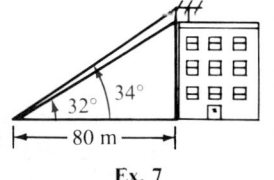

Ex. 7

Trigonometry and Vectors **599**

Additional A Exercises

In Exercises 1 and 2, find:
a. the length of $\overline{OB}$
b. sin A c. cos A d. tan A

1.

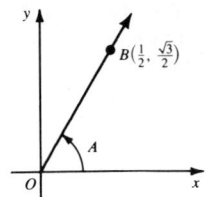

a. $OB = 1$ b. $\sin A = \dfrac{\sqrt{3}}{2}$

c. $\cos A = \dfrac{1}{2}$

d. $\tan A = \sqrt{3}$

2.

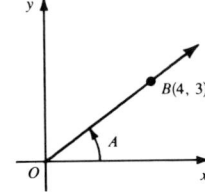

a. $OB = 5$ b. $\sin A = \dfrac{3}{5}$

c. $\cos A = \dfrac{4}{5}$

d. $\tan A = \dfrac{3}{4}$

In Exercises 3 and 4, find sin A, cos A, and tan A for an angle A in standard position whose terminal side contains the given point.

3. (1, 1)

$\sin A = \dfrac{\sqrt{2}}{2}$, $\cos A = \dfrac{\sqrt{2}}{2}$,

$\tan A = 1$

4. (1, 2)

$\sin A = \dfrac{2\sqrt{5}}{5}$, $\cos A = \dfrac{\sqrt{5}}{5}$,

$\tan A = 2$

(continued)

5. Find the approximate length x to the nearest tenth of a unit. Use the table on page 684 as necessary.

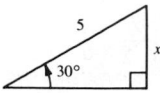

2.5

6. Find the approximate measure of $\angle A$ to the nearest degree. Use the table on page 684 as necessary.

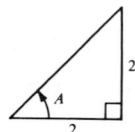

45°

Mixed Review

In Exercises 1–3 find the approximation in the table on page 684 for the given item.

1. sin 69° 0.9336
2. cos 15° 0.9659
3. tan 43° 0.9325

In Exercises 4–6 find the measure of angle A, $1° \le m \angle A \le 90°$, to the nearest degree.

4. sin A = 0.2500   14°
5. cos A = 0.3333   71°
6. tan A = 0.1429   8°

---

8. Two surveying transits are located 250 m apart. Both transits are sighted on the same rock. For each transit, the angle between the line of sight to the rock and the line of sight to the other transit is 45°. What is the distance x from the rock to the line connecting the two transits? 125 m

9. If the transits described in Problem 8 are 120 m apart, and if each angle described in Problem 8 measures 65°, how far is the rock from the line connecting the transits? 128.7 m

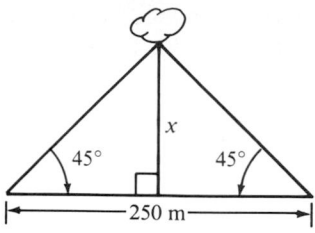

Exs. 8, 9

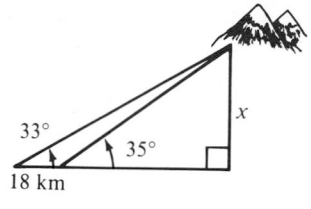

Ex. 10

10. At one point along a straight road the direction toward Mount Tanner makes an angle of 33° with the direction of the road. At another point 18 km farther along the road, the angle is 35°. Find the perpendicular distance x of Mount Tanner from the road. 161.1 km

C 11. A radar set at a point A sights a UFO at point B and tracks it to a point C along a straight and level path $\overline{BC}$. The distance from A to B is 12 km and the distance from A to C is 15 km. If $m\angle BAC$ is 6°, what is the distance from B to C? How fast was the UFO traveling (in km/h) if it took 10 s to move from B to C? (*Hint*: Use the dashed segment in the diagram.) 3.3 km; 1188 km/h

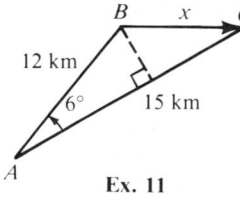

Ex. 11

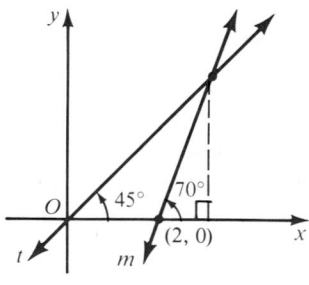

Ex. 12

12. On a coordinate plane, line t passes through the origin and forms a 45° angle with the positive x-axis. Line m passes through the point (2, 0) and forms a 70° angle with the positive x-axis. Find the coordinates of the point of intersection of lines t and m. (3.1, 3.1)

600   *Chapter 12*

Self-Test 1

VOCABULARY   angle (p. 579)
                       initial side (p. 579)
                       terminal side (p. 579)
                       directed angle (p. 579)
                       vertex of an angle (p. 579)
                       coterminal angles (p. 580)
                       measure of an angle (p. 580)
                       standard position (p. 580)

quadrantal angle (p. 580)
unit circle (p. 585)
sine of an angle (p. 585)
cosine of an angle (p. 585)
tangent of an angle (p. 585)
trigonometric functions (p. 586)
solve a triangle (p. 595)

Find the measure of $\angle A$, if $\angle A$ is coterminal with the given angle and $0° \leq m\angle A < 360°$.   *Obj. 1, p. 579*

1. 750°   30°
2. −240°   120°
3. 395°   35°

If $\angle A$ is in standard position, and the terminal side of $\angle A$ contains the point (−3, −4), determine the following.   *Obj. 2, p. 579*

4. $\sin A$ $-\frac{4}{5}$
5. $\cos A$ $-\frac{3}{5}$
6. $\tan A$ $\frac{4}{3}$

Use the table on page 684 to find the following.   *Obj. 3, p. 579*

7. $\sin 78°$ 0.9781
8. $\cos 51°$ 0.6293
9. $\tan 27°$ 0.5095

10. Solve the right triangle ABC given that $m\angle A = 48°$ and $b = 8$.   *Obj. 4, p. 579*
       Give lengths to the nearest tenth of a unit. $m\angle B = 42°$
                                               $a \approx 8.9$

Check your answers with those at the back of the book. $c \approx 12.0$

ON THE CALCULATOR

Many calculators have sine, cosine, and tangent keys that can be used in evaluating expressions involving trigonometric functions. Such calculators often have features that allow you to work with angles measured either in degrees or radians. (One *radian* is equal to the quotient when 180° is divided by π.)

Use a calculator to evaluate each expression to the nearest ten-thousandth. (Use the calculator's *degree mode*.)

1. $\sin 45°$ 0.7071
2. $\cos 71°$ 0.3256
3. $\tan 3°$ 0.0524
4. $2 \cos 22°$ 1.8544
5. $\frac{1}{2} \tan 84°$ 4.7572
6. $(\sin 51°)^2$ 0.6040
7. $(\sin 40°)^2 + (\cos 40°)^2$ 1
8. $(\cos 76°)^2 + (\sin 76°)^2$ 1

Trigonometry and Vectors   **601**

1. Find the measure of $\angle A$, if $\angle A$ is coterminal with the given angle and $0° \leq m\angle A < 360°$.
   a. 640°   280°
   b. −280°   80°
   c. 430°   70°

2. If $\angle A$ is in standard position, and the terminal side of $\angle A$ contains the point (12, −5), find:
   a. $\sin A$ $-\frac{5}{13}$
   b. $\cos A$ $\frac{12}{13}$
   c. $\tan A$ $-\frac{5}{12}$

3. Use the table on page 684 to find:
   a. $\sin 49°$ 0.7547
   b. $\cos 26°$ 0.8988
   c. $\tan 67°$ 2.3559

4. Solve the right triangle ABC given that $m\angle A = 41°$ and $b = 12$. Give lengths to the nearest tenth of a unit.
$m\angle B = 49°$; $a \approx 10.4$; $c \approx 15.9$

PROGRAMMING IN BASIC

The program given on these two pages can be used in solving right triangles. The variables A, B, C, A1, B1, and C1 refer to the angles and sides of a right triangle as labeled on the triangle at the right. The program will output values for all the angles and sides, except the right angle, when any of the following pairs are input.

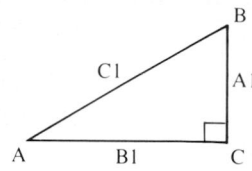

A, A1   A, B1   A, C1   A1, B1   A1, C1

Note that the trigonometric functions used in the BASIC language are defined for angles measured in *radians*, where

$$1 \text{ radian} = \frac{180}{\pi} \text{ degrees,} \qquad \text{or} \qquad 1 \text{ degree} = \frac{\pi}{180} \text{ radians.}$$

Thus, the program allows you to input an angle measure in degrees, but it uses $\pi \approx 3.14159$ to express this angle measure in radians.

Notice, too, that the program makes use of the BASIC function ATN(X). This function gives the radian measure of the angle between $-90°$ and $90°$ whose tangent is X.

```
10    PRINT "TO SOLVE A RIGHT TRIANGLE"
20    PRINT " WITH RIGHT ANGLE C AND"
30    PRINT " GIVEN SET (1) OR SET (2):"
40    PRINT " (1) A, A1 OR A, B1 OR A, C1"
50    PRINT " (2) A1, B1 OR A1, C1"
60    PRINT "SELECT SET NO. (1) OR (2)";
70    INPUT N
80    LET P = 3.14159
90    IF N = 2 THEN 320
95    REM *FOR ANGLE A AND A SIDE
100   PRINT "INPUT A";
110   INPUT A
120   LET B = 90 - A
130   LET A0 = A * P/180
140   PRINT "SELECT (1) A1 OR (2) B1 OR (3) C1";
150   INPUT N
160   ON N GOTO 170, 220, 270
165   REM *FOR A1
170   PRINT "INPUT A1";
180   INPUT A1
190   LET B1 = A1/TAN(A0)
200   LET C1 = A1/SIN(A0)
210   GOTO 480
215   REM *FOR B1
```

If the computer has a built-in value for π, use that.

```
220   PRINT "INPUT B1";
230   INPUT B1
240   LET A1 = B1 * TAN(A0)
250   LET C1 = B1/COS(A0)
260   GOTO 480
265   REM *FOR C1
270   PRINT "INPUT C1";
280   INPUT C1
290   LET A1 = C1 * SIN(A0)
300   LET B1 = C1 * COS(A0)
310   GOTO 480
315   REM *FOR TWO SIDES
320   PRINT "INPUT A1";
330   INPUT A1
340   PRINT "SELECT (1) B1 OR (2) C1";
350   INPUT N
360   ON N GOTO 370, 430
365   REM *FOR B1
370   PRINT "INPUT B1";
380   INPUT B1
390   LET A = (180/P) * ATN(A1/B1)
400   LET B = 90 - A
410   LET C1 = SQR(A1*A1 + B1*B1)
420   GOTO 480
425   REM *FOR C1
430   PRINT "INPUT C1";
440   INPUT C1
450   LET B1 = SQR(C1*C1 - A1*A1)
460   LET A = (180/P) * ATN(A1/B1)
470   LET B = 90 - A
475   REM *OUTPUT
480   PRINT
490   PRINT "ALL PARTS OF THE TRIANGLE:"
500   PRINT "A = ";INT(100*A + 0.5)/100  ⎤
510   PRINT "B = ";INT(100*B + 0.5)/100  ⎥  Values are
520   PRINT "A1 = ";INT(100*A1 + 0.5)/100 ⎬ rounded to
530   PRINT "B1 = ";INT(100*B1 + 0.5)/100 ⎥  hundredths.
540   PRINT "C1 = ";INT(100*C1 + 0.5)/100 ⎦
550   END
```

Exercises

Type in the program as given. Then RUN the program for the following pairs of parts.

1. $A = 30°$; $A1 = 0.5$ 2. $A = 60°$; $B1 = 5$

3. $A = 30°$; $C1 = 20$ 4. $A1 = 3$; $B1 = 4$

5. $A1 = 3$; $C1 = 5$ 6. $A1 = 1$; $B1 = 1$

Additional Answers
Programming in BASIC

1. $A = 30$
  $B = 60$
  $A1 = .5$
  $B1 = .87$
  $C1 = 1$

2. $A = 60$
  $B = 30$
  $A1 = 8.66$
  $B1 = 5$
  $C1 = 10$

3. $A = 30$
  $B = 60$
  $A1 = 10$
  $B1 = 17.32$
  $C1 = 20$

4. $A = 36.87$
  $B = 53.13$
  $A1 = 3$
  $B1 = 4$
  $C1 = 5$

5. $A = 36.87$
  $B = 53.13$
  $A1 = 3$
  $B1 = 4$
  $C1 = 5$

6. $A = 45$
  $B = 45$
  $A1 = 1$
  $B1 = 1$
  $C1 = 1.41$

Vectors

OBJECTIVES for Sections 12-5 through 12-7:
1. To sketch a vector in standard position on a coordinate plane.
2. To use the distance formula to find the norm of a vector.
3. To find the norm of the resultant of two given vectors.
4. To use vectors to solve problems.

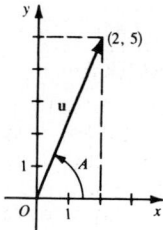
12–5 Vectors in the Plane

The red arrow, or *directed line segment*, in Figure 12 represents a *vector quantity*, namely, the displacement of a boat that has traveled 9 km northeast from port. Such an arrow indicating both a *magnitude* and a *direction* is called a **vector**. Force and velocity are other examples of vector quantities.

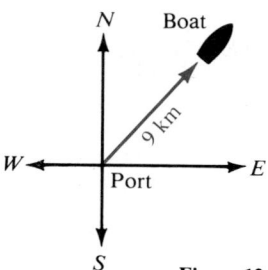

Figure 12   **Figure 13**

Vectors are usually designated by lowercase letters in bold type. For example, Figure 13 shows a vector **v** in *standard position* on a coordinate plane, that is, with **initial point** at the origin. The **terminal point** *P* is at the tip of the arrowhead.

The magnitude, or **norm**, of a vector **v** is represented by the symbol ‖**v**‖. The **direction** of a vector in standard position is the measure of the angle of rotation from the positive *x*-axis to the vector. The direction of vector **v** in Figure 13 is *m* ∠ *A*.

EXAMPLE 1   For the vector **v** shown at the right, find ‖**v**‖ to the nearest tenth and its direction to the nearest degree.

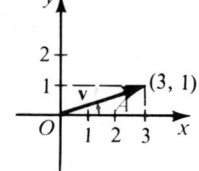

SOLUTION   From the Pythagorean Theorem:

$$‖\mathbf{v}‖^2 = 3^2 + 1^2 = 10$$
$$‖\mathbf{v}‖ = \sqrt{10} ≈ 3.2$$

Then to find the direction of **v**, you observe that $\tan A = \frac{1}{3} ≈ 0.3333$.

604   *Chapter 12*

From the table on page 684 and the fact that $\angle A$ is in the first quadrant, you find that $m \angle A \approx 18°$.

$\therefore \|\mathbf{v}\| \approx 3.2$ and the direction of $\mathbf{v} \approx 18°$.

You can find the norm and direction of a vector located anywhere in the plane if you know the coordinates of its initial and terminal points.

EXAMPLE 2   For the vector $\mathbf{u}$ shown at the right, find $\|\mathbf{u}\|$ to the nearest tenth and its direction to the nearest degree.

SOLUTION   Using the distance formula:

$$\|\mathbf{u}\| = \sqrt{(x_2 - x_1)^2 + (y_2 - y_1)^2}$$
$$= \sqrt{(4 - 2)^2 + (4 - 1)^2}$$
$$= \sqrt{2^2 + 3^2} = \sqrt{4 + 9} = \sqrt{13}$$
$$\approx 3.6$$

Further, to find the direction:

$$\tan A = \frac{y_2 - y_1}{x_2 - x_1} = \frac{3}{2} = 1.5$$

Thus, by the table on page 684:

$$m\angle A \approx 56°$$

$\therefore \|\mathbf{u}\| \approx 3.6$ and the direction of $\mathbf{u} \approx 56°$.

For any vector $\mathbf{v}$, the length of the *horizontal* displacement from the initial to the terminal point is called the *x-component* of $\mathbf{v}$. Similarly, the length of the *vertical* displacement is called the the *y-component* of $\mathbf{v}$. For example, in Figure 14 the x- and y-components of $\mathbf{v}$ are 1 and 3, respectively. For the vector $\mathbf{u}$, the x-component is $4 - 2$, or 2, and the y-component is $4 - 1$, or 3.

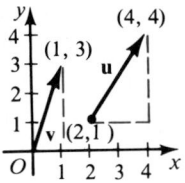

Figure 14

Note that a navigator gives the *bearing*, or *heading*, of a vector by measuring *clockwise* the angle from north to the vector. Notice that this is different from the definition of *direction* of a vector on the preceding page. Thus, in the diagram at the right, the bearing and magnitude of the ship's displacement from port are 135° and 12, respectively.

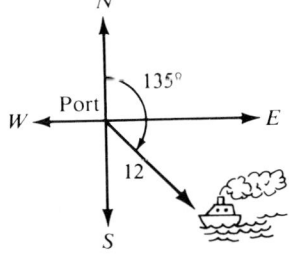

Trigonometry and Vectors   **605**

---

2. For the vector $\mathbf{w}$, find $\|\mathbf{w}\|$ to the nearest tenth and its direction to the nearest degree.

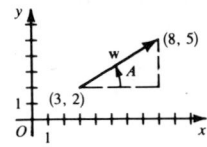

Using the distance formula, you have $\|\mathbf{w}\|$
$$= \sqrt{(x_2 - x_1)^2 + (y_2 - y_1)^2}$$
$$= \sqrt{(8 - 3)^2 + (5 - 2)^2}$$
$$= \sqrt{34} \approx 5.8.$$

$$\tan A = \frac{y_2 - y_1}{x_2 - x_1} = \frac{3}{5} = 0.6$$

so that $m \angle A \approx 31°$.

$\therefore \|\mathbf{w}\| \approx 5.8$ and the direction of $\mathbf{w} \approx 31°$.

3. Sketch a vector $\mathbf{t}$ given that $\|\mathbf{t}\| = 6$ and the direction of $\mathbf{t} = 225°$.

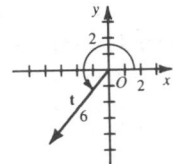

Common Errors

Students often confuse the direction of a vector with the bearing. Remind students that direction is measured counterclockwise from the positive x-axis to the vector. Bearing, on the other hand, is measured clockwise from the positive y-axis to the vector.

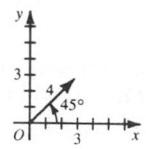
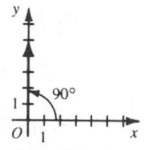
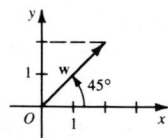
Oral Exercises

**Exercises 1–4 refer to the figure at the right. Give the
bearing of each vector.**

1. v 130°   **2. t** 340°   **3. s** 50°   **4. u** 225°

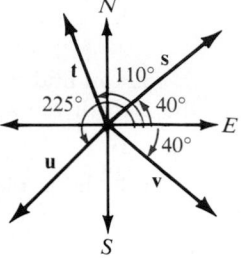

Exs. 1–4

**State the norm if the terminal point of a vector in standard
position has the given coordinates.**

5. (5, 5) $5\sqrt{2} \approx 7.1$ **6.** (1, 4) $\sqrt{17} \approx 4.1$ **7.** (0, −6) 6

8. (−2, −5) $\sqrt{29} \approx 5.4$ **9.** (3, *x*) $\sqrt{x^2 + 9}$ **10.** (*j*, *k*) $\sqrt{j^2 + k^2}$

Written Exercises

**Sketch a vector of the given magnitude and direction in
standard position on a coordinate plane.**

A   **1.** ‖**v**‖ = 3; direction 60°   **2.** ‖**v**‖ = 5; direction 140°
      3. ‖**v**‖ = 6; direction 260°   **4.** ‖**v**‖ = 2; direction 300°

**Exercises 5–8 refer to the figure at the right. Find the
required value.**

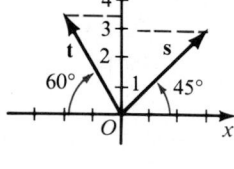

5. the norm of **s** $3\sqrt{2} \approx 4.2$ **6.** the norm of **t** $\frac{\sqrt{65}}{2} \approx 4.0$

7. the *x*-component of **s** 3   **8.** the *x*-component of **t** −2

Exs. 5–8

Find the norm and direction of the vector u.

9.

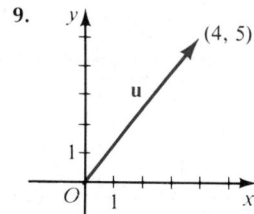

(4, 5) $\sqrt{41} \approx 6.4$;
51°

10.

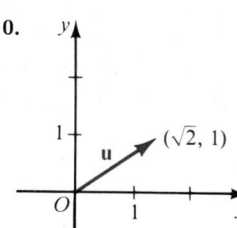

(√2, 1) $\sqrt{3} \approx 1.7$;
35°

11.

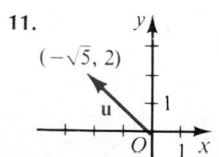

(−√5, 2)   3; 138°

12.

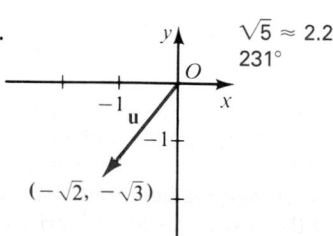

(−√2, −√3) $\sqrt{5} \approx 2.2$;
231°

13. A ship travels 8 km due east and then 8 km due north.
    a. What is the bearing of the ship from its initial point? 45°
    b. If the ship next travels 16 km due west, what is the final bearing from its initial point? 315°

14. An airplane flies 400 km due south from an airport and then 400 km due east.
    a. What is the bearing of the airplane from its starting point? 135°
    b. If the airplane next flies 1100 km due north, what is the final bearing from its starting point? 30°

Find the norm of u to the nearest tenth and the direction of u to the nearest degree.

B 15. **u** is in standard position, terminal point is $(-3, 5)$. 5.8; 121°
16. **u** is in standard position, terminal point is $(-3, -3)$. 4.2; 225°
17. **u** has initial point $(3, 2)$ and terminal point $(5, 6)$. 4.5; 63°
18. **u** has initial point $(2, 6)$ and terminal point $(6, 1)$. 6.4; 309°

C 19–22. What are the navigational bearings of the vectors described in Exercises 15–18?

19. 329°   **20.** 225°   **21.** 27°   **22.** 141°

Norbert Rillieux

1806–1894

Born in New Orleans, Norbert Rillieux was an engineer and inventor. He was educated in France, where he later taught applied mechanics.

In 1840, Rillieux returned to Louisiana, where he worked as an engineer in a sugar refinery. There he invented and patented a *multiple effect vacuum pan evaporator*, with which water could be removed from sugar cane without damaging the sugar. This invention increased the yield of sugar and reduced the cost of production. Today, the principle of multiple effect evaporation is still used in the manufacture of sugar, soap, glue, and condensed milk.

When Rillieux tried to interest New Orleans officials in advanced sewage-disposal methods, his ideas were rejected. Disheartened, Rillieux left the United States in 1854. He spent the rest of his life in Paris, where he adapted his evaporator for use in the sugar beet industry in France.

Trigonometry and Vectors **607**

In Exercises 5–6, find the norm of **v** to the nearest tenth and its direction to the nearest degree.

5.
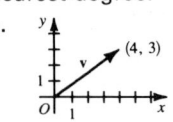

$\|\mathbf{v}\| = 5.0$
direction **v** ≈ 37°

6.
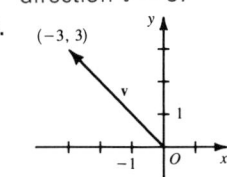

$\|\mathbf{v}\| \approx 4.2$
direction **v** = 135°

Mixed Review

1. Find the distance between $(0, -7)$ and $(2, 5)$. $2\sqrt{37}$

2. Find the distance between $(-1, -1)$ and $(-7, -7)$. $6\sqrt{2}$

3. Janet ran 100 m due north, then turned east and ran another 100 m. To the nearest tenth of a meter, how far was her finishing point from her starting point?   141.4 m

In Exercises 4–5, find the measure of ∠A to the nearest degree.

4.
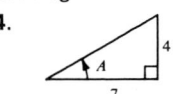

30°

5.
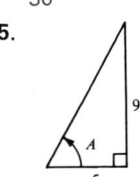

61°

Key Ideas

Find the sum of two vectors.
Find the x- and y-compo-
nents of the resultant of two
vectors.
Find the norm of the resultant of two vectors.

Chalkboard Examples

1. Draw a vector diagram
showing the sum **w** of **u**
and **v**, and find the x- and
y-components and the
norm of **w**.

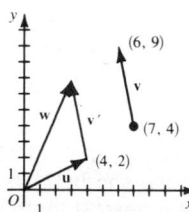

The vector **w** represents
u + **v**. The x-component
of **u** is 4, and that of **v** is
6 − 7, or −1. Hence, the
x-component of **w** is
4 + (−1), or 3.
The y-component of **u** is
2, and that of **v** is 9 − 4,
or 5. Hence, the y-compo-
nent of **w** is 2 + 5, or 7.
Then
$\|\mathbf{w}\| = \sqrt{3^2 + 7^2} = \sqrt{58}$.
Thus, the x-component of
w is 3, the y-component is
7, and the norm is $\sqrt{58}$.

12–6 The Sum of Two Vectors

Figure 15 pictures two successive displacements in the
plane, **s** and **t**. From the figure you can see that the x- and
y-components of **s** are 3 and 2, and those of **t** are 1 and 4.
The total resulting displacement from O is represented by
the vector **v**, with initial point O(0, 0) and terminal point
P(4, 6).

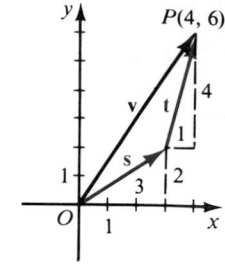

Figure 15

The x-component of **v** is the sum of the x-components of
s and **t**, namely, 3 + 1, or 4. The y-component of **v** is the
sum of the y-components of **s** and **t**, namely, 2 + 4, or 6.
The vector, **v** is called the **sum,** or **resultant,** of the vectors **s**
and **t**.

Vectors that have the same magnitude and direction are
called **equivalent vectors.** Figure 16 shows five equivalent
vectors, each having a norm of 2 units and directed 60°
counterclockwise from the positive x-direction.

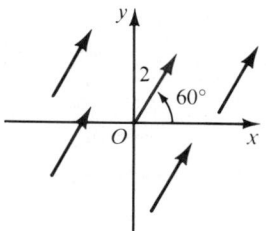

Figure 16

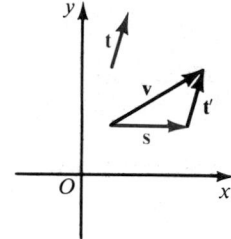

Figure 17

Figure 17 shows two vectors **s** and **t** in the plane. In order to find the
sum of **s** and **t**, you would first draw a vector **t'** equivalent to **t** and
positioned as shown, with its tail at the head of vector **s**. Then **v**
represents the sum **s** + **t**.

EXAMPLE   Draw a vector diagram showing the sum **v** of **s** and **t**, and find the x- and
y-components and the norm of **v**.

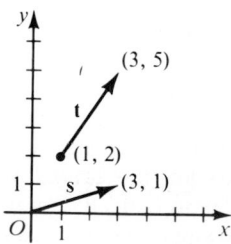

SOLUTION The red vector, **v**, represents **s** + **t**. The x-component of **s** is 3, and that of **t** is 3 − 1, or 2. Hence the x-component of **v** is 3 + 2, or 5.

The y-component of **s** is 1, and that of **t** is 5 − 2, or 3. Hence the y-component of **v** is 1 + 3, or 4.

Then

$$\|\mathbf{v}\| = \sqrt{5^2 + 4^2} = \sqrt{41}.$$

∴ the x-component of **v** is 5, the y-component is 4, and the norm is $\sqrt{41}$.

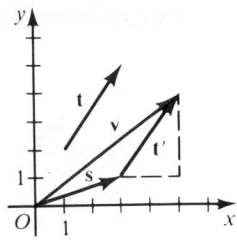

Oral Exercises

Tell which of the vectors p, q, or s is the resultant of the other two.

1.
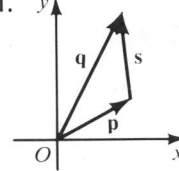
q

2.
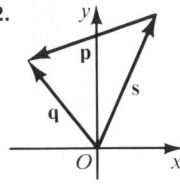
q

3.
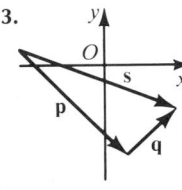
s

4.
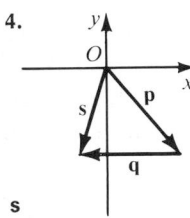
s

In Exercises 5–8, v is the sum of s and t. Name the following.
a. the x- and y-components of t   **b. ‖v‖**

5.
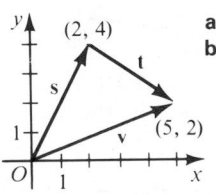
   a. 3; −2
   b. $\sqrt{29}$

6.
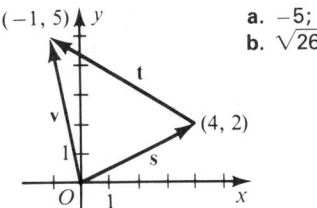
   a. −5; 3
   b. $\sqrt{26}$

7.
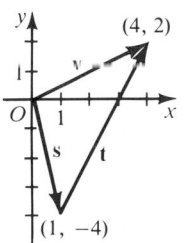
   a. 3; 6
   b. $2\sqrt{5}$

8.
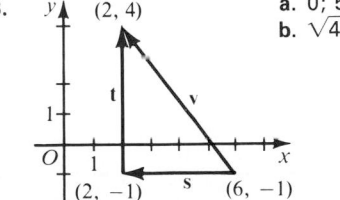
   a. 0; 5
   b. $\sqrt{41}$

2. Draw a vector diagram showing the sum **v** of **s** and **t**, and find the x- and y-components and the norm of **v**.

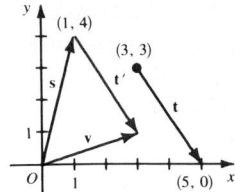

The vector **v** represents **s** + **t**. The x-component of **s** is 1, and that of **t** is 5 − 3, or 2. Hence, the x-component of **v** is 1 + 2, or 3.

The y-component of **s** is 4, and that of **t** is 0 − 3, or −3. Hence, the y-component of **v** is 4 + (−3), or 1. Then
$\|\mathbf{v}\| = \sqrt{3^2 + 1^2} = \sqrt{10}$.
Thus, the x-component of **v** is 3, the y-component is 1, and the norm is 10.

Reading Algebra

Direct students' attention to the definition of equivalent vectors. Emphasize that the definition of equivalent vectors only requires that they have the same magnitude and direction; there are no additional requirements. Since this is so, vectors equivalent to other vectors in the coordinate plane can be drawn in any desired location. Thus, to find the sum of two vectors, students should not hesitate to draw an equivalent vector at the head of another vector.

Trigonometry and Vectors   **609**

Additional A Exercises

In Exercises 1–2, **v** is the sum of **s** and **t**. Find the *x*- and *y*-components of **t** and the norm of the resultant vector **v**.

1.

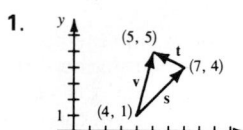

The *x*-component of **t** is −2, the *y*-component is 1, and the norm of **v** is $\sqrt{17}$.

2.

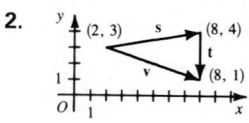

The *x*-component of **t** is 0, the *y*-component is −3, and the norm of **v** is $\sqrt{40}$, or $2\sqrt{10}$.

In Exercises 3–4, the vectors **v** and **u** are equivalent. Find the coordinates of the terminal point *P* of the vector **u**.

3.

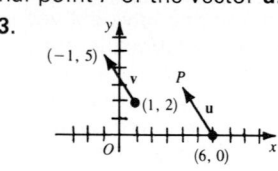

(4, 3)

4.

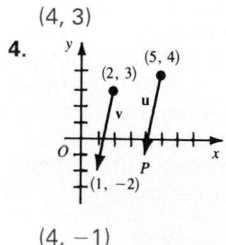

(4, −1)

Exercises 9 and 10 refer to the figure at the right.

9. Tell which vectors appear to be equivalent, and give a reason for your answer.

10. If you were to add **p** and **q**, what would be the magnitude of the vector for the sum? 0

11. What are the coordinates of the terminal point of the vector in standard position that is equivalent to the vector from (2, 3) to (5, 1)? (3, −2)

9. None. No two vectors appear to have the same magnitude and direction.

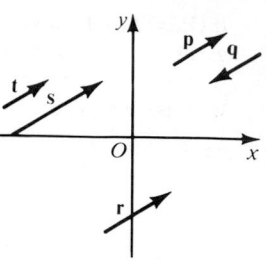

Exs. 9, 10

Written Exercises

Determine the following.

a. the *x*- and *y*-components of t   **b. the norm of the resultant vector v**

A **1.**

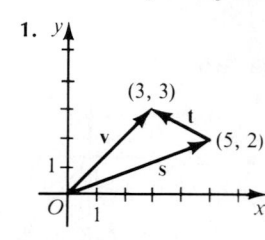
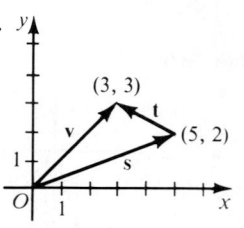

a. −2; 1
b. $3\sqrt{2} \approx 4.2$

2.

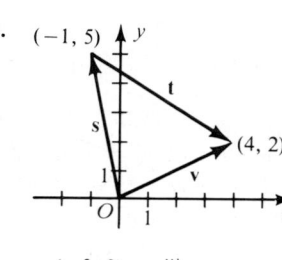
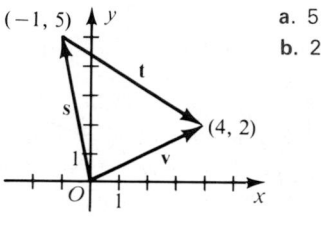

a. 5; −3
b. $2\sqrt{5} \approx 4.5$

3.

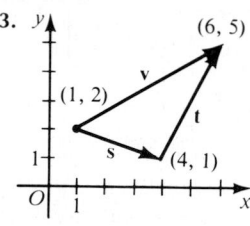
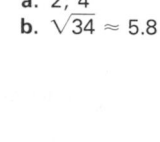

a. 2; 4
b. $\sqrt{34} \approx 5.8$

4.

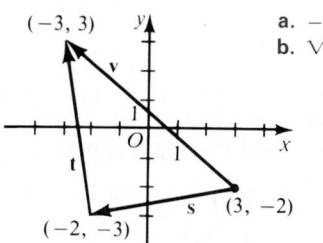

a. −1; 6
b. $\sqrt{61} \approx 7.8$

5–8. Find the direction of vector **v** in each of Exercises 1–4, to the nearest degree.
  5. 45°   **6.** 27°   **7.** 31°   **8.** 140°

In Exercises 9 and 10, the vectors u and v are equivalent. Find the coordinates of the terminal point *P* of the vector u.

9.

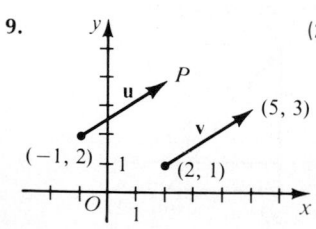

(2, 4)

10.

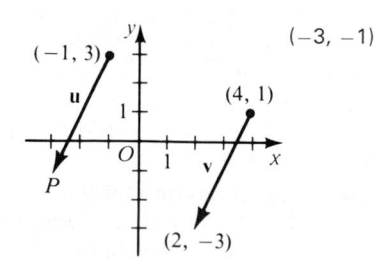

(−3, −1)

In Exercises 11 and 12:
a. Make a copy of the given vector diagram and on it show the sum
u = s + t and the sum v = t + s.
b. What can you conclude about u and v?

11.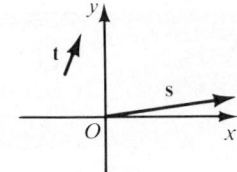

12.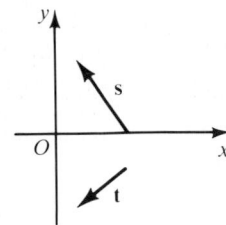

Determine *p* and *q* so that the given points will be the initial and terminal
points, respectively, of a vector equivalent to the vector from (5, 2) to
(1, 5).

B 13. $(-3, 1)$, (p, q) $p = -7$; $q = 4$
    15. $(-1, q)$, $(p, 7)$ $p = -5$; $q = 4$

14. (p, q), $(-3, 1)$ $p = 1$; $q = -2$
16. $(p, 3)$, $(-5, q)$ $p = -1$; $q = 6$

Using the norms and directions of vectors u and v as given, sketch u, v, and
their resultant u + v. Then find the norm of u + v and its direction to the
nearest degree. See Additional Answers for sketches.

C 17. **u**: 10; 0° $\|u + v\| = 2\sqrt{29}$ 18. **u**: 6; 45° $\|u + v\| = \sqrt{61}$ 19. **u**: 8; 30° $\|u + v\| = 10$
        v: 4; 90°   ≈ 10.8   **v**: 5; 315°   ≈ 7.8   **v**: 6; 120° direction: 67°
                 direction: 22°   direction: 5°

20. Is vector addition commutative or associative? Support your answer
with vector diagrams. Use vectors **r**, **s**, and **t**, with *x*- and *y*-
components of *a*, *b*; *c*, *d*; and *e*, *f* respectively.
Vector addition is both commutative and associative.

Computer Exercises For students with computer experience

1. Write a program that will determine the *x*- and *y*-components of a
vector when you input a positive number that represents its norm
and a positive degree measure that represents its direction with
respect to the positive *x*-axis. (*Hint*: Use the sine and cosine
functions. Remember to use a statement of the form LET A =
A * 3.14159/180, as discussed on page 592, before using the sine or
cosine of an angle.)

2. Modify the program that you wrote for Exercise 1 so that it allows
you to input a heading in degrees rather than the direction of a
vector in standard position.

Trigonometry and Vectors **611**

 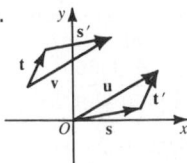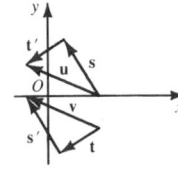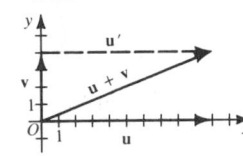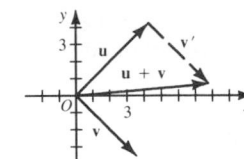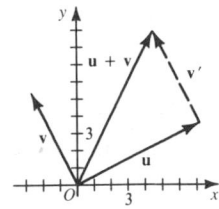

In Exercises 1–2, assume that $\angle A$ is in standard position and that
$0° \le m \angle A < 360°$.

1. If $\angle A$ is in Quadrant I, then $\underline{\ ?\ } < m \angle A < \underline{\ ?\ }$.
   $0° < m \angle A < 90°$

2. If $270° < m \angle A < 360°$, then $\angle A$ is in Quadrant $\underline{\ ?\ }$. IV

Find the norm and direction of the following vectors.

3.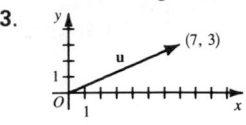

$\|\mathbf{u}\| = \sqrt{58} \approx 7.6$;
direction $\approx 23°$

4.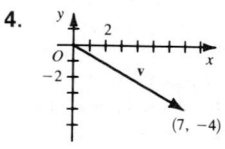

$\|\mathbf{v}\| = \sqrt{65} \approx 8.1$;
direction $\approx 330°$

Careers

Astronomy

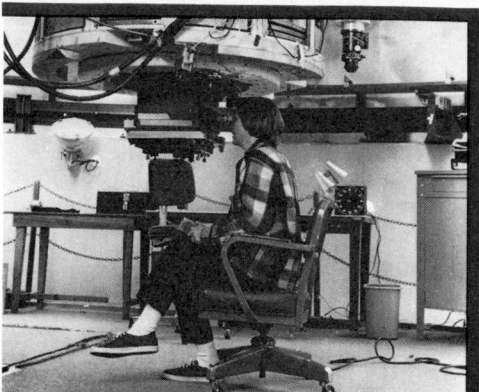

Astronomers are scientists who collect and analyze data pertaining to stars, planets, and other celestial bodies. They study the motions, chemical composition, sizes, and shapes of these bodies.

Astronomers use a variety of equipment. For example, *spectroscopes* are used to analyze the light from stars, while *radio telescopes* are used to measure x-rays and radio waves. Large telescopes such as the one at the right are used to observe the heavens and are often equipped with photographic devices to record these observations.

To find the distance from Earth to a star, astronomers use *parallax*, which is the apparent change in position of an object when it is viewed from two different points. To see an example of parallax, try this experiment: Hold your finger a few centimeters from your face. Now look at it first with one eye closed, then the other. Your eyes act as two different *observation points*, and your finger seems to change position.

In measuring the distance to a star, astronomers photograph it from two observation points that are as far apart as possible, that is, from opposite ends of Earth's orbit. When viewed against the background of other, very distant stars, the star being observed appears to change position. Astronomers measure this change in position in terms of the *parallax angle*, P. When $m \angle P$ is given in degrees, you can use the following proportion to find d, the distance from Earth to the star.

$$\frac{m \angle P}{360} = \frac{\text{radius of Earth's orbit}}{2 \pi d}$$

EXAMPLE Suppose that an astronomer determines that the parallax angle of the star Alpha Centauri is $0.0021°$. Given that the radius of Earth's orbit is approximately 1.5×10^8 km, find the distance from Earth to Alpha Centauri.

SOLUTION Substitute the given values into the parallax proportion, then solve for d.

$$\frac{0.0021}{360} = \frac{1.5 \times 10^8}{2 \pi d}$$

Using $\pi \approx 3.14$, $d \approx \dfrac{(1.5 \times 10^8)(360)}{2(3.14)(0.0021)} \approx 4.1 \times 10^{12}$

$\therefore$ the distance to Alpha Centauri is approximately 4.1×10^{12} km.

612 *Chapter 12*

12–7 Applications of Vectors

Figure 18 pictures the resultant force **v** when two different forces, **s** and **t**, are operating on an object at a point P. Physicists refer to the fact that the resultant force can be represented in this way as the *parallelogram law of forces*, because the resultant force **v** can be represented as the diagonal of the parallelogram formed by **s** and **t** and the equivalent vectors **s'** and **t'**. You can see from the figure that addition of vectors is a commutative operation since

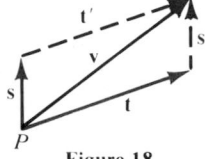

Figure 18

$$\mathbf{v} = \mathbf{s} + \mathbf{t'} = \mathbf{t} + \mathbf{s'},$$

or

$$\mathbf{v} = \mathbf{s} + \mathbf{t} = \mathbf{t} + \mathbf{s}.$$

EXAMPLE 1 Find the magnitude and the direction of the resultant force when an object is acted on at the point A by two forces at right angles to each other, each of magnitude 30 N (newtons).

SOLUTION First draw a *vector diagram*, such as the one shown at the right.

Since the two forces acting at A are at right angles to each other, you can use the Pythagorean Theorem to determine $\|\mathbf{v}\|$.

$$\|\mathbf{v}\| = \sqrt{30^2 + 30^2}$$
$$= \sqrt{1800} = 30\sqrt{2}$$

Since the two forces are equal and at right angles, triangle ABC in the diagram is an isosceles right triangle.

∴ the magnitude of **v** is $30\sqrt{2}$, and $m\angle CAB = 45°$.

EXAMPLE 2 A tractor pulls with a force of 800 N on a cable attached to an old tree stump. If the cable makes an angle of 15° with the ground, what are the horizontal component x and the vertical component y of the force, to the nearest tenth of a newton?

SOLUTION From the diagram, you can see that

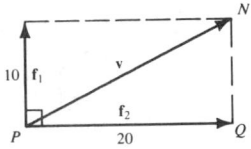

$$\cos 15° = \frac{x}{800} \quad \text{and} \quad \sin 15° = \frac{y}{800}$$

so that

$$x = 800 \cos 15° \approx 800(0.9659) = 772.72,$$

and

$$y = 800 \sin 15° \approx 800(0.2588) = 207.04.$$

∴ $x \approx 772.7$ N and $y \approx 207.0$ N

Teaching Suggestions
p. T113

Related Activities p. T113

Supplementary Material
Test 49

Key Ideas
Use vectors in solving problems.

Chalkboard Examples

1. Find **(a)** the magnitude and **(b)** the direction of the resultant force when an object is acted on at the point P by two forces at right angles to each other, one of magnitude 10 N and the other of magnitude 20 N.
    Draw a vector diagram.

    a. Since the forces are at right angles to each other, use the Pythagorean Theorem to determine $\|\mathbf{v}\|$.
    $$\|\mathbf{v}\| = \sqrt{10^2 + 20^2}$$
    $$= \sqrt{500} = 10\sqrt{5}$$

    b. $\tan \angle NPQ = \frac{1}{2}$
    ∴ $m\angle NPQ \approx 27°$

Additional A Exercises

Give magnitudes to the nearest tenth and angle measures to the nearest degree.

1. Two people push an object across a smooth surface. One pushes with a force of 25 N due north and the other with a force of 35 N due east. Describe the force acting on the object.
   magnitude ≈ 43.0 N;
   direction ≈ 36°

2. The captain of a boat sets a course due east at a speed of 9 km/h in a current flowing due south at a rate of 4 km/h. Describe the speed and bearing of the boat.
   speed ≈ 9.8 km/h;
   bearing ≈ 114°

3. If a balloon rises at the rate of 9 m/s and is blown horizontally at a speed of 4 m/s, what angle does the path of the balloon form with the level ground? 66°

Mixed Review

In Exercises 1–2, find the length x to the nearest tenth of a unit.

1.

38°
7

8.9

2.
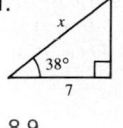

x
70°
4

11.7

Problems

Give magnitudes to the nearest tenth and angle measures to the nearest degree.

4. magnitude: 153.7 N; direction: 219°

A 1. A weather balloon is released vertically at a speed of 6 m/s in a wind blowing horizontally at a speed of 3 m/s. What angle does the path of the balloon form with the level ground? 63°

2. Laurie walks across fields from her home, first 2 km due north and then 1 km due east. To return home by the shortest path, how far and on what heading must she walk? 2.2 km; 207°

3. A boat moves due east at a speed of 12 km/h in a current flowing due south at a rate of 2 km/h. Describe the speed and bearing of the boat over the surface of the earth. 12.2 km/h; 99°

4. Two people push an object across a smooth surface. One pushes with a force of 96 N due south and the other with a force of 120 N due west. Describe the force acting on the object.

5. From a tractor equipped with a winch, a cable is attached to a large rock. If the cable makes a 40° angle with the ground, what is the magnitude of the force applied vertically to the rock when a force of magnitude 4500 N is applied along the cable? 2892.5 N

6. For the tractor in Problem 5, suppose that the cable makes a 35° angle with the ground and a force of 5000 N is applied along the cable. What is the magnitude of the force applied horizontally to the rock? 4095.8 N

B 7. A ship sails 90 km due north from a harbor, and then turns 30° toward the east and sails 80 km. How far from the harbor is the ship at that time, and what is the bearing from the harbor to the ship? 164.2 km; 14°

8. A kite string makes an angle of 48° with the horizontal. If the string is going out at the rate of 2 m/s, how fast is the kite rising? 1.5 m/s

9. An airplane flies at an air speed of 400 km/h on a bearing of 60° through a wind blowing due north at 50 km/h. Find the speed and bearing of the plane with respect to the ground.

10. By sitting in the center of a hammock, Mark exerts a force of 880 N downward. Find the pull on each supporting rope if each is 45° from the horizontal. 622.3 N

9. 427.2 km/h; 54°

C 11. A river flows from north to south. To cross from the east bank directly to the west bank a boat captain must keep on a course of 275°. If the trip takes 15 min when the boat travels 14.8 km/h, how wide is the river? What is the speed of the current? 3.7 km; 1.3 km/h

Self-Test 2

VOCABULARY vector (p. 604)
initial point (p. 604)
terminal point (p. 604)
norm of a vector (p. 604)

direction of a vector (p. 604)
sum or resultant of vectors
(p. 608)
equivalent vectors (p. 608)

1. Sketch the vector **v** in standard position with $\|\mathbf{v}\| = 5$ and direction   *Obj. 1, p. 604*
195°.

2. Find the norm to the nearest tenth and direction to the nearest degree   *Obj. 2, p. 604*
of a vector **v** whose initial and terminal points have the coordinates
(3, 2) and (6, 7), respectively. $\|\mathbf{v}\| = \sqrt{34} \approx 5.8$; direction: 59°

3. Find the norm of the resultant of **s** and   *Obj. 3, p. 604*
t in the diagram at the right.
$\|\mathbf{s} + \mathbf{t}\| = \sqrt{37} \approx 6.1$

4. A swimmer heads due east at a speed   *Obj. 4, p. 604*
of 3 km/h in a current that flows due
south at a speed of 2 km/h. Describe
the speed and bearing of the swimmer
over the surface of the earth.
3.6 km/h; 124°

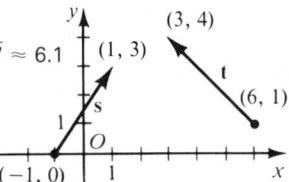

Check your answers with those at the back of the book.

Chapter Summary

1. A *directed angle* is the union of two ordered rays with the same
endpoint, together with a rotation from one ray, called the initial side,
to the other ray, called the terminal side.

2. The number of degrees through which a ray rotates from the initial
side to the terminal side is the *measure* of an angle.

3. An angle with its vertex at the origin and with the positive x-axis as
its initial side is in *standard position*.

4. If the terminal side of $\angle A$ in standard position intersects the unit
circle at (p, q), then the *cosine* of $\angle A$ (cos A) is p, the *sine* of $\angle A$
(sin A) is q, and the *tangent* of $\angle A$ (tan A) is $\frac{q}{p}$. The sine, cosine, and
tangent of an angle in standard position can be determined from any
point on the terminal side.

5. Given the measures of certain sides and angles of a triangle, *solving a
triangle* means to find its remaining sides and angles.

Trigonometry and Vectors **615**

In Exercises 3–4, find the magnitude and direction of the vectors.

3.

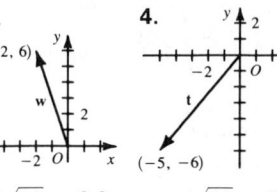

4.

$2\sqrt{10} \approx 6.3$;
108°

$\sqrt{61} \approx 7.8$;
230°

Quick Quiz

1. Sketch the vector **w** in standard position with $\|\mathbf{w}\| = 6$ and direction 210°.

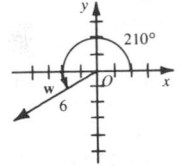

2. Find the norm and direction of a vector **t** whose initial and terminal points have the coordinates (2, 4) and (7, 8), respectively.
$\|\mathbf{t}\| = \sqrt{41}$; direction $\approx 39°$

3. Find the norm of the resultant of **u** and **v** in the diagram.

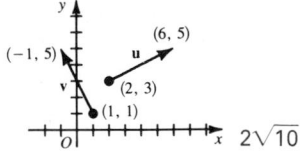

$2\sqrt{10}$

4. A ship sails 11 km due north from a harbor, and then turns due east and sails 12 km. How far from the harbor is the ship at that time, and what is the bearing from the harbor to the ship?
distance ≈ 16.3 km;
bearing $\approx 47°$

**Additional Answers
Written Exercises**
(continued from p. 589)

12. $\sin A = -\dfrac{\sqrt{3}}{2}$; $\cos A =$
$\dfrac{1}{2}$; $\tan A = -\sqrt{3}$

13. $\sin A = \dfrac{1}{2}$; $\tan A = \dfrac{\sqrt{3}}{3}$

14. $\cos A = \dfrac{4}{5}$; $\tan A = \dfrac{3}{4}$

15. $\cos A = \dfrac{\sqrt{2}}{2}$; $\tan A = -1$

16. $\sin A = \dfrac{3}{5}$; $\tan A = -\dfrac{3}{4}$

17. $\sin A = -\dfrac{\sqrt{2}}{2}$; $\cos A =$
$-\dfrac{\sqrt{2}}{2}$

18. $\sin A = -\dfrac{\sqrt{3}}{2}$; $\cos A = \dfrac{1}{2}$

19. $\sin A = \dfrac{5}{13}$; $\tan A = -\dfrac{5}{12}$

20. $\cos A = -\dfrac{12}{13}$; $\tan A = \dfrac{5}{12}$

24. If $\sin A = 1$, then $\cos A = 0$, since $\sin^2 A + \cos^2 A = 1$.

25. If $\sin A = 0$, then $\cos A = \pm 1$, since $\sin^2 A + \cos^2 A = 1$.

26. If $\sin A < 0$ and $\cos A > 0$, then $\tan A$ must be $<$ 0 since $\tan A = \dfrac{\sin A}{\cos A}$.

27. If $\sin A < 0$ and $\cos A < 0$, then $\tan A$ must be $>$ 0 since $\tan A = \dfrac{\sin A}{\cos A}$.

28. If $\tan A > 0$ and $\sin A > 0$, then $\angle A$ is in Quadrant I and $\cos A$ is positive and < 1.
Therefore, $\tan A = \dfrac{\sin A}{\cos A}$ must be greater than $\sin A$.

6. A *vector* is a directed line segment having both magnitude and direction. The *magnitude*, or *norm*, of a vector, $\|\mathbf{v}\|$, is its length. The *direction* of a vector in standard position is the measure of the angle of rotation from the positive x-axis to the vector.

7. *Equivalent vectors* have the same magnitude and direction.

8. The *sum*, or *resultant*, of two vectors is found by adding the x- and y-components of the vectors.

Chapter Review

Write the letter of the correct answer.

1. Which angle is *not* coterminal with an angle of 160° in standard position?   *12-1*
      a. −200°   **(b.)** 660°   **c.** 880°   **d.** −560°

2. If angle A is *not* a quadrantal angle, and $600° < m\angle A < 700°$, then the terminal side of angle A is in Quadrant ___?___.
      a. I or II   **b.** II or III   **(c.)** III or IV   **d.** I or IV

3. If $\sin A < 0$ and $\cos A < 0$ and angle A is in standard position, then the terminal side of angle A is in Quadrant ___?___.   *12-2*
      a. I   **b.** II   **(c.)** III   **d.** IV

4. The terminal side of $\angle A$ in standard position contains the point $\left(-\dfrac{\sqrt{2}}{2}, \dfrac{\sqrt{2}}{2}\right)$ on the unit circle. Then $\tan A = $ ___?___.
      a. 1   **(b.)** −1   **c.** $\sqrt{2}$ **d.** $-\sqrt{2}$

Use the table on page 684 to complete Exercises 5 and 6.

5. $\cos 75° = $ ___?___   *12-3*
      (a.) 0.2588   **b.** 0.9659   **c.** 3.7321   **d.** 0.9695

6. $\sin$ ___?___ $= 0.5592$
      a. 56°   **b.** 33°   **(c.)** 34°   **d.** 35°

7. The terminal side of $\angle A$ is in Quadrant II and $\cos A = -\frac{4}{5}$. Then $\sin A = $ ___?___.
      a. $-\frac{3}{5}$ **(b.)** $\frac{3}{5}$ **c.** $\frac{3}{4}$ **d.** $-\frac{3}{4}$

8. The terminal side of $\angle A$ in standard position contains the point $(-5, -12)$. Then $\cos A = $ ___?___.   *12-4*
      (a.) $-\frac{5}{13}$ **b.** $-\frac{12}{13}$ **c.** $\frac{5}{13}$ **d.** $\frac{12}{13}$

9. In right triangle ABC, $m\angle A = 70°$ and $c = 18$. Find b to the nearest tenth.
      a. 49.5   **b.** 16.9   **c.** 11.8   **(d.)** 6.2

616 *Chapter 12*

10. If a vector has initial point $(2, 1)$ and terminal point $(4, 5)$, find its direction to the nearest degree.   12-5
    a. 27°   (b.) 63°   c. 60°   d. 56°

11. If vector **v** is in standard position, with terminal point $(-8, 15)$, then $\|\mathbf{v}\| = \underline{\ ?\ }$.
    a. 12   b. $\sqrt{161}$ c. 7   (d.) 17

12. If vectors **s** and **t** are in standard position with terminal points   12-6
    $(-2, 2)$ and $(5, 2)$ respectively, and $\mathbf{v} = \mathbf{s} + \mathbf{t}$, then $\|\mathbf{v}\| = \underline{\ ?\ }$.
    a. $\sqrt{3}$ b. 4   (c.) 5   d. $\sqrt{10}$

13. If vector **p** has initial point $(1, 3)$ and terminal point $(-1, 5)$, vector **q** has initial point $(2, 4)$ and terminal point $(6, 6)$, and $\mathbf{v} = \mathbf{p} + \mathbf{q}$, then the x-component of vector **v** is $\underline{\ ?\ }$.
    (a.) 2   b. -2 c. 4   d. 6

14. A boat moves due west at a speed of 10 km/h in a current flowing   12-7
    south at the rate of 2 km/h. Then the bearing of the boat over the surface of the earth is $\underline{\ ?\ }$.
    a. 11°   b. 79°   c. 191°   (d.) 259°

Chapter Test

1. If A is a quadrantal angle and $-200° \le m\angle A \le -100°$, find $m\angle A$. $-180°$ 12-1

2. If $\angle A$ is in standard position, $180° \le m\angle A \le 270°$, and $\angle A$ is coterminal with an angle of $-465°$, find $m\angle A$. 255°

3. Evaluate $\tan 270°$. undefined   12-2

4. Angle A is in standard position and its terminal side contains the point $\left(-\frac{\sqrt{3}}{2}, \frac{1}{2}\right)$. Find $\tan A$. $-\frac{\sqrt{3}}{3}$

5. If $1° \le m\angle A \le 90°$ and $\cos A = 0.7777$, find $m\angle A$ to the nearest degree. 39°   12-3

6. In right triangle ABC, $c = 25$ and $b = 16$. Find $m\angle A$ to the nearest degree. 50°   12-4

7. At a point 25 m from a building, the angle of elevation of the top of the building is 72°. How tall is the building? 76.9 m

8. If a vector in standard position has direction 140° and y-component 30, find its norm. 46.7   12-5

9. If vectors **s** and **t** are in standard position with terminal points $(4, 2)$ and $(5, 2)$ respectively, and $\mathbf{v} = \mathbf{s} + \mathbf{t}$, find $\|\mathbf{v}\|$. $\|\mathbf{v}\| = \sqrt{97} \approx 9.8$ 12-6

10. In order to cross a river, a canoeist leaves a dock and paddles due north at a speed of 6 km/h. The river current is flowing due east at a speed of 5 km/h. Describe the speed and bearing of the canoeist over the surface of the earth. 7.8 km/h; 40°   12-7

29. If $\sin A < 0$ and $\tan A < 0$, then $\angle A$ is in Quadrant IV and $\cos A$ is positive and < 1. Therefore, $\tan A = \frac{\sin A}{\cos A}$ must be less than $\sin A$.

30. $\sin A = \frac{4}{5}$; $\cos A = \frac{3}{5}$; $\tan A = \frac{4}{3}$

31. $\sin A = \frac{12}{13}$; $\cos A = \frac{5}{13}$; $\tan A = \frac{12}{5}$

32. $\sin A = \frac{\sqrt{3}}{2}$; $\cos A = \frac{1}{2}$; $\tan A = \sqrt{3}$

33. $\sin A = \frac{\sqrt{2}}{2}$; $\cos A = \frac{\sqrt{2}}{2}$; $\tan A = 1$

38. Sine function graph:

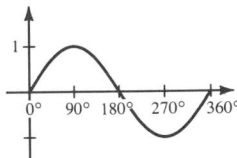

Cosine function graph:

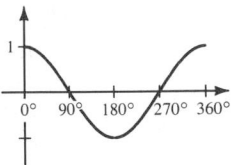

(Note that the unit circle must be rotated 90° counterclockwise in order to plot the lengths of line segments for $\cos A$ along the vertical axis of the cosine function graph.)

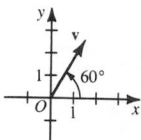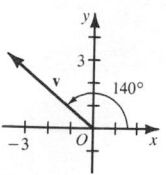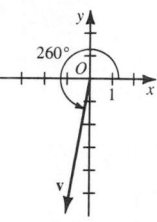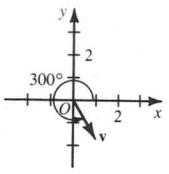
Mixed Review

Simplify.

1. $\dfrac{5x}{x^2 - 25} - \dfrac{x^2}{x^2 - 25} - \dfrac{x}{x + 5}$

2. $\dfrac{4a}{a^2 - 4a - 5} + \dfrac{1}{a^2 + 2a + 1}$ $\dfrac{4a^2 + 5a - 5}{(a - 5)(a + 1)^2}$

3. $\dfrac{3n}{n + 4} + 2$ $\dfrac{5n + 8}{n + 4}$

4. $\dfrac{6}{6 - m} - \dfrac{m}{m - 6}$ $\dfrac{6 + m}{6 - m}$

5. $\dfrac{x^{-4}y^3z^{-1}}{x^2y^{-1}}$ $\dfrac{y^4}{x^6z}$

6. $\left(\dfrac{2^{-3}a^{-1}b^3}{2^{-1}a^2b^{-3}}\right)^{-2}$ $\dfrac{16a^6}{b^{12}}$

7. $\dfrac{\dfrac{3}{m} - \dfrac{1}{n}}{\dfrac{1}{m} - \dfrac{3}{n}}$ $\dfrac{3n - m}{n - 3m}$

8. $\dfrac{\dfrac{r}{r + s} + \dfrac{s}{r - s}}{\dfrac{r}{r - s} - \dfrac{s}{r + s}}$ 1

9. $\sqrt{188}$ $2\sqrt{47}$

10. $\sqrt{x^6}$ $|x^3|$

11. $\sqrt{(y - 4)^2}$ $|y - 4|$

12. $\sqrt[4]{243}$ $3\sqrt[4]{3}$

13. $\sqrt{\dfrac{4}{5}}$ $\dfrac{2\sqrt{5}}{5}$

14. $\sqrt[3]{\dfrac{2}{3}}$ $\dfrac{\sqrt[3]{18}}{3}$

15. $\sqrt{18} + \sqrt{20} - \sqrt{50}$ $2\sqrt{5} - 2\sqrt{2}$

16. $\sqrt{2x^3} \cdot \sqrt{14x^3}$ $2|x^3|\sqrt{7}$

17. $\sqrt{18}(\sqrt{2} - \sqrt{10})$ $6 - 6\sqrt{5}$

18. $\dfrac{2}{1 - \sqrt{5}}$ $\dfrac{1 + \sqrt{5}}{-2}$

Express in scientific notation.

19. 0.00000531 5.31×10^{-6}

20. $37{,}000{,}000{,}000{,}000$ 3.7×10^{13}

21. $3000 \times 0.000003 \times 3$ 2.7×10^{-2}

22. $49{,}000{,}000 \div 0.0007$ 7×10^{10}

23. $\dfrac{20{,}000 \times 0.00015}{300}$ 1×10^{-2}

24. $\dfrac{0.00054}{0.0000012 \times 9000}$ 5×10^{-2}

Solve each system of equations.

25. $r - 4s = 3$ $\{(-5, -2)\}$
    $-5r + 3s = 19$

26. $m = 15 - 2n$ $\{(5, 5)\}$
    $3m - 2n = 5$

27. $x - y = 1$ $\{(-7, -8)\}$
    $-8x + 3y = 32$

28. $2a - 3b = 6$ $\{(4, \frac{2}{3})\}$
    $a = 4$

29. $3j = k$ $\emptyset$
    $3j - k = 4$

30. $v - w = 3$ $\{(v, w): v - w = 3\}$
    $2v = 2w + 6$

31. $15a - 4b = 8$ $\{(1, \frac{7}{4})\}$
    $-7a + 8b = 7$

32. $6r = 5s - 8$ $\{(-3, -2)\}$
    $2r + 5s = -16$

33. Express $\dfrac{11}{12}$ as a repeating or terminating decimal. $0.91\overline{6}$

34. Express $0.34\overline{3}$ as a fraction in simplest form. $\dfrac{103}{300}$

618   *Chapter 12*

In Exercises 35–40, $\angle A$ and $\angle B$ are the acute angles of a right triangle; a and b are the lengths of the sides opposite $\angle A$ and $\angle B$, respectively; and c is the length of the hypotenuse. Find the required measure, using a calculator or the tables on pages 683 and 684 as necessary. Give angle measures to the nearest degree and lengths to the nearest tenth of a unit.

35. If $a = 12$ and $b = 16$, find c. 20
36. If $b = 9$ and $c = 12$, find a. 7.9
37. If $m\angle A = 26°$ and $b = 10$, find a. 4.9
38. If $a = 3$ and $c = 5$, find $m\angle A$. 37°
39. If $b = 5$ and $c = 8$, find $m\angle A$. 51°
40. If $m\angle B = 28°$ and $a = 5$, find c. 5.7

Solve.

41. $a^2 - 14 = 0$ $\{\sqrt{14}, -\sqrt{14}\}$
42. $5 - 2b^2 = 1$ $\{\sqrt{2}, -\sqrt{2}\}$
43. $x^2 - 3x = 10$ $\{5, -2\}$
44. $3y^2 - 2y = 12$ $\{\frac{1 + \sqrt{37}}{3}, \frac{1 - \sqrt{37}}{3}\}$
45. $\sqrt{m} - 9 = 0$ $\{81\}$
46. $\sqrt{n - 4} = 2$ $\{8\}$
47. $\sqrt{r} + r = 2$ $\{1\}$
48. $\sqrt[3]{8 - s} = -3$ $\{35\}$

49. Alicia has $1.15 in change in her wallet, consisting entirely of quarters, dimes, and nickels. She has one more nickel than quarters and three more dimes than quarters. How many of each type of coin does she have? 2 quarters, 5 dimes, 3 nickels

50. Jen and Kevin cycled 14 km in 30 min with the wind. They made the return trip in 35 min cycling against the same wind. Find their speed of cycling without the wind. 26 km/h

51. The measure of the diagonal of a certain rectangle is 22 cm long. If the length of the rectangle is 16 cm, find its width. 15.1 cm

52. The measure in degrees of one of two supplementary angles is 20 more than half the measure of the other angle. Find the measures of the angles. $73\frac{1}{3}°$ and $106\frac{2}{3}°$

53. The sum of the digits of a two-digit number is 15. If 27 were subtracted from the number, the result would be the original number with its digits reversed. Find the original number. 96

54. When the dimensions of a rectangle that measures 2 cm by 5 cm are each increased by the same amount, the result is a rectangle whose area is 18 cm² greater than that of the original rectangle. Find the dimensions of the new rectangle. width: 4 cm; length: 7 cm

55. The sum of the squares of two consecutive negative odd integers is 202. Find the integers. −9 and −11

56. The cost of a ticket to the school play was $3 for adults and $2 for students. If 900 tickets were sold and the total receipts from tickets were $2300, how many of each type of ticket was sold? 500 adult tickets; 400 student tickets

PREPARING FOR ▮▮▮▮▮▮▮▮
COLLEGE ENTRANCE EXAMS

Strategy for Success: In solving problems such as those involving length, width, area, perimeter, or relative position, you may find it helpful to draw a sketch. Use any available space in the test booklet. Be careful to make no assumptions in drawing the sketch, using only information that is specifically given in the problem.

Decide which is the best of the choices and write the corresponding letter on your answer sheet.

1. Which of the following are roots of the equation $x^2 - 2kx = -9$? E
   I. $k + \sqrt{k - 3}$ II. $k + \sqrt{k^2 - 9}$ III. $k - \sqrt{k^2 - 9}$
   (A) I only   **(B)** II only   **(C)** III only   **(D)** I and II only   **(E)** II and III only

2. Sheila paddled a canoe 16 km up a stream and then returned. The round trip took 10 h. The current flowed at a rate of 3 km/h. How fast can Sheila paddle in still water? C
   (A) 2 km/h   **(B)** 4 km/h   **(C)** 5 km/h   **(D)** 6 km/h   **(E)** 8 km/h

3. Which of the following is true for the function $f: x \rightarrow 6 + x - x^2$? E
   I. The graph of the function opens downward.
   II. The function has no zeros.
   III. The function has two real roots.
   (A) I only   **(B)** II only   **(C)** III only   **(D)** I and II only   **(E)** I and III only

4. Which of the following is true if $(n - 3)^2 = 8$? A
   I. $n = 3 + \sqrt{8}$ II. $n = -3 + \sqrt{8}$ III. $n = 8 + \sqrt{3}$
   (A) I only   **(B)** II only   **(C)** III only   **(D)** I, II, and III   **(E)** none of these

5. Find $\|\mathbf{u}\|$ if the vector $\mathbf{u}$ is in standard position with terminal point $(-3, 4)$. A
   (A) 5   **(B)** −5   **(C)** 25   **(D)** $\sqrt{6}$ **(E)** $-\sqrt{6}$

6. Which of the following is true for $A = \{(x, y): x = -2y^2\}$? D
   (A) A is a function.
   (B) The minimum point of its graph is $(3, -4)$.
   (C) Its graph has a maximum point.
   (D) The graph opens to the left.
   (E) None of the above statements is true.

7. Find $\cos A$ for an angle A in standard position whose terminal side contains the point $(12, -5)$. B
   (A) $\frac{13}{5}$ **(B)** $\frac{12}{13}$ **(C)** $\frac{5}{13}$ **(D)** $\frac{5}{12}$ **(E)** $-\frac{12}{13}$

8. Solve $\dfrac{9}{x^2} - \dfrac{12}{x} + 4 = 0$. D
   (A) $\frac{2}{3}$ **(B)** $\frac{1}{2}$ **(C)** $2 + \sqrt{3}$ **(D)** $\frac{3}{2}$ **(E)** $2 - \sqrt{3}$

620 *Chapter 12*

Contest Problems

1. Suppose that you are allowed to make five vertical cuts through a large circular wheel of cheese, like the one that is shown at the right. What is the greatest number of pieces of cheese that you can get? 16

2. Solve the system: $\begin{array}{c} x^2 + y^2 = 25 \\ 4y^2 - 9x = 0 \end{array}$ {(4, 3), (4, −3)}

3. Solve the inequality $\dfrac{2m - 8}{m} < \dfrac{4}{m}$. {m: 0 < m < 6}

4. What is the next number in the following pattern?

   3, 5, 8, 13, 22, 39, 72, __?__   137

5. Find the least integral value of C for which the following equation has no real root.
   $$2x^2 + 6x + C = 0 \quad 5$$

6. Determine how many pairs of prime numbers have a difference of 3. 1

The science of statistics is useful in the study of the distribution of characteristics such as weight and life span among animal populations, such as the school of dolphins shown in the photo.

622 *Chapter 13*

Chapter 13

Statistics and Probability

Graphical Representation of Data

OBJECTIVES for Sections 13-1 through 13-3:
1. *To draw and interpret a dot frequency diagram and to determine relative frequencies for a given set of data.*
2. *To draw and interpret a histogram and a frequency polygon for a given set of data.*
3. *To draw and interpret a cumulative frequency polygon for a given set of data.*

13–1 Dot Frequency Diagrams and Relative Frequency

Statistics is the science of organizing and analyzing a set of numerical facts, or *data*, so that probable conclusions can be drawn from the data.

Suppose that 25 thirteen-year-old girls are measured and that their heights to the nearest centimeter are given in the following array:

158	160	155	159	160
162	161	158	163	157
156	165	160	157	165
161	157	163	160	160
158	161	159	161	158

Teaching Suggestions
p. T114

Key Ideas

Draw and interpret a dot frequency diagram for a given set of data.
Determine the relative frequencies for a given set of data.

1. A total of 20 patients admitted to a hospital have blood sugar levels as given in the following array:

67 69 74 73 70
70 71 67 73 74
73 75 69 72 70
70 72 70 73 74

Make a dot frequency diagram for the given data.

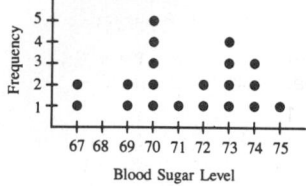

2. Make a table showing the frequencies and the relative frequencies for the data given in Exercise 1. Express each relative frequency as a fraction and as a percent.

BSL	Freq.	Rel. Freq. Frac.	%
67	2	$\frac{2}{20}$	10
68	0	0	0
69	2	$\frac{2}{20}$	10
70	5	$\frac{5}{20}$	25
71	1	$\frac{1}{20}$	5
72	2	$\frac{2}{20}$	10
73	4	$\frac{4}{20}$	20
74	3	$\frac{3}{20}$	15
75	1	$\frac{1}{20}$	5

In trying to analyze such a distribution of heights, it may be helpful to display the data as a **dot frequency diagram,** as shown in Figure 1.

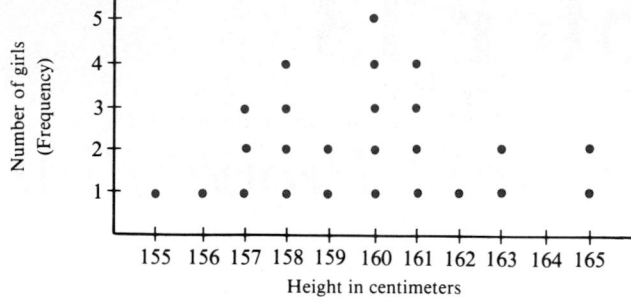

Figure 1

By observing the dots you can quickly see the **frequency,** or number of occurrences, of each measurement. You can also see, for example, that 21 of the 25 students have heights between 157 cm and 163 cm, inclusive.

You can obtain the **relative frequency** of each measurement if you divide its particular frequency by the total number of measurements. Converting these ratios to percents makes the data easier for most people to interpret.

Height	Frequency	Relative frequency Fraction	Percent
155	1	$\frac{1}{25}$	4
156	1	$\frac{1}{25}$	4
157	3	$\frac{3}{25}$	12
158	4	$\frac{4}{25}$	16
159	2	$\frac{2}{25}$	8
160	5	$\frac{5}{25}$	20
161	4	$\frac{4}{25}$	16
162	1	$\frac{1}{25}$	4
163	2	$\frac{2}{25}$	8
165	2	$\frac{2}{25}$	8
Total:	25	$\frac{25}{25} = 1$	100

From this table you can see that the heights of 72%, or about three fourths of the class, are between 157 cm and 161 cm, inclusive.

624 *Chapter 13*

Oral Exercises

In Exercises 1–3 state the frequency and the relative frequency, as a fraction and a percent, of the given letter from the word REPEATER.

1. T 1; $\frac{1}{8}$; $12\frac{1}{2}\%$ **2.** R 2; $\frac{2}{8}$; 25% **3.** E 3; $\frac{3}{8}$; $37\frac{1}{2}\%$

In Exercises 4–8 state the frequency and the relative frequency, as a fraction and a percent, of the given letter from the word INDEPENDENCE.

4. E 4; $\frac{4}{12}$; $33\frac{1}{3}\%$ **5.** N 3; $\frac{3}{12}$; 25% **6.** P 1; $\frac{1}{12}$; $8\frac{1}{3}\%$ **7.** D 2; $\frac{2}{12}$; $16\frac{2}{3}\%$ **8.** C 1; $\frac{1}{12}$; $8\frac{1}{3}\%$

9. Do you think that the set of arm length measurements of all the members of your algebra class would show the same pattern of distribution as the set of heights? That is, would most of the measurements cluster near the middle measurement? Yes

10. Name at least one word in which the letter s has a frequency of 2 and a relative frequency of $\frac{2}{5}$, or 40%. Answers will vary. Example: SENSE

Written Exercises

Make a dot frequency diagram for the given data. Label the axes "Number" and "Frequency."

A

1.
2	5	3	2	3	4
4	2	4	5	7	7
4	6	7	3	4	3

2.
32	30	33	36	30	36
36	34	35	30	30	36
30	31	36	35	32	33

Make a table showing the frequencies and relative frequencies, as fractions and percents, for the data in the given exercise.

3. Exercise 1   **4.** Exercise 2

5. Make a dot frequency diagram for the letters in the word ENTERTAINING.

6. Make a dot frequency diagram for the letters in the word MONOTONOUS.

Make a table for the data in the given exercise, showing frequencies and relative frequencies, as fractions and percents.

7. Exercise 5   **8.** Exercise 6

B

9. In order to express $\frac{1}{17}$ as a decimal, write $\frac{1}{17} = 0.\overline{0588235294117647}$. (Recall that the fraction $\frac{1}{n}$ can have at most $n - 1$ digits in the repeating block of digits in its decimal equivalent.)

  a. Make a table showing the frequency and relative frequency of each of the ten digits.
  b. Make a dot frequency diagram. Label the horizontal axis "Digit" and the vertical axis "Frequency."

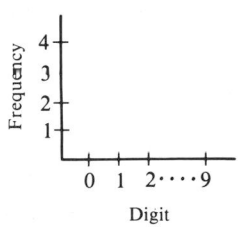

Statistics and Probability **625**

Suggested Assignments
Average
  625/1–10
Maximum
  625/1–11

Additional Answers
Written Exercises
(See p. 654.)

Additional A Exercises
Make a dot frequency diagram for the given data.

1.
15	14	11	16	12
17	15	15	18	16
13	12	14	16	15
18	16	15	14	13

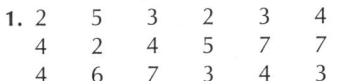

2. Make a table showing the frequencies and the relative frequencies for the data given in Exercise 1. Express each relative frequency as a fraction and as a percent.

No.	Freq.	Rel. Freq.	
		Frac.	%
11	1	$\frac{1}{20}$	5
12	2	$\frac{2}{20}$	10
13	2	$\frac{2}{20}$	10
14	3	$\frac{3}{20}$	15
15	5	$\frac{5}{20}$,	25
16	4	$\frac{4}{20}$	20
17	1	$\frac{1}{20}$	5
18	2	$\frac{2}{20}$	10

Simplify each ratio.

1. Boys to total number of students in a school containing 250 boys and 325 girls. $\frac{10}{23}$

2. Red pencils to total number of pencils in a box containing 12 red pencils and 38 blue pencils. $\frac{6}{25}$

Express each fraction as a percent and as a decimal.

3. a. $\frac{1}{10}$ 10%; 0.1

   b. $\frac{3}{25}$ 12%; 0.12

   c. $\frac{3}{200}$ 1.5%; 0.015

4. a. $\frac{1}{3}$ $33\frac{1}{3}$%; $0.\overline{3}$

   b. $\frac{3}{8}$ 37.5%; 0.375

   c. $4\frac{1}{2}$ 450%; 4.5

Teaching Suggestions
p. T114

Key Ideas

Draw and interpret a histogram for a given set of data. Draw and interpret a frequency polygon for a given set of data.

---

10. The birth dates of the first 20 United States presidents are shown.
    a. Make a table showing the frequency and relative frequency, as a fraction and percent, of each month.
    b. Make a dot frequency diagram for the data. Label the horizontal axis "Month" and the vertical axis "Frequency."

No.	Name	Date of Birth
1	George Washington	Feb. 22, 1732
2	John Adams	Oct. 30, 1735
3	Thomas Jefferson	Apr. 13, 1743
4	James Madison	Mar. 16, 1751
5	James Monroe	Apr. 28, 1753
6	John Quincy Adams	July 11, 1767
7	Andrew Jackson	Mar. 15, 1767
8	Martin Van Buren	Dec. 5, 1782
9	William Henry Harrison	Feb. 9, 1773
10	John Tyler	Mar. 29, 1790
11	James K. Polk	Nov. 2, 1795
12	Zachary Taylor	Nov. 24, 1784
13	Millard Fillmore	Jan. 17, 1800
14	Franklin Pierce	Nov. 23, 1804
15	James Buchanan	Apr. 23, 1791
16	Abraham Lincoln	Feb. 12, 1809
17	Andrew Johnson	Dec. 29, 1808
18	Ulysses S. Grant	Apr. 27, 1822
19	Rutherford B. Hayes	Oct. 4, 1822
20	James A. Garfield	Nov. 19, 1831

C 11. Record the first letter of the last name of all the students in your algebra class. Make a table showing the frequencies and relative frequencies, as fractions and percents, for the data. Answers will vary.

13–2 Histograms and Frequency Polygons

For a large set of data with a wide range of values, a dot frequency diagram is not a practical device for visualizing frequencies. Instead you can make a table showing a frequency distribution in which the data are grouped in equal intervals, and the frequency is shown for each interval. From this you can make a type of *bar graph* called a **histogram** to help visualize the distribution.

The following table shows the distribution of the weights in kilograms of 100 high-school sophomore boys.

626   *Chapter 13*

Interval	Frequency	Relative frequency Fraction	Percent
45–50	2	$\frac{2}{100}$	2
50–55	7	$\frac{7}{100}$	7
55–60	12	$\frac{12}{100}$	12
60–65	28	$\frac{28}{100}$	28
65–70	23	$\frac{23}{100}$	23
70–75	16	$\frac{16}{100}$	16
75–80	8	$\frac{8}{100}$	8
80–85	3	$\frac{3}{100}$	3
85–90	1	$\frac{1}{100}$	1
Total: 100		$\frac{100}{100} = 1$	100

The weights here are grouped into nine intervals, each of length 5. A large collection of data is usually compressed into anywhere from 10 to 20 such *class intervals*, depending on the number and range of the measurements.

A boundary value ordinarily is included in the interval on its *left*. For example, in Figure 2 a weight of 50 kg would be included in the 45–50 interval, *not* in the 50–55 interval.

The histogram of the given distribution (Figure 2) indicates that the greatest clustering of weights occurs between 60 and 70 kg (51%).

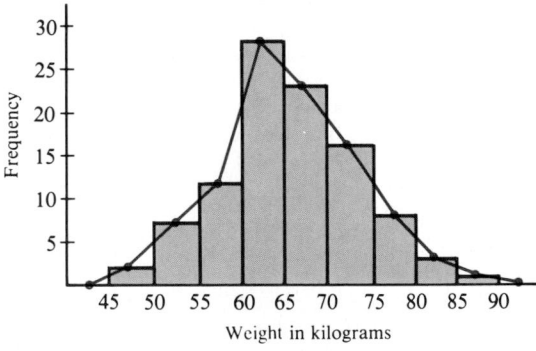

Figure 2

The red broken-line graph joining the midpoints of the intervals is called a **frequency polygon**.

Notice that the frequency polygon extends a half-interval beyond the histogram at each end, starting and ending on the horizontal axis. For a reason why this is done, see Oral Exercises 9 and 10.

A frequency distribution can be displayed by using either a frequency polygon or a histogram.

Statistics and Probability **627**

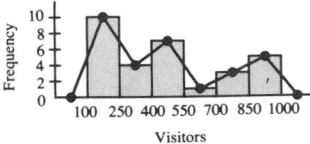
627

Common Errors

When drawing a frequency polygon, a misleading graph may result if the vertical scale is not set at zero. Consider the following frequency polygon showing the percentage of defective radios as detected by a company's testing policy over several years.

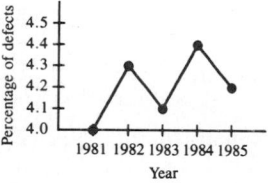

It would appear that the company's quality control is not doing its job. If the vertical scale were proportioned correctly, however, the correct inference, that the quality was relatively stable, could be made.

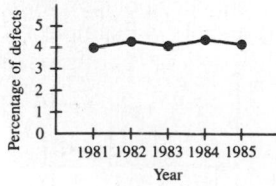

Oral Exercises

The histogram at the right shows a frequency distribution of the head size measurements in two algebra classes.

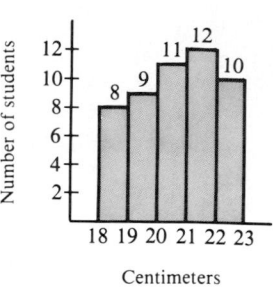

1. How many had head size measurements between 19 cm and 20 cm? 9
2. How many measured between 22 cm and 23 cm? 10
3. How many measured between 20 cm and 22 cm? 23
4. How many measured greater than 20 cm? 33
5. How many students were measured? 50
6. What was the relative frequency (as a fraction) of measurements between 19 cm and 20 cm? $\frac{9}{50}$
7. What percent of the measurements were between 21 cm and 22 cm? 24%
8. What percent of the measurements were between 18 cm and 20 cm? 34%

Exercises 9 and 10 refer to the figure at the right.

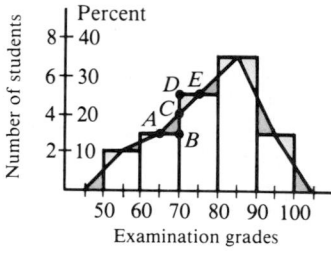

9. Triangles ABC and EDC are congruent. Do they have equal areas? How many other pairs of triangles in the figure have equal areas? Yes; 5
10. Explain why the area between the horizontal axis and the frequency polygon is equal to the sum of the areas of the rectangles in the histogram. For each triangle included in the area of the histogram but not in the area of the frequency polygon, such as triangle EDC, there is a triangle of equal area included in the frequency polygon area but not in the histogram area, such as triangle ABC.

Written Exercises

A

1. a. Make a frequency distribution table showing the data at the right. Group the data in the intervals 5–35, 35–65, 65–95, and show relative frequencies as fractions and percents.
   b. Make a histogram from the table in part (a).
   c. Draw a frequency polygon for the distribution in part (a), using the histogram in part (b).

42	8	54	93	38
27	61	84	33	30
60	21	24	91	52
31	39	37	68	39
77	64	36	47	44

2. Repeat Exercise 1 using the data at the right.

41	70	27	94	38
8	60	91	52	52
62	14	88	40	32
60	42	48	90	36
55	43	30	51	66

628 *Chapter 13*

3. Make a histogram and a frequency polygon for the frequency distribution of Scholastic Aptitude Test—Mathematics scores shown below.

Class Interval	Frequency
450–500	15
500–550	8
550–600	8
600–650	5
650–700	4
700–750	2

4. The speeds of cars passing a point on a highway were measured for fifteen minutes. Make a histogram and frequency polygon for the frequency distribution of the speeds shown below. Speed measurements are in miles per hour.

Class Interval	Frequency
40–45	12
45–50	24
50–55	30
55–60	9
60–65	6

B 5. For the set of measurements shown below:
a. Make a table, grouping measurements in the intervals 4.5–34.5, 34.5–64.5, 64.5–94.5, and showing frequencies and relative frequencies, as fractions and percents.
b. Draw a histogram for the data shown in the table of part (a).
c. Draw a frequency polygon for the data shown in the table of part (a).

48	91	10	36	70
39	51	29	62	57
20	27	42	81	41
25	42	18	56	83

6. Repeat Exercise 5 using these measurements:

59	65	16	39	90
8	35	42	84	36
73	7	34	48	61
26	19	28	43	58
88	43	48	38	11

Statistics and Probability **629**

Suggested Assignments
Average
628/1–7
Maximum
628/1–8

**Additional Answers
Written Exercises**
(See p. 658.)

Additional A Exercises

1. a. Make a frequency distribution table showing the data given below. Express each relative frequency as a fraction and as a percent. Group the data in the intervals (0–3), (3–6), (6–9), (9–12).

9	0	7	7	10	4	6
11	12	12	3	9	11	10
	4	8	12	5	10	8

		Rel. Freq.	
Int.	Freq.	Frac.	%
0–3	2	$\frac{2}{20}$	10
3–6	4	$\frac{4}{20}$	20
6–9	6	$\frac{6}{20}$	30
9–12	8	$\frac{8}{20}$	40

b. Make a histogram and a frequency polygon from the table in part (a).

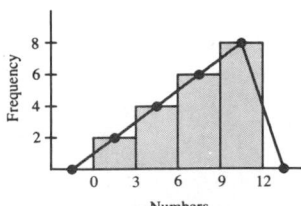

629

1. Make a dot frequency diagram for the letters in the
sentence, "I think, therefore I am."

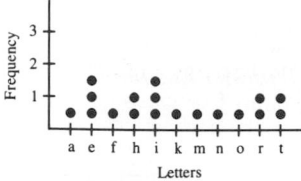

2. Record the month of birth
of every student in the
class. Make a table show-
ing the frequencies and
the relative frequencies.
Express each relative frequency as a fraction and
as a percent.

Answers will vary.

Express the following fractions as percents. Give answers to the nearest tenth.

3. a. $\frac{1}{30}$ $3\frac{1}{3}\% \approx 3.3\%$

   b. $\frac{1}{15}$ $6\frac{2}{3}\% \approx 6.7\%$

4. a. $\frac{4}{15}$ $26\frac{2}{3}\% \approx 26.7\%$

   b. $\frac{3}{35}$ $8\frac{4}{7}\% \approx 8.6\%$

7. The lengths in kilometers of some of the major rivers of the world
are listed below.

River	Length (km)	River	Length (km)
Albany	976	Loire	1014
Amazon	6400	Mekong	4160
Amur	4320	Mississippi	3757
Brahmaputra	2880	Nile	6632
Colorado	2320	Orinoco	2560
Columbia	1989	Ottawa	1264
Congo	4349	Po	648
Dnieper	2272	Rio Grande	3016
Elbe	1158	St. Lawrence	1280
Euphrates	3576	Thames	344
Garonne	571	Yangtze	5440
Indus	2880	Zambezi	2720
Jordan	320		

a. Make a table of frequencies and relative frequencies, as fractions
and percents, grouping the lengths in the intervals 0–1000,
1000–2000, 2000–3000, and so on.

b. Draw a histogram from the data in your table.

c. Use your histogram to draw a frequency polygon for the data.

C 8. Listed below are the numbers of immigrants in a recent year
entering the United States from selected countries (recorded as the
country of birth).

Country	Number of Immigrants	Country	Number of Immigrants
Austria	507	Italy	5,969
Canada	20,181	Japan	4,496
China	2,944	Lithuania	23
Denmark	378	Mexico	52,479
Finland	284	Norway	438
France	2,905	Poland	3,863
Germany	7,166	Spain	3,285
Greece	5,942	Turkey	1,306
Hungary	528	U.S.S.R.	1,919
India	18,625	Wales	128

a. Select appropriate grouping intervals and make a table of frequencies and relative frequencies, as fractions and percents.

b. Draw a histogram from the data in your table.

c. Use your histogram to draw a frequency polygon for the data.

d. Obtain a copy of a recent almanac. Compare the immigration
statistics for these countries for the two most recent years given.
What changes do you notice?

630 *Chapter 13*

13–3 Cumulative Frequency

In analyzing a set of numerical data, it is often helpful to tabulate **cumulative frequencies** and **cumulative percents,** that is, the number and percent of measurements that are *less than or equal to* a given value.

The table below shows the frequency distribution for the set of 25 heights given in Section 13-1, along with the *cumulative frequency* and the *cumulative percent.* (The second column gives the frequency as a percent.) Notice that any measurement that falls on an interval boundary is included in the interval to its left; for example, 157 is included in the 155–157 interval.

Interval	Frequency	Percent	Cumulative frequency	Cumulative percent
153–155	1	4	1	4
155–157	4	16	5	20
157–159	6	24	11	44
159–161	9	36	20	80
161–163	3	12	23	92
163–165	2	8	25	100

The table tells us, for example, that 80% of the students have heights equal to or less than 161 cm. In the **cumulative frequency polygon** (Figure 3) displaying the facts in the table, the ordinates of the red dots are cumulative frequencies, and the abscissas are the right-hand endpoints of the corresponding intervals.

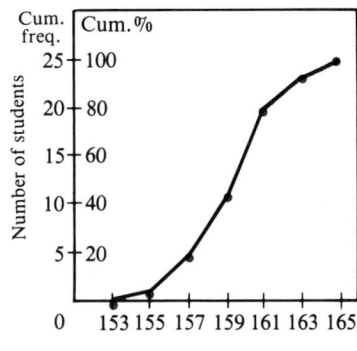

Figure 3

Because of rounding in the process of computation, the final value obtained in using cumulative percents sometimes does not appear to be exactly 100%.

Key Ideas

Determine the cumulative frequencies and cumulative percents for a given set of data.

Chalkboard Examples

1. Make a table showing the frequencies, percents, cumulative frequencies, and cumulative percents for the following set of measures given in meters. 10, 9, 11, 7, 9, 7, 10, 10, 6, 7, 10, 10, 5, 5, 10, 9, 6, 11, 8, 11

No.	Freq.	%	Cum. Freq.	Cum. %
5	2	10	2	10
6	2	10	4	20
7	3	15	7	35
8	1	5	8	40
9	3	15	11	55
10	6	30	17	85
11	3	15	20	100

2. Draw the cumulative frequency polygon for the data in Exercise 1.

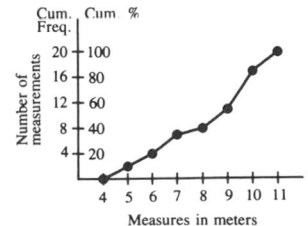

EXAMPLE Make a table showing the frequencies, percents, cumulative frequencies, and cumulative percents for the following set of measures:

$$7, 8, 8, 6, 7, 8, 7, 8$$

SOLUTION Count the number of times each measure occurs. Then complete the table.

Measure	Frequency	Percent	Cumulative frequency	Cumulative percent
6	1	$12\frac{1}{2}$	1	$12\frac{1}{2}$
7	3	$37\frac{1}{2}$	4	50
8	4	50	8	100
Total:	8	100		

Oral Exercises

Exercises 1–9 refer to the table of heights at the right.

1. What number does the entry *a* represent? 5
2. Explain how the third entry in the cumulative frequency column, 9, was determined, and then interpret what it means.
3. What number does *b* represent? 6
4. What number does *c* represent? 35
5. What number does *d* represent? 5
6. How many people were 1.55 m tall? 9
7. How many were at most 1.57 m tall? 40
8. How many people were measured in all? 50
9. How many people were more than 1.54 m but less than 1.57 m tall? State two ways of obtaining the answers to this question.

2. $9 = 3 + 2 + 4$; there are nine heights less than or equal to 1.52 m.

Height (m)	Frequency	Cumulative frequency
1.50	3	3
1.51	2	a
1.52	4	9
1.53	b	15
1.54	4	19
1.55	9	28
1.56	7	c
1.57	d	40
1.58	5	45
1.59	3	48
1.60	2	50

9. 16; add the frequencies for 1.55 m and 1.56 m (9 + 7) or subtract the cum. frequency for 1.54 m from the cum. frequency for 1.56 m (35 − 19).

Written Exercises

Make a table showing the frequencies, percents, cumulative frequencies, and cumulative percents for each set of data.

A 1. 7, 9, 2, 3, 4, 3, 6, 3, 8, 4, 8, 6
2. 23, 21, 25, 22, 23, 23, 25, 21, 23, 22, 25, 23, 21, 23, 25

3. Copy and complete the table.

Cost in dollars	Frequency	Percent	Cumulative frequency	Cumulative percent
20	2	8	_?_ 2	_?_ 8
21	3	_?_ 12	_?_ 5	_?_ 20
22	4	_?_ 16	_?_ 9	_?_ 36
23	9	_?_ 36	_?_ 18	_?_ 72
24	6	_?_ 24	_?_ 24	_?_ 96
25	_?_ 1	_?_ 4	_?_ 25	_?_ 100
Total:	_?_ 25	_?_ 100		

4. Copy and complete the table.

Ages of students	Frequency	Percent	Cumulative frequency	Cumulative percent
13	3	6	_?_ 3	_?_ 6
14	8	_?_ 16	_?_ 11	_?_ 22
15	12	_?_ 24	_?_ 23	_?_ 46
16	9	_?_ 18	_?_ 32	_?_ 64
17	10	_?_ 20	_?_ 42	_?_ 84
18	6	_?_ 12	_?_ 48	_?_ 96
19	_?_ 2	_?_ 4	_?_ 50	_?_ 100
Total:	_?_ 50	_?_ 100		

The cumulative frequency polygon shown at the right concerns the annual sales in each of 20 companies during a recent year.

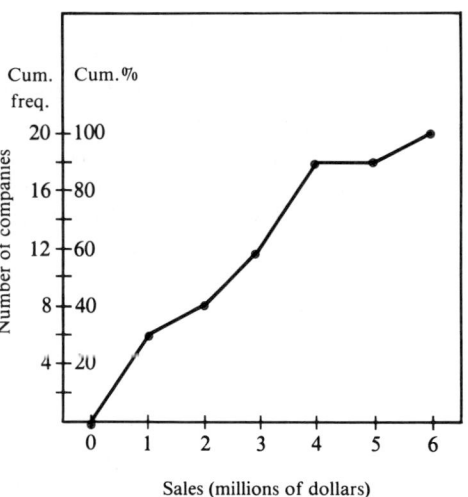

Sales (millions of dollars)

5. How many companies had sales no greater than $1,000,000? 6

6. How many companies had sales up to $3,000,000, inclusive? 12

7. How many companies had sales between $4,000,000 and $5,000,000? none

8. What percent of the companies had sales of at most $4,000,000? 90%

9. What percent of the companies had sales over $3,000,000? 40%

Make a table showing the frequencies, percents, cumulative frequencies, and cumulative percents for the set of data.

1. 1, 3, 4, 5, 5, 2, 1, 8, 5, 3

No.	Freq.	%	Cum. Freq.	Cum. %
1	2	20	2	20
2	1	10	3	30
3	2	20	5	50
4	1	10	6	60
5	3	30	9	90
8	1	10	10	100

The cumulative frequency polygon shown below concerns the heights in centimeters of 50 children.

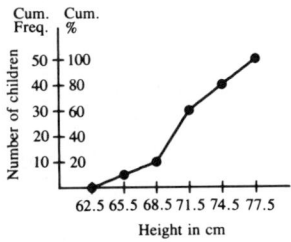

Height in cm

2. a. How many children had heights no greater than 68.5 cm? 10

b. How many children had heights up to 74.5 cm inclusive? 40

c. How many children had heights between 65.5 cm and 71.5 cm? 25

d. What percent of the children had heights at most 71.5 cm? 60%

e. What percent of the children had heights over 74.5 cm? 20%

10.

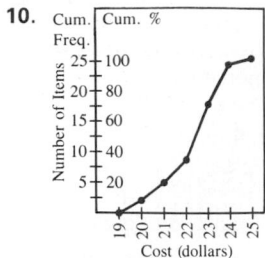

12.

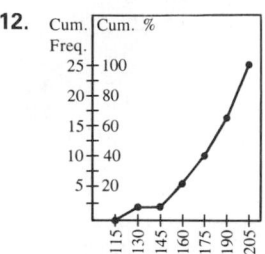

14.

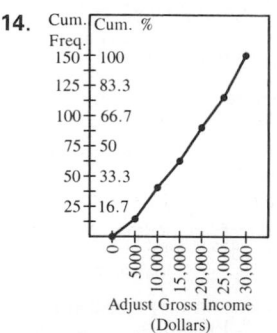

16.

Income (1000 $)	Freq.	Cum. %	Cum. Freq.	%
0– 5	3	10	3	10
5–10	4	13.3	7	23.3
10–15	4	13.3	11	36.6
15–20	3	10	14	46.6
20–25	4	13.3	18	59.9
25–30	3	10	21	69.9
30–35	2	6.7	23	76.6
35–40	2	6.7	25	83.3
40–45	1	3.3	26	86.6
over 45	4	13.3	30	99.9
Total:	30	99.9		

Draw a cumulative frequency polygon for the data in the given exercise.

10. Exercise 3

11. Exercise 4

For Exercises 12 and 13, copy and complete the table. Then draw a cumulative frequency polygon. See Additional Answers for frequency polygons.

B **12.**

Interval	Frequency	Percent	Cumulative frequency	Cumulative percent
115–130	2	? 8	? 2	? 8
130–145	0	? 0	? 2	? 8
145–160	4	? 16	? 6	? 24
160–175	4	? 16	? 10	? 40
175–190	7	? 28	? 17	? 68
190–205	8	? 32	? 25	? 100
Total:	? 25	? 100		

13.

Interval	Frequency	Percent	Cumulative frequency	Cumulative percent
0–100	7	14	? 7	? 14
100–200	10	? 20	? 17	? 34
200–300	12	? 24	? 29	? 58
300–400	? 15	? 30	? 44	? 88
400–500	6	? 12	? 50	? 100
Total:	? 50	? 100		

14. Following is a table showing the number of annual tax returns filed in a certain year with adjusted gross incomes of $30,000 or less. The number of returns in each interval has been recorded to the nearest million. Copy and complete the table, then draw a cumulative frequency polygon. Express all percents to the nearest tenth of a percent. See Additional Answers for frequency polygon.

Adjusted gross income	Frequency	Percent	Cumulative frequency	Cumulative percent
0–$5000	18	? 12	? 18	? 12
$5000–$10,000	22	? 14.7	? 40	? 26.7
$10,000–$15,000	23	? 15.3	? 63	? 42
$15,000–$20,000	25	? 16.7	? 88	? 58.7
$20,000–$25,000	30	? 20	? 118	? 78.7
$25,000–$30,000	32	? 21.3	? 150	? 100

C **15.** Toss two dice 100 times and record each total. Make a table showing the frequencies, percents, cumulative frequencies, and cumulative percents. Which total occurred most often? least often? Answers will vary.

16. A sociology professor found the following family incomes in a survey of an urban neighborhood.

$36,600	$18,000
$34,800	$21,000
$49,000	$12,000
$3,980	$43,900
$18,500	$37,540
$7,400	$53,270
$29,150	$13,700
$32,200	$4,500
$9,680	$24,430
$10,710	$26,000
$9,400	$4,780
$20,750	$8,670
$64,000	$56,350
$21,650	$15,880
$29,000	$11,940

Divide the incomes into ranges of $5000, with the final range being "over $45,000." Show them in a table with frequencies, percents, cumulative frequencies, and cumulative percents.

Computer Exercises For students with computer experience

1. The following data represent the number of household members in a sample of 20 households.

5	1	3	4	4	2	3	7	2	1
2	4	5	1	3	8	4	3	8	4

Write a program that will read these numbers from a DATA list and will display in table form each number of household members from 1 through 8 and its frequency in the sample.

2. Modify the program that you wrote for Exercise 1 so that it also displays the *relative* frequency of each number of household members, both as a fraction and as a percent.

3. Modify the program that you wrote for Exercise 1 so that it displays only the *cumulative* frequency of each number of household members. (*Hint:* Test whether or not each data value is equal to 8. If it is, branch to the first of a series of steps that increment the counters of the frequencies. If it is not, test whether or not it is equal to 7. If it is, branch to the *second* in a series of such steps. If it is not, continue testing against 6, 5, 4, 3, 2, and 1 in a similar manner.)

1. Make a frequency distribution table showing the data below. Group the data in the intervals (36–38), (38–40), (40–42), (42–44). Show relative frequencies as fractions and as percents.

38	40	41	37	39
43	39	43	39	39
43	39	38	37	37
41	42	37	39	39

Int.	Freq.	Rel. Freq. Frac.	%
36–38	6	$\frac{6}{20}$	30
38–40	8	$\frac{8}{20}$	40
40–42	3	$\frac{3}{20}$	15
42–44	3	$\frac{3}{20}$	15

2. Make a histogram and a frequency polygon from the table in Exercise 1.

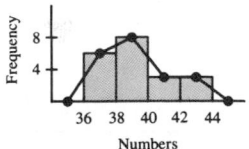

3. Make a histogram to describe the data given in Exercise 1 if it is grouped in the intervals (36–40) and (40–44).

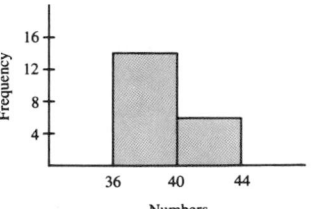

1. Make a dot frequency diagram for the given data and find the relative frequency of the measurement 12. $\frac{1}{4}$

```
14  10  12  13
13  11  11  12
14  14  14  14
12  13  14  12
```

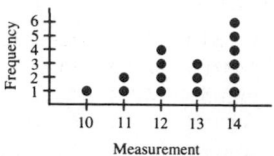

2. Complete: In the histogram below, _?_ people were surveyed, and _?_ of them own between 20 and 60 books. 26, 15

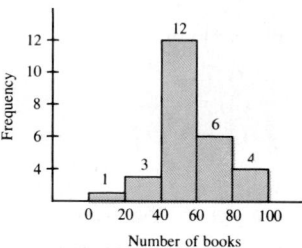

3. Make a table showing the frequency, percent, cumulative frequency, and cumulative percent for the following set of data: 9, 8, 7, 10, 8, 9, 9, 10, 8, 9 Draw a cumulative frequency polygon.

No.	Freq.	%	Cum. Freq.	Cum. %
7	1	10	1	10
8	3	30	4	40
9	4	40	8	80
10	2	20	10	100

636

Self-Test 1

VOCABULARY dot frequency diagram (p. 624)
frequency of a measurement (p. 624)
relative frequency of a measurement (p. 624)
histogram (p. 626)
frequency polygon (p. 627)
cumulative frequency (p. 631)
cumulative percent (p. 631)
cumulative frequency polygon (p. 631)

1. Make a dot frequency diagram for the given data and find the relative frequency of the measurement 5.   *Obj. 1, p. 623*

```
1   2   4   2   5
4   5   1   4   2
5   6   3   3   4
```

2. Complete: In the histogram shown below, _15?_ students received grades between 60 and 90, and _?_ students in all took the test. 20   *Obj. 2, p. 623*

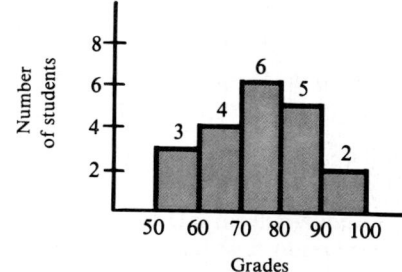

Grades

3. Make a table showing the frequency, percent, cumulative frequency, and cumulative percent for the following set of data: *Obj. 3, p. 623*

2, 3, 4, 3, 1, 2, 5, 4, 1, 5

Then draw a cumulative frequency polygon for these data.

Check your answers with those at the back of the book.

Arithmetical Description of Data

OBJECTIVES for Sections 13-4 and 13-5:
1. To find the median, mode, and arithmetic mean of a frequency distribution.
2. To find the range, variance, and standard deviation of a frequency distribution.

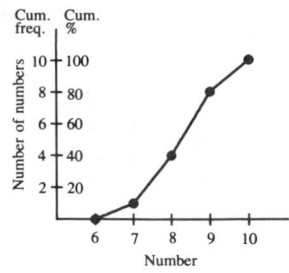

13–4 Statistical Averages

You have seen how a graph can give a quick visual summary of the distribution of values in a large collection of data. A compact description of how the data are "centered" can be given by obtaining certain "averages" from the data: the *median*, the *mode*, and the *arithmetic mean*.

To obtain the median from a set of data, such as the 25 heights in Section 13-1, the data must first be arranged in order, as at the right.

The **median** in an ordered set of *n* values is the *middle* entry if *n* is odd. If *n* is even, then there are two middle measurements, and the median is half their sum. In this example, the median is 160.

The **mode** is the value that occurs with the *greatest frequency*. In this example, the mode happens to be the same as the median, 160. There may be more than one mode in a list of data. For example, in 1, 3, 3, 4, 4, 6, there are *two* modes: 3 and 4.

The **arithmetic mean**, often called the **average**, or simply the **mean**, is the sum of the *n* values divided by *n*. Here the mean is $\frac{3994}{25}$, or 159.76.

In some cases the median may give a more accurate picture of a distribution than the mean because one or two extreme values can greatly affect the mean. For example, the annual income of the owner of a small business might be $35,000, while the earnings of the four employees might be $9000, $9000, $9000, and $10,000. The owner would perhaps want to point to the *mean income* of the entire group, which is

$$\frac{\$35,000 + \$9000 + \$9000 + \$9000 + \$10,000}{5} = \frac{\$72,000}{5}, \text{ or } \$14,400.$$

The employees, on the other hand, would probably feel that the *median income*,

$$\$9000,$$

is a more meaningful figure as far as they are concerned.

155	
156	
157	
157	
157	
158	
158	
158	
158	
159	
159	
160	
160	← Median
Mode { 160	
160	
160	
161	
161	
161	
161	
162	
163	
163	
165	
165	
Sum: 3994	

Teaching Suggestions
p. T115

Key Ideas

Find the median, mode and arithmetic mean of a frequency distribution.

Chalkboard Examples

1. For the list of data 1, 1, 2, 3, 4, 4, 4, 5, state:
   a. the mode(s)
   b. the median
   c. the mean.
   a. Mode: 4
   b. Median: $\frac{3+4}{2} = 3\frac{1}{2}$
   c. Mean: $\frac{24}{8} = 3$

2. Find the mean of the 20 weights in the table.

Weight	Frequency
158	1
159	2
160	3
161	5
162	6
163	2
164	1

158(1) + 159(2) + 160(3) + 161(5) + 162(6) + 163(2) + 164(1) = 3223

Mean = $\frac{3223}{20}$ = 161.15 ≈ 161.2

In most cases, however, the arithmetic mean is the most reliable measure, and also the most useful one for computational work. The mode is of very limited value.

EXAMPLE 1 For the list of data 1, 7, 8, 2, 3, 6, 8, state:

a. the mode(s)   b. the median   c. the mean

SOLUTION Order the data from least to greatest:

1, 2, 3, 6, 7, 8, 8

a. There is one mode, 8.

b. The number of values is odd, so the median is the middle number, 6.

c. $\dfrac{1 + 2 + 3 + 6 + 7 + 8 + 8}{7} = \dfrac{35}{7} = 5$

∴ the mean is 5.

EXAMPLE 2 Find the mean of the 25 heights in the table in Section 13-1 by using their frequency count.

SOLUTION

Height	Frequency
155	1
156	1
157	3
158	4
159	2
160	5
161	4
162	1
163	2
165	2

To find the sum, multiply each height by its frequency. Then add the products.

$155(1) + 156(1) + 157(3) + 158(4) + 159(2)$
$\quad + 160(5) + 161(4) + 162(1) + 163(2) + 165(2)$
$= 155 + 156 + 471 + 632 + 318 + 800 + 644 + 162 + 326 + 330$
$= 3994$

$$\text{Mean} = \frac{3994}{25} = 159.76 \approx 159.8$$

Oral Exercises

For the list of data 4, 4, 5, 6, 6, 7, 8, 8, 8, 9, state the following.

1. the mode(s) 8   2. the median 6.5   3. the mean 6.5

For the data in each dot frequency diagram, state the mode, median, and mean.

4.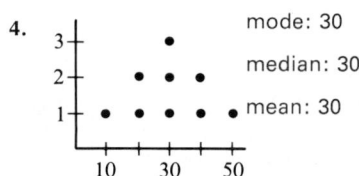

mode: 30
median: 30
mean: 30

5.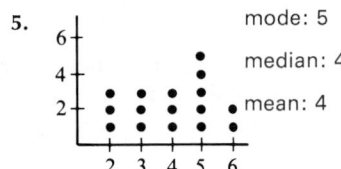

mode: 5
median: 4
mean: 4

6. Does the median have to be a member of a set of data? Does the mode? No; Yes

7. Does every set of data have a mode? Explain.

8. Suppose you received the following grades on six quizzes: 78, 91, 79, 41, 84, 83. Which average, the mean or the median, appears to provide the better description of your typical performance? Why?

Written Exercises

In Exercises 1–4, find the following for each list of data.

a. the mode(s)   b. the median   c. the mean

A   1. 4, 5, 5, 6, 7, 10, 12 **a.** 5   **b.** 6   **c.** 7   2. 15, 16, 19, 19, 20, 21, 21, 21 **a.** 21 **b.** 19.5 **c.** 19

3. 9, 13, 13, 23, 25, 12, 13, 18, 24, 25   4. 14, 26, 28, 17, 31, 23, 26, 21, 28, 26
   a. 13   **b.** 15.5   **c.** 17.5   **a.** 26   **b.** 26   **c.** 24

In Exercises 5 and 6, the data represent scores on an algebra test given to two classes. Find the mean from each table of scores. If necessary, round to the nearest tenth.

5.
Score	Frequency	67.3
90	8	
80	14	
70	11	
60	8	
50	8	
40	7	

6.
Score	Frequency	66.5
90	2	
80	12	
70	20	
60	7	
50	6	
40	5	

B   7. Find the median in Exercise 5. 70   8. Find the median in Exercise 6. 70

9. Determine the mode, median, and mean of the following four distributions of measurements. What do they have in common?

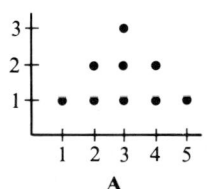
A

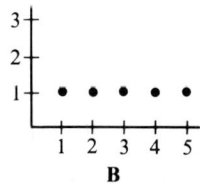
B

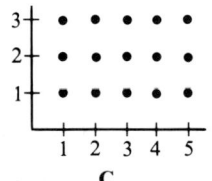
C

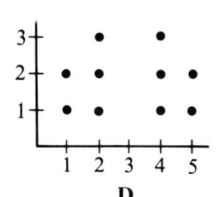
D

Statistics and Probability   **639**

Suggested Assignments

Average
  639/1–16
Maximum
  639/1–20

Additional Answers
Oral Exercises

7. No. If each measurement occurs with equal frequency, there is no mode.

8. The median. Most of the grades are clustered near the median, which is 81, whereas the one low grade brings the mean down to 76.

Mixed Review

1. Stephanie's grades on her first-term algebra tests were: 85, 73, 88, 95, 97. Find her average grade to the nearest tenth.   87.6

2. Simplify.
   a. $\dfrac{(-2a)^2 - (-4)^2}{a + 2}$ $4a - 8$

   b. $\dfrac{a^2b^{-3}c^0}{a^4b^{-5}c}$ $\dfrac{b^2}{a^2c}$

Solve.

3. a. $|x - 4| = 2$ $\{2, 6\}$
   b. $|x + 4| = 2$ $\{-6, -2\}$
   c. $3|x - 4| < -12$ $\emptyset$
   d. $-3|x + 4| < 12$ $\mathcal{R}$

4. a. $5 - 3x > 2$ $\{x: x < 1\}$

   b. $\dfrac{3x}{4} + 2 \le \dfrac{2x}{3} + 1$
   $\{x: x \le -12\}$

Additional Answers
Written Exercises

9. **A**: 3; 3; 3
   B: none; 3; 3
   C: none; 3; 3
   D: 2, 4; 3; 3
   They all have a median of 3 and a mean of 3.

18. If z is the mean of x and

y, $z = \dfrac{x + y}{2} = \dfrac{x}{2} + \dfrac{y}{2}$.

Since $x < y$, $\dfrac{x}{2} < \dfrac{y}{2}$. Add-

ing $\dfrac{x}{2}$ to each side of this

inequality, $\dfrac{x}{2} + \dfrac{x}{2} < \dfrac{x}{2} +$

$\dfrac{y}{2}$, and so $x < z$; adding $\dfrac{y}{2}$

to each side, $\dfrac{x}{2} + \dfrac{y}{2} < \dfrac{y}{2} +$

$\dfrac{y}{2}$, and so $z < y$. Then,

combining these
inequalities, $x < z < y$.

20. If y^2 is the mean of x^2
and z^2, then $y^2 =$

$\dfrac{x^2 + z^2}{2}$, and so $2y^2 =$

$x^2 + z^2$. Then mean of

$\dfrac{1}{y + z}$ and $\dfrac{1}{x + y} =$

$\dfrac{\dfrac{1}{y + z} + \dfrac{1}{x + y}}{2} =$

$\dfrac{\dfrac{x + 2y + z}{2y^2 + 2zy + 2xy + 2xz}}{} =$

$\dfrac{x + 2y + z}{x^2 + z^2 + 2zy + 2xy + 2xz}$

$= \dfrac{x + 2y + z}{(x + 2y + z)(z + x)} =$

$\dfrac{1}{z + x}$.

10. When the owner of a shoe store is putting in an order for more shoes, which descriptive measure of the foot sizes of the regular customers would be most useful: the mean, median, or mode? Explain. The mode. The owner would want to order the sizes most often sold.

11. If r_1, r_2, r_3, . . . , r_n is a list of n measurements, write a formula for their mean, M. $M = (r_1 + r_2 + r_3 + \ldots + r_n) / n$

12. If the list in Exercise 11 is ordered from least to greatest, and $n = 25$, what symbol would you use to denote the median measurement? r_{13}

13. If the sum of n measurements is 410 and their mean is 5, how many measurements are there? 82

14. If the mean of 6, 22, 40, x, 53, 14, 26, and 31 is 26, find the value of x. $x = 16$

15. The mean of 12 items is 7. If the mean of the first four items is 4, the mean of the next three items is 11, and the mean of the remaining 5 items is x, find the value of x. $x = 7$

16. Sean has scored an average of 172 after 27 games of bowling. What average must he achieve over the next 9 games to finish the season with a 173 average? 176

C **17.** Show that if x and y are integers, and one is odd and one is even, then their mean is not an integer.

18. Show that if $x < y$, and if the mean of x and y is z, then $x < z < y$.

19. a. If there are an equal number of measurements one unit less than some value n as there are one unit greater than n, show that the mean of the set of measurements is n.

b. If the measurements are as described in part (a), and there are also an equal number of measurements two units less than n as there are two units greater than n, show that the mean is n.

c. What is the generalization of this pattern?

20. Show that if y^2 is the mean of x^2 and z^2 and $x^2 \neq z^2$, then $\dfrac{1}{z + x}$ is the mean of $\dfrac{1}{y + z}$ and $\dfrac{1}{x + y}$.

13–5 Measures of Variation

The **range** of a collection of data is the difference between the greatest and least values in the set. It is the simplest *measure of the variation* in the data. The range tells us nothing, however, about how the data are scattered or clustered together, in particular, around the mean. For example, consider these two lists of data, A and B.

$$A:\ 2,\ 5,\ 6,\ 7,\ 10$$
$$B:\ 2,\ 2,\ 6,\ 10,\ 10$$

They have the same range, 8, and the same mean, 6. But in list A the numbers are clustered much more closely about the mean than in list B.

640 *Chapter 13*

You can compare the degree of scattering in two data samples by finding the *variance* of each one. The **variance** is the average of the squares of the *deviations,* or differences, of *all* the measurements from their mean.

EXAMPLE 1   Find the variance for each of lists A and B on page 640, and interpret the results.

SOLUTION   For list A, the deviations from the mean are: $2 - 6$, or -4; $5 - 6$, or -1; $6 - 6$, or 0; $7 - 6$, or 1; $10 - 6$, or 4. Then the variance is:

$$\frac{(-4)^2 + (-1)^2 + 0^2 + 1^2 + 4^2}{5} = \frac{34}{5}$$

$$= 6.8$$

For list B, the deviations are: $2 - 6$, or -4; $2 - 6$, or -4; $6 - 6$, or 0; $10 - 6$, or 4; $10 - 6$, or 4.
The variance is:

$$\frac{(-4)^2 + (-4)^2 + 0^2 + (4)^2 + (4)^2}{5} = \frac{64}{5}$$

$$= 12.8$$

Since the variance in A is 6.8 whereas in B it is 12.8, the data in B are considerably more scattered from their mean than are the data in A.

If the observations are measured in units, such as meters, then the variance is given in terms of square units. When you take the *principal square root* of the variance, you then have a measure of variation that is in the same units as those of the given data. This measure is called the **standard deviation,** usually denoted by s.

EXAMPLE 2   Find the standard deviation s for lists A and B.

SOLUTION   For A, the variance s^2 is 6.8, so $s = \sqrt{6.8} \approx 2.6$
For B, the variance s^2 is 12.8, so $s = \sqrt{12.8} \approx 3.6$

Oral Exercises

For Exercises 1–3, use the list of data 3, 4, 6, 7, 10, and find the following.

1. the range /   2. the mean 6   3. the deviation of each value from the mean
4. Find the variance for the list 3, 4, 6, 7, 10. 6 $-3; -2; 0; 1; 4$
5. The mean of each of the following lists of data is 8. Which one do you think has the greater variance? **b.**
   a. 5, 7, 8, 9, 11   **b.** 4, 5, 5, 6, 7, 9, 10, 11, 15
6. Check your answer to Exercise 5 by finding the variance of each list. **a.** 4   **b.** $11\frac{1}{3}$

Statistics and Probability   **641**

Teaching Suggestions
p. T115

Related Activities p. T115

Supplementary Material
Test 52

Key Ideas
Find the range, variance and standard deviation of a frequency distribution.

Chalkboard Examples

1. Find the variance for each of lists A and B below, and interpret the results.

   A: 3, 4, 4, 5, 6, 8
   B: 3, 4, 5, 5, 6, 7

   For list A, the variance is:
   $$\frac{4 + 1 + 1 + 0 + 1 + 9}{6}$$

   $$= \frac{16}{6} \approx 2.7$$

   For list B, the variance is:
   $$\frac{4 + 1 + 0 + 0 + 1 + 4}{6}$$

   $$= \frac{10}{6} \approx 1.7$$

   Since the variance in A is 2.7 whereas in B it is 1.7, the data in A are more scattered from their mean than are the data in B.

2. Find the standard deviation s for lists A and B.
   For A, the variance s^2 is 2.7, so $s = \sqrt{2.7} \approx 1.6$.
   For B, the variance s^2 is 1.7, so $s = \sqrt{1.7} \approx 1.3$.

Suggested Assignments

Average
First day: 642/1–6
              S 634/15
Second day: 642/7–11
              R 643/Self-Test 2
Maximum
First day: 642/1–6
              S 635/16
Second day: 642/7–12
              R 643/Self-Test 2

Written Exercises

Find the range for each list of data.

A **1.** 3, 14, 16, 29, 32, 50 47   **2.** 2, 3, 13, 14, 14, 15, 20, 23 21

3. For the list 2, 4, 7, 8, 9, find to the nearest tenth:
a. the variance, s^2 6.8   **b.** the standard deviation, s 2.6

4. For the list 10, 11, 13, 14, 17, find to the nearest tenth:
a. the variance, s^2 6   **b.** the standard deviation, s 2.4

In Exercises 5 and 6, the data represent hourly wages of employees in two small businesses. Use the frequency count and find the following to the nearest tenth.

a. the mean   **b.** the variance   **c.** the standard deviation

B **5.**

Hourly wage	Frequency	
$ 6	2	
$ 9	2	**a.** 11
$12	2	**b.** 18.8
$14	1	
$20	1	**c.** 4.3

6.

Hourly wage	Frequency	
$ 8	1	
$10	4	**a.** 13
$14	2	**b.** 14.4
$17	2	
$20	1	**c.** 3.8

7. Refer to the data lists A and B on page 640. Find the sum of the deviations for each list. How do the two sums compare? Do they tell you anything about how the data are scattered? A: 0, B: 0; They are equal; No

If $r_1, r_2, r_3, \ldots, r_n$ denotes a list of n observations and m denotes their mean, write an expression for each of the following.

8. the deviation of r_1 from the mean

9. the average of the set of all the deviations from the mean

10. the variance, s^2

11. the standard deviation, s

C **12. a.** If the deviations of data from the mean of the data are added, positive and negative quantities will offset each other, resulting in a sum of zero. An early statistical attempt to overcome this effect was the introduction of the absolute values of deviations in the addition process. Thus the *average deviation* from the mean was defined to be the following:

$$\frac{|r_1 - m| + |r_2 - m| + \cdots + |r_n - m|}{n}$$

Calculate this quantity for the data in Exercises 3 and 4. Compare this description of "scattering" with the standard deviation s found in part (b) of Exercises 3 and 4. Do the average deviations also appear to indicate which data are more scattered? **3.** 2.4; **4.** 2; Yes

b. The average deviation is seldom used except as a percent of the mean. For the data in Exercises 3 and 4, express the average deviation as a percent of the mean. **3.** 40%; **4.** 15.4%

Computer Exercises For students with computer experience

1. Write a program that will allow you to input a list of positive numbers and will compute the arithmetic mean of the numbers. The program should accept as many positive numbers as you want to input and should have you signal the end of your list by inputting one negative number. (Be careful that the program does not use this negative number in computing the mean.)

2. Modify the program that you wrote for Exercise 1 so that it will allow you to input a list of positive numbers *each followed by a frequency*.

3. Write a program that will allow you to input a set of numerical data and will compute the variance and the standard deviation of the data. The program should first ask that you input the *number* of data values to be entered, then ask you to input the data values themselves. (*Note*: This program requires the use of an *array* of values, and you will need to work with *subscripted variables*.)

4. Suppose that you are given a set of 25 quiz grades consisting of whole numbers from 1 through 10. Write a program that will compute the mode(s) of this set of data. (*Hint*: This program requires the use of an array of values. For each value of N from 1 through 10, find the frequency of the grade N and store this number as the value of G(N). Then, for each value of N, test whether G(N) is greater than or equal to *all* the values G(J), where J is a whole number from 1 through 10. If it is, then N is a mode of the set of data.)

▮ Self-Test 2

VOCABULARY   median (p. 637)   range (p. 640)
              mode (p. 637)   variance (p. 641)
              arithmetic mean (p. 637) standard deviation (p. 641)

1. Find the median, mode, and mean of the following data: *Obj. 1, p. 637*
   7, 2, 2, 6, 4, 2, 7, 2, 4, 14. 4; 2; 5

2. Find the range, variance, and standard deviation of the   *Obj. 2, p. 637*
   data in Exercise 1. 12; 12.8; 3.6

Check your answers with those at the back of the book.

Statistics and Probability   **643**

Mixed Review

1. Approximate each square root to the nearest hundredth. Use a calculator or the Table of Square Roots as necessary.
   a. $\sqrt{37}$ 6.08
   b. $\sqrt{481.5}$ 21.94
   c. $\sqrt{800}$ 28.28

2. A hockey team's mean score after 40 games is 4 goals per game. What mean must the team achieve over the next 20 games to finish the season with a mean of 5 goals per game?   7

Simplify.

3. $\dfrac{2x - 3}{5} - \dfrac{x - 1}{3} - \dfrac{x - 4}{15}$

4. $\dfrac{2a^2 + a - 6}{2a^2 + 5a + 3} \div \dfrac{a^2 + a - 2}{1 - a^2}$

   $-\dfrac{2a - 3}{2a + 3}$, or $\dfrac{3 - 2a}{3 + 2a}$

Quick Quiz

1. Find the median, mode, and mean of the following data: 10, 7, 8, 10, 11, 11, 14, 9, 11, 9   10, 11, 10

2. Find the range, variance, and standard deviation of the data in Exercise 1.   7, 3.4, 1.8

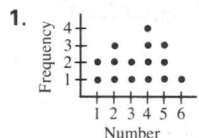
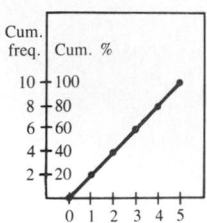
PROGRAMMING IN BASIC

The following program illustrates how a computer can be used in statistical analysis. Note that the program uses the READ and DATA statements in lines 60 and 410–450.

```
10    PRINT "ANALYSIS OF A SET OF DATA:"
20    LET S = 0
30    LET A = 0
40    LET B = 0
50    LET C = 0
60    READ M
70    IF M = −1 THEN 170
80    LET S = S + M
90    IF M < 40 THEN 150
100   IF M < 70 THEN 130
110   LET A = A + 1
120   GOTO 60
130   LET B = B + 1
140   GOTO 60
150   LET C = C + 1
160   GOTO 60
170   LET N = A + B + C
180   PRINT
190   PRINT "N = ";N,"AVERAGE = ";S/N
200   PRINT "INTERVAL      FREQ.      REL. FREQ."
210   PRINT "10–39          ";C;"          ";C/N
220   PRINT "40–69          ";B;"          ";B/N
230   PRINT "70–99          ";A;"          ";A/N
240   PRINT
250   PRINT "10–39:     ";
260   LET X = C
270   GOSUB 360
280   PRINT "40–69:     ";
290   LET X = B
300   GOSUB 360
310   PRINT "70–99:     ";
320   LET X = A
330   GOSUB 360
340   GOTO 460
350   REM *SUBROUTINE
360   FOR I = 1 TO X
370   PRINT "*";
380   NEXT I
390   PRINT
400   RETURN
```

644   *Chapter 13*

```
410    DATA 29,72,17,53,84,55,14,93
420    DATA 65,29,25,82,16,84,92,39
430    DATA 48,72,71,35,52,68,44,16
440    DATA 64,86,67,86,46,67,86,63
450    DATA 46,48,67,25,45,72,88,91,−1
460    END
```

Exercises

1. Type in and RUN the program as given.

2. Modify the program so that the intervals are 0–19, 20–39, 40–59, 60–79, and 80–99. Then RUN the revised program.

Probability

OBJECTIVES for Sections 13-6 and 13-7:
1. To find the probability of a specified event in an experiment with equally likely outcomes.
2. To calculate an experimental probability.

13–6 The Probability of an Event

Teaching Suggestions
p. T116

Related Activities p. T116

Key Ideas
Find the probability of an event in a random experiment.

The main use for statistical theory is to help people make decisions on the basis of incomplete information. For example, business people and politicians often use marketing surveys and opinion polls to assess the attitude of the *population* (the customers or voters) by studying a *sample* of it. In choosing a sample of data and then trying to draw conclusions about the population, the statistician must consider the theory of *probability*.

Simple games of chance offer ideal situations for understanding the notion of probability. Suppose that you perform the experiment of tossing a die and observing the number of dots on the top face. The six possible outcomes are the following:

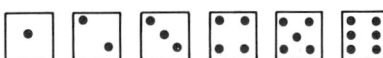

Figure 4

If the die is not loaded you can assume that each outcome is equally likely. That is, each outcome has an equal chance of occurring, namely, 1 out of 6. Then the *measure* of chance, or **probability,** that a particular outcome will occur is $\frac{1}{6}$.

1. There are 12 face cards in a standard bridge deck of 52 cards. What is the probability of the event that a card drawn at random is a face card?
   Since there are 12 face cards, and 52 cards in the deck, $e = 12$ and $n = 52$.

   Then $P(E) = \dfrac{e}{n} = \dfrac{12}{52} = \dfrac{3}{13}$.

2. A box contains 3 yellow, 4 green, and 2 white marbles. If a marble is drawn at random from the box, what is the probability of the event "the marble is not green"?
   Of the 9 marbles in the box, 5 are not green.

   Hence $P(E) = \dfrac{e}{n} = \dfrac{5}{9}$.

Reading Algebra

Using a 52-card bridge deck, have students state an event and calculate the probability of that event. For example:

"What is the probability that a card drawn is red and is a 3?" $\dfrac{2}{52}$, or $\dfrac{1}{26}$

"What is the probability that a card drawn is red or is a 3?" $\dfrac{28}{52}$, or $\dfrac{7}{13}$

"What is the probability that a card drawn is a black face card?" $\dfrac{6}{52}$, or $\dfrac{3}{26}$

"What is the probability that a card drawn is either black or a face card?" $\dfrac{32}{52}$, or $\dfrac{8}{13}$

Students should be able to differentiate events as described in each statement although the statements ap-

---

If an experiment has n possible, equally likely outcomes, then the probability that any one of them will occur is $\dfrac{1}{n}$.

An **event** is a *specified subset* of the set of all possible outcomes in an experiment. When you toss a die, the probability P of the event "the top face shows *either* 3 *or* 4 dots" is $\frac{2}{6}$. That is, there are 2 out of 6 chances that either one or the other of these outcomes will occur. If any *one* of the outcomes in an event occurs, you say that the event *occurs*.

If an experiment has n possible, equally likely outcomes and an event E consists of e of these outcomes, then the probability P that E will occur is $\dfrac{e}{n}$. That is:

$$P(E) = \frac{\text{number of outcomes in } E}{\text{number of possible outcomes}} = \frac{e}{n}$$

EXAMPLE 1 There are 4 jacks in a standard bridge deck of 52 cards. What is the probability of the event that a card drawn at random is a jack?

SOLUTION Since there are 4 jacks and 52 cards in the deck, $e = 4$ and $n = 52$. Then

$$P(E) = \frac{e}{n} = \frac{4}{52} = \frac{1}{13}.$$

EXAMPLE 2 A jar contains 2 red, 4 blue, and 5 white marbles. If a marble is drawn from the jar at random, what is the probability of the event "the marble is not red"?

SOLUTION Of the 11 marbles in the jar, 9 are not red. Hence

$$P(E) = \frac{9}{11}.$$

All the experiments described in this chapter are assumed to be *random* ones, that is, experiments conducted in such a way that the outcomes are strictly a matter of chance.

Oral Exercises

1. When a die is tossed, what is the probability that the side with just two dots will be on top? $\frac{1}{6}$
2. When you toss a coin, what is the probability that it will land with the head showing? $\frac{1}{2}$

3. When you draw a marble, while blindfolded, from a jar containing 4 marbles—1 red, 1 blue, 1 white, 1 green—what is the probability of drawing a blue marble? $\frac{1}{4}$

4. In Exercise 3, what is the probability of the event "either a red or a green marble is drawn"? $\frac{2}{4}$, or $\frac{1}{2}$

5. When a coin is tossed, what is the probability of the event "it will land either heads or tails"? 1

Exercises 6-8 refer to the spinner at the right. Assume that the arrow will neither stop on a dividing line nor favor a particular numbered region.

6. What are the possible outcomes of a spin? Are they all equally likely?

7. What is the probability that the arrow will stop on the region labeled 1? 2? 4? $\frac{1}{4}$; $\frac{1}{4}$; $\frac{1}{4}$

8. What is the probability that the arrow will not stop on the region labeled 3? $\frac{3}{4}$

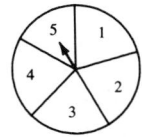

9. Suppose that in the city raffle to benefit the library, 5000 tickets are sold and you buy a single ticket.
   a. What is the probability that your ticket wins the grand prize, if only one grand prize winner is drawn at random? $\frac{1}{5000}$
   b. If your friend also has a single ticket, how does your chance of winning the grand prize compare to your friend's? It is the same.
   c. What can you do to "double your chances"? If you do this, what is the probability that you will be a grand prize winner? Buy two tickets; $\frac{1}{2500}$

6. The arrow will stop on the region labeled 1, 2, 3, or 4; Yes

Written Exercises

Exercises 1–14 refer to the spinner at the right. Assume that the arrow will neither stop on a dividing line nor favor a particular numbered region.

A
1. List the possible outcomes of a spin. 1, 2, 3, 4, 5, 6
2. List the outcomes that are prime numbers. 2, 3, 5

What is the probability that after a spin the arrow will stop on:

3. region 3? $\frac{1}{6}$

4. region 4? $\frac{1}{6}$

5. region 1 or 2? $\frac{1}{3}$

6. region 2 or 6? $\frac{1}{3}$

7. an even-numbered region? $\frac{1}{2}$

8. a region with a prime number? $\frac{1}{2}$

9. a region with a number less than 3? $\frac{1}{3}$

10. a region with a number greater than 2? $\frac{2}{3}$

11. a region with a number greater than 3? $\frac{1}{2}$

12. a region with a number less than 2? $\frac{1}{6}$

13. a region with a number greater than 6? 0

14. a region with a number less than 7? 1

___

Suggested Assignments

Average
First day: 647/1–26
Second day: 648/27–46
Maximum
First day: 647/1–26
Second day: 648/27–48

___

Additional A Exercises

Exercises 1–3 refer to the spinner below. Assume that the arrow will neither stop on a dividing line nor favor a particular numbered region.

What is the probability that after a spin the arrow will stop on the following?

1. region 4 $\frac{1}{5}$

2. an even-numbered region $\frac{2}{5}$

3. a region with a number less than 4 $\frac{3}{5}$

A box contains tickets numbered 1 to 30. If a ticket is drawn at random, what is the probability that the ticket number is the following?

4. 7 $\frac{1}{30}$

5. a multiple of 4 $\frac{7}{30}$

6. less than 35 1

647

Solve.

1. $6x^2 - 9x - 1 = 5$

$\left\{-\dfrac{1}{2}, 2\right\}$

2. $3x^2 - 7x + 4 = 0$

$\left\{1, \dfrac{4}{3}\right\}$

3. $x^2 - 1 = x$

$\left\{\dfrac{1 + \sqrt{5}}{2}, \dfrac{1 - \sqrt{5}}{2}\right\}$

Divide the first polynomial by the second.

4. $8x^3 - 4x + 3;\ 2x - 1$

$4x^2 + 2x - 1 + \dfrac{2}{2x - 1}$

5. $x^8 - 1;\ x^5 + x^4 + x + 1$

$x^3 - x^2 + x - 1$

A jar contains 2 white, 3 red, 1 green, and 4 blue marbles. If a marble is drawn at random from the jar, what is the probability that the marble is:

15. white? $\frac{1}{5}$ **16.** red? $\frac{3}{10}$

17. green? $\frac{1}{10}$ **18.** blue? $\frac{2}{5}$

19. not white? $\frac{4}{5}$ **20.** not blue? $\frac{3}{5}$

21. white or red? $\frac{1}{2}$ **22.** red or green? $\frac{2}{5}$

23. red, green, or white? $\frac{3}{5}$ **24.** red, green, or blue? $\frac{4}{5}$

25. yellow? 0   **26.** not yellow? 1

In a standard deck of 52 cards, there are four of each kind of card: ace, king, queen, jack, ten, nine, . . . , two. Half the cards are black and half are red, with two black and two red aces, two black and two red kings, and so on. If a card is drawn at random from the deck, what is the probability that it will be:

27. a black card? $\frac{1}{2}$ **28.** a ten? $\frac{1}{13}$

29. a king? $\frac{1}{13}$ **30.** a red card? $\frac{1}{2}$

31. a four? $\frac{1}{13}$ **32.** an ace? $\frac{1}{13}$

33. a red jack? $\frac{1}{26}$ **34.** a black ace? $\frac{1}{26}$

35. not a red queen? $\frac{25}{26}$ **36.** not an ace? $\frac{12}{13}$

37. not a queen? $\frac{12}{13}$ **38.** not a black ace? $\frac{25}{26}$

What is the probability that after a spin the arrow will stop on:

B   **39.** an unshaded region? $\frac{1}{2}$ **40.** a region shaded red? $\frac{1}{4}$

41. a region corresponding to a multiple of 3? $\frac{1}{3}$

42. a region corresponding to a multiple of 2? $\frac{1}{2}$

43. a region corresponding to a factor of 12? $\frac{1}{2}$

44. a region corresponding to a factor of 10? $\frac{1}{3}$

45. List the set of possible outcomes when a penny and a nickel are tossed, using the symbols H and T for heads and tails. (For example, for the outcome when the penny lands heads and the nickel tails, you would write HT; but for the outcome when the penny lands tails and the nickel heads, you would write TH.) HH, HT, TH, TT

46. In Exercise 45 do the events "they both land heads" and "they do not both land heads" have the same probability? What is the probability of each of these events? No; $\frac{1}{4}, \frac{3}{4}$

C   **47.** A jar contains four white tickets numbered 1–4, and three red tickets numbered 1–3. What is the probability that one ticket drawn randomly has either an even number or is red? $\frac{5}{7}$

48. What is the probability that a card drawn at random from a standard deck of 52 cards is:

a. either a king or a red card? $\frac{7}{13}$ **b.** neither a king nor a red card? $\frac{6}{13}$

648   *Chapter 13*

13–7 Experimental Probability

Teaching Suggestions
p. T116

Supplementary Material
Test 53

Key Ideas
Calculate an experimental probability.

Suppose that in a particular year, a school with 1000 students has 40 who are left-handed. Then there are 40 chances in 1000, or 1 in 25, that a student chosen at random will be left-handed. If there are 25 seats in each classroom, would you therefore conclude that every classroom should have exactly one seat with an arm for left-handed writers? Your intuition and experience tell you, "Of course not! For example there might be three left-handed students in one room and none in another."

What is the probability that exactly 40 of the school's 1000 students the next year will be left-handed? Again, experience based on repeated observations would indicate a relatively small probability of this event. Nevertheless, in the absence of further information regarding the percent of left-handed people in the population, you would have to give $\frac{1}{25}$ as your best estimate of the probability that a randomly chosen student in that school next year will be left-handed.

A certain baseball player may have a *batting average* (that is, a ratio of safe hits to times at bat) of 0.400 for a particular season. Then the **experimental probability** of the player making a safe hit the next time at bat is $\frac{4}{10}$, or $\frac{2}{5}$. This does not mean, however, that the player will *surely* make 2 hits in any given 5 times at bat.

In many real-life situations such as these, you have only experimental probabilities from which to predict future occurrences of events.

If an experiment is conducted n times, and an event E occurs e of these times, then the experimental probability P that E will occur in another trial is $\frac{e}{n}$. That is:

$$P(E) = \frac{\text{number of occurrences of } E}{\text{number of trials}} = \frac{e}{n}$$

EXAMPLE A thumbtack was tossed 100 times. It landed "point up" 60 times and "point down" 40 times.

Point up   Point down

  a. What is the experimental probability that on the next toss it will land point up?
  b. If it were tossed 50 additional times, about how many times would you expect it to land point down?
  c. If on 50 additional tosses it actually landed point down 25 times, what would be the experimental probability, based on all 150 tosses, that on the next toss it would land point down?

1. The Montgomery Cougars soccer team has won 25 of their last 40 soccer games.
   a. What is the experimental probability that they will win their next game?
   b. If they were to play an additional 20 games, how many would you expect them to lose?
   c. If the Cougars won 23 of their next 24 games, what would be the experimental probability based on all 64 games that the Cougars would win their 65th game?
   a. For the event E that the Cougars would win their next game, $e = 25$ and $n = 40$. Therefore,
   $$P(E) = \frac{25}{40} = 0.625.$$
   b. Since the probability that the team will lose is $\frac{15}{40}$ or 0.375, you would expect them to lose 24×0.375, or 9 games.
   c. If the Cougars win 25 of the first 40 games, and 23 of the next 24 games, then $e = 25 + 23 = 48$ and $n = 40 + 24 = 64$. Therefore, the experimental probability that the Cougars will win their 65th game is
   $$P(E) = \frac{e}{n} = \frac{48}{64} = 0.75.$$

SOLUTION a. For the event E that the thumbtack lands point up, you have $e = 60$ and $n = 100$. Therefore,

$$P(E) = \frac{60}{100} = 0.6.$$

b. Since the experimental probability that the thumbtack will land point down is $\frac{40}{100}$, or 0.4, you would expect it to land point down about 50×0.4, or 20, times out of the additional 50 tosses.

c. If the thumbtack lands point down 40 times out of the first 100 throws, and 25 times out of the next 50 throws, then in all you have $e = 40 + 25 = 65$ and $n = 100 + 50 = 150$. Therefore the experimental probability that the thumbtack will land point down is now

$$P(E) = \frac{e}{n} = \frac{65}{150} \approx 0.43.$$

Oral Exercises

1. A baseball player's batting average is 0.250.
   a. What is the experimental probability that the player will get a hit the next time at bat? $\frac{1}{4} = 0.25$
   b. What is the experimental probability that the player will not get a hit? $\frac{3}{4} = 0.75$
   c. In the next 8 times at bat, how many times would you expect the player to get a hit? 2

2. If the experimental probability is $\frac{5}{8}$ that a softball player will get a hit the next time at bat, does this mean that the player is sure to get 5 hits in the next 8 times at bat? No

3. The dietician in a company cafeteria made an informal survey one day, finding that 130 employees bought lunch in the cafeteria, 50 brought their lunches with them and ate in the cafeteria, and 20 did not come to the cafeteria at all. What is the experimental probability that an employee will buy lunch in the cafeteria? What is the experimental probability that an employee will not bring a lunch to the cafeteria? $\frac{130}{200} = 0.65$; $\frac{150}{200} = 0.75$

4. If two carriers of a certain trait marry, the experimental probability of any child of theirs having the trait is $\frac{1}{4}$. If they have 4 children, is it true that:
   a. one of them is certain to have the trait? No
   b. none of them might have the trait? Yes
   c. all of them might have the trait? Yes

Written Exercises

A

1. If a football quarterback is completing passes at the rate of 55%, what is the experimental probability that:
   a. the quarterback will complete the next pass? $\frac{55}{100} = 0.55$
   b. the quarterback will not complete the next pass? $\frac{45}{100} = 0.45$

2. A dart player has hit the bull's eye 8 times out of 40 throws. What is the experimental probability that:
   a. the player will hit the bull's eye in the next throw? $\frac{8}{40} = 0.2$
   b. the player will not hit the bull's eye in the next throw? $\frac{32}{40} = 0.8$

3. During a baseball season, a baseball player gets 10 hits out of 25 times at bat.
   a. What is the experimental probability that the player will get a hit the next time at bat? $\frac{10}{25} = 0.4$
   b. About how many hits would you expect the player to get the next 15 times at bat? 6
   c. Suppose that the player actually gets 2 hits the next 15 times at bat. What would be the experimental probability that the player would get a hit the next time at bat? $\frac{12}{40} = 0.3$

4. In a basketball game, a player makes the basket on 8 free throws out of 12.
   a. What is the experimental probability that the player will make a basket on the next free throw? $\frac{8}{12} \approx 0.67$
   b. How many times would you expect the player to make the basket on the next 3 free throws? 2
   c. If the player actually makes the basket all 3 times in the next 3 free throws, what is the experimental probability that the player will make a basket the next time? $\frac{11}{15} \approx 0.73$

Suppose that the results of 36 successive random drawings of one marble from a jar containing an unknown number of colored marbles are as follows: 4 blue, 14 red, 12 yellow, 6 green. What is the experimental probability that the next marble drawn will be:

5. red? $\frac{14}{36} \approx 0.39$ 6. blue? $\frac{4}{36} \approx 0.11$ 7. green? $\frac{6}{36} \approx 0.17$ 8. yellow? $\frac{12}{36} \approx 0.33$

9. yellow or green? $\frac{18}{36} = 0.5$ 10. red or blue? $\frac{18}{36} = 0.5$

11. not purple $\frac{36}{36} = 1$ 12. purple $\frac{0}{36} = 0$

B

13. Refer to the drawing described in Exercises 5–12. Assume that the ratios of colored marbles found in the drawing are representative of the total number of marbles that were in the jar. Using that fact, can you tell how many marbles were in the jar? Explain.

14. If it is found that there are 90 yellow marbles in the jar described in Exercises 5–12, about how many green ones would you expect to find? About how many blue ones would you expect to find? 45; 30

Statistics and Probability **651**

Suggested Assignments

Average
First day: 651/1–8
Second day: 651/9–15
          R 652/Self-Test 3
Maximum
   651/1–15
R 652/Self-Test 3

Additional Answers
Written Exercises

13. No. There can be any number of marbles that yields the same ratios. The contents of the jar must be $\frac{1}{9}$ blue, $\frac{7}{18}$ red, $\frac{1}{3}$ yellow, and $\frac{1}{6}$ green.

Additional A Exercises

Solve.

1. A softball player got 6 hits out of 10 times at bat.
   a. What is the experimental probability that the player will get a hit next time at bat?   0.6
   b. In the next 5 times at bat, how many times would you expect the player to get a hit?   3

2. A store orders 1000 watches. Of 200 watches randomly selected for testing, 1 was defective.
   a. What is the probability that the next watch will be defective?   0.005
   b. Of the first 600 of these watches, how many would they expect to be defective?   3
   c. If only 2 of the first 600 watches were actually defective, what is the experimental probability that the next watch tested will be defective?   0.003

Mixed Review

Exercises 1–3 refer to the spinner below. Assume that the arrow will not stop on a dividing line.

What is the probability that the spinner arrow will stop on the following regions?

1. an odd-numbered region $\frac{1}{2}$

2. region 7 0

3. a region with a number greater than or equal to 2 $\frac{5}{6}$

4. List the set of outcomes of tossing 3 coins at once. HHH, HHT, HTH, THH, TTH, THT, HTT, TTT

5. If 3 coins are tossed at once, what is the probability of getting the following?

   a. exactly two tails $\frac{3}{8}$

   b. at least two tails $\frac{1}{2}$

6. If 1 coin is tossed three times, what is the probability of getting the following?

   a. no heads $\frac{1}{8}$

   b. at least one head $\frac{7}{8}$

C 15. United States mortality tables show that about 95% of people born now can expect to be alive at age 40, but only about 75% of those born now will still be alive at age 65. What is the experimental probability that a person born now who reaches age 40 will still be alive at 65? (Assume that life expectancies do not change in the next 65 years.) $\frac{75}{95} \approx 0.79$

Computer Exercises For students with computer experience

You will need to use the computer's random-number function for the following exercises. The format of this function varies from computer to computer, so you may need to consult your teacher or the computer manual before you proceed. If you are programming in BASIC, on many computers either RND(1) or RND(0) will generate a random number greater than or equal to 0 and less than 1.

1. Write a program that will generate a random *integer* from 1 through 10 one hundred times and will determine how many times the integer 3 was generated.

2. Modify the program that you wrote for Exercise 1 so that it will determine how many times an integer greater than 7 was generated.

3. Modify the program that you wrote for Exercise 1 so that it will generate a random integer from 1 through 100 one hundred times and will determine how many times an integer divisible by 5 was generated.

4. Write a program that will simulate one hundred rolls of two dice and will determine how many times the total number of dots on the two faces that turn up is 7.

5. Modify the program that you wrote for Exercise 4 so that it will determine how many times the number of dots on each of the two faces that turn up is the same.

▌ Self-Test 3

VOCABULARY probability (p. 645) experimental probability
event (p. 646) (p. 649)

When you toss a die, what is the probability that:

1. the top face will show 4 dots? $\frac{1}{6}$

   Obj. 1, p. 645

2. the top face will show 7 dots? 0

652 *Chapter 13*

When you toss a die, what is the probability that:

3. the top face will show an odd number of dots? $\frac{1}{2}$
4. the top face will show either an even number or an odd number of dots? 1
5. Assume that it has rained on Thanksgiving Day in New York City for 35 out of the past 65 years. What would be the experimental probability that it will rain there next Thanksgiving Day? $\frac{35}{65} \approx 0.54$ *Obj. 2, p. 645*

Suppose the results, when you drew 40 slips of paper at random from a hat containing 100 slips, were as follows: 6 purple, 10 red, 8 yellow, 16 white. What is the experimental probability that the next slip you draw will be:

6. purple? $\frac{6}{40} = 0.15$ 7. red or yellow? $\frac{18}{40} = 0.45$ 8. yellow or white? $\frac{24}{40} = 0.6$
9. How many red slips would you expect to find in the hat? 25
10. How many purple slips would you expect to find in the hat? 15

Check your answers with those at the back of the book.

Chapter Summary

1. *Statistics* is the science of organizing and analyzing a set of facts, or data, so that probable conclusions can be drawn from the data.
2. The number of occurrences of a particular measurement is called the *frequency* of the measurement. The *relative frequency* is the ratio of the frequency of a particular measurement to the total number of measurements.
3. *Histograms* and *frequency polygons* are used to help visualize frequency distributions for large sets of data.
4. The number and percent of measurements that are less than or equal to a given value are called the *cumulative frequency* and the *cumulative percent*.
5. The *median*, the *mode*, and the *arithmetic mean* are three descriptions of the "center" of a set of data.
6. The *range*, the *variance*, and the *standard deviation* are three measures of the variation in a set of data.
7. If an experiment has n possible, equally likely outcomes, then the *probability* that any one of them will occur is $\frac{1}{n}$.
8. An *event* is a specified subset of the set of all possible outcomes in an experiment.

Statistics and Probability   **653**

Exercises 1–4 refer to the spinner pictured below. Assume that the arrow will not stop on a dividing line.

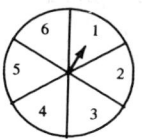

When you spin the arrow, what is the probability that the arrow will do the following?

1. stop on 3 $\frac{1}{6}$
2. stop on 0   0
3. stop on an even number $\frac{1}{2}$
4. stop on a prime number $\frac{1}{2}$
5. If the Bulldogs beat the Hornets 32 out of the last 50 times they played, what is the experimental probability that the Bulldogs will win the next game against the Hornets?   0.64

In Exercises 6–8, suppose the results, when you pick 20 marbles at random from a box containing 100 marbles, were as follows: 3 red, 8 blue, 4 green, 5 yellow.

6. What is the experimental probability that the next marble you pick will be the following?
   a. red   0.15
   b. green or blue   0.6
   c. green or red   0.35
7. How many blue marbles would you expect to find in the box?   40
8. How many yellow marbles would you expect to find in the box?   25

Additional Answers
Written Exercises (p. 625)

2.

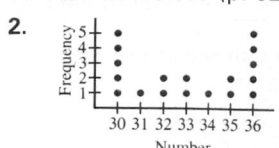

4.

No.	Freq.	Rel. Freq. Frac.	%
30	5	$\frac{5}{18}$	$27\frac{7}{9}$
31	1	$\frac{1}{18}$	$5\frac{5}{9}$
32	2	$\frac{2}{18}$	$11\frac{1}{9}$
33	2	$\frac{2}{18}$	$11\frac{1}{9}$
34	1	$\frac{1}{18}$	$5\frac{5}{9}$
35	2	$\frac{2}{18}$	$11\frac{1}{9}$
36	5	$\frac{5}{18}$	$27\frac{7}{9}$
Total:	18	1	100

6.

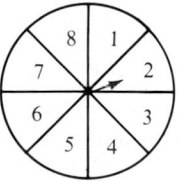

8.

Letter	Freq.	Rel. Freq. Frac.	%
M	1	$\frac{1}{10}$	10
O	4	$\frac{4}{10}$	40
N	2	$\frac{2}{10}$	20
T	1	$\frac{1}{10}$	10
U	1	$\frac{1}{10}$	10
S	1	$\frac{1}{10}$	10
Total:	10	1	100

Chapter Review

Write the letter of the correct answer.

1. What is the frequency of the letter I in the word IMITATION?   *13-1*

  a. 2   **(b.)** 3   **c.** $\frac{1}{3}$ **d.** $\frac{1}{6}$

2. What is the relative frequency of the letter N in the word DINNER?

  a. $\frac{2}{5}$ **b.** 2   **(c.)** $\frac{1}{3}$ **d.** $\frac{1}{6}$

3. For the list of data 46, 5, 14, 16, 21, 9, 19, 23, 31, 13, what is the relative frequency of the data in the interval 15–20?   *13-2*

  a. $\frac{1}{10}$ **(b.)** $\frac{2}{10}$ **c.** $\frac{3}{10}$ **d.** $\frac{4}{10}$

4. What percent of the data in Exercise 3 is in the interval 10–15?

  a. 10%   **(b.)** 20%   **c.** 30%   **d.** 40%

5. What is the cumulative frequency for the interval 45–50, given the following list of data: 26, 63, 34, 51, 40, 19, 62, 88, 53, 54, 28, 48?   *13-3*

  a. 1   **b.** 3   **c.** 5   **(d.)** 6

6. Find the mean for the following list of data: 16, 33, 25, 40, 38, 42, 19, 19.   *13-4*

  a. 19   **b.** 25   **(c.)** 29   **d.** 30

7. Find the median for the data in Exercise 6.

  a. 19   **b.** 25   **c.** 26   **(d.)** 29

8. Find the variance for the following list of data: 7, 9, 11, 15, 18.   *13-5*

  a. 4   **b.** 12   **(c.)** 16   **d.** 80

9. Find the standard deviation for the following list of data: 5, 6, 8, 9, 9, 11.

  a. 1.41   **(b.)** 2.00   **c.** 2.43   **d.** 4.90

10. A jar contains 5 white, 3 red, and 4 blue marbles. If a marble is drawn at random from the jar, what is the probability that it is not blue?   *13-6*

  a. $\frac{1}{4}$ **b.** $\frac{1}{3}$ **(c.)** $\frac{2}{3}$ **d.** $\frac{5}{12}$

11. What is the probability that after a spin the arrow will stop on region 3 or 8?

  (a.) $\frac{1}{4}$ **b.** $\frac{1}{2}$ **c.** $\frac{3}{4}$ **d.** $\frac{3}{8}$

12. Refer to the spinner in Exercise 11. What is the probability that after a spin the arrow will stop on a region corresponding to a prime number?

    a. $\frac{1}{4}$ (b.) $\frac{1}{2}$ c. $\frac{3}{4}$ d. $\frac{5}{8}$

13. A lacrosse player has hit the goal 7 out of 55 times. Approximately what is the experimental probability that the next attempt will be successful?

    (a.) 0.13   b. 0.07   c. 0.14   d. 0.21

        13-7

14. A baseball player's batting average is 0.167. In the next 12 times at bat, how many times would you expect the player to get a hit?

    a. 1   (b.) 2   c. 3   d. 4

Chapter Test

1. Make a table showing the frequencies and relative frequencies, as fractions and percents for the following set of data: 22, 20, 23, 21, 25, 21, 23, 25, 24, 23.

        13-1

2. Make a histogram and a frequency polygon for the data at the right. Group them in the intervals 10–15, 15–20, 20–25, 25–30.

13	18	23	27	29
24	16	14	12	19
21	30	19	24	22

        13-2

3. Make a table showing the frequencies, percents, cumulative frequencies, and cumulative percents for the data in Exercise 2.

        13-3

4. Find the median, mode, and mean for the following data: 9, 11, 13, 6, 22, 10, 8, 11, 16, 11, 9, 11, 10, 15, 18. 11; 11; 12

        13-4

For the list of data 15, 8, 18, 9, 20, find to the nearest tenth:

5. the variance 22.8

6. the standard deviation 4.8

        13-5

7. What is the probability of drawing at random a red ace from a standard deck of 52 cards? $\frac{1}{26}$

        13-6

8. A box contains three red tickets numbered 1–3, five green tickets numbered 4–8, and twelve blue tickets numbered 9–20. If a ticket is drawn at random from the box, what is the probability that it will be an even-numbered blue ticket? $\frac{3}{10}$

9. Claire predicted correctly 13 out of 35 times the day for each week's surprise algebra quiz. What is the experimental probability that her next prediction will be correct? $\frac{13}{35} \approx 0.37$

        13-7

10. A thumbtack was tossed 200 times. It landed "point up" 140 times and "point down" 60 times. If it is tossed 30 additional times, about how many times would you expect it to land point up? 21

Statistics and Probability   **655**

---

10. a.

Mo.	Freq.	Rel. Freq. Frac.	%
Jan.	1	$\frac{1}{20}$	5
Feb.	3	$\frac{3}{20}$	15
Mar.	3	$\frac{3}{20}$	15
Apr.	4	$\frac{4}{20}$	20
May	0	$\frac{0}{20}$	0
June	0	$\frac{0}{20}$	0
July	1	$\frac{1}{20}$	5
Aug.	0	$\frac{0}{20}$	0
Sept.	0	$\frac{0}{20}$	0
Oct.	2	$\frac{2}{20}$	10
Nov.	4	$\frac{4}{20}$	20
Dec.	2	$\frac{2}{20}$	10
Total:	20	1	100

b.

Additional Answers Chapter Test
(See p. 656.)

Additional Answers
Chapter Test (p. 655)

1.

No.	Freq.	Rel. Freq. Frac.	%
20	1	$\frac{1}{10}$	10
21	2	$\frac{2}{10}$	20
22	1	$\frac{1}{10}$	10
23	3	$\frac{3}{10}$	30
24	1	$\frac{1}{10}$	10
25	2	$\frac{2}{10}$	20
Total:	10	1	100

2.

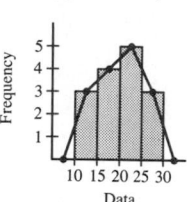

3.

Int.	Freq.	%	Cum. Freq.	Cum. %
10–15	3	20	3	20
15–20	4	$26\frac{2}{3}$	7	$46\frac{2}{3}$
20–25	5	$33\frac{1}{3}$	12	80
25–30	3	20	15	100
Total:	15	100		

Chapter 9

Solve.

1. $\frac{3x}{2} + \frac{8-4x}{7} = 3$ {2}

2. $\frac{a+1}{4} - \frac{3}{2} \geq \frac{2a-9}{10}$ {a: a ≥ 7}

3. $\frac{m}{m+2} = \frac{3}{5}$ {3}

4. $\frac{6}{n-8} - 1 = \frac{2}{n-3}$ {2, 13}

5. 74 is 5% of what number? 1480

6. 24 is what percent of 64? $37\frac{1}{2}\%$

7. What number is $\frac{3}{4}\%$ of 500? 3.75

8. If y varies directly as x^2, and if $y = 36$ when $x = 2$, find y when $x = 10$. 900

9. If c varies inversely as d, and if $c = 12$ when $d = \frac{1}{4}$, find c when $d = 9$. $\frac{1}{3}$

10. If y varies jointly as x and z, and if $y = 24$ when $x = 2$ and $z = 3$, find y when $x = 3$ and $z = 4$. 48

11. One pipe can empty a tank in 5 h, while a second pipe can empty the same tank in 7 h. How long would it take to empty the tank if both pipes are used at the same time? $2\frac{11}{12}$ h

12. The interest rate required to yield a given income is inversely proportional to the amount of money invested. Bonnie receives income from $20,000 that she has invested at an annual interest rate of 5.5%. How much money must Carl invest at an annual interest rate of 8% in order to receive the same income as Bonnie? $13,750

Chapter 10

13. Express $\frac{25}{16}$ as a terminating or repeating decimal. 1.5625

14. Express $4.\overline{09}$ as a fraction in simplest form. $\frac{45}{11}$

Simplify.

17. $-11\sqrt{3}$ **20.** $2 + 11\sqrt{2}$

15. $-\sqrt{(35)^2}$ -35

16. $\sqrt{120}$ $2\sqrt{30}$

17. $5\sqrt{12} - 7\sqrt{27}$

18. $\sqrt{14} \cdot \sqrt{21}$ $7\sqrt{6}$

19. $(4\sqrt{7})(2\sqrt{5})$ $8\sqrt{35}$

20. $(3\sqrt{2} - 1)(\sqrt{2} + 4)$

21. $\frac{\sqrt{40}}{\sqrt{5}}$ $2\sqrt{2}$

22. $\sqrt{\frac{1}{12}}$ $\frac{\sqrt{3}}{6}$

23. $\frac{3}{\sqrt{2}+1}$ $3\sqrt{2} - 3$

24. $\sqrt[3]{-32}$ $-2\sqrt[3]{4}$

25. $\sqrt{100x^2y^4}$ $10|x|y^2$

26. $\sqrt[5]{a^5b^{10}c^{15}}$ ab^2c^3

Solve.

27. $5 - 3x^2 = -22$ {3, −3}

28. $4y^2 - 9 = 127$ {$\sqrt{34}$, $-\sqrt{34}$}

29. $\sqrt{m} - 5 = 2$ {49}

30. $\sqrt[3]{x} - 5 = 3$ {32}

Chapter 11

Solve. Express irrational solutions in simplest form. **33.** $\{\frac{-7 + \sqrt{41}}{2}, \frac{-7 - \sqrt{41}}{2}\}$

31. $m^2 - 6m + 8 = 0$ $\{4, 2\}$

32. $n^2 + 6n + 4 = 0$ $\{-3 + \sqrt{5}, -3 - \sqrt{5}\}$

33. $j^2 + 7j = -2$

34. $k^2 = 20k - 19$ $\{19, 1\}$

35. $2x^2 = 3x - 1$ $\{\frac{1}{2}, 1\}$

36. $8y + 2 = 5y^2$ $\{\frac{4 + \sqrt{26}}{5}, \frac{4 - \sqrt{26}}{5}\}$

For each function, determine its zeros, if any, and its maximum or minimum value.

37. $f: x \longrightarrow -2x^2$ 0; maximum: 0

38. $g: x \longrightarrow \frac{1}{3}x^2$ 0; minimum: 0

39. $G: x \longrightarrow 2x + x^2$ 0, -2; minimum: -1

40. $H: x \longrightarrow 3 - x^2$ $\sqrt{3}, -\sqrt{3}$; maximum: 3

Sketch the graph of each function and estimate its zeros, if any.

41. $g: x \longrightarrow x^3 - 2$

42. $h: x \longrightarrow -x^4$

Chapter 12

43. If A is a quadrantal angle and $-350° \leq m\angle A \leq -250°$, find $m\angle A$. $-270°$

44. If $\angle A$ is in standard position, $90° \leq m\angle A \leq 180°$, and $\angle A$ is coterminal with an angle of $-585°$, find $m\angle A$. 135°

45. Use the table on page 684 to find sin 76°. 0.9703

46. If $0° \leq m\angle A \leq 90°$ and tan $A = 0.2250$, find $m\angle A$ to the nearest degree. 13°

47. In right triangle ABC, $c = 10$ and $a = 8$. Find $m\angle B$ to the nearest degree. 37°

48. In right triangle ABC, $c = 10$ and $m\angle A = 21°$. Find b to the nearest tenth of a unit. 9.3

49. If a vector in standard position has its terminal point at $(-2, 5)$, find the norm and direction of the vector. 5.4; 112°

50. If vectors **s** and **v** are in standard position with terminal points $(-1, 5)$ and $(2, 7)$, respectively, and $\mathbf{v} = \mathbf{s} + \mathbf{t}$, find $\|\mathbf{t}\|$. 3.6

Chapter 13

51. Make a table that shows the frequencies, the relative frequencies as fractions, and the relative frequencies as percents for the following set of data: 8, 9, 5, 7, 7, 10, 4, 8, 7, 9.

52. Find the median, mode, and mean of the following set of data: 25, 20, 22, 25, 27, 21, 25, 22, 16, 24. 23; 25; 22.7

53. When you toss a die, what is the probability that the top face will show 6 dots? $\frac{1}{6}$

54. An archer hits the bull's eye in 6 out of 50 tries. What is the experimental probability that the next arrow this archer shoots will miss the bull's eye? $\frac{44}{50} = 0.88$

Statistics and Probability **657**

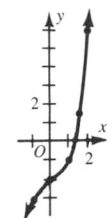
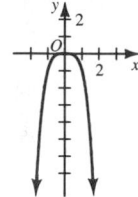

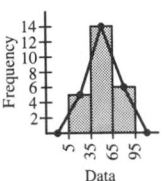
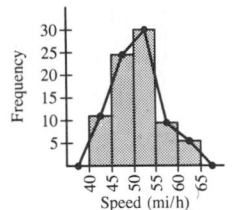
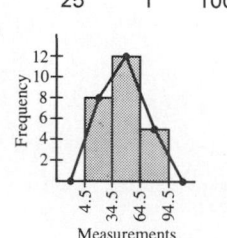
APPLICATION

Estimating Wildlife Populations

Ecologists often need to know how many animals of a certain species live in a particular area. It may not be possible to get an exact count. There is no practical way to be sure that, for example, all the whales in an ocean have been counted.

One way to estimate the number of animals is to catch some animals, tag them, and release them. Call the number of tagged animals t_1.

After the tagged animals have had time to mix with the rest of the population, more animals are caught or observed. Call this number c. Of these animals, a number t_2 have tags.

Assume that the ratio $\frac{t_2}{c}$ is equal to the fraction of all the animals that are tagged. If the total number of animals is n, then the following proportion can be used to calculate n.

$$\frac{t_2}{c} = \frac{t_1}{n}$$

$$t_2 n = c t_1$$

$$n = \frac{c t_1}{t_2}$$

For example, suppose that 50 tigers are tagged within a particular area. Later, 100 tigers are observed within the same area. Of these, 10 have tags. Then the total number of tigers can be estimated as follows.

$$n = \frac{c t_1}{t_2}$$

$$= \frac{100 \times 50}{10}$$

$$= 500$$

There are several possible sources of error in this procedure. Some animals' tags might come off. The tagged animals might not mix completely with the rest of the population. Animals may join or leave the population before the second capture. However, other counting methods such as aerial surveys confirm that this procedure produces good results.

Exercises

1. Ecologists tag 1000 whales. Later they observe 2500 whales. Of these, 400 have tags. Estimate the total number of whales. 6250 whales

2. To count fur seal pups, 5000 pups are tagged. Two weeks later, 1000 pups are caught. Of these, 200 have tags. Estimate the total number of fur seal pups. 25,000 seal pups

3. To count bass in a lake, 215 bass are tagged. Later 99 bass are caught, of which 15 have tags. About how many bass are in the lake? 1419 bass

4. In the same lake 240 pickerel are tagged. Later 342 pickerel are caught, of which 18 have tags. About how many pickerel are in the lake? 4560 pickerel

5. Observers in airplanes estimate that there are 1500 moose in a park. To check this estimate, 100 moose are tagged. Later 180 moose are observed. How many of these would be expected to have tags? 12 moose

8. **a.** Answers will vary.
Example:

Int. (Thousands)	Freq.	Rel. Freq. Frac.	%
0– 6	16	$\frac{16}{20}$	80
6–12	1	$\frac{1}{20}$	5
12–18	0	$\frac{0}{20}$	0
18–24	2	$\frac{2}{20}$	10
24–30	0	$\frac{0}{20}$	0
30–36	0	$\frac{0}{20}$	0
36–42	0	$\frac{0}{20}$	0
42–48	0	$\frac{0}{20}$	0
48–54	1	$\frac{1}{20}$	5
Total:	20	1	100

b., c. Answers will vary.
Example:

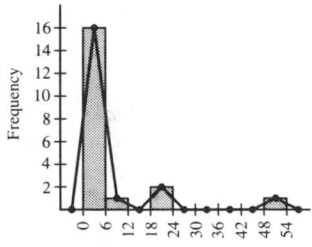

No. of Immigrants (Thousands)

d. Answers will vary.

1.

2.

8.

9.

10.

33.

34.

35.

Extra Practice
Skills

For use after Chapter 1.

Graph the given numbers on a horizontal number line. Construct a separate number line for each exercise.

1. $^-2$, $^-0.5$, 1, 2.5, 4

2. $^-3$, $^-1.5$, 0, 1.5, 3   *1-1*

Write the given numbers in order from least to greatest.

3. 4, $^-7$, 6, 0, $^-3$
   $^-7$, $^-3$, 0, 4, 6

4. 1.75, $^-2.25$, 3.5, $^-1.25$, 0.5
   $^-2.25$, $^-1.25$, 0.5, 1.75, 3.5

Replace __?__ with one of the symbols $\in$ or $\subset$ to make a true statement.

5. {0} __?__ {0} $\subset$

6. $\emptyset$ __?__ {1, 3, 5} $\subset$

7. 1 __?__ {1, 2} $\in$ *1-2*

Graph each set of numbers.

8. $\left\{ -2\frac{1}{2},\ 1, 0, 3 \right\}$

9. $\left\{ -5, -3\frac{1}{2}, -1\frac{1}{2}, 2 \right\}$

10. {the negative real numbers greater than $^-4$}

Specify each set by roster.

11. {the even integers less than 7} $\{\dots\ ^-4,\ ^-2, 0, 2, 4, 6\}$ *1-3*

12. {the natural numbers between $^-5$ and 5} {1, 2, 3, 4}

Simplify.

13. $189 + 556$ 745

14. 47×31 1457

15. $4896 \div 8$ 612   *1-4*

Evaluate each expression when $a = 12$ and $c = 9$.

16. $29 - a$ 17

17. $\dfrac{7 + c}{2}$ 8

18. $\dfrac{15}{a - c}$ 5

Simplify.

19. $7(10 + 68)$ 546

20. $[4(9) - 2(3)] \div 5$ 6

21. $\dfrac{8(12 \div 4)}{1 + 5}$ 4   *1-5*

Evaluate each expression when $r = 2$, $s = 5$, and $t = 8$.

22. $3(9 + t) - 4$ 47

23. $4r + 2s$ 18

24. $rt - s$ 11

Simplify.

25. $6(7^2 - 1^3)$ 288

26. $6^2 \div (2^3 + 1)$ 4

27. $(7 + 4^2)^2$ 529   *1-6*

Evaluate each expression when the variable has the given value.

28. $x^2 + 2x + 1$; $x = 3$ 16

29. $4(5w^2 - 7w)$; $w = 2$ 24

Solve each open sentence if $x \in \{0,2,4\}$.

30. $2x + 1 = 9$ {4}   **31.** $15 - 3x > 7$ {0, 2}   **32.** $x^2 \leq 4x$ {0, 2, 4}   *1-7*

Solve each open sentence if $y \in$ {the positive real numbers} and graph the solution set. See Additional Answers for graphs.

33. $3 \leq y$ {$y: y \geq 3$}   **34.** $3 < 5y - 7$ {$y: y > 2$}   **35.** $y^2 - 1 = 0$ {1}

Write a variable expression for each word phrase.

36. the cube of a number t, decreased by forty $t^3 - 40$ *1-8*

37. the quotient when nine times a number x is divided by the product of ten and a number y $\frac{9x}{10y}$

Write a mathematical sentence for each word sentence.

38. The product of seven and b is seventy. $7b = 70$ *1-9*

39. Jim is j years old and his age in six years will be three times his present age. $j + 6 = 3j$

40. The total value of n nickels and d dimes is not fifty-five cents. $5n + 10d \neq 55$

For use after Chapter 2.

Find a value of the variable that makes each statement true.

1. For some integer n, $2n - 1 > 8$. Answers may vary. Example: $n = 5$ *2-1*

2. There is a whole number x such that $x^2 = 9$. $x = 3$

Find a value of the variable that makes each statement false.

3. For all integers n, $3n + 1 \geq 4$. Answers may vary. Example: $n = 0$

4. For every real number a, $a^3 > a^2$. Answers may vary. Example: $a = 0$

Simplify.

5. $14 + 8 + 6 + 22$ 50   **6.** $2.3 + 1.9 + 3.7 + 2.6$ 10.5   *2-2*

7. $\frac{1}{2} \cdot 27 \cdot 8 \cdot \frac{2}{3}$ 72   **8.** $\frac{1}{8} \cdot \frac{1}{5} \cdot 16 \cdot 35$ 14

9. $t + 8t^2 + 3t + 2t^2$ $10t^2 + 4t$ **10.** $8h + 2d + 2d + 7h$ $15h + 4d$ *2-3*

11. $2\frac{1}{2} \times 9 + \frac{1}{2} \times 9$ 27   **12.** $1\frac{5}{6} \times 8 + 2\frac{1}{6} \times 8$ 32

13. $5 + 3(1 + k) + 6k$ $9k + 8$ **14.** $4(a + b) + (a + 2b)$ $5a + 6b$

15. $-[-(26 - 17)]$ 9   **16.** $7|-4| - |13 + 5|$ 10   *2-4*

Solve.

17. $-x = -7$ {7}　　　　　　　　　　　　　　**18.** $|y| = 2$ {−2, 2}

Simplify.

19. $[-2 + (-9)] + 5$ −6　　　　　　　　**20.** $(-6.5 + 4) + (-9.5 + 7)$ −5　　　　2-5
21. $-(-7 + 8) + [-17 + (-9)]$ −27　　**22.** $(-9)k + 21k^2 + (-9)k^2 + 9k$ 12k²　2-6

Give the reason that justifies each statement.

23. $a + (-a) = 0$　　　**24.** $0 + b = b$　　　　**25.** $c + d = d + c$　　　2-7
　　Axiom of additive　　　　Identity axiom　　　　　Commutative axiom
Simplify. inverses　　　　　for addition　　　　　for addition

26. $-52 - (-31 + 19)$ −40　　　　　　　**27.** $9a - (7a - 5) - (3 - 4a)$ 6a + 2　　2-8
28. $(-3)^2(-4)(-5)$ 180　　　　　　　　**29.** $-2(-13) - 8(-15)$ 146　　　　　　2-9
30. $(-5t + 7)(-4)$ 20t − 28　　　　　　**31.** $(-2x)(8x)(-x)$ 16x³

32. $\dfrac{1}{-6}\left(-\dfrac{1}{2}\right)(36)$ 3　　**33.** $\dfrac{1}{15}(-15xy)$ −xy　　**34.** $\left(-\dfrac{1}{3}\right)\left(\dfrac{1}{-7}\right)(-210)$ −10　2-10

35. $\dfrac{1}{y}(-3xyz), y \neq 0$　　**36.** $12pt\left(-\dfrac{1}{p}\right), p \neq 0$　　**37.** $-\dfrac{1}{5}(10x - 25y)$ −2x + 5y
　　　−3xz　　　　　　　　　　−12t

38. $10 \div \left(-\dfrac{1}{5}\right)$ −50　　**39.** $-\dfrac{7}{8}a \div \dfrac{1}{40}$ −35a　　**40.** $\dfrac{-72s^2}{-12}$ 6s²　　2-11

For use after Chapter 3.

Give the reason that justifies each statement.

1. If $-8t = 96$, then $\dfrac{-8t}{-8} = \dfrac{96}{-8}$. Division property of equality　　　3-1

2. If $\dfrac{x}{4} = 10$, then $4\left(\dfrac{x}{4}\right) = 4 \cdot 10$. Multiplication property of equality

3. If $r + 3 = 5$, then $(r + 3) - 3 = 5 - 3$. Subtraction property of equality

Solve.

4. $2 = 4 + b$ {−2}　　　**5.** $4 - y = 8$ {−4}　　　**6.** $-8 + |m| = -5$ {−3, 3}　3-2

7. $-3 = \dfrac{1}{5}b$ {−15}　　**8.** $29r = -29$ {−1}　　　**9.** $\dfrac{1}{7}s = -\dfrac{20}{7}$ {−20}　3-3

10. $\dfrac{a}{0.6} = -0.3$ {−0.18}　**11.** $\dfrac{x}{-4} = \dfrac{2}{-5}$ {$\frac{8}{5}$}　　**12.** $6|y| = 24$ {−4, 4}

13. $3y - 8 = 7$ {5}　　　　　　　　　**14.** $-4 + 2t - 3t = 9$ {−13}　　3-4

15. $-6(5 - r) = 18$ {8}　　　　　　　**16.** $\dfrac{1}{2}x - \dfrac{1}{3} = -\dfrac{5}{6}$ {−1}

17. $8m - (3m - 5) = 80$ {15}　　　　**18.** $-b - 4(2b - 3) + 5 = 8$ {1}

662　　*Extra Practice*

19. On a trip to the state capital, Debbie paid a toll on the turnpike and stopped for gasoline. The toll was $7.85 less than the cost of the gasoline. Together the toll and the gasoline cost Debbie $17.65. How much was the turnpike toll? $4.90   *3-5*

20. The length of a rectangular parking lot is 10 m more than twice its width. Find the dimensions of the lot if its perimeter is 320 m.   width: 50 m length: 110 m

21. $12 + 7r = 5r$ $\{-6\}$ 22. $1 + 2z = 16 - z$ $\{5\}$ *3-6*

23. $5(2 - b) = -3(b + 2)$ $\{8\}$ 24. $5(t - 1) + 4t = 9(t + 2)$ $\varnothing$

25. Kim is eight years older than Terry. In seven years, Kim will be 1.5 times as old as Terry. Find the present age of each. Kim: 17 years; Terry: 9 years   *3-7*

26. Find the number that is three greater than five times its additive inverse. $\frac{1}{2}$

Solve for a. Assume the variables represent real numbers that do not result in division by zero.

27. $m = c(a + 1.5)$ $a = \frac{m}{c} - 1.5$ 28. $t = \frac{r}{a - m}$ $a = \frac{r}{t} + m$ *3-8*
    or $a = \frac{m - 1.5c}{c}$ or $a = \frac{r + mt}{t}$

For use after Chapter 4.

1. Give the reason that justifies each statement in the given proof.   *4-1*

   Prove: If x, y, and z are real numbers and $x + y < z$, then $x < z - y$.

   1. x, y, and z are real numbers; $x + y < z$.   Hypothesis
   2. $(x + y) + (-y) < z + (-y)$ Addition property of order
   3. $x + [y + (-y)] < z + (-y)$ Associative axiom for addition
   4. $\quad\quad x + 0 < z + (-y)$ Axiom of additive inverses
   5. $x + 0 = x$ Identity axiom for addition
   6. $z + (-y) = z - y$ Definition of subtraction
   7. $\therefore x < z - y$ Substitution principle

Solve each inequality and graph its solution set. See Additional Answers for graphs.

2. $3 - 2a > 7$ $\{a: a < -2\}$ 3. $-3 + \frac{t}{5} > -2$ $\{t: t > 5\}$ *4-2*

4. $6r - 5 > 2 - r$ $\{r: r > 1\}$ 5. $4(m - 2) < 3m - (m + 8)$ $\{m: m < 0\}$

Specify the intersection and the union of the given sets.

6. $\{-3, -1, 0, 1\}$, $\{0, 1, 2\}$ $\{0, 1\}$; $\{-3, -1, 0, 1, 2\}$ *4-3*

7. $\{1, 3\}$, $\{2, 4\}$ $\varnothing$; $\{1, 2, 3, 4\}$

8. {the integers between -2 and 2}, {the whole numbers less than 3} $\{0, 1\}$; $\{-1, 0, 1, 2\}$

9. {the real numbers greater than -2}, {the real numbers greater than -2 and {the real numbers less than 3} less than 3}; $\mathcal{R}$

Additional Answers Chapter 4

2.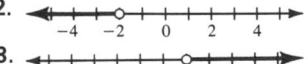

3.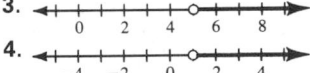

4.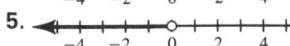

5.

10. $\{w: w < 4 \text{ or } w > 6\}$

11. $\{x: -3 < x \le 2\}$

12. $\{t: -1 < t < 2\}$

13. $\{y: y \le -1 \text{ or } y \ge 3\}$

14. $\{-5, 1\}$

15. $\{y: y < 3 \text{ or } y > 3\}$

16. $\{z: 2 \le z \le 6\}$

17. $\{0, 5\}$

18. $\left\{t: t \le -\dfrac{1}{2} \text{ or } t \ge \dfrac{1}{2}\right\}$

19. $\left\{d: -1 < d < \dfrac{13}{3}\right\}$

1.–4.

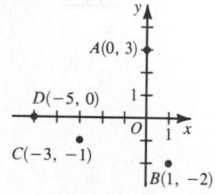

Solve each open sentence and graph its solution set.

10. $w - 5 > 1$ or $w - 5 < -1$ **11.** $-1 < x + 2 \le 4$ 4-4

12. $3t + 5 > 2$ and $2t < 4$ **13.** $3y - 1 \ge 8$ or $3 - 2y \ge 5$

14. $|x + 2| = 3$ **15.** $|y - 3| > 0$ **16.** $|4 - z| \le 2$ 4-5

17. $|2w - 5| = 5$ **18.** $|-8t| \ge 4$ **19.** $|5 - 3d| < 8$

Solve.

20. Find three consecutive odd integers whose sum is -3. -3, -1, and 1   4-6

21. Find the greatest two consecutive integers such that three times the lesser integer is at most 11 more than the greater integer. 6 and 7

22. The degree measures of the angles of a triangle are consecutive even integers. What are the measures? 58°, 60°, and 62°   4-7

23. Find the measure of an angle for which three times the measure of its complement is 50° more than the measure of its supplement. 20°

24. A disabled freight train is traveling toward a station 140 km away at a speed of 32 km/h. Another train, traveling 80 km/h, leaves the station to pick up the freight. When will the two trains meet? after $1\frac{1}{4}$ h   4-8

25. A car traveling at 87 km/h and a bus leave a toll booth at the same time. Twenty minutes later, the bus is 3 km farther from the toll booth than the car. What is the average speed of the bus? 96 km/h

26. At the planetarium, adult tickets cost $2.50 each and children's tickets cost $1.50 each. One day 408 tickets were sold in all, and $733 was collected. How many of each type of ticket were sold? adult: 121   4-9
children's: 287

27. A 30 L solution is 50% antifreeze. How many liters of antifreeze must be added to produce a solution that is 70% antifreeze? 20 L

28. Explain why the following problem has no solution: 4-10
Find four consecutive even integers whose sum is 224. The only solution of an equation that represents the relationships in the problem is an odd integer.

For use after Chapter 5.

Plot the given points on a coordinate plane.

1. $A(0, 3)$ **2.** $B(1, -2)$ **3.** $C(-3, -1)$ **4.** $D(-5, 0)$ 5-1

Exercises 5–7 refer to the relation $\{(0, 0), (-1, -1), (-1, 1), (2, 2)\}$.

5. State the domain D and the range R. $D = \{-1, 0, 2\}$; $R = \{-1, 0, 1, 2\}$ 5-2

6. Draw a mapping diagram that represents the relation.

7. Is the relation a function? No

Graph each of the following.

8. $\{(-1, 2), (0, 3), (0, -2), (1, -1)\}$ 5-3

9. $f: x \rightarrow 3x - 2$; $D = \{-2, -1, 0, 1, 2\}$

10. $g: x \rightarrow 3 - x^2$; $D = \{-2, -1, 0, 1, 2\}$

664   *Extra Practice*

Solve when $x \in \{-1, 0, 1\}$ and $y \in \{-2, -1, 0, 1, 2\}$.

11. $y = 1 - 2x$ $\{(0, 1), (1, -1)\}$ **12.** $y - 1 < 2x$ $\{(-1, -2), (0, -2), (0, -1),$ 5-4
$(0, 0), (1, -2), (1, -1),$

Graph each open sentence on a coordinate plane. $(1, 0), (1, 1), (1, 2)\}$

13. $y = -3$ **14.** $3x + 5y = 15$ **15.** $3x = 12 + 2y$ 5-5

16. $x \le -y$ **17.** $y \le 2x + 3$ **18.** $x + 3y > 6$ 5-6

Determine the slope of the line that passes through the given points.

19. $(5, 2), (-3, 4)$ $-\frac{1}{4}$ **20.** $(7, -4), (-3, -4)$ 0 **21.** $(1, 5), (1, -2)$ no slope   5-7

22. Graph the line that passes through the point $(-2, -3)$ and has slope $\frac{1}{3}$.

23. Determine the slope and the y-intercept of the line with equation   5-8
$3x - 2y = 6$. slope: $\frac{3}{2}$; y-intercept: -3

Write an equation of the line that satisfies the given requirements. The equation should be in the form $ax + by = c$, where a, b, and c are integers.

24. slope -2; y-intercept 3 $2x + y = 3$

25. y-intercept 5; parallel to the graph of $y = 2x - 3$ $2x - y = -5$

26. passes through $(-10, -3)$ and has slope $\frac{4}{5}$ $4x - 5y = -25$ 5-9

27. passes through the points $(4, -2)$ and $(8, 1)$ $3x - 4y = 20$

For use after Chapter 6.

Solve each system of equations using graphs. See Additional Answers for graphs.

1. $-x + y = 3$ **2.** $y = -2x + 1$ **3.** $y = -x$ 6-1
     $x + y = 5$ $\{(1, 4)\}$ $2x + y = 3$ $\emptyset$ $3x + y = -6$ $\{(-3, 3)\}$

4. Which (if any) of the systems in Exercises 1–3 are inconsistent?
                    the system in Exercise 2

Solve each system of equations using addition or subtraction.

5. $2x + 7y = -24$ **6.** $x - 3y = 18$ **7.** $5x - 2y = 12$ 6-2
    $2x + 5y = -20$ $8x + 3y = -18$ $3x - 2y = 4$
       $\{(-5, -2)\}$ $\{(0, -6)\}$ $\{(4, 4)\}$

Solve each system of equations using the linear-combination method.

8. $4x + 9y = 22$ **9.** $6x + 11y = -12$ **10.** $7x - 5y = 1$ 6-3
    $3x + 5y = 6$ $-4x + 7y = 8$ $5x + 2y = -16$
      $\{(8, 6)\}$ $\{(-2, 0)\}$ $\{(-2, -3)\}$

Solve each system of equations using the substitution method.

11. $2x + 5y = -13$ **12.** $4x + 3y = 43$ **13.** $12x + y = -20$ 6-4
    $x - 3y = 21$ $y = 2x + 1$ $11x - 4y = 21$
     $\{(6, -5)\}$ $\{(4, 9)\}$ $\{(-1, -8)\}$

Extra Practice   **665**

6.

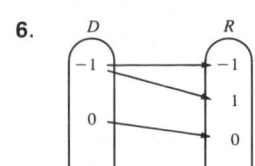

8.

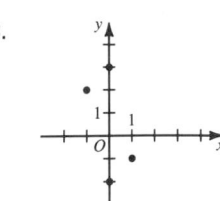

(continued on p. 678)

Additional Answers
Chapter 6

1.

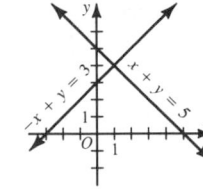

2.

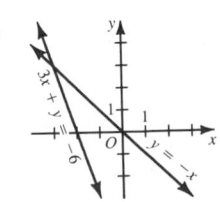

3.

665

Solve.

14. Three years ago Alice was twice as old as Bob was then. Nine years from now, she will be 1.5 times as old as Bob will be. How old are Alice and Bob now? Alice: 27 years; Bob: 15 years

6-5

15. The cost of 8 calculators and 24 protractors is $65.04. The cost of 7 calculators and 36 protractors of the same kind is $65.16. How much does each calculator cost? $6.48

16. With a tail wind, a plane can fly 1050 km in 84 min. Going against the wind, the plane can fly the same distance in 90 min. What are the wind speed and the air speed of the plane? wind speed: 25 km/h; air speed: 725 km/h

6-6

17. Tom rowed 6 km in 1.5 h against the current. He then rowed 10 km with the current in just 1.25 h. Find Tom's rowing speed in still water and the speed of the current. rowing speed: 6 km/h; speed of current: 2 km/h

18. The units' digit of a two-digit number is 1 more than twice the tens' digit. The sum of the digits is 10. Find the number. 37

6-7

19. The sum of the digits of a three-digit number is 13. The hundreds' digit is three times the tens' digit. The number formed by reversing the hundreds' and units' digits is 99 less than the original number. Find the original number. 625

Graph each system of inequalities.

20. $y > -x$
    $x + y \le 4$

21. $5x - 2y < 10$
    $4x + y > -5$

22. $x \le 3$
    $y \le -2$
    $y - x \le -1$

6-8

23. a. Graph the solution set of the following system of inequalities and label the corner points with their coordinates.
    $$x \ge 0, \quad y \ge 0, \quad y \le 3x + 6, \quad 3x + 2y \ge 8, \quad x + 2y \ge 4$$

6-9

  b. Find the minimum value of the expression $2x + 3y$ over the region graphed in part (a). 7

  c. Find the maximum value of the expression $5x - y$ over the region graphed in part (a). no maximum value

Solve each system of equations.

24. $x + y + z = 1$
    $x + 2y - z = -1$
    $2x - y - 3z = 8$ {(3, -2, 0)}

25. $-x - y + z = 7$
    $x + 2y + 3z = 7$
    $4x - 3y - 2z = -6$ {(-1, -2, 4)}

6-10

For use after Chapter 7.

Simplify.

1. $(3x^4 - 2x^2 + x) + (x^2 + 5x - x^3)$ $3x^4 - x^3 - x^2 + 6x$

7-1

2. $(a^2 + 4ab + 2b^2) - (a^2 - b^2)$ $4ab + 3b^2$

666 *Extra Practice*

Additional Answers
Chapter 6

20.
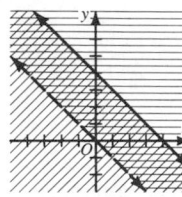

21.
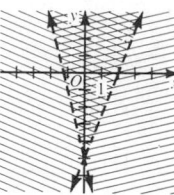

22.
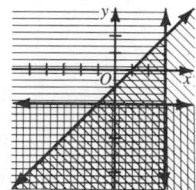

23. a.
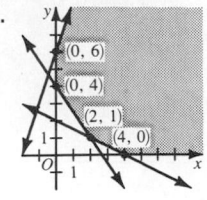
(0, 6)
(0, 4)
(2, 1)
(4, 0)

3. $(2s^2t - 3st - st^2) + (2s^2t + st + st^2)$ $4s^2t - 2st$

4. $(-z^4 + 6z^2 - 1) - (5z^3 - z^2 + 7z)$ $-z^4 - 5z^3 + 7z^2 - 7z - 1$

7-2

5. $(-3x^2)(-9x^3)$ $27x^5$

6. $(2r^4w)(-rw^3)$ $-2r^5w^4$

7. $2c(5c^2)(-3c^2)$ $-30c^5$

8. $(8d^3)(3d)^2$ $72d^5$

9. $(-m^2n)^2(mn)^3(-m)$ $-m^8n^5$

10. $(5y^2)^3(-3y^5)$ $-375y^{11}$

11. $f^5(2f^2 - f - 5)$ $2f^7 - f^6 - 5f^5$

12. $(h^3 + 4h + 2)g^2h$ $g^2h^4 + 4g^2h^2 + 2g^2h$

7-3

13. $-4m^2n(3m^4 - m^2n^2 + 7n^2)$

14. $(q + 4)(q - 3)$

15. $(r - 8)(2r - 6)$

16. $(a + 5)(2a^2 - 5a + 3)$

17. $(2x - 1)(3x + 4)$

18. $(7c + 2d)(3c + 5d)$

7-4

19. $(3t + 2y)^2$ $9t^2 + 12ty + 4y^2$

20. $(5 + 2z^3)(2z^3 - 5)$ $4z^6 - 25$

21. $(ab - c)^2$ $a^2b^2 - 2abc + c^2$

22. $(2a + 1)^3$ $8a^3 + 12a^2 + 6a + 1$

23. A rectangle is 3 cm longer and 2 cm less wide than a certain square. The area of the rectangle is 16 cm² greater than the area of the square. What are the dimensions of the rectangle? width: 20 cm; length: 25 cm

Find the GCF and the LCM of each pair of monomials.

24. 28, 42 GCF: 14; LCM: 84

25. 52, 91 GCF: 13; LCM: 364

7-5

26. $36c^2d^3$, $-27c^3d$ GCF: $9c^2d$; LCM: $108c^3d^3$

27. $64y^3z^4$, $-72y^5z^3$ GCF: $8y^3z^3$; LCM: $576y^5z^4$

Write each polynomial as the product of its greatest monomial factor and another polynomial.

28. $16a^3b^2 + 24a^2b^4$ $8a^2b^2(2a + 3b^2)$

29. $15x^3y^2 + 35x^4y^3 - 25x^2y^2$ $5x^2y^2(3x + 7x^2y - 5)$

7-6

Write each polynomial as the product of two binomials.

30. $t(t - 8) + 4(8 - t)$ $(t - 4)(t - 8)$

31. $3(n + 7) + 4(n + 7)^2$ $(n + 7)(4n + 31)$

Factor.

32. $36c^2 - 25d^2$ $(6c + 5d)(6c - 5d)$

33. $x^2 + 4xy + 4y^2$ $(x + 2y)^2$

7-7

34. $1 - 64a^3$ $(1 - 4a)(1 + 4a + 16a^2)$

35. $2x^2 + xy + 6x + 3y$ $(2x + y)(x + 3)$

36. $t^2 + 8t + 12$ $(t + 2)(t + 6)$

37. $m^2 + 3m - 28$ $(m + 7)(m - 4)$

7-8

Factor each trinomial in two steps. The first step should be to write the trinomial as the product of its greatest monomial factor and another trinomial.

38. $p^3 - 11p^2 + 18p$ $p(p^2 - 11p + 18) = p(p - 2)(p - 9)$

39. $-4a^2 + 4a + 48$ $-4(a^2 - a - 12) = -4(a - 4)(a + 3)$

Factor completely.

40. $3x^2 + 11x + 6$ $(3x + 2)(x + 3)$

41. $5a^2 - 38a - 63$ $(5a + 7)(a - 9)$

7-9

42. $12n^2 - 7n - 10$ $(4n - 5)(3n + 2)$

43. $20d^5 - 5d^3$ $5d^3(2d + 1)(2d - 1)$

44. $3t^2 + 21t + 36$ $3(t + 3)(t + 4)$

45. $30s^2 + 5s - 60$ $5(3s - 4)(2s + 3)$

13. $-12m^6n + 4m^4n^3 - 28m^2n^3$

14. $q^2 + q - 12$

15. $2r^2 - 22r + 48$

16. $2a^3 + 5a^2 - 22a + 15$

17. $6x^2 + 5x - 4$

18. $21c^2 + 41cd + 10d^2$

Solve.

50. $\{0, -\frac{1}{4}, -3\}$

46. $5x^2 = 35x$ $\{0, 7\}$ **47.** $r^2 - 6r + 5 = 0$ $\{1, 5\}$ 7-10

48. $q^2 - 2q = 24$ $\{-4, 6\}$ **49.** $25z^2 - 9 = 0$ $\{-\frac{3}{5}, \frac{3}{5}\}$

50. $4m^3 + 13m^2 + 3m = 0$ **51.** $(2x - 5)(x + 2) = 18$ $\{-\frac{7}{2}, 4\}$

52. The base of a triangle is 5 cm longer than the height. The area is   7-11
12 cm². Find the dimensions of the triangle. base: 8 cm; height: 3 cm

53. Find all pairs of consecutive integers such that twice their product
is 12. $-3, -2$ and $2, 3$

For use after Chapter 8.

Simplify. Assume that no denominator equals zero.

1. $\dfrac{24a^5b^2}{4ab^3}$ $\dfrac{6a^4}{b}$ **2.** $\dfrac{-18t^2z}{6t^2z^2}$ $-\dfrac{3}{z}$ **3.** $\dfrac{-7r^3s^4}{-35r^4s^7}$ $\dfrac{1}{5rs^3}$ 8-1

4. $\dfrac{30x^8y^{11}}{-18x^2y^3}$ $-\dfrac{5x^6y^8}{3}$ **5.** $\dfrac{72w^3p^8}{40w^9p^6}$ $\dfrac{9p^2}{5w^6}$ **6.** $\dfrac{20a^4b^2c^3}{0.4a^2b^5c}$ $\dfrac{50a^2c^2}{b^3}$

Express each quotient as a sum. Assume that no denominator equals zero.

7. $\dfrac{-28d^3 + 7d^2}{7d^2}$ $-4d + 1$ **8.** $\dfrac{6x^2y^2 + 8xy - 14y}{2y}$ $3x^2y + 4x - 7$ 8-2

9. $\dfrac{27m^4 - 30m^3 - 12m^2}{-3m}$ $-9m^3 + 10m^2 + 4m$ **10.** $\dfrac{8r^5s^5 + 5r^3s^7 + 9rs^9}{r^2s^3}$ $8r^3s^2 + 5rs^4 + \dfrac{9s^6}{r}$

Divide the first polynomial by the second. Assume that no divisor equals
zero.

11. $8y^2 + 6y - 5; 4y + 5$ $2y - 1$ **12.** $3t^2 + 5t - 12; t + 3$ $3t - 4$ 8-3

13. $9y^2 - 36y + 32; 3y - 8$ $3y - 4$ **14.** $2p^2 + 9p + 3; 2p - 7$ $p + 8 + \dfrac{59}{2p - 7}$

15. $z^3 - 3z^2 - z + 3; z - 3$ $z^2 - 1$ **16.** $8n^3 - 12n + 9; 2n + 3$ $4n^2 - 6n + 3$

Simplify.

17. $\dfrac{4a + 12}{a^2 + 3a}$ $\dfrac{4}{a}$ **18.** $\dfrac{3x}{2x^2 - 5x}$ $\dfrac{3}{2x - 5}$ **19.** $\dfrac{z^2 + 2z}{z^2 - 4}$ $\dfrac{z}{z - 2}$ 8-4

20. $\dfrac{6a - 1}{12a^2 + 16a - 3}$ $\dfrac{1}{2a + 3}$ **21.** $\dfrac{y^2 - 25}{2y^2 + 15y + 25}$ $\dfrac{y - 5}{2y + 5}$ **22.** $\dfrac{m^2 + 5m - 24}{m^2 + m - 12}$ $\dfrac{m + 8}{m + 4}$

Simplify using only positive exponents. Assume that no variable has zero as
a value.

23. $a^{-3} \cdot a^{-2}$ $\dfrac{1}{a^5}$ **24.** $b^{-4} \cdot b^2$ $\dfrac{1}{b^2}$ **25.** $(5ax^2)^{-2}$ $\dfrac{1}{25a^2x^4}$ **26.** $(r^{-3}s^0t^2)^{-3}$ $\dfrac{r^9}{t^6}$ 8-5

27. $\dfrac{7z^{-1}}{42z^{-4}}$ $\dfrac{z^3}{6}$ **28.** $\left(\dfrac{2m^3}{3n^{-2}}\right)^{-1}$ $\dfrac{3}{2m^3n^2}$ **29.** $\dfrac{b^{-3}c^3d^4}{b^{-2}cd^0}$ $\dfrac{c^2d^4}{b}$ **30.** $\dfrac{3^{-4}r^{-1}s^2t^{-5}}{9^{-1}rs^{-3}t^2}$ $\dfrac{s^5}{9r^2t^7}$

Express each number in scientific notation. Then simplify.

31. $600,000 \times 0.0007$ **32.** $0.00625 \div 5000$
$6 \times 10^5; 7 \times 10^{-4}; 420$ $6.25 \times 10^{-3}; 5 \times 10^3; 0.00000125$

668   *Extra Practice*

668

33. $(400,000)^3$ $\quad 4 \times 10^5$;
64,000,000,000,000,000

34. $(0.0005)^4$ $\quad 5 \times 10^{-4}$;
0.0000000000000625

Simplify. Assume that no variable has a value that results in division by zero.

35. $\dfrac{-15a}{4} \cdot \dfrac{2}{a^2}$ $\quad -\dfrac{15}{2a}$

36. $\dfrac{2y - 14}{-5y} \div \dfrac{3y - 21}{5y^3}$ $\quad -\dfrac{2y^2}{3}$ $\qquad$ 8-6

37. $\dfrac{z^2 + 3z}{-z^2 + 2z} \cdot \dfrac{7z - 14}{7z + 21}$ $\quad -1$

38. $(2r^2 + 3rs) \div \dfrac{2rs + 3s^2}{rs^2 - s^3}$ $\quad rs(r - s)$, or $r^2s - rs^2$

39. $\dfrac{16c^2 - d^2}{6c + 3d} \cdot \dfrac{2cd + d^2}{4cd - d^2}$ $\quad \dfrac{4c + d}{3}$

40. $\dfrac{x^2 + x}{-x^2 - 2x - 1} \div \dfrac{x^2 - 3x}{2x^2 - 2}$ $\quad -\dfrac{2(x - 1)}{x - 3}$

41. $\dfrac{4t - 1}{3t} - \dfrac{t - 7}{3t}$ $\quad \dfrac{t + 2}{t}$

42. $\dfrac{m + 1}{m - 5} - \dfrac{m - 6}{m - 5}$ $\quad \dfrac{7}{m - 5}$ $\qquad$ 8-7

43. $\dfrac{-a + 5b}{3a + 3b} + \dfrac{2a - 4b}{3a + 3b}$ $\quad \dfrac{1}{3}$

44. $\dfrac{-7}{8n - 4} + \dfrac{-4n + 9}{8n - 4}$ $\quad -\dfrac{1}{2}$

45. $\dfrac{5a}{b^2} + \dfrac{3b}{a^2}$ $\quad \dfrac{5a^3 + 3b^3}{a^2b^2}$

46. $\dfrac{d - k}{2d} - \dfrac{d + 4k}{3k}$

47. $\dfrac{p + 1}{4p} + \dfrac{2 - p}{4p^2}$ $\qquad$ 8-8

48. $\dfrac{9}{9r + 3s} - \dfrac{2}{12r + 4s}$

49. $\dfrac{1}{g - 4h} + \dfrac{1}{20h - 5g}$

50. $\dfrac{3}{m - n} - \dfrac{1}{m + n}$

51. $\dfrac{\dfrac{a}{4}}{\dfrac{3a}{8}}$ $\quad \dfrac{2}{3}$

52. $(a^{-2} + b^{-2})^{-1}$ $\quad \dfrac{a^2b^2}{a^2 + b^2}$

53. $\dfrac{2t + 3t^{-1}}{t + 4^{-1}}$ $\quad \dfrac{8t^2 + 12}{4t^2 + t}$ $\qquad$ 8-9

For use after Chapter 9.

Solve.

1. $\dfrac{3z}{4} + \dfrac{z}{7} = 1$ $\quad \left\{\dfrac{28}{25}\right\}$

2. $\dfrac{a}{6} + \dfrac{5a}{2} > \dfrac{1}{3}$ $\quad \left\{a\colon a > \dfrac{1}{8}\right\}$

3. $\dfrac{x - 7}{8} < \dfrac{x + 3}{6}$ $\quad \{x\colon x > -33\}$ $\qquad$ 9-1

4. $\dfrac{3b}{5} - 2 = \dfrac{b}{2}$ $\quad \{20\}$

5. $\dfrac{6r + 5}{9} - \dfrac{r - 3}{6} = \dfrac{r}{2}$ $\quad \varnothing$

6. $\dfrac{2m}{3} + \dfrac{7}{10} \geq \dfrac{11}{15} + \dfrac{5m}{6}$ $\quad \left\{m\colon m \leq -\dfrac{1}{5}\right\}$

Replace each ? with a real number to make a true statement.

7. 60% of 25 = __?__ 15 **8.** __?__% of 75 = 3 4 **9.** 24% of __?__ = 108 450 $\qquad$ 9-2

10. Jane invested $5000, part at an annual interest rate of 9% and the rest at an annual interest rate of 6%. Her income for the year from these investments was $345. How much did she invest at 6%? $3500

Solve.

11. $\dfrac{7}{x} - \dfrac{5}{2x} = 3$ $\quad \left\{\dfrac{3}{2}\right\}$

12. $\dfrac{2}{z} = \dfrac{9}{z + 1} + \dfrac{11}{2z}$ $\quad \left\{-\dfrac{7}{25}\right\}$ $\qquad$ 9-3

13. $\dfrac{2}{z + 11} = \dfrac{7}{-9z + 1}$ $\quad \{-3\}$

14. $\dfrac{4}{2x + 11} = \dfrac{2}{x - 8}$ $\quad \varnothing$

15. $\dfrac{3x}{x + 4} - \dfrac{2}{x - 3} = 3$ $\quad \{2\}$

16. $\dfrac{a}{a - 2} + \dfrac{a - 1}{a + 2} = \dfrac{3}{a^2 - 4}$ $\quad \left\{-\dfrac{1}{2}, 1\right\}$

46. $-\dfrac{2d^2 + 5dk + 3k^2}{6dk}$

47. $\dfrac{p^2 + 2}{4p^2}$

48. $\dfrac{5}{6r + 2s}$

49. $\dfrac{4}{5g - 20h}$

50. $\dfrac{2m + 4n}{(m + n)(m - n)}$, or
$\dfrac{2m + 4n}{m^2 - n^2}$

17. The sum of two numbers is 48 and their quotient is $\frac{5}{7}$. Find the numbers. 20 and 28

9-4

18. The denominator of a fraction is 7 less than three times the numerator, and the fraction is equal to $\frac{2}{5}$. Find the fraction. $\frac{14}{35}$

19. It takes Janet 4 h to take inventory of the stock in her store. Her assistant needs 6 h to do the same job. How long would it take to do the job if they work together? $2\frac{2}{5}$ h

9-5

20. A bank machine requires 45 s to process each transaction. A newer model can process each transaction in 30 s. How long will it take for the machines working together to process 40 transactions? 720 s, or 12 min

21. Jogging at a constant speed, Miguel can jog 7.5 km in 6 min less than twice as long as it takes him to jog 4 km. Find Miguel's speed in kilometers per hour. 5 km/h

9-6

22. The O'Keefe family drove 126 km at a certain speed and then took a scenic route of 28 km at two thirds their original speed. If the entire trip was 2 h long, what was the original speed? 84 km/h

Solve.

23. $\dfrac{2t + 5}{-2} = \dfrac{-t}{6}$ {-3}

24. $\dfrac{x + 8}{3x - 3} = \dfrac{5}{6}$ {7}

9-7

25. The measures of two complementary angles are in the ratio 2:3. Find the measure of each angle in degrees. 36° and 54°

26. If t varies directly as s, and if $t = 6$ when $s = -2$, find t when $s = 6$. -18

9-8

27. If x is directly proportional to y, and if $x = r$ when $y = s$, find x when $y = -2s$. $-2r$

If a varies inversely as b, and if $a = 15$ when $b = 12$, find b when a has the given value.

28. 18 10

29. 4 45

30. 22.5 8

31. $5\frac{5}{8}$ 32

9-9

32. If x varies directly as y and inversely as z, and if $x = 8$ when $y = 12$ and $z = 5$, find x when $y = 9$ and $z = 2$. 15

9-10

33. If r varies jointly as s and t^3, and if $r = 54$ when $s = 7$ and $t = 3$, find r when $s = 21$ and $t = 2$. 48

For use after Chapter 10.

1. Compare $\dfrac{35}{91}$ and $\dfrac{70}{190}$. $\frac{35}{91} > \frac{70}{190}$

10-1

2. Write $-\dfrac{21}{56}$, $-\dfrac{1}{3}$, and $-\dfrac{5}{12}$ in order from least to greatest. $-\frac{5}{12}, -\frac{21}{56}, -\frac{1}{3}$

3. Find the rational number that is one half of the way from $-\dfrac{5}{8}$ to $\dfrac{2}{5}$. $-\frac{9}{80}$

Express each rational number as a terminating or repeating decimal.

4. $\frac{13}{25}$ 0.52   5. $\frac{1}{18}$ $0.0\overline{5}$ 6. $-\frac{31}{36}$ $-0.86\overline{1}$ 7. $-\frac{53}{80}$ -0.6625 *10-2*

Express each rational number as a fraction in simplest form.

8. 9.36 $\frac{234}{25}$ 9. $0.\overline{8}$ $\frac{8}{9}$ 10. $3.41\overline{6}$ $\frac{41}{12}$ 11. $-1.\overline{30}$ $-\frac{43}{33}$

Simplify.

12. $\sqrt{121}$ 11   13. $-\sqrt{13^2}$ -13 14. $\sqrt{\frac{1}{100}}$ $\frac{1}{10}$ 15. $-\sqrt{(-25)^2}$ -25 *10-3*

Solve.

16. $y^2 - 12 = 37$ $\{-7, 7\}$ 17. $4x^2 + 9 = 90$ $\{-\frac{9}{2}, \frac{9}{2}\}$

Simplify.

18. $\sqrt{54}$ $3\sqrt{6}$ 19. $2\sqrt{48}$ $8\sqrt{3}$ 20. $\sqrt{\frac{7}{16}}$ $\frac{\sqrt{7}}{4}$ *10-4*

Solve. Simplify irrational solutions.

21. $y^2 - 3 = 29$ $\{4\sqrt{2}, -4\sqrt{2}\}$ 22. $4z^2 + 30 = 330$ $\{5\sqrt{3}, -5\sqrt{3}\}$

Approximate each square root to the nearest thousandth. Use the Table of Square Roots as necessary.

23. $\sqrt{95}$ 9.747   24. $\sqrt{162}$ 12.728   25. $\sqrt{294}$ 17.146

Determine whether or not each set of numbers could represent the lengths of the sides of a right triangle.

26. $\{9, 12, 15\}$ Yes   27. $\{8, 15, 18\}$ No   *10-5*

28. $\{20, 21, 29\}$ Yes   29. $\{2a, 3a, 4a\}$ No

Find the distance between the two points having the given coordinates. Simplify irrational distances.

30. $(0, -7), (1, -3)$ $\sqrt{17}$ 31. $(-6, -9), (-1, 4)$ $\sqrt{194}$ *10-6*

32. $(3, 1), (5, 7)$ $2\sqrt{10}$ 33. $(2, -8), (-5, 3)$ $\sqrt{170}$

Simplify. Assume that all variables denote positive real numbers.

34. $\sqrt{5} \cdot \sqrt{35}$ $5\sqrt{7}$ 35. $\frac{\sqrt{39}}{\sqrt{12}}$ $\frac{\sqrt{13}}{2}$ 36. $\frac{\sqrt{30}}{\sqrt{75}}$ $\frac{\sqrt{10}}{5}$ *10-7*

37. $\sqrt{2x} \cdot \sqrt{32x}$ $8x$ 38. $\frac{8}{\sqrt{a}}$ $\frac{8\sqrt{a}}{a}$ 39. $\sqrt{\frac{40}{3y}}$ $\frac{2\sqrt{30y}}{3y}$

Extra Practice **671**

13. The zero is 0.

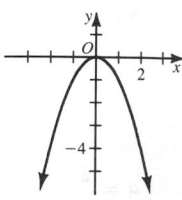

14. The zeros are -3 and 0.

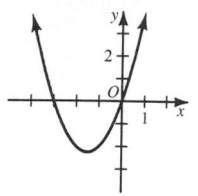

15. The zero is 3.

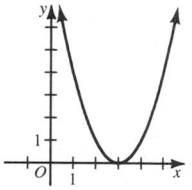

16. The zeros are -2 and 6.

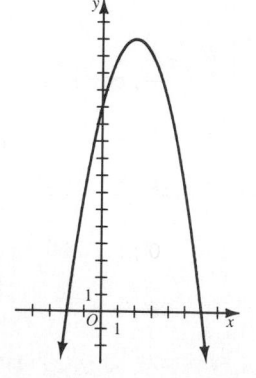

(continued on p. A1)

Simplify.

40. $5\sqrt{48} - 9\sqrt{3} + \sqrt{75}$ $16\sqrt{3}$ **41.** $2\sqrt{5}(7 + 3\sqrt{30})$ $14\sqrt{5} + 30\sqrt{6}$ 10-8

42. $(6\sqrt{2} + 1)(3\sqrt{2} - 4)$ $32 - 21\sqrt{2}$ **43.** $(4\sqrt{3} - \sqrt{7})(4\sqrt{3} + \sqrt{7})$ 41

44. $\dfrac{1}{5 - \sqrt{5}}$ $\dfrac{5 + \sqrt{5}}{20}$ **45.** $\dfrac{\sqrt{2} + 3}{\sqrt{2} - 1}$ $5 + 4\sqrt{2}$

46. $\sqrt[3]{-125}$ -5 **47.** $\sqrt[4]{48}$ $2\sqrt[4]{3}$ **48.** $-\sqrt[4]{81}$ -3 **49.** $\dfrac{5}{\sqrt[3]{2}}$ $\dfrac{5\sqrt[3]{4}}{2}$ 10-9

Solve.

50. $\sqrt{a} + 11 = 14$ $\{9\}$ **51.** $5\sqrt{x} = 4$ $\{\frac{16}{25}\}$ **52.** $\sqrt[3]{4x + 25} = 5$ $\{25\}$ 10-10

For use after Chapter 11.

Solve by completing the square. Give only the exact solutions, expressing any irrational roots in simplest form.

1. $x^2 + 2x - 35 = 0$ $\{5, -7\}$ **2.** $a^2 + 6a + 7 = 0$ $\{-3 + \sqrt{2}, -3 - \sqrt{2}\}$ 11-1

3. $m^2 + 5m + 1 = 0$ $\{\frac{-5 + \sqrt{21}}{2}, \frac{-5 - \sqrt{21}}{2}\}$ **4.** $y^2 - 12y - 7 = 0$ $\{6 + \sqrt{43}, 6 - \sqrt{43}\}$

Solve each equation using the quadratic formula. Give only the exact solutions, expressing any irrational roots in simplest form.

5. $2r^2 + r - 5 = 0$ **6.** $5g^2 - 3g = 0$ $\{0, \frac{3}{5}\}$ 11-2

7. $x^2 - 7x + 4 = 0$ **8.** $4y^2 + 9y + 2 = 0$ $\{-\frac{1}{4}, -2\}$

9. $5w^2 + 3w - 1 = 0$ **10.** $3c^2 + 20c - 32 = 0$ $\{\frac{4}{3}, -8\}$

Solve. **5.** $\{\frac{-1 + \sqrt{41}}{4}, \frac{-1 - \sqrt{41}}{4}\}$ **7.** $\{\frac{7 + \sqrt{33}}{2}, \frac{7 - \sqrt{33}}{2}\}$ **9.** $\{\frac{-3 + \sqrt{29}}{10}, \frac{-3 - \sqrt{29}}{10}\}$

11. The sum of a number and four times its reciprocal is 5. Find the number. 1 or 4   11-3

12. A picture is 30 cm by 40 cm. When it is placed in a frame of constant width, the area of the framed picture is 800 cm² more than the area of the picture itself. Find the width of the frame. 5 cm

Graph each function and determine its zeros, if any.

13. $f: x \longrightarrow -x^2$ **14.** $h: x \longrightarrow x^2 + 3x$ 11-4

15. $F: x \longrightarrow (x - 3)^2$ **16.** $G: x \longrightarrow 12 + 4x - x^2$

Sketch the graph of each function and estimate the zeros, if any.

17. $g: x \longrightarrow x^2 - x^3$ **18.** $h: x \longrightarrow x^3 - 9x$ 11-5

19. $f: x \longrightarrow \frac{1}{2}x^4$ **20.** $g: x \longrightarrow x^3 - 7x - 6$

Graph each inequality on a coordinate plane.

21. $y \geq x^2 - 4$ **22.** $y < -x^2 - 4x - 4$ 11-6

672 *Extra Practice*

23. Graph the solution set of the system: $y > x^2 + 4x + 4$
$$y < 3$$

For use after Chapter 12.

Find the measure of $\angle A$, if $\angle A$ is coterminal with the given angle and
$0° \le m\angle A < 360°$.

1. $440°$ $80°$ **2.** $-130°$ $230°$ **3.** $820°$ $100°$ **4.** $-210°$ $150°$ *12-1*

Give the sine, cosine, and tangent of an angle A in standard position whose
terminal side contains the given point on the unit circle.

5. $\left(\dfrac{8}{17}, -\dfrac{15}{17}\right)$ **6.** $\left(-\dfrac{21}{29}, -\dfrac{20}{29}\right)$ **7.** $\left(-\dfrac{\sqrt{2}}{2}, \dfrac{\sqrt{2}}{2}\right)$ *12-2*

The quadrant of $\angle A$ is given, and also the value of $\sin A$, $\cos A$, or $\tan A$.
Find the remaining two values.

8. $\sin A = \dfrac{4}{5}$, I   **9.** $\cos A = -\dfrac{1}{2}$, III   **10.** $\tan A = \sqrt{3}$; III

11. $\cos A = -\dfrac{3}{5}$; II   **12.** $\sin A = -\dfrac{1}{2}$; IV   **13.** $\tan A = -1$; II

Use the table on page 684 to find the required value.

14. $\sin 83°$ 0.9925   **15.** $\cos 12°$ 0.9781   **16.** $\tan 65°$ 2.1445   **17.** $\cos 48°$ 0.6691   *12-3*

Use the table on page 684 to find, to the nearest degree, the measure of
$\angle A$, $1° \le m\angle A < 90°$.

18. $\sin A = 0.5909$ 36°   **19.** $\cos A = 0.3611$ 69°   **20.** $\tan A = 0.4837$ 26°
21. $\sin A = 0.1405$ 8°   **22.** $\cos A = 0.2950$ 73°   **23.** $\tan A = 5.6378$ 80°

Solve the right triangle ABC using the given information and the table on
page 684. Give angle measures to the nearest degree and lengths to the
nearest tenth of a unit.

24. $m\angle B = 25°$; $c = 4$ *12-4*

25. $m\angle A = 50°$; $b = 3$

26. $m\angle A = 65°$; $c = 10$

27. $a = 2$; $c = 8$

28. $a = 3$; $b = 5$

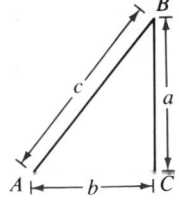

Find the norm to the nearest tenth and its direction to the nearest degree if
the terminal point of a vector in standard position has the given coordinates.

29. $(-3, 0)$ -3;   **30.** $(-4, -4)$ 5.7;   **31.** $(12, -5)$ 13;   **32.** $(-3, 5)$ 5.8;   *12-5*
        180°   225°   337°   121°

Extra Practice **673**

35.

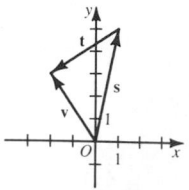

36.

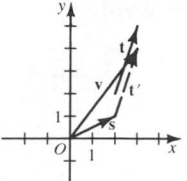

37.

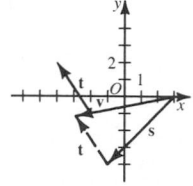

1.

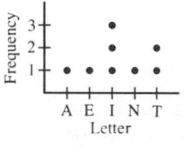

Letter	Freq.	Rel. Freq. Frac.	%
A	1	$\frac{1}{8}$	12.5
E	1	$\frac{1}{8}$	12.5
I	3	$\frac{3}{8}$	37.5
N	1	$\frac{1}{8}$	12.5
T	2	$\frac{2}{8}$	25
Total:	8	1	100

Find the norm to the nearest tenth and its direction to the nearest degree for the given vector.

33. initial point (2, 1) and terminal point (3, 4) $\|\mathbf{v}\| \approx 3.2$; 72°
34. initial point (2, 4) and terminal point (4, 7) $\|\mathbf{v}\| \approx 3.6$; 56°

The initial and terminal points of vectors s and t are given. Draw a vector diagram showing the sum v of s and t, and find the following.

a. the x- and y-components of v
b. the norm of v
c. the direction of v, to the nearest degree See Additional Answers for diagrams.

35. s: (0, 0), (1, 5)   t: (1, 5), (−2, 3) a. −2; 3   b. $\sqrt{13}$ c. 124°   12-6
36. s: (0, 0), (2, 1)   t: (2, 2), (3, 5) a. 3; 4   b. 5   c. 53°
37. s: (3, 0), (−1, −4)   t: (−2, −1), (−4, 2) a. −6; −1   b. $\sqrt{37}$ c. 189°

Give magnitudes to the nearest tenth and angle measures to the nearest degree.

38. Greg hiked 8 km east and 4 km northeast. Where was he then in relation to his starting point? 11.2 km away at a bearing of 75° from north   12-7

39. Find the magnitude and the direction of the resultant force when an object is acted on by two forces at right angles to each other, one of 60 N and the other of 175 N. 185 N, at an angle of 71° from the 60 N force

For use after Chapter 13.

Make a dot frequency diagram for the letters in each word. (Put the letters in alphabetical order.) Then make a table showing frequencies and relative frequencies, as fractions and percents.

1. initiate   2. dissension   3. successful   13-1

Use the data in the frequency distribution at the right. The data give the maximum speeds in kilometers per hour of some electric cars.

Speed	Frequency
70–90	2
90–110	5
110–130	3
130–150	1

4. Draw a histogram for the data.
5. Draw a frequency polygon for the data.   13-2

6-8. Make a table showing the cumulative frequency and cumulative percent for the letters in the words given in Exercises 1–3.   13-3

Find the mode(s), the median, and the mean for the data given.

9. 3, 4, 6, 6, 7, 9, 12, 12 6 and 12; 6.5; 7.4   13-4
10. 2, 5, 6, 8, 9, 12, 15, 16, 17 no mode; 9; 10

674 *Extra Practice*

11.	Number	Frequency
	2	5
	3	2
	4	1
	5	4

2; 3; 3.3

12.	Number	Frequency
	1	2
	3	3
	5	1
	7	3
	9	1

3 and 7; 4; 4.6

Find the range, the variance, and the standard deviation for the data given.

13-5

13. 2, 6, 7, 9
7; 6.5; 2.5

14. 1, 2, 5, 9, 13
12; 20; 4.5

15. 4, 5, 5, 6
2; 0.5; 0.7

A jar contains 8 pennies, 5 nickels, 10 dimes, and 7 quarters. If a coin is drawn at random from the jar, what is the probability that the coin is:

13-6

16. a penny? $\frac{4}{15}$

17. a nickel? $\frac{1}{6}$

18. not a dime? $\frac{2}{3}$

19. not a quarter? $\frac{23}{30}$

20. a nickel or a quarter? $\frac{2}{5}$

21. a nickel or a dime? $\frac{1}{2}$

An examination of 300 sample bicycles showed that 8 were defective, 285 were acceptable, and 7 exceeded standards. What is the experimental probability that a bicycle:

13-7

22. is defective?
$\frac{8}{300} \approx 0.03$

23. is not defective?
$\frac{292}{300} \approx 0.97$

24. exceeds standards?
$\frac{7}{300} \approx 0.02$

Problem Solving

For use after Chapter 1.

Choose a variable to represent one of the unknown quantities. Describe the quantity that the variable represents, then write a mathematical sentence that represents the given information.

1-9

1. The sum of a positive integer and the next greater positive integer is three hundred eighty-six.
2. The Stars played twenty-seven games and won twice as many games as they lost.
3. When twenty-eight votes were cast for the two candidates, Carol received six more votes than Brian.
4. Ed has two more dimes than nickels. The total value of the coins is one dollar fifty-five cents.
5. The length of a rectangle with a perimeter of one hundred centimeters is two centimeters more than twice its width.
6. In four years, Jack's age will be twice as great as his age fourteen years ago.

2.

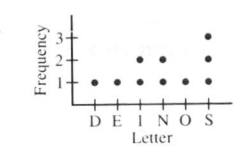

Letter	Freq.	Rel. Freq. Frac.	%
D	1	$\frac{1}{10}$	10
E	1	$\frac{1}{10}$	10
I	2	$\frac{2}{10}$	20
N	2	$\frac{2}{10}$	20
O	1	$\frac{1}{10}$	10
S	3	$\frac{3}{10}$	30
Total:	10	1	100

(continued on p. A1)

Additional Answers
Chapter 1

1. Let x = the first positive integer. $x + (x + 1) = 386$
2. Let l = the number of games that the Stars lost. $l + 2l = 27$
3. Let b = the number of votes that Brian received. $b + (b + 6) = 28$
4. Let n = the number of nickels that Ed has. $0.05n + 0.10(n + 2) = 1.55$
5. Let w = the width of the rectangle. $2w + 2(2w + 2) = 100$
6. Let j = Jack's present age in years. $j + 4 = 2(j - 14)$

For use after Chapter 2.

a. **Name a positive or a negative number to represent each measurement in the given problem.**
b. **Compute the sum of the numbers.**
c. **Answer the question.**

1. A bush pilot flew 285 km directly north from a landing strip. Then the pilot flew 360 km directly south. Where was the pilot then located relative to the landing strip?   2-6

2. At noon one day the temperature was 11°C. The temperature rose 2°C during the next six hours, fell 18°C during the next six hours, and then rose 6°C during the following twelve hours. What was the temperature at noon of the second day?

a. **Express the answer to each question as the difference between two real numbers and compute this difference.**
b. **Interpret the sign of the difference (positive or negative) and answer the question.**

3. When the wind is blowing at 32 km/h, an actual temperature of 2°C can feel like a temperature of −10°C. What is the difference between the temperatures?   2-8

4. In one day the temperature in Browning, Montana, dropped from 7°C to −31°C. How many degrees did the temperature drop?

5. A cargo plane took off from an airport 760 km south of the equator and flew due north to an airport 535 km north of the equator. How many kilometers did the plane fly?

For use after Chapter 3.

Solve.

1. Ella has eight fewer nickels than dimes. The total value of the nickels and dimes is $3.35. How many of each kind of coin does she have?   17 nickels; 25 dimes   3-5

2. The cost of a tie is $21 less than the cost of two shirts. A sweater costs $8.50 more than a shirt. If two shirts, a tie, and a sweater cost $55, find the cost of a sweater. $22

3. Art's age is two thirds Cal's age, and Bob is three years older than Art. The sum of their three ages is 45. Find the age of each.   Art: 12 years; Bob: 15 years; Cal: 18 years   3-7

4. The St. Lawrence River is 120 km longer than 3 times the length of the Albany River. The Yukon River is 130 km longer than the St. Lawrence River. The combined lengths of the Albany River and the St. Lawrence River is 850 km greater than the length of the Yukon River. How long is the Yukon River?   3190 km

5. Ed is 13 years old and his mother is 35. In how many years will Ed be half as old as his mother?   9 years

6. The formula $V = \frac{\pi r^2 h}{3}$ gives the volume V of a cone in terms of the   3-8
   radius of the base r and the height of the cone h.
   a. Express h in terms of the other variables. $h = \frac{3V}{\pi r^2}$
   b. Find the height of a cone whose radius is 3 cm if its volume is
      47.1 cm³. Use $\pi = 3.14$. 5 cm

For use after Chapter 4.

Solve. If a problem has no solution, explain why.

1. Find the least two consecutive even integers whose sum is more than   4-6
   50. 26 and 28

2. The measure of the supplement of a certain angle is 2° less than three   4-7
   times the measure of its complement. Find the measure of the angle. 44°

3. The degree measures of two angles of a triangle are consecutive
   multiples of 6. The measure of the third angle is 54° less than twice
   the sum of the other angles. Find the measure of each angle. 36°; 42°; 102°

4. Two bicyclists starting at the same time from towns 120 km apart rode   4-8
   toward each other at average speeds of 28 km/h and 44 km/h. How
   long did it take them to meet? $1\frac{2}{3}$ h

5. Karen drove a certain distance in 5 h and returned in 4.5 h. Her
   average speed on the return trip was 8 km/h more than her original
   speed. At what speed did she return? 80 km/h

6. The round-trip fare from Beachview to Warwick's Island is $39 for   4-9
   adults, $19 for children 2 through 12, and $5 for infants under 2. On
   Sunday there were 49 passengers who paid a total fare of $1475. If
   there were twice as many adult passengers as children, how many
   infants were there? 4

7. One solution is 10% acid and another is 25% acid. How many liters of 10% acid: 14 L
   each should be mixed to form 30 L of a solution that is 18% acid? 25% acid: 16 L

8. Pack Four hiked 14 km at an average rate of 3.5 km/h. How fast must   4-10
   they go on the return trip in order that their average speed for the
   round trip will be 7 km/h? No solution, because the return trip
   would have to be made in zero hours.

For use after Chapter 6.

Solve.   twenty-five 22¢ stamps; fifteen 14¢ stamps

1. Carol bought some 22¢ stamps and some 14¢ stamps. If 40 stamps   6-5
   cost $7.60, how many of each kind did she purchase?

2. Pearl is four years older than Jose. Eight years ago she was twice as
   old as Jose was then. How old are Pearl and Jose now? Pearl: 16 years; Jose: 12 years

3. Flying with the wind, a plane flew 700 km in 2.5 h. With no change in   6-6
   the wind, the return trip took 1.75 h. Find the wind speed and the air
   speed of the plane. wind speed: 60 km/h; air speed: 340 km/h

9.
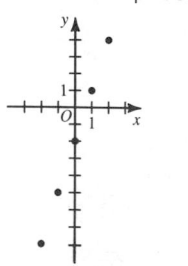

10.
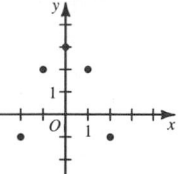

13.
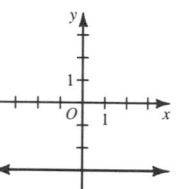

14.
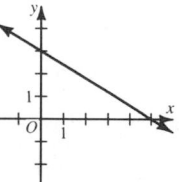

15.
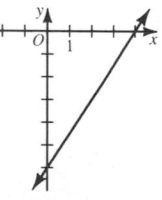

16.
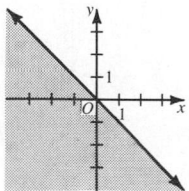

4. A boat traveled 9 km downstream in 36 min. It then traveled 21 km upstream in 2 h 20 min. Find the speed of the boat in still water and the speed of the current. speed of boat: 12 km/h; speed of current: 3 km/h

5. The sum of the digits of a two-digit number is 7. The number with its digits reversed is 27 more than the original number. Find the original number. 25 *6-7*

6. The hundreds' digit of a three-digit number is 2. The tens' digit is equal to the sum of the other two digits. If the order of the digits is reversed, the number formed is 396 greater than the original number. Find the original number. 286

For use after Chapter 7.

Solve.

1. The side of one square is 7 cm longer than the side of another square, and its area is 483 cm² greater than that of the second square. Find the length of a side of each square. 38 cm; 31 cm *7-4*

2. The product of the first and last of four consecutive integers is 12 greater than the square of the second of the integers. Find the integers. 13, 14, 15, and 16

3. The sum of two numbers is 20. The sum of their squares is 208. Find the two numbers. 8 and 12 *7-11*

4. Find two positive numbers whose difference is 5 and whose product is 84. 7 and 12

5. A rectangular playground is 24 m long and 18 m wide. It has been decided to increase each dimension of the playground by the same number of meters. This will increase the area of the playground by 400 m². By how many meters will each dimension of the playground be increased? 8 m

6. The Science Club launched a model rocket upward with an initial speed of 34.3 m/s. Use the formula $h = rt - 4.9t^2$ to determine how many seconds after launch the rocket will be 58.8 m above ground. after 3 s and after 4 s

For use after Chapter 9.

Solve.

1. Kim paid $10,269 for a new car, including a 5% sales tax. What was the price of the car? $9780 *9-2*

2. Maria invested a sum of money at an annual rate of 8%, the same sum at an annual rate of 11%, and $5000 less than each of these sums at an annual rate of 9%. If her income in one year from these investments was $3750, how much had she invested at each rate?
$15,000 at 8%; $15,000 at 11%; $10,000 at 9%

678   *Extra Practice*

3. The numerator of a fraction is 14 less than the denominator, and the fraction is equal to $\frac{5}{12}$. Find the fraction. $\frac{10}{24}$ 9-4

4. It takes Eddie 1.5 times as long to paint the garage as it takes his father to do the same job. Working together, they can complete the garage in 3.6 h. How long would it take Eddie to do the job alone? 9 h   9-5

5. Kim walked 12 km to her grandmother's house, and then rode her bicycle home. If her speed riding was 6 km/h faster than her speed walking, and the round-trip took 3 h, how fast did she walk? 6 km/h   9-6

6. Lee has $7.80 worth of nickels and dimes. The ratio of the number of dimes to the number of nickels is 7:12. How many of each kind of coin does Lee have? 72 nickels; 42 dimes   9-7

7. Property tax varies directly as assessed value. The tax on a certain house assessed at $80,000 is $1400. What is the assessed value of a house in the same community for which the property tax is $1610? $92,000   9-8

8. The height of a cylinder of fixed volume varies inversely as the square of the radius of a base of the cylinder. A certain cylinder has radius 3 m and height 8 m. What is the height of a second cylinder with radius 4 m and the same volume as the first cylinder? 4.5 m   9-9

9. The cost of operating an appliance varies jointly as the number of watts, hours of operation, and the cost per kilowatt-hour. If it cost 9¢ to operate a 1200-watt dishwasher for 40 min at a rate of 5¢ per kilowatt-hour, how much does it cost to operate a 3000-watt air conditioner for 4 h at the same rate? $1.35   9-10

For use after Chapter 10.

Make a sketch for each problem. Approximate each square root to the nearest hundredth. Use a calculator or the Table of Square Roots as necessary. See Additional Answers for sketches.

1. Find the length of a diagonal of a rectangular field whose dimensions are 54 m by 72 m. 90 m   10-5

2. A cruise ship sailed due south for 48 km and then due west for 14 km. At that point, how far was the cruise ship from its starting point? 50 km

3. A diagonal of a square is 14 cm long. Find the length of a side of the square. 9.90 cm

4. Find the dimensions of a rectangle whose length is seven times its width if one of its diagonals has length 40 cm. width: 5.66 cm; length: 39.60 cm

5. The altitude drawn to the base of any isosceles triangle divides the base into two line segments of equal length. If the base of a certain isosceles triangle is 24 cm and the length of the altitude drawn to the base is 16 cm, find the equal lengths of the other two sides of the triangle. 20 cm

17.

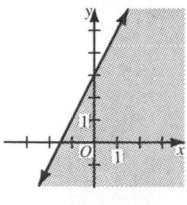

18.

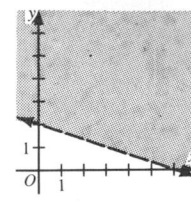

22.

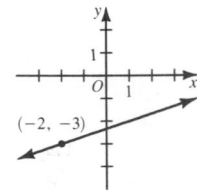

Additional Answers
Chapter 10

1.

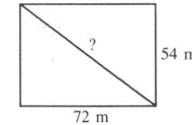

2.

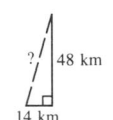

3.

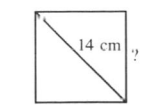

4.

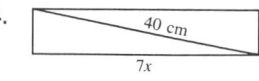

5.

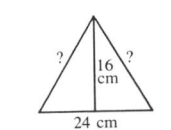

For use after Chapter 11.

Solve.

1. The sum of a number and three times its reciprocal is $\frac{7}{2}$. Find the number. 2 or $\frac{3}{2}$

2. The length of a rectangle is 4 cm less than twice its width. If the length is increased by 2 cm and the width is decreased by 3 cm, the area of the new rectangle is 646 cm². What are the dimensions of the original rectangle? width: 20 cm; length: 36 cm

3. It takes Diane 3 h less than it takes her brother Don to trim a hedge. If they work together, they can trim the hedge in 2 h. How many hours would each take, working alone, to trim the hedge? Diane: 3 h; Don: 6 h

4. The altitude of a triangle is 4 cm longer than its base. The area of the triangle is 198 cm². Find the dimensions of the triangle. altitude: 22 cm base: 18 cm

11-3

For use after Chapter 12.

Solve each problem using the table on page 684 as necessary. Give angle measures to the nearest degree and lengths or magnitudes to the nearest tenth.

1. The ruins of an ancient monument are 35 m tall. When the angle of elevation of the sun is 62°, how long is the shadow of the ruins? 18.6 m

12-4

2. An 8 m ladder is placed against a building, touching the building 7.5 m above the ground. What is the degree measure of the angle that the ladder makes with the level ground? 70°

3. Lucille, who can swim 5 km/h in still water, is swimming across a river. The river current carries her away from her original heading at a speed of 3 km/h. At what angle is she deviating from her original heading? 31°

12-7

4. An airplane flies due west at a speed of 288 km/h through a wind blowing south at a constant 84 km/h. Describe the speed and bearing of the plane. 300 km/h; 254°

Tables

Table 1 Formulas

Circle	$A = \pi r^2, C = 2\pi r$	Cube	$V = s^3$
Parallelogram	$A = bh$	Rectangular Box	$V = lwh$
Right Triangle	$A = \frac{1}{2}bh, c^2 = a^2 + b^2$	Cylinder	$V = \pi r^2 h$
Square	$A = s^2$	Pyramid	$V = \frac{1}{3}Bh$
Trapezoid	$A = \frac{1}{2}h(b + b')$	Cone	$V = \frac{1}{3}\pi r^2 h$
Triangle	$A = \frac{1}{2}bh$	Sphere	$V = \frac{4}{3}\pi r^3$
Sphere	$A = 4\pi r^2$		

Table 2 Metric Units of Measure

Base Units	Time	Temperature
Length: meter (m)	second (s), minute (min), hour (h)	degree Celsius (°C)
Mass: kilogram (kg)*	day (d), month (mo), year (yr)	Kelvin (K)

Capacity	Force	Pressure
liter (L)	Newton (N)	Pascal (Pa)
1 L = 1000 cm^3		

Prefixes

Factor	Prefix	Symbol	Factor	Prefix	Symbol
10^{18}	exa	E	10^{-1}	deci	d
10^{15}	peta	P	10^{-2}	centi	c
10^{12}	tera	T	10^{-3}	milli	m
10^9	giga	G	10^{-6}	micro	μ
10^6	mega	M	10^{-9}	nano	n
10^3	kilo	k	10^{-12}	pico	p
10^2	hecto	h	10^{-15}	femto	f
10	deka	da	10^{-18}	atto	a

A prefix multiplies a unit by the factor given in the table.

Examples   gigameter: 1 Gm = 10^9 m = 1,000,000,000 m
milligram: 1 mg = 10^{-3} g = 0.001 g*

Compound units may be formed by division or multiplication.

Examples   kilometers per hour: km/h   square centimeters: cm^2 cubic meters: m^3

*Although the kilogram is defined as the base unit, the gram (g) is used with the prefixes to name other units of mass.

Table 3   Squares of Integers from 1 to 100

Number	Square	Number	Square	Number	Square	Number	Square
1	1	26	676	51	2601	76	5776
2	4	27	729	52	2704	77	5929
3	9	28	784	53	2809	78	6084
4	16	29	841	54	2916	79	6241
5	25	30	900	55	3025	80	6400
6	36	31	961	56	3136	81	6561
7	49	32	1024	57	3249	82	6724
8	64	33	1089	58	3364	83	6889
9	81	34	1156	59	3481	84	7056
10	100	35	1225	60	3600	85	7225
11	121	36	1296	61	3721	86	7396
12	144	37	1369	62	3844	87	7569
13	169	38	1444	63	3969	88	7744
14	196	39	1521	64	4096	89	7921
15	225	40	1600	65	4225	90	8100
16	256	41	1681	66	4356	91	8281
17	289	42	1764	67	4489	92	8464
18	324	43	1849	68	4624	93	8649
19	361	44	1936	69	4761	94	8836
20	400	45	2025	70	4900	95	9025
21	441	46	2116	71	5041	96	9216
22	484	47	2209	72	5184	97	9409
23	529	48	2304	73	5329	98	9604
24	576	49	2401	74	5476	99	9801
25	625	50	2500	75	5625	100	10,000

Table 4   Square Roots of Integers from 1 to 100

Exact square roots are shown in red. For the others, rational approximations are given correct to three decimal places.

Number	Positive Square Root	Number	Positive Square Root	Number	Positive Square Root	Number	Positive Square Root
N	$\sqrt{N}$	N	$\sqrt{N}$	N	$\sqrt{N}$	N	$\sqrt{N}$
1	1	26	5.099	51	7.141	76	8.718
2	1.414	27	5.196	52	7.211	77	8.775
3	1.732	28	5.292	53	7.280	78	8.832
4	2	29	5.385	54	7.348	79	8.888
5	2.236	30	5.477	55	7.416	80	8.944
6	2.449	31	5.568	56	7.483	81	9
7	2.646	32	5.657	57	7.550	82	9.055
8	2.828	33	5.745	58	7.616	83	9.110
9	3	34	5.831	59	7.681	84	9.165
10	3.162	35	5.916	60	7.746	85	9.220
11	3.317	36	6	61	7.810	86	9.274
12	3.464	37	6.083	62	7.874	87	9.327
13	3.606	38	6.164	63	7.937	88	9.381
14	3.742	39	6.245	64	8	89	9.434
15	3.873	40	6.325	65	8.062	90	9.487
16	4	41	6.403	66	8.124	91	9.539
17	4.123	42	6.481	67	8.185	92	9.592
18	4.243	43	6.557	68	8.246	93	9.644
19	4.359	44	6.633	69	8.307	94	9.695
20	4.472	45	6.708	70	8.367	95	9.747
21	4.583	46	6.782	71	8.426	96	9.798
22	4.690	47	6.856	72	8.485	97	9.849
23	4.796	48	6.928	73	8.544	98	9.899
24	4.899	49	7	74	8.602	99	9.950
25	5	50	7.071	75	8.660	100	10

Table 5 Values of Trigonometric Functions for Angles in Degrees

$m\angle A$	sin A	cos A	tan A	$m\angle A$	sin A	cos A	tan A
1°	0.0175	0.9998	0.0175	46°	0.7193	0.6947	1.0355
2°	0.0349	0.9994	0.0349	47°	0.7314	0.6820	1.0724
3°	0.0523	0.9986	0.0524	48°	0.7431	0.6691	1.1106
4°	0.0698	0.9976	0.0699	49°	0.7547	0.6561	1.1504
5°	0.0872	0.9962	0.0875	50°	0.7660	0.6428	1.1918
6°	0.1045	0.9945	0.1051	51°	0.7771	0.6293	1.2349
7°	0.1219	0.9925	0.1228	52°	0.7880	0.6157	1.2799
8°	0.1392	0.9903	0.1405	53°	0.7986	0.6018	1.3270
9°	0.1564	0.9877	0.1584	54°	0.8090	0.5878	1.3764
10°	0.1736	0.9848	0.1763	55°	0.8192	0.5736	1.4281
11°	0.1908	0.9816	0.1944	56°	0.8290	0.5592	1.4826
12°	0.2079	0.9781	0.2126	57°	0.8387	0.5446	1.5399
13°	0.2250	0.9744	0.2309	58°	0.8480	0.5299	1.6003
14°	0.2419	0.9703	0.2493	59°	0.8572	0.5150	1.6643
15°	0.2588	0.9659	0.2679	60°	0.8660	0.50	1.7321
16°	0.2756	0.9613	0.2867	61°	0.8746	0.4848	1.8040
17°	0.2924	0.9563	0.3057	62°	0.8829	0.4695	1.8807
18°	0.3090	0.9511	0.3249	63°	0.8910	0.4540	1.9626
19°	0.3256	0.9455	0.3443	64°	0.8988	0.4384	2.0503
20°	0.3420	0.9397	0.3640	65°	0.9063	0.4226	2.1445
21°	0.3584	0.9336	0.3839	66°	0.9135	0.4067	2.2460
22°	0.3746	0.9272	0.4040	67°	0.9205	0.3907	2.3559
23°	0.3907	0.9205	0.4245	68°	0.9272	0.3746	2.4751
24°	0.4067	0.9135	0.4452	69°	0.9336	0.3584	2.6051
25°	0.4226	0.9063	0.4663	70°	0.9397	0.3420	2.7475
26°	0.4384	0.8988	0.4877	71°	0.9455	0.3256	2.9042
27°	0.4540	0.8910	0.5095	72°	0.9511	0.3090	3.0777
28°	0.4695	0.8829	0.5317	73°	0.9563	0.2924	3.2709
29°	0.4848	0.8746	0.5543	74°	0.9613	0.2756	3.4874
30°	0.5̄0	0.8660	0.5774	75°	0.9659	0.2588	3.7321
31°	0.5150	0.8572	0.6009	76°	0.9703	0.2419	4.0108
32°	0.5299	0.8480	0.6249	77°	0.9744	0.2250	4.3315
33°	0.5446	0.8387	0.6494	78°	0.9781	0.2079	4.7046
34°	0.5592	0.8290	0.6745	79°	0.9816	0.1908	5.1446
35°	0.5736	0.8192	0.7002	80°	0.9848	0.1736	5.6713
36°	0.5878	0.8090	0.7265	81°	0.9877	0.1564	6.3138
37°	0.6018	0.7986	0.7536	82°	0.9903	0.1392	7.1154
38°	0.6157	0.7880	0.7813	83°	0.9925	0.1219	8.1443
39°	0.6293	0.7771	0.8098	84°	0.9945	0.1045	9.5144
40°	0.6428	0.7660	0.8391	85°	0.9962	0.0872	11.4301
41°	0.6561	0.7547	0.8693	86°	0.9976	0.0698	14.3007
42°	0.6691	0.7431	0.9004	87°	0.9986	0.0523	19.0811
43°	0.6820	0.7314	0.9325	88°	0.9994	0.0349	28.6363
44°	0.6947	0.7193	0.9657	89°	0.9998	0.0175	57.2900
45°	0.7071	0.7071	1	90°	1	0	Undefined

Appendix A

The Divide-and-Average Method

If you do not have access to a calculator or a table of square roots, you can still approximate an irrational square root by using the *divide-and-average method*. This method is based upon the fact that, if you divide a positive number by a positive number that is *less than* the positive square root of the number, the quotient will be *greater than* the square root.

EXAMPLE   Approximate $\sqrt{51}$ to the nearest ten-thousandth.

SOLUTION
1. Select the integer whose square is nearest 51 as the first approximation, a. Since $7^2 = 49$, let $a = 7$.

2. Divide 51 by a, carrying out the division to two more digits than are in the divisor: $51 \div 7 \approx 7.28$

3. Find the average of a and $\frac{51}{a}$: $\frac{1}{2}(7 + 7.28) = 7.14$

4. Use the average as the new value for a. Continue repeating steps 2 and 3 as often as necessary.

   $51 \div 7.14 \approx 7.1428 \longrightarrow \frac{1}{2}(7.14 + 7.1428) = 7.1414$

   $51 \div 7.1414 \approx 7.141456$

5. The approximation is accurate to at least as many digits as match in a and $51 \div a$.

   $\therefore \ \sqrt{51} \approx 7.1414$

Exercises
11. 3.8730   12. -4.7958 13. -6.2450 14. 6.7082   15. -16.5227
16. 13.7840   17. 20.1990   18. 3.4496   19. -6.3797 20. 5.6480

Use the divide-and-average method to approximate each square root to the nearest hundredth.

1. $\sqrt{29}$ 5.39   2. $\sqrt{13}$ 3.61   3. $-\sqrt{47}$ -6.86 4. $-\sqrt{83}$ -9.11 5. $\sqrt{350}$ 18.71

6. $\sqrt{223}$ 14.93   7. $-\sqrt{167}$ -12.92 8. $-\sqrt{42.3}$ -6.50 9. $\sqrt{23.8}$ 4.88   10. $\sqrt{18.6}$ 4.31

Use the divide-and-average method to approximate each square root to the nearest ten-thousandth.

11. $\sqrt{15}$ 12. $-\sqrt{23}$ 13. $-\sqrt{39}$ 14. $\sqrt{45}$ 15. $-\sqrt{273}$

16. $\sqrt{190}$ 17. $\sqrt{408}$ 18. $\sqrt{11.9}$ 19. $-\sqrt{40.7}$ 20. $\sqrt{31.9}$

Appendix B

Summary of BASIC

BASIC is one of the so-called "higher-level" languages in which *programs* may be written for computers. A program written in BASIC is translated by a *compiler*, or interpreter, inside the computer into a *machine language* that tells the computer what to do.

BASIC is essentially linear in character. Each *statement* of a program is numbered in succession, often 10, 20, 30, and so on. (See page 34.) Such numbering allows statements to be inserted into a program as desired. (See page 77.) The computer follows through the statements in numerical order and carries out the instructions in each statement as it comes to it.

Some versions of BASIC require an

END statement

for each program. (See page 34.) This statement lets the computer know that no more statements follow in the program.

BASIC also includes several *commands*, which are not parts of programs and do not have line numbers.

The command RUN

This command tells the computer to RUN, or *execute*, the program.

The command LIST

This command tells the computer to list the statements of a program in numerical order. This is especially useful when changes have been made in a program and you want to see a clean copy of it. For example, if you were able to save the program on page 34 and then make the changes in it as shown on pages 76–77, you should type LIST to see and check the revised program.

A program generally includes both *input* and *output*. In the program on page 34, for example, the input is the domain given in line 60. The output is given by the PRINT statements in lines 90, 120, 140, and 150. (See pages 688-689 of this Appendix for more information about PRINT statements.)

The symbols for *operations* in BASIC are as follows.

$+$ addition	$*$ multiplication
$-$ subtraction	$/$ division
$(\ )$ grouping	$\uparrow$ or $\wedge$ raising to a power

The same rules are followed for *order of operations* as are followed in algebra. (See page 25.)

Other symbols used in BASIC are the following.

=	equals	<	less than
>	greater than	<>	not equal
<=	less than or equal	>=	greater than or equal

In BASIC, very large and very small numbers are stated in *E-notation*, which is a form of scientific notation. (See page 399.) For example:

1,200,000,000, or 1.2×10^9, is written as 1.2E+09.

0.0000012, or 1.2×10^{-6}, is written as 1.2E−06.

BASIC handles variables much as they are handled in algebra. A variable may be denoted by a letter, a letter followed by a digit, or possibly other combinations of symbols as allowed by the various versions of BASIC.

The input for a program involves giving values to variables. There are several ways to do this.

(1)   The INPUT statement

INPUT M

This statement will cause the computer to print a question mark and wait for you to type a value for M. (See page 195.)

INPUT A, B, C

This statement will print a question mark, and then you are to type in three values separated by commas. (See page 362.) If you type in only one or two values, the computer will continue to print question marks until you have entered three values.

(2)   The LET statement

LET S = 0

This statement will give the value 0 to the variable S. (See the program on page 690 of this Appendix.)

LET A2 = A/A1

This statement will find the quotient of A and $A1$ and give this value to $A2$. (See page 362.)

LET S = S + N

This statement will take the value of S, add to it the value of N, and give S this new value. (See the program on page 690 of this Appendix.)

Notice that the LET statement is *not* symmetric like an algebraic equation. When the computer encounters a LET statement, it finds the value of the expression to the *right* of the equals sign and gives that value to the single variable to the *left* of the equals sign.

(3)   READ and DATA statements

DATA 2, 5, 25, −10
READ M

The DATA statement contains a list of values that are to be given to one or more variables by a READ statement. In this case, the READ statement will read one value from the list of DATA and give that value to the variable M. (See pages 644–645.)

DATA 2, 5, 25, −10
READ M, N

In this case, the READ statement will read two values in succession from the list of DATA and give the first value to M and the second value to N.

A large list of DATA may be separated into several DATA statements, but the DATA will be read as if there were one continuous list. (See page 645.) DATA statements may be put anywhere in a program. However, they are usually put near the beginning or near the end for convenience.

If a list of DATA is to read more than once, a

RESTORE statement

is needed to direct the computer to start again at the beginning of the list.

Output is handled by

PRINT statements.

PRINT A

This statement will print the value of A. (See the program that follows on page 689.)

PRINT "A"

This statement will print the letter A. Any expression within quotation marks will be printed exactly as it appears, including spaces. (See page 34.)

PRINT "A = ";A

In this statement, the *semicolon* will cause the value of A to be printed immediately following the expression in quotation marks. (See the program that follows on page 689.) Note that a semicolon at the end of a PRINT statement preceding an INPUT statement will cause the question mark to be printed on that line.

PRINT A, B

In this statement, the *comma* will cause the values of A and B to be printed spaced apart. (See the program that follows on page 689.)

PRINT

(a) By itself, this statement will print a blank line. (See the program that follows on page 689.)

(b) This statement will cancel the effect of a semicolon or a comma at the end of the preceding line. (See page 138.)

To study the various forms of the PRINT statement, type in and RUN the following program.

```
10   PRINT "TO FIND AREA AND PERIMETER"
20   PRINT " OF A RECTANGLE:"
30   PRINT "INPUT LENGTH, WIDTH";
40   INPUT L, W
50   PRINT
60   LET A = L*W
70   LET P = 2*(L + W)
80   PRINT "AREA = ";A,"PERIMETER = ";P
90   END
```

Sometimes it is helpful to separate blocks of a program with

REM statements.

REM stands for *remark*. A REM statement may contain any comment that you wish to use to clarify the program. (See page 34.) It will appear in the LISTing of the program, but it will have no effect on the program itself. A long comment may run over several lines, but each line must begin with REM. Any letters or symbols may follow REM.

Sometimes in the construction of a program it is necessary to interrupt the line-by-line flow of the program. This is done by using

Transfer statements.

(1) GOTO 00

This is an *unconditional transfer* statement. It transfers the execution of the program to line 00. (See the program that follows on page 690.)

(2) IF (expression or relation) THEN 00

This is a *conditional transfer* statement. That is, IF the expression or relation is true, THEN the execution will be transferred to line 00. Otherwise (ELSE), the execution will proceed to the following line. (See page 34.) Unconditional and conditional transfers are often used together to form a *branched* program. (See page 34.)

(3) ON N GOTO 000, 000, 000 . . .

This is a *computed*, or *multibranch*, GOTO statement, with N equal to 1, 2, 3, and so on. Thus, in line 160 of the program on pages 602–603:

If $N = 1$, the execution is transferred to line 170.

If $N = 2$, the execution is transferred to line 220.

If $N = 3$, the execution is transferred to line 270.

An especially useful construction in BASIC is the *loop*, that is, passing through the same block of statements several times. One form of loop simply uses a GOTO statement to return the execution of the program to the beginning of the block of statements. (See the program below.) When using such a loop, some provision must be made for ending it. One way is to provide an ending number, such as the -1 in this program. Sometimes such a loop requires the use of a *counter* such as:

LET C = 0
. . . .
LET C = C + 1

This type of counter can be used to record the number of passes through the loop, as in the program that follows.

To study the different types of loops, type in and RUN the following program.

```
10    PRINT "TO FIND THE AVERAGE OF"
20    PRINT "SEVERAL POSITIVE NUMBERS:"
30    LET C = 0
40    LET S = 0
50    PRINT "INPUT N (−1 TO END)";
60    INPUT N
70    IF N = −1 THEN 110
80    LET C = C + 1
90    LET S = S + N
100    GOTO 50
110    PRINT "AVERAGE = ";S/C
120    END
```

When the number of passes through a loop is known in advance, a

FOR-NEXT loop

may be used. For a simple counter, you use:

FOR I = 1 TO N
. . . .
NEXT I

This loop will go through a block of statements N times. (See page 138.)

If a sequence of values is to be given to a variable X, the following form may be used.

FOR X = M TO N STEP S
. . . .
NEXT X

This construction will give values to X as follows:

$M, M + S, M + 2*S, M + 3*S$, and so on, up to N

If $M < N$, then $S > 0$. However, if $M > N$, then $S < 0$. (See page 247.) If $S = 1$, the STEP portion may be omitted. (See page 34.)

When one loop is contained within another loop, the loops are said to be *nested*. (See page 138.)

When a block of statements is to be used in several places in a program, it is often convenient to set off that block of statements as a *subroutine*. This is done by using

GOSUB . . . RETURN statements.

GOSUB 00

This statement will transfer the execution of the program to line 00. (See pages 362–363.) The block of statements forming the subroutine must end with:

RETURN

This statement will transfer the execution back to the line following GOSUB 00. (See pages 362–363.)

BASIC has several special built-in functions. The following are three of these.

ABS(X)

This function gives the absolute value of X, or $|X|$. (See pages 76–77.)

SQR(X)

This function gives the square root of X, or $\sqrt{X}$, for $X \geq 0$. (See page 566.)

INT(X)

This function gives the greatest integer less than or equal to X. It is often used in factoring programs (pages 362–363) and in rounding numbers (pages 602–603).

The TAB function

is often used in PRINT statements to adjust the spacing. (See pages 195–196.)

PRINT TAB(00);"A"

This statement will cause the computer to print the letter A in column 00 of the display. (Some computers assign the number 0 to the first column at the left, while others assign the number 1 to this column. You may need to check the computer manual to determine exactly how this function works on the computer that you are using.)

The TAB function can also be used to print out a table with computed spacing, as shown in the program that follows.

```
5    REM *PRINTS COLUMN HEADINGS
10   FOR I = 0 TO 9
20   PRINT TAB(I*3 + 3);I;
30   NEXT I
40   PRINT
45   REM *PRINTS HORIZONTAL LINE
50   FOR I = 1 TO 34
60   PRINT "−";
70   NEXT I
80   PRINT
85   REM *PRINTS ROW HEADINGS
86   REM *AND TABLE ENTRIES
90   FOR I = 0 TO 9
100   PRINT I;"!";
110   FOR J = 0 TO 9
120   PRINT TAB(J*3 + 3);I*J;
130   NEXT J
140   PRINT
150   NEXT I
160   END
RUN
```

	0	1	2	3	4	5	6	7	8	9
0!	0	0	0	0	0	0	0	0	0	0
1!	0	1	2	3	4	5	6	7	8	9
2!	0	2	4	6	8	10	12	14	16	18
3!	0	3	6	9	12	15	18	21	24	27
4!	0	4	8	12	16	20	24	28	32	36
5!	0	5	10	15	20	25	30	35	40	45
6!	0	6	12	18	24	30	36	42	48	54
7!	0	7	14	21	28	35	42	49	56	63
8!	0	8	16	24	32	40	48	56	64	72
9!	0	9	18	27	36	45	54	63	72	81

PRINT TAB statements may also be using in plotting graphs of functions. In doing this, the domain of the function is plotted *vertically* and the range of the function is plotted *horizontally*. (See pages 466–467.)

Subscripted variables are used to represent lists, or *arrays*, of values. For example, the variables

$$A(1), \ A(2), \ \text{and} \ A(3)$$

are often more convenient to use than A, B, and C. (See pages 414–415.) In some versions of BASIC a zero subscript is allowed, and so you may use

$$A(0), \ A(1), \ A(2), \ . \ . \ .$$

If more than 10, or in some cases 11, values are to be used, a

DIMENSION statement

is needed. For example,

DIM A(15)

will allow you to use values $A(1)$, $A(2)$, . . . , $A(15)$ or, in some cases, $A(0)$, $A(1)$, . . . , $A(15)$. (See pages 491–492.)

A *string* is a set of characters, including spaces, generally enclosed within quotation marks. A *string variable* is a variable having strings as values. Its name must end with $.

A string variable with values that are single-character strings may be used directly in a program. However, if the strings are to have more than one character, a DIMension statement is required. For example,

DIM N$(20)

will allow the values of N to have up to twenty characters, as in the program below.

Values may be given to string variables by INPUT, LET, or READ . . . DATA statements.

The following is a fragment of a program that shows how strings may be used in programs that are "chatty." Notice that the two string variables are DIMensioned in the same line, line 10. This statement allows the name that is input in line 30 to have up to twenty characters and the reply in line 80 to have up to three characters.

```
10    DIM N$(20), A$(3)
20    PRINT "WHAT IS YOUR NAME";
30    INPUT N$
40    PRINT "HELLO ";N$;"!"
50    PRINT "I AM GLAD TO MEET YOU."
60    PRINT "DO YOU ENJOY WORKING WITH ";
70    PRINT "ME (YES/NO)";
80    INPUT A$
90    IF A$ = "YES" THEN 120
100    PRINT "I AM VERY SORRY, ";N$,"."
110    GOTO 130
120    PRINT "I AM VERY GLAD, ";N$;"."
130    END
```

Appendix C

Preparing for College Entrance Exams

If you plan to attend college, you will most likely be required to take college entrance examinations. Some of these exams attempt to measure the extent to which your verbal and mathematical reasoning skills have been developed. Others test your knowledge of specific subject areas. Usually, the best preparation for college entrance examinations is to follow a strong academic program in school, to study, and to read as extensively as possible.

The following are test-taking strategies that may prove useful.

- Familiarize yourself with the test you will be taking well in advance of the test date. Sample tests, with accompanying explanatory material, are available for many standardized tests. By working through this sample material, you become comfortable with the types of questions and directions that will appear on the test and you develop a feeling for the pace at which you must work in order to complete the test.

- Find out how the test is scored so that you know whether it is advantageous to guess.

- Skim sections of the test before starting to answer the questions in order to get an overview of the questions. You may wish to answer the easiest questions first. In any case, do not waste time on questions you do not understand; go on to those that you do.

- Mark your answer sheet carefully, checking the numbering on the answer sheet about every five questions to avoid errors caused by misplaced answer markings.

- Write in the test booklet if it is helpful; for example, cross out incorrect alternatives and do mathematical calculations.

- Work carefully, but do not take time to double-check your answers unless you finish before the deadline and have extra time.

- Arrive at the test center early and come well prepared with any necessary supplies such as sharpened pencils and a watch.

College entrance examinations that test general reasoning abilities, such as the Scholastic Aptitude Test, usually include questions dealing with basic algebraic concepts and skills. The College Board Achievement Tests in mathematics (Level I and Level II) include many questions on algebra. The following first-year algebra topics often appear on these exams. For each of the topics listed on pages 695–697, a page reference to the place in your textbook where this topic is discussed has been provided. As you prepare for college entrance exams, you may wish to review the topics on these pages.

Types of Numbers (pages 1–2, 11, 179–181, 341)

Positive integers	$\{1, 2, 3, 4, \ldots\}$
Negative integers	$\{-1, -2, -3, -4, \ldots\}$
Integers	$\{\ldots, -4, -3, -2, -1, 0, 1, 2, 3, 4, \ldots\}$
Odd numbers	$\{1, 3, 5, 7, 9, \ldots\}$
Even numbers	$\{0, 2, 4, 6, 8, \ldots\}$
Consecutive integers	$\{n, n + 1, n + 2, \ldots\}$ (n = an integer)
Consecutive even integers	$\{n, n + 2, n + 4, \ldots\}$ (n = even integer)
Consecutive odd integers	$\{n, n + 2, n + 4, \ldots\}$ (n = odd integer)
Prime numbers	$\{2, 3, 5, 7, 11, 13, \ldots\}$

Properties and Axioms (pages 49–100)

Closure Axioms (p. 53)
Commutative Axioms (p. 53)
Associative Axioms (p. 54)
Reflexive Axiom (p. 55)
Symmetric Axiom (p. 55)
Transitive Axiom (p. 55)
Distributive Axiom (pp. 57–58)
Cancellation Property of Opposites (p. 62)
Identity Axioms (pp. 68, 88)
Axiom of Additive Inverses (p. 69)
Property of the Opposite of a Sum (p. 71)
Multiplicative Property of Zero (p. 88)
Multiplicative Property of -1 (p. 89)
Property of Opposites in Products (p. 90)
Axiom of Multiplicative Inverses (p. 94)
Property of the Reciprocal of a Product (p. 95)

Rules for Operations on Positive and Negative Numbers
(pages 72, 90)

If $a > 0$, $b > 0$, then $a + b = |a| + |b|$.
If $a < 0$, $b < 0$, then $a + b = -(|a| + |b|)$.
If $a > 0$, $b < 0$, and $|a| > |b|$, then $a + b = |a| - |b|$.
If $a > 0$, $b < 0$, and $|a| < |b|$, then $a + b = -(|b| - |a|)$.
If $a > 0$, $b < 0$, then $ab < 0$.
If $a > 0$, $b > 0$, then $ab > 0$.

Percents (pages 431–433)

Converting decimals and fractions to percents
Percents greater than 100
Percents less than 1
Percent problems

Solving Equations (pages 31, 118–119, 121–122, 124–126, 134–135, 363–365, 493–494, 499, 548–550)

Transformation by substitution (p. 119)
Transformation by addition (p. 119)
Transformation by subtraction (p. 119)
Transformation by multiplication (p. 122)
Transformation by division (p. 122)
Factoring (pp. 363–365)
$x^2 = k$ (pp. 493, 499)
Quadratic formula (p. 549)
Discriminant (p. 549)

Graphing (pages 3, 6–7, 32, 215–216, 232–238, 557–561, 567–568)

Points on a number line
Inequalities on a number line
Points and lines on a number plane
Inequalities on a number plane
Quadratic functions

Factoring (pages 341–360)

Integers
$a^2 - b^2$
$ax^2 + bx$
$a^2 + 2ab + b^2$
$a^2 - 2ab + b^2$
$ax^2 + bx + c$

Variation (pages 458–469)

Direct variation
Inverse variation
Direct variation involving powers
Inverse variation involving powers
Joint variation
Combined variation

Simultaneous Equations (pages 267–281)

The graphing method
The addition-or-subtraction method
The linear combination method
The substitution method

Rational Expressions (pages 392–417)

Simplification
Addition
Subtraction
Multiplication
Division

Word Problems (pages 35–36, 40–41, 130–132, 139, 144, 179–181, 184–185, 189–191, 196–198, 201, 284–285, 289, 292–295, 302–303, 337, 368–369, 431–433, 440–441, 443–447, 450, 452, 508–509, 552–553)

Age
Angle
Area
Consecutive integers
Cost and value
Digit
Fraction
Investment
Mixture
Motion
Percent
Proportion
Rate-of-work
Ratio
Uniform motion
Wind and Water Current
Without solutions

Glossary

abscissa (p. 216). The first coordinate of an ordered pair of real numbers that is assigned to a point on a coordinate plane. Also called *x-coordinate*.

absolute value (p. 63). The positive number of any pair of opposite nonzero real numbers. The absolute value of 0 is defined to be 0.

addition property of equality (p. 114). For all real numbers a, b, and c: if $a = b$, then $a + c = b + c$ and $c + a = c + b$.

addition property of order (p. 158). For all real numbers a, b, and c:
1. If $a < b$, then $a + c < b + c$ and $c + a < c + b$.
2. If $a > b$, then $a + c > b + c$ and $c + a > c + b$.

additive inverse (pp. 68, 69). For every real number a, the unique real number $-a$ such that $a + (-a) = 0$ and $(-a) + a = 0$. Also called the *opposite of a*.

angle (p. 579). The union of two rays that have the same endpoint. *See also* directed angle.

arithmetic mean (p. 637). The sum of n values divided by n. Also called the *average* or the *mean*.

associated equation (p. 237). The equation of the boundary of the graph of an inequality.

associative axioms (p. 54). For all real numbers a, b, and c:
Addition: $(a + b) + c = a + (b + c)$
Multiplication: $(ab)c = a(bc)$

average (p. 637). *See* arithmetic mean.

axes (singular: **axis**) (p. 215). On a coordinate plane, two perpendicular number lines that intersect at the origin of each. These lines are used for reference in locating points on the plane.

axiom (p. 53). A statement that is assumed to be true. Also called *postulate*.

axiom of comparison (p. 157). For all real numbers a and b, one and only one of the following statements is true: $a < b$, $a = b$, $b < a$

axioms of closure (p. 53). For all real numbers a and b: $a + b$ is a unique real number; ab is a unique real number.

axioms of equality (p. 55). *See* reflexive axiom, symmetric axiom, and transitive axiom.

axis of symmetry (p. 558, 561). A line that divides a curve into two matching parts. For a parabola with equation $y = ax^2 + bx + c$, $a \neq 0$, the axis of symmetry is the line $x = -\dfrac{b}{2a}$.

base (in a power) (p. 24). In an expression such as 3^4, 3 is the base.

basic property of quotients (p. 381). For all real numbers r and s and all nonzero real numbers t and u:
$$\frac{rs}{tu} = \frac{r}{t} \cdot \frac{s}{u}$$

binary operation (p. 52). An operation that pairs any *two* real numbers with a third real number.

binomial (p. 324). A polynomial of two terms.

Boolean algebra (p. 204). An "algebra of logic" in which letters such as p and q are used to represent statements. Truth values are assigned to compound statements produced from operations on the given statements p and q.

boundary of a half-plane (p. 236). *See under* open half-plane.

cancellation property of opposites (p. 62). For all real numbers a, $-(-a) = a$.

Cartesian coordinate system (p. 217). A one-to-one correspondence between the set of all points on a coordinate plane and the set of all ordered pairs of real numbers. Also called *plane rectangular coordinate system*.

closed half-plane (p. 237). The union of an open half-plane and its boundary.

coefficient of a monomial (p. 324). The numerical factor of a monomial that contains one or more variables. Also called *numerical coefficient*.

coefficients of a polynomial (p. 324). The coefficients of the terms of the polynomial.

coincident lines (p. 268). Two lines that have all their points in common.

collinear points (p. 244). Points that lie on the same line.

combined variation (p. 468). A relationship in which a variable varies *directly as* one variable (or a power of the variable) and *inversely as* another variable (or a power of the variable).

commutative axioms (p. 53). For all real numbers a and b:
Addition: $a + b = b + a$
Multiplication: $ab = ba$

complementary angles (p. 184). Two angles the sum of whose measures is 90°. Each angle is called a *complement* of the other.

complete factorization (p. 359). The factorization of a polynomial in which each factor is either a monomial or an irreducible polynomial whose greatest monomial factor is 1.

completing the square (p. 545). Transforming a quadratic expression into a trinomial square.

complex fraction (p. 415). A fraction with a numerator or denominator that contains a fraction or a term with a negative exponent.

conclusion (p. 78). That part of a theorem which states what follows logically from the hypothesis.

conjugates (p. 521). Two binomials that are of the form $a\sqrt{b} + c\sqrt{d}$ and $a\sqrt{b} - c\sqrt{d}$, where b and d are nonnegative real numbers, are called conjugates.

conjunction (p. 170). A sentence formed by joining two sentences with the word *and*.

consecutive even integers (p. 181). The numbers obtained by counting by twos from any given even integer.

consecutive integers (p. 179). The numbers obtained by counting by ones from any given integer.

consecutive multiples of a real number (p. 180). The

numbers obtained by multiplying the given number by consecutive integers.

consecutive odd integers (p. 181). The numbers obtained by counting by twos from any given odd integer.

consistent system of equations (p. 269). A system of equations that has at least one solution.

constant monomial, or **constant** (p. 323). *See under* monomial.

constant of variation (p. 458). *See under* direct variation *and under* indirect variation. Also called *constant of proportionality.*

constant term (p. 353). The numerical term of a polynomial in simplest form.

converse (p. 364). A statement formed by interchanging the hypothesis and conclusion of a given statement.

convex polygonal region (p. 303). A plane region that is the intersection of a finite number of closed half-planes.

convex region (p. 303). A region that contains the line segment drawn between any two points of the region.

coordinate of a point (p. 3). The number that corresponds to a point on a number line.

coordinate axes (p. 216). The x- and y-axes on a coordinate plane.

coordinate plane (p. 216). A plane on which a coordinate system has been set up.

coordinates of a point (p. 216). The pair of real numbers assigned to a point on a coordinate plane.

corner point (p. 303). A point where the boundary lines in a system of inequalities intersect.

cosine of an angle (pp. 585, 594). For an angle in standard position, the abscissa of the point at which the terminal side intersects the unit circle. For an acute angle of a right triangle, the ratio of the length of the side adjacent to the angle to the length of the hypotenuse.

coterminal angles (p. 580). Two or more angles that have the same initial side and the same terminal side.

counting number (p. 11). *See* natural number.

cube root (p. 524). A number b is a cube root of a if $b^3 = a$. The cube root of a is denoted by $\sqrt[3]{a}$.

cubic equation (p. 365). Any equation of the form $ax^3 + bx^2 + cx + d = 0$, where $a \neq 0$.

cumulative frequency (p. 631). In a set of data, the number of measurements that are less than or equal to a given value.

cumulative frequency polygon (p. 631). A broken-line graph joining dots whose ordinates are cumulative frequencies and whose abscissas are the right-hand endpoints of the corresponding intervals.

cumulative percent (p. 631). In a set of data, the percent of measurements that are less than or equal to a given value.

degree of a monomial (p. 323). The total number of times that the variables in a monomial occur as factors. A nonzero constant monomial has degree 0. The monomial 0 has no degree.

degree of a polynomial (p. 324). The greatest of the degrees of the terms of the polynomial after it has been simplified.

degree of a variable in a monomial (p. 323). The number of times that the variable occurs as a factor in the monomial.

density property of rational numbers (p. 484). Between any two different rational numbers, there is another rational number.

difference (p. 82). *See under* subtraction.

difference of cubes (p. 350). A polynomial that can be written in the form $a^3 - b^3$.

difference of squares (p. 336). A polynomial that can be written in the form $a^2 - b^2$.

direct proof (p. 78). The process of starting with the hypothesis of a theorem and arriving at its conclusion through a logical chain of statements.

direct variation (p. 458). Any function defined by an equation of the form $y = kx$, where k is a nonzero constant that is called the *constant of variation.*

directed angle (p. 579). The union of two ordered rays with a common endpoint, called the *vertex* of the angle, together with a rotation from the initial side to the terminal side.

directed numbers (p. 3). Positive and negative numbers.

direction of a vector (p. 604). The measure of the angle of rotation from the positive x-axis to a vector in standard position.

discriminant of a quadratic equation (p. 549). The value of $b^2 - 4ac$ is called the discriminant of the quadratic equation $ax^2 + bx + c = 0$.

disjoint sets (p. 167). Sets that have no members in common.

disjunction (p. 171). A sentence formed by joining two sentences with the word *or*.

distance formula (p. 514). Given any points $P_1(x_1, y_1)$ and $P_2(x_2, y_2)$: $P_1P_2 = \sqrt{(x_2 - x_1)^2 + (y_2 - y_1)^2}$

distributive axiom of multiplication with respect to addition (pp. 57, 58). For all real numbers a, b, and c, $a(b + c) = ab + ac$ and $(b + c)a = ba + ca$.

divisible (p. 386). One polynomial is said to be divisible by another polynomial if their quotient is also a polynomial.

division (p. 99). For all real numbers a and all nonzero real numbers b, the *quotient* denoted as $a \div b$ is defined by:

$$a \div b = a \cdot \frac{1}{b}$$

That is, to divide a by b, multiply a by the reciprocal of b.

division property of equality (p. 115). For all real numbers a and b and all nonzero real numbers c: if $a = b$, then $\frac{a}{c} = \frac{b}{c}$.

domain of a function (p. 221). *See under* function.

domain of a relation (p. 219). The set of all first coordinates of the ordered pairs that form the relation.

domain of a variable (p. 18). The set of numbers that a variable may represent. Also called *replacement set.*

dot frequency diagram (p. 624). A graph of the frequency of measurements that uses a dot to represent each measurement.

element of a set (p. 6). *See* member of a set.

empty set (p. 8). The set that contains no members. Also called *null set*.

equal sets (p. 6). Sets that contain exactly the same members.

equation (p. 30). A mathematical sentence which states that two expressions name the same number.

equation of a line (p. 232). An equation whose solutions give the coordinates of all the points on the line.

equivalent equations (p. 118). Equations that have the same solution set over a given domain.

equivalent expressions (p. 58). Expressions that represent the same number for all values of the variables that they contain.

equivalent inequalities (p. 163). Inequalities that have the same solution set over a given domain.

equivalent systems (p. 274). Systems that have the same solution set over a given domain.

equivalent vectors (p. 608). Vectors that have the same norm and same direction.

evaluate a variable expression (p. 18). To replace each variable in the expression with one of its values and simplify the resulting numerical expression. Also called *finding the value of the expression*.

event (p. 646). A specified subset of the set of all possible outcomes in an experiment.

experimental probability (p. 649). If an experiment is conducted n times, and an event E occurs e of these times, then the experimental probability that E will occur in another trial is $\frac{e}{n}$.

exponent (p. 24). In a power, the number of times that the base occurs as a factor.

exponential form of a power (p. 24). The expression x^n is the exponential form of the nth power of x.

extremes of a proportion (p. 450). In the proportion $\frac{a}{b} = \frac{c}{d}$, a and d are the extremes.

factor (p. 52). In a product, the numbers that are multiplied. For example, a and b are the factors of the product ab.

factor a number (p. 341). To express the number as the product of two or more members of a given set.

factor set (p. 341). The set from which the factors of a given number or expression are selected.

find the value of a variable expression (p. 18). *See* evaluate a variable expression.

finite set (p. 12). A set that is not infinite.

formula (p. 144). An equation that states a relationship among quantities represented by variables.

fractional equation (p. 437). An equation in which a variable appears in the denominator of one or more terms.

frequency of a measurement (p. 624). The number of occurrences of the particular measurement in a set of data.

frequency polygon (p. 627). A broken-line graph that

represents the frequency of measurements over given intervals.

function (pp. 219, 221). A relation in which no two ordered pairs have the same first coordinate. More formally, a function is a pairing that assigns to each member of one set, called the *domain*, exactly one member of a second set, called the *range*, and that assigns each member of the range to at least one member of the domain.

graph of an equation (p. 232). The set of all those points and only those points whose coordinates satisfy the equation.

graph of a number (p. 3). The point that corresponds to a number on a number line.

graph of an open sentence (p. 43). The graph of the solution set of the open sentence.

graph of an ordered pair (p. 216). The point that corresponds to the ordered pair on a coordinate plane.

graph of a relation (p. 220). The set of points on a coordinate plane that correspond to the ordered pairs that form the relation.

graph of a set of numbers (p. 6). The set of points on a number line corresponding to the set of numbers.

greatest common factor (GCF) of monomials (p. 343). The monomial with the greatest degree and the greatest numerical coefficient that is a factor of each of two or more given monomials.

greatest integer function (p. 227). The function $f: x \rightarrow [x]$, where the symbol $[x]$ is used to represent the greatest integer that is less than or equal to the real number x.

greatest monomial factor (p. 346). The greatest common factor of the terms of a polynomial that is in simplest form.

grouping symbol (p. 21). A device that is used to enclose an expression. Examples of grouping symbols include parentheses, brackets, braces, and fraction bars.

histogram (p. 626). A bar graph that represents the frequency of measurements over given intervals.

hypotenuse (p. 505). In a right triangle, the side opposite the right angle.

hypothesis (p. 78). That part of a theorem which states what is assumed to be true.

identity (p. 135). An equation that is true for all values of the variable(s).

identity element for addition (p. 68). Zero. When zero is added to any given real number, the sum is identical to the given number.

identity element for multiplication (p. 88). One. When a given real number is multiplied by one, the product is the given real number.

imaginary numbers (p. 530). Numbers that involve the imaginary unit, i.

imaginary unit, i (p. 530). A square root of -1. That is, $i = \sqrt{-1}$.

inconsistent system of equations (p. 269). A system of equations whose solution set is the empty set.

index (p. 525). In the symbol $\sqrt[n]{a}$, the positive integer n is called the index.

indirect proof (pp. 149, 150). The process of indirect reasoning used to prove a theorem by assuming that the conclusion of the theorem is false until a statement is obtained that contradicts an accepted fact. As a result of this contradiction, the assumption must be incorrect and the conclusion of the theorem must be true.

inequality (p. 30). A mathematical sentence that is formed by placing an inequality symbol between two mathematical expressions.

infinite set (p. 12). A set for which the process of counting its members would never end.

initial side (p. 579). The ray of an angle that is at the starting position of the generating ray.

integer (p. 11). A member of the set $\{\ldots, ^-3, ^-2, ^-1, 0, 1, 2, 3, \ldots\}$.

intersection (p. 167). The set consisting of the members belonging to both of two given sets.

inverse operations (p. 124). Operations that undo each other, such as addition and subtraction.

inverse variation (p. 472). Any function defined by an equation of the form $xy = k$, or $y = \dfrac{k}{x}$, where k is a nonzero constant that is called the *constant of variation*.

irrational number (p. 12). Any number that cannot be expressed as the quotient of two integers. Irrational numbers may be represented by nonterminating, nonrepeating decimals.

irreducible polynomial (p. 355). A polynomial that is not reducible over a given factor set.

joint variation (p. 467). A relationship in which one variable varies directly as the product of two or more other variables.

laws of exponents for division (p. 383). For all positive integers m and n and every nonzero real number a:

1. If $m = n$, then $\dfrac{a^m}{a^n} = 1$.

2. If $m > n$, then $\dfrac{a^m}{a^n} = a^{m-n}$.

3. If $m < n$, then $\dfrac{a^m}{a^n} = \dfrac{1}{a^{n-m}}$.

laws of exponents for multiplication (p. 329). For all real numbers a and b, if m and n are positive integers:
1. $a^m \cdot a^n = a^{m+n}$
2. $(a^m)^n = a^{mn}$
3. $(ab)^m = a^m b^m$

least common denominator (LCD) (p. 410). The least common multiple of the denominators of two or more fractions.

least common multiple (LCM) of integers (p. 342). The least positive integer that is a multiple of each of two or more given integers.

least common multiple (LCM) of monomials (p. 343). The monomial with the least degree and the least positive numerical coefficient that is a multiple of each of two or more given monomials.

least common multiple (LCM) of polynomials (p. 410). The polynomial of least degree and least positive constant factor that has each of two or more given polynomials as a factor.

like monomials (p. 324). *See* similar monomials.

linear combination (p. 278). The equation obtained by multiplying one equation of a system by a nonzero constant and another equation of the system by another nonzero constant and adding or subtracting the two resulting equations.

linear equation in three variables (p. 307). For the variables x, y, and z, any equation that is of the form $ax + by + cz = d$, where a, b, c, and d are real numbers and a, b, and c are not all zero. In space, the graph of any equation of this form is a plane.

linear equation in two variables (p. 232). For the variables x and y, any equation that is of the form $ax + by = c$, where a, b, and c are real numbers and a and b are not both zero. The graph of any equation of this form is a line.

linear function (p. 234). A function whose ordered pairs satisfy a linear equation.

linear inequality in two variables (p. 237). An inequality whose associated equation is a linear equation in two variables.

linear programming (p. 302). A method of solving problems in which a quantity represented by a linear equation is to be maximized or minimized subject to conditions expressed by a system of linear inequalities.

linear term (p. 353). A term of degree one in a quadratic polynomial in simplest form.

mathematical expression (p. 18). A numerical or variable expression.

mathematical sentence (p. 30). A statement that indicates a relationship between two mathematical expressions.

maximum point of a curve (p. 557). A point on the curve whose y-coordinate is greater than or equal to the y-coordinate of every other point on the curve.

maximum value of a function (p. 557). The y-coordinate of the maximum point of the graph of the function.

mean (p. 637). *See* arithmetic mean.

means of a proportion (p. 450). In the proportion $\dfrac{a}{b} = \dfrac{c}{d}$, b and c are the means.

measure of an angle (p. 580). The number of degrees through which a ray rotates from the initial side to the terminal side of the angle.

median (p. 637). In an ordered set of n values, the median is the middle entry if n is odd, and is half the sum of the two middle entries if n is even.

member of a set (p. 6). Any object in the set. Also called *element*.

minimum point of a curve (p. 557). A point on the curve whose y-coordinate is less than or equal to the y-coordinate of every other point on the curve.

minimum value of a function (p. 557). The y-coordinate of the minimum point of the graph of the function.

mode (p. 637). In a set of data, the value that occurs with greatest frequency.

monomial (p. 323). An expression that is either a numeral, a variable, or an indicated product of a numeral and one or more variables. A *constant monomial*, or *constant*, contains no variable.

multiple (p. 180). The product of any real number and an integer is called a multiple of that real number.

multiplication property of equality (p. 114). For all real numbers a and b: if $a = b$, then $ac = bc$ and $ca = cb$.

multiplication property of order (p. 159). For all real numbers a, b, and c:
1. If $a < b$ and $c > 0$, then $ac < bc$ and $ca < cb$.
   If $a > b$ and $c > 0$, then $ac > bc$ and $ca > cb$.
2. If $a < b$ and $c < 0$, then $ac > bc$ and $ca > cb$.
   If $a > b$ and $c < 0$, then $ac < bc$ and $ca < cb$.

multiplicative inverse (pp. 93, 94). For all nonzero real numbers a, the unique real number $\frac{1}{a}$ such that $a \cdot \frac{1}{a} = 1$ and $\frac{1}{a} \cdot a = 1$. Also called *reciprocal*.

natural number (p. 11). A member of the set $\{1, 2, 3, \ldots\}$. Also called *counting number*.

negative number (p. 2). A number that corresponds to a point on the negative side of a number line.

norm of a vector (p. 604). The length, or magnitude, of a vector.

nth root (p. 524). If n is a positive integer and a is a real number, any real number whose nth power equals a is called an nth root of a.

null set (p. 8). *See* empty set.

number line (p. 1). A line for which numbers have been placed in correspondence with points on the line.

numerical coefficient (p. 324). *See* coefficient of a monomial.

numerical expression (p. 17). A name for a number. Also called *numeral*.

one-to-one correspondence (p. 6). A pairing of the members of two sets that assigns to each member of each set one and only one member of the other set.

open half-plane (p. 236). Either of two regions into which a line separates a coordinate plane. The line is called the *boundary* of each half-plane.

open sentence (p. 31). A mathematical sentence that contains at least one variable. Also called *open mathematical sentence*.

opposite of a number (p. 62). *See* additive inverse.

order of operations (p. 25). The steps to be followed in order when simplifying expressions.

ordered pair (p. 216). A pair of numbers in which the order of the numbers is important.

ordered triple (p. 306). A group of three numbers in which the order of the numbers is important.

ordinate (p. 216). The second coordinate of an ordered pair of real numbers that is assigned to a point on a coordinate plane. Also called *y-coordinate*.

origin (pp. 2, 215). On a number line, the point that corresponds to the number zero. On a coordinate plane, the point of intersection, labeled O, of the axes.

parabola (p. 557). A smooth curve that is the graph of a

quadratic function with domain $\mathcal{R}$.

parallel lines (p. 249). Lines in a plane that do not intersect.

percent (p. 431). Percent means "per 100" or "divided by 100." The symbol for percent is %.

perfect square (p. 494). Any number that can be expressed as the square of a rational number.

plane rectangular coordinate system (p. 217). *See* Cartesian coordinate system.

plotting a point (p. 216). To find the point on a coordinate plane that corresponds to an ordered pair. Also called *graphing an ordered pair*.

point of intersection (p. 268). The one point that is common to two or more lines.

polynomial (p. 324). A monomial or a sum of monomials.

polynomial equation (p. 365). Any equation that can be written equivalently with 0 as one side and a polynomial as the other side.

polynomial function (p. 564). Any function of the form $x \to P(x)$, where $P(x)$ represents a polynomial.

positive number (p. 2). A number that corresponds to a point on the positive side of a number line.

postulate (p. 53). *See* axiom.

power (p. 24). The number named by an expression of the form a^n which represents the product of n factors equal to a. For example, $3^4 = 3 \times 3 \times 3 \times 3 = 81$.

prime factorization (p. 341). The expression of a positive integer as the product of prime factors.

prime number (p. 341). An integer greater than 1 that has no positive integral factor other than itself and 1. Also called *prime*.

principal nth root (p. 524). If n is an even integer, the positive nth root of a number is called the principal nth root.

principal square root (p. 493). The positive square root.

probability (p. 645). The measure of the chance that a particular outcome will occur.

product (p. 52). The number that is paired with two real numbers a and b by the operation of multiplication. This product is denoted by ab.

property of completeness (p. 489). Each decimal represents a real number, and every real number can be represented as a decimal.

proportion (p. 450). An equation which states that two ratios are equal.

Pythagorean numbers (p. 508). Any set of positive integers that satisfies the equation $c^2 = a^2 + b^2$. Also called a *Pythagorean triple*.

Pythagorean Theorem (p. 505). In any right triangle, the square of the length of the hypotenuse is equal to the sum of the squares of the lengths of the other two sides.

quadrant (p. 216). One of the four regions into which a coordinate plane is separated by the coordinate axes.

quadrantal angle (p. 580). An angle in standard position whose terminal side coincides with a coordinate axis.

quadratic equation (p. 363). Any equation that can be written equivalently in the form $ax^2 + bx + c = 0$, where $a \neq 0$.

quadratic formula (pp. 549–550). The solutions of the equation $ax^2 + bx + c = 0$, $a \neq 0$, are given by the formula

$$x = \frac{-b \pm \sqrt{b^2 - 4ac}}{2a}.$$

quadratic function (p. 557). Any function of the form $f: x \rightarrow ax^2 + bx + c$, $a \neq 0$.

quadratic inequality (p. 567). An inequality whose associated equation is a quadratic equation.

quadratic polynomial (p. 353). A polynomial of the form $ax^2 + bx + c$, $a \neq 0$.

quadratic term (p. 353). A term of degree two in a quadratic polynomial that is expressed in simplest form.

quadratic trinomial (p. 354). A polynomial that can be expressed in the form $ax^2 + bx + c$, $a, b, c \neq 0$.

quantifier (p. 50). A word or phrase that is used in combination with a variable in an open sentence to convey the idea of how many, or quantity. For example, *any*, *some*, and *at least one* are quantifiers.

quotient (p. 99). *See under* division.

radical (p. 493). When a mathematical expression appears under a radical sign, the entire expression is called a radical.

radical sign (p. 493). The symbol $\sqrt[n]{}$ in an expression like $\sqrt[n]{a}$.

radicand (p. 493). Any numeral or expression under a radical sign.

range of a function (p. 221). *See under* function.

range of a relation (p. 219). The set of all the second coordinates of the ordered pairs in the relation.

range of a set of data (p. 640). The difference between the greatest and least values in the set.

ratio (p. 449). The quotient of one number divided by another, provided that the divisor is not zero.

rational expression (p. 390). Any expression that can be written as the quotient of two polynomials, provided that the denominator is not zero.

rational number (p. 12). Any number that can be expressed as the quotient of two integers, provided that the denominator is not zero.

rationalizing the denominator (p. 518). The process of expressing a fraction with an irrational denominator as an equivalent fraction with a rational denominator.

real number (p. 3). A number that is either a positive number, a negative number, or zero.

reciprocal (p. 93). *See* multiplicative inverse.

reducible polynomial (p. 355). A polynomial that can be expressed as a product of two or more polynomials of lower positive degree taken from a given factor set.

reflection (p. 372). A transformation, or mapping, of a plane in which each point on the plane is transformed into its mirror image across an *axis of reflection*.

reflexive axiom of equality (p. 55). For all real numbers a, $a = a$.

relation (p. 219). Any set of ordered pairs.

relative frequency of a measurement (p. 624). In a set of data, the frequency of the measurement divided by the total number of measurements.

repeating decimal (p. 486). A nonterminating decimal in which the same block of digits *repeats* without end.

replacement set (p. 18). *See* domain of a variable.

resultant of vectors (p. 608). The vector whose x-component is the sum of the x-components of two or more given vectors and whose y-component is the sum of the y-components of the given vectors. Also called *sum of vectors*.

right angle (p. 184). An angle whose measure is 90°.

root of an open sentence (p. 31). *See* solution of an open sentence.

roster (p. 6). A list of all the members of a set.

rule (p. 6). A description that identifies the members of a set.

satisfy (p. 31). Each member of the solution set of an open sentence is said to satisfy the open sentence.

scientific notation (p. 399). A notation used to express any positive number in the form $k \times 10^n$, where $1 \leq k < 10$ and n is an integer.

set (p. 6). A collection of objects.

set-builder notation (p. 32). A notation used in describing a set. For example, $\{x: x < 5\}$ is written in set-builder notation.

side of an equation (p. 30). One of two mathematical expressions that are joined by an equals sign.

side of an inequality (p. 30). One of two mathematical expressions that are joined by an inequality symbol.

similar monomials (p. 324). Monomials that are identical or that differ only in their numerical coefficients. Also called *like monomials*.

simple fraction (p. 415). A fraction free of negative exponents and fractions in the denominator or numerator.

simplest form of a fraction (p. 393). That form of the fraction in which the greatest common factor (GCF) of the numerator and denominator is 1.

simplest form of a polynomial (p. 324). That form of the polynomial in which no two of its terms are similar.

simplest form of a radical (p. 518). An expression that contains a radical is in simplest form when:
1. no integral radicand has a perfect square factor other than 1,
2. no fractions are under a radical, and
3. no radicals are in a denominator.

simplify a numerical expression (p. 17). To replace a numerical expression with the simplest, or most common, name of its value.

simplify a variable expression (p. 58). To replace a variable expression with an equivalent expression that has as few terms as possible.

simultaneous linear equations (p. 267). *See* system of linear equations.

sine of an angle (pp. 585, 594). For an angle in standard position, the ordinate of the point at which the terminal side intersects the unit circle. For an acute angle of a right triangle, the ratio of the length of the side opposite the angle to the length of the hypotenuse.

slope of a line (p. 243). The steepness of a nonvertical line as defined by the ratio:
$$\frac{\text{difference of ordinates}}{\text{difference of abscissas}}.$$
More formally, if (x_1, y_1) and (x_2, y_2) are any two different points on the line, then
$$\text{slope} = \frac{y_2 - y_1}{x_2 - x_1} \quad (x_1 \neq x_2).$$
A horizontal line has slope 0; a vertical line has no slope.

slope-intercept form of a linear equation (p. 249). $y = mx + b$, where m is the slope of the line represented by the equation and b is its y-intercept.

solution of an open sentence (p. 31). Any value of the variables for which the open sentence is true. A solution of an open sentence in two variables is an ordered pair; a solution of an open sentence in three variables is an ordered triple. Also called *root*.

solution of a system (p. 267). A solution of a system of linear equations in two variables is an ordered pair that satisfies each equation in the system. A solution of a system of linear equations in three variables is an ordered triple.

solution set of an open sentence (p. 31). The set of all solutions of the open sentence.

solution set of a system (p. 267). The set of all solutions of the system.

solve an open sentence (p. 31). To determine the solution set of the open sentence over a given domain.

solve a triangle (p. 595). To find measures, or approximations to measures, of angles or sides of a triangle when measures of other angles or sides of the triangle are given.

square root (p. 493). A number b is a square root of a number a if $b^2 = a$. The positive square root of a is denoted by $\sqrt{a}$, the negative square root by $-\sqrt{a}$.

standard deviation (p. 641). In a set of data, the nonnegative square root of the variance.

standard position of an angle (p. 580). On a coordinate plane, the position of the angle with its vertex at the origin and the positive x-axis as its initial side.

subset (p. 7). If every member of a set S is also a member of a set T, then S is a subset of T.

substitution method (p. 281). A method for solving a system of linear equations in two variables by:
(1) solving one equation for one of the variables,
(2) substituting this expression in the other equation and solving this equation,
(3) finding the corresponding value of the other variable, and
(4) checking the solution in both original equations.

substitution principle (p. 17). Changing the numeral by which a number is named in an expression does not change the value of the expression.

subtraction (p. 82). For all real numbers a and b, the *difference* denoted as $a - b$ is defined by
$$a - b = a + (-b).$$
That is, to subtract b from a, add the opposite of b to a.

subtraction property of equality (p. 114). For all real numbers a, b, and c: if $a = b$, then $a - c = b - c$.

supplementary angles (p. 184). Two angles the sum of whose measures is 180°. Each angle is called a *supplement* of the other.

sum (p. 52). The number that is paired with two real numbers a and b by the operation of addition. This sum is denoted as $a + b$.

sum of cubes (p. 350). A polynomial that can be written in the form $a^3 + b^3$.

sum of vectors (p. 608). *See* resultant of vectors.

symmetric axiom of equality (p. 55). For all real numbers a and b, if $a = b$, then $b = a$.

system of linear equations (p. 267). A set of two or more linear equations in the same variables. Also called a set of *simultaneous linear equations*.

tangent of an angle (pp. 585, 594). For an angle in standard position whose terminal side intersects the unit circle at the point (p, q), with $p \neq 0$, the ratio $\frac{q}{p}$.
For an acute angle of a right triangle, the ratio of the length of the side opposite the angle to the length of the side adjacent to the angle.

tautology (p. 206). A compound statement that is true for all truth values of its component statements.

terminal side (p. 579). The ray of an angle that is at the final position of the generating ray.

terminating decimal (p. 486). A decimal in which the division process *terminates* when a final remainder of zero is reached.

term of a polynomial (p. 324). A monomial in the expression for the polynomial.

terms (p. 52). In a sum, the numbers that are added. For example, a and b are the terms of the sum $a + b$.

theorem (p. 78). A statement that can be proved to be true.

transform an equation (p. 118). To change an equation into an equivalent equation.

transformation Each of the following always produces an equation equivalent to the original equation:
by addition (p. 119): Adding the same real number to each side of a given equation.
by division (p. 122): Dividing each side of a given equation by the same nonzero real number.
by multiplication (p. 122): Multiplying each side of a given equation by the same nonzero number.
by substitution (p. 119). Substituting for any expression in a given equation an equivalent expression.
by subtraction (p. 119). Subtracting the same real number from each side of a given equation.

transitive axiom of equality (p. 55). For all real numbers a, b, and c, if $a = b$ and $b = c$, then $a = c$.

transitive property of order (p. 158). For all real numbers a, b, and c:
1. If $a < b$ and $b < c$, then $a < c$.
2. If $a > b$ and $b > c$, then $a > c$.

translation of a plane (p. 256). A *transformation*, or *mapping*, of the plane in which the axes remain fixed while the points "slide" to new locations.

trigonometric functions (p. 586). A set of functions that includes the sine, cosine, and tangent functions. The domain of each trigonometric function is a set of angles, and the range is a set of real numbers. For example, $\sin 90° = 0$.

trinomial (p. 324). A polynomial of three terms.

trinomial square (p. 335). A trinomial obtained by squaring a binomial.

uniform motion (p. 189). An object that moves at a constant speed, or rate, is said to be in uniform motion.

union (p. 167). The set consisting of the members belonging to at least one of two given sets.

unit circle (p. 585). A circle whose radius is one unit long and whose center is at the origin of a coordinate plane.

value of a function (p. 225). A member of the range of the function.

value of a numerical expression (p. 17). The number named by the expression.

value of a variable (p. 18). A number in the domain of the variable.

variable (p. 18). A symbol that is used to represent one or more numbers.

variable expression (p. 18). An expression that contains a variable.

variance (p. 641). In a set of data, the average of the squares of the deviations, or differences, of all the measurements from their arithmetic mean.

vector (p. 604). An arrow indicating both magnitude and direction, with an initial point at the origin and a terminal point at the tip of the arrowhead.

Venn diagram (p. 8). A diagram that shows how certain sets are related.

vertex of an angle (p. 579). *See under* directed angle.

vertex of a parabola (p. 558). The maximum or minimum point of the parabola.

whole number (p. 11). A member of the set $\{0, 1, 2, 3, \ldots\}$.

x-**axis** (p. 216). Usually, the horizontal axis on a coordinate plane.

x-**component of a vector** (p. 605). The length of the horizontal displacement of the vector from its initial point to its terminal point.

x-**coordinate** (p. 216). *See* abscissa.

y-**axis** (p. 216). Usually, the vertical axis on a coordinate plane.

y-**component of a vector** (p. 605). The length of the vertical displacement of the vector from its initial point to its terminal point.

y-**coordinate** (p. 216). *See* ordinate.

y-**intercept** (p. 249). The ordinate of the point at which a line intersects the *y*-axis.

zero of a function (p. 559). Any member of the domain of the function for which the value of the function is zero.

zero-product property (p. 364). For all real numbers a and b, $ab = 0$ if and only if $a = 0$ or $b = 0$.

Index

Abscissa, 216
Absolute value(s), 63
  in open sentences, 174–176
Addition
  associative axiom for, 54
  axiom of closure, 53
  commutative axiom for, 53
  distributive axiom of multiplication with respect to, 57
  of expressions containing radicals, 520–521, 525
  identity axiom for, 68
  identity element for, 68
  on a number line, 67–69
  of polynomials, 323–325
  of rational expressions, 407–413
  rules for, 71–72
  used to solve systems of linear equations, 274–275
  theorems about, 78–80
  transformation by, 119
Addition property of equality, 114
Addition property of order, 158
Additive inverse(s), 68
  axiom of, 69
Angle(s), 579
  adjacent right, 184
  complementary, 184
  coterminal, 580
  directed, 579
  initial side of an, 579
  measure of, 580
  problems about, 184–185
  quadrantal, 580
  right, 184
  standard position of, 580
  supplementary, 184
  terminal side of an, 579
  of a triangle, 184
  vertex of an, 579
Applications
  acceleration, 212–213
  days of the week, 264–265
  estimating wildlife populations, 658–659
  gravity, 479
  metric prefixes, 425
  raising your sights, 540–541
  temperature scales, 110–111
Approximations
  of decimals, 488
  of irrational square roots, 497–498
Area problems, 339–340
Arithmetic mean, 637
Arrow notation for a function, 224

Associated equation, 237
Associative axioms of addition and multiplication, 54
Augmented matrix, 313
Average, 637
Average rate, 189
Axes, 215
  coordinate, 216
  x-, 216
  y-, 216
Axiom(s), 53
  of additive inverses, 69
  associative, 54
  of closure, 53
  commutative, 53
  of comparison, 157
  distributive, of multiplication with respect to addition, 57
  of equality
    reflexive, 55
    symmetric, 55
    transitive, 55
  identity
    for addition, 68
    for multiplication, 88
  of multiplicative inverses, 94
Axis of reflection, 372
Axis of symmetry, 558

Bar graph, 626
Base
  in a percentage, 431
  in a power, 24
BASIC
  programming: *See* Programming in BASIC
  summary of, 686–993
Bearing of a vector, 605
Between, 483
Binary operation, 52
Binomials, 324
  containing radicals, 521–522
  multiplying mentally, 335–337
  squaring, 335
Boole, George, 204
Boolean algebra, 204–206
Boundary(ies)
  of a half-plane, 236–238
  of a set, 7
Braces, 6, 21
Brackets, 21
BROWNE, MARJORIE LEE, 10

Calculator, *see* On the Calculator
Cancellation property of opposites, 62

Careers
  accounting, 142
  aerospace engineering, 312
  architecture, 457
  astronomy, 612
  broadcasting, 188
  oceanography, 16
  pharmacy, 388
  transportation engineering, 503
Cartesian coordinate system, 217
Chapter Review, *see* Reviews
Chapter Summary, *see* Summaries
Chapter Test, *see* Tests
Charts used in problem solving, 139
CHU SHIH-CHIEH, 556
Closed half-plane, 237
Closure, axioms of
  addition, 53
  multiplication, 53
Coefficient(s)
  of a monomial, 324
  numerical, 324
  of a polynomial, 324
  of x, 232
  of y, 232
Coefficient matrix, 313
Coincident lines, 268
College Entrance Exams, *see* Preparing for College Entrance Exams
Collinear points, 244
Combined variation, 468
Commutative axioms of addition and multiplication, 53
Comparison, axiom of, 157
Complementary angles, 184
Complete factorization, 359
Completeness, property of, 489
Completing the square, 545
Complex fraction, 415
Components of a vector, 605
Computer Exercises, 24, 39, 66, 87, 93, 103, 129, 137, 163, 166, 183, 200, 223, 231, 252, 273, 277, 331, 345, 353, 396, 402, 413, 430, 435, 456, 501, 512, 515, 548, 551, 564, 584, 592, 611, 635, 643, 652
Conclusion, 78
Conditional, 204–205
Conjugates, 521
Conjunction, 170
Consecutive integers, 179
  even, 181; odd, 181
Consecutive multiples, 180

Order
addition property of, 158
multiplication property of, 159
of operations, 25
of real numbers, 3
transitive property of, 158
Ordered pair(s), 215, 216
graphing of, 216–217
as a solution of an open sentence in two variables, 228
Ordered triple, 306
Ordinate, 216
Origin
of a coordinate plane, 216
on a number line, 2
of a number plane, 215

Parabola, 557
axis of symmetry of, 558, 560–561
vertex of, 558
Parallel lines, 249
Parallelogram, 184, 613
Parentheses, 21
Percent, 431
cumulative, 631
problems involving, 431–433
Percentage, 431
Perfect square, 494
Plane(s)
coordinate, 216
mapping of, 256
number, 215
reflection of, 372–373
rotation of, 373
transformation of, 256
translation of, 256
Plane rectangular coordinate system, 217
Plotting of points, 216–217
Point(s)
collinear, 244
coordinate(s) of, 3, 216
corner, 303
determining an equation of a line through two, 253
distance between two, 512–514
of intersection, 268
plotting of, 216–217
Polygon
cumulative frequency, 631
frequency, 627
Polygonal region, convex, 303
Polynomial(s), 324
addition of, 323–325
coefficients of, 324
degree of, 324
divisible, by another polynomial, 386
division of, by monomial, 386
division of, by polynomial, 389–390
factoring of, 346, 350–351, 353–

356, 358–360
greatest monomial factor of, 346
irreducible, 355
multiplication of, 332–333
quadratic, 353
quotients of, 389–390
reducible, 355
simplest form of, 324
subtraction of, 323–325
term of, 324
Polynomial equation, 365
Polynomial function, 564
graphing of, 564–565
Positive number, 1, 2
Positive side of a number line, 2
Postulate, 53
Power(s), 24
exponential form of, 24
of a monomial, 329
of a power, 328
of a product, 329
product of, 328
quotient of, 383
Preparing for College Entrance Exams, 109, 211, 320, 424, 539, 620
Appendix, 694–697
Prime, 341
Prime factor, 341
Prime factorization, 341
Prime number, 341
Principal nth root, 524
Principal square root, 493
Probability, 645
experimental, 649
Problem(s)
addition, 74–76
angle, 184–185
area, 339–340
digit, 292–295
involving factoring, 369–371
formulas in, 144
involving inequalities, 179–183
integer, 179–181
linear programming, 302–304
mixture, 196–198
motion, 446–447
number, 440–441
percent, 431–433
perimeter, 140
involving Pythagorean Theorem, 508–509
involving quadratic equations, 552–553
involving ratios and proportions, 454–456
subtraction, 86–87
involving trigonometry, 599–600
uniform motion, 189–191
involving variation, 461, 465, 470

involving vectors, 614
wind and water current, 289–291
without solutions, 201
word, solving of, 130–132
work, 443–445
Problem solving
addition in, 74–76
using charts in, 139
using equations, 130–132
by factoring, 368–369
formulas used for, 144
using inequalities, 179–180
procedure for, 504
using quadratic equations, 552–553
subtraction in, 86–87
using three variables, 310
using trigonometry, 599–600
using two variables, 284–285
Product(s), 52
of binomials, 335–336
of expressions containing radicals, 521
of monomials, 328–329
opposites in, 90
of polynomials, 332
of powers, 328
power of a, 329
of rational expressions, 403–404
reciprocal of a, 95
Product property of square roots, 498, 516–517
Programming in BASIC, 34, 76, 138, 195–196, 247, 301, 362–363, 414–415, 466–467, 491–492, 566, 602–603, 644–645
Proof
direct, 78
indirect, 148–150
Property(ies)
cancellation, of opposites, 62
of completeness, 489
density, of rational numbers, 484
of equality, 113–115
multiplicative, of −1, 89
multiplicative, of zero, 88
of opposites in products, 90
of the opposite of a sum, 71
of order, 158–159
product, of square roots, 498
quotient, of square roots, 517
of quotients, 381
of the reciprocal of a product, 95
zero-product, 363–365
Proportion, 450
extremes of, 450
means of, 450
Proportionality, constant of, 458
Proving statements, 78–80

of an inequality, 30
of a triangle, 505, 594–595
Similar monomials, 324
Simple fraction, 415
Simple interest, 432
Simplest form
of a fraction, 393
of a polynomial, 324
of a radical, 518
Simplest form of a radical, 518
Simplifying
numerical expressions, 17
polynomials, 324
expressions containing radicals, 494, 498–499, 517–518, 520–521, 522, 525
rational expressions, 392–394
variable expressions, 58
Simultaneous equations, 267
Sine of an angle, 585
Slope, 243
Slope-intercept form of a linear equation, 248–250
Solution set
of a linear equation in three variables, 306
of an open sentence
in one variable, 31
in two variables, 228
of a system of inequalities, 298, 568
of a system of linear equations
in three variables, 307
in two variables, 267
Solving equations
by factoring, 363–365
fractional, 437–438
by inspection, 118
quadratic equations, 362–365, 543–550
system of linear equations
algebraically, 274, 277, 280, 307
graphically, 267–270
by matrices, 314
by transformations, 118–119, 121–122, 281
having the variable in both sides, 134–135
having whole-number denominators, 427–428
Solving inequalities, 165
quadratic, 567–568
systems, 298–299, 568
with whole-number denominators, 427–428
Solving open sentences
in one variable, 31
in two variables, 228–229
with whole-number denominators, 427–428
Solving triangles, 593–595
Square(s)

of a binomial, 335
completing the, 545
difference of, 336
perfect, 494
table of, 682
trinomial, 335
Square root(s), 493
irrational, 497–499
principal, 493
product property of, 498
quotient property of, 517
rational, 493–494
rational approximations of irrational, 497–498
table of, 683
Standard deviation, 641
Standard position
of an angle, 580
of a vector, 604
Statistics, 623
Subset, 7
Substitution, transformation by, 119
Substitution method, 280–281
Substitution principle, 17
Subtraction
definition of, 82
of expressions containing radicals, 520–521, 525
of polynomials, 323–325
of rational expressions, 407–413
used to solve systems of linear equations, 274–275
transformation by, 119
Subtraction property of equality, 114
Sum, 52
of cubes, 350
of expressions containing radicals, 520–521, 525
opposite of a, 71
of polynomials, 325
of rational expressions, 407–413
of vectors, 608
Summaries, Chapter, 44–45, 104–105, 150–151, 207, 258–259, 315, 374, 419–420, 472–473, 533, 572, 615–616, 653
Summary of BASIC, 686–693
Supplement, 185
Supplementary angles, 184
Symbol(s)
equality, 6, 30
grouping, 21–22
inequality, 3, 30
list of, x
Symbolic logic, 204
Symmetric axiom of equality, 55
Symmetry, axis of, 558
System(s)
consistent, 269

coordinate, 217
equivalent, 274
inconsistent, 269
of linear equations, 267
graphing of, 267–270
in three variables, 306–308
solution of, 267–269, 274–275, 277–278, 280–281, 306–308
solution set of, 267
of linear inequalities, graphs of, 298–299

Tables
of metric units of measure, 681
of formulas, 681
of square roots, 683
of squares, 682
of symbols, x
of trigonometric functions, 684
Tangent of an angle, 585
Tautology, 206
Term, 52
constant, 353
linear, 353
of a polynomial, 324
quadratic, 353
Terminal side of an angle, 579
Terminating decimal, 486
Tests
Chapter, 46–47, 107, 153, 209–210, 261–262, 318, 376–377, 422, 476, 536, 574–575, 617, 655
Self-, 15, 27, 43, 66, 87, 103, 129, 148, 178, 203, 227, 240, 255, 283, 297, 311, 340, 361, 371, 392, 402, 419, 436, 449, 472, 502, 516, 530, 556, 569, 601, 615, 636, 643, 652–653
See also Extra Practice Reviews
Theorem(s), 78
about addition, 78–80
converse of, 364
Pythagorean, 505
Transformation(s)
by addition, 119
by division, 122
of a formula, 144
by multiplication, 122
of a plane, 256
that produce an equivalent inequality, 163
that produce an equivalent system of linear equations, 281
rigid, 373
by substitution, 119
by subtraction, 119
Transitive axiom of equality, 55
Transitive property of order, 158
Translation, 256
equations of, 256

Credits

Book designed by The Quarasan Group, Inc.
Cover concept and design by Lehman Millet, Inc.
Mechanical art by ANCO
Portrait art created by Nathan Goldstein (pages 10, 61, 240, 348, 396, 436, 556, 607)

Photos

page xii: © David Cupp 1983 / THE STOCK BROKER
page 16: C. B. Jones / TAURUS PHOTOS
page 43: S. J. Krasemann / PETER ARNOLD, INC.
page 48: Horst Schafer / PETER ARNOLD, INC.
page 86: Jerome Friar / SOUTHERN LIGHT
page 110: © Peter Menzel MCMLXXX
page 112: NASA
page 140: Russell A. Thompson / TAURUS PHOTOS
page 142: © Joseph Nettis / PHOTO RESEARCHERS, INC.
page 156: Hugh Rogers / Monkmeyer Press Photo Service, Inc.
page 188: © Peter Menzel
page 192: Jeff Dunn / THE PICTURE CUBE
page 212: © Melanie Wall / SOUTHERN LIGHT
page 214: © Frank Siteman / THE PICTURE CUBE
page 264: NASA photo, from Carlin / THE PICTURE CUBE
page 266: © Peter Menzel MCMLXXXIV
page 287: © Melanie Wall / SOUTHERN LIGHT
page 291: © Richard Wood 1981 / THE PICTURE CUBE
page 312: © Peter Menzel / STOCK BOSTON
page 322: Courtesy of The German Information Center
page 370: Harold M. Lambert / E. P. JONES, INC.
page 380: Harvard College Observatory
page 388: © Michal Heron / Monkmeyer Press Photo Service, Inc.

page 425: Courtesy of The California Institute of Technology
page 426: © Tom McHugh / PHOTO RESEARCHERS, INC.
page 446: John W. Manos / THE PICTURE CUBE
page 457: Mimi Forsyth / Monkmeyer Press Photo Service, Inc.
page 479: © Abraham Menashe 1980 / PHOTO RESEARCHERS, INC.
page 480: © James Nachtwey 1981 / BLACK STAR
page 503: © Richard Wood / THE PICTURE CUBE
page 511: © Ulrike Welsch / STOCK BOSTON
page 541: GRANT HEILMAN PHOTOGRAPHY
page 542: © David Plowden 1974 / PHOTO RESEARCHERS, INC.
page 555: © Frank Siteman MCMLXXXIII / THE PICTURE CUBE
page 578: © Frank Siteman MCMLXXX / THE PICTURE CUBE
page 612: Courtesy of Dr. Martha L. Hazen / Harvard College Observatory
page 614: © Catherine Noren / PHOTO RESEARCHERS, INC.
page 622: © J. J. Languepin 1977 / PHOTO RESEARCHERS, INC.
page 635: © Ellis Herwig / THE PICTURE CUBE
page 659: © Arthur Grace / STOCK BOSTON

Answers to Selected Exercises

Chapter 1 Numbers and Variables

Written Exercises, pages 4–5
1. 30, ⁻30, thirty losses **3.** 5000, ⁻5000, a loss of $5000 **5.** 3, ⁻3, three floors down **7.** 400, ⁻400, 400 m below sea level **9.** 90, ⁻90, 90 km west

11.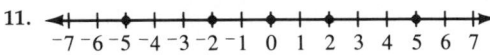

13.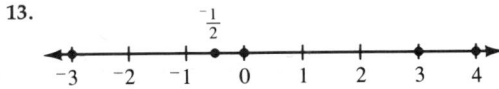

15.
(line graph with −1.5, 0.5, 3.5 marked)

17.
(line graph with $-2\frac{1}{4}$, $\frac{-3}{4}$, $\frac{1}{4}$, $3\frac{1}{4}$ marked)

19. < **21.** > **23.** > **25.** > **27.** <, < **29.** >, >
31. <, < **33.** ⁻9, ⁻7, ⁻3, 0, 4, 5 **35.** $\frac{-1}{2}, \frac{-1}{3}, \frac{-1}{4}, \frac{-1}{5},$
$\frac{-1}{6}, \frac{-1}{10}$ **37.** ⁻2.5, ⁻2, ⁻1.5, 0, 4.5, 5.5 **39.** 5
41. $1\frac{1}{2}$ **43.** $\frac{-1}{2}$ **45.** $5\frac{1}{3}$ **47.** $4\frac{2}{5}$ **49.** $\frac{1}{3}$ **51.** 4
53. ⁻3

Written Exercises, pages 9–10
1. ∈ **3.** ⊂ **5.** ∈ **7.** ⊂

9.
(line graph −7 to 7)

11.
(line graph with $-3\frac{1}{2}$, $\frac{-1}{2}$, $2\frac{1}{2}$, $4\frac{1}{2}$ marked; −6 to 6)

13.
(line graph 0 to 14)

15.
(line graph −12 to 0)

17.
(line graph −12 to 12)

19.
(line graph −4 to 10)

21.
(line graph −2 to 5)

23.
(line graph −4 to 10)

25.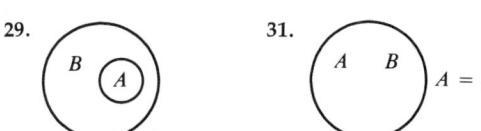
(line graph 0 to 6)

27.
(line graph −3 to 3)

29.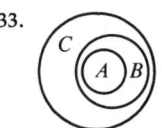

31.

33.
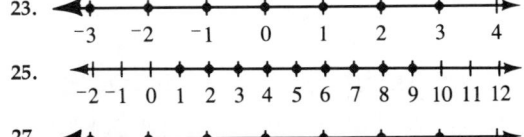

35. {a}, {b}, {c} **37.** {a, b, c} **39.** 8 **41.** {a}, {b}, {c}, {d}, {e}, {a, b}, {a, c}, {a, d}, {a, e}, {b, c}, {b, d}, {b, e}, {c, d}, {c, e}, {d, e}, {a, b, c}, {a, b, d}, {a, b, e}, {a, c, d}, {a, c, e}, {a, d, e}, {b, c, d}, {b, c, e}, {b, d, e}, {c, d, e}, {a, b, c, d}, {a, b, c, e}, {a, b, d, e}, {a, c, d, e}, {b, c, d, e}, {a, b, c, d, e}, Ø; 32 **43.** 256

Written Exercises, pages 13–15
1. {⁻1, 0, 1, 2, 3, . . .} **3.** {1, 2, 3, 4, 5, 6} **5.** {⁻9, ⁻8, ⁻7, . . . , 1} **7.** {⁻2, ⁻1} **9.** {10, 11, 12, . . . , 99} **11.** {the natural numbers less than 6} **13.** {the integers greater than ⁻7} **15.** {the even integers greater than 1 and less than 101} **17.** {the integers that are multiples of 3} **19.** {the integers greater than 1} **21.** {the natural numbers less than 2}

23.
(line graph −3 to 4)

25.
(line graph −2 to 12)

27.
(line graph −5 to 2)

29.
(line graph −7 to 7)

31. All **33.** No **35.** Some **37.** All **39.** All
41. No **43.** {⁻3, 0, 3} **45.** {⁻3, ⁻2, 0, 1, 2, 3}
47. $\{⁻3, ⁻2, \frac{-1}{2}, 0, \frac{1}{2}, 1, \frac{3}{2}, 2, 3\}$ **49.** Ø
51. {4, 5, 6, . . .}

Self-Test 1, page 15
1.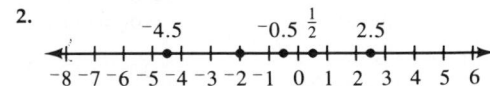
(line graph −7 to 7)

2.
(line graph with −4.5, −0.5, $\frac{1}{2}$, 2.5 marked; −8 to 6)

3. > **4.** < **5.** >,> **6.** false **7.** false **8.** true

9.

10.

Written Exercises, pages 19–20

1. 4665 **3.** 900 **5.** 11.51 **7.** 50 **9.** $150\frac{1}{4}$ **11.** $9\frac{1}{3}$

13. 23 **15.** 80 **17.** 21 **19.** $\frac{1}{5}$ **21.** 22 **23.** 144

25. 20 **27.** $\frac{1}{2}$ **29.** 14 **31.** 576 **33.** 36 **35.** 16

37. $\frac{5}{6}$ **39.** 2.16 **41.** $1\frac{3}{8}$ **43.** $1\frac{1}{2}$ **45.** 17 **47.** $1\frac{1}{2}$

49. 5 **51.** $\frac{1}{4}$ **53.** Examples: $z - 2$, $\frac{z}{3}$, $2z - 5$

55. Examples: $a = 1$, $b = 0$ **57.** Examples: $a = 2$, $b = 1$ **59.** $\frac{1}{n}$ decreases toward 0; $n + \frac{1}{n}$ increases

Written Exercises, pages 22–23

1. 109 **3.** 14 **5.** 176 **7.** 4 **9.** 54 **11.** $\frac{1}{2}$ **13.** 12

15. 240 **17.** 18 **19.** 21 **21.** 4 **23.** 1 **25.** 7

27. 3 **29.** $5\frac{4}{5}$ **31.** 32 **33.** 60 **35.** 0 **37.** 70

39. $\neq$ **41.** = **43.** = **45.** = **47.** 37; 1 **49.** 190; 1

51. 5; $3\frac{1}{5}$ **53.** 0; $\frac{9}{100}$

Computer Exercises, page 24

1. 14 **3.** 3 **5.** 0.5 **7.** 1.4 **9.** $\frac{2n - 1}{2n + 1}$ increases to 1

Written Exercises, pages 26–27

1. x^5 **3.** $7a^4$ **5.** $(x - y)^3$ **7.** m^2 **9.** $p + q^2$

11. $\frac{1}{2}t^5$ **13.** 243 **15.** 2401 **17.** 43 **19.** 287

21. 94 **23.** 152 **25.** 10 **27.** 25 **29.** 35 **31.** 8
33. 0 **35.** 1 **37.** 41 **39.** 78 **41.** 54 **43.** 2169

45. 0 **47.** $\frac{9}{10}$ **49.** 0.75 **51.** 7 **53.** 0.25 **55.** ÷;

+; ×; − **57.** ×; ÷; ×; −; ÷ **59.** −; ÷; ×; ×; ÷

Self-Test 2, pages 27–28

1. 11 **2.** 7 **3.** 12 **4.** 37 **5.** 8 **6.** 2 **7.** 625
8. 88 **9.** 4

On the Calculator, page 28

1. 55 **3.** 163 **5.** 216 **7.** 58 **9.** 82 **11.** 2316

Reading Algebra, page 29

1. Examples: fourteen minus eight, the difference when eight is subtracted from fourteen **3.** Examples: four multiplied by fifteen, the product of four and fifteen **5.** 6×20 **7.** $\frac{1}{10} \times 90$ **9.** $12 + 4$ **11.** $\frac{9}{3}$

Written Exercises, pages 32–33

1. {4} **3.** {7} **5.** {2} **7.** Ø **9.** {4} **11.** {0, 1, 2, 3}
13. {0, 1} **15.** {3, 4, 5, 6, 7, 8}

17. $\{y: y \geq 2\}$

19. $\{y: y > 0.5\}$

21. $\{y: 1 < y < 6\}$

23. $\{y: 3 < y < 6\frac{1}{2}\}$

25. {6}

27. {5.5}

29. {4}

31. {1}

33. $\{y: 2 \leq y \leq 6\}$

35. $\{y: 6 < y < 10\}$

37. $\{y: y > 0\}$

39. Ø

41. $1 + 3 = 6$, False; $2 + 3 = 6$, False; $1 + 4 = 6$, False; $2 + 4 = 6$, True. **43.** $7 - 3 = 2(1)$, False; $7 - 4 = 2(1)$, False; $7 - 3 = 2(2)$, True; $7 - 4 = 2(2)$, False **45.** $3 < 3(1)$, False; $4 < 3(1)$, False; $3 < 3(2)$, True; $4 < 3(2)$, True **47.** $3(1) + 3 > 7$, False; $3(1) + 4 > 7$, False; $3(2) + 3 > 7$, True; $3(2) + 4 > 7$, True **49.** Examples: $x - 1 = 4$, $x + 1 = 6$ **51.** Examples: $x + 1 > x$, $x - 2 < x$ **53.** Examples: $2x \neq 0$, $x^2 > 0$

On the Calculator, page 33

1. 5.58 **3.** 6.48 **5.** 0.54 **7.** 7.8 **9.** 1.5 **11.** 8.7

Written Exercises, pages 37–39

1. $2x + 5$ **3.** $45 - \frac{1}{4}z$ **5.** $17(m - 7)^2$

7. $(p + 11)(q - 14)$ **9.** $\frac{a + 2b}{c + d^5}$ **11.** $7w$ **13.** $\frac{h}{24}$

15. $5n + 10d$ **17.** $0.05n + 0.25q$ **19.** $j + (j + 1)$
21. $x(x + 2)$ **23.** $24 - a$ **25.** $2(m - 6)$
27. $48 - x$ **29.** $a + 10$ **31.** $5n + 25(n + 6)$
33. $9d + 10d$ **35.** $3(m - 5)$ **37.** $y + y + (y + 5)$
39. $vh - (v - x)h$ **41.** $gn + r(n + z)$

On the Calculator, page 39
1. $\{1\}$ **3.** $\{3\}$ **5.** $\{10\}$

Written Exercises, pages 41–43
1. $4x = 76$ **3.** $20 - z < 4$ **5.** $21 = 1 + \frac{2}{3}v$
7. $m + 8 = 3m$ **9.** $6w = 21$ **11.** $10d + 25q \neq 90$
13. Let $x =$ the unknown number. $2x + 2 = 42$
15. Let $z =$ the second number. $3z + z = 32$
17. Let $b =$ the cost in dollars of a new basketball.
$\frac{b}{5} < 6$ **19.** Let $s =$ the length in centimeters of one
side of the square. $4s = 96$ **21.** Let $n =$ the
number of nickels. $5n < 90$ **23.** Let $n =$ the first
integer. $n + (n + 1) = 139$ **25.** Let $x =$ the
length in centimeters of the shorter piece of balsa
wood. $x + (x + 4) = 14$
27. Let $w =$ the width in centimeters of the rectangle.
$w(w + 3) < 45$ **29.** Let $a =$ the measure in
centimeters of the adjacent side of the parallelogram.
$2a + 2\left(\frac{1}{3}a - 1\right) = 22$ **31.** Let $j =$ Jack's age now.
$\frac{1}{2}(j + 20) = 2(j - 1)$ **33.** Let $q =$ the number of
quarters Carlos has. $0.10(q - 5) + 0.25q = 7.55$
35. Let $g =$ the number of girls. $g + (g + 3) = 25$;
$\{11\}$ **37.** Let $n =$ the number. $5n - 4 = n + 8$; $\{3\}$

Self-Test 3, page 43
1. $\{4\}$

2. $\{x: x > 6\}$

3. $\{x: x > 0\}$

4. $\frac{(m + 1)^3}{m^6}$ **5.** $10d + 100x$ **6.** $\left(\frac{n}{3}\right)^2 < \frac{n^2}{3}$

7. $\frac{1}{2}(b + 14) = 2(b - 10)$

Chapter Review, pages 45–47
1. d **2.** c **3.** b **4.** d **5.** a **6.** c **7.** d **8.** b
9. b **10.** c **11.** a **12.** d **13.** c **14.** d **15.** a
16. d

Contest Problems, page 47
1. 2 values **3.** 576

Chapter 2   Working with Real Numbers

Written Exercises, pages 51–52
1. $u = 5$ **3.** $j = 1$ **5.** $s = 0$ **7.** $p = 0$ **9.** $a = 1$
11. $m = 0$ **13.** For any real number x, there is a real
number y such that $y = 3 + x$. **15.** For every
whole number x, there exists a whole number y such
that $y = 5 + 3x$. **17.** For all real numbers x and y,
$x + y = y + x$. **19.** There exists a nonzero real
number y such that, for each real number x, $\frac{x}{y} = x$.
21. Examples: 7. For some real number p, $p^2 > 0$.
8. There is a positive integer q such that $q < q^3$.
9. There exists a natural number a such that
$a^2 + a > 2$.   10. For at least one whole number b,
$0 < b^2 - b$.   11. For some integer m, $5m \neq \frac{m}{5}$
12. There is a real number n such that $2n \neq n^2$.

Written Exercises, pages 56–57
1. 680 **3.** 690 **5.** 108 **7.** 16 **9.** 16 **11.** 21
13. $6z + 117$ **15.** $52m$ **17.** $91pq$ **19.** $\{47\}$
21. $\mathcal{R}$ **23.** $\mathcal{R}$ **25.** $\emptyset$ **27.** $\emptyset$ **29. a.** 9 **b.** True
c. Commutative **d.** Associative **31. a.** 13 **b.** True
c. Not commutative **d.** Not associative **33. a.** 49
b. True **c.** Commutative **d.** Not associative
35. a. $5\frac{1}{2}$ **b.** False **c.** Not commutative **d.** Not
associative **37.** Transitive **39.** Symmetric

Written Exercises, pages 60–61
1. 44 **3.** 312 **5.** 540 **7.** $81x$ **9.** $100z^2$ **11.** $9g$
13. $29s + 43$ **15.** $17p + 9q$ **17.** $11a^2 + 2a$
19. $23c^3 + 2c^2$ **21.** $6m + 25$ **23.** $22x + 18y$
25. $25p^4 + 10$ **27.** $9u + 15$ **29.** $11c^2 + 42$
31. $20a + 28$ **33.** $19y^2 + 31y$ **35.** $18j + 15k$
37. $6g^2 + 12g + 9$ **39.** $8a^3 + 11a^2 + 9a$
41. $12p^3 + 10p^2 + 17p + 11$
43. $6y^3 + 13y^2 + 14y + 5$
45. $19m^2 + 5m + 12$ **47.** $33r + 12s + 51$
49. $3(u + v) + 2(u + 2v)$; $5u + 7v$
51. $5(3 + w^2) + 2(w^2 + 5)$; $7w^2 + 25$
53. $2y^3 + 6(3y^2 + 2y^3)$; $14y^3 + 18y^2$
55. $70a + 42$ **57.** $53x + 53$
59. $48m + 34n + 44$ **61.** $49p^2 + 27p + 34$

Written Exercises, page 65
1. 3.5 **3.** -85 **5.** 57 **7.** $6\frac{1}{7}$ **9.** -6.2 **11.** 84

13. 4 **15.** 60 **17.** 0 **19.** $\{-18\}$ **21.** $\{4\}$
23. $\{5, -5\}$ **25.** $\{0\}$ **27.** $\{10, -10\}$ **29.** $\{6, -6\}$
31. $\{-4\}$ **33.** $\{-15\}$ **35.** $\{6, -6\}$ **37.** $\{25, -25\}$
39. $\{1, -1\}$ **41.** $\{6, -6\}$

43.

45.

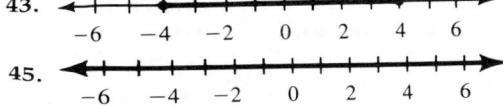

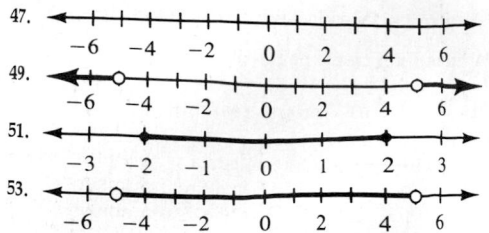

47.

49.

51.

53.

55. {negative real numbers} **57.** {nonpositive real numbers}

Computer Exercises, page 66
1. 25; 25 **3.** 25; 25 **5.** 27; 27 **7.** 50; 50

Self-Test 1, page 66
1. $n = 7$ **2.** $w = 0$ **3.** 20 **4.** 77 **5.** $5x^2 + 15$
6. $21y + 28$ **7.** 15 **8.** 35 **9.** $\{-8\}$ **10.** $\{19, -19\}$

Written Exercises, page 70
1. -8 **3.** 10 **5.** -3 **7.** -18 **9.** -25 **11.** $\frac{5}{7}$
13. 2 **15.** -18 **17.** 6 **19.** -3 **21.** -23 **23.** 0
25. $\{-4\}$ **27.** $\{-8\}$ **29.** $\{11\}$ **31.** $\{-1\}$ **33.** $\{0\}$
35. $\{20\}$ **37.** $\{-17\}$ **39.** $\{-9\}$ **41.** $\{14\}$ **43.** $\{0\}$
45. $\{9\}$ **47.** $\{16\}$ **49.** $\{-4\}$ **51.** $\left\{-2\frac{1}{2}\right\}$ **53.** $\{-4\}$
55. -7 **57.** -5 **59.** -7

Written Exercises, pages 73–74
1. 6 **3.** -72 **5.** 40 **7.** 9 **9.** -315 **11.** 3 **13.** $\frac{1}{9}$
15. -4 **17.** -15 **19.** $(-8)x$ **21.** $(-17)m + (-9)n$
23. $(-4)r^2 + 8r$ **25.** $(-21)a + 15b + 5c$
27. $(-15)j^3 + (-9)j^2 + 4j$ **29.** $3p + (-17)$
31. a. -1 **b.** 3 **c.** -1 **d.** -3 **33. a.** 0 **b.** 8
c. 0 **d.** -8 **35. a.** 0.4 **b.** 1 **c.** -0.4 **d.** -1
37. $=$ **39.** $<$

Problems, pages 74–76
1. a. 43, -51 **b.** $43 + (-51) = -8$ **c.** The helicopter was 8 miles directly south of its base.
3. a. 3200, -340, 75, -800 **b.** $3200 + (-340) + 75 + (-800) = 2135$ **c.** The glider's new altitude was 2135 m. **5. a.** -12.05, -13.15, -13.45, -12.05, 12.25, 12.85, 13.75, 12.95 **b.** $-12.05 + (-13.15) + (-13.45) + (-12.05) + 12.25 + 12.85 + 13.75 + 12.95 = 1.10$ **c.** She made $1.10 in selling the coins. **7. a.** 8, -15, 9, -2 **b.** $8 + (-15) + 9 + (-2) = 0$ **c.** The temperature was 0°C at midnight of the second day. **9. a.** -38.50, -220.95, 84, 84, -6.55 **b.** $-38.50 + (-220.95) + 84 + 84 + (-6.55) = -98$ **c.** Alice still owed $98 on her account.
11. 13 m below **13.** $255 **15.** at the 6th floor
17. at his starting point

Written Exercises, pages 81–82
1. 82 **3.** 104 **5.** -52 **7.** -49 **9.** -53 **11.** 81

13. 1. Hypothesis 2. Axiom of additive inverses 3. Commutative axiom for addition 4. Associative axiom for addition 5. Axiom of additive inverses 6. Identity axiom for addition 7. Transitive axiom of equality **15.** 1. Hypothesis 2. Property of the opposite of a sum 3. Commutative axiom for addition 4. Associative axiom for addition 5. Axiom of additive inverses 6. Identity axiom for addition 7. Transitive axiom of equality

17.
1. a and b are real numbers.	Hypothesis
2. $-a$ is a real number.	Ax. of add. inv.
3. $-a + (a + b)$	Assoc. ax. for add.
$\quad = (-a + a) + b$	
4. $\quad = 0 + b$	Ax. of add. inv.
5. $\quad = b$	Iden. ax. for add.
6. $\therefore -a + (a + b) = b$	Trans. ax. of $=$

19.
1. a and b are real numbers.	Hypothesis
2. $-a$ and $-b$ are real numbers.	Ax. of add. inv.
3. $[a + (-b)] + (-a)$	Comm. ax. for add.
$\quad = (-b + a) + (-a)$	
4. $\quad = -b + [a + (-a)]$	Assoc. ax. for add.
5. $\quad = -b + 0$	Ax. of add. inv.
6. $\quad = -b$	Iden. ax. for add.
7. $\therefore [a + (-b)] + (-a) = -b$	Trans. ax. of $=$

21.
1. a and b are real numbers.	Hypothesis
2. $-a$ and $-b$ are real numbers.	Ax. of add. inv.
3. $-(a + b) + b$	Prop. of opp. of a sum
$\quad = [-a + (-b)] + b$	
4. $\quad = -a + (-b + b)$	Assoc. ax. for add.
5. $\quad = -a + 0$	Ax. of add. inv.
6. $\quad = -a$	Iden. ax. for add.
7. $\therefore -(a + b) + b = -a$	Trans. ax. of $=$

23.
1. a and b are real numbers.	Hypothesis
2. $-a$ and $-b$ are real numbers.	Ax. of add. inv.
3. $-[a + (-b)] + (-b)$	Prop. of opp. of a sum
$\quad = [-a + (-(-b))] + (-b)$	
4. $\quad = (-a + b) + (-b)$	Canc. prop. of add.
5. $\quad = -a + [b + (-b)]$	Assoc. ax. for add.
6. $\quad = -a + 0$	Ax. of add. inv.
7. $\quad = -a$	Iden. ax. for add.
8. $\therefore -[a + (-b)] + (-b) = -a$	Trans. ax. of $=$

25.
1. a and b are real numbers.	Hypothesis
2. $-a$ and $-b$ are real numbers.	Ax. of add. inv.
3. $(a + b) + [-a + (-b)]$	Prop. of opp. of a sum
$\quad = (a + b) + [-(a + b)]$	
4. $\quad = 0$	Ax. of add. inv.
5. $\therefore (a + b) + [-a + (-b)] = 0$	Trans. ax. of $=$

27. 1. a, b, and c are real numbers. Hypothesis
  2. $-a$, $-b$, and $-c$ are real Ax. of add. inv.
   numbers.
  3. $-[(a + b) + c]$ Prop. of opp. of
    $= -(a + b) + (-c)$ a sum
  4. $= [-a + (-b)] + (-c)$ Prop. of opp. of
         a sum
  5. $\therefore -[(a + b) + c]$ Trans. ax. of $=$
    $= [-a + (-b)] + (-c)$

Written Exercises, pages 84–85

1. -23 3. -53 5. 44 7. 100 9. -10.6

11. 4.4 13. $\dfrac{5}{8}$ 15. $6\dfrac{4}{7}$ 17. -69 19. 24 21. 23

23. 25 25. 133 27. 95 29. -140 31. $-16x - 3$
33. $-21z^2 - 8$ 35. $-23m^2 - 9m - 6$ 37. $-15a - 6$
39. $-21p - 3$ 41. $3m + 13n$ 43. $-11x^3 + 7x$
45. $-22z^2 + 4z + 7$ 47. 1. Hypothesis 2.
Definition of subtraction 3. Associative axiom for
addition 4. Axiom of additive inverses 5. Identity
axiom for addition 6. Transitive axiom of equality

49. 1. a and b are real numbers. Hypothesis
  2. $(a - b) + b = [a + (-b)] + b$ Def. of subt.
  3. $= a + (-b + b)$ Assoc. ax. for
         add.
  4. $= a + 0$ Ax. of add. inv.
  5. $= a$ Iden. ax. for add.
  6. $\therefore (a - b) + b = a$ Trans. ax. of $=$

51. 1. a, b, and c are real numbers. Hypothesis
  2. $-[(a + b) + c]$ Prop. of opp.
    $= -(a + b) + (-c)$ of a sum
  3. $= [-a + (-b)] + (-c)$ Prop. of opp.
         of a sum
  4. $= (-a - b) - c$ Def. of subt.
  5. $\therefore -[(a + b) + c]$ Trans. ax. of $=$
    $= (-a - b) - c$

53. closed 55. not closed 57. not closed
59. closed 61. closed

Problems, pages 86–87

1. **a.** $-2 - 29 = -31$ **b.** Carlos rode 31 floors
down. 3. **a.** $1935 - 1882 = 53$ **b.** She was 53
years old. 5. **a.** $58 - (-88.3) = 146.3$ **b.** The record high is 146.3°C higher than the record low.
7. **a.** $-17 - 2 = -19$ **b.** The temperature feels 19°C
colder (or lower). 9. 10,915 m below sea level
11. 71° north latitude

Computer Exercises, page 87
1. $|a - b| = |b - a|$ for all a, $b \in \mathcal{R}$

Self-Test 2, page 87
1. -5 2. -9 3. 0 4. 39 5. $(-5)a + (-7)b$
6. Identity axiom for addition 7. Axiom of
additive inverses 8. Cancellation property of
opposites 9. -54 10. $6n + 10$

Written Exercises, pages 91–93
1. 595 3. 0 5. -540 7. 150 9. -792 11. 250
13. -270 15. 0 17. 560 19. -350 21. -660
23. -890 25. $-x^3$ 27. $72a^3$ 29. $-80pqr$
31. $-10m - 6$ 33. $-70j - 28k$ 35. $3x - 6y$
37. $w - 2x$ 39. $-11j^2 - j$ 41. $-3a^2 - a - 9$
43. $-2m + n$ 45. $r + 3s$ 47. $11c - 9d$
49. $-10a - b$ 51. $-x - 28y$ 53. $8p + q$
55. -7 57. -4 59. 2 61. 1. Hypothesis
2. Axiom of additive inverses 3. Multiplicative
property of -1 4. Associative axiom for
multiplication 5. Multiplicative property of -1
6. Transitive axiom of equality
63. 1. a and b are real numbers. Hypothesis
  2. $-a$ is a real number. Ax. of add. inv.
  3. $(-a)b = [(-1)a]b$ Mult. prop. of -1
  4. $= (-1)(ab)$ Assoc. ax. for
        mult.
  5. $= -ab$ Mult. prop. of -1
  6. $\therefore (-a)b = -ab$ Trans. ax. of $=$
65. 1. a, b, and c are real numbers. Hypothesis
  2. $-a$ is a real number. Ax. of add. inv.
  3. $-a(b + c) = (-a)b + (-a)c$ Dist. ax.
  4. $= -ab + (-ac)$ Prop. of opp.
        in products
  5. $= -ab - ac$ Def. of subt.
  6. $\therefore -a(b + c) = -ab - ac$ Trans. ax. of $=$
67. 1. a, b, c and d are real numbers. Hypothesis
  2. $a[(b - c) - d]$ Exercise 62,
    $= a(b - c) - ad$ page 92
  3. $= (ab - ac) - ad$ Exercise 62,
        page 92
  4. $= [ab + (-ac)] + (-ad)$ Def. of
        subt.
  5. $= ab + [-ac + (-ad)]$ Assoc. ax.
        for add.
  6. $= ab + [(-1)(ac) + (-1)(ad)]$ Mult. prop.
        of -1
  7. $= ab + [(-1)(ac + ad)]$ Dist. ax.
  8. $= ab + [-(ac + ad)]$ Mult. prop.
        of -1
  9. $= ab - (ac + ad)$ Def. of
        subt.
  10. $\therefore a[(b - c) - d]$ Trans. ax.
    $= ab - (ac + ad)$ of $=$

Computer Exercises, page 93
1. positive 3. negative 5. zero 7. positive
9. positive 11. positive

Written Exercises, pages 97–98
1. $\dfrac{1}{18}$ 3. $-\dfrac{1}{28}$ 5. -15 7. 24 9. 3 11. -1
13. -3 15. jk 17. $-27a^3$ 19. $-18m$ 21. $-4q$
23. $5h^2$ 25. $-5a + 11$ 27. $-4c + 6d$
29. $-7j - 5k$ 31. $6w - 9z$ 33. $8s - 11t$
35. $-5m^2 - 6m - 7$ 37. $-6w^2$ 39. $-2r - 10s$

41. $4a - 50b$ **43.** $\{-4\}$ **45.** $\left\{\dfrac{2}{5}\right\}$ **47.** $\emptyset$ **49.** $\emptyset$

51.
1. a and b are real numbers such that $b \neq 0$.   Hypothesis
2. $\dfrac{1}{b}$ is a real number.   Ax. of mult. inv.
3. $(ab)\left(\dfrac{1}{b}\right) = a\left[b\left(\dfrac{1}{b}\right)\right]$ Assoc. ax. for mult.
4. $\phantom{(ab)\left(\dfrac{1}{b}\right)} = a \cdot 1$ Ax. of mult. inv.
5. $\phantom{(ab)\left(\dfrac{1}{b}\right)} = a$ Iden. ax. for mult.
6. $\therefore (ab)\left(\dfrac{1}{b}\right) = a$ Trans. ax. of $=$

53.
1. a and b are real numbers such that $b \neq 0$.   Hypothesis
2. $\dfrac{1}{b}$ is a real number.   Ax. of mult. inv.
3. $-\dfrac{1}{b}(ab) = -\left[\dfrac{1}{b}(ab)\right]$ Prop. of opp. in products
4. $\phantom{-\dfrac{1}{b}(ab)} = -\left[\left(\dfrac{1}{b}\cdot b\right)a\right]$ Comm. and assoc. ax. for mult.
5. $\phantom{-\dfrac{1}{b}(ab)} = -(1 \cdot a)$ Ax. of mult. inv.
6. $\phantom{-\dfrac{1}{b}(ab)} = -a$ Iden. ax. for mult.
7. $\therefore -\dfrac{1}{b}(ab) = -a$ Trans. ax. of $=$

55.
1. a and b are real numbers such that $a, b, \neq 0$.   Hypothesis
2. $\dfrac{1}{a}$ and $\dfrac{1}{b}$ are real numbers.   Ax. of mult. inv.
3. $-\dfrac{1}{a}\left(-\dfrac{1}{b}\right) = \dfrac{1}{a}\cdot\dfrac{1}{b}$ Prop. of opp. in products
4. $\phantom{-\dfrac{1}{a}\left(-\dfrac{1}{b}\right)} = \dfrac{1}{ab}$ Prop. of recip. of a product
5. $-\dfrac{1}{a}\left(-\dfrac{1}{b}\right) = \dfrac{1}{ab}$ Trans. ax. of $=$

Written Exercises, pages 101–103

1. -32 **3.** -25 **5.** 75 **7.** $-\dfrac{1}{3}$ **9.** $-29jk$
11. -12 **13.** $-272w^3$ **15.** $14r$ **17.** 39 **19.** 0
21. a. $\dfrac{1}{2}$ **b.** $\dfrac{1}{2}$ **23. a.** 1 **b.** 1 **25. a.** -1 **b.** 1
27. a. -1 **b.** -1 **29.** 16 **31.** -2 **33.** $-\dfrac{3}{2}$
35. 4 **37.** -7 **39.** $\dfrac{1}{2}$ **41.** 1. Hypothesis

2. Definition of division   3. Associative axiom for multiplication   4. Axiom of multiplicative inverses 5. Identity axiom for multiplication   6. Transitive axiom of equality

43.
1. a and b are real numbers such that $b \neq 0$.   Hypothesis

2. $(a \div b)b = \left(a \cdot \dfrac{1}{b}\right)b$ Def. of div.
3. $ = a\left(\dfrac{1}{b}\cdot b\right)$ Assoc. ax. for mult.
4. $ = a \cdot 1$ Ax. of mult. inv.
5. $ = a$ Iden. ax. for mult.
6. $\therefore (a \div b)b = a$ Trans. ax. of $=$

45.
1. a is a real number such that $a \neq 0$ Hypothesis
2. $\dfrac{a}{a} = a \cdot \dfrac{1}{a}$ Def. of div.
3. $\phantom{\dfrac{a}{a}} = 1$ Ax. of mult. inv.
4. $\therefore \dfrac{a}{a} = 1$ Trans. ax. of $=$

47.
1. a and b are real numbers such that $b \neq 0$.   Hypothesis
2. $\dfrac{a + b}{b} = (a + b)\dfrac{1}{b}$ Def. of div.
3. $\phantom{\dfrac{a + b}{b}} = a \cdot \dfrac{1}{b} + b \cdot \dfrac{1}{b}$ Dist. ax.
4. $\phantom{\dfrac{a + b}{b}} = a \cdot \dfrac{1}{b} + 1$ Ax. of mult. inv.
5. $\phantom{\dfrac{a + b}{b}} = \dfrac{a}{b} + 1$ Def. of div.
6. $\phantom{\dfrac{a + b}{b}} = 1 + \dfrac{a}{b}$ Comm. ax. for add.
7. $\therefore \dfrac{a + b}{b} = 1 + \dfrac{a}{b}$ Trans. ax. of $=$

49. closed   **51.** closed   **53.** not closed   **55.** closed
57. closed

Self-Test 3, page 103

1. -44 **2.** $63x^2$ **3.** $-12y + 7$ **4.** $z^3 - 9z$ **5.** $\dfrac{1}{21}$
6. $-15w$ **7.** $-7p + q$ **8.** -13 **9.** $\dfrac{8}{5}$ **10.** $-4n$

Chapter Review, pages 105–106

1. c   **2.** c   **3.** d   **4.** c   **5.** a   **6.** d   **7.** b   **8.** a
9. c   **10.** b   **11.** c   **12.** d   **13.** d   **14.** c   **15.** d
16. d   **17.** a   **18.** b   **19.** c   **20.** a

Mixed Review, pages 107–108

1. -18 **3.** 140 **5.** $-16a$ **7.** $36c$ **9.** 17 **11.** 10
13. $\emptyset$ **15.** $\emptyset$ **17.** $\{-4, -3\}$ **19.** $\{-4, -3, -2, -1, 0\}$ **21.** -14 **23.** -56

25.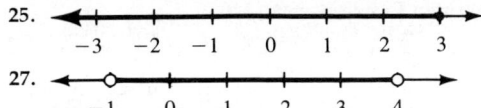

27.
$$-1 \quad 0 \quad 1 \quad 2 \quad 3 \quad 4$$

29. True **31.** True **33.** $5n + 25q + 100x$
35. 9000 m

On the Calculator, page 108
1. 0.0625 **3.** -16 **5.** 0.025 **7.** 0.25

Preparing for College Entrance Exams, page 109
1. E **3.** D **5.** C **7.** C

Application, page 111
1. 25°C **3.** -39.15°C to 355.85°C

Chapter 3 Solving Equations and Problems

Written Exercises, pages 116–118
1. (a) Addition property of equality **(b)** Associative axiom for addition **(c)** Axiom of additive inverses **(d)** Identity axiom for addition **3. (a)** Multiplication property of equality **(b)** Associative axiom for multiplication **(c)** Axiom of multiplicative inverses **(d)** Identity axiom for multiplication **5. (a)** Definition of subtraction **(b)** Addition property of equality **(c)** Associative axiom for addition **(d)** Axiom of additive inverses **(e)** Identity axiom for addition
7. 1. Hypothesis 2. Axiom of additive inverses 3. Hypothesis 4. Addition property of equality 5. Definition of subtraction

9. 1. a, b, and c are real numbers.   Hypothesis
2. $-c$ is a real number.   Ax. of add. inv.
3. $a + c = b + c$ Hypothesis
4. $(a + c) + (-c)$ Add. prop. of =
$\quad = (b + c) + (-c)$
5. $a + [c + (-c)]$ Assoc. ax. for
$\quad = b + [c + (-c)]$ add.
6. $a + 0 = b + 0$ Ax. of add. inv.
7. $\therefore a = b$ Iden. ax. for add.

11. 1. a, b, and c are real numbers.   Hypothesis
$\quad a - c = b - c$
2. $a + (-c) = b + (-c)$ Def. of subt.
3. $[a + (-c)] + c$ Add. prop.
$\quad = [b + (-c)] + c$ of =
4. $a + [(-c) + c]$ Assoc. ax. for
$\quad = b + [(-c) + c]$ add.
5. $a + 0 = b + 0$ Ax. of add. inv.
6. $\therefore a = b$ Iden. ax. for add.

13. 1. a and b are real numbers.   Hypothesis
2. $-b$ is a real number.   Ax. of add. inv.
3. $a + b = 0$ Hypothesis
4. $(a + b) + (-b) = 0 + (-b)$ Add. prop. of =
5. $a + [b + (-b)] = 0 + (-b)$ Assoc. ax. for add.

6. $a + 0 = 0 + (-b)$ Ax. of add. inv.
7. $\therefore a = -b$ Iden. ax. for add.

15. 1. a, b, and x are real numbers; $a - b = x$ Hypothesis
2. $x = a - b$ Symm. ax. of =
3. $x = a + (-b)$ Def. of subt.
4. $x + b = [a + (-b)] + b$ Add. prop. of =
5. $x + b = a + [(-b) + b]$ Assoc. ax. for add.
6. $x + b = a + 0$ Ax. of add. inv.
7. $x + b = a$ Iden. ax. for add.
8. $\therefore b + x = a$ Comm. ax. for add.

17. 1. a, b, c, and d are real numbers; $a = b$ Hypothesis
2. $a + c = b + c$ Add. prop. of =
3. $c = d$ Hypothesis
4. $\therefore a + c = b + d$ Subs. prin.

19. 1. a, b, c, and d are real numbers; $a = b$ Hypothesis
2. $ac = bc$ Mult. prop. of =
3. $c = d$ Hypothesis
4. $\therefore ac = bd$ Subs. prin.

21. 1. a, b, c, and d are real numbers; $a = b$ Hypothesis
2. $a - c = b - c$ Subt. prop. of =
3. $c = d$ Hypothesis
4. $\therefore a - c = b - d$ Subs. prin.

Written Exercises, pages 120–121
1. $\{-2\}$ **3.** $\{12\}$ **5.** $\{1\}$ **7.** $\{43\}$ **9.** $\{5\}$
11. $\{-21\}$ **13.** $\{20\}$ **15.** $\{0\}$ **17.** $\{-3\}$ **19.** $\left\{-\dfrac{1}{6}\right\}$
21. $\{0.22\}$ **23.** $\{7\}$ **25.** $\{-11\}$ **27.** $\{-1\}$
29. $\{-0.72\}$ **31.** $\{14, -14\}$ **33.** $\{1, -1\}$ **35.** $\emptyset$
37. $\{-11\}$ **39.** $\{23\}$ **41.** $\{-3\}$ **43.** $\{-14\}$
45. $\left\{-\dfrac{8}{3}\right\}$ **47.** $\left\{-\dfrac{11}{6}\right\}$ **49.** $\{0.1\}$ **51.** $\{8, -8\}$
53. $\emptyset$ **55.** $\{0\}$ **57.** $\{-13\}$ **59.** $\left\{\dfrac{2}{3}\right\}$ **61.** $t > 0$
63. $t < -1$ **65.** $t < 2$

Written Exercises, pages 123–124
1. $\{-11\}$ **3.** $\{-21\}$ **5.** $\{14\}$ **7.** $\{60\}$ **9.** $\{0\}$
11. $\{90\}$ **13.** $\left\{-\dfrac{7}{6}\right\}$ **15.** $\{11\}$ **17.** $\left\{-\dfrac{4}{49}\right\}$ **19.** $\{1\}$
21. $\{-0.1\}$ **23.** $\{1.5\}$ **25.** $\{-0.8\}$ **27.** $\left\{\dfrac{8}{3}\right\}$
29. $\{2, -2\}$ **31.** $\{15, -15\}$ **33.** $\left\{\dfrac{9}{8}, -\dfrac{9}{8}\right\}$ **35.** $\emptyset$
37. $\{0\}$ **39.** $\{400, -400\}$ **41.** $\emptyset$ **43.** $\{6\}$
45. $\left\{\dfrac{3}{4}\right\}$ **47.** $\left\{\dfrac{3}{2}, -\dfrac{3}{2}\right\}$ **49.** $\{-2\}$

1. $\{1\}$ 3. $\{2.4\}$ 5. $\{6\}$ 7. $\{-8\}$ 9. $\left\{-\dfrac{17}{2}\right\}$

11. $\left\{\dfrac{16}{3}\right\}$ 13. $\{9\}$ 15. $\left\{\dfrac{7}{3}\right\}$ 17. $\left\{\dfrac{41}{5}\right\}$ 19. $\left\{\dfrac{5}{6}\right\}$

21. $\{0\}$ 23. $\{3\}$ 25. $\left\{-\dfrac{14}{3}\right\}$ 27. $\{2\}$ 29. $\{12\}$

31. $\{-14\}$ 33. $\{3\}$ 35. $\{4\}$ 37. $\{-6\}$ 39. $\{6\}$

41. $\{17\}$ 43. $\{15\}$ 45. $\left\{\dfrac{17}{4}\right\}$ 47. $\left\{\dfrac{15}{4}\right\}$ 49. $\{4\}$

51. $\{0\}$ 53. $\{-6\}$ 55. $\{3\}$ 57. $\left\{\dfrac{1}{4}\right\}$ 59. $\left\{-\dfrac{2}{3}\right\}$

Computer Exercises, page 129
1. $\{-2\}$ 3. $\{-12\}$ 5. $\{26\}$ 7. $\{1\}$ 9. $\{14, -14\}$
11. $\{7, -7\}$

Self-Test 1, page 129
1. Addition property of equality   2. Multiplication property of equality   3. $\{-26\}$ 4. $\{16\}$ 5. $\{-75\}$
6. $\{9, -9\}$ 7. $\{5\}$ 8. $\{12\}$ 9. $\{-1\}$ 10. $\{-1\}$

Problems, pages 132–134
1. 19, 25   3. 22 min   5. Darcy's: 1.2 m; Damon's: 1.6 m   7. width: 1 m; length: 9 m   9. 9 atoms
11. 52, 17   13. 38 quarters   15. Aardvark: $1860; Afghan: $950; Armadillo: $1020.   17. Debating: 24; Math: 32; Ecology: 12   19. 18 nickels, 22 quarters
21. 273 m²   23. 4550 m²   25. boxer: $200; Dalmation: $220; Airedale: $250; miniature schnauzer: $200

Written Exercises, page 136–137
1. $\{7\}$ 3. $\{11\}$ 5. $\varnothing$ 7. $\{-9\}$ 9. $\left\{\dfrac{18}{25}\right\}$ 11. $\mathscr{R}$

13. $\{0\}$ 15. $\left\{\dfrac{3}{10}\right\}$ 17. $\{3\}$ 19. $\left\{\dfrac{66}{35}\right\}$ 21. $\{9\}$

23. $\left\{-\dfrac{20}{7}\right\}$ 25. $\{1\}$ 27. $\left\{-\dfrac{29}{11}\right\}$ 29. $\{-4\}$

31. $\left\{\dfrac{7}{4}\right\}$ 33. $\left\{\dfrac{5}{9}\right\}$ 35. $\left\{\dfrac{1}{2}\right\}$ 37. $\left\{\dfrac{23}{16}\right\}$ 39. $\{34\}$

41. $\left\{-\dfrac{13}{4}\right\}$ 43. $\left\{-\dfrac{3}{2}\right\}$ 45. $\varnothing$ 47. $\{5\}$ 49. $\{-2.2\}$
51. $\{-5\}$ 53. a. 7   b. 3   c. $2p + 3$ d. Example: $2p + 2$ e. $2p + 3$ 55. a. 7   b. 1   c. $3k + 1$
d. Example: $3k + 2$ e. $3k + 1$

Computer Exercises, page 137
1. $\{1\}$ 3. $\{3\}$ 5. $\varnothing$

Problems, pages 139–141
1. -6 3. 11   5. 3, 13   7. 19 years   9. in 1992
11. 6400 km   13. Panama Canal: 160 km; Suez Canal: 81 km   15. 35 books   17. width: 30 m; length: 50 m   19. $17.10   21. 2013   23. a. 1954
b. 2010

Reading Algebra, page 143
1. $3d$ 3. $d + 3d + (3d + 4)$ 5. No   7. Yes
9. a. 8 is the number of *dimes*.   b. $7.00 is the *value* of the quarters.

Written Exercises, page 146

1. $I = \dfrac{V}{R}$ 3. $v = \dfrac{-4xy}{3w}$ 5. $c = \dfrac{a^2b}{3x}$ 7. $h = \dfrac{-xy}{3g^2}$

9. $x = \dfrac{b - 2c}{-2a}$ 11. $c = \dfrac{-2ax - b}{-2}$, or $\dfrac{2ax + b}{2}$

13. $a = \dfrac{-4d}{b - c}$ 15. $b = \dfrac{ac - 4d}{a}$, or $c - \dfrac{4d}{a}$

17. $c = \dfrac{ab + 4d}{a}$, or $b + \dfrac{4d}{a}$ 19. $x = \dfrac{9a^2 + 4c}{3a^2}$, or

$3 + \dfrac{4c}{3a^2}$ 21. $b = \dfrac{2A}{h}$ 23. $m = \dfrac{2E}{c^2}$ 25. $c = 3n - t$

27. $s = \dfrac{-0.06k}{t^2}$ 29. $h = \dfrac{2A}{a + b}$ 31. $a = \dfrac{2(A - wt)}{t^2}$

33. $b = \dfrac{3}{2}C - a$ 35. $y = 3x - \dfrac{2A}{5h}$, or $\dfrac{15hx - 2A}{5h}$

37. $P = \dfrac{A}{1 + rt}$ 39. $d = \dfrac{kA}{4\pi C}$ 41. $t = \dfrac{c_1 - c_2}{2C}$

43. $c_2 = c_1 - 2Ct$ 45. $E = I(R + r)$

47. $r = \dfrac{E - IR}{I}$, or $\dfrac{E}{I} - R$ 49. $b = \dfrac{at + 3}{t - 4}$

51. $t = \dfrac{-4b - 3}{a - b}$ 53. $f_1 = \dfrac{-Ff_2}{F - f_2}$ 55. $\{0\}$ 57. $\{6\}$

Problems, page 147
1. a. $h = \dfrac{V}{lw}$, $l, w \neq 0$ b. 14 cm   3. a. $h = \dfrac{SA}{2\pi r} - r$,
$r \neq 0$ b. 11 cm   5. 24 m   7. 48 ft

Self-Test 2, page 148
1. $x + (x + 6) = 42$; $\{18\}$; 18, 24
2. $10d + 5(d - 5) = 110$; $\{9\}$; 9 dimes   3. $\{2\}$
4. $\{-10\}$ 5. 17 years   6. width: 1 cm; length: 4 cm
7. $b = \dfrac{2A}{a}$, $a \neq 0$ 8. $b = \dfrac{2A}{h} - a$, $h \neq 0$

Extra, page 150
1. *Plan:* Assume $a = b$.
  1. $a + c = b + c$ Add. prop. of $=$
  2. $a + c \neq b + c$ Hypothesis
Statement (1) contradicts the hypothesis $a + c \neq b + c$. $\therefore$ the assumption $a = b$ is incorrect and the conclusion $a \neq b$ is true.

3. *Plan:* Assume $a = b$.
  1. $ac = bc$ Mult. prop. of $=$
  2. $ac \neq bc$ Hypothesis
Statement (1) contradicts the hypothesis $ac \neq bc$ $\therefore$ the assumption $a = b$ is incorrect and the conclusion $a \neq b$ is true.

5. *Plan:* Assume $-a = -b$.
  1. $-1(-a) = -1(-b)$ Mult. prop. of $=$
  2. $1 \cdot a = 1 \cdot b$ Prop. of opp. in prod.
  3. $a = b$ Iden. ax. of mult.
  4. $a \neq b$ Hypothesis
Statement (3) contradicts the hypothesis $a \neq b$.

∴ the assumption $-a = -b$ is incorrect and the conclusion $-a \neq -b$ is true.

7. *Plan*: Assume $1 = 2$.

1. $1 + (-1) = 2 + (-1)$ Add. prop. of $=$
2. $0 = 2 + (-1)$ Ax. of add. inv.
3. $0 = 1$ Subst. prin.

Statement (3) contradicts the iden. ax. for mult. which states $1 \neq 0$.

∴ the assumption $1 = 2$ is incorrect and the conclusion $1 \neq 2$ is true.

Chapter Review, pages 151–152

1. b **2.** a **3.** c **4.** a **5.** b **6.** c **7.** d **8.** b
9. d **10.** d **11.** b **12.** c **13.** a **14.** c **15.** b
16. b **17.** c **18.** a.

Cumulative Review, pages 154–155

1. True **3.** True **5.** $7\frac{13}{24}$ **7.** 26 **9.** 26 **11.** 12

13. 3 **15.** $\frac{13}{2}$ **17.** -83 **19.** 253 **21.** -28

23. $-4a^2 + 7$ **25.** $105c^3$ **27.** $21u - 12v$
29. Axiom of additive inverses **31.** Property of the opposite of a sum **33.** Property of opposites in products **35.** $\{20\}$ **37.** $\left\{-\frac{1}{2}\right\}$ **39.** $\{-11\}$ **41.** $\{-2\}$
43. $\{7\}$ **45.** $\{-4\}$ **47.** width: 6 cm; length: 10 cm

Contest Problems, page 155

1. $-\frac{1}{6}$ **3.** $-\frac{1}{5}$ **5. a.** Yes **b.** u

3. $x^2 \cdot x > 0 \cdot x$ Mult. prop. of order
4. $x^2 \cdot x = x \cdot x \cdot x$ Def. of x^2
5. $x \cdot x \cdot x = x^3$ Def. of x^3
6. $x^2 \cdot x = x^3$ Trans. ax. of $=$
7. $0 \cdot x = 0$ Mult. prop. of 0
8. ∴ $x^3 > 0$ Subs. prin.

37. 1. x and y are real numbers;   Hypothesis
    $x < 0$ and $y < 0$
2. $xy > 0 \cdot y$ Mult. prop. of order
3. ∴ $xy > 0$ Mult. prop. of 0

39. 1. m and n are real numbers;   Hypothesis
    $m > 0$ and $n > 0$;
    $m < n$
2. $m \cdot m < m \cdot n$ Mult. prop. of order
3. $m \cdot n < n \cdot n$ Mult. prop. of order
4. $m \cdot m < n \cdot n$ Trans. prop. of order
5. $m \cdot m = m^2$ Def. of m^2
6. $n \cdot n = n^2$ Def. of n^2
7. ∴ $m^2 < n^2$ Subs. prin.

41. 1. x and y are real numbers;   Hypothesis
    $x < y$
2. $x + x < x + y$ Add. prop. of order
3. $2x < x + y$ Subs. prin.
4. $\frac{2x}{2} < \frac{x + y}{2}$ Mult. prop. of order
5. ∴ $x < \frac{x + y}{2}$ Subs. prin.

43. False; $a = \frac{1}{2}$, $b = \frac{1}{3}$ **45.** False; $a = \frac{1}{2}$, $b = \frac{1}{3}$

Chapter 4   Solving Inequalities and Problems

Written Exercises, pages 161–163

1. $<$ **3.** $<$ **5.** $>$ **7.** $<$ **9.** $<$ **11.** $=$ **13.** $>$
15. $>$ **17.** $<$ **19.** $<$ **21.** 1. Hypothesis 2. Addition property of order 3. Hypothesis 4. Addition property of order 5. Transitive property of order
23. 1. Hypothesis 2. Addition property of order 3. Definition of subtraction 4. Axiom of additive inverses **25.** $\{0\}$ **27.** $\{t: t \neq 0\}$ **29.** $\{t: t < 0\}$
31. $\{x, y: |x| < |y|\}$
33. 1. a is a real number;   Hypothesis
    $a < 0$
2. $-1(a) > -1(0)$ Mult. prop. of order
3. $-1(a) = -a$ Mult. prop. of -1
4. $-1(0) = 0$ Mult. prop. of 0
5. ∴ $-a > 0$ Subs. prin.
35. 1. x is a real number;   Hypothesis
    $x > 0$
2. $x^2 > 0$ Exercise 34

Written Exercises, page 166

1. $\{a: a > -1\}$
3. $\{c: c > -6\}$
5. $\{b: b < -11\}$
7. $\{m: m < 12\}$
9. $\{t: t > -6\}$
11. $\{w: w > 0\}$
13. $\{c: c > 7\}$
15. $\{d: d < -13\}$
17. $\{e: e < -6\}$

19. $\{f: f > -12\}$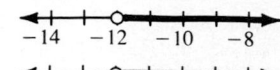

21. $\{u: u > 15\}$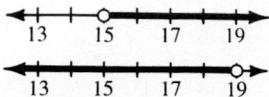

23. $\{h: h < 19\}$

25. $\{i: i < 2\}$ **27.** $\{j: j < 12\}$ **29.** $\{k: k > -10\}$

31. $\left\{t: t > \dfrac{1}{9}\right\}$ **33.** $\{a: a > -6\}$ **35.** Ø

37. $\{s: s < -2\}$ **39.** $\{b: b < 4\}$

Computer Exercises, pages 166–167
1. $\{x: x < 3\}$ **3.** $\{z: z > -6\}$ **5.** $\{a: a < 4\}$
7. $\{b: b > -3\}$ **9.** $\{m: m < 5\}$ **11.** $\{r: r > 3\}$
13. Ø

Written Exercises, pages 169–170
1. $\{-1, 1\}$; $\{-3, -1, 0, 1, 3\}$ **3.** Ø; $\{-6, -4, 1, 2, 4,$
$6\}$ **5.** $\{-2, 1, 8, 10\}$; $\{-2, 1, 8, 10\}$ **7.** Ø; $\{1, 2, 3, 4\}$
9. $\{2, 4\}$; $\{1, 3$, the even whole numbers$\}$ **11.** Ø;$\{$ the
integers less than 4 and the integers greater than 6$\}$
13. $\{2, 3, 4\}$; $\{1, 2, 3, 4, 5\}$ **15.** $\{0, 1, 2, 3, 4, 5, 6\}$;
$\{0, 1, 2, 3, 4, 5, 6, 7\}$

17. a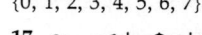

   b.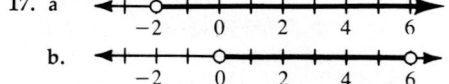

19. a.

   b.

21. a.

   b.

23. a.

   b.

25.

27.

29.

31.

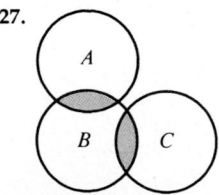
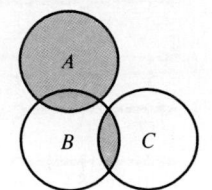
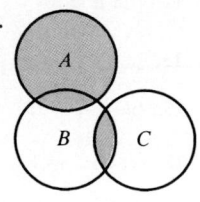

33.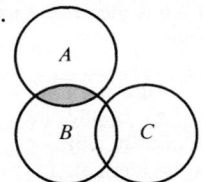

35. $\mathcal{R}$

37. Ø

39. $\{1, 3, 6\}$

41. $\{0, 1, 2, 3, 4, 5\}$

43. $\{0, 3, 5, 6\}$

Written Exercises, pages 173–174
1. $\{a: a < -3 \text{ or } a > 1\}$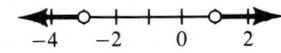

3. Ø

5. $\mathcal{R}$

7. $\{z: -1 < z < 2\}$

9. $\{w: w > 0\}$

11. $\{x: -12 \leq x < -3\}$

13. $\{z: z > 1 \text{ or } z < -7\}$

15. $\{a: -3 < a \leq 5\}$

17. $\{t: -5 \leq t \leq 6\}$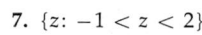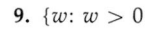

19. $\{x: -2 < x < 3\}$

21. $\{d: -4 < d \leq 8\}$

23. $\{r: -4 < r < 2\}$

25. $\left\{k: -1 \leq k < \dfrac{2}{3}\right\}$ **27.** $\{h: 2 < h < 6\}$

29. $\left\{y: y < \dfrac{5}{2}\right\}$ **31.** $\left\{g: -\dfrac{8}{3} < g < \dfrac{10}{3}\right\}$
33. $\{x: -6 \leq x \leq 10\}$ **35.** $\{d: 2 < d < 5\}$ **37.** $\mathcal{R}$
39. $\{e: -4 < e < 1\}$ **41.** $\left\{x: \dfrac{1}{2} < x \leq 2\right\}$ **43.** $\mathcal{R}$
45. $\{x: -2 \leq x < 3\}$ **47.** $\{x: x \geq -3\}$

Written Exercises, pages 177–178
1. $|x - 5| = 2$ **3.** $|x + 1| < 3$ **5.** $|x - 1| \leq 6$
7. $|x - 10| \geq 5$ **9.** $|x - 5| > b$

11. $\{-2, 10\}$

13. $\{-3, 1\}$

15. $\{v: 0 \leq v \leq 12\}$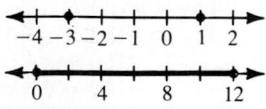

17. $\mathcal{R}$

19. $\{v: 8 \leq v \leq 12\}$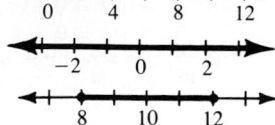

21. $\{x: x < -6 \text{ or } x > 8\}$

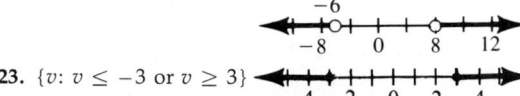

23. $\{v: v \le -3 \text{ or } v \ge 3\}$

25. $\{w: -6 \le w \le 6\}$

27. $\{w: w \le 2 \text{ or } w \ge 5\}$

29. $\{c: c \le 1.3 \text{ or } c \ge 2.7\}$

31. $\mathcal{R}$

33. $\{x: -11 < x < -1\}$

35. $\{c: c \le 0 \text{ or } c \ge 6\}$ **37.** $\emptyset$ **39.** $\left\{c: -\frac{13}{2} < c < \frac{7}{2}\right\}$
41. $\{x: 1 \le x \le 11\}$ **43.** $\{d: d < -9 \text{ or } d > -1\}$
Examples are given for Exercises 45–55.
45. $|x| = 2$ **47.** $|x| < -1$ **49.** $|x - 7| = 2$
51. $|x + 10| = 2$ **53.** $|x| < 4$ **55.** $|x - 2| > 3$
57. a. $\emptyset$ **b.** $\{x: x \ge -3\}$ **c.** $\{x: x < -3\}$
59. a. $\mathcal{R}$ **b.** $\mathcal{R}$ **c.** $\{x: x \le 0\}$ **61. a.** $\{x: x > 0\}$
b. $\{x: x \le 0\}$ **c.** $\emptyset$
63. $\{d: -5 < d < -3 \text{ or } 3 < d < 5\}$ **65.** $\emptyset$

Self-Test 1, pages 178–179
1. Addition property of order **2.** Multiplication property of order **3.** $\{x: x > -4\}$ **4.** $\{y: y \ge -1\}$
5. {the whole numbers less than 5} **6.** {the integers greater than -5 and less than or equal to 6}

7. $\{t: -10 \le t < 2\}$

8. $\{m: m > 5 \text{ or } m < -11\}$

9. $\{-5, 7\}$

10. $\{-6, -2\}$

11. $\{p: p < 2 \text{ or } p > 8\}$

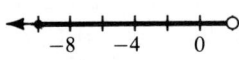

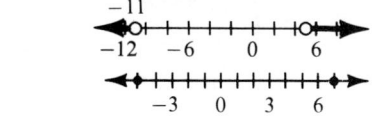

12. $\{x: -\frac{3}{2} \le x \le \frac{9}{2}\}$

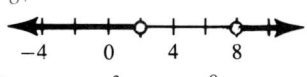

Problems, pages 182–183
1. 191, 192 **3.** $-81, -79$ **5.** 21, 28, 35 **7.** 94, 96, 98
9. 23, 25 **11.** 7 **13.** 360 m^2 **15.** 12 years, 15 years, 18 years, 21 years **17.** $\{3, 4, 5\}, \{4, 5, 6\}, \{5, 6, 7\},$

$\{6, 7\,8\}$ **19.** $\{-45, -42, -39\}, \{-42, -39, -36\},$
$\{-39, -36, -33\}$ **21.** 4, 6; 0, 2 **23.** The sum of an odd integer x and the next odd integer $x + 2$ is $2x + 2$ or $2(x + 1)$, which is even because it is a multiple of 2.

Problems, pages 186–187
1. $61°$ **3.** $74°, 16°$ **5.** $4°$ **7.** $72°, 24°, 84°$ **9.** $50°$, $130°, 130°$ **11.** $30°$ **13.** $52°, 54°, 74°$ **15.** 15 **17.** 36
19. 22 **21. a.** $75°$ **b.** $89°$ **23.** measure of $\angle A$ is $80°$, measure of $\angle B$ is $60°$, measure of $\angle C$ is $40°$

Problems, pages 192–194
1. van, 84 km/h; truck, 60 km/h **3.** $3\frac{1}{2}$ h **5.** 4 km
7. 4.5 m/s **9.** 56 km/h; 20 km/h **11.** 100 km/h
13. 80 km/h; 240 km/h **15.** 1560 km **17.** 52.2 km/h; 84.4 km/h **19.** 27 km/h **21.** 2 km/h **23.** 30 km
25. 12 m **27.** 0.8 km; 21 s

Problems, pages 199–200
1. 7.5 g **3.** 100 g **5.** 15 dimes, 30 quarters **7.** 19 students **9.** 55 km/h **11.** 100 yellow, 105 pink, 410 red **13.** 28 students **15.** 1 h **17.** 40 kg; $6\frac{2}{3}\%$

Problems, pages 202–203
1. No solution; the sum of two consecutive integers must be odd because the sum of an odd integer and an even integer is odd. **3.** 42 m, 44 m, 46 m
5. No solution, $\angle B$ and $\angle C$ are complementary angles so they are acute angles whose measures total $90°$; the measure of $\angle A$ is $90°$, and so the measure of $\angle B$ cannot be greater than the measure of $\angle A$. **7.** Yes, the clerk received 20 nickels, 40 dimes, and 60 quarters. **9.** 5 dimes **11.** No solution; not enough information is given; the time traveled by bus is needed. **13.** 1440 km **15.** 45 kg
17. 60 km/h; 150 km/h; no solution because the return trip would have to be made in zero hours.

Self-Test 2, page 203
1. 76, 78, 80 **2.** the measure of $\angle A$ is $50°$, the measure of $\angle B$ is $62°$, and the measure of $\angle C$ is $68°$ **3.** 7.5 km **4.** 3 kg **5.** 13 quarters

Extra, page 206
1. T **3.** T **5.** T **7.** T **9.** T **11.** $p \to q$ is false implies that p is true and q is false. $\sim q$ is false, so $p \wedge \sim q$ is true.
13. No;

p	q	$p \vee q$	$(p \vee q) \to p$
T	T	T	T
T	F	T	T
F	T	T	F
F	F	F	T

15. Yes;

p	q	$p \vee q$	$q \vee p$	$(p \vee q) \leftrightarrow (q \vee p)$
T	T	T	T	T
T	F	T	T	T
F	T	T	T	T
F	F	F	F	T

17. Yes

Chapter Review, pages 208–209

1. c **2.** d **3.** a **4.** c **5.** c **6.** c **7.** a **8.** c **9.** b
10. a **11.** b **12.** d **13.** b **14.** a

Mixed Review, page 210

1. $\frac{8}{3}$ **3.** $-\frac{3}{2}$ **5.** $-3a^2$ **7.** $4s^2$ **9.** -6 **11.** 113
13. -5

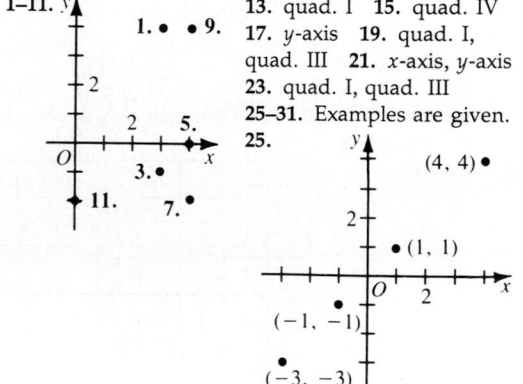

15.
17.
19.

21. $\left\{-\frac{1}{2}\right\}$ **23.** $\{-1\}$ **25.** $\{m: m < 3\}$ **27.** $\emptyset$

29. $\{3, 7\}$ **31.** $\left\{j: -4 < j < \frac{3}{2}\right\}$ **33.** 11 nickels,
9 dimes, 5 quarters **35.** 6 years

Preparing for College Entrance Exams, page 211
1. A **3.** B **5.** D **7.** A

Application, page 213
1. a. 17 km/h/s **b.** 42.5 km/h **c.** approximately 0.06
km **3.** approximately 0.16 km

Chapter 5   Graphs and Functions

Written Exercises, page 218
1–11.
13. quad. I **15.** quad. IV
17. y-axis **19.** quad. I,
quad. III **21.** x-axis, y-axis
23. quad. I, quad. III
25–31. Examples are given.
25.

27.

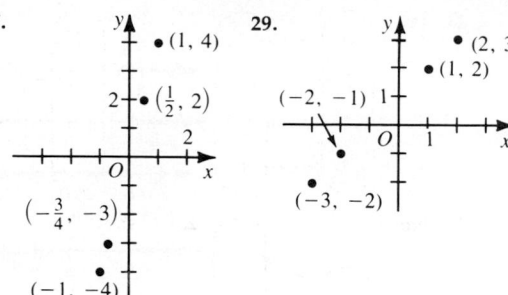

29.

31.

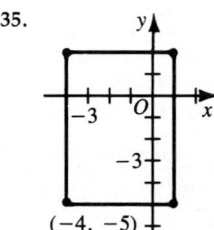

33.

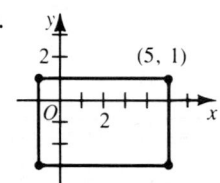

35.

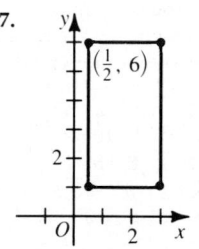

37.

39. (1, 0), (9, 2), (3, -6) **41.** (-2, -2), (6, -6)
43. $|b - c|$

Written Exercises, pages 222–223

1. $\left\{\left(\frac{3}{4}, -\frac{3}{4}\right), \left(\frac{1}{2}, -\frac{1}{2}\right), \left(\frac{1}{4}, -\frac{1}{4}\right)\right\}$

3. $\{(0, 0), (1, -1), (1, 1)\}$

5. $\{(-1, 1), (0, 0), (1, 1), (2, 4)\}$

7. $\{(1, 10), (2, 10), (5, 10), (10, 10)\}$

9. Yes;

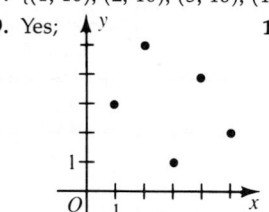

11. No;

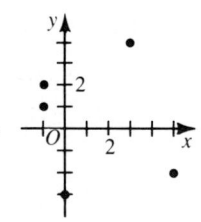

13. Yes;

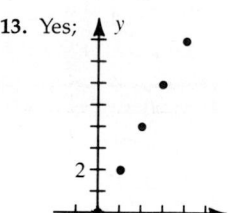

15. No;

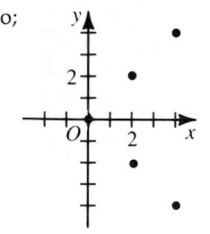

17. Yes;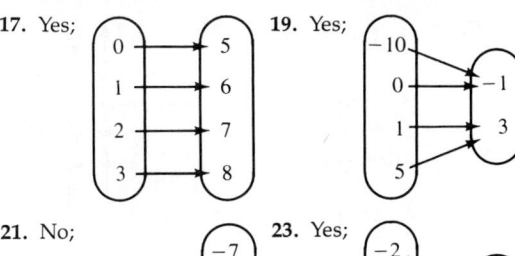

19. Yes;

21. No;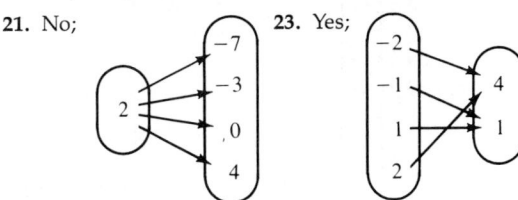

23. Yes;

25. {((2, 1), 1), ((3, 1), 2), ((5, 2), 3), ((5, 1), 4), ((10, 5), 5)}; Yes; subtract the second coordinate from the first **27.** 2 **29.** 5 **31.** No, at least one element of the domain must be mapped onto at least two elements of the range. ∴ Two ordered pairs will have the same first coordinate.

Computer Exercises, page 223
1. quad. I; quad. III; quad. IV; quad. II; *x*-axis; *y*-axis; quad. IV; quad. III; quad. I; quad. III; *y*-axis; *x*-axis, *y*-axis

Written Exercises, pages 226–227
1. {1, 2, 3, 4, 5} **3.** {−4, −2, 0, 2, 4} **5.** {−7, −4, −1, 2, 5} **7.** {2, 1, 0} **9.** {4, 1, 0} **11.** {9, 4, 1, 0}

13.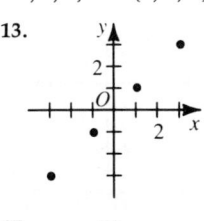

15.

17.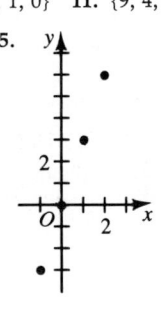

19.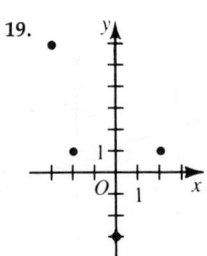

21.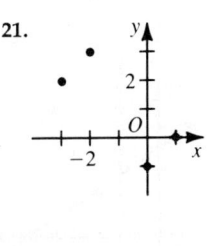

23.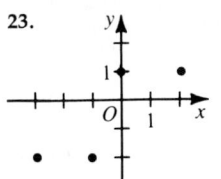

25. a. 5 **b.** $x = -5$
27. a. 0 **b.** $x = 0$
29. a. 6 **b.** $x = -3$
31. a. −1 **b.** $x = 1$;
$x = -1$ **33. a.** −9
b. $x = 3$; $x = -3$
35. a. 0 **b.** $x = 0$

37. 9 **39.** 36 **41.** 17 **43.** 5 **45.** 0 **47.** 0 **49.** 3
51. −1 **53.** $x + 2$ **55. a.** 3 **b.** 3 **c.** −3 **d.** −4

Self-Test 1, page 227
1–3. **3.**

2.

1.

4. $D = \{0, -1, 5, -3\}$,
$R = \{2, 3, -2, 6\}$

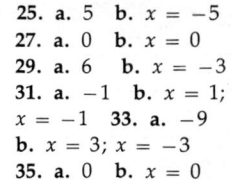

5. $D = \{2, -1, 0, 5\}$,
$R = \{1, 0, 4\}$

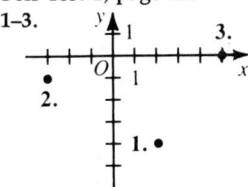

6. Yes **7.** No

8.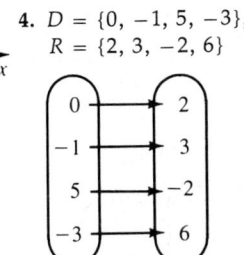

9. **10.**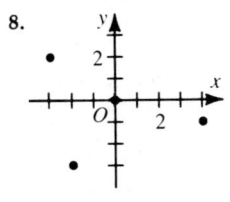

Written Exercises, pages 230–231
1. (5, −5), (0, 0), (−3, 3) **3.** (9, 4), (5, 0), (−7, −12) **5.** (1, 1), (0, −1), (−3, −7)
7. (2, 1), (0, 9), (−5, 29) **9.** (1, 3), $\left(\frac{11}{2}, 0\right)$, (13, −5) **11.** (3, 4), (2, −1), (−2, −1)
13. {(−2, −2), (−1, −1), (0, 0), (1, 1), (2, 2)}

Exercises 15–35: graphs follow.
15. {(−2, −5), (−1, −2), (0, 1), (1, 4), (2, 7)}
17. {(−2, 9), (−1, 7), (0, 5), (1, 3), (2, 1)}
19. {(−2, −8), (−1, −5), (0, −2), (1, 1), (2, 4)}
21. {(−1, −2), (1, 1)}   **23.** ∅
25. {(−1, −1), (−1, 0), (−1, 1), (−1, 2), (0, 0), (0, 1), (0, 2), (1, 1), (1, 2)}
27. {(−1, −1), (−1, 0), (−1, 1), (−1, 2), (0, 0), (0, 1), (0, 2), (1, 1), (1, 2)}
29. {(−1, −2), (−1, −1), (−1, 0), (−1, 1), (0, −2), (0, −1), (0, 0), (1, −2), (1, −1)}
31. {(0, −2), (0, −1), (1, −2), (1, −1), (1, 0), (1, 1)}
33. {(−1, −2), (0, −2), (0, −1), (0, 0), (1, −2), (1, −1), (1, 0), (1, 1), (1, 2)}
35. {(−1, 0), (−1, 1), (−1, 2), (0, 0), (0, 1), (0, 2), (1, 0), (1, 1), (1, 2)}

31.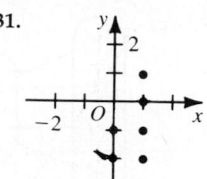

33.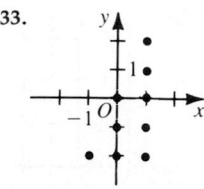

35.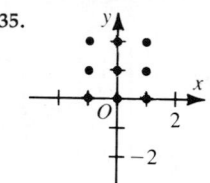

37. {(2, 1), (2, 2), (3, 1)}
39. {(−2, −3), (−2, −2), (−2, −1), (−1, −1)}
41. {(3, 0), (3, 1), (4, 0), (5, 0)}
43. {(−3, 2), (−2, 3), (2, 3), (3, 2)}
45. {(−5, −3), (−5, 3), (−3, −1), (−3, 1), (3, −1), (3, 1), (5, −3), (5, 3)}
47. {(−1, 0), (0, −1), (1, 0)}

15.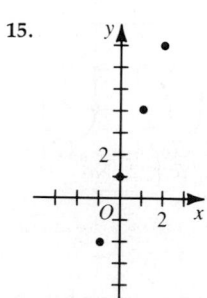

17.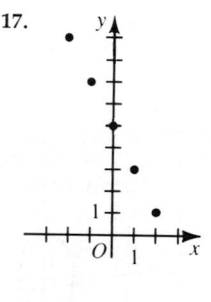

Computer Exercises, page 231
1. Yes **3.** No **5.** Yes **7.** Yes **11.** (−6, 12); (−3, 10); (0, 8); (3, 6); (6, 4); (9, 2); (12, 0)

19.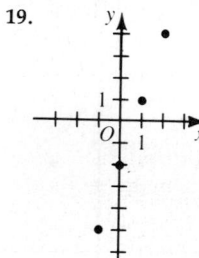

21.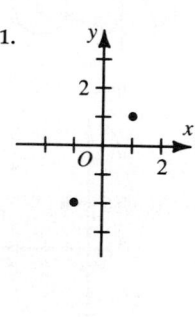

Written Exercises, pages 235–236
1. $3x − 4y = −5$ **3.** $5x − 3y = 8$ **5.** $x + 2y = 6$

7.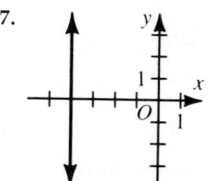

9.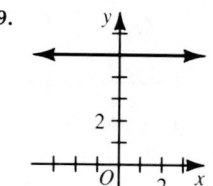

23.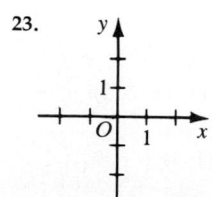

25.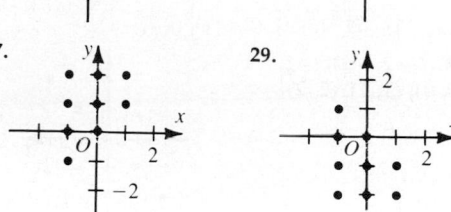

11.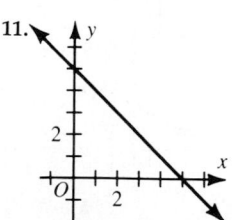

13.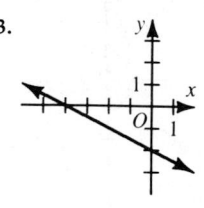

27.

29.

15.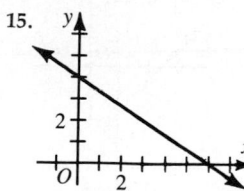

17.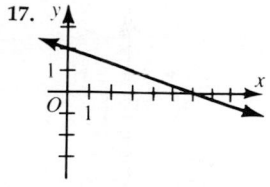

19.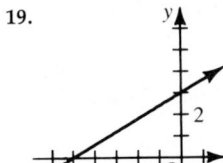

21.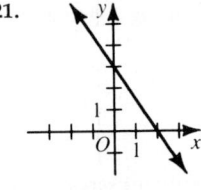

23.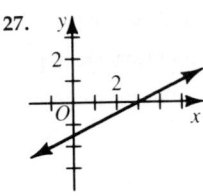

25.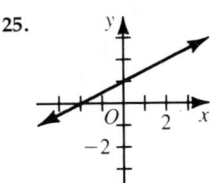

27.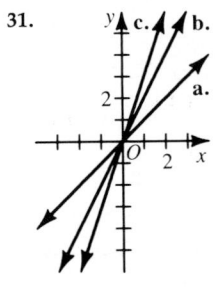

29.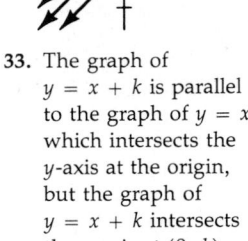

31.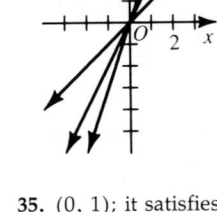

33. The graph of $y = x + k$ is parallel to the graph of $y = x$, which intersects the y-axis at the origin, but the graph of $y = x + k$ intersects the y-axis at $(0, k)$.

35. $(0, 1)$; it satisfies both equations.

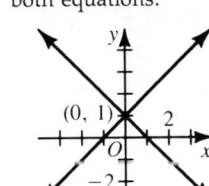

37. $(2, 3)$; it satisfies both equations.

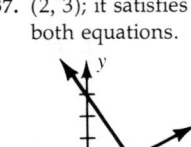

39. not linear

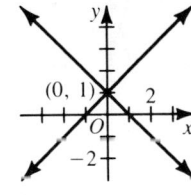

41. not linear

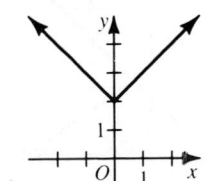

43. not linear

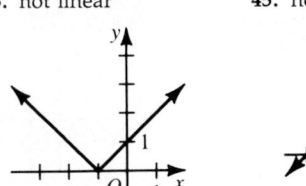

45. not linear

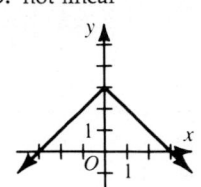

47. The graph of $y = |x| + k$ is the graph of $y = |x|$ translated (slid) vertically k units (up if $k > 0$ and down if $k < 0$).

Written Exercises, page 239

1.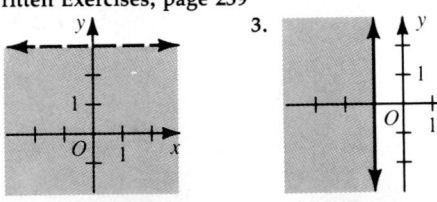

3.

5.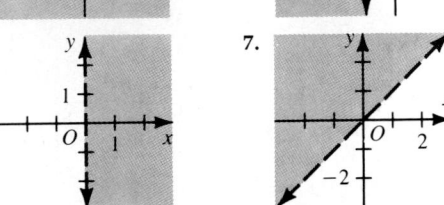

7.

9.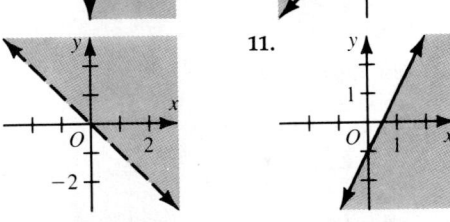

11.

13.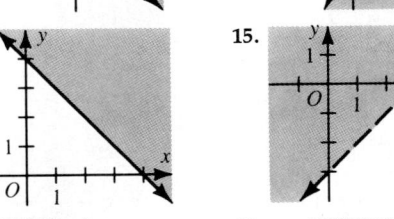

15.

17.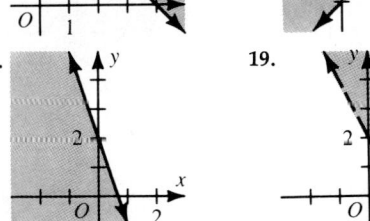

19.

21.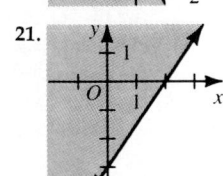

23.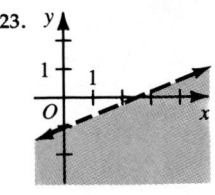

25. 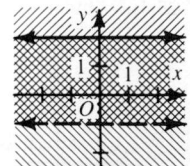 **27.**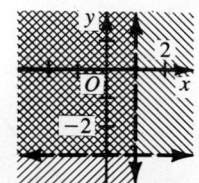

29.  **31.**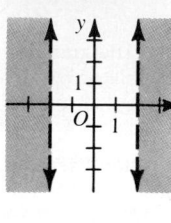

33. 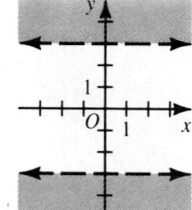 **35.**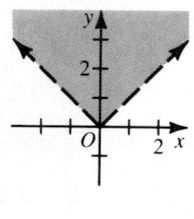

1. {(2, −2)} **2.** {(2, −4)} **3.** {(−2, −4), (−2, −2), (0, −4), (0, −2), (2, −4), (2, −2)}

4. 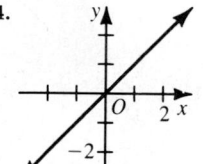 **5.**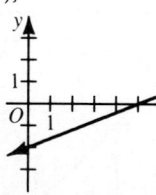

6. 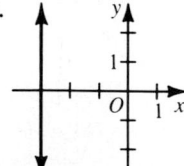 **7.**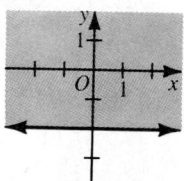

8. 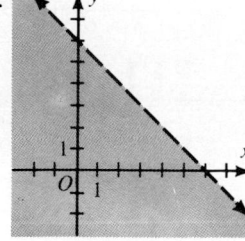 **9.**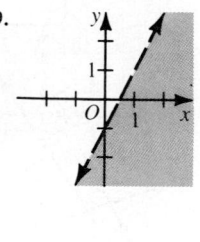

1. Obj. 1, p. 228 **3.** Obj. 2, p. 228 **5.** Sec. 5-5
7. The opposite of the sum of a and b is equal to the sum of the opposite of a and the opposite of b.

1. 2 **3.** 0 **5.** −1 **7.** −2 **9.** $\frac{7}{2}$ **11.** no slope

13. 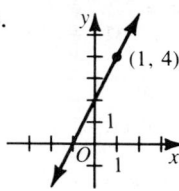 **15.**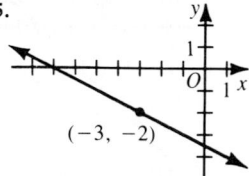

17.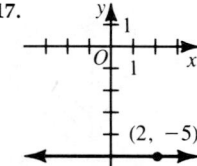

19. collinear; slope = 2
21. collinear; slope = $-\frac{1}{3}$
23. collinear; slope = 0
25. −17 **27.** 2 **29.** $\frac{1}{2}$
31. $\overline{AB}:-2;\ \overline{AC}:-\frac{1}{3};\ \overline{BC}:\frac{1}{2}$

Exercises 1-11: graphs follow.
1. slope = 2; y-intercept = 1
3. slope = −3; y-intercept = 2
5. slope = 1; y-intercept = −4
7. slope = 3; y-intercept = 5
9. slope = $\frac{1}{2}$; y-intercept = 3
11. slope = −3; y-intercept = $\frac{1}{2}$

1. 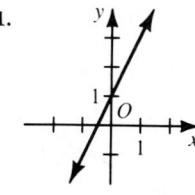 **3.**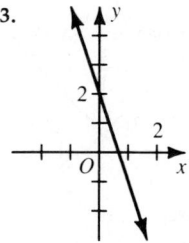

5. 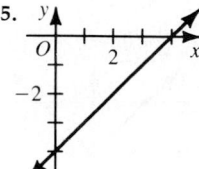 **7.**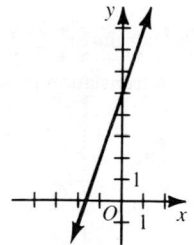

9. 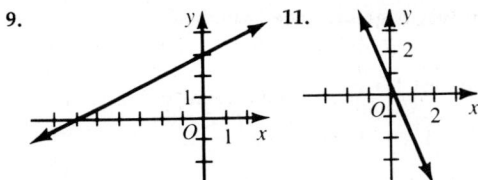 **11.**

13. $5x - y = 1$ **15.** $2x + y = 0$ **17.** $y = 2$
19. $x - 2y = 8$ **21.** $3x - 2y = -4$
23. $6x + 10y = 5$ **25.** $5x + y = 3$
27. $3x - y = 11$ **29.** $x - y = 7$ **31.** $x = -3$
33. $\frac{3}{2}$ **35.** -4 **37.** $y = -\frac{a}{b}x + \frac{c}{b}, b \neq 0$

39. If $by - c = 0$, $y = \frac{c}{b}$. From Exercise 37, the

y-intercept of the graph of $ax + by = c$ is $\frac{c}{b}$.

Replacing y with $\frac{c}{b}$ in the equation $by - c = 0$,

you obtain $b\left(\frac{c}{b}\right) - c = 0$, or $0 = 0$. Thus, $\frac{c}{b}$ is a

root of $by - c = 0$.

Written Exercises, pages 254–255
1. $2x - y = 1$ **3.** $2x + y = 4$ **5.** $x - 2y = -6$
7. $x + 5y = -13$ **9.** $y = 5$ **11.** $x = -2$
13. $x + y = 6$ **15.** $3x + y = 0$ **17.** $5x + y = 24$
19. $x - 3y = 8$ **21.** $y = 7$ **23.** $x = -3$ **25.** 3

27. -3 **29.** 0 **31.** $f: x \to -x + 3$

33. $f: x \to \frac{5}{2}x + \frac{1}{2}$ **35.** $f: x \to \frac{4}{3}x + \frac{20}{3}$

37. $r = -9, s = 4$ **39.** The equation of the line
with slope m and y-intercept b is $y = mx + b$.
Since point (x_1, y_1) is on the line, the coordinates
satisfy the equation. So $y_1 = mx_1 + b$. $\therefore b = y_1 - mx_1$

Self-Test 3, page 255
1. 3 **2.** $-\frac{3}{2}$ **3.** $m = 3, b = -5$ **4.** $m = -\frac{3}{2}$,

$b = \frac{7}{2}$ **5.** $4x - y = 1$ **6.** $2x + 5y = 15$

7. $y - 3x$ **8.** $4x - 3y = 6$

Extra, page 257
1. a. $x' = x + 1, y' = y - 3$ **b.** $(-6, 0)$
3. a. $x' = x + 13, y' = y - 3$ **b.** $(6, 0)$
5. a. $x' = x + 7, y' = y + 2$ **b.** $(0, 5)$
7. a. $x' = x + 9, y' = y - 5$ **b.** $(2, -2)$
9. A translation in which $h = 0$;
a translation in which $k = 0$

Chapter Review, pages 259–261
1. c **2.** c **3.** d **4.** b **5.** b **6.** a **7.** d **8.** c
9. a **10.** c **11.** d **12.** b **13.** c **14.** b **15.** c
16. c **17.** b **18.** d **19.** b **20.** a

Cumulative Review, pages 262–263
1.

3.   **5.**

7. 54 **9.** 2 **11.** $\frac{1}{2}$ **13.** Subtraction property of

equality **15.** $\{39\}$ **17.** $\{-96\}$ **19.** $\{6\}$ **21.** $\emptyset$
23. $\{2\}$ **25.** 52 years

27. $\{v: v \leq -8\}$ **29.** $\{a: a > 9\}$

31. $\{p: p < -8 \text{ or } p > 4\}$ **33.** $\{g: -1 < g < 6\}$

35. $\{x: x \leq 1 \text{ or } x \geq 7\}$

37. $\{-2, -1, 0, 1, 2, 3\}; \{-1\}$ **39.** $\{2, 4, \text{the odd}$
whole numbers$\}; \{1, 3\}$ **41.** $\mathcal{R}$; {the real numbers

between $-\frac{1}{3}$ and $2\frac{2}{3}$}

Application, page 265
1. Thursday **3.** Tuesday

Chapter 6 Systems of Open Sentences

Written Exercises, pages 272–273
1. $\{(-7, 2)\}$ **3.** $\{(1, 1)\}$

5. $\emptyset$ **7.** $\{(x, y): y = 2x + 3\}$
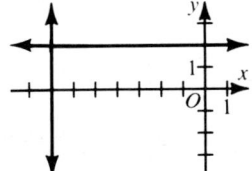 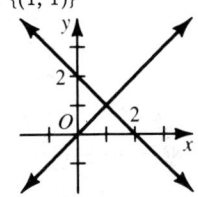

9. $\{(-3, -4)\}$ **11.** $\{(x, y): x = 4y - 12\}$
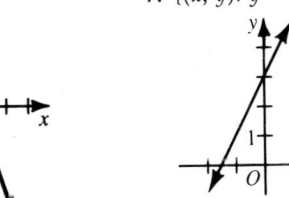

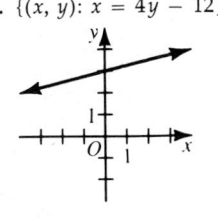

Answers to Selected Exercises **17**

13. exactly one member; consistent   **15.** infinite set; consistent   **17.** exactly one member; consistent   **19.** infinite set; consistent   **21.** empty set; inconsistent   **23.** empty set; inconsistent

25. $\left\{\left(3\frac{1}{2}, 1\frac{1}{2}\right)\right\}$ **27.** $\{(-3, 2)\}$

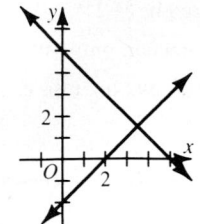 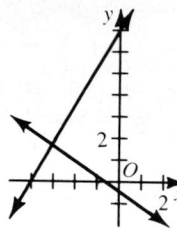

29. $(2, 2)$ **31.** $(5, -5)$
33. a. $(-2, -1), (3, 4), (3, -6)$ **b.** 25
35. a. $(1, 7), (-3, 8), (-3, -9)$ **b.** 34
37. a. $(6, 4), (6, -4), (-3, -7), (-3, 1)$ **b.** 72

Computer Exercises, page 273
1. No   **3.** Yes

Written Exercises, page 276
1. $\{(6, -11)\}$ **3.** $\left\{\left(-\frac{1}{3}, 4\right)\right\}$ **5.** $\left\{\left(\frac{1}{2}, -2\right)\right\}$
7. $\left\{\left(\frac{1}{2}, -\frac{2}{3}\right)\right\}$ **9.** $\left\{\left(-4, \frac{3}{4}\right)\right\}$ **11.** $\varnothing$ **13.** $\left\{\left(-\frac{2}{3}, 0\right)\right\}$
15. $\left\{\left(-\frac{11}{2}, -\frac{3}{2}\right)\right\}$ **17.** $\left\{\left(\frac{7}{3}, \frac{3}{5}\right)\right\}$ **19.** $\{(-2, 5)\}$
21. $\{(-5, 7)\}$ **23.** $\varnothing$ **25.** $\left\{\left(1, \frac{14 - c}{3}\right)\right\}$ **27.** $a = -b$

Computer Exercises, page 277
1. $(10, 3)$ **3.** $\varnothing$

Written Exercises, pages 279–280
1. $\{(1, 1)\}$ **3.** $\{(4, 0)\}$ **5.** $\{(0, -3)\}$
7. $\left\{\left(-3, \frac{1}{3}\right)\right\}$ **9.** $\varnothing$ **11.** $\{(0, 2)\}$ **13.** $\{(n, 2n)\}$
15. $\left\{\left(\frac{k}{3}, -\frac{k}{2}\right)\right\}$ **17.** $\{(b, -4a)\}$
19. $(0, 3), (-3, 0), (3, 0)$

Written Exercises, pages 282–283
1. $\{(-2, 5)\}$ **3.** $\{(8, 1)\}$ **5.** $\left\{\left(6, \frac{1}{3}\right)\right\}$ **7.** $\{(7, 1)\}$
9. $\{(-1, -2)\}$ **11.** $\varnothing$ **13.** $\left\{\left(\frac{1}{2}, 5\right)\right\}$ **15.** $\left\{\left(\frac{3}{5}, \frac{2}{5}\right)\right\}$
17. $\{(3, 1)\}$ **19.** $\{(2, -3)\}$ **21.** $\left\{\left(-\frac{2}{3}, \frac{5}{3}\right)\right\}$
23. $\left\{\left(\frac{50}{31}, \frac{15}{31}\right)\right\}$ **25.** $\varnothing$ **27.** $\left\{\left(-\frac{1}{2}, -\frac{3}{2}\right)\right\}$
29. $\{(8, 6)\}$ **31.** $a = \frac{1}{2}, b = -1$ **33.** $a = 3$,
$b = -4$ **35.** Assume that (x_1, y_1) is a solution of the system. Then $m_1x_1 + b_1 = m_2x_1 + b_2$. By hypothesis $m_1 = m_2$, so $m_1x_1 + b_1 = m_1x_1 + b_2$. Then $b_1 = b_2$,

but this contradicts the hypothesis $b_1 \neq b_2$.
$\therefore$ the system has no solution.

Self-Test 1, page 283
1. inconsistent   **2.** $\{(2, 3)\}$ **3.** $\{(2, -1)\}$
4. $\{(-3, 4)\}$ **5.** $\{(5, -1)\}$

Problems, pages 286–288
1. Susan has 25 fish, Tammy has 18 fish.   **3.** $105°$
5. \$3.25   **7.** 12 m by 4 m   **9.** Charles read 21 books, Frank read 8 books.   **11.** 6 nickels, 34 quarters
13. twenty 20¢ stamps, fifteen 25¢ stamps
15. Lee's mother: 47 years; her aunt: 36 years
17. 3 bath towels, 5 hand towels   **19.** 40 cm by 80 cm   **21.** 40 nickels, 6 dimes, 4 quarters

On the Calculator, page 288
1. $\{(1.2, -0.6)\}$ **3.** $\{(3.3, 1.9)\}$

Problems, pages 290–291
1. wind speed: 50 km/h; air speed: 550 km/h   **3.** air speed: 310 km/h; wind speed: 10 km/h   **5.** rowing rate: 4 km/h; speed of current: 2 km/h   **7.** 15 km/h
9. rate in still water: 4 km/h; rate of current: 2 km/h.
11. $s = 2c$ **13.** 0.6 km/h; 25 min   **15.** original wind speed: 25 km/h; distance from Reno: 1125 km

Problems, pages 296–297
1. 64   **3.** 13   **5.** 75   **7.** 19   **9.** 386   **11.** 824
13. $100h + 10t + u$ may be rewritten $99h + h + 9t + t + u$ or $9(11h + t) + h + t + u$. Since $9(11h + t)$ is divisible by 9, the entire number is divisible by 9 if and only if $h + t + u$ is divisible by 9.   **15.** $10t + u + 10u + t = 11t + 11u = 11(t + u)$, which is divisible by 11.   **17.** 832
19. $10t + u - [10(t - 1) + (u + 1)] = 10t + u - (10t - 10 + u + 1) = 9$

Self-Test 2, page 297
1. 8 years   **2.** rowing rate: 6 km/h; rate of current: 2 km/h   **3.** 81

Written Exercises, page 300
1. 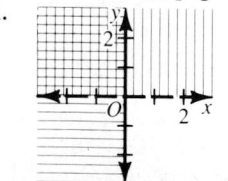   **3.**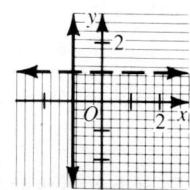

5. 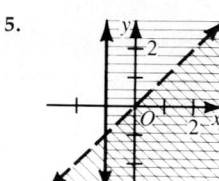   **7.**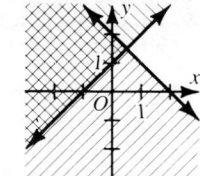

9.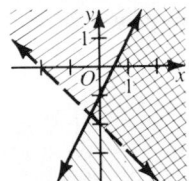

11.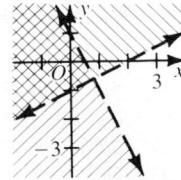

13.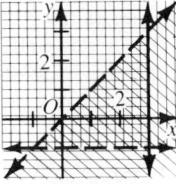

15.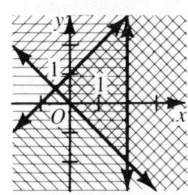

17.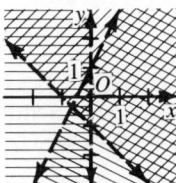

19.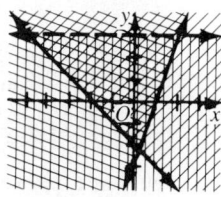

21.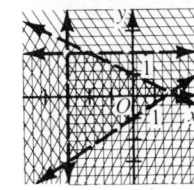

Written Exercises, pages 304–306

1.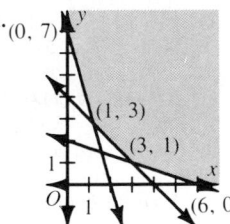

3. a. 11 **b.** 14 **c.** 5 **5.** 8

7. a.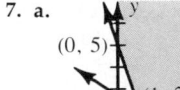

b. 8

9. a.

b. 9

11. Let x = number of hamsters $x \geq 0$
Let y = number of rabbits $y \geq 0$
$$x + y \geq 8$$
$$5x + 25y \geq 100$$

13. 5 hamsters and 3 rabbits **15.** Shop A: 5 hours; Shop B: 6 hours

Written Exercises, pages 309–310

1. $\{(1, -1, 2)\}$ **3.** $\{(2, 3, 1)\}$ **5.** $\{(2, -3, -2)\}$

7. $\left\{\left(\frac{1}{2}, 3, -2\right)\right\}$ **9.** $\left\{\left(\frac{104}{19}, \frac{35}{19}, \frac{120}{19}\right)\right\}$

11. $\{(x, y, z): 4x - 10z = -8\}$ **13.** $a = 4$, $b = -3$, $c = 1$ **15.** $a = 3$, $b = -2$, $c = 4$
17. $a = 1$, $b = -1$, $c = 3$

Problems, page 310
1. length: 3 cm; width: 5 cm; height: 9 cm
3. 1 nickel, 4 dimes, 6 quarters **5.** 564

Self-Test 3, page 311
1.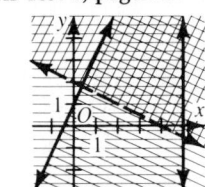

2. 480 **3.** 1 order from Provincial and 5 orders from Rocky Mountain
4. $\{(5, -4, 2)\}$

Extra, page 315
1. $\{(1, 2)\}$ **3.** $\{(4, 2)\}$ **5.** $\{(4, -1)\}$

Chapter Review, pages 316–317
1. d **2.** a **3.** d **4.** c **5.** c **6.** a **7.** c **8.** c
9. b **10.** c **11.** b **12.** b **13.** d **14.** c **15.** a

Mixed Review, page 319
1. 8 **3.** −7 **5.** $\frac{1}{6}$ **7.** $\frac{1}{2}$ **9.** $4x + 10y$ **11.** $7a - 1$
13. $11m + 34$ **15.** $-15c - 12d$ **17.** $60pqr$

19.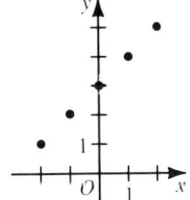

21.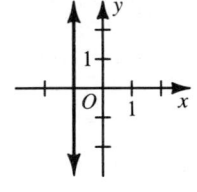

23.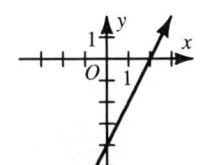

25. 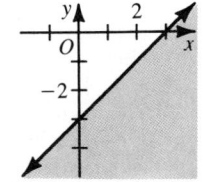 **27.**

29. **31.**

33. 0 **35.** −12 **37.** $16x + 20y = 35$ **39.** $y = 6$

Preparing for College Entrance Exams, page 320
1. E **3.** D **5.** D

Contest Problems, page 321
1. 15 3. $7c + 9$ 5. 27

Chapter 7 Polynomials and Factoring

Written Exercises, pages 326–327
1. 3; 5 3. 3; j 5. 2 7. $\underline{3rs} + \underline{\underline{2}} - \underline{2rs} - \underline{\underline{3}} = rs - 1$ 9. $b^2 - \underline{7b} - 6b^2 + \underline{b} = -5b^2 - 6b$
11. $-\underline{3t} + \underline{\underline{t^3}} - \underline{5t} - \underline{\underline{t^3}} - 1 = -8t - 1$
13. $\underline{9ab} + \underline{\underline{bc}} + \underline{2ab} + \underline{\underline{bc}} + 7 = 11ab + 2bc + 7$
15. $\underline{4p^3q^2} - \underline{\underline{5p^2q^3}} - 9pq + \underline{\underline{p^2q^3}} - \underline{2p^3q^2} = 2p^3q^2 - 4p^2q^3 - 9pq$
17. $\underline{9rs} - \underline{\underline{3st}} + \underline{6rt} - \underline{\underline{5st}} - \underline{6rt} - \underline{rs} = 8rs - 8st$
19. $7a + 8b + c$ 21. $12s^2 + 4s - 2$
23. $4m^4 - 4m^2 - 1$ 25. $-3a + 2b - 3c$
27. $6s^2 + 4s - 8$ 29. $8m^4 - 6m^2 + 1$
31. $5p + q + 6r - s$ 33. $-x^2 - 6x - 5$
35. $ab - 3b + 12a$ 37. $-5m + 8$
39. $3b^2c + 3bc - 3bc^2$ 41. $7x^3 + 3x^2 + 2x - 4$
43. {2} 45. {3} 47. {−4}
49. $-3c + 6cd - c^2d^2 - 2d^2$
51. $7rs - 2r^2 + 3r^2s^2 - 3s^2$
53. No, a term of one that is the additive inverse of a term of the other may change the degree of the sum.

Written Exercises, pages 330–331
1. $-10b^3$ 3. $-192m^6n$ 5. $9a^3z^4$ 7. $-8y^7$
9. $-15u^3v^4$ 11. $4c^9$ 13. $72g^7$ 15. $a^{10}b^{11}$
17. $-4a^{16}b^{11}$ 19. x^5y^5 21. $-12m^8n^4$ 23. $r^5s^7t^4$
25. $11z^4$ 27. $-6b^4c^4$ 29. $14x^7y^5$
31. $-3y^9 - 32y^7$ 33. $b^7 - 11b^4$
35. $5c^4d^2 + 125c^3d^6$ 37. $3r^{11}s^{11}t^{10}$ 39. x^{3n}
41. z^{2n} 43. r^{3n} 45. {6} 47. {4} 49. {7}

Written Exercises, page 334
1. $2m^3 + 10m^2 - 18m$ 3. $z^4 - 5z^3 + 2z^2$
5. $-3a^3 + 7a^4 + 4a^5$ 7. $12s^5 - 8s^4 + 4s^3 - 20s^2$
9. $2p^4q - 2p^3q^2 + 2p^2q^3 - 2pq^4$
11. $-10u^2v^7 + 4u^4v^5 - 6u^6v^3$ 13. $2x^3 + 3x^2 - 10x$
15. $2a^3 - 5a^2$ 17. $7j^3 + 5j^2$ 19. $a^2 + 4a + 3$
21. $m^2 - 2m - 8$ 23. $2b^2 + 7b + 3$
25. $8t^2 + 6t - 9$ 27. $9z^2 - 49$ 29. $4c^2 + 20c + 25$
31. $r^3 + r^2 - 8r - 6$ 33. $2c^3 - 9c^2 + 10c - 3$
35. $u^3 - 2uv^2 + v^3$ 37. $6c^3 - c^2d - 12cd^2 - 5d^3$
39. {5} 41. {−5} 43. {3} 45. $x^3 + 2x^2 - 5x - 6$
47. $6m^3 + 11m^2 - 3m - 2$ 49. $r^3 + 6r^2 + 12r + 8$
51. $x^4 + 4x + 3$ 53. $x^{2n} + x^n$ 55. $x^{2n+1} + x^{n+1}$
57. The degree of the product is the degree of the product of the terms of greatest degree of the two polynomials, which is the sum of the degrees of the polynomials.

Written Exercises, pages 338–339
1. $j^2 + 2j - 15$ 3. $21 + 10u + u^2$ 5. $s^2 - 121$

7. $m^2 - 14m + 49$ 9. $3y^2 + 5y - 2$
11. $4c^2 + 12c + 9$ 13. $6p^2 - p - 2$ 15. $16a^2 - 49$
17. $12r^2 - rs - 6s^2$ 19. $12c^2 - 28cd + 15d^2$
21. $25b^2 + 20bc + 4c^2$ 23. $9a^4 - 1$
25. $16m^4 + 24m^2 + 9$ 27. $j^4 - 2k^2j^2 + k^4$
29. $25c^6 - 10c^3d^3 + d^6$ 31. $x^3 + 6x^2 + 12x + 8$
33. $27m^3 - 27m^2 + 9m - 1$ 35. $a^6 + 3a^4 + 3a^2 + 1$
37. $100^2 - 1^2 = 9999$ 39. $30^2 - 2^2 = 896$
41. $90^2 - 3^2 = 8091$ 43. $9^2 - 0.1^2 = 80.99$
45. {1} 47. {2} 49. {−1} 51. {7} 53. $\left\{-\dfrac{4}{3}\right\}$
55. $x^{2n} + 2x^ny^n + y^{2n}$
57. $x^{3n} - 3x^{2n}y^n + 3x^ny^{2n} - y^{3n}$
59. $|(x + 1)^2 - x^2| = |x^2 + 2x + 1 - x^2| = |2x + 1| = |x + x + 1|$

Problems, pages 339–340
1. 7 cm 3. 40 and 42 5. 17 and 19 7. $h = 5$ cm; $w = 7$ cm; $l = 12$ cm 9. 9 cm 11. Any three consecutive integers have this property.

Self-Test 1, page 340
1. $4r^2 - r + 4$ 2. $-4c^3 + 3c^2 + c + 1$
3. $2u^2 + 5u^2v - 3uv^2 + v^2$ 4. p^7 5. $-8a^5$
6. $3x^3y^4$ 7. m^{10} 8. $9s^2t^2$ 9. $8u^6v^9$ 10. $5y^3 - 10y^2$
11. $3a^4b - 2a^3b^2 + a^2b^3$ 12. $c^2 + c - 6$
13. $2d^2 + 3d - 20$ 14. $9x^2 - 12x + 4$
15. $4m^2 - 25n^2$

Written Exercises, pages 344–345
1. $3 \cdot 5^2$ 3. $2 \cdot 7 \cdot 11$ 5. $2^4 \cdot 3^2$ 7. $2^4 \cdot 3^4$
9. 1, 2, 3, 6, 9, 18 11. 1, 2, 3, 6, 9, 18, 27, 54
13. 1, 101 15. 1, 2, 3, 4, 6, 8, 12, 16, 24, 32, 48, 96 17. $1, -1, 2, -2, 4, -4, 8, -8, a, -a, 2a, -2a, 4a, -4a, 8a, -8a$ 19. $1, -1, 3, -3, r, -r, rs, -rs, rs^2, -rs^2, s, -s, s^2, -s^2, 3r, -3r, 3rs, -3rs, 3rs^2, -3rs^2, 3s, -3s, 3s^2, -3s^2$ 21. 11, 66 23. 15, 225
25. 1, 210 27. 45, 270 29. $3ab, 90a^3b^2$
31. $48r^3s^2, 144qr^5s^2$ 33. $22abc^3, 770a^2b^2c^5$ 35. $3k$
37. $-5r^2s$ 39. $13s^3t$ 41. $-5uvw$ 43. $6b^3c^3$ 45. 0
47. $3x$ 49. $-9a^n$ 51. $6j^{n+1}$ 53. $3a^7b^4$

Written Exercises, pages 347–348
1. $7(r + 2s)$ 3. $m(5m - 6)$ 5. $3b(5b + 2)$
7. $2pq(5 - 6p)$ 9. $12r^2s^2(3r - 5s)$
11. $16x^2y^2(x^2y^3 + 1)$ 13. $5b(b^2 - 7b + 2)$
15. $7z(3z^4 - 11z^2 - 7)$ 17. $2bc(3b - 2c + 2)$
19. $5y^3z(3y + 2z - 4z^2)$ 21. $9a^2b^3(5b + 2a^2 - 9ab)$
23. $6r^2s^2t(7s^3t^2 + 9r)$ 25. $(z - 1)(z + 2)$
27. $(m + 5)(2m - 3)$ 29. $(d - 5)(d - 7)$
31. $(1 - q)(9 + q)$ 33. $(x + 7)(x + 12)$
35. $(a - 6)(a - 2)$ 37. $(m - 8)(m - 7)$
39. $(z + 1)(4z + 7)$ 41. $(p - 3)(5p - 13)$
43. $(a - 1)(3a - 5)$ 45. $1 + x$ 47. $1 - j^n$
49. $r^{4n}s - s^3$ 51. $w^3 + v$ 53. $a^2b - ab^4$

Reading Algebra, page 349
1. an equation 3. coincident lines 5. a coordinate

Written Exercises, pages 352–353

1. $(8z + 5)(8z - 5)$ 3. $(11c^2 + 1)(11c^2 - 1)$
5. $(8u + 11v)(8u - 11v)$ 7. $(a + 7)^2$ 9. $(6 - r)^2$
11. $(m + n)^2$ 13. $(2c + 1)^2$ 15. $(2g - 5)^2$
17. $(5y - 4z)^2$ 19. $(r + 2)(r^2 - 2r + 4)$
21. $(4u - 1)(16u^2 + 4u + 1)$
23. $(3j - k)(9j^2 + 3jk + k^2)$
25. $3(c^2 - 25) = 3(c + 5)(c - 5)$
27. $5(16 + 8x + x^2) = 5(4 + x)^2$
29. $5(9v^2 - 24v + 16) = 5(3v - 4)^2$
31. $4m(4m^2 - 9) = 4m(2m + 3)(2m - 3)$
33. $3x(x^2 + 2x + 1) = 3x(x + 1)^2$
35. $2j(25j^2 + 20j + 4) = 2j(5j + 2)^2$
37. $(a + 6 + b)(a + 6 - b)$
39. $(5x - 1 + y)(5x - 1 - y)$ 41. $(a - b)(3a + 2)$
43. $(5 + r)(r - s)$ 45. $(2a + 3b)(5c + 2d)$
47. $(2p + 3r)(p + 5q)$ 49. $(x^n + 1)(x^n - 1)$
(Note: If n is an even integer, the second factor
may be factored further.)   51. $(x^n - 1)^2$
53. $(x^n - y)(x^{2n} + x^n y + y^2)$

Written Exercises, page 357

1. $(a + 3)(a + 4)$ 3. $(r + 5)(r - 1)$ 5. irreducible
7. $(g - 8)(g + 3)$ 9. $(m - 5)(m - 4)$
11. $-(x + 8)(x + 3)$ 13. $(6 + x)(1 + x)$
15. $-(j - 12)(j + 3)$ 17. $(m + 2n)(m + n)$
19. $4(c^2 - 11c + 30) = 4(c - 5)(c - 6)$
21. $k(k^2 - 8k + 15) = k(k - 3)(k - 5)$
23. $m^2(m^2 - 7m - 18) = m^2(m - 9)(m + 2)$
25. $3v(v^2 - v - 20) = 3v(v - 5)(v + 4)$
27. $5b^2(b^2 - 3b + 2) = 5b^2(b - 2)(b - 1)$
29. $-6x^3(x^2 - 4x + 3) = -6x^3(x - 3)(x - 1)$
31. $2x^2(y^2 - 4y - 45) = 2x^2(y - 9)(y + 5)$
33. $st(15 + 8s + s^2) = st(3 + s)(5 + s)$
35. $2u^3(20 + 8v - v^2) = 2u^3(10 - v)(2 + v)$
37. $s(s^2 - st - 12t^2) = s(s - 4t)(s + 3t)$
39. $m^2(m^2 - 5mn + 6n^2) = m^2(m - 3n)(m - 2n)$
41. $2j^2(1 - 5k + 4k^2) = 2j^2(1 - 4k)(1 - k)$
43. a. $3, -3$ b. $1, -1$ 45. a. $7, -7, 8, -8, 13,$
-13 b. $1, -1, 4, -4, 11, -11$
47. $(x^2 - 2)(x^2 + 1)$ 49. $(x^4 + 3)(x^4 + 2)$
51. $(x^n + 5)(x^n - 2)$

Written Exercises, pages 360–361

1. $(5a + 2)(a + 1)$ 3. $(2c - 5)(c + 1)$
5. $(3x - 8)(x + 1)$ 7. $(3r - 2)(2r + 1)$
9. $(2z - 5)(2z - 3)$ 11. $(4m - 5)(2m + 3)$
13. $(j - 4)(10j + 3)$ 15. $(3p + 2)(4p + 5)$
17. $(5u - 2)(4u + 5)$ 19. $4(x + 1)(x + 5)$
21. $2(2m + 1)(m + 3)$ 23. $2(2a - 1)(3a + 2)$
25. $6m(2m - 1)(m + 1)$ 27. $2z^2(3z + 1)^2$
29. $4a^3(2a + 1)(2a - 1)$
31. $2x(x + 2)(x - 2)(3x^2 + 1)$
33. $3z^2(z + 2)(z - 2)(2z^2 + 1)$
35. $b(1 - 2b)(1 + 2b)(1 + 3b^2)$

37. $u(3u - 2v)(2u + 9v)$
39. $w^2(3w - 8z)(4w - z)$ 41. $2yz(3y - z)(u + 2z)$
43. $(x^2 + y^2)(x + y)(x - y)$
45. $(x + y)(x^2 - xy + y^2)(x - y)(x^2 + xy + y^2)$
47. $(y^2 + y + 1)(y^2 - y + 1)$
49. $a = 2, b = 1, c = -3$

Self-Test 2, page 361

1. $2^4 \cdot 3 \cdot 5$ 2. $16c^2d^5$; $96c^3d^6$ 3. $5m(5m - 8)$
4. $2x^4y(3y - 5xy^2 + 7x^2)$ 5. $(s + 3)^2$
6. $(9m + 7)(9m - 7)$ 7. $(x - y)^2$
8. $(n - 1)(n^2 + n + 1)$ 9. $(t - 3)(t - 4)$
10. $(u - 4)(2u + 5)$ 11. $3g(g + 3)(g - 3)$
12. $2h(3h^2 - 5h + 4)$

Written Exercises, pages 366–367

1. $\{4, -1\}$ 3. $\left\{\dfrac{1}{5}, \dfrac{3}{2}\right\}$ 5. $\{0, 2, -5\}$ 7. $\left\{0, -7, \dfrac{3}{5}\right\}$
9. $\{-5, 2\}$ 11. $\{3, -3\}$ 13. $\left\{\dfrac{4}{3}, -3\right\}$ 15. $\left\{\dfrac{1}{3}, -\dfrac{3}{2}\right\}$
17. $\left\{-\dfrac{1}{4}, -\dfrac{4}{5}\right\}$ 19. $\left\{\dfrac{5}{2}, \dfrac{2}{5}\right\}$ 21. $\left\{\dfrac{5}{2}, -\dfrac{5}{2}\right\}$
23. $\{0, 4\}$ 25. $\{0, 9, -4\}$ 27. $\{0, 3, 1\}$
29. $\left\{0, 1, -\dfrac{3}{2}\right\}$ 31. $\left\{0, \dfrac{5}{3}, -\dfrac{5}{3}\right\}$ 33. $\{2, -2\}$
35. $\{1, -1, 2, -2\}$ 37. $\left\{5, -\dfrac{3}{2}\right\}$ 39. $\left\{\dfrac{1}{3}, -\dfrac{5}{3}\right\}$
41. $\{-6, 1\}$ 43. $\left\{\dfrac{5}{3}, -\dfrac{4}{5}\right\}$ 45. $x^2 - 10x + 21 = 0$
47. $x^2 - x - 6 = 0$ 49. $8x^2 - 10x + 3 = 0$
51. $10x^2 + 9x - 9 = 0$ 53. 1. Axiom of multiplicative inverses   2. Hypothesis   3. Multiplication property of equality   4. Identity axiom for multiplication
5. Multiplicative property of zero

Problems, pages 369–371

1. 11 and 12   3. 8   5. -8 and -7 7. 1 m   9. 6 m
11. 2 and 18   13. after 3 s and after 2 s   15. 30 s
17. 6 cm and 16 cm   19. 4 m by 18 m or 9 m by 8 m
21. 4 years and 16 years   23. 6 m   25. The sum
of the squares of two consecutive integers, $x^2 +$
$(x + 1)^2$, may be represented as $2(x^2 + x) + 1$. If x
is even, x, may be represented as $2y$ and the sum
above as $2[(2y)^2 + 2y] + 1$ or $4(2y^2 + y) + 1$. If x is
odd, x may be represented as $2y + 1$ and the sum as
$2[(2y + 1)^2 + 2y + 1] + 1$ or $4(2y^2 + 3y + 1) + 1$.
Both expressions are one greater than a multiple of
four.

Self-Test 3, page 371

1. $\{-2, 7\}$ 2. $\left\{0, -\dfrac{5}{3}\right\}$ 3. $\{6, -3\}$
4. $\left\{0, 5, -\dfrac{5}{2}\right\}$ 5. $\left\{\dfrac{2}{3}, -\dfrac{3}{2}\right\}$ 6. $\left\{0, 2, -\dfrac{3}{2}\right\}$
7. 3 m by 16 m   8. 8 and 9

Extra, page 373

1.

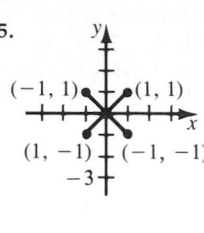
(−3, 1) (3, 1)
(5, 0)
(−5, 0)
(3, −1)

3.

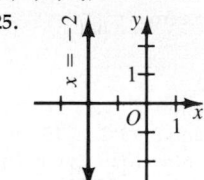
(−1, 4) (1, 4)
(−2, 3) (2, 3)
(−2, −3)
(1, −4)

5.

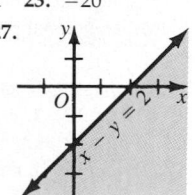
(−1, 1) (1, 1)
(1, −1) (−1, −1)
−3

9. It is perpendicular to the y-axis; it is perpendicular to the x-axis.   11. The line segment with endpoints (1, 2) and (2, 5)

Chapter Review, pages 374–376
1. d 2. c 3. d 4. d 5. c 6. d 7. d 8. b
9. d 10. c 11. d 12. b 13. d 14. d 15. c
16. a 17. d 18. a 19. b 20. c 21. c 22. c
23. c

Cumulative Review, pages 377–379
1. False 3. False 5. Associative axiom for multiplication 7. Identity axiom for addition
9. 60 11. $16x + 2$ 13. $x + 8y$ 15. $-6cd$
17. Range: {1, 3, 5}; not a function 19. Range: {0, 1, 2, 3}; a function 21. -1 23. -20
25.

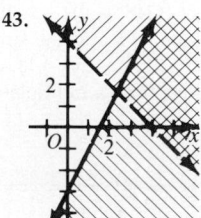

27.

29. $x - 4y = -4$ 31. $x = -2$ 33. {(5, −2)}
35. {(5, −1)} 37. Ø 39. $\left\{\left(\frac{9}{4}, \frac{1}{2}\right)\right\}$ 41. {(−2, 3)}
43.

45. 13 blocks
47. 153

Contest Problems, page 379
1. -24 3. 3 5. 31

Chapter 8 Polynomials and Rational Expressions

Written Exercises, pages 384–385

1. x^4 3. $\frac{1}{b^4}$ 5. $2c$ 7. $-\frac{1}{11n^2}$ 9. $8e^4f$ 11. $\frac{1}{3v}$

13. $-\frac{3b^3}{2c}$ 15. $2p$ 17. $\frac{4a^2}{b^2c^2}$ 19. $\frac{20m^2}{n^2}$ 21. xy^2

23. $-a^4$ 25. t^4 27. $-\frac{1}{d}$ 29. $\frac{1}{3a^6}$ 31. $\frac{(3t - 1)^2}{s}$

33. d^n 35. $\frac{1}{3p^{3n-1}}$ 37. $\frac{1}{f^2}$ 39. u^{m+3}

41. a. $m \in$ {all integers greater than 5} b. {4}
c. {5} 43. x^{1-m} when $m \in$ {all integers less than 1}; $\frac{1}{x^{m-1}}$ when $m \in$ {2, 3}; 1 when $m = 1$

45. a^{-1-2t} when $t \in$ {−2, −1}; $\frac{1}{a^{1+2t}}$ when $t \in$ {0, 1}

Written Exercises, pages 386–387

1. $2m + 3n$ 3. $3e - 1$ 5. $\frac{7}{2h^2} + \frac{3}{2h^3}$ 7. $\frac{1}{2ab^2} -$
$\frac{3b}{4a^3}$ 9. $\frac{-1}{4t^2} + \frac{t}{2}$ 11. $-1 - 3v^4$ 13. $3r^2 + 4r - 7$

15. $-6z^2 + 3z - 8$ 17. $\frac{5uv^2}{3} + \frac{2v}{3} - \frac{4u^2}{3}$

19. $\frac{-5e^2}{t^2} + \frac{5}{2} - \frac{2t^2}{e^2}$ 21. $4x^4 - 3x^2 + 5$

23. $4x^4 - x^2 + 3 - \frac{1}{8x^2}$ 25. $b^{2n} - b^n + b$

27. $-a^{2n}b^n - 2a^nb + \frac{3b^{n-1}}{a^{n-1}}$ 29. 34 31. 0.6

33. a. Yes, the solution set for each equation is {2}.
b. No, the solution set of the first equation is Ø; the solution set of the second equation is {0}.

Written Exercises, page 391

1. $m + 3$ 3. $3s + 2$ 5. $5t - 3 - \frac{2}{7t - 6}$

7. $-2x^2 + 5x - 4 + \frac{3}{x + 1}$ 9. $d^2 + 2d + 4 + \frac{133}{d - 2}$

11. $6x^2 + 3x - 1 - \frac{3x - 1}{3x^2 + 1}$ 13. $t^4 + t^2 + 1 +$

$\frac{2}{t^2 - 1}$ 15. $r^2 - 2r + 9 - \frac{24r - 26}{r^2 + 2r - 3}$ 17. Yes
19. No 21. $(k - 3)^2(k + 5)$
23. $(w + 1)(w - 1)(w + 2)(w^2 - 2w + 4)$
25. $p = -12$ 27. a. $a^{99} + a^{98} + a^{97} + \cdots + 1$
b. $a^{n-1} + a^{n-2} + a^{n-3} + \cdots + 1$
29. No; the quotient is a polynomial only if the dividend is divisible by the divisor.

Self-Test 1, page 392

1. $7m^2$ 2. $\frac{10}{s}$ 3. $3b^2 - 4b + 1$ 4. $2e^3f - 3e^2 + \frac{4}{ef}$
5. $c + 7$ 6. $3r - 5 + \frac{5}{2r + 3}$

Written Exercises, pages 395–396

1. $\frac{5ab}{7}$ 3. $\frac{b-4}{3}$ 5. $\frac{3w+2}{3w-2}$ 7. $-\frac{3b^2}{5}$ 9. $\frac{2}{f-5}$

11. $\frac{c+3}{-2(c+2)}$ 13. $\frac{d+3}{3d+4}$ 15. $\frac{a-4b}{a-3b}$

17. $\frac{3(2e-1)}{4(e+1)}$ 19. $x=a+4;\ a\neq\frac{3}{2}$

21. $x=a+3;\ a\neq\frac{5}{2}$ 23. $\frac{3(a^2+2a+4)}{a-2}$

25. $\frac{a^4-a^2+1}{a^2-1}$ 27. $\frac{3x-14}{x-2}$ 29. $\frac{2a+3b}{3b}$

31. $x=\frac{4}{7}$ 33. a. $a=-2$ b. $a=2$ c. $x=\frac{a+3}{a-2}$

35. 1. r, s, and t are real numbers with s and t ≠ 0 — Hypothesis

2. $\frac{r}{s}=\frac{r}{s}\cdot 1$ — Ident. ax. for mult.

3. $=\frac{r}{s}\cdot\frac{t}{t}$ — Subs. prin.

4. $=\frac{rt}{st}$ — Basic prop. of quotients

5. $\therefore\ \frac{r}{s}=\frac{r\cdot t}{s\cdot t}$ — Trans. ax. of eq.

Written Exercises, pages 400–401

1. 10 3. $-\frac{1}{32}$ 5. $\frac{5}{z^3}$ 7. $\frac{6b^2}{c^3}$ 9. $\frac{m^3}{n^3}$ 11. $\frac{1}{a^5}$

13. $\frac{r^4}{2}$ 15. $\frac{1}{u^2v^4}$ 17. 1 19. $\frac{1}{e+f}$ 21. $\frac{s}{5}$ 23. $\frac{d^3}{2c^2}$

25. 2500 27. 0.000016 29. 0.0000000144

31. 30 33. $-\frac{h^4}{2c}$ 35. $-\frac{1}{2}$ 37. $\frac{s^2r^2}{625}$ 39. $3m^{-1}n^{-5}$

41. $2d^{-1}e^{-6}f^7$ 43. $\frac{2x^4z^{-1}}{5}$ 45. 30 47. $-\frac{5}{13}$

49. $\frac{8a^3}{b^2}=\frac{8(2^m)^3}{(2^{m+1})^2}=\frac{8(2^{3m})}{2^{2m+2}}=\frac{2^3(2^{3m})}{2^{2m+2}}=\frac{2^{3m+3}}{2^{2m+2}}=2^{m+1}$

Self-Test 2, page 402

1. $-\frac{2b}{3d}$ 2. $\frac{3}{5}$ 3. $2a+1$ 4. $\frac{1}{c^3}$ 5. x^4 6. $\frac{b^{10}}{9a^4}$

7. 1 8. $\frac{3b}{2a^2}$ 9. r^5pm 10. 2.7×10^5

11. 2.7×10^{-3}

Written Exercises, pages 406–407

1. m 3. $\frac{15a}{2}$ 5. $-\frac{5}{4}$ 7. $-6t$ 9. $\frac{2p^2}{5q}$

11. $\frac{s(r+s)}{4}$ 13. 1 15. $\frac{t+2}{3t-1}$ 17. $\frac{3}{2}$ 19. $\frac{2}{v^3}$

21. 15 23. $\frac{1}{6x}$ 25. $\frac{8a^6c^6}{27b^{12}d^6}$ 27. 1 29. $-\frac{72b^3}{a^9c^4}$

31. 3 33. $\frac{t-5}{t-2}$ 35. $\frac{(m+2)(m+3)}{(m-2)(m+1)}$

37. $\frac{x^2+2xy+4y^2}{x-2y}$ 39. $\frac{2t+3}{3t-1}$ 41. $\frac{3x+2}{x+1}$

43. $-\frac{(j+5)}{12(j+1)}$ 45. $-\frac{h}{h+6}$ 47. Yes

49. If you multiply any two rational numbers, the product is a rational number.

51. 1. a, b real numbers, n a positive integer, $b\neq0$ (Hypothesis);

2. $\left(\frac{a}{b}\right)^n=\left(\frac{a_1}{b_1}\right)\left(\frac{a_2}{b_2}\right)\left(\frac{a_3}{b_3}\right)\cdots\left(\frac{a_n}{b_n}\right)$ (Definition of a power);

3. $\frac{a_1}{b_1}\cdot\frac{a_2}{b_2}\cdot\frac{a_3}{b_3}\cdots\frac{a_n}{b_n}=\frac{(a_1\cdot a_2\cdot a_3\cdots a_n)}{(b_1\cdot b_2\cdot b_3\cdots b_n)}$ (Rule for multiplying rational numbers);

4. $\frac{(a_1\cdot a_2\cdot a_3\cdots a_n)}{(b_1\cdot b_2\cdot b_3\cdots b_n)}=\frac{a^n}{b^n}$ (Definition of a power)

5. $\therefore\left(\frac{a}{b}\right)^n=\frac{a^n}{b^n}$ (Transitive axiom of equality)

Written Exercises, page 409

1. $\frac{9}{a}$ 3. $\frac{3n-1}{n+1}$ 5. $\frac{2t+1}{4}$ 7. 1 9. $\frac{3}{2}$

11. $\frac{x}{x-3}$ 13. -1 15. $2(a-1)$ 17. $\frac{-2}{d-2}$

19. $\frac{1}{c+1}$ 21. $\frac{4}{3d+4}$ 23. $\frac{-1}{4n+3}$ 25. $\frac{e}{e-5}$

27. $\frac{1}{v+3}$ 29. $\frac{3r+2}{3r-5}$ 31. $\frac{1}{x-y}+\frac{1}{y-x}=\frac{1}{x-y}+\frac{-1}{x-y}=\frac{1}{x-y}-\frac{1}{x-y}=0$

Written Exercises, pages 412–413

1. $\frac{10x}{3}$ 3. $\frac{5d+c}{cd}$ 5. $\frac{3t+1}{t}$ 7. $\frac{h+2}{2h^2}$

9. $\frac{u^2+2uv-v^2}{uv}$ 11. $\frac{-7}{3(a-2)}$

13. $\frac{2t^2-t+3}{(t+3)(t-3)}$ 15. $\frac{b^2-2b-3}{b-2}$ 17. $\frac{3-5a}{a(a-3)}$

19. $\frac{3h}{g-3}$ 21. $\frac{-6}{(w+2)(w-2)}$ 23. $\frac{5x-3}{6x(x-2)}$

25. $\frac{2m+3n}{(m-n)^2}$ 27. $\frac{10b}{(b+5)(b-5)(b+4)}$

29. $\frac{c^2}{c-5}$ 31. $\frac{2a^2+10a+16}{(a+2)(a+5)(a-3)}$

33. $\frac{q^2-p^2}{pq}$ 35. $\frac{s^3-2s^2-41s+12}{s^2-36}$ 37. $\frac{t+s}{t-s}$

39. $\frac{a^2+ab+b^2}{a^2-ab+b^2}$ 41. $\frac{-a(a-b)^2}{2b(2a-3b)}$ 43. a. Yes
b. Yes

Computer Exercises, page 415

1. $9x^4+12x^3+10x^2+4x+1$ 3. $6x^8+23x^7+19x^6-18x^5-9x^4+9x^3-6x^2+4x-1$
5. $2x^3-x^2-7x+2$

Written Exercises, page 418

1. $\frac{3}{2}$ 3. $\frac{3}{2}$ 5. cd^2 7. $\frac{4e+3f}{6e-4f}$ 9. $\frac{5w-9z}{2w+z}$

11. $u-v$ 13. $\frac{3d+1}{d-1}$ 15. $\frac{x(x+2y)}{y(y-2x)}$ 17. $3c+1$

19. $\frac{1}{x-2}$ 21. $\frac{rs}{s-r}$ 23. $\frac{u^2+v^2}{u^2v^2}$ 25. $\frac{c^2+d^2}{cd}$

27. $\dfrac{e}{e+1}$ **29.** $\dfrac{-1}{z-3}$ **31.** $\dfrac{5-2t}{2-t}$

33. $r = \dfrac{1 - \dfrac{1+t}{1-t}}{1 + \dfrac{1+t}{1-t}} = \dfrac{1-t-(1+t)}{1-t+(1+t)}$

$= \dfrac{-2t}{2} = -t; \; \therefore r + t = -t + t = 0$

Self-Test, page 419

1. $\dfrac{3(x-1)}{2x}$ **2.** $\dfrac{2(c-4)}{3}$ **3.** $\dfrac{1}{2d-1}$

4. $\dfrac{9e^2+4}{6e^3}$ **5.** $\dfrac{6b^2+4}{2b^2+b}$ **6.** $\dfrac{a^2+1}{a^2-a}$

Chapter Review, pages 420–422
1. b **2.** a **3.** c **4.** a **5.** c **6.** b **7.** a **8.** a **9.** c
10. d **11.** b **12.** c **13.** d **14.** b **15.** a **16.** d

Mixed Review, page 423
1. $6c^2 - c - 15$ **3.** $-36x^6y^3$ **5.** $4x^2 - 12x + 9$
7. $49x^2 - 4y^2$ **9.** $2ax(4x+3)(4x-3)$
11. $x^2(x-6)(x+4)$ **13.** $(6+x)(3-4x)$
15. $(y^2+4)(y+2)(y-2)$ **17.** $3a(2a^2+3a-1)$
19. $5(r-2s)(r^2+2rs+4s^2)$ **21.** $w = -2$
23. $z > -9$ **25.** $\dfrac{5}{3} < u < 3$ **27.** $m = -3, 5$

29. $(-5, 2)$ **31.** $n = \dfrac{m}{t+v}$ **33.** $b = \dfrac{3a-T}{4}$

35. $c = \dfrac{ef}{2bd}$ **37.** 20 km/h

Preparing for College Entrance Exams, page 424
1. B **3.** B **5.** E **7.** C **9.** C

Application, page 425
1. less **3. a.** 5.9 terameters (5.9 Tm) **b.** 20 kilo-
seconds (20 ks) **c.** 4 femtograms (4 fg)

Chapter 9 Rational Expressions in Open Sentences
Written Exercises, pages 429–430
1. $\{6\}$ **3.** $\{6\}$ **5.** $\left\{f : f \leq \dfrac{10}{3}\right\}$ **7.** $\{n : n < 8\}$
9. $\{7\}$ **11.** $\{p : p \leq 9\}$ **13.** $\{s : s > 10\}$
15. $\{s : s > -2\}$ **17.** $\left\{-\dfrac{9}{4}\right\}$ **19.** $\left\{\dfrac{3}{5}\right\}$ **21.** $\left\{\dfrac{11}{2}\right\}$
23. $\{a : a > -15\}$ **25.** $\left\{\dfrac{9}{50}\right\}$ **27.** $\{4\}$
29. $y = -\dfrac{2}{3}x + \dfrac{11}{36}; \; m = -\dfrac{2}{3}$. **31.** $y = \dfrac{7}{4}x; \; m = \dfrac{7}{4}$
33. $\{(15, 14)\}$ **35.** $\left\{\left(\dfrac{17}{48}, -\dfrac{133}{72}\right)\right\}$ **37.** $\left\{\dfrac{\sqrt{2}}{2}, -\dfrac{\sqrt{2}}{2}, \dfrac{2}{3}\right\}$

Computer Exercises, page 430
1. $\{1\}$ **3.** $\{8\}$ **5.** $\{3\}$ **7.** $\left\{-\dfrac{24}{11}\right\}$

Problems, pages 433–435
1. 1100 **3.** 35% **5.** 5 L **7.** $32 **9.** $2,000 at 5%;
$1,000 at 12% **11.** 25 **13.** 12,500 **15.** $2000 at 9%;

$1500 at 8%; $1200 at $6\dfrac{1}{2}$% **17.** $3000 **19.** 12 L

Self-Test 1, page 435
1. $\{48\}$ **2.** $\{1\}$ **3.** $\{b : b \geq -4\}$ **4.** $\{x : x > 10\}$
5. $44.90 **6.** $2300 at 8%; $2700 at $10\dfrac{1}{2}$%

Written Exercises, pages 438–439
1. $\{14\}$ **3.** $\{12\}$ **5.** $\{8\}$ **7.** $\{2\}$ **9.** $\emptyset$
11. $\left\{3, -\dfrac{11}{2}\right\}$ **13.** $\{-5, 3\}$ **15.** $\left\{\dfrac{20}{9}, 3\right\}$ **17.** $\left\{\dfrac{3}{5}\right\}$
19. $\{-4, 3\}$ **21.** $\{-2\}$ **23.** $\left\{\dfrac{1}{3}, 2\right\}$ **25.** $\{10\}$

27. $\{-2\}$ **29.** $x = \dfrac{by}{1-y}$ **31.** $x = \dfrac{cn-m}{cP}$

33. $\left\{\dfrac{9}{11}, 3\right\}$ **35.** $\{-4\}$ **37.** $\{-3\}$

39. a. No; the solution set for the first equation is
$\{x : x \neq \pm 1\}$ and the solution set for the second
equation is $\mathcal{R}$. **b.** Yes; the solution set for each equation is $\mathcal{R}$. **41.** $k = 20$ **43.** $\{x : 1 < x < 4\}$
45. $\{x : -6 < x < 1\}$

Problems, pages 442–443
1. 12, 18 **3.** 42 **5.** 8, 9 **7.** 32, 18 **9.** $\dfrac{77}{33}$ **11.** 6,
8 **13.** $-3, -12$ or 3, 12 **15.** $-8, -12$ or 8, 12
17. $\dfrac{5}{4}, \dfrac{4}{5}$ or $-\dfrac{5}{4}, -\dfrac{4}{5}$ **19.** 12, 48 **21.** 12, 88
23. 1. $a < b; \; a, b > 0$ Hypothesis
   2. $\dfrac{a}{b} < \dfrac{b}{b}$ Mult. prop. of order
   3. $\dfrac{a}{b} < 1$ Subs. prin.
   4. $\dfrac{1}{b} < \dfrac{1}{a}$ Mult. prop. of order
   $\left(\text{or } \dfrac{1}{a} > \dfrac{1}{b}\right)$

Problems, pages 445–446
1. $3\dfrac{3}{7}$ h **3.** $1\dfrac{1}{11}$ h **5.** 20 h **7.** 16 min, 48 min.
9. $4\dfrac{1}{2}$ h **11.** 9 min **13.** $2\dfrac{3}{16}$ h **15.** 16 h

Problems, pages 447–448
1. 80 km/h, 120 km/h **3.** 30 km/h **5.** 100 km/h
7. 255 km/h **9.** 5 km/h **11.** 42 km/h on bicycle;
105 km/h on motorcycle **13.** 30 km

Self-Test 2, page 449
1. $\{3\}$ **2.** $\{13\}$ **3.** $\{7\}$ **4.** $\{-2\}$ **5.** 16, 80
6. $3\dfrac{1}{3}$ h **7.** 40 km/h

Written Exercises, pages 453–454
1. $\dfrac{6}{5}$ **3.** $\dfrac{2}{9}$ **5.** $\dfrac{14}{17}$ **7.** $\dfrac{3}{2}$ **9.** $\left\{\dfrac{10}{3}\right\}$ **11.** $\{0\}$ **13.** $\{25\}$
15. $\{3\}$ **17.** $\{10\}$ **19.** $\{12\}$ **21.** 6 to 1 **23.** -1 to 2

25. 3 to 4 **27.** c to d **29.** 9 to 2 **31.** -5 to 4

33. $\left\{\dfrac{R}{R^2 - 1}\right\}$ **35.** $\{y, -6y\}$ **37.** $\{6\}$

39. 1. $\dfrac{a}{b} = \dfrac{c}{d}$; $b, d \neq 0$ Hypothesis

2. $\dfrac{a}{b}(bd) = \dfrac{c}{d}(bd)$ Mult. prop. of eq.

3. $\left(\dfrac{a}{b} \cdot b\right)d = \left(\dfrac{c}{d} \cdot d\right)b$ Commut. and assoc. ax. for mult.

4. $ad = cb$ Ex. 43, page 102

5. $ad = bc$ Commut. ax. for mult.

41. 1. $\dfrac{a}{b} = \dfrac{c}{d}$; $b, d \neq 0$ Hypothesis

2. $\dfrac{a}{b} + 1 = \dfrac{c}{d} + 1$ Add. prop. of eq.

3. $\dfrac{a}{b} + \dfrac{b}{b} = \dfrac{c}{d} + \dfrac{d}{d}$ Subs. prin.

4. $\dfrac{a + b}{b} = \dfrac{c + d}{d}$ Theorem, page 407

Problems, pages 454–456

1. 240 acres, 300 acres **3.** 42°, 48° **5.** 12 cm, 12 cm, 20 cm **7.** 10 kg **9.** 8 cubes **11.** $84 **13.** 19.5 m²
15. 48
17. Profit (P) = Selling Price (S) − Cost (C)

$$P = S - C$$

$\dfrac{P}{S} = \dfrac{S - C}{S} = \dfrac{1}{4}$ $\dfrac{P}{C} = \dfrac{S - C}{C} = \dfrac{1}{3}$

$4(S - C) = S$ $3(S - C) = C$

$4S - 4C = S$ $3S - 3C = C$

$3S = 4C$ $3S = 4C$

19. $\dfrac{an}{a + b}, \dfrac{bn}{a + b}$ **21.** $2h$ **23.** 10 g of calcium, 8 g of sulfur, 16 g of oxygen **25.** 168 g of iron, 40 g of nickel, 30 g of copper

Computer Exercises, page 456

1. 21 **3.** -11.25 **5.** 2.67 **7.** -12.5

Problems, pages 461–462

1. $\dfrac{28}{3}$ **3.** $3a$ **5.** 6 cm **7.** 144 cm **9.** 28

11. $e = 500f^3$ **13.** 384 cm² **15.** $35,000 **17.** No; the graph of the function must go through the origin. **19.** $r_1 = kt$; $r_2 = k(2t) = 2(kt)$; $\therefore r_2 = 2r_1$
21. $u = 6v$

Problems, pages 465–466

1. 20 **3.** 14 **5.** 400 kHz **7.** 125 cm **9.** 64 rpm

11. 5.6 **13.** $a = \dfrac{0.18}{t^2}$ **15.** $12,800 **17.** y is halved.

19. a will be quadrupled.

Problems, pages 470–471

1. 24 **3.** 0.15 **5.** 14 **7.** 4 **9.** No change **11.** Q is divided by 4. **13.** 75π cm³ **15.** $\dfrac{16}{3}$ h **17.** 96 cm³

19. x is multiplied by 12 **21.** Halve the resistance.

1. 3.5 m, 4.9 m **2.** $2350 **3.** 6 **4.** 264 cm³ **5.** 2

Chapter Review, pages 473–475

1. b **2.** a **3.** c **4.** b **5.** c **6.** b **7.** c **8.** a
9. b **10.** a **11.** d **12.** c **13.** b **14.** b **15.** d
16. d **17.** a **18.** c **19.** a **20.** c

Cumulative Review, pages 477–478

1. $\{x: 2 \leq x < 7\}$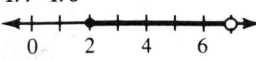

3. $\{a: a > 5 \text{ or } a < -1\}$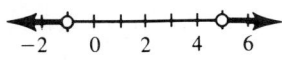

5. $\{p: -3 \leq p \leq -1\}$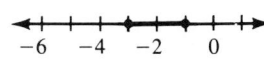

7. 130° **9.** 2.8 km **11.** False **13.** False
15. $\{(-1, -1)\}$ **17.** $\{(-1, 2), (0, 2), (1, 2)\}$ **19.** 17
21. 7
23. 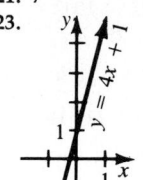 **25.**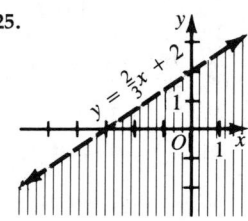

27. $y = -\dfrac{3}{2}x + 1$ **29.** $y = -2x - 6$

31. $x^4 + x^3 + 3x - 6$ **33.** $24y^9$ **35.** $-x^6y^2 + 4x^5y^4 - 2x^3y^5$ **37.** $2x^2 - x - 3$ **39.** $16c^2 + 24c + 9$ **41.** $8x^3 + 12x^2 + 6x + 1$ **43.** $(4x - 3)^2$
45. $10(n + 1)(n^2 - n + 1)$ **47.** $(3x + 5y)(3x - 5y)$
49. $4x(3x - 1)(x + 3)$ **51.** $\left\{0, 3, -\dfrac{1}{3}\right\}$
53. $\left\{-\dfrac{3}{2}, -1\right\}$ **55.** $-9, -11$ **57.** $\dfrac{e^8h^4}{k^2}$
59. $-5m^4 - 7m^2 + 9$ **61.** $\dfrac{4m + 3}{2m - 5}$ **63.** $\dfrac{3}{5c - 3}$
65. $\dfrac{(4x - y)(x + y)}{6(4x + y)}$ **67.** $\dfrac{x - y}{4}$
69. $a + 7 + \dfrac{3}{2a - 3}$

Application, page 479

1. 9.8 N **3.** 1.5×10^{-10}

Chapter 10 Irrational Numbers and Radicals

Written Exercises, pages 484–485

1. $<$ **3.** $<$ **5.** $=$ **7.** $>$ **9.** $<$ **11.** $=$

13. $\dfrac{3}{16}, \dfrac{8}{29}, \dfrac{5}{14}$ **15.** $-\dfrac{15}{40}, -\dfrac{27}{81}, -\dfrac{14}{63}$

17. $\dfrac{7}{43}, \dfrac{4}{24}, \dfrac{13}{65}, \dfrac{5}{16}$ **19.** $\dfrac{9}{40}$ **21.** $-\dfrac{37}{60}$ **23.** $2\dfrac{7}{8}$

25. $\dfrac{58}{45}$ 27. $-\dfrac{77}{72}$ 29. 9 31. $\dfrac{59a}{60}$ 33. $\dfrac{99}{70}$

Written Exercises, pages 490–491

1. 0.25 3. $0.\overline{5}$ 5. $0.41\overline{6}$ 7. -0.525 9. $-6.\overline{2}$
11. $4.\overline{428571}$ 13. $\dfrac{21}{25}$ 15. $\dfrac{2627}{50}$ 17. $\dfrac{4}{9}$ 19. $\dfrac{25}{99}$
21. $\dfrac{229}{99}$ 23. $\dfrac{5155}{999}$ 25. $-\dfrac{31}{30}$ 27. $-\dfrac{10604}{909}$
29. $\dfrac{2}{9}$ 31. $\dfrac{34}{15}$ 33. $\dfrac{5}{27}$ 35. $-\dfrac{20}{121}$ 37. $\dfrac{9}{8}$ 39. $\dfrac{19}{45}$
41. $\dfrac{10}{9}$ 43. $\dfrac{29}{33}$ 45. $\dfrac{31}{10}$ 47. 6 49. a. $\dfrac{36}{9}, \dfrac{72}{9}, \dfrac{144}{9}$
b. $x + 1$

Written Exercises, pages 495–496

1. 9 3. -3 5. 17 7. 4 9. 25 11. $-\dfrac{4}{9}$
13. $\{5, -5\}$ 15. $\{9, -9\}$ 17. $\left\{\dfrac{9}{10}, -\dfrac{9}{10}\right\}$
19. $\{4, -4\}$ 21. $8|c|$ 23. $3|mn|$ 25. $4u^2$
27. $\left|\dfrac{c}{e}\right|$ 29. $5a^2|b|$ 31. $\dfrac{x^2}{3|y|}$ 33. $-1.1g^2h^2$
35. $5|ef|$ 37. $|r + s|$ 39. $(f + g)^4$ 41. $14x^2y^8$
43. $\dfrac{13|j^{15}|}{8k^{10}}$ 45. $\sqrt{25a^2 + 30ab + 96^2} =$
$\sqrt{(5a + 3b)^2} = |5a + 3b|$ 47. c^{2m} 49. $|q^{5m}|$
51. No 53. Yes; domain: $\{x: x \geq 0\}$, range:
$\{y: y \geq 0\}$ 55. $\sqrt{a^2} = \sqrt{(-a)^2} = |a|$. You must
know if a is negative or nonnegative before you can
decide if $|a| = a$ or $-a$.

Written Exercises, pages 500–501

1. $4\sqrt{3}$ 3. $5\sqrt{3}$ 5. $8\sqrt{5}$ 7. $8\sqrt{6}$ 9. $\dfrac{\sqrt{5}}{4}$
11. $0.6\sqrt{2}$ 13. $\{\sqrt{11}, -\sqrt{11}\}$ 15. $\{\sqrt{7}, -\sqrt{7}\}$
17. $\{2\sqrt{5}, -2\sqrt{5}\}$ 19. $\{\sqrt{30}, -\sqrt{30}\}$
21. $\{2\sqrt{15}, -2\sqrt{15}\}$ 23. $\left\{\dfrac{3\sqrt{5}}{2}, -\dfrac{3\sqrt{5}}{2}\right\}$
25. 4.461 27. 57.295 29. 16.662 31. 0.779
33. 29.226 35. 6.261 37. $20 - \sqrt{85}$
39. $2 + \sqrt{8}$ 41. $\sqrt{204} - 1$
43. $(3 - \sqrt{5})^2 < 3 - \sqrt{5}$; $(3 - 2.24)^2 < 3 - 2.24$;
$(0.76)^2 < 0.76$; $0.58 < 0.76$
45. $\{1.79\}$ 47. $x = -y$; for example $x = 1$,
$y = -1$ 49. -5 51. -9 53. 15 55. $a - b$
57. No; $\sqrt{2} - 1$ 59. Yes; if $r = s$ then $\dfrac{r}{s} = 1$ is
rational.
61.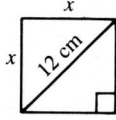

Self-Test 1, page 502

1. $>$ 2. $<$ 3. $2\dfrac{5}{6}$ 4. 0.625 5. $0.458\overline{3}$ 6. $\dfrac{48}{125}$
7. $\dfrac{24}{55}$ 8. 8 9. 36 10. 18 11. $2\sqrt{6}$ 12. $4\sqrt{21}$
13. $8\sqrt{5}$ 14. 9.381 15. 8.307 16. 15.652

26 *Answers to Selected Exercises*

1. 2.65 3. 11.18 5. 1.58 7. 0.79

Written Exercises, page 510

1. 5.39 3. 7.48 5. 11.18 7. 5.10 9. 8.25
11. 5.66 13. Yes 15. No 17. No 19. Yes
21. No 23. No 25. Yes 27. Yes 29. Yes
31. Yes 33. $0.625 \approx 0.63$ 35. 8.94 37. 40
35. $a = 8.94; b = 4.47$ 41. $a = 4.47; c = 6.71$

Problems, pages 511–512

1. $x = 20$ cm 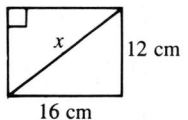 3. $x = 14.14$ m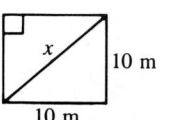

5. $x = 17$ km 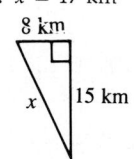 7. $h = 43.59$ m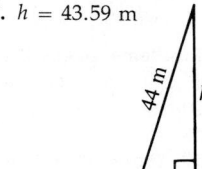

9. $x = 8.49$ cm   11. $b = 8.66$ m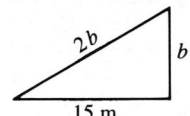

13. $x = 6.24$ m or $y = 9.43$ m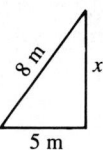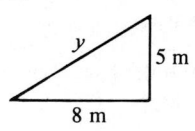

15. diagonal $= 11.87$ cm 17. $d = k\sqrt{3}$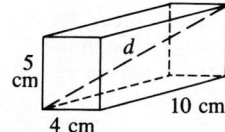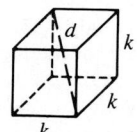

Written Exercises, pages 514–515

1. $\sqrt{5}$ 3. 5 5. $3\sqrt{5}$ 7. $3\sqrt{2}$ 9. 17 11. $18\sqrt{2}$
13. $2\sqrt{a^2 + b^2}$ 15. $|a - b|\sqrt{2}$ 17. $|a - b|\sqrt{2}$
19. Yes 21. Yes 23. No 25. No
27. $y = -7x + 5$ 29. $A = (18, 4)$, $B = (12, 12)$,
$C = (8, 4)$; $AB = AC = 10$
31. $d = \sqrt{(x - 0)^2 + (y - 0)^2} = \sqrt{x^2 + y^2}$
33. $m = -6.5$ or $m = -1.5$

Self-Test 2, page 516
1. Yes 2. No 3. No 4. Yes 5. 30 6. 6.24
7. 14.07 8. 10.58 9. $\sqrt{41}$ 10. 13
11. $6\sqrt{29}$ 12. $2\sqrt{73}$

Written Exercises, pages 519–520
1. $5\sqrt{2}$ 3. $6\sqrt{2}$ 5. $6\sqrt{5}$ 7. $3\sqrt{2}$ 9. $\dfrac{\sqrt{30}}{3}$
11. 2 13. $\sqrt{2}$ 15. $4\sqrt{10}$ 17. 4.47 19. 16.97
21. 0.45 23. 0.76 25. $c\sqrt{6}$ 27. $10e\sqrt{e}$ 29. $\dfrac{\sqrt{3t}}{t}$
31. $16r^4$ 33. $\dfrac{5\sqrt{2}}{e}$ 35. $\dfrac{3\sqrt{10c}}{4c}$ 37. $m\sqrt{n}$
39. $\dfrac{\sqrt{15k}}{3}$ 41. $\dfrac{g\sqrt{3gh}}{h}$ 43. $2\sqrt{2}$ 45. $\dfrac{\sqrt{a+b}}{a+b}$
47. $\dfrac{\sqrt{2x+1}}{2x+1}$ 49. $\dfrac{\sqrt{x+3}}{x+3}$ 51. $\dfrac{(r+s)\sqrt{r-s}}{(r-s)}$

Written Exercises, pages 523–524
1. $4\sqrt{6}$ 3. $10\sqrt{2}$ 5. $-20\sqrt{2}$ 7. $-4\sqrt{c}$
9. $15\sqrt{30}$ 11. $3\sqrt{2}-6$ 13. 2 15. $-5+2\sqrt{13}$
17. $1+\sqrt{2}$ 19. $26+7\sqrt{3}$ 21. -5 23. 17
25. $a-2\sqrt{ac}+c$ 27. $9+6\sqrt{2}$ 29. $23+4\sqrt{15}$
31. $1+\sqrt{3}$ 33. $4\sqrt{6}-4\sqrt{3}$ 35. $\dfrac{5+3\sqrt{3}}{2}$
37. $-4-3\sqrt{2}-2\sqrt{3}-2\sqrt{6}$
39. $\dfrac{10+\sqrt{6}+2\sqrt{10}+\sqrt{15}}{17}$
41. $\dfrac{5+6\sqrt{6}-3\sqrt{10}-2\sqrt{15}}{-13}$ 43. $6\sqrt{6}$
45. $-34j\sqrt{2k}$ 47. $\dfrac{5}{6}u$ 49. $(82+8\sqrt{10})$ m^2
51. $10-3\sqrt{3}$ 53. $13-9\sqrt{2}$ 55. $(2+\sqrt{5})^2-$
$4(2+\sqrt{5})-1=4+4\sqrt{5}+5-8-4\sqrt{5}$
$-1=0$; $(2-\sqrt{5})^2-4(2-\sqrt{5})-1=$
$4-4\sqrt{5}+5-8+4\sqrt{5}-1=0$

Written Exercises, page 526
1. 3 3. 3 5. -128 7. $2\sqrt[4]{5}$ 9. $5\sqrt[3]{3}$ 11. $\dfrac{3\sqrt[3]{4}}{2}$
13. $3\sqrt[4]{2}$ 15. $\dfrac{\sqrt[3]{5}}{2}$ 17. $\dfrac{\sqrt[5]{18}}{2}$ 19. $\dfrac{\sqrt[3]{10}}{4}$ 21. x^3
23. x^2y^3z 25. $x^2y^3z^2\sqrt[3]{xy^2z^2}$ 27. $\dfrac{y^2\sqrt[4]{x^2y}}{x}$
29. $\dfrac{\sqrt[3]{x(x-1)}}{x-1}$ 31. $\dfrac{\sqrt[6]{(x+1)^5}}{x+1}$

On the Calculator, page 526
1. 54.96 3. 1.21 5. 1.17 7. 0.89

Written Exercises, page 529
1. $\{64\}$ 3. $\{4\}$ 5. $\varnothing$ 7. $\left\{\dfrac{1}{9}\right\}$ 9. $\{100\}$ 11. $\{13\}$
13. $\left\{-\dfrac{3}{2}\right\}$ 15. $\left\{\dfrac{81}{4}\right\}$ 17. $\left\{\dfrac{9}{25}\right\}$ 19. $\left\{\dfrac{7}{2}\right\}$ 21. $\{12\}$

23. $\{3\sqrt{2},\ -3\sqrt{2}\}$ 25. $\{4\}$ 27. $\{14\}$ 29. $\{-3\}$
31. $\{2\}$ 33. $\left\{-\dfrac{1}{3}\right\}$ 35. $\{0,\ 4\}$ 37. $\{(4,\ 16)\}$

Self-Test 3, page 530
1. $18\sqrt{6}$ 2. $\dfrac{4\sqrt{10}}{15}$ 3. $9\sqrt{3}-6\sqrt{2}$
4. $3\sqrt{2}+2\sqrt{3}$ 5. $\dfrac{\sqrt{35}-2\sqrt{7}-3\sqrt{5}+6}{-2}$
6. $18-\sqrt{2}$ 7. 3 8. $\dfrac{\sqrt[3]{20}}{6}$ 9. $7\sqrt[4]{2}$ 10. $\{63\}$
11. $\{11\}$ 12. $\{3\}$

Extra, page 532
1. 1 3. -1 5. i 7. $2i\sqrt{7}$ 9. $2i\sqrt{15}$ 11. $15i$
13. $-24\sqrt{2}$ 15. $-\dfrac{2i}{7}$ 17. $-\dfrac{4i\sqrt{5}}{15}$ 19. $\{7i,\ -7i\}$
21. $\{4i\sqrt{5},\ -4i\sqrt{5}\}$ 23. $\{4i,\ -4i\}$

Chapter Review, pages 534–535
1. c 2. c 3. d 4. d 5. b 6. c 7. c 8. c
9. a 10. b 11. c 12. a 13. d 14. b 15. b
16. b 17. c 18. b 19. c 20. d 21. b 22. c
23. d 24. a

Mixed Review, pages 537–538
1. $3n^3$ 3. $4x^2-25$ 5. $x^3-15x^2+75x-125$
7. x^4z^3 9. $-\dfrac{m+2}{2}$ 11. $9b^2-14ab-1$
13. $\dfrac{c^2+6cd+d^2}{3c^2+10cd+3d^2}$ 15. $-\dfrac{5}{2}$ 17. $\left\{\dfrac{5}{2},\ -3\right\}$
19. $\{1\}$ 21. $\{k: k<12\}$ 23. $\{40\}$ 25. $\{1,\ -2\}$
27. $\left\{\dfrac{5}{3}\right\}$ 29. $(x-6)(x+3)$ 31. $(5z+2)(4z-1)$
33. $3(m+4)(m-1)$ 35. $2ab(b-3)(b+3)$
37. 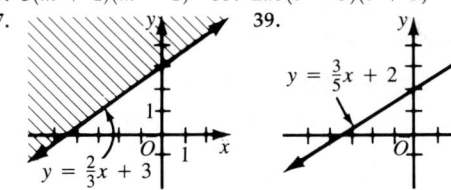 39.
41. $x=-1$ 43. $y=\dfrac{2}{3}x-5$, or $2x-3y=15$
45. 9.07×10^{13} 47. \$2000 at $6\dfrac{1}{2}$%; \$4500 at 8%
49. 12 and 20 or -12 and -20 51. 8.75 cm
53. 15 m long and 8 m wide 55. 24 cm^3

Preparing for College Entrance Exams, page 539
1. B 3. C 5. B 7. E

Application, page 541
1. 80 km 3. 226 km 5. 2 m 7. a. 106,189 km
b. 35,721 km c. Since h is large, the h^2 term is
significant and must not be omitted in the calculation.

Chapter 11 Quadratic Equations and Functions

Written Exercises, page 547

1. $\{-1, 3\}$ **3.** $\{-2, -4\}$

5. $\{-4 + \sqrt{6}, -4 - \sqrt{6}\}$; $\{-1.55, -6.45\}$

7. $\{2 + 2\sqrt{3}, 2 - 2\sqrt{3}\}$; $\{5.46, -1.46\}$

9. $\{1, 3\}$ **11.** $\{1, 7\}$

13. $\left\{\dfrac{1 + \sqrt{17}}{2}, \dfrac{1 - \sqrt{17}}{2}\right\}$; $\{2.56, -1.56\}$

15. $\left\{\dfrac{-3 + \sqrt{13}}{2}, \dfrac{-3 - \sqrt{13}}{2}\right\}$; $\{0.30, -3.30\}$

17. $\left\{\dfrac{9 + \sqrt{41}}{2}, \dfrac{9 - \sqrt{41}}{2}\right\}$; $\{7.70, 1.30\}$

19. $\left\{-\dfrac{1}{2}, 1\right\}$ **21.** $\left\{\dfrac{1 + \sqrt{13}}{3}, \dfrac{1 - \sqrt{13}}{3}\right\}$

23. $\{-3 + \sqrt{14}, -3 - \sqrt{14}\}$ **25.** no real roots

27. $\left\{\dfrac{-1 + \sqrt{61}}{10}, \dfrac{-1 - \sqrt{61}}{10}\right\}$

29. $\left\{\dfrac{-2 + \sqrt{3}}{5}, \dfrac{-2 - \sqrt{3}}{5}\right\}$ **31.** no real roots

33. $\left\{\dfrac{1}{2}, 2\right\}$ **35.** $\left\{\dfrac{4 + \sqrt{10}}{3}, \dfrac{4 - \sqrt{10}}{3}\right\}$

37. $\left\{\dfrac{5 + \sqrt{41}}{8}, \dfrac{5 - \sqrt{41}}{8}\right\}$

39. $\left\{\dfrac{-\sqrt{3} + \sqrt{15}}{2}, \dfrac{-\sqrt{3} - \sqrt{15}}{2}\right\}$

41. $\left\{\dfrac{1 + \sqrt{6}}{5k}, \dfrac{1 - \sqrt{6}}{5k}\right\}$

43. $\left\{\dfrac{-b + \sqrt{b^2 - 4c}}{2}, \dfrac{-b - \sqrt{b^2 - 4c}}{2}\right\}$

Written Exercises, pages 550–551

1. $\{5, -1\}$ **3.** $\{-4, 1\}$ **5.** $\left\{4, -\dfrac{2}{3}\right\}$ **7.** $\{-1 + \sqrt{2},$

$-1 - \sqrt{2}\}$; $\{0.41, -2.41\}$ **9.** $\left\{\dfrac{3 + \sqrt{2}}{7}, \dfrac{3 - \sqrt{2}}{7}\right\}$;

$\{0.63, 0.23\}$ **11.** no real roots

13. two **15.** none **17.** two **19.** one **21.** none

23. two **25.** $\left\{-3, \dfrac{1}{2}\right\}$ **27.** $\{1, 5\}$

29. $\left\{\dfrac{13 + \sqrt{89}}{8}, \dfrac{13 - \sqrt{89}}{8}\right\}$ **31.** $\left\{\dfrac{1}{2}, \dfrac{3}{2}\right\}$

33. $\{2 + 2\sqrt{3}, 2 - 2\sqrt{3}\}$ **35.** $\{7 + 2\sqrt{7}, 7 - 2\sqrt{7}\}$

37. $\left\{\dfrac{\sqrt{6} + 3\sqrt{2}}{2}, \dfrac{\sqrt{6} - 3\sqrt{2}}{2}\right\}$

39. $\{\sqrt{5} + 2\sqrt{2}, \sqrt{5} - 2\sqrt{2}\}$

41. $\left\{\dfrac{-4 + \sqrt{10}}{3}, \dfrac{-4 - \sqrt{10}}{3}\right\}$ **43.** 9

45. By the quadratic formula, the roots are

$\dfrac{-b + \sqrt{b^2 - 4ac}}{2a}$ and $\dfrac{-b - \sqrt{b^2 - 4ac}}{2a}$. Thus,

$\dfrac{-b + \sqrt{b^2 - 4ac}}{2a} + \dfrac{-b - \sqrt{b^2 - 4ac}}{2a} = \dfrac{-2b}{2a} = -\dfrac{b}{a}$.

47. $x^2 - 4x - 1 = 0$

Problems, pages 554–555

1. 70 cm; 40 cm **3.** $\dfrac{4}{3}$ or $\dfrac{3}{4}$ **5.** width; 2.2 m; length:

10.8 m **7.** 8 km; 15 km **9.** 8 arrangements

11. 7.2 km/h **13.** Ron: 8 h; Russ: 10 h

15. 40 shares **17.** 60 km/h; 80 km/h

Self-Test 1, page 556

1. $\{-2 + \sqrt{7}, -2 - \sqrt{7}\}$ **2.** $\{3 + 2\sqrt{2},$

$3 - 2\sqrt{2}\}$ **3.** $\left\{\dfrac{-5 + \sqrt{17}}{4}, \dfrac{-5 - \sqrt{17}}{4}\right\}$

4. $\left\{\dfrac{2 + \sqrt{10}}{3}, \dfrac{2 - \sqrt{10}}{3}\right\}$ **5.** 8 cm; 12 cm

Written Exercises, pages 562–563

1. The zero is 0. **3.** The zero is 0.

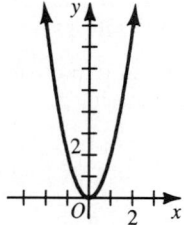 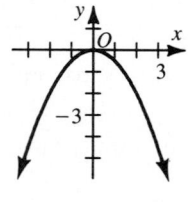

5. The zeros are $\sqrt{3}$ and $-\sqrt{3}$. **7.** The zeros are 0 and -4.

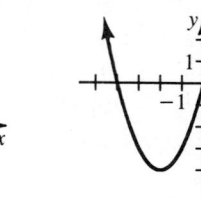

9. The zero is -1. **11.** The zeros are $-3 + \sqrt{2}$ and $-3 - \sqrt{2}$.

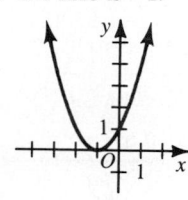 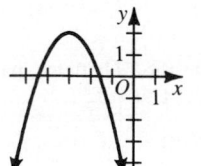

13. $-\dfrac{9}{4}$ **15.** $-\dfrac{25}{4}$ **17.** -6 **19.** $\dfrac{9}{4}$ **21.** -1 **23.** $\dfrac{1}{8}$

25.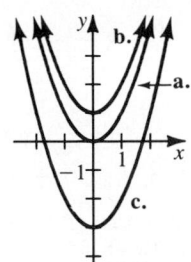

27.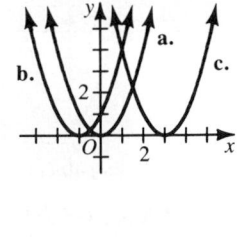

29. The graph of $y = ax^2 + k$ is the graph of $y = ax^2$ translated (slid) vertically k units (up if k is positive and down if k is negative).

31. 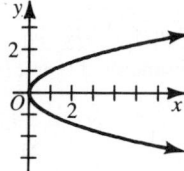 **a.** no; no **b.** no

33. Answers may vary. (The graph is any non-horizontal line.) **35.** Least positive integral value of x_1: 500,000

37. Let $x = -\dfrac{b}{a} - u$.

$$ax^2 + bx + c = a\left(-\frac{b}{a} - u\right)^2 + b\left(-\frac{b}{a} - u\right) + c$$

$$= a\left(\frac{b^2}{a^2} + \frac{2ub}{a} + u^2\right) - \frac{b^2}{a} - bu + c$$

$$= \frac{b^2}{a} + 2ub + au^2 - \frac{b^2}{a} - bu + c$$

$$= au^2 + bu + c$$

Since $P(u, v)$ lies on the graph,
$au^2 + bu + c = v$.

Thus, $a\left(-\dfrac{b}{a} - u\right)^2 + b\left(-\dfrac{b}{a} - u\right) + c = v$, and

so $Q\left(-\dfrac{b}{a} - u, v\right)$ lies on the graph.

Written Exercises, pages 565–566

1. The zero is 0.   **3.** The zero is approx. $1\frac{1}{4}$.

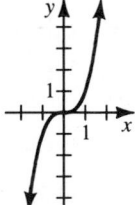 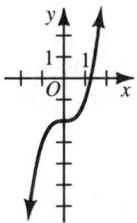

5. The zeros are 0 and approx. $-1\frac{3}{4}$ and $1\frac{3}{4}$.

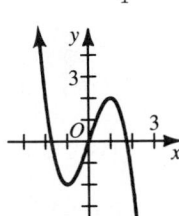

7. The zeros are approx. $-2\frac{1}{8}, \frac{1}{4}$, and $1\frac{7}{8}$.

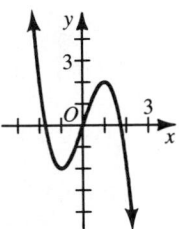

9. The zeros are -1 and 1.

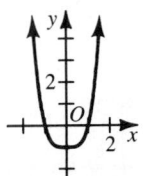

11. The zero is 0.

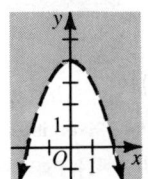

13. a. -1010000 **b.** 1010000
15. a. -1000100000000 **b.** 1000100000000
17. a. 999000000 **b.** -999000000 **19. a.** The values of f get smaller; the values of f get larger.
b. The values of f get larger; the values of f get smaller. **21.** No. For example, $Q(0) = 0$.
Thus, Q has at least one zero between -1 and 0.

Written Exercises, page 569

1. 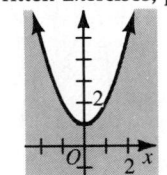   **3.**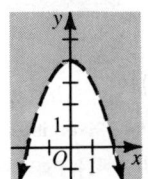

5. 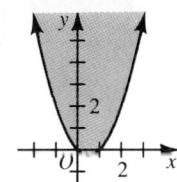   **7.**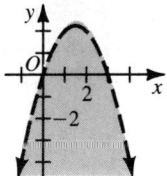

9. 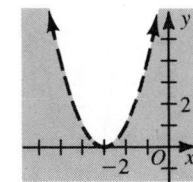   **11.**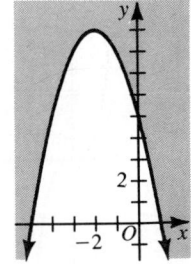

13.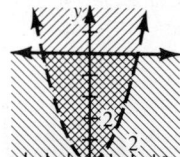

15.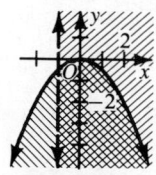

17.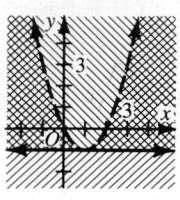

19.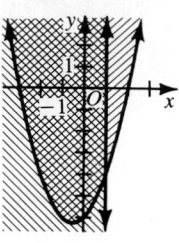

21.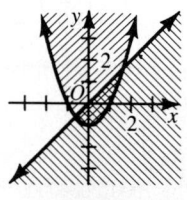

23.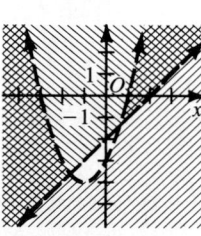

25.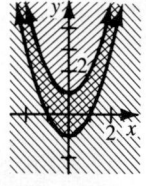

27.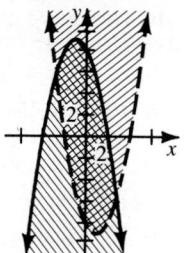

Self-Test 2, page 569

1.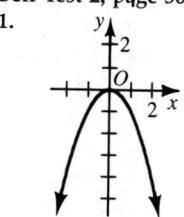

2.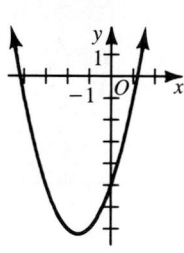

3.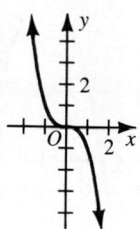

4.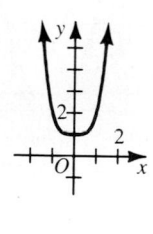

5.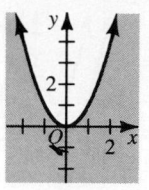

6.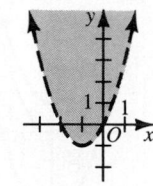

Extra, page 572

1. 2 **3.** −2 **5.** not a real number **7.** 8 **9.** $\frac{1}{6}$
11. 9 **13.** $4y^4$ **15.** $4x^2$

Chapter Review, pages 573–574

1. b **2.** c **3.** d **4.** a **5.** c **6.** b **7.** b **8.** b
9. a **10.** a **11.** a **12.** a **13.** d **14.** b

Cumulative Review, pages 575–577

1. {(2, 1)} **3.** ∅ **5.** {(0, 2)} **7.** {(5, 2)}
9. {(−4, 6)} **11.** 15 sophomores; 13 juniors
13. $-3uv^2$ **15.** $-r^3s + r^2s^2 - rs^3$
17. $16x^2 - 24x + 9$ **19.** $-12cd^2e(1 + 4c^2d)$
21. $(1 + 6w)^2$ **23.** $(2x - 5)(8x - 5)$ **25.** 32, 4
27. {−30} **29.** {a: a ≥ −4} **31.** {−1} **33.** 24
35. 40 **37.** $3\frac{3}{5}$ h **39.** 15 **41.** 150 **43.** 40

45. $\frac{\sqrt{14}}{6}$ **47.** $\frac{\sqrt[3]{15}}{3}$ **49.** $\frac{11 - 6\sqrt{3}}{-13}$,

or $\frac{-11 + 6\sqrt{3}}{13}$ **51.** {−6, 6} **53.** {9}

Contest Problems, page 577
1. 30 **3.** 256 **5.** 91 **7.** $n + 1$

Chapter 12   Trigonometry and Vectors

Written Exercises, pages 582–584
1. 660° **3.** −390° **5.** −240° **7.** −420° **9.** III; IV
11. 0°; 90° **13.** 270° **15.** 320° **17.** 560° **19.** III; II
21. 0° **23.** −45° **25.** II; IV **27.** 120° in 1 s;
600° in 5 s; 14,400° in 2 min **29.** 18,000°

Written Exercises, pages 588–590

1. $\sin A = \frac{4}{5}$; $\cos A = \frac{3}{5}$; $\tan A = \frac{4}{3}$ **3.** $\sin A = \frac{3}{5}$;

$\cos A = -\frac{4}{5}$; $\tan A = -\frac{3}{4}$ **5.** $\sin A = \frac{12}{13}$; $\cos A =$

$\frac{5}{13}$; $\tan A = \frac{12}{5}$ **7.** $\sin A = -\frac{\sqrt{2}}{2}$; $\cos A = \frac{\sqrt{2}}{2}$;

$\tan A = -1$ **9.** $\sin A = \frac{\sqrt{2}}{2}$; $\cos A = \frac{\sqrt{2}}{2}$; $\tan A =$

1 **11.** $\sin A = \frac{1}{2}$; $\cos A = \frac{\sqrt{3}}{2}$; $\tan A = \frac{\sqrt{3}}{3}$

13. $\sin A = \frac{1}{2}$; $\tan A = \frac{\sqrt{3}}{3}$ **15.** $\cos A = \frac{\sqrt{2}}{2}$;

$\tan A = -1$ **17.** $\sin A = -\frac{\sqrt{2}}{2}$; $\cos A = -\frac{\sqrt{2}}{2}$

19. $\sin A = \frac{5}{13}$; $\tan A = -\frac{5}{12}$

21.

	I	II	III	IV
sin A	+	+	−	−
cos A	+	−	−	+
tan A	+	−	+	−

23. $\sin A = 0$; $\cos A = 1$; $\tan A = 0$
25. If $\sin A = 0$, then $\cos A = \pm 1$, since $\sin^2 A + \cos^2 A = 1$. **27.** If $\sin A < 0$ and $\cos A < 0$, then $\tan A$ must be > 0 since $\tan A = \dfrac{\sin A}{\cos A}$.

29. If $\sin A < 0$ and $\tan A < 0$, then A is in quadrant IV and $\cos A$ is positive and < 1. Therefore, $\tan A = \dfrac{\sin A}{\cos A}$ must be less than $\sin A$.

31. $\sin A = \dfrac{12}{13}$; $\cos A = \dfrac{5}{13}$; $\tan A = \dfrac{12}{5}$
33. $\sin A = \dfrac{\sqrt{2}}{2}$; $\cos A = \dfrac{\sqrt{2}}{2}$; $\tan A = 1$
35. $k = 1\frac{1}{2}$ **37.** $k = 1\frac{1}{4}$

Written Exercises, page 592
1. 3.7321 **3.** 0.7071 **5.** 2.1445 **7.** 0.5736
9. 0.9877 **11.** 0.2924 **13.** 14° **15.** 84° **17.** 56°
19. 71° **21.** 76° **23.** 27° **25.** sin 45° **27.** sin 20°
29. sin 30° **31.** 18° **33.** $\tan 10° = \dfrac{\sin 10°}{\sin 80°}$

Written Exercises, pages 596–598
1. a. $OB = 1$ **b.** $\sin A = \dfrac{\sqrt{2}}{2}$ **c.** $\cos A = \dfrac{\sqrt{2}}{2}$
d. $\tan A = 1$ **3. a.** $OB = 13$ **b.** $\sin A = \dfrac{12}{13}$
c. $\cos A = \dfrac{5}{13}$ **d.** $\tan A = \dfrac{12}{5}$ **5. a.** $OB = 3$
b. $\sin A = -\dfrac{2}{3}$ **c.** $\cos A = \dfrac{\sqrt{5}}{3}$ **d.** $\tan A = -\dfrac{2\sqrt{5}}{5}$ **7. a.** $\sin A = \dfrac{\sqrt{2}}{2}$ **b.** $\cos A = \dfrac{\sqrt{2}}{2}$
c. $\tan A = 1$ **9. a.** $\sin A = \dfrac{1}{2}$ **b.** $\cos A = -\dfrac{\sqrt{3}}{2}$
c. $\tan A = -\dfrac{\sqrt{3}}{3}$ **11. a.** $\sin A = \dfrac{\sqrt{2}}{2}$ **b.** $\cos A = -\dfrac{\sqrt{2}}{2}$ **c.** $\tan A = -1$ **13. a.** $\sin A = \dfrac{5}{13}$
b. $\cos A = \dfrac{12}{13}$ **c.** $\tan A = \dfrac{5}{12}$ **15. a.** $\sin A = 0$
b. $\cos A = -1$ **c.** $\tan A = 0$ **17. a.** $\sin A = -\dfrac{3}{5}$
b. $\cos A = -\dfrac{4}{5}$ **c.** $\tan A = \dfrac{3}{4}$ **19.** $x = 4.4$
21. $x = 7.2$ **23.** $m \angle A = 64°$ **25.** $m \angle A = 36°$
27. $m \angle A = 62°$; $m \angle B = 28°$; $m \angle C = 90°$;

$AB = 9$; $BC \approx 7.9$; $AC \approx 4.2$ **29.** $m \angle A = 30°$; $m \angle B = 60°$; $m \angle C = 90°$; $AB = 12$; $BC = 6$; $AC \approx 10.4$ **31.** $m \angle A = 54°$; $m \angle B = 36°$; $m \angle C = 90°$; $AB = 50$; $BC \approx 40.5$; $AC \approx 29.4$
33. $m \angle A = 66°$; $m \angle B = 24°$; $m \angle C = 90°$; $AB \approx 9.8$; $BC = 9$; $AC = 4$
35. $x \approx 11.8$

37. $x \approx 11.7$

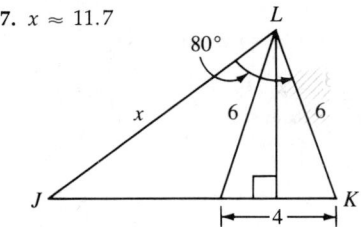

39. $x \approx 103.2$

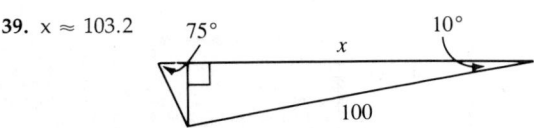

Problems, pages 599–600
1. $x = 168.6$ m **3.** $m \angle A = 12°$ **5.** 29.7 m
7. 4.0 m **9.** 128.7 m **11.** 3.3 km; 1188 km/h

On the Calculator, page 601
1. 0.7071 **3.** 0.0524 **5.** 4.7572 **7.** 1

Self-Test 1, page 601
1. 30° **2.** 120° **3.** 35° **4.** $-\dfrac{4}{5}$ **5.** $-\dfrac{3}{5}$ **6.** $\dfrac{4}{3}$
7. 0.9781 **8.** 0.6293 **9.** 0.5095 **10.** $m \angle A = 48°$; $m \angle B = 42°$; $m \angle C = 90°$; $a \approx 8.9$; $b = 8$; $c \approx 12.0$

Written Exercises, pages 606–607
1.

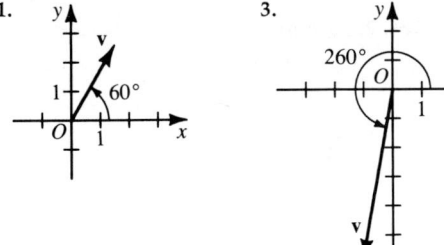

5. $\|\mathbf{s}\| = 3\sqrt{2} \approx 4.2$ **7.** 3 **9.** 6.4; 51° **11.** 3; 138°
13. a. 45° **b.** 315° **15.** 5.8; 121° **17.** 4.5; 63°
19. 329° **21.** 27°

Written Exercises, pages 610–611
1. a. x-component, −2; y-component, 1
b. $\|\mathbf{v}\| = 4.2$ **3. a.** x-component, 2; y-component, 4
b. $\|\mathbf{v}\| = 5.8$ **5.** 45° **7.** 31° **9.** (2, 4)

11. a.

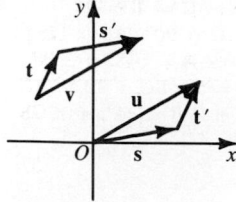

b. **u** and **v** are equivalent

13. $p = -7; q = 4$ **15.** $p = -5; q = 4$
17. $\|\mathbf{u} + \mathbf{v}\| = 10.8$; direction $22°$

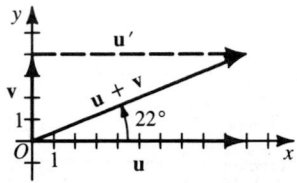

19. $\|\mathbf{u} + \mathbf{v}\| = 10$; direction, $67°$

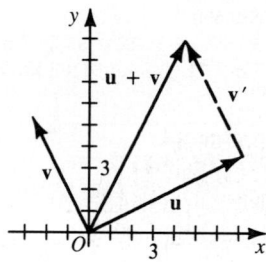

Problems, page 614
1. $63°$ **3.** 12.2 km/h; $99°$ **5.** 2892.5 N **7.** 164.2 km;
$14°$ **9.** 427.2 km/h; $54°$ **11.** 3.7 km; 1.3 km/h

Self-Test 2, page 615
1.

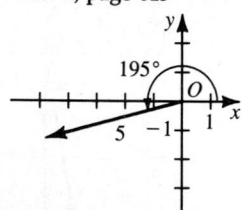

2. $\|\mathbf{v}\| = 5.8$; direction, $59°$
3. $\|\mathbf{s} + \mathbf{t}\| = 6.1$
4. 3.6 km/h; $124°$

Chapter Review, pages 616–617
1. b **2.** c **3.** c **4.** b **5.** a **6.** c **7.** b **8.** a **9.** d
10. b **11.** d **12.** c **13.** a **14.** d

1. $-\dfrac{x}{5 + x}$ **3.** $\dfrac{5n + 8}{n + 4}$ **5.** $\dfrac{y^4}{x^6 z}$ **7.** $\dfrac{3n - m}{n - 3m}$

9. $2\sqrt{47}$ **11.** $|y - 4|$ **13.** $\dfrac{2\sqrt{5}}{5}$ **15.** $2\sqrt{5} - 2\sqrt{2}$

17. $6 - 6\sqrt{5}$ **19.** 5.31×10^{-6} **21.** 2.7×10^{-2}
23. 1×10^{-2} **25.** $\{(-5, -2)\}$ **27.** $\{(-7, -8)\}$

29. $\emptyset$ **31.** $\left\{\left(1, \dfrac{7}{4}\right)\right\}$ **33.** $0.91\overline{6}$ **35.** 20 **37.** 4.9

39. $51°$ **41.** $\{\sqrt{14}, -\sqrt{14}\}$ **43.** $\{5, -2\}$ **45.** $\{81\}$
47. $\{1\}$ **49.** 2 quarters; 5 dimes; 3 nickels
51. 15.1 cm **53.** 96 **55.** -9 and -11

Preparing for College Entrance Exams, page 620
1. E **3.** E **5.** A **7.** B

Contest Problems, page 621
1. 16 **3.** $\{m: 0 < m < 6\}$ **5.** 5

Chapter 13 Statistics and Probability

Written Exercises, pages 625–626
1.

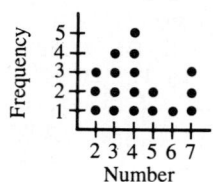

3.

Number	Freq.	Relative Freq.	
		Frac.	%
2	3	$\dfrac{3}{18}$	$16\dfrac{2}{3}$
3	4	$\dfrac{4}{18}$	$22\dfrac{2}{9}$
4	5	$\dfrac{5}{18}$	$27\dfrac{7}{9}$
5	2	$\dfrac{2}{18}$	$11\dfrac{1}{9}$
6	1	$\dfrac{1}{18}$	$5\dfrac{5}{9}$
7	3	$\dfrac{3}{18}$	$16\dfrac{2}{3}$
	18	$\dfrac{18}{18} = 1$	100

5.

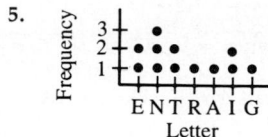

7.

Letter	Freq.	Relative Freq. Frac.	%
E	2	$\frac{2}{12}$	$16\frac{2}{3}$
N	3	$\frac{3}{12}$	25
T	2	$\frac{2}{12}$	$16\frac{2}{3}$
R	1	$\frac{1}{12}$	$8\frac{1}{3}$
A	1	$\frac{1}{12}$	$8\frac{1}{3}$
I	2	$\frac{2}{12}$	$16\frac{2}{3}$
G	1	$\frac{1}{12}$	$8\frac{1}{3}$
Total:	12	$\frac{12}{12} = 1$	100

9. a.

Digit	Freq.	Relative Freq. Frac.	%
0	1	$\frac{1}{16}$	$6\frac{1}{4}$
1	2	$\frac{2}{16}$	$12\frac{1}{2}$
2	2	$\frac{2}{16}$	$12\frac{1}{2}$
3	1	$\frac{1}{16}$	$6\frac{1}{4}$
4	2	$\frac{2}{16}$	$12\frac{1}{2}$
5	2	$\frac{2}{16}$	$12\frac{1}{2}$
6	1	$\frac{1}{16}$	$6\frac{1}{4}$
7	2	$\frac{2}{16}$	$12\frac{1}{2}$
8	2	$\frac{2}{16}$	$12\frac{1}{2}$
9	1	$\frac{1}{16}$	$6\frac{1}{4}$
Total:	16	$\frac{16}{16} = 1$	100

b.

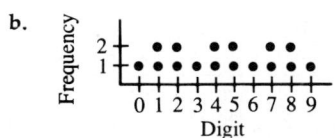

11. Answers will vary.

1. a.

Interval	Freq.	Relative Freq. Frac.	%
5–35	7	$\frac{7}{25}$	28
35–65	13	$\frac{13}{25}$	52
65–95	5	$\frac{5}{25}$	20
Total:	25	$\frac{25}{25} = 1$	100

b., c.

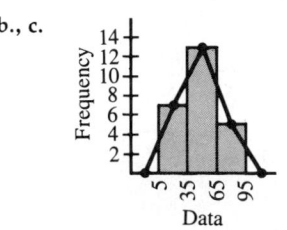

3.

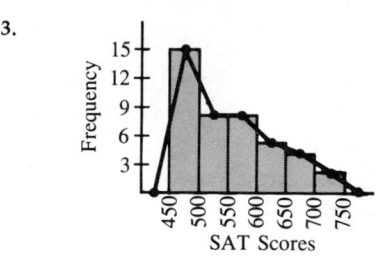

5. a.

Interval	Freq.	Relative Freq. Frac.	%
4.5–34.5	6	$\frac{6}{20}$	30
34.5–64.5	10	$\frac{10}{20}$	50
64.5–94.5	4	$\frac{4}{20}$	20
Total:	20	$\frac{20}{20} = 1$	100

b., c.

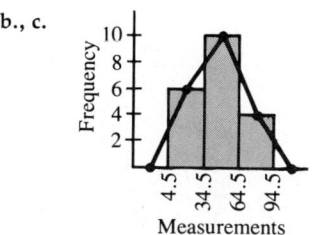

7. a.

Interval	Freq.	Relative Freq. Frac.	%
0–1000	5	$\frac{5}{25}$	20
1000–2000	5	$\frac{5}{25}$	20
2000–3000	6	$\frac{6}{25}$	24
3000–4000	3	$\frac{3}{25}$	12
4000–5000	3	$\frac{3}{25}$	12
5000–6000	1	$\frac{1}{25}$	4
6000–7000	2	$\frac{2}{25}$	8
Total: 25		$\frac{25}{25} = 1$	100

b., c.

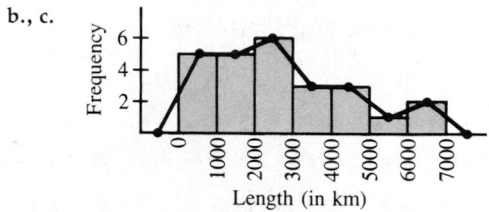

Length (in km)

Written Exercises, pages 632–635

1.

Number	Freq.	%	Cum. Freq.	Cum. %
2	1	$8\frac{1}{3}$	1	$8\frac{1}{3}$
3	3	25	4	$33\frac{1}{3}$
4	2	$16\frac{2}{3}$	6	50
6	2	$16\frac{2}{3}$	8	$66\frac{2}{3}$
7	1	$8\frac{1}{3}$	9	75
8	2	$16\frac{2}{3}$	11	$91\frac{2}{3}$
9	1	$8\frac{1}{3}$	12	100
Total:	12	100		

3.

Cost ($)	Freq.	%	Cum. Freq.	Cum. %
20	2	8	2	8
21	3	12	5	20
22	4	16	9	36
23	9	36	18	72
24	6	24	24	96
25	1	4	25	100
Total:	25	100		

5. 6 companies   **7.** no companies   **9.** 40%

11.

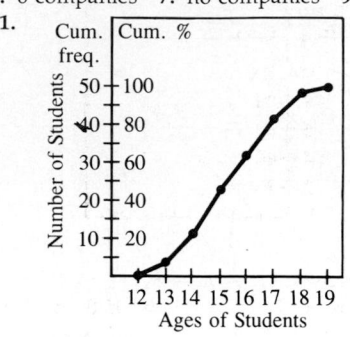

Ages of Students

13.

Interval	Freq.	%	Cum. Freq.	Cum. %
0–100	7	14	7	14
100–200	10	20	17	34
200–300	12	24	29	58
300–400	15	30	44	88
400–500	6	12	50	100
Total:	50	100		

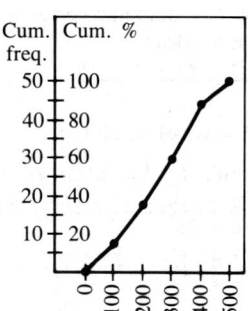

15. Answers will vary.

Self-Test 1, page 636

1.

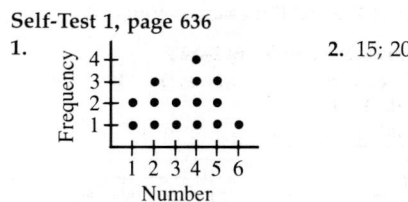

Number

2. 15; 20

Relative frequency of 5 is $\frac{3}{15}$, or 20%.

3.

Number	Freq.	%	Cum. Freq.	Cum. %
1	2	20	2	20
2	2	20	4	40
3	2	20	6	60
4	2	20	8	80
5	2	20	10	100
Total:	10	100		

3. (cont.)

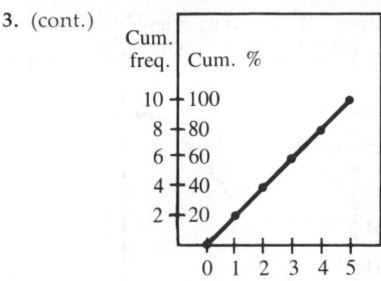

Written Exercises, pages 639–640

1. a. 5 **b.** 6 **c.** 7 **3. a.** 13 **b.** 15.5 **c.** 17.5
5. 67.3 **7.** 70 **9.** A: 3; 3; 3 B: 1, 2, 3, 4, 5; 3; 3
C: 1, 2, 3, 4, 5; 3; 3 D: 2, 4; 3; 3 They all have a
median and a mean of 3.

11. $M = \dfrac{r_1 + r_2 + r_3 + \cdots + r_n}{n}$ **13.** 82 **15.** $x = 7$

17. Let $x = 2m$ and let $y = 2n + 1$, where m and n
are integers. Mean of x and $y = \dfrac{x + y}{2} =$

$\dfrac{2m + 2n + 1}{2} = m + n + \dfrac{1}{2}$, which is not an

integer. **19. a.** Let a = equal number of measurements one unit from n. Mean =
$\dfrac{a(n - 1) + a(n + 1)}{2a} = \dfrac{an - a + an + a}{2a} =$

$\dfrac{2an}{2a} = n.$ **b.** Let b = equal number of
measurements two units from n. Mean =
$\dfrac{b(n - 2) + a(n - 1) + a(n + 1) + b(n + 2)}{2a + 2b} =$

$\dfrac{bn - 2b + an - a + an + a + bn + 2b}{2a + 2b} =$

$\dfrac{2an + 2bn}{2a + 2b} = \dfrac{2n(a + b)}{2(a + b)} = n.$ **c.** If there are an
equal number of measurements that are the same
amount less than and greater than some value, then
the mean of the set of measurements is that value.

Written Exercises, pages 642–643

1. 47 **3. a.** 6.8 **b.** 2.6 **5. a.** 11 **b.** 18.8 **c.** 4.3
7. A: 0; B: 0. No.

9. $\dfrac{(r_1 - m) + (r_2 - m) + \cdots + (r_n - m)}{n}$

11. $\sqrt{\dfrac{(r_1 - m)^2 + (r_2 - m)^2 + \cdots + (r_n - m)^2}{n}}$

Self-Test 2, page 643

1. 4; 2; 5 **2.** 12; 12.8; 3.6

Written Exercises, pages 647–648

1. 1, 2, 3, 4, 5, 6 **3.** $\dfrac{1}{6}$ **5.** $\dfrac{1}{3}$ **7.** $\dfrac{1}{2}$ **9.** $\dfrac{1}{3}$ **11.** $\dfrac{1}{2}$

13. 0 **15.** $\dfrac{1}{5}$ **17.** $\dfrac{1}{10}$ **19.** $\dfrac{4}{5}$ **21.** $\dfrac{1}{2}$ **23.** $\dfrac{3}{5}$ **25.** 0

27. $\dfrac{1}{2}$ **29.** $\dfrac{1}{13}$ **31.** $\dfrac{1}{13}$ **33.** $\dfrac{1}{26}$ **35.** $\dfrac{25}{26}$ **37.** $\dfrac{12}{13}$

39. $\dfrac{1}{2}$ **41.** $\dfrac{1}{3}$ **43.** $\dfrac{1}{2}$ **45.** HH, HT, TH, TT **47.** $\dfrac{5}{7}$

Written Exercises, pages 651–652

1. a. $\dfrac{11}{20} = 0.55$ **b.** $\dfrac{9}{20} = 0.45$ **3. a.** $\dfrac{2}{5} = 0.4$ **b.** 6

c. $\dfrac{3}{10} = 0.3$ **5.** $\dfrac{7}{18} \approx 0.39$ **7.** $\dfrac{1}{6} \approx 0.17$ **9.** $\dfrac{1}{2} = 0.5$

11. $\dfrac{36}{36} = 1$ **13.** No. There can be any number of
marbles that yields the same ratios. The contents of
the jar must be $\dfrac{1}{9}$ blue, $\dfrac{7}{18}$ red, $\dfrac{1}{3}$ yellow, and $\dfrac{1}{6}$

green. **15.** $\dfrac{75}{95} = \dfrac{15}{19} \approx 0.79$

Self-Test 3, pages 652–653

1. $\dfrac{1}{6}$ **2.** 0 **3.** $\dfrac{1}{2}$ **4.** 1 **5.** $\dfrac{7}{13} \approx 0.54$ **6.** $\dfrac{3}{20} = 0.15$

7. $\dfrac{9}{20} = 0.45$ **8.** $\dfrac{3}{5} = 0.6$ **9.** 25 **10.** 15

Chapter Review, pages 654–655

1. b **2.** c **3.** b **4.** b **5.** d **6.** c **7.** d **8.** c **9.** b
10. c **11.** a **12.** b **13.** a **14.** b

Cumulative Review, pages 656–657

1. 2 **3.** 3 **5.** 1480 **7.** 3.75 **9.** $\dfrac{1}{3}$ **11.** 2 h 55 min

13. 1.5625 **15.** -35 **17.** $-11\sqrt{3}$ **19.** $8\sqrt{35}$
21. $2\sqrt{2}$ **23.** $3\sqrt{2} - 3$ **25.** $10|x|y^2$ **27.** $\{3, -3\}$

29. $\{49\}$ **31.** $\{4, 2\}$ **33.** $\left\{\dfrac{-7 + \sqrt{41}}{2}, \dfrac{-7 - \sqrt{41}}{2}\right\}$

35. $\left\{\dfrac{1}{2}, 1\right\}$ **37.** 0; max: 0 **39.** 0, -2;

min: -1 **41.** Estimated zero: $1\dfrac{1}{4}$

43. $-270°$ **45.** 0.9703 **47.** 37° **49.** 5.4; 112°

51.

Number	Freq.	Relative Freq. Frac.	%
4	1	$\dfrac{1}{10}$	10
5	1	$\dfrac{1}{10}$	10
7	3	$\dfrac{3}{10}$	30
8	2	$\dfrac{2}{10}$	20
9	2	$\dfrac{2}{10}$	20
10	1	$\dfrac{1}{10}$	10
Total:	10	$\dfrac{10}{10} = 1$	100

53. $\dfrac{1}{6}$

Application, page 659

1. 6250 whales **3.** 1419 bass **5.** 12 moose

17. The zeros are 0 and 1.

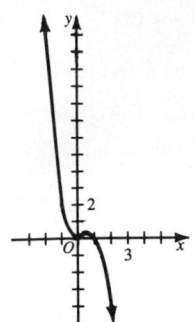

18. The zeros are −3, 0, and 3.

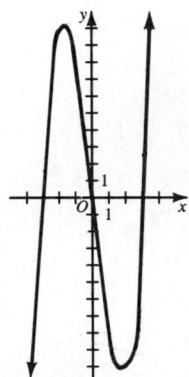

19. The zero is 0.

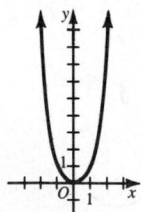

20. The zeros are −2, −1, and 3.

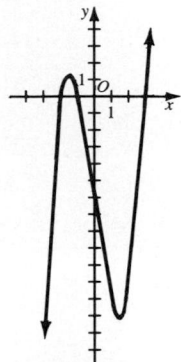

21.

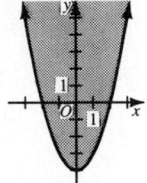

22.

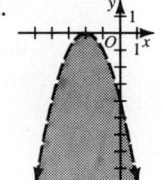

23.

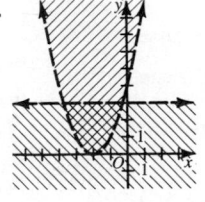

3.

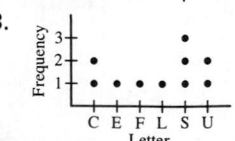

Letter	Freq.	Rel. Freq. Frac.	Rel. Freq. %
C	2	$\frac{2}{10}$	20
E	1	$\frac{1}{10}$	10
F	1	$\frac{1}{10}$	10
L	1	$\frac{1}{10}$	10
S	3	$\frac{3}{10}$	30
U	2	$\frac{2}{10}$	20
Total:	10	1	100

4., 5.

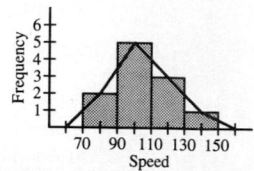

6.

Letter	Cum. Freq.	Cum. %
A	1	12.5
E	2	25
I	5	62.5
N	6	75
T	8	100

7.

Letter	Cum. Freq.	Cum. %
D	1	10
E	2	20
I	4	40
N	6	60
O	7	70
S	10	100

8.

Letter	Cum. Freq.	Cum. %
C	2	20
E	3	30
F	4	40
L	5	50
S	8	80
U	10	100